ingenious ire!and

About the Author

Mary Mulvihill is a science writer and broadcaster. A former editor of *Technology Ireland*, she has a degree in genetics and an MSc in statistics from Trinity College Dublin, and has written widely about science. She was instrumental in founding WITS (Women in Technology & Science), and edited *Stars, Shells & Bluebells – biographies of Irish women scientists and pioneers* (WITS, 1997)

ingenious ireland

A county-by-county exploration
of Irish mysteries and marvels

Mary Mulvihill

TOWN
HOUSE
DUBLIN

First published in 2002 by
TownHouse & CountryHouse Ltd
Trinity House, Charleston Road
Ranelagh, Dublin 6, Ireland

1 2 3 4 5 6 7 8 9 10

A CIP catalogue record for this book is available from the British Library.

ISBN: 1-86059-145-0

Cover and text design by Wendy Williams Design, Dublin
Maps by Christine Warner, Dublin
Printed by WS Bookwell, Finland

Contents

List of Maps

for my parents

Maureen and in loving memory of Joe

Acknowledgements

The island of Ireland has a wonderfully rich heritage of ingenuity and inventiveness, and a unique natural history. Sadly, much of this is not well known and, unlike our sporting and artistic achievements, seldom celebrated. I hope this book will heighten awareness of and pride in this often-overlooked aspect of our history, and give readers a new insight into the fascinating Irish countryside.

When this project began six years ago, it was planned as a small directory of visitor centres. It soon became clear that a larger volume was needed, giving readable background information and the relevant historical context. It was something of a shock to discover, halfway through, that much of this had been done before. But since *The Scientific Tourist through Ireland* – a "guide to the principal objects of antiquity, art, science and the picturesque" – was published in 1818, it was surely time for an update. Alongside places of interest, however, I have also included biographies of inventors, engineers, philosophers and scientists, both those born in Ireland, and others who, though not Irish, worked here – and making one exception for Robert Fulton, who was born in the US but of Irish parents.

Along the route to publication, I have been assisted by a great many organisations and people. Part of the early research work was grant-aided by the Science Technology & Innovation Awareness Programme run by Forfás (Dublin), and the Innovation Awareness Programme run by Invest Northern Ireland (Belfast). I am very grateful to Noel Gillatt and Dermot O'Doherty in Dublin, and Harry Cherry in Belfast for supporting the project from the outset.

It would be impossible to reference all the information gleaned from primary and secondary sources, and even unpublished theses, but key ones are footnoted in the relevant chapter, and a bibliography is also appended. Numerous researchers and local historians kindly shared their knowledge with me, and I have tried to credit them where appropriate in the text. I hope I have done justice to their work, and any errors that have crept in are my own.

Countless people helped me with information, images, and comments on the text, and I am indebted to them all. It is a pleasure to thank in particular: Don Bagley (Chester Beatty Library), Gerard Barrett (Patents Office), David Bedloe (Commissioners of Irish Lights), Dr Philip Boland, Dr Máire Brück, John Callanan (Institution of Engineers), Jimmy Carolan (Dublin Port Company), Prof Davis Coakley, Dr Tim Collins, Dr Stephen Coonan, Dr Owen Corrigan, Dr Ron Cox, Dr Pete Coxon, Nick Coy, Dr Bill Davis, William Dick, Prof Jim Dooge, Prof Ian Elliott, Dr John Feehan (UCD), Dr Billy Flynn (Irish Wildlife Trust), Ann Goldsmith (DIAS), Prof Gerry Harrington, Máiréad Johnston, Dr Roy Johnston, Dr John Joyce (Marine Institute), Dr Eugene Kelly, Seamus Kelly, Brian Lacy (Tower Museum, Derry), Joe Langtry (Mount Jerome Cemetery), Martin Lyes, Cóilín MacLochlainn (BirdWatch Ireland), Ciara McMahon (Physics, UCD), Mary McMahon (Industrial Heritage Association of Ireland), Paul Mohr, Nathalie Morgado (Musée des Arts et Métiers, Paris), Austin O'Sullivan (Teagasc), Grace Pasley (National Botanic Gardens), Susan Poole, Jim Rees, Ian Roberts (Ove Arup), Dr Colin Rynne, Dr John Sweeney, Pat Sweeney (Maritime Institute), Darina Tully, Tom Wall (Eircom), Michael Walsh (Ove Arup), and Prof Alastair Wood (DCU). Una MacConville (*Archaeology Ireland*) kindly supplied me with a database of heritage centres, and Ita Murphy helped trawl for information.

Thanks to the many geologists who patiently explained rocks and landscapes to me, and sourced images, especially Dr Matthew Parkes (GSI), Dr Nigel Monaghan (National Museum), Dr Patrick Wyse Jackson (TCD), Dr Patrick McKeever (GSNI), Dr Barry Long (GSI) and, at the Ulster Museum, Dr Ken James, Dr Mike Simms and Dr Peter Crowther. Many visitor centres, tourist offices and local history societies provided other useful information and images. I must also single out the ever-helpful staff of the National Library in Dublin.

An Post kindly allowed us to reproduce several of its stamps, and I am most grateful to Anna McHugh and Brenda Dwyer for facilitating this. For assistance with permissions and sourcing other images, I am indebted to the various visitor centres and institutions, especially to: Marie Kearns (Enterprise Ireland), Gráinne MacLochlainn (National Library of Ireland), Michael Kelly and Gerry Hampson (ESB), Kenneth Anderson (Ulster Folk & Transport Museum), Katherine Lindsay (The National Trust, Northern Ireland), Pamela Yeh (Northern Bank), Kieran O'Higgins (Commisisoners of Irish Lights), Gerry Camplisson (Northern Ireland Tourist Board), Stephen Curran (Ordnance Survey of Ireland), Katherine Meenan (Bord na Móna), Clíodhna Ní Fhatharta (Iarnród Éireann), John Kelly (TCD Physics Dept), Gráinne Kilcoyne (Mayo Naturally), Tony Roche (Dúchas), Patricia McLean (Ulster Museum), and Alasdair Smeaton (Dublin Central Library/Dublin Civic Museum). Scans of most of the stamps were prepared by David Davison; most other scans were prepared by CPL Ltd.

Several people took early versions for a 'test drive' and provided valuable feedback, notably Simon Brooke, Louise Buffini, Patricia Deevy, Anna Heussaff, Danaë Maguire, Frank Miller, Fiachra Ó Cairbre, Ena Prosser, Marion Palmer and Faith White.

I thank my former colleagues at *Technology Ireland* for their help with an earlier, unpublished version of this book: in particular Tom Kennedy, whose contribution was above and beyond the call of duty, and Marie Kearns, who toured the country taking photographs for me; also Duncan Black, June O'Reilly and Maureen Egan. I am indebted to Jonathan Williams for much good counsel about the publishing process then, and Therese Caherty who proofread that earlier text.

Thanks to my editor at TownHouse, Claire Rourke, for caring about this project as much as I did, to Wendy Williams, whose careful design work enlivened the text and images, and to Gay Needham for her expert typesetting.

A heartfelt thanks to all the friends who buoyed me up throughout this long project; to the Hemingways, the Bryants, Steve Bevan, Marion Gerry, Moira Bruce, Alan Mould and the Lindsays for the Scottish respite care; and to Harry Robinson who drove me around Northern Ireland. I would never have finished this without Brian Dolan, who accompanied me on research forays, corrected my physics, provided technical support at critical moments and, most of all, believed in the book.

Finally, it is a pleasure to dedicate this book to my parents, who introduced me to the wonderful world of books, libraries and learning; I am only sorry it took so long that Joe, though he had read every word of it, did not live to see it in print.

MARY MULVIHILL
Dublin, August 2002

Sponsorship

The author gratefully acknowledges the receipt of grants from the Science Technology & Innovation Awareness Programme run by Forfás (Dublin), and the Innovation Awareness Programme run by Invest Northern Ireland (Belfast), during the research for this book.

How to use this book

Ingenious Ireland is designed for dipping into, and each short story stands alone, but the material is also indexed and cross-referenced – an asterisk (*) means there is further information elsewhere in the book – so you can browse at random or pursue particular subjects across the country via the index. The bigger picture for the island's geology, natural history and climate is in the introductory chapter, *In the Beginning*. The tour moves from Dublin anticlockwise around to Co. Wicklow, taking, as much as possible, a sequential route across one county and into the next.

Outline maps at the start of each chapter show the location of the visitor facilities and the main places of interest referred to in the text; further information on the visitor facilities is in the Directory Section *(page 274)*. The OSI Discovery and OSNI Discover maps (1:50,000 or 1.25 inches to the mile), are ideal for exploring in greater detail the places featured in this book. Please observe any access restrictions to sites, seek permission where this is required, leave habitats as you find them, and do not disturb wildlife or birds' nests, or interfere with sites or artefacts.

Dislaimer

Every effort has been made to ensure that the information in this book is accurate and up to date. However, the author accepts no responsibility for any loss, injury or inconvenience sustained by anyone as a result of the information contained in this guide, or of visiting a location described here. This guide aims only to provide information about a place, and including a place in this book should not be taken to mean that there is necessarily any right of public access. Where a site is on private property, permission to enter should always be sought from the owner.

Some abbreviations used

BAAS	British Association for the Advancement of Science
CIÉ	Coras Iompair Éireann (Irish State transport authority)
DATI	Department of Agriculture and Technical Instruction (forerunner to the current departments of agriculture and education)
ESB	Electricity Supply Board
FÁS	Foras Áiseanna Saothair (Irish State training and employment agency)
GSI	Geological Survey of Ireland
GSNI	Geological Survey of Northern Ireland
NUI	National University of Ireland
OPW	Office of Public Works
OSI	Ordnance Survey of Ireland
OSNI	Ordnance Survey of Northern Ireland
QUB	Queen's University Belfast
RCSI	Royal College of Surgeons in Ireland
RDS	Royal Dublin Society
RIA	Royal Irish Academy
RIC	Royal Irish Constabulary
RSPB	Royal Society for the Protection of Birds
RTÉ	Radio Telefís Éireann (Irish State television service)
TCD	Trinity College Dublin
UCC	University College Cork
UCD	University College Dublin
UCG	University College Galway

Bord na Móna is the National Turf Authority, Teagasc is the National Agricultural Research & Advisory Institute.

IN THE BEGINNING

Ireland's long and varied geological past includes a previous existence in two pieces separated by a prehistoric ocean, and time spent living in the southern hemisphere (the slow crawl northwards continues). The island of Ireland has relatively few plants and animals but they include many rare and unusual species. And its notoriously unpredictable weather can also surprise. The result is a rich legacy of wonders and treasures.

Ireland, as we know it, is relatively young – a mere few thousand years old – yet it is made of ancient rocks that, in turn, are composed entirely of primordial stardust. The island's familiar teddy-bear shape emerged only when sea and land levels stabilised after the last Ice Age, and when rising sea levels flooded coastal areas, severed the land bridges connecting us to Britain and Europe and, some 7,000 years ago, finally made an island of Ireland. But the

Vital statistics

Area:	83,000sq km Republic of Ireland 69,000sq km Northern Ireland 14,000sq km
Latitude:	51°25'–55°25' north
Longitude:	5°25'–10°35' west
Population:	5.6 million (Republic of Ireland 3.9 million; Northern Ireland 1.7 million)
Highlands:	5% (more than 300m above sea level)
Lowlands:	75% (less than 150m above sea level)
Longest stretch:	485km
Widest stretch:	304km
Direction:	drifting northwards with the rest of Europe, at about 10km per million years

Source: Central Statistics Office (www.cso.ie)

oldest Irish rocks are nearly 2 billion years old, reaching back into what geologists call 'deep time'. Fortunately, stones can talk, and they reveal a fantastic geological story, that includes scorching deserts, tropical seas and thick ice sheets, mountains higher than the Himalayas, and violent earthquakes and volcanoes. This geological history, which can be read from Irish rocks and landscapes like pages in a book, shapes the island's appearance, even its very nature: the soils we use for farming, the peat bogs we cut for fuel, the minerals we mine, the rocks we quarry, the scenery we admire… are all determined by our geological history.

The **oldest-known rocks in Ireland**, at Inishtrahull* (Co. Donegal) and Annagh Head* (Co. Mayo), are some 1.75 billion years old. Over the millions of years since they were formed, they have been shaken, baked, melted, crushed and considerably altered, so that today their original nature is unrecognisable. All the other rocks of Ireland are younger, and formed of minerals recycled from previous formations, for rocks are constantly being recycled by geological processes. The constituent elements of the various rocks were themselves formed in supernova explosions and in nuclear reactions at the heart of extinct stars, long before even the Solar System came into being – truly, this world is made entirely of stardust.

Thanks to continental drift, the Earth's landmasses are constantly moving. Some 600 million years ago Ireland – or at least, the bedrock it lies on – was located as far south as South Africa is today (the evidence is in magnetic grains in the rocks), and it has since been drifting northwards, at about 10km every million years. Moreover, this prehistoric Ireland was split in two: the northern half was at the edge of the ancient continent of Laurentia, along with North America and Greenland; the southern half was in Avalonia, with Wales and Brittany; and in between lay the Iapetus Ocean (Iapetus was the father of Atlantis). Over the ensuing 200 million years, Laurentia and Avalonia inched closer together, squeezing the Iapetus Ocean, just as the Pacific is being squeezed today. Sandy areas at the edges of these ancient continents were later petrified into the distinctive quartzite mountains of Errigal and Croagh Patrick on the Laurentian side, and Wicklow's Sugar Loaf and Bray Head across the ocean in Avalonia. Rocks at Bray Head contain

Drifting continents

When Alfred Wegener proposed his theory of continental drift in 1912, it was ridiculed. But evidence in its favour, from evolution and geology, mounted over the years and now plate tectonics is a cornerstone of geological science. The Earth has a solid nickel-iron core, but the surrounding mantle is hot, and flows much like sticky toffee can; floating on top is the outer crust, a jigsaw of landmasses, or 'plates'. The plates are constantly jostling, and moving at about 2–5cm a year (or 5–10km per million years) – the rate fingernails grow – but sensitive satellite instruments can detect these movements. Ireland is currently drifting north with the Eurasian plate; the Atlantic Ocean is widening as North America and Europe move apart; the Himalayas are being pushed up as the Indian plate rams into Asia; the Pacific Ocean is being squeezed by the surrounding plates; and a new ocean is appearing in the Great African Rift Valley where the Somali plate is tearing away from the rest of Africa. Two million years ago, North and South America became joined at Panama, separating the Pacific and Atlantic oceans and initiating the warm Gulf Stream so important for Ireland's climate.

A rough guide to the world some 470 million years ago: Ireland was split in two pieces on opposite sides of the Iapetus Ocean, and was still in the southern hemisphere.

Ireland's oldest fossils: frond-like traces called *Oldhamia*,* dating from 550 million years ago. By 420 million years ago, simple plants had evolved to live on land, and their fossils (*Cooksonia*)* have been found on the Devil's Bit mountain. Eventually, 400 million years ago, the two continents collided to form a single landmass, Pangaea, and the Iapetus Ocean disappeared. The collision created an impact zone 60km wide, and in Ireland the 'seam' runs from Limerick to Clogher Head* in Co. Louth. The collision and the attendant volcanic activity deformed the adjacent rocks; minerals melted in the Earth's crust, and seeped into the rocks to form veins of ore, including the gold deposits* of Tyrone and Mayo. Molten magma also welled up under the Earth's surface, where it cooled to form the granite that is today exposed in the mountains of Wicklow and Donegal.

The collision pushed up a massive mountain range, extending from Scandinavia through Scotland, northwest Ireland and Newfoundland to the Appalachians, and trending northeast–southwest. The mountains were eroded so quickly that geologists conclude they were higher than today's Himalayas, though all that now remains are modest hills. The erosion continued for millions of years, producing massive amounts of sand which ultimately became the **Old Red Sandstone** (ORS) rocks found all over Ireland, but especially in Munster. (Kerry's ORS mountains are, in turn, being eroded today, and the sand is accumulating on the beaches of the Dingle peninsula, where it may one day form sandstone again.) The ORS formed 370 million years ago, much of it in a desert – the deep-red colour comes from iron oxide, a compound that forms in hot, dry conditions. 'Ireland' was still south of the equator and a sea occasionally

Ireland in outer space

Minor Planet Ireland is a small asteroid or minor planet (specifically, MP 5029) lying in the asteroid belt between Mars and Jupiter. It was named, in 1995, in honour of the 150th anniversaries of the GSI and the University Colleges in Cork, Galway and Belfast, by the two US astronomers who discovered it, Carolyn and the late Eugene Shoemaker. Made from material left over after the Sun formed 5 billion years ago, MP 5029 is much older than our Ireland, but just 25km long.

washed over the land. The first fish had evolved by then, while on land leafy plants were starting to contribute oxygen to the atmosphere. Fossils from this era are preserved in the yellow sandstones of Kiltorcan* (Co. Kilkenny).

By 350 million years ago, Ireland had reached the equator and resembled the Bahamas today, with limey muds washed by warm seas that teemed with life. The explosion of life on land and sea included amphibians – the first animals that could breathe air and survive on land – and one left its prints on a mudflat now preserved at Valentia.* Other amphibian fossils were found at Jarrow* (Co. Kilkenny). The limey muds and the copious animal shells settled in sediments thousands of metres deep, which became the thick **limestone beds** that still cover much of Ireland. Limestone is easily eroded, producing at the coast the wide bays of Galway, Sligo and Dublin and the flooded valleys of the south, west and north, and, in the midlands,

Ireland's top ten treasures

The Burren: for its unique assortment of arctic, alpine and Mediterranean plants, its lunar landscape, disappearing rivers, vanishing turloughs, caves, archaeological remains and more.
Birr telescope: built in 1845, and for 75 years the world's largest telescope, it revealed for the first time the spiral nature of some distant nebulae.
Our raised and blanket bogs: we urgently need to conserve and treasure what little is left of these ecosystems.
Three beautiful bridges: the elegant engineering gems of Lisdoonvarna's Spectacle Bridge, Bunlahinch Clapper Bridge, and the Boyne railway viaduct at Drogheda.
The National Botanic Gardens and Natural History Museum: irreplaceable botanical and zoological archives, compiled over the centuries and cared for by generations of dedicated naturalists.
Killarney National Park: for its ancient oak woods, red deer and Kerry cows, and the 700-year-old Muckross yew tree – probably the oldest living thing in Ireland.
Valentia Island's fossil footprints: made 385 million years ago by a primitive amphibian, they are the oldest-known footprints in the northern hemisphere.
Marsh's Library: for its priceless collection of rare and historic books.
The Antrim Coast Road: this engineering triumph is packed with fascinating sites and geological gems.
The Céide Fields: a 5,000-year-old neolithic farming landscape hidden beneath Mayo's blanket bog.

the broad depressions that contain the Shannon lakes and the extensive raised bogs. The same calcium-rich limestone also underpins the bone-building pastures of Kildare and Meath.

Next came the Carboniferous period, when 'Ireland' and the rest of what would be today's northern hemisphere was blanketed with swampy tropical forests of tree ferns and giant horsetails, the putrid remains of which became **coal**. In Britain, younger rock subsequently covered and protected the coal, and it was mined to fuel the Industrial Revolution, but not in Ireland, where the coal was quickly lost to erosion, apart from small deposits at Castlecomer,* Arigna* and the Kish Bank.*

Some 300 million years ago the European and African landmasses jostled together, pushing up new mountains in central Europe. The shock waves were felt in Munster, where the land was crumpled into parallel east–west folds, forming the Cork, Kerry and Galtee mountains, and the Bandon, Lee and Blackwater* valleys. The same events brought up fresh minerals, notably the lead and zinc deposits at Tynagh, Navan* and Galmoy–Lisheen.* Desert conditions returned 250 million years ago, by which time Ireland had reached the latitude of the Canaries. Huge sand dunes formed at Scrabo Hill* (Co. Down) and elsewhere, and geologists can tell the direction of the prevailing wind from the shape of the fossil dunes. These deposits formed the New Red Sandstone which is quarried for Belfast's construction industry. Salt lakes, or 'dead seas', occasionally formed, and **salt** and **gypsum** deposits from these are mined today at Carrickfergus* and Kingscourt* respectively. The world was again rocked by violent events 200 million years ago, when the Pangaean super-continent, which had existed for 200 millions years, began breaking up. The Atlantic Ocean opened as South America pulled away from Africa, and North America from Europe (though North America and Europe stayed connected through Greenland for a further 150 million years). Lava still pours out from an open wound in the mid-Atlantic and pushes apart the landmasses a few centimetres a year.

The North American connection

For centuries, people have been trying to bridge the ever-widening Atlantic divide. Early voyages, like those of Saint Brendan, could take months but, in 1833, the steamship *Sirius** crossed from Cobh to New York in a record 18 days. By 1865, a telegraphic cable linked Valentia* to Newfoundland, and messages could cross in seconds. In 1905, the turbine-powered *Victorian* crossed from Moville in Co. Donegal to New York in eight days and, in 1919, Alcock and Brown* flew the 3,040km in 16 hours 12 minutes. Yet you do not have to leave Ireland to visit North America: just step north of that 400-million-year-old seam running from Limerick to Clogher Head.

By the Jurassic era 150 million years ago, when the **dinosaurs** were at their peak, Ireland was at the latitude of Gibraltar. Dinosaurs were land creatures, however, and Ireland then was covered by the sea, so we are unlikely to find dinosaur fossils here, though fragments of a Skelidosaurus* were found in Co. Antrim, where its body was probably washed into the Jurassic sea. The seas that covered Ireland, especially around 100 million years ago, were very pure, and the limestone deposits they left are a brilliant **white chalk**.* Most of the chalk was quickly lost to the relentless erosion, along with the underlying coal and much of the limestone; only Co. Antrim's chalk survived, protected under the later basalt, and a small deposit concealed in an ancient sinkhole near Killarney. Ireland was again rocked violently 65 million years ago, as Greenland finally broke away and the northern Atlantic Ocean opened. The break came not along the original Limerick–Louth join, but a little to the west, leaving part of North America stuck to Ireland and Scotland. Volcanic activity continued for 15 million years, producing Co. Antrim's great basalt deposits and the **Giant's Causeway**; related events gave rise to the igneous mountains of the Mournes and Slieve Gullion.* So much lava was extruded then that the Earth's surface eventually collapsed, leaving a deep depression that flooded to form Lough Neagh.* The lough — much bigger than today — was surrounded by sub-tropical redwood swamps, which later formed thick lignite* deposits.

When Ireland had more or less reached its current latitude, the African plate ploughed into Europe and pushed up the Alps. A shockwave hit Ireland, and uplifted Antrim's basalt and chalk layers, producing a plateau and cliffs. Only the northeast was affected, so Ireland falls gently towards the southwest. The late Prof Frank Mitchell believed this explained why the southwest coast has so many flooded valleys, why the tide penetrates far up the Barrow estuary (which is tidal to St Mullins), and why there are more waterfalls in the north, such as Cathaleen Falls on the River Erne, now harnessed for hydroelectricity at Ballyshannon (Co. Donegal).

500,000 years of climate change

For the past 500,000 years, the world's climate has oscillated between warm spells and **ice ages**. Ice ages, first proposed in the mid-1800s, were not fully accepted until the 1920s, but geologists now believe there were four glaciations during the past 500,000 years, as the ice advanced and retreated approximately every 100,000 years. Sea levels* also fluctuated wildly: much land that today is

Waxing hot and cold – the last four ice ages

Timescale (years ago)	Climate
500,000–480,000	Warm (Ballylinian)
480,000–428,000	Cool
428,000–302,000	Warm (Gortian)
302,000–132,000	Cold – polar conditions over Ireland
132,000–122,000	Warm (Eemian)
122,000–100,000	Mild (Fenitian)
100,000–80,000	
80,000–65,000	Cold – but Munster is ice-free
65,000–35,000	Mild (Aghnadarraghian)
35,000–15,000	Cold – northern drumlins laid
15,000–10,500	Late-glacial – ice melting, tundra conditions
10,500–10,000	Nahanagan cold snap
10,000–present	Mild – apart from the Little Ice Age AD 1350–1850

submerged was, at times, above sea level; at other times, land that today is dry has been submerged. During each cold period, great ice sheets, sometimes over a kilometre thick, swept over and reshaped the land, ironing smooth the surface, gouging out corries, dumping massive sand and gravel deposits, changing the course of rivers, and laying clutches of drumlins* and cords of eskers.* Each time, the land was swept clean and, after the ice melted, soil began forming and the land was recolonised from the south: first tundra species and grasses, then shrubs, conifers and finally broad-leaved trees, until the climate flipped again, temperatures dropped, and the process reversed as the ice returned.

Pollen grains trapped in sediments reveal how Ireland's climate fluctuated and its vegetation changed. A 450,000-year-old sample from **Ballyline** (Co. Kilkenny) contains pollen from the wing nut tree; this species grows in Turkey today, suggesting it was once much warmer in Ireland (see: **Waxing hot and cold**). It was warm again 350,000 years ago when box and rhododendron grew at **Gort** (neither species is native to Ireland now). Ice dominated much of the subsequent 200,000 years: thick ice sheets covered the country, and glaciers flowed down the Irish Sea from Scotland bringing bits of Ailsa Craig* as far as Ballycotton. Milder weather returned at times, especially 120,000–100,000 years ago, when pine trees grew at **Fenit** (Co. Kerry), and the rising sea cut new beaches that are now stranded above the current sea level. During the next, 'Midlandian', Ice Age, 80,000–65,000 years ago, the Scottish sea ice reached the Saltee Islands off Co. Wexford; ice covered Ireland north of a line from Kilrush to Wicklow Head, and deposited extensive gravel moraines around Cork that are still quarried. Alternating mild and cold spells ensued and, during one cool period 30,000 years ago, a mammoth* died at Dungarvan when conditions there were tundra-like. The drumlin swarms over northern Ireland were laid under the ice sheets then. This landscape of small hills is thus younger than the more rugged terrain to the south which escaped the most recent ice. By 15,000 years ago the climate was again warming, as the world entered the current mild spell. The glaciers stopped growing, slowly melted and retreated, and within 2,000 years the ice was gone, apart from a 500-year-long cold snap (the [Lough] Nahanagan* period), when the Gulf Stream temporarily disappeared around 10,500 years ago, temperatures plummeted by nearly 7°C, small glaciers formed on the Irish mountains and icebergs floated off the coast.

Natural history

The first plants colonising the bare ground were tundra lichens and grasses, followed in an almost orderly succession, as soil developed and temperatures warmed, by rich meadow grasses, scrubby juniper, birch and aspen, and hazel and pine. When early hunters arrived, perhaps 9,000 years ago, Ireland was richly forested: oak and elm on the heavy soils, alder on wetland, ash and yew on limestone, and pine in sandy areas. Evidence for this succession comes from pollen analysis, which also reveals that the elm trees disappeared from Ireland and northern Europe around 5,000 years ago, probably killed by a fungal disease similar to Dutch elm disease.

The plants and animals that colonised Ireland were racing against the tide. Sea levels were rising – 12,000 years ago they were 50 metres lower than today – and about 7,500 years ago Ireland became an island when the last land bridge* to Britain was submerged. Snakes and moles were just some of the many species that did not make it across in time. Britain's bridge to Europe lasted longer and, consequently, Britain has more native species than Ireland, though still considerably fewer than France. Even counting

Ireland's impoverished flora and fauna

Native species	Ireland	Britain
Wild plants	815	1,172
Birds	354	456
Reptiles	1	4
Amphibians	2	6
Land mammals	14	32
(Extinct mammals)	(3)	(4)

alien species, Ireland's total comes to just 28 species of land mammal, 420 birds, one frog, one toad, one lizard and one newt, 1,300 vascular plants (of which only two-thirds are native) and some 12,000 invertebrates. The seas remain open to immigrants, however, and at the last count, 25 species of marine mammal – whales, dolphins and seals – had been found in Irish waters. Ireland's flora and fauna, though small and ecologically impoverished, can still surprise: several species and races are unique to Ireland, as is the Burren's cosmopolitan botanical assembly, and then there are the intriguing **Lusitanians**.* These dozen species, among them the Kerry slug and the strawberry tree, have an unusual distribution, being apparently native to Ireland, Spain and Portugal yet not found in Britain. They mainly occur in the southwest, where they probably survived in sheltered pockets while the rest of the country was under ice.

After the great thaw, salmon, eel and trout – three fish species that migrate between river and sea – recolonised Irish rivers. Ireland's only other **native freshwater fish** are

The seven wonders of Ireland

The Giant's Causeway: its basalt columns are so perfectly formed that early visitors thought they were man-made.

The Blackwater bend: the river takes a sharp right-hand turn at Cappoquin, where a new route was imposed on an ancient river – but when and how did this happen?

Lough Hyne: it is a near-perfect rectangle, but no one knows how it formed.

The mysterious frog: did it come of its own accord, or was it introduced by the Anglo-Normans, or by a zoological zealot?

The disappearing springs at Fore: this puzzling fossil landscape also includes Lough Lene that, perversely, drains both east and west.

Newgrange: the 5,000-year-old tomb and astronomical observatory is a feat of precision engineering, pre-dating the pyramids, and built without the aid of metal tools or wheels – but by whom, how and why?

Our intriguing 'Lusitanians': a dozen species, found in Spain and Portugal ('Lusitania') but not in Britain – so how come they also occur in southwest Ireland?

the pollan, char and Killarney shad; originally, these were also migratory but, during the Ice Age, they became isolated by the ice in a few lakes and gave up migrating to the sea. All the other freshwater fish here were introduced: pike in the 1500s; carp and tench in the 1600s; and roach and dace in 1889. Other exotic species deliberately introduced by people include: flax* and horses in the Bronze Age; hens in the Iron Age; frogs* and rabbits* with the Normans; the potato* in the 1500s; and mink in the 1960s. Accidental arrivals include the black rat,* New Zealand flatworm* and zebra mussel.* Exotic trees planted in 18th-century gardens included beech, chestnut and the monkey puzzle; later a growing posse of international plant collectors brought us buddleia, rhododendron and giant hogweed,* all now spread beyond the garden gate. Some people welcome this new diversity, others fear the consequences of what has been called the 'McDonaldisation' of the natural world.

Just as we can gain new species, we run the risk of losing others. Mammoth* and giant deer* have already disappeared, victims of climate change, and the wolf* and wild boar* were hunted to extinction here; the corncrake* and meadow saffron are being pushed out by intensive farming, while turf cutting and drainage schemes threaten bog plants such as marsh saxifrage and serrated wintergreen. Recent legislation provides some protection: all wild birds are protected under the Wildlife Act, 1976, for example, and Ireland's offshore waters are a whale and dolphin sanctuary. There are also numerous wildfowl sanctuaries and national nature reserves, in addition to the national parks at Wicklow, Killarney, the Burren, Connemara, Mayo and Glenveagh. Individual species deemed to be 'protected' include the pine marten,* common lizard and pearl mussel.* But legislative protection is no guarantee that these species will survive the onslaught of 'development'.

Birdlife: Ireland's many islands, extensive coastline, bogs and wetlands, mild climate and location at the edge of both

Europe and the Atlantic Ocean, make it ideal for birds. Over 420 species have been recorded here, though some were rarities spotted only once, and the list is continually growing. Unique Irish races include the Irish jay and Irish red grouse. Many migrating birds stop over, whether to re-fuel or spend the winter, and occasional strays arrive, blown in from Europe, Africa or America. Aliens that arrived and stayed include the magpie* and collared dove. Some native species disappeared in recent times, including the great auk (1834) and golden eagle (1931), though the latter is now being reintroduced to Glenveagh National Park. Internationally important sites include Skellig Beag's gannet colony,* Rockabill's roseate terns,* and the Greenland white-fronted geese on the Wexford Slobs.* Cape Clear,* Mizen Head,* the Wexford Slobs and Rathlin Island* are noted for bird-watching; further information can be had from BirdWatch Ireland (Dublin) and the RSPB (Belfast).

Invertebrates: Ireland is home to 12,000 invertebrate species, including 360 types of non-biting aquatic midge (chironomid), 28 butterflies, 22 ants, 59 species of wasp and almost 100 kinds of bee. Though often treated with disgust and disdain, these are economically and ecologically vital: without them, plants would not be pollinated, crops would fail, insect-eating animals would starve, rotting vegetation pile up, pests multiply out of control, and the finely tuned web of life would fall asunder. Some native invertebrates are rarities, including a few 'glacial relics' – alpine and arctic species that remained after the ice retreated. The list continues to grow, thanks especially to pioneering work by the Natural History Museum's entomologist, Dr James O'Connor. Exotic arrivals include: the honeybee, introduced by early Christians for its honey and wax; the oriental rat flea, allegedly the cause of the Black Death;* the bark beetle that carries Dutch elm disease; and the varroa mite, now decimating bee colonies. Many insects pose health, economic or ecological problems, and growing international imports of produce, timber and plants mean these hazards can only increase.

Woodlands and trees: Ireland's post-Ice Age forests were well established 7,000 years ago. An ancient dugout canoe at the National Museum, measuring 18 metres long by 45cm wide, conveys something of the majestic size of those prehistoric trees – modern ones are puny by comparison. Forest clearances began 5,000 years ago with the first farmers, and continued relentlessly. Timber has been called the 'oil of the Middle Ages', and Ireland had it in abundance: with timber you could build houses, boxes, barrels, carts, wheels and boats; and make charcoal, weapons, tools, musical instruments, and even paper. Destruction of the Irish woodlands began in earnest in the 1600s, when they were felled to lay waste the country and flush out the rebels, to make charcoal for the iron smelters, to make wooden goods and barrels (the cardboard boxes of the day), to rebuild London after the Great Fire, and to cask the wines of France and Spain. Within a hundred years, Ireland had become a net importer of timber. No original wild wood survives intact and Ireland is now the least wooded country in Europe. Even after decades of commercial State-sponsored forestry, only 9 per cent of the island is forested – compared with 30 per cent on average in Europe – and most of that is alien conifer species, such as Sitka spruce.

Native tree species include oak,* holly, ash, rowan, yew,* hazel* (more shrub than tree), elm, willow (sally) and the unusual strawberry tree. All had their uses: pigs were

Special Irish trees

Significant tree collections: Castlewellan, Rowallane and Mount Stewart Gardens, all in Co. Down; Belfast Botanic Gardens; Glenveagh Gardens in Co. Donegal; Muckross House and Killarney National Park in Co. Kerry; Illnacullin (Garinish) and Fota in Co. Cork; JFK Arboretum, Co. Wexford; National Botanic Gardens, Dublin.

Tallest broad-leaved: a 45-metre tall poplar at Birr Castle.

Tallest conifer: a 60-metre tall western hemlock at Castlewellan.

Oldest tree: yew tree planted at Muckross Abbey, Killarney, reputedly in 1344.

Source: The Tree Council of Ireland (www.treecouncil.ie)

fattened on acorns, for instance, oak bark was used in tanning, and even the soft upper leaves of holly were fed to cattle in winter in the days before the Anglo-Normans introduced haymaking.* Traditionally, trees were venerated and revered; for example, the rowan or mountain ash, considered a druid's tree, was planted beside a house to protect the home. And the Brehon Laws set down penalties for those who damaged trees – the fine for felling a 'noble of the forest', such as an oak, was two milch cows. The many place names that derive from trees reflect how widespread woodlands once were – for instance, over 1,300 townland names incorporate the element 'derry', from *doire*, meaning an oak wood.

Boglands

After the last Ice Age much of northern Europe was covered by lakes – the weight of the ice had compacted the ground, leaving poorly drained land, and there was no shortage of meltwater. Many of these primordial lakes subsequently developed into **raised bogs**, as in the central plain of Ireland. The transition began around 8,000 years ago, when marsh plants and sedges started encroaching on the lakes. As the colonisation continued, watery **fens** formed; eventually, a bed of fen peat accumulated which was dry enough to support sphagnum moss, and raised bog took over. The tussocks of moss trapped water, impeded drainage and, in a self-perpetuating cycle, created the conditions they needed to develop. Dead and dying vegetation accumulated in the waterlogged bog to form peat, and a growing dome of moss and peat rose in place of the original lake.

Raised bogs began in the midlands as a natural post-glacial development, but **blanket bogs** only began about 5,000 years ago, after early farmers had cleared the forests, and then only on the rain-soaked mountain sides and the wet west coast. In these treeless wastes, rain leached minerals from the soil, producing a watertight 'iron pan' below the surface and a waterlogged soil. Sphagnum moss colonised this, and created its own acidic environment where vegetation accumulated, and peat began to form a blanket over the land.

Bogs may appear bleak, but they are sophisticated ecosystems, home to specialised plants and animals that have evolved to cope with the harsh conditions. Vegetation does not decay in a bog, so mineral nutrients are in short supply – although bogs along the west coast have relatively high levels of chlorine, sodium and magnesium salts, thanks to the sea spray and salty rain – and minerals must be conserved. Heathers do this by not shedding their leaves, for instance, while bog cotton shunts nutrients from its dead leaves to underground storage organs (which are an important winter food for grazing birds). Most bog plants take their nutrients from rainwater, but some are carnivorous and get their minerals from eating insects: the sundews have sticky leaves that act like fly traps; the butterworts have scented leaves that act like flypaper; and the bladderworts trap aquatic insects in their bladder-like leaves.

Nutrients are in short supply in bogs, so some plants, such as this sundew, have evolved to trap insects; others get their mineral nutrients from the rain.

Peat accumulates slowly, about 10–100cm every thousand years, though some raised bogs can grow by 10cm a year. The peat can eventually cover everything from old tree stumps, to prehistoric fields walls and standing stones. Thanks to their acidic waterlogged conditions, bogs also

The sine qua non *of bogs*

Sphagnum moss is the ultimate bog plant. It can absorb 20 times its weight in water, storing the liquid in special cells, and is perfectly adapted to poor, waterlogged conditions, absorbing its nutrients from rain. Significantly, sphagnum secretes hydrogen ions, which make the surrounding environment acidic, preventing vegetation from decaying, and rendering the bog unattractive to most other plant species. About 25 types of sphagnum grow in Irish bogs, but a hand lens is needed to reveal the fine detail of their structure. The moss also secretes its own antibiotics and, during World War I, volunteers collected Irish sphagnum for use as an antiseptic dressing in military hospitals.

preserve objects, such as bog bodies, wooden trackways and firkins of butter, and are important archives: bog oak can be tree-ring dated, while pollen grains and volcanic dust (tephra*) reveal information about prehistoric vegetation and ancient eruptions.

In the 19th-century, bog reclamation began in earnest, with drainage schemes and mechanised briquette factories; large-scale turf cutting continues in the midlands, mainly to fuel electricity-generating stations. Originally, nearly one-fifth of Ireland was under bog and fen (about 1.1 million hectares), but over 80 per cent has gone or is going: only 7 per cent of the raised bog (23,000 hectares) and 18 per cent of the blanket bog (143,000 hectares) survives "relatively intact". Yet what remains is internationally important: Ireland has 8 per cent of the world's blanket bog resource, and half of all oceanic raised bog. In the northern hemisphere, only Canada and Finland have more bog than Ireland. The main threats remain overgrazing, turf cutting and drainage. The Irish Peatland Conservation Council, a non-governmental organisation founded in 1982, believes 220,000 hectares are still worth conserving. In 1987, the government committed itself to conserving 40,000 hectares of blanket bog and 10,000 hectares of raised bog. By 2002, the target for blanket bog has nearly been met, mostly through national parks, but only half the target for raised bog has been met.

Irish weather

In the long term, Ireland's climatic history is as varied as its geological one, thanks to continental drift. In the medium term, climate is influenced by cyclical phenomena in the Solar System, while, in the short term, weather can be affected by events such as volcanic eruptions (as when dust blocks out sunlight for several years and the world cools). These short-term changes are recorded in tree-ring patterns (a narrow ring indicates a poor growing season), pollen profiles, sediment layers and even the growth rings of cave stalagmites.

After the end of the last Ice Age, the climate warmed fairly steadily, and by 6,500 years ago temperatures were 2°C warmer than today. Conditions became cooler and wetter for a time, and Ireland's raised bogs began forming, then temperatures rose again until, by AD 1200, farmers in northern Europe were enjoying bumper harvests and grapes grew in northern England. That ended with the **Little Ice Age** (c.1300–1850) – paintings from that time depict freezing winters, with Londoners skating on the Thames – and Irish winter temperatures regularly dropped to -12°C. Poor harvests and bad weather frequently led to famine, food riots and disease, while watermills – the power stations of the day – regularly froze, bringing industrial activities to a halt. Now, industries spew out gases that could be warming the planet; certainly, weather patterns worldwide appear to be changing and sea levels rising. In the past, the climate has regularly flipped between warm and cold, sometimes in under a decade. We could be about to experience another flip, though no one knows if we are headed for a greenhouse or an ice house, nor when.

Ireland's weather is determined by its location at the edge of the north Atlantic, where westerly winds arrive laden with moisture. These westerlies moderate the climate, so that it avoids the extremes experienced on the continent. On the downside, Ireland is consequently cloudy and wet, especially in the west which receives 1,200mm of rain a year, compared with 800mm in the east. The Gulf Stream current keeps sea temperatures warmer than one would

Record-breaking Irish weather

Hottest day: 33.3°C recorded at Kilkenny, June 26th, 1887
Coldest day: -19.4°C recorded at Omagh, January 23rd, 1881
Driest year: 1887, only 357mm of rain fell in Dublin (45% of annual average)
Worst wind: Night of the Big Wind, January 1839, estimated wind speed of 52m/s (180 kmph)
Heaviest rain: (i) August 25th 1986, Hurricane Charlie: 270mm fell in 24 hours at Kippure, Co. Wicklow; and (ii) August 26th 1949: 38mm fell in 12 minutes at Waterford

Source: Met Éireann (www.met.ie)

expect for this latitude, but that current could weaken or disappear as climate change progresses, so that perversely, even if the world warms, Ireland could become colder.

The prevailing weather systems develop over the Atlantic, where cold, polar air meets warm, moist tropical air, producing vigorous, changeable phenomena that make Irish weather notoriously variable and difficult to forecast. Forecasting depends on knowing what conditions are like upwind but, in the days before the telegraph and satellites and with no observers in the Atlantic, Irish people had to rely on a folklore of beliefs to predict the weather. The scientific approach to weather began with the development of instruments like the barometer and thermometer; telegraphy meant that, by 1860, weather information could be sent rapidly over long distances, and people could at last know in advance what was coming their way. Today, satellites record weather information over wide areas – Irish forecasts take account of conditions from Russia across Europe and the Atlantic to North America – while conventional ground-based weather stations record local information. High-powered computers then crunch the many numbers to produce a forecast for the coming days.

Mineral wealth

Geology is crucial to economies, since everything that cannot be grown must be dug from the ground. Fortunately Ireland, though small and lacking coal, is otherwise blessed with numerous rock types, diverse mineral ores – from andalusite to zinc – and a wealth of building materials, all of which people have exploited since the earliest times. Stone

> ## Geological riches – from A to Z
> Ireland has deposits of, among other things: andalusite, antimony, barytes, basalt, bauxite, chalk, coal, cobalt, copper, diatomite clay, flagstone, gold, granite, gypsum, iron (and bog iron), lignite, lead, limestone, maërl, marble, marl clay, phosphates, pottery clay, quartz, salt, sandstone, silica sand, silver, slate, sulphur, talc, uranium, zeolite and zinc. However, not all these deposits are commercial.

Age people used the flint nodules found in Antrim chalk,* and made axes from Lambay porphory* and Rathlin porcellanite;* Ireland's oldest, recognised metal-working site is a 4,000-year-old copper mine at Mount Gabriel* (Co. Cork); gold* was also mined in prehistoric Ireland. Iron ore* is more plentiful than copper or gold, but requires a more complex smelting process, and ironwork arrived here only about 500 BC.

Irish mining has long been an international industry; there was a flourishing overseas trade in prehistoric Irish stone axes and gold, Italian entrepreneurs opened a mine at Silvermines* in 1289, and skilled mine workers frequently came here from Cornwall and Wales. The past 1,000 years saw considerable mining activity in Ireland: Wicklow had a gold rush* in 1795, remote corners such as Allihies,* that today seem quiet, were once thriving sites employing thousands of people, and every small village had its stone quarry, and a lime kiln for mortar and fertiliser. Widespread iron smelting in the 17th and 18th centuries contributed to the destruction of Ireland's forests, as oak trees were felled for charcoal to fuel the smelters.

The Industrial Revolution increased the demand for metals. By then, steam-powered pumps had made it possible to sink deep mine shafts and, later, railways made it easier to transport the ore. But mine fortunes frequently fluctuated, depending on international prices, trade restrictions, wars and, more recently, competition from big American and African mines. By the 1880s, production at Irish mines had peaked; a global recession meant prices and demand had slumped, the workforce had been decimated by famine and emigration, and political unrest discouraged would-be investors. The result was 75 years of near inactivity, and the widely held belief that turf was Ireland's only natural resource. All that changed in the 1960s, when modern prospecting began in earnest, and sizeable deposits of lead and zinc were discovered at Tynagh, Gortdrum, Silvermines, Navan* and Galmoy–Lisheen.* Today, Navan mine is one of the world's largest zinc mines. More recently, there has been talk of diamonds in Donegal, and gold mining in Tyrone and Monaghan. And all made possible by ancient processes and geological accidents. (The Mining Heritage Trust of Ireland is compiling a database of mine sites at www.mhti.com.)

Marine resources

The island of Ireland has an area of 83,000sq km, but, under the 'law of the sea', can claim a seabed territory of over 900,000sq km – hence the Marine Institute's slogan that "90 per cent of Ireland is undiscovered, undeveloped and underwater". Some 86 per cent of the population lives within 50km of the coast, and 90 per cent of international trade is still conducted by sea. **Fishing rights** have long been important: in 1465, England banned strangers from fishing in Irish waters without a licence. In 1556, Philip of Spain paid England £20,000 a year to allow 600 Spanish ships to fish here and, by 1600, the Dutch were paying £30,000 for similar access. During the 1700s, the English fisheries claimed they had been ruined by Irish successes, and restrictions followed so that, by the Great Famine, Ireland had no fishing industry to speak of, though foreign fleets still fished Irish waters. Various measures were introduced post-Famine to resurrect Irish fishing; the Clifden–Galway railway line,* for example, was intended to bring Clifden's catch quickly to market, numerous harbours and piers were built, fishermen were given loans to buy boats, and, by 1887, Ireland was competing with Norway in the herring market. Today, Ireland's 200-mile fishery zone encompasses 465,000sq km, half its seabed territory.

Thanks to a geological accident, Ireland has an extensive **coastal shelf**; when the north Atlantic Ocean was opening 100 million years ago, as Europe and North America moved apart, initial false starts stretched the coastal shelf, before breaking it. Consequently, the west coast has a wide and shallow 'shelf' stretching hundreds of kilometres, before plunging to the abyssal depths of the Atlantic Ocean rift – the Spanish shelf, by comparison, is very narrow. Ireland is also a safe distance from the volcanic activity of the mid-Atlantic ridge, where lava erupts and forms new seabed. Scientists now study these remote places using sophisticated surveying techniques, exploring the vast canyons of the Rockall Trough and the rocky formations of the Porcupine Bank.

Marine scientific surveys began in earnest in the 1860s, prompted by the unusual specimens trawled from the ocean floor during surveys for the first transatlantic telegraph cable, and spurred on by the Victorian desire to collect as many species as possible. Surveys organised then by the Royal Dublin Society and Royal Irish Academy yielded new information on ocean currents and discovered numerous species new to science, including some dredged from a then-record 4,289 metres. The best-known Irish **research vessel** was *Helga II*, which had an onboard laboratory. Frequently requisitioned for fisheries protection and military work, it was infamously brought in during the 1916 Easter Rising to shell Dublin from the Liffey quays. In 1947, it became the first vessel in the modern Irish naval service, renamed LE (*Long Éireannach*) *Muirchú*. The latest research vessel, the Marine Institute's *Celtic Voyager*, launched in 1997, carries six crew and eight scientists and is equipped with laboratories and sophisticated instruments for seismic and fisheries research, and for studying the marine environment and geology.

DUBLIN CITY

Dublin has, for centuries, been a busy port, an administrative capital and, for most of its history, Ireland's largest city. It is home to the oldest Irish university – Trinity College, founded 1592 – and several newer universities and third-level colleges. It is also home to the Royal Irish Academy and Royal Dublin Society, which were founded in the 18th century, and to the National Museum, National Gallery, National Library and Natural History Museum, which opened during the second half of the 1800s. Dublin has a distinguished medical history too and, in the 19th century, the Meath and Rotunda hospitals were world renowned. Like all cities, Dublin also has a long and varied industrial tradition: from the Viking shipyards of 1,000 years ago, to Richard Turner's ironworks (which, in the 1840s, produced the elegant curvilinear glasshouses for Dublin, Belfast and Kew Gardens), to Thomas and Howard Grubb, who made some of the greatest of the 19th century's telescopes. Significantly, by 1845, Dublin was also the headquarters of the Ordnance and Geological Surveys of Ireland, which attracted professional scientists and engineers to Ireland, at a time when British science was still dominated by gentlemen amateurs. Dublin's name derives from its old Irish name *Dubh Linn* (the Vikings used Dyfflin), for a 'black pool'. In AD 140, a Greek geographer Ptolemy referred to the settlement as Eblana. In modern Irish, the city is *Baile Átha Cliath* ('town at the ford of the wattles').

Chapter 1: Dublin – the Liffey to the Bay

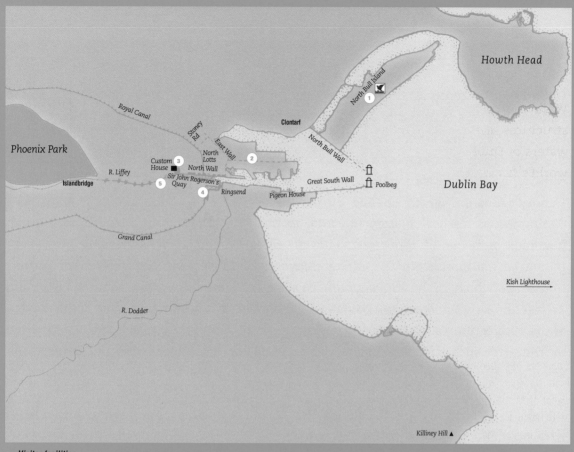

Visitor facilities

1. North Bull Island Interpretive Centre
2. Dublin Port, Alexandra Road
3. Custom House Visitor Centre
4. Waterways Visitor Centre, Grand Canal Quay
5. ENFO, Andrew Street

♀ woodland
🦆 nature reserve or wildfowl sanctuary
🏛 lighthouse

*Visitor facilities are numbered in
order of appearance in the chapter,
other places in alphabetical order.
How to use this book: page 9;
Directory: page 474.*

Walk the great wall of Dublin, inspect a Victorian diving bell, see Dublin's tallest structures, meet a Dublin Bay fraud, and visit an island that is still emerging from the waters of Dublin Bay. It was at Dublin Bay, in 1663, that a novel catamaran became the world's fastest ship.

Anna Livia Plurabella

Every major city is built on a river. Dublin grew up around the estuary of the **River Liffey** and, for over 1,000 years, Dubliners have been using the river for water and energy, food, transport and waste disposal. The Liffey, sometimes called Anna Liffey, from the Irish *Abhann na Liffe*, or Anna Livia Plurabella, as James Joyce dubbed her in *Finnegans Wake,* rises in the Dublin Hills on the slopes of Kippure Mountain, just metres from the source of the River Dodder.* But while the Dodder begins on Kippure's northern flank and flows directly to the sea, reaching Ringsend in 29km, the Liffey starts on Kippure's southern side, and must take a circuitous route via counties Wicklow and Kildare, finally entering Dublin Bay after a journey of 130km – yet only 19km from its source. En route, the Liffey still provides water and power, as it has done for centuries: historically, it filled storage tanks and turned mill wheels; today it fills a reservoir at Blessington,* and drives hydroelectricity turbines at Pollaphuca* and Leixlip.* The last mill powered by the Liffey, the 19th-century **Anna Liffey Mill** in Lucan, closed only in the 1990s; it was also the last water-powered roller mill in Ireland, and had been grinding semolina for the pasta industry. Its three turbines are virtually intact, and Fingal County Council is converting it into a milling museum, similar to that at Skerries.*

Early and Viking Dublin developed on high (and dry) ground above the Liffey estuary, around what is now Christ Church Cathedral. The estuary was long and muddy then, and probably 1km wide. It could be forded only at one shallow point, near modern-day Church Street, and then only at low tide. *Slíghe Cualainn*, the main road south from Tara,* crossed the river there, and the ford, which was reinforced with wattles, gave Dublin its Irish name: *Baile Átha Cliath,* the 'town at the ford of the wattles'. In 1214, the ford was replaced by a bridge, the Liffey's first.[1] The Liffey is still tidal inland to **Islandbridge Weir,** which was built in 1220 by monks from Kilmainham Priory. The weir created a sharp line between the fresh water upstream and the brackish water downstream, and meant that, at last, the river there could be used for drinking water.

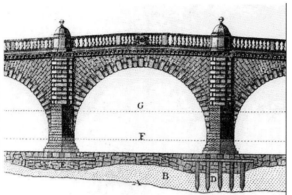

The former Essex [Grattan or Capel Street] Bridge, built in the 1750s by pioneering Dublin engineer George Semple,* who is buried in Drumcondra churchyard.

Over the centuries, Dublin expanded on both sides of the river and, as it grew, so did the need for bridges – though these could be erected only after quay walls had been built to corset the river into a manageable width. Today, 16 bridges cross the Liffey between Islandbridge and the sea, including one rail and two pedestrian bridges, with two further road bridges planned (at the docks, and near Heuston Station). Several bridges replaced ferries, including King's [Heuston] Bridge, which replaced a ferry operated by Dr Steevens' Hospital,* and the East Link toll bridge which replaced the last Liffey ferry in 1984. There is also a tunnel under the river at Ringsend: built in 1928 to carry water pipes and electricity cables, it was to have included a pedestrian way, but this was never built.

Dublin's oldest-surviving bridge is Queen's (or Mellowes) Bridge, which dates from 1764. Sadly, little survives of the bridge built at Capel Street in the 1750s by the self-taught Dublin engineer **George Semple** (1700–82). A builder's son from Dublin, Semple had just six weeks of formal schooling, yet he became an architect and master builder and his bridges were renowned among 18th-century engineers. He built his first bridge in 10 days in 1751, when he put a temporary structure across the Liffey at Capel Street after a storm destroyed the Essex Bridge. This was so successful that Semple was asked to rebuild the main bridge. Knowing little about bridges, he bought some engineering books and travelled to examine London's bridges. The one book he found on bridges was in French, which he could not read, so he developed his own techniques. To excavate foundations for the new Essex Bridge, Semple erected timber walls (cofferdams) around the foundation site, then pumped the water out. He later wrote a book, *A Treatise on Building in Water* (1776), which became a classic and is said to be one of the first true civil-engineering books in English (civil engineering as a discipline only began in the 1760s). Semple's contract for the Essex Bridge was for two years and £20,000. It took two years and eight days, and cost £20,661 – not bad for a beginner. The bridge was widened for traffic in 1873, and only Semple's foundations were retained.

When Semple built the bridge, it was at the heart of the city, linking Capel Street, then the main thoroughfare, and Dublin Castle. It was also the last bridge on the river before the sea; boats could sail up to it, and the Custom House was located nearby. As part of his design, Semple proposed widening the street (now Parliament Street) that led from his bridge to the castle… and so began the Wide Streets Commission, which over the next 100 years dramatically changed the city. Significantly, by developing Sackville [O'Connell] Street, the commission shifted the city's centre of gravity downstream to the east. And with the new Carlisle [O'Connell] Bridge blocking river traffic, the Custom House was forced to shift downstream too, moving to its current location in 1790.

The city's most distinctive bridge is the Liffey or **Ha'penny Bridge,** so-called because of the halfpenny toll pedestrians had to pay until 1916. The oldest iron bridge in Ireland, it was built in 1816 to replace a ferry and was initially considered an eyesore, but is now a popular landmark. The cast-iron structure of three parallel arched ribs spans 42 metres; the sections were cast at a Shropshire foundry, brought to Dublin by boat, and assembled on-site. The bridge was renovated in 2001, retaining most of the original metalwork, and painted off-white – the original colour decreed by George III to assist river navigation.

The hazardous gateway to Dublin

Dublin Bay is a broad natural bay that stretches 10km across at its widest point. The rocky headlands of Howth Head and Killiney Hill protrude to form the mouth of the bay, and lighthouses at both points guard the approaches. The bay owes its existence to local geology: the Howth and Killiney headlands are made of hard rock that resists erosion; and between them lies soft limestone that has been slowly eroded over thousands of years by the sea and rivers flowing into the bay, removing a large bite out of the coastline. Howth Head, made of a 500-million-year-old, golden-coloured quartzite, is the oldest part of greater Dublin. Killiney Hill is a little younger: the granite there is 400 million years old. By contrast, the youngest natural place in the bay (not counting artificially reclaimed land in the port), is North Bull Island, a mere infant not yet 200 years old.

Dublin Bay is choked with sand and gravel: vast deposits were dumped there by glaciers flowing down the Irish Sea during the four most recent ice ages; and Dublin's rivers, 'captured' and channelled into the bay by the low-lying limestone, still bring a continuous stream of sand and soil. The resulting sandbanks have long been a danger to shipping, not helped by the fact that they are constantly moving with the currents and tides. The greatest hazard is the **bar**, a large sandbank lying 1km east of Poolbeg Lighthouse. It forms where the Liffey, hitting the currents of the Irish Sea, is forced to dump its sediment load. A lightship was anchored there in 1735, equipped with a coal fire in a brazier, and the bar is still dredged regularly to maintain a safe channel. But rather than navigate the shifting, treacherous shallows, most ships preferred to dock at Howth* or Dalkey,* at least until new docks were built in Dublin in the 1700s. Even then, the mailboat continued to use Howth, switching to Dún Laoghaire* after a large harbour was built there in the 1820s. Today, navigational lights and buoys mark the channel through the various

The Liffey river god – original wax model for a carving on the Custom House. *
© Enterprise Ireland

hazards to Dublin port. With 20 ferries passing in and out each day, plus considerable freight traffic, the port authority has to operate a water traffic control service, and what amounts to a roundabout for boats at the mouth of the bay.

In earlier times, occasional deep pools provided a valuable anchorage among the shifting sandbanks; some pools had names – the Poolbeg, the Salmon Pool, the Iron Pool. These pools disappeared, however, after the Bull Wall was built in the 1820s and the currents in the bay altered. There were also vast sandy areas at Clontarf and Sandymount that were dry at low tide, the **North and South Bulls**, respectively. Their name may derive from *bølge*, an old Viking word for a wave. To complete the etymological tangle, Clontarf comes from the Irish *Cluain Tarbh*, meaning 'meadow of the bull', supposedly from the bull-like sound of waves thundering in across the flat sands.

A gift from the sea

For centuries Clontarf had a large sandbank that was dry at low tide. When the Bull Wall was completed in 1825, however, the currents in the bay altered and sand began to accumulate on the sandbank. Soon, it was high enough to trap windblown sand, and the sandbank became a sand dune. New land was materialising from the waters of the bay. A decade or so later and a second line of sand dunes had grown parallel to the first, beginning a cyclical process that continues today, for **North Bull Island,** as it is now called, is still growing and evolving. It is also reaching out to join the mainland: this began when the channel on the island's landward side became sluggish and brackish, enabling mud flats and salt marshes to develop there; alder and willow subsequently colonised the marsh; and the next phase, when nature is ready, will probably see the arrival of Scots pine.

North Bull Island now measures 5km by 1km and is an important bird sanctuary and nature reserve that, despite its young age, has a range of habitats, including mud flats, salt marshes and grasslands which, in spring, are dotted with purple orchids. The island is on a major bird migration route and, no sooner had the first dunes stabilised, than birds began flocking there, drawn to the new hotel on their migration motorway. Thousands congregate there each winter – over 200,000 on one occasion, the highest concentration of birds ever counted in Ireland. Many are stopping off to re-fuel en route to warmer parts, but others, such as the brent geese, spend all winter there. Two other nature reserves in North Co. Dublin that are internationally important for brent geese are Rogerstown estuary and Baldoyle estuary. North Bull Island, which was used as a rifle range during World War I, is now home to two golf courses and an interpretive centre run by Dublin City Council. ①

The Dublin Bay fraud

The **Dublin Bay prawn,** that frequent star of the seafood menu, is a fraud. And doubly so – it does not come from Dublin Bay, and it is not a prawn. Dr Paul Hillis, a marine biologist who has made a special study of the creature scientists call *Norphops norvegicus,* says true prawns have a soft and transparent polythene-like shell, but the Dublin Bay prawn is more like a small lobster, with claws and a hard pink shell which it sheds regularly as it grows. "If we were American, we'd probably call it a lobsterette," says

Hillis. The DBP has five pairs of legs, including its claws, plus another eight pairs of swimming paddles under its tail. The creatures are gregarious but territorial, living in burrows in mud or clay, and in water that is 10–400 metres deep. According to Hillis, the crowded sites "can look like Aero chocolate". Left undisturbed, the animals could live to be 25 years old, though this would be rare, and reach 35cm long. Despite its name, the DBP is not found in Dublin Bay, though it is found in the Irish Sea north of Lambay, along the west coast and Aran Islands, also around Scotland, and in the Atlantic from Iceland to Morocco and east into the Mediterranean – in fact, almost anywhere except Dublin Bay, where the sediment seems unsuited to its burrows. One story has it that the name arose from the custom of allowing fishermen to eat the 'prawns' as their boats entered the bay on their return to port. Today, the DPB are caught either in pots, or nets dragged along the sea floor; the catch is subject to EU quotas.

1663 – the fastest ship in the world

January 6th, 1663, and a most unusual boat is racing in Dublin Bay. Called *Invention,* it has two hulls and is the brainchild of Sir William Petty,* the surveyor and scientist who masterminded the Down Survey.* Petty had gone to sea as a boy, and his life's ambition was to design a fast boat. Boat building then was a traditional craft, the knowledge handed down from father to son, but Petty, a keen advocate of the new scientific revolution, decided instead to conduct experiments. Working with small-scale models, he found that the narrower the hull, the faster it cut through the water. A single narrow hull was unstable, so Petty opted for two narrow hulls, stabilised with a connecting deck. The same principle had long been used in Polynesian catamaran canoes, though these had not yet been seen by westerners, and Petty was the first to use the design for a large vessel. *Invention*, built with backing from the Royal Society in London, had hulls measuring 6 metres

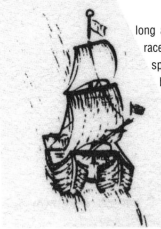

long and 0.6 metres across. It won its first race in Dublin that January day, reaching speeds of 16 knots against the tide, becoming what was then the world's fastest ship. Petty's next vessel, *Experiment,* could carry 13 men and 10 guns and, in July 1663, it raced the Holyhead mailboat, winning by a full 15 hours. Unfortunately, *Experiment* sank two years later while on a voyage from Portugal, and Petty's third and final boat, *St Michael* (1684), proved too unstable. Boat design has improved since then, and Petty would surely be pleased that large high-speed catamarans now regularly ply the Dublin– Holyhead ferry route. In 1991, the Irish Nautical Trust built a replica of *Invention* and raced it on the Liffey.

Petty's invention: this sketch of a 2-hulled ship appears on the Irish Sea in Petty's 1685 map of Ireland.

'The Great Wall of Dublin'

Dublin's docklands grew out of literally nothing over the past 350 years, built on reclaimed land and created by engineers who devised ingenious new ways of building sea walls, winning land from the water, and tackling the severe silting in the estuary. The port's origins lie at Wood Quay, which the Vikings built c.AD 900. The city and bay looked very different then: the mouth of the Liffey was more or less at Wood Quay; there was no North Bull Island; high tides washed in as far as Merrion Square in the south and Ballybough to the north; the north, east and south dock walls had not been built; and much of the North Lotts and Ringsend was still underwater. Over the next thousand years, the city would push slowly eastward, building quay walls to constrain the Liffey, reclaiming new land and creating a large industrial dockland.[2]

Initial development was slow. By 1300, only Merchant's Quay had been added, and when the Black Death* hit Dublin in 1348, the city came to a standstill. Not until the 1660s did

port development begin again, when William Hawkins built a wall on the Liffey's south bank from Temple Bar to Hawkins Street; the shore behind the wall was reclaimed, including what is now Westmoreland Street. The last bridge on the Liffey was then at Capel–Parliament Street, making Wellington Quay a good location for a custom house. And when Dublin Corporation opened one there in 1707, tall-masted ships could sail almost to its door.

The next major development was the construction, in the early 1700s, of the north and south port walls. In a bid to hold back the encroaching sands, the new Ballast Office (see: **Taking on ballast**) extended Hawkins's wall out to Poolbeg, a deep pool among the sandbanks. First, rows of wooden piles were erected, these were then lined with wattles, and stuffed with kishes – willow baskets filled with stones, and named from *cis*, the Irish for basket. Several Dutch labourers, experienced at building sea defences and dykes, were employed and, in 1731, the piles reached Poolbeg where a new lightship was anchored. A shorter kish wall was built on the north side enclosing an area of shore east of where O'Connell Street is today. By 1728, this wall

Taking on ballast

Empty ships need ballast for stability at sea. Originally this was sand and gravel excavated from the nearest shore but, as shipping traffic increased, port authorities needed to regulate where ballast was excavated, and where unwanted ballast was dumped. In 1707, the Ballast Office Act established a board responsible for "cleansing the port, harbour and rivers of Dublin"; in addition to ballast, it was responsible for dredging the shipping channels and maintaining the quay walls. In 1786, the board was replaced with the Corporation for Preserving and Improving the Port of Dublin, known as the **Ballast Board**. Initially, its offices were on Sackville [O'Connell] Street but, in 1801, a new Ballast Office opened on Westmoreland Street overlooking Carlisle [O'Connell] Bridge. There, in 1865, it erected a time ball that, for decades, provided ships with an accurate time signal: a large 1.2-metre copper ball dropped down a 4.6-metre mast, controlled by a telegraphic time signal from Dunsink Observatory.* Several changes later, the Ballast Board has become Dublin Port Company, and moved to Alexandra Road in the north docks, where it hosts an open day each autumn. ②

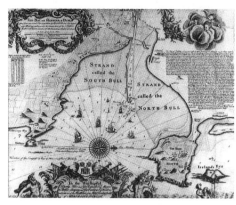

George Gibson's 1756 chart of Dublin Bay: the Great South Wall is half-built, as the old pile-and-kish construction is replaced with a granite wall.
Courtesy Dublin Port Company

had reached a point opposite Ringsend, where it turned north to form the East Wall; the shore within the L-shaped wall was reclaimed and sold off as the North Lotts. The new walls dramatically altered the city and port, and triggered the dockland developments. Ships could now anchor safely in deeper water, and not risk sailing upstream, and the city's centre of gravity began shifting eastward. This was reinforced when Sackville [O'Connell] Street was built in the 1740s, and, in 1791, the erection of a new bridge at Sackville Street forced the **Custom House** to move downstream to elegant new premises designed by James Gandon. The façade of the Custom House bears carvings of 13 Irish river gods. Burned by the IRA in 1921 (intense heat melted the copper dome, and the ruins took weeks to cool) and later rebuilt, it has been home at various times to government bodies, such as the Board of Works and the Department of the Environment, whose work features in a visitor centre. ③

The piles and kishes needed constant repair, however, and were later replaced with substantial granite and rubble walls, although it is said the remains of the original wooden piles can be seen at low tide. The long southern wall, now called the **Great South Wall**, took longest to build – 1748–90 – though, by 1768, a lighthouse had been built at Poolbeg, replacing the lightship. The wall stretches 7km from O'Connell Bridge to Poolbeg, and is reputedly Europe's longest sea wall. It was a feat of engineering and the final 2km walk out into the sea to Poolbeg lighthouse is the perfect place to inspect the structure: originally no mortar was used in its construction, and the massive granite blocks were brought by barge to the building site from Dalkey quarry. Ships in the bay finally had much-needed shelter from the prevailing southwesterly winds,

and the sands encroaching from Sandymount were held back. Following the 1798 Rising, a 'half-moon' gun battery was inserted midway along the wall and a fort built at Pigeon House to guard the port approaches (it was later a store for the city's bullion and archives). Cannon from those turbulent times now guard the ESB's Poolbeg power station,* while the 'half-moon' is a bathing place.

Despite the new wall, problems remained, notably the lack of a sizeable harbour, the treacherous bar,* and the silting of the estuary. Numerous consultants were brought in, including **Capt William Bligh** of HMS *Bounty* fame. Bligh, a skilled maritime surveyor, produced the first accurate navigational chart of Dublin Bay and, in 1801, recommended building a **North Bull Wall** from the existing North Wall almost to Poolbeg. The entrance to the port would then be a narrow gap between the two sea walls; the resulting funnel-shape would, Bligh hoped, increase the Liffey's flow and scour the channel clean, taking silt out but, in the reverse of a lobster pot, not letting it in again.

Other consultants suggested moving the mouth of the Dodder river in a bid to control the silting, or developing Dalkey as Dublin's port, with a canal link to the city. After years of technical arguments, work began on a North Bull Wall in 1819, but starting from Clontarf and not from the North Wall as Bligh had suggested. Completed in 1825, its first 1.7km are above high water; the final kilometre is a submerged breakwater, the end of which is marked by a lighthouse. The strategy worked: in 1819 there was often less than 2 metres above the bar at low tide; 50 years later clearance had improved to almost 5 metres, and today it is nearly 8 metres, though it still needs occasional dredging. The North Bull Wall had an unexpected side effect: by altering the currents within the bay it inadvertently created the North Bull Island.*

The North Bull Wall was part of a major programme to develop Dublin's docklands, fuelled by increasing overseas trade – all of which necessarily went by sea – and the arrival of big steamships needing larger berths. The engineer who did most to develop the docks was **Bindon Blood Stoney** (1828–1909), who devised an ingenious new way to build dock walls, and pioneered the use of pre-cast concrete as a building material. Stoney was born at Clareen in Co. Offaly, to a family that had come to Ireland from 'stoney' Yorkshire during the 17th-century Plantations (Blood was his mother's family name). He and his brother George Johnstone Stoney* (who 'invented' the electron), worked as astronomical assistants on the great telescope at nearby Birr Castle.* Bindon studied engineering at TCD and made his name in the 1850s working on the Boyne railway viaduct at Drogheda* – the elegant metal viaduct was an engineering triumph and is still in use. He moved to Dublin as a port engineer and, under his care, the docks were rebuilt and extended, especially on the northern side. Several kilometres of new quays were built, as well as new graving (dry) docks and the Alexandra Basin, named after the princess of Wales, who opened it in 1885.

The system Stoney invented for building dock walls was to use massive pre-cast concrete blocks instead of stone and rubble. First, the ground had to be levelled, then the blocks made on-site and lowered into place. Instead of men working inside a wooden dam, from which the water had been pumped out, Stoney designed a diving bell (see: **A Victorian diving bell**) so they could work underwater. The concrete blocks used were veritable monoliths, weighing an unprecedented 350 tonnes; each took four weeks to make and a further 10 to 'cure' before they could be moved. The design called for precision engineering to ensure a tight fit between the blocks. Once part of the new quay wall was built, the blocks for the next section could be made there, and so the wall inched forward. The technique was first used on the North Quay extension beginning in 1870; only when that was complete, in 1884, did attention shift to the southern side, where the docks were by then in bad repair. The resulting berths were independent of the tides, with depths that were unsurpassed anywhere, and international interest was considerable – when the British Association for the Advancement of Science* met in Dublin in 1878, an expedition visited the docks to inspect the technique. The docks are an enduring monument to Bindon Blood Stoney – 'the father of Irish concrete' – who won international fame for his innovative work. Stoney Road near the East Wall is named after him.

A Victorian diving bell

This odd metal structure on Sir John Rogerson's Quay is the diving bell designed in 1860 by Bindon Blood Stoney, and used in building the docks. The lower section is hollow and bottomless, with just enough room for six men to work. Once the 80-tonne bell had been lowered into position in the river, the crew entered through an access tunnel from the surface. Compressed air was fed in from an adjacent barge but, even though the air was cooled, the temperature inside quickly became unbearably hot, and shifts lasted only 30 minutes. The bell was crucial to Stoney's innovative way of building dock walls: the men inside the bell worked on the riverbed exposed at their feet, excavating the site where a massive concrete block would later go; all the excavated soil was stashed in baskets hanging inside the bell, and brought up when the bell was lifted. The bell was still in use in the 1960s to repair dock walls, by which time it had a telephone link to the surface. It was renovated in 2000, when holes were cut to provide a view of the interior.

An alternative diving technology can be seen at the Waterways Centre: a canvas diving suit, tightly woven to keep water out, and worn with a metal helmet; air was supplied via a tube connected to a pump on the surface. The suit was bought in 1905, and worn by Denis Madigan and later by his son when repairing canal locks and inspecting sea-planes at Foynes.* Despite being patched, it was still used in the 1970s. ④

Wildlife and nightlife

Like all ports, Dublin is home to a diverse assemblage of plants and animals – most are native, but several are blow-ins that arrived as stowaways on ships. Nearly 350 species of plants have been found growing in the city streets, seals have been seen on the Liffey as far upstream as Islandbridge, and O'Connell Street, where you might expect to find only traffic and people, has an unusual bird colony. One of the first people to study the city's plantlife was **Caleb Threlkeld** (1676–1728), an English doctor and Dissenting minister who moved to Dublin in 1713. Threlkeld identified 268 different wildflowers in Dublin, and he includes them in his landmark book, *Synopsis Stirpium Hibernicarum* (1727), listing their Latin, English and Irish names, where each was found (helpful for those revising his work later), and their herbal uses – snapdragon would, he wrote, protect against witchcraft and necromancers. This was the first reasonably reliable account of Irish plantlife, though Nathaniel Colgan, a naturalist who wrote a comprehensive *Flora of County Dublin* (1904), described it as "a piquant medley of herbal and homily in which this medical missionary from Cumberland delivers himself of his opinions on botany, medicine, morals, theology, witchcraft and the Irish Question". In 1998, the Dublin Naturalists' Field Club published a revised *Flora of County Dublin,* the culmination of a decade's work by dedicated volunteers, with details of over 1,300 plant species, not just flowering plants, but also ferns, mosses and stoneworts.

Only two copies of Threlkeld's book survive (one is in the National Library*), while TCD's herbarium has the plant specimens that he collected and dried. Some 250 years after Threlkeld's survey, two Dublin botanists repeated the exercise: Peter Wyse Jackson and Micheline Sheehy Skeffington found 345 wildflower species, nearly 100 more than Threlkeld.[3] Among the newcomers were buddleia, now common on waste ground having escaped from ornamental gardens, and **Oxford ragwort** *(Senecio squalidus)*. This ragwort originates on the sulphurous slopes of volcanic Mount Etna and has larger flowers than native ragwort. Its fondness for sulphur means it can tolerate urban pollution and, for over 200 years, it has been spreading from city to city: it reached Oxford in 1794, disembarked at Cork in 1800, and with the coming of the railways, spread north

along the tracks, reaching the Inchicore* railway works in 1890. Now widespread in Dublin, it is just one of the many exotic species that was brought here by trade, traffic and human activity.

City residents must be able to tolerate crowds, noise, traffic and pollution. London plane trees are particularly resilient, and so are often planted in cities, and there is a line of them down O'Connell Street. A flock of pied wagtails roost in those trees in winter, proving remarkably at home amidst the hustle and bustle. **Pied wagtails** are gregarious little birds that feed on insects and are more usually associated with riverbanks and beaches than city streets. They arrived in O'Connell Street in the winter of 1929, presumably attracted by the trees and the warmth of the lights, and spent the winter nights sleeping there, settling at dusk and leaving at dawn. Their numbers increased yearly, peaking at several thousand in the 1950s, when their twittering caused quite a racket. Today there are only a few hundred, possibly because, naturalist Christopher Moriarty suggests, as the city's cattle markets, piggeries and stables closed, there are fewer insects for them to feast on.

Exotic insects and other wee creatures are occasionally found in the city's fruit, vegetable and flower markets. Expert entomologists from the Natural History Museum* are on call to identify anything suspicious, such as the large Egyptian grasshoppers that sometimes arrive with Mediterranean imports. Most worrying is the Colorado beetle: this yellow-and-black striped beetle is a major agricultural pest, and would destroy the potato crop if it ever became established here.

County Dublin's coastal location and many estuaries make it attractive to **birds**. Rockabill, an islet off Skerries,* is home to Europe's largest colony of **roseate terns** and over 500 pairs breed there each summer when they visit from Africa, along with smaller numbers of common and arctic terns. The roseate tern, a darting, swallow-like bird that nests on the ground, is Europe's rarest breeding seabird, making Rockabill internationally important. BirdWatch Ireland (BWI) manages the colony, with a full-time warden all summer, and tern numbers at Rockabill are now increasing. BWI also manages the east coast's other tern colonies on Maiden Rock (Dalkey) – you can watch the terns in summer from the road – Newcastle (Co. Wicklow), and Lady's Island* (Co. Wexford); also the nature sanctuaries at Balbriggan, Shenwick's Island (frequented by seals and seabirds), and Rogerstown estuary (where brent geese, herons and falcons can be seen). ENFO, the environmental resource centre, has information on Dublin's, and Ireland's, nature reserves. ⑤

RINGSEND

A gas place

The elegant wrought-iron **gasometer** in Barrow Street is the only visible reminder that this was once a major industrial area. For over a century, several companies made gas there, originally by distilling coal to make coal-gas, and later by 'cracking' petroleum compounds such as naphtha. Three telescopic gasometers, or gas holders, stored the gas on-site until customers needed it, with sections that rose and fell guided by an iron skeleton.

Two gasometers were demolished in 1995: the Clayton (built in 1871, and named after John Clayton*) and the

Barrow Street's gasometer: destined to be incorporated into an unusual apartment complex.

© Enterprise Ireland

Dickens (1925). Only the 10-storey Alliance, built in 1884, survives. It measures 26 metres across, and has 24 ornate, cast-iron columns linked by girders and cross-bracing. Now a protected structure, it is to form part of an innovative housing development: the iron skeleton will be filled with a 9-storey apartment complex, and the underground gas storage drum will become a car park. Much of the surrounding land is 'made ground', reclaimed from the sea in the 1700s by Sir John Rogerson. Survey work for the new housing development recently uncovered a thick layer of old brick, cinder and rubbish, which had been used to reclaim the site.

The area between Ringsend and Poolbeg, and a twin district on the north quays, was also associated with shipbuilding and maritime work such as rope-making.* There were several **ropewalks** – long sheds where workers walked backwards twisting strands together to make new rope – as at Ropewalk Place beside Ringsend Park. **Windmill Lane** nearby, now famous as U2's recording studio, had one of the many windmills* that powered the port's industries in the days before steam.

PIGEON HOUSE

1903 – a powerful place

In 1903, Dublin Corporation built Ireland's first major power plant, and significantly the first in the world to generate **3-phase electricity,** at Pigeon House. The corporation had opened a small plant in the city at Fleet Street in 1892, having decided that electricity generating should be in public hands. The steam turbines and dynamos there produced electricity for a limited network of electric street lights,* which proved popular, being brighter, cleaner, simpler and safer than gas lamps. By 1899, demand was outstripping the 900kW that Fleet Street could generate, and the corporation decided to build a larger 3,000kW (3MW) station at Pigeon House. The remote location was initially ridiculed, but electricity is still generated there a century later. Pigeon House was the first station ever to generate 3-phase electricity – basically, using a more efficient generator, with three wire coils instead of one – and it was a model for stations around the world. Over the next 25 years, demand continued to grow, fuelled by the arrival of cheap and efficient filament bulbs; the station's capacity was increased, the network was developed, and the corporation built transmission centres and substations around the city. Then, in 1929, as Ardnacrusha* came on stream, the station was taken over by the new Electricity Supply Board and joined the infant national grid.

In 1971, the Pigeon House station closed, replaced with the new **Poolbeg station** which had been built alongside it on 36 hectares of newly reclaimed land (750,000 tonnes of sand were dredged from the bay to fill the site). The new plant's twin chimneys are Dublin's most distinctive landmark: 207 metres tall and visible from miles around. They were the tallest structures in Ireland until two 225-metre chimneys were built at Moneypoint power plant in Co. Clare in the 1980s, but they remain Dublin's tallest structures. Where the old station burned coal, the new one can burn oil and natural gas, but it mostly burns gas, increasingly from the North Sea via the pipeline from Scotland. It generates 900MW every year – 300 times what the original station produced in 1903. That station survives relatively intact, complete with ornate ironwork and the

remains of early generating equipment; a local group has ambitious plans to make it into a technology museum.

Pigeon House was named after John Pigeon who was caretaker of the port buildings there in the 1730s. A hotel was built in 1795 when the area was briefly a popular resort. It is now home to the Bolton Trust, an organisation that helps young entrepreneurs. The city's first **sewage treatment** plant was also built at Pigeon House; it opened in 1896 – before then raw sewage was pumped into the bay – and is still in use. The treatment plant was simple: sewage was allowed to settle in sludge beds; the liquid effluent was then discharged into the estuary, and the solids taken in sludge boats and dumped at sea. Sludge dumping ceased only in 2000, with the opening of a new plant that dries the sludge to produce compost.

Spot heights
– the Poolbeg datum

On April 8th, 1837, surveyors from the Ordnance Survey* recorded the low-water mark of the spring tide by chiselling four notches onto the outside of Poolbeg lighthouse. Their crow's foot mark – shaped so as to fit with surveying equipment – became the reference point against which all the heights for the rest of the country were measured in the survey's massive mapping project, which was then underway. Crow's foot benchmarks were hammered discreetly into walls every few hundred metres, enabling surveyors to map the terrain. They have been superseded by modern instruments and satellite technology, but can still be spotted on old walls, and on Poolbeg lighthouse at very low water. (They are no longer protected, and many were moved from their original position when walls were rebuilt.) The Poolbeg low-water mark, or Ordnance Datum (OD) Poolbeg, remained in use

The tallest structures in Dublin: the Poolbeg power plant's twin chimneys, erected in 1969–77, stand 207 metres high.
© ESB

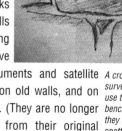

A crow's foot: surveyors no longer use these discreet benchmarks, but they can still be spotted on walls all over the country.
© Ordnance Survey of Ireland

until 1959 when the OSI switched to a new reference point, OD Malin Head.*

A telescopic tower

The Kish Bank is a great sandbank that lies just outside Dublin Bay. It has always been a major shipping hazard and, in 1811, the Ballast Board* anchored a lightship there. It was equipped with a gong that was sounded in fog, also a gun was fired in fog when the mailboat was due, and then at 15-minute intervals until an answering shot indicated the mailboat had safely passed. In 1841, the Ballast Board decided to erect a permanent lighthouse there, but this was a considerable engineering challenge because of the strong current, the fact that the bedrock lies buried under almost 100 metres of sand, and because the depth of water above the sandbank varies with the tide (from 16 metres at low tide to 21 metres at high tide). The board opted to use a new 'screw pile' technique developed for use in soft ground by Irish engineer, Alexander Mitchell.* Mitchell's screw piles had already been used successfully in several estuaries, but the open sea proved too much: the structure was only part built when a gale flattened it in November 1842. The project was abandoned – though the submerged wreckage is apparently still there, and occasionally snags a trawler's gear – and the lightship was retained.

In the 1960s, the development of new building techniques for offshore oil rigs prompted a fresh look at the idea of a lighthouse on the Kish Bank. Eventually, a novel Scandinavian design was chosen for a telescopic structure made in two parts: the base was a large tube or caisson, 13 metres in diameter; the tower was formed from a smaller caisson designed to telescope up. They were built in the shelter of Dún Laoghaire harbour,* starting in August 1963, and attracted much public interest during the 15 months they took to complete – there was even a viewing platform for day trippers. When ready, the caissons were floated out to a prepared site on the Kish Bank, where the base was sunk, the tower was then raised by flooding the inner caisson. Finally, the water was pumped out and the base filled with concrete and sand. One of the last Irish lighthouses built, it was equipped with the most modern equipment and a powerful light that can shine at up to 3 million candelas in fog. Unusually, it rises sheer out of the sea, and the only place where keepers could stroll was the helipad. The strength of the current at the site had earlier been measured by a simple, yet effective technique: molten jelly was poured into a bottle that was tethered to a brick and dropped to the seabed; from the angle at which the jelly set in the bottle, engineers could calculate the strength of the current.

The unusual telescopic Kish light, erected on-site in 1965, was the first of its type used in the open sea.
© Commissioners of Irish Lights

The Kish area also has Ireland's largest coal* deposit, discovered in the 1970s after test drills and seismic studies. There are 200 million tonnes of coal at two locations, the first about 8km off Greystones (45 million tonnes), and the second a little distance off to the northeast (155 million tonnes). The seams, some of them 4 metres thick, are in bedrock that is buried under 70 metres of sand. Underwater coal is not unusual and has been mined in Japan for over a century. In the 1980s, the Geological Survey of Ireland* considered mining the Kish coal, either via tunnels from the shore or from floating artificial islands; or gasifying the coal on-site and piping the gas ashore. The mines would have been productive for 20 years, but none of the options was economic at the time and, for the moment, the coal remains where it is. Feasibility studies are under way for a large wind farm on the Kish Bank, similar to that planned for the Arklow Bank.*

Chapter 1: between the canals

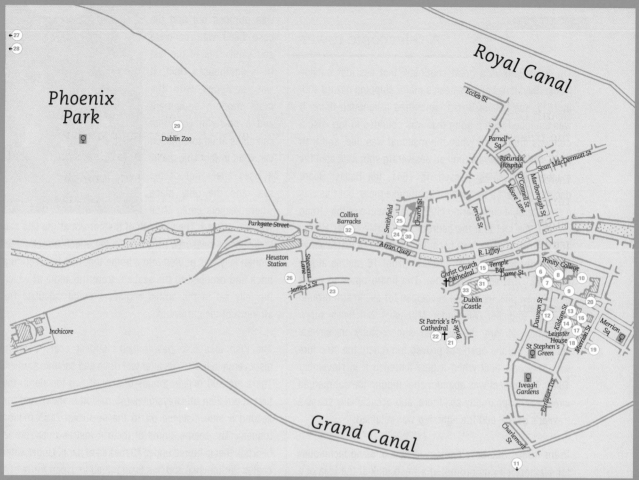

Visitor facilities

6 TCD Library
7 TCD Geology Museum
8 Centre for Civil Engineering Heritage
9 TCD Zoology Museum
10 TCD Chemistry Department
11 TCD Botanic Gardens, Dartry
12 Royal Irish Academy, Dawson St
13 Royal College of Physicians of Ireland, Kildare St
14 National Library of Ireland, Kildare St
15 National Photographic Archive, Temple Bar
16 National Gallery of Ireland, Merrion Sq
17 Natural History Museum, Merrion St
18 Government Buildings, Merrion St
19 Irish Architectural Archive, Merrion Sq
20 Apothecaries Hall, Merrion Sq

21 Marsh's Library, Patrick's Close
22 St Patrick's Cathedral
23 Guinness Storehouse, James's St
24 Dublin Brewing Company, Smithfield
25 The Old Jameson Distillery, Bow St
26 Dr Steevens' Hospital, Steevens' Lane
27 Phoenix Park Visitor Centre
28 Ordnance Survey of Ireland
29 Dublin Zoological Gardens
30 Michan's Church, Church St
31 City Hall, Dame St
32 National Museum, Collins Barracks
33 Chester Beatty Library, Dublin Castle

♀ woodland
🦆 nature reserve or wildfowl sanctuary

Visitor facilities are numbered in order of appearance in the chapter, other places in alphabetical order. How to use this book: page 9; Directory: page 474.

SOUTH DUBLIN CITY: FROM THE LIFFEY TO THE GRAND CANAL

South Dublin city is blessed with riches: a geology museum hidden in an attic (at TCD), the treasures of Marsh's Library and of the Natural History Museum, a mediaeval gold laboratory (at Dublin Castle), a large Victorian engineering works (at Inchicore), and institutions such as Trinity College Dublin, the Royal Irish Academy and the world-renowned (but little-known at home) Dublin Institute for Advanced Studies. Leinster House, former headquarters of the Royal Dublin Society, was the scene of world trade fairs, centre of a scholarly campus and home to a pioneering cancer treatment laboratory. The hypodermic syringe was invented at the Meath Hospital by Francis Rynd and George Francis FitzGerald, a TCD physicist, helped pave the way for Einstein's theories.

TRINITY COLLEGE DUBLIN
1592 – Ireland's first university

There were repeated attempts to start a college in Dublin from as early as 1311. Had they succeeded, Irish history might have been different: there would, for instance, have been contact with European centres of Renaissance thought – a movement that largely passed Ireland by – and a formal Catholic educational system would have been established. But it was not until after the Reformation that Ireland gained its first college: Dublin University, founded by Elizabeth I in 1592, and built beyond the city walls on lands confiscated from a disbanded monastery. Modelled on Oxford and Cambridge, it was to have had several colleges, but only one, Trinity College, was ever established.

The aim was to provide an educated, Protestant clergy for the established church in Ireland, and admission was restricted to Protestants and to Catholics who renounced their faith. The latter restriction was lifted in 1793 after the French Revolution, amid fears that Catholics forced to go to Europe for their education would acquire revolutionary ideas. (The Catholic hierarchy retained its ban on attendance until 1970, however, and Catholics wishing to attend needed their bishop's dispensation.)

TCD came into being during the relatively quiet years between the Munster Rising and O'Neill's Ulster Rebellion, but political events regularly disrupted its first century. Its revenue was the rent from the several confiscated estates it had been given, but the rent roll dried up during the 1641

The architecturally flamboyant Museum Building, with its colourful use of Irish marbles and building stones. Skeletons of two giant Irish deer* guard the hallway. Home to TCD's Geology Museum and the Centre for Civil Engineering Heritage.
© Trinity College Dublin

Kane,* the brothers George Johnstone and Bindon Blood Stoney,* and the physicist Ernest Walton,* who won a Nobel prize for his work on splitting the atom.

The college has the oldest and largest **library** in Ireland, with over 4 million volumes and an internationally important collection of early manuscripts, including the famous 9th-century *Book of Kells*. Other treasures include James Ussher's personal library of 10,000 early printed books, and numerous early atlases and astronomical, botanical and scientific books. The library also has a high-tech conservation laboratory. ⑥

Like many universities, TCD has several small museums and a herbarium of dried plants (see: **How sap rises**). The **Geology Museum** is a particular treasure with 50,000 fossils, 15,000 rock specimens and 7,000 minerals. The collection was assembled over centuries, much of it by Samuel Haughton and John Joly, and features all the major Irish fossils – notably Oldhamia* (at 540 million years old, the oldest fossils in Ireland); 355-million-year-old tree ferns

Rising, and the college stopped taking students until the Cromwellian regime a decade later. Things came to a standstill again in the 1680s, with the rise of the Catholic King James II, when most Protestant settlers left, returning only after William of Orange won the Battle of the Boyne.

From the outset TCD offered courses in subjects as varied as divinity, Hebrew, mathematics and chemistry – the latter to would-be physicians. In 1711, it opened Ireland's first school of physic (medicine) and in 1783, founded Dunsink Observatory.* Among the many great scientists and thinkers associated with TCD were James Ussher* (who famously calculated the age of the Earth from an analysis of the Old Testament); the philosopher George Berkeley;* geophysicist John Joly,* who invented an early technique for colour photography; Samuel Haughton,* who invented the humane hangman's drop; and the great 19th-century physicists and mathematicians William Rowan Hamilton,* George Gabriel Stokes* and Robert Ball.* The innovative engineers Robert Mallet,* Howard Grubb* and Charles Parsons* all studied there, as did the chemist Sir Robert

Protestant and Catholic science

Trinity College always had a strong scientific tradition and most of the big names in Irish science over the centuries were associated with the college in one way or another. Restrictions on Catholic education, and the fact that the Royal Irish Academy* and Royal Dublin Society* were strongly associated with the Protestant Ascendancy tradition, meant Irish science was, for almost 300 years dominated by a Protestant elite, many of whom were English born or of English extraction. (There were a few notable exceptions, such as Nicholas Callan* and Sir Robert Kane.*) Even in the late 1800s, only 10 per cent of Irish scientists were Catholic; the rest were Anglo-Irish, or British-born professional scientists who came to work for government institutions such as the Ordnance* and Geological Surveys.*[4] Various theories have been suggested to account for this imbalance: that the Irish were incapable of rational scientific thought, for example; that the Catholic Church actively discouraged science; that the colonial powers wanted it that way; and that Irish nationalism was a Romantic movement and anti-science. The main reason, however, is probably that Catholics were given little opportunity or encouragement. For similar reasons, there were also few women scientists then – TCD did not open its doors to women until 1904.

from Kiltorcan;* 320-million-year-old amphibians from Jarrow* coalfield; and footprints left by an early reptile that walked across mud at Scrabo* 230 million years ago. There are also dinosaur eggs from the Gobi desert; lead, zinc and copper ores from Irish mines; a lump of a meteorite* that fell in Co. Tipperary in 1865; zeolites* from Co. Antrim; and (industrial) diamonds* from Shannon. ⑦

The Museum Building also houses Ireland's **Centre for Civil Engineering Heritage**. ⑧ Civil engineering dates from 1760 – before that engineering* was a military discipline – coinciding with the start of canal building in these islands. The centre's database features nearly 300 sites of Irish engineering heritage, from aqueducts and viaducts, to windmills and reservoirs; there are photographs, plans and detailed information for each site. A small collection of historic surveying instruments and computational equipment, dating from 1700, includes a spiral slide rule patented in 1912 by Englishman Walter Lilly, who joined the college as a lecturer in 1891; being spiral, it was longer and therefore more precise than a conventional rule, but no more cumbersome. There are also various devices designed by another college engineering professor, Scotsman Thomas Alexander (1847-1933), whose elliptographs enabled engineers to draw more accurate diagrams. The foyer of the Parsons Building, home to the department of mechanical engineering, has more engineering heritage: a small exhibition of steam engines, including the world's third-oldest, steam-powered turbo-generator. Designed by Charles Parsons,* who invented the steam turbine, it is a small portable generator that was used for demonstrations.

Extinct species, such as the Tasmanian wolf, passenger pigeon, giant deer,* and the last great auk found in Ireland, can be seen in the **Zoology Museum**. The collection of over 30,000 specimens, started in 1876, also features 12,000 Irish and tropical insects, marsupials and platypuses, and exquisite glass models of marine animals handmade in the 1860s by the famous German Blaschka* family. ⑨

The first modern 'scientists' in Ireland?

You could say that the Dublin-born Molyneux brothers, William and Thomas, were the first modern scientists in Ireland. The word 'scientist' was not coined until the 1840s, derived from the Latin *scientia* meaning knowledge, and replacing 'natural philosopher'. Yet, in the late 1600s, William and Thomas Molyneux eagerly embraced the new ideas emerging from Europe, and the emphasis on scientific experiments. **William Molyneux** (1656–98) was multi-talented: engineer, surveyor, civil servant, barrister, author, meteorologist (the detailed weather records he kept, including barometric pressure readings, are the earliest known for Ireland), inventor (he designed a sundial-mounted telescope), politician (he was MP for TCD in the days when the college was represented at parliament), and scientist. Significantly, he was instrumental in founding the Dublin Philosophical Society,* which introduced modern scientific ideas to Ireland.

When Molyneux's young wife Lucy became blind shortly after they married, it prompted him to propose a philosophical question that became celebrated as **Molyneux's problem**: in 1693, he asked his friend, Scottish philosopher John Locke, whether a man who was born blind and had learned to distinguish a sphere and a cube by touch, would, if he gained his sight, be able to identify the objects without touching them. The question raised issues about perception, knowledge and the brain. Empiricists like Locke and Molyneux said No, arguing that all our knowledge of the external world comes from our experience. Rationalists, believing that we are born capable of reason, said Yes. Molyneux's problem was not resolved until 1980 when an unusual patient at Bristol University was able to prove the Rationalists right – and Molyneux wrong. Physicists remember Molyneux as author of the first English treatise on optics, *Dioptrica Nova* (1692). Irish historians remember him for arguing that England had no right to legislate for Ireland: his book *The Case of Ireland...* *Stated* was denounced in parliament as dangerous and burned by the hangman outside Parliament House (now Bank of Ireland, College Green). William Molyneux is buried at St Audoen's Church in Dublin.

Thomas Molyneux (1661–1733), the 'father of Irish medicine', studied at TCD and on the continent, before returning to TCD as professor of physic. He was interested in medicine, natural history, geology and antiquities. Research techniques then were relatively primitive, however – he studied the 1693 flu epidemic, for instance, by monitoring coughing among church congregations. He wrote the first scientific report on the Giant's Causeway* – from the absence of mortar he concluded the formation must be natural – and the first detailed description of the giant deer.* Thomas Molyneux also wrote on round towers and coal mines, the mammoth* unearthed in Co. Monaghan in 1715, and the human remains found at Dunmore* Cave in Co. Kilkenny.

Humphrey Lloyd's magnetic observatory: opened in 1838 in TCD, it was iron-free. To make way for a new arts block, the building was moved to UCD's Belfield campus in 1974.

Measuring the Earth's magnetic field

In the 1830s, a network of observatories around the world began monitoring the Earth's magnetic field, and the standard instrument they used was an Irish one: invented by Dublin scientist **Humphrey Lloyd** (1800–81), who was professor of natural philosophy (science) in TCD. Little was known then about the puzzling phenomenon that is the Earth's magnetic field – some people even thought there were four magnetic poles. Interest was sparked by scientific curiosity, as the magnetic field is constantly shifting as the Earth's iron core gyrates, and by practical necessity, as the changes have implications for navigators. By 1832, Sir John Ross had discovered the north magnetic pole in Arctic Canada, and a German scientist, Carl Friedrich Gauss, had predicted the south magnetic pole would be found in Antarctica at 66° south. Gauss had also

invented an instrument for measuring the strength of the Earth's magnetic field and laid the groundwork for scientific studies of the phenomenon.

In 1833, Humphrey Lloyd invented an instrument that was a considerable improvement on Gauss's. Using a suspended magnetic needle that was free to rotate, it was very sensitive, accurate and, significantly, could be used to measure both the strength of the Earth's magnetism and its dip (an aspect of its direction).[5] In 1834, Lloyd, the Irish naturalist–explorer **Edward Sabine** (1788–1883) and the polar explorer James Clark Ross, used the new instrument to survey Ireland, and subsequently Scotland, England and Wales. Measurements had to be taken away from any large source of iron and, so as not to disturb it, the instrument was read from a distance through a telescope. Lloyd invented six further instruments, including a sizeable device called a balance magnetometer for monitoring changes in the magnetic field's dip, which became the international standard.

In 1838, Lloyd opened a magnetic observatory in TCD, where measurements were taken round the clock in a special iron-free building. Russia and several European countries had established similar observatories and, in 1839, Britain joined this growing network, opening observatories in its colonies in India, Singapore, Africa and Australia. It was a major international collaboration with 33 observatories in all. Lloyd equipped the British ones, with instruments made in Dublin by Thomas Grubb,* and trained the observers at TCD; many of the other observatories used Irish instruments too. James Clark Ross also led an expedition to survey Antarctica (1839–43), but could not reach the inaccessible south magnetic pole – it was not reached until 1909, when it was at 72° south on Victoria Land; since then it has moved over 800km.

Lloyd's other great success was a delicate optical experiment he performed in 1832 using a special biaxial crystal (with two optical axes), which confirmed William Rowan Hamilton's* prediction of conic refraction; the result was a major boost to the wave theory of light. He is also remembered for a classic mirror experiment in 1834, which showed that reflected light could be made to interfere with light from its source. He died of dysentery and is buried in Mount Jerome.*

Physicist-powered flight: George Francis FitzGerald trying (unsuccessfully) to fly a glider in the TCD sports grounds. Note the gentlemanly top hat. A crater on the Moon is named after him.
Courtesy Trinity College Dublin

Radio waves, ether and dirty snowballs

The first person to suggest the possibility of radio waves was Dublin-born **George Francis FitzGerald** (1851–1901), who was professor of physics at TCD from 1881 until his death. FitzGerald was one of the great ideas men of the late-19th century, who influenced the development of electromagnetism and radio, as well as thinking in several areas: he suggested one of the cornerstones of Einstein's theory of relativity (the FitzGerald contraction), was the first to propose that nothing could travel faster than the speed of light, and speculated that comets were dirty snowballs, made of ice and gravel.

FitzGerald was immersed in science from an early age: the physicist George Johnstone Stoney* was his maternal uncle; and he was tutored at home by a sister of George Boole.* FitzGerald was an early champion of James Clerk Maxwell's radical ideas about electricity and magnetism, and it was in correspondence with Maxwell that FitzGerald

suggested an oscillating electric current in a Leyden jar would produce radio waves. In 1888, he was proved right when Heinrich Hertz in Germany generated radio waves in this way. FitzGerald brought Hertz's results to public attention, and Irish-Italian inventor Guglielmo Marconi* quickly exploited the new discovery for his wireless telegraphy. Marconi and FitzGerald later collaborated on the first sports radio report – a commentary on the 1898 Kingstown [Dún Laoghaire] regatta.* A great scientific debate then was about the existence of 'the ether'. Just as sound waves need air to travel, so it was thought light waves need an ether, though no one had ever detected it. But if the ether existed, then the Earth moving through it would generate an 'ether wind' and alter the speed of light. Light coming towards the Earth from the direction in which the ether wind is blowing, for instance, would appear to be travelling faster than light moving away. In 1887, two US physicists, Albert Michelson and Edward Morley, tried to measure this effect. Puzzlingly, they found that the speed of light was constant in every direction. In 1889, FitzGerald wrote to the US journal *Science* with a radical proposal to explain their result: that objects moving through the ether would shrink in the direction they were moving – a bit like a boat pushing through water is compressed by the water – and the faster they moved the more they would shrink. FitzGerald knew that electrical forces were affected by motion, and thought the same could happen to matter.

Measuring instruments would themselves shrink and so would never detect the effect, and the speed of light would appear constant. A Dutch physicist Hendrik Lorentz had the same idea independently in 1892, and worked out the possible effect in greater detail. FitzGerald was pleased to learn of Lorentz's contribution, telling him in a letter: "I have been rather laughed at for my view over here." Their idea is now called the Lorentz-FitzGerald contraction, and was proposed as a way of 'rescuing' the ether. Ironically, in 1905, Albert Einstein's theory of relativity did away with the need for an ether, but not with the Lorentz-FitzGerald contraction: in the relativistic world, moving objects really do contract in the direction of their motion, though the change is detectable only at speeds approaching the speed of light. It is like saying that a moving boat shrinks even in a vacuum, purely because it is moving.

A great frustration with FitzGerald is that he published very little, and many of his insights were never developed. For instance, he was probably the first person to suggest that nothing could travel faster than the speed of light, which he proposed in a letter to physicist Oliver Heaviside in February 1889: "What if the velocity be greater than that of light? I have often asked myself that but got no satisfactory answer. The most obvious thing to ask in reply is, is it possible?" FitzGerald was also first to explain why a comet's tail always points away from the Sun – solar radiation blows it back – and, in 1892, he helped conduct the first-ever electrical measurement of starlight* at Earlsfort Terrace. Generous in championing other people's ideas, FitzGerald advocated women's education and technical education, helping to establish Kevin Street College of Technology (now part of the Dublin Institute of Technology). He was also Unionist, anti-Home Rule and hostile to Catholics, once writing that he thought their beliefs and superstitions prevented them "from rational activity almost as much as the fetid worship of an African savage". FitzGerald died of a perforated ulcer at 49, and is buried in Mount Jerome.*

How to dry alcohol

Distilling is an ancient Arabic technique for separating and purifying liquids, but much of our modern understanding of the process is due to **Sydney Young** (1857–1937), an Englishman who was professor of chemistry at TCD from 1903–28. By 1900, there was growing industrial interest in distillation, particularly to extract paraffins from raw petroleum, and Young, who had studied the effects of heat on solids and liquids, became a world authority on distillation. Significantly, he invented a novel and efficient way of separating water and alcohol (ethanol) that, bizarrely, entailed adding a third substance to the mix: Young found that adding a small amount of benzene to the water and ethanol – creating what chemists call an azeotropic solution – made it easier to distil the ethanol, yielding a nearly pure 'absolute alcohol'. Young also discovered that ice kept at low pressure will not melt when it is heated; this phenomenon is now exploited in freeze-drying techniques. The chemistry department, founded in 1711 as part of TCD's school of medicine, has a small collection of chemical instruments, including some of Young's intricate hand-blown glass distillation columns. ⑩ One of Young's predecessors there was Limerick man **James Apjohn** (1796–1886) who devised a formula (Apjohn's formula) for calculating the dew point – the temperature at which a given sample of air will deposit dew if it is in contact with the ground. As part of this work, Apjohn invented the wet and dry bulb thermometer, for calculating the relative humidity of air; it was made in Dublin by Mason's* and sold around the world.

How sap rises

The first person to explain how sap rises in tall trees was Dubliner **Henry Horatio Dixon** (1869–1953), who was TCD's professor of botany for nearly 50 years. Some trees are over 100 metres tall and, until Dixon's work, it was believed cells living in the tree trunk pumped the sap up from root to leaf. But, in the 1930s, Dixon saw that sap rose even when these stem cells were dead. With his colleague John Joly,* Dixon suggested that the sap was actually pulled up from the roots, drawn by the evaporation of water from the leaves. Using elegant apparatus, which they made themselves, and several ingenious experiments, Dixon and Joly showed that the cohesion or stickiness of a column of water (or sap) was greater than previously thought, and could indeed withstand being stretched to the top of the highest tree. It took 10 years to convince the sceptics, but Dixon's theory is now accepted. Dixon may have inherited his ingenuity from his mother, Rebecca Yeates,* of the noted Dublin family of instrument makers.

The botany department has important collections of dried plants (its herbarium), living plants (its botanic garden), and frozen plants (the national seed bank). The **herbarium**, begun in 1840, was developed by noted Limerick botanist **William Henry Harvey*** (1811–66). The dried specimens, collected around the world and carefully preserved in special files, form a botanical library, archive, research resource and museum all in one. Thanks to Harvey's explorations in Australia, North America and South Africa, his hard work – starting in the herbarium at 5am each day – and friendship with scientists such as Charles Darwin, the collection quickly grew. It now boasts over 250,000 specimens from every continent, including Antarctica, with more deposited each year. It is Ireland's second-largest herbarium, after the National Botanic Gardens.* A highlight is the specimens collected by Caleb Threlkeld,* which are the oldest Irish plant specimens known.

In 1687, the college planted a 'physic garden' for medical students, growing medicinal herbs on what is now the college tennis courts. In 1806, the garden, by then a general **botanic garden**, was transplanted to Ballsbridge, a site now occupied by Jury's Hotel, where some of the garden's trees survive. In 1967, the garden was again transplanted, to its current location in Dartry. ⑪ It is primarily a research facility, but it also hosts the **national seed bank collection**. Run in conjunction with the Wildlife Service, Department of Agriculture and Irish Genetic Resources Conservation Trust, this stores the seeds of endangered species. Already, seed from 60 species has been collected, dried and frozen; some is kept at Dartry, and some is stored overseas as a backup. The team is concentrating initially on rare species and on unique Irish strains of crops, such as the thatching grasses of the Aran Islands. The seed bank may ultimately be used to re-introduce species that become extinct here.

A cure for leprosy

In the 1940s, a team of Irish scientists began looking for a chemical cure for tuberculosis. They failed in their task, but they did find a cure for leprosy, one that is still recommended by the World Health Organization. The notion of treating diseases with chemicals really began when German chemist Paul Ehrlich (1854–1915) discovered arsenic compounds that killed the syphilis parasite. In 1943, the Irish **Medical Research Council** (MRC) wondered if Ehrlich's 'magic bullet' approach would work against TB. A team led by Cork-born **Vincent Barry** (1908–75), spent years making and testing thousands of chemicals, working initially in laboratories at UCD and later at TCD. Many compounds looked hopeful initially, but none was effective against TB. Fortunately, antibiotics had become available by then. One compound, however, a red dye they labelled B663, was tested on the bacterium that causes leprosy, since it is closely related to the tuberculosis bacterium. The results were spectacular, and 22 countries asked to join the clinical trial. In 1981, the new Irish drug, then called Clofazimine, won the UNESCO prize for science. A modified form of a dye originally extracted from a lichen, it is now being tested as a cancer treatment and to promote wound healing. During the 1980s, the MRC was replaced by the Health Research Board, which funds medical research, including acclaimed studies on using folic acid to prevent

A touch of the Renaissance

Ireland's first real brush with the Renaissance came in 1683, when several Protestant clerics and Trinity College academics founded the Dublin Society for the Improving of Naturall Knowledge, Mathematicks and Mechanicks or, in short, the **Dublin Philosophical Society**. The prime mover was Sir William Petty* who, in 1660, had helped Robert Boyle* set up the Royal Society in London. The Dublin society, Ireland's first scientific institution, was designed along similar lines, and was influenced by Francis Bacon and the new ideas about scientific experimentation. Others associated with it were William Molyneux,* Narcissus Marsh* and George Berkeley.* They opened a laboratory above an apothecary's premises on Crow Street, assembled a collection of scientific instruments and books, and planted a herb garden, and Petty, a great believer in the new scientific method, encouraged members to conduct experiments. One of its first activities was to survey Ireland's natural history and resources, with contributions from around the country – Roderic O'Flaherty,* for instance, surveyed west Connaught – but the work was never completed. For these were troubled times, and the society disbanded in 1687 with the rise of the Jacobites, when many prominent Protestants fled to England. A second phase in 1693–98 also ended abruptly, when Molyneux's controversial book *The Case of Ireland... Stated* was publicly burned by the hangman; the society re-formed in 1706, but finally disbanded two years later. It was, perhaps, ahead of its time. Despite a disrupted existence, it inspired the formation of the **Royal Dublin Society** (RDS)* and the **Royal Irish Academy** (RIA) which both proved more enduring.

The academy was established in 1785 to promote the study of "science, polite literature and antiquities", under the motto: "We Will Endeavour". English engineer Richard Lovell Edgeworth,* who had just inherited a Longford Estate, was instrumental in starting it; members were drawn from among the academics, nobility and parliamentarians; and the Earl of Charlemont was elected first president. One of its first activities was to spend £30 on barometers and thermometers for a network of weather stations around the country. Projects organised since then included dredging expeditions to Rockall* in 1896, and a comprehensive bio-geographical survey of Clare Island* in 1909–12, which is currently being revisited (1992–). The academy quickly established a reputation, and its roll call of members included the top scientists and thinkers in 19th-century Ireland – William Rowan Hamilton,* Richard Kirwan,* Thomas Romney Robinson* and Sir William Wilde* were among those taking an active part. It was in the academy's proceedings that Hamilton published his invention of quaternions, for instance, and where Robert Mallet* outlined his ground-breaking 1846 thesis on seismology.

The academy met at 114 Grafton Street until 1851, when it moved to its present home on Dawson Street. Thanks to its many benefactors and activities, it amassed an important library and collection of antiquities. The antiquities were given to the nation in 1890 to start what would become the National Museum of Ireland,* but the academy retained its library of ancient manuscripts, early books, pamphlets and other documents. ⑫

Its antiquities work attracted Nationalists, and so the academy was not associated as strongly with the

Ireland's oldest scientific instrument?

Around 1415, Aedh Buidhe, of the Fermoy branch of the scholarly O'Leighlin clan, produced an Irish translation of an astronomical text written by an 8th-century Arabic scientist Masha'-allah of Alexandria. In addition to the original 14 chapters on heavenly bodies, O'Leighlin added another 13 explaining why the Earth is not flat, and dealing with the origin of fossils, earthquakes and tides. His 12 tightly written vellum pages are now in the RIA's library. On the front of the manuscript is a rotula: a circular paper device, 14cm across, which was used to demonstrate the movement of heavenly bodies and to calculate the Sun's position in the zodiac for any day of the year. It is arguably Ireland's oldest scientific instrument.

Ascendancy tradition as was the case with the RDS that, with its emphasis on agricultural improvement, was dominated by Protestant landowners. Nevertheless, the academy's survival was threatened in 1922 by a split between Nationalist and Unionist academics, mirroring the contemporary pro- and anti-Treaty divide that was driving the country to civil war. This culminated, in May 1922, with the launch of a rival **National Academy of Ireland** (*Dámh Ioldánach Éireann*), which had 100 members, a grant from the Free State government and five divisions: Irish language and literature; mathematics and physics; philosophy and history; anthropology and biology; and fine arts. Among the principals were two UCD professors, Eoin MacNeill (history) and Arthur Conway (mathematics). The start of the Civil War prevented any real development, however, and the NAI disbanded in November 1923.

The academy, now funded by the Higher Education Authority, still publishes journals and books, organises symposia and conferences, convenes 20 scientific committees, and liaises with sister institutions abroad, such as Edinburgh's Royal Society which dates from the same time. Honorary membership has been awarded to people such as Einstein, Pavlov, Heisenberg, Humboldt and Niels Bohr; Scottish science writer Mary Somerville was elected an honorary member in 1834, but it was not until 1949 that botanist Phyllis Clinch became the first woman elected to ordinary membership. Currently there are some 300 members – eight new ones are elected every St Patrick's Day, four each in the sciences and the humanities.

DAWSON STREET

"Number, please"

Ireland's **first telephone line** connected the Gresham Hotel in O'Connell Street and Maguire's Hardware in Dawson Street. It opened in January 1878 – just 22 months after Alexander Graham Bell made the first voice transmission – and both companies invited customers to witness the spectacle of the newfangled device. Theirs was a dedicated line linking the two premises yet, within a year, switching exchanges had been developed, enabling telephone companies to connect all subscribers to each other. The next 100 years would bring more changes: the telephone is no longer an expensive service reserved for the head of the household; automatic systems replaced operators and manual exchanges; fibre optic cables are replacing copper wire; analog transmission is giving way to digital (the information coded not as current, but as on/off pulses of laser light); microwave distribution and satellite systems circle the Earth; mobile phones are everywhere; call charges are a fraction of what they were; and emigrants have a direct link to home.

The United Telephone Company opened Ireland's first switching exchange in Dame Street in 1880 with five customers. Business was slow, and the youth employed to operate the switch played marbles out of boredom and was quickly dismissed. Women, it was decided, would be more pleasant, careful and easier to train and, in 1881, Agnes Duggan became Ireland's first female telephone operator. Business increased 50-fold in the next year, to 271 subscribers – enough to merit publishing a list of their

An early Irish telephone, c.1900: turn the handle to summon the operator. Part of the Eircom collection of apparatus, now held by the National Museum at Collins Barracks; * *the National Photographic Archive (National Library of Ireland) has the telecom photographic archive. © Eircom/National Library of Ireland.*

names, subscriber numbers and addresses. The first Irish telephone directory, it came complete with instructions:

"(1) To call the Exchange, press the button three times and wait until your bell rings.
(2) Take your Telephone off the hook and place it firmly against your ear. Tell the Exchange the number of the Subscriber you wish to speak to…"

In 1884, the first Irish trunk line was laid, linking Dublin and Belfast; by 1893 Dublin, Belfast, Cork, Limerick, Dundalk, Drogheda and Derry had a telephone service between 9am and midnight. Subscribers could now also phone Britain, thanks to a submarine cable between Donaghadee* in Co. Down and Portpatrick in Scotland. By 1900, Ireland had 56 exchanges, though only Dublin, Belfast and Cork had a 24-hour service. Telecommunications were nationalised in 1912 and taken over by the Post Office. By the end of World War I, there were 12,000 telephones in Ireland – though there were none as yet in counties Mayo, Leitrim or Roscommon.

The country's first public telephone box was installed in Dublin's St Stephen's Green in 1925 (an original box was re-erected on Dawson Street and is still in use), and direct dialling for local calls became possible in 1927 with a new automatic switch at Dublin's Ship Street exchange. Subscribers were assured there was no extra cost and that connections would take just five seconds:

Pipes of light

Beams of laser light transmitted down fibre optic cables are crucial to modern digital telecommunications. But the first person to beam light down a pipe was noted Carlow-born scientist **John Tyndall**.* In 1851, he shone light into a stream of water as it left a tap, and demonstrated that the light was 'trapped' within the water stream, emerging unaltered at the far end (the beam of light is internally reflected within the pipe of water). With great foresight, Tyndall suggested this seemingly simple phenomenon might one day be useful, though he would surely be surprised to know it is used to send information and conversations along very thin, glass fibres.

"You first lift the receiver and listen. You will hear almost at once a sound, this is known as the dialling tone and is the automatic language for 'Number, please'. You can dial a five-digit number in about seven seconds. Please do not to try to improve on the speed by forcing the dial back. It does not like it and will give you a wrong number."

Party lines were used in thinly populated rural areas: each subscriber sharing the line was assigned a ringing signal; but this did not necessarily stop other parties from eavesdropping.

Larger towns gained subscriber trunk dialling (STD) during the 1950s enabling people to dial some long-distance numbers directly. The first transatlantic telephone cable was laid in 1956 (100 years after the first transatlantic telegraphic cable*) and people could now phone North America – until then, all transatlantic communication was by telegraph, radio or mail. Five per cent of Irish people had a telephone by 1960, compared with 40 per cent in the USA. Irish people also used their phone less: making on average two calls per day, compared with four in the USA. During the 1980s, Ireland's telephone network was modernised and, in 1987, the country's last manual exchange closed at Mountshannon, Co. Clare. The next decade saw the country's networks and telecom services deregulated and privatised, passing from a State company, Telecom Éireann, to private companies, such as Eircom and Esat. After a century of developments, some things came full circle.

1654 – the start of Dublin's modern medical system

The first half of the 17th century was a bad time to be sick in Ireland. The monasteries, which traditionally cared for the sick, had been closed by Henry VIII, and the once powerful medical clans had disappeared. The only medical institutions left were leper colonies – such as the one outside Dublin's city walls near modern-day St Stephen's Green. The situation began to improve in the 1650s, with the arrival of a few college-educated physicians. Unlike the old order, these were Protestants who had studied at Dublin's new Trinity College, or come to Ireland with Cromwell; most had also trained abroad in at least one European medical school. In 1654, these men formed a fraternity of physicians, later the **Royal College of Physicians of Ireland** (RCPI), an elite body, initially open only to graduates of Trinity, Oxford and Cambridge, that was given the right to control the city's medical practitioners – even today Irish hospital consultants must join and pass its exams. The founders included **John Stearne** (1624–69), professor of medicine, Hebrew and law at TCD; Dubliner Thomas Molyneux;* the ubiquitous William Petty,* physician to Cromwell's army in Ireland and champion of the collection of mortality statistics, such as cause of death; and Scotsman **Sir Patrick Dun** (1642–1713). Dun came to Ireland in 1657 as physician to the Duke of Ormond, and later served William of Orange, attending the wounded at the Battle of the Boyne. He left his considerable fortune to support medical training and, in 1816, TCD used part of this to fund the first purpose-built Irish teaching hospital, named Sir Patrick Dun's in his honour.

All those associated with the golden years of Irish medicine in the 19th century – men like Robert Graves* and William Stokes* who pioneered radical new approaches to medical care – were members of the RCPI. Today, the college's concerns are the same as in 1654 – medical standards and doctor education – and from its elegant Kildare Street headquarters (built in 1864) it organises postgraduate courses in occupational medicine, public health, paediatrics, and obstetrics and gynaecology. ⑬

'For the gratuitous vaccination of the poor'

For centuries, **smallpox** was a major scourge. The early Irish annals record outbreaks in the 6th century, and in the late 1600s the disease accounted for 20 per cent of all deaths in Dublin, according to mortality statistics compiled at William Petty's* suggestion. Victims not killed by the disease were left disfigured and frequently blind. Inoculation, introduced to England from Constantinople in 1721 by Lady Mary Montague, may have been brought to Ireland by itinerant healers. In 1784, the practice was sufficiently established for the city's foundling hospital in St James's Street to inoculate its young charges. Vaccination – a more refined form of inoculation developed by English doctor Edward Jenner – was introduced in 1800. Four years later, the **Cow-pock Institution** was established at Sackville [O'Connell] Street for "the gratuitous vaccination of the poor, and for supplying all parts of the country with genuine matter of infection" (for use in preparing vaccines). Within 10 years, it had treated 30,000 people at its Dublin clinics, and distributed 27,000 "packets of infection" to physicians and others around the country. The widespread vaccination programme helped to tame smallpox: by 1841 the disease accounted for only 5 per cent of deaths, possibly contributing to the dramatic rise in Ireland's population pre-Famine. The Cow-pock Institution closed in 1877, when its activities were taken over by hospitals and dispensary doctors; its records are held by the RCPI. The World Health Organization eradicated smallpox in the 1970s, the first infectious disease ever eradicated.

Stamp issued in 1978 to mark the global eradication of smallpox.
© An Post

During the 1700s, the RCPI's members established private medical schools and several hospitals, some of which survive, laying the foundations for Dublin's modern health system. This culminated in the golden age of Irish medicine in the 1830s, centred on the Rotunda* and Meath* hospitals. Among the first of the new hospitals was the City Workhouse and Foundling Hospital which opened in 1704 on the site of today's St James's Hospital. Over the next 50 years, nearly a dozen more opened, all Protestant run and voluntary, dependent on donations and fundraising. Among them were: the Charitable Infirmary, Jervis Street (1718–1987); Dr Steevens'* (1733–1985); Mary Mercer's Hospital, on the site of Stephen's Green leper colony (1734–1983), which now houses the RCSI* library; the Rotunda (1745–); St Patrick's* Hospital for the Insane (1749–); and the Meath, which opened in 1753 in the Coombe district, moved in 1822 to Heytesbury Street, and is now part of Tallaght Hospital. A number of regional infirmaries were founded towards the end of the 1700s, as in Cork and Limerick. During crises, such as the 1826 typhoid outbreak and the deadly cholera* epidemic of 1832, these city hospitals were in the firing line.

Not until the rise of a Catholic middle-class in the 19th century were Dublin's first Catholic hospitals begun, notably St Vincent's, by the Sisters of Charity (1834–), and the Mater, by the Sisters of Mercy (1861–). Outside of the cities, there was little access to medical care, however, and during the fever epidemics of the Famine years in the late 1840s, the only medical care available to most was at the workhouse. But this prompted the creation of a national dispensary system and, by 1852, there were 700 district dispensaries around the country, each with a salaried midwife and medical officer. It was the start of a national health service.

RDS – a great improving society

A cluster of scholarly institutions encircle Leinster House because, from 1815–1922, it was the headquarters of the **Royal Dublin Society**. This private society undertook initiatives that in other countries were often the responsibility of government, and its great legacy is the institutions it founded or helped establish: the National Botanic Gardens,* National Library, Natural History

'The Dublin Method'

From 1914–22, Leinster House was home to an unusual scientific laboratory, the **Dublin Radium Institute**. As soon as radioactivity was discovered in the 1890s, it was realised the phenomenon could both cause and kill tumours. Early cancer treatments used radium itself, but this was expensive, highly radioactive – and therefore unsafe – and ineffective. A safer, cheaper, more effective technique was invented in 1914 by John Joly,* professor of geology at TCD, and Walter Stevenson, a cancer expert at Dr Steevens' Hospital.* Instead of using radium, they used the radioactive radon* gas, or emanation, which is given off by the radium, collecting the gas in sealed needles so it could be injected directly into a tumour. Joly provided the scientific theory, and Stevenson the practical techniques, and the pair probably began experimenting with the procedure around 1904. Their technique, the basis for the modern radium needle treatment, became known worldwide as 'the Dublin method'. It was safer for both doctor and patient, and more effective because the gas could be injected at several sites in a tumour and percolate through it. It was also cheaper because it did not consume the original radium source, which could be re-used. At Joly's suggestion, the RDS established the Radium Institute at Leinster House. A 200mg radium source was bought at a cost of £3,000 with public and RDS funds, and used to provide radon for cancer units around the country. In 1920, the Medical Research Council in England loaned 200mg of precious radium – originally intended for use in painting luminous gunsights on World War I weapons – in recognition of Stevenson's work in treating war wounds; in 1930, additional radium was bought after an appeal raised £1,000 from the public, and donations of £1,000 each from Lord Iveagh and Sir John Purser Griffith.* The facility moved to Ballsbridge with the RDS in 1922, and continued until 1952 when the Radiological Institute was established at St Luke's cancer hospital, at which point the radium source on loan from England was returned.

Leinster House: the Royal Dublin Society's headquarters from 1815–1922, around which were established the National Gallery, National College of Art, Natural History Museum, National Library and National Museum. A statue of Science inside Leinster House is a relic from the RDS era.

Phoenix Park. A year later, it acquired various agricultural implements, which it kept in a room at the House of Parliament (now Bank of Ireland at College Green), where the public could inspect them, and from this small beginning its museum grew. By the 1740s, the society was offering grants for various activities, such as reclaiming bogs, and had started a drawing school (later the National College of Art and Design) to foster improved product and industrial design. Eminent scientists were employed to conduct research and give public lectures, among them Richard Kirwan,* William Higgins,* Robert Kane* and Edmund Davy.* Various projects were funded, sometimes in association with the Royal Irish Academy,* such as marine research expeditions.

Museum, National Gallery, National College of Art and Design, National Museum, Geological Survey of Ireland,* Veterinary College, UCD's science faculty, even the craft council, cancer radiation units, and marine research programme. Most would eventually have been established anyway, but thanks to the RDS they came into being sooner rather than later.

The society began in 1731 when a dozen eminent citizens founded the 'Dublin Society for Improving Husbandry, Manufactures and Other Useful Arts'. A week later, they had second thoughts, adding 'and Sciences' to the title. (The society was not 'Royal' until George IV became its patron in 1821.) The founders were prompted by what they saw as the relatively primitive agricultural and industrial practices then in use and, from the outset, organised award schemes, lectures and other activities aimed at disseminating useful information. The society is still in existence and little changed, making it one of the oldest of its kind in the world – the 18th-century English agrarian reformer Arthur Young described it as "the father of all similar societies now existing in Europe".

By its second year, it was conducting experiments on flax and grasses, and organising trials of new ploughs in the

In 1790, the society began developing its botanic gardens at Glasnevin* and, in 1792, it bought a substantial collection of over 7,000 minerals, plus numerous shells, fossils, insects and dried botanical specimens. Known as the Leskean collection, after its German owner Dr Leske, it cost the then princely sum of £1,200, which was provided by parliament. This was the foundation of its natural history museum, a collection that quickly grew: Richard Griffith* was paid to collect Irish geological specimens, travellers donated objects from around the world – Leopold McClintock,* for instance, brought arctic fossils and plants from his polar expeditions – and the Royal College of Surgeons* presented a stuffed giraffe.

In 1815, the society bought the Earl of Kildare's elegant residence at Leinster House – previously it held its meetings, museum collections and lectures in smaller premises in Hawkins Street, and before that at 112 Grafton Street. Part of Leinster House was converted into a lecture theatre and laboratories; six rooms were given over to the museum collections, and opened daily to members, and to the public twice a week. In 1831, the RDS held its first agricultural spring show in the extensive grounds in front of Leinster House, and a few years later held its first industrial show. Its biggest such undertaking was the first Dublin International Exhibition,* which took place at Leinster House in 1853 and was part-funded by railway contractor William Dargan.* Prompted by the success of the exhibition's fine art hall – where a substantial collection of Old Masters was on show – Dargan helped establish a National Gallery the following year in the grounds of Leinster House (see: **A repository of knowledge**). The natural history collection was increasingly cramped and many specimens were perishing. With a grant from parliament, work began on a dedicated Natural History Museum, which opened in 1857 across the lawn from the gallery (see: '**The dead zoo**').

During the 1850s, there was increased competition from the growing number of new State bodies, such as the Geological Survey of Ireland* and the Museum of Irish Industry.* In 1857, the society's teaching activities were ceded to the Museum of Irish Industry (which later became the Royal College of Science). Then, in 1877, the society handed its institutions over to the government – for which it was compensated – namely the Natural History Museum, National Botanic Gardens and National Gallery. When, 10 years later, the government established the National Library and the Science and Art Museum (later the National Museum of Ireland), it drew substantially from the society's library and its scientific, industrial and agricultural collections. By the 1880s, the society's agricultural shows had grown so big that it bought a 16-hectare site at Ballsbridge as a showground. Then, in 1922, the new Free State government took over Leinster House, and the society was forced to move (it eventually received £68,000 in compensation). New headquarters were built at Ballsbridge, where the RDS still organises events, lectures, recitals, youth science and arts weeks, spring livestock and horse shows, publishes occasional books on the history of Irish science, and keeps the herd book for the endangered Kerry cow.*

A repository of knowledge

The **National Library** is a vast repository of learning spanning almost 1,000 years. It has over 1 million books, plus extensive archives, photographs, newspapers, maps and manuscripts, needing over a dozen miles of shelving. [14] The collection, started by the RDS, has grown thanks to acquisitions, donations and the fact that as a copyright library it is entitled to a copy of every book published in Britain and Ireland. The oldest map in the library is a coloured sketch from Giraldus Cambrensis's* *Topographica Hibernica* (1188). Among the other 150,000 maps and atlases are 18th-century copies of the Down Survey* maps, and 1,906 of the original Ordnance Survey* 6-in (15-cm) sheets (1830–1846) – each sheet covers 24 square miles (62sq km) of the country and can be examined on microfiche. Among the manuscripts is a draft of Captain Cook's journal of his voyages (1772–75), plus samples of bark cloth he brought from the South Seas. The photographic collections, at the **National Photographic Archive**, include photographs taken *c*.1860 by the countess of Rosse,* and early colour images made by John Joly* using a technique he invented. [15]

The **National Gallery** was established by the RDS with substantial funding from William Dargan,* following the success of the 1853 Industrial Exhibition; the main hall is named after Dargan and his statue stands on the front lawn. [16] When it opened in 1854, it was the first public gallery lit by gas jets – 2,000 lamps in all. The collection includes a portrait of Dargan, and a watercolour by James Mahoney depicting Queen Victoria's visit to the 1853 exhibition. There are landscapes by geological artist George du Noyer,* and botanical works by Rosalind Praeger (sister of Robert Lloyd Praeger*). The gallery can be an unusual source of information on times past: 17th- and 18th-century works, for instance, frequently depict the frozen winters that were characteristic of the Little Ice Age, and James Malton's watercolours from the late 1700s accurately record Dublin city's development. The collection also includes depictions of early Irish balloon* ascents, and even industrial scenes – a sketch of Rathgar quarry in the 1800s shows a windmill pumping the quarry dry. Behind the scenes, skilled scientists and conservators work in the gallery's laboratories, conserving and restoring paintings, drawings and sculptures.

The fossil of a giant deer: just one of the delights awaiting visitors to the Natural History Museum.
© National Museum of Ireland

'The dead zoo'

The **Natural History Museum** is a treasure. Virtually unchanged since the Royal Dublin Society* opened it in 1857, this museum of museums is a cross between a hunter's trophy room and a taxidermist's paradise, and affectionately known as 'the dead zoo'. Rows of specimen jars hold cuttlefish, parasitic worms, even a tick engorged by its last blood meal, stuffed animals gaze down on you, and two 11,000-year-old skeletons of giant deer* greet you at the entrance. Top of the bill for the opening celebrations was a public lecture by explorer David Livingstone about his African adventurers. Now restored to its original splendour, it has none of the gee-whizzery of modern museums, but where else can you safely inspect a piranha's teeth at close quarters? ⑰

The museum has over 2 million zoological and geological specimens from around the world. The Irish collection features fossils of animals that once lived here, such as brown bear,* lynx, lemming and hyena, case upon case of Irish birds – can there really be so many? – and native mammals such as the fin whale, pine marten* and pygmy shrew. There is a basking shark* caught in Galway in 1870, a massive sunfish from Lough Swilly, and an unfortunate sturgeon that once ventured up the Liffey. The international collection includes a king penguin that stands waist high, a weaver bird's nest, and skeletons of extinct birds such as the dodo and the roc or elephant bird. A turtle's skeleton reveals how its shell is an extension of the backbone, while in another case, a series of chicks in their eggs show how the young embryo develops from Day 6 until it hatches on Day 21. At the top of the house is an exquisite collection of over 500 delicate glass models of soft-bodied marine animals such as anemones and jellyfish, which were difficult to preserve or otherwise display. Intricate and zoologically correct, they were made by **Leopold Blaschka** and his son Rudolph, a famous team of Dresden glassmakers, and bought by the museum in 1878–86. The Blaschkas subsequently worked exclusively for Harvard University, and Ireland is fortunate to have this collection.

The geological material includes the historic Leskean cabinet of minerals, with which the RDS began its museums in 1792, and Sir Philip Crampton's* plesiosaur. Most is in storage pending a new earth sciences gallery at Collins Barracks,* but you can still see several fascinating fossils and a tiny piece of Moon rock – black and unremarkable looking, it was a gift to the nation from the USA. The museum's scientific staff continues to deepen our understanding of Ireland's natural history – Dr Jim O'Connor, for instance, specialises in Irish insects and is forever finding new species here. They also help identify specimens sent in by the public, and the more exotic creatures occasionally found among imports in Dublin's fruit and vegetable market.

1867 – A college of science

By the early 1840s, there was a mounting campaign for a technical college in Dublin to kick-start Ireland's industrial development. The RDS* had, for decades, organised public lectures on technical subjects, and several mechanics institutes had started around the country – transport entrepreneur Charles Bianconi* began one in Clonmel – but people still had to travel to England or Europe for technical training. The indefatigable Sir Robert Kane (see: **A man of science**) argued that Ireland had all the resources necessary for industrial development, apart from official support and a formal, government-funded, third-level technical college providing appropriate courses.

In 1845, the government and the Board of Education in London agreed, establishing the **Museum of Irish Industry** (MII) with Kane in charge. This was a kind of state laboratory-cum research institute-cum college, with departments of mechanical arts, mining, engineering, and manufacture; the curriculum was drawn up along English lines, however, rather than tailored for Irish conditions. Now largely forgotten, its collection of industrial equipment would later form part of the Science and Art Museum (now the National Museum of Ireland); and its educational wing, though slow to start, later became the Royal College of Science which, in 1926, merged with University College Dublin.

The MII was established at the same time as the Geological Survey of Ireland* and they shared premises at No 51 St Stephen's Green. There was considerable rivalry, however, between the MII and the RDS, and the society was forced to withdraw from most of its lecturing activities and, in 1853, saw most of its professors transferred to the MII. In 1867, the MII's training division was reorganised and renamed as the **Royal College of Science**, though still at St Stephen's Green. It offered three-year courses, trained science teachers and, significantly, was the first Irish third-level college to admit women. About half the students came from England. Ironically, instead of fostering industrial development, the college created a brain drain: there was little demand for trained science teachers in Irish schools and few industrial jobs for graduates.

In 1900, largely at Sir Horace Plunkett's* instigation, the government established a new department in Ireland, the **Department of Agriculture and Technical Instruction** (DATI), forerunner of today's Departments of Education

A man of science

Sir Robert Kane (1809–90), who was instrumental in founding the Museum of Irish Industry and the Geological Survey of Ireland,* was an important figure in 19th-century Ireland. He was one of the first Catholics to study at TCD, though it took him seven years to obtain a degree. Admittedly, he was delayed by taking time out to train as a medical doctor and found a journal *(Dublin Journal of Medical and Chemical Science)*. Chemistry was Kane's main interest, and his many chemical firsts included being the first person to convert a straight-chain chemical into a ring compound (tri-methyl benzene), and to analyse a mineral he had discovered in Germany, arsenide of manganese, which is named kaneite after him. His textbooks were international bestsellers – the massive 1,200-page *Elements of Chemistry* (1841) was a standard text in British and US universities – and his work on natural dyes, which he extracted from plants, won international acclaim. In 1844, he published his seminal study, *The Industrial Resources of Ireland*. This assessed the commercial potential of everything from potatoes and peat, to wind power and limestone quarrying, and was a rallying call to action. But it was fated to become one of the great might have beens of Irish history, for no sooner was it published than the Great Famine intervened in Ireland's economic and industrial development. When Prime Minister Robert Peel established a commission in 1845 to investigate the cause of the puzzling potato blight, Kane was one of those appointed. Kane also found time to be president of Queen's College Cork* and of the Royal Irish Academy*, and a government adviser on education, food distribution and public health.

and Agriculture. One of its first tasks was to reorganise and expand the college as a polytechnicum. Funding was increased, new courses were added – including electrical technology and physics – and work began on a purpose-built college at Merrion Street, adjacent to the RDS's Leinster House* campus, and with space for several laboratories. Designed on a grand scale, the new building was the height of modernity, with concrete floors, electric power and even elevators. George V opened the initial phase in 1911 and the first students moved in. Construction work continued until 1922, at which time the Executive Council of the new Free State seized part of the building. Ironically, the college was now at its peak, with student numbers at their highest ever and, thanks in part to returning ex-servicemen, over 200 new students each year. But once in occupation, the government never left, and the college's days were numbered. In 1926, it was merged with the newly established **University College of Dublin** (UCD), forming its faculties of science and agriculture.

UCD can trace its origins to the Catholic University of Ireland, begun in 1854 at what is now Newman House on St Stephen's Green (named after its first rector, English theologian John Henry Newman). The college became moribund in the 1880s, until the Irish Universities Act of 1908 revived it in the form of University College Dublin (the former Queen's Colleges* in Galway and Cork simultaneously became university colleges, and all three were incorporated into the new National University of Ireland). UCD retained Newman House, acquired a concert hall on Earlsfort Terrace, which had been built for Dublin's 1865 Industrial Exhibition* and, in 1926, when it merged with the Royal College of Science, gained space at Merrion Street, albeit cohabiting with government departments. In the 1960s, a modern campus was built at Belfield and, in 1989, UCD – and science – finally left Merrion Street. The government took over the entire complex for its senior offices, though statues of two famous Irish scientists, Robert Boyle* and William Rowan Hamilton,* still guard the entrance, a relic from those earlier days. (18)

Hamilton country – de Valera's vision of Ireland

Behind the Georgian façade of 5 Merrion Square, lurk professional stargazers. They watch galaxies evolve, count cosmic rays and, with the aid of the Hubble Space Telescope, observe new stars forming. They also watch planet Earth: mapping gravitational fields, listening for earthquakes, surveying Ireland's offshore seabed, even monitoring the African Rift Valley. For No 5 is home to the astrophysicists and geophysicists of the **Dublin Institute for Advanced Studies** (DIAS). The building itself is historic: former home of noted Dublin doctor William Stokes,* it was where **Éamon de Valera** secretly met Lloyd George's emissary, Jan Smuts, in 1921 in a prelude to the Treaty negotiations.

De Valera, one-time mathematics teacher and later taoiseach and then president of the Republic, is often remembered for his vision of an Ireland with "comely maidens" and crossroad dances. But he also saw Ireland as "the country of [William Rowan Hamilton], a country of great mathematicians". Hamilton* had been astronomer royal for Ireland at Dunsink Observatory,* a position that de Valera's former professor, Sir Edmund Whittaker, took up in 1906. The observatory, owned by Trinity College Dublin, had closed in 1921, and de Valera decided to save it for the nation.

After a Meteorological Service* was established in 1936, there was discussion about other scientific areas the new Irish State could support. De Valera believed theoretical physics suited a small country with limited resources, since it did not call for elaborate equipment, only "an adequate library, the brains and the men [sic] and [some] paper" and, in 1939, he proposed a Dublin Institute for Advanced Studies, similar to that in Princeton, USA. Fusing his twin visions of Ireland, it would have two schools: Celtic Studies and Theoretical Physics. There was considerable political opposition to a school of theoretical physics, but de Valera's Dáil majority ensured the Bill was passed. In 1940, DIAS was established, with offices at 65 Merrion Square.

Austrian Nobel physicist **Erwin Schrodinger** (1887–1961) was invited to head the school of theoretical physics, after Whittaker suggested to de Valera that neutral Ireland might accept scholars fleeing persecution in war-torn Europe. Schrodinger, who invented wave mechanics – using mathematical equations to describe the motion of sub-atomic particles – was to stay 17 years in Ireland, returning to Vienna in 1956. Noted German physicist **Walter Heitler** (1904–81) joined him in 1941. Schrodinger quickly put the new institute on the scientific map and, despite the war, organised international conferences and attracted renowned speakers, among them Max Born, Wolfgang Pauli and Paul Dirac. The public lectures Schrodinger gave at TCD in 1943 were published as a book, *What Is Life?*, that famously inspired many young scientists to study biology after the war, notably Jim Watson and Francis Crick, who later discovered DNA's double helix. Even today, DIAS retains its international reputation, and is better known abroad than at home.

The School of Cosmic Physics opened at 5 Merrion Square in 1947 and, significantly, incorporated Dunsink Observatory. More scientific refugees now came to Dublin: the astronomer Herman Brück, who fled Potsdam in 1936 and was Dunsink's director (1947–57); Leo Pollak, who fled Prague, and headed up the new school of cosmic physics, where he was joined by Hungarian physicist Lanos Janossy. One Irish scientist who worked at the institute was Dublin-born **John Lighton Synge** (1897–1995), remembered for his work on the geometry of relativity. Synge, uncle of playwright John Millington Synge, studied at TCD. He had an international career, including ballistics research for the US army, and founded the mathematics department at Carnegie Mellon University in Pittsburgh, before returning to DIAS in 1948. His students included Nobel prize-winning economist John Nash and his books reputedly spurred

Stephen Hawking to study relativity. The institute's theoretical physicists and Celtic scholars moved to Burlington Road in 1971, but the cosmic physicists continue at Merrion Square and Dunsink.

MERRION SQUARE

Buildings on file

The Irish Architectural Archive, which opened in 1976, holds extensive material on all manner of Irish buildings, from early Christian times to modern day. The collection of photographs, plans, models, maps and manuscripts includes a file on the work of Irish engineer James Waller,* who invented a cheap and novel way of erecting concrete buildings. One of Waller's unusual jelly mould buildings can be seen at Locke's Distillery in Kilbeggan.* The archive undertakes research and counselling work. It is a charitable organisation and welcomes gifts, loans and copies of architectural material. [19]

ST STEPHEN'S GREEN

Marble halls and geological curiosities

St Stephen's Green nearly became a university campus in 1658, when some Cromwellians proposed building Oliver the Lord Protector's College there, as a sister to Trinity College.* It was intended as a Protestant seminary "for the advancement of ingenious learning", but failed to attract funding and came to naught. At the time, the area was open ground and used as common grazing by Dublin citizens. In 1663, however, when Dublin Corporation needed to raise funds after the ravages of the Cromwellian rebellion, it fenced off part as a public park, and sold the surrounding land as building lots; one site was bought by William Petty,* who built himself a city residence on what is now the Shelbourne Hotel. Traces of gold were reputedly found in the gravel in the park pathways – the gravel was dredged from the River Dodder* – and, more prosaically, there is lead in the bedrock beneath the park.[6]

The foyer of **51 St Stephen's Green** is adorned with 40 large slabs of beautiful stone. This unusual geological gallery was erected in 1845 by Sir Robert Kane* to promote the commercial potential of Ireland's quarries and natural resources (Kane was director of both the Museum of Economic Geology* and the Museum of Irish Industry,* which were then at No 51). Six of the slabs are beautiful samples of green Connemara marble;* some have stunning zigzag patterns, a souvenir of the powerful geological forces that altered the original limestone 600 million years ago. The other 34 are not true marble, but polished limestone, and include Cork red, Galway black, Clonmacnoise grey and Pallaskenry purple 'marble'; many are festooned with fossils, some are stained red-brown by iron in the rock. The foyer is open during office hours and is worth including in any geological tour of Dublin.

Cities are excellent places for a **geological walkabout** – nowhere else will you see so much naked stone.[7] The oldest buildings were constructed for convenience with local stone, in Dublin's case a dark and muddy limestone called calp. Dublin's oldest-surviving stone building, the 12th-century Christ Church Cathedral, has walls of local calp, though a harder limestone imported from Somerset was used to frame the doors and windows. Calp, quarried at Donnybrook, Crumlin and Rathgar, was also used in the city walls and the old library at TCD.* Wicklow granite is also common in older Dublin buildings – the Rotunda Hospital,* for instance, completed in 1757, is clad in granite quarried at Ballybrew in Co. Wicklow. Many 18th-century buildings are dressed in creamy Portland stone – a 150-million-year-old Jurassic limestone from Dorset, which Sir Christopher Wren popularised when he rebuilt London after the Great Fire of 1666. The stone is still quarried, and was used in the 1990s to re-clad Brown Thomas's shop on Grafton Street. Dublin's Georgian squares and the 19th-century suburbs of Donnybrook and Rathmines were built of red brick, some of which was imported from the English potteries, and some brought by canal from the Athy brickworks. Modern buildings often use exotic stones imported from far-flung places, such as the zany pink Brazilian gneiss at Thomas Cook's on Grafton Street, and the pale-orange Norwegian marble on Hickey's fabric shop in the Stephen's Green Shopping Centre.

Slate* for the city's roofs came from Irish and Welsh slate quarries, and copper for the many domes may have been mined at Allihies.* Many of the city's footpaths were paved with hard-wearing Wicklow granite, though, increasingly, imported Chinese granite is used. Some new pavements are made from concrete, or a false granite made from concrete mixed with flecks of mica. Less common is Liscannor flagstone,* though it can be seen in the Peace Garden opposite Christ Church. Many streets were originally paved with setts, such as those that survive at Foster Place and were quarried from Charles Stewart Parnell's* Wicklow estate. TCD's front square is about the only place in Dublin you will see true cobblestones – unlike setts, which are cut, cobblestones are rounded, having been collected from beaches and riverbeds. While on walkabout, do not miss the wonderful fossils in the otherwise non-descript wall of No 4 Kildare Street, and the sparkling gemstones at Lapis Jewellers in Nassau Street.

Barbers and saw-bones no longer

Surgeons are respected members of the medical profession today, but it was not always so. For centuries, physicians were the recognised medics: they treated disturbances of a patient's bodily 'humours', yet often had little knowledge of anatomy. They were not to be confused with the barber-surgeons: men skilled with a razor who could shave you, bleed you, even amputate gangrenous limbs, but were looked down on as butchers and 'saw-bones'. Things began to change in the mid-1700s, as our understanding of the human body improved and, in the early 1800s, the Napoleonic wars brought a growing need for medically trained surgeons. However, there was still public disquiet about the work of the anatomists and their shady dealings with body snatchers,* and surgical operations remained crude, painful and dangerous until the development of antiseptics and anaesthetics.*

The **Royal College of Surgeons in Ireland** was established in 1784 as surgeons were gaining in respectability here. The college was first proposed in 1765 by a noted Limerick eye surgeon **Sylvester O'Halloran** (1728–1807), who also started Limerick's County Infirmary. O'Halloran studied medicine in London, Paris and Leyden, devised new cataract treatments, and wrote a widely read book on cataracts. He once studied the blood supply to a dog's eye by hanging an unfortunate hound upside down to increase the blood flow to its head, and regularly dissected fresh calf eyes.

The RCSI, initially a fraternity of surgeons, held its first meetings at the Rotunda hospital.* By 1810, it had built a school on a disused Quaker graveyard at St Stephen's Green, and quickly flourished as the demand for skilled surgeons increased (there were also several private anatomy schools then). Noted scientists and medics associated with the college include **Walter Hartley** (1846–1913), an English chemist who helped develop the new science of spectroscopy during his 30 years there as chemistry professor. Spectroscopy is a technique for studying the chemical composition of something by examining the spectrum of light it emits. Each element has its own characteristic pattern, and Hartley was the first to realise this depended on the element's position in the periodic table. Hartley invented several spectroscopic instruments (one is in London's Science Museum, and another at Maynooth College's science museum*).

1865 – Dublin's Crystal Palace

The great **industrial exhibitions** of the 19th century were the Expos and the World Trade Fairs of their day. There were pleasure gardens to amuse the visitors, activities to exercise them, recitals to entertain them, and exhibitions to educate and inform them. Leading companies from around the world came to show high-tech gadgetry, the latest domestic appliances, sophisticated scientific instruments, modern industrial machines and exotic curiosities. These exhibitions, starting in London in 1851 when the original Crystal Palace was built, were massive events that took

Dublin's Crystal Palace: built at the rear of the concert hall for the 1865 International Exhibition.

years of careful planning, and cost (and often lost) millions to mount. Many Irish firms attended – Yeates & Son* won an award for its instruments at London in 1851, for instance, and Thomas Grubb* brought his telescopes to the 1855 Paris exhibition.

The first Irish exhibition took place at Cork in 1852, and the following year Dublin held its first exhibition in temporary buildings at Leinster House, then the RDS headquarters.* The railway contractor William Dargan* bankrolled that project with £20,000 of his own money, but the small Irish market failed to attract enough international companies and, though Queen Victoria visited four times while on a one-week visit to Dublin, it was a commercial disaster. Undaunted, planning soon began for a second exhibition, which eventually opened in the spring of 1865 in a green space leased from Lord Iveagh (of the Guinness family). A concert hall seating 3,000 people was built on Earlsfort Terrace (it is now the National Concert Hall), and behind it, running the length of the terrace, was an enormous crystal palace: a vast confection of cast-iron and glass housing a winter garden and exhibition galleries. Its construction was an engineering feat: 14-tonne weights were hung from the cast-iron components, for instance, to test their strength; 1,000 cannon balls were rolled along the floors to ensure they were level; and, in March 1865, before it opened to the public, 600 men from the 78th Highlanders marched in formation through the galleries to test their robustness. In case of fire, there were telegraphic connections to the city's five fire stations, plus fire hydrants and pumps on each floor.[8] The outdoor gardens included a maze, an archery arena (sunken, to protect spectators), fountains, woodlands and wildernesses. Water cascaded over a substantial grotto that was made from boulders selected to represent Ireland's geological resources.

The buildings and gardens were reused for a similar exhibition in 1872. The crystal palace was demolished soon after, but the concert hall was taken over in the 1880s by the Royal University* (later University College Dublin). Further Dublin exhibitions followed in 1895 and 1907. The last exhibition, which took place in Herbert Park, had pavilions (subsequently demolished) devoted to gas appliances, "mechanical Art", industry and exhibitions from abroad. There was even an early big dipper: a massive water chute that carried people aloft in small boats before plunging them into a pool below. Iveagh House is now the Department of Foreign Affairs; the Iveagh Gardens are open daily.

EARLSFORT TERRACE

1892 – measuring starlight

The garden of No 16 Earlsfort Terrace was the scene of an historic experiment in 1892, when starlight was measured electrically for the first time. Until then, there was no way to measure a star's brightness – it came down to a person's individual judgement as to whether one star was brighter than another, making comparative studies difficult. Then Irishman **George Minchin,** professor of engineering at what is now Brunel University in England, invented a device (a selenium cell) that produced a voltage when light was shone on it. The more light there was, the higher the voltage. Minchin wondered if his cell was sensitive enough to detect starlight, and contacted his friend William Monck, an amateur astronomer living at Earlsfort Terrace. Monck supplied a telescope and venue for the experiment, and TCD physicist George Francis FitzGerald* supplied a sensitive electro-motor and helped connect the selenium cell to the telescope. After several unsuccessful attempts and a run of bad weather, Minchin returned to England. Then, early on August 28th, 1892, Monck, FitzGerald and a neighbour Stephen Dixon pointed the telescope at the Moon, Venus and Jupiter, and at the stars Vega and Capella, and successfully detected the light coming from them. It was the first absolute measurement of the light from heavenly bodies and enabled them to conclude, for instance, that Venus was 2.5 times brighter than Jupiter. It was the start of the important science of stellar photometry and, in 1992, an international conference was held in Dublin to mark the experiment's centenary. Minchin went on to design more sensitive selenium cells, and one was used at Daramona Observatory* in 1895 to measure the starlight from other stars. A plaque at No 16 commemorates the historic experiment.

ESSEX STREET

1681 – the elephant man

In June 1681, a fire in Essex Street killed an elephant that had been in a show there, and a young physician **Allan Mullen** (or Molines, 1654–90) jumped at the opportunity to dissect the unfortunate animal. Working by candlelight and assisted by butchers, Mullen operated in front of a large crowd that fought for souvenirs. His account of the dissection, published the following year, is the first modern description of the elephant's anatomy; detailed and accurate, it remained the standard source for 300 years and can still be read at Marsh's Library.* Intriguingly, Mullen discovered that, unlike other land mammals, the elephant has no pleural cavity (the fluid-filled space between the lungs and chest walls); some zoologists now think this is because the elephant evolved from an aquatic mammal, such as the African dugong or sea cow. Born in Ballycoulter, Co. Down, Mullen studied at TCD, befriended Narcissus Marsh* and Robert Boyle,* and helped found the Dublin Philosophical Society* and the Royal College of Physicians.* Not all his experiments would win approval today: he injected mercury into condemned men to investigate its effects on their lungs, and drained the blood from animals to see how much they had. Mullen died while on a fortune-hunting trip to Barbados, but whether of a fever or drink is disputed.

The hallmark of quality

There is a sophisticated laboratory at Dublin Castle, one which, strangely, is run by a mediaeval guild. This is the **Assay Office,** responsible for testing and hallmarking precious metals, and run by the Company of Goldsmiths, an organisation that received its royal charter in 1637 and traces its origins to the ancient City Guild of Gold and Silversmiths. In charge is the assay master or 'touch warden', a title that derives from the days when gold was tested by rubbing it on a touchstone of basaltic rock. For nearly 400 years, the company has been testing the purity of the gold and silver (and since 1983 the platinum) used and sold in Ireland. It is an offence to sell non-hallmarked objects, so, each year, it assays millions of objects of precious metal, made here or imported. The laboratory uses a mix of ancient techniques and high-tech equipment and can detect impurities in a matter of hours in samples the size of a pinhead – there is no logjam of valuable items waiting to be tested, and only a tiny sample is taken for testing (items are usually submitted semi-finished, with part of the casting left for sampling).

Platinum is assayed by sophisticated atomic absorption; gold, however, is merely heated to 1,100°C to separate the impurities from the precious metal – and all that has changed in 350 years is that natural gas heats the furnace in place of charcoal. Items that pass the test are hallmarked, those that fail are smashed and the pieces returned to the owner. The term hallmark derives from when goldsmiths took their wares to their hall in London for marking. The Irish mark is the figure of Hibernia, together with a letter denoting the year of manufacture. The standards for the precious metals are set by an international convention: 18 carat gold, for instance, is defined as 18 parts pure gold in 24, or 750 parts per 1,000 (equivalent to 75 per cent pure metal); silver must be 925 parts per 1,000 (92.5 per cent pure metal).

Purges and potions – Dublin's drug trade

For centuries Bride Street was at the heart of Dublin's drug trade, and it has been said that **apothecaries** were mixing medicines there when Grafton Street was still a muddy lane. Their guild met regularly in a room above the city's Pole Gate at the junction of Bride and Werburgh streets, and the last pharmaceutical company – Boileau & Boyd – left the area only in 1975, when it moved to an industrial estate.

The first recorded apothecary in Ireland was an Englishman, Thomas Smith, who arrived in Dublin in 1554. Smith became a major figure in the Dublin business scene, rising to the rank of Lord Mayor, in which capacity he laid the foundation stone for TCD* in 1592. Apothecaries were by then a recognised profession and could join the city guild of "barbers, chirurgeons [surgeons], apothecaries and periwig makers". They were not, however, held in the same esteem as today, and "gentlemen physicians", who often competed with them for patients, regarded them as tradesmen. A royal charter in 1745 granted apothecaries their own corporation – the Guild of St Luke the Evangelist

(Luke was a physician) – and, in 1791, in a bid to control the quality of the medicines being supplied, the government established the **Apothecaries Hall**. This statutory body ran a medical school in Cecilia Street, offered courses in subjects such as botany, anatomy and chemistry, held its own examinations and awarded diplomas; the noted chemists William Higgins* and Robert Kane* were among its professors. From 1860–1970, it also had permission to qualify apothecaries as medical doctors, and the few remaining members of the Apothecaries Hall are medical doctors. The hall, on Merrion Square, has the records of the apothecaries guild. ⑳ Training in pharmacy is now provided by the School of Pharmacy at TCD.

Boileau & Boyd is Ireland's oldest scientific company, tracing its origins to "Robert Wilson, Apothecary", who commenced trading in Bride Street in 1700. The firm was renamed when J T Boileau, descendant of a Huguenot immigrant, married a Wilson in 1781.

The company moved from Bride Street in 1975 to Walkinstown, but still operates under its historic sign of the unicorn, though it no longer manufactures its own products. It owns Wilson's original 18th-century *materia medica* jars, for storing exotic items such as 'Treacle of Andromachus'; some are on loan to the National Museum, others are kept by the Pharmaceutical Society of Ireland, the modern-day guild representing pharmacists.

Taking the pulse – the golden age of Irish medicine

The practice of taking a patient's pulse as a simple way of assessing their condition was pioneered at the **Meath Hospital** in the 1830s. The Meath (then in Heytesbury Street, now part of Tallaght Hospital) was then a great centre for medical training and research, attracting students from Britain, Europe and North America. Other innovations introduced there included the hypodermic syringe,*

invented in 1844. This was the golden era of Irish medicine: Robert Collins* was introducing radical new practices at the Rotunda;* and Robert Graves and William Stokes were leading the revolution at the Meath, where their ideas about patient care and medical training became internationally renowned as 'the Dublin School of Medicine'.[9]

Robert Graves (1796–1853) was from a prominent family that came to Ireland with Cromwell (the poet Robert Graves was from the same family). He studied at TCD, then at Edinburgh, Berlin, Vienna, Paris and Italy, acquiring a facility for languages that once landed him in an Austrian jail on suspicion of being a German spy. At the age of 25, Graves was appointed to the Meath, and immediately set about reforming practices – in his first lecture he bravely denounced indifferent medical treatment which, he said, often killed patients. Graves promoted bedside training for medical students, and urged students to walk the wards in search of medical knowledge and not entertainment. Doctors and students were exhorted to take detailed observations in order to make accurate diagnoses, tailor prescriptions to

Two medicine men

The first person to realise that the heart can affect the mind was **Robert Adams** (1791–1875), a Dubliner and surgeon to Queen Victoria. In 1827, Adams studied an elderly man who suffered from blackouts, and realised that the loss of consciousness was caused by a slow pulse rate. William Stokes* studied similar conditions 20 years later, and the condition is named Stokes-Adams syndrome after them. On New Year's Day 1847, Adams was part of the medical team that performed the first operation under anaesthesia (ether) in Ireland, when they amputated a girl's arm at the Richmond Hospital, just weeks after the first such operation had taken place in Boston.

The medical use of the microscope was pioneered in the 1840s by **John Houston** (1802–45), son of a northern Presbyterian clergyman, who worked at Baggot Street Hospital. People were starting to examine cells under a microscope, and Houston realised the instrument could be used to reveal changes in diseased cells, such as cells taken from a cancer, and he proposed categorising tumours based on the microscopic appearance of the cells. Folds in the wall of the rectum are named Houston valves, as he was the first to describe them in 1830.

individual patients, communicate clearly with patients, treat rich and poor alike, be sympathetic, and not practice some inhumane clinical detachment. Just as importantly, Graves encouraged a spirit of inquiry and research, and the names of many of his students and colleagues are immortalised in the conditions and techniques named after them, such as Graves disease, Cheyne-Stokes syndrome, Wilde's snare and Stokes-Adams syndrome.

The 1826 typhus epidemic sparked Graves's lifelong interest in fevers. He revolutionised fever treatment, advocating that patients be fed instead of starved, the better to fight infection, and suggested his epitaph be: "He fed fevers." Despite the dangers – 25 per cent of doctors died of infections caught at work – he urged doctors to treat fever patients. During a cholera epidemic after the Great Famine in the late 1840s, the official view was that cholera was not contagious, and patients need not be isolated. Graves proved otherwise, tracing the spread of the disease along lines of communication, but failed to change the official view. In 1830, Graves established the practice of pulse taking; he was also the first to describe a goitre condition, now called Graves disease (1835), peripheral neuritis (in 1828) and angioneurotic oedema (in 1843); and he was ahead of his time in cautioning against the use of mercury as a medicine, and of bleeding, then a popular practice. His lectures, published in 1838, and his book *Systems of Classical Medicine* brought him international renown, and his many students included Sir William Wilde,* Sir Robert Kane* and William Stokes.

William Stokes (1804–77) was also from a noted Cromwellian family and related to George Gabriel Stokes.* He joined the Meath in 1826, succeeding his physician father, Whitley Stokes (1763–1845), and with Graves, made it into a great teaching hospital. While still a student, Stokes wrote the first book in English on the stethoscope,* a relatively new device that many physicians still ridiculed. At the Meath, Stokes pioneered the stethoscope's use in medicine, becoming an early heart specialist. He was the

first to identify a breathing problem now called Cheyne-Stokes respiration, and the first to explain that seizures associated with a slow pulse were caused by lack of oxygen in the brain (first identified by Robert Adams (see: **Two medicine men**) and now called Stokes-Adams syndrome). He pioneered the use of iodine to treat goitre and thyroid disorders and, in 1852, published a major book on heart disease. A keen antiquarian, Stokes and his friend and colleague William Wilde were early movers in the post-Famine Celtic revival. Stokes's Swiss-born grandson **Adrian Stokes** (1887–1927) also went into medicine and, in 1927 in Africa, discovered the cause of the mysterious yellow fever, even as he lay dying of the disease. He proved it was caused by a virus, and that blood-sucking insects could spread the infection. His selfless work paved the way for the first yellow-fever vaccine.

1844 – the first hypodermic syringe

The world owes a debt of gratitude to **Francis Rynd** (1801–61) a Dublin doctor who, on June 3rd, 1844, performed the world's first subcutaneous injection, and thus invented the hypodermic syringe. Rynd was treating a woman at the Meath Hospital in Heytesbury Street, who had a severe pain in her face for years. She had been taking morphine by mouth to no avail, so Rynd decided to place

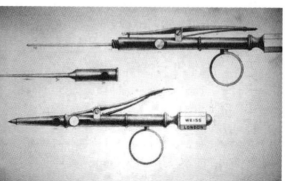

Commercial version of Francis Rynd's syringe. There is no plunger (the lever is to facilitate the injection), and the solution simply flowed in under gravity. Courtesy Davis Coakley

morphine directly under her skin near the facial nerves – essentially a local anaesthetic. He made an improvised syringe from a tube, or cannula, and a cutting implement called a trocar. Having made four puncture holes in the woman's face, he allowed some morphine solution to flow in through the cannula. The operation was relatively painless and the woman later slept well for the first time in months. Rynd's technique was soon widely used to treat pain and championed as "the greatest boon to medicine since the discovery of chloroform". Florence Nightingale benefited from it during an illness and wrote: "Nothing did me any good, but a curious little newfangled operation of putting opium under the skin, which relieved one for twenty-four hours." In 1853, the plunger syringe was invented, enabling doctors to inject solutions faster and into a vein against blood pressure.

"For all graduates and gentlemen"

Marsh's Library and its priceless collection of early books is a national treasure. Opened in 1701 "for all graduates and gentlemen", it is Ireland's oldest public library and one of the few 18th-century buildings in Dublin still used for its original purpose – the beautiful oak interior is virtually unchanged and the books are shelved where they were 300 years ago. The library was founded by **Narcissus Marsh** (1638–1713), a noted English scholar who came to Ireland in 1679 as provost to TCD. Discovering that there was no public library – TCD's library was open only to staff and students – Marsh set about starting one. In 1701, when he was Archbishop of Dublin, he built his library in the garden of the bishop's residence (now the site of Kevin Street Garda Station). ㉑

Like most early scientists, Marsh had diverse interests: he studied acoustics, and coined the word 'microphone'; invented a lamp capable of lighting large churches; suggested classifying insects according to their appendages (which was novel at the time); and enjoyed mathematical problems, once declaring: "Thy name be praised O Lord for all thy mercies, this evening I invented a way to find out the Moon's distance from the centre of the Earth without the help of its parallax." He was a founder of Ireland's first scientific association, the Dublin Philosophical Society,* and published an Irish-language edition of the Old Testament – he himself organising the translation, and Robert Boyle* paying for the printing.

The library's 25,000 books reflect Marsh's own interests in science, medicine and botany – indeed, many of the books were annotated by him. His friends, such as William Petty* and William Molyneux,* presented copies of their publications, but Marsh acquired 10,000 of the books in 1705, when he bought the library of Bishop Edward Stillingfleet, who had died in 1699. A Huguenot refugee, Elias Bouhéreau, who was the first librarian at Marsh's Library, contributed an important collection of 17th-century French Calvinist books. Early scientific works in the library include a first edition of Newton's *Principia* (1687), a French edition of *Euclid* (1516), Linnaeus's *Systema Plantarum* (1762), and works by Kepler, Galileo, Danish astronomer Tycho Brahe, and Robert Boyle.* There are also maps by Mercator, the first maps from the Ordnance Survey of Ireland* in the 1840s, and a pamphlet by Allan Mullen* on the elephant's anatomy. Narcissus Marsh is buried in St Patrick's Cathedral.

Swift, satire and science

You could say that **Jonathan Swift** (1667–1745) predicted the existence of the two **moons of Mars** 150 years before they were discovered. For in *Gulliver's Travels*, a satirical tale published in 1726, he announces that astronomers in Laputa had detected two satellites orbiting Mars – yet the real martian moons, Phobos and Demos, were not discovered until 1877. Ironically, Swift had no great love of scientists and he was ridiculing astronomers and the Royal

Society,* with its fascination with measuring devices and experiments. In Laputa, Swift wrote, was a man who spent eight years trying to extract sunbeams from cucumbers. *Gulliver's Travels* also includes unkind references to the mathematician Sir Isaac Newton, whom Swift held responsible, as Comptroller of England's mint, for the debased coins then circulating in Ireland.

Jonathan Swift is buried at St Patrick's Cathedral.

Swift, who was Dean of St Patrick's Cathedral for many years, was plagued by unexplained deafness and dizziness in his later years. Convinced he was going mad, and concerned at the conditions in existing institutions for mentally ill people, he left his personal fortune of £11,000 to found a hospital for the insane – St Patrick's in Steevens' Lane, still there today – which, when it opened in 1749, was one of Europe's first psychiatric hospitals. A century after his death, Swift was himself the subject of scientific research: his body was exhumed when the crypts flooded in 1835, and two Dublin doctors took the opportunity to perform late post mortems. John Houston, a phrenologist, was interested in the bumps on Swift's skull; Sir William Wilde* diagnosed Ménière's disease, a disorder of the inner ear, and not insanity. [22]

Brought to book

Gulliver's Travels is not the only novel by an Irish writer, or set in Ireland, to feature science. Flann O'Brien used science satirically, notably in *The Third Policeman,* which features de Selby's molecular theory. English physicist Fred Hoyle wrote a 20th-century science-fiction novel *Oisín's Ride,* set in an Ireland overtaken by aliens. The main character in C P Snow's *The Search* is based on the Tipperary scientist J D Bernal* and James Joyce's *Ulysses* includes numerous references to scientific and medical institutions, such as the Rotunda Hospital* and Dunsink Observatory.* There are plays too: Dublin paediatrician Robert Collis* wrote a play about social conditions in the Dublin slums – *Marrowbone Lane* – and George Bernard Shaw's *The Doctor's Dilemma* was prompted by a remark made by his friend, Dr Almroth Wright:* "The time is coming when we shall have to decide whether this man or that is most worth saving."

CHARLEMONT STREET

Teach Ultan – a radical children's clinic

In 1919, a committee of women nationalists and suffragists, many of them members of Sinn Féin's Women's League, founded the first Irish hospital dedicated to children. Teach Ultan, or St Ultan's Clinic, opened in Charlemont Street with two cots and £100. It was managed entirely by women, led by **Dr Kathleen Lynn** (1874–1955), and with Madeleine french-Mullen as administrator. They were concerned about children's health, social conditions and the deadly Spanish flu epidemic, but especially an outbreak of syphilis after soldiers returned from World War I, resulting in huge numbers of infected war babies. Lynn was one of Ireland's first women doctors, graduating in 1899 from the Royal University (later UCD). As a Dublin GP she attended hunger-striking suffragists in 1912 and workers during the 1913 lockout, and was chief medical officer to the 1916 Rising, assisting the wounded James Connolly and earning a spell in Kilmainham Gaol.

St Ultan's was staffed and run predominantly by Protestants, which became an issue in the 1930s when the Catholic hierarchy, then campaigning for its own Catholic Children's Hospital at Crumlin, objected to any enlargement of St Ultan's. Catholic children attending the clinic, it argued, might be exposed to inappropriate sex education and contraception. Ironically, it was then that St Ultan's pioneered a new vaccine to fight tuberculosis, a killer disease then endemic in Dublin's tenement slums. The vaccination campaign was led by Dublin doctor **Dorothy Price** (*née* Stopford, 1890–1954), who had visited European TB clinics and learned of a new French vaccine, the BCG (Bacille Calmette et Guerin). With social conditions slow to change, Price hoped a preventative vaccine might help control the disease. In 1936, she vaccinated 35 children at St Ultan's, and the success of the approach eventually persuaded Dublin Corporation to introduce a free BCG vaccination programme for children in 1948. In 1950, however, Dorothy Price contracted TB and died four years later. St Ultan's closed in 1980 and the site is now a private clinic.

JAMES'S STREET

Pints and measures

The Irish have been brewing and distilling* for centuries. The ancient Irish drank *coirm*, a fermented malt drink (whence *coirm cheoil*, the Irish for a musical party); St Patrick employed a brewer, St Brigid performed brewing miracles, and most monasteries had an ale house. In 1672, Dublin had, according to an economical survey commissioned by William Petty,* 5,000 drinking houses, 1,180 ale houses and 91 public breweries. Over the next 300 years, all those breweries disappeared bar one: a brewery in James's Street that Dublin businessman Arthur Guinness leased in 1759. **Guinness's Brewery** still produces 4 million pints of porter a year and, in recent years, has been joined by some new small breweries, notably the Dublin Brewing Company.

The James's Street district was the centre of Dublin's drinks industry in the 19th century: there was Guinness's, **Power's distillery** at John's Lane and the **Roe distillery** on Thomas Street. Power's closed in 1976 and the site is now the National College of Art and Design, though it retains some of the distillery's big, steam-powered beam engines; Irish Distillers now owns the Power's brand, along with Jameson, Midleton* and Bushmills.* The Roe distillery opened in 1757, when there was already a big **windmill** on the site. There would have been several windmills in the city then but only this one survives. Guinness's bought the site in 1949 and have preserved the windmill tower which, at 50 metres, is reputedly the tallest ever built in these islands. With its elegant copper cap, it is a prominent landmark.

Although Guinness's popularised porter as an Irish drink, it was actually invented in London in 1722, when a brewer discovered that extra hops improved a beer's shelf life. That one small recipe change dramatically altered the industry: beer could now be produced on a larger scale, and breweries expanded accordingly. The new 'stout' beer proved popular with porters at Covent Garden markets, whence the 'porter' label. The Irish porter industry developed in the early 1800s in Cork where Beamish & Crawford* employed 500 men in what was then Ireland's largest brewery, a position it lost in 1833 to Guinness's. By the early 1900s, nearly 10,000 people worked at Guinness's turning yeast, water, barley and hops into porter. The James's Street plant was a veritable town with its own fire brigade and medical service, canal harbour (now filled in), timber yard, blacksmith, carpentry workshop, metal foundry, and cooperage yard holding 250,000 barrels. Automated production techniques mean fewer than 800 people work there now, most of whom are engaged not in producing beer, but in administration and marketing.

Sampling Guinness: Student's t-test

If you wanted to know the average height of the Irish population you could get a reasonably good estimate by measuring a random sample of 1,000 people. But how good would your estimate be if you could measure only 10 people? In 1900, brewers at Guinness's faced just this kind of problem: they were testing new barley varieties that had been grown in trials by the Department of Agriculture, but many of the trials were small and could not easily be repeated – for instance, it would take another year to grow more and the weather then would be different. Trying to evaluate some of the varieties was like trying to measure the Irish population by studying 10 people. Fortunately, Guinness's had just recruited a new brewer, a mathematically minded Englishman, **William Sealy Gosset** (1873–1937), who devised a statistical way of evaluating the results produced from small samples. Guinness staff could not publish under their own name then, so Gosset published his technique in an international journal under the pseudonym 'Student'.[10] Called *Student's t-test*, it is still used by statisticians.

Guinness is a global brand, with 3.5 billion pints of the black stuff drunk each year in 151 countries. The brewery's stylish visitor centre is housed in a 1904 seven-storey fermenting- and store-house. ㉓ Or, you can tour a working brewery at the Dublin Brewing Company in Smithfield. ㉔ The Old Jameson distillery in Bow Street is also now a visitor centre. ㉕ Founded in 1780, and for years a rival to the Power's distillery in John's Lane, it closed in 1970 when production moved to Midleton.*

Hospital; the buildings are now the headquarters of the Eastern Health Shared Services. Neighbouring **St Patrick's Psychiatric Hospital** was founded in 1749 at Jonathan Swift's* bequest and is still going. Swift had been a trustee of Dr Steevens', which prompted him to fund a similar institution for the insane, stipulating that it be near Dr Steevens', in open country as "the cries and exclamations of the outrageous would reach a great way and ought not to disturb neighbours".

STEEVENS' LANE

Curing distempers and wounds

INCHICORE

'The Works'

When **Richard Steevens**, professor of medicine at TCD, died in 1710, he left his considerable wealth to fund somewhere "for maintaining and curing from time to time such sick and wounded persons whose Distempers and Wounds are curable". **Dr Steevens' Hospital**, founded by his sister Grizel, was the first, successful voluntary hospital in these islands.[11] Eight men and two women were admitted when it opened in 1733, and were cared for by a matron, four nurses, a cook and a porter. The trustees raised funds where they could, even running a ferry across the Liffey until King's [Heuston] Bridge was built in 1827. One benefactor was **Dr Edward Worth** (1678–1733), who donated his library of medical and scientific books, including early printed books (pre-1500), and works by Galileo, Newton and Paracelsus. A small library, built for the collection, survives unchanged 250 years later. ㉖ In the 1820s, Dr Steevens opened what was probably the country's first ward for venereal diseases – a radical move at the time – and, in the 1860s, started a nursing school, just six years after Florence Nightingale's school opened in London. The first surgical use of X-rays* in Ireland took place at Dr Steevens' in 1896 when Richard McCausland removed a needle from a woman's hand, having first located it using a 'shadowgraph', or X-ray, taken at the Royal College of Surgeons.* Dr Steevens' Hospital closed in 1987, and its services transferred to nearby St James's

A vast Victorian engineering works is still operating in Dublin. It is the Inchicore railway works, Ireland's only surviving railway works and one of the world's oldest, where some 400 people maintain and repair Iarnród Éireann's rolling stock. 'The Works', as it is affectionately known, has produced trains and locomotives, as well as buses, lorries, armoured vehicles, armour-plated trains, experimental battery trains, turf-burning locomotives, even munitions. It was opened in 1846 on a 28-hectare green-field site in rural Inchicore by the Great Southern & Western Railway (GS&WR). Each railway company then had its own works, and Inchicore was built to serve the GS&WR's Dublin–Cork line, the first stretch of which opened later in 1846. Soon, several hundred men were working a 54-hour

A busy Inchicore workshop during the Emergency years.
© Iarnród Éireann

Ireland's largest-ever steam locomotive Maedb: built at Inchicore in the 1930s, it now has pride of place at the Ulster Folk & Transport Museum.* © Ulster Folk & Transport Museum

week there, and a community developed at Inchicore, as schools, accommodation and recreation facilities were added.[12] At its peak in the 1950s, Inchicore employed 2,000 people.

The works were self-contained and self-sufficient: there were workshops, engine running sheds, a sawmill, timber stores, creosote works, lime kilns, coal banks, gas and coke plants and an iron foundry; engine cleaners worked there, along with fitters, coach builders, carpenters, metalworkers, painters, body and boiler makers, and upholsterers, making everything a railway company needed, from shovels and wheelbarrows to locomotives. After the 1916 Rising, Inchicore made an armoured Daimler for the British army (protected behind a spare boiler casing) and, during the Civil War, produced armoured Lancias that could travel along the railways. During World War I, some workshops became munitions factories, turning out shell cases and grenade parts; and during the Emergency (1940–47) they made submersible mines for the Irish navy. Inchicore

innovations include the revolutionary Drumm* battery train built in the 1930s; less successful turf-burning locomotives developed in the 1870s and again in the 1950s when coal was scarce; and the continuous vacuum brake, still widely used, which was invented by English engineer John Aspinall, who joined Inchicore in 1877.

As rail traffic grew, so did the weight a locomotive had to pull – from under 100 tonnes in 1860 to over 250 tonnes by 1900. This culminated in Inchicore building the three heaviest locomotives ever used in Ireland – *Táilte*, *Macha* and *Maedb*. Launched in 1939, and weighing 120 tonnes, only the Dublin–Cork line could take them. *Maedb* ran for 20 years, clocking a record 155kmph, and now has pride of place at the Ulster Folk & Transport Museum.* In 1925, following partition, all the Free State railway companies were amalgamated into the Great Southern Railway, with Inchicore as its main works, the only sizeable heavy engineering works in the 26 counties. There are now high-tech laboratories there, for fuel quality testing, and metallurgy testing in the event of equipment failure. Train travellers pass its castellated stone buildings just before entering **Heuston station**. That station was built in the 1840s during the golden age of Irish railway development. The roof over the train shed – where trains pull up to the platforms – is an engineering feat in itself, covering 1 hectare, and held in place by 72 cast-iron columns. All the ironwork was supplied by Robert Mallet's* foundry. Only platform No 1, the 'troop platform', is outside the roof's umbrella, soldiers being expected to wait in the elements. The Irish Railway Records Society keeps its archive at Heuston. Part of a freight line linking Heuston and Connolly stations runs in a tunnel under the Phoenix Park.*

North Dublin city can boast the wild and uncultivated Phoenix Park, Dublin Zoo, the world's oldest-surviving maternity hospital, and the birthplace of Lucien Bull, who pioneered high-speed cinematography. The first Irish blood transfusion took place at Jervis Street Hospital in 1865, while that same year a Mater Hospital physician, Francis Cruise, invented the first practical endoscope.

PHOENIX PARK

A wild and uncultivated land

It was Henry VIII who began the Phoenix Park, when he dissolved the monasteries and confiscated land belonging to Kilmainham Priory. The lands, which spanned both sides of the Liffey, subsequently became a royal deer park, with a herd of fallow deer. In 1680, the land on the Liffey's south bank was given to the military's Royal Hospital in Kilmainham (now the Irish Museum of Modern Art), and the remainder was enclosed behind a high wall. Then, in 1745, the lord lieutenant, Philip Stanhope, Earl of Chesterfield, opened the deer park to the public, ordering that "this wild and uncultivated land… be ornamented for the pleasure of the citizens". It became known as the Phoenix Park, from *fionn uisce* (meaning 'clear water') after a popular mineral spring near the zoo and, despite the name change and the intervening centuries, the herd of fallow deer survives and now numbers over 500.

The Phoenix Park, at 700 hectares, is one of the world's largest enclosed public parks, and its main road, named Chesterfield Avenue in the earl's honour, stretches 4km from Conyngham Road to the Castleknock Gate. It was the scene of two great scientific jamborees – in 1835 and 1854 – when the British Association for the Advancement of Science* (BAAS) held its annual summer festival there. The 1835 meeting was a gala occasion – among the huge crowd was songwriter Thomas Moore, who reputedly prowled the proceedings in search of pretty women – and a big attraction in 1854 was the colossal fossil of a plesiosaur that had recently been presented to the surgeon general **Sir Philip Crampton** (1777–1858). Crampton donated the fossil to

Crampton's plesiosaur: this fossil of a Nessie-like reptile, Plesiosaur cramptoni, found in Yorkshire in 1848, was first displayed at a scientific gala in the Phoenix Park in 1854; now in a dozen crates, it will be on show again shortly in the new geology gallery at the National Museum, Collins Barracks.

the Royal Zoological Society (Dublin Zoo), where it was displayed under a large tent.

Despite being close to the city, the park has a rich diversity of habitats, thanks to its size and the fact that it was never disturbed or built on. Caleb Threlkeld* surveyed the flora in 1726 and many of the species he found are still growing where he recorded them; more recently, the Dublin Naturalists' Field Club recorded over 300 species of flowering plants and grasses there. One-third of the area is planted with trees, including many specimens that were planted centuries ago. Some 2,000 mature elms were felled in the 1980s to control Dutch elm disease, but 20,000 saplings are now being planted in a woodland restoration programme. ㉗ Among the organisations located in the park are the Ordnance Survey of Ireland,* ㉘ Dublin Zoo,* and An Garda Síochána (Irish police force). A vice-regal lodge, built there for the crown's representative in Ireland, is now Áras an Uachtaráin (the residency of the President of Ireland).

PHOENIX PARK

"Useful information and innocent amusement"

Numerous gentlemen, earls and other Dublin worthies met at the Rotunda Gardens* in 1830, and agreed to establish a zoo in Dublin to "provide useful information and innocent amusement… for those who have not the benefit of travel". The following year, Dublin's **Zoological Gardens** opened in the Phoenix Park on a site donated by the lord lieutenant, and with just one wild boar. By 1832, however, there were 47 mammals (15 of them monkeys), 72 birds and four reptiles; today there are over 1,000 animals in the park. However, during the 1916 Rising and the Emergency years of World War II, food and fuel shortages meant the donkeys and goats were sacrificed to feed the lions and tigers, and park trees felled to feed furnaces in the tropical houses. In 1927, the lion (called Cairbre) that was to become MGM's trademark roaring lion was born at Dublin Zoo. ㉙

There have been private menageries since ancient times, when rulers and wealthy people kept and exchanged captive animals – witness, for instance, the 2,000-year-old skeleton of a Barbary ape found at Armagh's *Emain Macha*/Navan Fort.* The modern notion of a public zoo began in 1765 with the founding of Vienna's Imperial Menagerie; Madrid followed suit in 1775, Paris in 1794, London in 1828, and Dublin in 1831, becoming only the fifth city with a public zoo. Many of Dublin's initial acquisitions came from London Zoo, including the first giraffe ever seen in Ireland, which attracted huge crowds in 1845 despite growing fears about the failure of the potato* crop.

Today the emphasis is less on information and amusement, and more on conservation and education: Dublin is part of 30 international breeding programmes for endangered species, and manages the European studbook for the golden lion tamarin and the Moluccan cockatoo. Endangered species bred at Dublin include snow leopards, cheetahs and orangutans. Zoologists at the zoo are also researching ways to enrich the conditions for its captives: animals are no longer simply given their food, for instance, but have to forage for it; this reduces boredom and stress, and prepares them for possible introduction to the wild. Following concerns about cramped and inappropriate conditions during the 1980s and 1990s, the zoo spent €20 million on improvements, doubling its size and creating an open grassland for the larger African mammals that is more wildlife park than zoo.

PARKGATE STREET

Muspratt and the alkalis

The founder of Britain's modern chemical industry was a colourful Dubliner **James Muspratt** (1793–1886). Muspratt was born in Dublin of English parents and, after fighting in the Napoleonic wars, returned to open a dye factory at 14 Parkgate Street. In 1822, however, he moved to Liverpool and started a chemical empire. The year 1822 was significant, for that was when the government abolished the hefty £30 tax on salt, giving the infant chemical industry a much-needed boost. Salt was an important raw material, and crucial to a process devised by French chemist Nicolas Leblanc for making soda (sodium ash or sodium carbonate), an alkali that was vital to the glass, paper, soap and chemical industries. (Alkali comes from *al-kali*, meaning 'ashes', as soda and potash were originally made from the ashes of certain plants.) With the salt tax gone, Leblanc's process was at last economic, and Muspratt was quick to cash in, moving to Liverpool to be closer to the Cheshire salt mines. He was joined initially by another Irish entrepreneur Josias Gamble,* but their partnership was short-lived.

The Leblanc process produces noxious fumes and acidic by-products, and for decades Muspratt's business was embroiled in legal wrangles over pollution (eventually resolved with the erection of tall absorbing chimneys). Despite the problems, Muspratt was so successful he has been called 'the father of the British alkali industry'. By 1840, he was producing nearly 20 per cent of Britain's soda and exporting to North America. Perhaps if the Carrickfergus salt deposits* had been known in the 1820s, Gamble and Muspratt would have stayed at home, and Ireland's industrial development would have been different?

CHURCH STREET

Coal-gas and streetlights

St Michan's Church, founded in 1095 by the Vikings, is famous for the mummified bodies in its vaults and for its 12th-century tower, which is the oldest-surviving building on the northside. ㉚ But it is also believed to be the final resting place of the **Rev John Clayton** (*d.*1725), the man who discovered coal-gas and thus made possible the gas streetlighting and gas plants of the 19th century. Clayton was a vicar at Crofton in England before becoming rector at St Michan's in 1700. While at Crofton, he heard of a stream that burned "like brandy" and how people even boiled eggs over the flaming water. On investigation, Clayton discovered

coal in the ground and realised that the gas came not from the water, but from the coal. Back home, he experimented by heating some coal in an oven; this gave off a flammable gas that he collected in a bladder, and then lit as it escaped. Clayton described his discovery in a letter to Robert Boyle,* which was later published by the Royal Society. Yet his "distillation of coal" was not exploited until the early 1800s when companies began making coal-gas, selling it for lighting and later for cooking and heating. Coal-gas was eventually replaced by town gas (made by distilling, or 'cracking', petroleum products, such as naphtha), which in turn was replaced by natural gas.

Streetlighting began in Dublin in 1616 when, in a bid to fight crime, the corporation decreed that on winter nights every fifth house must display a lantern and candle – previously, people setting out at night carried their own torch. Public streetlighting began in 1697 when the corporation allocated money for tallow lamps. By 1800, Dublin had 6,000 lamps – today there are 46,000 – and teams of lamplighters to maintain, fill, light and extinguish each lamp every day. Tallow, whale, fish and rapeseed oils were all burned at various times. The corporation initially rejected coal-gas as 'noxious', but relented as the quality of gas lamps improved – the development of an automatic filling system was a major plus – and gas streetlighting began in Dublin in 1825. It lasted well into the 1900s, and indeed the Phoenix Park* is still lit by gas. (Acetylene* was also popular for lighting in the early 1900s, and one of the city's acetylene manufacturers was the **Sunlight Gas Company** on Essex Quay; the firm also made soap, and its former premises at the corner of Parliament Street still carries a terracotta frieze depicting the story of soap.)

Ireland's first electric light was an arc lamp erected outside the Dublin offices of the *Freeman's Journal* in 1880. By 1892, Dublin Corporation had erected 80 arc lamps on the main shopping streets, powered by its new electricity generating station in Fleet Street, and some department shops installed the new carbon filament incandescent electric bulbs. These were cleaner and safer than smoky gas lamps, but were very expensive. Only when cheap, metallic filament bulbs became available in 1909 did electric lighting really become popular – and demand for electricity shot up so much that Dublin Corporation had to double the capacity of its new Pigeon House* power station. Over 500 electric street lamps were erected in Dublin, each with a dazzling 500-watt bulb equivalent to 70 gas lamps. The history of Dublin streetlighting features in a City Hall exhibition. ㉛ A Dublin gasometer* was named after Clayton: built in 1871 at Ringsend, and demolished in 1995. See also Carrickfergus* gasworks, and the Argory's acetylene plant.*

ARRAN QUAY

An instrumental district

Galileo's telescope, van Leeuwenhoek's microscope, the spectroscopes that revealed the composition of the Sun, Moon and stars, the X-ray photographs that exposed the hidden structure of DNA. Our understanding of the world around us has been shaped by a few great ideas, but mostly as a result of new and improved instruments.

Ireland had a thriving scientific instrument trade by the 1700s. Trades then often clustered into a particular district: Dublin apothecaries* gathered around Bride Street, while the instrument makers congregated around Capel Street, and Essex and Arran quays. Most instrument makers began as opticians, and when microscopes, telescopes and other optical instruments were developed, they naturally added these to their business. Many firms went on to invent, design and make their own instruments, frequently to international acclaim, and as their trade flourished in the 19th century, several moved to more upmarket locations around Grafton Street, close to the scientific establishments of TCD and the RIA.* These firms included Spear & Company of Capel Street (and later College Green), Clarke's of Sackville [O'Connell] Street, and Spencer & Company of Aungier Street. The greatest of all Irish scientific firms

was Grubb & Son,* makers of some of the world's finest telescopes at their Rathmines workshop. Two other companies stand out: **Yeates & Son**, which ceased trading only in 2001, and **Mason Technology** which is still in business after 220 years.

Sundial made by the Dublin instrument-making firm of Mason, and now at the National Museum, Collins Barracks. © National Museum of Ireland

Yeates & Son was founded in 1790 by Samuel Yeates, an optician. His was an inventive family, frequently improving on existing designs, and the firm exhibited at London's exhibition at the Crystal Palace in 1851, winning honourable mention for its surveying instruments. Many of the instruments Yeates made for universities survive in college collections. The firm later reverted to its original business as an optician, but closed in 2001. Mason Technology was begun in 1780 by Seacome Mason, grandson of an English tanner who came to Dublin in 1712. He opened an optician's practice at No 8 Arran Quay and, from the outset, sold a range of products – from reading and opera glasses

Business card for Yeates & Son, a Dublin firm of instrument makers founded in 1790, which ceased trading in 2001. Courtesy Yeates Opticians

to silk bathing caps and umbrellas, microscopes, pantographs (for scaling up technical drawings), and linen provers (magnifiers for counting threads and assessing fabric quality). Over the years, other instruments were added to the catalogue: thermometers, barometers, quadrants and theodolites and, later, photographic equipment. In the 1830s, the firm began making the wet and dry bulb thermometer invented by Professor James Apjohn,* which was sold around the world as Mason's hygrometer. Mason Technology survives as an equipment supply company, run by the seventh generation of the Mason family from offices on the South Circular Road.

The oldest-surviving, Irish-made scientific instrument is a surveying compass made in 1667 by W. R. Dublin and now at Oxford University. The oldest scientific instrument in Ireland is thought to be a rotula contained in a 15th-century astronomical manuscript owned by the RIA.* The **National Museum** has a fine collection of Irish and international scientific instruments, some of which are on show at Collins Barracks, including a Grubb telescope and a Mason sundial. (32) The oldest object in the collection is an ornate 400-year-old astrolabe (navigational aid), made in Prague *c.*1600 and designed for calculating the positions of heavenly bodies. Also on display are a linen prover (*c.*1730) and, in the furniture gallery, an enormous old studio camera, microscopes, barometers (including one made by Mason's), globes and an orrery. The latter, a clockwork model of the solar system, was named after **Charles Boyle** (1676–1731), Earl of Orrery in Cork (and grandnephew of the chemist Robert Boyle*), for whom one of the first such devices was made.

Elsewhere in the museum, you can see a souvenir slice from the 1866 transatlantic telegraph cable,* the 30,000-year-old leg bone of a mammoth found 100 years ago in Shandon Cave,* and the massive meteorite* which fell at Brasky in Co. Limerick in 1813. You can also examine the inner workings of the barracks clock: made in 1849, it has a complex timekeeping mechanism of cogs and weights, which tick off the seconds and drive the hands on the clock faces outside. In 2001, the museum acquired an extensive collection of old telecom equipment and apparatus from Eircom. The **Chester Beatty Library**, donated to the State by US mining magnate **Chester Beatty** (1875–1968), is renowned for its manuscripts and rare books, including early works on science and mathematics, but also has several historic Arab and European navigational instruments. (33)

JERVIS STREET

Mixing humours – Ireland's first blood transfusion

On March 27th, 1865, 14-year-old Mary Ann Dooley mangled her arm when she caught her hand in a roller at the paper mill where she worked. She was admitted to Jervis Street Charitable Infirmary, but later developed serious lockjaw (tetanus). Violent spasms prevented her from eating or drinking, and treatments with tobacco, chloroform, valerian and deadly nightshade were to no avail. In a last-ditch attempt to save her life, **Dr Robert McDonnell** decided to give her some of his own blood, and on April 20th he performed Ireland's first blood transfusion. He took 12fl oz (0.35 litres) of blood from his left arm; stirred the blood, strained it through muslin, then pumped it "into the corresponding vein in the patient's left arm" using a syringe and piston.[13] The young girl, conscious throughout, is said to have described "feeling an agreeable sensation, an undefined sensation of warmth". The spasms continued, however, and she died the following day. Undeterred, McDonnell conducted a dozen transfusions over the next decade, and recommended the procedure for treating various illnesses including cholera. Transfusing was safe and simple, he said, and the blood donor could be a patient's relative or a willing medical student. (Incidentally, McDonnell's experiment was probably the basis for the blood transfusion described in Bram Stoker's *Dracula*; Stoker could have heard of the experiment from his brothers who were surgeons in Dublin at the time.)

Blood's life-giving properties had long been known and in Leviticus it is written that "the life of the flesh is in the blood". Ancient Romans drank the blood of dying gladiators and, in 1492, Pope Innocent drank the blood of young boys in a vain attempt to stave off death. Transfusion was first suggested in the 1600s, after William Harvey discovered the circulation system, but blood was still believed to be one of the body's humours, central to a person's personality – people could be cold-blooded or hot-blooded, for example – and early transfusions were concerned with mixing humours and treating what were perceived as personality disorders, rather than blood loss.

Richard Lower at Oxford gets the credit for the first transfusion when, in 1665, he transfused blood from one dog to another. Two years later the first transfusion involving a human was performed in Paris by Louis XIV's physician, Jean-Baptiste Denis, who gave a small volume of lamb's blood to a feverish youth (amazingly, the boy recovered). Some months later, before an audience at the Royal Society in London, Richard Lower took half a pint of blood from a docile sheep and gave it to 22-year-old Arthur Coga, described as a harmless lunatic and eccentric scholar, who was paid 20 shillings for volunteering. Coga also survived, but the third recipient was less lucky: Antoine Mauroy died in Paris after receiving two transfusions of calf's blood, though whether due to the operations or because his wife had poisoned him was never clear. Nonetheless, countries rushed to ban the procedure, and it was nearly 200 years before doctors risked it again.

By the late-19th century, however, blood transfusion was again tried to treat various problems, including severe blood loss in women who haemorrhaged after giving birth. Some transfusions were successful, but many failed inexplicably and the patient died. The reason became clear in 1900, when a Viennese doctor Karl Landsteiner discovered the first blood groups (the rhesus factors were not discovered until 1940). Landsteiner showed that blood from some donors will 'clump' when mixed with blood from certain recipients. In all, he found four blood groups: A, B, AB (both A and B), and O (neither A nor B). O-type blood could be given to anyone (the universal donor), and AB people could receive any blood (the universal recipient). The most common Irish group is O (55 per cent of people), the rarest is AB (3 per cent).

Once donor and recipient were carefully matched, the success rate of transfusions dramatically increased. The procedure was still used only in emergencies, however, and because there was as yet no way to store the blood, it was taken directly from donor to recipient, their veins connected by a tube. By the 1920s, large hospitals had panels of volunteers who could be called at any time to donate. Belfast's donor panel, for example, comprising policemen and former soldiers, was begun in 1924 by Sir Thomas Houston. It was an Irish-Swedish doctor Almroth Wright (see: **The Doctor's Dilemma**), who made the breakthrough

The Doctor's Dilemma

The colourful and often controversial **Sir Almroth Wright** (1861–1947) was half-Irish, half-Swedish and English born.[14] He made many important contributions to science, and has been ranked on a par with the great bacteriologists Pasteur and Koch, but he was also a lawyer and gifted writer, and his friend George Bernard Shaw once said that he could "handle a pen as well as I". Wright studied in Dublin and Belfast, and worked in Europe and Australia before settling in England. In 1896, working as an army doctor, he developed the first successful vaccine against typhoid. Vaccines were still controversial, and it was not until World War I that his typhoid vaccine was widely used and all soldiers were vaccinated – saving, it is estimated, over 100,000 lives. In 1902, Wright started an inoculation department at St Mary's Hospital in Paddington that quickly gained an international reputation. Alexander Fleming, who later discovered penicillin there, was one of those who joined the centre that, today, is called the Wright-Fleming Institute.

Wright was one of the first to explain how immune cells (phagocytes) fight infection by attacking microbes present in the blood. An explanation of this phagocytosis appears in *The Doctor's Dilemma* (1906), Shaw's satirical play about the medical profession. Sir Colenso Ridgeon, the central character, is based on Wright and the 'dilemma' – the doctor is forced to decide between treating a mediocre doctor and a brilliant artist – was prompted by Wright's remark to Shaw, that, with scarce resources, doctors were increasingly forced to decide who to save. During World War I, Wright and Fleming studied why so many wounded soldiers developed gangrene. Controversially, they concluded that part of the problem was swabbing the wounds with antiseptics – the antiseptics could not penetrate deep wounds where infection lurked, and also killed off the immune cells that were fighting the infection.

that paved the way for blood banks, when he showed that calcium plays an important role in clotting. In 1914, it was discovered that adding sodium citrate would sequester the calcium, preventing the blood from clotting and keeping it fresh for three weeks. This, coupled with improvements in refrigeration, led to the setting up of blood banks in the 1930s, and donors could now donate at their own convenience. Transfusion and blood storage techniques improved during World War II and, in 1946, the Northern Ireland Blood Transfusion Service was established in Belfast; two years later Dr Noel Browne, then minister for health, launched the Republic's service.

Like all potentially life-saving medical procedures, blood transfusions come with health warnings: blood typing removed some of the dangers, but problems remain, particularly transmitting diseases via contaminated blood products. Consequently the search is on for artificial substitutes, but blood is complex and not easily replaced and, with no alternative yet in sight, donors are still needed. Jervis Street Infirmary, one of the hospitals that benefited from the 1742 charity première of Handel's *Messiah*, closed in the 1980s to make way for a shopping centre.

The 'club orange' bottle?

In the days before screw tops, several types of bottle were invented for storing carbonated drinks and mineral waters. One bottle, designed by Englishman Hiram Codd, had an internal glass marble that was held in place at the bottle's mouth by the pressure of the carbon dioxide gas in the drink. Dublin inventor **Francis Hamilton** patented his glass bottle in 1809: shaped like a wooden club with a pointed bottom, it had to be sat upright in a special stand when it was brought to table; otherwise it was stored on its side, and closed with a cork held in place by wire, just as champagne bottles are. According to Hamilton, this arrangement prevented the gas escaping. His bottles were used by Thwaites & Co, which had a mineral water factory in Moore Lane for over 100 years. Thwaites, founded by Dublin apothecary Augustine Thwaites, began making mineral waters in the mid-1700s. Its Moore Lane factory closed in 1927, and the company was later taken over by the Irish drinks firm of Cantrell & Cochrane. C&C concocted the popular 'club orange' soft drink in the 1930s – it was named, not after the Hamilton bottle, but after the Kildare Street Gentlemen's Club, where many of the company's directors were members. When Hamilton invented his bottle there were several 'glasshouses' or glass factories in Dublin city; today, AGB Scientific (Associated Glass Blowers) makes specialised laboratory glassware in Glasnevin, but the city's last glass bottle factory, Ardagh Glass at Ringsend, closed in 2002.

Hamilton's patented club bottle: used for 100 years by Thwaites Mineral Waters of Dublin.
© Cantrell & Cochrane

"Accouching the Rotundities"

Dublin's **Rotunda** is the world's oldest-surviving maternity hospital. It was founded in 1745 by **Dr Bartholomew Mosse** (1712–59), a surgeon and midwife who was born in Maryborough [Portlaoise]. Mosse trained abroad – Ireland having no midwifery school then – and returned determined to establish a lying-in and teaching hospital to treat Dublin women, rich and poor. His hospital opened in a small premises on South Great George's Street, which it quickly outgrew, so Mosse began campaigning for a larger premises: running lotteries (for which he was arrested), seeking donations, and holding fundraising events (his fund benefited from the charity première of Handel's *Messiah*). In 1757, his New Lying-in Hospital, as it was called, opened in fine purpose-built premises, popularly known as the Rotunda, on account of its tower and cupola, or "Accouching the Rotundities", as James Joyce dubbed it in *Ulysses*. Sadly, Mosse never lived to enjoy its success: he died two years later, penniless and exhausted.

The hospital set out to provide for "destitute females in their confinement, [to ensure] a supply of well-qualified male and female practitioners throughout the country, and [to prevent] child murder". Designed by noted architect Richard Cassells, it had a pleasure garden, theatre and concert hall where the fundraising continued, to enable the hospital to treat poor women for free. The hospital's theatre is now the Gate Theatre, its concert hall the Ambassador Cinema. The Rotunda still relies on donations and volunteers. From the outset, the Rotunda had an international reputation for its midwifery training – classes for male and female midwives began in 1774 and attracted students from Britain, Europe, Russia and North America – and for its novel attitude to hygiene, and success in reducing deaths from childbirth fever.

Childbirth fever was a constant problem in every maternity hospital – at one point in the early 1800s, it was so bad that the Rotunda nearly closed. Until Pasteur's germ theory of disease was established in the 1860s, most doctors pooh-poohed the suggestion that they were responsible for spreading infection by examining patients without washing their hands. Yet, as early as 1790, **Joseph Clark** at the Rotunda realised that cleanliness was important in controlling childbirth fever, and he instituted hygiene regulations for doctors and midwives aimed at reducing infection rates. His son-in-law, Scottish-born **Robert Collins** (1800–68), succeeded him as Master at the Rotunda and took the practice further, though many medics ridiculed his ideas. Collins insisted that wards be filled and used in strict rotation, and disinfected and fumigated with chlorine afterward; straw from used mattresses was burned, and linen thoroughly washed; staff had to wash their hands before seeing a patient; ventilation was improved; and women with a contagious disease were quarantined in separate isolation units. Collins succeeded in bringing down the death rate among mothers at the Rotunda to 1 per cent – compared with 4–5 per cent in many continental hospitals.[15] His book, *A Practical Treatise on Midwifery* (London, 1835), attracted international attention and some European hospitals began experimenting with his revolutionary ideas. But childbirth fever remained a problem, and was not eradicated until the discovery of antibiotics in the 1930s.

One of the Rotunda's 20th-century innovators was Dublin doctor **Robert Collis** (1900–75) who pioneered the technique of feeding premature infants via a tube through the nose – previously, they were spoon fed, usually unsuccessfully, as they had not yet learned to suck – and designed an incubator for premature infants that was cheap to make and which less well-off countries could afford.[14] Collis exposed the dreadful living conditions in Dublin slums, and his play on the topic – *Marrowbone Lane* – was premièred, appropriately, at the Gate Theatre. Working with the Red Cross at Belsen concentration camp, Collis helped revive emaciated children. In Dublin, he started a cerebral palsy clinic, where he treated the young Christy Brown, later assisting him to publish his autobiography, *My Left Foot*. Brown returned the favour, writing a foreword to Collis's autobiography, which was published after Collis's death from a riding accident.

MARLBOROUGH STREET

The *Cabinet Encyclopaedia*

There was a tremendous thirst for practical knowledge in the 19th century, witness the many mechanics institutes, libraries, societies "for the diffusion of useful knowledge" and suchlike that started then. A great and prolific writer supplying that market was **Dionysius Lardner** (1793–1859), who was born in Marlborough Street and earned an international reputation, particularly for his 133-volume *Cabinet Encyclopaedia*. Lardner took degrees in mathematics, philosophy and science at TCD, where he later worked as a 'grind', helping students cram before exams. Some of his first books were aimed at that market. He also taught at the Dublin Mechanics Institute and the RDS* and was an innovative, popular lecturer – for an acclaimed series on steam engines, he used cutaway demonstration models of the engines. His early books on algebraic geometry and calculus won him an international reputation and, in 1828, after assiduously campaigning on his own behalf, he was appointed as a science professor at the new University College London. Five years later, he left to concentrate on his *Cabinet Encyclopaedia*. Lardner wrote most of this himself over a 20-year period, though he also took contributions from various experts. He produced a phenomenal number of other popular books and schoolbooks on a tremendous range of subjects, from silk manufacture, heat and railways, to the electric telegraph. And he found time for a complicated personal life: one of his extra-marital children was the playwright Dion Boucicault, and there was a scandal when, in 1840, he left his wife for a married woman. Lardner died while visiting Naples in 1859.

SEÁN MACDERMOTT STREET

In slow motion

A bullet smashing through glass, a drop of water falling, an insect in full flight – we can watch all these in slow motion thanks to Dublin inventor **Lucien Bull** (1876–1972). Bull was born at Gloucester [Seán McDermott] Street Upper, where his father was a merchant and 'Parisian agent'. He was educated in France (his mother was French) and, in 1896, went to work with a famous French biologist Étienne Jules Marey. Marey had a small laboratory at his Paris home, where he was using high-speed photography to study human and animal movement. Bull pioneered ultra-rapid cinematography, striving to capture as many images a second as possible, so that fast events could be viewed in slow motion. The technical difficulties were considerable: how to handle delicate film moving at speed, how to light the moving subject, and with apparatus that initially shot one frame every 15 minutes. Yet, by 1902, Bull had achieved 500 images per second, by 1910 he had clocked 5,000 frames, and 40 years later he recorded 1 million frames a second. The man who gave us a glimpse of a bullet-in-flight and an insect's wing beating, spent his life at the Institut Marey,

where he was later the director. A prolific inventor, his many patents included cameras, an improved version of the electrocardiogram (1938), and various optical, acoustic and scientific instruments, all of which won him numerous honours in France and elsewhere. Paris's Musée des Arts et Métiers and the Cinémathèque Française have a wealth of information on this Irish inventor.

ECCLES STREET

1865 – the first practical endoscope

It was a Mater Hospital physician, **Francis Cruise** (1834–1912), who invented the first practical endoscope, which enabled doctors to see inside a patient's body. A primitive endoscope had been invented in the USA in 1827 – a cystoscope, for looking inside the bladder – but its light source was poor and it was not effective. Cruise's version had a powerful paraffin lamp, and mirrors to reflect the light into the patient's bladder. At its first trial in 1865, Cruise used it to identify correctly mystery objects (a bullet and screw) concealed in a cadaver's bladder.[15] His device received considerable international attention and commercial versions came with various attachments for peering into nearly every orifice – bladder, ear, nose, throat, stomach, rectum and womb. It was widely used across Europe, until the invention in Germany of an improved lighting source in 1877. Cruise, who was born in Mountjoy Square and is buried at Glasnevin Cemetery, was a gifted cellist, crack marksman and classical scholar, and an authority on Thomas à Kempis (1380–1471), author of the devotional *The Imitation of Christ*. The Mater Hospital, founded by the Sisters of Mercy in 1861, is still at Eccles Street. The Royal College of Surgeons in Ireland* has Cruise's original endoscope.

The first practical endoscope, invented by Francis Cruise. The box holds a paraffin lamp; the attachment is a cystoscope to view inside the bladder. Courtesy Davis Coakley

Chapter 1: County Dublin

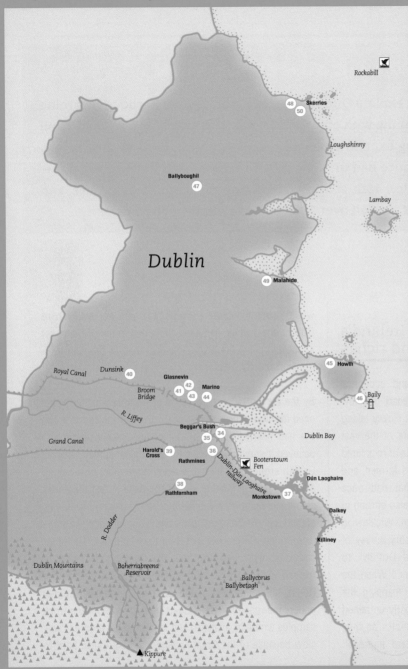

Rockabill

Skerries 48 50

Loughshinny

Lambay

Ballyboughil 47

Dublin

Malahide 49

Royal Canal   Dunsink 40

Glasnevin

Broom Bridge   41 42 43   Marino 44

R. Liffey

Grand Canal

Beggar's Bush

Harold's 39   35 34

Cross   36

Rathmines

Rathfarnham 38

R. Dodder

Howth 45

Baily 46

Dublin Bay

Booterstown Fen

Dún Laoghaire

Monkstown   37

Dalkey

Killiney

Dublin Mountains

Bohernabreena Reservoir

Ballycorus Ballybetagh

▲ Kippure

Visitor facilities

34 Geological Survey of Ireland, Beggar's Bush
35 National Print Museum, Beggar's Bush
36 Institution of Engineers of Ireland, Clyde Rd
37 National Maritime Museum, Dún Laoghaire
38 Rathfarnham Castle
39 Mount Jerome Cemetery, Harold's Cross
40 Dunsink Observatory
41 National Botanic Gardens, Glasnevin
42 Met Éireann, Glasnevin
43 National Metrology Laboratory
44 Dublin Fire Brigade Museum, Marino
45 Howth Transport Museum
46 Baily Lighthouse
47 Ballyboughil Hedgerow Society
48 Ardgillan House, Balbriggan
49 Malahide Demesne
50 Skerries Mills

♀ woodland
⚐ nature reserve or wildfowl sanctuary
⌂ lighthouse

Visitor facilities are numbered in order of appearance in the chapter, other places in alphabetical order. How to use this book: page 9; Directory: page 474.

Travel on Ireland's first railway, explore Dún Laoghaire's vast 'asylum harbour', examine early printing technology, commune with eminent Victorians at Mount Jerome, and stroll Killiney beach where pioneering experiments with explosives gave birth to the modern science of seismology. Dalkey once had an atmospheric railway that 'sucked' the trains uphill, some of the world's greatest 19th-century telescopes were built at Rathmines, and a boggy field at Ballybetagh has yielded over 100 fossil skeletons of giant deer.

BEGGAR'S BUSH

Mapping Ireland's underground riches

By 1800, there was growing scientific and economic interest in geology. Mining companies needed to know where to dig, and geology as a science was slowly taking shape, as people became increasingly aware of the vast timescale involved, saw how the various strata of rock fitted together, learned how the various rock types formed, and uncovered fossils of extinct creatures that had once inhabited the Earth. Geological interest was also stirring in Ireland. There had been a gold rush* in Co. Wicklow in 1795 and, in 1809, the RDS* began a cursory survey of certain areas, and commissioned Richard Griffith* to investigate the coalfields around Castlecomer.* When the Ordnance Survey of Ireland (OSI)* began mapping the country in the 1820s, the field workers (mostly untrained foot soldiers) were told to collect rock samples as they worked, though this random and haphazard approach

would never constitute a useful geological survey. Richard Griffith, meanwhile, had been appointed to the OSI, and set about compiling his own geological map of the country – much of it on OSI time.

By the early 1840s, there was a growing realisation of the need for an organised geological survey of the country. The OSI had nearly finished producing its detailed maps and believed it should undertake the geological survey. Richard Griffith argued that he had already done most of the work and, with a little funding, could complete it. But there was also a campaign to establish a separate geological survey, staffed with trained geologists and using the new OS maps to produce a more detailed map than Griffith's. The latter campaign won, and, in 1845, the **Geological Survey of Ireland** (GSI) was established, with offices at the Custom House,* together with a **Museum of Economic Geology** under the chemist Sir Robert Kane,* where the rock samples would be analysed and stored. Initial progress was slow because of the Great Famine; staff were also hard to

Part of the Geological Survey of Ireland field sheet for Co. Wexford, where the survey began in 1845, and bearing notes and remarks added by surveyors over the subsequent 50 years.
© Geological Survey of Ireland

detailed maps and accompanying memoirs. These are often consulted still, a lasting testament to the work.

After this first flurry of activity, however, the survey entered a dormant phase, during which time Ireland was dismissed as having little or no mineral wealth. It began awakening from this slumber in the early 1950s with the arrival of a new generation of trained geologists, the development of new survey techniques, and a growing interest in Ireland's geological resources. Substantial mineral deposits were discovered at Avoca* and Abbeytown,* and new coalfields at Castlecomer, and a search began for uranium* deposits. In 1958, the government drafted its first *Programme for Economic Expansion.* Then, in 1960, there was confirmation of a substantial lead-zinc deposit at Tynagh in Co. Galway, and the country's first sizeable modern mine opened there some years later. Ireland's economic expansion had begun: during the 1950s, just 23 prospecting licences were awarded but, in 1962 alone, nearly 120 were issued; and the GSI, which employed five geologists in 1964, recruited 13 in 1968. More discoveries followed: Silvermines,* Gortdrum… culminating in 1970 with Navan,* one of the world's largest zinc deposits of recent times, and the Kinsale gasfield* which was confirmed in 1974.

come by, and, after a year, the survey had to make way for famine relief agencies, so it joined the Museum of Irish Industry* at No 51 St Stephen's Green.

Fossil of Vestinautilus formosus from 340-million-year-old limestone at Lisbane, Co. Limerick; part of the GSI's heritage collection of Irish fossils.
© Geological Survey of Ireland

Initially, it was thought the work would take 10 years and, in the summer of 1845, a small team of surveyors began at Hook Head* in Co. Wexford. In the end, it took 40 years to survey and map the whole island.[16] Some of the best geologists of the day joined the GSI, among them Thomas Oldham* and Joseph Beete Jukes,* and they had the benefit of the detailed new OS maps, which were at an unprecedented scale. But the surveyors worked in difficult conditions, spending eight months of each year in the field and four months in Dublin analysing their finds. They examined bedrock, fossils, rock formations, landscapes, and the quaternary deposits of soil, sand and gravel. Artists, notably George Victor du Noyer,* recorded the findings and formations. The final result, completed in the late 1880s, was some 200 wonderfully

Today, the GSI employs over 50 geologists at its Beggar's Bush headquarters, including experts in bedrock, quaternary deposits, groundwater, land use and soils, minerals and offshore deposits. They continue to collect information about Ireland's geology, produce detailed maps and even popular guides to places of geological interest, and provide expert advice to local authorities and private companies on everything from rehabilitating mine wastes to groundwater reserves. With the Marine Institute, the GSI is also analysing the structure and composition of Ireland's seabed. The GSI has extensive archives dating from 1845, 20,000 boxes of drill cores from around the country, and 24,000 Irish fossils (recently catalogued for a heritage project). It hosts occasional exhibitions about Irish geology and landscape, and has a map sales office. ㉞

BEGGAR'S BUSH

The printed word

Printing* is one of civilisation's greatest and most powerful inventions. The first mass media, it replaced the monks in their scriptoria, made the written word cheap and plentiful, and fanned the flames of political revolution and religious reformation. Printing was initially a slow process demanding great skill: the printing blocks for each letter or character had to be individually carved; each word was manually pieced together, or composited, letter by letter; and the sheets of paper were fed one by one into a printing press. In the 19th century, books became cheaper and more widely available as steam-powered machines made long print-runs possible, and hand-composition gave way to linotype machines with keyboards that could set a full line of text in one go. These were the steamy, smelly, noisy days of hot metal and men in aprons: the type was cast from molten lead which, being soft, could be used for only one print-run before the quality deteriorated; the cast was then melted down and the lead reused. *Irish Independent* newspapers was still using linotype machines in the late 1980s, but changed then to computerised typesetting and page-making systems. Today, anyone with a PC and printer can be their own publisher, choosing from a myriad of typefaces and font sizes at the push of a button.

The many developments and changes from hand-composition to desktop publishing can be seen at the **National Print Museum**. The collection includes a 17th-century compositor's stick used in hand setting, linotype machines and lead melting pots, setting rules, Gestetners, and a device used to print the red and blue rules in school copybooks. Printers often donate equipment when they upgrade, so the collection continues to grow. It is housed in the garrison chapel of the former Beggar's Bush barracks; framed copies of historic editions of Irish newspapers line the walls. ㉟

CLYDE ROAD

Ingenious Ireland

Engineer, engine and ingenious... the three words share a common root in the Latin term *ingenium*, meaning a clever device. Down through history, engineers have provided us with ingenious devices and solutions, from engines and other machines of various kinds, to the infrastructure that makes travel possible, and the public health works that bring us power and clean water and take our dirt away. Often their contribution is unseen, unknown and unacknowledged. In the early 1800s, the government planned the Shannon navigation,* arterial drainage schemes, the railway network, new harbours and the commercial exploitation of bogs, and all the groundwork was done by engineers, especially two Scotsmen, Alexander Nimmo* and John Rennie.* Noted Irish engineers included Bindon Blood Stoney,* who developed Dublin's docklands; lighthouse engineer Alexander Mitchell;* Thomas Mulvany,* who developed Germany's Ruhr coalfield; and mathematical engineer Peter Rice,* who helped shape the roof of Sydney Opera House. The Institution of Engineers of Ireland (founded 1835) has an extensive engineering archive. ㊱

BOOTERSTOWN

A fragment of the old coast

Booterstown Fen, a small piece of suburban wilderness sandwiched between the Rock Road and the DART railway, is the last-surviving remnant of un-reclaimed shoreline in South Co. Dublin. Ironically, it owes its continued existence to the railway. The original shoreline was at the Rock Road, fringed with salt marshes that were covered at high tide.

But, in 1834, the Dublin–Kingstown [Dún Laoghaire] railway* embankment was built along the shore, enclosing 19 hectares of salt marsh between Merrion Gates and Blackrock. Much of the enclosure was drained for development, apart from 4 hectares that, fed by streams and a spring, became Booterstown Fen. A sluice gate and drains were added during World War I so that crops could be grown there and, in 1971, An Taisce (the heritage group) acquired it to manage it as a nature reserve. During the 1990s, however, conditions there deteriorated: there was no longer enough oxygen circulating in the water as local building work had cut the fen off from its essential groundwater sources. An invasive rush began colonising the area, choking the waterways and threatening a rare grass (*Puccinellia fasciculata*) growing there. Fortunately, the remedy was simple: remove the 1914 sluice gate. The tide now floods in twice a day, flushing in much-needed oxygen; the saltwater also killed off the unwanted rush and the birds have returned in huge numbers.

MONKSTOWN

A marriage of heaven and earth

It was an astronomical love affair and it laid the foundation for modern astrophysics – the 20-year collaboration of Dublin-born **Margaret Lindsay Murray** (1848–1915) and her husband, noted English astronomer **Sir William Huggins** (1824–1910).[17] Murray, later Lady Huggins, had no formal scientific training, but had long been interested in astronomy. The pair met through her neighbour and friend, Sir Howard Grubb,* who had made telescopes for Huggins, and they married in 1875. Huggins was already known for his pioneering work in astronomical spectroscopy: using the newly invented spectroscope to analyse the light from heavenly bodies, he had examined stars, comets and nebulae, proving they were made of chemical elements that were also found on Earth. When the 1866 nova was brought to his attention by an amateur Galway astronomer John Birmingham,* Huggins discovered that the bright display was caused by a shell of hot hydrogen gas surrounding the dying star.

After marrying, Margaret Huggins worked as her husband's assistant, and later full collaborator, at his private observatory in London. They were the first to use the newly invented dry plate photographic technique in astronomical spectroscopy, which enabled them to accumulate light and to study faint objects for the first time. They studied several planets and the Orion Nebula, in whose spectrum they discovered puzzling violet lines (identified years later as ionised oxygen), and published a *Photographic Atlas of Stellar Sources* in 1889. The couple collaborated equally and jointly published their scientific papers, but the conventions of the day meant Sir William was accorded the many honours. In 1903, though, the Royal Astronomical Society did elect Margaret Huggins and her friend Agnes Clerke* as honorary members (women were not admitted on the same terms as men until 1915). An advocate of women's education, Lady Huggins bequeathed her scientific and personal papers to Wellesley Women's College in the USA. A commemorative plaque at 23 Longford Terrace, Monkstown, marks her former home.

DÚN LAOGHAIRE

1848 – the world's largest asylum harbour

In 1817, some 1,000 men began building a huge harbour at the small fishing village of Dún Laoghaire. When it was finished 25 years later, it was the world's biggest asylum harbour, providing shelter from the treacherous waters of Dublin Bay. Together with the arrival in 1834 of Ireland's first railway (see: **1834 – the coming of the railways**), it was the making of Dún Laoghaire, which is now a busy ferry port and Ireland's premier maritime centre. Yet no one would have predicted this in the early 1800s, when Dublin Bay was rated among the world's most treacherous on account of its shifting sandbanks, and ships preferred to dock outside the bay at Howth* or Dalkey.* There were

One of Thomas Romney Robinson's* anemometers: erected in 1852 on Dún Laoghaire's east pier, it is now part of a computerised station where you can watch weather data being collected.
© Enterprise Ireland

frequent tragedies, the worst during a storm in November 1807 when two troop ships were wrecked, and locals watched helplessly from the shore as 400 people drowned. The disaster prompted Captain Richard Toutcher, a Norwegian mariner who had settled in Dalkey, to campaign for an 'asylum harbour' at Dún Laoghaire where ships could shelter. It was 1817, though, before permission was granted and work began.[18]

The harbour, designed by George Rennie,* had a 1km east and a 1.5km west pier, embracing 102 hectares of sheltered water, and forming the world's largest artificial harbour. In 1821, George IV came to inspect developments, and Dun Leary (as it was then known) was renamed Kingstown in his honour. Massive granite blocks for the piers – philanthropically paid for by Toutcher – were quarried at nearby Dalkey, and transported on a gravity-fed tramway: trucks laden with stone fell from Dalkey, pulling the empty trucks up. (The route of the quarry tramway was later followed by the Dalkey railway.) The harbour was never intended as a landing port, but the new steamships of the day were too big for Howth, so, in 1826, they began docking at Kingstown's half-built piers. The piers were completed in 1848, but Toutcher never saw his dream complete, having died bankrupt in 1841.

Kingstown (renamed Dún Laoghaire in 1920), quickly became an important maritime centre, with a small naval base, fishing fleet, yacht clubs, and later the headquarters of institutions such as the Royal National Lifeboat Institution, Bord Iascaigh Mhara (the Fisheries Board) and the Sea Scouts. The **National Maritime Museum** was begun there in 1959, in the Mariners' Church, which was built in the 1830s for seamen whose ships were sheltering in the harbour. The core collection belonged to Robert Halpin,* who commanded the steamship *Great Eastern* on its epic telegraph cable-laying voyages, and the museum has samples of the transatlantic cables. (37)

In July 1898, Kingstown was the venue for an historic **sports radio transmission** during the annual regatta. Thick fog obscured the view for spectators on the shore, but Irish-Italian radio pioneer Guglielmo Marconi* was in a boat out in the bay, with his newfangled transmission equipment; receiving equipment, provided by George Francis FitzGerald,* was temporarily installed in the harbour house. Marconi radioed details of the race ashore in Morse Code, and they were relayed to the *Dublin Express* newspaper, which was able to print a full report. It was the first radio coverage of a sporting event, and earned Marconi and FitzGerald considerable publicity. A plaque at Moran Park (formerly Harbour) House, commemorates the transmission; the equipment used is at Maynooth Science Museum.*

1834 – the coming of the railways

Ireland's first railway opened on December 17th, 1834, and ran between Westland Row and the pleasant seaside resort of Kingstown [Dún Laoghaire]. It was the world's first suburban and commuter line, and it is still heavily used, a testament to the wisdom of the route chosen. The world's first railway was an industrial one in north England that opened in 1825 and linked Darlington coal mine to Stockton

The first Dublin–Kingstown train, taken from an illustrative book published in 1835 to mark the railway's opening.

port. The second was an inter-city route between Liverpool and Manchester that opened in 1830. Ireland's first railway company had begun planning a Limerick–Waterford line in 1825, but the government gave the go-ahead instead to the Dublin–Kingstown option. It was the perfect choice: a short route, serving a sizeable catchment population, and linking the city to the new port town. Not everyone was pleased, however. Kingstown residents particularly resented the "intrusion of such a vulgar and democratic mode of conveyance", and, to ease local objections, the line initially stopped short of Kingstown.

The 9-km line was built in 18 months, including substantial embankments, several stations and bridges, and a short section of tunnel; the embankments were laid along the shore, and tamed the hitherto wild coastline. The work was done by hand by nearly 2,000 men and was overseen by William Dargan,* who later built most of Ireland's railways. Two locomotives were commissioned – *Hibernia* and *Vauxhall*, the first locomotives ever named – and trials began in October 1834. Straffan Steam Museum* has a small-scale model of the railway's first locomotive. The line opened in December and proved tremendously popular, carrying 350,000 passengers in its first six months. "Hurried by the agency of steam," according to a contemporary report, "the astonished passenger glides [over houses and streets] to the amusements of a fashionable watering place." There were first, second, third and fourth-class carriages – the last was uncovered – and a man rode outside atop the first carriage, a hangover from

the days of the stagecoach. In August 1835, several hundred scientists from Britain took the train, having arrived at Kingstown for the British Association for the Advancement of Science* annual meeting, which was happening in Dublin for the first time.

The line was initially laid with granite sleepers, but the vibrations shattered them, and they were replaced with wooden ones; some original granite sleepers were incorporated into the sea walls and can be seen at Seapoint train station. In 1837, the railway was finally brought into Kingstown: a causeway was built across part of the harbour and a new station was opened; the portion of the harbour that was cut off by the causeway was filled in and later used as a gasworks site. The line was extended to Dalkey in 1844 with an 'atmospheric' or pneumatic railway (see: **1884 – an 'atmospheric' railway**); passengers travelling from Dublin to Dalkey changed platform at Kingstown.

The coming of the railways, which coincided with the arrival of the telegraph, marked the end of one era and the start of another. The country was changing, and quickly. Until then people could travel only as fast as a horse could carry them, now they could achieve previously unimagined speeds; long journeys became more feasible and, in 1850, you could travel from Dublin to Cork in six hours. The railways put an end to stage and mail coaches and introduced mass transport, especially with the sale of affordable third-class tickets. They fostered emigration and increased urbanisation, the development of seaside resorts and, increasingly, a standard time* zone across the country. National markets developed, with national newspapers and brand-name products replacing local ones. Indeed, whole empires were built on the back of railways. Soon after the successful Dublin–Kingstown line opened, work began on a

The Colossus, a small-scale prototype of the locomotives commissioned in 1834 for the Dublin–Kingstown railway, and now at Straffan Steam Museum. © Straffan Steam Museum

Belfast–Lisburn line to aid northern industry and, by 1859, the country had 1,500km of mainline rail. Elegant stations and railway hotels were also built during this golden railway era. Next came an extensive network of local narrow-gauge railways,* which reached their peak in Co. Donegal. The railway era ended with World War I, when interest switched to roads and motor cars, and many railway lines fell into disrepair and closed.

DALKEY

1844 – an 'atmospheric' railway

For their first 25 years, railways were marred by the 'war of the propulsion systems'. On one side were George and Robert Stephenson, who pioneered the railway locomotive; opposing them was Isambard Kingdom Brunel* who favoured engine-less trains. Brunel thought attaching a heavy locomotive to a train was crazy: the engine had to haul its own weight, a substantial railway was needed to bear the combined weight of train and engine, and the steam engine made for a noisy, smoky and rough ride. It would be faster, cheaper, safer and more comfortable to generate the power elsewhere with a fixed engine, transmit the power to the railway, and run a lightweight train of carriages on a correspondingly lightweight track.

Various engine-less systems were devised – some, such as 'rack and pinion' trains and cable cars, are still used – but the main contender was Brunel's 'atmospheric' or pneumatic railway, so-called because the fixed steam engine generated suction power to move the train. Following successful experiments with small-scale models, the developers of the new Kingstown [Dún Laoghaire]–Dalkey railway opted for Brunel's idea and, in July 1844, opened the world's first commercial atmospheric railway to considerable international attention. (Brunel's South Devon atmospheric railway opened some months later on an experimental basis, was not fully operational until 1847 and closed a year later.) A steam engine in Dalkey generated the power to pull the trains uphill from Kingstown; for the return journey they simply fell slowly downhill under gravity – and if the momentum was not enough to carry the train into Kingstown station, third-class passengers were expected to get out and push.

The pneumatic system was intricate: a cast-iron pipe was laid between the railway tracks, and an airtight piston in the pipe was connected to the train; the steam engine at Dalkey pumped air out of the pipe ahead of the train, creating a vacuum; and the atmospheric pressure of the air behind the piston pushed the train along. The pipe had a narrow slot along its top through which the piston arm moved; a complex flap and valve system let the piston arm pass, but otherwise kept the slot closed; and wheels and rollers on the underside of the train manoeuvred the flap open, and pressed it back in place afterwards. To ensure a tight seal, the flap was also greased. But maintaining an airtight seal was difficult: the grease attracted rats that ate the leather; in summer, the grease melted away; and in winter the leather froze. Running the pumping station intermittently was also costly.

Nevertheless, the Kingstown–Dalkey railway operated for 10 years, following the old tramway cutting between Dalkey quarry and Kingstown. Trains ran every half-hour between 8am and 6pm, averaging 48kmph uphill to Dalkey, and 32 kmph when falling to Kingstown. However, in 1848, the South Devon Railway Company, believing they were losing money (they were actually making a handsome profit), pulled the plug on their atmospheric railway, handing victory to Stephenson and his locomotives. A Parisian atmospheric railway continued for several years, and Dalkey's survived until 1854, when it was converted to a conventional track and gauge, though part of the line survives, known locally as 'the metals'.

When the Earth quakes

Ireland is relatively free of earthquakes, yet modern seismology was invented here in 1846, the year **Robert Mallet** (1810–81) invented a seismometer and coined the word 'seismology'. Over the next 30 years, Mallet pioneered the science of seismology, detonating explosions at Killiney beach, travelling to Naples – where he conducted the first thorough scientific analysis of an earthquake site – analysing the speed of shock waves travelling through various rock types, and compiling a catalogue of every recorded earthquake in history.

Robert Mallet, an inventive engineer, was born in Ryder Row, Dublin (a plaque marks the site of his birthplace), where his father, an Englishman, had established a brass foundry. After studying classics and mathematics at TCD, Mallet joined the business. Thanks to his scientific approach, the Victoria Foundry became Ireland's leading engineering works. They built steam barrel-washing plants

Ireland's underground movements

About 250 million years ago, as the continents moved apart and the Atlantic Ocean opened up, Ireland was as seismically active as California is today. And after the end of the last Ice Age, as the land rebounded after the weight of ice was lifted, large earthquakes were again common here, hitting over 7 on the Richter scale – geologists find traces of these quakes in sandy layers: telltale patterns where the sand, shaken by an earthquake, flowed like a liquid. From 1917–67, there was a seismic observatory in Dublin at Rathfarnham Castle;* and now the DIAS* maintains seven monitoring stations around Ireland, which send data continuously to an international network. They detect about one event a year, mostly small ones that otherwise pass unnoticed. The largest recent quake occurred on July 19th, 1984, and measured 5.4 on the Richter scale; the epicentre in north Wales was fortunately 23km below ground, and it caused only minor damage. Ireland's two most active regions are the southeast and the northwest on account of their geological history – Donegal, for instance, is geologically allied to Scotland's Great Glen fault, and several faults in Donegal still judder occasionally, even 400 million years after they formed.

for the Guinness brewery,* cast-iron swivel bridges for the Shannon navigation,* a metal lighthouse tower for the Fastnet Rock* and mortar guns for the Crimean War. The railway years of the 1840s and 1850s were a boom time: the foundry supplied tracks for Ireland's mainline routes, the ironwork for railway stations and train sheds, and equipment for Dalkey's experimental atmospheric railway.* In 1852, Mallet invented a metal plate for flooring bridges; thin yet strong, it was first used on Westminster Bridge and then widely adopted.

Mallet also indulged his wide-ranging scientific interests: experimenting with electricity, testing for pollution in sea water, investigating corrosion in iron ships, and exploring geology – how glaciers move, how slate formed, and what happens when the Earth quakes. Arab scientists of the 10th century already understood much about earthquakes, but their knowledge was lost, until Robert Mallet's classic paper to the Royal Irish Academy in 1846.[19] Mallet was attempting "to bring the phenomena of the earthquake within the range of exact science… and deducing from… the enormous mass of disconnected and often ill-observed facts… a theory of earthquake motion". He coined nearly a dozen technical terms that are still in use, such as epicentre, seismic focus and angle of emergence, and gave details of his invention of the first completely self-registering seismometer. This consisted of five glass tubes filled with mercury – four held horizontally, and the fifth vertically; a tremor would break the mercury's contact with an electrical circuit, producing a pencil mark, the length of which depended on the duration of the disturbance.

To investigate how seismic waves travel through the Earth's surface, Mallet detonated 11kg of gunpowder at Killiney beach in 1849, and measured the time taken for the shock wave to travel through the sand. He subsequently experimented with explosions at Dalkey quarry and elsewhere, and calculated that the transit time in sand was 825ft (251 metres) per second, for instance, but 1,165ft (385 metres) in granite. Today, prospecting companies exploit the same principle to identify the underlying geology in a region. After Naples was rocked by a catastrophic quake in December 1857, Mallet travelled there to conduct the first scientific analysis of an earthquake site. By analysing how buildings fell and objects were thrown, and other patterns, Mallet deduced that the cause was a series of shock waves from a deep focus or epicentre.

In 1858, he published the first global distribution map of earthquakes, and with his son John (who was later professor of chemistry at the University of Louisiana), he compiled *On the Facts and Theory of Earthquake Prevention*. This catalogued every recorded earthquake, including one in 1606 BC at Mount Sinai that is recorded in Exodus, together with details of timing and impact. From this wealth of data, Mallet looked for patterns that might be used to predict an earthquake – animal behaviour, seasonal or tidal patterns, some distribution in time or space, rivers being diverted or springs disappearing – but in vain. Robert Mallet, the father of modern seismology, was awarded numerous international honours. He moved to England in 1861 and worked as a consulting engineer, on projects such as Thomas Mulvany's* new coal mine in Germany's Ruhr valley. He frequently returned to Ireland, but died in England in 1881. A crater on the Moon is named after him and TCD's Geology Museum* has the volcanic samples Mallet collected in Naples.

BALLYBETAGH

Mystery and misnomer – the giant deer

The most majestic animal ever to roam Ireland was surely the giant deer (*Megalocerus giganteus*). It stood 2 metres tall at its shoulders and the males carried an impressive rack of antlers, weighing up to 35kg and spanning nearly 4 metres, which they used for displays and fighting. In Europe, where Stone Age people probably hunted them to extinction, 25,000-year-old cave paintings depict them with a patterned coat and a pronounced shoulder hump –

presumably the thick muscle needed to hold the antlers. These great creatures re-colonised Ireland after the last Ice Age, along with mammoth* and brown bear,* but disappeared 10,500 years ago – long before humans arrived – and for centuries no one knew why.

Giant deer roamed across prehistoric Europe and Russia, but most of the fossils come from Ireland, where they have been turning up in bogs and lake sediments for centuries; by far the greatest number, over 100, comes from Ballybetagh. Not surprisingly, the enormous antlers always attracted attention. The earliest Irish record is a skull and antlers found in Co. Meath in 1588, and one of the first scientific descriptions was written in 1677 by Dublin naturalist and philosopher Thomas Molyneux:* "A discourse concerning the Large Horns frequently found underground in Ireland, concluding from them that the Great American deer, called a Moose, was common in that Island" (whence, presumably, the notion that the deer was an elk or moose).

Fossil of a misnomer: often called the 'great Irish elk', the giant deer was neither Irish (being found across Europe and Russia), nor an elk, but a true deer, and closely related to the fallow deer.

But why did they disappear from here? Many blamed the enormous antlers, and pictured the males fatally snagged in shrubs or trapped in mud under the excess weight. Some suggested severe epileptic fits, caused by a rush of blood to the brain when the antler velvet was shed in winter. Yet comparative studies reveal the antlers were not disproportionate, merely what you would expect for such a large animal, and unlikely to be a problem. More recently, scientists have blamed malnutrition and climate change.

Much of the evidence comes from the deer's jaws and teeth, and from Ballybetagh. At the end of the last Ice Age, Ballybetagh was a narrow gorge with a small lake where male deer gathered each winter. In the early 1800s, it was still damp enough to be called Killegar Lake; today it is a boggy field. Drainage work in the 1840s uncovered the first fossil, and, over the next century, fossil hunters found dozens more, plus the scattered bones of many more deer that died by the lakeside and were later trampled by their herd mates. Modern investigations and pollen studies reveal that, when the deer were there, the vegetation was shrubby grassland. We know from their jaws and teeth that the deer were grass-eaters and, thanks to Ice Age debris and crushed limestone, the grass then would have been rich in calcium – perfect for deer that needed phenomenal amounts of calcium each summer to grow their annual rack of antlers. But 10,500 years ago, Ireland was plummeted back into a local Ice Age – the (Lough) Nahanagan* period. It lasted 500 years, when the warm Gulf Stream temporarily disappeared and temperatures dropped by 7°C. Only a sparse tundra vegetation grew then and, in the absence of grass, the giant deer probably starved to death. Ballybetagh Bog is on private land and part of a farm. Fossil giant deer can be seen at the Natural History Museum* or TCD'S Geology Museum.*

BALLYCORUS

The lead mine's chimney

Ballycorus chimney: the flue entered by the small 'door'. An upper brick layer was removed for safety after the smelter closed.
© Enterprise Ireland

Throughout the 19th century, lead was in strong demand for use in roofing and pipes. There were several lead mines around the country, and most of the ore was smelted at Ballycorus. A small lead mine had opened there in 1807, which also produced some silver and, in 1824, the newly formed Mining Company of Ireland bought the mine and the associated works. By 1836, they had established a sizeable smelter and, to conduct away the noxious fumes, ran a long flue to the top of a nearby hill and erected a tall chimney there. That chimney was replaced in 1862 by a taller granite and brick one, which quickly became a prominent landmark. The flue, most of which has since disappeared, ran for over 1km and was large enough to walk through – children were regularly sent in to recover the soot, which contained significant traces of silver. The small lead veins at Ballycorus were quickly exhausted, but the smelter continued to process ore shipped in from mines in counties Wicklow and Donegal, and later from the Isle of Man. The ore was brought by train to Shankill and from there by horse and cart to Ballycorus, until the smelter closed in 1913.

RATHFARNHAM CASTLE

Earthquakes and 'the big O'Leary'

In the early 1900s, the Jesuit Society, prompted by the catastrophic 1906 earthquake in San Francisco, established several stations to monitor earth tremors. For 50 years, one such observatory was at the Jesuit-owned **Rathfarnham Castle**, managed by Dublin-born **Fr William O'Leary** (1869–1939), who even designed his own seismographs. O'Leary had studied science at Louvain University, and exhibited his first seismograph at London's Coronation Exhibition in 1911. Five years later, he designed the one which would be dubbed 'the big O'Leary' – built with the help of Howard Grubb,* it needed its own small building in the castle grounds. (38) The device was essentially a finely balanced inverted pendulum: 1,800kg of iron sat atop a 2-metre rod; the rod was suspended in a pit from three steel cables, so that the weight was just above ground level. The assembly was almost unstable, and sensitive to even slight tremors, but after any movement, electromagnets restored the pendulum in time to record the next tremor. Levers attached to the top magnified any movement, and guided a needle that drew a trace in sooty paper.[20]

A wavy trace indicated a tremor, though it took skill to diagnose how big and how far away the earthquake was. The seismograph recorded a continuous trace, and monthly summaries were sent to similar observatories around the world. In 1929, William O'Leary moved to Australia's Riverview Astronomical Observatory, where he invented several astronomical instruments, but the Rathfarnham Observatory continued under Fr Richard Ingram SJ, who also lectured in mathematics at UCD. A second seismograph was later installed, this time a commercial instrument with a horizontal pendulum. In a 1953 experiment, the observatory recorded an explosion at a Roadstone quarry, but interest slowly waned and, in 1967, the instruments were unhooked. Seismic monitoring in Ireland is now done by geophysicists at the DIAS,* and Maynooth Science Museum* has what is left of 'the big O'Leary'.

1244 – drinking the Dodder

The Liffey* is tidal in Dublin and therefore undrinkable, so the earliest settlers took their water from the Poddle. Now almost entirely culverted, the Poddle rises in Tallaght and joins the Liffey near Parliament Street, but is little more than a stream. By 1244, the sheriff was forced to look to the Dodder for drinking water, a fine river that rises south of Tallaght in Glenasmole. Monks at Balrothery near Firhouse had already built a weir on the Dodder for their own water supply, and the sheriff exploited this to divert some of the river into the Poddle. The enhanced Poddle was then split at Mount Argus by a stone wedge: two-thirds continued in the Poddle to the Liberties, and one-third was diverted into a city watercourse. Part canal, part wooden conduit, part lead pipe and running atop an earthen embankment, this was the city's first **water supply scheme**, bringing water to a public cistern at St James's Gate near Christ Church, and to a few private houses and industries. The arrangement was vulnerable to attack – when besieging Dublin in 1534,

Silken Thomas easily cut the water supply – but it sufficed for over 400 years.

In 1670, to improve the supply in dry weather, the corporation replaced the cistern with a small reservoir near James's Street; this was extended, in 1721, to store 60 days' reserve (500,000 hogsheads), and lead mains were laid to 90 streets. Trees were planted at the basin, which became a popular amenity (until, in 1869, the Roundwood* reservoir made it redundant and it was filled in). Demand for water continued to grow so, in the 1740s, the Liffey was finally tapped, and a pump was installed above Islandbridge weir to lift water to the city. But not for long – in 1777 the Grand Canal introduced a much-needed new water source, filling both the city basin and a new basin at Portobello, and the Islandbridge pumps were stopped. From 1803, the Royal Canal also filled a small reservoir at Blessington Street on the city's growing northside.

The corporation started replacing the old wooden and lead pipes with cast-iron mains in 1809, levying a special tax to pay for the improvements. But the city's modern water supply begins in the 1860s with the Roundwood reservoir on the Vartry. This was the grand era of public health engineering, when there was growing awareness of the need to supply not just sufficient water, but clean water. In 1888, the Dodder was dammed to supply the new suburbs of Rathmines and Rathgar: two reservoirs were built near the river's source at Bohernabreena – one for drinking water (which was carried to the city in pipes), and a second to supply water in dry weather to the many mills relying on the Dodder for power. For the Dodder was also an important industrial river and, in the 1840s, powered at least two distilleries and 27 mills – 10 flour and oil mills, four paper mills, three woollen mills, one cutlery mill, five iron foundries, one cotton spinning factory, two calico printing plants and one sawmill. (There is still a paper mill on the Dodder – the Smurfit mill at Clonskeagh, which was established in 1938.)

The Liffey was tapped in the 1940s with the flooding of Blessington reservoir,* which doubled the city's water supply, and again, in 1966, when a reservoir was created at Leixlip to supply the towns of North Co. Dublin. Dubliners no longer drink canal and Poddle water, only rainwater falling on the Dublin and Wicklow mountains and brought to the city by the Vartry, Dodder and Liffey schemes. (The black siphon pipe across the Grand Canal at Leeson Street Bridge is the Vartry aqueduct.) The water is cleaned in treatment plants at Leixlip, Ballymore Eustace, Ballyboden and Vartry, and pumped to homes and businesses via 3,000km of water mains.

ENFO publishes an information sheet on Dublin's water supply. Blessington Basin (recently renovated as a city park), Balrothery's mediaeval weir (part of Firhouse Dodder Park) and Bohernabreena Reservoir are all open to the public.

Casting the mirror for the Great Melbourne Telescope in 1866 at the Grubb telescope 'factory' in Rathmines (now Leinster Cricket Club). A crucible of molten metal is being moved across, to be poured into the circular mould.

RATHMINES

Telescope makers to the world

For nearly a century, a Dublin firm led the world in making telescopes. The Grubb family's "optical and mechanical works" was one of only a handful of firms working in this competitive field of precision engineering, and observatories around the world – Austria, Australia, India, South Africa, South America, Crimea, not to mention Ireland – had instruments bearing the plaque "Grubb Dublin". The business was founded by **Thomas Grubb** (1800–78), a self-taught Quaker mechanic who was born in Waterford. The family, which came to Ireland during the Cromwellian settlements, originated in northern Germany and Denmark, and its ancestors included the famous astronomer Tycho Brahe. Thomas Grubb had an inventive turn of mind and, by the 1830s, ran a small machine business in Dublin making, among other things, cast-iron billiard tables.[21] He had several successful inventions, including a camera lens and an automated machine for printing and numbering banknotes – Grubb had found a way to ensure that every banknote printed was identical, thus cutting down on forgeries and earning himself the title 'engineer to the Bank of Ireland'. A keen amateur astronomer, Grubb also built himself a small telescope and came to know Thomas Romney Robinson* of Armagh Observatory. Robinson was clearly impressed, for he recommended Grubb to several astronomers and, in so doing, decided the company's fate for the next century.

Thomas Grubb's first astronomical contract was to repair a telescope in 1834 at Markree* in Co. Sligo. Then, in quick succession, came contracts to design and build telescopes for Armagh Observatory and for the Royal Greenwich Observatory in London, followed by other orders. Each instrument was unique, designed afresh by Grubb. The parts were made and assembled at his workshop, then shipped and erected on site and adjusted over several nights of observations. Significantly, Robinson introduced Grubb to William Parsons,* the Earl of Rosse, who was building the largest reflecting telescope the world had then seen, with a 6-ft (1.8-metre) mirror. Grubb designed a mounting that distributed the mirror's weight evenly, preventing it distorting or breaking under its own weight. Though Grubb was increasingly known as a telescope maker, he also designed and made surveying equipment, seismographs and other scientific instruments.

In 1866, Thomas Grubb won his largest contract yet: to build the Great Melbourne Telescope, a 122-cm reflector equatorial telescope that would be the largest of its kind in the world. His son **Howard Grubb** (1844–1931) gave up his engineering studies at TCD to join the project, and the company built an extensive works at the rear of No 57 Rathmines Road. The Melbourne telescope was delivered in 1868 and, two years later, Thomas Grubb retired. Howard then took over, and grew the company into a formidable international export business, producing increasingly sophisticated instruments for the new fields of astrophysics and astronomical photography equipped with electrical controls. Howard Grubb introduced numerous refinements to telescope design, which other companies often adopted.

In 1878, Grubb designed and made a 68.5-cm refracting telescope, then the world's most powerful telescope, for the Vienna Royal Observatory. Over the next 20 years, the Rathmines workshop produced countless scientific instruments and telescopes, including those used in observing the 1882 transit of Venus, and seven of the 10 telescopes for the international Carte du Ciel project to photograph the skies in the 1890s. By the early 1900s, Howard Grubb had been knighted for his work, and turned to military optics: he invented the first successful periscope for one of John Philip Holland's submarines;* and, by 1904, held 61 patents for gunsights, though most were kept secret by the British government. The Dublin company supplied most of the periscopes for World War I, and many of the gunsights, but its very success made it a security worry: the essential products had to be shipped across a channel that was increasingly patrolled by German submarines and, after the 1916 Rising, it was feared the Rathmines workshop would become a Republican target. Under government pressure, the company transferred to England. But the post-war chaos, the Russian Revolution – two telescopes built for the Russian government could not be shipped – and the fact that no astronomical work had been undertaken for four years precipitated a financial crisis. Grubb & Son went into receivership in 1924, and Sir

Howard returned to Ireland. The business was rescued by Charles Parsons,* son of the Earl of Rosse and inventor of the steam turbine, who founded Grubb Parsons & Company at Newcastle-upon-Tyne. It made telescopes until 1985, including some for the international observatory at La Palma in the Canary Islands, notably the 249-cm reflector named after Isaac Newton.

Many Grubb telescopes are still in use, although several have been modified over the years. At least three are still in Ireland: at the observatories of Armagh,* Dunsink* and UCC.* The Rathmines street name, Observatory Lane, is the only trace of the renowned Grubb works. A plaque at 13 Longford Terrace, Monkstown, marks Howard Grubb's former home.

A city of the dead

Fancy communing with some of the great scientists of the 19th century? Then get thee to Mount Jerome where, among others, you will find Sir William Rowan Hamilton,* George Francis FitzGerald,* John Joly,* Thomas Grubb,* Mary Ball* and Sir William Wilde.* Mount Jerome began in 1834 after a group of Dublin businessmen was refused planning permission for a cemetery in the Phoenix Park, and instead bought Mount Jerome, the Earl of Meath's estate in Harold's Cross. Harold's Cross had a sizeable cotton factory then and several mills, yet was mostly a rural village and popular with Dubliners escaping the city's dirt and bustle. Local residents were none too pleased with the planned cemetery, fearing it would destroy their tourist trade but, ironically, the cemetery is now something of a tourist attraction in its own right, on account of the many distinguished Dubliners buried there.

Burials started in 1836, and there have been over 300,000 since (compared with 1 million at Glasnevin Cemetery, which opened in 1832). Mount Jerome quickly became the main cemetery for Dublin's Protestant community and

Last resting place of Thomas Drummond,* Scottish engineer, inventor of limelight, and later under-secretary for Ireland at Dublin Castle. His dying words were: "Bury me in Ireland, the land of my adoption. I have loved her well, and served her faithfully."
© Enterprise Ireland

of TCD, are buried there. The early registers recorded the cause of death – "decline of life" was common, and accounted for Richard Griffith;* Humphrey Lloyd* succumbed to dysentery; and medical man Abraham Colles* died of "disease of the heart".

If visiting Mount Jerome, take time to admire the lettering on the inscriptions, enjoy the varied stones and memorial designs, inspect the mort safes (designed to deter body snatchers*) and a bell-and-chain device installed for a woman who feared being buried alive. Mount Jerome is also something of a wildlife refuge – with fox, pheasant and squirrel – and some of the original estate's elegant tree-lined avenues survive. Specimen trees include redwoods, the upright Irish yew* and the unusual Christ-thorn, its lower trunk circled with thorns. The River Poddle, which once powered the cotton and flour mills of Harold's Cross, now flows in a culvert beneath the cemetery gates. The cemetery office may be able to direct visitors to some of the more well-known graves. ㊴

consequently most of 19th-century Ireland's scientists, engineers and businessmen, who were drawn from the educated ranks of the Anglo-Irish Ascendancy and the staff

The delights of North Co. Dublin include the National Botanic Gardens, the mediaeval hedgerows around Ballyboughil, the Skerries water and windmills, and Dunsink Observatory, for many years Ireland's official time keeper and where William Rowan Hamilton invented quaternions in 1843. In 1865, in a pioneering experiment, Howth's Baily lighthouse acquired the world's most powerful gaslight. Howth was also one of the first places in Ireland hit by the Black Death in 1348; coincidentally, the black rat, traditionally blamed for bringing that plague, now survives only on the nearby Lambay Island.

BROOM BRIDGE

"A spark flashed forth"

In the 1830s and 1840s, the Irish mathematician **Sir William Rowan Hamilton** (1805–65) bestrode the scientific world as Einstein would a century later. Hamilton made several important contributions, including a new type of algebra called quaternions, and he hoped his work would earn respect for Ireland and "remove the prejudice which supposes Irish men to be incapable of perseverance".[22] He was showered with international honours – the American National Academy of Science, for instance, voted by an overwhelming majority to elect him their first honorary overseas fellow in 1865 and, more recently, a crater on the Moon was named in his honour. Hamilton was born in Dublin but, as a toddler, was fostered out to his uncle, the Rev James Hamilton, a curate and schoolmaster in Trim, Co. Meath, under whose tutelage this child prodigy

flourished: he could read Latin, Greek and Hebrew before he was five and, by 17, had mastered a dozen languages, was writing poetry, and reading the great mathematical classic works of Newton and Laplace – he even spotted a minor error in one of Laplace's equations. Hamilton soon came to the attention of Dublin's scientific community and, while still an undergraduate at TCD, was appointed professor of astronomy at Dunsink Observatory.* Hamilton was no astronomer, but the position was deliberately chosen because, with no teaching duties, he was free to pursue mathematics, and he held the professorship until his death.

Mathematicians then were struggling to accept such artificial and seemingly illogical concepts as imaginary numbers (involving i, the square root of -1), and even simple negative numbers – mathematics was, after all, supposedly allied to Logic. Hamilton was able to make the transition into the 'illogical' world of modern mathematics.

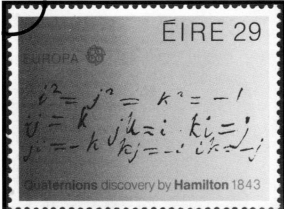

EIRE 29

EUROPA

$i^2 = j^2 = k^2 = -1$
$ij = k$ $ji = i$ $ki = j$
$ji = -k$ $kj = -i$ $ik = -j$

Quaternions discovery by **Hamilton** 1843

Hamilton's quaternion equations.
© An Post

He even threw away some of the old rules, and once attempted to describe time mathematically, with negative numbers representing a step back in time. He also believed that, through the language of mathematics, it would be possible to understand a system without conducting experiments on it.

His major contributions were in mechanics, optics, geometry and algebra. First, he developed a geometric formulation for optics, which put this new science on a firm footing. Then, in 1832, he made a daring prediction, based purely on a mathematical analysis of the laws of light: that in certain circumstances, when a ray of light was refracted in a biaxial crystal, it would produce an infinite number of refracted rays, which would take the shape of a cone. Such a theoretical prediction was unprecedented and, two months later, when his TCD colleague Humphrey Lloyd* proved the phenomenon of conic refraction experimentally, the news shook the scientific world. Hamilton was granted a royal pension of £200 and, in 1835, was knighted for his discovery. That same year, he produced what is probably his greatest contribution, his general theory of dynamics. He devised a generalised and, it later emerged, very powerful way of writing Newton's laws of motion: his 'Hamilton equations' are derived by expressing the energy of a mechanical system in terms of special variables; when written in this way the energy is called the Hamiltonian. This approach was later crucial to the development of quantum mechanics in the 20th century, where Hamiltonians are central in describing the energy and quantum motion of particles.

Hamilton is often remembered for quaternions, which he invented in 1843. He had been struggling to describe rotations in three dimensions, and the solution came to him in one of those classic eureka moments, as he was walking by Broom (Hamilton called it Brougham) Bridge over the Royal Canal. Years later, writing to his son, he described how "an electric circuit seemed to close, and a spark flashed forth". Taking his pocket-knife, he scratched his new equation on the bridge: $i^2=j^2=k^2=ijk=-1$. Because there were four parts – one real (-1), and three imaginary (i, j, k, corresponding to the x, y and z axes in space) – he called them quaternions. Significantly, he had dispensed with a cornerstone of arithmetic – the commutative law, which says that 'A times B' is the same as 'B times A' (AB=BA). In three dimensions, and in quaternions, the order (direction) of actions is important: rotating something around its x-axis first, and then its y-axis, brings it to a different position than if it is rotated first around the y, and then the x-axis. So, in quaternions, $ij \neq ji$. This creative leap paved the way for the concept of vectors. Hamilton spent the last 20 years of his life looking for a practical application for quaternions, but they were clumsy to use and vectors proved more popular, though some computer graphics applications now employ quaternions.

Hamilton could lose himself in work all day, often forgetting to eat, and he had two other unusual traits: double vision and an odd speaking voice that alternated between a high treble and a deep bass. Though his professional life was successful, his personal life was less so: a board game he invented failed commercially; and, after he was spurned by his first two loves, he married a friend, Helen Bayly, but it was not an easy marriage and he later sought refuge in drink. Hamilton never forgot his early love of poetry, writing sonnets at every opportunity for his many friends and correspondents, who included the writers Wordsworth, Coleridge and Maria Edgeworth,* and scientists such as Herschel and Thomas Romney Robinson.* Plagued by gout, Hamilton died in 1865 at Dunsink after a severe epileptic fit and was buried at Mount Jerome.*

A plaque at Broom Bridge commemorates the spot where Hamilton invented quaternions; each October 16th, scientists retrace his walk from Dunsink Observatory to the canal. A statue of Hamilton stands at the entrance to Government Buildings, formerly the Royal College of Science,* and TCD's mathematics department has put much of his work on the internet.

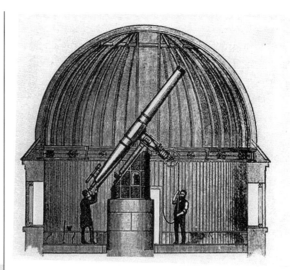

Dunsink's South telescope, in a dome designed by Thomas Grubb. On open nights visitors may try the telescope.*

Ireland's oldest scientific institute

Dunsink Observatory, founded by TCD* in 1783, and now part of the Dublin Institute for Advanced Studies,* is the oldest scientific institution in Ireland. It gained a measure of immortality when James Joyce mentioned it in *Ulysses*, especially the time service it provided for decades (see: **The Dunsink Meridian and Dunsink mean time**). The observatory was built on a hill outside the city, far from smoke and pollution, the better to see the stars. The design was state-of-the-art for its time (the telescope had its own foundation, independent of the building, so that it would not rock in strong winds), and noted instrument makers were brought in from London to fit out the meridian room. This long room, where astronomers conducted nightly observations, originally had shutters in its ceiling that opened to reveal the heavens. In 1868, the observatory gained a new telescope: Sir James South, an English astronomer, donated a 30.5-cm lens which, when made in Paris in 1829 was the largest then made; this was set in a telescope mount that Thomas Grubb* had built in 1853. The 'South telescope' has a clockwork mechanism that enables it to track an object for over two hours, and it was installed in a dome (also designed by Grubb) that is one of the oldest of its kind still functioning. It is no longer used for research – today Dunsink's astronomers time-share powerful telescopes in the Canary Islands and elsewhere.

The observatory had a staff of two: a professor of astronomy (who was accorded the honorary title, Astronomer Royal of Ireland) and an assistant. Sir William Rowan Hamilton* was professor there from 1826–65, and his friend, the poet William Wordsworth, visited Dunsink several times during his tenure. **Sir Robert Stawell Ball** (1840–1913), Dunsink's professor from 1874–92, became famous for his work on screws, wrote several science bestsellers, and was popular on the American lecture circuit. He was born in Dublin, where his father, Robert Ball,* was a prominent figure in scientific circles. His father's early death left the family in straightened circumstances so, after graduating from TCD, Robert Jr worked at Birr Castle* as tutor to the Parsons children and assistant on the great telescope. In 1867, he was appointed to the new Royal College of Science* in Dublin, where he taught mechanics and began developing his theory of screws, finding ways to describe mathematically what happens when a rigid body is twisted and wrenched. He first published his ideas in 1876, expanding on them in his

Treatise on the Theory of Screws (1900), which became a classic. His work was rediscovered in the 1970s by robotics engineers and republished in 1999.[23] Ball was a gifted, if sometimes pompous speaker, who gave several lecture tours. In diaries of his US tours, he mentions his encounters with various new technologies, such as "electric lighted streets", typewriting machines, and Edison's incandescent electric light bulbs and "loud-speaking telephone".[24]

In 1892, Ball moved to Cambridge University. His successors at Dunsink included **Sir Edmund Whittaker**, who was director from 1906–12, and who, while a lecturer

The Dunsink Meridian and Dunsink mean time

There is an intimate link between observatories, longitude and time. We need astronomers to tell us our time and longitude, for time and longitude are determined by our position relative to the Sun, Moon and stars. Thus London's Greenwich Observatory established the Greenwich Meridian (longitude) and Greenwich mean time (GMT). Similarly Dunsink Observatory gave Ireland the Dunsink Meridian and Dunsink mean time – at least until 1916 when, as a wartime measure, Ireland adopted GMT. For over 60 years, Dunsink also supplied a time signal that regulated several Dublin clocks and helped to keep the city and the railways punctual.[24]

The first public clock in Ireland was erected in 1466 at Dublin city's Tholsel (the mediaeval corporation toll booth near Christ Church). Over the next 200 years, public and private clocks became cheaper, reliable and more commonplace. They were regulated, and set to the local time, by referring occasionally to a sundial, or possibly corrected by an amateur astronomer – in the 1830s, for instance, Skibbereen's clock was set by the bank manager, Agnes Clerke's* father, based on observations he made with a telescope in his garden.

When Dunsink Observatory opened, its meridian effectively became Dublin's meridian, and Dublin adopted Dunsink time, which is 25 minutes behind GMT. The mail coaches brought this Dublin time along their routes, thanks to portable and reasonably affordable timepieces that staff carried from stage to stage. Each place still kept local time: Belfast, for instance, was 1 minute 19 seconds ahead of Dublin, while Cork was 11 minutes behind. All this changed in the 1840s with the coming of the railways* and the telegraph*: the trains ran to Dublin time, punctual to the minute, and the telegraph wires brought Dublin time to the railway stations. Galway station had two public clocks, one for Dublin (and train) time, one for Galway time. But to avoid confusion, railway towns increasingly switched from local to Dublin time, and surrounding areas eventually followed suit.

In 1865, the General Post Office in Dublin began receiving Greenwich Observatory's 10am telegraph time signal, as an aid to setting the GPO clock. That same year, Dublin's port authority erected a time ball outside the Ballast Office beside Carlisle

Heaven and Earth: Dunsink Observatory, which provided a time service for Dublin city from 1865–1937.
© An Post

[O'Connell] Bridge, so that ships could set their clock before leaving port. The copper ball, 1 metre in diameter, was held aloft on a tall pole, and an electrical mechanism controlled its descent each midday, the time signal coming from Dunsink on a portable clock. However, when the RDS surveyed Dublin clocks in 1873, they found tremendous variation – worst was the Alliance Gas Company clock, at over seven minutes fast. So, the following year, two time-distribution systems began in Dublin: the RDS, then at Leinster House,* subscribed to the Greenwich telegraphic signal (at £6.12s per clock per annum); and Dunsink began its formal time service – a master clock at Dunsink was connected by telegraph to a series of slave clocks (at the Ballast Office, GPO, TCD and Bank of Ireland); and a pulsed signal from the master clock controlled the slave pendulums. The RDS/Greenwich system was reliable but expensive, and the RDS stopped in 1921 because of costs. Dunsink's system ran into frequent difficulties: it depended on a small rural telegraph line, and had few staff to make the observations, calculate and send the time signal. Nevertheless, it continued until 1937 when it was replaced by a radio time signal from Britain. Still, in the 1950s, Dunsink staff were occasionally asked for a time check by CIÉ and they were sometimes asked for an expert opinion in legal cases. Ireland's official timekeeper now is the National Metrology Laboratory,* which takes its time signal from an atomic clock in England.

at the Royal University (a forerunner of UCD), had taught the young Éamon de Valera. In 1921, Dunsink Observatory, like many other scientific instutitions in the new Free State, was in trouble. Always a small institution, and physically isolated from TCD, it needed a massive injection of funds, but these were difficult times, financially and politically, and the observatory closed. A caretaker maintained the time signal until 1937, by which time de Valera, advised by Whittaker, had begun to establish the DIAS. In 1947, the State bought the historic Dunsink Observatory and it became part of the institute's School of Cosmic Physics. It hosts open nights every first and third Wednesday of winter, when, weather permitting, visitors may try the South telescope. (40)

Where there's muck, there's brass

When Dunsink Observatory was founded in 1783, it was built atop a hill overlooking Dublin. The hill has since grown, however, and now overlooks the observatory – it was for many years a landfill site and, unusually, it now generates electricity. Since the 1990s, Irish Power Systems has been harvesting methane gas from the mound of rotting waste; the gas is burned on-site to generate electricity, which is then sold to the national grid. The initiative doubly helps to cut Ireland's greenhouse emissions: by reducing the amount of fossil fuels which might otherwise be burned to generate the electricity; and by trapping methane, a powerful greenhouse gas, which would otherwise be released into the atmosphere. The methane, produced by bacteria digesting rotting waste, is collected in pipes laid through the tiphead (a thick layer of soil seals the mound and traps the gas). There are similar plants at five other Irish landfills, generating in all 20MW of power each year, enough to supply 15,000 homes. Not all landfill sites can be tapped, and test drills are needed to establish the volume of gas being produced and the presence of any contaminants that might make it unusable.

A blooming jewel

When Dublin surgeon Walter Wade campaigned, in 1790, for a public **botanic garden** in the city, he picked the right time. For this was the Age of Reason, when people believed that education would civilise the world, when the (not yet Royal) Dublin Society* was actively working to improve Irish agriculture and horticulture, when the mood of the Irish Parliament was favourable, and the speaker there, John Foster, had his own botanic collection in Cavan. These happy coincidences ensured Wade's petition was successful. Parliament granted funds and, in 1795, the Dublin Society acquired 11 hectares of land along the south bank of the Tolka River at Glasnevin, part of the Delville Estate formerly owned by Dean and Mrs Mary Delany. Donations of plants arrived from Irish gardens, from Kew Gardens in London and from collectors around the world. In 1877, the RDS handed the gardens into government care, and today the 'Bots', as they are affectionately known, are renowned throughout the botanical world. Now a flourishing 19 hectares, they are home to 20,000 species and cultivars collected from diverse environments – from desert to rainforest, and tundra to tropic – a peaceful haven, a training ground for 150 horticultural students every year, a research centre, and a source of information for expert gardener and lay person alike. Each year, the garden replaces thousands of plants, but it retains a yew avenue that dates from the 1740s, and the design laid down in the 1830s by the landscape gardener Sir Ninian Niven. (41)

The Dublin gardens introduced several new species to cultivation, including pampas grass, collected in Argentina in the 1840s, the giant Himalayan Lily and Lady Cuffe's rhododendron from Burma. During the mid-1800s, under a Scottish curator, **Sir David Moore**, the gardens had two major achievements where everyone else failed: growing orchids from seed to flowering stage and raising insectivorous pitcher plants. In 1845, when a new potato

Part of Turner's famous curvilinear glasshouse at Glasnevin; faithfully restored in 1995 to its original glory.
© An Post

ballad "The Last Rose of Summer". Glasnevin also has an important library and an irreplaceable herbarium (the botanical equivalent of a museum) that contains 500,000 dead, dried and pressed plants – everything from seaweeds and mushrooms, to mosses and conifers and specimens of cocoa, coffee and rubber brought from the Congo in 1904 by Roger Casement. Pride of place goes to a collection compiled, in 1661, by a Dutch pharmacist Antoni Gaymans and brought to Ireland by Thomas Molyneux.*

Glasnevin has rock, rose, herb and vegetable gardens, an arboretum, and glasshouses for its alpines, succulents, palm trees and orchids. Plant collectors around the world regularly send donations and, though the garden never had funds to mount its own expeditions, it occasionally sponsors expeditions organised by others. As part of its conservation work, Glasnevin grows several rare Irish plants, and partners TCD's botanic gardens in the Irish seed bank project, collecting and freezing seeds from endangered Irish species. The garden's 'acidic wing' is at Kilmacurragh* in Co. Wicklow.

disease appeared in Ireland, Moore experimented at Glasnevin to investigate the cause. He became convinced, as had several other botanists here and in Europe, that a fungus was to blame, and even came close to discovering a treatment for the blight* – but the germ theory of disease had not yet been accepted, and it was 20 years before the blight's fungal nature was confirmed.

In the 1840s, work began on a curvilinear range of glasshouses to shelter tender exotic plants. Made of cast and wrought iron, the buildings have a distinctive curved roof which was novel at the time. They were designed and built by **Richard Turner** (1798–1881), a Dublin ironmaster who could make glazing bars that were slender and thin yet strong. Turner experimented with shapes and designs, and built elegant, towering glasshouses that let in as much light as possible. He designed the great curvilinear glasshouses at Kew Gardens and Belfast,* as well as Glasnevin, and made glasshouses, gates and railings for many of the big Irish demesnes.

Glasnevin's oldest plant is a venerable 400-year-old fern, a gift from Australia to TCD* in 1892 for the college's tercentenary; it was given to Glasnevin in 1968, though TCD's botanic garden* kept an offshoot. Other unusual plants at Glasnevin include: the rare Killarney fern;* the giant Amazon water lily (grown from seed each year, it reaches a spectacular size); and a rose raised as a cutting from the shrub that inspired Thomas Moore's famous

Whither the weather?

The first weather service in these islands began in 1860 when the Royal Navy gathered data by telegraph from a network of coastal stations, including Valentia Observatory.* The network passed to the British Meteorological Service (BMS) which was set up in 1867 and, in 1936, when the **Irish Meteorological Service** was established, it took over in the Republic from the BMS. The fledgling Irish service had a small staff, most of whom had worked for the BMS at Valentia, but the timing was fortuitous, and two international experts were recruited – Mariano Doporto, a Basque fleeing Franco's Spain, and Leo Pollak, a refugee from Czechoslovakia – who might otherwise not have joined a small Irish agency. In addition to Valentia, there were four part-time telegraphic stations (Malin Head,* Blacksod, Birr Castle* and Roche's Point), 18 climatological stations (which recorded a variety of measurements) and 172 rainfall stations.

The new service developed partly in response to the growth in aviation and the need to provide pilots with weather information. In 1939, a weather station opened at Foynes* to serve the new transatlantic flying-boat service and, by 1941, there were stations at Baldonnell Aerodrome and Dublin airport. Wartime restrictions made it difficult to recruit staff, however, and censorship meant the service could not broadcast public forecasts, and had to encrypt its weather transmissions to aircraft. Yet the network of climatic and rainfall stations grew and, by 1944, there were 530 rainfall stations (as against 650 today), many of the new ones located at garda stations and lighthouses.

Weather reports were first broadcast on Radio Éireann after World War II, bringing up-to-date information to a wider audience; a recorded telephone service started some years later. Data gathering also expanded: atmospheric conditions were monitored from radiosonde balloons released each week at Valentia, and the Irish shipping fleet carried instruments to record conditions at sea. With the atmospheric testing of atomic weapons in the 1950s, the Meteorological Service began measuring the radioactivity of the air over Ireland; subsequently it also tested for air pollution and acid rain. The service, which moved from its O'Connell Street offices to purpose-built headquarters at Glasnevin in 1980, is continually modernising, keeping abreast of improvements in computing power, mathematical modelling, telecommunications and radar and satellite technology. Despite a name change – to Met Éireann – it still issues weather reports on the radio, provides recorded telephone forecasts, and produces blight warnings for farmers and wave-height forecasts for coastal workers, among other services. (42)

Standards in public life

Ireland's official standard-bearer is the **National Metrology Laboratory**, a high-tech facility that safeguards the national kilogram and national metre. The laboratory is also our timekeeper – its clock is regulated by an accurate radio signal received from an atomic clock in Britain – and it maintains the national standards for temperature (based on the 'triple point' of water), and for electricity (volt and ohm). These are all used to calibrate industrial and commercial equipment, such as scales, thermometers and electricity meters, and ensure that, when you buy a litre of petrol, unit of electricity, kilogram of sugar or metre of timber, you get precisely what you paid for. (The Assay Office* at Dublin Castle is responsible for precious metal testing.)

For much of the past 800 years, Ireland used the imperial system, introduced to Dublin in 1217 by King John, but though traffic speed limits are still quoted in miles per hour, Ireland is now mostly metric. The metric system began during the French Revolution, as a rational decimal alternative to the diverse national weights and measurements then in use. The gram (unit of mass) was defined as the mass of $1cm^3$ of pure water measured at 4°C (water's maximum density). The litre (unit of volume) was defined as the volume of a cube having sides of 10cm. And the metre was 1 ten-millionth of the distance between the North Pole and the Equator. The new system spread across Europe on the back of Napoleon's campaigns, helped by scientists who appreciated the benefits of a common system – though many in Britain opposed it, seeing it as a revolutionary French invention. William Thomson (Lord Kelvin)* helped pioneer it during his work on the transatlantic telegraph cable,* and George Johnstone Stoney* first proposed the electron as a metric unit of electricity. In 1875, a Treaty of the Metre, signed in Paris, added several other measurements to create the International System of Units. This spelt the death knell for idiosyncratic local measurements, such as pecks and bushels, furlongs and chains, not to mention the Irish mile – at 6,721ft (2,048.6 metres) it was 27 per cent longer than the English statutory mile of 5,280ft (1,609.3 metres).

Ireland's national kilogram is a highly polished ingot of very pure stainless steel. The legal Irish unit of mass, it is checked every five years by the International Bureau of Weights and Measures in Paris. Exposing it to the atmosphere could affect its composition and thus its mass, so it is kept in a special sealed container and removed only occasionally to calibrate reference-weighing equipment. The national metre, a 1-metre long bar of pure stainless steel, is no longer a legal unit – after the metre was redefined, in 1983, to be the distance travelled by light in a vacuum during a time interval of 1/299 792 458 of a second – but it too is kept carefully at Glasnevin. (43)

MARINO

To prevent calamities

Dublin got its first **fire-brigade service** in 1711 when, "to prevent the calamities that may happen by fire", the corporation paid John Oates £6 to supply two pumps and attend any outbreak of fire. (In an advertisement from the time, Oates was described as a "water ingineer" who made pumps for reasonable rates at his Dame Street premises.) Up until then, fires were fought by bucket brigades as they had been for centuries – people with buckets forming human chains to ferry water from a river. Oates could use pumps partly because of improvements in the city's water supply that, by then, was circulating in wooden conduits, albeit at low pressure: he simply cut into the wooden water main, and plugged it afterward with a bung carried specially for that purpose. Insurance companies also began offering a fire-fighting service to clients: they issued plaques that customers mounted on their building, and their fire brigade attended only premises displaying their plaque. This could mean that the first team on the scene left if the building was insured by another company so, in the late 1800s, local

authorities around the country took over responsibility for fire fighting.

Several fire stations were established around Dublin, together with a network of alarm cables for summoning the brigade (there being few telephones* then): anyone needing to call the brigade simply went to the nearest crossroads and pulled on the alarm cable. This rang a bell at the fire station on a special control panel that also revealed the call's location. The control panel installed in Tara Street station when it opened in 1907 can be seen at the **Dublin Fire Brigade Museum**. (44) The small museum, at the brigade's training centre, relates the development of fire fighting from the mediaeval days of bucket brigades. A highlight is the **turntable ladder**, invented in 1911 by Capt Purcell of the Dublin brigade, at a time when fire tenders were being mechanised. Purcell's ladder was easier to manipulate and more stable than stand-alone ones, and his design was subsequently adopted worldwide. The museum also has equipment, early breathing apparatus and samples of Purcell's designs. Howth Transport Museum has a fire engine with an early Purcell ladder and other fire tenders. (45) The city's first official fire station was in the basement of Dublin Corporation's Assembly House, now home to the corporation's Civic Museum.

HOWTH

1865 – the world's most powerful lamp

Seafarers the world over owe their lives to **John Richardson Wigham** (1829–1906), a Quaker engineer who invented powerful new lamps and signalling devices for lighthouses, harbours and navigational buoys. Wigham's experiments began in earnest in 1865 at Howth's Baily lighthouse, where he installed a revolutionary new gaslight. Despite a bitter controversy, Wigham's innovations were ultimately adopted the world over. Born in Scotland, Wigham came to Dublin at 14 to work with his brother-in-law, Joshua Edmundson, whose company designed and supplied gas plants and

Lighting the way

Lighthouses have had to experiment with various fuels over the years, as no single energy source suited every location. The earliest beacons were coal fires, candles came later, and oil lamps were introduced in the late 1700s. Skellig lighthouse,* for example, burned sperm oil until the 1840s, then changed to rapeseed oil. Burning oil generated a considerable amount of moisture, however, and the resulting condensation, particularly if it froze on the windows, could dim the light. Oil lamps also had to be carefully tended, and the heavy oil carried to the top of the lighthouse. Wigham's gas lamps were a major improvement and his burners were installed at several Irish lighthouses, notably Baily, Wicklow Head, Rockabill and Hook Head.* Acetylene,* discovered in Dublin in 1836, was tried at several locations, including Carlingford lighthouse, which burned acetylene until 1922. Propane and paraffin vapour were used elsewhere. Electric lighting, such as arc- and battery-powered lights, was available from the 1840s, but the devices were unreliable, expensive and tricky to maintain. It was not until the mid-1930s, when electricity was more widely available and electric bulbs more reliable, that lighthouses began switching to this relatively clean and easy energy source. By then, electric lamps were also more powerful, and easier to see, particularly against the increasing night glow from city lights, although Rathlin West's paraffin lamp was not electrified until 1983. The Commissioners of Irish Lights even experimented with nuclear power in the 1970s at Rathlin O'Birne,* in Co. Donegal. These days, solar and wind power are increasingly popular, especially for remote and offshore lighthouses.

A lighthouse optic: the huge lens focuses the light beam. Various lighting technologies have been tested at Irish lighthouses down the centuries.
© Commissioners of Irish Lights

appliances. When Wigham inherited the company in 1848, after Edmundson died, he expanded into Scotland, the Channel Islands and England, and diversified into the new electricity market.

His initial foray into marine navigation was when he patented the first gas-lit navigational buoy – this was successfully installed on Scotland's River Clyde in 1861. The gas burner he designed in 1865 for the Baily had 108 jets and four settings: 'low' for clear weather, when only 28 jets were lit, through to 'high' for heavy fog, using all 108 jets and producing a light equivalent to 3,000 candles. It was the world's most powerful gaslight, and 12 times more brilliant than the previous 240-candle oil lamp; it was also easier to use and maintain. The gas was made on-site by burning coal and shale at a small gas plant and pumped up to the lantern room. Numerous innovations followed. In 1868, at Wicklow Head lighthouse, Wigham installed a gas burner with a clockwork valve, producing the first-ever intermittent light. The system installed three years later at Rockabill* off Skerries had a revolving lens and an intermittent light; this combination produced a novel flashing pattern, with a series of quick flashes, followed by an eclipse, which was then repeated. Now called group flashing, this was an improvement on existing flashing patterns, which were used to give each lighthouse its own recognisable signal. Wigham's approach was eventually deployed at most of the world's lighthouses. By 1885, Wigham was designing burners with beams equivalent to 2 million candles, and testimonials in his praise flowed in from ships' captains and from the noted Irish scientist John Tyndall,* who was an adviser to Trinity House, the authority then overseeing all Irish and British lighthouses. Yet Trinity House preferred the oil burners designed by its own engineers so, after a bitter row, Tyndall resigned. Trinity House later used Wigham's innovations without his permission, and eventually was forced to pay him £2,500 compensation.

Among Wigham's other inventions were sirens, fog signals, and an oil-lit navigational buoy that could burn unattended for three months. His company also pioneered the use of acetylene* in lighthouses, and supplied lamps, beacons and buoys around the world. Unionist in his politics, Wigham was a major figure in Dublin business circles, becoming a director of the Alliance & Dublin Gas Company and Dublin United Tramways. He was twice offered a knighthood for his services to marine navigation, but declined, as a title was at odds with his Quaker principles. His company is still in business, still helping to make the seas safe, and still in Quaker hands, as Barrett & Company of Dublin, designing, making and supplying marine and navigational lamps.

There had been a beacon on Howth Head since at least the 9th century, and a lighthouse cottage since 1667, but that light was often obscured in fog so, in 1814, a tower was built closer to the water at the Baily. Howth, a rugged and rocky headland, was once an island, but 5,000 years ago when the sea levels dropped, it became joined to the mainland by a sand and shingle bank – Sutton, where Howth joins the mainland, is a raised beach. For centuries, Howth was the main landing and departure point for cross-channel boats, and a new harbour was begun there in 1807. A much bigger harbour was begun at Dún Laoghaire* 10 years later, however, and all the traffic eventually moved across the bay.

The Irish lighthouse service's heritage collection, with artefacts dating from the service's inception in 1768, is held at the Baily; ㊻ the National Maritime Museum* has the massive optic used at the Baily 1902–72.

HOWTH

1348 – Black Death and black rats

The **Black Death** hit Ireland in the summer of 1348, arriving at the main east coast ports of Howth, Dalkey* and Drogheda.* It had begun devastating Europe the previous year, and now it was Ireland's turn: by Christmas it had killed 14,000 people in Dublin alone, where people were buried in mass graves, such as the Black Pits, and it had reached Kilkenny, 100km away. Victims died in agony, vomiting blood, with swollen glands, and covered in painful sores. Fear was rife, pilgrimages were popular, food was in short supply, building work came to a halt, and civic life came to a standstill as public places were closed, meetings were banned, and the courts and parliament adjourned. This was not Europe's first plague epidemic – there were sporadic outbreaks every few hundred years – nor was it the last – London, for instance, was badly hit again in 1664–5 – but it was the worst on record, killing between one-third and one-half of Europe's population. Dublin took 200 years to recover to its pre-plague level. Towns, being relatively densely populated, were worst hit, and in Ireland this meant the English and Anglo-Norman communities suffered most, the predominantly rural native Irish less so. Some settlements, such as Cadamstown* in Co. Offaly, were destroyed or deserted. In the immediate aftermath, though, survivors benefited as rents fell and wages rose.

The epidemic was blamed on various causes, including contaminated goods, foul vapours or 'miasma' exhaled by the Earth, and divine retribution for the sins of man. But, in the 19th century, it was diagnosed as bubonic plague, after French scientist Alexandre Yersin noticed similarities between the bubonic plague and descriptions of the Black Death, in particular the presence of swollen glands or buboes. If so, the Black Death was probably spread by the **black rat** – for the black rat carries the flea that carries the germ that causes bubonic plague, a germ (bacterium) now called *Yersinia pestis*. The black rat (*Rattus rattus*, more accurately called the ship's rat, since colour is not a good guide, and 'black rats' can be brown), originated in Malaysia, and spread across Europe during the first centuries AD. The earliest known Irish reference to it is an illustration in the 8th-century *Book of Kells*. It was the only rat in Western Europe until the arrival of the much larger brown rat (*Rattus norvegicus*), which originated in China, and reached Britain and Ireland in the 1720s. The brown rat quickly ousted its smaller cousin, so much so that the black rat, once widespread, is now Ireland's rarest mammal. Occasionally, one is caught escaping from a ship arriving at a port, but the only surviving Irish colony lives on the island of Lambay, coincidentally near Howth, where their ancestors are thought to have brought the plague in 1348. Or did they?

Some scientists now believe the Black Death was not bubonic plague, but a highly infectious virus similar to Ebola (though bubonic plague was no doubt present at the time).[25] They argue that the Black Death spread quickly, at several kilometres a day, while animal-borne diseases generally move slowly – some modern-day outbreaks of bubonic plague have spread at a rate of only 20km a year. That some towns in the Middle Ages controlled the epidemic by curtailing travel, which would not have been possible if rats carried the disease. And that, tellingly, Iceland had no black rats, yet was devastated by the Black Death. Scientists are now taking samples from various Black Death burial sites in a bid to identify the disease. We may yet have to rehabilitate the black rat.

LAMBAY

Animals odd, new and rare

Lambay, home to Ireland's last few black rats, is a small volcanic island with a rich history. The Vikings used it as a safe lambing site (the name means 'lamb island' in Old Norse), and it was later owned by Christ Church Cathedral,* and by the Talbots of Malahide, who sold the island in 1903 to Cecil Baring of the merchant bank family. Baring, later Baron Revelstoke, commissioned the noted architect Lutyens to renovate Lambay's 15th-century castle, and stocked his island with exotic creatures – Corsican mountain sheep, kinkajous, even snakes. But his menagerie was ill-matched and ill-suited to the Irish environment, and none survived long. While researching the island's history,

however, Baring met librarian and naturalist Robert Lloyd Praeger,* at whose suggestion several naturalists spent months scouring the island for plants and animals. Their study, a forerunner of the Clare Island survey,* uncovered five small creatures that had never been discovered before – three worms, a mite and a bristletail – plus 90 species that had not previously been known in Ireland. Geologically, Lambay is a complex mix of sandstones, limestone, clay-slates and green porphyry, reflecting its volcanic origins 500 million years ago. The porphyry was quarried in prehistoric times to make stone axes,* and the quarry site and axe 'factory' survive.

BALLYBOUGHIL

Hedgerows – mediaeval demarcation lines

The ancient Gaelic clans held their land tribally and had no need for enclosures, but the 12th-century Anglo-Norman settlers had a feudal, manorial system of land ownership and, in North Co. Dublin and elsewhere, they divided their land into a network of small fields, planting the boundaries with hedgerows of fruit trees and shrubs. Many of these mediaeval fields and hedgerows survive at Ballyboughil, notably around the ruined Gracedieu nunnery. They are living heritage, rich in species and wild fruit trees, and at Balrothery they shelter at least one rare plant, the short-styled field rose *Rosa stylos*.

An estimated 2 per cent of Ireland's area is given over to **enclosures**, including hedgerows, stone walls and ditches. Some are prehistoric, such as the Neolithic stone walls of the Céide Fields* and the hedges that circle ring-forts; some are mediaeval, as around Ballyboughil and the Pale villages of Lusk and Swords; but most date from the 1700s and 1800s when farmers were encouraged, and sometimes obliged, to enclose fields. The expanding road* network led to a parallel increase in the number of miles of hedges, and likewise when the Congested District Board and later the Land Commission required farmers to plant hedge boundaries. Since the 1950s, however, many hedgerows have been replaced with wire fences.

Enclosures can serve several functions: shelter from the elements, marking boundaries, corralling stock (as at Céide), and as vital wildlife corridors and refuges. The type of enclosure varies from region to region, stone walls being most common in the west of Ireland and hedges elsewhere. The species growing in a hedge will depend on the region, climate and soil: fuchsia and montbretia thrive in the southwest, for instance; hawthorn, blackthorn and ash in the rich Golden Vale; and willow, alder and birch in the wetter parts of the northern drumlin belt. Ballyboughil Hedgerow Society organises occasional walks and talks. ㊼

SKERRIES

1656 – the Down Survey of Ireland

When Cromwell conquered Ireland, one of his first priorities was to confiscate the estates of the rebellious Catholics and give them – 3 million hectares, or about 40 per cent of the island – to those who had funded his campaign. Since no accurate maps existed, Cromwell ordered a survey of the country, which came to be known as the 'Down Survey', because the information was written down. Despite the scale of the undertaking and the inaccessible terrain, it was completed in little over a year, thanks to **Sir William Petty*** – inventor, political scientist and physician to Cromwell's army in Ireland – who planned the survey like the military campaign it was.

The 1650s were troubled times, so the survey had to be executed quickly, which meant employing large numbers of fieldworkers. It was probably the biggest project the country had yet seen. Rather than engage skilled surveyors and engineers – of whom there were only a few – Petty recruited 1,000 foot soldiers. He also devised a division of labour: simple measuring chains and compasses were used for the survey work, and each man was trained in only one

William Petty's 1685 map: the first faithful portrait of Ireland, it was not bettered for nearly 200 years.

skill, such as adjusting the compasses. The end product was several hundred maps of baronies and parishes around the country, which Petty collated 30 years later into the first printed atlas of Ireland, *Hiberniae Delineatio* (1685).[26] It included the first truly accurate and recognisable map of the island, and formed the basis for all subsequent maps until the Ordnance Survey* of Ireland in the mid-1800s.

There had been previous maps, but all were relatively rudimentary. *Topographica Hiberniae*, written in 1188 by Giraldus Cambrensis,* has the earliest-known map of Ireland: a shapeless oblong island with four towns (Dublin, Wexford, Waterford and Limerick) and four rivers (*Sinnenus* [combined Shannon and Erne], *Slana* [Slaney], *Suir* and *Auenliff* [Liffey]). Ireland also featured in a "survey of the known world" or *Geographia* compiled in the second century AD by a Greek astronomer, Claudius Ptolemy. He listed the longitude and latitude co-ordinates for several rivers, headlands, settlements and tribes, but the information, gathered from travellers and seafarers, is often ambiguous: the Oboka river, for instance, might be the Avoca, or the Liffey – Ptolemy suggests it flows into Dublin Bay near the settlement of Ebdana [Eblana/Dublin]. Mediaeval cartographers used Ptolemy's data to draw maps of Ireland, but all are necessarily sketchy, and some border on the fantastical with 'here be dragons'. As mapping techniques improved, however, and England's political interest in Ireland grew, so too did the accuracy of the maps, culminating in Petty's survey. (Marsh's Library and the National Library have editions of Petty's maps. The latter sells a wallet of 16 historic maps of Ireland, including Petty's.)

There is an exhibition about the Down Survey at **Ardgillan House**, which was built in 1737 by Robert Taylor, grandson of Thomas Taylor, who served on Petty's survey. The Taylor family lived at Ardgillan until 1962; the demesne was later bought by the local council. ⓽ The council also runs Malahide Demesne, home of the **Talbot Botanic Gardens**. Malahide Castle was owned by the Talbot family from 1185–1976, interrupted only in 1654–65, when Cromwell confiscated the estate. It was noted for its gardens since the early 1800s, though the current planting scheme dates from the mid-1900s. The soil borders on alkaline, so planting is restricted primarily to lime-loving plants. Over 5,000 species and varieties grow there, mostly from Chile, Australia, Tasmania, New Zealand, also coastal California and Africa, and seed is exchanged with botanic gardens around the world. ⓽

Old mills and copper mines

Most of Ireland's millers were men, but a few were women and among them was Mrs Bridget Flynn, who, for 20 years, was miller for Skerries. By the mid-1800s, this small port and fishing town had a substantial watermill, a 5-storey, 5-sail windmill, and a smaller 4-sail windmill. All were located in the town's park, alongside a threshing barn, grain-drying kiln, bakery and forge, forming a self-contained milling complex. Mrs Flynn inherited the business when her husband Richard died in 1849, and ran it until she died in 1869. The mills closed in the early 1900s, but are now restored to full working order by Fingal council, and grind local organic grain; the flour is used in the mill bakery and café. ⓾

There is copper in nearby **Loughshinny** headland. However, the rock is heavily contorted and folded and hence the ore veins are small and difficult to follow. Nevertheless, the copper was probably mined in prehistoric times, and various mining companies worked it in the late 1700s and again during the Napoleonic wars, producing in all about a thousand tonnes.[27] The mine closed in 1844, but traces of the workings survive. The extremely contorted rock can be seen on the beach below, which is also the landfall for a 200-km **natural gas pipeline** from Scotland that opened in 1999. The supply complements Ireland's indigenous gas sources – Cork's near-depleted Kinsale gasfield,* and Mayo's new Corrib gasfield.* **Rockabill Island** off Skerries is home to Europe's largest colony of roseate terns.*

Chapter 1 sources

Dublin City

1. de Courcy, John, *The Liffey in Dublin* (Gill & Macmillan, 1996).

2. Gilligan, H A, *A History of the Port of Dublin* (Gill & Macmillan, 1988).

3. Wyse Jackson, Peter & Micheline Sheehy Skeffington, *The Flora of Inner Dublin* (Royal Dublin Society, 1984).

4. Whyte, Nicholas, *Science, Colonialism & Ireland* (Cork University Press, 1999).

5. O'Hara, James "Humphrey Lloyd: ambassador of Irish science and technology" in Nudds *et al* (eds), *Science in Ireland 1800–1930: Tradition and Reform* (Dublin, 1988).

6. Rutty, John, *An Essay Towards a Natural History of the County of Dublin* (Dublin, 1772).

7. Jackson Wyse, Patrick, *The Building Stones of Dublin – A Walking Guide* (TownHouse & CountryHouse Ltd, 1993).

8. Johnston, Máiréad, "Dublin's Crystal Palaces" *Technology Ireland* (October 1990).

9. Coakley, Davis, *The Irish School of Medicine: Outstanding Practitioners of the 19th Century* (TownHouse & CountryHouse Ltd, 1988).

10. Student, "The Probable Error of a Mean" *Biometrika* 6: 1-25 (1908).

11. Coakley, Davis, *Dr Steevens' Hospital: a brief history* (Dr Steevens' Hospital, 1992).

12. Ryan, Gregg, *The Works: Celebrating 150 Years of Inchicore Works* (Iarnród Éireann, 1996).

13. McDonnell, Robert, "Remarks on the Operation of Transfusion and the Apparatus for Its Performance" *Dublin Quarterly Journal of Medical Science* (November 1870).

14. Coakley, Davis, *Irish Masters of Medicine* (TownHouse & CountryHouse Ltd, 1992).

15. Coakley, Davis, *The Irish School of Medicine: Outstanding Practitioners of the 19th Century* (TownHouse & CountryHouse Ltd, 1988).

County Dublin

16. Davies, Gordon L H, *North from the Hook* (Geological Survey of Ireland, 1995).

17. Brück, Máire, "An Astronomical Love Affair" in Mary Mulvihill (ed.), *Stars, Shells & Bluebells* (WITS, 1997).

18. de Courcy Ireland, John, *History of Dún Laoghaire Harbour* (Edmund Burke, 2001).

19. *Transactions Royal Irish Academy* XXXI (June 1846).

20. "Earthquake Recording at Rathfarnham Castle" *Irish Jesuit Directory & Yearbook* (1938).

21. Glass, Ian, *Victorian Telescope Makers: the lives and letters of Thomas and Howard Grubb* (Institute of Physics, 1997).

22. Attis, David, "The Social Context of W R Hamilton's Prediction of Conical Refraction" in P J Bowler & N Whyte (eds), *Science & Society in Ireland 1800–1950* (Institute of Irish Studies, 1997).

23. Ball, Sir Robert Stawell, *Treatise on the Theory of Screws* (Cambridge University Press, 1999).

24. Wayman, Patrick, *Dunsink Observatory 1785–1985* (RDS/DIAS, 1987).

25. Scott, Susan & Christopher Duncan, *Biology of Plagues* (Cambridge University Press, 2001).

26. Andrews, J H, *Shapes of Ireland: Maps and their Makers 1564–1839* (Geography Publications, Dublin, 1997).

27. Cole, Grenville, *Memoir of Localities of Minerals of Economic Importance and Metalliferous Mines in Ireland* (Geological Survey of Ireland Memoir, 1922).

Chapter 2: Westmeath • Meath • Louth

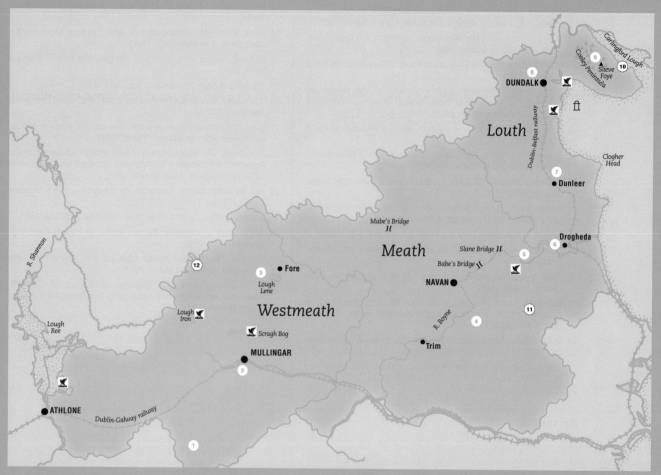

Visitor facilities

1 Locke's Distillery, Kilbeggan
2 Tullynally Castle, Castlepollard
3 Belvedere House, Mullingar
4 Hill of Tara
5 Brú na Boinne (Newgrange)
6 Millmount Museum
7 White River Mills
8 Louth County Museum
9 Táin Holiday Village

Other places of interest

10 Carlingford village
11 Rathfeigh
12 Street
♀ woodland
🖾 nature reserve or wildfowl sanctuary
🏛 lighthouse

Visitor facilities are numbered in order of appearance in the chapter, other places in alphabetical order. How to use this book: page 9; Directory: page 474.

This is a county of strange and well-kept secrets – like the unusual jelly-mould building at Locke's Distillery in Kilbeggan, the unique Lough Lene that drains both east and west and the many wonders of Fore. The county was created during the Plantations when Meath was divided in two.

KILBEGGAN

The jelly-mould building

Locke's Distillery is reputedly Ireland's oldest distillery site, dating from 1757. It has a most unusual whiskey warehouse shaped like a large, inverted jelly-mould, with a distinctive corrugated surface. It was built of concrete using a cheap and ingenious technique devised by Irish engineer **Jim Waller** (1884–1968).

Waller was born in Tasmania, but his family came from Nenagh and he went to college in Ireland. During World War I, he watched soldiers camouflage their tents by daubing them with concrete, and realised the approach could be used for buildings. A shortage of steel in the 1940s, coupled with his own humanitarian outlook, prompted Waller to develop his 'Ctesiphon technique', which he hoped could be used in developing countries and to rebuild Europe's bombed cities.

First, a skeleton of timber arches was erected, this was then covered with a sheet of hessian. Two or three layers of mortar were applied to the hessian, which sagged between the ribs under the weight of mortar, giving the structure a corrugated form. Finally, the wooden arches were removed to leave a self-supporting concrete shell. The arched shape Waller devised was inspired by Baghdad's ancient Great Arch of Ctesiphon (pronounced 'tessifon'), which he saw while working in Iraq as a surveyor. That arch, built of crude brick, has lasted over 1,600 years thanks to its design: taller than it is wide, it is a catenary arch, and the weight is evenly distributed across all points.

Waller's approach enjoyed a certain vogue in the 1940s and 1950s: it was cheap and flexible, did not require steel reinforcing, was quick to erect and used local materials. It proved especially popular in Africa, where it was used for everything from small hen houses to large aircraft hangars. The British Ministry of Works also built and tested various Ctesiphon structures. Interest in Waller's technique waned in the 1950s, though with the resurgence of traditional building techniques, we may yet see more jelly-mould buildings.

Whiskey maturing inside Kilbeggan's unusual 'Ctesiphon' warehouse.
© Cooley Distillery

Locke's warehouse is probably Ireland's only surviving Ctesiphon building in good condition. A low structure beside the distillery, and visible from the Dublin road, it is still 'bonded' – used by Louth-based Cooley whiskey to store whiskey – and not open to the public. The Kilbeggan distillery closed in 1957 but was left virtually intact – only the copper piping was scrapped. A community group has developed an industrial museum there, complete with millwheels and steam engines. ①
More information on Waller is available at the Irish Architectural Archive, Dublin.*

ATHLONE

Navigating the Shannon

The broad reaches of the mid- and lower-Shannon river were, for centuries, a natural barrier to movement (east–west), yet also a major trade route (north–south). Witness the remains of ancient fords and bridges along the banks, and the monastic and other settlements on many of the river's islands. In the 1750s, as the volume of traffic and the size of boats using the Shannon grew, the Inland Navigation Board built locks and canals at Meelick and Athlone to bypass the falls there and make the river more navigable.

Steamer services for passengers and freight began in the 1820s between Killaloe and Athlone and, to a lesser extent,

Carrick-on-Shannon in the north, Thomas Rhodes, the Shannon Commission engineer, drew up plans for major drainage and navigational improvements, replacing earlier canals and trying to tackle the annual problem of winter floods. Steam-powered buckets were used to dredge channels through shallow stretches, and new locks and canals were built along the river where needed. In Athlone, a large weir and lock were built, bypassing the earlier canal, and a new road bridge was installed with a central span that opened for the steamer traffic (sadly this was replaced in the 1960s by a fixed slab). No passenger or freight services ply the Shannon now but, thanks to the old navigation and the recently restored Ballinamore–Ballyconnell canal, flotillas of leisure craft can again journey from Limerick to the Erne. [Next stop on the river: **The magical Shannon Pot.***]

MULLINGAR

The railway on the bog

Railways running across boggy ground are usually lightweight narrow-gauge systems, like that still used by Bord na Móna.* But the Midland & Great Western Railway Company (MGWR), which built the Dublin–Galway line in the late 1840s, wanted to run its heavy standard-gauge main line across part of the great midland bog.

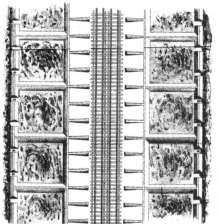

George Hemans devised an ingenious way of laying the railway across the bog.
Institution of Engineers of Ireland

The MGWR, set up in 1844 to rival the Great Southern & Western Railway, bought the Royal Canal* for a bargain price and began laying the track for its Galway line on the level ground prepared alongside the canal. For a stretch between Enfield and Mullingar, however, the company left the canal route, opting for a risky shortcut across bog that in places was 20 metres deep. Building a railway across deep bog posed many problems. First, to consolidate the ground, numerous drains were cut into the bog. This drained and helped stabilise it and also lowered the height of the surface. Then the company's engineer, Englishman George Hemans, devised an ingenious way of laying timbers so as to spread the weight of the railway across a larger than usual area. When they were first used, the tracks still deformed considerably but, eventually, after tests showed that the deformation had reduced to just 5cm, the line was deemed safe for public use. Opened in October 1848, Hemans's route is still used today.

To span the broad Shannon at Athlone, Hemans designed a railway bridge 180 metres long, with a central swivel portion to accommodate the Shannon steamers. Six pairs of cast-iron legs carried the wrought-iron latticed girders. The dozen legs, each hollow and 3 metres in diameter, were cast in Limerick and floated up the Shannon on barges and sunk into the river bed using, for the first time in Ireland, compressed air to exclude the water, they were then filled with masonry to anchor them in place. The Board of Trade tested and approved the bridge's load-bearing capacity (at 200 tonnes) and the *Westmeath Independent* welcomed it as "a beautiful and permanent piece of Art".

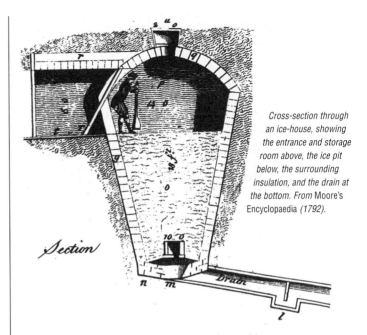

Cross-section through an ice-house, showing the entrance and storage room above, the ice pit below, the surrounding insulation, and the drain at the bottom. From Moore's Encyclopaedia (1792).

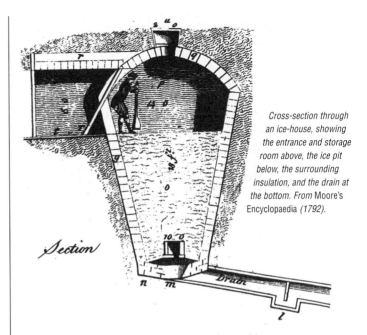

CASTLEPOLLARD

Historic fridge-freezers

Ice houses were the fridge-freezers of the 18th century, and Westmeath has some fine examples, including those at Tullynally Castle and Belvedere House (see below), where the estates are open to the public. People have long known that chilling preserves food: caves, and the souterrains of the Late Bronze Age may have been used as cold rooms, and butter was often stored in cool bog pools. During the 18th century, most of Ireland's large demesnes had ice houses. These were filled in winter with ice that was harvested from a nearby pool, or bought in, perhaps from ice makers on the Dublin Mountains, or even transported from Norway or Canada by special ships. During the summer, food could be stored in the ice house and the ice itself used in cold drinks.

The ice was kept in an underground storage pit to insulate it; a drain at its base carried away any meltwater and kept the ice dry. The cold room was in a domed space above this pit, often brick lined and insulated on the outside with sods. Two doors, an outer and an inner one, separated by an air trap, gave access to the room and the ice. Máiréad Johnston, an archaeologist who has studied Irish ice houses, says good design was crucial.[1] Anything that might make the place damp, such as its orientation, the soil it was built on and whether it was overshadowed by trees, detracted from its performance. Some ice houses were large structures – that at the former Vice-Regal Lodge in

Dublin's Phoenix Park,* now the Zoological Gardens, is 6 metres deep.

For many years, ice was made at the Featherbeds ice works in the Dublin Mountains. (Europe was still in the grip of the Little Ice Age until *c*.1850.) The works consisted of a shallow reservoir; when the water in it froze, blocks of ice were cut and harvested and stored nearby in stone-lined pits under an insulating straw thatch. Horse carts took the ice to the city's merchants and markets as required. All this changed in the late-19th century with the arrival of ice chests and ice-making machines, but ice houses remain on many Irish estates, leftover from an earlier time. Tullynally Castle, home to the Pakenhams, the earls of Longford since the 1600s, has gardens, woodlands and an avenue of 200-year-old Irish yews. ②

Exploring Everest

The first ever expedition to Everest was led in 1921 by a Mullingar man, **Lt Col Charles Howard-Bury** (1881–1960), a British army officer based in India who had a passion for exploring. His Everest expedition did not reach the summit, but the climbing team (which included George Mallory who died on Everest in 1924), had many notable achievements: reaching the North Col (at 7,000 metres, still 1.5km from the top), and climbing several 7,000-metre peaks in the range. More important was the mapping done by the reconnaissance team who, in three intense months, surveyed several thousand square kilometres of what was until then a *terra incognita*. Their charts and geological surveys laid the basis for all subsequent Everest maps. As well as climbing, Howard-Bury was a keen hunter and plant collector, and the formal gardens at Belvedere, his Mullingar estate, contain Himalayan plants he collected. Belvedere House, itself a 1740s fishing lodge, now includes an angling museum. ③

Walking on water

Scragh Bog is a rare and special place and, despite the name, it is in fact a fen.* It is in the process of changing from an alkaline fen into an acidic raised bog* – part of that slow progression whereby a small freshwater lake becomes overgrown and develops first into a fen and later a raised bog. In truth, Scragh Bog is more like a lake that is now so covered by floating vegetation you could walk on the water. It is an excellent example of the transition from fen to bog and, consequently, of international importance.

The unusual habitat means it is also home to many rare species of plant and animal, including the round-leaved wintergreen and a tiny, hairy fly called *Telmatoscopus maynei*. Both of these are glacial relict species – a hangover from when the climate was colder at the end of the last Ice Age – and, in Ireland, both are found only at Scragh Bog. Thanks to money donated by the Dutch Foundation for the Conservation of Irish Bogs, the 23-hectare nature reserve was bought by the Irish Peatland Conservation Council in 1992 and it is now a national nature reserve.

The Edgeworth-Kuiper belt?

Beyond the planet Neptune, at the outer edges of the Solar System, is a vast ring of comets and would-be comets, dust and debris left over after the planets formed. It is called the Kuiper Belt after Gerard Kuiper who suggested its existence in 1951. However, **Kenneth Edgeworth**, an amateur Irish astronomer, who first studied astronomy in his uncle's private observatory at Daramona, near Street, made the same suggestion two years earlier.

In the second half of the 19th century, a number of wealthy, Irish, amateur astronomers built their own observatories. The best known is probably the telescope at Birr,* but **William Wilson**'s (1851–1908) observatory on his

Daramona Estate was also important. When Wilson was 19, he joined an expedition to Iran to observe a total eclipse of the Sun. The following year, he bought a 12-inch (30.5-cm) reflecting telescope from Howard Grubb,* the renowned Dublin telescope maker, and started observing at Daramona. Later, using a larger telescope, he took many fine early photographs of nebulae and star clusters. In 1896, the Dublin scientists who, four years earlier, had made the first-ever electrical measurements of starlight,* came to Daramona with a more sensitive version of their electrical equipment. With Wilson, they measured the light of over a dozen more stars, as well as making the first accurate estimate of the temperature of the Sun's surface.

Wilson's nephew, Kenneth Edgeworth (1880–1972), related to the noted Edgeworth family of Edgeworthstown,* was born at Daramona and learned his astronomy there, though he grew up at nearby Kilshrewly.[2] After working as an engineer with the British army, he retired to Dublin in 1931 where he began writing and publishing books, initially on economics and later on astronomy. Edgeworth was a regular visitor to Dunsink Observatory* where he discussed astronomy at length, being particularly interested in how the planets in our solar system formed.

In an article he wrote for the *British Royal Astronomical Journal* in 1949, Kenneth Edgeworth speculated about the existence of unseen bodies in the outer reaches of the solar system and suggested: "The material out of which the embryo planets were first formed existed originally in the form of a rotating disk of small solid particles extending beyond the orbit of Neptune." The residual material left after the planets formed continued to orbit as a flat ring and, he thought, some objects could leave the ring to become comets.

Edgeworth has since been proved right but, while his insight came two years before Kuiper's, the Irishman's work was not well known and Kuiper gained the credit. Edgeworth's contribution has at least been acknowledged in recent years, though it is unlikely that astronomers will start referring to the Edgeworth–Kuiper belt. It was 1992 before astronomers actually saw the first object in the Kuiper Belt: a tiny, faint object some 250km in diameter, and labelled 1992KB1, it was seen from Mauna Kea observatory in Hawaii.

The seven wonders

The village of Fore claims seven historic wonders, including water that will not boil and a stream that apparently runs uphill. But the countryside here is also a wonder, a puzzling landscape that scientists are still trying to understand. Two of Fore's seven wonders – the Anchorite's Cell and the Rock of Fore – are by the ruins of the Church of St Fechin, while the others are in and around the ruined 13th-century Benedictine Priory (information about all seven is available locally). The priory is one of the wonders, for being built on the 'quaking bog'. Nearby are the tree that will not blossom, the well water that will not boil – be warned, illness apparently befalls those who try! – the mill without a millrace and the stream that runs uphill. The latter leaves Lough Lene, disappears underground, and seems to emerge as a spring higher up and across the road…

But nothing is quite what it seems in this unusual place and, in fact, the stream and spring are probably not connected at all, though they are a clue to understanding the region's puzzling and deceptive terrain. Like much of the midlands, the rock here is limestone. But, unlike the flat plain to the west, there is a swarm of three-dozen small hills here, and no obvious explanation as to how they formed. There are also numerous lakes, but some, such as Lough Bane, have no visible inlet or outlet.

David Drew, a geographer at Trinity College Dublin who has studied the area, believes this is a **fossil landscape**: the remains of an ancient limestone landscape that, 2 million years ago, was as impressive as that seen today in China and Vietnam, with great rocky towers and pinnacles, deep chasms and gorges, immense caves and a sub-tropical climate.[3] Fore's limestone landscape has since been severely eroded, and what is left is now mostly hidden under gravel left by later glaciers, though there are a few clues to its past. The small hills in the area may be the remains of limestone 'towers', for example, and the many lakes are possibly flooded depressions (depressions being common in a mature limestone landscape).

Clues to the amazing world that exists below ground level come from two very different springs in Fore, just 300 metres apart. At Toberfaonagh, the water is always 10.5°C and the flow, which comes underground from Lough Bane 3km to the north, is constant. This suggests the water travels a slow and ancient route, and spends a long time underground. The temperature at the second spring changes with the weather and the flow varies with the season, usually drying up in summer. That is because this spring is fed from Lough Lene to the south where the inlet, which is on the lake side, is left dry and exposed when the water level drops. This stream does not spend much time on its underground journey – perhaps a day at most – and this fast route suggests it is younger than its neighbour. Another confirmation of a fossil landscape here is the presence of many caves but these, like the many wonders of Fore, have yet to be fully explored.

Pigeon pie

The circular pigeon house or dovecote at Fore's Benedictine Priory dates from mediaeval times when the monks kept birds as a source of winter meat. There was limited winter fodder then: haymaking,* introduced to Ireland by the Normans, was still not widespread and, in a cold spell, cattle often died for want of grass. So rabbits* and pigeons were a handy source of winter meat. The pigeons were kept in flocks of up to 1,000 birds in a dovecote that had nesting spaces in the walls. They produced eggs, guano for fertiliser, and meat. Pigeons are especially prolific, producing two chicks a month, and tender month-old chicks were a delicacy. Large estates might also have had an artificial fish and duck pond for similar reasons. In the 18th century, however, plant breeders introduced new crop varieties, such as the swede turnip, that could be used to feed animals over winter, and farming changed forever.

Lough Lene is unique, for it drains both west to the Atlantic and east to the Irish Sea. The flat central plain means Ireland has no watershed, unlike say North America, which the Rocky Mountains divide in two: rivers west of the Rockies drain into the Pacific, those to the east into the Atlantic. But Lough Lene has it both ways: one stream drains into the River Deel and thence to the Boyne and Irish Sea; an underground channel supplies a spring at Fore, which drains west to the Inny river and thence to the Shannon and Atlantic Ocean. This unusual behaviour was confirmed in the 1980s when geographers from TCD traced Fore's underground springs. Science, far from demolishing the myths at Fore, is adding to its list of wonders.

COUNTY MEATH

Meath is rich in ancient remains: such as the world's oldest astronomical observatory at Newgrange, the Hill of Tara, which was once at the hub of Ireland's historic road network, and the many mediaeval bridges across the Boyne. Navan, birthplace of the man who invented the Beaufort Wind Scale, is now home to one of the world's largest zinc mines. Meath, from *an Mhí*, meaning 'the middle', refers to the original fifth province of Ireland that once occupied the centre of the country and incorporated neighbouring Westmeath, until that was created as a separate county during the Plantations.

Crossing the Boyne

Some of the oldest-surviving **bridges** in Ireland are those that cross the Boyne and its tributaries in Co. Meath, a relic perhaps of the ancient road network that once converged on Tara. Ireland has an estimated 30,000 masonry-arch bridges with a span of at least 1.8 metres (6ft – anything less is considered a culvert). Nearly all date from after 1775, but a handful of mediaeval bridges survive, at least in part. Sadly, the records for many were destroyed during the fire at the Four Courts in 1922.

The oldest-surviving and unaltered bridge arch is at **Babe's Bridge** over the Boyne near the round tower of Donaghmore. Named after an Anglo-Norman landowner John le Baube, the bridge was originally 90 metres long and had a dozen arches. Only one arch survives, though traces remain of the piers that supported the rest of the bridge.

The bridge is mentioned in *Annales Hiberniae* (1330), but engineering historian Peter O'Keeffe believes it may have been built 100 years earlier by a Norman stonemason sent by King John, who crossed the Boyne at Trim in 1210.[4]

Ireland's oldest-surviving and unaltered complete bridge is probably that at the historic town of Trim. Thought to date from 1393, it is still in use. The bridge is 6.5 metres wide, with four pointed-arch spans; the piers were reinforced in the 1970s during drainage work. Another old bridge in the area is Mabe's Bridge on the Blackwater a mile north of Ceanannas (Kells). Peter O'Keeffe believes this was built in the 15th century by the Mapes, a local Norman family. Restored and partly rebuilt in the 1980s, it is again open to traffic.

Newtown Bridge on the Boyne just east of Trim is a gentle humpbacked bridge of five arches, reputedly built about 1470 by William Sherwood, Bishop of Meath. Bective

Bridge, also on the Boyne, is a long, elegant structure with 11 arches and probably dating to the mid-1600s. Finally, part of Slane Bridge may date to the 1300s, though most of the 13 arches were built later. This narrow bridge, less than 7 metres wide, still carries all the traffic on the main Dublin–Monaghan road.

Ireland's two oldest-surviving documented bridges are over the Liffey at Kilcullen (1319) and over the Barrow at Leighlinbridge (1320). Both were built by Canon Maurice Jakes and sadly were much altered down the centuries and widened for traffic. Other notable bridges are Mayo's unusual Clapper Bridge;* the engineering triumph that is Drogheda's Boyne Viaduct;* Dublin's Ha'penny Bridge* (Ireland's oldest metal bridge) and Ennistymon's Spectacle Bridge.*

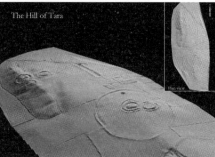

The Hill of Tara

A computer-generated, 3D view of the Hill of Tara, made with new data from the Discovery Programme survey. The image reveals archaeological remains and highlights the local terrain, making analysis easy.

© Discovery Programme

TARA

A hidden landscape

The Hill of Tara, seat of the high kings of Ireland, was an important political and religious site for over 5,000 years. At least 30 archaeological structures are visible in the area and there are copious references to Tara in the ancient manuscripts. The site was also the hub for the five major roads that radiated out across the country (see: **Old roads and milestones**). Yet not much was known about Tara until a major study was undertaken in the 1990s using the latest archaeological techniques.

Tara is built on one of the many small, sandy hills that occur in a band sweeping south into Co. Kildare. Those hills are a glacial moraine, produced during the last Ice Age as the ice sheets melted and retreated north. The area around Tara and to the south was then part of a vast delta flooded with glacial meltwater. The band of sandy hillocks marks the edge of that delta, while the rich farmland to the south owes its fertility to the well-drained base it gained from the ancient delta's outwash.

Reading the landscape

Science is providing archaeology with powerful new tools that can reveal hidden information about landscapes, monuments and artefacts. The tools, many of which were used at Tara, can help identify the best place to dig or even enable some sites to be studied without turning a sod. Thus time and money are not wasted and disruption is minimised.

Aerial photography: can reveal features not visible from the ground.

Carbon-14 dating: dates many objects reasonably accurately and precisely.

Tree-ring dating: (dendrochronology) this amazing technique can date oak timbers to precise calendar years. The oak logs used at Corlea trackway,* for example, were felled in 148 BC.

Pollen analysis: the presence and relative abundance of pollen from trees, grasses and crops reveals much about the climate, vegetation and agriculture of the time.

Crop marks: structures hidden under the soil can affect crop growth. Where soil is shallow (perhaps a stone wall lies under the surface), plants will be stunted and ripen early; deeper soil (e.g. over a filled-in ditch), produces tall plants that ripen late. The variations are often visible on a summer's day and in aerial photographs.

Magnetometry: heat from a fire makes atoms slightly magnetic; a magnetometer can detect this and locate burnt or fired objects, e.g. pottery, bricks, hearths and kilns.

Soil resistivity: instead of digging, a site can be mapped by running an electric current between probes placed in the ground. Damp soil conducts electricity well; dry and well-drained soil (for example, around a buried stone wall) does not. A computer can plot the results to produce a map of the area. This technique was used at Navan Fort,* Armagh.

Computers: 3D graphics, electronic theodolites, GPS and geographical information systems… just some of the tools used to store and analyse archaeological data. Archaeology is not so much about objects and artefacts as about information, and computers are good at converting archaeological data into 3D pictures of what the site might have looked like. This approach was used extensively in the Tara project.

There has been some digging at Tara over the last 200 years, mostly at the Rath of Synods. Much of it was treasure hunting: two gold torcs found there in 1810 prompted a gold rush. In 1899, the British Israelites arrived, convinced that the Ark of the Covenant was buried at the rath. They unearthed an underground passage, 15 Roman coins, a brooch, and considerable public outcry – WB Yeats was among those calling for an end to their activities – but sadly no Ark. In the 1950s, Prof Seán Ó Ríordáin led a small archaeological excavation, again at the Rath of Synods, since the site had already been disturbed by the earlier activities. An intensive 5-year survey at Tara began in 1992, led by Conor Newman of the Office of Public Works. Using the latest techniques (see: **Reading the landscape**), the archaeologists probed the site without disturbing it. Hidden under the landscape, they discovered some 30 previously unknown features, mostly barrows and ring ditches.[5] The Hill of Tara, which is overseen by Dúchas, is always open. ④

TARA

Old roads and milestones

In the year AD 123, on the night of the birth of Conn, first high king of Ireland, the *Annals of the Four Masters* record that five *slígheanna* or primary roads appeared connecting Tara with the north, west and south of Ireland. The Romans

Tara was at the centre of an ancient road network, with five major slígheanna *radiating out to the north, west and south of Ireland, and often passing through important monastic settlements.*

A wheeled chariot, depicted on a high cross at Clonmacnoise, a monastic site on the Slíghe Mhór/Eiscir Ríada.

never brought their road-building engineers here, but clearly the ancient Irish had their own techniques.

The *slígheanna* (*slíghe* from *sligid*, 'to fell', suggesting a way cut through a forest), were the motorways of their day, though they were probably little more than dirt tracks, wide enough to enable two chariots to pass. Where possible, the routes followed dry esker ridges;* elsewhere, rivers were forded and timber tracks were laid across bogs. (Many timber trackways are preserved in the bog, as at Corlea,* Co. Longford.)

Up to the 1600s, most journeys were short and on foot – even invading armies walked – and routes followed a line of least resistance. Wheeled traffic was mostly confined to towns, where it caused problems for road maintenance – in 1480, Dublin Corporation levied a cart tax of 2s.2d on each iron-clad cart to fund city roadworks.[6] Local authority responsibility for roads started with the Highway Act of 1615, which obliged parishes to appoint surveyors, and required farmers to supply men and materials to mend the roads each spring. Ireland's modern road network began to develop in 1710–65, following a series of acts that gave Grand Juries (forerunners of the county councils) power to acquire land for local road building and, in 1765, set down the first road standards – the standard width was to be 21 feet (6.4 metres), for example, with the central 14 feet (4.5 metres) gravelled.

A network of **turnpike** (**toll**) **roads** developed in parallel starting in 1729. Most radiated out from Dublin and, to a lesser extent, from Cork, Limerick and Belfast. They were authorised by government, and any profits were to be spent on repairs. The introduction of **mail coaches** in 1789 gave a new impetus to road development. Roads were straightened and levelled, and stringent standards were laid down, including, for example, the steepest gradient allowed (bearing in mind the limits of horse-drawn coaches).

Between 1805–11, the Post Office surveyed over 3,000km of road, and many new routes were built, including the Dublin–Slane road. Those designing and building the roads included three noted Scottish engineers: Alexander Nimmo,* who is associated in particular with roads in Connemara; William Bald,* who built the Kenmare–Glengarriff road and the spectacular Antrim Coast Road; and Alexander Taylor, who masterminded Wicklow's Military Road* (1800–03). Consequently, Ireland had a good road network, while England was less well served. In 1813, Richard Lovell Edgeworth,* reporting on road engineering to the House of Commons, wrote: "I have visited England and have found… scarcely twenty miles of well-made road. In many parts… the roads are in a shameful condition."

Ironically, the most intense period of Irish road development came immediately after the Famine and, during 1846, an estimated 350,000 people were employed in famine-relief work on roads. In 1834, Ireland had 23,000km of road; by 1884 that had quadrupled to 83,000km. Since then, the total length of road has increased by only 40 per cent to 118,000km. The second half of the 19th century also saw a dramatic increase in road transport, as more people journeyed further afield – whether to reach towns, railway stations, emigration ports or seaside resorts. Demand for train services lead to new feeder roads linking the smaller towns with the railways.

In 1888, John Dunlop* designed the first successful pneumatic tyre. Intended to make bicycle riding smoother,

it had a greater impact on the motor car. By 1909, there were 3,790 motor cars licensed in Ireland, plus 3,245 motor cycles and 71 motor wagons. People could now travel further and faster than ever. They also demanded smoother surfaces and so the tarring of roads began in earnest.

RATHFEIGH

The body snatchers

The small hut built into the wall at Rathfeigh graveyard is a watch-house, dating from when fresh corpses were a valuable commodity much sought by anatomy schools, when metal grilles and heavy slabs were used to protect graves from the 'sack-'em-ups', and 'security men' would stand watch over each fresh burial. Today, people can donate their body to science, but society's attitude to **dissection** has not always been this accepting. The Christian churches historically considered the body sacred and opposed the practice, believing a body must remain intact for resurrection on the Last Day. Despite this opposition, European anatomy schools began dissecting cadavers in the mid-16th century, leading to many important scientific discoveries, notably the circulation of blood.

The growing number of medical schools in the 18th century created a strong demand for fresh corpses. The bodies of hanged convicts were sometimes donated to the anatomists – this was termed "a fate worse than death" – but the main source was bodies dug up (usually without permission) by medical students, who helped pay for their studies with the proceeds of a night's digging, or by professional resurrection men. Winter was the best time, as the cold preserved the body and discouraged mourners from staying in the cemetery. Pauper graveyards were often unguarded and offered rich pickings; elsewhere, unscrupulous gravediggers might be persuaded to turn a blind eye.

Builders demolishing a house in Dublin's Liffey Street in 1999 discovered a dozen skeletons buried in the basement. Worse, the remains seemed to have been butchered: small holes were drilled in the bones, and skulls had been sliced open as if to reveal the brain. When it became clear this was not a recent burial, the Gardaí called in TCD anatomist Dr Maura Delaney who specialises in historic remains and bog bodies. The premises, it emerged, was a medical school in the early 1800s, and the holes in the bones were where copper wire had been threaded to hold the skeletons together for display. Strict guidelines now govern the burial of bodies donated for dissection, but the sites of older anatomy schools may still hold secrets such as that found at Liffey Street.

The Napoleonic wars brought increased demand for trained surgeons, and a corresponding demand for bodies. By then, Dublin had several anatomy schools, as well as the Royal College of Surgeons in Ireland.* Most relied on the shady underworld of body snatchers, and prices soared as schools competed for scarce subjects – bodies sold for 2 guineas in the early 1800s in Dublin, and for £13 in 1830. Ireland was also an important source of bodies for the great London and Edinburgh medical schools, though the remains were hardly fresh on arrival. Understandably, there was considerable public unease about the anatomists' activities. Some anatomy schools were attacked by angry mobs, and there are reports of pitched battles at cemeteries between body snatchers and those guarding graves.

The most notorious Irish body snatchers were **William Burke** (1792–1829) and **William Hare** ("flourished 1820s"). In 1827, when someone at their Edinburgh lodging house died, the pair sold the body for £7.10s to a renowned Scottish medic, Robert Knox, rather than pay for a burial. Attracted by this easy money, they murdered 15 people over the next year, selling the bodies to Knox. When caught, Hare turned Crown witness, Burke was hanged, and Knox had to flee the city.

To counter fears about dissection, 100 prominent surgeons pledged to donate their own bodies for dissection, but it was the crimes of Burke and Hare that eventually forced the introduction of the 1832 Anatomy Act. This allowed surgeons to take the "unclaimed bodies" of workhouse paupers, provided they or a close relative had not previously objected; and 48 hours had to elapse after death before dissection. However, the main source of cadavers remained the poor and those in State care, until the 1920s, when attitudes to life, death and dissection again changed, and voluntary donation was introduced.

Our mineral wealth

During the 1950s and 1960s, thousands of Irish children learned from a school geography book that "Ireland has no natural or mineral resources". Ironically, at that same time, and thanks to new exploration techniques, geologists were learning that Ireland is relatively rich in minerals, especially zinc. Indeed the list is now quite long: uranium* and possibly even diamonds* in Donegal; lignite* at Lough Neagh; coal* in the Kish Bank; gold* at various locations; and enormous lead-zinc deposits at Navan, Lisheen,* Galmoy* and elsewhere. One of the largest zinc mines in the world is Tara Mines, on the Kells side of Navan, where each week thousands of tonnes of ore are mined. The crushed and washed ore, which contains zinc and some lead, is shipped to European smelters for refining. In recent years, Navan's annual output was 200,000 tonnes of zinc

Working underground at Tara Mines, one of the largest zinc mines in the world.
© Outokumpu/ Tara Mines

metal, which is mostly used to galvanise steel; and 40,000 tonnes of lead, used in lead-acid batteries.

Mineral exploration began at Navan in the early 1960s. By then, mining companies had identified new commercial deposits around Ireland: notably the lead-zinc and silver mines at Tynagh in Co. Galway, and at Silvermines* in Co. Tipperary, and a new copper mine at Gortdrum, also in Tipperary. But these were small compared with the ore body discovered at Navan in 1970 which, at 80 million tonnes (10 per cent zinc and 2.6 per cent lead), was the largest in Europe. Mining began in 1977 and there are still 30 million tonnes of ore left – enough for another decade (though the mine nearly closed in the late 1990s when the world zinc price dropped). The mines at Tynagh, Silvermines and Gortdrum are all exhausted and closed.

The metals mined at Navan were spewed out by super-hot 'black smoker' vents on the sea floor 350 million years ago, when this part of Ireland was covered by a tropical sea. The minerals were later covered by layers of limestone and sea floor. The ore veins, some of them 80 metres thick, lie 60–600 metres under the surface. The ore is mined underground by blasting out the rock. It is then transported to the surface where it is crushed to a fine sand. This is mixed with water and additives that help concentrate the ore and remove much of the unwanted limestone. The end products are a lead concentrate containing 67 per cent lead metal, and a zinc concentrate with 56 per cent zinc metal. In 1989, the Finnish company Outokumpu bought out the government's shareholding in Tara Mines, and it continues to explore for new deposits at over 30 sites in Westmeath, Longford, Roscommon and Galway.

NAVAN

When the wind blows

The Beaufort Wind Scale was invented by Navan-born **Francis Beaufort** (1774–1857), one of two Irish people who made important contributions to the study of wind.

The other was Thomas Romney Robinson* who, in 1846, made the first instrument capable of measuring actual wind speed. Both the Beaufort Scale and Robinson's anemometer are still used.

Admiral Sir Francis Beaufort was born in Navan where his father, Rev Daniel Augustus Beaufort, was rector. Rev Beaufort, born in England of Huguenot parents, was a learned man, a founder member of the Royal Irish Academy* and a skilled map-maker. His *New Map of Ireland* (1792) accurately depicted Irish coasts, rivers and place names and was not bettered until the Ordnance Survey of Ireland* 30 years later.

Francis Beaufort, despite his inland home, had a passion for the sea and joined the Royal Navy at 13. The following year, his father arranged for him to spend five months studying astronomy and meteorology at Dunsink Observatory.* There he learned to calculate latitude and longitude from the stars, a training that served him well in his later career as the Royal Navy's hydrographer.

Beaufort always kept a weather journal and, in 1806, realised the need for a standard system to describe wind and weather conditions. Until then, people used subjective descriptions and vague terms like 'light breeze'. The system Beaufort devised classified the wind according to the amount of sail a man-o'-war battleship could safely hoist.

The air over Ireland is seldom calm! We have still weather less than 5 per cent of the time, winter being the windiest season, and so we are well placed for wind farms. The north, west and south coasts are the windiest: Malin Head weather station records 80 gale days a year on average (winds over 18 metres/second), compared with 10 at Rosslare.

Wind is air rushing from an area of high pressure to one of low pressure. The pressure differences are due to unequal heating of the Earth's atmosphere; and the greater the pressure difference, the faster the wind blows. The predominant winds over Ireland are southwesterlies, which bring in mild, rain-laden air from the Atlantic and greatly influence our weather.

His 13-point scale, allied with the terms commonly used by sailors, graded the wind from Force 0 (calm), through Force 5 (fresh breeze), to Gale Force 8 and finally Hurricane Force 12, "which no canvas can withstand" and when the air at sea "is filled with foam and spray". Beaufort also invented a system of coding weather conditions using letters and dots to indicate intensity, which some meteorologists still use. His wind scale was later modified for the post-sail era and to include actual wind speeds once these could be measured: Force 0 corresponds to a wind speed of at most 0.3 metres/second; and a Hurricane Force 12 wind travels at over 30.2 metres/second.

The first recorded use of Beaufort's wind scale was not until 1831, in the log of HMS *Beagle* under Capt Robert Fitzroy. (Fitzroy later organised a telegraphic network of weather stations around these islands, including one at Valentia,* to provide early storm warnings.) Beaufort's other involvement in that historic voyage of the *Beagle* was to arrange for a young naturalist called Charles Darwin to travel as the captain's companion.

In 1829, Beaufort was appointed Hydrographer of the Admiralty. The appointment reflected his tremendous talents, for the Irishman would have had little patronage in the Royal Navy. He set about mapping uncharted coasts at home and abroad, and changed the hydrographer's office into a major scientific survey department. The charts he produced – 1,500 in all, equivalent to a rate of over one each week – were renowned for their accuracy and must surely have prevented numerous shipwrecks. Some of his charts were not bettered until the 1970s.

Beaufort had many scientific interests: he helped found the Royal Geographical Society and organised polar expeditions and, in 1803, he joined Richard Lovell Edgeworth,* the Longford-based engineer, to experiment on Edgeworth's design for a semaphore telegraph system.

The two men were already brothers-in-law – Beaufort's sister Frances was Edgeworth's fourth wife – when, in 1838, Beaufort married Edgeworth's daughter Honora. (It was Edgeworth who, in the 1780s, first suggested building a device to measure wind speed; but it was not until 1846 that Thomas Romney Robinson, who was also related to Edgeworth by marriage, succeeded in building a practical anemometer.) The Beaufort Sea off Alaska is named after this talented Irishman, and a commemorative plaque at St Ultan's School in Navan marks his birthplace.

Ireland's first aeronaut?

The first manned balloon flight in these islands reputedly ascended from Navan town on Thursday April 15th, 1784. Just a few months earlier, in November 1783, the French Montgolfier brothers had proven that hot-air ballooning was possible. An unmanned balloon was sent up from Dublin's Rotunda Gardens as early as February 1784, but the Navan flight (if it really happened) was more daring. *Faulkner's Dublin Journal* reported that at 2.30pm, the rope was cut and a Mr Rousseau, accompanied by a 10-year-old drummer boy, took off in a wicker basket attached to a balloon. The balloon disappeared from sight after half-an-hour but the drum could still be heard. The pair landed safely near Ratoath at 4pm (though the drummer boy apparently sustained some bruising in his rush to leave the basket), and were back in Navan two hours later.

Sadly, Rousseau's flight is unconfirmed, and Ireland's first manned balloon flight probably took place the following year on January 19th, 1785, when Wicklow-born **Richard Crosbie** (*b.*1755), watched by a large crowd, took to the air for a short flight from Leinster House* lawn to Clontarf. He generated the hydrogen gas for the balloon by pouring acid over iron and zinc filings, and raised funds by exhibiting his balloon beforehand in the Rotunda* rooms. Crosbie

First flight by an Irishman 1785

undertook several further flights in 1785, on one occasion carrying a prototype device for measuring wind speed devised by Richard Lovell Edgeworth.* Ballooning soon became a craze, with numerous ascents from Dublin and around the country, so much so that Dublin's mayor wanted to ban balloons altogether. He was not without reason: in May 1785, Tullamore* town was destroyed by fire after an unmanned balloon crashed and set fire to a thatch. In 1817, **Windham Sadler** (1796–1824) became the first person to cross the Irish Sea by balloon: he took off from Portobello Barracks at lunchtime and landed in Holyhead in time for tea, returning home on the Howth Packet (mailboat) the following day. (For more on Irish aviation history see: Godwin Swifte,* Harry Ferguson,* Alcock and Brown,* Amelia Earhart* and Sophie Pierce.*)

THE BEND IN THE BOYNE

River of wisdom

The River Boyne follows a convoluted route to the sea and, at Knowth, takes a long detour south, forming a lazy loop before eventually swinging north again towards Tullyallen. That lazy loop encircles *Brú na Boinne* ('the dwelling place on the Boyne'), the most important neolithic site in the country. It has numerous monuments, ritual ponds and enclosures, and nearly 40 tombs, among them Newgrange with its celebrated astronomical alignment (see: **The first astronomers**). It is easy to see why this site appealed to neolithic people: the soil is fertile, thanks to the underlying limestone and the thick covering of rich, well-draining glacial till; the river is navigable and provided a highway to the coast, as well as an important salmon fishery.

The Boyne has been important since the earliest times. Named after a Celtic goddess and hunter, Boann, it is marked in Ptolemy's 2nd-century map of Ireland as Buvinda. It is the local geology that forces the river to bend. From Trim, the Boyne takes a northeasterly route among gentle limestone hills and hollows. A wall of rock at Oldbridge forces the river east, however, and down a gorge to Slane. This wall of rock is the face of a geological fault that thrust up the land on the northern bank by some tens of metres. Past Slane, the river hits another fault and turns south along this line, skirting the higher ground to the north that forms *Brú na Boinne*. Along the Bend in the Boyne, the river digs itself ever deeper into the thick beds of glacial debris that carpet the area. Eventually the river reaches an old gorge, carved out of the limestone 15,000 years ago by tremendous torrents of glacial meltwater, and it follows this easier route north, completing its bend and resuming its journey to the sea.

The Boyne has long been an important **salmon** river, and a Boyne salmon eaten at the Celtic festival of *Lughnasa* on August 1st was said to ensure long life and good fortune. Fionn McCumhaill tasted the salmon of knowledge, *an bradán feasa*, when he accidentally burned his thumb on the magical fish as the druid Finn Eigeas roasted it after catching it in the Boyne. Fionn gained his wisdom from the Boyne fishery, but he also met his end there, killed by Athach, a Boyne fishermen who speared the legendary warrior with a salmon gaffe.

The earliest evidence for fishing on the river comes from neolithic stone flakes found at Newgrange that are thought

This stretch of the Boyne boasts four natural heritage areas: Crewbane Marsh, Dowth wetlands, Rossnaree riverbank, and the Boyne islands. The Bend in the Boyne also has an industrial past: the river here was repeatedly harnessed in historic times to power mills, and the 12th-century millrace for Broe Mill, built by Cistercians, can be seen below Knowth.

An bradán feasa: the salmon has featured on Irish coins (the old florin or two-shilling piece and the now defunct decimal 10p piece) and on stamps.
© An Post

to have been used to cut and skin fish. Salmon were traditionally caught by spearing or in nets. The Boyne salmon gaffe or spear had, like Mercury's trident, three barbed prongs. It could be thrown or thrust at the fish and was used until it was outlawed in 1716. Netting was traditionally done from a **Boyne coracle**, a small flat-bottomed and nearly circular craft that is unique to the river (one can be inspected at Millmount Museum,* Drogheda). In the Middle Ages, the Boyne salmon fishery was managed by the Cistercians at Mellifont who built weirs to control the water flow, and fish traps to catch salmon and eel. The **mediaeval weirs** at Oldbridge and Rossnaree are still in place and virtually intact.

The first astronomers

The elaborate burial mound at Newgrange is arguably the world's oldest astronomical observatory. The tomb is built in such a way that, at each winter solstice, the first rays of the rising Sun shine through a special opening (called the roof-box) and light up the inner chamber. The construction is a feat of precision engineering, made 5,000 years ago by Stone Age people who possessed neither metal tools nor the wheel. Yet their observatory marks the turning point of the solar year and it was built 500 years before the pyramids and 1,000 years before the astronomical alignment at Stonehenge. (On the *evening* of the winter solstice, the Sun shines into the chamber of a tomb at Knockroe, Co. Kilkenny.*)

Like Knowth and Dowth, the two other major burial mounds in the area, Newgrange is built in a prominent position on top of a rocky outcrop, and its entrance passage climbs uphill for 15 metres into the burial chamber. The roof-box is set over the entrance, but level with the floor of the inner chamber. This chamber has an elegant corbelled roof made of overlapping slabs angled so that the water runs off the outside. No mortar was used and the edifice holds together under its own weight. The stone structure was topped with layers of stones and sods to form a mound 10 metres high.

In truth, Newgrange no longer marks the actual moment of the mid-winter dawn: Prof Tom Ray, an astronomer with the Dublin Institute of Advanced Studies,* found that because the Earth's axis has shifted a few degrees in the 5,000 years since Newgrange was built, the Sun now has to climb above the horizon for 4½ minutes before it reaches the tomb's opening. Despite this shift, the alignment still functions incredibly well: seen from inside the chamber, the winter solstice Sun appears first in the bottom left corner of the roof-box; it rises diagonally to fill the opening, which is just large enough to frame the Sun, and a beam of light creeps slowly up the passage to the inner chamber; then the Sun passes from view in the top right-hand corner and the chamber is left in darkness for another year.

The design was the result of a precise 3D arrangement of passage, chamber and roof-box, all aligned with the point on the horizon that marks dawn on the winter solstice. Scratches on the underside of the stones used in the roof-box suggest that various positions were tried before the construction was perfected. The winter solstice on December 21st–22nd (the date varies from year to year), marks the end of the longest night, and would be an important date in a farming calendar. And it was farming people who built Newgrange: pollen studies reveal that the woods roundabout had been cleared by then and crops were being grown. But we can only guess at the full significance of this tomb-observatory, and who, if anyone, would be present in the chamber to witness the event.

Sunrise on the winter solstice at Newgrange: the Sun shines through the roof-box above the entrance to illuminate the inner chamber. Many of the large stones in Newgrange and the other Boyne tombs are beautifully carved; some say the frequent crescent and circle designs represent the phases of the Moon. Other designs may represent the Moon's face.
© Dúchas

Building Newgrange was a major undertaking. It probably took many years and involved a significant percentage of the small population living in Ireland at the time. The enormous stones lining the passage and the outer edge of the mound all had to be quarried, ferried, hauled uphill to the building site and then set in place. Some of the stones used were probably quarried from the sea cliffs at Clogher Head* in Co. Louth and ferried by sea and river to the neolithic site 30km away on the banks of the Boyne; the rounded granite cobbles may have come from beaches near the Mournes; and the angular white quartz was probably quarried in Wicklow.

Newgrange had long been known locally and local lore held that the Sun shone into the chamber on a special day. Over the centuries the mound's soil cover slipped, however, and hid the entrance. The tomb was 'discovered' in the late 1600s but was not fully excavated until the 1960s. The archaeological team, led by Prof Michael O'Kelly, found the locals had been right: at 8.58am on December 21st, 1969 they became the first modern people to witness Newgrange's astronomical phenomenon. The megalithic cemetery of nearly 800 hectares has been accorded World Heritage Status by UNESCO, and Dúchas runs a visitor centre at Donore. ⑤

This eastern county boasts two engineering gems: the cast-iron railway bridge over the Boyne at Drogheda, which was a world triumph when it opened in 1855; and Dundalk's leggy lighthouse, which is anchored to the sand by screw piles. The explorer Francis McClintock, who solved the mystery of the Franklin expedition, which disappeared in the Arctic in 1845, was also from Dundalk. Louth or *Lú*, after the village of the same name, derives from *lubhadh*, meaning 'a plain'.

The Boyne Viaduct

The Boyne has cut a deep channel for itself at Drogheda, where the river's long estuary starts. The drop is steepest and most impressive on the southern side, whereas the northern bank is gentler and streets there have less of a climb. The Boyne estuary is 6km long from port to sea, but, despite being this far inland, Drogheda has a fine harbour and has long been an important trading centre. The first settlement there was associated with an ancient ford across the river. The Norsemen had a trading post there, and the Anglo-Normans built the first bridge across the river, the *droichead átha* ('ford bridge') from which the town is named. In the Middle Ages, Drogheda was one of the busiest ports on Ireland's east coast, and it was there, and at the ports of Howth and Dalkey, that the Black Death* first reached Ireland in 1348.

While there was a road bridge over the river for centuries, the steep gorge and broad estuary were a major obstacle when the Dublin–Belfast railway was being built in the 1840s. Initially the Dublin–Drogheda line, completed by 1844, ended on the Boyne's southern shore; by 1852 the line down from Belfast had reached the river's northern bank. Passengers wanting to continue their journey had to cross the river by carriage or on foot from one railway terminus to the other.

Building a bridge capable of carrying a fully laden train proved a major undertaking, especially securing the foundations in the soft river bed, as the bridge's piers had

Sir John Benjamin MacNeill (*c*.1793–1880) initially worked on roads and bridges in Connemara under Scottish engineer Alexander Nimmo* and, in England, with Thomas Telford. There he was responsible for the London–Shrewsbury section of the road to Holyhead that, via the mailboat, linked London and Dublin. In the 1840s, MacNeill worked on Irish railways and, according to engineering historian Ron Cox, pioneered the use in these islands of latticed cast-iron girders in bridge building. His greatest achievement was the Drogheda Viaduct. The first professor of civil engineering at TCD, MacNeill died in London and is buried in an unmarked grave at Brompton churchyard.

to be massive enough to carry the weight of both the bridge and a train. Hence the engineer, Dundalk-born **John Benjamin MacNeill**, suggested using a lattice of cast-iron girders to minimise the bridge's weight. Work began on building a double-track viaduct in 1852, but such were the problems encountered – the foundations, for example, had to be dug 13 metres down into the soft river bed – that one firm went bankrupt. But there was so much rail traffic travelling to the 1852 Great International Exhibition* in Dublin that a temporary single-track wooden bridge was built.

That same year a young Irish engineer, **Bindon Blood Stoney**,* joined the design team. Stoney used the latest mathematical and experimental techniques to test various designs and components, and helped to realise MacNeill's vision of a lattice bridge. This experimental approach was new to bridge building as, previously, builders relied mostly on trial and error to get their design right. For some of the experiments, Stoney teamed up with his uncle, William Bindon Blood,* professor of civil engineering at Queen's College Galway,* who found a way of calculating the stresses for each component.

In April 1855, Stoney hammered in the final rivet and the viaduct was officially opened. It was a major engineering triumph and the longest lattice construction of its kind in the world. The elegant cast-iron section of the bridge consists of three spans and is 155 metres long (the masonry approaches on either side bring the bridge's full length to 540 metres). Stoney and MacNeill's work greatly increased our understanding of cast-iron design and construction.

A wonder of the engineering world: Drogheda's lattice railway viaduct, opened in 1855, was the longest, cast-iron structure of its kind at the time. When it was renovated in the 1930s, the historic cast-iron lattice was replaced by a steel girder version, though the original masonry piers were retained; the line was also converted to single track.
© National Library of Ireland

"Merchandise our Glory"

Drogheda has long valued its trades and industries. The town motto proclaims, "God our Strength" and "Merchandise our Glory". That commercial tradition is celebrated at Millmount Museum, especially the historic guild and trade banners. The bricklayers' elaborate banner features the barbican at the town's St Laurence Gate, which still stands as a testament to their workmanship. One of 10 original town gates, it is the best preserved of its type in the country.

The collection is housed in an old military barracks beside a Martello tower and atop a substantial mound said to mark the 3,000-year-old grave of Amergin, one of the sons of the legendary Milesius. A good geological section features maps, photographs and over 300 local and international specimens, while the industrial room highlights past local enterprises including pipe making, ironworks and shipbuilding. Among the many curiosities is a Boyne coracle reputedly made in 1943 from hazel twigs and the hide of a prize bull. ⑥

Hide-covered coracles are no longer used, but local fishermen do still use another ancient craft, the **Boyne canoe**. Darina Tully, an expert on traditional Irish boats, describes the canoe as "very, very unusual, and pure Norse in design". The canoe is a wooden, pram-type, clinker-built boat, about 5 metres long. Round hulled, with no keel and flat at the front, its shape makes it ideal for estuary work. Tully estimates that about 100 of these boats are still used on the Boyne estuary, as they have been for centuries, for drift-net salmon fishing and, in winter, for mussel collecting. The mussels are gathered following an age-old technique using a hand-rake to drag the shellfish from the estuary bed. The rake also sifts the shells so that only the larger mussels are removed.

Where Europe and North America meet

There is a seam running across Ireland from Limerick to Louth. It marks the join where, 400 million years ago, two continents collided and an ocean disappeared. There is still some debate about the precise location of the seam as much of the evidence for it has been destroyed or buried during the intervening 400 million years. Geologists have pieced together a general picture, however, from the fossil and geological records, especially around Clogher Head, which is the only place in Ireland where the seam is exposed and visible.

The half of Ireland that lies north of the seam was originally joined with Scotland, Greenland and North America in the ancient continent of **Laurentia**. The southern half of Ireland, along with Wales, England and Brittany, formed the smaller continent of **Avalonia**. Some 500 million years ago, Laurentia lay at the Equator, Avalonia lay further south, and between them was the **Iapetus Ocean**. The fossil records for the Laurentian ('American') half of Ireland and the Avalonian ('European') half are very different and distinct, reflecting the fact that they come from distant parts of the world once separated by an ocean.

Laurentian Ireland

Avalonian Ireland

The seam that joins North America and Europe runs approximately from Limerick to Louth. Clogher Head is the only place in Ireland where it can be seen in the surface rocks.

For 100 million years, the two continents inched together, squeezing the Iapetus Ocean, just as the Pacific Ocean is today squeezed between Asia and North America. Volcanoes and mountain ranges formed around the ancient ocean's rim as the two continents pressed in. The ocean grew smaller and shallower until eventually the same fossils are found in the rocks on both sides. Finally, 400 million years ago, Laurentia and Avalonia collided and the ocean vanished. The result was a massive continent (**Pangaea**) and an impact zone 60km wide stretching across Ireland and Britain. During these violent events, minerals were deposited (as at Wicklow, Mayo and Tyrone), volcanoes erupted (as at Lambay and Vinegar Hill), and rocks were crumpled – as can be seen at Clogher Head.

These crumpled, complex, beautifully folded rocks are visible at Port Oriel 400–800 metres east of the harbour. Nearby, geologists have found fossil evidence for a meeting of Europe and America: 'European' fossils at Clogher Head, but 'American' ones at Annagassan, 6km to the north. And the same picture inland: 'European' fossils at Collon quarry, but 'American' ones at Mullaghash quarry 4km to the north.

The great landmass of Pangaea lasted for 200 million years, before the restless wandering continents of North America and Europe separated again. A rift valley cracked open between them, which has since widened to form the Atlantic Ocean. The continents did not separate along the same line as before, but instead a little to the west, and so small pieces of North America were left behind, tacked on to Scotland and the northwest half of Ireland.

White River mills

This working mill and farm complex, part of which dates to 1698, has been in the same family for four generations and is now open to the public. The watermill is well preserved and much of the original mechanism survives, including the gearing and millstones, which are still used to grind grain.

The complex also includes a millpond and grain-drying kiln, as well as a stable and piggery. The mill, originally part of the local Bellew Estate, was leased in 1722 to Anthony Foster, grandfather of John 'Speaker' Foster of the Irish Parliament. ⑦

Ireland's bubble car

A most unusual vehicle was made in Dundalk in the early 1960s, in a strange convergence of road, rail and air transport. This was the Heinkel bubble car, made from the modified cockpit of Heinkel III fighter planes, and produced at the former Great Northern Railway (GNR) works. One of these cars can be seen at Louth County Museum, which spotlights the industrial heritage of Louth and of Dundalk in particular. That diverse heritage includes brewing (Dundalk once had 32 breweries, though now there are just two, Macardles and Harp); cigarette making (P J Carroll's is still there and the museum is in one of the firm's tobacco warehouses); linen processing (a major industry in the area during the late-18th and early-19th centuries); and railway engineering (the GNR works opened in 1881, making and repairing steam locomotives, and at one time employed 4,000 people before closing in the 1950s). The museum's other displays feature local geology, agriculture and transport. ⑧

The Heinkel bubble car can be seen at Louth County Museum, Dundalk.
© Louth County Museum

Kerley lines and TB

Thin lines often show up in chest X-rays taken of people who have high blood pressure or heart failure. The lines, which reveal the damage done to the patient's lung, are called Kerley lines, after the Dundalk-born doctor who first discovered them.[7] **Peter Kerley** (1900–79) was an eminent radiologist whose patients included Winston Churchill and George VI, and he was a key architect of the mass X-ray programmes used in Ireland and Britain to diagnose TB. A nephew of the medic and botanist Augustine Henry,* Kerley studied medicine at ucd before joining the British army's medical corps. In the early 1940s, he was released, however, at the Ministry of Health's request, to help develop a national screening programme for TB.

X-rays had been in use since 1896, but it took 30 years to develop the necessary techniques and equipment for mass TB screening, especially small portable machines. Shortages of photographic film and plates during World War II delayed the introduction of mass screening, yet, in 1943, the British government was able to begin a civilian programme, which Kerley helped develop. The X-rays were used to diagnose various chest infections, but especially early TB and long-term TB masquerading as bronchitis. It was a major step in the fight against tuberculosis. Not everyone was pleased though: being diagnosed with TB brought its own problems, not least for patients was having to give up work while being treated. Following the success of the British programme, Kerley advised the Irish authorities on establishing a similar scheme here. In 1947, he also worked for the International Refugee Organization, helping to control TB in the former concentration camps and refugee centres of post-war Europe.

Solving the Franklin mystery

Ireland is not famous for its tradition of polar exploration, yet a number of Irishmen were noted polar pioneers and explorers. The best celebrated today is Ernest Shackleton,* but Dundalk-born **Capt Francis Leopold McClintock** (1819–1907) won international fame when he solved the mystery of the Franklin expedition. That ill-fated expedition sailed from England in 1845 in search of a northwest passage to the Pacific around Canada's Arctic coast. The ship never returned. Among those on board was Co. Down man, Francis Crozier* who, it later emerged, played an important role in the expedition's attempts to reach safety.

In the decade after the expedition disappeared, a dozen search parties were dispatched but no trace was found. When the British government refused further help, Franklin's wife commissioned a private expedition under Capt McClintock. Born at Century House in Seatown Place, McClintock was an experienced explorer who had already taken part in some of the earlier searches. In 1859, his team reached King William Island in the Canadian Arctic and found graves, belongings and most importantly, a journal. The diary, preserved under a cairn and now at London's

Capt Francis Leopold McClintock from Dundalk, who won international fame in 1859 when his expedition to the Arctic solved the mystery of the missing Franklin expedition.

Greenwich Museum, revealed how Franklin's crew had had to abandon their ship and, led by Capt Crozier and with minimal provisions, made their way across the ice to the remote island. The last survivors died there in 1848, three years after their expedition began.

As well as solving the Franklin mystery, McClintock discovered Prince Patrick Island, drew the first geological map of the region, and named 70 features in the landscape. He later published popular accounts of his voyages. A keen naturalist, he also brought home numerous fossils and a polar bear (now in the Natural History Museum,* Dublin). Other fossils he buried on Melville Island were uncovered in 1960 and are now with the Canadian geological survey.

The screw-pile light

Dundalk Lighthouse, built in 1849, stands on nine cast-iron legs that are screwed firmly into the sandy floor of Dundalk Bay. This unusual way of securing a platform in soft terrain was invented by the blind engineer **Alexander Mitchell** (1780–1868). So successful was his technique, that Mitchell's screw piles were used around the world to build lighthouses, harbour piers and even railways. Mitchell was born in Dublin, but spent most of his life in Belfast. Despite

Still standing: Alexander Mitchell's screw-pile lights are a testament to the ingenuity of this blind Irish engineer.
© Commissioners of Irish Lights

becoming totally blind in his early twenties as a result of childhood smallpox, he was a prodigious inventor and in his first business venture, a brick-making enterprise, he introduced numerous innovations to improve the manufacturing processes.

He patented his underwater screw-pile system in 1833. Before then it was well-nigh impossible to secure a lighthouse platform in sandy areas. Mitchell used cast-iron rods some 6–12in (15–30cm) across and with a threaded end that could be firmly screwed into soft or sandy ground to provide a secure mooring. The approach was first used in 1838 at Maplin Sands lighthouse in the Thames estuary. Other screw-pile lighthouses quickly followed in Britain and Ireland, including ones in Belfast Lough, Cobh Harbour and Dundalk Bay. Mitchell had a rare failure attempting to secure a light on the Kish Bank* in Dublin Bay when his piles were prostrated by a gale. Once screw piles had been successfully used at sea, they were adapted for other structures, especially in India where they were used to anchor the Indian telegraph system and the viaduct and bridges of the Bombay–Baroda railway. Mitchell's other inventions included an improved ship's propeller. He is buried at Belfast's Clifton Street cemetery.

An epic landscape

Carlingford Lough and the Cooley peninsula have long been of strategic importance. The cargo-handling port at Greenore was also an important passenger port in the last century with its own train line. After the Anglo-Norman invasion, King John built a castle to guard Carlingford Harbour. Centuries before, the Norsemen chose the area for a settlement. But long before even the Norsemen, this was the scene of the epic *Táin Bó Cuailgne*, where warring Maeve led a cattle raid and Cúchulainn single-handedly defended Ulster against the rest of Ireland. This strategic importance is mostly due to the local terrain: the Cooley peninsula is surrounded by mountains (the Cooleys, Slieve

Gullion and the Mournes); the long inlet at Carlingford Lough facilitates boat access; and it is flanked on both sides by fertile land that sweeps down to the sea.

Though not as dramatic as Killary in Co. Mayo, Carlingford Lough is also a fjord,* scoured out by glaciers that flowed down from the surrounding mountains and cut through the soft limestone. As in all true fjords, the mouth of Carlingford Lough is barred by rock and a heaped moraine of debris that the glaciers left behind, creating a shallow stretch at the entrance that is obvious from the hills above. The same glacial moraine forms the fertile lowland at the end of the Cooley and Mourne headlands. Today, oysters are farmed on the lough's shallower shores and can be sampled, in season, at pubs in the attractive mediaeval village of Carlingford.

The Cooley Mountains behind the village are made of a dark igneous rock called gabbro. They formed 60 million years ago when powerful volcanic events shaped and reshaped the northern corner of Ireland and formed not just the Cooleys, but also the Giant's Causeway,* Lough Neagh,* the Mournes* and the unusual Ring of Gullion* in neighbouring Co. Armagh. In the Science Centre at Táin Holiday Village, visitors can study local geology and history as well as experiment with electricity and learn about planetary orbits. ⑨

Chapter 2 sources

Co. Westmeath sources

1. Johnston, Máiréad, *Ice & Cold Storage* (published by Dublin ice firm Autozero, 1988).

2. Edgeworth, K, *Jack of all Trades – The Story of My Life* (Allen Figgis, 1965).

3. Drew, David, "Karst Land Forms and Hydrology of the County Westmeath 'Lakeland' Area" *Irish Speleology* 16, 17–21, 1997.

Co. Meath sources

4. O'Keeffe Peter & Tom Simington *Irish Stone Bridges* (Irish Academic Press, 1991).

5. Newman, Conor *Tara: An Archaeological Survey* Discovery Programme Monographs 2 (Royal Irish Academy, 1997).

6. O'Keeffe, Peter *Development of Ireland's Road Network* (Institution of Engineers of Ireland, 1973).

Co. Louth sources

7. Davis, Coakley *Irish Masters of Medicine* (TownHouse & CountryHouse Ltd, 1992).

Chapter 3: Armagh • Down • Antrim

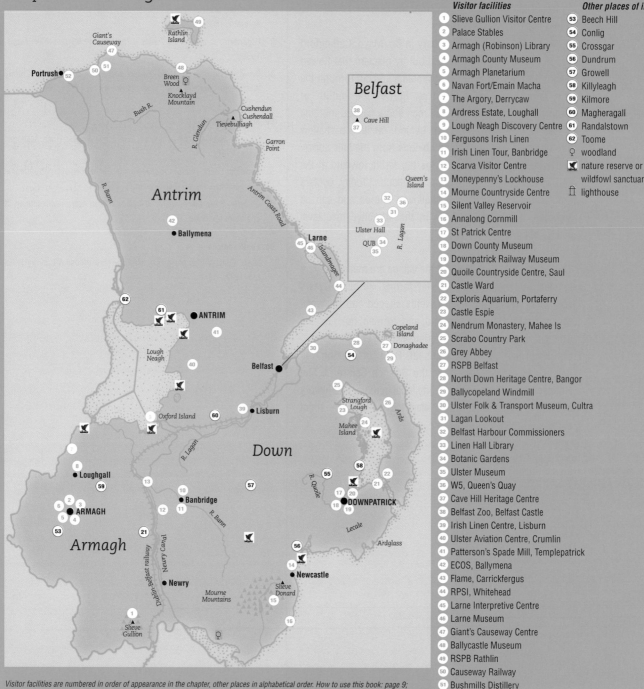

Visitor facilities

1. Slieve Gullion Visitor Centre
2. Palace Stables
3. Armagh (Robinson) Library
4. Armagh County Museum
5. Armagh Planetarium
6. Navan Fort/Emain Macha
7. The Argory, Derrycaw
8. Ardress Estate, Loughall
9. Lough Neagh Discovery Centre
10. Fergusons Irish Linen
11. Irish Linen Tour, Banbridge
12. Scarva Visitor Centre
13. Moneypenny's Lockhouse
14. Mourne Countryside Centre
15. Silent Valley Reservoir
16. Annalong Cornmill
17. St Patrick Centre
18. Down County Museum
19. Downpatrick Railway Museum
20. Quoile Countryside Centre, Saul
21. Castle Ward
22. Exploris Aquarium, Portaferry
23. Castle Espie
24. Nendrum Monastery, Mahee Is
25. Scrabo Country Park
26. Grey Abbey
27. RSPB Belfast
28. North Down Heritage Centre, Bangor
29. Ballycopeland Windmill
30. Ulster Folk & Transport Museum, Cultra
31. Lagan Lookout
32. Belfast Harbour Commissioners
33. Linen Hall Library
34. Botanic Gardens
35. Ulster Museum
36. W5, Queen's Quay
37. Cave Hill Heritage Centre
38. Belfast Zoo, Belfast Castle
39. Irish Linen Centre, Lisburn
40. Ulster Aviation Centre, Crumlin
41. Patterson's Spade Mill, Templepatrick
42. ECOS, Ballymena
43. Flame, Carrickfergus
44. RPSI, Whitehead
45. Larne Interpretive Centre
46. Larne Museum
47. Giant's Causeway Centre
48. Ballycastle Museum
49. RSPB Rathlin
50. Causeway Railway
51. Bushmills Distillery
52. Portrush Countryside Centre

Other places of interest

53. Beech Hill
54. Conlig
55. Crossgar
56. Dundrum
57. Growell
58. Killyleagh
59. Kilmore
60. Magheragall
61. Randalstown
62. Toome
♀ woodland
🦅 nature reserve or wildfowl sanctuary
🗼 lighthouse

Visitor facilities are numbered in order of appearance in the chapter, other places in alphabetical order. How to use this book: page 9; Directory: page 474.

Armagh's delights include the unusual volcanic crater that is the Ring of Gullion, the old orchard country around Loughgall, and Lough Neagh, a lake that formed where the Earth collapsed. At Armagh city, astronomer Thomas Robinson invented the first device to measure wind speed and Archbishop James Ussher famously calculated that the world was created in 4004 BC. Armagh (*Ard Mhacha*, 'Macha's height'), is the hill where St Patrick subsequently founded a cathedral town; nearby is prehistoric Queen Macha's settlement of *Emain Macha* (Navan Fort)

SLIEVE GULLION

A geological gem

Slieve Gullion sits at the centre of a large crater of volcanic origin, a unique landscape within an unusual ring of hills. This is the legendary **Ring of Gullion** – the warrior Cúchulainn reputedly strode these hills and the small lake on top of Slieve Gullion is said to be bottomless. The ring of hills, which measures 12km across and resembles an ancient crater, is technically a ring dyke. It formed when a circular fracture developed in the Earth's surface, enabling molten magma to rise, during the volcanic era some 60 million years ago. Other volcanic events around the same time – all triggered when Greenland and North America broke finally from Europe – gave rise to the Giant's Causeway,* Lough Neagh* and the Mountains of Mourne.*

The Ring of Gullion was built up over time and after several events: first, a circular fracture developed in the Earth's surface around the remains of a volcano that is now Slieve

Gullion; next, the circular crater, or caldera, collapsed and magma rose into the ring (basalt also seeped out through fissures in the Earth's surface); then small volcanoes around the ring erupted violently, spewing out lava, as at Forkhill; after the volcanic material was extruded, the centre of the ring subsided forming a large crater, this later filled with more lava, before subsiding again. In all, the crater collapsed and filled several times so that the rocks here are a complex mix of basalt, lava, gabbro and igneous conglomerate. Millions of years later, erosion has removed the softer, surrounding rock leaving the hard ring-dyke hills and central Slieve Gullion.

Some of the complex rock structure can be seen in a disused quarry by **Cam Lough**. The lough sits in a long, deep trench that glaciers cut in the ring during the last Ice Age. The glaciers breached the ring at various points, notably the **Gap of Moiry** (or Moyry), also known as the Gap of the North. For centuries, this was an important defence point and route into Ulster: the ancient highway of

Slíghe Mídluachra, from Tara* in Co. Meath to Dunseverick in Co. Antrim, ran through the gap, and the route is still used by the Dublin–Belfast train. Hugh O'Neill held the gap in the 1590s until Elizabeth I's forces captured it and built a castle to guard the pass.

The Ring of Gullion's sheltered floor has a fertile soil and is well farmed, contrasting starkly with the poor soil and bogland of the encircling hills. Good views of the ring and surrounding landscape – Ireland's central limestone plain to the south and drumlin swarms to the north – can be had from Moiry Castle and Slieve Gullion's summit. This varied landscape can also be seen from the train. Travelling north from Dundalk, you leave the central plain and enter the Ring of Gullion through a cutting in the igneous rock; the train crosses the ring in under five minutes, then exits through another rocky cutting to emerge among the drumlins surrounding Newry. (Just after Newry station, the train crosses Bessbrook river on the impressive 18-arch Craigmore Viaduct.) There are cairns on Slieve Gullion, early Christian churches, wild goats, a forest park and a visitor centre. ①

1838 – a longitude experiment

In May 1838, the British army sent up fireworks each night from the summit of Slieve Gullion as part of an unusual scientific experiment. Thomas Romney Robinson* at Armagh Observatory, and Sir William Rowan Hamilton* at Dunsink Observatory in Dublin, were calculating the longitude for the two observatories. (Longitude essentially gives the relative position on the globe of various places.) Earlier that year, in a more conventional longitude calculation, the pair had used 15 different chronometers (accurate timepieces) to time astronomical events, such as the passing of a particular star or planet, at Greenwich, Dunsink and Armagh, ferrying the instruments from one observatory to the next by boat, train and stagecoach.

To verify their results and to check their chronometers, the astronomers agreed to simultaneously time a series of fireworks set off on Slieve Gullion, the highest point between Armagh and Dunsink. Explosives (and guards) were provided by the army, which was using similar flares to triangulate mountain tops during its ordnance survey* of Ireland. From the timing of the fireworks, as noted at Armagh and Dunsink, Robinson calculated that the longitude for Armagh Observatory was 0hrs 26min 35.2sec west of Greenwich, whereas Dunsink was 0hrs 25mins 21.1sec west, a difference of 1min 14.1sec. The Slieve Gullion measurement was more accurate than the previous chronometer ones, which had given a difference of 1min 14.2sec between the two observatories. So the position of Armagh Observatory was 'moved' 50cm west, relative to Dunsink and Greenwich. The longitude values Robinson calculated are still the official ones quoted for Armagh and Dunsink (though they are now quoted in degrees of longitude).

The chemistry of life

In the second half of the 19th century, scientists made great strides in analysing and understanding the chemicals found in living things. This organic chemistry was concerned, in particular, with compounds containing carbon (C), but also nitrogen (N), hydrogen (H) and oxygen (O), and one of its founding fathers was **Maxwell Simpson** (1815–1902), who was born at Beech Hill House. Simpson, who worked for many years with some of the great European chemists, devised a powerful analytical technique for determining the amount of nitrogen in a compound; he was also the first to synthesise certain organic compounds, notably succinic acid, and thus create some of the chemicals of life; and it was he who realised that all organic acids contain a particular arrangement of atoms, -COOH, now called the carboxyl group.

At school in Newry, Simpson met John Martin and John Mitchell, later famous as Young Irelanders.* The three remained friends and, in 1845, Simpson married John Martin's sister, Mary. After studying medicine and chemistry at TCD, Simpson taught at various Dublin medical schools. Then, in 1851, he moved to the renowned Marburg institute in Germany, where he devised his technique for determining the nitrogen content of organic compounds, a method that was still widely used a century later. Simpson's international reputation grew and, in 1856, he moved to Paris.

By 1860, he was back in Dublin where he held various positions. It was there, over the next 10 years and in a small laboratory at his home, that Maxwell Simpson investigated organic acids and realised that their acidity was due to the presence of the -COOH group. In 1872, he was made professor of chemistry at Queen's [University] College Cork* and stopped research to concentrate on teaching and administration. By then, however, his reputation was assured and he had been awarded numerous international honours. He retired to London in 1892 where he died 10 years later. A commemorative plaque has been erected at Simpson's Cork home at 11 Dyke Parade.

ARMAGH

4004 BC – the date of the Creation

In 1650, the Archbishop of Armagh **James Ussher** (1581–1656) carefully counted the 'begats' in the Old Testament, studied numerous ancient Egyptian and Hebrew texts, analysed the various ways in which ancient calendars were calculated, and concluded that the world began the evening before October 23rd, 4004 BC. He chose October 23rd as this would have been the date, under the old calendar, of the autumn equinox, and a traditional start to the year; he specified the previous evening, as traditionally this was when each day began; and he believed the 23rd would have been a Sunday, since time would surely have

begun on the first day of the week. Ussher also calculated that the Flood occurred in 2348 BC and King David's accession in 1056 BC.

James Ussher: Archbishop of Armagh and classical scholar. Armagh Public Library

We may laugh now at these calculations of the **age of the Earth**, but the technique Ussher used was the only one possible at the time. Other scholars also used the scriptures as an accurate historical record and computed similar sums – the Venerable Bede, for instance, put the year of Creation at 3952 BC, and Isaac Newton estimated 3998 BC. Most agreed the world was about 4,000 years old, but the dates were hotly disputed. John Lightfoot, for instance, an eminent Hebrew scholar at Cambridge, believed the Creation took place at 9 o'clock on the morning of the autumn equinox and not, as Ussher suggested, the previous evening. Ussher's chronology was the one chosen for use in printed English bibles, however, and accorded the same respect as the scriptures; it was widely accepted until the late-19th century.

By then, scientists were experimenting with other ways of calculating the Earth's age – based on the amount of salt that had accumulated in the oceans, for instance, or the time it had taken the Earth to cool from a molten ball to a solid planet. Some techniques provided useful estimates, while others were seriously flawed. In the early-20th century, and thanks to the discovery of radioactivity, scientists found new ways of dating rocks accurately and we now know the Earth is 4.6 billion years old. Other Irish people who made important contributions to calculating the age of the Earth were Samuel Haughton,* William Thomson (Lord Kelvin)* and John Joly.*

James Ussher was born in Dublin and, in the 1590s, became one of the first students to attend Trinity College

Dublin* (his extensive library later formed the nucleus of the college's collection). He was a renowned scholar, a professor of theology at TCD, and a keen astronomer who used a telescope to verify for himself the theories of Galileo, Kepler and Copernicus. He was in England in 1642 when the Civil War broke out and remained there, managing to be both a Royalist and a friend to Oliver Cromwell. He attended Charles I at the scaffold, and when he himself died he was buried, at Cromwell's request, in Westminster Cathedral.

The Heavens declare the Glory of God

In 1765, Richard Robinson, Archbishop of Armagh, set about transforming his small ecclesiastical seat into a centre of learning. He planned to establish a university and bring the town to international prominence, as in the early days of Christianity when a noted college flourished there around St Patrick's first church. Robinson built an elegant Georgian town and several notable buildings: a Bishop's Palace (1770), now home to Armagh council; a library (1771), which today houses Robinson's extensive book collection; a Royal School (1773); and, in 1791, Armagh Observatory, the first building in his planned college. Robinson died in 1794 and his university never happened, though 200 years later Queen's University Belfast* established an outreach campus in the town.

Armagh Observatory, which opened five years after Dublin's Dunsink Observatory,* is the oldest scientific institution in Northern Ireland, and its board is still chaired by Richard Robinson's successor, the Church of Ireland primate. The purpose-built observatory was set on a drumlin on high ground at the edge of town. A large pillar in the centre of the building supports both the stairs and the telescope above. The telescope, installed in 1795, is still there, the oldest telescope still in its original dome in the world. It was made in London and its brass surface is badly pitted, caused by London's sulphurous air pollution at the time.

Armagh's international reputation began with Thomas Romney Robinson (no relation of the founder), who headed the observatory from 1823–82. Robinson compiled the weighty *Places of 5,345 Stars Observed from 1828–1854 at Armagh Observatory* – a mammoth undertaking – and invented a device to measure wind speed (see: **Catching the wind**). He was succeeded by a Danish astronomer who had worked at Birr,* **Emil Dreyer**, and who compiled another classic reference work, *The New General Catalogue of Nebulae*. In the early part of the 20th century, the observatory, along with most Irish scientific institutions, went into decline, until the arrival in the 1930s of an enthusiastic director **Eric Lindsay**. Lindsay brought over Estonian astronomer Ernst Opik (noted for his pioneering work on comets, asteroids and the origins of the solar system), helped set up the first major joint international observatory, based at Bloemfontein in South Africa (a new collaborative approach, replacing smaller, national observatories) and developed Armagh planetarium.* Nearly 30 astronomers now work at Armagh, researching areas such as the solar system, interplanetary dust, sunspots and climate change.

Weather has always been important to astronomers – clouds can obscure the skies, and air temperature and pressure can affect observations – so, in 1795, the Armagh astronomers began taking weather measurements. Their records, which continue today, are the longest continuous set of weather records in Ireland and they are now being used to study **climate change**. Dr John Butler of Armagh Observatory has found evidence in the records of a relationship between sunspot cycles and weather: the average sunspot cycle lasts 11 years; but there are fast cycles, of about nine years, when the Sun may be brighter

Armagh's two cathedrals sit atop neighbouring drumlins. The Palace Stables Heritage Centre has an ice house* and a sensory garden. ② Armagh's Robinson Library, the first Irish public library outside of Dublin, has Archbishop Robinson's book collection and mediaeval manuscripts, etc. ③

and hotter, and these are associated with warmer weather; while slow cycles of 13 years are associated with colder weather.[1] Butler found a similar pattern in tree rings:* wide rings, indicating good growing years, are seen in the warmer decades associated with fast sunspot cycles; while narrow rings are seen during the colder years associated with slow sunspot cycles.

ARMAGH

Catching the wind

Sunday night, January 6th, 1839, was the **Night of the Big Wind**. The worst storm to hit Ireland in recent history, it destroyed thousands of houses – at least 5,000 in Dublin alone – and uprooted thousands of trees. Thirty-seven people were lost at sea and as many as 300 were killed by falling masonry and trees, or swept away in the floods. Afterwards, cattle starved because their fodder had

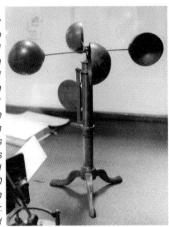

Thomas Romney Robinson's 4-cup anemometer. One still spins on Armagh Observatory's roof, and another at Dún Laoghaire's east pier.* Anemometers are generally placed in open sites, and 10 metres above ground level, as friction from the ground slows wind speed by 10 per cent, compared with its speed over water.
© Enterprise Ireland

been destroyed and the timber market collapsed as people exploited the ready supply of fallen trees. Based on eyewitness descriptions of the 1839 Big Wind, it is thought the wind speeds reached at least Hurricane Force, a frightening 193 kmph (120 mph). In 1909, when the old-age pension was introduced, one technique reputedly used to identify people over 70 was to check their memory of that storm. The storm may have prompted the **Rev Thomas Romney Robinson** (1792–1882) director of Armagh Observatory, to devise an instrument to measure wind speed. This would put numbers on the Beaufort Wind Scale, a descriptive scale devised in 1806 by Francis Beaufort.*

Robinson, whose father was a successful portrait painter, was born in Dublin. A child prodigy, he entered TCD at 12 and published a book of poetry at 13. He was appointed director of Armagh Observatory in 1823 (see: **The Heavens declare the Glory of God**), a position he held for 60 years. His second wife, whom he married in 1843, was Lucy Edgeworth, daughter of engineer Richard Lovell Edgeworth* and sister of novelist Maria. It was R L Edgeworth who, as early as the 1780s, thought of making an instrument to measure wind speed. Edgeworth even made a rudimentary anemometer which was tested at Dublin in 1786 during a balloon ascent.* But 60 years passed before Robinson found a way of making an accurate and effective instrument. His anemometer, which he demonstrated to the Royal Irish Academy* in 1850, had four cups which could turn freely in the wind around a central spindle. The faster the wind blew, the faster the cups spun. Crucially, Robinson devised a mechanism to count the number of times the cups spun each minute, and from this he could calculate the wind speed. In 1852, Robinson began recording wind speed and direction at Armagh every hour, and this monitoring continues at the observatory, which is an international reference site for the World Meteorological Organisation. Armagh County Museum has a portrait of Robinson painted by his father; the museum was begun in 1850 by the (still extant) Armagh Natural History & Philosophical Society. ④

ARMAGH

An astronomical family

In the 19th century, four members of an Armagh family worked at the Royal Greenwich Observatory, three of them at the same time, in what is surely a record. They were **Hugh Breen** (1791–1848) and his sons **Hugh Jr** (*b*.1824), **James** (1826–66) and **John William** (1832–71). They rose to positions of some responsibility at Greenwich but, being Catholic, were barred from more senior appointments, which would have required them to subscribe to the Anglican Church. Their story has now been pieced together by astronomers Máire Brück and Allan Chapman, and Armagh historian Sheelagh Green.[2]

Hugh Breen senior, a talented mathematician and teacher, was head of science at the Mechanics Institute, which opened in Armagh in 1826. The institute was never successful and, after it closed in 1831, Breen moved his family to London and found work as a computer at the Royal Greenwich Observatory. (Computers did the routine calculations that were essential to 19th-century astronomy, though many of them also undertook observations.) Hugh Breen was responsible for overseeing a major analysis of measurements of the Moon's orbit made at Greenwich over the previous century. In 1848, the year he died, he wrote an official report for the project and, with two of his sons, also published a paper on the mathematics involved in computing the orbit of Venus.

Hugh Jr worked for 20 years as a computer at Greenwich, retiring in 1858 possibly because of mental illness. While at the observatory, he wrote a substantial chapter on astronomy for a popular encyclopaedia, *The Natural History of Inanimate Creation* (1856). Later he tried to open a school "for Catholic gentlemen" in Ireland and, when that failed, returned to work occasionally at Greenwich. James Breen started work at Greenwich in 1840, moving to Cambridge Observatory six years later on the recommendation of the astronomer royal, Sir George Airy, who described him as "a rough genius". James spent 12 years at Cambridge, where he rose to the rank of senior observer. He published independent observations of double stars and comets, but resigned unexpectedly in 1858, the same year as Hugh retired from Greenwich. James also took to writing and while at Cambridge wrote a popular science book, *The Planetary Worlds* (1854). There was heated debate at the time about the possibility of life on other planets and Breen's book brought together all the information then known about the planets. He studied in Europe for a while, before returning to London to prepare a book on nebulae, but died of tuberculosis before finishing

it. The fourth member of the family to work at Greenwich was John William, who joined the staff as a computer at 14 but left at 28 to work in business. Hugh Breen and his sons are buried at Nunhead Cemetery in southeast London.

1969 – meteorite hits Antrim and Derry

On the evening of April 25th, 1969, a meteorite* fell to Earth in Northern Ireland. It broke up just before impact into at least two fragments, the first of which hit the roof of a Royal Ulster Constabulary (RUC) building at Sprucefield, Co. Antrim, and the second fell in a field at Bovedy, Co. Derry. The Sprucefield fragment of what is now called the Bovedy meteorite, can be seen at Armagh Planetarium. The planetarium also has a piece of the planet Mars: part of a meteorite that landed in Nigeria in 1962, and which has since been identified, thanks to its distinctive chemical composition, as coming from Mars. Armagh Planetarium opened in 1967, with well-known television astronomer Patrick Moore as its first director, and its collection also includes dinosaur fossils and a flag that flew on the space shuttle in 1993. The wide-ranging exhibition covers both astronomy and earth science – the Sun, the Moon and stars, as well as planet Earth. You can explore the Universe or learn about volcanoes and the local weather. The grounds surrounding the planetarium and the neighbouring observatory have a stone calendar, a sundial,* a large-scale model of the solar system and a hazel coppice.* ④

A prehistoric mystery

In the year 94 BC, or shortly after, the people of *Emain Macha* built a massive timber temple. Thirty metres in diameter, it is the largest prehistoric ritual building known in Ireland or Britain. Bizarrely, once the structure was finished, it was filled with limestone boulders and then, for reasons we can only guess at, it was burned down and the

Infrared photograph of Navan Fort's royal mound. 'Navan' is an anglicisation of Emain Macha (the twins of Macha). Navan Fort/© Chris Hill/Jill Jennings

remains covered with soil. From the remains of the mysterious structure, and using sophisticated scientific techniques, archaeologists have pieced together a more detailed picture. Tree-ring dating,* for instance, revealed that an oak tree used in the temple was felled in the year 94 BC; soil resistivity measurements enabled archaeologists to probe the site without actually turning a sod; and plant remains have shed light on the climate and vegetation of the time.

For 700 years, *Emain Macha* was the seat of power in Ulster, home to the Knights of the Red Branch and Cúchulainn and a rival to Tara* in Co. Meath. The main site is a large mound set among drumlins* near Armagh town. Other significant monuments include a rare sacrificial pool thought to date from 1000 BC, one of only a few such known in these islands. A limestone quarry at the edge of Navan Fort provided stone for many of Armagh's fine buildings for over 150 years; it was closed in 1986 to protect the prehistoric site. The site is an open access; the visitor centre at Navan Fort has an exhibition that describes scientific techniques used to explore the site, and finds include the skull of a Barbary ape, presumably an ancient royal pet. ⑥

Preston's puzzling phenomenon

It is tempting to speculate that, if **Thomas Preston** (1860–1900) had lived beyond 40, he might have won a Nobel prize for physics. For, in 1897, Preston discovered a puzzling phenomenon, known as the 'anomalous Zeeman effect'. It was later important in the development of quantum mechanics, a revolutionary new theory about atomic structure that was first proposed in the 1920s. Nobel prizes were awarded to the physicists who discovered and explained the normal Zeeman effect, and to those who developed the theory of quantum mechanics but, by then, Preston was dead, and there are no posthumous Nobels. Thomas Preston was born near Kilmore and studied physics at TCD.* In 1891, he was appointed professor of natural philosophy [science] at the newly established University College Dublin.* The college was then based at Earlsfort Terrace, in what is now the National Concert Hall, and it was there Preston did his research.

Just as sunlight can be split by a prism into a rainbow, or continuous spectrum of colours of differing wavelength, so the light emitted when chemical elements are heated can be split into a spectrum, but a discontinuous one, with sharply defined spectral lines corresponding to specific wavelengths. In 1896, Dutch physicist Pieter Zeeman discovered that the spectral line for sodium will, in a magnetic field, split into two lines of light with slightly different wavelengths, a phenomenon known as the Zeeman effect. Then, using a new spectroscope and a powerful magnet (40,000 Gauss), Thomas Preston discovered that when the magnetic field is very strong, the spectral lines for nearly every element will separate into many more lines. He published his findings in the *Proceedings of the Royal Dublin Society* in December 1897. Classical physics could explain the normal

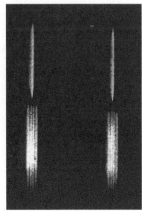

Part of the spectra Preston observed for cadmium (left) and zinc (right). The top row was taken with no magnetic field; the lower line, taken in a strong magnetic field, shows the anomalous splitting into four lines.

Zeeman effect, and Preston's 'anomalous effect' was initially seen as a curiosity. Its significance was only realised later when it became clear that existing theories about the atom could not explain the multiple spectral lines. It was 25 years before scientists explained Preston's discovery, when, in the 1920s, they developed new theories about electron spin and quantum mechanics.

Preston made other contributions to spectroscopy, and wrote two textbooks – *The Theory of Light* (1890) and *The Theory of Heat* (1894) – that became international classics and were used until the mid-20th century. In 1900, at the height of his career, he died of a perforated ulcer. His work was subsequently overlooked, however, and the anomalous Zeeman effect is usually credited to the French scientist Cornu. There are plans for a commemorative plaque at Earlsfort Terrace, where Preston worked.

Lit by acetylene gas

There is no electric lighting at the Argory. Instead, this elegant estate house is lit by acetylene gas. In the days before cheap and reliable electric light bulbs, acetylene lighting was a popular alternative to coal-gas and oil. The gas plants were reliable and simple to run, the gas was easy

The Argory's acetylene plant, installed in 1906 and still used to make acetylene. The estate is run by the National Trust; there is also a sundial in the garden.
© National Trust

to generate and the light could be controlled. Acetylene was discovered in Dublin in 1836 by Edmund Davy,* when he accidentally dissolved calcium carbide in water; the reaction gave off a gas that burned with a bright flame. Early bicycle lights were carbide lamps; and potholers still use them because the system is simple and reliable. The Argory uses the original plant installed in the stable yard in 1906 by the Sunbeam Company (the gas is pumped from there around the house). The system is one of only a few surviving in these islands. ⑦

Apple blossom time

This is the heart of Ireland's orchard country. In Loughgall even the lampposts have an apple motif, and each May–June, the town hosts an apple-blossom festival. The surrounding drumlin hills are planted with neat orchards and the area has a long tradition of apple growing and cider making – St Patrick is said to have planted an apple tree near Armagh town; Macan [McCann], a local chieftain who died in 1155, was noted for his strong cider and, in 1690, King William used Armagh cider to fortify his troops. Today, the **Orchard Trust** is working to save old Irish apple varieties and orchards; it has a temporary nursery at the 17th-century Ardress Estate, near Loughgall, ⑧ and hopes to establish a permanent home at Drumilly Bawn, in the old walled-garden at Loughgall demesne, where the Department of Agriculture has a horticultural and apple research unit. It would be a fitting home for, in the 1540s, Art Óg O'Neill (grandfather of Owen Roe O'Neill) kept a bountiful orchard there.

All domestic apples are cultivated varieties (or 'cultivars') of the wild crab apple (*Malus*). The crab is native to Ireland and remains found at Mount Sandel in Co. Derry reveal that hunter–gatherers were eating apples there 9,000 years ago. Apples feature in the great Ulster legends, and the ancient Irish Brehon Laws classed the tree as a "noble of the wood", along with oak and ash. Apples were also recommended to keep the doctor at bay: in 1735 the Co. Roscommon herbalist Rev John Keogh* prescribed them to "comfort and cool the heat of the stomach".

Place names containing the elements *úll* ('apple') or *úllghort* ('apple field') – such as Oulart* in Co. Wexford – are a reminder that orchards were once common in the Irish landscape. Each area had its own local variety and, at one time, over 1,000 varieties were grown in these islands, differing in characteristics such as colour, taste, flowering season and keeping quality. Irish cultivars included Blood Butcher from Co. Meath, The Smeller from Co. Tyrone, Co. Sligo's Irish Pitcher, and Co. Armagh's Dockney. Most have disappeared over the past 150 years, replaced by modern commercial varieties, such as Bramley, a cooking apple now widely grown in Armagh for the food and drink industry. (Dessert apples need a warmer climate and are generally found in more southerly orchards, as around counties Dublin, Kilkenny, Wexford and Tipperary.)

In the 1940s, a Dublin horticulturalist, Dr J G D Lamb, collected as many as possible of Ireland's remaining varieties, and planted cuttings at the Albert Agricultural College on Dublin's northside.[3] This important heritage collection was bulldozed in the 1970s to make way for Dublin City University but, fortunately, 40 of the varieties had also been planted in Kent. These backups were rescued and now form the nucleus of a National Historic Apple Collection, maintained by UCD.*

Collecting continues – old hedgerows often shelter forgotten apple trees – and UCD, the Seed Savers' Association* (Co. Clare) and the Orchard Trust have together rescued 120 Irish varieties, including some mediaeval ones and others previously presumed lost. Peadar MacNeice, an Armagh pommologist and prime mover in the Orchard Trust, says the three groups exchange cuttings to ensure there are copies of all conserved varieties. Mediaeval rescues include Cockagee, a renowned Irish cider variety that was grown in the 1400s, and

Caledon, an Armagh cider apple developed by the monks at Tynan Abbey near Killylea. These could one day be a source of new cultivars, since they are suited to Irish conditions and, because they developed in the days before pesticides, are probably more disease-resistant than modern varieties. The Seed Savers' Association sells apple trees, including 38 old Irish varieties.

The lake where the Earth collapsed

Lough Neagh is the biggest and oldest lake in these islands. It formed, so the legend goes, after Fionn McCumhaill picked up a lump of earth from the north of Ireland and threw it into the sea. Depending on where you come from, the lump formed either the Isle of Man or Ailsa Craig,* while the hole that was left in the ground became Lough Neagh. If anything, the truth is more intriguing: scientists believe this great lake formed when the Earth's surface subsided there 40 million years ago, partly under the weight of basalt rock above, and partly because the ground below was undermined when vast quantities of magma were removed during the volcanic era 20–25 million years before. The result was an enormous lake basin, about twice the size of modern Lough Neagh.

Those were warmer times and lush tropical swamps with redwoods and palm trees grew by the prehistoric lakeshore. Their fossil remains are preserved today as thick beds of **lignite**,* especially on the eastern side of the modern lake where some of the lignite has been mined. Over time, the lake filled with sediment, becoming both shallower and smaller. Old shorelines or raised beaches were left high and dry, as at Randalstown Forest on the northern side. Surrounding land that was formerly part of the lake is covered with a heavy, waterlogged clay-rock and, unless the land is drained, it makes for poor farming. (This Lough Neagh clay, less than 40 million years old, is the youngest rock found in Ireland.)

Modern Lough Neagh is more or less rectangular, covers 400sq km, has 125km of shoreline and is the primary water source for one-third of Northern Ireland's population. Because it formed in a large depression, many rivers flow into it, but only one flows out – the Bann, to the north. During the last Ice Age, however, Scottish ice sheets blocked that exit and Lough Neagh drained south through Newry into Carlingford Lough. The eight major rivers flowing into Lough Neagh today bring in vast quantities of sand and silt, and in places the sand deposits have been thick enough to quarry. The sediment, coupled with the lake's shallow nature and low-lying surroundings, also made Lough Neagh prone to serious **flooding**. During one wet winter in the 1870s, the water level rose 4 metres at Toome,* Co. Antrim. Flood relief and drainage schemes in the 1840s, 1930s and again in the 1950s, reduced the lake level by 3 metres and helped control flooding. The former shoreline left dry by these measures can still be seen. The lake's deepest point now is a 30-metre deep trench in the northwest corner; otherwise, Lough Neagh is on average just 9 metres deep. As a result, sunlight penetrates to the bottom and algal growth is vigorous.

Pieces of fossilised or **petrified wood** occasionally wash up on the shore. These gave rise to the belief that the lake water could petrify things – as one local rhyme had it: "Lough Neagh hones, Lough Neagh hones, you put in sticks and you take out stones." In fact, the trees were petrified 30 million years ago when the lake water was particularly rich in silica. The fossils preserve the fine detail of the original trees, most of them conifers that were growing in the swamps around the prehistoric lake. Some trees are only partly petrified and resemble lignite more than stone. The rock-hard "Lough Neagh hones" were traditionally used for sharpening knives and blades.

Lough Neagh is an important **bird sanctuary**, thanks to its size, rich waters and the varied habitats around its shores, such as fen, bog, reed bed, meadow and scrub. In winter, the lake teems with enormous flocks of birds, including the largest concentration in these islands of diving duck; it is also noted for swans, herons and great crested grebe. The lough's islands are important nesting sites, and are protected as a nature reserve. In summer, the air above the water is thick with half a dozen species of midge, collectively known as the 'Lough Neagh fly'. Lough Neagh is home to two **unique fish** species: the dollaghan, a type of trout unique to the lake; and the pollan, related to herring, which gave up the sea during the last Ice Age and is now confined to Lough Neagh and its rivers. Lough Neagh is also the only site in Ireland for some unusual aquatic plants, among them river water crowfoot and holy grass.

There are several **nature reserves** around the lough shore: Randalstown, Farr's Bay and Rea's Wood are wetland woodland and species-rich fen sites at the northeastern corner of the lough in Co. Antrim; there are important bog and woodland sites at Mullenakill and Annagarriff in the southwest corner. The wet meadows and scrub around Portmore Lough and Lough Beg are managed as a reserve by the RSPB. The Lough Neagh discovery centre at Oxford Island has a natural history exhibition, bird hides, ecolabs for school groups, and information about the various nature reserves (access to some is restricted). ⑨ Oxford Island became a peninsula when the lake level dropped.

In the 19th century, the lake was an important part of Ireland's **canal and navigation** networks. Barges and boats ferried coal and flax across the lake to Ulster's linen and mill towns, and brought the finished fabrics to market. A number of small industries also developed around the lake: turf cutting, reeds for thatching, willows for basketry, eel fishing, and cutting and quarrying the fossil or diatomite Bann clay.* Today, Lough Neagh supports a leisure/tourism industry, and the largest eel fishery* in Europe.

Down's delights include the lovely landscapes of Strangford Lough and the Lecale Peninsula. The county produced several famous inventors, among them: William Clanny, who designed the first miner's safety lamp; James Drumm, who pioneered battery-powered transport; James Martin, who invented the ejector seat; and Harry 'mad mechanic' Ferguson, whose lightweight tractor revolutionised farming. Down (*an Dún*, 'the fort') is named after a prehistoric fort at Downpatrick.

SCARVA

1741 – barging uphill

It is often said that the canal era in these islands began in 1761, the year the Duke of Bridgewater built a canal and locks, linking his Worsley coal mine with Manchester port. In fact, Ireland had a modern canal 20 years before that. The **Newry Canal**, which runs from Lough Neagh* to Newry port, opened in 1741 and linked the newly discovered Tyrone coalfields* with the Dublin market. It was the first summit canal in Ireland or Britain: thanks to nine locks, barges could climb from Newry to the summit at Poyntzpass, 25 metres above sea level; six locks on the other side stepped down to the River Bann at Portadown and thence to Lough Neagh. Lough Shark at the summit provided the water supply.

Simple navigational waterways had been used for centuries – the earliest Irish one was probably the channel known as Friar's Cut, made in 1178, that enabled boats to pass from Lough Corrib* to the sea at Galway city. It was not until the mid-1500s, and the invention of the mitre lock-gate, that canals could be made to climb hills. (Mitre gates, reputedly invented by Leonardo da Vinci, close into a 'V' that points upstream when shut; water pressing against the gate keeps it closed.) The Newry Canal was first proposed in 1703, however, work did not begin until 1731 and the canal opened 10 years later. A ship canal built in 1769 between Newry and Carlingford Lough,* further improved access, and Newry became for a time the largest port in the north of Ireland. New villages sprang up along the Newry canal, notably at Scarva, Poyntzpass and Jerrettspass. A coal dock was built at Scarva, which became a coal-supply centre for the surrounding area. The improved coal supply led in turn to the development of several linen* mills – one survivor is Ferguson's of Banbridge. ⑩ ⑪

In 1813, a canal passenger service began. An alternative to the stagecoach, it covered the 24km between Newry and Portadown in four hours. Traffic on the canal peaked in 1840, by which time there were five more canals in the region, though their popularity rapidly declined with the arrival of the railways. Newry Canal was last used in 1937, but the channel survives and there are plans to reopen it. The superb engineering works also survive – locks, lock-gates, lock-keeper's cottages, quays, docks and warehouses... some now serve as canal interpretive centres. ⑫ ⑬ The old towpath is part of the Ulster Way. As the canal is connected to Lough Neagh and Carlingford Lough, it is relatively rich in wildlife: fish and aquatic plants thrive in the slow waters, birds and insects frequent the banks, and there are bird observation points at Lough Shark and at Victoria Sea Lock on the Newry ship canal.

BANBRIDGE

1848 – "A last bold struggle for life"

Four stone polar bears guard a statue in Banbridge town square. It commemorates a 19th-century Irish polar explorer, **Capt Francis Crozier** (1796–1848), who was born in the town. Crozier joined the Royal Navy at 14, took part in several Arctic voyages and, in 1839, was second-in-command on an expedition to Antarctica in search of the magnetic south pole.* His final voyage, in which he played an important role, was the ill-fated Franklin expedition that was searching for a Northwest Passage.*

For centuries, sailors had sought a route to the Pacific Ocean through the tangled Arctic maze of Canada's Elizabeth Islands. In 1845, the Royal Navy launched its most ambitious expedition, with 129 men led by Capt Thomas Franklin, and with Crozier as second-in-command. No expense was spared: the two expedition ships, *Erebus* and *Terror*, had hot-water central-heating systems; they were sheathed in iron to protect them from the ice; and each had a powerful steam engine that could be used to propel it through the ice (although, coal supplies were limited). The expedition also carried a library of 3,000 books, and three years' provisions, including 8,000 tins of canned food.

The ships left England in May 1845 but, by 1847, had still not appeared in the Pacific. The loss was a major blow to the navy, which sent 39 search-and-rescue expeditions, and offered a £10,000 reward. Initially, there were hopes of finding the men alive but, in the early 1850s, contact was made with local Inuit people who reported seeing the expedition's ships crushed in the ice, and survivors who staggered and died as they walked. The mystery was finally resolved by Capt McClintock* from Co. Louth, who found the only written record from the Franklin expedition, a message scribbled on a scrap of paper and concealed in a cairn on one of the islands. Dated April 1848, it revealed that Franklin and 21 others had died in 1847; that Capt Crozier had taken command, and that the remaining 100 men intended to walk to the Canadian mainland on "a last bold struggle for life". More clues were found on other islands, including graves, a skeleton dressed in a steward's uniform, a lifeboat intriguingly stuffed with luxury goods but little food, and finally, several mutilated corpses that bore the marks of cannibalism. Ironically, some of Franklin's crew may unwittingly have discovered the Northwest Passage, as their bodies were found at Simpson's Strait, the final leg of the passage. (The route was officially discovered in 1850–53 by Wexford man Robert McClure.*)

Archaeological studies in the 1980s and 1990s shed further light on the expedition's fate. Autopsies conducted on corpses, preserved in the permafrost, revealed severe lead

Capt McClintock's expedition opens the cairn at Point Victory, which contains the record of captains Crozier and FitzJames, ending the mystery of the Franklin expedition. Illustrated London News, 1859

poisoning, probably from the lead solder used to seal their tinned food. Lead poisoning hampers decision-making and this may explain why exhausted and hungry men hauled a lifeboat laden with useless luxury goods; and why Capt Crozier led his men towards the mainland, rather than Lancaster Sound where they were more likely to be rescued. Traces of the expedition are still occasionally found, and the search continues to locate the wrecks of Franklin's two ships.

Of the heart and haemoglobin

Our understanding of blood and the circulation system was greatly advanced by two world-famous physiologists from Newry: **Henry Newell Martin** (1848–96) was the first to reveal the workings of a live mammalian heart and **Sir Joseph Barcroft** (1872–1947) studied the red-blood pigment, haemoglobin. Henry Martin, son of a clergyman, won a scholarship to the University of London when he was 15, and seven years later won a scholarship to Cambridge University. There he worked with Thomas Huxley, famous scientist and champion of Charles Darwin's theory of evolution. In the 1870s, the president of the newly established Johns Hopkins University in Baltimore, Maryland, came to Europe in search of talented academics for his new college. Huxley recommended his brilliant young assistant for the professorship of biology and head of the medical school, and Martin moved to Baltimore. Not much was known then about how the mammalian heart worked, as it was difficult to study a living, beating heart. But Henry Martin devised a way of isolating a dog's heart so that it could be studied while still beating. It was an important breakthrough in the emerging field of cardiology, and Martin soon became the country's leading heart physiologist. Anti-vivisectionists denounced his experiments, however, as "barbarous". Martin denied inflicting unwarranted pain and always believed the experiments were for the greater human good, but the protests upset him and he turned to drink. His health deteriorated after his wife died in 1892, and he became addicted to morphine. In 1894, he left Baltimore and returned to work with Huxley, but died two years later.

Joseph Barcroft, a Quaker, also studied at Cambridge, where he later became professor of physiology, specialising in blood and haemoglobin, and in foetal development. He showed, among other things, that oxygen passes into the blood by simply diffusing across the alveolar membrane in the lungs, and invented a device for measuring the pressure of oxygen in the blood. Barcroft took part in mountaineering expeditions to study how altitude affects the blood, which was useful in aviation medicine, as pilots had begun flying at higher altitudes. In the 1930s, to study how the body responds to extreme cold, he would lie naked on a couch for up to an hour in freezing temperatures, and record his reactions. Though a pacifist, he undertook military research during both world wars: on the effects of gas poisoning during World War I, and animal physiology and food production during World War II. Barcroft wrote several books that became classic texts, including *The Respiratory Function of the Blood* (1914) and *Researches on Pre-Natal Death* (1946). Mount Barcroft in California, where the University of California has a high-altitude research laboratory, is named after him. His son **Henry Barcroft** (1904–98) was also a world authority on blood function. Professor of physiology at Queen's University Belfast in the 1930s, and later at St Thomas's Hospital in London, he studied how blood flow is affected by adrenalin, and devised simple techniques to measure peripheral blood flow in skin, muscle and the limbs.

The Kingdom of Mourne

Thanks to their young age, the Mountains of Mourne are rough, rugged and crowned by craggy tors of shattered granite. The Wicklow Mountains, which are made of similar granite, are nearly 400 million years older, and so in contrast are more eroded and smoother. The Mournes formed 55 million years ago, at the end of the volcanic era that shaped so much of this corner of Ireland. It was the last major geological event in Ireland – since then things have been relatively quiescent. The neighbouring Cooley Mountains* and Ring of Gullion* formed at the same time, but are petite by comparison with the Mournes which, on their longest axis, stretch 25km from Newcastle to Rostrevor, and, at their highest point, Slieve Donard, rise to 850 metres. The National Trust acquired part of the Mournes in 1991, including Slieve Donard.

The Mournes formed after part of the Earth's surface fractured and subsided, enabling molten magma to rise and come close to the surface. There, cocooned under older rock, the magma cooled slowly to form granite. At least five of these 'intrusion' events happened at the Mournes, three in the higher, eastern section, and two in the lower, western section of the range. Each differs slightly from the others in terms of mineral make-up, and the type and size of the granite crystals. Heat from the molten magma also baked the surrounding rock up to a distance of several hundred metres, and this, together with various lava flows and subsidences, created a complex mix of old, young and altered rock. Semi-precious gemstones occur in places – at Hare's Gap you can find beryl, topaz and emeralds – but never in economic quantities. The Mourne granites, however, especially the early ones that were hardened by the heat from subsequent intrusions, have been much quarried, notably at Newcastle and Kilkeel. One track up Slieve Donard starts at the old Newcastle quarry tramline.

There is an **ancient oak wood** in the steep, wooded valley above Rostrevor. This mature oak forest, with a holly understorey, re-grew after the area was clear-felled in the 1730s. The valley was not disturbed in the interim, and the trees growing there today are direct descendants of the ancient oaks that once covered most of the country. Rostrevor Wood is now a nature reserve. Across the hills at Newcastle, another nature reserve protects the 6,000-year-old **sand dunes*** at Murlough, where the Mountains of

Mourne sweep down to the sea. Ireland's last wild boar was reputedly killed in the 1600s on Slieve Croob, an outlying hill north of the main Mourne range. ⑭

The dam builders

These days it is easy to take tap water for granted, but securing a reliable water supply could sometimes be an epic undertaking. So it was with the building of the **Silent Valley reservoir** in the Mourne Mountains in the 1920s. For centuries, Belfast had taken its drinking water from the local Farset river but, by the late 1800s, the demand for drinking, domestic and industrial water greatly exceeded supply. In 1890, the city council decided to tap the Kilkeel river in what it termed the "uninhabited Mourne catchment", where the water was free of industrial pollution, and could fall to Belfast under gravity. A 60-km pipeline was laid to Belfast's Knockbracken reservoir – including long sections tunnelled through the mountains – and the water came on stream in 1901. This was the grand era of public-health engineering when, in a bid to prevent the return of the cholera and fever epidemics of previous decades, drinking water and sewerage schemes were inaugurated for Ireland's growing urban population.

Digging the dam trench in pressurised shafts. Note the steel plates of the shafts, built to withstand the pressure of the deep, waterlogged bog.
Public Records Office of Northern Ireland

To define the Mourne scheme's catchment area, the council next built a massive wall. It stands 3 metres high, is 1 metre thick and over 30km long, encircles the Annalong and Kilkeel valleys, passes through the summits of 15 mountains, and took 19 years to complete. No sooner was it finished, than Belfast's water needs again exceeded supply. The council decided to dam the Kilkeel river, and create a reservoir that could hold 3 billion gallons in the long and deeply glaciated Happy Valley (renamed the Silent Valley because, it is said, construction work scared away all the birds).

The valley turned out to have a very deep bed of glacial clay mixed with enormous boulders – some the size of a cottage – and it took 2,000 men 10 years to excavate and build the dam trench and reservoir.[4] The waterlogged bog was nearly 60 metres deep so, to excavate the trench, men worked in pressurised shafts beneath the bog, a building technique more usually associated with underground tunnelling. At the start of each shift, the men spent an hour acclimatising, and two hours after work decompressing to avoid the 'bends'. For this hazardous work they were paid 30 per cent extra. Inevitably with such a large building project there were accidents, and eight men died while the dam was being built – even water does not come cheap.

In the 1950s, the Silent Valley scheme was expanded again, this time to include the neighbouring Annalong river. The rock in Annalong Valley is deeply fissured, however, and unsuited to a reservoir, so the river was diverted into the Silent Valley via a tunnel driven through Slieve Binnian. A second reservoir was built in the Silent Valley, but this time in three years and by 200 men, thanks to technical developments in the intervening decades. Since the 1990s, Belfast has also been drawing water from Lough Neagh.* There are lakeside walks and an information centre at the Silent Valley. ⑮

"Father of the Suez Canal"

The first practical proposal for a Suez canal was made by a military surveyor from Annalong, **Francis Rawdon Chesney** (1789–1872). Governments and engineers had long dreamed of linking the Mediterranean and Red seas, which are separated by just 160km. A canal would mean ships could reach the Indian Ocean without having to sail around Africa, but when Napoleon's engineers surveyed the route in the early 1800s, they concluded that a canal was impossible. Francis Chesney proved them wrong.

Chesney, who came from a distinguished military family of Scottish origin, joined the artillery in 1805 and eventually rose to the rank of general. In 1829, the army commissioned him to survey possible routes overland to India across Ottoman territory. Chesney investigated two options: a canal cut across Egypt's Suways [Suez] peninsula; and an overland route into Syria, then down the River Euphrates to the Persian Gulf. Not wanting to attract attention, Chesney surveyed the Euphrates by rafting down the river, and taking surreptitious soundings of the water's depth through a hole cut in the raft. Despite the precautions, he was frequently shot at from the riverbank.

Chesney's surveys showed that a Suez canal was possible, but the British government preferred the Syria–Euphrates option, and asked him to test that route. In 1835, he led an overland expedition that left Antioch on the Mediterranean coast, carrying two dismantled paddle steamboats across the desert into central Syria. They joined the River Euphrates there, and assembled the boats for the journey down river. One ship sank in a storm, but the other completed the journey. The government never developed the route, although, in the 1850s and again in the 1860s, Chesney was asked to re-survey the route for a possible railway line. Chesney later commanded the British artillery in China during the Opium Wars, and published several accounts of his travels, before retiring to Co. Down. He lived to see the French build his Suez Canal, however: Ferdinand de Lesseps used Chesney's survey and credited the Irishman as the "father of the Suez Canal". Today, the canal is one of the world's busiest shipping lanes, thronged with oil tankers serving the oilfields of the Persian Gulf and Saudi Arabia.

In Annalong, the village's old, water-powered cornmill has been restored. The mill, built in the early 1800s, was in use until the 1960s, by which time it was one of Ulster's last working watermills. Three pairs of millstones are powered by a 5-metre waterwheel (or a 20hp engine, when water levels are low). ⑯

The ingenious Drumm battery train

In the 1930s, Ireland led the world in the field of battery-powered trains. Thanks to a novel battery invented by **James Drumm** (1896–1974), who was born in Dundrum, electric trains ran on one of the main Dublin suburban rail lines during the 1930s and 1940s. A fleet of battery-powered bread vans was also used around the city. Despite these successes, Drumm's technology was never widely commercialised, due in part to the intervention of World War II. In 1949, when the original batteries were exhausted, Dublin's innovative electric trains were replaced by diesel engines.

James Drumm studied chemistry at UCD,* then worked for some years as an industrial chemist in England and Ireland. In 1926, he became interested in developing a battery that could be used for traction. This echoed work done a century before by another Irish chemist, Nicholas Callan,* who invented powerful batteries and dreamed of running a battery-powered train on the then new Dublin–Kingstown [Dún Laoghaire] railway.* The only storage batteries available at the time, however, were the lead-sulphuric acid accumulator and Edison's nickel-iron alkaline battery, and

neither was suited to traction – they were heavy, slow to charge and discharge, had a short working life, and the lead battery disintegrated with vibration.

After much experimenting, some of it funded by Dublin County Council, in 1930 Drumm invented the nickel-zinc rechargeable battery that was ideal for traction: mechanically robust, cheap to run, with a long life, relatively lightweight – to shift the combined weight of the vehicle and itself – quick to charge and, in order to give rapid acceleration, quick to discharge as well. Worldwide patents were taken out, the scientific journal *Nature* described the battery as "remarkable", and the Edison Company expressed interest in commercialising it. It quickly became the subject of a major political debate, however. The ambitious Shannon hydroelectricity station at Ardnacrusha* had just come on stream, and the government was anxious to develop projects that would use the electricity. It would not have been economic to electrify the railway lines, but a rechargeable train battery seemed attractive, with the added advantage of reducing the country's dependence on imported fuel. The opposition, however, considered the whole Shannon scheme to be a white elephant and opposed the project.

Nonetheless, in late 1931, Drumm's battery was tried on a train. It successfully pulled an 85-tonne load, going from 0–80 kmph in under a minute, and travelled 130km on a single charge. Two battery trains were built at the Inchicore

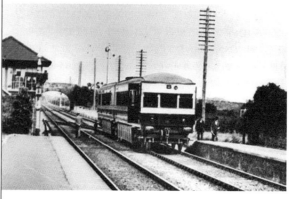

A Drumm battery train: note the large batteries underneath the carriages. Drumm invented the nickel-zinc battery: its positive plate was nickel oxide; the negative plate was pure nickel gauze; and the electrolyte was zinc oxide in potassium hydroxide.
© Iarnród Éireann

railway works* in Dublin and, in 1932, they went into service on the Dublin–Bray route. Each train could carry 130 passengers and the batteries took just a few minutes to recharge at the terminus stations. Uniquely, the trains incorporated regenerative braking: when they braked, the kinetic energy of the wheels was transmitted to small spinner wheels; when the train restarted, this energy was transmitted back to the train wheels. This arrangement, which helped conserve energy and improve acceleration, is used on modern buses and electric trains.

The two Drumm trains ran successfully for years, and, in the late 1930s, two more were added. They continued running during the war years, when other trains were hampered by coal and diesel shortages. In the 1930s, Drumm batteries were also used on a fleet of Dublin bread vans and on at least one commercial truck. There was never much political support for the project, however, nor funding to develop the battery. By 1949, when the train batteries had reached the end of their life, cheap diesel had become available and the batteries were scrapped in favour of diesel engines. So ended a chapter of innovative Irish transport. Modern versions of Drumm's invention, however, such as the nickel-zinc batteries made by US-based Evercel Corporation, are used to power small electric vehicles.

AD 540 – St Patrick, tree rings and the Dark Ages

Why is so little known about St Patrick? Despite his profound influence on Ireland, his life is shrouded in mist and there is little agreement about when he died – some say he died in AD 461, others 491 or 493. Some even believe there were two Patricks. Similar uncertainty surrounds England's King Arthur, who lived at the same time. Yet there are no doubts about Julius Caesar's dates, even though he lived 500 years earlier. The reason, according to Prof Mike Baillie of QUB, is that in AD 540, not long after Patrick and Arthur died, the world was plunged into a state of chaos. It was the start of the Dark Ages, and Mike Baillie believes it was caused by a series of collisions with heavenly bodies.

Prof Baillie has made a special study of tree-ring* patterns in Irish oaks, including prehistoric bog oaks dating back 7,000 years. The patterns of narrow and wide rings resemble a bar code, and can be used to date timber objects to a precise calendar year, as at the Corlea* bog trackway. They can also be used to study climate change: a narrow ring indicates poor growing conditions and bad weather; a wide ring indicates a good growing season. According to Baillie, the Irish oak rings reveal that the worst weather in the last 2,000 years began in AD 540, and the dark skies and chilly conditions lasted several years. It was a global catastrophe: tree-ring patterns from North and South America, Europe and Russia, all record the same bad weather; while conventional historical documents report major famines in China and the Mediterranean region, devastating floods and the Justinian Plague which killed one-third of Europe's population.

The narrow tree rings suggest that the Sun was obscured for years by a thick veil of dust in the atmosphere. There is no evidence of a sizeable volcanic eruption in AD 540, or of a major meteorite impact, to account for a dust cloud. So Mike Baillie believes the events were triggered by a series of smaller meteorite collisions, similar perhaps to the string of cometary fragments that collided with Jupiter in 1994. One result of the apocalyptic start to the Dark Ages, Mike Baillie suggests, was that the Irish converted in droves to the Christian religion that Patrick had preached a century before. Patrick is associated with various sites in the Lecale Peninsula, notably Saul (from *sabhal*, 'a barn'), Loughinish and Downpatrick. [17]

Lecale – the island plain

After the end of the last Ice Age, when sea levels* were higher, Lecale was an island. Even after sea levels dropped, the area remained virtually islanded, cut off by the marshy estuaries of the Blackstaff and Quoile rivers, until 1745 when the marshes were drained. Originally called *Magh Inis* ('island plain'), the Lecale peninsula stretches around the coast from Clough to Killough, Ardglass and Strangford, and inland to Downpatrick, the peninsula's capital. [18] It is a low-lying wedge of land; the only high ground are drumlins,* and the Castlemahon ridge near Downpatrick, which rises to 125 metres. Despite its small size, Lecale has four nature reserves and a rich diversity of habitats – saltwater and freshwater marshes, tidal mudflats, small lakes, sandy beaches, sea caves and rocky shores.

The bedrock is shale and slate, but these surface only at the rocky southeast coastline. Otherwise, the land is covered with a thick blanket of glacial till. Lecale's fertile soil, good fishing, gentle aspect, accessible creeks and natural harbours have always attracted people: there are Mesolithic, Neolithic and Iron Age sites; numerous Early Christian remains – the region has strong associations with St Patrick – and Viking influences. The Anglo-Normans built Ardglass port and the mediaeval walled town of Downpatrick. In the 18th and 19th centuries, improving landlords, notably the Wards of Castle Ward,* converted the area to tillage, built mills and created new towns, such as Killough, with its harbour, quays and salt pan. Small farm buildings of that era were made in a unique local style of corbelled dry-stone walls. Many have been preserved and the pigsties at Murphystown are unusual among pigsties in being listed for conservation.

Quoile served as Downpatrick's port for nearly 200 years. A quay was built there in 1717, and, in the 1830s, a paddle-steamer service operated to Liverpool, but the quay's importance declined after the Belfast–Downpatrick train line opened (part of the line has been restored for steam excursions). [19] Flooding was a perennial problem at Quoile and Downpatrick, so floodgates were built across the estuary at Quoile bridge in 1745. One result of this was that the tidal marshes and mudflats were slowly reclaimed, and

Lecale finally gave up its insular nature. There were further drainage works in 1802 and 1934. When Downpatrick was again flooded in the 1950s, a new barrier was built at Castle Island. This created the **Quoile Pondage**, to hold flood waters back until the next low tide. The pondage has slowly been desalinating since the barrier was built. Now a nutrient-rich lake and nature reserve, it continues to evolve. It is an important wildfowl site in winter, and is noted for swans, waders and duck. [20]

Upriver at **Hollymount**, the marshes, which were drained in 1745, are still in the process of becoming dry land. This nature reserve has ancient alder carr (a wetland wood), also fen and reed swamp. Lecale's other nature reserves are on the coast near Strangford: **Killard** has rocky and sandy shores, dune-grassland growing on glacial clays, and a very rich flora with several types of orchid; **Cloghy rocks** are popular with Common and Grey seals. (There is open access to the two coastal nature reserves; access to Hollymount is restricted.)

Sea beans and drinking chocolate

The British Museum owes its existence to **Sir Hans Sloane** (1660–1753), a physician and naturalist who was born in Killyleagh. Sloane also introduced drinking chocolate to Europe, and he was the first to realise that the 'sea beans' found on Irish and British beaches are the seeds of Caribbean trees, carried across the Atlantic on ocean currents.

Sloane's father came to Ireland as Receiver-General of Taxes during the Plantation of Ulster, and the family settled at Killyleagh. Hans Sloane had been seriously ill in his teens, possibly with tuberculosis and smallpox. As a result, he became something of a health fanatic, but he also resolved to train as a physician. He studied in London and European universities, before returning to London to work in 1685. He quickly established a reputation, and later became physician to the king and queen, but, in 1687, he was appointed surgeon to the West Indies fleet, and left to spend two years in the Caribbean. While there, Sloane studied the islands' natural history and climate, and acquired 800 plant specimens, which formed the start of his collection. He was also introduced to the local cocoa drink, but added milk to make it more palatable. London apothecaries sold his **drinking chocolate** as a remedy, and Cadbury's popularised it in the 1800s.

Sir Hans Sloane invented the milk-chocolate drink when he was in the Caribbean.

In 1696, Sloane spotted **sea beans** on a Scottish beach, and recognised them as the seeds of Caribbean trees. Sea beans were a mystery then: some said that they came from seaweeds, others that they had been cast overboard from ships. Not everyone believed Sloane's theory at first, but sea beans are naturally buoyant, with a thick water-resistant shell, and can survive years in the sea. The transatlantic voyage can last 14 months, and many are viable when they arrive. Nine types of sea bean have been found on Irish beaches, including sea kidney, nickar nut and even coconut.

Hans Sloane settled in London, married a wealthy widow, Mrs Elizabeth Rose, and became president of the Royal Society and the Royal College of Physicians. He helped establish smallpox vaccination* in 1718, when he vaccinated the royal family, and advised the Royal Navy on how to prevent scurvy. His circle included great writers, scientists and composers, among them Newton, Liebnitz, Linnaeus and Handel. Sloane was an enthusiastic collector and, thanks to his investments in Caribbean sugar and quinine, his wife's fortune, and his successful practice, he amassed a vast collection of books, manuscripts, gemstones, plant and animal specimens and antiquities from around the world. He kept all these at his Bloomsbury home, and employed a full-time curator. Hans Sloane left his entire collection to the nation, provided it was kept together and held in London, where as many people as possible might see it. His terms were accepted, funds were raised to build somewhere to house the collection – and so began the British Museum.

1946 – the first ejector seat

On July 24th, 1946, Bernard Lynch willingly sat on top of a controlled explosion. It was the first live test of an ejector seat invented by an engineer from Crossgar, **Sir James Martin** (1893–1981). Lynch was flying at an altitude of 2,438 metres, and a speed of 515 kmph, when an initial explosion blew away the cockpit covering him; then a second explosion propelled him and his seat along guide rails and out of the plane, enabling him to parachute to safety. The test was a success, and James Martin's invention won the RAF's approval. Within 12 months, every new British military aircraft was fitted with an ejector seat and by the time Martin died, in 1981, his invention had saved nearly 5,000 lives.

James Martin was a talented, self-taught engineer who enjoyed inventing things. He had a small business in

Sir James Martin features on the Northern Bank's £100 note. The Ulster Folk & Transport Museum has one of his early ejector-seat designs.*
© The Northern Bank

London making various machines, then he switched to building aeroplanes, and, in 1934, went into partnership with his former flying instructor, Capt Valentine Baker. The Martin-Baker Aircraft Company designed several civilian and military aircraft, but despite successful test flights, failed to win any orders. They had more success, however, with other Martin inventions, such as a novel cartridge-feed system for cannon, and a device to cut the cables tethering barrage balloons – 250,000 cable cutters, fixed to aircraft wings, were used during World War II.

In 1942, after Capt Baker was killed while attempting an emergency landing, Martin started designing an ejector seat. The government, concerned at the number of pilots being killed in action, asked him to develop a prototype. In the early days, when planes flew at relatively slow speeds, pilots had time to bale out without injuring themselves, but as flight altitudes and speeds increased, this became more difficult. Martin intended to use explosives to expel the pilot and his seat out of the aeroplane, but no one knew if the human body could withstand the forces involved. The first experiments were conducted on a test rig with sandbags and dummies, and then human volunteers, before the live test in July 1946 proved the concept. Martin's ejector seat soon won international approval, and it is now used by 90 air forces. Modern ejector seats are increasingly sophisticated: for example, if the pilot is unconscious they can take care of the parachute descent, inflate if they land on water, and flash a beacon to aid rescue. The Martin-Baker Company is still on the go and, in the 1980s, designed crash-resistant seating for passenger planes.

Mistress of the microscope

The first person killed in an automobile accident in Ireland was a bestselling science writer **Mrs Mary Ward** (1827–69). Ward was also a noted illustrator and she encouraged a whole generation of readers to experiment with microscopes and telescopes. Born Mary King in Ferbane, Co. Offaly, she was a cousin of William Parsons,* 3rd Earl of Rosse and builder of Birr's monster telescope, which she often visited. Even as a young girl, she was interested in natural history, and, for her 18th birthday, was given a 50-piece microscope, then the best in Ireland. She taught herself how to mount samples on slides for viewing – improvising with thin paper rather than use the thick glass slides of the day – and studied everything from feathers to tadpole jaws and insect eyes.

In 1854, Mary King married Capt Henry Ward from Castle Ward, with whom she had eight children. The family struggled to survive on her small dowry, so she turned to writing and illustrating science books. With no access to a science library, she collected articles where she could, and corresponded with scientists she had met at Birr. Critical of existing books about microscopes, Mary Ward wrote and illustrated her own: *Sketches with the Microscope* (1857), written in an engaging style, described how a microscope could be used to study the natural world, and went to five editions. Her second book, *Telescope Teachings* (1859), prompted by the Birr telescope, was equally successful.

Mary Ward photographed by her friend Mary, Countess of Rosse. Ward's microscope, telescope and other memorabilia are displayed at Castle Ward.

Despite her lack of formal education, Mary Ward's accurate descriptions won the respect of scientists. Her books were shown at the 1862 Crystal Palace exhibition, she wrote for publications such as *The Intellectual Observer*, and illustrated books written by Scottish astronomer Sir David Brewster, whom she first met at Birr. Sir William Rowan Hamilton* of Dunsink Observatory arranged for her to be one of only three women to receive the proceedings of the Royal Astronomy Society (the others were Queen Victoria and Mary Somerville, another science writer).

In 1869, Mary Ward was travelling in a steam carriage at Birr, when she was thrown from the carriage and killed. Ironically, the vehicle had been designed by her late cousin, the Earl of Rosse. Her husband later inherited Castle Ward as fifth Viscount Bangor, and moved his late wife's belongings to the house, where they are now exhibited (Birr Castle* Museum also has a microscope of hers). Castle Ward also has a restored water-powered sawmill and is home to the Strangford Lough wildlife centre. (21)

GROWELL

The man who modernised farming

The Industrial Revolution left the world of farming largely untouched. Admittedly, by 1900, there were large, steam-powered threshing machines, but farmers were still using horse-drawn ploughs as they had for centuries. **Harry Ferguson** (1884–1960) changed all that. A mechanical genius, he designed a safe and efficient tractor and plough system that dramatically improved farm productivity. Ferguson also built his own plane, became the first Irish man to fly, designed an early 4-wheel-drive system, and inadvertently started the Airfix model-kit business.

The fourth of 11 children born to a Plymouth Brethren farming family at Growell, Ferguson left school at 14 to work on the family farm but, ironically, never took to farming. At 17, he joined his elder brother's car and bicycle repair shop in Belfast and, by 1903, had built his own motorcycle, which he raced all over Ireland. In 1909, he was bitten with the flying bug, after visiting an English air show and, within months, had built a monoplane, despite having no design to work from. On December 31st, 1909, he became the **first Irish man to fly**, when his plane flew for 30 metres at an altitude of 4 metres. Within a year, he managed a 5-km flight, at a of height of 30 metres over the beach at Newcastle, Co. Down. In 1912, the plane was proudly shown at a Sinn Féin exhibition in Dublin, as Irish made. The plane was unstable, however, and regularly

turned over in flight – on one occasion ejecting both pilot and engine. It earned Ferguson the nickname of the 'mad mechanic', but also brought him to national attention. After one lucky escape, Ferguson gave up flying, and turned his attention to racing cars. By then he had his own business in Belfast, selling cars and imported US tractors.

Tractors at the time were heavy, clumsy, inefficient, fuel-thirsty and expensive. They were also used simply as mechanical horses, to drag a heavy plough across the ground. Despite weighing 4 tonnes, they delivered less than 3hp. They were unsuited to the wet Irish conditions, and worse, were inherently unsafe: if the plough hit an obstacle, the tractor's back wheels would keep spinning and the vehicle would turn over, crushing the driver. During World War I, the Irish Board of Agriculture faced mounting food shortages and, in 1917, asked farmers to till 20 per cent more land. The board imported 6,000 Ford tractors from the USA, and commissioned the 'mad mechanic' to design a more efficient plough. It was to prove an inspired move.

Ferguson designed a lightweight plough and made it part of the tractor, so that it did not need its own wheels. Crucially, he also designed a new 3-point linkage system to couple the plough to the tractor. The plough was suspended from the system, rather than dragged behind the tractor, which meant the tractor and plough could now be easily reversed.

Harry Ferguson with one of his revolutionary tractors. Ferguson features on the Northern Bank's £20 note, and there are commemorative plaques at Growell, and at Donegall Square in Belfast, site of his former business. The Ulster Folk & Transport Museum has early Ferguson tractors and a replica of his historic monoplane.*
Ulster Folk & Transport Museum

A fixed bar held the plough, and this prevented the tractor turning over if it hit an obstacle. And the drill depth could be adjusted from the driver's seat – for a deeper drill, you lengthened the bar and this dropped the plough more. The new plough was designed to work with Ford tractors and, in 1922, Ferguson opened a plough factory in Indiana, making implements for the large North American market.

Ferguson continued refining his design. Realising that his lighter plough did not need to be pulled by a heavy tractor, he designed a small vehicle weighing just 0.8 tonne, and an hydraulic linkage system so the plough could be lifted clear of the ground when not in use. A British company, David Brown, began making the new Ferguson tractor in 1936, but European farmers were suspicious of the lightweight vehicle, so Ferguson turned again to North America. In 1939, he and Henry Ford shook hands on a deal that allowed the Ford Company to make Ferguson tractors and implements. Over 300,000 tractors were sold but, in 1947, when Henry Ford died, his successor reneged on the deal. Ferguson began a protracted court case, but also opened his own tractor factories in the USA and England, producing his latest design, the TE 20 or 'wee Fergie'. Later he signed with a Canadian firm, Massey-Harris, to produce the famous Massey-Ferguson tractor. The TE 20 reflected Ferguson's attention to detail: the nuts and bolts came in only two sizes, $^{11}/_{16}$ in (1.75cm) and $1^1/_{16}$ in (2.7cm); the tractor tool set was one spanner: $^{11}/_{16}$ inch at one end, and $1^1/_{16}$ inch at the other; the spanner also served as a dipstick (there was no fuel gauge) and a measuring stick for drill width. Ferguson's attention to detail was legendary and bordered on eccentricity: everything in his workshop had to be arranged just so, and every mechanic had to carry a notebook in their left pocket, so they could jot down ideas while they worked.

In 1934, he inadvertently triggered the model-kit business.[5] He ordered small-scale models of his prototype tractor from Nicolas Kove's Airfix company, which made plastic items using the then new technique of injection moulding.

Kove produced the necessary pieces, but lacking the resources to assemble them, sent them out in kit form.

Harry Ferguson resigned as chairman of Massey-Ferguson in 1954, by which time the firm had sold 500,000 of his tractors. He continued inventing, despite ill health and persistent insomnia. He designed a 4-wheel-drive racing car, but failed to find a backer, though the system was later used with conventional cars; he also invented an anti-skid braking system and designed a family car to rival the Volkswagen Beetle. In 1960, he died at his home in the Cotswolds.

A liquid landscape

The sinuous coastline of Strangford Lough stretches for almost 200km, from Strangford village around to Portaferry, making it the longest sea lough in either Britain or Ireland. Over 100 islands roll in and out of the water, like a pod of humpback whales, blurring the line between land and liquid. The lough is held in a firm embrace by the long arm of the Ards Peninsula, and is almost a lake – its only connection to the sea is through a rocky gorge called the Narrows, which is just 800 metres wide. The tide rips through there at 10 knots, which prompted the Vikings to name this the violent, or *strang*, fjord. Strong currents carry 400 million tonnes of water through the gorge every day – which is bridged by a small car ferry – and there have been proposals to harness that energy with a hydroelectricity-generating barrage.

Strangford Lough is one of Europe's prime marine nature reserves. It has a tremendous diversity of habitats – many more than you would find on an open coastline – from fast currents and whirlpools in the south to calm mudflats in the north. Most of the world's population of pale-bellied brent geese graze beside the shore in winter; and Ireland's largest common seal colony congregates at the lough's rocky entrance. The lough is not a true fjord,* being neither narrow nor deep. Rather it is a shallow drowned glacial basin, a land form geomorphologists call fjardic. The ancient Irish annals state that the lough formed in 1654 BC, when the sea flooded the Kingdom of Brena. The flooding probably happened thousands of years earlier, immediately after the last Ice Age, when sea levels* rose dramatically and while the land in the north of Ireland was still compressed from the weight of the ice. A clutch of drumlins,* which had been 'laid' by the glaciers, was half-drowned in the flood. These drumlin islands are most noticeable in the lough's calmer western half; along the exposed eastern shore, wind and waves have reduced most of them to flat piles of rubble known locally as **pladdies**. Gerard Boate* wrote in the 1640s that the lough had 250 islands, but modern estimates put the number at 120. Several are inhabited, and many more of them are farmed. Raised, or 'fossil', beaches,* etched high onto the sides of the islands, reveal that the water level was originally much higher and the lough larger. As sea levels dropped and the land rebounded, however, the lough has slowly shrunk.

People have been exploiting the lough's resources for thousands of years. In the 18th and 19th centuries, for example, there was a thriving kelp* industry around the coast. Fishing has long been the main activity, and archaeologists have found mediaeval **fish traps** dotted around the shore. Fish traps are V-shaped stone walls or wooden fences, perhaps 200 metres long. Erected on the lower shore, they trap fish drifting down on the falling tide, and were used up to the 1600s. Remains of seven traps have been found near Grey Abbey Monastery (see: **A pharmacy garden**), and others can be seen at Ogilvie and Chapel islands.

Strangford Lough has much in common with Lough Hyne,* Ireland's other marine nature reserve, not least the fact that both have been heavily studied by naturalists for over a century. In Strangford Lough, zoologists have found over 2,000 species of marine animal, among them the horse mussel, the Norway lobster, the burrowing sea-cucumber, numerous sponges and over 300 species of mollusc. The Lough's marine life features at the Explorer's Aquarium. ㉒

In 1994, an invasive new arrival began colonising the lough's northern shores: **japweed** (*Sargassum muticus*), an extremely fast-growing Pacific seaweed, with brown-green fronds up to 5 metres long. It may have hitched a ride on boats coming from England, where the weed is now firmly established. Japweed can foul boats and mooring sites but, on the plus side, it provides a habitat for anemones and small fish. If you spot the weed, do not attempt to remove it as small fragments can break off and become established elsewhere.

There are seven **nature reserves** around the lough and its approaches: at Ballyquintin Point, on the southern tip of the Ards Peninsula, wind-dwarfed scrub grows on raised shingle beaches, surrounded by salt marshes; Cloghy rocks support an important seal colony; the shore and tidal islands at Granagh Bay within the Narrows are rich in marine life; wintering wildfowl and waders can be seen among the rocks and pladdies at Dorn; and the extensive mudflats at the lough's northern end are an important site for invertebrates and birds. The final two reserves are the Quoile Pondage* and Killard* on the Lecale Peninsula. Good bird-watching sites include Grey Abbey, Hill Island (just east of Comber), Castle Ward Bay and the National Trust bird hide on Reagh Island near Castle Espie. Flooded disused quarries at Castle Espie provide important habitats for plants, insects, mammals and birds. ㉓ Details about the various nature reserves can be had from the Quoile Centre* and the wildlife centre at Castle Ward.*

Ancient sundials and tidal mills

In continental Europe, no village square was complete without a public sundial fixed to the south-facing wall of the town hall or church. In Ireland, sundials were rare, perhaps because the climate was less conducive to this type of chronometer. Nevertheless, at least 10 ancient Irish

Nendrum sundial: the main rays indicate the three times for service:– terce (9am), sext (noon) and nones (3pm). Environment & Heritage Service NI

sundials have been found. All were associated with monastic cemeteries, and three were found in Co. Down: at Nendrum on Mahee Island (which is connected to the mainland by a causeway); at Bangor on the Ards Peninsula; and at Saul on the Lecale peninsula. Pieces of the Nendrum sundial were found during excavations in the 1920s, and the stone has since been reconstructed, but the Saul sundial, which was recorded in the 19th century, has since disappeared. Other sundial sites include Clogher,* Monasterboice and Kilmalkedar. It is difficult to date the stone sundials, but they are probably Early Christian.

The ancient Irish sundials are distinctive: all are standing stones, rather than wall-mounted sundials; many incorporate a cross and/or carved designs; and the time-lines divide the daylight hours into the classical Christian prayers times, rather than the civilian 12-hour day.[6] They were all found in graveyards, but may originally have stood elsewhere on the monastery grounds.

In the 1990s, archaeologists led by Tom McErlean from the University of Ulster uncovered the remains of three rare 7th- and 8th-century **tidal mills** built one on top of the other by the lough shore at Nendrum.[7] Amazingly, much of the structure survives, including the remains of millstones, a wheelhouse, a tailrace, stone dams and even wooden water wheels. Tree-ring dating* of oak timbers suggests that the first mill was built in AD 619, and the latest in AD 788, making these the oldest known tidal mills in Europe. With no river on Mahee Island, the monks at Nendrum had to harness the tide to turn their millstone: the incoming tide would have filled a small pond behind the stone dam; the stored water was later released through a stone flume or tunnel to drive the mill wheel and grind the monastery's grain. This was a horizontal mill, the oldest type of watermill known – the paddle wheel lay on its side – and a fascinating example of early Irish industrial technology. There is open access to Nendrum's monastic site and a small museum. (24)

Fossil footprints in the sandstone

At the head of Strangford Lough stands the prominent rocky mound of Scrabo Hill. The core of the hill is New Red Sandstone, which formed 220 million years ago when the area was an arid sandy desert. It is topped with a basalt cap, which protected the softer sandstone from erosion, especially during ice ages. The sandstone is rich in fossils – worms and scorpions, mud cracks and ripple marks, are all preserved in the now petrified sand. The most interesting, however, are footprints that were made when a small reptile walked across a wet patch in the sand 220 million years ago. This type of fossil is called *cheirotherium*, Greek for 'like a hand', because of the shape of the impressions preserved in the rock. The tracks, discovered in 1938, are the only *cheirotherium* fossils found in Ireland – Valentia's fossil footprints* were made 385 million years ago by an early amphibian. There is an exhibition with a plaster cast of the fossil footprints at Scrabo Country Park (the original is at the Ulster Museum*). (25)

Scrabo sandstone has been quarried for 1,000 years – 12th-century Grey Abbey* is built with the stone – but extensive quarrying began in 1826. The stone was used at Belfast, especially in the 1880s when the city was expanding rapidly, but it was also shipped as far away as New York. Despite all the quarrying, Scrabo Hill remains a substantial presence on the landscape, and from the summit there are fine views over Strangford Lough. The hill is crowned by a tall basalt tower, built in 1857, a post-Famine tribute from the Marquess of Londonderry's grateful tenants. The tidal mudflats south of the hill are a nature reserve, popular in winter with flocks of brent geese.

A pharmacy garden

A link with Scotland

Herb gardens and hedgerows were a valuable source of medicines in the days before pharmacies. The common name of many plants reflects this ancient practice – balm (for soothing), self-heal (for wounds), feverfew (for a temperature) – and gave illiterate people access to herbal wisdom. Mediaeval Irish monasteries provided something of a health service, with an infirmary, a **physic garden** of medicinal plants (such as the one recreated at Grey Abbey ㉖) and, in the library, a herbal book of remedies. The herbals used in mediaeval Ireland were generally copies of European texts, and it was not until the early 1700s that botanists compiled details of Irish remedies. Caleb Threlkeld* wrote one of the earliest: a catalogue of plants he found growing in parts of Ireland, published with details of their medicinal uses in 1726; snapdragon, for instance, could be used against necromancers. (Remedies to ward off evil date from when diseases were poorly understood and bacteria and viruses were unknown.)

Herbals were still being compiled in the 19th century – a manuscript written by Aodha Mhic Dhomhnaill (1802–1867), for example, now part of the McAdam Collection in Belfast Public Library, gives the medicinal uses for some 50 plants. Local plant lore persisted into the 20th century: the Clare Island survey* (1910) recorded seven plants that islanders used medicinally, including self-heal; even in the 1930s the Folklore Commission collected detailed plant lore around the country. There was often merit in the traditional remedies, and many modern pharmaceutical drugs derived from them. Salicylic acid, for example, the active ingredient in aspirin, is found in plants traditionally taken to treat fever, headache and inflammation. Other remedies and plants are being studied as a possible source of new drugs. Ingrid Hook at the School of Pharmacy in Trinity College Dublin is analysing sundew (*Drosera*), an insect-eating bog plant that features in traditional cough mixtures. She has found that sundews contain several anti-bacterial and anti-viral chemicals.

The first **telegraph** cable between Ireland and Britain was laid in 1853 between Donaghadee and Portpatrick in Scotland. The Irish and British telegraph networks were now linked, and messages could go from Dublin to London in hours if not minutes. The English & Irish Magnetic Telegraph Company used a heavy-duty cable, weighing 7 tonnes per mile, for the submarine connection. At 35km, Donaghadee–Portpatrick is one of the shorter routes between Ireland and Britain – the 20-km crossing at Torr Head, Co. Antrim was too remote to be useful – and it had long been used for trade and communications between the north of Ireland and Scotland. A packet, or mail-boat, service began in the 1690s and provided a valuable link for Belfast merchants, and by the early 1800s, was carrying 300,000 letters a year. Passengers landed at Donaghadee too, among them Peter the Great of Russia, who came to buy horses for his army, and Daniel Defoe, who came on a spying mission. Donaghadee Harbour was built in the early 1800s as part of a major government programme to develop the Irish fishing industry. New piers and harbours were built around the coast then, many of them, including Donaghadee, designed by Scottish engineer **John Rennie** (1761–1821). When Larne gained its harbour in 1849, however, the Scottish traffic moved there from Donaghadee. The RSPB runs bird-watching trips in summer from Donaghadee to Copeland Island. ㉗

1813 – the first miner's safety lamp

Safety was a major concern in collieries for centuries. Up to the early 1800s, naked candles were widely used, but if there was any methane in the air, the flame could ignite it. Explosions were frequent and tragic, and many people tried designing a safety lamp for coal miners. The best-known was probably English chemist Sir Humphry Davy, who

William Clanny from Bangor, with two of the safety lamps he designed.

unveiled his safety lamp in 1815. But the first safety lamp had been invented two years previously by a doctor from Bangor, **William Clanny** (1776–1850). Educated at Edinburgh, Clanny practised as a physician in Sunderland. He began working on a safety lamp in 1811, and presented his first design to the Royal Philosophical Society in May 1813 in a paper entitled "On the means of procuring a steady light in coal mines without danger of explosion". His was an oil lamp, shielded within a glass lantern, and isolated from the surrounding atmosphere by water seals; a bellows supplied air to the lamp. It was used in several mines in 1816, and Humphry Davy pronounced it "an ingenious arrangement". Davy's lamp took a different approach, using a metal gauze chimney around the flame to absorb the heat.

There were drawbacks to both lamps: Clanny's had to be continuously pumped, and Davy's had to be constantly watched. Indeed, Davy's lamp was not so much a light source, as an instrument for detecting the presence of methane in the air, since the size of the flame varied depending on the amount of methane present. Improved safety lamps were introduced over the next few years, the best of which combined the principles devised by Clanny and Davy. Clanny himself designed five further safety lamps and won numerous honours. In his final lamp, invented in 1840 and known as the 'Clanny', a glass cylinder surrounded the flame, and air was supplied through a wire gauze.

In the 19th century, Bangor had a large cotton mill and was a popular sea-bathing resort and busy port – ore from the nearby **Conlig lead mines** was shipped from there for smelting in Wales. Conlig mines opened in the 1780s after a man discovered lead ore while ploughing a field. The lode, which contained a high-quality ore averaging 76 per cent lead, ran for over 1km, from Whitespots to Conlig. It was softer and more pure than most English ores and during its best years Conlig produced 30 per cent of Ireland's lead output. The ore was worked until the 1860s at various locations, notably Conlig itself. Large steam engines were used there to pump the shafts dry and to hoist the ore from underground; a windmill that formerly ground grain was converted to crushing ore; and a marsh was dammed to create a lake where the crushed ore could be washed, before being carted to Bangor. The period 1845–65 was the mine's most productive, with 150 miners and another 200 workers above ground, including women and children who picked over the crushed ore. Deeper shafts were sunk – some are 400 metres deep – and, in 1850, production peaked at 1,500 tonnes per year. In the 1860s, output fell to under 200 tonnes a year, and the mines became uneconomic, though some 1,500 tonnes of good ore remained in the ground.

In 1910, a German company mined the colossal spoil heaps, using oil-fired furnaces to extract any remaining lead. The venture was short-lived, however, and some believed it was a front for German spies – one employee, who was still in the area when war broke out in 1914, was arrested on suspicion of spying. The mines have not been worked since. The spoil heaps are gone – used as in-fill for building projects – and the underground shafts are flooded, but the artificial lake and the ruined engine houses and windmill survive. Ask locally about access, as some is on private land and some in Whitespots Country Park. North Down Heritage Centre in Bangor has information about the mines and the local history and geology. 28

Ulster's last windmill

At the end of the 18th century, the government introduced measures to encourage cereal growing in Ireland. Tillage farming became more common – Co. Down was noted for its oats – and grain mills were built around the country. By 1840, there were over 2,500 mills in Ireland. Most were water powered, but 15 per cent were **windmills**. These were most popular along the east coast, where damaging westerly gales were less of a problem. Windmills were never powerful enough for heavy industrial applications, such as paper milling or metalwork, but they were used in many other applications, including wool carding, grinding bones and crushing potatoes to make starch.

Of the many 19th-century windmills in Co. Down, only Ballycopeland's survives intact. This 4-storey stone tower was built in 1790 at the heart of the county's grain belt and was in use until 1915. Like all Irish mills, it has a kiln to dry the grain and reduce its moisture content before grinding it. The mill's four sail-arms are carried on the tower's revolving cap; a fantail at the opposite side is used to turn the cap and move the sails into the wind. The area of sailcloth exposed on the arms can be varied from inside the mill, according to wind conditions. The mill has two opposing doors, so that when one door is blocked by the sails the miller could still get in and out. There are three pairs of millstones: one for grinding wheat, another to remove the husks from oats and a third to grind the oats into oatmeal. ㉙

Step back in time

Spread across 60 hectares, and straddling the Belfast–Bangor road, is one of Ireland's finest cultural institutions, the **Ulster Folk & Transport Museum**. It has its own farm, and railway stop – Cultra Halt on the Bangor line – and has deservedly been voted one of Europe's top transport museums. The vast transport halls feature every

The first Irish man to fly: Harry Ferguson,* inventor of the modern tractor, with his plane and a rare photograph of the plane in flight. Courtesy Ulster Folk & Transport Museum

conceivable mode of locomotion, from man-hauled sledges to vertical take-off aircraft. Many of the items have a local connection, such as a replica of Harry Ferguson's* 1909 monoplane, a Shorts Bros* Skyvan, an early Martin-Baker ejector seat,* penny-farthing bicycles with Dunlop pneumatic tyres,* a gull-winged Delorean car, a 'toast rack' carriage from the Bushmills electric tramway* and a schooner built across the bay at Carrickfergus in 1893. A special exhibition is devoted to the *Titanic*, which was launched at Belfast in 1911. Pride of place in the railway hall goes to 155-tonne *Maedb*, the largest steam locomotive ever built in Ireland: made at CIÉ's Inchicore* works in Dublin in 1939, *Maedb* ran for 20 years on the Dublin–Cork route, and clocked a record 155kmph.

The extensive open-air folk museum recreates rural Ulster landscape and small-town life of *c.*1900. Dozens of buildings have been carefully moved from their original location and rebuilt in a facsimile of their former environs. They include a Coalisland spade mill* (which was in operation *c.*1840–1950), a linen bleach green, a flax mill, a Fermanagh sawmill and a printer's workshop. The museum's large farm is worked as living history, using century-old methods and equipment, and traditional crops and breeds. There are exhibitions on the history of farming and the 1847 Famine, and demonstrations and courses on traditional skills and crafts. ㉚

BELFAST

By 1914, Belfast was Ireland's newest, biggest and most industrialised city. It was at centre of the world's linen trade, had the world's biggest ropeworks and shipyards and produced several world famous inventors and scientists – including one who proved Einstein wrong. The city straddles the Down–Antrim border, nestling beneath the southern edge of Antrim's basalt plateau, and is named after the River Farset (*Béal Feirste*, 'the mouth of the Farset').

Ireland's youngest city

Had you come to Belfast 7,000 years ago, you would have found it under water. For the sea level was higher then, and the land was also lower, depressed with the still-remembered weight of ice from the last Ice Age. Over the millennia, the sea slowly retreated and the ground slowly rebounded. (The north of Ireland is still rebounding, though the rate has slowed in modern times to about 1mm a year.) The result was a muddy slobland at what is now Belfast. In places, the mud – known locally as 'sleech' – is 15 metres deep, and buildings there have to be erected on deep piles. Despite these precautions, the Albert memorial clock on High Street has subsided into the sleech and developed a noticeable tilt. High Street is in the oldest part of Belfast, and built on the original seabed – some of the prehistoric shoreline can be seen at Royal Avenue and York Street, where the higher ground is a fossil beach.*

The Lagan and Farset rivers converge at Belfast; the Farset is now culverted and the Lagan corseted by embankments.

Historically, the mudflats and broad estuary were difficult to cross, and there was only one ford at firm ground, near the present-day Queen's Bridge and new Lagan weir. As a result, development was slow and, even 1,000 years ago, Belfast was little more than a defensive fort at the crossing. By 1600, some 500 people had settled there, although ship access was still difficult and Carrickfergus* was the main port. A small linen* industry began and received a major boost in 1685 with the arrival of French Huguenots, so that, by 1700, Belfast's population had swelled to 2,000.

Over the next century, Belfast grew an astonishing 12-fold, fuelled by the expanding linen trade and associated industries – such as the iron foundries that made steam engines and machines for the mills – and aided by the **Lagan navigation**, which improved access to the hinterland. Work on the navigation began in 1756, and a dozen locks were built on the river between Belfast and Lisburn, giving industries in the interior easier access to the coast. In the 1790s, a canal was added linking the navigation to Lough Neagh; this would eventually connect Belfast to Limerick via Ireland's canal network. Shipbuilding

The Lagan river god: swathed in linen fabric and crowned with two swans.
© Enterprise Ireland

started at Belfast in 1791, when Scottish brothers William and Hugh Ritchie opened a shipyard on the banks of the Lagan. This, and Ulster's thriving machinery sector was the foundation of Belfast's heavy-engineering industry.

In 1800, Belfast had a population of 25,000. Over the subsequent century, although the population of rural Ulster fell by one-third, Belfast expanded rapidly. In 1842, it was made a borough, in 1888 it was granted city status and, in 1914, it became, for a time, Ireland's largest city, with a population of 400,000, compared to 300,000 in Dublin. It was Ireland's newest, yet most-industrialised, city, with the world's largest shipyards and ropeworks,* and serving Europe's most extensive linen* industry. This development began in earnest in 1839 when William Dargan,* the engineering contractor who built most of Ireland's mainline railways, was commissioned to dredge a channel through the estuary mudflats. The channel greatly improved ship access and Belfast port at last began developing. Significantly, Dargan dumped the dredged material by the riverbank, creating a new island that proved an ideal location for shipbuilding. Dargan's Island, renamed Queen's Island in 1849, is still at the heart of the Harland & Wolff* shipyards, and the new railway bridge at the quays is named after Dargan, who kick-started the city's development.

Belfast prospered during World War I. The Harland & Wolff shipyards produced nearly 10 per cent of the world's shipping output, for example, and the linen industry turned to making uniforms. During World War II, even more of Belfast's and Northern Ireland's industry was channelled into the **war effort** – Shorts* changed to making fighter planes, a factory by Lough Neagh made torpedoes, many small foundries produced artillery shells, linen firms again made uniforms and soldiers' kits, a mill at Carrickfergus became a parachute factory, a company that used to make farm gates changed to making intricate submarine valves,

and Gallagher's tobacco factory in Belfast, among the largest in the world, made cigarettes for the troops. But this strategic importance made Belfast an obvious target and, in 1941, the Luftwaffe bombed the city twice.

Belfast's rivers and streams continuously wash red clay and sandy silt into the Lagan estuary. This provided a convenient raw material for brickworks, and much of the city is built of local redbrick, but the silting also threatened to choke the port. So, in 1924, the council decided to narrow the Lagan, the aim being to improve the river's flushing action, and thus reduce silting. Five kilometres of embankment were built (now topped by a road), from Annadale through Stranmillis to Ormeau, and a weir was added in 1937 to ensure the fetid sloblands remained flooded at low tide. The scheme was never fully successful so, in the 1980s, a **new weir** was added. It has five large flood gates, made by Harland & Wolff, which hold back the tide and create a lagoon that covers the mudflats at low water. The weir helps control silting and improves the channel for river traffic, as well as rejuvenating the waterfront and protecting the area against high tides. The control tower for the new weir incorporates a visitor centre, and you can watch the flood gates in operation from a pedestrian bridge. ㉛

There are several good **bird-watching** sites along the surrounding coast, thanks to the rivers, the estuary mudflats, and the many small bays of Belfast Lough. The Lagan estuary is noted for wintering populations of great-crested grebe and godwit; Victoria Park for roosting waders; Cultra for turnstone and sandpiper; Helen's Bay for red-breasted merganser and eider duck; Carrickfergus Harbour for divers; and Lagan Valley Park for kingfishers. Belfast Harbour Commissioners has port archives dating from 1600. ㉜ The Linen Hall Library has a collection of historic newspapers, material relating to local industries, meteorological records (from 1796) and the minutes of the Belfast Natural History and Philosophical Society. ㉝

Samson, Goliath and *Titanic*

The earliest record of **shipbuilding** in Belfast is 1636 but, given the sheltered coastal location, there was probably shipbuilding activity in Belfast Lough long before then. The first substantial Lagan shipyard was started in 1791 by Scottish brothers Hugh and William Ritchie, who already had a yard in Ayrshire. They arrived in Belfast with 10 workmen, launched their first ship a year later and, by 1812, were employing over 120 men (Corporation Street is on the former site of their dock). Other companies joined them by the Lagan, such as Coates & Young which, in 1838, built Ireland's first iron steamboat – named the *Countess of Caledon*, it served on Lough Neagh* towing barges. In 1851, a new shipyard was built on Queen's Island, an artificial island created from material dredged during harbour improvements in the 1840s. Queen's Island, now joined to the mainland, is still at the centre of Belfast's shipbuilding industry.

By 1853, the shipyard had dry docks and slipways, and a 23-year-old English engineer, **Edward Harland** (1830–95), arrived to manage the works. Five years later he bought the business for £5,000, and took on a partner, **Gustav Wolff** (1834–1913) a German engineer. Harland & Wolff experimented with ship design, in search of ever-larger and faster vessels. They made boats longer but not wider, thus capable of carrying more cargo without compromising speed; they dispensed with bowsprits and figureheads, and tried various rigs and designs. In the 1890s, they began making truly enormous vessels, such as *Majestic* for the White Star Line, a 10,000-tonne, 177-metre ship. The first of many company firsts, it was the largest thing then afloat – but though powered by steam, it was also equipped with sails.

By the early 1900s, the company had become the world's leading shipbuilder, and the Belfast yard was the Cape Canaveral of its day. In 1911, it launched the world's largest liners, the ill-fated *Titanic* and its sister ship *Olympic*, each weighing an unprecedented 77,000 tonnes. The company built naval ships and merchant vessels during World War I, and afterwards won orders to replace the vessels lost in the conflict. In the depressed 1930s, Harland & Wolff partnered Shorts in making aircraft. World War II brought another boom time, when the company built no fewer than 139 naval vessels, including six aircraft carriers, as well as 130 other merchant ships. Inevitably this made it a Luftwaffe target, and 60 per cent of the yard was destroyed during air raids in 1941.

After the war, the company turned to making bulk carriers and offshore rigs for the oil and gas industry. In the 1960s, it began making boats in sections that were later welded together. The yard's two enormous cranes – affectionately called **Samson** and **Goliath** – were installed then, and each can lift sections weighing 1,000 tonnes. The yard continued to build record-breaking vessels, such as the 200,000-tonne oil tanker *Myrina* (1967) but, in the 1970s, hit troubled times and was nationalised. Following a management buyout in 1989, Harland & Wolff now specialises in offshore rigs.

The world's oldest aircraft maker

Short Brothers is the oldest aircraft company in the world. It began in 1901 in England making aerial balloons, and secured the first-ever contract to produce aeroplanes, in 1909, when the pioneering Wright brothers commissioned them to build six 'Flyers'. Shorts achieved several firsts, including the first aircraft with folding wings and, by the 1920s, was specialising in flying boats, such as those used at Foynes.* In 1936, the company opened an aeroplane factory on Queen's Island, in partnership with Harland & Wolff, and initially called Short & Wolff. When war broke out, the factory changed to making bombers. It became an important part of Britain's war industry and, in 1943, was nationalised. After the war, Shorts turned to making passenger planes and guided missiles, and supplying other aircraft manufacturers, including Boeing, Lockheed and Fokker. In 1989, the firm was bought by the Canadian company Bombardier, and currently employs over 6,000 people at five sites around Belfast, making it Northern Ireland's largest manufacturer. It also owns Belfast City Airport.

BELFAST DOCKS

Walking the rope

For many years, the largest rope factory in the world was **Belfast Ropework Company**. At its peak in the 1930s, it employed 4,000 people (men and women – unlike the shipyards which were nearly all male). Working in three enormous mills on a sprawling 16-hectare site at Belfast docks by the Connswater river, they produced 20,000 tonnes of rope a year for customers around the world. The works were founded in 1873 by an Englishman **William Smiles** (1846–1945), who had moved to Belfast that year. Thanks to the Harland & Wolff shipyard, demand for rope there was growing. The rope factory started with 50 workers, using local flax, as well as soft cotton and jute, and rougher sisal and coir, which were all imported through the now thriving Belfast port. In one early marketing campaign, the company persuaded famous French tightrope walker Blondin to use its products.

The traditional way to make rope was first to braid the raw fibre and then, walking backwards down a long shed, or 'ropewalk', to twist the stands into a finished rope. Ropewalks were a common feature of 19th-century docklands and Belfast's longest one stretched for 150 fathoms (275 metres). By 1900, there were three large mills on the site and the factory's product range included ships' halliards, agricultural baler twine, fishing nets, sash cords, clotheslines and domestic twine. The ropeworks were destroyed during the air raids of 1941 and, despite massive post-war investment, never really recovered. Synthetic yarns and Sellotape were replacing natural ropes and twines and, in 1970, the Belfast ropeworks closed.

SHORT STRAND

Where the Sirocco blows

One of Belfast's most prolific inventors was **Sir Samuel Davidson** (1846–1921), who held over 120 patents. He began by drying tea, but went on to invent a very efficient fan for ventilation systems, and his company is still in business, albeit now owned by a multinational group. Davidson was born in Belfast where his family had a mill. They also had an interest in the Indian tea industry and, at 18, Davidson left to work on the plantations. He realised that to produce a consistently high-quality tea, he needed to improve the drying process and, in 1877, he patented his first dryer design, using it to produce a quality tea that fetched higher prices. Orders for the new dryer rolled in and, in 1879, Davidson returned to Belfast to set up a company on the Short Strand making his Sirocco tea dryers.

For air flow, Davidson initially relied on the draught from the dryer's fire, but this was not enough for large commercial machines, where the hot air had to be forced through many trays of tea leaves. So Davidson experimented with fan design, eventually inventing the 'forward bladed centrifugal fan'. It was more efficient than existing fans and was soon widely used in industrial systems and to ventilate tunnels and mines and, at its peak, the Sirocco works employed 1,500 people. Davidson's other inventions included a hand-held Howitzer gun. He was knighted in 1921, but died shortly afterwards. In 1988, his business was bought by the

A Sirocco fan – designed by Sir Samuel Davidson, who features on the Northern Bank's £50 note. Howden-Sirocco

Howden Group, an international ventilation company; in 1999, the Sirocco site was sold for redevelopment, but parts of the old factory will be conserved.

The 'Inst' and the godless college

From 1592 until 1845, Trinity College Dublin* was the only university in Ireland. The situation was less than satisfactory: only the wealthy could afford the costs of attending college and living in Dublin for several years, and the college was controlled by the established Church of Ireland, which effectively excluded Catholics and, to a lesser extent, Presbyterians. By 1800, there were growing demands for more universities, notably at Cork and Belfast. In Cork, wealthy businessmen funded the Royal Cork Institution,* which opened in 1802, and a similar campaign in Belfast led to the establishment, in 1810, of the **Belfast Academical Institution**. The 'Inst', as it came to be known, enrolled students, ran a medical school and provided courses of a university standard, although it had no power to award degrees. One of the founders was **William Drennan** (1754–1820), a poet and a leader of the United Irishmen, who became the institution's first president. Many of the staff came from Belfast and the surrounding counties, among them the physician and chemist Thomas Andrews,* and the mathematician James Thomson from Ballynahinch, Co. Down, who was father of William Thomson* (aka Lord Kelvin) and James Thomson Jr.*

Thanks to the institution's success, Belfast was chosen over Armagh when, in 1845, the government set up the Queen's University of Ireland. The three Queen's Colleges – at Cork,* Belfast and Galway* – were non-sectarian, with staff appointed without regard to religion. They were denounced by the Roman Catholic Church, however, and when blight destroyed the potato crop that year, many blamed the tragedy on these "godless colleges". In Belfast though, Presbyterians and Protestants gave the new college a guarded welcome. The academical institution closed and its students and staff transferred to the new university.

When Queen's College Belfast opened its doors in 1849, it had 20 professors, 100 students and four faculties – Arts, Science (including engineering), Law and Medicine. By 1860, student numbers had doubled. Over the next 40 years, however, the three Queen's Colleges struggled for funds as, once the government had completed the initial building programme, it turned its attention to devising a university structure to satisfy the Roman Catholic Church. So, in 1901, Queen's College Belfast began fundraising among local businesses – Harland & Wolff, for example, donated £7,500 for new engineering laboratories. Then, in 1908, the government reorganised the universities: Belfast was granted independent status as Queen's University Belfast, while its sister colleges at Cork and Galway joined University College Dublin* in the new National University of Ireland. Today, QUB has over 15,000 students and an academic staff of 1,200. Its archives include the papers of both Thomas Andrews and James Thomson, two of the men who helped found the college.

From a gas to a liquid

Until 1861, it was thought that the gaseous elements, such as hydrogen, nitrogen and oxygen, were permanent gases, and that this was their only possible state. But Belfast-born **Thomas Andrews** (1813–85), who was QUB's first professor of chemistry, proved that, under the right conditions, any gas can be made into a liquid. Liquids and gases were, therefore, merely different phases which a substance could take, depending on the pressure and temperature. Andrews' work paved the way for liquefying gases, and led

Thomas Andrews, who proved it was possible to make a gas into a liquid.

to a whole new industry, with applications such as oxygen therapy in medicine. He was also the chemist who discovered that a strong smelling gas called ozone is a form of oxygen.

Thomas Andrews studied at some of the great 19th-century centres of learning – chemistry in Glasgow, Paris and Giessen in Germany; and medicine in Edinburgh and in Dublin, then at the height of its fame as a medical centre. In 1835, he returned to Belfast to work at the city's general hospital and to head the medical school at Belfast Academical Institution. Thanks to his success with the medical school, the government chose Belfast in 1845 for one of the three new college's in the Queen's University. When Queen's College Belfast opened, Andrews was appointed vice-president. He proved a skilled administrator and among his innovations was the introduction of a school of Celtic languages.

Andrews is famous for his work on gases. In one crucial series of experiments, he trapped samples of carbon-dioxide gas in heavy-duty glass tubing; he put the gas under pressure, and carefully monitored how it changed. His experiments echo the work done nearly 200 years earlier by another great Irish chemist, Robert Boyle.* Andrews, however, worked at higher pressures and temperatures and, indeed, he proved that Boyle's Law does not apply under extreme conditions. Andrews showed that for carbon dioxide the temperature of 31°C was critical: below that temperature, by increasing the pressure, he could compress the gas and eventually liquefy it; but above that temperature no amount of pressure would liquefy the gas. Consequently, this temperature is known as the **critical temperature**. This finding suggested that any gas could be liquefied, provided the temperature was low enough and, in 1871, Andrews wrote that "we may yet live to see… such bodies as oxygen and hydrogen in the liquid, perhaps even in the solid-state". This would have been heresy to many chemists then, yet six years later a French chemist Louis Cailletet made liquid oxygen, the first gaseous element to be liquefied. Physics are still investigating critical phenomena today. QUB's chemistry department has some of Andrews's experimental apparatus and his manuscripts are in the college archives.

From turbines to alcoholic 'legs'

Belfast-born **James Thomson** (1822–92) has been largely overshadowed by his younger brother William (see: **The ingenious baron**). Yet he was a pioneering engineer with an international reputation, who invented a new type of turbine, and improved fan and pump designs. He was also the first to explain why, when you swirl an alcoholic drink around a glass, it drains back as rivulets or 'legs'. James Thomson graduated from Glasgow University with a science degree when he was 17, by which time he had suggested design improvements to the paddle wheels on steamships. He was professor of civil engineering at QUB when, in the 1840s, he experimented with making waterwheels more efficient and, in 1850, designed his vortex turbine, which has at its core a waterwheel laid horizontally. This was more efficient than other turbines of the time, and was soon widely used in place of waterwheels.

In 1855, Thomson explained the "curious motions observable at the surfaces of wine and other alcoholic liquors". Alcoholic drinks contain a mix of ethanol (alcohol) and water, which evaporate at different rates, this creates a tension in the surface of the liquid, which generates evaporation and convection currents, even if the glass is held still. His insight was overlooked, however, and the phenomenon came to be called the Marangoni effect, after an Italian scientist who studied it in the 1870s. James Thomson also invented a successful planimeter, or device to measure the area under a curve, which his brother William used in his tide predictor. QUB's archive has some of James Thomson's papers.

What are the stars?

It was a Greek philosopher Anaxagoras who first suggested, in the 5th century BC, that stars were made of the same elements as are found on Earth. And for nearly 2,500 years people accepted this, believing the Sun was predominantly made of iron. By the 1920s, however, spectroscopic studies were suggesting otherwise: from analyses of the light emitted by the Sun, it seemed our star was made of hydrogen – not a metal, but a lightweight gas. In 1928, a young Dublin-born astrophysicist, **Sir William Hunter McCrea** (1904–99), did the calculations that proved conclusively that the Sun was three-quarters hydrogen and one-quarter helium, with only traces of a few other elements. Soon it was clear that hydrogen is the main constituent in all stars. McCrea, who was then completing his PhD at Cambridge University, was professor at QUB from 1936–44. His findings laid the basis for our modern understanding of how stars form, and the origins of the Solar System, galaxies and even the Universe. The preponderance of hydrogen, which is the simplest of all chemical elements, is also evidence for the Big Bang theory, though McCrea preferred to think that the Universe formed from several 'small bangs'. McCrea, awarded numerous international honours, is also known for his contributions to relativity and cosmology and, in addition to QUB, worked at Edinburgh, Imperial College London and Sussex University.

The man who proved Einstein wrong

The quantum world of particle physics is weird. Very weird. So weird, in fact, that even Einstein could not accept it. One of the people instrumental in uncovering that weirdness was a brilliant Belfast scientist **John Bell** (1928–90). Bell studied science at QUB, but spent most of his working life at CERN, the particle physics laboratory in Geneva. He made many important contributions, often collaborating with his wife Mary Ross, who was also a physicist, but he is best-known for an idea he proposed in 1964.[8]

First, a little history. In 1935, Albert Einstein, Boris Podolsky and Nathan Rosen drew attention to a paradox in quantum theory. If twin particles, say for example two particles of light (photons), were to be emitted from a single source, and fly off in opposite directions, the properties of one would still depend on the properties of the other – they are 'entangled'. In the bizarre world of quantum mechanics some of the properties are not concretely realised until they are actually measured – until then they do not have a concrete existence – yet they are still entangled. Measure the properties of one particle, and quantum theory predicts you will instantly know the properties of the other, as if information had passed between them faster than the speed of light. Einstein called this "spooky action at a distance", and thought it most unlikely.

John Bell proposed a way to explore this apparent paradox, but it was 20 years before physicists, led by Alain Aspect at Orsay in Paris, could attempt the experiment: they created pairs of photons, separated the twins, and sent each flying on its separate way; but when they measured the polarisation of one photon, they instantly knew the polarisation of its twin. This proved that John Bell was right, and Einstein was wrong – the world really is spooky, weird and entangled.

Bell is also known for pioneering work in elementary particle physics where he helped explain (along with two US physicists, Stephen Adler and Roman Jackiw) why a particular sub-atomic particle, a neutral pion, spontaneously explodes into two photons – though the prevailing ideas said this should not happen so quickly. Their explanation uncovered a subtle quantum effect (the Adler-Bell-Jackiw anomaly), which is central to all modern attempts to unify the forces of nature. Bell was awarded many international honours, and had been nominated for a Nobel prize, but died suddenly of a stroke in 1990.

Noted northern naturalists

In the early 1800s, naturalists began rigorously cataloguing species, recording when and where they were found. Two naturalists who contributed to this important early groundwork in Ireland came from Belfast: **John Templeton** (1766–1825), an expert on Irish plants, and **William Thompson** (1805–52), whose special interest was Irish birds.

Templeton, son of a prosperous Belfast merchant, was the first to compile a detailed study of Ireland's flora and fauna. Robert Lloyd Praeger,* another great northern naturalist, described him as the most eminent naturalist Ireland ever produced. Templeton's *magnum opus* was a comprehensive catalogue of native Irish plants, which he finished in 1801. He next prepared two books, *Hibernian Flora* and *Hibernian Zoology*, for which he also made numerous illustrations. His books and catalogue were never published, but the manuscripts survive (at the Ulster Museum* and the RIA*) and are a valuable source of information. Templeton corresponded with eminent European naturalists of his day, and his garden at Cranmore, off Belfast's Malone Road, was renowned. (The house at Cranmore, reputedly the oldest in Belfast, and where William of Orange rested on his march to the Battle of the Boyne, is now a ruin.) A genus of Australian legumes, the *Templetonia*, was named after Templeton, who was also a founder of the Belfast Academical Institution.*

William Thompson was another self-taught naturalist from a successful Belfast linen family. Like Templeton, he forsook the family business, preferring to study natural history. Thompson made a detailed study of Irish plants and animals, carefully recording which ones were found here, and which were absent. He discovered several species that were new to science, added some 1,000 species to the lists for Irish fauna, and published over 70 scientific papers about his findings. He began a major book on Ireland's natural history, which incorporated observations made around Ireland by other naturalists. The first three volumes on birds were published before his untimely death; a fourth volume on the remaining animal species was published posthumously by his friends.

John Templeton had dreamed of starting a botanic garden in Belfast, similar to those in Dublin and Cork. But it was not until 1827, two years after his death, that the Belfast Botanic and Horticultural Society was formed with the aim of establishing a **Belfast Botanic Garden**. ㉞ The society sold 500 shares at 7 guineas each to finance the project and, in 1828, bought 5.5 hectares outside the city, at the Malone and Stranmillis roads. When the garden opened later that year, shareholders were admitted free, others had to pay a shilling. In 1839, Richard Turner* built the garden's palm house. Turner, who had an iron foundry in Dublin, had developed a way of building attractive curved conservatories. In the 1840s, he designed and built the great curvilinear houses at Kew Gardens in London and at Dublin's Botanic Garden.* Belfast's glasshouse, however, is the earliest known example of his curvilinear work. The east wing is heated and tropical plants are grown there, such as coffee, cinnamon, rubber and banana. In 1852, a high central dome for tall palm trees was inserted between the two wings. In the late-1970s, the glasshouses were renovated; some of Turner's original cast iron metalwork was reused, but badly corroded sections were replaced with freshly cast ribs.

The palm house at Belfast Botanic Gardens: the earliest known example of Richard Turner's curvilinear work.
© Belfast City Council

For years, the gardens were a popular venue for balloon* ascents and a gas pipeline was laid down for fuelling the balloons. Harland & Wolff, which occasionally built naval gunships in secrecy in large sheds, donated one such shed that served for a time as an exhibition hall. The gardens are run by Belfast City Council, which bought them in 1895, after the horticultural society ran into financial problems. Other Belfast gardens of note are **Barnett Demesne** (daffodil gardens), **Grovelands** (heathers) and **Sir Thomas and Lady Dixon Park** (Japanese gardens).

Made in Belfast

Early pneumatic Dunlop bicycles, aeroplanes, ropes and steam engines built in Belfast for the linen mills – just some of the local products and industries featured at the **Ulster Museum**. The eclectic collection ranges from art to industry, from prehistory to modern times, and from American dinosaurs to Irish plants. There are waterwheels, steam engines, artefacts recovered from an Armada galleon, and Neolithic stone axes made from Antrim porcellanite.* The geological section boasts 3,000 rocks, 25,000 minerals and 200,000 fossils, including Ireland's only dinosaur – fragments of a 200-million-year-old scelidosaurus found at Islandmagee* in 1989 – and some 200-million-year-old fossil footprints from Scrabo Hill.* The mineral collection features Antrim zeolites,* an Australian gold nugget, and an aquamarine from the Mourne Mountains.

Many of the geological specimens were collected by a Belfast quarryman and gem dealer, **'Diamond' Pat Doran** (1784–1880). The earliest known accurate depiction of the Giant's Causeway,* painted in 1739 by Dublin artist Susanna Drury, is in the art gallery. The museum opened in 1831 at College Square and moved to the Botanic Gardens in 1929. ㉟

1874 – Darwinian storm hits Belfast

In August 1874, the **British Association for the Advancement of Science** (BAAS) held its annual gathering in Belfast. It had met there once, 21 years earlier and, as that meeting had been successful, the *Northern Whig* welcomed the association's return. The 5-day scientific festival, the newspaper told its readers, promised a pleasant respite from Orange riots, striking linen workers and ecclesiastical squabbles. The Belfast Naturalists' Field Club even published a guidebook for the visitors.[9] However, the meeting put the scientific cat squarely among the ecclesiastical pigeons, generating a storm that raged for years.

The BAAS was started, in 1831, to promote research, help the growing numbers of scientists to meet and keep in touch, and improve public understanding and support for science. Its annual summer gathering is held in a different place each August – the first were at Oxford, Cambridge and Edinburgh, then, in 1835, the BAAS came to Dublin. These gatherings were, and are, great scientific jamborees, with dozens of public lectures, distinguished speakers, gala dinners and considerable media attention. In the early days, the attendance would include MPs, local celebrities, members of the public, gentlemen interested in science, as well as professional scientists, though the latter were initially in a minority as science was still largely a pursuit of amateurs.

The 1835 Dublin meeting was a festive occasion. Over 1,000 people attended, 700 of whom travelled over from Britain. Marquees were erected in the Phoenix Park,* and there were

Brilliant Irish scientist John Tyndall, whose address to the 1874 BAAS meeting in Belfast put the scientific cat squarely among the ecclesiastical pigeons.

lectures about comets, tides, steam engines, map-making, magnetic engines, and much more, and considerable press coverage of the events. In 1843, the scientists gathered in Cork, but though many Irish people attended, few came from Britain, and the meeting was a financial failure. As a result, subsequent meetings in Ireland were confined to Dublin and Belfast. The BAAS first came to Belfast in 1853. Clergymen, MPs, successful businessmen and noted medics attended from around the country, alongside members of the ascendancy such as the Earl of Rosse* from Birr, and a growing number of professional scientists – from the three new Queen's Colleges,* for example, also the Royal Dublin Society,* Trinity College Dublin,* the Museum of Irish Industry,* and British universities and institutions. There were lectures on growing and processing flax, on railways and telegraphs, on boilers and bridges and on Belfast's sanitation. James Thomson* described his new water turbine, and John Locke from the London Statistical Society discussed aspects of the 1847 Famine. The association gathered in Dublin again in 1857, but its next visit to Ireland was not until 1874, when it came to Belfast.

The 1874 meeting proved to be one of the BAAS's most controversial. On Wednesday night, August 19th, the association's president John Tyndall* gave his presidential address to a packed Ulster Hall. Tyndall, born in Co. Carlow, was, by then, a director of the Royal Institution in London, and an important figure in the British scientific establishment. Pugnacious by nature, he was regularly embroiled in controversy. The theme for his talk – a history of atomic theory from ancient Greece to the modern day – seemed inoffensive, but Tyndall used the occasion to unleash a major attack on religion. Scientific knowledge, he claimed, was the highest level of human experience, and religion must take second place. "We claim, and we shall wrest from theology, the entire domain of cosmological theory. All schemes and systems which thus infringe upon the domain of science must, insofar as they do this, submit to its control."

Tyndall was backed by various colleagues, notably Thomas Henry Huxley, the man who some years before had championed Charles Darwin. But their views were anathema to much of the audience, many of them clergymen for whom science was a way of exploring the beauty of God's creation. A storm erupted that pitched science against theology, the first time the Darwinian controversy had really hit Ireland. On the following Sunday, the "atomists" and the "blasphemous professors of materialism" were denounced from church pulpits. The Catholic bishops issued a pastoral letter, and Belfast's Presbyterian community organised public lectures in defence of religion. Many of the sermons were later published as pamphlets on both sides of the Atlantic, and the debate ran for years. The BAAS's gathering had one calming influence, however: a long-running strike by Belfast linen workers was resolved during the week, thanks, it is said, to interventions made by economists attending the meeting. The BAAS still holds a science festival each August. It last visited Ireland in 1987 when it again came to Belfast.

1888 – the first successful pneumatic tyre

Small things can sometimes be significant. Like the rubber tubes filled with air which **John Boyd Dunlop** (1840–1921) fixed around the wheel rims on his son's tricycle. The tubes made cycling more comfortable and faster, but their greatest impact was on the new automobile. Without Dunlop's pneumatic tyre, the motor car might never have become popular, the internal combustion engine might not have been commercialised and aeroplanes might never have taken off. John Dunlop was born in Ayrshire, Scotland, where his family were farmers. He trained as a veterinary surgeon and, in 1867, opened a practice in Belfast at Gloucester Street. He was a sound businessman and, within 15 years, had the largest veterinary practice in Ireland, with a staff that included 12 farriers, and a branch in Kirkpatrick,

Co. Down. He also had an inventive streak, and devised various veterinary medicines and implements.

In 1887, his 9-year-old son Johnnie asked if there was some way to make cycling more comfortable. At the time, all wheels were solid, whether on cycling machines or carriages, and most roads were rough or cobbled, so that cycling was a bone-shaking experience. Dunlop realised cushioned wheels would be more pleasant and, according to one report, initially thought of using tubes filled with water.[10] But the family doctor, John Fagan (founder of Belfast Children's Hospital), was familiar with air cushions and suggested using air. Dunlop made a tube from a sheet of rubber, inflated it with a football pump, and tacked it in place around the rim of a wheel. When he rolled the new wheel across his yard he discovered that it travelled much faster than a conventional wheel. Over the next few months, Dunlop fashioned pneumatic wheels for his son's tricycle; the tubes had 1-way valves, and the tube and tyre were tied to the wheel using copper wire. Johnnie tested it in January 1888, and returned ecstatic – the new tyres were both fast and comfortable. In December 1888, Dunlop was granted a patent for his invention. He also coined the word 'pneumatic'. But, contrary to popular wisdom, he did not invent the pneumatic tyre: unknown to Dunlop, that had been invented in 1845 by another Scotsman, Robert Thompson, as an improvement to carriage wheels. Thompson's 'air tyre' was never commercialised, however.

John Boyd Dunlop, who invented the first practical pneumatic tyre and features on the Northern Bank's £10 note. His home on Ailesbury Road, Dublin (now an ambassador's residence) bears a commemorative plaque. Ulster Folk & Transport Museum

Dunlop's first tyres were too bulky for a conventional bicycle, so he commissioned a local company, Edlin, to make special frames. The first official outing was in May 1889 at a Belfast cycling race when, to considerable derision, local racer Willie Hume appeared on a bike with thick and clumsy-looking pneumatic tyres. He won every race, however, beating the Irish champion Arthur du Cros, whose father, Dublin businessman **Harvey du Cros**, was watching.

John Dunlop planned to set up a tyre business in Coventry, the centre of the British bicycle industry. He was persuaded to keep the jobs in Ireland, however, and, in late 1889, agreed to go into business with Harvey du Cros and the Dublin bicycle firm of Booth Bros (though Dunlop had wanted to involve the Edlins). The world's first pneumatic-tyre factory was in Stephen Street, Dublin. News of the new tyre spread – especially when it was banned from some races as being unfair to the competition. But there were technical problems: mending a puncture was tricky and could only be done by an expert; the 1-way valve and the way the tyre was fixed to the wheel were also unsatisfactory. Within a couple of years, other inventors solved these problems – with the beaded tyre, for example, and the 2-way valve. Du Cros quickly bought these inventions, as he set about building an industrial empire. He also acquired a rubber factory, so as to control rubber supplies, and thus began the Dunlop Rubber Company Ltd. In 1892, Dublin residents objected to the fumes from his tyre factory, and the company relocated to Coventry.

John Dunlop resigned from the firm soon after, his gentle manner at odds with du Cros's hard business style. He sold most of his shares in Dunlop Rubber, and became chairman of a Dublin textile company, but retained an interest in cycling, and, up to his death, led the annual Dublin 'old-timers' bicycle outing. The Dunlop Company is now owned by Goodyear Tires. and the Ulster Folk & Transport Museum* have early Edlin and pneumatic bicycles. The Ulster Museum* and the Ulster Folk & Transport Museum* have early Edlin and pneumatic bicycles. Young would-be inventors and entrepreneurs can experiment and even build a robot at the W5 interative science centre. ㊱

The man who discovered carbon dioxide

In 1750, gases were still thought of as mysterious, incorporeal, untamable and quite unlike other substances. **Joseph Black** (1728–99) changed all that. Through a series of careful experiments, he discovered the first gas, carbon dioxide; and he proved that it was present in the atmosphere, that it behaved like an acid, and could take part in chemical reactions, just like solids and liquids. There was no longer anything special about gases, and Black's work paved the way for the discovery of other gases, such as nitrogen, chlorine and ozone (see: **From a gas to a liquid**), and especially Lavoisier's discovery of oxygen. Black also made important contributions to the science of heat, inventing the concepts of specific heat and latent heat, and helping James Watt invent a better steam engine.

Black was born in Bordeaux where his father, an Ulster-Scot, was a wine trader. At 12, he was sent home to attend the Latin School at Ann Street in Belfast, and, four years later, went to university in Glasgow, as Belfast still had no higher college. He was a talented artist and musician and initially studied Arts, until his father persuaded him to choose something more practical and he switched to medicine and chemistry. In 1750, Black began experimenting with magnesium carbonate and calcium carbonate. By rigorously weighing everything, he found that when he heated 1oz (28g) of magnesium carbonate for an hour, it lost $7/12$ of its weight. Something must have been emitted, and Black spent two years chasing this volatile substance. After a series of experiments, he showed that it was a gas, which he called 'fixed air', and which we know as carbon dioxide. Moreover, he proved that the gas was present in the air, and could take part in chemical reactions: a lump of calcium oxide exposed to the atmosphere absorbed carbon dioxide from the air, to become calcium carbonate; and when this was heated, the carbon dioxide could be driven off again, leaving calcium oxide. Black published his study in a thesis that is a classic of early-modern chemistry.

Joseph Black, the man who discovered carbon dioxide, and whose ideas about heat helped James Watt to design an improved steam engine. He is buried at Greyfriars in Edinburgh.

Black also experimented with heating water and ice, and is said to have "waited with impatience for the winter" so that he could freeze water. He found that, as ice melts, it absorbs heat, yet its temperature does not rise; he concluded that the heat mixed with the ice and became hidden or 'latent'. This **latent heat** is used to change the state from a solid to a liquid; a similar phenomenon happens when a liquid boils to become a gaseous vapour. Black also discovered that, for every substance, there is a specific amount of heat, or a **specific heat**, that is needed to increase its temperature by 1 degree. James Watt, who was an instrument maker at Glasgow University at the time, was friendly with Black and their discussions about heating water helped Watt to design an improved steam engine. Black was professor of chemistry at Glasgow University, and later Edinburgh University, and was famous as a popular lecturer – his talks were illustrated by ingenious demonstrations and attracted students from around the world. He was a noted physician, who reputedly did not charge poor patients, and his friends included the philosopher David Hume, the economist Adam Smith, the geologist James Hutton and others of the Scottish Enlightenment.

A turbulent engineer

In the late 1800s, engineers were able to design better boilers, propellers, turbines and pumps – indeed, better machinery all round – thanks to the work of Belfast-born **George Osborne Reynolds** (1842–1912). Reynolds, one of the great theoretical engineers of the 19th century, is remembered for his work on heat, turbulent fluids, and

George Osborne Reynolds: his work on lubrication and fluids helped engineers to design better machines.

lubrication. Though he studied mathematics at Cambridge, Reynolds always had a keen interest in mechanical things and, at 26, was appointed professor of engineering at Owens College (later the University of Manchester), a position he held until he retired in 1905.

In the 1870s, screw propellers were taking over from paddlewheels on steamships. Often, although a propeller might race, the ship would go no faster. Reynolds was the first to explain that this was because of cavitation: a fast spinning propeller creates a cavity in the fluid that, ironically, slows the ship down.[11] Reynolds pioneered the technique of tank-testing, with small-scale models of ships, and using dyes to reveal the flow patterns in the water. Significantly, he developed equations that enabled engineers to predict the full-scale performance, based on these tests, and his cheap and effective technique is still used. (Charles Parsons* developed cavitation studies in the 1890s when he designed an ultra-fast, turbine-powered ship.)

Reynolds also pioneered the study of fluid mechanics, and especially the complex area of turbulent or unsteady flow. He formulated a relationship, the **Reynolds number**, to predict the precise point where a flow would become turbulent. His number relates the velocity, density and viscosity of the fluid, and the diameter of the channel or pipe; if the Reynolds number is greater than 2,000, the flow becomes turbulent, but under 2,000 the flow remains steady. Reynolds's other contributions include a theory of lubrication, and a mathematical representation of what is happening when a thin film of gas or liquid lubricates a surface. This greatly improved the design of moving parts in machinery, and helped drive Britain's industrial development in the late 19th century.

The ingenious Baron Kelvin

An early photograph of the great 19th-century scientist and inventor William Thomson. There is a commemorative statue to him in Belfast Botanic Gardens; the plaque reads: "He elucidated the laws of Nature for the service of Man." He is buried beside Sir Isaac Newton in Westminster Abbey.*

William Thomson (1824–1906) was an ingenious inventor, a theoretician and a towering figure in 19th-century science. His great success as an inventor was designing the equipment that made it possible to send and receive telegraphic signals over very long submarine cables; this spawned a telecommunications revolution, and brought him fame and fortune. As a scientist, he is remembered for his work on the nature of heat and energy. Thomson was born at College Square, opposite the Belfast Academical Institution where his father was the mathematics professor; William and his brother James, who became a pioneering engineer (see: **From turbines to alcoholic 'legs'**), were taught at home by their father. William was especially precocious, and at the age of eight reputedly sat in on some of his father's lectures. In 1832, his father was appointed to Glasgow University, and the family moved to Scotland. Thomson began attending lectures at the university when he was just 10, later studying at Cambridge and Paris. At 22, he returned to Glasgow and, thanks to his father's intervention, was appointed professor of physics. He established the first physics laboratory in Britain and stayed at Glasgow until his retirement.

In the 1840s, phenomena such as heat, mechanical movement, and light and electro-magnetism were all thought to be separate. One of Thomson's great insights was to realise that these were all different forms of energy. Collaborating with James Joule, and subsequently as a mentor to James Clark Maxwell, Thomson helped formulate theories to unify these phenomena, and pave the way for the 20th-century theories of Einstein and others. With Joule, for example, he explained that, below a certain temperature, a gas will cool as it expands, because the gas molecules expend energy moving apart. This Joule-Thomson effect was later used to cool and liquefy gases. In 1848, noting that a gas loses $1/273$ of its 0°C-volume with every 1°C drop in temperature, Thomson suggested that it reaches a thermodynamic absolute at -273°C, and that this point be called **absolute zero**. When his idea was eventually adopted, the scale was named kelvin in his honour (after the title he took in 1892, from the Kelvin river in Glasgow); 1 kelvin (K) is equivalent to 1°C, and 0°C (the freezing point of water) corresponds to 273K. Thomson's book *On the Dynamical Theory of Heat* (1851) is a classic, and includes his formulation of the Second Law of Thermodynamics – essentially that to get useful work (energy) out of a system, there must be a temperature difference.

Studying electricity and magnetism, Thomson realised the current flowing through a wire was similar to heat energy flowing through a conductor. His work brought him to the attention of the consortium planning the first long-distance submarine telegraph cable,* from Valentia, Co. Kerry to Newfoundland. After the first cable failed in 1857, Thomson proved that the problem was the high-voltage current, which had burned the cable's insulation. He redesigned the cable, switched to a very low-voltage system, and invented a sensitive receiver (a mirror galvanometer) capable of detecting the tiny current. This was crucial to the success of the second cable in 1866, and Thomson was knighted for his contribution. He became a consultant for subsequent submarine cables and made a fortune from his invention. He patented over 70 other inventions and set up a company in Glasgow to make his instruments. Interested in sailing since his student days, he produced several nautical devices, notably an improved compass,* a depth-sounding device, and a calculating machine that could predict the tide for any time and place. The tide-predictor was a complex engine, however, and only a few were made,

though one was used in 1944 to predict the Normandy tides for the D-Day landings.*

Thomson was an opinionated man with strong views. An ardent imperialist, he relished his role in creating the telegraph network that cemented the British Empire; he was firmly opposed to Home Rule for Ireland, to admitting women to Cambridge and to Darwin. He once calculated that the Earth was at most 100 million years old, based on the rate at which the molten planet would have cooled. This was nowhere near long enough for evolution, and his result was a major blow for Darwin. Thomson also found it hard to believe in radioactivity when it was discovered in 1896. Radioactivity, however, is what generates the heat at the Earth's core. Had Thomson been willing to take it into account, he might have calculated a more accurate age for the Earth, of 4.6 billion years.

1963 – the arrival of the killer worm

An unusual worm was spotted for the first time in a Belfast garden in 1963. Irish earthworms have a segmented body, but the stranger did not. It was also slightly flattened, could stretch to nearly a foot long, and was distinctively coloured with a purple-brown back and pale yellow belly. The creature was identified as a **New Zealand flatworm**, *Artioposthia triangulata*. It had not been seen outside New Zealand before, but soon it was also spotted in England and Scotland, where it may have been before reaching Northern Ireland. It has since spread to elsewhere, including Iceland, and across Ireland, having recently reached Co. Kerry. The worm presumably arrived with imported plant material, possibly as early as the 1950s. It was initially a curiosity, until it emerged that the New Zealand flatworm eats earthworms.

Ireland has over a dozen different types of earthworm. They digest and bury rotten vegetation, and aerate the earth, and soil fertility depends on them. There can be over 400 earthworms in a square metre of ground, but researchers at Northern Ireland's Department of Agriculture found that six New Zealand flatworms could dispose of all 400 in just one year. The flatworm wraps itself round an earthworm and secretes digestive enzymes, then eats the resulting soup; the whole process takes less than an hour. In New Zealand, the flatworm numbers are controlled by natural predators and the warm weather – it cannot survive temperatures above 20°C – but nothing in Ireland preys on the flatworm and it thrives in our mild climate.

During the 1990s, there was growing concern about the threat to native earthworms and farming. Gardeners in badly affected areas had to aerate their lawns mechanically, and initial studies suggested the flatworm could cut farm and grassland productivity by over 30 per cent. Recent studies, however, indicate that earthworm populations recover from a flatworm attack, though never to their original levels, so perhaps some balance will eventually be struck. Meanwhile, three more flatworm species have arrived from the southern hemisphere: *Australoplana sanguinea*, *Kontikia andersoni* and *Rhynchodemus sylvaticus*.

BELFAST

1896 – Ireland's first medical X-ray

Two months after X-rays were discovered, they were used by a Belfast doctor to diagnose an injury. Wilhelm Röntgen discovered X-rays in late December 1895, and the news soon spread around the world. On January 17th, 1896, a Dublin paper, the *Freeman's Journal*, reported: "Newly discovered light: some striking experiments." The first Irish demonstration of the powerful new technique came a few weeks later, when Prof John Joly* gave a public lecture in Dublin, and showed X-ray images he had taken revealing spectacles hidden in a case, and the bones of his hand. The first diagnostic use of X-rays in Ireland took place some days later in Belfast, when Dr Cecil Shaw used the technique to locate a bullet that had lodged in a man's finger when he shot himself during revolver practice. Dr Shaw presented a public lecture, reported in the *Lancet* of February 28th, 1896, at which he showed a 'radiograph' of the man's injury.

BELFAST CASTLE

The caves on the hill

To the northwest, Belfast is hemmed in and overlooked by high ground and a line of hills. Collin, White Hill, Black Mountain, Divis, Wolf Hill and Cave Hill… these are the southern edge of the Antrim basalt plateau. Beneath the basalt lie older beds of chalk and limestone and the hillsides are pockmarked with limestone quarries. Most of the quarries are closed, but at Limestone Road on **Cave Hill** you can see the remains of an old quarry railway. Cave Hill is the most distinctive of the Belfast hills, and its brooding, craggy peak is known as Napoleon's Nose. Numerous paths lead up the hill and, from the top, there are panoramic views over Belfast city and lough. Below the craggy summit are five small caves, or openings, the lowest of which is accessible. There is no evidence of any occupation in the caves, and all are at least partly artificial – possibly prehistoric mines. Cave Hill is part of Belfast Castle demesne, where the varied habitats include parkland, moorland and wooded hillside. (37) The woods are home to what may be **Ireland's rarest plant**: *Adoxa moschatellina* grows in just a few patches on Cave Hill and produces small green flowers in April.

Exotic creatures, such as Falkland penguins and Malayan tapirs, can be seen at neighbouring Belfast Zoo. The zoo is part of an international network that co-operates to breed rare and endangered species and, in the 1990s, Belfast reported successes with lion-tailed macaques, bongo antelope and cheetah. (38)

Antrim is rich in geological wonders: Ireland's one and only dinosaur; an amazing collection of rare minerals; brilliant-white chalk deposits; Ireland's only salt mine; and an exhilarating volcanic landscape, that includes the world-famous Giant's Causeway. The ideal approach to the causeway is via the breathtaking Antrim Coast Road, which is both an engineering *tour de force,* and a stunning geological excursion, skirting the coast and connecting the scenic nine glens of Antrim. Antrim (*Aontroim*) has been variously translated as the 'solitary farm/home/tribe'.

LISBURN

Fine linen and damask

When the French revoked the Edict of Nantes in 1685, little did they realise the effect it would have on Irish industry. Many of the Huguenots, forced to flee religious persecution in France after that date, settled in Ireland, especially after the Protestant William of Orange's victory. Among the various commercial enterprises they started was the Ulster linen industry. People had been making linen in Ireland since prehistoric times, but it was a Huguenot refugee, **Louis Crommelin**, who started a linen industry proper at Lisburn in the late 1680s. He was given government grants to develop his business and to induce skilled Flemish linen workers to settle at Lisburn. (At the same time the Irish woollen industry, seen as a major threat to England's woollen industry, was killed off by an Act of Parliament that imposed severe taxes on Irish wool exports.)

Crommelin's Flemish workers introduced several new and improved techniques, notably the spinning wheel –

previously, linen was spun in Ireland using the distaff and spindle. Throughout the 1700s, the Irish linen industry developed and, by 1750, the world's largest linen sailcloth factory was at Douglas in Co. Cork. But the industry became increasingly concentrated in Ulster, thanks in part to the Tyrone coalfields, which supplied coal for the mill steam engines. By 1800, 80 per cent of Irish linen was made in Ulster, mostly in the 'linen triangle' of Belfast–Dungannon–Armagh; moreover, it accounted for over half of all Irish exports. There were three broad classes of linen: natural (brown colour), dyed (black) and bleached (white). In Ulster, each district also produced a distinct type of fabric: Armagh, for example, was known for its coarse linen, and Ballymena for its "fine yard-wide plain" linen.

The linen industry changed the fabric of Irish life. Several new linen **mill towns** were founded, among them Fintona in Co. Tyrone, and Emyvale in Co. Monaghan, and bleach greens and mills became a common feature. As the industry became increasingly mechanised, it spawned a heavy engineering sector in Ulster. Likewise, the linen-bleach works fostered a chemical industry, and Josias Gamble,* who was among those who kick-started the English chemical industry, began by supplying chemicals for Ulster bleach works. Country markets and fairs prospered, thanks to the linen trade, and rural society became increasingly 'monetised'.

Yet even in 1820, when output was at 55 million yards, linen was largely a cottage industry, and much of the work was done by hand. The first Irish machine-spun linen was produced in Co. Down in 1805, but manual labour was still cheap and plentiful and the machinery was not widely adopted for some time. By 1850, though, nearly all production had moved to factories and, in 1873, there were 20,000 linen power looms in Ireland. By then, the Irish flax crop had peaked at 112,500 hectares, and Belfast had become the largest linen-making centre in the world. The industry received a major fillip when the American Civil War cut the supply of cotton to Europe.

Flax remained a crop for small farmers, however, despite the best efforts of organisations like the Linen Trade Board and the RDS,* which regularly looked to improve the quality of seed used, and brought in expert growers from abroad to instruct Irish farmers. It was a labour-intensive crop that had to be pulled by hand, and it suffered in bad weather. By the time pulling machines became available in the 1940s, the Irish flax crop had virtually disappeared. Ulster still has a vibrant linen industry, and several factories offer tours to visitors, but they all now use imported yarn. The Irish Linen Museum is in Lisburn. ㊴ See also: Wellbrook Beetling Mill* and Benburb Heritage Centre* (both in Co. Tyrone); William Clarke & Sons* linen factory (Co. Derry); and Ferguson's Linen Centre* (Co. Down).

Why a compass needle works

The first person to explain why a compass needle works was a noted physicist from Magheragall, **Sir Joseph Larmor** (1857–1942). Larmor bridged the 19th-century world of classical physics and the new 20th-century world of relativity and quantum mechanics. He worked on electricity, dynamics and thermodynamics, helped to develop a scientific understanding of the electron,* even before the particle's existence was widely accepted, and paved the way for the theories of relativity and quantum mechanics. Larmor was educated at QUB* and Cambridge, and was professor of natural philosophy (science) at Queen's (University) College Galway* for five years, before moving to Cambridge in 1885. There he developed his theories about matter, electricity and the ether.

Since Aristotle, scientists had believed in an ether, a transparent, weightless substance, thought to permeate all space and matter. In the 19th century, the ether was invoked to explain how light and other electromagnetic waves were transmitted through air and space. By the end of the 19th century, however, there was growing evidence that the ether did not exist. Larmor continued to believe in the ether, but rather than take a classical mechanical view, he saw it as electrical particles in motion, an idea he developed in his classic book *Aether and Matter* (1900). Despite adhering to an increasingly out-of-date idea,

Making linen

Converting flax plants into finished linen fabric calls for several processing steps. The flax seed is sown in spring and the plants harvested three months later. The next stage is **retting** – soaking the flax plants in ponds for two weeks to loosen the fibres – followed by **scutching**, where the fibres are separated from the rest of the plant material, so that only the useful fibres (about 10 per cent of the plant mass) are transported to the mill for **spinning**. Mill spinning entails several processes, such as **carding**. In the 1820s, **wet spinning** was introduced: the fibres are first passed through hot water, to remove any gummy plant residue; the result is a finer yarn, and this new technique was important in fostering a large-scale linen industry. Next, the yarn is **woven** and then, if desired, the cloth is **bleached**. Initially, bleaching was a time-consuming process, entailing several cycles of soaking, treating, boiling and drying but, in the 19th century, this was increasingly handled by specialised bleach works. Finally, comes **beetling**,* where the linen is hammered repeatedly to give it a smooth sheen.

Larmor made several significant contributions, and he gave his name to Larmor's precession, Larmor's frequency, Larmor's theorem and Larmor's formula, and was the first to calculate the rate at which energy is radiated by an accelerating electron.

In 1919, Larmor suggested that the Earth contains a self-exciting dynamo: that the Earth's inner core is electrically conducting and in constant motion, and that this generates both a current and a magnetic field. Larmor's explanation was the first to account for the behaviour of a compass needle, which essentially aligns itself along the Earth's magnetic field. In 1903, Larmor succeeded another noted Irish scientist George Gabriel Stokes* as Lucasian professor of mathematics at Cambridge University, a prestigious position once held by Isaac Newton and today held by Stephen Hawking. Larmor also found time to be the university's MP at Westminster from 1911–22, and to edit the works of several great 19th-century scientists, among them Kelvin,* Stokes and Cavendish. A crater on the Moon is named in his honour.

CRUMLIN

Flying history

The first Irish man to fly a plane was Harry Ferguson* from Co. Down, later famous as the man who revolutionised tractor design. His plane is not at the Ulster Aviation Society's heritage centre – though the Ulster Folk & Transport Museum* has a full-scale replica – but the aviation society has several other historic aircraft, including a Buccaneer, a Wildcat and a Seahawk, along with other aviation vehicles and aircraft engines. The heritage centre is at Langford Lodge Airfield, a former World War II airfield and depot. ④ Built in 1940, Langford was used by the US Airforce, and by the Lockheed Corporation which had an engineering research centre there until the airfield closed in 1945. The site was bought in 1958, by the Martin-Baker company – which was begun by another Co. Down entrepreneur, James Martin,* who invented the ejector seat. The company leases some buildings to the aviation society, including the original World War II control tower.

CRUMLIN

Antrim's brown coal

There are vast deposits of **lignite** around the shores of Lough Neagh.* The lignite formed from the rotting remains of swampy vegetation that grew by the lough shore 25 million years ago. Fossil spores in the lignite reveal the presence of ferns, cypresses and palm trees, and suggest that the climate then was warm and frost free. Lignite, or 'brown coal', is intermediate between peat and black coal, and partially petrified. It is used as a fuel, and as a raw material for the chemical industry (to make polishes and waxes, such as those in carbon paper). The Lough Neagh lignite can be seen in places, notably around Crumlin and Glenavy, but the full extent of the deposit only became known during commercial prospecting in the 1980s. There are over 500 million tonnes of recoverable lignite in the area, making it the largest lignite deposit in these islands. Nevertheless, a lignite mine was considered uneconomic at the time, likewise a proposal to build a lignite-burning power station on the site. There are smaller lignite deposits trapped within the basalt along the north Antrim coast. This lignite formed over 60 million years ago from the rotting remains of vegetation that grew on the basalt during the quiet intervals between volcanic eruptions. These deposits were mined sporadically, notably at Ballintoy, where old mine shafts can be seen in the sea cliffs. The Ballintoy mine closed in the mid-1700s after the lignite there caught fire.

RANDALSTOWN

Icelandic ash in Irish bogs

At Sluggan's Bog near Randalstown, scientists have found evidence of volcanic eruptions that happened on Iceland thousands of years ago. The evidence, first discovered in the early 1990s, is being combined with pollen profiles* and tree-ring patterns* to shed light on Ireland's prehistoric environment. Since Irish bogs started growing some 7,000 years ago, there have been several major volcanic eruptions on Iceland. These eruptions shot microscopic ash particles (tephra or pumice) into the atmosphere. Tephra is full of gas bubbles and is so light it can be carried long distances on the wind, until it is washed down by rain. Any tephra that falls on a bog will become trapped within the peat.

Bogs are wonderful archives. Anything that rains down on them – pollen grains, insects, atmospheric dust – is stored in neat layers, and in chronological order. Taken together, this can provide valuable information about the environment of the time. Valerie Hall and John Pilcher from QUB, who study Ireland's prehistoric climate, have found tephra layers in bogs at Randalstown and in the Lower Bann valley. By analysing the chemical composition of the tephra, and the gases trapped in the ash particles, they have identified some of the volcanic eruptions: the eruption of Veidivotn in the 9th century AD, for example, and of Hekla in the 2nd and 3rd millennia BC.

A separate source of information on Ireland's prehistoric climate comes from tree-ring patterns compiled by Prof Mike Baillie, also from QUB. Significantly, the tree rings reveal a period of poor growth starting in 2354 BC, and again in 1159 BC, which Baillie suggests were due to the Hekla eruptions: a major eruption can throw massive amounts of dust into the Earth's atmosphere, blocking sunlight, and resulting in cold, dark conditions, until the dust is eventually washed down. So, pollen grains trapped in the bog confirm the poor growing conditions, the tephra particles reveal the culprit and the tree rings provide the precise calendar date for the eruption.

TEMPLEPATRICK

Calling a spade a sophisticated agricultural implement

For centuries, the spade was the main tool on Irish farms. The small size of most Irish holdings, and the ready availability of cheap labour, meant spades remained popular well into the 20th century, even when farm machinery became more widely available. But not just any spade would do: there was a tremendous diversity of spades, each designed for a particular soil type and job. Spades for digging trenches, for turning sods and for making cultivation ridges. A lightweight foot slane to cut turf on deep raised bogs; but a breast slane on blanket bog, where it was easier to get a firm footing. The Office of Public Works also designed spades for its drainage programmes, including "a narrow spade for bottoming small drains" and an "earth scoop for cleaning drains". There were regional variations too, such as the 2-pronged *gabhal gob* which was unique to Bangor-Erris in Co. Mayo. In all, there were over 1,000 different spades in use in 19th-century Ireland.

They are making spades again at Patterson's spade mill in Templepatrick.
© National Trust

Spades were made by local blacksmiths and, increasingly during the 19th century, at specialist **spade mills**. The mills called for powerful equipment to drive the massive hammers, rollers, cutters and sharpening blades used in working the metal, so most such mills were in Ulster, Dublin and Cork, where there was a tradition of heavy engineering. Each mill and forge kept patterns for the spades they made, also the designs circulated by the OPW. One Co. Tyrone spade mill had 230 different spades in its portfolio, according to Dr Jonathan Bell, a farming historian with the Ulster Folk & Transport Museum.* This kind of product range was possible only in the days before mass production. Ireland's last traditional spade mill was Patterson's of Templepatrick, which was working until 1990. Restored by the National Trust, it is again in production making turf and garden spades with a water turbine, furnace, massive hammer and blades – the last place in the world where you can buy a hand-crafted spade.
(41) (See also: Monard spade mill,* Co. Cork.)

Sketching Ireland

Bats – and then there were nine

A talented artist and master draughtsman, who produced thousands of sketches and paintings for the Ordnance* and Geological* surveys of Ireland in the mid-19th century, is buried at All Saints' Parish Church in Antrim town. **George Victor du Noyer** (1817–69) died of scarlet fever, while working in Antrim for the Geological Survey. His family were with him at the time, and his 5-year-old daughter Fanny died of the fever within a day of her father.

Du Noyer, born in Dublin of Huguenot descent, was an inveterate sketcher, even as a child. He studied art under George Petrie, a noted Irish antiquarian who later introduced him to the Ordnance Survey. From 1835, du Noyer worked for the survey, until the antiquities division was closed in 1842. He produced detailed drawings of a tremendous range of subjects including landscapes, antiquities and fossils. In 1847, du Noyer joined the Geological Survey and spent the next 20 years making maps and detailed geological drawings, including the geological structures that were exposed as the new railway lines were laid. He mapped all of counties Wexford, Waterford, Cork and Kerry, and the maps he produced are still of use because of the fine detail they record.

Du Noyer exhibited at the Royal Hibernian Academy during his lifetime and, in 1995, a special exhibition of his watercolours was held at the National Gallery in Dublin. Several fossils discovered in Ireland were named after him, including a redwood tree, *Sequoia du noyeri*. Various institutions own work by du Noyer, among them the National Gallery of Ireland,* the National Botanic Gardens* in Dublin and the Royal Society of Antiquarians. There is a commemorative plaque to him at All Saints' Church. Illustrations in this book by du Noyer include an esker at Clonmacnoise,* the Cloughlowrish erratic* and Glendalough valley.*

You might have thought that, by the end of the 20th century, scientists would have discovered all of the mammals living in Ireland. Not so. Until 1992, seven species of bat were known in Ireland. Then zoologists found two more, bringing the total to nine, and there are unconfirmed recordings of a tenth. The first discovery was unusual: in 1993 a zoologist at Bristol University discovered that the common pipistrelle bat – a tiny creature smaller than your thumb – was in fact two separate species. One – *Pipistrellus pipistrellus* – produces echo location signals at 45kHz, and the other – *P. pygamaeus* – at 55kHz. Although they can be distinguished only by sonar detection equipment and they do not interbreed.

Then, in May 1997, zoologists from QUB discovered a colony of Nathusius pipistrelle breeding in an old stable on the outskirts of Antrim town. It was the first time this bat had been found in Ireland, and it brought the species tally to nine. Nathusius pipistrelle have since been found at other sites around Lough Neagh, but not, so far, in the Republic of Ireland. Also in 1997, Scandinavian zoologists, who were in Ireland recording bat signals, detected what they believed was the barbastelle bat. Bat signals are notoriously difficult to study, however, and until a barbastelle bat is seen here, the Scandinavian recording does not count.

Ireland also has internationally important populations of Leisler's bat and the lesser horseshoe bat. The other species found here are the brown long-eared, the whiskered, Daubenton's and the natterer. All are relatively small and insectivorous. A common pipistrelle can eat 600 midges in an hour, Leisler's bats feast on dung flies, while Daubenton's bats fly low over water and catch aquatic

The lesser horseshoe bat, with its characteristic horseshoe-shaped nose plate. Bats are protected in Ireland and should not be disturbed if found.
© Guy Troughton/ Vincent Wildlife Trust

insects with their large feet. The Irish bats are nocturnal, and use echo location to navigate and locate insects. They hibernate in winter, in quiet roosts such as old ruins where they are unlikely to be disturbed. They are often greeted with fear and loathing, but these reactions are unwarranted, and there is no truth in the myth that they will fly into your hair or bite you.

Bann clay

For over a century, an unusual white clay was dug from around the shores of Lough Beg and in the Lower Bann valley, and processed at factories in Toome and Newferry for a range of industrial uses. This Bann clay is made almost entirely of glassy skeletons of microscopic fossil plants, or **diatoms**, which lived in vast numbers in the fresh water there 7,000–4,000 years ago. The water level was higher then, and Lough Neagh* was much bigger than today. As well as being commercially useful, the clay is scientifically and archaeologically important: it records changing water levels and climate; and hidden in its layers are artefacts left by early hunter–gatherers.

While they are alive, diatoms contain a green gel rich in chlorophyll (the molecule which plants use to convert sunlight into energy). When they die, however, the diatoms turn white. Dr David Jewson from Ulster University's

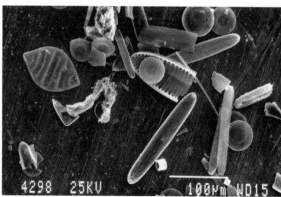

Fossil skeletons of the tiny plants found in diatomite Bann clay (this sample from Toome was photographed under an electron microsope).
© Jewson & Lowry/Ulster University

freshwater laboratory, who has been studying Bann clay, says the diatomite layers were laid down each winter when the river flooded. The thickest and most commercial deposits were in the old flood plains, notably west of Toome and between Newferry and Portna. In summer, when the floods retreated, prehistoric hunters came onto the mudflats to trap fish; artefacts they left behind were later buried in the clay. The water levels and flood patterns of the region changed dramatically about 4,000 years ago, at the same time as Ireland's bogs started to grow, and very little diatomite clay formed after that. Diatoms still live in the water, but drainage works in the 1860s and 1950s mean the river seldom floods now, and pure diatomite clay is no longer forming; instead, the diatoms join the general lake-bottom mud.

Diatomite clays are abrasive, porous, highly absorbent and heat resistant, and were used in various industries: as an abrasive powder (in polish, face powder and toothpaste); in firebricks and to line furnaces; as a filter (in beer and sugar production); and as a stabilising absorbent in explosive artillery shells. Factories drying and processing the clay operated for many years in Toome and Newferry, where drying pans and factory sheds can still be seen. Clay dug from Lough Beg was processed at Newferry and used in cosmetics. Toome clay was mostly used to pack artillery shells and production there peaked during World War II at nearly 16,000 tonnes per annum, when 80 people were employed cutting the clay by hand (in later years a mechanical digger was used and fewer people were employed). The last factory closed in 1997 and the remaining deposits are now protected, because of their archaeological and scientific importance, and because freshwater diatomite clay from this period is rare: most commercial deposits around the world are marine in origin and millions of years old. Toome is also home to Europe's largest eel fishery (see: Kinturk,* Co. Tyrone).

The rash on the rocks

Scraith chloch, the Irish for **lichen**, literally means 'rock rash' or 'rock scab'. It is a fitting description for at least some Irish lichens, which form colourful crusts on boulders, gravestones and rocky beaches. Not all lichens are found on rocks, though, and many grow on trees, festooning the branches with grey-green whiskers. Lichens are unusual plants, for they are a co-operative – a symbiotic relationship – involving an alga and a fungus. They are slow growing, and the diameter of a lichen crust can be used to estimate how long a stone has been exposed to the elements. Lichens are also sensitive to air pollution and their presence or absence is a useful indicator of air quality. For years, however, they were greatly overlooked, on account of their low profile and because they are difficult to study – a microscope is needed to reveal the detail of their structures.

Matilda Knowles, who added greatly to our understanding of Irish lichens. National Botanic Gardens

Matilda Knowles (1864–1933) from Ballymena greatly added to our understanding of Irish lichens.[12] After studying at the Royal College of Science* in Dublin, Knowles joined the staff of the National Museum, where she worked on its herbarium, or collection of dried botanical specimens (the herbarium is now maintained by the National Botanic Gardens* in Dublin). Realising that lichens had long been ignored, Knowles undertook a detailed study of marine lichens at Howth in Co. Dublin. She discovered three species there that were new to science, and 25 that had not previously been found in Ireland. The next 15 years were spent comprehensively surveying all Irish lichens. She added over 20 species to the known list, bringing the total to over 800. When the RIA* published her findings in 1929, her study was described as "one of the finest pieces of work ever carried out in any section of the Irish flora".

Ballymena is also home to Northern Ireland's new environment information centre which is built on a 60-hectare site that floods several times a year. But, instead of draining the site, **ECOS** exploits its damp setting: wet meadows and a lake have been created, along with reed beds to treat the centre's waste water and sewage. As befits a centre designed to demonstrate the principles of sustainable living, nothing is wasted: sludge from the reed bed is spread as a fertiliser on the centre's willow fields; and the willow is then coppiced* at 3-year intervals, and burned on-site to generate electricity. ECOS plans to generate two-thirds of its energy needs, from a wind turbine, a photovoltaic system, solar water heating and the willow biomass. Set on the River Braid flood plain, the state-of-the-art centre is built on the site's highest ground, and constructed from materials chosen for their environmental impact. There is a rich mosaic of habitats now, and the number of bird species has doubled since the park was created. ㊷

Ireland's only salt mine

This corner of Ireland was once a sandy red desert with a salty 'dead sea'. Eventually, the desert dunes turned to sandstone and, 200 million years ago, the sea began to evaporate. Initially it formed a chain of salt lakes similar to those found today in Utah, USA. Later, the lakes evaporated leaving thick deposits of rock salt and gypsum. Subsequent geological events covered the salt with a protective cap of impervious rock, and it was not discovered until 1839 by miners prospecting for coal. It has since been mined at various sites, first at Duncrue and Eden near Carrickfergus, where miners chipped out the rock salt; this was then washed and purified to make table salt. During the first half of the 20th century, the salt was solution mined at Red Hall, by the shore of Larne Lough: hot water was forced down boreholes, and the resulting brine pumped out; the brine was evaporated, again to make table salt. Red Hall mine closed in 1958.

Since 1965, the Irish Salt Mining Company has been drilling and blasting at Kilroot, producing 500,000 tonnes of salt each year; it is left unwashed and used to de-ice roads. The mine is 300 metres underground and stretches inland for nearly 4km, following a deposit that continues at least to Larne and probably further under Antrim's basalt plateau. The rock salt is crushed underground then carried by conveyor belt from the mine straight to a deepwater jetty in Belfast Lough. The deposit consists of five stacked beds of salt, separated by thin layers of red mudstone. Each mudstone layer represents an ancient flooding event, when sediment was washed into the prehistoric salt lake. The rock salt is also stained red by sand blown onto the lake from the surrounding desert. At Kilroot, the salt layers are 5–50 metres thick, but at Larne they are up to 400 metres thick. There is probably a considerable salt deposit under the Antrim plateau, but it is too deep to be economic. The gypsum deposit mined at Kingscourt,* Co. Cavan, was part of the original salt-lake chain.

CARRICKFERGUS

Making gas from coal

In these all-electric days, it is hard to imagine that, 50 years ago, people in towns around Ireland used gas to power fridges, irons and even hair dryers, as well as for the more conventional uses in lighting, cooking and heating. The gas was made from coal at local chemical works. During the 1970s, however, oil, electricity and natural gas began to replace coal-gas, and the gasworks either switched to making bottled gas from oil, or closed. Ireland's only surviving coal-gas plant is at Carrickfergus, a unique industrial heritage site with many artefacts from this vanished industry.

The Carrickfergus retorts, where coal was heated to generate gas. This historic chemical plant has been restored and is open to the public, with an exhibition that includes gas-powered domestic appliances.
© Brian McKee

Coal-gas was discovered in the late 1600s by the Rev John Clayton.* A former minister at St Michan's parish, Dublin, Clayton found that when he heated coal, flammable gases were given off. The discovery was not commercialised until 1812, when it was used in London to make gas for street lighting. By 1821, there were four coal-gas plants in Ireland. The plant at Carrickfergus opened in 1855 and, by 1900, there were hundreds of similar works around the country, some private, others publicly owned. All used imported low-ash coal as their starting raw material. This was placed in special airtight metal furnaces, or retorts; fireboxes beneath the retorts were stoked with coke, and the whole assembly was heated to 1,000°C. As there was no air in the retorts, the coal could not ignite; instead, over about five hours, it gave off a dirty mix of gases, notably hydrogen, methane and carbon monoxide. The gases were collected, treated and purified, and the methane piped to customers.

The process generated a variety of waste products and, during the 19th century, this spawned a veritable chemical industry. The waste gases, for example, included hydrogen sulphide, which could be used to make sulphuric acid, and ammonia, which was used in making fertiliser. The tar-like residue was collected and distilled, to yield tar and pitch for road making, creosote for wood preservation, naphthalene for mothballs and various dye stuffs, even some early drugs were derived from coal-gas waste. Finally, the heating had converted the starting coal to coke; this was used to heat the next batch of coal, so the process was economic and simple, albeit dirty and smelly.

At Carrickfergus, according to industrial archaeologist Fred Hammond, every 200kg of coal generated 85 cubic metres of gas, enough for 15 customers for a day. Initially, the gas was used for lighting – gaslights were cleaner, cheaper and safer than oil lamps or candles, and remained popular long after the electric light bulb was invented in the 1880s. In 1900, Carrickfergus had 172 customers, only four of whom had gas stoves; by 1913, however, 580 customers were using the gas and 45 had stoves. In 1960, Carrickfergus had 2,000 customers, but the following year the plant closed – labour costs had risen, and coal was now expensive relative to oil. The site was used until 1987, however, to store and distribute gas made at an oil-gas plant in Belfast. Now maintained by a preservation society, the site boasts three retort houses, with 36 retorts; a 1940s gasholder, or gasometer; and a World War II air-raid shelter. [43]

The Gobbins path

In the early 1900s, a path along the sea cliffs of Islandmagee was a more popular tourist attraction than even the Giant's Causeway.* The path, blasted through the basalt at the Gobbins, had the advantage of being close to Whitehead train station, and it became a popular excursion from Belfast. Indeed, the local railway company developed Whitehead as a resort and commuter town, and built the path as a tourist attraction. The Gobbins Path was the brainchild of Wexford-born railway engineer **Berkeley Deane Wise** (1853–1909), who joined the Belfast & Northern Counties Railway (BNCR) in 1888. Under Wise, the BNCR built tea rooms, promenades, bandstands and hotels along its network, and even had plans for golf courses. The company also leased Glenariff, one of the Antrim glens, and built a tourist railway up the valley in 1890.

Whitehead had been a small village until the railway arrived in the 1870s, when it became popular as a seaside resort. In the 1890s, the BNCR added a promenade, and granted seven years' free travel to anyone buying a house in the town. Then Wise conceived of an even more daring plan: a 5-km walkway along the spectacular sea cliffs and caves at the Gobbins, just north of the town. Until then, Islandmagee had chiefly been known for a massacre in 1641, when local Catholics were allegedly chased over the cliffs and, in 1711, for Ireland's last witchcraft trial. Work on the path began in 1901: tunnels were blasted through the basalt; and iron footbridges, including one unusual tubular section, were built in Belfast, brought out by barge and lifted into place to span vertiginous gullies and sea stacks. For the 1902 opening, it was advertised as: "New cliff path along the Gobbins Cliffs, with its ravines, bore caves, natural aquariums… no parallel in Europe (admission 6d)". The British Association for the Advancement of Science* reported that "there is, in short, nothing like the Gobbins Path anywhere in the world". The path was never finished, however – only the first 3km were constructed – and BNCR's interest waned after Wise retired in 1906 due to ill health. By the 1940s, the iron footbridges had corroded and sections became unusable. Boat trips can bring visitors to see the sea cliffs and the remains of the path. Gobbinsite, a rare zeolite* mineral, was discovered in basalt at the Gobbins in 1982.

Whitehead railway station is today the headquarters of the **Railway Preservation Society of Ireland**. The society has a workshop there with nine vintage locomotives of various classes and ages. All are in running order and many have featured as extras in movies. In summer, the society runs frequent short steam trips at Whitehead, and 1-day excursions to Portrush, Rosslare and elsewhere. [44]

Ireland's only dinosaur

In 1989, at a secret location on the coast at Islandmagee, a fossil hunter struck lucky: he found three small, black pebbles that, subsequently, turned out to be the only dinosaur fossils ever found in Ireland. Dinosaurs were land animals, and they roamed the Earth for 150 million years, until they became extinct 65 million years ago. Unfortunately, Ireland has no surviving land rocks from this era – all have long since been eroded. Any rocks we have from the dinosaur era are of marine and aquatic origin. So we find ichthyosaurs and other marine fossils here – such as the one found at the Bald memorial near Larne* – but no dinosaurs. Or so it was thought, until 1989. Dr Mike Simms, a geologist with the Ulster Museum,* which now displays the fossils, says the Islandmagee pebbles were found in Lias clay, a rock of marine origin and 200 million years old. When the fossils were examined at QUB under a powerful microscope, however, they were found to be weight-bearing bones, and therefore from a land animal, rather than a marine animal; significantly, the bone marrow cavity resembles that of a dinosaur. The animal has tentatively been identified as **scelidosaurus**, a dinosaur of the appropriate size and which lived at that time. Scelidosaurus was a plant-eating dinosaur, about 4 metres long, and one of the earliest armoured dinosaurs. Its body was presumably washed into the sea after it died, where it subsequently fossilised.

A pioneering anaesthetist

Adam had it easy. Before the Lord God removed his rib, according to Genesis, "He caused a deep sleep to fall on Adam." For centuries, other patients awaiting surgery were not so lucky, and the best they could hope for was alcohol or opium to numb the pain. Even in the late 1800s, by which time scientists had discovered the anaesthetic effects of chloroform, ether and nitrous oxide (laughing gas),

surgeons continued to rely on "surprise and speed" – aiming to amputate a leg in under 2 minutes – and strong straps to restrain the patient. Moreover, there was a fine line between giving the patient enough anaesthetic to put them to sleep, and giving them a fatal overdose. Someone who helped change all that, and made anaesthetics safe and effective, was **Sir Ivan Magill** (1888–1986), a physician from Larne.

Magill invented various ingenious techniques that enabled the patient to breathe, the anaesthetist to deliver the anaesthetic, and the surgeon to operate. When he started working with anaesthetics at the end of World War I, there was no such thing as a specialist anaesthetist, and the ether was as likely to be administered by a physician as by a porter wielding a bottle and rag. Magill worked in particular with Harold Gillies, a surgeon who reconstructed the shattered faces of soldiers injured in the war. Magill's first major contribution was to devise a way of placing a breathing tube into a patient's windpipe through their neck. This endotracheal intubation, which is still used, made the facial surgery easier, and meant the surgeon was less likely be overcome by the anaesthetic. Magill also developed a suction technique to clear phlegm from the lungs of TB patients, making lung surgery easier and, in the 1930s, he invented a sophisticated breathing and anaesthetic delivery system that made chest and heart operations possible. Ivan Magill, who was awarded numerous international honours, is acknowledged as a father of modern anaesthesia and the department where he worked at Westminster Hospital, London, is named in his honour.

Petrified potatoes and star stones

There is much of geological and archaeological interest in Larne, including the earliest Irish arms factory. The local chalk, which has been quarried for centuries – the massive Ballylig quarry has been worked since at least 1796 –

contains abundant flint nodules. Known locally as 'petrified potatoes', the nodules were used 8,000 years ago by Mesolithic people to make **arrowheads** and other stone implements. One excavation at Larne uncovered over 15,000 unfinished flint pieces, at what was presumably an early arms factory. The 'Larnian' people who lived in the area 8,000 years ago may have crossed from Scotland via a land bridge.* For years it was thought they were the first people in Ireland, but remains found subsequently at Mount Sandel by the River Bann, are 1,000 years older. Larne is probably the only Irish town with a rare mineral named after it – **larnite** was discovered along with similar minerals in the 1930s; all formed in chalk that was baked by hot lava 60 million years ago. Larne also has raised beaches,* swallows holes and beautiful star stones. The star stones are small starfish-like fossils, called crinoid ossicles; the size of a fingernail, they can be found on Bank Quays beach. ㊺ Nearby Swan Island is home to an important tern colony; the reserve is managed by the RSPB, and access is restricted.

LARNE

A must see – the Antrim Coast Road

The Ring of Kerry may be scenic, but it cannot hold a candle to the Antrim Coast Road. For the coast road hugs the chalk and basalt cliffs, and winds along by the water's edge for over 30km. It offers breathtaking views of a stunning coastline, takes you to the scenic glens of Antrim, shows you Scotland on a clear day and, on top of all that, brings you along a fossil beach, takes you through (literally) an ancient sea cave, and past amazing fossil beds, brilliant-white chalk quarries, ancient volcanoes, and much more besides. If you are even remotely interested in geology or engineering – the road is also an engineering feat – or simply enjoy varied scenery, then the coast road is a must, as well as a natural prelude to the Causeway Coast.* Fortunately, there are frequent stopping places, making it easy to inspect the many features along the way.

The road was proposed in the early 1800s to improve access to the coastal villages that had developed at the foot of the Antrim glens. New postal and policing services meant there was growing emphasis on accessibility, but the coastal villages could be reached only by sea, or by steep mountain tracks, which were impassable in wet weather. A Scottish engineer **William Bald** (1789–1857) was commissioned to design and build a coast road, for the then-sizeable sum of £25,000, and work began at Larne in 1832. Bald had been in Ireland since 1809, working as a surveyor and engineer, and he spent most of the next 35 years in this country. His first major project was a detailed survey of Co. Mayo. The map Bald produced is a classic, but it cost a staggering £6,372 to produce, which caused controversy at the time. Bald also surveyed large tracts of western bogland for drainage programmes, and built several new coach roads, including Antrim–Coleraine and Bantry–Kenmare, but the Antrim Coast Road was his masterpiece.

The road took 10 years to build – five years for the coastal stretch to Cushendun, and a further five for the inland mountainous stretch to Ballycastle. The starting point, just north of Larne, is marked by a memorial to Bald and the 'Men of the Glynnes'. English writer William Thackeray, who witnessed the final stages in 1842, described it as "one of the most noble and gallant works of art that is to be seen in any country… torn sheer through the rock… immense work leveling, shuffling, picking, blasting, filling, is going on along the whole line". For the cross-country stretch, Bald first had to drain a deep bog, but the coastal stretch posed even greater challenges. First, there was the proximity to the sea and the threat of storm damage. Fortunately, there are raised, or 'fossil', beaches* in places along this coast; they appear as stepped platforms above the modern beach, and Bald could sometimes use one of the higher ones as a bed for his road. For the most part, however, the road had to be blasted out of the cliffside. The local rock structure created unique geological difficulties, however, which can occasionally close the road even today.

The cliffs that tower above the road are actually the escarpment or edge of the **Antrim plateau**, which is composed of layers of several very different rock types. Topmost is basalt, and below that a thick bed of pure white chalk. Then comes the problem layer: blue Lias clay, a rock that dates from the early Jurassic era and is rich in fossils, but, also impermeable. Rain percolating down through the basalt and chalk collects in the Lias. Eventually, the clay becomes waterlogged and unstable; at this point, it can start to buckle, or move under the weight of the overlying rock, or act as a lubricant. Catastrophic landslides can occur, as massive chunks of chalk and basalt lurch seaward. Some **spectacular landslides** happened at the end of last Ice Age at Garron Point: passing ice sheets had eaten away at the cliffside but, so long as the ice was there, it supported the cliffs; once the ice melted, however, the cliffside collapsed in a massive landslide, the jumbled remains of which can still be seen. Landslides still happen along this coast, notably at Minnis North, where they occasionally close the road temporarily, but fortunately on a smaller scale than the prehistoric one at Garron Point. William Bald was aware of this problem when he designed the coast road, and he incorporated extensive drainage facilities to minimise the dangers. The road has been patched in places, but it remains largely unaltered – a testament to Bald's design and the men who built the road. The building of the road features in Larne Museum. ㊻

A natural starting point for an excursion is the Bald memorial, which stands in one of Larne's old chalk quarries. The Lias clay on the beach below is good for fossils – a large **ichthyosaur** (200 million-year-old fish-reptile) was found there in 1991. **Ballygalley Head**, 4km up the road, is the plug of an **ancient volcano** which, 60 million years ago, produced much of the lava that covers the surrounding land. Some of the lava is in rough columns, akin to those of the Giant's Causeway, and some of the chalk was turned to a marble by the heat. **Scawt Hill** nearby is the original source of scawtite, one of the many **rare minerals** found on the Antrim plateau; it formed where chalk was baked by the hot lava. Minnis North, at Drumnagreagh Bay, is a site of occasional landslides and, in 1967, after some very wet weather, 400 tonnes of rock fell there and closed the road. There was talk then of closing the road permanently but, after several attempts, engineers stabilised the cliff and the road was secured. The quarry-like pit by the road was excavated at the time to trap any subsequent rock falls.

At **Madman's Window** car park, just before Glenarm, there are basalt boulders rich in zeolites.* **Glenarm** is home to one of Ireland's biggest **chalk quarries**. The chalk there is a brilliant white, more pure than any English chalk, and it is used as a pigment in white paint for road markings across Europe. In the 19th century, the chalk was also quarried for whitewash and agricultural lime. Iron and bauxite (aluminium ore) were also mined at Glenarm in the 19th century. Neighbouring **Carnlough**, at the foot of Glencloy,

The Antrim glens: reading from south to north, the nine glens are Glenarm ('the glen of the army'), Glencloy ('wall' or 'fence'), Glenariff ('ploughman'), Glenballyeamon ('Eamon's town'), Glenaan ('little fords'), Glencorp ('body' or 'slaughter'), Glendun ('brown'), Glenshesk ('sedges') and Glentaisie ('Taisie's glen'). The first seven open on to the east coast, the last two on to the northern Causeway Coast.

was likewise home to **iron** and **bauxite mines** and a large chalk quarry in the 19th century. The stone archway spanning the road is part of a truck way that connected the quarries to the harbour. The system was gravity-fed: laden trucks, falling under their own weight, pulled empty trucks up from the harbour.

There were similar mines and a railway in the 19th century at **Garron Point**, the massive headland that separates Glencloy and Glenariff; bits of an old bridge flanking the road are left over from the railway. The headland is the largest tract of ground without a trackway or road in Northern Ireland. It is also one of the geologically unstable points on the road, and scene of some spectacular landslides in the past. Significantly, it is one of the few places in Ireland with rocks from the **Jurassic era**, and it was there, in the 1980s, that bits of an ichthyosaur and a **plesiosaur** (such as the Loch Ness monster) were found. In 1993, despite a major outcry from geologists, the site was covered in concrete to stabilise the cliff against rock falls.

Scenic **Glenariff**, the most accessible of the Antrim glens, has been described as "Switzerland in miniature" and, in the 1890s, a tourist train ran up the glen. This was a small river glen 60,000 years ago, but glaciers flowing off the Antrim plateau during the Ice Age greatly enlarged it. Glenariff river runs through a deep gorge with numerous waterfalls, and the glen is a nature reserve noted for luxuriant mosses, ferns and liverworts. In Waterfoot, at the foot of the glen, the coast road passes through **Red Arch**, a natural stone arch that is the remains of a sea cave. There are other fossil caves in the village – all were carved thousands of years ago when sea levels were higher than today, and one was used for a classroom in the 18th century. **Cushendall** village, at the confluence of three of the Antrim glens, is overshadowed by Tievebulliagh Mountain, site of a prehistoric stone axe* factory.

The road next turns inland to follow Glencorp to the heritage village of **Cushendun**. Cushendun has fossil sea caves and raised beaches and, behind the Bay Hotel, some amazing red-sandstone conglomerate. This distinctive rock, formed from a jumble of sand and boulders, is the petrified remains of a massive flash flood that crashed through there 400 million years ago, when this area was a hot sandy desert. Cushendun village was designed for Lord Cushendun by Welsh architect Clough Williams-Ellis. Built in 1912–25, it is now owned by the National Trust. Cushendun marks the end of the coastal stretch of Bald's road. A steep, narrow and twisting road – distinctly unsuited to heavy traffic and long vehicles, but offering fine views of Scotland on a clear day – continues around the coast to Torr Head and Ballycastle.

The main road goes inland over Cushleake Mountain and the Antrim plateau to Ballycastle, but first must cross the Glendun River. For this, Bald built a fine **stone viaduct**; though it is hard to appreciate the massive construction while you are on it. On the northern side of the glen, the road cuts into the hillside; massive retaining stone walls were built and, in an elegant piece of engineering, the whole assembly is held in place by the weight of the road. There are gold-bearing rocks in Glendun, and the river has been panned in the past. **Vanishing Lake**, or Loughareema, at the top of Cushleake, is an unusual kind of turlough* or transient lake: the underlying rock is chalk, through which rain water can drain, to emerge as springs amid the trees lower down in the valley; occasionally the drainage channel blocks with peaty debris from the surrounding bog and, in wet weather, a substantial volume of water quickly accumulates.

During the last Ice Age, this glen was blocked to the north by Scottish sea ice, and a large, deep lake of meltwater accumulated in the valley. For ages, the only outlet was over the top of the mountain, where the torrents of water cut a deep incision through the rock. That cut, now called Atldorragha, is the route taken by Bald's road. As the ice receded, the glacial lake drained slowly in stages, leaving behind flat, sandy terraces on the hillside, and a thick bed of sand and gravel on the valley floor below. The terraces can still be seen, and each marks an intermediate stage in the draining of the glacial lake.

Land of fire, water and ice

"It looks like the beginning of the world", was how William Makepeace Thackeray described the Giant's Causeway in 1842. "The sea looks older than in other places, the hills and rocks strange, and formed differently from other rocks… shattered into a thousand fantastical shades… when the world was moulded and fashioned out of formless chaos, this must have been the bit over – a remnant of chaos!" It is easy to see why people thought the Giant's Causeway was not natural. It has to be Ireland's most spectacular coastline, made all the more special by the dramatic contrast between the underlying white chalk and the black basalt. There is nothing quite like it anywhere else in these islands, and the basalt columns are so unusual and so regular they were surely man-made. Admittedly, there are small patches of roughly columnar basalt elsewhere in Ireland – in counties Limerick and Waterford, for example – but nothing that compares with the audacity, perfection and scale of the Giant's Causeway. Such a landscape needs explaining.

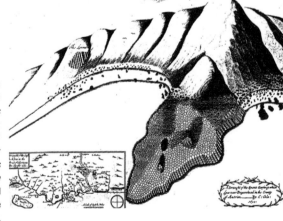

The earliest-known representation of the Giant's Causeway: published in 1694 to accompany Thomas Molyneux's article in the Philosophical Transactions of the Royal Society.

For centuries, legend attributed it to Fionn McCumhaill, a giant who reputedly built it to get to Scotland. The first scientific description was in 1694, when Thomas Molyneux,* a Dublin medic and philosopher, wrote about the rock formations in the *Philosophical Transactions of the Royal Society*. Molyneux believed the columns were natural – there was no evidence of any chisel marks or mortar and so, he reasoned, it could not be the work of man. International scientific interest was aroused in the 1740s, after a Dublin artist, **Susanna Drury** (flourished 1733–70), produced the first accurate paintings of the site, and copies of her work circulated across Europe. For the next 75 years, the rocks of the Causeway Coast were at the centre of a major scientific and religious debate, between the Vulcanists and Neptunists,* concerning how the Earth was formed. By the early 1800s, however, it was widely accepted that basalt was a volcanic rock, and that the causeway columns formed when a vast lake of molten lava cooled slowly.

The origins of the Giant's Causeway lie 60 million years ago. At that time, 'Ireland' was at the latitude where Spain is today. The weather was warm, and there was a rolling mature chalk landscape, with hills, valleys, caves and chasms, and pockets of vegetation. But then all hell broke loose, as the continental plates went on the move again, and the North Atlantic Ocean started opening up (the South Atlantic had begun opening 100 million years before, when South America broke away from Africa). Now, North America and Greenland started to pull away from Europe. Across northern Ireland and the Scottish Hebrides, the Earth's crust was stretched taut and thin. Molten magma rose from deep within the Earth. Eventually, fissures opened in the surface, and lava poured out and covered the chalk landscape with a thick blanket of molten rock, baking the chalk to form a myriad of unusual minerals. The basalt had arrived.

The north of Ireland experienced several major **volcanic episodes** on and off for 5 million years. They were separated by quiet intervals lasting tens of thousands of years, during

One of many engravings made of Susanna Drury's celebrated paintings of the Giant's Causeway, which aroused international interest due to their unusual formations. This engraving was published in the 1740s.

which time the lava weathered to clay, and trees re-colonised the landscape. The volcanic events varied in their violence and intensity. The first lava flows, for example, were relatively quiet extrusions. Next came an explosive phase, when several volcanoes erupted, violently ejecting lava. Telltale signs can be read from the landscape: layers of ash and tuff, which are only produced by violent eruptions; the remains of the volcanoes themselves, as at Knocklayd, Slemish, Ballygalley Head and Carrick-a-rede; and the shattered rock around the myriad smaller volcanic vents, where lava was forced through the chalk. When there was not enough energy to force the magma to the surface, it was inserted between layers in the existing sedimentary rocks. This produced hard dolerite sills, and when the surrounding softer rock was eroded away, the sills were left as rocky headlines, as at Portrush. Swarms of dolerite dykes formed where molten rock rose up through fissures; these dark dykes survive as conspicuous wall-like features in places.

There followed a long interval, during which time most of the region was quiet. In the hot, humid weather, the top of the basalt decayed to form a **red laterite clay**, like that seen in tropical Africa today, and rich in iron and aluminum. Indeed, the Antrim plateau has the only substantial **aluminium** deposit in these islands, and it was mined at various sites around the coast until the 1920s. There were river valleys and lakes; redwood and eucalyptus trees grew there, and their fossil remains reveal much about the climate and vegetation of the time. Some of the vegetation was turned to coal and some to lignite;* these were mined in the past, notably at Ballintoy in the 1700s.

Along the Causeway Coast, however, there was a major geological fault. Even while the rest of northern Ireland was quiet, this region remained active, and several times the valleys there filled with molten lava. The ponded lava took months to cool slowly from its initial temperature of 1,000°C but, as it cooled, it crystallised into neat and many-sided columns of basalt. A second period of widespread volcanic activity followed, shaking the whole region and blanketing the area with more thick layers of basalt. The Mourne* and Cooley Mountains and Slieve Gullion* formed around this time, from massive amounts of molten igneous rock that rose close to the Earth's surface. Eventually, however, the Atlantic rift drifted away from Ireland – it is now at Iceland – and the volcanic activity died away. By 50 million years ago it was all over, and apart from the occasional earthquake* since then, Ireland has been geologically quiet.

The volcanic era dramatically changed the landscape of northeast Ireland, and produced a broad basalt plateau that is the largest of its kind in these islands. So much basalt was extruded that the Earth's crust was undermined, and it eventually collapsed to form a depression that is now filled by Lough Neagh.* In places the basalt is nearly 1km thick, even though much of it was removed by erosion in the

Chalk – a fossil sea

Most of Ireland was once covered by a thick layer of chalk. The chalk, a pure type of limestone, was deposited 140 million years ago when Ireland was covered by a warm and relatively shallow sea. The water was thick with tiny, single-celled plants called coccoliths, which are unusual in having a hard shell. When they died, their shells accumulated on the sea floor, and were later compressed to form a pure, fine-grained chalk. (The fossil shells can be seen under a microscope.) Around 100 million years ago the sea withdrew and 'Ireland' was again dry land. Erosion began to eat away at the chalk, but, in the northeast, this process stopped when basalt was spread over the chalk. As well as preserving the chalk, the basalt baked and compressed it. As a result, Antrim chalk is harder and more impervious then British chalk of the same age, and it has been quarried for centuries. The Antrim plateau has also been mined at various times for flint, basalt, iron ore, bauxite, coal, lignite and porcellanite.*

where lignite, iron ore and bauxite were mined from between the basalt layers. The chalk and basalt quarries at **Larrybane** are now owned by the National Trust. **Carrick-a-rede**, noted for its rope bridge, is the best example of a volcanic crater in Ireland, although the sea has broken through. This was the scene of one of the north's most explosive volcanic eruptions – witness the layers of tuff and ash in the surrounding rock, and the volcanic bombs and boulders of basalt that were thrown high before falling back.

An enlightened entrepreneur

intervening millions of years. The hill of Slemish, for instance, is merely a plug that was once at the core of a large volcano. And the plateau's southern edge today ends north of Belfast, but the basalt once stretched to Armagh.

The best place to explore this volcanic landscape (which is a UNESCO heritage site and a national nature reserve) is not on the plateau, which is relatively featureless, but at the edges, particularly the Antrim Coast Road* and the Causeway Coast and, to a lesser extent, Cave Hill* above Belfast. One option is to follow the causeway path from Portrush to Ballycastle, which is surely Ireland's most interesting walk.[13] At the **Giant's Causeway**, as well as the world-famous columnar formations, you can see layers of red laterite clay and, embedded in the clay, occasional kernels of uneroded basalt, known locally as 'giant eyes'. [47] At **White Rocks**, you can see the remains of explosive volcanic vents (basalt agglomerate, which plugged the vent, is surrounded by shattered chalk); there are also fine chalk cliffs, sea stacks and caves on the beach. **Ballintoy Harbour** lies on a fault line, and the rock has shifted so that now the basalt on the northwest of the fault lies alongside the chalk to the southeast. There are also fine raised beaches and fossil sea stacks there, which were left high and dry after sea levels fell. Ballintoy is one of the many places along this coast where chalk was quarried, and

In the early 1700s, **Colonel Hugh Boyd** (1690–1765) ran a pioneering industrial estate at Ballycastle, employing 300 people. Boyd, who was something of a visionary, even provided pensions and free housing for retired workers. He began by opening a small colliery in the 1720s, to mine the local coal and ironstone; he built a harbour, with an inner and outer dock (the inner dock is now a tennis court), from where the coal was exported; and, in the 1730s, added a tramway – possibly the first tramway in Ireland – linking the colliery and harbour. He also had a fleet of ships to ferry the coal to Dublin and England. Realising that the coal could be used to fuel other local industries, and that there was ready availability of raw materials, he built salt pans by the coast, and evaporated the brine over coal fires. Later, he switched to heating the sea water in iron pans made at his ironworks. One by one, he added a bleachworks, soap factory, tanneries and a glassworks. They all used local materials, and each was chosen to exploit products and waste material produced by the others. At the end of the chain came a brewery and a distillery – to fill the bottles made at the glassworks. According to Ballycastle historian Cathal Dallat, the colliery underpinned everything, and, in his will, Boyd specified that coal be delivered each week to the glassworks at 5s.6d a ton. By the time he died, however, coal cost twice that to produce, and Boyd's grandson and heir refused to honour the agreement. The whole project foundered, and little now remains of this pioneering industrial experiment. ⑱ South of Ballycastle, at the head of Glenshesk, is **Breen Wood** nature reserve, an ancient oak wood that survives from the original prehistoric wild woods. The wood is on steep ground, which may have helped ensure its survival.

1788 – Ireland's last volcano?

In May 1788, the top of Knocklayd burst open in a violent volcanic eruption. According, that is, to *Faulkner's Dublin Journal*. The paper reported tragic events in which villages were lost under the lava, and those killed apparently included both the Dissenting minister and the parish priest. The *Dublin Courant* carried further details a fortnight later, in a report filed from Ballycastle by Pliny the Younger, who reported further eruptions and general destitution, though on a positive note he claimed that there were now so many dead fish they were being used as fertiliser. The story ran for several years in various periodicals, before being put to sleep. Knocklayd is a volcano, and it did smother the surrounding area in hot lava – but 60 million years ago. Reports of its re-awakening in 1788 were, happily, greatly exaggerated, and probably something of a hoax. Historians have suggested it might have been prompted by a bog slide,* or by the debate about the volcanic origins of basalt, which were being hotly argued then by Vulcanists and Neptunists.*

There are no snakes in Ireland

In 1617, the Laird of Lisnorris claimed Rathlin for Scotland. The island, thanks to its strategic location, had been claimed by Ireland and Scotland at various times but, in 1617, a protracted legal case ensued. Randal MacDonnell, first Earl of Antrim, won the argument – and the island – by resorting to natural history: there were no snakes on Rathlin, he pointed out, so it was obviously Irish. The absence of snakes in Ireland had been noted centuries before: St Donatus, a 9th-century bishop, had written that "no snakes are creeping there with venomed guile"; and in the 12th century, Giraldus Cambrensis* wrote that "the land and air of Ireland are so sacred that nothing poisonous could live there". A separate 12th-century story claimed that St Patrick rid Ireland of snakes. But there were never any snakes to start with: snakes were just some of the many creatures that had not reached Ireland before rising sea levels 10,000 years ago drowned the land bridges* linking us to Britain and Europe.

Rathlin Island may not have snakes, but it does have buzzard, kestrel and peregrine falcon and, at its western end, Northern Ireland's largest seabird colony. Tens of thousands of fulmars, kittiwakes and auks crowd the spectacular basalt and chalk cliffs and sea stacks, which are a RSPB nature reserve. ⑲ The nesting sites are busiest in June – only the fulmars are year-round residents. Rathlin's varied habitats include sea caves, maritime heath, freshwater marshes, rough pasture and a small wooded ravine near Church Bay, which has most of the island's few trees. A sea whirlpool, *Slough na Morra*, can be seen off Rue Point.

In 1898, Rathlin Island was the scene of an early **radio experiment** conducted by Guglielmo Marconi* and two assistants, John Kemp and Edward Granville. The previous year in Europe, Marconi had deployed his wireless telegraph to send signals over distances of 20km, but failed to convince companies of the system's commercial merits. So, in 1898, Marconi arranged a demonstration for Lloyd's of London: he chose Rathlin Island, which was beside a major shipping approach to England; at the time, a carrier-pigeon service was used there to send information from passing ships to Ballycastle on the mainland, and thence by telegraph to London. Marconi's team was able to radio information even in foggy conditions to Whitehead, and from there to Ballycastle. The experiments convinced Lloyd's to erect a radio mast on Rathlin, the remains of which can be seen near East Lighthouse. Sadly, Edward Granville fell to his death while on the island. A plaque at Whitehead commemorates Marconi's experiment.

A prehistoric Irish arms trade

Stone axes are the archetypal Stone Age tool. They were widely used between 7000–2000 BC, and nearly 20,000 have been found in Ireland. Some were imported, but nearly half were made at two Antrim axe factories – Tievebulliagh near Cushendun,* and **Brockley** on Rathlin Island. Stone axe heads need to be made from a rock that is hard enough to withstand use – more than likely felling trees and not people – yet one which can be easily fashioned into shape. Irish axe makers used various rock types: shale cobbles, as at Fisherstreet (Doolin) in Co. Clare; porphyry from Lambay,* a volcanic island in Dublin Bay; but the ideal material was **porcellanite** from Tievebulliagh and Brockley.

Porcellanite is a fine-grained rock, glassy blue in appearance, and with the smooth texture of porcelain. It forms where basalt rock has weathered in a warm climate to produce a mineral-rich clay; and where that clay was subsequently baked by hot lava, much as china clay is baked in a kiln to form porcelain. This particular set of circumstances occurred at Tievebulliagh and Brockley. At Brockley, you can still see the neolithic quarries where the porcellanite was extracted, while the hillside at Tievebulliagh is littered with unfinished axe heads, the remains of a neolithic arms factory.

Archaeologists from UCD, led by Prof Gabriel Cooney, are using sophisticated techniques, such as X-ray fluorescence, to identify the source of the rock used in stone axe heads, and thus reveal prehistoric trade routes. Some of the axes found in Ireland are now known to have been made of Cornish gabbro, others of Cumbrian tuff; yet others are of semi-precious jadeite from the French Alps, although these were probably for ceremonial use. Similarly, Irish stone axes have been found in Britain and across Europe, suggesting a highly developed trade route between Ireland and Europe. Antrim stone axes can be seen at the Ulster Museum.*

1882 – the world's first water-powered tramway

Tramways are usually associated with cities, but the north of Ireland had a number of scenic services, the most unusual being the hydro-powered causeway tram that ran from Portrush to Bushmills village. Powered by a small generating station built at the Salmon Leap waterfall on the Bush river, it was the world's first hydro-powered tramway, and the first electrified railway in Ireland or Britain. The service ran from 1882 until 1949, and was something of a tourist attraction in its own right, while it also helped develop the Giant's Causeway as an attraction in the days before motor cars were common. Day trippers from Belfast could take the train to Portrush, catch the tram to Bushmills, then take a jaunting car the final 3km to the causeway (though, by 1887, the tramline had been extended almost to the causeway, and the jarveys were out of a job).

The tramway was the brainchild of **William Traill** (1844–1933), who was born near the Giant's Causeway and studied engineering at TCD. By 1882, a small turbine house had been built on the river, with two 45-hp turbines and a Siemens dynamo. Traill designed and managed the

William Traill, inventor and engineer, features on the Northern Bank's £5 note. The Ulster Folk & Transport Museum* has one of his 'toast rack' trams.
© Northern Bank

tramway, and invented the technique of supplying the power in a third conducting rail (rather than in an overhead line), a system later adopted by other electric trams and railways. In 1900, however, the causeway tram changed to using overhead lines and trolley cars. The tramway service is probably unique in being disrupted by submarine activity: during World War I the line was damaged by shells fired when a German submarine was attacked by an armed coaster. With the advent of the motor car, the tram's popularity declined and it closed in 1949. In 2001, the route was re-laid with narrow gauge tracks and steam train excursions run in summer. ⑤⓪

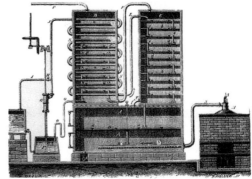

Aeneas Coffey's ingenious still.

BUSHMILLS

Distilling the essence

In 1830, a Dublin man **Aeneas Coffey** (1780–1852) invented a highly efficient distilling apparatus. His patent still was also the world's first heat-exchange device, and it would have saved distilleries a fortune in fuel costs alone. Unfortunately, the alcohol it produced was so pure it lacked flavour and individuality, and most distilleries continued to use pot stills. Coffey's apparatus proved ideal for the chemical industry, however, and the company he set up to make it is still in business. Aeneas Coffey joined the Customs and Excise service in 1800. Ten years later he was in charge of a major operation against illicit distillers in Co. Donegal, and on one occasion he and his militia were beaten and left for dead. Coffey had an inventive turn of mind, he resigned from the Customs in 1824, presumably to concentrate on his inventions, and, by 1830, had patented his new still.

Distilling apparatus had changed little in the previous 1,000 years, since Arab alchemists invented the alembic (pot still) around AD 700. The still was designed to extract essential oils for perfumes, but was soon used to concentrate alcoholic solutions. (Missionaries returning from Arabia brought the principle of the pot still to Ireland, where it was used in the 8th century to make the world's

first whiskey.) Not understanding the chemistry involved, the Arabs attributed the still's success to 'spirits', a term which has stuck. The alembic consisted of a container topped with a condensing tube; the starting liquid was placed in the vessel and, as it was heated, vapour rose into the condensing tube, where it cooled and dripped into a receptacle. The process exploited the fact that different liquids have different boiling points and, by carefully controlling the temperature, it was possible to separate alcohol from water. The pot still was, however, wasteful of fuel, and could only be run in batches, starting each time with a fresh volume of cold liquid. Moreover, to produce a reasonably concentrated spirit, the process had to be repeated several times.

Aeneas Coffey's invention was a significant improvement. It consisted of two tall columns, connected by pipes, and filled with stacks of perforated metal plates. Each plate acted

Black Bush

On April 20th, 1608, King James I granted a seven-year distilling licence to a Bushmills landowner, Sir Thomas Phillipps. This, **Bushmills Distillery** claims, makes it the world's oldest whiskey distillery – although the company was not founded until 1783 and, no doubt, people were making whiskey there long before 1608. Bushmills, now owned by Irish Distillers, is the only working Irish distillery that is open to the public. ⑤① All the others – the Old Jameson Distillery in Co. Dublin,* Midleton Distillery in Co. Cork,* and Locke's at Kilbeggan* – are visitor centres in distilleries that no longer make whiskey.

as a small distillation chamber; and steam could rise, and liquid fall, through the perforations. Significantly, Coffey used the heat from the outgoing vapour to warm the incoming liquid, the first known heat exchanger. His patent still was more efficient than the old pot still, could be run continuously, used less fuel, and produced a near-pure spirit that was at least 95 per cent alcohol. The principles Coffey developed are still used in the chemical industry and refineries.

1803 – the Firemen *versus* the Watermen

A major geological controversy raged throughout the 1700s, but fossils found at Portrush in 1803 played a crucial role in resolving the issue. On one side were the **Neptunists**, aka the 'Watermen'. Aligned with Noah, the Flood and a biblical interpretation of geology and the creation of the world, they were led by Abraham Werner (1749–1812) from the University of Freiburg. Neptunists believed that all rocks were sedimentary and had crystallised from sea water – apart from a few modern lava formations that, they admitted, were the result of volcanic eruptions. Opposing them was a group of heretical geologists, the **Vulcanists** or 'Firemen', led by, among others, James Hutton (1726–97), the 'father of modern geology'. The Vulcanists took a less literal reading of the biblical story. They believed that some rocks were sedimentary, but they also held that certain rocks such as basalt were igneous, and that these had formed during volcanic eruptions aeons ago.

In 1803, an Irish geologist, **Rev William Richardson** (1740–1820), discovered fossils in a hard, dark, basalt-like rock at Portrush. This Portrush rock appeared to clinch the argument for the Neptunists. Richardson sent samples to geologists around Europe, and many came to Portrush to see the fossils for themselves. By 1816, however, a detailed examination of the rock revealed that it was not basalt, but hornfels – originally Lias clay, a sedimentary mudstone, that had been baked by molten rock intruded into the area during the volcanic era. Together with other evidence from ancient and modern volcanic sites around Europe – including an eruption of Vesuvius in 1790 – this eventually won the argument for the Vulcanists. The fossil site is now protected as a nature reserve. ⑤²

Portrush town sits on a prominent headland, Ramore Head. This is made of dolerite, a hard igneous rock that was intruded between layers of the older Lias mudstone 60 million years ago.

Chapter 3 sources

Co. Armagh

1. Zhou, Keqian, & C J Butler, "Statistical study of the relationship between solar cycle length and tree-ring index values" *Journal Atmos. & Solar Terrestrial Phys* **60** 1711–1718 (1998); and Butler, C J, *et al.*, "Precipitation at Armagh Observatory 1838–1997" *Proc. RIA* **96B** 123–140 (1998).
2. Brück, M T & S Grew, "A family of astronomers – the Breens of Armagh" *Irish Astr J* **26** 121–128 (1999).
3. Lamb, J G D "The Apple in Ireland: its history and varieties" *Econ Proc RDS* **4** 1–23 (1951).

Co. Down

4. Carson, W H, *The Dam Builders* (Mourne Observer Press, 1980).
5. Hart-Davis, Adam & Paul Bader, *Local Heroes* (Sutton, 1997).
6. Arnaldi, Mario, "Ancient sundials of Ireland" *British Sundial Society Bulletin* **4**:13–20 (1997).
7. McErlean, Thomas & Norman Crothers, "Tidal power in the seventh and eight centuries AD" *Archaeology Ireland* (Summer 2001).

Belfast

8. Gribbin, John, "The Man Who Proved Einstein Was Wrong" *New Scientist* Vol 128 November 24th: 43–45 (1990).
9. *A Guide to Belfast and Adjacent Counties* (Belfast Naturalists Field Club, 1874).
10. Garvin, Wilbert & Des O'Rawe, *Northern Ireland Scientists and Inventors* (Blackstaff Press, 1993).
11. O'Donnell, Sean, "Thomson and Reynolds – The Physicists of Flow" *Technology Ireland* Vol 9 February: 64–65 (1978).

Co. Antrim

12. Scannell, Maura, "Inspired by lichens" in Mary Mulvihill (ed.), *Stars, Shells & Bluebells* (WITS, 1997).
13. Lyle, Paul, *A Geological Excursion Guide to the Causeway Coast* (Environment and Heritage Service NI, 1996).

Chapter 4: Derry • Donegal • Tyrone

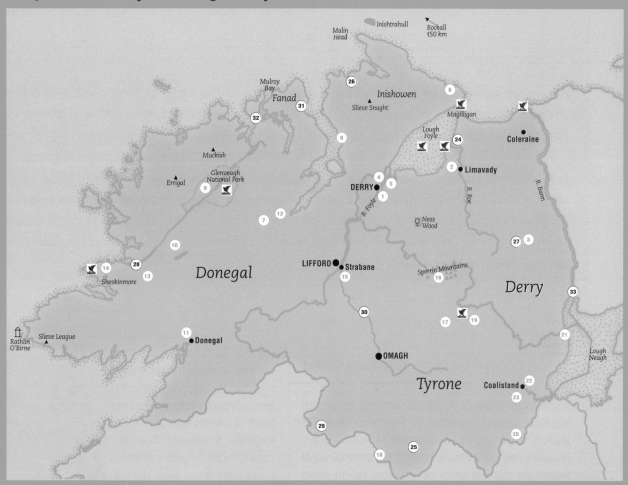

Visitor facilities

1. Derry Tower Museum
2. Roe Valley Country Park
3. Wm Clarke & Sons Ltd, Upperlands
4. Derry Harbour Museum
5. Amelia Earhart Cottage, Ballyarnet
6. Greencastle Maritime Museum
7. Newmills
8. Tullyarvan Museum
9. Glenveagh National Park
10. An Mhuc Dhúbh Railway, Fintown
11. CDR Railway Heritage Centre, Donegal
12. Donegal County Museum, Letterkenny
13. St Connell's Museum, Glenties
14. Dolmen Centre, Kilclooney
15. Gray's Printing Press
16. Sperrin Heritage Centre
17. An Creagán / Creggan Centre
18. Fivemiletown Library
19. Wellbrook Beetling Mill
20. Benburb Valley Linen Museum
21. Kinturk Cultural Centre
22. Cornmill Heritage Centre
23. Tyrone Crystal

Other places of interest

24. Broighter
25. Clogher
26. Clonmany
27. Culnady
28. Gweebarra
29. Lisnahanna
30. Newtownstewart
31. Portsalon
32. Rosapenna
33. Toome

♀ woodland
🦆 nature reserve or wildfowl sanct
🏛 lighthouse

*Visitor facilities are numbered in
appearance in the chapter, other
alphabetical order. How to use th
page 9; Directory: page 474.*

All of Ireland was measured against a line drawn in the sand at Magilligan when the Ordnance Survey began mapping the country in the 1820s. The start and end points of that historic baseline can still be seen. The extensive sand dunes at Magilligan are among the most impressive in the country and this coastline was and is an important landfall, both for aviators such as Amelia Earhart and for countless migrating birds. Milk of magnesia and artificial fertilisers were invented by a Derry physician, James Murray and, in 1842, the largest screw-propelled ship the world had seen was built at Derry shipyards. Derry comes from *doire* ('oak wood'), and remnants of the Derry wildwoods survive today in Ness Wood.

Where Ireland was measured

A straight line nearly 9km long was laid out in the sand at Magilligan between September 1827 and November 1828 and its length was measured with, literally, military precision. This was the baseline for the **Ordnance Survey of Ireland**, the rule with which the rest of the country would be measured. The Magilligan baseline was the starting point for **triangulating Ireland**, the first step in mapping the country. Triangulation uses simple geometry to calculate long distances: if you know the length of one side of a triangle (the baseline), and the size of two of its angles, you can calculate the length of the other two sides. Angles are relatively easy to measure, generally using theodolites to take sightings to distant mountain tops. And once one triangle has been set up and calculated, the process can be readily extended to include neighbouring triangles. But first

you must select and measure your baseline. Colonel Thomas Colby, who had already overseen the army's Ordnance Survey of Britain, chose to site his Irish baseline at Magilligan as the land there was remarkably flat for quite a distance, it was close to sea level, and not far from Scotland, making it easy to triangulate across to Scotland and connect the Irish and British mapping systems.

The baseline had to be surveyed as accurately as possible as any error would be amplified when the survey spread out across the country. Surveyors measured distances then using a known length of metal chain, a kind of flexible ruler. But links in a chain wear and stretch, and the length of the chain also varies with temperature as the metal expands

One of the six compensation bars designed specially to measure the Irish baseline. The Tower Museum in Derry has one, as does the OSI in Dublin; the remaining four were sent abroad for other ordnance surveys.

Measuring the baseline: the manoeuvres at Magilligan attracted considerable attention and among the visitors were English astronomer John Herschel, who drew this sketch, William Rowan Hamilton* from Dunsink Observatory and computing pioneer Charles Babbage.

themselves with the new instruments. On average they were able to measure 60ft (18 metres) of line in an hour.

The baseline started at a point called North Station and, by late October, the team had reached Minearny, where they broke for winter. The work resumed the following summer, broke for August– September to avoid damaging crops, and finally reached South Station at Ballykelly on the outskirts of Derry city in November 1828, after measuring a line precisely 30,533ft long (9,306 metres).[1] Permanent enclosures were built around the crossed wires that marked the precise start and end points of the line so that, if it was ever needed, the identical line could be re-measured. Finally, the co-ordinates for longitude and latitude were calculated from the stars, to locate the baseline on the actual globe. Only then could the mapping begin in earnest.

The main reason to map Ireland was tax purposes: to provide accurate townland measurements so that land could be valued fairly, as the existing valuations were old, unreliable and inequitable. The Ordnance Survey of Britain had just been completed by the army and the government decided the same team should map Ireland because, as the lord-lieutenant, Richard Wellesley declared, Irish engineers "had [n]either, science, nor skill, nor diligence, nor discipline, nor integrity" for the work.[2] The survey established its Irish headquarters at Mountjoy House in Dublin's Phoenix Park* (still the headquarters of the OSI). Only one Irish engineer was employed: Richard Griffith* was put in charge of the townland division and was ultimately responsible for calculating the new land valuations, what came to be known as the Griffith valuations.

Triangulating Lough Foyle around the Magilligan baseline (from Yolland, 1847).

and contracts. So Colby and his assistant Lieut Thomas Drummond* devised a radical new way of measuring distance using a series of metal compensation bars which they designed. Each bar (in essence a sophisticated ruler) was precisely 10ft (3.04 metres) long and made from two rods, one made of brass and the other of iron, which compensated for differences in the way each metal expanded and contracted with changing temperatures, such that any change in the overall length of the bar could be calculated precisely.

The survey team at Magilligan used microscopes to align the starting point of one compensation bar with the end point of the previous one; theodolites and telescopes to check the orientation of the baseline; trestles and tripods to support the bars and instruments; spirit levels to ensure the bars were horizontal; and a tent to shade the whole assembly and minimise temperature changes. The complex compensation bars took two years to design and make – six in all were made – and so, while the Ordnance Survey of Ireland was approved in 1824, the baseline survey did not begin until September 1827. The work was slow and tedious, especially until the men had familiarised

The Magilligan baseline's precise start and end are protected by small enclosures. In 1960, the OSNI re-measured the baseline, taking one hour, where Colby's team took four months. The two measurements differed by just 2.5cm, a testament to the accuracy of the original survey.

Once the Magilligan baseline had been measured, the primary triangulation of Ireland – a network of large-scale triangles spanning the country – was conducted. Next came a network of smaller triangles at parish level. These provided fixed control points for the detailed local mapping, which was done by teams of field workers (sappers). Using chains and theodolites they recorded and mapped "everything attached to the ground". The survey team started in the north and moved south county by county; the budget for the entire country was £300,000, or 1½d per acre. At its peak 2,000 men were employed in the 1830s, including skilled map-makers and engravers, and artists whose job was to sketch in the hills and topography. Colby's instructions were detailed – he even exhorted those working in rural areas not to appear out of the ordinary by sporting beards or moustaches.

Spot heights were included on the maps. Initially the heights in each county were measured relative to a local reference point but, in 1840, the whole of the country was levelled relative to one point: the low-water mark at Poolbeg lighthouse* in Dublin. In 1841, contour lines replaced the way hills and valleys were drawn by the artists. The first maps were published in 1833 for Co. Derry. They were at a generous scale of 6in to the mile, and it took numerous large maps (or sheets) to cover the county. Thirteen years later, in 1846, the last of the maps was published, for Co. Kerry. In all, the survey produced 1,939 sheets, recording every house, road and field. It was the first full portrait of Ireland – the most-detailed map until then, the Down Survey,* had surveyed only the lands confiscated by Cromwell's government. It was also the first time ever that any country had been mapped in such fine detail, and the procedures developed by Colby and his team, including their compensation bars, were adapted by other surveys abroad. The original plan was to produce a detailed biography or memoir to accompany each map, but these proved too costly and only one was ever published, for the Derry parish of Templemore (see: **The Templemore memoir**).

The new 6-in maps proved invaluable. As well as being the basis for Griffith's land valuation, they were used to plan the population census, the developing railway network, arterial drainage, and a new geological survey* of Ireland. This last project, begun in 1845, mapped not what was on the surface of the ground, but what lay hidden below. After the initial townland maps came even more detailed maps of towns and cities – Derry at 24in to the mile, for example. Other series were published over the subsequent decades – a 1-in series in the 1850s, for instance and a ¼-in series in the 1880s. In the early 1900s, the country was re-surveyed and the maps updated, and colour maps were produced for the growing tourist trade. In the 1950s and 1960s, the country was re-surveyed, this time with a new triangulation. Concrete trigonometric ('trig') point pillars were built on 57 mountain tops as part of this exercise. The country was also re-levelled: the old Poolbeg reference, against which every height had been measured, was replaced in the Republic with a new reference taken at Malin Head, and in Northern Ireland with a new Belfast one.

These days Ireland is continuously being surveyed, driven by the need to keep up with changes, and aided by the arrival of new technologies. Modern mapping uses aerial photography, global-positioning satellites and high-powered computers, though field surveyors still walk the land, much as they did in the 1830s, to ensure no detail is missed. In 1996, during one survey, Ireland became 3.977 metres longer and 2.042 metres wider, almost overnight: accurate global-positioning satellites were used for the first time to fix the positions of 95 stations around the country, and revealed that the distance from Malin Head to Skibbereen was 443,820.424 metres; and from Achill to Howth 275,278.459 metres.

Part of Magilligan headland is still occupied by the UK Ministry of Defence. The Tower Museum in Derry has an exhibition about the survey and one of Colby's compensation bars. ① The original field notes for the survey are at the National Archives (Dublin); various places have copies of the maps, including the National Library (Dublin).* The OSI has its headquarters in Dublin, and OSNI in Belfast. Their *Discovery* maps (1:50,000) are ideal for exploring the places featured in this book.

Sea, sand – and ice!

Most of Ireland's **sand dunes** owe their existence to the glaciers of the last Ice Age. For it was the glaciers that made the sand, grinding down mountains and rocks to fine grains. The same glaciers also transported the sand, dumping it in vast quantities in sand banks off the coast. Over the ensuing thousands of years, sea currents, tides and winds brought the sand back onshore to form sandy beaches, sand hills and, eventually, complex systems of dunes. The oldest dunes in Ireland date from 6,000 years ago, and most developed during the subsequent 2,000 years.

About 10 per cent of Ireland's coastline is sandy and most of this is in the windier northwest, from Mayo to Antrim. Dunes form where sand is blown onshore and has a chance to accumulate, and where marram grass can stabilise it. Once a first row of dunes has formed, a second row can grow on the seaward side, and this process continues with parallel rows of dunes forming in line with the shore. Dunes are dynamic systems, constantly changing, but also vulnerable to erosion by the very wind that helped form them in the first place. Cutting the marram grass for thatching and rabbit burrows can also damage dunes, rapidly destroying what took thousands of years to develop and sometimes triggering a disaster: the villages of Rosapenna* in Donegal and Bannow in Wexford were buried under sand when the dunes beside them were destabilised by rabbits.

The best-studied sand dunes in Ireland are at **Magilligan**, a vast flat triangle of sand 32sq km in area. The dunes formed in three phases 6,000–4,000 years ago.

The Templemore memoir

The most-detailed study of an Irish parish ever published is surely the memoir of Templemore produced by the Ordnance Survey in 1837. Initially, the survey intended to produce not just maps of the country, but also scholarly memoirs detailing each parish's natural state (features and natural history), its artificial state (modern settlements and ancient remains), and its general state (including commerce and industry). A team was recruited to compile this information, including the historian and antiquarian **John O'Donovan** (1809–61), who worked on place names; and the noted Celtic scholars **George Petrie** (1789–1866) and **Eugene O'Curry** (1796–1862).

The 400-page Templemore memoir records the parish's flora, fauna and geology, industries and economy, history and settlements, the sites of old churches, the town's rope-making industry, even the extent of charitable works.[3] The volume cost a staggering £1,700 to produce and appeared to hostile reviews – the cost for this one parish alone was more than the amount allocated for three whole counties. The scheme was cancelled, and no further memoirs were published, though data had been collected for other northern counties. This detailed information is now a valuable archive for historians, and a facsimile edition of the Templemore memoir was published in 1990.

Occasionally, a layer of peat occurs in the sand, marking an interlude when the climate changed, the sand stopped accumulating and vegetation grew there for a while. These peat layers can be carbon dated, providing approximate dates for when the dunes formed. Neolithic remains found in the sands provide further dates for the dunes. By dating Magilligan's topmost layer of peat, we know that the most recent dunes there began forming 1,000 years ago.

Throughout its history Magilligan has grown outwards, closing off Lough Foyle and changing the area from open sea to a more lough-like estuary. This infill continues today as more dunes form at the point. A good place to view the extent of Magilligan is from the basalt table top of **Binevenagh**. The basalt cliffs at Binevenagh are a nature reserve and home to several rare species, including purple saxifrage. Another nature reserve noted for plantlife is the steep wooded glens at **Banagher**. The sandy soil of Magilligan is well suited to tillage, and rye was once extensively grown there for thatching. The flat expanse was ideal for the Ordnance Survey's baseline – measured there in 1827–28 – and for the **first Irish flight**. In 1909, Harry Ferguson* became the first person in Ireland to fly, when he flew his home-made aeroplane at Magilligan; on his first successful attempt he managed a distance of 40 metres, at a height of 3 metres.

On the defence

A small proportion of the sand on Irish beaches comes from present-day erosion of our mountains, and from crushed shells and coral seaweeds, but most has not been replenished since the glaciers melted. Sand removal needs to be controlled, because beaches and dunes are important **coastal defences**: beaches absorb much of the sea's energy, while dunes form a windbreak and act as a sand reservoir. In winter, energetic waves often remove sand from a beach and dump it offshore; but the resulting sandbank reduces the energy of the waves before they hit the beach; in summer the sand is returned to the beach. Interfering with this natural cycle can destroy a vital beach or dune system.

The case of the gold hoard

A most unusual court case was heard in London in 1903 over who might rightfully own the fabulous Broighter gold hoard. Among the expert witnesses was the noted Irish naturalist Robert Lloyd Praeger,* who was called to give evidence about changing sea levels in Ireland over the past 5,000 years. The hoard, thought to date from 100 BC, was found by a man ploughing a field at Broighter near the Roe Estuary. The objects included a beautiful model boat, complete with a mast, oars and benches for a crew of eight. The find passed to the man's employer, who sold it to a collector, who finally sold it to the British Museum in 1893 for £600.

Treasure that is found is generally considered either to have been voluntarily abandoned, in which case 'finders keepers', or else to be 'trove' – accidentally lost, or hidden with a view to later recovery, in which case, if the rightful owner is not found, the Crown or State can claim the property on behalf of any next of kin. The Royal Irish Academy* (RIA) heard that the British Museum had acquired the Broighter collection and, acting for the Crown in Ireland, began legal proceedings to have the hoard declared treasure trove and Crown property and thus returned to Ireland. The museum refused to hand over the collection, however, arguing that the hoard had been a votive offering to the Celtic sea-god, Mananan mac Lír, and that, in this case, finders could be keepers. In the ensuing court case, part of the legal argument centred on whether the field at Broighter, now 1.5 metres above the high-tide mark, could have been underwater 2,000 years previously, when the offering was supposedly made.

For its case, the RIA asked Praeger and George Coffey to investigate the evidence for changing sea levels around Lough Foyle. Excavating near Portstewart, they discovered an undisturbed Neolithic site that was clearly much older

than the hoard, and only about 60cm above the modern-day high-tide mark. If the site at Port-stewart had not been disturbed by the sea, they argued, the Broighter field must also have been above sea level, and therefore an unlikely site for an offering to any sea-god. The court found in the RIA's favour: the hoard was declared treasure trove and Crown property, and the museum must hand it over. The judge said in his report that the British Museum's argument about sea-gods was "a fanciful suggestion more suited to the poem of a Celtic bard than any English law report". The museum returned the objects to the academy, which presented them to the National Museum of Ireland. Ironically, many archaeologists now accept that large hoards found at coastal locations are very probably votive offerings to Mananan. But if so, how did the Broighter hoard end up in a field high above sea level? Perhaps the Roe estuary once covered the field but later changed course? Or perhaps the objects were washed up in a storm? Or deliberately buried in the field? A century after the court case, the Broighter hoard remains a golden mystery.

Roe Valley power house

For over 70 years, the town of Limavady received its electricity thanks to an innovative hydro-scheme that harnessed the local River Roe. This municipal hydroelectricity project, the brainchild of local landowner **John Ritter** (1853–1901), was one of the earliest public electricity schemes in the North and among the first in Ireland to use alternating current.

Downstream of Dogleap (*léim a'mhadaidh*, whence Limavady), the River Roe falls 30 metres in 3km. This substantial drop has been used since early times to power waterwheels and mills. There have been cornmills there, a distillery and, in the 18th and 19th centuries, a bustling linen industry with water-powered scutching mills, looms and beetling* engines. In 1893, John Ritter experimented using the mill's waterwheel to power a generator and provide lighting for his home at Roe Park House. Three years later, he began a larger project to supply electricity to the local town. A new power station was built by the river and an intake pond, tunnel, turbine pit and tailrace were blasted out of the rock. Three turbines, made in Belfast by Robert MacAdam, were installed in the pit and overhead power lines were strung along the estate to a substation and transformer at The Lodge, where Miss Lancy became the first customer. But not everyone was pleased. The power station diverted water from a neighbouring Dogleap farm and a legal battle ensued. And the local gas company, quick to spot the competition, objected to Ritter running his supply lines along public streets. Ritter in turn accused the gas company of poisoning the local air with its leaking pipes.

Ritter's electricity was expensive: the minimum charge per quarter in 1897 was 13s.4d, when the weekly wage for a labourer was about 10 shillings. Yet the charges were probably realistic – the Ritter family profited little from the business and the power station was efficiently run. The arrival of efficient light bulbs, and later electric heaters and cookers, increased the demand for electricity. During World War I, a shortage of coal, from which the town's gas supply was made, gave the hydro-scheme an extra boost. By 1918, even the local town hall was wired for the electricity, though the council owned the gas company then!

Improved turbines and other equipment were installed as demand grew, and branch lines were laid to connect local farms. In 1929, a "hire-purchase wiring scheme" was offered to local cottagers: to wire in one 40w lamp cost 7d a week over two years, a hefty £3 all told. The Limavady Electricity Supply Company was taken over by Northern Ireland's electricity board in 1946, by which time the company was supplying 1,095 customers. The Roe Valley station closed in 1965 in favour of a new oil-fired station at **Coolakeeragh**. The visitor centre in Roe Valley Country Park tells the story of Ritter's power station and the local linen industry. ②

A host of golden daffodils

The world's largest collection of daffodils grows in a special garden at the University of Ulster, Coleraine. Daffodils (*Narcissus pseudonarcissus*) are not native to Ireland – those you see growing 'wild' probably escaped from a garden – but Irish plant breeders have led the world for over a century in developing new varieties. One of the most successful breeders of all time was **Guy Wilson** (1885–1962), who was especially noted for his white-coloured varieties. Wilson was from Broughshane, Co. Antrim, and bulbs from his original stock go on sale there each August.

© An Post

The Coleraine garden, created in Wilson's memory, was begun in 1971 in an old quarry, and now has over 1,800 types of daffodil. Most were developed by Irish plant breeders, and they include some unusual pink cultivars bred in Co. Waterford by Lionel Richardson. Mid-March–April is the best time to visit.

The man who made Milk of Magnesia

Sir James Murray (1788–1871), a physician and chemist born in Culnady, discovered ways of transforming hard and insoluble minerals into soluble forms that could be taken up and used by plants and animals. His inventions gave rise to two successful businesses: Milk of Magnesia and the first artificial phosphate fertilisers. Magnesium and phosphorous are essential minerals, but the forms present in rock and soil are insoluble and of little use. In 1812, Murray discovered that solid magnesium carbonate would dissolve in water if carbon dioxide was bubbled through, producing a milky-white liquid solution of magnesia. Murray patented his new process, advocated its medicinal use for treating stomach upset and opened a factory in Belfast that proved very successful.

Sir James Murray: he invented the first artificial fertilisers and Milk of Magnesia

The waste from that factory included insoluble phosphate and, in 1817, Murray experimented at converting this to a soluble form, which he believed could be used as an **artificial fertiliser**. He found that sulphuric acid dissolved the phosphate, producing a soluble 'super-phosphate'. Guano and ordinary manure were added to make a rich fertiliser, and the resulting paste was mixed with sawdust to yield a powder that was easy and safe to handle. Murray patented his process, and crop trials were set up in Belfast and Dublin to demonstrate the power of his new fertiliser compared with ordinary compost. Belfast Botanic Gardens, for example, reported that Murray's fertiliser doubled the grass growth there. Murray's new fertiliser caused quite a stir: courses were organised around the country to promote its use and crowds attended public lectures on the topic.

But despite the obvious benefits and public interest, Murray failed to commercialise his discovery. Instead, the patent was bought by Englishman John Lawes, founder of the Rothamstead agricultural research centre, who used Murray's process as the basis for a successful factory making the first commercial artificial fertilisers. Murray returned to his medical career and his Milk of Magnesia business and, in 1831, he was appointed physician to the lord-lieutenant in Dublin. When cholera* broke out in Ireland the following year, there were many ideas about what caused the terrible disease. Some thought wearing poplin would keep it at bay. Murray believed cholera was caused by electricity and recommended that houses be insulated. His idea seems far-fetched now, but no one knew then what caused infectious diseases.

The linen factory tour

Ireland's oldest linen* factory, **Wm Clarke & Sons Ltd**, was established at Upperlands in 1726. The Clarke family still runs the factory, employing 200 people and producing over 1 million square yards of fine damask linen every year. Little has changed in three centuries and most of the intricate processing required to turn flax into linen is still done on-site: the 11-hectare complex includes scutching mills, where the flax is battered and the fibre removed; weaving sheds; and beetling mills,* where the woven fabric is hammered to give it a glossy sheen. The firm's beetling engines may well be the oldest in the world still in place and still in everyday use. The factory continues to draw power from the Clady river, which now drives three turbines instead of the original waterwheels. Very little flax is grown locally now, and most of the factory's raw material is imported. ③

Ancient oaks

Remnants of the original wild oak woods that once covered Ireland, and from which Derry takes its name, survive today in the steep gorges of the Burntollet river valley at Ness

Wood. Two oak species are native to Ireland: the pedunculate oak (*Quercus robur*), so called because its acorns are attached to the tree by a long stalk or peduncle; and the sessile oak (*Q. petraea*), whose acorns are attached directly to the twig with little or no stalk. Oak trees appeared in Ireland about 8,000 years ago. They are our natural climax vegetation – the final phase in the recolonisation of land that had been wiped clean by glaciers during the last Ice Age. This recolonisation began nearly 12,000 years ago with grasses and heathers, which gave way gradually to juniper, then birch and hazel, followed by pine, and finally oak and elm woods. For over 2,000 years, great oak woods dominated the Irish landscape, until the climate started getting wetter, bogs started developing and farming began. Yet Ireland was still heavily wooded even in the 1500s, witness the fact that Derry (*doire*, 'oak wood') features in over 1,600 townland names.

Oak leaves: dugout canoes and bog trackways were made from oak trees, which archaeologists can date precisely from the tree-ring pattern in the timber.*

The oak woods of Derry were finally felled during the Plantation of Ulster, when various London guilds were granted land in the county – Ness Wood, for example, was allotted to the Grocers' Company of London. The woods were cleared to provide timber for shipbuilding and barrel staves, and for charcoal to fuel the many furnaces around the country. The steep-sided valley at Ness, however, made felling and re-planting difficult. The natural wood had a chance to regenerate, and today boasts a fine mix of oak, holly, birch and rowan, all descended from Derry's ancient wild woods.

Coppin's *Great Northern*

On July 23rd, 1842, a public holiday was declared in Derry so that people could attend the launch of the *Great Northern*, the largest screw-propelled ship the world had

When it was launched in Derry in 1842, The Great Northern was the largest screw-propelled ship the world had seen, but it made only one voyage before being sold for scrap.

yet seen. The enormous ship was the work of a prolific marine inventor **William Coppin** (1805–95), who single-handedly kick-started Derry's development as a port and shipbuilding yard. Coppin was born in Kinsale, Co. Cork, and left for Canada as a teenager to study shipbuilding. In 1825, he designed a sledge-boat to run on frozen rivers, but did not build his first boat until 1829. The following year, in the West Indies, he met a group of Derry merchants who commissioned him to build a vessel for their transatlantic trade. The boat was delivered to Derry in 1831 and Coppin followed it, settling in the city and taking over the local shipbuilding yard. He built steamers for cross-channel traffic and sailing ships for the transatlantic routes, and, at one stage, employed 500 men.

The *Great Northern*, begun in 1841, was his most ambitious project. The first of the really large iron ships of the 1840s, it was a 3-masted sloop, 76 metres long. A 360-hp steam engine drove the two huge screw propellers on either side of the ship's rudder. Some 20,000 people gathered for the launch and, afterwards, hoping to find a buyer, Coppin took the ship to London. The enormous steamer created a stir when it docked and the *London Illustrated News* called it a "monument of marine architecture… a huge monster, but pleasing in her mould and trim". The voyage to London was the vessel's only trip. Coppin failed to find a buyer and, unable to pay the costly harbour fees, sold the vessel for scrap at a considerable loss. Undaunted, he returned to Derry to build a replacement, but this was destroyed by fire in 1846 before it could be launched.

Coppin designed a triple-hulled vessel in the 1880s, the *Tripod Express*, which sailed the Atlantic, but mostly he engaged in salvage work. He devised an ingenious technique for raising sunken vessels by pumping in air to refloat them, and invented an improved diving suit that had a novel breathing system: as well as a tube supplying fresh air pumped from the surface, Coppin added a second tube to remove the exhaled air. Previously, divers breathed out directly into the water, though with difficulty because of the pressure difference. Coppin claimed his suit enabled divers to spend an hour at depths up to 35 metres. Derry's Harbour Museum celebrates the city's maritime history, including Coppin, who is buried at St Augustine's Church in the town. ④

The Bogside

The walled city of Derry was built on a perfect defensive site, atop a steep hill at the mouth of the Foyle. This hill was often isolated as an island in prehistoric times – at the end of the last Ice Age, for example, torrents of glacial meltwater flowed around the hill, cutting a deep channel in places. And when sea levels were higher than today, the hill was again cut off from the surrounding land. The channel on the northern side of the hill was particularly deep and marshy, whence its name: the Bogside.

Crossing the Atlantic – first landfall

The shores of Lough Foyle are a convenient landfall for transatlantic flyers. Kansas-born aviator **Amelia Earhart** (1898–1937) became the first woman to fly the Atlantic solo when, on May 20th, 1932, she landed unexpectedly in a field at Ballyarnet. Her original destination was Paris, but she had been blown off course. Yet she predicted that "such crossings will become commonplace… [possibly even] regular scheduled transatlantic crossings". The journey from Newfoundland had taken her 13 hours and 15 minutes, twice as fast as Charles Lindbergh's crossing five years previously. On July 3rd, 1987, English businessman **Richard Branson** and his partner **Per Lindstrand** landed near Limavady, becoming the first people to cross the Atlantic by hot-air balloon. Their 5,000-km journey from Maine took nearly 32 hours. The Ballyarnet field where Amelia Earhart landed is now a park; a small display in a cottage there includes a one-quarter-scale working model of her aeroplane. (5)

Every autumn thousands of **migrating birds** arrive at Lough Foyle, their first landfall since leaving Greenland. Some will spend the winter there, for others this is a short bed-and-breakfast stop: they will feed for a while on the lough's lush eel grasses, then head south for warmer weather. **Lough Foyle** is wide and shallow, bordered by mudflats, shingle banks, salt marshes and reclaimed polder lands, offering a tasty variety of food for migrating birds. The area is famous for its birdlife, especially swans, and the RSPB holds a 1,300-hectare reserve between Longfield and the Roe estuary. Species to look for include whooper and Bewick's swans; brent, light-bellied and Greenland white-fronted geese; wigeon, whimbrel and snow bunting; and birds of prey such as kestrel, buzzard and, occasionally, gyrfalcon. Access points for the reserve are at Longfield, Faughanvale and Ballykelly Marsh.

There are other important sites around Lough Foyle's shores, notably the salt marshes, tidal flats and mussel beds of the **Roe estuary**, which attract wintering curlew and various divers. There is public access to part of the nature reserve at **Magilligan Point**, where red-throated divers, snow bunting and skua can be seen from the dunes in winter. (Neighbouring **Ballymaclary** nature reserve falls within the Ministry of Defence's firing range and access there is restricted.) Elsewhere, **Castlerock Strand** on the Bann estuary attracts duck; and **Ballyronan Point** on Lough Neagh boasts the largest concentration in Europe of scaup, alongside goldeneye and tufted duck.

The pearl makers

Pearls are usually associated with oysters and Pacific islands. Yet an Irish freshwater mussel also produces these lustrous stones and, historically, was a valuable source of pearls. In 1094, the bishop of Limerick presented an Irish pearl to the archbishop of Canterbury and, in the 1600s, Charles I used Derry's rich pearl beds to entice settlers to the new plantation by the Foyle. The **pearl mussel** (*Margaritifera margaritifera*) is one of several Irish freshwater mussels. A slow-growing animal that can reach 8cm long, it reputedly lives for up to 120 years, making it one of Europe's longest-lived animals. It lives in burrows in sand or mud, and likes big fast-flowing rivers. Its shell is yellow-brown when young, but darkens with age almost to black. According to a 17th-century account, 1 per cent of mussels contained a pearl, and 1 per cent of the pearls were "tolerably clear". The mussel shells were used to make mother-of-pearl and, in some places, the shells themselves were used as spoons.

The Foyle river god: crowned with Derry's walled city and dated 1689, the year of the Siege of Derry.
© Enterprise Ireland

Overfishing in the past, and now water pollution, mean the pearl mussel is no longer common and it is now a protected species, so the days of Irish pearl fishing are gone. Ireland's 20 other freshwater mussel species are mostly small and short-lived and none produces a pearl. They include the pea

The Derry punt

There are many native styles of **boat** in Ireland. Each evolved a unique form to suit its particular environment and function, like the Shannon ganglow, the Waterford prong and the Erne cot.* Many of those used in river estuaries are flat-bottomed, such as the Derry punt. The punt is a simple boat that evolved from a primitive raft-like craft. It is short (under 5 metres) but broad (at nearly 2 metres across), with a flat bottom made from half a dozen planks, and is rowed using oars. Small numbers of the Derry punt are still used for salmon draft-net fishing on the river Foyle, according to boat historian Darina Tully, and one is on display at Greencastle Museum, Co. Donegal.

mussel, the swan and duck mussels, which are found in slow-moving waters and canals, and the recently arrived zebra mussel.*

A rock and hard place

The tiny islet of Rockall is probably the most remote piece of dry land in the world. Located 250km west of St Kilda in the Outer Hebrides, it is just 18 metres high and 30 metres wide and, in high seas, probably disappears under the Atlantic. Bereft of plantlife, incapable of supporting habitation, too small and exposed for a lighthouse, well-nigh impossible to land on, barren save for a covering of bird droppings – and yet this speck of granite is a most interesting place. Rockall appears to be the highest point of a submerged landmass called the Rockall Plateau, which measures 900km by 450km. This is separated from the main European continental shelf by the deep Rockall Trough. Geologists now believe the Rockall Plateau formed 100 million years ago when the North Atlantic Ocean was opening up, and Greenland and western Europe were moving apart. There are two other reefs near Rockall, but neither breaks the surface, though they may have done in the past when sea levels were lower and when, presumably, more of Rockall would also have been exposed.

Rich fishing in the area attracts numerous birds and the resulting guano cap gives Rockall a distinctive appearance making it a useful navigational beacon, at least on the days when it can be seen. The Vikings probably knew of it; an Armada galleon spotted it on its storm-tossed journey home in 1588; and it is marked as a significant island on a French chart of 1640, leading some to suggest it is the basis for the

The legendary island of Hy Brasil? Rockall, sketched by Rev Green in 1896, is reputedly "further out to sea than any other rock of the same diminutive size in all the world" and "the smallest piece of new territory over which a flag has ever flown"

legendary island of Hy Brasil. People first set foot on Rockall in 1810: HMS *Endymion* chased what it thought was the white sails of a ship, only to discover it was Rockall and, in exceptionally calm weather, landed a party. Fifty years later, during seabed surveys for the first transatlantic telegraph cable,* HMS *Porcupine* landed a man who returned with rock samples which revealed that Rockall was made of granite.

In June 1896, the RIA mounted a major scientific expedition to Rockall, led by Robert Lloyd Praeger* and Rev William Spotswood Green (see panel). Their steamer left Killybegs with seven scientists plus dredges, trawls, harpoon guns and sounding machines but, in atrocious weather, the expedition had trouble even finding the rock. No one landed, yet, despite the weather, they gathered considerable scientific information, dredging samples from the shore, including three species of sponge never seen before. They also dispelled the myth that Rockall was the breeding ground for the great shearwater, a seabird now known to breed on Tristan da Cunha in the South Atlantic.

In the 1950s, Rockall's remote location took on a certain strategic importance and the rock was annexed by Britain when a party landed by helicopter and cemented a flagpole on the rock. The territory was soon the subject of numerous discussions and Sea Law conferences, but international treaties now recognise it as part of the British zone of the continental shelf. Because Rockall cannot support habitation, however, UK territorial limits are measured from St Kilda.

Rev William Spotswood Green (1847–1919), the Youghal-born mariner, cleric – he preferred a coastal parish – fishery inspector and mountaineer, founded Irish marine research. He often explored deep waters, once dredging samples from a record 2.3km, and a sea anemone discovered on one such trip was named *Paraphellia greenii* in his honour. Green wrote books about his climbing trips to Canada and New Zealand and is buried at Sneem, Co. Kerry.

Slieve Snaght in Co. Donegal was the first place in the world to bask in the limelight. The county's many riches include: diamonds (perhaps) and uranium; the majestic scenery of Glenveagh, the Poisoned Glen and Errigal; and the fossil beaches of Malin Head. The small island of Inishtrahull, at 1,778 million years old, is probably the oldest place in Ireland. County Donegal (*Dún na nGall*, 'the foreigner's fort'), is named after the town, and probably refers to the Vikings who took over a fort there in the 10th century.

SLIEVE SNAGHT

The first use of limelight

The first place in the world to bask in the limelight was Slieve Snaght on the Inishowen peninsula. Limelight is usually associated with theatres, yet it was first used on this Donegal mountain during the Ordnance Survey of Ireland* in November 1825. This bright form of lighting, produced by burning a block of lime in a flame of hydrogen and oxygen gases, was invented in 1816 by a talented Scottish engineer, **Lieut Thomas Drummond** (1787–1840). Drummond, who also invented an improved mirror heliostat (a device that reflects sunlight and which surveyors use when taking sightings by day), served with the Royal Engineers on the British Ordnance Survey and

Cartographers such as Drummond were not always welcome in Donegal: in 1602 the locals beheaded Richard Bartlett, a map-maker with Elizabeth I's army, "for they would not have their country discovered".

came to Derry with them in 1824 for the mapping of Ireland. Part of that task was triangulation (see *page 190*), which entailed taking sightings to poles erected atop distant mountains.

But the survey pole on top of Slieve Snaght was obscured by bad weather for months. The Argand lamps of the time were not powerful enough to be visible at 60 miles (96.5km), so Drummond proposed using his limelight. On the night of November 9th, 1825, a survey team carried Drummond's apparatus to the top of Slieve Snaght. The lime was set alight, and the brilliant beam was clearly visible against the night sky from the other side of the country.

Thomas Drummond: engineer, administrator and inventor of limelight.
Courtesy
Ordnance Survey
of Ireland

It was another 12 years before limelight, or 'Drummond light' as it was first known, was used in a theatre. It quickly proved popular, even though each lamp needed constant minding, as the block of lime had to be turned as it burned, and the two gas canisters tended. Drummond later fell out with his senior officer and left the army for a career in administration, becoming under-secretary to the lord-lieutenant at Dublin Castle. A conscientious and reforming administrator, he famously told landlords that "property has its duties as well as rights". He died young and is buried at Mount Jerome,* Dublin.

found in other geological formations, including some that occur among the older rocks of northern Donegal. In 1997, Cambridge Mineral Resources found tiny sapphires and rubies in samples from rivers at Carndonagh, Culdaff and Glentogher. These gems are indicators, often found in association with diamonds. More extensive tests are now under way to determine if Donegal has any diamonds. The Brookeborough diamond* is the only diamond ever found in Ireland, but its origins are a mystery. Ireland does produce masses of synthetic diamonds,* made at Shannon by De Beers and, in Dublin, by General Electric.

Digging for diamonds

The oldest place in Ireland

Ireland is not noted for its rare gems, but an English exploration company, Cambridge Mineral Resources, is looking for diamonds on the Inishowen peninsula. It acquired the first-ever Irish licence for diamond prospecting in 1998, and has since been drilling at several sites in the area. Diamond, the hardest substance known, forms deep within the Earth at depths of over 100km and at extreme temperatures and pressures. Subsequently, molten magma rising through the rock can carry the crystals closer to the surface. Diamond deposits (and mines) are thus associated with volcanic rocks but, if the volcanic material becomes exposed and eroded, the diamonds can be released and end up in river gravel. Diamonds also occasionally occur in meteorite impact craters, where they probably formed during the extreme conditions of the impact. Most diamonds are over 500 million years old, much older than the rocks that carry them. Sadly, they are not forever: the crystals are unstable and most decay to graphite before they reach the Earth's surface.

Most natural diamonds are found in a volcanic mineral called kimberlite, named after South Africa's Kimberley mine. (The Irish company, Jacob's Biscuits,* invented its Kimberley biscuit in 1893 when the South African mines were in the news.) In recent years, diamonds have been

The small, rocky island of Inishtrahull is the oldest place in Ireland. The rocks there have been scientifically dated to 1,778 million years old – and they show every trace of their venerable age. They are gnarled, much faulted and folded, and have been greatly altered by geological events throughout their existence (a type of rock geologists call gneiss). The minerals in the rock have also separated out over time and now form colourful bands. Geologists date ancient rocks by measuring the radioactivity of their zircon crystals, similar to the carbon-dating of archaeological artefacts.

Even before we knew its tremendous age, Inishtrahull was suspect, for its rocks are not like anything else in Ireland, and certainly unlike neighbouring Malin Head. This raises the question of how the island came to be there, and only in recent years have geologists solved the puzzle: Inishtrahull, it turns out, is actually part of Greenland. Its ancient rocks are nearly identical with those of Greenland, though also strangely similar to those of Islay in the Hebrides. Geologists now believe that millions of years ago a small piece of Greenland broke off from its motherland, and thanks to plate tectonics, floated away from its parent, anchoring first at Islay and finally off Malin Head. Over the past century, botanists have surveyed Inishtrahull thoroughly and, despite its barren surface, recorded over 100 species there. They include some plants usually associated with Scotland, among them Scots lovage and Scottish scurvy grass. The second-oldest rocks in Ireland, at Annagh Head* in Co. Mayo, are 1,753 million years old.

An Armada cannon

In 1556, Remigy de Halut, master gun-maker at the Malines foundry near Antwerp, then the centre of Europe's gun-making industry, cast an ornate 3-tonne cannon for King Philip of Spain. The gun was installed on a galleon, *La Trinidad Valencera*, which later joined the Armada fleet. In September 1588, *La Trinidad*, caught in the storm that scattered the invading Spanish forces, sought shelter in Kinnagoe Bay, near Greencastle. Kinnagoe is a most inviting beach in summer, but the Spanish vessel ran aground on a hidden reef 150 metres from the shore. Fortunately all on board were saved and most of the valuables removed before the ship broke up. Then it was forgotten, until a Derry subaqua club explored the wreck in 1971. The divers recovered various artefacts, most of which are now in the Ulster Museum,* but de Halut's gun is at **Greencastle Maritime Museum**. ⑥ The gun is of interest because de Halut was an innovative armourer: among the first to

The Greencastle yawl or drontheim, native to these waters of the northwest and one of the many local boats on show at Greencastle Museum. Sketch by Dónal Mac Poilín. Greencastle Museum

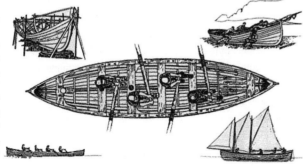

standardise gun types, he laid down strict formulae for calibre, barrel length, metal type, etc. He also introduced an early form of quality control, test firing each of his guns three times before they left his foundry.

Fossil beaches, flooding seas

Malin Head, the most northerly point in Ireland, is one of the best places in the country to see fossil beaches. These days we worry about rising sea levels brought on by climate change, but sea levels have changed continuously throughout geological history. During a warm period 120,000 years ago, sea levels were 6 metres higher than today. At the height of the last Ice Age, 20,000 years ago, seas were 60 metres lower than today, with land bridges linking Ireland, Britain and the continent. One reason so few hunter–gatherer sites have been found in Ireland is probably because most are now underwater.

Sea level is influenced by the amount of water present in the oceans (mostly determined by how much water is locked up in glaciers and continental ice sheets) and by the water's temperature. Land levels also vary, however. The rocks that make up the continents are relatively buoyant and, as well as moving sideways (continental drift), can also 'bob' up and down. During an ice age, the weight of the ice compresses the land below; but once the ice melts the land can start to rebound. Even 14,000 years after the last ice melted, the land in the north of Ireland is still rising. Initially it rose at about 6mm a year, but the rate has now slowed to about 1mm on average a year. The south of Ireland, which was under less ice, has stabilised and may even be sinking slightly to compensate for the lift in the north.

The net effect, taking account of changes in both the land *and* sea levels, is called Relative Sea Level (RSL) and is of most practical importance. All around Ireland's coast, there is evidence of changes in Relative Sea Level. Submerged peat deposits, as at Ballycotton in Co. Cork, were once dry land and fossil (or raised) beaches, now left high and dry, were once underwater. At Malin Head, three main terraces of fossil beach can be seen. They formed at various times over the past 15,000 years and record the complex interplay between changing land and sea levels. Fossil sea cliffs, with fossil shingle beaches at their feet, can also be seen in the area. Ballyhillin village, the most northerly village in Ireland, sits on the crest of the highest raised beach, at 25 metres above the present sea level. The other fossil levels are at 16 metres and at 8 metres above sea level. They can also be seen at Lough Foyle – though the lowest and newest shoreline at Lough Foyle is not a fossil beach but polder land that was reclaimed for farming in the 19th century – and along the stunning Antrim Coast Road,* which in places runs on a fossil shoreline.

Tide gauges worldwide have recorded a rise in relative sea level of nearly 15cm since 1880, and climatologists are predicting a possible sea level rise of up to 6 metres by 2100. This would be a return to the levels that existed 120,000 years ago and have major implications for coastal and low-lying areas – anywhere that lies close to sea level, including, for example, parts of Dublin, Belfast, Cork and Galway cities would be increasingly prone to flooding.

The Malin datum

Since 1959, all the spot heights recorded on OSI* maps are measured relative to mean [average] sea level at Malin Head, based on a decade of observations and known as the Malin datum. Previously, the reference point was one recorded at low tide at Poolbeg lighthouse* in Dublin Bay in 1837. Malin was chosen because it is open sea and not a channel, which might have distorted the tidal measurements. The difference between the old and new systems is 2.7 metres – heights on all pre-1960 maps are 2.7 metres higher than modern ones. Northern Ireland maps use a Belfast datum, which differs from the Malin one by only 2cm.

"This new way of reasoning"

The pioneering free thinker **John Toland** (1670–1722) was born at Clonmany into a Catholic and Gaelic environment. A controversial figure and prolific writer, he coined the word 'pantheist' and helped shape debates about the nature of God and the Universe in the new Enlightenment of the 18th century. Toland's ideas also influenced the scientist Isaac Newton, though the two never met. Toland, allegedly the illegitimate son of the local priest, was a radical from an early age, converting to Protestantism at 16 and later becoming a Dissenter. He studied divinity and philosophy at Edinburgh, Leyden and Prague and became fluent in a dozen languages.

The only known portrait of John Toland, radical Irish free thinker and deist.

In the late-17th century, questions about the nature of religion, the Universe, matter and even politics were intimately connected. Hotly debated issues included whether God had any sense organs and how a remote God could influence proceedings on Earth. Many philosophers invoked the idea of a cosmic ether as a medium for God's action. Toland believed in pantheism, however: God was everywhere, and so there was no need for an ether (the concept of the ether was eventually blown away 200 years later by Einstein). When it came to religion, Toland argued that people should be allowed to think for themselves. Conviction, and not popish authority, should guide their beliefs. Toland's ideas were seen as dangerous and seditious for, if the Universe had no need of a grand ruler, then neither did any country. And unlike Newton's regular clockwork universe, Toland envisaged a world where matter – and politics – were in constant flux.

After the censorship laws were relaxed in 1695, Toland published his most controversial book, *Christianity Not Mysterious* (1696). It met with a hostile reception and was burned by the common hangman at the gates of the Irish parliament. Toland, who was visiting Dublin, fled and never returned. Denounced from pulpits, vain and arrogant, in debt and dangerous to know, Toland now sought work and patronage around Europe, developing his ideas and publishing widely. He came to be part of a philosophical circle at the Prague court, and with Locke was described as the "gentlemen of this new way of reasoning". At odds with Newton on many points, he nevertheless played a major part in shaping that great scientist's arguments, especially his work on *Optics* (1706). Toland left a rich legacy of writings, publications and ideas, but died a pauper in London.

NEWMILLS

Food and fabrics

There is a cornmill side by side with a flax-scutching mill at Newmills, an unusual combination of food and textile processing that reflects the historic importance of the linen industry* to Ulster farming. Both mills are powered by the River Swilly (*súileach*, 'of the eyes or eddies'), and there has been a mill there for nearly 400 years. The 3-storey cornmill is driven by an enormous wheel over 7 metres in diameter. Made at Stevenson's foundry in Strabane in 1867, it drives the millstones, sieves, elevators, fans and sack hoists. The operation was highly mechanised and could easily be managed by one person, who controlled the waterwheel from inside with a small sluice. The mill, in use until the mid-20th century, worked year round, grinding grain until May, and mixing animal feed in the summer. At the smaller scutching mill, the flax was beaten to extract the linen fibre; the woody residue could be burned as a fuel. This mill survived because of a revival in flax growing in the area during World War II, but it closed in 1955. The mill complex includes a grain-drying kiln, a public house, grocery, miller's residence and forge. ⑦

Compared with the industrial-scale enterprises of the other Ulster counties, Donegal's linen* industry was small. The county's **woollen industry** was extensive, however, and ranged from small cottage work, to large commercial mills, providing jobs and additional income in a rural region. In 1783, a weaving centre opened in a 4-storey mill at Tullyarvan and, by 1805, it was employing 600 people. The mill has been restored as a textile museum. ⑧

PORTSALON

The flora of Donegal

The person who surveyed the flora of Donegal more than any other was the explorer, athlete, naturalist and scholar, **Henry Chichester Hart** (1847–1908). In his *Flora of Northwestern Donegal* (1879), which covers just one part of the county, he listed nearly 500 species. The family estate was on the Fanad peninsula where, according to Hart, his people "had been settled since Elizabethan times". A keen botanist and explorer, he spent two years in the Arctic as a naturalist with the *Discovery* expedition of 1875, and later travelled to Palestine. Some 20 years after his flora of northwest Donegal was published, Hart compiled a botanical survey of the whole county. This comprehensive 400-page book includes an appendix of plant names and plant lore.

Hart was a prolific scholar: he published numerous books on flora, including one of Howth and another of the Aran Islands, and a study of the Ulster dialect, and was an editor for the Arden editions of Shake-speare. He is buried on the family estate near Portsalon, at a site he chose for himself.

ÉIRE 28

Saxifraga hartii

This rare saxifrage was discovered in northwest Donegal by Hart and later named after him. Many of the plant specimens Hart collected in Donegal and around the world are in the herbarium at the National Botanic Gardens. *
© An Post

Overwhelmed by sand and rabbits

Ireland's great glen

There is no mention of rabbits in ancient Irish tales or the *Book of Kells* and for good reason. The rabbit is an alien species, brought here in the 12th century by the Anglo-Normans. They kept rabbits for meat and fur, farming large numbers in warrens, or *cunicularia*, usually in sandy areas as at Rosapenna, for example, and at Magilligan* in Co. Derry. The earliest reference to an Irish warren is at Lambay* in 1191. Rabbits quickly became commonplace and place names containing the element 'warren' or 'burrow', such as Ballyteigue Burrow in Co. Wexford, date from that era. In the 1600s, Ireland exported 2 million rabbit skins annually. Even in the 1800s, Magilligan Warren was exporting 50,000 skins a year, and most of the rabbit meat at Manchester market came from Ireland.

A downside of these enormous warrens was that the many burrows could destroy the fragile sand dunes and trigger rapid erosion. When that happened, the sand would be blown away and deposited elsewhere, often burying a neighbouring village or farm. Rosapenna disappeared under the sand several times over the centuries and, in 1784, a dozen farms were buried. On that occasion, the damage was attributed to foxhunting: with fewer foxes to control the rabbit population, their numbers had dramatically increased.

Shifting sands also shaped nearby **Mulroy Bay**, with its numerous quiet bays and small lakes, great stretches of sand, dunes and machair,* and a consequent confusion of tides and currents. The Rossguill and Fanad peninsulas were probably isolated as islands at the end of the last Ice Age. But 6,000 years ago, the winds and tides began to deposit sand and gravel in the area, and the two islands gradually became linked to the mainland by tracts of new land (Mayo's Mullet Peninsula* formed in a similar way).

Glenveagh National Park has some of Ireland's most majestic landscape. Thanks to a varied geological past, this corner of Ireland has the great glen of the Gweebarra fault line, the towering Derryveagh and Glendown mountains, the awesome Poisoned Glen and the distinctive cone of Errigal. In scenery and habitat, this part of Ireland has much in common with the Scottish Highlands.

Scotland's great glen, the long narrow glen of Loch Ness, is a major geological fault that nearly cuts the country in two. A smaller version of this is the **Gweebarra fault** that cuts across Donegal in a straight line from Lettermacaward in the southwest, up through the valleys of Gweebarra and Glenveagh, to Glen in the northwest. Though not as impressive as Loch Ness, it forms an obvious feature on the map: a blue line of inlet and river, dotted with a string of lakes that is broken only by the high ground at Ballaghgeeha Gap, which forms the watershed between Gweebarra and Glenveagh. This fault is one of many in Donegal, and all have a similar southwest–northeast orientation. They formed 60 million years ago as Greenland pulled away from Europe, and the North Atlantic Ocean opened up, and the rock walls on either side of the Gweebarra fault were wrenched past each other. Earthquakes then also generated numerous fractures and zones of weakened rock, which were later eroded and subsequently scoured out by glaciers, producing deep valleys, as at Glenveagh.

The granite mountains that border the Gweebarra fault formed 400 million years ago when the two ancient continents of Laurentia and Avalonia collided, closing the Iapetus Ocean.* Rock melted around the collision zone – which, in Ireland, runs from Limerick to Louth – and this molten magma later rose towards the Earth's surface. In Donegal, it came within 5km of the surface, before slowly solidifying, forming a granite intrusion. The cap covering this rocky mass was eroded over millions of years so that now the granite is exposed. Heat and pressure from the molten granite cooked and compressed the neighbouring sandstones, converting them to quartzite. Any pebbles in the sandstone were compressed and elongated by the powerful forces. Originally these sandstones were flat sediments, but powerful upheavals later pushed them up to form mountains as at **Errigal** and **Muckish**. The sand in the Muckish quartzite is pure silica. On the mountain's northern face, weathering has leached out the cement that once bound the grains together, releasing a fine white silica sand that has been quarried on occasion. Its purity makes it ideal for use in optical glass.

Errigal and Muckish are part of a chain of quartzite mountains that stretches as far as Croagh Patrick in Co. Mayo. All were originally sandy dunes at the edge of the ancient continent of Laurentia. Their quartzite reacts differently to weathering than the neighbouring granite: granite forms smooth summits, but quartzite shatters, producing scree piles and mountains with a distinctive sharp profile. During the last Ice Age, glaciers flowing north from Derryveagh breached the Errigal–Muckish ridge at three points, creating the passes of Lough Altan, Lough Aluirg and especially Muckish Gap, where the ice gouged a pass that is 100 metres deep, and is exploited today by the Falcarragh Road.

The other rock type in this region is basalt – this dates from 60 million years ago and the same series of events that created the Giant's Causeway.* In Donegal, the basalt was injected into fractures in the granite to produce basalt dykes. Being softer than granite, these dykes erode faster, leaving steep gulleys on the glen sides. (Hill walkers should note that this basalt contains a magnetic mineral called ilmenite, which can interfere with a compass.)

The **Poisoned Glen** is an awe-inspiring, ice-carved corrie. It once held a glacier that flowed north towards Dunlewy, and debris left by the glacier still covers the valley's marshy floor. The glen is named for a poisonous spurge that grew there. Other evidence for glaciers in the vicinity includes the perfect U-shaped valleys of Glenveagh and Gweebarra, with their ice-carved cliffs, hanging valleys, truncated spurs and moraines. Glenveagh was excavated by a glacier that was at least 250 metres thick and flowing north, while Gweebarra was sculpted by a separate glacier flowing south.

The climate in this part of Ireland can be harsh, averaging only five hours of sunshine a day in summer and just one hour a day in December, and the habitats are similar to those of the Scottish Highlands. The short growing season and poor soils mean only specialised plants and animals survive there. Arctic-alpine species grow on the exposed mountain summits, including rose-root and starry saxifrage; Arctic birds overwinter there, among them snow bunting and ring ouzel. The majestic **golden eagle**, shot to extinction in Ireland by the 1930s, is being reintroduced at Glenveagh, with chicks from Scotland.

The national park boasts a number of important habitats: extensive tracts of intact lowland blanket bog that are of national importance; **Derrybeg Bog**, on the flat ground at the head of Lough Veagh, is an old quaking bog with 4 metres of accumulated sediment and peat, recording 8,000 years of climate and vegetation; and **Mullangore Woodland**, rich in birch and oak, is a remnant of the native wild wood for which Glenveagh is named. **Lough Veagh** has a unique population of arctic char: landlocked there since the end of the last Ice Age, the fish have evolved separately and are now smaller than those found in neighbouring Dunlewy Lake.

Glenveagh Estate, at the heart of the national park, was established in the 1850s by John Adair. All the tenants were evicted in 1860, although Glenveagh valley itself, being remote, was probably never inhabited. Native red deer had disappeared from the area by 1845, but a fresh herd was established there in the 1880s; a 45-km fence built to contain the deer is reputedly Ireland's longest. The estate was bought in 1936 by Henry McIlhenny, whose grandfather John, born at Carrigart in 1830, made his fortune in Philadelphia having invented the first successful **gas meter**. In 1981, McIlhenny presented the estate to Ireland. Conservation work in the park today includes controlling deer numbers, protecting old woodlands and removing invasive rhododendron. ⑨

Prospecting for uranium

In the mid-1970s, exploration companies travelled Ireland in search of uranium. Nuclear power stations were being built around the world – Ireland's ESB was even considering one at Carnsore Point* in Co. Wexford – and the search was on for new uranium sources. Gweebarra Valley was one of the areas surveyed. Uranium is a heavy whitish metal that occurs naturally in various rocks, typically in granite, such as that found in Donegal. The heaviest naturally occurring element, it is unstable and decays slowly, releasing energy in the form of radioactivity, though this was not realised until 1900. It was the 1930s before any use was found for uranium, or at least its radioactivity, in atomic bombs and nuclear power stations and, more recently, to coat armour-piercing weapons. Uranium occurs in the Earth's crust at about four parts per million, making it more common than gold and silver. It seldom forms commercial deposits, but this is fortunate, for if the concentration in any one place was high it could form a natural nuclear reactor.

In the 1970s, uranium exploration began in Ireland in earnest. Radiation sensors were mounted on cars and planes that toured the country to find radiation hot spots. Three potential mine sites were identified: around Tullow in Co. Carlow, Thomastown in Co. Kilkenny and, especially, at Gweebarra. Trenches were dug at Gweebarra, bore-holes drilled and rocks samples removed for analysis. In the end, the exploration came to naught: the economic and political climates changed, the ESB dropped its controversial plans for Carnsore, the exploration trenches at Gweebarra were filled in and the mining companies withdrew. The 1970s uranium survey found what may be the most radioactive place in Ireland, in a boggy field near Liscolman,* Co. Wicklow.

The black pig returns

The county that once had the most extensive network of narrow-gauge railway in these islands now has no railway service at all, and just one short functioning stretch of track: 2km of line that has been restored along Lough Finn. **Narrow-gauge railway** has a track width of 3ft (0.9 metres), compared with the Irish mainline gauge of 5ft 3in (1.6 metres). It is lightweight, cheap to build and ideal for the boggy and often difficult terrain of rural Ireland, especially as these regions were often lightly populated. Extensive local networks were laid in the second half of the 19th century that helped bring agricultural produce and fresh fish to market, and the benefits of the railways to remote parts of the country. At their peak a century ago, there was 800km of narrow-gauge railway in Ireland, and 40 per cent of that was in Donegal. (Narrow-gauge railway is also ideal for traversing bogs and Bord na Móna* has over 1,000km of narrow-gauge track, probably the largest industrial railway in the world.)

The southern part of Co. Donegal was served by the County Donegal Railway (CDR) from the twin towns of Ballybofey and Stranorlar. The County Londonderry & Lough Swilly Railway (CL&LSR) served the northern half from its headquarters at Derry city. Both lines opened in 1863 and, over the next 50 years, a veritable web of branch lines was laid down to Letterkenny, Carndonagh, Killybegs, Glenties and Ballyshannon. Connections between the two networks could be made at Letterkenny and Derry. Perhaps the most ambitious undertaking was a branch line linking Derry and Burtonport that opened in 1903. Powerful locomotives were needed to tackle the steep gradients encountered, but the 100-km journey still took five hours.

Partition badly affected Donegal's railway network, especially the CL&LSR, which had its headquarters in Northern Ireland but most of its track in the new Republic. Duties and border customs were just some of the obstacles it faced. In the 1930s, the Donegal companies pioneered the use of diesel railcars – small, tram-like vehicles. But the declining fishing industry and the increasing move to road transport spelled the end. Most of the rail traffic had ceased by the 1950s and the last vestiges of this once great network closed in 1960.

At Fintown, once an important trading post on the CDR route, railway enthusiasts have renovated 2km of track and excursions are run in summer using a former mining locomotive. ⑩ When the railway arrived in Fintown in 1895, belching and snorting, it was christened *an mhuc dhúbh* ('the black pig'). Now the black pig has returned, at least for a short stretch. Enthusiasts undertake engine and coach restoration work at the Railway Heritage Centre in Donegal town. ⑪ Railway memorabilia can be seen there, and also at Donegal County Museum in Letterkenny ⑫ and St Connell's Museum in Glenties. ⑬

SHESKINMORE

A hidden lagoon

They say that, if you are not familiar with the area, it can take days to find Sheskinmore. But the effort amply repays those interested in birds and plantlife. The rich machair* grasslands are internationally important: in summer they are alive with insects and wild flowers, especially limestone-loving orchids; in winter they provide safe grazing for flocks of barnacle* and Greenland white-fronted geese. A ridge of sand dunes separates the lough from the sea, though these have suffered severe erosion in recent years. The lough itself is a freshwater coastal lagoon, noted for its rare aquatic species, including the **slender naiad** (*Naias flexilis*). This is an elegant rush-like plant, with slender, toothed leaves and an angular branching stem. It likes a warm climate, and its remains have been found around Ireland in sediments dating to the warm interval before the Munsterian Ice Age (300,000–130,000 years ago). The slender naiad survives as a relict species,* growing in just a few places along the west coast; it is most common now in North America. BirdWatch Ireland and the National Parks and Wildlife Services jointly manage Sheskinmore. They are restoring the water level in the lake, which was lowered by drainage works, and restricting grazing by farm animals to protect the machair grasslands.

Other **nature reserves** in Donegal include: Pettigo Plateau (an excellent blanket bog and wet heath); the rich alder woodland at Duntally Wood; an old oak wood at Rathmullan on the shore of Lough Swilly; and an old coppiced oak wood at Ballyarr Wood. **Wildfowl sanctuaries** in the county are at the Ards, Blanket Nook, Lough Fern, Trawbreaga Bay and Dunfanaghy Lake. **Dunfanaghy** had a second lake for a while 80 years ago. During World War I, marram grass in the dunes there was cut for thatching, but this destabilised the dunes. The sand was blown away and deposited in the bay, blocking the estuary. Farmland flooded, and a new lake was created. Today, controlled planting of marram grass helps stabilise the dunes and ensure Dunfanaghy has just one lake.

KILCLOONEY

An ancient meteorite impact crater?

About 400 million years ago, some geologists think that a large meteorite,* possibly measuring 1km across, crashed to Earth somewhere between Kilclooney and Ardara. It would have created a big crater, fracturing the Earth's surface and leaving lots of debris; rock deep beneath the crater would have melted with the heat, risen closer to the surface (forming what geologists call a granite pluton) and filled the cracks. The impact crater – if it existed – has now disappeared, removed by erosion over the intervening millennia. But a few traces survive, enabling geologists to piece together a story. Some believe that what we can see today is the crater's root which, 400 million years ago, lay

10km below the surface. According to Malcolm McClure, who prepared the Dolmen Centre's geological exhibition, one vital piece of evidence is a circular granite pluton, measuring 10km across and encompassing Kilclooney and Ardara; this is surrounded by a ring of metamorphic rock, material that was altered by the heat and pressure of the impact; there are also breccias (rocks made from broken gravel) and other structures suggesting that tremendous damage occurred there long ago. The Dolmen Centre, which has an exhibition about the local geology, is powered by a wind turbine, solar panels and geo-thermal heat pump, and has a reed sewage bed. (14)

RATHLIN O'BIRNE

A nuclear lighthouse

The channel between the mainland and the small island of Rathlin O'Birne is deep and subject to dangerous currents – there has been a light on the island since 1856. Finding an appropriate and reliable energy source to power lighthouses* in remote or awkward locations has always been a problem. Coal, gas and various oils were used in the 19th century; wind, wave and solar power are increasingly the choices of today. But, in 1974, the lighthouse on Rathlin O'Birne was the location for an international experiment, when it became one of the world's first nuclear-powered lighthouses (there had been at least one earlier nuclear trial at a Scottish lighthouse in the 1960s). The pioneering project was a joint venture between the Commissioners of Irish Lights and Britain's Atomic Energy Research Establishment (AERE). The AERE had developed a mini-nuclear generator, a small version of the type used on nuclear submarines. In August 1974, one of these generators was installed at Rathlin O'Birne, the only lighthouse it was ever tested at. In the end, the generator was deemed inappropriate for a lighthouse and, when it ran down after 10 years, it was replaced by a wind turbine. Today, Rathlin O'Birne is solar powered, as are an increasing number of Irish lights.

SLIEVE LEAGUE

The flagstone cliffs

The tallest sea cliffs in Europe are said to be those at Slieve League, which drop a vertiginous 595 metres from the summit to the waves below – the Cliffs of Moher are the sea cliffs with the longest *sheer* drop in Europe, a vertical fall of 205 metres. The ridge that runs along Slieve League's summit has just enough room for an exhilarating track known as One Man's Path, which is as wide as its name suggests. This path is probably the only true arrête in Ireland: it separates two corries, each of which was formed by a glacier biting deep into opposite sides of the mountain; the outer corrie is drowned by the sea, the inland one contains a small lake. The cliff above that lake is an important habitat for arctic and alpine species, including alpine clubmoss and dwarf willow, and was a popular haunt of noted Donegal botanist, Henry Chichester Hart.* The rocks of Slieve League (*Sliabh Leic*, 'flagstone mountain') are colourful quartzites: originally sandstones, they have been much altered by geological processes over millions of years.

DONEGAL TOWN

A treatise on plague

A noted 17th-century medical doctor, **Niall Ó Glacan**, was born near Donegal town in 1600. Medical knowledge in Ireland then was still mostly the preserve of powerful families like the O'Lees (from *liaig*, 'leech', the blood-sucking animal often used in medical treatment). By the late 1600s, however, this old Gaelic system had given way to a new breed of university-educated medical doctor or physician, many of whom trained on the continent. Niall Ó Glacan spans that changing era: born to a Gaelic family, he studied abroad and became professor of medicine at the University of Toulouse. His *Tractatus de Peste*, an early treatise on the plague,* earned him an international reputation, and his later 13-volume *Cursus Medicus* became a standard textbook and was widely published.

A unique linen beetling mill at Wellbrook, a beautifully preserved printing works at Strabane and the chance to pan for gold in the Sperrins, are just some of the county's attractions. Tyrone (*Tír Eoghan*, 'Eoghan's land') is named after a son of Niall of the Nine Hostages.

The first 'lady computers'

In 1890, the Royal Observatory in Greenwich recruited a small team of women graduates to work as 'lady computers'. It was a pioneering attempt at positive discrimination, designed to employ some of the new university-educated women. Those recruited were the first women to work professionally in astronomy in these islands. Among them were two women from Northern Ireland: **Annie Scott Dill Russell** (1868–1947), who was born in Strabane, where her father was a Presbyterian minister; and the inventor **Alice Everett** (1865–1949), who was born in Glasgow but grew up in Belfast, where her father Joseph Everett was professor of science at the Queen's University.[4] Unusually for women at the time, Alice Everett and Annie Russell went to university, entering Girton College together in 1886. Girton, established at Cambridge in 1873, was the first British women's college of university rank; but, while women could sit examinations, it was 1948 before the college awarded them degrees.

In 1888, William Christie, head of Greenwich Observatory, sought permission to recruit lady computers (computers being people who performed routine but often complicated calculations). Permission was given for eight recruits, but civil service rules allowed women to hold only lowly positions, so the salary offered was £50 a year. This proved a major problem in attracting qualified women, and the scheme was dropped after five years. Annie Russell was so interested in the astronomical work, however, that she left her job as a mathematics schoolteacher and took a 50 per cent cut in pay to go to Greenwich. There, she worked with Edward Walter Maunder, whom she later married, on a major project photographing sunspots and recording their position and size. Alice Everett worked on the international *Carte du Ciel* survey, photographing the entire sky and cataloguing all stars brighter than magnitude 11. Both women were keen astronomers, despite the nightwork, and the fact that their impractical clothes made it awkward to use the telescopes and other instruments. Their Greenwich (male) colleagues proposed them for membership of the Royal Astronomical Society, but that all-male society refused them.

In 1895, Alice Everett moved to the Astrophysics Observatory at Potsdam, becoming the first woman employed at a German observatory. Later that year, Annie Russell resigned to marry Maunder, for the civil service had a bar on married women. Russell was the last lady computer at Greenwich and the pioneering programme ended when she left. Annie Russell Maunder and Alice Everett continued their scientific work, however. Annie collaborated with her husband, becoming a noted solar observer, especially of eclipses, and later an authority on religious and ancient astronomies. In 1916, she was at last made a fellow of the Royal Astronomical Society, the ban on women having been lifted in 1915.

Alice Everett went on to a career in physics, including many years research at the National Physical Laboratory near Reading. On retiring aged 60, she began a second career in the new technology of wireless radio and television. In 1933, she designed and, with the Baird Television Company, jointly patented an improved version of Baird's mirror drum, the scanning device used in early TV transmissions. Everett had found a way to add more mirrors and improve the scanner without making the drum bigger, but her design was never commercialised as, by that time, newer scanning devices were in use.

Ireland's oldest-surviving print workshop – Gray's of Strabane, now run by the National Trust.
© National Trust

newspaper. Gray's, the oldest-surviving Irish print workshop, is on the town's Main Street. Located in the courtyard behind the printer's shop, it contains everything from an 18th-century press to 19th-century hand-printing machines; much of the equipment is still used for demonstrations. ⑮ Gray's also has one of the oldest, traditional shop fronts in the country: the elegant Georgian double-bow fronted façade is typical of the late 1700s. (See also: National Print Museum,* Dublin.)

God, gravel and bog

The Sperrin Mountains in Co. Tyrone may look bleak, but there is gold* in those hills. Indeed, **Ireland's first commercial gold mine** of modern times opened at nearby Cavanacaw in 2000. People have known about the gold for centuries, and Gerard Boate* recorded its occurrence "in nether Tyrone" in his *Irelands Naturall History* (1652). Cavanacaw, south of the Sperrin Mountains, is actually one of Tyrone's smaller gold deposits – the Sperrins have more substantial deposits, and mining companies prospected there for nearly 20 years, but that gold lies deep and extraction would require explosives. Official permission for explosives was not forthcoming in the recent past and so, for the moment, the Sperrin gold lies untouched. Small flecks occasionally wash out into local streams, however,

The power of print

Printing technology revolutionised human society. It enabled stories, information and documents to be shared, distributed, stored and archived. It fostered literacy and fomented revolutions. Yet the technology involved was relatively simple, although one of the principle prerequisites was a ready supply of paper.

Strabane was once an important Irish printing centre. It was a Strabane man, John Dunlap, who printed the American Declaration of Independence in Pennsylvania in 1776, and the *Pennsylvania Packet*, North America's first daily and at the Sperrin Heritage Centre you can try panning for gold in the Glenelly river.

Cavanacaw mine is a small opencast pit where Galantas Irish Gold digs 100 tonnes of earth each summer, yielding small but worthwhile amounts of gold and silver. The minerals occur in quartz veins in the schist rock, and each tonne of rock contains about 15g each of gold and silver. The ore is processed abroad, and the gold and silver are returned for making into Galantas's range of Irish-designed jewellery.

The landscape around the Sperrins holds plenty of evidence that the region was once glaciated, notably the kettle lakes at the Murrins nature reserve and around Gortin, and the extensive **sand** and **gravel** deposits throughout the area. Glaciers flowing down from the Donegal Mountains during the last Ice Age deposited the sand and gravel, both as sinuous eskers,* and as substantial moraines, some of which are still 10 metres high. These deposits are a feature of Boorin nature reserve, where heathland, birch and oak wood have developed on the well-drained, gravelly terrain. **Kettle lakes** occur in small depressions, or kettles, which formed after the Ice Age where large boulders of ice that were buried in the gravel melted, and the ground's surface

Cloudberry land

The cloudberry is a low-growing, strawberry-like plant that is usually found in Scandinavia. It also grows in Ireland, but only on the north-facing slopes of the Sperrin Mountains, where it was first discovered in the 1820s. It is probably a glacial relict,* a species that was widespread in Ireland when the climate was colder at the end of the last Ice Age, but which now clings on in just a small area.

The creeping cloudberry is a member of the rose family, with downy leaves and white flowers. The fruit is an eye-catching yellow, clustered berry, but conditions in Ireland are less than ideal now and the Sperrin plants seldom produce fruit. They do spread vegetatively, however, sending out runners, and flourish in their Tyrone niche.

subsided. The sand and gravel deposits were heavily quarried in the past; Tyrone also has significant deposits of fireclay and coal* near Coalisland, and lignite* by the shores of Lough Neagh.

Two educational centres introduce visitors to this often overlooked part of Ireland. At Sperrin Heritage Centre visitors can enter a tunnel that recreates conditions under a glacier, and take a guided walk over the post-glacial landscape and blanket bog; an exhibition explains the geological processes that formed the gold. ⑯ At An Creagán (Creggan) visitors can explore the **Black Bog**: some 12 metres deep in places, this 200-hectare site is one of the largest intact raised bogs* in Ireland and it is now protected as an area of special scientific interest. Pollen tests reveal that this extensive hummock-and-hollow complex of sphagnum sponge has been growing for 7,000 years. Special trails provide limited access to the fragile ecosystem. ⑰

Other features of note in the area are the Esker Road (built along a snaking esker ridge from Dunamore to Blackrock), old field systems exposed by turf cutting, and similar to those at the Céide Fields,* and the prehistoric monuments of Beaghmore (including seven stone circles, 10 stone alignments and a dozen cairns).

Tyrone's other **nature reserves** are at Killeter Forest and Meenadoan, which are noted for their bog. The county has several important **wildfowl sites**: Grange, on the River Foyle north of Strabane, is noted for winter flocks of Bewick's and whooper swans and greylag geese; Greenland white-fronted geese winter at Annaghroe on the Blackwater river. Kestrel can be seen in Strabane Glen, and Sawel Mountain in the Sperrins is noted for peregrine falcons and golden plover.

The astronomer of Good Hope

The astronomer Sir Thomas Maclear

When the explorer David Livingstone needed advice in 1850 on how best to calculate his position while on expedition, he turned to his friend **Sir Thomas Maclear** (1794–1879), a Tyrone man who for nearly 40 years was 'Her Majesty's Astronomer of the Royal Observatory at the Cape of Good Hope'. Maclear, born in Newtownstewart, trained as a doctor, but he was a keen amateur astronomer and in 1833, despite his relative inexperience, was appointed to the South African observatory. He arrived at the Cape in 1834 with his wife, five children and a governess. During his time there, Maclear came to Livingstone's aid, took part in Herschel's mammoth survey of the entire sky, and carried out a gruelling geodetic survey (published in 1866) that formed the basis for the later mapping of South Africa. Maclear is buried under southern skies in the grounds of the observatory. A crater on the Moon was named in his memory.

The Blacker 'hedgehog'

The 'hedgehog' was just one of many weapons devised by a military inventor from Lisnahanna, **Latham Valentine Blacker** (1887–1964). Blacker's experiments apparently began early, and he reputedly destroyed his teacher's glasshouse in England using a home-made mortar. He joined the British Imperial General Staff in the 1920s where his talents were appreciated – many inventors are never financially successful, but Blacker was paid £25,000 for his designs. His hedgehog showered mortar bombs at its target and was especially effective against submarines, claiming some 50 U-boats in World War II. Another of his mortar devices, the Blacker bombard, was used against helicopters and tanks and some 120,000 bombards were deployed in the early 1940s. Blacker was also a pioneering aviator and keen explorer and, in 1933, he led the first expedition to fly over Everest.*

A main-road railway

For over 50 years, the **Clogher Valley railway** chugged slowly to and fro between Clogher and Fivemiletown. Though the line closed in 1941, it is remembered with affection. The hilly terrain posed problems for the small engines and meant the line was never noted for speed or punctuality. On uphill stretches, it is said that first-class passengers were asked to get out and walk, while those travelling third class were asked to get out and push! For most of its 60-km route, this narrow-gauge line – in truth more like a tram than a train – travelled on the main road, and in Fivemiletown, the track ran along the main street. With a stop every couple of kilometres, it provided an essential local transport service in the days before cars were common. Fivemiletown Library has some of the railway memorabilia. ⑱ Clogher Cathedral has an early Christian sundial carved on a standing stone.* The face is partitioned for the monastic prayer schedule; on one side of the pedestal is an interlaced design and fish, and on the rear is a human face. The sundial formerly stood in the graveyard, but it is now inside the church.

The linen beetling mill

The characteristic sheen and smooth finish of damask **linen*** is produced by pounding the cloth for up to nine hours with heavy wooden hammers, or 'beetles'. It was part of the complex process of making linen and time was when every Ulster linen centre and village had its own beetling mill. Each mill had several beetling engines – each engine

Some of the wooden beetling hammers from Wellbrook Mill, used to pound and soften damask linen.

having perhaps 3-dozen hammers arranged in a row, a bit like large-scale versions of the hammers in an upright piano. One of Ireland's last working beetling mills is at Wellbrook, where seven beetling engines are powered by a large waterwheel. Beetling began in Wellbrook in 1764 and, within a few years, six mills were serving the local linen bleach works. This last mill, which closed in 1961 with all its original equipment, has been restored by the National Trust. Demonstrations are held when the mill is open. (19)

Tyrone, like most Ulster counties, had a strong linen industry in the late-18th and 19th centuries. The larger linen centres were usually located close to a port, or at least adjacent to a canal for easy access to the imported coal that was needed to power the heavy steam engines. A number of smaller mills, however, continued to rely on water power often into the 20th century, as at Wellbrook. Tyrone's linen industry also features at Benburb's Linen Museum which is in a renovated weaving factory by the Ulster Canal. (20)

KINTURK

The creature from the Sargasso Sea

You will find the **European brown eel** (*Anguilla anguilla*) in most Irish rivers and lakes. Lough Neagh* has 200 million of them, making it one of the largest eel centres in Europe.

At Toome, where the lough funnels into the Lower Bann river, a co-operative of 500 fishermen runs Europe's largest eel fishery. The eel is one of Ireland's oldest foods: our Mesolithic ancestors ate them at camps along the Bann 9,000 years ago, and they were still popular in the 19th century, when they were eaten fresh, smoked or salted. Ironically, few eels are eaten in Ireland now and most of Toome's catch is exported, packed on ice and shipped out live through Belfast airport.

For thousands of years, the European brown eel was a mystery. Adults were common in lakes and rivers, yet no young were ever seen – Aristotle believed eels emerged fully formed from river mud. Young elvers were often caught at sea, but no one realised these transparent, glassy creatures were related to the long, brown freshwater eels. The mystery was only solved in the early-20th century, when scientists discovered the eel's spawning grounds in the Sargasso Sea, near the Caribbean. The adults spawn there in March and, after the eggs hatch, the tiny young, measuring less than 7mm, drift north across the Atlantic on the warm Gulf Stream current. It is not known why eels cross the Atlantic. Perhaps, when the Atlantic Ocean began opening up over 100 million years ago, their spawning ground fell on the wrong side of the divide; and in the intervening millennia, as plate tectonics pulled the Sargasso Sea away at a rate of a few centimetres a year, the eels simply adapted to the change. One thing is sure: if climate change alters the Gulf Stream current, Ireland's eel population will be badly hit.

It takes eels a year to reach Ireland, by which time they are 70mm long and, when they enter fresh water, they begin to darken. The journey up the River Bann to Lough Neagh takes another year. The eels spend 8–50 years in the lake; when fully mature, they will be nearly 1 metre long, silver in colour and with thick fat deposits. At this point, they stop feeding and start living off their fat reserves, for the long journey back to the Sargasso Sea, where they will breed once before dying. Mature eels gather at Toome each autumn, waiting for the right conditions to journey *en masse*: preferably a dark night with a southwesterly wind, and a river in flood. Eels are carnivorous, eating insects, fish and even other eels. They are nocturnal, to avoid predation, and are most active in summer, spending the winter hiding in burrows. They are especially long-lived, and one female caught in Co. Mayo was 57 years old. (Scientists age eels by counting the growth rings in a small bone in the skull.)

Eels are found in most Irish lakes, apart from land-locked lakes and lakes cut off by high waterfalls. Eel fishing on Lough Neagh is tightly regulated and subject to a quota system: long lines and draft nets are used, and there is an eel weir or trap across the river at Toome; the season is from May to December. There is also an important eel fishery on the Shannon, managed by the ESB which controls the weir at Ardnacrusha.* The story of the Lough Neagh eel is told at Kinturk cultural centre. The centre, located on the lough's western shore, runs boat tours in summer. ㉑

Ducart's hurries

An unusual canal* was built in the 1770s between Coalisland and Drumglass to transport coal* from the local mines. The canal was designed by a Sardinian engineer and architect, **Daviso Ducart** (de Arcourt), who also designed Cork city's mayoralty house and Limerick's custom house. The Coalisland canal is in three sections and these are linked, not by the usual locks, but by a simpler and cheaper arrangement of dry slopes or hurries. The coal was transported in barrows on canal barges, and at the end of each stretch of canal, the barrows were wheeled down the hurry to a barge waiting in the next stretch of canal. Coal production here was always on a small scale, however, and the canal soon fell into disuse. Part of it has since been converted to carry a road, but some hurries survive, as at Farlough beside the Dungannon–Newmills road.

As well as coal, the region has commercial deposits of clay and sand that have been worked on and off down the centuries, giving rise to various local industries, including brickworks, and pottery and glass-making. The coal was mined for over 200 years; more recently the local Bann clay* was used in making fire-proof brickworks. The region's diverse industrial history is the subject of Coalisland's Cornmill Heritage Centre. ㉒ One of those industries is glass-making:* in 1771 Benjamin Edwards opened a 'glass manufactory' at Drumveagh, and precisely 200 years later, and just 2km from Edwards's original premises, the Killybrackey community opened **Tyrone Crystal** to create local employment. The crystal is made on-site from ingredients that include silica sand, potassium nitrate and lead oxide. All items are mouth-blown and hand-cut using techniques that have changed little since Edwards's day. ㉓

Chapter 4 sources

Co. Derry

1. Yolland, Capt William, *An account of the Measurement of the Lough Foyle Base in Ireland* (1847).

2. Andrews, J, *A Paper Landscape: the Ordnance Survey in 19th-century Ireland* (Oxford, 1975).

3. *The Memoir of the City and Northwestern Liberties of Londonderry: Parish of Templemore* (Ordnance Survey 1837, reprinted by North West Books, 1990).

Co. Tyrone

4. Brück, Máire, "Torch-bearing women astronomers" *Irish Astronomy Journal* **21** 281–290 (1994).

Chapter 5: Fermanagh • Cavan • Monaghan

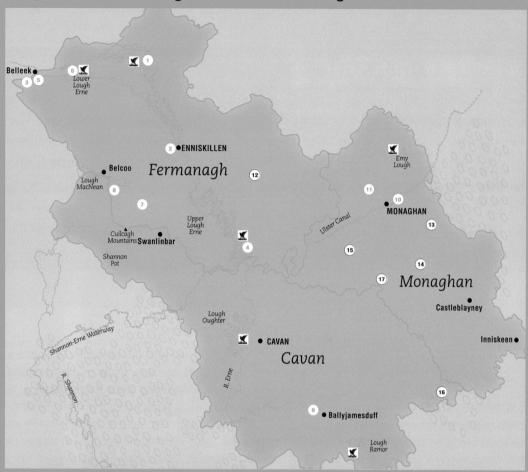

Belleek

Lower
Lough
Erne

ENNISKILLEN

Emy
Lough

Belcoo

Fermanagh

Lough
MacNean

MONAGHAN

Upper
Lough
Erne

Ulster Canal

Cuilcagh
Mountains Swanlinbar

Monaghan

Shannon
Pot

Castleblayney

Lough
Oughter

Shannon-Erne Waterway

Cavan

CAVAN

Inniskeen

R. Erne

R. Shannon

Ballyjamesduff

Lough
Ramor

Visitor facilities

1. Castle Archdale Country Park
2. Explore Erne
3. Enniskillen Castle
4. Crom Estate
5. Belleek Pottery
6. Castle Caldwell Nature Reserve
7. Florence Court
8. Marble Arch Cave
9. Cavan County Museum
10. Monaghan County Museum
11. Vintage Museum

Other places of interest

12. Brookeborough
13. Clontibret
14. Cornamucklagh
15. Drumurcher
16. Kingscourt
17. Rockcorry
♀ woodland
⚑ nature reserve or wildfowl sanctuary
◐ drumlins

*Visitor facilities are numbered in order of appearance in the chapter, other places
in alphabetical order. How to use this book: page 9; Directory: page 474.*

Fermanagh, a convoluted landscape of over 300 satellite lakes and as many islands, took decades to map accurately. At the heart of this lake district lies Lough Erne (from the Erni, a Fir Bolg tribe said to have lived there before the lake formed). Much of the county can be explored by boat and the area is noted for its birdlife. Fermanagh is home to the original Irish yew and what may be Ireland's only diamond. Denis Burkitt from Enniskillen greatly improved our understanding of cancer, and his father James pioneered the practice of ringing birds. Fermanagh, from *Fear Manach* ('Manach's men'), the tribe that settled there after fleeing their home in Leinster.

The Erne waterland

In Fermanagh, the boundary between dry land and open water is blurred and terrain can change seamlessly from lake to waterlogged bog to island. One-third of the county is underwater, dominated by a complex system of rivers and lakes and a veritable archipelago of islands – 157 in Upper Lough Erne and a further 97 in the lower lough, most of them drowned drumlins.* So convoluted is this watery region, that it was the last part of Ireland to be accurately mapped. In the past, the lakeshore clays were used by local industries, notably Belleek pottery. Today, the wetlands, flood plains and islands are important for wildlife and plants, especially birds and rare orchids.

The Erne occupies a shallow basin, with only a gentle gradient until it reaches Belleek, where it drops to the sea in a series of rapids and falls. This makes the Erne prone to severe flooding and, before the Upper Lough was drained, it could double in size during floods, hence the saying: "In summer the Erne is in Fermanagh, but in winter Fermanagh is in the Erne." Engineering works in the 19th century and again in the 1940s improved the drainage and reduced the flooding. Channels were deepened and widened and a control barrage was built at Enniskillen. The river was also harnessed for power: in 1863 at Belleek to drive the new pottery and, more recently, with two hydroelectricity schemes at the falls near Ballyshannon, Co. Donegal. These moves dropped the water levels by 3 metres, revealing new shore that is now being colonised by woodland.

The Erne has long been important for **navigation**, and one 15th-century report counted 1,500 boats at Enniskillen. The Erne and Shannon systems were connected when the Ballinamore–Ballyconnell canal* was built in Co. Cavan, but the route was never a commercial success, though the renovated canal is now popular with leisure craft. The great 19th-century builder William Dargan* developed a trade route linking the Erne and Liverpool: goods from Fermanagh could reach northern England in just five days using a combination of canal to Portadown, rail from there to Newry and onwards by ship. In 1842, Dargan also introduced the first paddle steamer to Lough Erne, starting a freight and passenger business that continued for some 50 years. During World War II, the flying boats that patrolled the Atlantic for U-boats were stationed at Castle Archdale, where the old docks can still be seen. ①

The Erne archipelago gave rise to an unusual tradition of **island farming**. Some two dozen islands in Upper Lough Erne are still farmed as they have been for centuries, the cattle being ferried from island to island using a special boat or 'cot' (*coite*) unique to the area. The cot is a shallow, flat-bottomed, blunt-ended craft that evolved from the dugout canoe. During the 1600s, there was a change to making them from larch planks; the modern cots are made of metal and can carry a tractor or 20 head of cattle. There is also a long tradition of **fishing**, especially eel.* In the 1580s, the Abbot of Assaroe in Co. Donegal owned 10 eel weirs along the Erne, but these disappeared during subsequent drainage and navigation works. There is still an eel fishery on the Erne, but it is smaller than the one on Lough Neagh. The heritage centre at Enniskillen Castle ② and Explore Erne, ③ tell the story of Lough Erne and its natural history.

The Erne area is renowned for its **birdlife**. Upper Lough Erne is noted for whooper swans, and 5 per cent of the world's whooper swans winter there. The Upper and Lower Macnean loughs are important for Greenland white-fronted geese, while a 400-year-old oak grove on Inishfendra, an island off the Crom Estate, is home to Ireland's largest heronry. ④ The Crom oak woods are descended from woodland that was managed historically, but which was itself descended from the ancient wild woods of Ireland.

The woodland is noted for a rare parasitic orchid, the bird's-nest orchid, which lives on other plants and, not needing much sunlight, can survive in the deep shade under the trees. Other unusual plants in the area include the moor pea, fen violet and, on Knockninny's limestone grassland, the dense-flowered orchid. The nine national **nature reserves** in Fermanagh are: Castle Archdale Islands, Castle Caldwell, Correl Glen, Crossmurrin, Hanging Rock and Rossa Wood, Lough Naman Bog, Marble Arch,* Reilly and Gole Woods and Ross Lough.

The geography of cancer

One of the great Irish medical scientists of the 20th century, **Denis Burkitt** (1911–1993), was the first person to identify a cancer caused by the environment, and later proved that diet played a major role in bowel cancer. Burkitt, a deeply religious man, was born near Enniskillen and studied medicine at TCD. His father, **James Burkitt** (1870–1959), was a surveyor and keen naturalist who pioneered the use of leg rings to identify birds, map their territories and plot their distributions. This greatly improved our understanding of bird populations and behaviour, and ringing is now standard ornithological practice. Denis Burkitt later used a similar technique to survey and map the distribution of cancer.

After World War II, Burkitt and his wife, Olive Rogers, a nurse, went to Africa on medical work. There, Burkitt studied a disfiguring swelling of the jaw, common in children living in certain parts of Africa, and fatal if untreated. He identified the swelling as a lymph tumour (now called Burkitt's lymphoma) and then set about mapping the incidence of the disease across the continent. In a ground-breaking study published in 1959, he showed that it

Burkitt recorded his lymphoma only in central African areas. Scientists later discovered the cancer was caused by a virus.

occurs in a central belt and, significantly, its distribution coincided with various insect-borne diseases.[1] Despite working with limited resources, Burkitt developed a cheap and effective cure for the disease using low doses of a chemotherapy compound, methotrexate, making Burkitt's lymphoma only the second cancer to be treated with drugs.

Next, an international team led by Tony Epstein, and including Irish geneticist Yvonne Barr, found a virus in samples taken from patients with Burkitt's lymphoma. In 1966, they showed that the Epstein-Barr virus, as it is now called, can cause cancer in people whose immune system has been weakened by malaria. Burkitt's lymphoma was thus the first cancer shown to have an environmental cause.

In 1966, after 20 years in Africa, Burkitt moved to the Medical Research Council at London, where he switched to studying diet and disease, then considered something of a fringe area. Prompted by South African research linking diet and bowel movements, Burkitt proposed in 1971 that a high-fibre diet helps prevent bowel cancer. His book *Don't Forget Fibre In Your Diet* (1979) was a bestseller and helped raise public awareness of the need for dietary fibre. Burkitt was awarded many international honours, including membership of the prestigious French Académie des Sciences. Active to the end, he gave a lecture the afternoon before he died.

ENNISKILLEN

Quite a gamble

The British chemical industry took off in the 1820s thanks, in part, to two Irishmen – Dubliner James Muspratt* and Enniskillen-born **Josias Gamble** (1776–1848). The Procter & Gamble Corporation was co-founded by another Northerner, **James Gamble** (*b*.1802), but it is not known if James and Josias were related. Josias Gamble, a Presbyterian minister who served at Enniskillen and Belfast, developed an interest in the linen industry, especially the

technologies of flax processing. He quickly left one 'cloth' for another, starting a small company making bleaching powder and sulphuric acid in Dublin, where he met Muspratt, who was engaged in a similar business.

In the 1820s, the British government repealed the Salt Tax, then a hefty £30 per tonne. Overnight, the Le Blanc process for producing soda (sodium carbonate) became economic. Soda was an important industrial chemical used in glass and soap-making, bleaching, and dyeing. When the Salt Tax was repealed, Gamble and Muspratt moved their businesses to Liverpool to be near the great Cheshire salt fields. (Irish industry would surely have developed differently had they chosen instead the salt mines at Carrickfergus.*) In England, they formed a short-lived partnership, but went on to establish separate chemical empires.

James Gamble, son of a travelling Methodist preacher, emigrated to Cincinnati, Ohio, in 1819 where he developed a soap factory. Cincinnati, or 'Porkopolis', was a major pig-butchering centre and produced plentiful animal fat, the chief ingredient, not just for Gamble's soap-making, but also for the candle-making trade run by his brother-in-law, Englishman William Procter. In 1837, the pair merged their businesses and, within 20 years, Procter & Gamble had a turnover of $1 million. The company made candles and soap for the Union army and its moon and stars logo symbolised the original 13 US colonies. Its innovations over the years included ivory soap (1878) and the first synthetic laundry detergent, Tide (1946). Procter & Gamble also introduced the first soap opera, sponsoring the radio programme *Puddle Family* in the 1930s.

BELLEEK

Fine china pottery

In 1849, John Caldwell Bloomfield inherited the run-down Castle Caldwell Estate from his father. Bloomfield was keen to develop a local industry and improve the lot of his

John Caldwell Bloomfield: he discovered china clay on his Castle Caldwell Estate and started Belleek Pottery.
Belleek Pottery

tenants in the bleak post-Famine years. He knew nothing about farming, but he did know something about geology and pottery, having lived in China. In 1853, he commissioned a geological survey of his estate, prospecting for commercial resources and mindful that, in the 1840s, the Earl of Enniskillen had established a brick factory near Florence Court* that used local clay.

Bloomfield hit pay dirt, discovering kaolin (or china clay), along with flint, clay and shale – all the ingredients needed to make pottery. Labour was plentiful, there was turf to fire the kilns and the River Erne was harnessed at Belleek village to provide water power. Bloomfield also persuaded the railways to bring a line to Belleek, so that coal could be imported to supplement the turf and the finished pottery be taken to market. Experienced potters were brought in from England to train local staff and the factory opened in 1857, specialising in high-quality ware including telegraph insulators, hospital pans and tableware. The Erne kaolin gave the finished ware a fine lustre. Belleek pottery developed a distinctive style, won numerous gold medals at international exhibitions and established an export market with sales in the USA, Australia and elsewhere.

The business has faced various crises over the past century, notably during the two world wars, when orders fell and coal was rationed. It also changed hands on occasion, most recently in 1990 when it was bought by Erne Heritage Investments, owned by US-based, Dundalk-born entrepreneur George Moore. In 1952, two electric kilns were installed and, for the first time, the pottery could achieve the constant high temperatures needed to make the fine parian china for which, today, it is world famous. Apart from electricity, Belleek's production techniques are little changed from Bloomfield's day, though the raw materials are now imported and no longer come from the local estate.

For moulded items, a master mould is first made for each model from the design drawing; plaster of Paris working moulds are made from this master. Each mould is filled with liquid slip – a slurry made by grinding and mixing china clay, feldspar, flint-glass and water. This is left to stand and dry and shrink a little before being removed from the mould, finished and fired. Finally, the unique vitreous Belleek glaze, made from borax and frit, is applied and dried under an infrared heater. ⑤ Bloomfield's former home at Castle Caldwell is now a nature reserve, noted for its terns and herons. The English writer Arthur Young described the view from the demesne in 1776 as "the most pleasing [I have] seen anywhere". ⑥

The Colebrook diamond

There may be diamonds* in Co. Donegal, if recent geological studies prove correct, and industrial diamonds* in Co. Clare, but so far only one diamond has ever been found in Ireland, and even its origins are questioned. The mystery Irish diamond was found in the early-19th century in the Colebrook river by a girl; the stone later passed to the local landowners, the Brookeborough family. Dr Patrick McKeever of Northern Ireland's Geological Survey says the gem is a true diamond, but no one knows how it came to be in a Fermanagh river. The Brookeborough family is thought to have known some diamond importers at the time and so it is possible the stone is not Irish at all. Modern scientific tests could reveal clues to its origins, but Dr McKeever says the owners were never interested in having their gem analysed.

The milling soke

Mullycovet cornmill, just outside Belcoo, was probably built in the late 1700s. It belonged to the earls of Erne, whose tenants were obliged to use the mill under the milling soke arrangement. This mediaeval system gave a miller the exclusive right to grind all the grain produced in the 'soke' or local jurisdiction. Typically, the mill was built by a local landlord and rented to a miller, who took a proportion of the grain or meal, sometimes as much as one-sixth, by way of payment. The landlord's tenants, as here at Mullycovet, were obliged to use the mill, an arrangement which ensured that the miller had a guaranteed business and the landlord a regular rental income. When tenants refused to co-operate, they could be punished, for example by losing their right to cut turf for fuel. The milling soke was gradually abandoned in the late-19th century with the rise of large milling companies.

Mullycovet mill probably stopped working in the late-19th century, though one report suggests it was used up to 1920. It was excavated and restored to working order by a local group. The mill was much rebuilt in the 1850s, as confirmed by recent tree-ring dating* of the oak beams installed then. The milling complex includes a grain-drying kiln (in Ireland's damp climate, grain needs to be dried before it can be ground), and a seed house where the oat husks were blown off using a system of fans and sieves.

The world's finest fossil fish

Florence Court, former home of the earls of Enniskillen, was, for much of the 19th century, also home to the world's finest collection of fossil fish. Leading geologists came to Fermanagh to study at this amazing private museum as they struggled to understand the nature of fossils and their relationship with living species. The collection was the work of the third earl, **William Willoughby Cole** (1807–1886). A man with a passion for fossils, Cole studied geology at Oxford, where he met Sir Philip de Malpas Grey Egerton. The two became lifelong friends, travelling Europe in search of fine fossils and regularly sharing specimens that had split in two, tossing a coin to decide who would have which side (the part and counterpart).

Fossil fish from William Cole's collection. The drawing was made at Florence Court by noted artist, Joseph Dinkel, paid for by Cole and later used in Louis Agassiz's publication on fossil fish. This fossil (Tetragonolepis speciosus) dates from Jurassic times and was found at Lyme Regis in Dorset.
© Kenneth James/ Ulster Museum

In 1830, on a geological grand tour of Europe, they met Swiss naturalist **Louis Agassiz** (1807–1873). Agassiz, who was the first person to propose the idea of an Ice Age, was busy comparing living and fossil fish species. Having no money to travel or buy specimens, he persuaded the two friends to help by concentrating their efforts on fish fossils. Their tour continued, including a visit to Monte Bolca quarry near Verona, still an important site for fossil fish. Back home, Cole converted the south pavilion at Florence Court into a museum. Glass cabinets were installed and, to protect the collection from fire, the ceiling of the kitchen below was lined with lead. Not everyone was pleased however: according to geologist Kenneth James of the Ulster Museum, Cole's father objected to his son's expensive hobby as "this damned nonsense" – one fossil plesiosaur had cost £200.[2]

In 1835, some of the world's leading geologists, including Agassiz, gathered at Florence Court to study what was then probably the world's finest collection of fossil fish. Scottish scientist Charles Lyell, Cole's friend and a founding father of modern geology, described it as a party of "jolly hammer bearers". Indeed, Cole was renowned for his conviviality and menus that included 'plesiosaur pie' (woodcock, alas).

Cole never published any scientific studies himself, being happy simply to collect, but he gladly allowed others to use his collection, which eventually ran to 10,000 items, mostly fossil fish. They were all the best specimens that could be had and many were scientifically important. To help

Agassiz's research, Cole and Egerton generously paid the celebrated artist Joseph Dinkel to draw various fossils in their collections. These illustrations later appeared in Agassiz's monumental 5-volume *Recherches sur les Poissons Fossiles* (1833–1844) in which he identified over 1,700 species.

Cole's interest in geology never waned. When archaeopteryx was discovered in 1861 he travelled to Germany to bid for the fossil, but lost to the British Museum, which paid £700 for it. He was also active in local politics, serving as MP (1831–1840), and later as the first Imperial Grandmaster of the Orange Order. In the 1840s, after inheriting the family estate, he established a brick and tile factory using local clay and also encouraged drainage schemes to improve the heavy clay soil. Before he died, Cole transferred his collection to the Natural History Museum in London, where it joined Egerton's, his old friend having died shortly before him. Sadly, there is no trace at Florence Court of the geological museum that once attracted the world's leading geologists to this corner of Fermanagh. However, the estate, now run by the National Trust, does have a water-powered sawmill and an ice house,* and the original Irish yew. ⑦

All Irish yews around the world are descended from the Florence Court mother plant.
© An Post

FLORENCE COURT

The original Irish yew

In 1740, George Willis found two unusual yew seedlings on the slopes of Cuilcagh Mountain. Normally yew grow low and horizontally, but these shrubs had tall upright branches. Willis transplanted one seedling to his own garden and gave the other to his landlord at Florence Court demesne. Cuttings from the Florence Court yew were gradually distributed far and wide; it was especially popular in graveyards and was listed in commercial plant catalogues from 1818. Initially classified as a new species, the Irish yew – *Taxus baccata* fastigiata – was later identified as merely a variant of the ordinary yew, but with an unusual upright or 'fastigiate' habit. Other upright yews have since been discovered but none is as elegant as the Florence Court variety. The original Florence Court tree is now only a shadow of its former self; the second one Willis planted died in 1865.

Yew trees have long been associated with Ireland. Giraldus Cambrensis* described the coniferous evergreen in 1188 as "the favourite of the Irish", popular in sacred places and cemeteries. It was commonly held that, because yew leaves are poisonous, it was safe to plant the trees only among the dead. Graveyard yews thrive, probably benefiting from the minerals released by decaying caskets and corpses, and the protection from grazing afforded by high cemetery walls.

Yews are extremely long-lived – some are over 1,000 years old – and they are the oldest living things in Europe (only in dry climates can trees, such as California's bristlecone pine, live for 4,000 years). The oldest Irish yews are probably those at Muckross Abbey, Co. Kerry, which were reputedly planted in 1344. Yews produce a strong alkaloid, taxine, from which a powerful anti-cancer drug, Taxol, has been developed. Taxine was first harvested from the Pacific yew, but Ingrid Hook of TCD has discovered taxine in the leaves of the Irish yew. Yew orchards could soon be planted around Ireland, to provide taxine for the drug industry.

A limestone underworld

The limestone landscape of the Marlbank and Marble Arch is second only to the Burren in Co. Clare.* Though relatively small, it is a complex area with a myriad sinkholes, caves, dry valleys and resurgences, and an amazing subterranean world that is still being formed by underground rivers. Among the caves is **Marble Arch Cave**, probably Ireland's most impressive show cave, and the only one that includes a boat trip as part of the tour. Other caves in the area are suited only to experts – Coolarkan, for example, has an 80-metre pothole – and all are liable to flood quickly in heavy rain. Marble Arch itself has an alarm system and flood monitors in the passageways, and visitors are advised to phone ahead, in case it is flooded. Despite these constraints, access has been thoughtfully planned and can even accommodate wheelchairs. ⑧

Marble Arch Cave, though known for centuries, was first properly surveyed in 1895, when a French caver, Edouard Martel, explored it in a small folding boat. A decade later, Robert Lloyd Praeger,* the noted Irish naturalist, took part in an expedition, swimming through the passages with a candle in his hat. The 6.5km cave system is formed by three rivers flowing off Cuilcagh Mountain, which straddles the Fermanagh–Cavan border to the south. The rivers combine inside the cave to form the Claddagh river, emerging into daylight at the Marble Arch, a limestone arch 10 metres high, which is the remains of an older, collapsed cave passage.

Edward Martel fires a magnesium flare to illuminate the junction where the three underground rivers meet to form the Claddagh. From Martel's account of the expedition in his book Irlande et Cavernes Anglaises *(1897). Marble Arch Cave*

The limestone landscape of the neighbouring **Marlbank nature reserve** is worth a visit. Claddagh Glen is an attractive wooded gorge that starts at the Marble Arch; the local geology can be seen in the riverbed there and in the surrounding rock sides, and in spring the woods are bright with primroses and bluebells. Killykeeghan is an area of fragile limestone pavement: the soil there is thin and the eco-system vulnerable, and visitors should keep to the tracks.

The countryside around **Cuilcagh Mountain** to the south includes important bog habitats, complex limestone landscapes and several nature reserves. Cuilcagh itself rises out of the bog to a height of 667 metres, and the Ulster Way climbs to the top, rewarding hill walkers with dramatic cliffs and panoramic views. The rocks of Cuilcagh formed 330 million years ago when this part of Ireland resembled the Bahamas today. At various times there were muddy deltas, silted river beds, swamps and coral reefs. Thus Cuilcagh is made of layers of muddy shales, siltstones, sandstones and limestone, capped by gritstone. This gives the mountain a step-like profile and, in places, a steep cliff face. Large blocks of rock, fractured by freeze–thaw action, regularly break off and fall to boulder fields below.

Cuilcagh's boggy slopes can seem bleak but the area is an important wildlife habitat, noted for its golden plover, Greenland white-fronted geese and merlin. Rare plants grow on the mountain's steep northern face and in the boulder fields below, including dwarf willow and starry saxifrage. The surrounding bogland is also 'active', and subject to sudden bog bursts.* Many go unnoticed but, in 1992, one major slide was large enough to flow downstream and into Marble Arch Cave, leaving a peaty tidemark high up on the passageways inside. To prevent similar catastrophes in future, **Cuilcagh Mountain Park** was formed: managed by Fermanagh Council, and with EU funding, the aim is to encourage less intensive farming, promote conservation and restore damaged peatland.

COUNTY CAVAN

Cavan, like neighbouring Fermanagh, is dotted with small hills and jewelled with numerous lakes, and land and water seem braided together here. Highlights include the Lifeforce Mill, scene of some 19th-century industrial espionage; Swanlinbar, once a famous spa town; and Ballyjamesduff, birthplace of the world's greatest copper tycoon.

GLANGEVLIN

The Shannon Pot

This small yet mystical pool on the southwest side of Cuilcagh Mountain is the legendary source of the longest river in these islands, the Shannon (*an tSionnainn*). Legend has it that the ancient druids gathered all their knowledge into seven hazel trees growing by the pool. A young woman called Sionnan coveted their power and picked a hazelnut after the druids had left, but the pool erupted and drowned her, then crashed down in an enormous wave and cut a channel south to form the great river. The tree-ringed pool is actually a deep natural well, and is the exit point of an underground stream that starts higher up on Cuilcagh. Geologist Patrick McKeever says the source of the water there is complex, but at least two springs flow into the pool, from where the river officially starts its journey.

The Shannon is 251km long from source to estuary; the estuary adds another 83km, and the 15 lakes and numerous tributaries a further 1,500km. This vast system drains 20 per cent of Ireland (14,000sq km) spread across 12 counties, but only with difficulty: the midlands are flat and, despite the river's size, its gradient is shallow, dropping just 17 metres between Lough Allen and Killaloe. (The river drops 30 metres between there and Limerick, where Ardnacrusha* hydro-scheme was built.) In essence, the Shannon lies across Ireland like an enormous gutter, and one that, despite numerous drainage programmes, is still prone to severe flooding. The very gentle gradient means the situation could probably be remedied only by costly and continuous pumping. On the plus side, the seasonal flood plains or callows* form an unusual and important wildlife habitat.

Originally, the river formed a vast lake that covered most of central Ireland after the end of the last Ice Age, stretching from Lough Allen in Co. Leitrim to Lough Boora in Co. Offaly. Gradually bog and sediment encroached, squeezing off individual lakes and reducing the Shannon to a thread of its original size. Yet the river is still a major force – for example, the grey squirrel,* though introduced nearly 100 years ago in Co. Longford, has yet to cross the Shannon to the west.

The Shannon river god; garlanded with acorns, oak leaves and cornucopias, and crowned with a trident. © Enterprise Ireland

Given that this is part of Ireland's lake district, it is not surprising to find three wildfowl sanctuaries there: in the wonderful lakeland area of Lough Oughter, at Lough Ramor, and at Dartrey/Fairfield. [Next stop on the Shannon: **Mid-river meanderings** in Co. Roscommon.]

SWANLINBAR

'The Harrogate of Ireland'

In the late 1600s, four entrepreneurs built a watermill to process iron ore from nearby Sliabh an Iarainn.* They called the mill Swanlinbar from their surnames – **Sw**ift, **Sa**nders, Dar**lin**g and **Bar**ry – though in Irish it was known as Muileann an Iarainn (Iron Mill). In the 1700s, it was other minerals that drew people there, as crowds came to drink the sulphur- and magnesium-rich waters from springs around the town. This was the boom era for **spa towns**. The mineral waters were thought to be medicinal and spas became popular resorts – indeed, probably the first modern tourist attractions, after religious shrines and trading fairs. Assembly rooms, pump rooms and hotels were built and the fashionable set gathered to take the waters. Just a few Irish spas survived into the 20th century and only Lisdoonvarna* remains today. The main Irish spa resorts were: Castleconnell in Co. Limerick; Kilmeadan in Co. Waterford; Leixlip in Co. Kildare; Lisdoonvarna* in Co. Clare; Lucan near Dublin; Mallow* in Co. Cork, where the thermal springs were said to cure consumption; and Ballyspellin and Spa, both in Co. Kerry.

Swanlinbar was a bustling 18th-century spa town attracting visitors from Britain and Europe, and renowned as 'the Harrogate of Ireland', after the famous English spa town. Little trace of this remains, though the foundations of the Spa Hotel can be seen near Uragh well. After the Napoleonic wars, Irish spa towns dwindled in popularity as transport improved and with it access to the larger continental resorts. The rising popularity of sea bathing and the emergence of seaside resorts in the mid-1800s was the final blow. The minerals found in the Swanlinbar springs come from the rocks of Cuilcagh Mountain. Water collects the minerals and percolates down through porous rock and shale layers, to surface at the various springs. Two wells are still accessible: Drumbrochas, a sulphur-rich spring about 1km southeast of the town, was said to cure rheumatism; the spring is behind a low wall by the road. Uragh well, near the site of the old hotel, is rich in magnesium and was reputedly good for stomach aches.

CAVAN TOWN

A tale of industrial espionage

The water turbine lovingly restored to working order at Lifeforce Mill may be evidence of 19th-century industrial intrigue, for it is probably a pirated copy of an original French design. For centuries, water was Irish industry's primary power source, but conventional waterwheels are relatively inefficient. Eighteenth-century English industry could use cheap coal to produce steam power, but the best that Irish industry could do was build ever bigger waterwheels. French mills also depended on waterwheels but, in 1837, Benoît Fourneyron invented the water turbine. A significant improvement on the wheel, it consisted of a fanned arrangement of blades, set horizontally at the bottom of a short shaft and connected to a spindle. Water entered at the top of the shaft and passed down over the blades, turning them and the spindle as it went, before flowing out at the bottom.

Sir Robert Kane* brought Fourneyron's invention to Irish attention in his book, *The Industrial Resources of Ireland* (1844), where he calculated that Irish rivers could generate 1.25 million horsepower. According to industrial archaeologist Colin Rynne, two Irish entrepreneurs hurried to France to negotiate a licence to manufacture the turbines in Ireland. Fourneyron refused, but the pair made covert sketches of the French equipment before returning to Ireland. By 1846, at least two Belfast engineers – James Thomson* and Robert MacAdam – had designed water turbines based on Fourneyron's original.

Lifeforce Mill: the millrace (right) from the Kennypottle river drives the MacAdam turbine, which turns the cogs and thence the millstones. Grain is delivered at an upper level (left), and dried on the heated floor above the kiln room, before being ground. A small engine, installed between the stairs and cog room during the renovations, is used when the water level in the river is insufficient to drive the historic turbine. © Lifeforce Mill

Ulster firms were interested in turbines to drive the heavy machinery of the linen* factories, but one of the earliest MacAdam turbines was installed at this Cavan cornmill when it opened in 1846. (A similar Irish turbine was installed in 1853 at Ballincollig Gunpowder Mills* in Cork, where it can be seen today.) The Cavan mill and turbine were recently restored by Lifeforce Foods and the company believes its turbine may be the oldest of its kind in the world that is in working order and still in its original location.

There has been a mill on the site since the 1300s. The present complex was built in 1846 by the Greene family of millwrights who ran the mill until it closed in the 1950s. The restored mill is again in production, grinding wholemeal flour on the 150-year-old millstones. The company claims its slow stone-grinding does not heat or damage the flour, unlike the faster industrial mills.

CAVAN TOWN

Place names – a local archive

The place names of Ireland are a valuable record of culture and tradition. They contain a wealth of information on local landscape and geology, natural history and even industry. Cavan town for example, *an Cabhán*, is named for the small, grassy hill beside the town or possibly the small hollow beneath the hill, as *cabhán* can mean hill or hollow.[3] Yet *cabhán* is just one of over a dozen Irish words for hill. Others include *ard, brí, céide, cnap, cnoc, droim, maol, mullach*. Each is subtly different: *droim*, a long low hill, usually refers to a drumlin;* *céide* means flat topped; *cnap* is a small round hill and *maol* a bare one.

Many places take their name from the dominant feature of the local landscape, perhaps a lake, river or bog. A curragh is a grassy plain; barren rocky places are often called burren (*boirinn*, 'stony place'), while *leic* or *leac* in a place name refers to flagstone, and carrick (*an charraig*) a rock, often a glacial erratic.* Fords, bridges, mills and churches often feature in place names, even if the original structure has long since vanished. Place names can also reveal much about a locality's natural history. Feltrim (*faeldruim*) in North Co. Dublin translates as wolf hill, presumably predating the last Irish wolf,* which was shot in the 1780s. A sizeable number of the country's 67,000 townlands are named after trees, reflecting Ireland's wooded past, even if it is now the least-forested country in Europe. The generic *coill* ('wood') is common; some 1,500 contain the element derry (*doire*) meaning oak wood; *iubhair/eó* ('yew') is also frequent, as in *an tlúiur* (Newry) and *Maigh Eo* (Mayo).

While most Irish place names are Gaelic some, like Wicklow and Lambay,* are Norse in origin. Names that are English probably date from the Pale or the Plantations, like Parsonstown (now Birr*) and Edgeworthstown.* Apart from mills, mines and furnaces, a handful of place names record industrial enterprises, notably Prosperous* in Co. Kildare. Interpreting place names is fraught with problems and many explanations are disputed. Most Gaelic names were in use for centuries and their original meaning had been lost before they were written down, and, moreover, which version was written down could depend on who had done the recording. The place names department of the Ordnance Survey of Ireland* continues to research Irish place names, and publishes an occasional gazetteer of them.[4]

BALLYJAMESDUFF

The world's copper king

One of the USA's greatest mining tycoons was **Marcus Daly** (1841–1900) from Ballyjamesduff. He ran the world's largest copper mine, founded the city of Anaconda, Montana, USA and supplied the copper that fuelled the new and emerging electrical technologies, especially the telephone and the electric motor. The young Daly left Cavan

Mining magnate Marcus Daly from Ballyjamesduff supplied the copper that fuelled the new electrical industries of the late-19th century. Cavan County Museum.

for America in 1856, finding work at mines in California and Nevada. A self-taught engineer, he earned a reputation for his uncanny ability to assess the commercial value of ore – reputedly able "to see farther into the ground than any mining engineer" – and quickly became a prosperous mine manager. In 1880, backed by George Hearst (father of newspaper magnate William Randolph Hearst), he bought a silver mine near Butte, Montana, for $30,000.

As he developed the mine, Daly regularly struck copper. In fact, his mine was sitting on the largest, richest copper bed then known, but, at the time, copper had little commercial value. Then, in 1882, Thomas Edison opened a power-generating station in New York and Daly realised that, if the newfangled electricity took off, copper, a cheap and excellent conductor, would be in great demand. In 1882, Daly shipped 37,000 tonnes of copper ore to Wales for smelting, then the nearest copper smelter. The following year he built his own smelter – the world's largest copper-processing plant – at Warm Springs Creek in Montana. That same year he laid out the town of Copperopolis, renamed Anaconda in 1888.

Daly, by then a millionaire, also became active in state politics and founded a newspaper, *Anaconda Standard*. A man of energy and vision, he supplied the copper that made the electrical revolution of the late-19th century possible. Copper smelting remained Anaconda's main industry until 1980, when the smelter's last owner, Atlantic Richfield Company, closed the plant. Daly's former mansion home in Hamilton, Montana, is now a state-owned national historic site and open to the public. Ballyjamesduff also remembers Daly, who helped fund the town's church, and there are plans to erect a plaque at the old family home (now in

ruins). Cavan County Museum, which is in the town, has information about Daly, and about mining in Cavan. ⑨

Desert, a dead sea and gypsum

There was once a desert in the northeast of Ireland, and a very salty 'dead sea'. But about 225 million years ago, the sea evaporated, leaving behind a rich bed of salts – white gypsum salt at Kingscourt–Carrickmacross and rock salt at Carrickfergus,* both of which are mined today. The gypsum deposit occurs in a block that is about 2.5km across and 14km long. It originally covered a much larger area, but it has been eroded over time, except at Kingscourt, where the beds survived because they were protected by overlying rock, thanks to a geological fault: at some point in time the ground around Kingscourt slipped below the level of the surrounding rock, taking the gypsum with it; subsequently, rock covered over the gypsum and preserved it. There are two gypsum beds in Kingscourt: the lower bed, which is over 20 metres thick, and a thinner upper bed. The two are separated by a band of mudstone 10 metres thick.

Gypsum (crystalline calcium sulphate) is used in making plaster of Paris and plaster products for the building industry. The Kingscourt deposit has been quarried on and off since the 1870s, but large-scale mining began in the 1940s when it was realised that in places the gypsum is tens of metres thick. There have been gypsum mines at Lisnaboe, Drumgoosat and Drumgill, and in Co. Meath at Ballynaclose. Today, the gypsum is mined at Knocknacran by Gypsum Industries (est. 1935), which uses the raw material at its Kingscourt plaster factory. The older underground mine became an opencast facility in 1990.

The modern opencast gypsum mine at Knocknacran. © Gypsum Industries

Monaghan *(Muineacháin)*, according to the *Annals of the Four Masters*, means "the place of the small hills", an appropriate description for a county covered by swarms of drumlins. These give Monaghan its distinctive "basket of eggs" landscape, and ensure that there is a different view at every bend in the road. There is gold buried beneath the drumlins, and Clontibret may one day have a gold mine. Noted Monaghan people include John Robert Gregg, who invented Gregg shorthand, and Peter Rice, one of the great design engineers of the 20th century.

INNISKEEN

Kavanagh country

Monaghan's stoney grey soil, small fields and black hills made poet Patrick Kavanagh into "the sort of man I am/A fellow who can never care a damn/For Everestic thrills". Monaghan is classic **drumlin country** and swarms of these small hills shape not just the poetry but also the terrain: creating a fragmented landscape, controlling the routes of roads and rivers, and damming small lakes. The whole surface is stitched over with a patterning of numerous hedgerows and small fields and farms. In 1744, a Mr Charles Smith, realising how this landscape must appear from above, described drumlin country as "a basket of eggs". But it was another hundred years before people realised that, amazingly, these clutches of hills were laid by glaciers.

Robbie Meehan, who has studied drumlins with the Geological Survey of Ireland,* says their internal structure varies: some are all rock, some all gravel, others all clay and yet others are a mix. Their size and the steepness of the slope also vary. Some are small hillocks, others are large enough to hold a cathedral, as at Armagh. But all have the same streamlined, tear-drop shape, because the ice sheet that laid them was moving. We can even tell the direction of the ice from the drumlin's orientation (long axis): at Castleblayney, for example, the glacier was moving northwest–southeast.

Drumlins traditionally make for poor farm land: the ground was compacted by the glacier and so drainage is poor; marshy areas often develop at the base of the hills; there are many small lakes; and the steep hillsides make cultivation difficult. Any dry meadow sites were, therefore, worth naming, and Clon/Cloon (*cluain*, 'a meadow'), is a common place name element in drumlin country.

The last Ice Age left large swarms of drumlins across Ireland in a belt stretching from Strangford Lough to Clew

Drumlin and other neologisms

When geologists around the world talk about the small hills deposited by glaciers, the technical term they use is the Irish word 'drumlin', which comes from *droim*, ('back' or 'hump'). Drumlin is just one of many neologisms we have given the world. Others include:

barometer: coined by Robert Boyle,* who invented the siphon-type barometer.

calp: name given by Richard Kirwan* to a muddy limestone found in Leinster and much used in Dublin buildings.

electron: named in 1891 by George Johnstone Stoney.*

esker:* another glacial feature from an Irish word *eiscir* (for example, the *Eiscir Ríada* in the midlands).

fluorescent: George Gabriel Stokes* named the phenomenon after the glowing mineral fluorite.

kyanise: a way of preserving wood, called after its Irish inventor John Kyan.*

microphone: Narcissus Marsh* coined the word for an instrument that amplifies a quiet sound (what we today call a megaphone).

Neanderthal: term invented in 1864 by William King,* professor of geology at Galway, after the German Neander Valley, where the first such fossil was found.

orrery: a clockwork model of the solar system named after Charles Boyle, Earl of Orrery* in Cork, who owned one of the first ones.

quarks: sub-atomic particles named by US physicist Murray Gell-Mann from a word invented by James Joyce in *Finnegans Wake*: "Three quarks for Muster Mark."

Yahoo: an internet company and web search-engine called after the race of brutish creatures in Jonathan Swift's* *Gulliver's Travels* (1726).

Bay, via counties Down, Armagh, Monaghan, Cavan, Leitrim, Sligo, south Donegal and west Mayo. There are also smaller clusters elsewhere, notably around Gort in Co. Clare and Bantry Bay* in Co. Cork. Drowned drumlins form island swarms at Clew Bay* and Strangford Lough,* where the waves cutting into a drumlin often expose a rare view of its interior.

INNISKEEN

An ingenious engineer

Inniskeen, best known as the home of poet **Patrick Kavanagh** (1905–67), was also the hometown of **Peter Rice** (1935–1992), one of the great design engineers of the 20th century. Rice, celebrated in architectural circles as 'the James Joyce of engineering', is not well known in Ireland. Yet he made possible some of the most famous buildings of the past 50 years, including the Sydney Opera House, the Lloyd's Building in London and, in Paris, the Pompidou Centre and Louvre glass pyramid. As an engineer, Rice helped architects to realise their designs, found elegant solutions to structural problems and pioneered the use of many new materials, including cast steel and glass, and particularly lightweight materials, such as fabric.

Sydney Opera House: the distinctive roof was made possible by Irish engineer Peter Rice.
© Ove Arup & Partners

Rice had a flair for mathematics and a talent for design and innovation. He went to school in Dundalk and studied civil engineering at QUB. On graduating, he joined the London office of engineering firm, Ove Arup, where he was later a director, specialising in lightweight materials research. His first major project was the Sydney Opera House roof (1958–1964), then a controversial building but now an icon of 20th-century architecture. Rice's contribution was to find the right geometrical curve for the shell-like structures the architect had designed. For the Pompidou Centre in Paris, Rice chose cast steel for the building's external skeleton of ducts and tubes, and devised a way of making the tubing so it was thickest at the bottom where the stresses were greatest. For the vast Parisian museum of science and industry in La Villette, the architect wanted to build broad façades from large panes of glass. Peter Rice devised a way to bind, support and tension 16 panes of glass at once, so that they formed a sturdy wall.

As a structural engineer interested in lightweight materials, Rice said he drew his inspiration from spiders. The architect he worked with most was the Italian Renzo Piano, and they even designed a concept car together for Fiat. Rice was also passionate about horse-racing and once wrote that the ultimate challenge would be "to design a car as flexible as a race horse". Rice's work was an unusual combination of art, mathematics, science, design and engineering. Just months before he died, he was awarded the Royal Gold Medal for Architecture by the Royal Institute of British Architects – a rare honour for a civil engineer.

Nahamagan* period) and small glaciers formed again in Irish mountains. The last of the giant deer* stalked Ireland then, and bear* were probably still here, but the experts were surprised to find mammoths were around too (presuming the carbon-date of 10,500 for the vegetation also applies to the mammoth). Several mammoth fossils were found in Ireland subsequently, including one in Awbeg Cave, Co. Cork, dating from 35,000 years ago, and another in Shandon Cave,* Co. Waterford, from 30,000 years ago. A leg bone from Shandon's mammoth is at the National Museum,* Collins Barracks, Dublin.

An Irish mammoth

An ill-fated canal

Workers building a new water-powered flax mill at Drumurcher in 1715 unearthed a most unusual find: the skeleton of an ancient woolly mammoth. It was the first mammoth found in Ireland and the news generated quite a stir. The Dublin physician and philosopher Thomas Molyneux* described the fossil in a letter to the Royal Society in London, that was published along with an account of the mammoth's teeth. The now-ruined site at Drumurcher was excavated in the 1940s by the late Frank Mitchell and again in the 1970s. Among the ancient vegetation, Mitchell found the remains of a hundred different species of arctic beetle, and seeds and pollen from

Mammoth (Mammuthus primigenius)

The mammoth (Mammothus primigenius): the largest mammal that ever lived in Ireland, weighed about 7 tonnes and probably had few enemies.
© An Post

ÉIRE 30

arctic plants that today grow in tundra regions, in particular an arctic poppy – Drumurcher is the only place in Ireland where this plant has so far been found.

The various plant remains were carbon-dated to c.10,500 years ago, a time when there was a short cold spell (the Lough

The **Ulster Canal**, intended to connect Lough Neagh* and Lough Erne,* and passing through Monaghan town, would have made it possible to travel by boat from Limerick to Belfast via the Shannon. It was the last major canal built in Ireland but, though much of the work was completed by the 1840s, the section between Ballinamore and Ballyconnell,* which would have linked the Erne and Shannon, was not finished until the 1860s. And, by then, the rest of the Ulster Canal was closed for repairs! The project was never a success: there were also problems with the water supply and the locks (which are inexplicably narrow – one is less than 3.65 metres wide) and, in any case, the railways were taking over. Some of the canal has since been filled in, including the stretch in Monaghan town. The successful restoration of the Ballinamore–Ballyconnell waterway, however, has prompted plans to rebuild the Ulster Canal. Monaghan County Museum, in the town, has a small exhibit on the Ulster Canal, and the county's flax and lace industries. ⑩

The lace-making, based at Carrickmacross, began in 1816 and became world famous, despite being mostly a cottage industry. It developed a distinctive style and was exhibited at international industrial fairs, including the Dublin Exhibition* of 1872.

Vintage steam

The Vintage Museum at Ballinode claims to have the largest collection of stationary steam engines* in Ireland and Britain. The oldest dates from 1915 and they include large engines once used to generate power in creameries, hospitals and laundries in the days before widespread electrification. These engines were typically bolted to the floor and generated electricity using steam piped from a nearby boiler. The exhibits are in working order and on open days (in April and July) are again put through their paces, albeit more slowly than their original working speed. The museum also has an extensive collection of agricultural implements including early Ferguson* and Ford* tractors and horse-drawn machinery. ⑪

Shorthand script

The shorthand system of speed writing was invented by **John Robert Gregg** (1868–1948), who was born at Rockcorry near Ballybay. His system of notation, first published in 1888 when he was just 20, and still in use, was developed after he had studied the way people move their hands when writing. The strokes he chose for the letters are a simplified form of those used when writing longhand; the shorthand stroke for *n*, for example, is a short and straight horizontal line, while the stroke for *m* is similar but twice as long. Gregg's shorthand is also phonetic, and any silent letters are dropped; thus *know* and *no* are written the same.

An example of the Gregg shorthand, invented by Monaghan man John Robert Gregg.

In search of gold

Mention Irish gold* and most people think of Wicklow, Croagh Patrick or maybe the Sperrins, yet Ireland's first large-scale modern gold mine might be at Clontibret. The rocks there contain various ores, especially lead-antimony. The lead was mined during the 18th and 19th centuries, notably at Lisglassan and Tullybuck, and then, in the 19th and early-20th centuries, the ore was mined for the antimony. In the 1950s, geologists discovered there was also gold in the Clontibret ore and over the past 40 years prospectors have drilled hundreds of bore holes in the area. In 1999, Conroy Gold & Diamond became the latest company to join the treasure hunt. So far, Monaghan's gold deposit seems uneconomic, but that might all change if gold mining becomes cheaper or more efficient or the price of gold rises. A small-scale gold mine opened near Omagh* in 2000. For the story of Ireland's prehistoric gold and the 1795 gold rush, see: Croghan Mountain, Co. Wicklow.

Besides Clontibret's lead mine, there were also two **lead mines** at Castleblayney: Lisdrumgormly, which was worked sporadically from the 1700s to 1956, and neighbouring Hope Mine (*c.*1850–1969). There are some ruined remains at both sites, including the Hope Mine's engine-house chimney. A major lead smelter was located at Cornamucklagh, east of Ballybay. Lead was an important building material in the 19th century, used for pipes, plumbing and even roofing. Gypsum salt is also found in Monaghan, part of the same deposit currently mined at Kingscourt,* Co. Cavan. The gypsum beds, which start near Carrickmacross and run south to Kingscourt and beyond, were left behind after an ancient 'dead sea' evaporated 200 million years ago.

Chapter 5 sources

Co. Fermanagh

1. Burkitt, Denis, "A Sarcoma involving the joints in African Children" *British Journal of Surgery* Vol 46 218–202 (1959).

2. James, Kenneth, *This Damned Nonsense – the geological career of the third Earl of Enniskillen* (Ulster Museum, 1986).

Co. Cavan

3. Room, Adrian, *A Dictionary of Irish Place Names* (Appletree Press, 1994).

4. *Gazetteer of Ireland* (Ordnance Survey of Ireland, 1989).

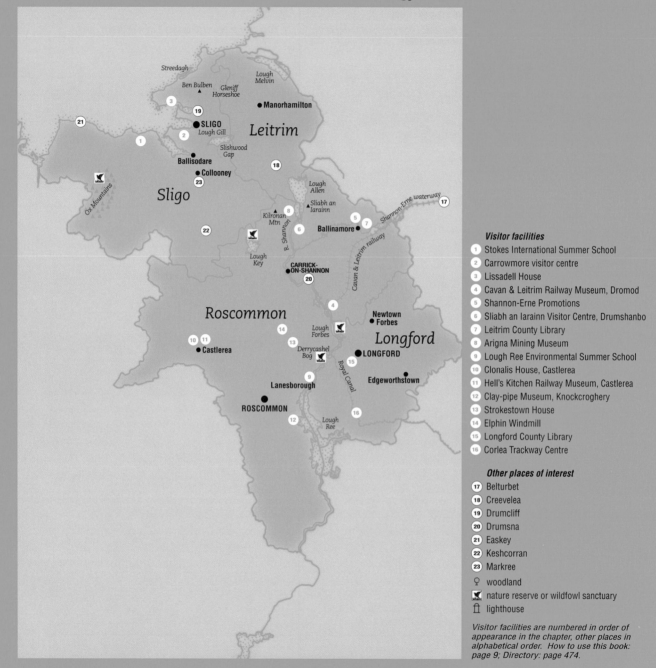

Visitor facilities

1. Stokes International Summer School
2. Carrowmore visitor centre
3. Lissadell House
4. Cavan & Leitrim Railway Museum, Dromod
5. Shannon-Erne Promotions
6. Sliabh an Iarainn Visitor Centre, Drumshanbo
7. Leitrim County Library
8. Arigna Mining Museum
9. Lough Ree Environmental Summer School
10. Clonalis House, Castlerea
11. Hell's Kitchen Railway Museum, Castlerea
12. Clay-pipe Museum, Knockcroghery
13. Strokestown House
14. Elphin Windmill
15. Longford County Library
16. Corlea Trackway Centre

Other places of interest

17. Belturbet
18. Creevelea
19. Drumcliff
20. Drumsna
21. Easkey
22. Keshcorran
23. Markree

♀ woodland
🗙 nature reserve or wildfowl sanctuary
🏛 lighthouse

Visitor facilities are numbered in order of appearance in the chapter, other places in alphabetical order. How to use this book: page 9; Directory: page 474.

COUNTY SLIGO

Sligo comes from *sligeach* ('a shelly place'), an apt description for a county with fine beaches, sandy bays and wonderful fossil shells. The county can claim one of the great scientists of the 19th century, George Gabriel Stokes, as well as interesting caves, exotic minerals and unusual landscapes.

SKREEN

A giant of science

Want to win $1 million? All you have to do is solve the Navier–Stokes equations. First proposed in 1821 by a French mathematician Claude Navier, and subsequently improved on by Sligo-born **George Gabriel Stokes** (1819–1903), the equations describe how a viscous fluid flows. Engineers use them when designing aeroplanes, for instance, but because the mathematics are so complex, they currently use only approximate solutions. To win the prize offered by the Clay Mathematics Institute in Massachusetts, you do not even have to solve the equations – merely prove that for fluids in 3-dimensional space, a solution exists that will never break down.

George Gabriel Stokes was one of the great scientists of the 19th century. A prodigious worker, he made valuable contributions to many fields, notably light, viscosity and how fluids behave, and was the first to explain fluorescence, and how the red-blood pigment haemoglobin works. Stokes was born at Skreen vicarage into something of a scientific dynasty: an ancestor, Gabriel Stokes (*b.*1682

in Dublin) had been Deputy Surveyor General of Ireland and a maker of mathematical instruments; and George Gabriel's brothers included a botanist (William) and an astronomer at Armagh (John Whitely), while other relatives included William Stokes,* the eminent Dublin medic. George Gabriel Stokes was a child prodigy at mathematics and, after studying at Cambridge, was appointed to the prestigious Lucasian Chair of Mathematics there – a position once held by Isaac Newton and now occupied by Stephen Hawking, and which Stokes held for over 50 years until his death.

Some of his most important work was in understanding, and mathematically describing, what happens to particles caught up in a moving fluid, work that is valuable today in studying sedimentation. Stokes had a practical bent and his work on water, waves and fluids led him to advise on aqueduct design (moving water) and railways (where wind and air currents are a concern). Researching the maximum

George Gabriel Stokes was one of the greats of 19th-century science, witness the many things named after him: Stokes's conjecture, Stokes's phenomenon, Stokes's law of hydrodynamics and his law of fluorescence... there is even a unit named after him: the stokes, the standard unit of kinematic viscosity, is equal to 1cm²/second.

height of ocean waves, he collected data from around the world, and laid the foundations of work that is still vital to those working at sea. Stokes experimented with quinine solutions, which have an unusual blue glow, a phenomenon he named 'fluorescence' after the mineral fluorite that has similar properties; and, in 1878, he was the first to explain both fluorescence and phosphorescence.

In 1859, two German scientists, Kirchoff and Bunsen, realised we could study the composition of substances and even distant stars by analysing the wavelength of the light they emitted and absorbed. This idea of spectral analysis revolutionised science, yet Stokes had suggested the same idea five years earlier in a letter to his friend William Thomson* (Lord Kelvin), another great Irish-born scientist. Turning this spectral analysis on haemoglobin, Stokes discovered it occurs in two states, with and without oxygen, and thus explained its vital role of carrying oxygen in the blood.

His work on friction in fluids led him to study air friction acting on a pendulum (useful in improving clock design). This led to research on gravity and, in 1849, Stokes pioneered the study of how gravity varies around the Earth, depending on the rocks and minerals present. Stokes became a consultant to the Royal School of Mines and his concept of gravitational variation is still used by prospectors looking for oil and gas. Stokes also pioneered the science of modern geodesy (the study of curved surfaces, including the shape of the Earth) and helped establish the Indian Geodetic Survey, echoing the work of his 17th-century ancestor.

George Gabriel Stokes married Mary Robinson, daughter of the astronomer Thomas Romney Robinson,* after he met her at Birr Castle where he was advising on telescope lenses. A reserved and deeply religious man, Stokes was also president of the Royal Society and an MP for Cambridge (1887–92), sitting with the Conservatives and voting against Home Rule. Skreen rectory where he was born has been demolished, but a commemorative plaque has been erected nearby. An international summer school is held at Skreen in Stokes's honour every second year. ①

Cytochrome discovered

Where George Gabriel Stokes discovered how haemoglobin functions, another Sligo man, **Charles MacMunn** (1825–1911), was the first to realise that similar pigments occur in other tissues, where they play an essential role in converting food to energy. These two Irish scientists made important contributions to our understanding of respiration and how the body functions.

MacMunn was born at Seafield House in Easkey, near Stokes's home in Skreen, and studied medicine with another member of that family, William Stokes.* MacMunn later moved to a general practice in Wolverhampton where, unusually perhaps for a GP, he set up a small private laboratory and conducted his own research using the latest in equipment, in particular one of the newly invented spectroscopes. In 1884, he discovered pigments in muscle cells and other cells, and showed they were similar to haemoglobin. MacMunn was not working at a recognised institution, however, and it was not until the 1920s that scientists realised the importance of his discovery. Instead of an oxygen atom (as in haemoglobin), MacMunn's pigments carry energised electrons, and these play a central role in extracting energy from food. MacMunn named the various pigments myohaematin (in muscle) and histohaematin (in other tissues) but today they are called the cytochromes.

The enigmatic frog

The frog is the most abundant of Ireland's three amphibians, the other two being the natterjack toad* and the smooth newt. But the frog is something of an Irish mystery. Is it a native species? Or was it introduced here, perhaps by the Normans? Happily, fossil bones found in a cave at Keshcorran may hold the answer.

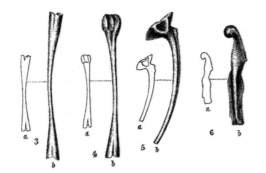

Some of the fossil frog bones found in a Kesh cave a century ago. Transactions of the RIA

According to one school of thought, the Anglo-Normans brought frogs to Ireland as a delicacy. This is at least plausible, since we know for a fact that they introduced the rabbit* here for its fur and meat. Moreover, there is no mention of any frogs in Ireland before the 12th century. For example, St Donatus wrote in the 9th century that "no raucous frogs disturb the rustling reeds". Then, in 1188, Giraldus Cambrensis* recorded that, while historically, there were no frogs in Ireland, because Ireland's air was sacred and nothing poisonous could breed there, "within my time one was found on a grassy meadow near Waterford". It was beheld with astonishment and, because it was classed with poisonous snakes, was seen as a portent of doom for the country.

A second theory holds that frogs were introduced in the 1690s by one Dr Gwithers, an over-zealous English naturalist who was taking part in a survey of Ireland's flora and fauna. The survey was organised by William Molyneux* of Trinity College Dublin, and Gwithers was responsible for surveying "the quadrupeds". Failing to find any frogs, he allegedly corrected this omission by placing imported English frog-spawn in a ditch in the college grounds.

In 1901, however, fossil frog bones were found on the floor of a cave at Keshcorran, along with the remains of other species, including arctic lemming, reindeer, wolf* and bear.* The fossils are currently being scientifically dated, but it now seems possible that frogs were present in Ireland over 10,000 years ago. If so, why were they not recorded until the 12th century? Did they, like the other arctic species, become extinct in Ireland at some stage and subsequently have to be re-introduced? Were the early naturalists simply mistaken when they said there were no frogs in Ireland? Or did some 'modern' frogs die in the Keshcorran Cave, where their bodies became mixed with more ancient ones?

If frogs were introduced here from Britain some time during the last 800 years, then DNA profiling could be used to investigate this, by measuring how closely related are the various frog populations of these islands. It might also be possible to compare the DNA profiles of our modern frogs and those of the fossil frogs found at Keshcorran. But until such tests are done, the Irish frog will remain a natural mystery. Keshcorran's caves are steeped in lore and Cormac MacAirt, king of Ireland at the time of Fionn McCumhaill, was reputedly raised by a she-wolf in one.

Two reactionary chemists

William Higgins (1763–1825) was a pioneering chemist, who invented a new form of chemical notation – some of which we still use today – helped promote the revolutionary new theory of oxygen, and claimed to be the first person to propose the modern atomic theory of the elements. The O'Higgins, formerly a great bardic clan, switched from poetry to the more prosaic world of medicine in the 17th century, and Higgins, born at Collooney, was a third-generation chemist.[1] He studied for a while at Oxford, but left without a degree and spent some years in London apprenticed to his uncle, Bryan Higgins, a successful medic running a chemistry school in Soho.

Two new, and not unrelated, scientific ideas were taking shape at the end of the 18th century, concerning the nature of the chemical elements and their reactions. One was Lavoisier's new theory of oxygen and combustion; the second was the modern atomic theory of the elements. Many people then, among them a senior Irish chemist, Richard Kirwan,* continued to believe that when a metal burned it gave off something they called **phlogiston**, although the great French scientist Antoine Lavoisier (1743–94) had recently proven that, on the contrary, oxygen is *taken up* when something burns, and that combustion is accompanied by a weight gain, and not weight loss. In 1789, provoked by Kirwan's work, Lavoisier published his influential *Elementary Treatise on Chemistry*, setting out his ideas about atoms, elements and how they react to form compounds.

That same year, the 26-year-old William Higgins published his first book, *The Comparative View of the Phlogistic and Antiphlogistic Theories*, in which he favoured Lavoisier and dismissed Kirwan. The book, an immediate success, significantly included Higgins's own ideas about the "ultimate particles of simple elementary matter" and how they react or combine with each other to form more complex compounds. When English scientist John Dalton (1766–1844) published his proposal for an atomic theory of elements in 1808, Higgins claimed he had had the idea first. But, in fairness, the Irishman had only ever stated his ideas in outline, whereas Dalton had calculated a table of atomic weights for various elements, and so is rightly credited with formulating the theory.

That said, Higgins pioneered two conventions still used today: he was the first to use letters to denote chemical elements, such as S for sulphur and O for oxygen, and lines linking these symbols to denote chemical bonds, or what he called 'affinities'. The fame brought by his book earned him the job of chemist at the Irish Apothecaries' Hall* and, later, that of chemistry professor at the RDS.* But Higgins, who has been described as spiteful, continued to argue against Dalton and died a bitter man in 1825.

William's uncle **Bryan Higgins** (1740–1818), also from Collooney, was an affable man who studied medicine on the continent and later ran a successful school in Soho, then the heart of London's chemical quarter. He was well travelled and, at Catherine the Great's invitation, pioneered chemistry teaching in Russia and he introduced German mineral waters to the British market. He patented various inventions including a cheap quick-setting cement in 1779, and a warm-air central-heating system in 1802, after he returned from five years in Jamaica where he had been advising on the industrial manufacture of sugar and rum.

Stars and salmon

For years, one of the best-equipped astronomical observatories in the world was at Markree Castle. It was established by **Edward Cooper** (1798–1863), noted as an astronomer and a pioneering aquaculturalist and, like Birr Castle, was one of the private Irish observatories that made important contributions to 19th-century astronomy. Cooper, who inherited his mother's interest in astronomy, spent the 1820s travelling the world. At each destination he would calculate his longitude and latitude from the heavens, and quickly became a proficient astronomical observer. He started Markree observatory in 1830 when he took over the family estate and spared no expense equipping the facility. His first purchase was a new 34-cm lens that had just been made in Paris and was then the largest lens in the world. More equipment was added over the years so that, in 1851, Markree was described by British astronomers as "the most richly furnished of private observatories". The first important astronomical news from Markree came in 1848 when Cooper's assistant, Andrew Graham, discovered a minor planet (a large asteroid). Graham named it *Metis*, from the Greek for meticulous, a

The telescope mounted at Markree in 1934 by Thomas Grubb. Markree Castle is now a hotel, and nothing remains of Cooper's observatory.*

reference to Cooper's careful planning of the observations that led to its discovery. Significantly, Graham and Cooper were the first people to discover a minor planet and also calculate and analyse its orbit. Graham and Cooper's most important work, however, was an amazing 8-year study measuring the positions of 60,000 stars, over 80 per cent of which were measured for the first time. Allowing for the fact that the sky over Ireland is cloudy two-thirds of the time, the two men must have studied at least 60 stars every possible working night from 1848–56. The Royal Society in London rightly praised their undertaking and the government paid for their results to be published in four volumes as *A catalogue of stars near the ecliptic observed at Markree*.

Cooper established a weather station at Markree Castle, taking daily readings from 1833 until his death. The observations resumed in 1875 and continue to this day, forming the longest continuous weather record for the west of Ireland. Astronomical research at Markree ceased when Cooper died, and the observatory finally closed in 1902, by which time no amateur facility could compete with the new high-tech professionals. Cooper also established a pioneering and very successful **fish farm** on the Owenmore estuary at Ballisodare. The falls there are too steep for salmon to climb, so Cooper installed fish ladders and later a hatchery and thus succeeded in introducing salmon upstream of the falls.

Abbeytown lead mine

People have mined minerals at Abbeytown, near Ballisodare, since the 1400s if not earlier, according to local geologist Dr Eamonn Grennan. Silver and lead were mined there in the 19th century and, again in the 1950s and 1960s, when 1 million tonnes of lead and zinc were extracted. Limestone quarrying in recent years, however, has removed any trace of the mine. The minerals had been deposited on an ancient sea floor 330 million years ago by 'black smokers', volcano-like chimneys that spew out hot plumes rich in metals.

Kames and kettles

The amazing megalithic cemetery at Carrowmore is set in a hummocky landscape but, with over 60 cairns and other monuments around, it is easy to miss the undulating terrain. Yet this hummocky ground is itself an ancient relic, a remnant of the last Ice Age. The mounds (kames) contain gravel and sand and they were made, not by people, but by streams that flowed out from under the glaciers and deposited the rocky debris. The depressions (kettles), some containing a pond, formed where large buried boulders of ice melted and the overlying gravel then subsided. Like other major glacial deposits of sand and gravel, Carrowmore has been much quarried over the centuries for building materials. Carrowmore's megalithic sites are on open access; Dúchas runs a seasonal exhibition there. ②

Ancient gneiss and green serpentinite

The **Ox Mountains** (Sliabh Ghámh, 'stoney hill'), may look bleak, but the rock that forms them is beautiful. It is ancient – 1 billion years older than neighbouring Ben Bulben – and having been around for so long has been melted, crushed and folded many times by the powerful forces that make mountains and continents. Now it is barely recognisable as the sandstone it once was. Geologists call this kind of ancient and much-altered rock a gneiss.

The minerals in the Ox Mountain gneiss form pretty bands of pink (potassium-rich feldspar), grey (quartz) and silvery-white mica. A handy place to see them is on the forest walk at Slish Wood above Lough Gill and "the lake isle of Innisfree", the attractive conjunction of rock, wood and water that inspired Yeats's famous poem. Another consequence of the gneiss's venerable age is that it is less fertile than the surrounding limestone, since much of its mineral content has been weathered away. The transition moving north across the county from ancient gneiss to 'young' limestone is evident in the vegetation, changing from bleak heathery slopes to green pasture.[2]

The unusual valley of **Slishwood Gap** is even more exotic, for here is found serpentinite, a dark green rock that has fibres of natural (chrysolite) asbestos, and was once buried 50km below the Earth's surface, where the temperature reaches 850°C. (Serpentinite is the same mineral that gives Connemara marble its characteristic colour.) Slishwood is a verdant valley that owes its existence, and its fertility, to this unusual green rock: the serpentinite is softer than the surrounding gneiss, and over time it has been eroded away to leave the gap. Moreover, the primary mineral in serpentinite is magnesium silicate, which enriches the valley soil, setting it apart from the more barren hillsides above. Look for the serpentinite in exposed rock at the southern end of the gap, before the main road bends east. The Geological Survey of Ireland's informative leaflet on *The Landscape of Yeats Country* is sold locally.

LISSADELL

Roe deer and wildlife

Ireland's native deer are the red deer,* the reindeer (no longer found here) and the giant deer* (now extinct). Over the past 800 years, there have been various attempts at introducing exotic deer into Irish estates. The only significant attempt to introduce roe deer was at Lissadell where, in the 1870s, the Gore-Booth family had a thriving herd of these shy, nocturnal animals. (One of the more famous members of the Gore-Booth family was Constance Gore-Booth, better known as Countess Markiewicz.) The roe is a small European deer, about 75cm high at the shoulder (two-thirds the size of native red deer); its antlers have, at most, three points. It was successfully introduced into Britain, but not in Ireland: for reasons unknown, the Lissadell herd died out in the 1920s – though some say a few of these timid animals could still be hiding in the surrounding woods. ③

On the shore near Lissadell, large numbers of barnacle geese* flock each winter to Ballygilgan nature reserve. The broad flat strand there is used for oyster farming and is a hive of activity during the short working shift permitted when the tide is out. Sligo's other nature reserve is at **Easkey Bog** on the northern slopes of the Ox Mountains, which has extensive tracts of three types of blanket bog* – lowland, highland and mountain – all in close proximity.

STREEDAGH

Stroll by a tropical sea

Ireland was once washed by a warm tropical sea teeming with corals and exotic creatures, just like the Bahamas is today. That was 330 million years ago, but the remains of these long-dead animals have been preserved as fossils in the limestone, thanks to some amazing chemistry, whereby hard quartz and other compounds replace the softer minerals in the original shells and skeletons. At Sligo, the fossils include coral colonies, stalked sea lilies, trilobites that look for all the world like enormous woodlice, and brachiopods – bivalve shells similar in appearance to mussels and clams. The limestone rocks at Streedagh and Lissadell beaches are especially rich in these fossils if you fancy a stroll on an ancient tropical sea floor.

An Irish ammonite fossil "drawn from nature" by George du Noyer.

DRUMCLIFF

To every cow its calf

Drumcliff Church is known as the burial place of poet W B Yeats, but it is also the site of an ancient conflict that established the important concept of **copyright** – something poets and other writers still depend on for their livelihood. Like the related concept of a patent for inventions,* copyright protects what lawyers call intellectual property.

The instigator in the ancient conflict was the munificent **Columba** (AD 521–597), later canonised as St Columcille. Columba would give as gifts copies that his monks had made of manuscripts that he had 'borrowed' from other monasteries. But Finnian, Abbot of Clonard, objected when Columba copied his psalter and, in AD 561, took his case to the High King at Tara. The Brehon Laws of the time ruled that, in the case of a wandering calf, the calf, regardless of where it was found, belonged with its mother. King Diarmuid, in the first recorded copyright ruling, declared: *Le gach bain a bainín, le gach leabhar a leabhrán* – to every cow its calf, to every book its copy. Columba refused to give Finnian back the copy, however, and 3,000 men died in the ensuing Battle of the Book at Cooldrumman, by Drumcliff. Columba won the battle, but ashamed perhaps at the bloodshed, fled in penitential exile to the Scottish island of Iona. King Diarmuid's judgment was especially apposite as the ancient manuscripts were inscribed on calf-skin vellum. Today, the same concept of copyright applies to paper and the new electronic media and web publishing.

BEN BULBEN

A limestone landscape

Ben Bulben towers above Drumcliff and Sligo's flat coastal plain, rising 526 metres in a series of layers and steps. Its cliff-edged profile is a well-known landmark, but Ben Bulben is no mountain – rather it is the edge of a broad plateau that falls gently away to the east. On a clear day, there are fine views from the summit over Sligo's sand-choked bays and estuaries.

In the midlands, the limestone plain forms a vast lowland region, but in Sligo–Leitrim it is a highland plateau, that was pushed up 100–200 million years ago when the North Atlantic Ocean was opening up. The Ben Bulben plateau once reached further west to a geological fault that runs between Lissadell and Mullaghmore. But the edge of the plateau is continually receding inland, as the soft limestone is eaten away by rain and weather.

Ben Bulben's glacial relics

Sometimes, plant and animal species are discovered that had not previously been recorded in Ireland. Most are unobtrusive species, such as the small alpine caddis-fly (*Tinodes dives*), discovered on Ben Bulben in 1983, and not found elsewhere in Ireland. This fly appears to be a **glacial relict**, that survived after the last Ice Age and adapted to the modern climate. It may have survived there because Ben Bulben's summit was a 'nunatak', high enough to protrude above the ice. Other glacial relicts in Ireland include four rare heathers* found in Connemara.

Layers of various rock types are visible in Ben Bulben's bare slopes: hard limestone and harder sandstone, interspersed with softer mudstone and shale. These reflect the way sea levels changed in these parts 330 million years ago. The land that is now 'Ireland' lay just south of the Equator then and was covered by a shallow tropical sea teeming with creatures whose shells form much of Ben Bulben's limestone. From time to time, the sea levels dropped and swamps grew there, or massive river deltas formed and brought in mud (which later formed the shales) and sand (sandstone).

Much of the limestone has since disappeared, dissolved by millions of years of erosion, but Ben Bulben reveals that what is left is still hundreds of metres thick. Despite the boggy covering atop Ben Bulben, you can tell there is limestone underfoot from the numerous sinkholes, springs and disappearing streams, and the many depressions, some deceptively bridged with insubstantial peat. **Truskmore** (647 metres), the highest point on the plateau, is 120 metres higher than Ben Bulben, partly because it has not yet lost its sandstone cap and partly because that cap protects the softer limestone underneath from erosion.

The remote plateau between Ben Bulben and Truskmore was once an industrial site where **barytes** was mined on and off from 1850 until as recently as 1980, and employing at its peak some 70 people. Barytes (barium sulphate) is a soft, heavy, pale-pink mineral, used as a drilling mud on oil rigs and in medicine for barium meals. The barytes vein at

Glencarbury mine is about 3 metres wide, quite conspicuous in places, and 2.5km long, cutting deep into the upper layers of the plateau. The older mine workings were underground (now fenced off), the later ones opencast. Local geologist Eamonn Grennan says some ore remains, if you want to keep an eye out for it. In the 19th century, the ore was processed at Gleniff mill in Ballaghnatrillick to the north, where a local group is restoring the sluice and channel system used to wash the impurities from the ore. In the 1950s, processing moved south to Glencar; the ore was carried down from the mine on an aerial rope-way, the remains of which can be seen in places on the southern hillside. The mine road is on private property and access to it from Gleniff has been denied locally in recent years.

Gleniff Horseshoe is a pleasant valley off the beaten track and bounded on the west by the unusual and dramatically sharp profile of **Benwiskin**, another of the plateau's edges. The old road to Glencarbury mine started at the bend in the horseshoe. From the bend you can also see the dramatic glacial cirque of Annacoona at the head of the valley and at least one of the caves on its steep cliff slopes. These north-facing cliffs are an important habitat for various arctic-alpine plants, the rarest of which is fringed sandwort, discovered there in 1699 by a Welsh naturalist, Edward Lhwyd. There are also attractive glens on the Leitrim side of the plateau, notably the Swiss Valley at Glencar*on the other side of Ben Bulben.

Leitrim is a county of contrasts: from the steep-sided glens around Ben Bulben in the west, to the lakelands of the east. Despite its reputation for quiet fishing and pleasant canals, it has a strong industrial heritage, notably the historic ironworks around Sliabh an Iarainn, and the coal-mining tradition shared with Arigna in Co. Roscommon. Leitrim (*Liatroim*, 'grey ridge') is named after the rising ground east of Leitrim village.

GLENCAR

The Leitrim glens

The cataract that smokes upon the mountain side, is how poet W B Yeats described Glencar waterfall. Waterfalls are common in the lovely glens around Ben Bulben, as are landslides. Both features are caused by the alternating layers of soft and hard rock that are sandwiched together to form this fine plateau. The landslides occur where a layer of soft shale low down on a cliff side is eaten away by erosion. This removes the support from under the harder layer of limestone above. Eventually, the limestone will break off under its own weight, collapsing into the valley below, sometimes as a huge slice of rock, more often as a jumble of scree. The landslips would have been most spectacular at the end of the last Ice Age: glaciers filled the valleys then and, though they eroded the shale, they also supported the limestone ledges. But once the glaciers melted, the limestone collapsed and slipped. Landslides still happen occasionally, albeit on a smaller scale, nibbling away at the plateau's edges. The fresh scree slopes and cliff sides produced in this way are an important habitat for alpine plants. The many waterfalls occur when streams tumbling off the high plateau hit a layer or lip of harder rock, and are tossed over the edge to cascade down. The prettiest and best-known waterfall is at Glencar, where the water drops 30 metres into an attractive wooded pool in the **Swiss Valley** above Glencar Lough. Manorhamilton lies at the eastern end of Glencar Valley, it is a pleasant spot where half a dozen Leitrim glens converge. A small clutch of drumlins* was laid there by a glacier during the last Ice Age, witness the many place names that contain the element *druim* ('low hill'), such as Drumnahan, Dromore, Drumman and Drumdillure.

LOUGH MELVIN

Three strange trout

Lough Melvin is unique and internationally important. For despite its small size (21sq km), it is home to three very different types of brown trout. It is also one of the few remaining lakes in northern Europe that has stayed relatively undisturbed since the end of the last Ice Age 13,000 years ago. The lake's trout are the strangely named

The ferox: largest of Lough Melvin's three trout species, it feeds on other fish and can live for 12 years.
© Andy Ferguson/QUB

sonaghen (occasionally called the black-finned), the **gillaroo** (*gille ruaidh* or red fellow, for its red spots), and the **ferox** (a term often used for large trout). Prof Andy Ferguson, a zoologist at QUB who has studied Lough Melvin's trout in detail, says the three have been known for at least 200 years, although it was once thought the small sonaghen were merely young trout.

The three look very different, lead different lifestyles and have different diets and, because they spawn in separate places, there is no crossbreeding or mixing. The three are classed as separate subspecies of the trout family: *Salmo trutta ferox*, *S. trutta nigripinnis* (for the black-finned sonaghen), and *S. trutta stomachius* (for the gillaroo's supposedly thick stomach).

The sonaghen is the smallest and most abundant of the three. It eats tiny insects, lives for five years at most, and an adult will weigh less than 500g. Sonaghen spawn in small rivers flowing into the lake, especially the Ballagh and Tullymore. The gillaroo, larger but less abundant, feeds mostly on snails and the like, reaching about 1–1.5kg, but nonetheless rarely lives past five years. Its preferred spawning ground is the out-flowing Drowes River and Lareen Bay. The ferox, least common of Lough Melvin's trout, is very different. It is a large, heavy, long-lived trout that dines on other fish: an adult ferox, on a diet of perch, charr and even other trout, can live to be 12 years old and weigh 6kg. Its spawning grounds are the deep downstream stretches of the Glenaniff river.

The three types of trout probably arrived in Lough Melvin once the ice had melted at the end of the Ice Age. Like all brown trout, they are descended from sea trout that adapted to spending their life in fresh water. Recent genetic studies suggest that the sonaghen and gillaroo are related, but also that they are unique to Lough Melvin and not found anywhere else. The ferox, however, seems to be related to the brown trout of Lough Erne.

From a conservation point of view, Lough Melvin is special: none of the three trout was introduced and the situation there evolved naturally over thousands of years. Indeed, there are no introduced fish in the lake at all, and little pollution, and Lough Melvin today is much as it was at the end of the Ice Age. But this also makes the lake vulnerable: if pike or roach were introduced, or the water quality dropped, then the delicate balance between sonaghen, gillaroo and ferox could be destroyed forever.

DROMOD

The Cavan & Leitrim Railway

The 54km Cavan & Leitrim Railway ran from Dromod to Belturbet, with a branch line linking the coal mines at Arigna* to the canal at Ballinamore. This narrow-gauge route, which opened in 1887, was crucial in sustaining Arigna coal mine until the 1920s. The maximum speed was

The Cavan & Leitrim Railway: narrow-gauge railway cars could travel alongside other road traffic and stop at crossroads to pick up passengers.
Leitrim County Library

a leisurely 19 kmph. In the 1920s, the route was taken over by the Great Southern Railway Company but it finally closed in 1959. Local enthusiasts have restored Dromod station, workshop, train shed and water tower, and 800 metres of track where vintage steam trains run every weekend in summer. There are plans to rebuild the line to Mohill station, which is also being restored. This museum is 'work in progress' and as well as old railway equipment from around the country, visitors can witness a railway being rebuilt or take a short excursion on a steam or diesel train. ④

BALLINAMORE TO BALLYCONNELL

From Leitrim to the Ruhr

The **Ballinamore and Ballyconnell Navigation** was part of the ill-fated Ulster Canal,* intended to link Limerick and Belfast via the River Shannon and Lough Erne. The Ballinamore–Ballyconnell stretch runs for 60km, from near Leitrim village to Lough Erne via Lough Scur. But this section of the Ulster Canal did not open until 1860 and was never a success, closing just nine years later. During that time, it carried only eight boats, earning £18 in tolls, compared with building costs of £250,000. It was restored, however, and reopened to leisure craft in 1994 as the **Shannon Erne Waterway**, and this second lease of life has been much more successful.

The canal route was selected and the canal designed in 1845 by a Dublin engineer, **William Thomas Mulvany** (1806–85), who masterminded the arterial drainage programme of 1842–52, and kick-started the industrial development of Germany's Ruhr Valley.[3] Mulvany worked on the new Ordnance Survey,* the Shannon navigation* and later as a commissioner for the Board of Public Works where he was the prime mover behind a national programme of publicly funded **arterial drainage** work which began in 1842. The aim was to design and re-engineer river and canal navigation, control floods, ensure

William Mulvany: the Irish engineer who designed the Ballinamore–Ballyconnell canal and kick-started industrial development in Germany's Ruhr valley.
James Dooge

Floods and flows

Drainage works must be able to cope with the worst downpours and flash floods. But, in the 1840s, there was no accurate way to predict the flooding that might occur at a particular river or lake after heavy rain. Then two Irish engineers working on the arterial drainage programme – **Thomas Mulvany** (1821–92, who was William Mulvany's youngest brother) and **Robert Manning** (1816–97) – devised ways to accurately predict flood levels and flow rates that are still used today.

At the time, rain gauges simply measured the total volume of rain that fell in a 24-hour period. Thus, William Mulvany noted that, on one November day in 1840, the water level in Lough Derg rose 30.5cm. Knowing the lake was 12,140 hecatres in area, with a catchment area of 930,760 hectares, he calculated that, over the 24 hours, the rain would have run off the surrounding land at a rate of 0.02 cubic metres per minute per hectare of catchment. Mulvany directed his drainage staff to make similar calculations for other Irish rivers and lakes, but they all got different results: there was no standard rate of run-off that could be universally used, and worse, the engineers often underestimated the actual flood levels.

In 1851, Thomas Mulvany realised that this 24-hour approach was inappropriate for small, hilly areas, where flood waters could run-off in just a few hours. He suggested calculating the time taken for the flood water to arrive from the most remote point, as by then the whole catchment area would be contributing to the run-off and the flood would have reached its peak. He called this "the time of concentration". Mulvany designed a clockwork rain gauge to measure rainfall continuously and record the measurements on paper. His approach, known as the 'rational method', is still used. In 1855, Thomas Mulvany followed his brother William to Germany, becoming director of the Shamrock and Hibernia coal mines. In 1878, he moved to New Zealand, where he is buried.

The second Irish contribution to predicting water-flow rates was made by Robert Manning, who came from Crooke parish in Waterford and was later chief engineer with the Office of Public Works. Manning wanted to devise a simple technique to predict flow rates in open channels, such as drains and rivers. Using measurements taken by other engineers on rivers around the world, including the Mississippi, Manning devised an equation to accurately predict flow rate, taking account of the channel's slope, size and shape.[4] He presented his equation at a meeting of the Institution of Civil Engineers in Dublin in 1889, and it was soon used internationally, appearing in textbooks from London to Cairo. In 1989, a major international engineering conference held in the USA commemorated the centenary of this disarmingly simple, yet still valuable, Irish equation.

consistent water supplies to watermills, drain wetlands, and generally improve access and transport. The Ballinamore–Ballyconnell canal was part of this ambitious programme. Landowners were initially keen on a scheme which promised to improve their estates at public expense, and Mulvany gathered a team of skilled engineers and surveyors for the work, including his brother Thomas and Robert Manning, both of whom went on to devise ways to predict flood and flow rates that are still used today (see: **Floods and flows**).

During the black Famine years, William Mulvany campaigned for his drainage works to be included in the government relief schemes, arguing that the work could provide useful employment, unlike the unproductive tasks usually assigned to the starving workers. But the landowners believed Mulvany was exceeding his powers and they convened a House of Lords inquiry under Lord Rosse* of Birr. In 1852, the Lords found in their own favour, the drainage works were stopped and Mulvany was retired on a pension. His skilled staff disbanded, many emigrating to work on irrigation schemes for the East India Company.

Within two years, Mulvany had also found a new career abroad: in 1854 he was asked to survey some coal deposits near the Ruhr village of Gelsenkirchen, by a Brussels-based Irishman, Michael Corr, whose parents had fled Ireland after Robert Emmet's 1803 Rising. The Ruhr was a prime farming region then, with just a few small, shallow coal mines supplying local needs; otherwise, Germany used coal imported from England. Mulvany examined a local geological map and realised there could be deep, rich coal deposits there. On St Patrick's Day 1856, with funding from four Irish Quakers – James Perry from Dublin and three Malcolmson* brothers from Waterford – Mulvany opened 'Shamrock', the Ruhr's first deep coal mine. The company used deep-mining techniques imported from England, notably cast-iron tubing in place of the brick-lined shafts used in the neighbouring German-owned shallow mines.

More Irish coal mines soon followed – Erin, Zollern, Hansa – all using the latest techniques to reach the deep, rich deposits. The mines were all successful, many continuing until the coal was exhausted well into the 20th century – the Erin mine in Castrop closed only in 1988.

Prof Jim Dooge, who has written about Mulvany, says the Irish engineer also helped develop the area's industrial infrastructure, including a railway and canal network to ship the coal, but especially the Rhine as a major shipping route. He also founded the Dusseldörf stock exchange and introduced steeplechasing to Germany. The mining empire Mulvaney founded became VEBA, now a major German industrial group. Perhaps, not surprisingly, Mulvany is better known in Germany than in Ireland: a biography of him was published there in 1922, and Mulvanystrasse in Dusseldörf commemorates this Irish engineering entrepreneur. Information about the Ballinamore–Ballyconnell canal is available from Shannon Erne Promotions. ⑤

Sliabh an Iarainn's ironworks

Sliabh an Iarainn (iron mountain) is aptly named: the land to the east of Lough Allen holds Ireland's richest iron deposit; streams in the region are often stained a rust colour; and nearby Drumshambo was once the site of a major industrial ironworks. Iron is a common element and widely distributed throughout the Earth's surface, but always as a rocky ore containing earthy materials and other minerals. So making iron always entails some processing, and its appeal as a metal began only with the development of reasonable smelting techniques, though we will never

There were many small iron smelters in 17th- and 18th-century Ireland, and some provided iron for the United Irishmen's pikes in the 1798 Rebellion.* Most used bog iron, as at Tuamgraney* in east Clare. True iron ore was also mined at Sliabh an Iarainn, the Antrim Coast* and Castlecomer.*

know how people discovered that a purple-brown rock can be transformed into a shiny, useful metal. Iron smelting originated in the Middle East 4,000 years ago and spread slowly across Europe, arriving in Ireland probably with the Celts about 400 BC.

Early smelters were simple clay ovens where lumps of ore were placed on top of hot charcoal; the charcoal acted as a fuel, but also absorbed impurities from the ore. A manual bellows might be used to encourage the fire, and after some hours the ore would have reduced in size to a 'bloom' of spongy **wrought iron** that could be hammered into shape. By the Middle Ages, ironworks were using waterwheels to operate more powerful bellows and hammers, and hence had to be built near a water source.

In the 15th-century, Belgian ironworkers discovered they could make a purer iron if they smelted the ore *with* the charcoal, rather than on top of it. This led to the blast furnace, an example of which can be seen at Creevelea. It was a tall oven with an opening at the top where the charcoal and ore were introduced, and a facility at the base where the molten **cast iron** was let out. The next improvement came when it was discovered that adding limestone to the furnace produced an even cleaner iron, because the limestone combined with the earthy ore to form a slag from which the molten metal readily separated.

Blast furnaces changed iron smelting: they were larger than the old bloom furnaces; they operated round the clock; and consumed vast amounts of charcoal. Some of the larger British ironworks overcame this last problem by planting and coppicing* oak woods: cutting the trees at the base, then allowing a flush of shoots to re-grow before cropping them again at intervals of 16–24 years. But coppicing was seldom practised in Ireland. Instead, once the trees around an ironworks were consumed, the company simply moved further downstream, leaving behind the cleared land, abandoned furnaces and slagheaps and, sometimes, a succession of places called Furnace. Iron smelting was a major industry in 17th-century Ireland, and it contributed to the rapid disappearance of what was left of the country's native woodlands – it took an estimated 0.8 hectares of mature oak wood to produce enough charcoal to smelt just 1 tonne of iron ore.

The final historic development in iron smelting came in 1709 when Englishman Abraham Darby discovered he could smelt iron successfully using coke, a kind of charcoal-like fuel made from coal (raw coal cannot be used because of its impurities). Iron could now be made wherever coal, iron ore and water occured together and, as geological luck would have it, coal and iron ore are often found in close proximity, as at Lough Allen.

The iron occurs there as seams and nodules in shale that formed when this area was covered with iron-rich swamps and lagoons 325 million years ago. (The coal at Arigna on the other side of Lough Allen formed in similar swamps about the same time.) There have been ironworks around Sliabh an Iarainn since at least the 15th century, especially at Drumshambo, according to local historian Breda Wynne. The Planter Sir Charles Coote reputedly employed 3,000 Dutch and Englishmen there in the 1600s – no Irish were recruited for fear they would learn the secrets of the smelting trade. Gerard Boate,* in *Irelands Naturall History* (1652), writes that over 18 different trades were employed

The ruined works at Creevelea include this enormous 19th-century blast furnace with a complex 2-tiered structure and chimney above the smelting oven.
Leitrim County Library

at each Irish ironworks, from blacksmiths to charcoal makers, and "as for the charcoal, it is incredible what quantity thereof is consumed".

Most Irish ironworks were destroyed during the rebellion of 1641. But, in the 1690s, after the Battle of the Boyne, works reopened in Leitrim at Drumshambo, Dromod, Ballinamore and **Creevelea**. Drumshambo finally closed in 1765 and Creevelea a few years later, though it reopened and even employed 600 men again for a short while in the 1830s. In 1851, a Scottish firm worked Creevelea again, initially using coal carted from Arigna over 13km away. The business continually struggled, though during the 1860s it had some success with an experimental charcoal made from peat. By then, however, Ireland was importing almost all its iron from Britain and Sweden. Belfast's new shipbuilding industry and the developing agricultural sector were major

users, also the numerous spade mills* and foundries begun at this time, including Pierce* of Wexford. There are still significant iron deposits at Sliabh an Iarainn, but they are no longer economic.

The region around Lough Allen and north Co. Leitrim also has commercial deposits of an unusual clay called **attapulgite**. All clays are formed from weathered rock of one kind or another, and this clay comes from the weathered shale of the surrounding countryside. Attapulgite is rich in fibrous minerals, notably magnesium-aluminium-silicates, which are highly absorbent, and the clay can be used for cat litter, among other things. The Sliabh an Iarainn visitor centre at Drumshambo features the local iron and coal mining, and the Cavan & Leitrim Railway. ⑥ Leitrim County Library in Ballinamore has archive reports on the local mining. ⑦

Fringed with furze

In Leitrim, furze was once a common crop, sown in "a plot or garden of a quarter acre in rough strong land". We are more likely to think of it as a troublesome if colourful weed, but furze – gorse, as it is called in the north, or whin, its common name in parts of Fermanagh and Tyrone – once had many uses and was an important crop. In the 1950s, with a view to reviving furze as a crop, the Folklore Commission and the National Museum of Ireland surveyed its traditional uses. The resulting report is probably the most comprehensive study of any Irish plant.[5]

The Irish for furze is aiteann (Welsh: eithin, Basque öte). It occurs in many place names, like Mullachanaitinn, reflecting the plant's widespread distribution.

Furze is a legume, related to clover and peas but, in furze, only the young shoots have the characteristic trefoil leaves; otherwise furze leaves are reduced to sharp thorns. These protect the plant from grazing animals and parching winds and furze can, therefore, survive in exposed places. And because legumes have root nodules that 'fix' nitrogen and produce their own fertiliser, furze can also thrive in poor soil, where it out-competes other species. Two types of furze grow in Ireland: common furze

(*Ulex europaeus*), found over most of the country, can grow to about 1.5 metres in height and flowers from March–May. Dwarf furze (*U. gallii*) is smaller and hardier and can survive on poorer soil and higher up mountains. It flowers from August–September and is common in the southern half of the country and in counties Armagh and Down.

The 17th-century Down Survey* recorded that furze could "yield as much profit as the arable land". The Folklore Commission's survey 200 years later found it was valued for hedging and numerous other uses. Bakers used 'fyrris faggots' in their ovens because they lit quickly, gave an intense heat and produced little ash. Mashed furze tops were fed to horses and cattle, while the wood was used in roofing, in foundations for paths and to make hurleys and walking sticks. Chimneys could be cleaned with a bundle of furze branches, the yellow flowers were used in dyeing, and some herbal remedies recommended furze to treat ringworm. In the 19th-century, homesick Irish and Scots emigrants introduced furze to New Zealand and the Caribbean where it is now a major pest, and furze was recently introduced to some of Ireland's western islands in soil imported from the mainland, where it may yet become a problem.

Highlights include one of Ireland's last coal mines, the Famine Museum at Strokestown, and the varied Shannon shoreline that forms the county's eastern border. Roscommon (*Ros Comáin*, 'St Comán's wood') is named after the 8th-century priory by Roscommon town.

Ireland's coal mines

There are various coalfields around Ireland – including Ballycastle,* Coalisland,* Castlecomer,* and even the Kish Bank* in Dublin Bay. Aside from the Kish, most are small and the coal was often of poor quality, and though they were mined on and off over the centuries, it was usually easier to use local wood or turf as a fuel or even to import coal from abroad. The Arigna coalfield however was worked until 1990, making it the last large-scale coal mine in Ireland (a small mine employing six men at Castlecomer was still being worked in the spring of 2000). Arigna's coal formed from swamps that grew there about 325 million years ago, and it lies in horizontal seams in the limestone hills around Lough Allen. The best of the deposits is found to the west of the lake, on either side of the Arigna river between counties Leitrim and Roscommon, although even that coal produces nearly 40 per cent ash.

For centuries, locals probably worked small outcrops of this coal for their domestic needs as others might cut turf. But, in the 1770s, an extensive seam was discovered and commercial mining began. For most of the next 200 years, the mines were worked on and off at over a dozen sites, sometimes employing over 600 men. Horizontal tunnels were dug into the hillsides, but, more recently, large excavators were used in opencast mining. A small canal at Drumshambo, opened in 1817 as part of the upper Shannon navigation,* helped to improve access in the area. The Cavan & Leitrim Railway,* which opened in 1887, made access even easier and also used the local coal. Within the Arigna Valley and on the surrounding hills, a complex network developed of narrow gauge tramways, railways and ropeways, operated by horses and pulleys, and connecting the tunnel openings with the access routes below. For the past 100 years, the Leyden family was important in developing the Arigna mine, and during the Emergency of the 1940s, employed some 600 men. The coalfield was also the scene of a workers' soviet in 1919–20.

The **Arigna Miner's Way** is a 50-km walking route that circles Kilronan mountain, starting at Arigna village, where maps and leaflets are available. The trail passes through villages whose inhabitants have mined the local coal and iron for over 400 years. Ruins of ancient **sweat houses** – the Irish equivalent of the sauna – also dot the route.

Arigna coal has almost no sulphur, making it a clean fuel, but it is friable and so must be handled carefully, produces a lot of ash, and has a low volatile content making it difficult to ignite and slow to burn. All of which pose problems for any user. For a long time, the main customer was the local railway but the Great Southern Railway took over the line in the 1920s and changed to using imported Welsh coal. When the railway closed in 1959, a new coal-burning power station on the western shore of Lough Allen gave the mine a further lease of life. The mine closed in 1990, however, and the power station followed shortly after. Most of Arigna's coal is now exhausted, and what is left there and elsewhere around Lough Allen is of poor quality. The hills around Arigna and Lough Allen still bear traces of their long industrial history – in place names, ancient slag heaps and furnaces, old mine shafts and tracks. Not everything has stopped though: local shale is used in making cement, Arigna flagstones are quarried for the building trade, and a local firm is making smokeless briquettes, but from imported ingredients. A new visitor centre at Arigna celebrates the valley's mining heritage. ⑧ Sliabh an Iarainn* centre in Co. Leitrim also features the Arigna coal mines.

RIVER SHANNON

Mid-river meanderings

From Knockateen to Shannonbridge is a distance of some 170km as the Shannon flows. This long stretch of the river forms Roscommon's eastern border and it is especially

varied: from the expansive basins of loughs Allen, Boderg, Forbes and Ree, to the confined channels at Roosky and Cloondara, and the broad flood plains or callows* around Shannonbridge, all of which are probably best explored by boat.

Like all rivers, the Shannon's route and form are determined by the local geology and topography. At Lough Bofin, for example, the rock is a soft sandstone and the river has been able to create enough space for a lake. Immediately to the south is harder shale where the Shannon has managed to excavate only a narrow channel. Soft limestone around Lough Forbes permits another lake, before the river is again confined, until soft limestone is reached once more at Lough Ree. The second-largest lake on the Shannon, **Lough Ree**, has a convoluted shape and a very irregular floor. There are frequent shallows, numerous deeps, and many places where the limestone still rises high enough to break the lake surface and form islands. Clearly, the limestone has been much eroded and weathered – sufficient even to form clays, as at Knockcroghery (see: **Clay-pipe factory**) – but also very irregularly. South of Athlone, the Shannon enters esker* country: the river contours around some of these glacial gravel ridges (as at Clonmacnoise), elsewhere it cuts through them (as at Athlone and Shannonbridge).

During the Napoleonic era, Lough Ree and Lough Derg were charted by the British Admiralty fearful of a French invasion as, by then, the Shannon was navigable. For the same reason, gun batteries and barracks were installed at various points on the river, including Shannonbridge and Keelogue, and a Martello tower (a fortification normally associated with the coast) was built on Cromwell's Island.

The Shannon river has not always followed the same route from Lough Allen to Lough Ree: a long rocky ridge bars the southern end of Lough Boderg and, before the last Ice Age, the original Shannon could not cross this barrier. Instead, it left the lake to the west, passed between Drumdaff and Roscommon town, and so on to the prehistoric Lough Ree.

Then some 20,000 years ago, glaciers deposited debris that blocked this original route; fortunately the ice also cut through the rocky ridge at Lough Boderg, opening a new passage for the river at the Derrycarne Narrows. When barge traffic began on the Shannon 200 years ago, these narrows were deepened by blasting. At Lanesborough, water discharged from the ESB's power station is 10°C warmer than the rest of the Shannon and much appreciated by local fish. Lanesborough is also the setting for the annual ESB Loughree environmental summer school. (9)
[Next stop on the Shannon: **The Shannon callows** in Co. Offaly.]

CASTLEREA

Wildean wisdom

Oscar Wilde is one of Ireland's best-known writers, but his father William also had an international reputation as a writer, physician and antiquarian, and students came from Europe and even North America to learn new techniques at his eye-and-ear hospital in Dublin. **Sir William Wilde** (1815–76) was born in Castlerea and studied medicine at various Dublin hospitals under some of the city's great physicians, including Robert Graves* and William Stokes.* Soon after gaining his licence from the Royal College of Surgeons* in 1837, Wilde contracted a fever and as part of his recuperation embarked on a Mediterranean cruise, working as medical attendant to a wealthy businessman. His first book was an account of this voyage, published in 1840, which proved successful and helped fund his medical career. Wilde went on to travel widely, developing an interest in archaeology and natural history, and publishing all the while. His book on Austrian medical, scientific and literary institutions (1843) is still a useful reference work.

Visiting Egypt, he was horrified to see so many people blinded by trachoma infection and resolved to specialise in eye surgery. He studied at leading European centres before returning to Dublin to establish his own eye-and-ear clinic, first at Westland Row then, from 1848, in Lincoln Place, as

the new St Mark's teaching hospital, with accommodation for 20 public patients. (In 1897, St Mark's and the National Eye & Ear Infirmary amalgamated to form the Royal Victoria, which is still at Adelaide Road.) Wilde was a great innovator, especially as an ear surgeon: he developed the first dressing forceps, an 'aural snare', and an operation to relieve inflammation of the mastoid bone, called Wilde's incision. His practical book on *Aural surgery & the nature & treatment of Diseases of the Ear* (1853) was the standard text for many years and widely translated.

William Wilde's gravestone, Mount Jerome,* Dublin: "Surgeon oculist to her Majesty Queen Victoria, chevalier of the Royal Swedish Order of the North Star, founder of St Mark's ophthalmic hospital, and the author of many works illustrative of the history and antiquities of Ireland."
© Enterprise Ireland

Appointed medical advisor and compiler for the 1841 census of Ireland, Wilde collected valuable medical and scientific information, much of it for the first time, including the cause of death. He repeated these surveys for the census of 1851, 1861 and 1871. The reports include valuable notes as well as tables of figures, providing important historic records on the health of the Irish people before, during and after the Great Famine.

Wilde maintained his interest in antiquities and natural history, exploring and writing about the Boyne Valley and Lough Corrib* (his country house was by the lake, at Moytura). He also wrote biographies of three important Dublin medical men: Thomas Molyneaux,* Bartholomew Mosse* and Jonathan Swift.* He was honoured with many awards at home and abroad but, late in life, was sued for libel by Mary Travers, following a separate court case in which he was accused of sexually assaulting Travers. Wilde lost the libel case and was broken by the trial, foreshadowing the fate that would later befall his son Oscar.

Clonalis concrete

Mass concrete was first used as a building material in the mid-19th century. One surviving early example of this type of construction is **Clonalis House**, built in 1880, and the ancestral home of the O'Conor clan, former kings of Connaught. The walls throughout the house are of concrete and even the architectural details were cast in concrete. ⑩ The house holds the important O'Conor archive of over 100,000 documents, including a copy of the last Brehon Law judgment, which was handed down in 1580. (The archive catalogue can be consulted at the National Library, Dublin.*). At the same time as Clonalis House was built, a new technique for building dock walls with massive pre-cast concrete blocks was being pioneered in Dublin by the innovative Irish engineer Bindon Blood Stoney.*

A varied collection of railway memorabilia fills the **Hell's Kitchen** pub in Castlerea, including an A-class steam locomotive, bought from CIÉ in 1955 when diesel was replacing steam. Other artefacts include railway staffs – the rods carried on a train to ensure only one vehicle was on a particular stretch of track at any one time; when the train passed through, the staff was handed in at the nearest station for use by the next train. ⑪

Clay-pipe factory

An unusual clay pit at Carnamaddy on the shore of Lough Ree gave rise to a clay-pipe industry at Knockcroghery where, at one stage in the 1890s, seven factories were working. The clay is weathered rock, from limestone that the lake eroded over thousands of years. Some of the clay was deposited in a depression or 'solution pipe' in the surrounding limestone that protected it from the elements until it was discovered in the 1700s.  Unlike the fine china clay found at Belleek,* the Carnamaddy clay is a low-quality earthenware clay, but it is ideal for making pipes. At Knockcroghery each pipe, or *dúidín,* was individually handmade using a metal mould and a hand-operated press. The moulded pipes were dried for five days, then finished and trimmed. A skilled worker could produce 5,000 pipes a week, a rate of about two a minute. During the 1890s, some 60 people were employed, but by then the factories were using clay imported from England.

The pipes, often stamped with advertising slogans, were popular with emigrants overseas. Knockcroghery: from Cnoc an Chrochaire, Hangman's Hill, after the small hill in the village once used for executions. Clay-pipe Museum

Kept carefully, a clay pipe would last a lifetime, but drop it and it would shatter. Fortunately they were never expensive, costing less than a ha'penny. The market for clay pipes declined in the early-20th century, however, with the growing availability of cigarettes and cheap wooden pipes. Production at Knockcroghery ceased abruptly when the village was burned by the Black and Tans in 1921. The craft was recently revived using the original moulds and techniques at a workshop on the site of A & PJ Curley's pipe factory. ⑫

The alien pitcher plant | Blight and the Big House

A colony of colourful insect-eating pitcher plants grows on the small raised bog at Derrycashel, south of the Feorish river. Insectivorous plants are common in Irish bogs, where the insects provide the plants with much-needed nutrients, but these pitcher plants are alien: they originate in Canada and were introduced here in 1904. Having successfully established themselves at Derrycashel, they have since been transplanted to a number of other bogs, though arguably it is not a good idea to help an alien species spread. The plants have fat bunches of lurid purple-green leaves that curl together to form a pitcher some 12cm high. Any insect falling in drowns in liquid at the bottom of the pitcher. Enzymes secreted by the plant can then digest the insect's body.

Roscommon has important areas of raised bog* and several turloughs,* many of which have been designated special areas of conservation, notably: the raised bogs of Bellanagare, Cloonchambers, Cloonshanville, Derrinea and Lough Forbes, and the turloughs at Lough Croan and Lisduff. The machair grassland* on the lime-rich Castlesampson Esker is also a designated SAC. Given Roscommon's numerous small lakes and extensive Shannon shoreline, it is not surprising the county has six wetland **wildfowl sanctuaries**: Lough Key, Lough Gara, Little Brosna, Annaghmore Lough, Lough Croan and Lough Funshinagh. Though much of Roscommon is low-lying, the land rises in the north at the Curlew Mountains and the coal-bearing hills of Arigna.*

Herbal remedies

The first true Irish herbal was probably *Botanalogica Universalis Hibernica*, written in 1735 by Roscommon man **John K'Eogh** (*c.*1681–1754), although herbal remedies had been included in earlier manuscripts and books. K'Eogh's cure for deafness was eardrops made from heron fat warmed with oil of amber.

Strokestown's Major Denis Mahon: an unpopular landlord during the Famine years. Strokestown House

In August 1845, potato fields around Ireland began to succumb to an unusual disease. No one knew what caused this blight,* nor where it came from. Potatoes had been successfully grown in Ireland for over 200 years and, apart from occasional problems caused by weather, there had been no major disease concerns. Meanwhile, in Co. Roscommon, Major Denis Mahon had just inherited the debt-ridden Strokestown Estate, and his agent, John Ross Mahon, planned to pay off the debts by clearing two-thirds of the tenants to make way for tillage farming. Instead of throwing the evicted tenants on the mercy of the local workhouse, where Major Mahon would have to pay for their keep, they would be sent abroad on specially chartered emigrant ships.

Over the next few years, these two developments would converge at Strokestown with tragic consequences for all concerned. Tenants, blight-stricken, starving and ill, who could not pay their rent, died or were evicted. Then, in November 1847, Major Mahon was assassinated. Two men were later hanged for his murder.

The story of these black years is told in an acclaimed Famine Museum housed in the stable yard at Major Mahon's old home of Strokestown House. At the core of the exhibition is the Mahon family archive, which includes estate papers, management documents, even letters written by tenants. The exhibition devotes a section to the potato* – varieties grown locally, their nutritional value, even the local spade* used to make the cultivation ridges: the massive Roscommon *loy*, described in an 1830s account as "a hand plough" – and a section to blight, especially theories popular in the 1840s as to what it was and how to treat it. The 18th-century house and estate have been restored by a local firm, the Westward Group. A guided tour spans the last 300 years of Irish history, and the gardens boast the longest herbaceous border in these islands. (13)

As the tragic events were unfolding at Strokestown, the **windmill** at neighbouring Elphin was closing. Built in the 1730s to grind grain, it was worked for over a century before being abandoned. A local group has restored it to working order, complete with thatched roof, a new timber interior, and oak-wood cogs similar to those used in the 18th century. The sails are set into the wind by turning the cap, which can rotate through 360 degrees; it was originally pulled by a horse or donkey, but is now turned by a tractor. The mill is again grinding grain and demonstrations can be arranged, wind permitting. (14)

Longford *(Long phort,* 'long fort') has two connections with historic road-making: the amazing prehistoric oak track at Corlea Bog, and engineer Richard Lovell Edgeworth – arguably he, and not John McAdam, should be credited with inventing the principles of modern road-making.

Squirrel squabbles

In 1913, the Earl of Granard released a dozen North American grey squirrels on his estate at Castle Forbes. This simple act sowed the seeds of a conflict that continues today: for wherever grey squirrels are introduced into Europe they threaten their smaller red cousins, and Ireland is no exception. Across Europe, everywhere red and grey squirrels meet, the native red loses out to its larger, more adaptable North American cousin. Zoologists do not know why this happens, especially as the two species prefer different habitats: red squirrels thrive best in coniferous forests, while grey squirrels are happiest in broad-leaved woodland. Most worrying in Ireland is that the red squirrel is now rare or absent in counties where the grey squirrel has been present the longest, notably Longford and Westmeath.

The red squirrel (*Sciurus vulgaris*), found throughout Europe, is native to Ireland but, by the 1700s, had become extinct here. In the early-19th century, it was reintroduced from England at a dozen sites around the country and by the time the grey squirrel (*S. carolinensis*) arrived in 1913, red squirrels were again found in all counties. Initially the grey squirrels remained in Longford. Then, in the late 1930s, they began spreading north, east and south and are now present from counties Derry and Tyrone through to Wicklow and Wexford. For the moment, the Shannon bars their movement west and so far they have also not made it southwest to Cork and Kerry. The grey squirrel is clearly here to stay, but scientists do not know what the future holds for the Irish red squirrel. Will the cousins find a way

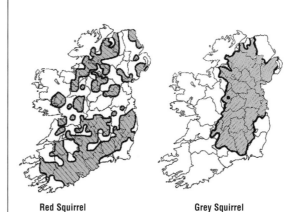

Red Squirrel   **Grey Squirrel**

Zoologists Shane Reilly (TCD) and Denis Tangney (QUB) studied the distribution of the two squirrels in the 1990s. The red squirrel's distribution is fragmented, with many small isolated populations that could be vulnerable to competition from their grey cousins.
© TCD Zoology Department

of co-existing? Will our large, commercial coniferous plantations favour the red? Or will it be driven to extinction in Ireland for a second time?

An ingenious squire

Edgeworthstown or Mostrim was home to the celebrated novelist **Maria Edgeworth** (1767–1849). But her father, **Richard Lovell Edgeworth** (1744–1817), though less well known, was a gifted inventor and educator and arguably he, and not John McAdam of tarmacadam fame, should be credited with inventing the principles of modern road-making. The Edgeworths came to Ireland in the 16th century and were granted 243 hectares in Longford. Richard Lovell was born in England but moved his family over in 1782 when he inherited the Irish estate. Edgeworth was a liberal thinker, sometimes ridiculed for his far-sighted ideas, and a respected engineer appointed to the Bog Commission investigating the best way to drain and exploit Ireland's vast bogs.

Longford's 'Lunatic': Richard Lovell Edgeworth. While still in England, Edgeworth was one of the 14 select members, along with the inventors Benjamin Franklin and James Watt, of Birmingham's Lunar Society, so called because their meetings coincided with each full Moon. In Ireland, Edgeworth was active in founding the Royal Irish Academy. Longford Library

Fascinated by mechanical contrivances, he invented an early bicycle and a sail-powered carriage designed for Longford's straight roads. He recommended using springs to make carriages more comfortable, built a horse-drawn railway on his estate for transporting peat, and designed a vehicle for crossing bogs equipped with primitive caterpillar treads – years before George Cayley, the English inventor traditionally credited with inventing the caterpillar tread in 1825. Edgeworth designed a central-heating system for his neighbours, the Pakenhams of Tullynally, and, in 1783, suggested a device for measuring wind speed to his son-in-law, the astronomer Thomas Romney Robinson* (40 years later Robinson succeeded in making it, the first instrument ever to measure the speed of the wind).

In the 1760s, while a student, Edgeworth designed what was probably the first aerial telegraph, prompted by a bet about the time taken to send racing results from Newmarket to London. His design for a mechanical semaphore relied on a network of towns built at 50-mile intervals; triangular pointers on top of the towers displayed the coded message to observers in the next tower. Edgeworth failed to attract any backing until 1803, however, when the British government, fearful of a French invasion, asked him to test his idea. Working with Admiral Francis Beaufort* (the man who devised the wind scale and who was both Edgeworth's brother- and son-in-law), Edgeworth succeeded in sending a message from Dublin to Galway in eight minutes. Clear weather was needed for the signals to be visible, however, and the system was never taken up, although a similar telegraph was established in France in the 1790s to send messages between Calais and Paris, and used until the

Edgeworth's road

At the end of the 18th century, road traffic began to increase and Richard Lovell Edgeworth, seeing the need for better roads and carriages, built a road on his Longford estate using new principles he devised himself.

The ancient Romans built fine roads paved with heavy stone slabs but, until the 1500s, European roads were mostly dirt tracks, muddy when wet and heavily rutted when dry. The first major improvement was using broken stone (road metal), introduced by the 16th-century Italian engineer Toglietta. In 1775, the French engineer Trésaguet made two changes: he used an underlying foundation to support the road, and he laid three layers of stone, starting with the largest stones at the bottom. But his road was built in a trench and drainage was a problem.

Edgeworth's design incorporated two significant new features, both now an integral part of modern roads. His road had a slightly arched profile or 'camber' so that rain could drain away; and a top 'blinding' layer 20-25cm thick of angular, broken stones. Finally, stones larger than 1½ in (3.8cm) were cleared from the surface as they would damage the underlying 'metal'. The small angular stones were crucial: as wheels passed over the surface, the flat angular sides were forced together; any intervening spaces then filled with smaller stones and sand, until the whole layer locked in place with a flat surface. Edgeworth published his road design in 1813,[6] three years before John McAdam built the Bristol road for which he is famous.

coming of the electrical telegraph. Ironically, Edgeworth spent some years in France in the 1770s, where he worked on plans to divert the Rhône and designed flourmills for the reclaimed land, as well as a new moving platform to load and unload boats. He was granted land there by the grateful company, but this was confiscated during the Revolution.

Edgeworth's large family – he was married four times and had 19 children who were all schooled at home – gave him a practical interest in education. With his eldest daughter Maria, he wrote numerous educational books expounding his radical beliefs, notably that girls and boys should receive the same education, and that science and engineering were as important as the arts and classics. He founded a school at Edgeworthstown for Protestants and Catholics to promote his ideas, discouraging learning by rote in favour of educational toys.

Maria inherited her father's love of science and her first book, *Letters to Literary Ladies* (1795), advocated science classes for women. She is best known for her novel *Castle Rackrent* (1800), but her many later works included science books for children, some of them collaborations with eminent scientists, such as the Scottish astronomer David Brewster, inventor of the kaleidoscope. Maria's friends included many scientists, and the astronomer John Herschel and Charles Babbage, 'the grandfather of computing', were visitors at Edgeworthstown.

If Richard Lovell Edgeworth is not well known it is perhaps because his interests were so diverse and, being wealthy, he never needed to commercialise his ideas. He and Maria are buried in the family vault at St John's Church, Edgeworthstown. Other family members include **Michael Pakenham Edgeworth** (1812–81), a botanist after whom the *Edgeworthia* genus is named, and Kenneth Edgeworth,* the first person to propose the existence of the Kuiper belt beyond Neptune. Longford County Library has material relating to the Edgeworth family. ⑮ The Edgeworth family house is now a private nursing home.

The old bog road

In the winter of 148 BC, the people of Corlea cut down 200 oak trees and used the trunks to make a trackway across Ringdong bog to the drumlin 'island' of Derryadd. Theirs is an impressive road running for about 2km. Birch branches form a layer of runners under the oak sleepers; and many of the logs are massive, some held in place with hazel pegs. The woodwork is skilful and the construction would have taken thousands of hours and required a communal effort. Indeed, it may have proved too big an undertaking for the community, as the track looks as if it was never used. This track, or 'togher', is just one of several dozen in the area, dating from Neolithic to Early Christian times, and which were investigated in the 1980s by the Irish Wetland Archaeology Unit under Prof Barry Raftery. In nearby Derryoghil, there are at least five tracks all about the same age and all just a few feet from each other.

Toghers were used throughout northern Europe often up to the late-Middle Ages and over 1,000 have been studied in Ireland alone. They were usually built across otherwise inaccessible bogs or wet areas, either to reach dry land or perhaps a source of bog iron* or a remote monastic site. Many were sturdy enough to carry animals and a few show signs of use by wheeled carts, but most were footpaths. Some were built with gravel, others used stone, but most were made of timber – and from the pattern of tree rings in the oak logs, the main Corlea track can be dated to 148 BC (see: **Tree-ring calendar**).

The Corlea road is the largest of its kind uncovered in Europe. After seven years of excavations, an 18-metre section was lifted and preserved by treating it in wax for several months before freeze-drying it. The timbers were re-laid in their original position and are now in a special centre where the humidity is controlled to prevent the timbers cracking. ⑯ The rest of the track was wrapped for protection and several small lakes were created to help

The tree-ring calendar

Amazingly, the growth rings of the oak timbers at Corlea record the exact year when the trees were felled, and also what the climate was like then, invaluable information for archaeologists and scientists. Each year, trees growing in temperate regions add a fresh growth ring to their trunk. In good growing years, the tree ring will be thick, in bad years thin, so that over the years a pattern of thick and thin rings is laid down. In the 1960s, scientists realised these patterns could be used to date the timbers and they began compiling tree-ring or dendro-chronologies.

An Irish chronology was pieced together by a team from QUB led by Prof Mike Baillie. They studied oak trees, as these are long-lived and widely used in buildings, and because there are ancient oaks preserved in bogs. Starting with modern oak trees, where they could easily assign a known year to each growth ring, they used computers to measure each ring accurately, and began compiling a sequence. Next they took timber samples from 17th- and 18th-century buildings around the country, looking for any overlap with the patterns in the oldest of the modern trees. In this way they worked backwards with overlapping samples, counting off the years.

The period 1350–1450, after the Black Death,* was difficult to bridge as little building took place then and few oak timbers were available, but that gap was eventually crossed and the sequence reached 13 BC. After that, the team switched to bog oak. Years of laborious work finally brought them to 5289 BC, a chronology of over 7,000 years. Their work means we can now confidently say, from the tree-ring patterns, that the Corlea oaks were felled in 148 BC, while those used at Navan Fort* were cut in 95 BC. There

By matching overlapping samples of timber, the QUB researchers could compile a calendar 7,000 years long. Courtesy Ulster Museum

are now other oak chronologies from England and Germany. The German series extends back to 7200 BC, as more oak trees grew on the continent then, but its tree-ring pattern is remarkably like the Irish one, suggesting that Ireland and Germany shared similar weather.

Scientists can also use the long tree-ring sequences to study climate change and to date catastrophic events like major volcanic eruptions. The Irish tree rings reveal a 20-year run of bad weather starting in 4370 BC that must have devastated Neolithic farmers. More narrow rings after 1628 BC may have been caused by an eruption at Santorini in the Mediterranean, which threw up so much dust that the skies over Ireland darkened for years. A decade of poor weather in AD 536–546 is seen in all the tree-ring chronologies of the northern hemisphere. It ushered in the Dark Ages and Prof Mike Baillie believes it might have been caused by a massive meteorite* impact.

conserve the boggy surroundings. Corlea bog is managed by the National Parks and Wildlife Service, which has erected some information boards there. (Co. Longord has no nature reserves or wildfowl sanctuaries, but Fortwilliam Turlough and a raised bog* at Lough Forbes have been designated as special areas of conservation.)

Chapter 6 sources

Co. Sligo

1. Wheeler, T S, & J R Partington, *The life and work of William Higgins* (Pergamon, 1960).

2. Tietzsch Tyler, Daniel, *The Landscape of Yeats Country* (Geological Survey of Ireland)

Co. Leitrim

3. Dooge, J, "Hibernia im Ruhrgebiet: William Thomas Mulvany & the industrialisation of the Ruhr" *History Ireland* 31–35 (Autumn 1997).

4. Dooge, J, "Manning and Mulvany: river improvement in 19th century Ireland" in Günther Gabrecht (ed.), *Hydraulics & Hydraulic Research* (Braunschweig Technical Univ, 1987).

5. Lucas, A T, "Furze: A Survey and History of Its Uses in Ireland" *Bealoideas* vol 26 (1958).

Co. Longford

6. Edgeworth, Richard Lovell, *An Essay on the Construction of Roads and Carriages* (London, 1813).

Chapter 7: Laois • Offaly • Kildare

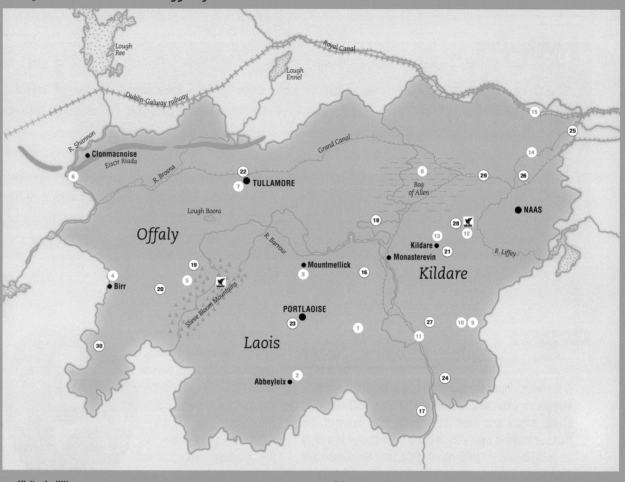

Visitor facilities

1. Stradbally Steam Museum
2. Abbeyleix Heritage Centre
3. Mountmellick Enterprise Centre
4. Birr Castle
5. Slieve Bloom Centre, Kinnitty
6. Bog Rail Tour, Shannonbridge
7. Tullamore Heritage Centre
8. Peatland Museum, Lullymore
9. Quaker Museum, Ballitore
10. Crookstown Mill
11. Athy Heritage Centre
12. County Library, Newbridge
13. National Stud
14. Straffan Steam Museum
15. Science Museum, Maynooth

Other places of interest

16. Ballybrittas
17. Ballyhide
18. Bracknagh
19. Cadamstown
20. Clareen
21. The Curragh
22. Durrow Abbey
23. The Great Heath
24. Kilkea
25. Leixlip
26. Millicent
27. Motte of Ardscull
28. Pollardstown Fen
29. Prosperous
30. Shinrone
♀ woodland
⚐ nature reserve or wildfowl sanctuary

Visitor facilities are numbered in order of appearance in the chapter, other places in alphabetical order. How to use this book: page 9; Directory: page 474.

COUNTY LAOIS

The Great Heath near Dunamase, a historic but little-known landscape, is one of Laois's hidden attractions. The face of 19th-century Ireland was changed by a Laois man, William Dargan, who built the country's main railways, and the latch-hook needle, used by rug and carpet-makers the world over, was invented by another Laois man, Robert Flower. Mountmellick was once home to a sugar factory that played a small but crucial role in the fight against slavery. Laois (or Leix) refers to the legendary Lugaid Laígne, who was granted land there after driving invaders out of Munster. From the Plantations of 1556 until 1920, Laois was known as Queen's County and Portlaoise as Maryborough.

BALLYHIDE

A philanthropic engineer

If anyone changed the face of Ireland in the 19th century, it was Laois man **William Dargan** (1799–1867). He built roads, canals and most of Ireland's main railways; his harbour improvement work at Belfast kick-started that city's shipping industry;* he bankrolled the 1853 Great Industrial Exhibition* in Dublin; and, with his personal fortune, helped establish the National Gallery.*

Dargan, son of a prosperous tenant farmer from Ballyhide, near Carlow town, had a natural talent for mathematics, engineering and business. He worked as a surveyor in Carlow, where his skill attracted the attention of his father's friend, the local MP Sir Henry Parnell, who recommended him to the great British engineer Thomas Telford. Under Telford, Dargan worked on various projects during the 1820s, including the Liverpool docks and parts of the London–Holyhead road. Back in Ireland, he took on various contracts, including Dublin's North Circular Road and the Howth–Dublin road (which completed the Dublin–Holyhead–London route, as the mailboat terminus was then at Howth*).

In 1831, Dargan won his first major public contract: to build Ireland's first railway, the Dublin–Kingstown [Dún Laoghaire] line.* The work entailed coastal embankments and much civil engineering, and the original line is still used. When the route was later extended south, Dargan built a novel vacuum-powered atmospheric railway* to (literally) suck the trains up Dalkey Hill. The great English engineer Brunel, who was building a similar railway in England, came to witness Dargan's work.

William Dargan was noted for his maxims, among them: "A spoonful of honey catches more flies than a gallon of vinegar" and "When a thing is put anyway right, it takes a great deal of mismanagement to make it go wrong." A statue of Dargan stands at the National Gallery in Dublin and there are commemorative plaques at Portlaoise and Carlow train stations; Belfast's new rail bridge over the Lagan is named after him. Iarnród Éireann

Dargan went on to build most of Ireland's rail network – 1,000km in all, including many of today's mainline routes. The new infrastructure dramatically changed Ireland, in often unforeseen ways: the rapid transport facilitated emigration, urbanisation and the movement of labour; access to a country-wide market of consumers fostered new national products, newspapers and brand names in place of local ones; telegraph lines crisscrossed the country following the rail lines; and local time differences gave way to a standard Irish time,* to simplify railway timetables.

Dargan also built canals, including the Ulster Canal.* In Belfast, he dredged a new shipping channel that dramatically improved port access and, with the dredged material, he created an artificial island (originally Dargan's Island, now Queen's Island), where a company called Harland & Wolff* opened a shipbuilding yard. Dargan was also renowned as a just employer: during the Famine years of the 1840s, he regularly employed more labour than was needed – overall he is thought to have employed 50,000 people – and paid generous wages. He also shunned the tied system that forced labourers to spend their wages buying over-priced goods from company shops.

In 1853, the now wealthy Dargan bankrolled Ireland's first Great Industrial Exhibition,* part of his grand vision for national industrial development. Held in the Royal Dublin Society's grounds at Leinster House, in a glass 'temple of industry' that was modelled on London's Crystal Palace, it cost the equivalent of €25 million today. Queen Victoria spent a week attending the exhibition, and visited Dargan at home, reputedly offering him a baronetcy, which he declined for political reasons. Dargan had plans for a technical institute in Dublin, but fundraising fell short of the targeted £100,000, so, instead, he established a National Gallery of Art on the site of the 1853 exhibition.

In addition to engineering and construction, Dargan invested in canal boats, cross-channel steamers, peat works in the midlands, the linen* industry, sugar beet* growing and slobland drainage. He helped develop the resort of Bray* (modelled on Brighton), and had a farm at his Mount Anville Estate (now a school) in South Co. Dublin, where he experimented with breeding cattle and sheep. Sadly, his later ventures proved costly – one of his firms went bust while building Dún Laoghaire's Royal Marine Hotel – and he died bankrupt. Dargan is buried in Glasnevin Cemetery, Dublin.

A turf entrepreneur

The second half of the 19th century saw tremendous activity across Ireland's vast midland bogs, as landlords, entrepreneurs and government agencies experimented with ways of converting peat into a fuel that industry could use. One such enterprise in the 1860s and 1870s was masterminded by **Robert Alloway** from Ballybrittas, who devised a way of producing an early type of peat briquette which he called 'peat coal'.

Alloway knew that the major problem with fresh peat is that over 90 per cent of its weight is water. Drying turf in an oven, however, consumes more energy than the finished peat can supply and is just not economic. Eschewing new industrial kilns and other machinery, Alloway exploited what he termed the "natural workers" of sun, wind and rain. His peat was first cut and soaked, macerated using a machine he designed, and then stored over winter in tanks. In spring, the resulting slurry was carted to tables where moulders (mostly women and children) shaped the mud into sods that were left to dry in the air. Skilled moulders could make 2,000 sods a day. No expensive machinery was needed, only barrows and tables that could be moved around the bog. Alloway claimed a tonne of his sods burned as long as a tonne of coal but cost less, and produced no black smoke. He later patented "an improved process for [drying] oak bark, flax, hemp, tobacco, peat and other substances by atmospheric evaporation".

Magh Rechet – the Great Heath

The Great Heath near Dunamase is visible from space – the vegetation of the plain is so different from the surrounding area that it shows up clearly on satellite images. This important and historic landscape, which has much in common with the more well-known Curragh* in Co. Kildare, was ruled for centuries from the **Rock of Dunamase**. At various times, it served as a burial and gathering place for the people of Leinster, as fair ground and prairie farm, as commonage and even as a sports arena. A race track was built there in the late 1700s, it was used for military manoeuvres in the 1800s, and today a golf course takes up one-third of the area. Traces of this varied history are still visible, including the remains of mediaeval cultivation ridges.

Furze (gorse) is threatening to swamp the Great Heath, so a management plan has been drawn up to protect the area.

Old manuscripts refer to the area as *Magh Rechet* (the plain of Rechet, daughter of Dian). In the early Iron Age, when it would have been much larger than today, it was the site of *Aenach Carmain*, a week-long fair hosted by the king of Leinster to mark the Celtic festival of *Lughnasa*. Numerous circular earthworks dating from this period dot the heath, including one which forms the golf course's 13th green. When the Anglo-Normans took over Dunamase, they transformed the area into one vast wheatfield, a mediaeval prairie. There were no hedgerows, and the ground was ploughed into countless cultivation ridges.* This period ended in the 1340s at a time of great social unrest across Europe: the years 1342–1347 were the coldest and wettest of the past millennium and, in 1348, the Black Death* arrived in Ireland. The plain has not been ploughed since, and the mediaeval cultivation ridges can still be seen in places.

In the intervening 600 years, the area has reverted to heathland. It has shrunk in size, however: by 1800 it was down to 250 hectares, today it covers just 170 hectares (including the 50-hectares Heath Golf Club). It is a true heath, a habitat that is rare in Ireland. The turf is shallow, and the soil is acidic and low in minerals. The main plants are sedges, although, where the land is grazed, there is a grassland rich in a tasty grass called sheep's fescue. Because the land is held as commonage and has been grazed for centuries without anyone adding fertiliser, it is low in nutrients. A special mix of plants has developed as a result and the site is also rich in toadstools, notably the colourful waxcaps. The most obvious plant on the heath today, however, is furze* (gorse) which, being a legume, makes its own fertiliser. A management plan is being implemented to control its spread. *Magh Rechet: The great Heath* by Dr John Feehan, an informative 30-page colour guide, is sold by UCD's Environmental Resource Management Unit.

Hissing dragon gods

Agricultural work, once all done by hand, has gradually been mechanised. Steam-powered farm machinery first appeared in the late-19th century. These were large and expensive machines – for instance, it could take 20 men to operate a steam-threshing machine – and it was feared these behemoths would mean the start of industrial-scale farming, and the end of small farmers and rural communities. If anything, the opposite happened: farmers came together to share the cost of hiring a machine, then used it collectively, and communities worked more closely than ever. **Stradbally Steam Museum**, run by the Irish Steam Preservation Society, has numerous agricultural steam engines, including reapers, binders, elevators and cornmills, and other artefacts from the steam age, notably a 100-year-old steam locomotive from Guinness's Brewery. ① A steam rally is held there each August bank-holiday weekend, when steam engines take over the village.

Home of the latch-hook needle

The latch-hook needle, used by craft workers the world over to knot yarn in handmade carpets, was invented in 1903 by **Robert Flower** (1836–1919), the eighth Viscount Ashbrook of Durrow. Flower's needle, and a handloom that he also designed, were the basis of a small carpet factory that flourished at Abbeyleix for a few years, making high-quality carpets for prestigious clients, including the ill-fated liner *Titanic*.[1]

Robert Flower was an enthusiastic inventor, though one early experiment with gunpowder left him with a permanent limp. His patents included designs for organ pipes, water-heating stoves and soldering aluminium, but his primary interest was textiles. He invented improved automatic looms, and easy-to-use handlooms designed to enable unskilled workers and people with disabilities to earn an income. His patented latch-hook needle had a hinged latchet that kept the yarn hooked so that, with one quick movement, the yarn could be drawn through the canvas foundation and knotted.

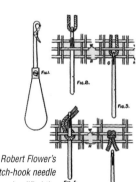

Robert Flower's latch-hook needle simplified the making of hand-tufted rugs and carpets. (Note the canvas foundation he also designed, with its double warp and single weft.) Abbeyleix Heritage Centre

Deep-pile, hand-tufted carpets attracted a premium price. They were traditionally made on large upright looms and heavy beams supported the fabric rolls; each tuft was manually woven around the warp and knotted in place. Flower patented a new system for use with his needle, in which the carpet was made on a small flat-bed loom that he also designed; and the carpet's canvas foundation was made on a separate automatic loom. Flower claimed this was simpler and faster than the traditional approach.

Ivo de Vesci, a philanthropic entrepreneur from nearby Abbeyleix, was keen to provide local employment and, in 1904, he opened a carpet factory in Abbeyleix using Flower's system. Similar ventures were being established around the country to create work for local women in a bid to slow emigration and save rural communities. The Abbeyleix factory commissioned designs from well-known artists, and earned an international reputation and numerous awards. At its peak, 24 women worked there, and a team could produce a carpet 12ft (3.5 metres) by 15ft (4.6 metres) in three months, with yarn tied at 20 knots to the square inch. Their carpets graced Ascot's grandstand and Dublin's Mansion House, and were sold in Harrods of London and Marshall Fields of Chicago.

In 1909, it merged with Naas carpet factory (begun by the Mercy Sisters in 1902). The Abbeyleix site was extended and now supplied the company's most notable order: carpets for RMS *Olympic* and *Titanic*, orders won through de Vesci's friend and White Star Line chairman, Lord Perrie. However, the merger with Naas had never been happy, and industrial problems in England disrupted the crucial yarn supplies. In December 1912, the Abbeyleix factory closed, and Naas followed a few years later. Robert Flower's innovative needle and the Abbeyleix factory are the subject of a special heritage museum that has original carpets. ②

Drifts and erratics

Laois, Kildare and Offaly form much of Ireland's large central plain. The region is often thought of as flat and boring yet there is great diversity there – from the broad Shannon flood plains, to the Slieve Bloom Mountains,* the Bog of Allen* and the Curragh.* The plain is floored by limestone, although this is now mostly hidden under deep peat. During the last Ice Age (and before the bogs grew), glaciers scraped up much of the limestone and carried it along before depositing the rubble further downstream. Consequently, much of the midlands is covered in glacial drifts, moraines and eskers.* The drifts are deepest in the east, producing prime, well-drained land, good for horses and fattening stock. In the west, where the glacial deposits are thinner, there is more bare rock and the farms are smaller. Though the bedrock is limestone, boulders of granite are occasionally found. These are glacial erratics:* during the Ice Age they were carried great distances by glaciers, before being dumped far from their original source. Cobbles of granite found on high ground east of Abbeyleix, for example, originated in Galway over 130km away.

Sugar, slaves and Quaker industry

Mountmellick, which is almost islanded by the River Owenass, was developed by the Quakers in the 1700s. At one stage, it had 27 manufacturing industries, and was home to an early Irish sugar* company. The **Royal Irish Beet-Root Sugar Factory**, founded in 1851 at a cost of £10,000, employed 160 people and extracted sugar from the then new sugar-beet plant.[2] The factory was one of the most efficient of the time: its sugar extraction rate of 7.5 per cent was 0.5 per cent better than any other European factory; and its manufacturing cost was £7.5s.0d per tonne (compared with a European average of £9). The venture won praise from local Quakers and other anti-slavery campaigners, including some Mormons from Utah who visited Mountmellick in 1856 while on a fact-finding tour of Europe's sugar-beet industry. Impressed by the Laois operation, the Mormons established a successful sugar-beet industry in Salt Lake City. Sadly, import taxes made the Mountmellick venture uneconomic, and it closed after 10 years.

The town's main industry was a cotton mill that, in the 1830s, employed 2,000 people, nearly half the town's population. There was also a woollen mill, a noted linen factory – Mountmellick linen was shipped on canal barges to Dublin and abroad – glue factory, soap and candle manufactory, potteries and iron foundries – Robert Welsh's foundry reputedly made steam and even train engines in the 1830s. The town also supported three breweries and a distillery and the Bank of Ireland established a branch office there in 1836. Today, an enterprise centre-cum-tourist office in a restored watermill has a small exhibition about the town's varied industrial tradition. ③

Mountmellick was the home of **Anne Jellicoe** (1823–1880), a Quaker who campaigned for women's education and employment.[3] She trained local woman as seamstresses and later investigated the working conditions for women and girls in Dublin factories. Jellicoe was also a founder of the pioneering **Queen's Technical Training Institute for Women** (1861–1881). Based in Dublin, this was the first women's technical college in these islands and it offered various courses including printwork, book-keeping, telegraphy and engraving. Jellicoe, who later founded Alexandra College for girls, also in Dublin, is buried in the Friend's burial ground at Mountmellick.

Birr Castle, famous for its giant telescope, was home to the ingenious Parsons family, including Charles, who invented the steam turbine and helped electrify the world. Offaly also produced George Johnstone Stoney who 'invented' the electron, and the brilliant John Joly who helped calculate the age of the Earth. Other Offaly delights include the Shannon callows, or flood meadows, the *Eiscir Ríada*, and the newly flooded Boora lakelands. Offaly takes its name from *Uíbh Fhailí*, the descendants of Failge. From 1556 until 1920, it was called King's County.

BIRR (PARSONSTOWN)

1845 – the world's largest telescope

The biggest scientific undertaking in 19th-century Ireland was when **William Parsons** (1800–1867), third Earl of Rosse, built the world's largest telescope in the grounds of Birr Castle. It was the Hubble telescope of its day, enabling Parsons to see further into space than ever before, and it was nearly all built with local labour. The third earl also headed an inventive family: his wife, Mary, was a pioneering photographer; their eldest son, Laurence, took some of the first accurate measurements of the Moon's temperature; and their youngest son, Charles, invented the steam turbine, which revolutionised marine transport and electricity generation.

William Parsons was born in Yorkshire and educated at Birr Castle (the family's Irish seat since the Plantation), TCD*

and Oxford. For 10 years he was MP for King's County (Offaly), but he resigned in 1832 to concentrate on astronomy. Though Parsons was an amateur astronomer, and his telescope was mostly a solo run, he was still part of the scientific establishment, at various times president of the Royal Society in London (1849–1854) and chancellor at TCD (1862–1867). In 1836, he married a Yorkshire heiress, Mary Field, who bankrolled his ambitious plans.

Until 1845, the world's largest telescope was one with a 49-in (124-cm) mirror, installed at Greenwich Observatory in the 1780s. A bigger telescope would collect more light and enable astronomers to see further and study fainter objects. It might also answer a burning question: Were the heavens fixed and immutable as God created them, or could they change? Interest focused in particular on faint, fuzzy patches of distant light called **nebulae**. These mysterious objects were neither stars nor planets – but were they thick clusters of distant stars, as William

Parsons believed? Or were they clouds of nebulous material, as others would have it and, if so, where had the dust come from? Or could they be regions of space where new stars and planets were still condensing?

To study nebulae, William Parsons resolved to build the world's largest telescope at Birr. He chose to build a Newtonian or reflecting telescope, which used mirrors (the alternative refracting-type uses lenses). Glass technology was not very advanced then, so most large mirrors were made from speculum, a hard, white alloy of copper and tin that can be polished to a mirror-like finish. The art of mixing and casting the metals was a trade secret, however, so William Parsons began his own experiments in the 1820s.[4] By 1839, after years of research, he had successfully cast a mirror 3ft (0.9 metres) in diameter, using an alloy of two parts copper to one of tin. He polished it using a large steam-powered grinding machine, which he also designed and made, and installed the mirror in a tube he had built earlier. This sizeable instrument, supported by a timber scaffolding, was blown down during the Night of the Big Wind,* but, fortunately, not seriously damaged. William determined, however, that his next telescope would be bigger and better protected.

Work began in 1842 on a 6ft (1.8 metre) mirror and a telescope tube to match. There were several logistical problems, not least was casting the largest metal mirror ever made: it needed 4 tonnes of metal, which took 26 hours to melt, and consumed an estimated 62.3 cubic metres of Offaly turf. After casting, the mirror had to be cooled slowly for four months in a special oven to

prevent cracks developing. Then it had to be carefully polished. Much of this precision work was done by a local mechanic, William Coughlan, who worked at the castle for 55 years until 1895.

The Leviathan: Birr telescope photographed in 1862 by Mary, Countess of Rosse. Her photographs were a valuable source of information when the telescope was being restored in the 1990s. Birr Castle Demesne

Six mirrors were made, but only two were good enough to use. They were so heavy they were moved on purpose-built trolleys that ran on tracks between the telescope and the castle workshops. The chief astronomer from Dunsink Observatory, William Rowan Hamilton,* introduced an engineer called Thomas Grubb* who designed a cradle to hold each mirror so it did not break under its own weight. Local coopers meanwhile made the 54-ft (16.5-metre) long telescope tube, probably Ireland's largest barrel. The tube and mirror together weighed 12 tonnes and could only be moved using a complex system of counter weights, pulleys and chains. Thick walls 56ft (17 metres) high and 72ft (22 metres) long were built to support and shelter the telescope, and hold the observing platforms and lifting mechanisms.

By March 1845, the telescope was ready, made by local tradesmen and at a cost of about €1 million in today's money. A month later, it made its one and only real discovery: in April 1845 William Parsons trained the telescope on a nebula known in the catalogues as M51, and saw that it was a spiral-shaped cluster of stars. He named it the **Whirlpool Nebula**. The discovery caused a sensation, as it seemed to confirm the theory of a fixed Universe. Over the ensuing decades, the Birr telescope was used to study numerous other nebulae. Many could not be resolved – which Parsons often attributed to poor weather and the mirror tarnishing – but those that could be resolved all seemed to be star clusters. Today, we know that while many nebulae, such as the Whirlpool, are indeed star clusters, others are dust clouds (such as the Crab nebula, which Parsons named, and which is the remains of the supernova explosion of AD 1054*); and yet others, such as the Orion Nebula, are breeding grounds where new stars are forming. The heavens can, and do, change.

But all this was in the future. In 1845, as the summer skies became too bright for astronomy, potato blight* arrived and, for the next few years, the Parsons concentrated on famine-relief works at Birr Castle, funded from the countess's fortune. Astronomy began again in 1848 when George Johnstone Stoney,* a young TCD graduate from Offaly, was appointed tutor to the Parsons children and assistant at the observatory. At one stage, Parsons tried to photograph the observations made with his telescope, but early photographic techniques were clumsy and slow, so all the observations were instead meticulously drawn by hand. However, his wife later used his darkroom and equipment.

Birr telescope functioned until the third earl's death, one mirror being polished while the other was in use. It remained the world's largest until 1917, but size is not everything in astronomy. The Leviathan's bulk made it difficult to manoeuvre and its view of the sky was restricted. It was too clumsy to use with new instruments like cameras and spectroscopes. Moreover, its location in Ireland's boggy midlands was not ideal: skies were seldom clear and the metal mirrors tarnished quickly. Other smaller telescopes with lenses or better mirrors and in a better location proved more useful. Professional astronomers were also taking over from amateurs. By 1914, the Birr telescope was derelict and the metal parts were gone for scrap. The timber tube was decrepit. One mirror had left for London's Science Museum, where it can still be seen; the second one vanished. Most importantly, knowledge regarding how the telescope was constructed and used also disappeared.

Nevertheless, for years Birr observatory was a valuable training ground for many Irish scientists, including the Stoney brothers and Robert Ball,* as well as Parsons' sons, Charles and Laurence. Astronomers came from around the world to see the amazing instrument. And the engineer Thomas Grubb went on to found a company that lasted until 1985 making telescopes for some of the world's great observatories. William Parsons also built various steam-powered machines, including an early steam motor carriage (tragically responsible for the country's first motoring fatality involving the earl's cousin, Mary Ward*), and he was awarded numerous international honours from as far away as St Petersburg. Parsons died at his Dublin home and is buried in Birr.

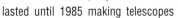

Parsons' wife was **Mary Field** (1813–1885), a remarkable Yorkshire woman who married William in 1836. Her personal fortune – an annual income of £50,000, equivalent perhaps to €5 million today – funded her husband's ambitious projects, but she was also a noted early photographer. In the 1850s, she took over her husband's darkroom and cameras, and experimented with various photographic processes including waxed paper, daguerreotypes, stereo photography and the wet collodion process. Her subjects included the telescope, castle, family and friends, and the scientists from around the world who visited Birr. Her work was exhibited in Ireland and Britain and, in 1859, she won the first ever medal awarded by the Photographic Society of Ireland. The world's first news photographs were premiered at a function in the Parsons' London home; they were taken by William Fox Talbot, a friend of the Parsons, and the man who invented the photographic negative, enabling an image to be printed

William Parsons' 1845 sketch of the Whirlpool nebula, a spiral-shaped cluster of stars, as seen through his giant telescope. The sensational discovery seemed to confirm theories of a fixed Universe.

many times. The castle darkroom survives, and is the world's oldest extant darkroom.

The great telescope was renovated in the 1990s; its new mirror, made in London of nickel-plated aluminium, is one-third the weight of the original. An accompanying exhibition celebrates the achievements of the Parsons family. ④ Birr Castle gardens are open to the public; reputedly Ireland's largest, they boast the world's tallest box hedge, Europe's oldest wire-suspension bridge (from the 1820s) and over 2,000 plant species from around the world.

Taking the Moon's temperature

Laurence Parsons (1840–1908), fourth Earl of Rosse, took after his father. He inherited his title, followed in his footsteps as chancellor of TCD (1888–1908), and shared his interest in astronomy and engineering. He improved his father's giant telescope by adding a water clock to move it at a constant rate, making it easier to handle and, in 1880, he harnessed the Camcor river to generate electricity for Birr Castle and town. He also devised a way of taking accurate measurements of the Moon's temperature.[5]

People had long speculated about the Moon's heat. Aristotle noted in 335 BC that nights when the Moon was full were warmer, but he thought the heat was generated by the Moon's movement through space. Early attempts to measure the Moon's radiant heat were hampered by poor telescopes, thermometers and techniques. People could confirm the Moon had a temperature, but they could not measure it. In 1868, Laurence Parsons devised a way around this, using his father's 3-ft telescope to focus the Moon's image onto one of a pair of thermocouples (wire circuits that generate electric current that varies with their temperature). Commercial thermocouples were not sensitive enough for the tiny temperature increase involved, so Parsons made his own.

Others were attempting something similar at the same time, but Parsons was the first to realise the need for an infrared filter (in his case, a piece of plate glass). This enabled him to isolate the relevant radiation that caused the heating effect. He also took account of the heat from the sky, his surroundings and even the telescope. Over a number of years, he and Dr Otto Boeddicker, an astronomer working at Birr, measured the Moon's radiant heat at various phases in the lunar cycle and when the Moon was at various heights above the horizon, refining their technique with time. From their thermocouple data, they calculated the Moon's surface temperature. In 1877, Parsons published his best estimate, 197°F (92°C), in the journal *Nature*, but, much to his dismay, it went largely unnoticed. He had, however, made some assumptions in his calculations that later proved wrong. In 1958, his original data were re-used in a more up-to-date calculation and the result was 70°C, close to the true value. Parsons' approach was vindicated. (The first ever electrical measurement of starlight* was made in a Dublin back garden in 1892.)

An invention that changed the world

Sir Charles Parsons (1854–1931), youngest son of the 'telescope earl', was a gifted engineer and entrepreneur who invented the steam turbine. Within a few years, his engines had revolutionised marine transport and naval warfare and, most importantly, made cheap and plentiful electricity a reality. Today, every power station in the world uses turbine generators based on Parsons' idea. Were it not for his turbine, we might still be using gas lighting and gas-powered appliances.

Charles, like his brother Laurence, was educated at home in Birr by a distinguished tutor (in his case the mathematician Robert Ball*). He was interested in engineering as a boy and his father encouraged him to design and make his own machines. After studying at Cambridge, Charles worked as an apprentice at Newcastle-upon-Tyne, England's great engineering city. For the previous 150 years, conventional steam engines had powered the industrial revolution, employing steam to move pistons and crankshafts. However, steam engines were inefficient, noisy, slow behemoths that consumed prodigious amounts of coal, were costly to maintain and vibrated violently. People were starting to change over to gas machines. Then, in 1884, 30-year-old Charles Parsons invented a new and more efficient steam engine, the **steam turbine**.

Instead of using steam to drive pistons, Parsons used it to turn a rotor directly. In his design, the steam was allowed to escape in a controlled manner – at 2,000kmph – into an enclosed space. Instead of harnessing the power all at once, Parsons had the ingenious idea of tapping it in stages as it expanded, by forcing it through a series of ever-larger, multi-vaned wheels. Steam, passing over the vanes, rotated each wheel in turn and thus the central axle rotor. In this way, he could extract as much of the energy as possible, and the turbine blades did not overspeed. His first model spun at an unprecedented 20,000rpm and generated 7KW. The design was ideal for turning the dynamos used in generating electricity, and power stations quickly spotted the potential. It was Parsons' turbine generator that facilitated global electrification.

A daring stunt: Turbinia's uninvited dash at the 1897 naval review. Charles Parsons's custom-built, turbine-powered boat, flames spewing from its funnel, clocked a record 34.5 knots. The restored Turbinia is the centrepiece of an exhibition about Parsons at Newcastle-upon-Tyne's Discovery Museum. TCD's engineering faculty has a small-scale Parsons turbine dating from the 1890s.
Newcastle-upon-Tyne Discovery Museum*

Parsons founded his own company (now the NEI Parsons Group) and refined his designs, improving speed and efficiency each time. Next he tried to interest the Royal Navy in turbine-powered ships, but they scoffed at his idea. (Coincidentally, the week he was born, his father had tried unsuccessfully to interest the Admiralty in a steam-powered fleet.) Undaunted, Charles built a demonstration boat, *Turbinia*, a sleek, lightweight vessel 30 metres long and 2 metres wide, that was designed for speed. But initial trials were disappointing until experiments revealed that, ironically, the boat's propeller was turning so fast it was creating vacuum cavities in the surrounding water that were slowing it down. Parsons built the world's first cavitation tunnel to investigate the phenomenon and improve the propeller's design. His cavitation work was later used to improve aeroplane propellers.

By 1897, *Turbinia* was ready and Parsons ventured a daring stunt: gate-crashing Queen Victoria's diamond jubilee naval review at Spithead, near the Isle of Wight, where 150 vessels were gathered. *Turbinia*, a small, uninvited guest, steamed around the party at a record-breaking 34.5 knots (64kmph). Nothing could catch it, and a suitably impressed navy ordered two turbine-powered destroyers. Parsons' turbine dramatically changed marine transport. Enormous ocean-going liners such as *Lusitania* now became possible – previously, a ship that size would have sunk under the weight of coal needed to drive conventional steam engines. HMS *Dreadnought* (1906), the first turbine-powered destroyer, was capable of 21 knots and rendered all previous warships obsolete. However, designs improved so fast that, by 1916, even *Dreadnought* was too slow to take part in the Battle of Jutland.

Parsons developed the turbine because he was concerned that reserves of fossil fuels would run out. For similar reasons, he studied renewable energy sources, including tidal, wave, geothermal and solar power (one scheme envisaged using mirrors in the Sahara to heat steam for electricity). Dubbed "the most innovative British engineer since James Watt", his many other patents included searchlight designs, non-skid chains for cars, and even a music amplifier. He also spent a fortune trying (unsuccessfully) to make synthetic diamonds* in high-pressure vessels. In 1925, he took over a telescope and periscope company owned by another Irishman, Howard Grubb,* and the Grubb-Parsons company made telescopes for the world's top observatories until 1985. Parsons' own company, NEI Parsons, is still making turbines.

CADAMSTOWN

The Silver river gorge

The compact **Slieve Blooms** – *Sliabh Bladhma*, 'the blazing hills', named presumably on account of their heathery slopes – are an island of ancient rock rising out of the central limestone plain. The only mountain range in the midlands, and straddling the Offaly–Laois border, these gentle, at times bleak, hills retain a certain beauty. The summits were originally jagged sandstone but they were worn smooth by glaciers during the various ice ages. A number of small fens* around the perimeter of the range are an important habitat and home to rare plants.

The lower slopes are now mostly forested, but the summit has Ireland's largest continuous upland blanket bog* (2,300 hectares of which is a nature reserve), with a luxuriant carpet of heather, bog moss and lichen, as well as sedges, crowberry and bilberry. In the middle of the ridge, near the source of the Blackstairs river, is the enigmatically circular **Clear Lake**. ⑤

A necklace of river valleys rings the uplands – Glenafelly, Glendossaun, Glenregan, Glenletter… and clockwise around to Glendine. A few, such as the Silver river gorge, cut deep into the mountain revealing the ancient core: a tough blue-grey Silurian rock that was deposited on an ocean floor 425 million years ago. This core is seldom seen outside the gorges, for it is usually hidden under younger, iron-rich red sandstones and siltstones. These 'Old Red Sandstones', the remains of ancient deserts, are 350–400 million years old. They are the main local building stone and are still quarried.

The Slieve Blooms bear many traces of **glaciation**: esker* gravel ridges snake across the terrain; erratics* lie here and there, including granite boulders brought from Galway by the glaciers; thick beds of gravel blanket the ground, especially on the northern flanks where the glaciers had to drop their rocky load as they struggled uphill. The gravel can be 50 metres thick in places, as at Glenbarrow where the river cuts deep into the glacial debris before carrying the sand and stone downstream.

You can explore the local geology at Cadamstown's **Silver river** nature trail and geological reserve, which is owned by Offaly County Council. The river is named for the tiny particles of silver found in the rock upstream and washed down in the river. About 800 metres upstream of the village, past the waterfall and bridge, is a break, or 'non-conformity', between the red sandstone and the older, underlying blue-grey rock. The steep gorge is wooded with oak, birch and hazel, and a rich underlayer of shrubs and plants. The fast waters attract dippers, while fox, deer, otter and red squirrel frequent the woodland.

Cadamstown itself is an interesting historical site: the original settlement, founded by the Anglo-Normans just north of the present village, was reputedly ravaged by plague during the Black Death* in 1348, and the village was destroyed to prevent the disease spreading. However, its mediaeval sandstone bridge survives: tall and narrow, with a single arch, **Ardara Bridge** leads to an unusual, walled, sunken way and thence to the site of the mediaeval village.

A sacred tree

The **hawthorn** or whitethorn (*Crataegus monogyna*, or *an sceach gheal* in Irish) is native to Ireland and common in hedgerows. In May, its masses of cheerful white-and-pink flowers brighten the countryside, giving it its third name, the May tree. A solitary hawthorn, however, was thought to be a fairy tree and considered sacrosanct: moving or cutting one risked disaster, and wood from even a dead one was never used for anything. This unease may stem from thorn trees being associated with forbidden and sacred places. Some also said Christ's crown of thorns was made from a hawthorn, and in many countries it was thought unlucky to bring May flowers into a house. Ireland has had many notable thorn trees. The Beggar's Bush area of Dublin is named after an elderly hawthorn that grew there until about 1900. And some blamed the collapse of Northern Ireland's Delorean car factory in the 1970s on a fairy tree that was cut down when the factory was built. In Clareen, they show more respect: when the new road was built there, it was made to skirt an old hawthorn tree known locally as **St Ciarán's bush**.

The hawthorn: a small tree or shrub with sharp thorns at intervals; small, deeply lobed leaves; orange-brown bark; clusters of flowers in May; clusters of indigestible red haws in autumn.

The man who 'invented' the electron

You could say that the electron, one of nature's fundamental particles, was invented by the Irish physicist **George Johnstone Stoney** (1826–1911). Stoney, born at Oakley Park, Clareen, was the first science professor at the new university of Queen's College Galway.* A great champion of the metric system, he believed science would benefit from a wise choice of standard units and this prompted him to invent the idea of a standard unit of electricity. Stoney initially called his unit an 'electrine' (he later changed it to

electron) and he presented his idea to the British Association for the Advancement of Science (BAAS)* at its 1874 annual meeting which was held in Belfast. Stoney also calculated that his electron's charge was 10^{-20} coulombs, based on his analysis of chemical reactions (we now know the charge is 16 times greater, but Stoney's estimate was reasonable for the time).

George Johnstone Stoney, with a heliostat he invented. A mirror with a clockwork device to track the Sun's movement, it provided a constant beam of sunlight. Scientists used it when analysing the composition of sunlight. Stoney's design, made commercially in Ireland and England, was cheaper than other versions.

Many international scientists liked his idea for an electron, including the German physicist Hermann Helmholtz. In Ireland, Stoney's nephew and noted scientist George Francis FitzGerald* also championed the concept. When, in 1897, an English physicist, J J Thomson, discovered that cathode rays were beams of negatively charged particles, FitzGerald immediately realised these particles were Stoney's electrons. Today, Thomson is credited with discovering the electron (though for years he persisted in calling them corpuscles) and Stoney is credited with inventing the concept and name.

George Johnstone Stoney had learned his science at TCD. In 1848, he went to work for Lord Rosse as an astronomical assistant on the Birr telescope* and, in 1852, thanks to Rosse's influence, he became, at 26, Galway's first professor of science (John Tyndall* also applied for the job but, though better qualified, had no one in Ireland to champion his cause). As well as conceiving of the electron, Stoney worked on optics, gases and extraterrestrial bodies, and calculated the amount of energy expended in riding a bicycle. He developed a musical shorthand, suggested new designs for pianos and organs, and persuaded the RDS* to hold public concerts. Later, as a university administrator with the Queen's colleges, he campaigned for higher education to be opened to women – thanks to him, women earned medical qualifications in Ireland long before their sisters in Britain. In 1893, Stoney moved to London where he felt his daughters would have better career prospects (one took after her father and became a noted physicist, the other a distinguished medical doctor). Stoney's younger brother was the innovative engineer Bindon Blood Stoney,* who achieved engineering fame by helping to build the Boyne viaduct* and much of Dublin's docklands.

An early glass factory

French Huguenot glass-makers came to Offaly c.1600 and opened 'glasshouses' or manufactories, making window glass and glass vessels at various sites in the county. The increased availability of glass would no doubt have helped bring down the local cost of, what was then, a very expensive product. Glass was valuable in 1600: when the Duke of Northumberland was away from home, the staff removed and stored the glass windows until his return. The Bigault family had operations near Birr and Lusmagh, while the Hennezells/Henseys established factories at Shinrone and later Portarlington. The remains of the Shinrone furnaces can be seen in a field near the village, in a townland still called Glasshouse.[6]

The Shinrone factory had at least two furnaces: one to melt the ingredients (at temperatures up to 1,200°C) and the other to slowly and carefully cool the finished item. The whole process probably took five days and nights of continuous hard, hot work, and careful watching and manipulating. The furnaces were stone-built arched structures with a stoking tunnel at either end and a central fire trench. The main ingredients used then (as now) were sand (75 per cent), ash (15 per cent) and lime (10 per cent). The Shinrone glasshouses probably made green glass as they would have used wood-ash, which is rich in potassium. The sand came from nearby eskers;* wood for the furnaces was also plentiful locally. (Coal was not used until after 1650; coal-fired glasshouses had a characteristic bottle shape and were located near a coalfield or a canal.)

Callows and corncrakes

Between Lough Ree and Lough Derg, the Shannon follows a series of gentle S-bends for over 40km. This stretch of river is the Shannon callows (from *caladh*, a river meadow). It has no houses, no hedges or walls, and no tarred roads. This is because, every winter, the river breaks its banks there, forming a shallow lake up to 4km wide. When the flood retreats in spring, it leaves a narrow strip of soft fertile ground that can be farmed as meadow. The annual floods and the soft ground mean callow farming has never been intensive. The result is a unique and internationally important ecosystem, with a rich assembly of birds, insects and plants, and covering 35sq km. Hundreds of plant species have been counted on the callows, including 17 rare species, among them the marsh pea and summer snowflake. In winter, the flooded fields support large numbers of wildfowl and waders, such as whooper and Bewick swans, Greenland white-fronted geese and 10 per cent of the European population of black-tailed godwit. In summer, the place is home to numerous waders including water rail, redshank and lapwing.[7]

The corncrake: protection measures have helped to stabilise the population, but the bird is still endangered.
© Eoin MacLochlainn

The area's most important bird is the **corncrake**, Ireland's only globally endangered bird. It is named in Latin and Irish (*Crex crex* and *an traonach* respectively) for the male's chainsaw-like crek-crek call (the females are the silent type). Formerly, Ireland had so many corncrakes their endless racket was a summer nuisance. However, by 1988, there were only 1,000 calling males here, and 10 years later there were just 149 males left – with 54 of those in the Shannon callows. The corncrake is a small bird, seldom seen, with yellow-grey plumage, chestnut-coloured wings and long legs that dangle in flight. It winters in Africa, coming to Ireland in April to breed. The males advertise their territory by calling from dawn to dusk throughout the summer, usually from low down in the grass. Nests are built on the ground in meadows, and broods are hatched in June and July. Hence, the major threat to the nests and the young is from mowing machines and silage cutting in early summer. An emergency rescue programme, run by BirdWatch Ireland, the RSPB and the National Parks and Wildlife Service, asks local farmers not to mow the callows before August 1st, and to cut their fields from the centre out. Thanks to these measures, corncrake numbers on the callows seem to have stabilised, though the bird is still endangered.

Numerous **arterial drainage** schemes over the past 200 years, much of it famine-relief work, help to control the Shannon floods. The most recent programme, the Brosna scheme of the 1940s, was a major engineering project that drained 125,000 hectares: 400km of channels were improved and 200 bridges underpinned. Similar schemes around the country removed many callows, and the only significant ones left are those on the Shannon and Brosna, and on the Blackwater at Lismore in Co. Waterford. The Shannon still floods, however, and each time there are renewed calls to drain the region further. But the area is so flat, the only effective solution would be expensive pumping. [Next stop on the Shannon: **Lough Derg**, in Co. Tipperary.]

CLONMACNOISE

Eskers and the Eiscir Ríada

The monastic university of Clonmacnoise developed at the historic crossroads where the River Shannon and the *Eiscir Ríada* meet. The *Eiscir Ríada* was one of the ancient Irish highways or *slíghe*. It started in Tara* and, to avoid boggy ground, ran along glacial gravel ridges wherever possible. Today, such glacial ridges are known worldwide as eskers, after their Irish name. (The small hills known as drumlins* are another glacial feature named after an Irish word.)

This esker from the Slieve Bloom area was sketched c.1860 by geologist and artist, George du Noyer.

Eskers, found wherever there were glaciers, are usually steep-sided, at least 5 metres high and 10 metres wide, and made of sand and gravel. They are stranded, or fossil, riverbeds that were deposited underneath glaciers. All glaciers have rivers of meltwater flowing under them through tunnels in the ice. The surrounding ice banks up the riverbeds; but when the glaciers melt and the rivers disappear, the beds are left as upstanding dry gravel ridges snaking their way across the country. Many Irish eskers are quarried for their sand and gravel; they were also exploited as natural causeways to cross boggy ground, and numerous small country roads still run along eskers in places. Esker slopes, too steep to farm, are often covered in furze* and, because they are rich in limestone, also support alkaline-loving plants, very different to the acidic vegetation of the surrounding bog.

The *Eiscir Ríada* highway ran along a series of eskers in Offaly, locally called the Arden Hills. At Clonmacnoise, the main ridge turns south then crosses the river at Shannonbridge. The dry esker ridge on either bank made this a natural fording place in an otherwise boggy flood plain. In 1995, archaeologists found the remains of a timber bridge near there that has been dated to AD 800. And, of the two bridges that today cross the Shannon between Lough Ree and Lough Derg, one is still at Shannonbridge, the other at Banagher.

Behind Clonmacnoise, and trapped between two esker ridges, is **Mongan Bog** nature reserve, a near-intact raised bog* that lies just inland of the Shannon's flood plain. A survey in the 1950s found over 150 species of moth and butterfly there. Beside it is **Fin Lough**, a tiny remnant of the vast ancient lake that covered the midlands immediately after the last Ice Age. Lime-rich water draining into the lough from the eskers keeps the acidic bog at bay and sustains the lough's fen-like vegetation.

A 10,000-year-old landscape

At the end of the last Ice Age 12,000 years ago, the midlands were covered by a vast lake. This inland sea, which some call Lough Shannon, was so big that waves threw up storm beaches of boulders above its shore. The modern Shannon and its loughs are surviving remnants of the larger ancient lake. Today, where the turf has been cut away, Bord na Móna is returning the land to the water. The result is a small version of the lake that existed before the bogs grew, a new wetland habitat. Traces of the vanished lake can be read in the landscape. The ancient lake bed, a thick layer of shelly marl clay, is now covered by bog, but can be glimpsed in drains and cutaway sections as a distinct white layer. Fossil storm beaches are occasionally found as rocky patches in the bog, and tall mushroom-shaped stones, which can be seen at Crancreagh, Creevagh, Drinagh and elsewhere, were eroded by the waves and mark the edge of the vanished lake.

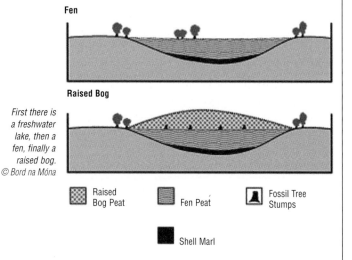

Fen

Raised Bog

First there is a freshwater lake, then a fen, finally a raised bog.
© Bord na Móna

| | | Raised Bog Peat | | | Fen Peat | | | Fossil Tree Stumps |

Shell Marl

A group of **Mesolithic hunter–gatherers** camped for a while beside the ancient lake 8,400 years ago. Their campsite was found by turf cutters at Boora Bog in the 1970s. Excavations revealed a dozen hearths and 1,500 artefacts including blades, arrowheads and stone axes. There is little to be seen there now, yet this site is the oldest archaeological site in southern Ireland – only one older site is known in the country, dating from about 9,000 years ago, at Mount Sandel near Derry.

Even as the hunters camped by Lough Shannon, the lake was shrinking. Most lakes are ultimately temporary, slowly disappearing as the surrounding vegetation encroaches. The transformation at Lough Shannon began 9,000 years ago, first from open water to fen,* which took 2,000 years to complete, and then from fen to raised bog.* The vast tracts of midland raised bog all grew out of the ancient lake. In all, 20 per cent of Offaly is covered by this bog and Bord na Móna has been extracting turf there for decades (**Clara Bog** and **Raheenmore Bog** are important tracts of Offaly raised bog conserved as nature reserves). Each year this generates 1,200 hectares of 'cutaway' where the peat has been removed leaving only a poor soil. Trials are now under way to find new uses for the cutaway bog. Where it is naturally well drained, Bord na Móna is experimenting with commercial forestry. Other areas naturally flood, as at Turraun, and there Bord na Móna has blocked the drainage channels to create experimental wetlands.

Turraun wetland, created in 1990, is a nature reserve with a mosaic of habitats including open water and reedbanks, fen and heath, and grassland and scrub. Ireland's only flock of grey partridge, comprising about 40 breeding pairs, inhabits part of the grassland. Partridges from Poland and France have been imported in a bid to secure the endangered Irish population, and the site is managed to provide nesting cover and food for the chicks, and to reduce predation by magpies and crows. Turraun's transition to wetland nature reserve is being monitored and recent surveys counted over 180 bird species and 270 plant species there. In summer, Turraun is rich in orchids, especially the rare bee orchid that recently arrived there. In winter, the wetlands attract swan and duck.

Turraun also has a place in **turf-cutting history**. In the 1850s, local man Kieran Farrelly ran a pioneering moss-peat business there on 100 hectares, with a horse-powered grinder and baling press. In the 1920s, an entrepreneurial Welsh engineer, Sir John Purser Griffith,* bought the bog, convinced it could be commercially developed. Griffith, who also championed the 1939 Liffey hydroelectricity scheme at Pollaphuca,* spent £70,000 of his personal fortune turning Turraun into an industrial showcase. He chose Turraun because of its size and proximity to the Grand Canal. Extensive drainage channels were dug, a narrow-gauge railway installed on the bog, and machine peat and peat moss were produced using imported German machinery. Significantly, Griffith believed the best use of the peat was in electricity-generating stations located beside the bog. He built a small station to generate electricity in summer for the peatworks, and in winter for local towns. Ironically, the Shannon electricity scheme at Ardnacrusha,* which Griffith had campaigned against, put paid to his plans for local peat-fired power stations. In 1935, however, the Turf Board (later to be Bord na Móna) bought Turraun, and subsequently adopted Griffith's idea of bog-based power stations. Turraun, the first bog to be intensively cut, was also the first to be 'bottomed', and it is now being returned to nature.

Cutting turf by hand was notoriously hard work, but taking heavy machinery on to soft bog brings its own problems. So, over the years, **Bord na Móna**, which works 40,000 hectares of Offaly bog, has had to invent new techniques and machines for the job. The Irish agency is a world leader in peat technology and its innovations are used around the world. The big problem is moisture content: bogs are 95 per cent water, and a raised bog that has not been drained could easily swallow a tractor. A first step in any turf cutting is to prepare the ground by digging a network of drains

The new wetland at Turraun, created from flooded cutaway bog, has bird hides and walking trails; Lough Boora is being developed as an amenity area.
© Bord na Móna

that, over time, will reduce the moisture content and stabilise the surface.

Even after peat has been cut, it must be dried before it can be used. Bord na Móna uses various types of harvester to cut moss peat, peat for briquettes and peat crumb for power stations. Crumbed, or milled, peat was pioneered by Bord na Móna: milling machines with blades and hooks cut and crumb just the top few millimetres of peat each time; the resulting crumbs dry in a few days to a moisture content of about 50 per cent, unlike thick sods which must be regularly turned and take weeks to dry. The dried crumb is then swept up and shipped to a power station, usually located as close as possible to the bog (Shannonbridge station, for example, is beside Bord na Móna's Blackwater Bog).

To move harvested peat around a working bog, Bord na Móna uses an extensive network of **narrow-gauge railway**,* probably the largest industrial railway in the world. The bog network comprises 1,000km of permanent track plus 40km of temporary way that can be moved around as needed. Narrow-gauge railway is ideal for bog transport: the tracks spread the weight; narrow-gauge vehicles are light; and it is relatively simple to lay tracks direct onto the bog surface. Part of the Shannonbridge Railway is open to the public. ⑥

A mediaeval observatory

1785 – great balloon disaster

In July 1054, the monks of Durrow Abbey observed something strange in the heavens – a bright star that literally appeared overnight. Their observation, recorded in the *Annals of the Four Masters*, is the only known European sighting of the **supernova of 1054**. A supernova is a violent explosion of a dying star. They are rarely seen, but this one, which first appeared on July 4th, 1054, was observed by Chinese and Japanese astronomers, and keen-eyed Irish monks. The enormous explosion was so bright it was visible by day for three weeks during July, and by night for two years afterwards, and the debris gave rise to the beautiful Crab Nebula.

Irish monks probably watched the skies for heavenly signs or portents. For centuries, their monasteries doubled as astronomical observatories and they recorded their observations in the annals. Daniel McCarthy of TCD* and Aidan Breen of the DIAS* examined annals dating from 442–1133 and found numerous astronomical references: solar and lunar eclipses, comets and auroral displays, in addition to the supernova of 1054, and the eruption of an Icelandic volcano, Eldgjá, in 939.[8] Old- and Middle-Irish terms used in the annals included *retlu* (similar to the modern *réalt*) for a star, *retlu mongach* (hairy star) for a comet, *erchra* for an eclipse and *gcaoir adtuaidh* (northern berries) for the *aurora borealis*. The Milky Way has been called *Buar Thethreac* (Tethrach's herd), after the Formorian giant whose cattle left hoof marks in the heavens, and sometimes *Bealach Bó Fhoinn* (the way of the white cow) that sprinkled milk from her udders as she crossed the skies.

Durrow Abbey was built on a 6th-century monastic site noted for its high crosses, which can still be seen, and the manuscript *Book of Durrow*, now in TCD's library. During the Plantation, the abbey was granted to Nicholas Herbert, who built an estate there; Tullamore town developed on former abbey lands.

On May 12th, 1785, a hot-air balloon was released during the celebrations when Charles Bury inherited Tullamore Estate. Ballooning* was something of a craze then and there had been a number of successful Irish flights that year. But the Tullamore balloon got caught in a chimney and, in the ensuing fire, one-third of the village burned down. It was Ireland's first air disaster. For Tullamore, it was a blessing in disguise: the houses were rebuilt and, when the Grand Canal arrived a few years later, development began in earnest. Breweries, distilleries and textile mills were soon shipping their output to Dublin, and Tullamore, having outgrown neighbouring Birr, adopted a phoenix rising from the ashes for its town crest.

Daly's Distillery, begun in Tullamore in 1829, created two world-famous brands: **Tullamore Dew** whiskey and **Irish Mist** liqueur. The 'Dew' was named from the initials of the distillery's later owner, Daniel Edmund Williams; the 'Mist' is reputedly made from a secret 17th-century Irish recipe involving whiskey, honey and herbs. (Both brands are now owned by Cantrell & Cochrane.) Tullamore (*Tulach Mhór*, the big mound), is named for the hilly ground at Moore Street, former site of the town's windmills. ⑦

How old is the Earth?

The age of the Earth was a burning question for centuries. Many Irish scientists were among those tackling the problem, starting with Archbishop Ussher* in 1650, who famously calculated that the Earth was created in 4004 BC. One of those who helped solve the puzzle was a brilliant polymath from Bracknagh, **John Joly** (1859–1933). Joly also invented numerous scientific instruments, an early technique for colour photography and a method of treating cancer with radiation. He had a multinational background:

his father, the local rector, was of Belgian origin and his mother, an Austrian contessa, was of English, German, Italian and Greek extraction. Joly studied science and English at TCD, where he stayed for the rest of his life, working in the engineering and physics departments before being appointed geology professor in 1897.

By then, the controversy about the age of the Earth was connected with the debate about Darwin and evolution. Creationists believed the planet was young, at most a few tens of millions of years old, but, if Darwin was right and evolution had happened, the Earth had to be at least 1 billion years old. In the 1840s, Belfast-born William Thomson* (Lord Kelvin), an anti-Darwinist who initially made his name for work on heat and temperature, estimated that the Earth was at most 40 million years old, based on a guess of how long the planet took to cool from a molten ball. John Joly, at his first of two attempts, used the rate at which salt accumulates in the ocean and estimated that the oceans were 100 million years old. Most other estimates lay somewhere between those of Kelvin and Joly.

By the early 1900s, Joly had become fascinated with the newly discovered phenomenon of radioactivity. He was the first to realise that Kelvin's estimate of 40 million years for the Earth's age ignored the fact that radioactive material in the Earth generated heat, and that, therefore, the Earth was older than Kelvin thought. (In related work, Joly also proposed an early version of plate tectonics to explain the origin of the Earth's crust, suggesting that the continents were floating on a molten layer.)

Joly's insight was to realise that radioactivity in rocks could be used to measure their age. In particular, that the 'haloes' surrounding radioactive particles in rocks were caused by the radiation and that, the larger the halo was, the older it was. Joly collaborated with Ernest Rutherford in England to estimate the age of these haloes: Rutherford conducted laboratory tests, while Joly measured haloes in a sample of rock from Co. Carlow. Their results, published in the prestigious journal *Nature* in 1913, proved their rock was 400 million years old, then the oldest authenticated object in the world. The result offered the first evidence of a sufficient timescale for evolution. It also laid the foundations for radiometric dating of rocks: using modern versions of Rutherford and Joly's technique, geologists have since shown that the world's oldest rocks are 3.9 billion years old; that the world itself is 4.5 billion years old; and that the oldest rocks in Ireland, at Inishtrahull* in Co. Donegal and Annagh Head* in Co. Mayo, are nearly 1.8 billion years old.

Meanwhile, in 1894, Joly found time to invent an early technique for **colour photography**. His 'Joly process' produced the first full-colour slides that could be viewed in the hand or shown on an ordinary projector without any additional equipment. Previously, colour photos could only be created using red, green and blue filters and three corresponding images. Joly's process used a glass sheet coated with a thin gelatine layer on which fine lines were ruled at the rate of 220 to the inch, and in alternating red, green and blue transparent dyes. This ruled screen was placed in front of the glass photographic plate when the photograph was taken, and in the same position when the resulting slide was projected on to the screen.

Although the lines of the screen could still be discerned, the final image had natural colours and a 3D quality. The technique was restricted to glass lantern slides, however, and the images could not be printed onto paper. Joly patented the process and the Natural Colour Photograph Company, set up in Dublin by a US corporation, commercialised the technique. It lasted for a decade, until the invention, in 1904, of autochrome film that could be used to produce colour prints.

Joly's many other inventions included: the steam calorimeter, to measure heat energy and the specific heat of materials; a meldometer, to detect the melting point of a tiny sample of material (useful with archaeological and valuable artefacts); and a simple photometer to measure the intensity of light sources, made with two transparent pieces of wax separated by a sheet of tinfoil. With the Dublin cancer surgeon Clegg Stevenson, Joly also developed an improved way of using radium to treat cancer. Their 'Dublin method' was used around the world and at the Dublin Radium Institute* (1914–1952). This ingenious polymath is buried at Mount Jerome* in Dublin.

COUNTY KILDARE

Kildare highlights include Pollardstown Fen, the Curragh, the great Bog of Allen and more canal miles than any other county. The Antarctic explorer Ernest Shackleton was born in Kildare, as was Richard Griffith, who made the first map of Ireland's underground. The ingenious Nicholas Callan, a priest who invented the induction coil among other things, was based at Maynooth. Kildare derives from *Cill Dara* (the church of the oak wood), after the site in a pagan oak grove near Kildare town, where St Brigid is believed to have established a church.

THE BOG OF ALLEN

Turf treasures, peat products

Postcards, horse bedding, mud baths, cattle feed, charcoal, mineral supplements, creosote, firelighters… just some of the many uses peat from the Bog of Allen has been put to over the last 250 years. The bog, which spans the Offaly–Kildare border, was once Ireland's greatest **raised bog**,* but now it is significantly reduced by drainage and commercial turf cutting. It began developing on the site of a post-glacial lake, when conditions became wetter and colder 7,000 years ago. As moss accumulated, the bog grew upwards to form a great but insubstantial dome of watery vegetation, smothering any trees growing in the area. These buried trees convinced Gerard Boate* in 1652 that the bogs of Ireland were due to the "wretchedness of the Irish who daily let more and more of their good land grow boggy through their carelessness".

For thousands of years raised bogs, and especially the vast Bog of Allen, were inhospitable wildernesses, obstacles to transport and progress. They did at least provide bogwood, in otherwise treeless areas, which was used in building and furniture, for torches and, shredded into fibres, for ropes and thatching. The bog also provided fuel, but only at great

Back-breaking work: stacking turf sods to dry on the bog, from a sketch c.1780.

trouble and then only in a good summer – in a wet year the hard-won sods might never dry.

From the late 1700s, however, bogs were seen as a commercial resource. The Grand Canal* was crucial in opening up the Bog of Allen to transport and industry, giving turf cutters access to the Dublin market. A government **Bog Commission** (1809–1813) engaged distinguished engineers, such as Richard Griffith,* Alexander Nimmo* and William Bald,* to survey the major bogs and recommend ways of draining and developing them. The commission investigated options, such as flax growing, and proposed various drainage schemes, but ultimately stimulated little economic growth. Nevertheless, as industry turned increasingly to steam power, and needed fuel, entrepreneurs devised new ways of drying and compressing sods to form briquettes, and turf-processing plants started up beside the more accessible bogs.

Others found new uses for the peat. In the 1840s, the world's largest **turf distillation** plant was run at Kilberry near Athy by a Welsh chemist, Rees Reece. Turf is younger and less altered than coal and, therefore, richer in chemical compounds. Reece's distillation produced gas, charcoal and tar; the gas and charcoal were used as fuel to further distil the tar, in a process that yielded paraffin, lubricating and petroleum oils, creosote and asphalt. The plant lasted 10 years, until cheaper coal tar became available.

There was an impressive **peat briquette** factory at Derrylea in the 1860s employing 200 workers. Run by Charles Hodgson, who patented an early briquette press, it operated as an assembly line: peat cut in the morning was dried in kilns, then compressed into nuggets at a rate of 40 a minute to produce 180 tonnes a week of Hodgson's 'brown coal'. This forerunner of Bord na Móna developments was a large industrial complex with its own workers' village, narrow-gauge railway and foundries – the latter used local bog-iron* to make the factory tools. The plant lost out to cheaper imported coal and closed around 1870. It was excavated in 1975 and some of the artefacts found are at Lullymore's Peatland Museum.[8] Derrylea bog was later harvested by Bord na Móna.

A peat factory at Umeras on the Grand Canal made **horse-bedding** in the 1880s for the British cavalry's Irish stables. In 1903, the Callendar Paper Company in Celbridge made brown paper and card from moss-peat mixed with waste paper pulp; the paper was never economic but the peaty **postcards** were popular with emigrants. Other factories made firelighters from a mix of peat and coal dust; cattle feed, by adding molasses; and a wood substitute, by adding plaster of Paris – all a tribute to peat's versatility.

In the 1930s, the government tried to promote a new awareness of turf as a cheap native alternative to coal. National **turf-cutting championships** were held at Allenwood and, like today's ploughing championships, these bog jamborees attracted hundreds of competitors and spectators.[9] The fastest slanesman ever recorded was at the 1945 competition when Christie Daly from Kerry clocked up 100 sods a minute. During the Emergency (1940–1947), the government ran summer turf-cutting camps on the Bog of Allen to produce fuel for priority industries and the railways. Hostels were built at Lullymore and elsewhere and 4,000 men, mainly from the west, were brought to Kildare and paid to cut turf.

In 1946, the Turf Development Act established a State body, **Bord na Móna**, to exploit Ireland's bogs on an industrial scale. Most of the Bord's output today is used to fuel electricity stations, but it also makes briquettes for domestic use. At its Newbridge headquarters, scientists study new uses for peat, especially as an industrial purifier or filter, to clean for example gas emissions and septic tank effluent.

By the 1970s, large areas of the Bog of Allen had been cut away and Bord na Móna experimented with uses for the left-over land. At Turraun* in Co. Offaly new wetlands have been created and, at Lullymore, the Bord tried growing trees and even grass pastures. Today, thanks to the Irish Peatland Conservation Council, there is growing awareness of the need to conserve what is left of our bogs, and a number of intact raised and blanket bogs and also cutaway bogs are now protected as nature reserves.

'Venice of the midlands'

Kildare is canal country. It has more waterways than any other county in Ireland – nearly 150km in all – including parts of the Grand and Royal canals (which link Dublin and the Shannon), the Barrow branch line to New Ross, and numerous small feeder canals, such as the Milltown channel that links Pollardstown Fen* and the Grand Canal. The canals, built for the most part between 1750–1800, were Ireland's first modern transport network and the motorways of their day. British canals of the same era were usually private affairs serving a coal mine. But the great Irish canals were publicly funded ventures, planned by government, ordained by Acts of Parliament and overseen by boards. Moreover, they were designed to open up the boggy midlands, provide a transport infrastructure, develop local industry and give the cities ready access to cheap turf (in return, 'scavenger's manure' from the city streets was brought back by the barges for use in the bog fields).

The first canal to come through Kildare was the 200-km **Grand Canal**, linking Dublin to the lower-Shannon and Limerick, with a branch line south to the Barrow. It began in 1756, fuelled by promises of cheap turf from the Bog of Allen, and was completed 50 years later. The full journey from Limerick to Dublin took 18 hours at a leisurely 8km an hour. It was cheaper than the rival stagecoach: in 1798 it cost 8 shillings to travel by coach from Dublin to Leixlip, but only 1 shilling by canal fly-boat. Feeder channels kept the water levels topped up, and also supplied water to Dublin city.

Spurred by the Grand Canal's commercial success, a rival line, the **Royal Canal**, began in 1790 connecting Dublin with the upper-Shannon. The influential Duke of Leinster wanted the canal to pass his Carton Estate at Maynooth, and forced the Royal into a difficult and costly route. First, it had to cross the River Rye at Leixlip, where hundreds of men spent six years digging and filling the foundations of a massive aqueduct to carry the canal across the Rye (the excavation pits can still be seen by the canal). The aqueduct cost £28,300 – nearly four times its budget – and almost bankrupted the project. Next, the canal had to cross deep and unstable bog. Drainage channels helped reduce the height of the bog by up to 6 metres, but substantial embankments were still needed in places. By the time the canal was finished, however, the railway era had begun, and the canal was never a commercial success. In 1844, it was sold at a bargain price to the Midland & Great Western Railway Company, which intended to fill in the channel and build the Dublin–Galway railway line on top. In the end, the railway was laid alongside, and the canal survived.

After 1850, the canals lost much of their traffic to the railways but, by then, they had had a major impact on Kildare's development, witness the many canal-side hotels, mills, industries and warehouses that can still be seen. A gasworks at Naas, for example, on the Grand Canal quays, used coal, brought by barge, to make gas for the town's lamps, cookers and heaters. Brickworks developed along the canal, exploiting the clay that lay under the peat. Monasterevin found itself at a convenient location thanks to the canals and quickly became a prosperous industrial town. It had an extensive distillery, brewing and malting business, oil mills and even a tobacco and pipe factory, and was known as the 'Venice of the midlands', because of its waterways, aqueduct and bridges, especially the elegant lifting bridge over the Barrow.

The Barrow river god: spewing fish and crowned with a sheep, signifying rich waters and pastures.
© Enterprise Ireland

Up to the 1950s, non-perishable goods like coal were still carried by barge, but by then the canals were falling into disrepair. The changing fortunes of the canal are mirrored in those of Robertstown Hotel: built as a hotel for canal passengers, it closed in 1844 when rail travel took over; later it was a Royal Irish Constabulary barracks and, in the 20th century, a hostel for turf workers. The canals have since been restored and the waterways and towpaths are now popular leisure ways, giving people a chance to admire again the fine engineering of the canal embankments, harbours, bridges, locks and aqueducts. A guide to Kildare's canal walks, *Towpath Trails*, is sold locally.

Quaker village and mill

In 1685, two English Quakers, John Barcroft and Abel Strettel, were travelling from Dublin to Cork when they rested their horses by the Griese Valley. So taken were they with the valley that they bought it, planted it with trees, orchards and hedgerows, and founded a neat settlement at Ballitore – Europe's first planned Quaker town. ⑨ Another English Quaker, Abraham Shackleton, founded a renowned village school there in 1726 which counted among its students the writer Edmund Burke, the botanist William Henry Harvey* and the anti-slavery campaigner James Haughton.* Shackleton's family came to prominence in the area: Richard Shackleton owned Ballitore mill (now in ruins), and his daughter Mary Leadbetter (1758–1826) was a noted author (her village home has been restored). The best-known family member, however, was the explorer Sir Ernest Shackleton (see: **"By endurance we conquer"**). Just outside Ballitore is Crookstown mill. Built in 1840 and powered by the River Griese, it only stopped milling in 1965. The mill has been restored as a home and museum, with millwheels and grain-drying kiln, though it no longer grinds grain. ⑩

"By endurance we conquer"

"Fortitudine vincimus". An appropriate family motto for **Sir Ernest Shackleton** (1874–1922), whose 1914–1916 Antarctic expedition on the aptly named *Endurance* is celebrated in the annals of exploration. Shackleton's every endeavour failed yet he remains a hero, a man who inspired loyalty and confidence among his men. Moreover, despite the hardships, no one died on any of his expeditions. The climber Sir Edmund Hillary, who crossed Antarctica in 1958, succeeding where Shackleton failed, once said: "For scientific discovery give me Scott, for speed and efficiency Amundsen; but when disaster strikes and all hope has gone, get down on your knees and pray for Shackleton." Business and management schools now study 'the Shackleton way', yet, ironically, the Irish explorer's attempts at business, journalism and politics all failed. He was happy only at sea.

Shackleton, descended from the Yorkshire Quaker who founded Ballitore's famous school, was one of 10 children born at Kilkea House where his father was the local doctor. The family moved to Dublin when Shackleton was six, and to London four years later. At 16, Shackleton enlisted in the merchant navy and, in 1901, in search of adventure, joined Scott's first Antarctic expedition that got within 750km of the South Pole. By 1907, Shackleton had raised funds for his own first expedition. Though ill-equipped – they brought ponies to pull sledges, and ate them when that failed – they came within 150km of the South Pole, closer than anyone before. Sick and short of rations, Shackleton reluctantly turned back rather than risk lives. It was his last time in Antarctica.

After Amundsen reached the South Pole in 1911, Shackleton decided to try crossing the continent. His *Endurance* expedition was five years in the planning; 8,000 volunteered from whom 27 were selected, including Kerryman Tom Crean* as second officer. They brought

Explorers and voyagers

Ireland has a strong tradition of exploring starting with **St Brendan the Navigator** (flourished 530–550) who may have been the first European to visit North America. The first European to see the frozen continent of Antarctica was a Cork man, **Edward Bransfield** (1781–1852): press-ganged into the British navy for the Napoleonic wars, Bransfield later took part in voyages to the South Atlantic where he mapped and named the South Shetland Islands and, in January 1820, saw the Antarctic mainland across the Bransfield Strait, named later in his honour.

Francis Crozier* from Co. Down and Dubliner **Edward Sabine** took part in scientific expeditions in the 1830s in search of the magnetic South Pole.* Crozier was later lost with the ill-fated 1845 Franklin expedition that sought a Northwest Passage through the Arctic to the Pacific. The mystery of that expedition's disappearance was solved by polar explorer, **Capt Francis McClintock*** from Dundalk, while **Admiral Robert McClure*** from Wexford, who also led a search party for Franklin's expedition, is credited with discovering the Northwest Passage. Donegal botanist **Henry Chichester Hart*** was the naturalist with the British Arctic expedition of 1875–76, which attempted to sledge to the North Pole; they reached a latitude of 83°North before turning back, the most northerly reached by any expedition then. Cork-born **Jerome Collins** was the meteorologist aboard the *Jeannette* (1879), an ill-fated expedition that perished while seeking a route through to the North Pole. The indomitable Kerryman **Tom Crean,*** one of Shackleton's *Endurance* party, was also on Scott's last expedition to the South Pole, though not selected for the fatal final leg.

Noted Irish mountaineers include **Anthony Reilly** (from Co. Westmeath) who, in 1865, made the first accurate map of Mont Blanc; **John Tyndall*** (from Co. Carlow) who was the first to climb the Weisshorn; **John Pallisser** (from Co. Waterford) who discovered important routes across the Rocky Mountains; and marine biologist **Rev William Green*** (from Co. Cork), who was known as 'Canada's first recreational mountaineer'.

Shackleton, Crean and four others left in one of the boats to get help at South Georgia whaling station, 1,200km away.

Their journey is spoken of with awe. Ill and unfit, the six spent 16 days in appalling weather and mountainous seas. They were able to take only a handful of navigational observations as the sun seldom shone, yet they reached South Georgia safely, albeit the inhospitable southern side. Shackleton, Crean and Frank Worsley then spent two further exhausting days crossing the island's unmapped and mountainous interior, to reach the whaling station in May 1916, nearly two years after they had last been seen. With the war still on, no boats could be spared and it was three months before the men left on Elephant Island were rescued. By then interest in Antarctic exploration had waned and, in any case, the expedition was deemed to have failed. It was the end of an era of polar exploration, although Shackleton's account of the expedition, *South* (1919), became a bestseller. Ernest Shackleton sailed again for the South Atlantic in 1922 but with no particular aim. He died of a heart attack while still at South Georgia, where he is buried, leaving a wife and three children in London. Mountains, glaciers and ice shelves have all been named in his honour. Athy Heritage Centre has one of Shackleton's sledges. [11] Kildare County Library has papers relating to the Shakleton family. [12]

1903 – Gordon Bennett road race

When the automobile speed limit was raised in the 1890s from 4mph to 14mph, road races were held to celebrate the new-found freedom. One such event was an international race organised by US millionaire Gordon Bennett. Countries could enter three-car teams, but every part of each car had to be made in that country. The 1903 race was to have been in England but closing the roads there caused problems, so instead it took place in Ireland over a complex 145-km, figure-of-eight route, starting at the Motte of Ardscull.

sledging dogs, Norwegian tents, a ski instructor, and a scientific team that included a physicist, biologist, geologist and meteorologist. They sailed in August 1914 as World War I broke out. *Endurance* became trapped in the ice, however, within sight of the Antarctic mainland and, after drifting for ten months, broke up and had to be abandoned. Then, in early 1916, when the ice melted, the crew spent six terrifying days in three small open lifeboats, before reaching the inhospitable Elephant Island. A week later

France, Germany, England and the USA entered, a large grandstand was erected across the road at the starting point, and a huge crowd gathered, brought by special trains and steamers from all over Europe. The 526-km race over four laps of the course began on July 2nd at 7am and ended at 5pm. At each town, the cars were chaperoned through the streets by a cyclist, making it difficult to compute the race times. Of the 12 cars that started, only five completed the gruelling course. One British car was disqualified for receiving help in Athy, where cold water was thrown over the hot tyres to help keep them on their rims. In the end, Herr Jenatzy was declared the winner, driving a Mercedes for Germany, in a time of 6 hours 39 minutes, and an average 49mph (79kmph).

This was a far cry from the 1860s when the British government, alarmed by the growing numbers of steam-powered carriages, set a speed limit of 4mph on country roads and 2mph in city streets. Moreover, each vehicle had to be 'driven' by two people and a third person had to walk in front carrying a red flag. From the 1820s on, there were probably several steam-powered carriages on Irish roads and one, made by the Earl of Rosse in Birr,* was involved in Ireland's first motoring fatality in 1869. The first petrol-driven car in Ireland was imported in 1899 by Dr John Colohan of Co. Dublin: a Benz Comfortable, registration number IK52; restored in the 1980s, it has taken part in the London–Brighton vintage races.

Motoring became a practical mode of transport when the speed limit was relaxed to 14mph in 1896. Roads were poor and dusty, however, and consequently cars needed constant servicing. Service stations, garages and car showrooms now appeared throughout the country and several Irish firms started making cars. They included NF Pearce of Wexford, which made three- and four-wheeled vehicles; Chambers Motors of Belfast, some of whose cars can be seen at the Ulster Folk & Transport Museum;* also Burke Motors of Clonmel and Ailesbury Motors of Edenderry, though both their factories had closed by 1908.

In the 1920s, as motorised road traffic developed, Ireland introduced a system to record road-traffic accidents and, in 1927, the first legislation in Europe for standard signposting. By 1928, 69 per cent of traffic was mechanically propelled. Today, there are 2 million registered motor vehicles in Ireland, including for example, hearses, tractors and boats, but mostly motor cars. And they have access to 115,000km of tarmac road. Each July, the vintage car club holds a Gordon Bennett fun run on the anniversary of the 1903 race.

Horoscopes for horses

In 1900, an eccentric Scotsman, **Colonel William Hall-Walker**, established a stud farm beside Kildare town. A keen astrologer, Hall-Walker believed that the heavens controlled the destiny of all living creatures. To ensure the planets could exert their maximum influence on his horses, he installed elaborate skylights in his stables. Certainly, his stud enjoyed considerable success, but he had also, despite his eccentricities, chosen the perfect location for a stud farm: Kildare's limestone-rich grasslands and waters are good for building bones, and the area has been at the centre of Irish horse-racing since ancient times. Today, the county is home to racing's Turf Board, and two of Ireland's best-known race courses – Punchestown (near Naas) and the Curragh* (by Newbridge). Hall-Walker also drained an area of bog and had Japanese designer Tassa Eida build a garden there on the theme of 'the path of life'. In 1915, the Colonel presented his 400-hectare stud farm to the Crown which, in 1943, passed it to the new Irish Republic. The government established a national stud farm there in 1945 to aid the bloodstock industry by providing top-quality stallions for breeding. Each season about eight stallions cover between them 250 mares. [13]

Aquifer and ancient arena

The Curragh of Kildare is one of Ireland's ancient landscapes. The name comes from an old Irish word meaning 'race course', and the area has been used for thousands of years as an arena, commonage and training ground. Legend has it that Fionn and the Fianna trained there in pre-Christian times, and that certain boundary stones were erected by St Brigid. The Curragh is a vast heathy plain, very different from neighbouring Bog of Allen* to the west, and Pollardstown Fen (see: **A land of water**) to the northeast. Its distinctive nature is due to the vast ice sheets that blanketed the midlands at the height of the last Ice Age over 20,000 years ago. This so-called Midlandian ice did not cover the Curragh, but stopped a little to the northwest along a line marked by the high ground of the Galtrim moraine. To the west of this line is the Bog of Allen, a water-logged region that developed where the enormous weight of ice compacted the ground. To the east of the line, meltwater streams flowed out from under the ice laden with sand and gravel. The gravel was deposited in a delta at the Curragh, creating a well-drained plain with a springy turf. Occasional depressions, such as Donnelly's Hollow, are glacial kettle holes,* which formed when large boulders of ice that were trapped in the gravel melted and the ground above subsided.

The enormous limestone gravel bed beneath the Curragh is 60 metres deep, and it forms Ireland's largest **aquifer** or underground reservoir. It filters, enriches and stores water, and supplies local drinking wells, the springs at Pollardstown – giving that fen its distinctive alkaline qualities – and from there, the Grand Canal. It is vulnerable to pollution, however, and any contamination at the Curragh would quickly percolate throughout the area.

Unlike the Great Heath of Portlaoise,* which was ploughed by the Anglo-Normans, the ground at the Curragh has never been disturbed. Instead, it was grazed as commonage for centuries without seeding or fertiliser. The result is an acidic heathland, low in minerals, but with a distinctive vegetation. The close-cropped grasses include purple moor grass, sheep's fescue and mat-grass; heathers like calluna and ling are common, as is furze,* which, being a legume, can make its own fertiliser. In 1837, Lewis's *Topographical Dictionary* described the Curragh as "a fine extensive sheepwalk of above 6,000 acres [2,428 hectares], forming a more beautiful lawn than the hand of art ever made… nothing exceeds the extreme softness and elasticity of the turf, which is of a verdure that charms the eye". Lewis also recorded that local cottagers collected the sheep dung to fertilise their cabbage plants.

The Curragh was never enclosed and the broad open space was often used for military manoeuvres. The British army established a temporary camp there during the Napoleonic era, and a permanent camp in 1853 during the Crimean War. The latter move precipitated a major campaign by local farmers and horse trainers to retain the ancient rights of way and commonage. After a lengthy inquiry, these rights were enshrined in the 1870 Act for the Better Management and Use of the Curragh. The army camp was built as a self-contained town with a hospital, church, school, bakery, abattoir, sports ground, even narrow-gauge railway. **Donnelly's Hollow** was the site of a celebrated boxing match where in 1815 Irishman Dan Donnelly defeated English champion George Cooper.

A land of water

Just 1.5km separates Pollardstown Fen from the Curragh, yet the two places could hardly be more different. The Curragh is a dry and acidic scrubland, while Pollardstown is a watery world, an alkaline fen of rare plants and insects. Strangely, it is the Curragh's mineral-rich aquifer that supplies the 30 springs that feed Pollardstown, and gives the fen its distinctive character. Pollardstown began developing 12,000 years ago at the end of the last Ice Age as a lake in a basin bounded by low ridges of glacial debris. With 130 hectares, it is Ireland's largest fen and now a nature reserve.

Fens are unusual places: they are 'bogs in embryo', an early step in the transition from mineral-rich freshwater lake to poor, acidic bog. First, plants at the shore trap silt so that the lake slowly fills in from the edge. Then dead plants fall to the lake bottom and form fen-peat. As this peat accumulates, plants at the surface become cut off from the mineral springs beneath and find it increasingly difficult to survive. In this nutrient-poor environment, bog plants eventually take over, notably sphagnum moss that can live on rainwater alone, by which time the fen is part way to becoming a raised bog.*

Because fens are still rich in minerals, their ecosystem is more diverse than that of a bog. Aquatic plants, such as duckweed, pondweed and water lily, colonise the open water. Black bog rush forms tussocks, orchids provide summer colour, bulrush and flag iris are common, as are

(A) Early Stage

(B) Middle Stage

(C) Final Stage

Three stages in the development of a fen. If the process continues, the fen will eventually become a raised bog.
© Bord na Móna

| | Glacial Deposits | | Shell Marl | | Reed Swamp Peat |
| | Woody Fen Peat | | Lake Waters | | |

brown fen mosses. Sedges are important fen plants and Pollardstown is rich in saw-tooth sedge (*cladium*) and great fen-sedge. In 1990, the entomologist Dr Jim O'Connor found a small caddis-fly at Pollardstown (*Leptocerus tineiformes*), the first time this insect was found in Ireland. (There is a bird-hide in the sourthern section of the fen.)

The transition to bog has begun at Pollardstown where the peat is up to 7 metres thick in places and acid-loving bog plants are colonising parts of the fen. We are lucky to have Pollardstown Fen at all: fens, being rich in minerals, are often reclaimed for farming and Pollardstown was drained in the 1960s for that reason. The OPW acquired the site in 1983, however, and blocked the drains, and the area has since reverted to wetland. In 1999, there was concern that the new Kildare bypass road would disconnect the fen from its crucial water source in the Curragh aquifer, but special drains are being installed to prevent this happening.

What shape is benzene?

Benzene was discovered in 1825 by the great English scientist, Michael Faraday, when he was distilling whale oil. But for over a century this important compound was a chemical mystery: each benzene molecule contains six carbon atoms and six hydrogens, and this broke all known chemical rules – there ought to have been more hydrogen. The mystery was partly solved in 1865 when German chemist Auguste Kekulé dreamed of a snake biting its tail and suggested that the carbon atoms were linked in a ring. But what kind of ring? Circular? Bent? And with what kind of chemical bond? The mystery was solved in 1929 by a brilliant young scientist from Newbridge, **Dame Kathleen Lonsdale** (1903–1971).

Lonsdale's father Harry Yardley was the Newbridge postmaster and she later said she inherited her love of facts and science from him. The family moved to England in 1908 and Lonsdale studied science at Bedford Women's

Dame Kathleen Lonsdale from Newbridge, who solved the mystery of benzene's shape.
Dublin Institute of Advanced Studies

Kathleen Lonsdale became professor of chemistry at University College London. She continued to work in crystallography, studying pharmaceutical compounds, and producing the first accurate measurements of the atomic structure of diamond.* In 1967, a newly discovered diamond-like form of carbon, variously known as delta-carbon or hexagonal diamond, was named lonsdaleite in her honour. Like her contemporary and countryman Bernal, Lonsdale was politically active. A Quaker and pacifist, she was interned as a conscientious objector in 1943, and later joined the distinguished Pugwash Conference on science and society. She drafted the first international guidelines to protect uranium miners from radiation and, in 1945, was the first woman to be elected a fellow of the Royal Society since its foundation in 1660. In 1968, she became the first woman president of the British Association for the Advancement of Science.* Dame Kathleen Lonsdale died of cancer in 1971, a few months before Bernal.

College in London. She then won a place with the distinguished scientist Sir William Bragg, who was experimenting with X-ray diffraction at London University. In X-ray diffraction, photographs are taken of crystals; the location of the atoms in the crystals can be calculated from the resulting patterns (in 1953 the technique famously revealed the double-helix structure of DNA).

In the 1920s and 1930s, two of the world's best X-ray diffraction laboratories were those of William Bragg and, at London's Birckbeck College, the brilliant Nenagh-born J D Bernal* who specialised in biological molecules. Significantly, Bragg and Bernal both supported women scientists and thanks partly to them, three of the big names in X-ray diffraction were women: Lonsdale, Rosalind Franklin (who worked on DNA) and Nobel laureate Dorothy Hodgkin. In 1929, Kathleen Lonsdale took and analysed photographs that revealed that benzene has a flat and hexagonal ring, with the six carbon atoms at the corners of the hexagon. Her discovery was a vital piece in the jigsaw of benzene chemistry. The picture was completed in 1931 when US Nobel laureate Linus Pauling suggested that the carbon bonds in benzene resonate between single and double bonds, thus explaining the limited amount of hydrogen in the molecule.

1839 – a map of the underground

For centuries map-makers were concerned with literally superficial detail – surveying and drawing what they saw on the surface around them. However, in the early 1800s, prospecting techniques improved, so that companies and governments began searching

Sir Richard Griffith with his geological map of Ireland. The Geological Survey* sells modern maps of Ireland's geology.

for mineral wealth, and people became increasingly interested in what lay beneath the Earth's surface. Map-makers responded with maps of the geological underground. The man who almost single-handedly produced the first geological map of Ireland was a prodigious engineer, **Sir Richard Griffith** (1784–1878). This 'father of Irish geology' was also responsible for the land valuations during the first Ordnance Survey of Ireland.* Griffith was born in Dublin but grew up at Millicent near Clane, where his father, a former MP, had bought an estate with money made trading in the East Indies. Richard Griffith was educated for the army and spent a year in the Irish artillery, but left in 1801 to study engineering, mining and geology in London, Edinburgh and Cornwall. He returned to Ireland in 1808.

Richard Griffith came to be one of the most important engineers in 19th-century Ireland. From 1808, as a member of the government's Bog Commission* assessing the economic potential of Irish bogs, he surveyed over 100,000 hectares of remote bogland. At the same time, he surveyed the Castlecomer coalfields for the Royal Dublin Society.* From 1812–1830, he was that society's professor of geology and mining, and the government's inspector of mines in Ireland. In 1822, Griffith was dispatched to Munster to organise famine-relief roadworks and, over the next decade, built numerous bridges and over 400km of roads, opening up remote parts of Cork and Kerry. His Listowel–Newmarket road, for example, halved the distance to Cork and greatly improved access to the Cork butter market.* In the 1830s, he joined the government's Railway Commission and, from 1850, as chair of the Board of Works, he also managed the country's extensive land improvement and drainage programmes.

Most importantly, in 1825 he was put in charge of the Boundary Commission for the Ordnance Survey of Ireland, which was compiling the first detailed map of the country. Griffith, the only Irish engineer in the whole survey, was responsible for mapping the boundaries of every townland and calculating the land valuations, based on soil quality and rock type, which would be used to assess taxes and rates. Griffith's valuation study took decades to complete and produced detailed information that is still used by historians.

Throughout all this varied work, Griffith collected geological data where he could, exploiting his many jobs to provide the necessary information and intending to produce Ireland's first geological map. The first edition, published in 1839 at a scale of one-quarter inch to the mile, was spread over six sheets and sold for £1 (uncoloured). Much of the important detail came from Patrick Ganly,* who was part of Griffith's land valuation team. As Ganly's geological work was unofficial, done on valuation office time, his contribution was never credited and only emerged in the 1940s when 600 of his letters to Griffith were discovered. The last edition of Griffith's geological map was issued in 1855, by which time the Geological Survey of Ireland* had been established and begun a detailed (and more official) assessment of the country's rocks and mineral wealth. Although Griffith is now remembered as the man responsible for compiling the land valuations, his geological map was an amazing private undertaking.

Cotton town

Prosperous was founded in 1780 by **Robert Brooke** as a purpose-built, industrial settlement providing work in a poor, rural area. It was one of several such ventures around the country made possible by the new steam engines. Previously, industrial power was provided by watermills and industry had to be located beside a river, but steam engines used coal power instead and broke that connection. Most of these industrial towns were founded around textile mills, usually linen* or cotton,* and often started by enterprising and public-spirited Quakers. They include the mill towns of Bessbroke (Co. Armagh), Sion Mills (Co. Tyrone) and Portlaw* (Co. Waterford). The government backed some of the projects, especially schemes that would cut the number of unemployed people arriving in the overcrowded cities.

At Prosperous, Robert Brooke invested his own money and £25,000 from the government to build steam-powered spinning and weaving factories, plus workshops to make and repair the machines. Within three years, there were 200 houses and 1,000 inhabitants, however the scheme went bankrupt in 1786. Manufacturing continued on a small scale until the 1798 Rebellion but, by 1837, Prosperous was that in name only: Lewis's *Topographical Dictionary* described it as "little more than a pile of ruins" with no advantages and no hope of revival. Yet the town is again thriving, thanks to the big electronics companies that have located nearby, a 21st-century equivalent of the large textile mills of the 1780s.

1797 – the world's oldest 'automobile'

The world's oldest 'automobile', or 4-wheeled self-propelled object: a small-scale working model dating from 1797 of a Trevithick steam engine.
© *Straffan Steam Museum*

The world's oldest automobile takes pride of place at Straffan Steam Museum. This 'automobile' – the original sense of the word meant a self-propelled object, rather than a motor car – was made in 1797, and it is a small-scale working model of a Trevithick steam engine. **Steam engines** were the workhorses of the Industrial Revolution. They burned coal to heat steam that pushed pistons, moved shafts and turned wheels, and thus did useful work. The first successful steam engine was the beam engine, invented in 1712 by Englishman **Thomas Newcomen**: as steam expanded inside a chamber, it pushed a piston; cold water was then injected into the chamber to condense the steam and, as the steam contracted, it created a vacuum that sucked the piston back. The net result was a beam that slowly nodded up and down. Newcomen's engine was inefficient, however: heating, cooling and re-heating the chamber for each stroke was wasteful, and converted just 1 per cent of the energy available in the coal. Nonetheless, mine owners loved it, and they used Newcomen engines to pump mines dry, replacing the horse power that had been used for centuries. The first Newcomen steam engine in Ireland was probably one installed in 1744 at Doonane in the Castlecomer* collieries.

In 1769, **James Watt** realised it was more efficient to do the heating and cooling in separate chambers – the steam expansion chamber could then be kept hot, and a separate condenser chamber kept cold. Watt also designed a way of converting the beam's nodding action into a rotary motion, which was much more useful. The result was a powerful, efficient and versatile engine, capable of operating at higher temperatures and pressures, and perfect for driving the wheels of the Industrial Revolution. Despite being expensive, Watt's engine proved even more popular than that of Newcomen. In the 1770s, a Dublin steel mill acquired what was probably Ireland's first Watt engine.

Watt thought that high-pressure steam was too dangerous, so his engines operated at atmospheric pressure. **Richard Trevithick** found a way of controlling high-pressure steam, by enabling it to expand within the cylinder. He was thus able to design engines that were as powerful as Watt's, yet smaller and 200 times lighter (5 tonnes versus 1,000 tonnes). In 1797, he built the first high-pressure stationary and locomotive engines. Small-scale models, such as the one at Straffan, were built as prototypes to be copied by engine makers who might not be able to follow mechanical drawings. Trevithick is best remembered as the man who, in 1801, built the first locomotive or 'steam carriage'. The cast-iron tramways of the day were too brittle for the weight of his engines, though, and he abandoned the project. Never a successful businessman, he died penniless, while other engineers, notably George Stephenson, profited from his inventions.

By the late-19th century, most Irish firms were using stationary steam engines to power their mills and machinery, and to drive pumps and lifts; some are still *in situ*, such as the beam engine at Midleton* distillery in Co. Cork. The Straffan Steam Museum has an excellent collection of steam engines, from locomotives to distillery beam engines. Most were used in Ireland at one time or another, and all are restored to their former glory. They include a small-scale model of *Colossus*, the locomotive used on Ireland's first railway, the Dublin–Kingstown [Dún Laoghaire] line.* The collection is housed in a church originally built at the Great Southern & Western Railway's Inchicore works* in Dublin, but rebuilt at Straffan in 1992. ⑭ The village hosts a steam rally each August bank-holiday weekend.

When the ESB built the dam at Leixlip, part of the larger Blessington–Pollaphuca* scheme, it created a narrow lake 3km long on the Liffey. Leixlip, at the confluence of the rivers Rye and Liffey, has a number of iron-rich mineral springs and was a noted spa* town in the late-1700s when people came to drink and bathe in the waters. The town later lost out to Lucan spa, which was closer to Dublin. Two of Leixlip's **spa wells** were restored in the 1970s: the deep bath-like structures near the canal have steps down and grilles to let water in and out.

A salmon lift

An ingenious showman-priest

The Vikings named this point on the River Liffey *Lax Hlaup* (salmon leap), because every autumn the fish jumped a waterfall there as they migrated upstream to spawn. In the 1950s, when the ESB built a hydro-generating station at Leixlip to harness that waterfall, the salmon had to be accommodated. The solution devised by two ESB engineers, Jim Dooge and Anthony Murphy, was the world's first hydraulic fish lift. Their design has since been used by hydro-stations around the world.

The lift operates like a canal lock, with an upstream and downstream gate and a lock in between, but instead of lifting boats, it lifts fish. The lower inlet is opened first to let in the salmon; it closes and the lock is flooded, then the upper gate is opened, releasing the fish upstream; each full cycle takes one hour. The main problem, according to Prof Jim Dooge, was attracting salmon to the lock to begin with. Studies of fish behaviour suggested they were attracted to white water, so the lower gate was made narrow, and when it opens this creates turbulence around which the salmon congregate. Before the lift was built, the design had to be approved by fisheries inspectors: a small-scale model was made as a test, and 12 salmon were counted into the lower pool; the design was approved when, according to Jim Dooge, not 12 but 13 fish emerged safely at the top.

The small priest must have seemed like an Irish Frankenstein, experimenting with electricity in his basement laboratory at Maynooth College, dishing out al-mighty electric shocks to unsuspecting volunteers and electrocuting turkeys. Yet the **Rev Nicholas Callan** (1799–1864) was one of Ireland's great inventors. He invented the induction coil and self-exciting dynamo – both of which are still used – built the most powerful batteries and electromagnets of his time, and made it into the *Encyclopaedia Britannica.*

The Rev Nicholas Callan, ingenious inventor and pioneer in the field of electricity. His exploits were recorded by the Encyclopaedia Britannica. Maynooth Museum

Callan, born to a middle-class family in Dromiskin, Co. Louth, was the first major Irish scientist from a Catholic background. For 200 years, Irish science had been dominated by Anglo-Irish Protestants who studied at TCD.* Callan, however, studied for the priesthood at the Irish College in Rome. There he learned about Alessandro Volta's battery, a novel chemical device that could store electricity. Ironically, Callan was later professor of science at St Patrick's College Maynooth, which the British government established in 1795 to remove the need for Catholic priests to study on the continent where, after the French Revolution, they might be exposed to dangerous ideas.

In Maynooth, and with funding from friends and family, Callan set about designing better batteries. Electricity was still something of a toy, but Callan realised that, with powerful batteries, it could be put to practical use. Where other batteries used costly platinum, Callan found a way of using cheap cast iron: his positive plate was a small cast-iron box filled with strong acid; in it sat a porous pot, also filled with acid, and containing a zinc sheet as the negative plate. Callan hoped it would be used to power lamps at lighthouses and remote navigational buoys, but electric lighting was still in its infancy. His most successful battery design sold commercially in London as 'the Maynooth battery'. The Duracell of its time, it was hard-wearing and cheap, long-lasting and powerful.

Callan often connected large numbers of these battery cells, and once joined 577 together, using 136 litres of acid, forming the world's largest battery. Since there were no instruments yet to measure current, Callan assessed his batteries by the weight an electromagnet could lift when connected to the battery. His best effort lifted 2 tonnes and made it into the *Encyclopaedia Britannica* (eighth edition) as the world's most powerful magnet. When Callan reported this in the *Annals of Electricity*, a London professor came to witness the spectacle and was said to be incredulous.

In 1836, Callan made his most important discovery – **the induction coil**. He wound two long wires around the end of a large electromagnet and connected the ends of 1 wire to a battery. But when he interrupted the current from the battery, he got a spectacular spark between the ends of the second, unconnected coil. It was the world's first transformer: he had induced a high-voltage in the second wire, starting with a low voltage in the first. And the faster he interrupted the current, the bigger the spark. In 1837, he produced his giant induction machine: using a clock escapement to cut the current 20 times a second, he generated 38-cm sparks – an estimated 600,000 volts – and the largest artificial bolt of electricity then seen.

Nicholas Callan's giant induction coil: incorporating miles of wiring, it could generate an estimated 600,000 volts – the largest artificial bolt of electricity then seen.
© *Maynooth Museum*

The coils were made from miles of hand-drawn metal wire that Callan laboriously insulated by hand with tape and wax. A model was sent to London for demonstrations and similar machines soon appeared in Europe and North America. The voltmeter had not yet been invented so, to measure the voltage from his induction coils, Callan improvised: he would electrocute large birds or shock chains of volunteers holding hands. One almighty shock rendered a future archbishop of Dublin unconscious, after which the college asked Callan to limit his experiments. Callan's induction coil is today used in car ignitions, for example, to generate powerful voltages from a low-voltage battery and produce sparks to cross the gap in the spark plugs.

In 1838, this intrepid priest stumbled on the principle of the self-exciting dynamo. Simply by moving his electromagnet in the Earth's magnetic field, he found he could produce electricity without a battery. The effect was feeble, however, so Callan never pursued it, and the discovery is generally credited to Werner Siemens in 1866. Callan probably also had one of the world's first electric vehicles: in 1837 he was using a primitive electric motor to drive a small trolley around his laboratory. With great foresight he predicted electric lighting, at a time when oil was still widely used and gas was the next new thing. In 1853, he patented an early form of galvanisation, using a lead–tin mix to protect iron from rusting. He even proposed using batteries instead of steam engines on the newfangled railways. Callan later realised his batteries were not powerful enough and it took a hundred years before battery-powered trains invented by another Irishman, James Drumm,* were used on Dublin railways. The Science Museum at Maynooth College has Callan's instruments and laboratory. ⑮ It also has the equipment Marconi* used in 1898 to broadcast the results of the Dún Laoghaire regatta.

———————

Chapter 7 sources

Co. Laois

1. Johnston, Máiréad, *Hidden in the Pile: the Abbeyleix Carpet Factory 1904–1912* (Abbeyleix Heritage Company, 1997).

2. Foy, Michael, *The Sugar Industry in Ireland* (Irish Sugar Co. Ltd, 1976).

3. Phillips, Patricia, "The Queen's Institute, Dublin (1861–1881): the First Technical College for Women in the British Isles" in Norman McMillan (ed.), *Prometheus's Fire* (Tyndall Publications, 2000).

Co. Offaly

4. Tubridy, Michael, *Reconstructing the Rosse six-foot telescope* (Birr Castle, 1998).

5. Taylor, David, & Mary McGuckian, "Lunar temperature measurements at Birr Castle" in Nudds *et al.*, *Science in Ireland: 1800-1930* (Dublin, 1988).

6. "Forest glass furnaces in County Offaly" *Archaeology Ireland* **11** No.4 (1997).

7. Heery, Stephen, *The Shannon Floodlands* (Tír Eolas, 1993).

8. McCarthy, D & A Breen, "Astronomical observations in the Irish annals and their motivation" *Journal of the Mediaeval Academy of Ireland* vol **11** (1997).

Co. Kildare

9. Feehan, John & Grace O'Donovan, *The Bogs of Ireland* (UCD, 1996).

Chapter 8: Mayo • Galway • Clare

Visitor facilities

1. Céide Fields
2. Mayo North Heritage Centre, Crossmolina
3. Belderrig Research Centre
4. Basking Shark Survey
5. Foxford Woollen Mills
6. Knock Folk Museum
7. Clew Bay Heritage Centre
8. Centre for Island Studies
9. Doon Nature Peninsula, Burriscarra
10. Killary Cruises
11. Leenane Cultural Centre
12. Ocean's Alive, Renvyle
13. Connemara National Park, Letterfrack
14. Walking Ireland Tours, Clifden
15. Connemara Environmental Education Centre
16. Atlantaquaria, Salthill
17. Glengowla Mine, Oughterard
18. Tuam Mill Museum
19. Galway City Museum
20. James Mitchell Museum NUIG
21. Coole Park
22. Ionad Árann/Aran Heritage Centre
23. Burren Exposure, Ballyvaughan
24. Burren Centre, Kilfenora
25. Burren Perfumery, Carran
26. Aillwee Cave
27. Spa Wells, Lisdoonvarna
28. Liscannor Stone Story
29. Clare County Museum
30. Dolphinwatch, Carrigaholt
31. Scattery Island Ferries, Kilrush
32. Shannon Dolphin & Wildlife Foundation
33. Bunratty Castle & Folk Park
34. Irish Seed Saver Association, Scariff
35. Ardnacrusha

Other places of interest

36. Aughrim
37. Ballynahinch
38. Burrishoole
39. Carna
40. Cill Chiaráin/Kilkieran
41. Corrandulla
42. Derrigimlagh
43. Doolin/Fisherstreet
44. Glenamoy
45. Gort
46. Killeen
47. Moycullen
48. Mulranny
49. Roundstone
50. Shannon
51. Six-Mile-Bridge

♀ woodland

🦌 nature reserve or wildfowl sanctuary

⌂ lighthouse

Visitor facilities are numbered in order of appearance in the chapter, other places in alphabetical order.
How to use this book: page 9; Directory: page 474.

Mayo's many riches include amazing porous limestone, several rare and unusual heathers, one of the world's most intensely studied islands, prehistoric farm landscapes, a spectacular coastline and gold in Croagh Patrick. This scenic county also produced the inventor of the world's first guided missile. Mayo (*Maigh Eo*, 'yew tree plain') refers to the flat terrain south of Castlebar and near Mayo village.

Neolithic cattle ranchers

Had you visited the Céide Fields 5,000 years ago, you would have seen, not today's vast expanse of blanket bog, but a pleasant pastoral scene. Regular stone walls ran inland from the coast, and herds of small Neolithic cattle (the ancestors of the Kerry cow*) grazed peacefully in the neat fields. The climate was 2°C warmer than today, so the animals could continue grazing through the winter. The cattle provided milk, butter and cheese, ox power and, after slaughter, meat, tallow, sinews, leather and bone – no wonder the early Irish valued them like gold. Some 500 people lived in Céide then, while at nearby Belderrig their neighbours farmed cereals (see: **Ireland's oldest plough marks**). Céide was a peaceful community – no defensive structures were found there – and it lasted for 500 years, until the climate turned colder and wetter. Then the farmers moved out, and the bog moved in, gradually burying the Neolithic farmscape, until no visible trace remained.

We know about these Neolithic farmers thanks to UCD archaeologist Dr Seamus Caulfield. His father, a schoolmaster at Belderrig, was the first to realise the significance of the piles of stones occasionally found under the peat and clearly older than the bog. In 1934, he alerted the National Museum to their existence, but it was not until the 1970s, when his son had trained as an archaeologist, that investigations began.

Excavating the whole area would have been a major undertaking, and would have destroyed the bog. So Seamus Caulfield mapped Céide's hidden stone structures by adapting a technique traditionally used to find bogwood – in treeless areas these fossilised tree stumps were an important source of timber, and were found by probing the peat with a rod. To locate and trace the stone walls, Caulfield probed the bog with long graduated rods. The technique is simple, cheap, fast and effective: students can learn it in minutes; it does not disturb the site; one person can map a kilometre of stone wall in a day; the results need no costly or time-consuming analysis; and, as no artefacts are unearthed, there is no call for expensive conservation work.

Caulfield's team uncovered a Neolithic farmscape that spanned 1,000 hectares, the most extensive of its kind known in the world. They found a well-planned field system: walls running inland divided the land into strips; and cross-walls divided the strips into fields. They also found circular enclosures, or 'paddocks', hut sites and occasional tombs. The walls were not big enough to restrain deer and too big for sheep, so the most likely stock were cattle; moreover, pollen studies reveal that grasses grew in the fields and not cereals. Other studies yielded additional detail: carbon-dating puts the site firmly in the Neolithic Stone Age, 5,000 years ago, while pollen profiles reveal that previously the area was heavily wooded; and a charcoal layer, between the trees and grasses, suggests the area was cleared by burning the trees. Similar pre-bog field systems have been found elsewhere in Ireland and Europe, but Céide's are the most extensive known, and among the oldest. Small parts of the site have since been excavated more thoroughly by conventional techniques, but, for the most part, the walls remain hidden beneath the bog. The Céide Fields visitor centre has a special 'totem pole', a large, bog pine stump, tree-ring dated to 4,500 years old. ①

Sod-busting cereal farmers

The Neolithic farmers at Céide were cattle ranchers, but their Belderrig neighbours were grain growers. Stone-walled fields found under the peat at Belderrig were too small for livestock, but the conclusive evidence comes from 5,000 year-old **cultivation ridges** preserved under the bog, and from scratch marks in the soil, which were probably made by a primitive plough. In 1780, an English agrarian reformer, Arthur Young, writing of his travels in Ireland, reported that cultivation ridges had been found under bog in the west of Ireland. Had he realised the truth of

these claims, one wonders would he have grasped the timescale involved?

Cultivation ridges, or lazy beds, are usually associated with potato growing in the congested years of the early-19th century. The technique is ancient, however, and was practised in Ireland 5,000 years before the potato* arrived here. Cultivation ridges may be simple – 19th-century agrarian reformers denounced them as crude – but they are also sophisticated: turning a sod in from either side creates the small ridge; and this one efficient move simultaneously deepens the soil where the crop will be grown, and improves the drainage. The technique can be used on untilled ground, whence perhaps the notion that it was a 'lazy' option. In fact, turning a perfect ridge calls for considerable skill and spadework. Farming historian Jonathan Bell, of the Ulster Folk & Transport Museum,* says there are countless ways to make cultivation ridges: the ridges could be 1–20 metres wide, and turned by spade,* horse-drawn plough or tractor.* Their shape and form could be varied depending on terrain, soil type, date of planting, the crop to be grown, and even its place in the crop rotation cycle. The slope and aspect could be tailored to minimise soil erosion, or an asymmetrical ridge made to shelter tender young plants. There were regional variations, too – in Kerry, for example, ridges were made with a special hoe or *grafáin*, while north Mayo farmers used a unique two-pronged spade, the *gabhal gob* or gowl gob, designed

Cultivation ridges, or lazy beds, from 19th-century Co. Mayo; the Neolithic farmers at Belderrig grew their oats on similar ridges, which can be seen preserved beneath the bog.

for the local sandy soil. ② Unfortunately, this tremendous diversity of techniques was ill-suited to mechanisation or automation.

Lazy beds were ideal for growing potatoes, especially on marginal land. A potato crop helped break up the soil and brought the land into cultivation, and, in the early-19th century, much of Ireland's burgeoning population was sustained by potatoes grown in this way on boggy ground. Many remote hillsides, which are now deserted, retain the corrugated traces of these undulating earthworks, relics from the densely populated pre-Famine era.

Ireland's oldest plough marks?

There are regular scratch marks visible in the soil beneath the bog at Belderrig which were probably made by a primitive plough or ard. **Ploughing** has been an important technique since the earliest days of farming: it loosens the soil, buries the remains of the previous crop, and controls weeds. Belderrig's Stone Age farmers might have ploughed with the shoulder blade of an ox, or with a digging stick, neither of which has a cutting point, so first the grass and soil surface had to be torn with a sharper pointed implement. Then an ard was dragged through the soil first in one direction, next in a second direction at 90 degrees to the first. It is these regular scratches that can be seen at Belderrig. This type of primitive ploughing calls for stone-free and square-shaped fields.

Ploughing improvements came slowly over centuries. The coulter, a sharp and heavy metal share that could cut the sod, was a Roman innovation; it had to be pulled by an ox. The mould board, which turns the furrow cut by the share, was not introduced until the 1700s. Horse-drawn ploughs were restricted to a single share because of the weight, but tractors could pull up to five shares, and plough five furrows simultaneously. Horse-drawn and early tractor-drawn ploughs can be seen at Teagasc's National Agricultural Museum* (Co. Wexford), Bunratty Folk Park* (Co. Clare), and the Ulster Folk and Transport Museum* (Co. Down).

*A single-furrow horse-drawn plough, made by Pierce & Company, * the celebrated Wexford farm machinery firm. This type of plough was still widely used in the 1930s.*
National Agricultural Museum, Wexford

At Belderrig, the Neolithic farmers grew oats on their cultivation ridges. Elsewhere on the site, there are Neolithic stone walls, the remains of a Bronze Age house with a quern stone, which was used to crush grain, and the charcoal remains of a hearth. The later Bronze Age remains date from 3,500 years ago, when a small group – possibly miners chasing copper in the hillside nearby – lived at Belderrig, though it was then part-covered by bog. Details about the self-guiding trail around Belderrig's prehistoric farm site are available at the Céide Fields centre. A new research centre for bog studies and archaeology opened at Belderrig in 2001. ③

"Europe's most spectacular coastline"

Dizzying, vertiginous views at every turn, prominent rocky headlands, teetering sea stacks and countless sea arches and caves… no wonder English geologist J B Whittow, in his book *Geology and Scenery in Ireland* (1974), described the coast between Downpatrick Head and Benwee Head as the most spectacular in Europe. Historically, most of the headlands here were guarded by a promontory fort, and the remains of two such forts can be seen at Downpatrick Head and Dún Briste (a substantial sea stack that became isolated from Downpatrick in 1393 when a storm brought down the connecting rock bridge). During World War II, several Mayo headlands also acquired a modern version of the promontory fort: a lookout tower built as a landmark to aid returning US pilots; alongside each tower the word *Éire* and an identifying number were written large in stones on the ground, some of which survive in places.

The cliffs at Glinsk plummet 300 metres to the sea, and Whittow rates them highly. Nearby, between the mainland and the small island of Illaunamaistir, is Moista or Maistir Sound; this is a 'geo' or deep cleft that formed where soft rock was eroded, and in stormy weather waves crash through spectacularly. Illaunamaistir is noted for puffins,

Manx shearwaters and storm petrels, which can be watched from the shore. The walk around Benwee Head from Portacloy is exhilarating, with views to the Stags of Broad Haven that rise 100 metres out of the sea 2km away. Sea caves at Broad Haven Bay shelter some unusual creatures, among them a rare anemone (*Parazoanthus anguicomus*) and a soft coral (*Alcyonium glomeratus*); the mouth of the bay is guarded by a dense forest of tall kelp. Dooncarton on the bay will be the landfall for the submarine supply pipe coming from the new **Corrib gasfield**. The gasfield, 70km offshore in the Slyne Trough, was discovered in 1996; its five wells will start supplying gas in 2004, just as the Kinsale gasfield* comes to the end of its commercial life. The Corrib gas reservoir is in sandstone rocks that lie 3.5km below the seabed, and in relatively deep water at 350 metres; the supply is expected to last for 15 years. (The Kinsale reservoir, which lasted for over 20 years, lies underwater that is only 100 metres deep.)

Held together by sand

The Mullet peninsula is a fascinating, complex and much underrated landscape. It is also, contrary to appearances, an island, albeit an artificial one: thanks to a navigational channel cut through the rock at Belmullet, you must cross a bridge to reach the peninsula. This treeless, low-lying finger of land is formed from a chain of rocky islets, or skerries, linked by deposits of glacial debris, and bound together by a covering of wind-blown sand. The sand – and the Mullet – are slowly but relentlessly moving eastward, driven by Atlantic storms. This is a peninsula of two halves: the northern end is top-heavy with sea cliffs and rocky headlands formed from hard quartzite; the land slopes down to Blacksod village at the southern end which is composed of granite; in between is a narrow strip of land which, at Annagh Head, has some of the oldest rock in Ireland. The **Annagh gneiss** is a beautiful rock with glistening bands of pink, white, grey and black minerals that are folded into wonderful, tortuous shapes; cobbles of the gneiss can be found on beaches south of Annagh Head. Geologists date the rock at 1,753 million years old – a shade younger than the gneiss of Inishtrahull* in Co. Donegal, which at 1,778 million years old is Ireland's oldest rock.

There are two noted **bird-watching sites** on the Mullet: Termoncarragh, a freshwater coastal lake with reed beds, is internationally important for its breeding waders and wintering wildfowl, among them whooper swan, and Greenland white-fronted and barnacle geese; Annagh Marsh is a brackish wetland set amid coastal machair, and noted for breeding snipe, dunlin and curlew, also ducks and swans. Access to both sites is restricted, but they can be watched from the road. Blacksod Bay was home in the past to Ireland's rarest breeding bird, the red-necked phalarope; sadly the species is no longer breeding there. (Weather information sent from Blacksod lighthouse in June 1944 was reputedly decisive in timing the Allied D-Day landings* in Normandy.)

In vegetation, topography and habitat, the Mullet closely resembles Scotland's Outer Hebrides, especially North and South Uist with their broad areas of machair* grassland. Along the Mullet's western edge, the wind-blown sand forms extensive carpets of machair, which is rich in calcium, and thick with wild flowers. Some of the beaches there also have layers of coastal peat that formed thousands of years ago when sea levels were lower. Most of the farming and habitation is on the more sheltered eastern side. A bicycle is a good way to investigate the Mullet, and the peninsula rewards the explorer.

1908 – Norwegian whaling stations

In the days before modern petrochemicals and plastics, an amazing range of valuable industrial materials were obtained from whale carcasses. Whale oil was used to lubricate industrial machines, to light oil lamps, and to make crayons, explosives, margarine, linoleum, soap, and even lipstick and ice cream; the horny baleen, or whalebone, was used in springs, bristles, corsets and as a fabric stiffener; spermaceti, a waxy substance taken from the head of sperm whales, was used in making candles; whale offal was often fed to pigs, and the dried meat and leftovers made into a manure; ambergris, a waxy substance found in whale intestines and once thought to be hardened faeces, was a valuable ingredient in perfumes... even the teeth went into bonemeal.

Whales are common off Ireland's west coast and, in the 1760s, a Donegal man **Thomas Nesbitt** invented the first effective harpoon gun. Mounted on a swivel gun, it could be launched with greater power and accuracy than the hand-thrown variety. Nesbitt and his brother operated a small whaling fleet out of Inver, killing several whales a year. Most Irish whaling was passive, however, taking advantage of beached whales, rather than venturing into dangerous waters. But in the early 1900s, two Norwegian companies ran a sizeable whaling business from Co. Mayo: at Rusheen, on Inishkea South (1908–14); and at Ardelly on the Mullet Peninsula, which operated from 1909 until World War I, and again in the early 1920s, employing 100 men, until an arson attack destroyed the station in 1923. By then, the North Atlantic whale catch was declining, whalers were switching to the Antarctic Ocean, and whaling in Ireland came to an end.

The Norwegians, prohibited from whaling in Norway at certain times of the year, had intended to open a station on the Donegal islands of Aranmore and Gola. Local objections forced them to look elsewhere and, eventually, they were granted permission to open on Blacksod Bay. They were supported by the Fisheries Inspectorate and the Local Government Board, although local landlords objected to the smell, and the damage that might be done to fishing grounds.

Norwegian sailors crewed the whaling boats, but the onshore processing work was done by locals. Whale carcasses were moored at the pier, before being winched up a slipway to the wooden factory sheds; the blubber was removed and boiled, and the resulting oil stored; the meat was dried, for making into fertiliser; the bones were ground at a crushing mill. In all, the Mayo factories processed 899 whales: including 592 fin whales, 124 blue whales, 97 sei whales, 63 sperm whales and five humpback whales. Irish waters are now a whale and dolphin sanctuary; for whale watching, see Kilrush, Co. Clare. Several basking shark* fisheries also produced oil, including one at Achill that closed only in 1975.

Barnacle goose – fish or fowl?

If a caterpillar can turn into a butterfly, and a tadpole into a frog, might not a barnacle become a goose? The goose barnacle is a small crustacean that attaches to driftwood, while the barnacle goose is a black-and-grey migratory bird that breeds in Greenland and Canada, but spends its winters grazing in Ireland and Britain. In the Middle Ages both creatures were at the centre of an unusual theological and zoological debate – for it was widely believed that the goose barnacle was an immature barnacle goose. At the time, the goose's breeding grounds in the Arctic were unknown, and naturalists believed that when migrating birds disappeared, they were hiding in river muds. The barnacle's body is shaped and coloured like a miniature goose, and a cluster of feathery tentacles protruding from its shell resemble the goose's tail; the shell, it was thought, protected the immature goose. This being so, the goose was clearly fish and not fowl, and could be eaten on Fridays by Catholics without breaking their fast. Those who subscribed to this theory included the Welsh, or Anglo-Norman, historian Giraldus Cambrensis,* who refers to the changeling in his *Topographica Hibernica* (1188).

The North and South Inishkea islands are the barnacle goose's main Irish haunt – indeed, the name Inishkea may derive from *Inis Gé* (goose island). The Inishkeas also have a large hauling-out ground, where hundreds of grey seals congregate. The islands, last inhabited in 1937, are a Mullet in miniature, being a series of low, rocky skerries connected by sand. Legend has it that the north and south islands were once connected both to each other and to the Mullet peninsula. Intriguingly, there was a substantial **purple dye factory** on Inishkea North around AD 500–700 which produced pigment for the emperors of Europe and Byzantium. The pigment, Tyrian purple or purpurin, was extracted from a gland in the dog whelk. It was costly to make, and took countless numbers of whelks to yield enough dye for one garment; consequently, before the arrival of modern synthetic dyes, only wealthy people could afford purple clothes. Despite being on the edge of Europe, Inishkea's dye industry was well known and the islands were included on early European sea charts, though conclusive evidence was discovered only in 1950 by a French archaeologist, Françoise Henry, who found substantial mounds of crushed whelk shells at Baily Mór.[1] There are boat trips in summer to Inishkea North from Belmullet.

MULRANNY

Bellacragher's rare heather

Bellacragher Bay cuts deep into the land – 1km further and it will break through to Mulranny and make an island of the Corraun peninsula. As it is, Bellacragher Bay is a long, narrow defile, protected to the east and west by the hills of Corraun and Claggan. It is a sheltered spot and one of the few places where you can find the rare **Irish heath** *(Erica erigena)* sometimes called Mediterranean heath. At Bellacragher the plants grow to great height, sometimes reaching 3 metres and, in the past, the long branches were cut and used as brushes (not now, though, as the heather is a protected species). The heath was first identified in Ireland in 1830 near Roundstone by a local naturalist William M'Calla,* and subsequently found at several places in Connemara and Mayo. Pollen grains from prehistoric sediments reveal that it grew in the thick forests that covered Ireland after the last Ice Age. The plant somehow survived the subsequent climatic and environmental changes, and it is still here, now growing in a warmer environment and on open blanket bog, albeit confined to a few places.

Apart from the west of Ireland, *Erica erigena* is also found in northwest Spain and Portugal but, strangely, not in Britain. Naturalists call this unusual pattern a Lusitanian* distribution (other Irish Lusitanians include the Kerry slug* and the strawberry tree*). Connemara and Mayo boast no fewer than four Lusitanian heathers, and botanists are still struggling to understand how they got there – birds do not carry heather seeds, and if the plants had crossed by land bridges* we would expect to find them in Britain and southeast Ireland. **Mackay's heath** *(Erica mackaiana)* is a rarity that grows only at a few places in counties Galway, Mayo and Donegal, and in Spain; it was first discovered by William M'Calla near Roundstone. **St Dabeoc's heath** *(E. daboecia)* is found only in the west of Ireland – it is common in Connemara – and in Spain, including the Pyrenees, where it can survive the winter buried under snow. **Dorset heath** *(E. ciliaris)* grows in Ireland at only one site near Clifden; it also grows on the continent but, unlike the other three, it does grow in Britain. Are these four heathers glacial relics that tenaciously survived from the end of the Ice Age? Perhaps they hitched a ride with Spanish wine traders? Or did pilgrims bring them from Santiago de Compostella in northern Spain? Further pollen studies, coupled perhaps with genetic

St Dabeoc's heath, one of the four rare heathers found in Mayo and Connemara.
© An Post

fingerprinting, may one day reveal the answer. Mulranny celebrates a Heather Festival every July.

Hunting the sunfish

It is as long as a bus, weighs 7 tonnes, and has a dorsal fin that is as tall as a man. It is *Cetorhinus maximus* or the **basking shark** – at 11 metres long, the biggest fish in Europe, and the second-biggest in the world (after the 12-metre long tropical whale shark). Basking sharks were once common off the Irish coast, where they were hunted for centuries for their valuable oil, and a shark fishery operated out of Keem Bay in Achill until 1975. In Irish, the basking shark has many names: *ainmhíde na seolta* (creature of the sails), on account of its large fin; *liamhán mór* (leviathan), on account of its size; and *liamhán gréine* (great sunfish), for its habit of lolling about on the surface.

The basking shark's enormous size can strike terror into observers, yet this is a peaceful giant that lives on tiny copepods (microscopic marine animals). It spends much of its time cruising just below the water surface where the copepods swarm, with its huge mouth agape; horny rakers on its gills strain the copepods from the water, and it can filter 1,500 tonnes of water every hour.

Basking sharks were once so common off the west coast that the sea from Slyne Head to Achill was known as the Sunfish Bank. The sharks were hunted for centuries, yet little is known about them: some zoologists think they hibernate in deep water, when copepod numbers decline in winter; others think they migrate, chasing seasonal copepod swarms. The sharks generally appear off Ireland around April, when the sea temperature warms to at least 12°C, and stay until November. In the 1700s and 1800s, these great creatures were hunted for their liver oil: the liver accounts for 15 per cent of a shark's weight, and a 400kg liver would yield a dozen barrels of oil (nearly 250kg). The sharks were harpooned, the liver removed at sea, and the carcass dumped; back on shore the liver was boiled to extract the oil. In the 18th century, the oil was prized as a lamp oil, as it was the least smoky oil available, and the streets of Dublin were lit with shark oil. The oil was also used to treat burns and, in the 19th century, to lubricate industrial machinery.

Shark hunting off the west of Ireland declined in the 1870s as shark numbers dropped. It was revived in Achill in 1947, partly because of an oil shortage after World War II, and because the sharks were thought to damage salmon nets. At its peak in the 1950s, the Achill fishery caught nearly 2,000 sharks a year. This time nothing was wasted – the fins were sold to Hong Kong restaurants and the carcasses were used to make fishmeal. The catch gradually declined and in its last year (1975), only 38 sharks were caught. Basking sharks are no longer as common as in the past, whether because of overfishing or because of changing sea currents and temperatures, is not known. Sightings of basking sharks should be reported to the Basking Shark Survey. ④

Bleak but beautiful – the Barony of Erris

Thousands of years ago, when most of Ireland was still densely wooded, there was probably a rainforest along the west coast – a green and mossy primeval forest, dripping wet and watered by the heavy rainfall coming off the Atlantic Ocean. Today, the high rainfall continues but the trees are gone, so instead of a west of Ireland rainforest we get **blanket bog**. And the Barony of Erris has blanket bog in abundance. This northwestern corner of Mayo is a vast, uninhabited wilderness with mile after mile of bleak and apparently barren bog. Robert Lloyd Praeger,* in *The Way That I Went* (1937), declared this to be the "wildest loneliest stretch of country in all Ireland", while Lewis's *Topographical Dictionary* (1837) warned that a traveller would need eight guides just to haul his horse across the swamp. Yet for naturalists this is a fragile heaven, a home to many rare plants, and a sanctuary for several endangered bird species. Despite decades of commercial turf cutting, much of it to feed a peat-burning power station at Bellacorick, large tracts of relatively intact blanket bog survive, and 10,000 hectares of this are now protected in **Ballycroy [Mayo] National Park**. Centred on the Nephin Beg Mountains and Owenduff Bog, this is Ireland's sixth, newest and largest national park. Ireland has already lost over 80 per cent of its blanket bog, yet what survives in north Mayo is among the largest intact areas of blanket bog in western Europe, so the new national park is internationally important.

The great raised bogs* of the midlands developed from the lakes that filled Ireland's central plain at the end of the last Ice Age. Vegetation gradually encroached on the lakes, producing first freshwater fens and, eventually, great mossy domes of raised bog. The blanket bog of the west is very different, however. It began developing around 5,000 years ago (much more recently than the raised bog), when the climate became cooler and wetter, and after Neolithic farmers had cleared the forests, and it developed on poorly drained soil in areas with high rainfall. The rain washes minerals in the soil down to form a watertight layer, or iron pan, beneath the surface; the earth above becomes waterlogged as a result; dead vegetation no longer decays; and peat begins to accumulate, often burying the stumps of the pine trees that previously grew in the area. Blanket bog that develops in lowland areas (at most 200 metres above sea level) is called Atlantic or oceanic blanket bog; above that level it is mountain blanket bog. The depth of peat in blanket bog can vary tremendously, from a thin skin to treacherous tracts that are many metres deep. Raised bog is dominated by sphagnum moss, but blanket bog has a grassy appearance – grasses and sedges predominate, especially purple moorgrass, black bog rush and bog cotton. The latter is a sedge that produces an underground food storage organ, which is much appreciated by the geese that graze the bog in winter.

Much of Mayo's blanket bog is threatened by turf cutting, afforestation and grazing, but two relatively intact areas around Bellacorick were set aside as **nature reserves** in 1986. One, Knockmoyle Sheskin, is an extensive lowland blanket bog that is studded with numerous small pools, many of them rich in minerals, which support an interesting combination of plants. One pool in particular, discovered in 1957 in the Oweniny valley, has many rare mosses, ferns and flowering plants, notably the marsh saxifrage (*Saxifraga hirculus*), growing alongside wild angelica, marsh arrowgrass and other bog plants. The second nature reserve, Owenboy, is a lowland blanket bog, but with several small raised-bog-like domes, and mineral-rich pools, or flushes, where rare mosses and ferns abound. This bog is an important sanctuary for the endangered red grouse and for flocks of Greenland white-fronted geese, which spend the winter grazing there. Owenduff bog at the foot of Nephin Beg, and now in Mayo National Park, is one of the largest river catchment areas in Ireland that is still relatively free of coniferous plantations; it too is noted for rare plants, among them ivy-leaved bell flower, marsh clubmoss and shining sickle-moss.

For most of the 20th century, the blanket bogs of Bangor–Erris were seen not as something to conserve, but as harsh places useful only for supplying fuel, growing conifers and grazing sheep. In the late 1950s, the Agricultural Research Institute (now Teagasc) set up a **peatland research centre** at Glenamoy to investigate the commercial potential of the bogs and various reclamation projects (see: **Captain Cook's lily**). In the early 1960s, a peat-burning power station opened in the middle of the bog at Bellacorick, supplied with machine-cut peat by Bord na Móna; peat is conveyed to the power station via an extensive narrow-gauge bog railway. In 1992, the power station site became Ireland's first windfarm, now with 21 turbines. Overgrazing by sheep has severely eroded the Mayo bog especially around Nephin Beg, leaving a vast expanse of naked, black peat. Silt and acidic water running off the exposed peat can pollute local rivers and streams,

Captain Cook's lily

Windbreaks and hedges around Glenamoy are often planted with phormium, a tall, distinctive lily which Captain Cook brought from **New Zealand** to Kew Gardens in 1775. Phormium leaves contain a flax-like fibre – hence the plant's other name, New Zealand flax. Unlike conventional linen* flax, which must be soaked for weeks in a smelly retting pond to loosen the plant fibres, phormium fibres can be removed easily and mechanically. In the 1960s, scientists from the Agricultural Research Institute (now Teagasc) tried growing phormium commercially on the Mayo bog, using plants that were reputedly descended from Captain Cook's original specimens. In the 1970s, synthetic fibres replaced natural ones, however, and commercial interest in an Irish fibre industry waned. Phormium remains popular in Mayo and other coastal areas, where it is often grown as a windbreak, as it can tolerate wind, frost and salty sea-spray. The plants can reach 5 metres, have long, evergreen sword-shaped leaves, and a tall spike of distinctive bronze flowers in summer.

and clog the spawning grounds of wild trout and salmon, and controlling the overgrazing and erosion is just one of challenges facing the new national park. Mayo National Park's visitor centre will be at Ballycroy. Bellacorick village has a musical bridge that resonates when a stone is run along it. Two walking trails (the Western Way and Bangor Trail) traverse the Erris bogs.

FURNACE

Salmon and science

Lough Furnace, together with neighbouring Lough Feeagh and several smaller lakes in the valley, make up the Burrishoole fishery, known to trout and salmon anglers far and wide. They are also one of Europe's best-studied river systems, thanks to a **salmon research centre** that was established at Furnace in 1955 by Guinness's, as part of the brewery's fish-farming subsidiary, to study water quality and fish stocks, and to train gillies and manage the fishery. Over the intervening decades, the centre has contributed greatly to our knowledge of salmon stocks in Ireland and, in 2001, it became part of the State's Marine Institute, with responsibility for the commercial fishery passing to the local fishery board. Scientists from the research centre

continue to monitor the whole Furnace catchment area – 50km of rivers and streams, two large lakes and five small ones – and the surrounding countryside, taking account of local geology, and the effects of forestry plantations and overgrazing on the water and the fish stocks. Sophisticated instruments monitor 23 aspects of weather and environment at Lough Feeagh, and the data are sent to research teams around the world as part of an international network. One important project at Furnace is an investigation into the decline in wild salmon numbers, with scientists tagging and releasing over 300,000 young salmon each year in several rivers.

Ireland has 135 main salmon rivers, each with its own genetically unique population of wild fish. Salmon numbers are dropping dramatically, however, as fewer fish return from the sea to breed. Previously, 20 per cent of wild salmon might return from the sea, now that figure is closer to 3 per cent, and some local salmon stocks are dangerously close to extinction. The problem could be caused by overfishing at sea, pollution, disease, infection, and/or changing sea currents and temperatures, all of which can affect fish survival. New fishing restrictions have been introduced to limit catches in a bid to control the problem. Sea trout are also affected, though in Mayo this species has been rescued thanks to the Furnace scientists: by 1990, only 30 sea trout were breeding at Lough Furnace; a few were captured for a breeding programme to restock the local waterways; now several thousand sea trout are there, all descended from the original population.

Furnace–Burrishoole glen is a broad, flat-bottomed glacial valley that runs approximately north–south. The valley's southern end is choked with gravel that was deposited by glaciers during the last Ice Age; today, the gravel impounds the valley's lakes, including Lough Furnace, which is partially tidal. In the 1500s, the lough's tidal flow was harnessed at Burrishoole to drive a tidal mill owned by Richard Burke, a Mayo chieftain and second husband of pirate queen Grace O'Malley (Granuaile). Furnace, like many other Irish places of the same name, takes its name from a small ironworks that operated there in the 17th century.

1892 – a nun, a mason, a mill and providence

It was an early, and ambitious, rural development programme. To alleviate poverty, unemployment and emigration, especially among young, single women and, at a time when the only local jobs were in seasonal farm work, the Sisters of Charity opened **Providence Woollen Mill** at Foxford in 1892. The nuns also organised primary education and vocational training, housekeeping and hygiene classes, as well as housing, a handball court and a brass band. The prime mover behind the project was Mother Agnes Morrogh Bernard (Mother Arsenius), and funding came from the Congested District Board. In setting up the woollen mill, however, technical advice was provided free, in what was considered an unusual move at the time, by John Charles Smith, an Ulster mill owner, Protestant and Freemason.

Foxford mill was small by the standards of the day and never employed more than 200. The building was bright and well ventilated, with windows and skylights, and warmed in winter by the heat from the machinery. At first the Moy river was harnessed to drive the looms, but steam power was later used. The nuns employed mostly single women – a woman was expected to leave when she married – and teenagers, and a few men for the heavier work in the warehouse. It was a six-day week, with the women working up to 10 hours a day, but the children never more than eight hours. The Sisters of Charity ran the woollen mill until it hit financial trouble in the 1980s, and was bought by a local business group. The working mill is now open to the public, and Foxford's brass band survives, proud monuments to Mother Arsenius's entrepreneurial spirit. (5)

1879 – investigating an apparition

An apparition appeared on the gable end of Knock parish church on August 21st, 1879. It started at about 8pm, continued for two hours, and was visible in daylight, dusk and at nightfall, and despite a steady downpour. Witnesses said they saw three figures, interpreted as a cloaked figure of the Virgin Mary, with St Joseph and St John the evangelist; also a lamb on an altar, a cross and a host of hovering angels. The apparition appeared in a bright light, yet cast no shadows, and the ground beneath was said to be dry despite the rain. In October, an ecclesiastical commission took statements from witnesses, while several scientists investigated the possibility that the apparition had been projected from a magic lantern, then the only image projection technology. Light sources for magic lanterns were usually hot, such as a burning block of lime (limelight*), and the scientists concluded that a slide would have melted during the two hours the apparition lasted; moreover, the light from a magic lantern would have thrown shadows. Whatever the source of the apparition, the scientists concluded, it was not a magic lantern. A second ecclesiastical commission re-interviewed the surviving witnesses in 1936 and again concluded they were reliable, and Knock was established as a site of Marian pilgrimage. (6) The lantern used in the scientific investigation is now in the science museum at Maynooth college.*

1877 – the world's first guided missile

The world's first guided missile was patented in 1877 by a talented engineer from Castlebar, **Louis Brennan** (1852–1932), who also designed an unusual monorail train, a two-wheeled car and an early helicopter. But it was his missile system – a directable torpedo controlled by guide wires and used at British coastal defence forts until the early 1900s – that made Brennan wealthy. In the days before aeroplanes, enemy attacks generally came by sea. Torpedoes were much in vogue then, as weapons of both offence and defence – John Philip Holland's submarine* was essentially an offensive torpedo delivery system, while Louis Brennan's guided torpedo was a coastal defence system.

When Brennan was nine, his family emigrated to Australia, where he later trained as a watchmaker. It was while watching a belt-driven machine at a workshop, that Brennan reputedly conceived of his torpedo control system. He produced his first design aged 25 and, with £110,000 from the British navy, spent 10 years refining the idea. In 1887, the Admiralty approved the system and opened a government factory making 'Brennans' at Gillingham in Kent. The Brennan was a sizeable missile, weighing 3 tonnes and carrying nearly 100kg of explosives. The guidance system consisted of two reels, each holding 3km of wire, and each connected to one of the torpedo's two propellers. When a missile was launched (from a guide rail and runway on shore), a steam-driven winding machine extracted the wires as the torpedo pulled away. An observer (generally a member of the Royal Engineers) followed the missile's progress by watching a small mast that protruded above the surface, and steered it by varying the rate at which the left and right guide wires were extracted. The first weapon that could be guided all the way to its target, it had a range of 3km and could travel at 40kmph, cruising 3 metres below the surface. Launch stations were built at strategic points around the British coast, and in Hong Kong and Malta; only one was built in Ireland, at Fort Camden near Crosshaven to defend Cork Harbour (where the launch rails can still be seen). For 20 years, Brennan's system was the ultimate in coastal defence technology, though, so far as is known, it was never fired in anger – until the early 1900s when the development of new guns and other military technologies rendered it obsolete.

Louis Brennan went on to design several novel transport systems, all using gyroscopic controls. In 1909, he demonstrated a working model of a gyroscopic monorail train: a single rail was laid on the ground, and gyroscopes kept the carriage balanced on it, even when all passengers sat on one side. (Contrast this with the Lartigue* monorail in Co. Kerry, where weight had to be carefully balanced on either side of an elevated rail.) Brennan believed his monorail would be quick and easy to lay, and that it would appeal to the military and the British colonies. Despite backing from Winston Churchill, however, it was never developed. During World War I, Brennan worked for the British government as an inventions consultant, and afterwards for the Royal Aircraft Establishment at Farnborough, where he developed a gyroscopically controlled helicopter. He achieved the first ever unmanned helicopter flight, but when his prototype crashed on a trial run in 1925, the government opted for a competing design. Louis Brennan was working on a two-wheeled gyro-motor car when, ironically, he was knocked down by a car in Switzerland and died shortly afterwards from his injuries. Louis Brennan had been a founder member of the National Academy of Ireland,* when it was established in 1922 as a rival to the Royal Irish Academy. London's Science Museum has a small working model of Brennan's monorail train; the Royal Engineers Museum at Kent has the only surviving Brennan torpedo.

KILLEEN

A rare clapper bridge

The delightful footbridge at the ford of Bunlahinch near Roonagh Lough could be Ireland's oldest bridge – there are no surviving records to date it, but the structure may well be mediaeval. It certainly has more arches than any other Irish bridge, with 37 in all. It is, however, no ordinary road or vehicle bridge, but rather an elevated footpath, about 15 metres long, and standing a metre above a stream that flows into Roonagh Lough. It is made from over 30 large limestone slabs supported on rough piers of rubble and stone; a low parapet stone wall on the downstream side has regular small openings that enable water to flow safely

A small section of the unusual clapper bridge near Roonagh Lough, a rare surviving Irish example of this ancient type of bridge.
© Mayo Naturally

through the footbridge during floods. A minor road alongside the bridge also fords the stream. This type of structure is the simplest and oldest bridge design, and is called a clapper bridge, from the Latin *claperius*, meaning 'pile of stones'. Surviving clapper bridges are rare in Ireland, but Ron Cox in *Civil Engineering Heritage of Ireland* records both this one and another small one at Ballybeg near Buttevant in Co. Cork, associated with an abbey founded there in 1229.

Gold, drumlins and old oaks

Beautiful Clew Bay with its hundred islands, and majestic Croagh Patrick with its near-perfect pyramidal peak, together form one of Ireland's finest landscapes. And, as if all that grandeur was not enough, the area also boasts two historic oak woods. **Croagh Patrick**, probably Ireland's most distinctive mountain, and known locally as The Reek, has been venerated as a holy mountain for 5,000 years, the traditional pilgrimage day being the last Sunday in July. It is a quartzite peak, as are neighbouring Nephin, Co. Donegal's Errigal* and Co. Wicklow's Sugar Loaf,* all of which have a conical shape, thanks to the nature of quartzite. Quartzite is an old rock that was originally sandstone, but which has been much altered and hardened by geological processes over time. Although it resists erosion, it also shatters easily to form shards of scree, which eventually release the original sand grains from which the rock formed. Consequently, quartzite mountains are typically conical peaks with flanks covered in scree slopes. The rock that forms Croagh Patrick was pushed up 400 million years ago, during the violent collision between the ancient continents of Laurentia and Avalonia,* when the Iapetus Ocean closed.

The sandstone of Croagh Patrick, Nephin and Errigal formed from coastal sands at the edge of Laurentia, while Wicklow's Sugar Loaf formed in Avalonia on the other side of the ancient ocean. There is gold in Croagh Patrick and minute flecks can be seen in quartz veins in the rock – those with patience can try panning in a local stream. Explorations in the early 1990s confirmed that the gold deposit was commercial, but plans for a mine on this holy mountain proved controversial and, eventually, the local authority banned all mining on Croagh Patrick.

Clew Bay is choked with nearly a hundred islands that break the surface like a large pod of humpback whales. These are drowned drumlins,* the tail end of a swarm that starts further inland and, indeed, is part of a drumlin belt that runs across the country to Strangford Lough.* Strangford Lough is similarly crowded with drowned drumlins, but at Clew Bay the drumlins are exposed to the Atlantic's full force, which easily erodes the sand and gravel mounds. There were once many more drumlins further out in the bay, but these have long since disappeared, and only the innermost drumlins retain anything like their original shape. In places, the resulting debris has been reshaped into shingle spits and tombolos, linking islands to islands, and islands to mainland. As the drumlins were originally laid on dry land, the sea levels 20,000 years ago were clearly much lower than today. Even 5,000 years ago much of Clew Bay was still dry – witness the Neolithic stone alignments along the modern day shore at Killadangan, 2km east of Murrisk, which were presumably well inland when they were erected. There is a wide range of habitats around Clew Bay's many islands and interlocking bays, including salt meadows, sheltered lagoons, stony banks and shifting dunes, which make this an important area for flora and fauna. ⑦

The area's two **historic oak woods** are Brackloon Wood, which shelters on the eastern side of Croagh Patrick, and Old Head Wood, which nestles on the eastern side of a hill by the seashore at Louisburgh. Brackloon Wood is the relic of an ancient oak wood, where the trees have managed to regenerate for centuries. Alongside the oaks, many of which are probably 200 years old, grow birch, ash, willow, hazel, elm and holly. The trees are extravagantly festooned with mosses and lichens, and there is a rich under-storey of ferns. At least one rare plant grows there, the narrow-leaved helleborine (*Cephalanthera longifolia*). The wood was once extensive, but, in the mid-1600s, it was drastically reduced by tree felling for timber and charcoal making (traces of charcoal-burning platforms have been found there). The State bought Brackloon in the 1940s, and under-planted conifers, but these have since been removed as part of a restoration programme. Since 1991, Brackloon has been intensively studied by UCD's forest ecosystem research group, led by Prof Ted Farrell, and with EU funding. On a first count, the team identified 775 fungi and plant species in the wood, including 30 slime moulds. This tremendous diversity is evidence of the wood's great age, and there are moves afoot to register the wood as a national nature reserve. Old Head Wood, which is a national nature reserve, is also a semi-natural oak wood, with birch, rowan and willow, and some introduced beech and sycamore.

CLARE ISLAND

When scientists "ransacked" an island

Hordes of scientists from around Europe descended on Clare Island in 1909 and, over the next three years, systematically "ransacked" the place. At least, that was how their leader, Robert Lloyd Praeger (see: **A great Irish naturalist**) described their activities. In fact, the Clare Island project was highly organised, and nothing less than a comprehensive inventory of all the plants and animals living on the island, in the sea around and on the neighbouring mainland. When it was over, the scientists had identified a grand total of 8,488 species, including 120 that were new to science. In all, they found 3,219 plant species, of which 585 had not been known in Ireland before, and 11 were new to science. And of the 5,269 animal species found, 1,253 were new to Ireland, and an amazing 109 were new to science. Most of the newly discovered species were small animals, such as mites and insects, and unobtrusive plants, such as mosses and tiny seaweeds. Clare Island may look barren, but this ground-breaking study revealed that there is tremendous diversity on a small Atlantic island.

Islands are interesting places, and island studies were much in vogue in the late-19th century, thanks partly to Darwin's pioneering work on the Galapagos. Scientists were keen to know which plants and animals were found on islands, how they had reached their island home and whether they differed from the species found on the nearest mainland. In the 1890s, several Irish and British biologists surveyed parts of Valentia Island* and, in 1905, Robert Lloyd Praeger compiled a detailed inventory of Lambay,* a small island in Dublin Bay. By 1908, the main Irish and

A great Irish naturalist

Robert Lloyd Praeger (1865–1953), who organised the Clare Island survey, was one of the great Irish naturalists. Born in Holywood, Co. Down, to a prosperous Presbyterian family, his paternal grandfather was of Dutch (and ultimately Czech) extraction – Praeger means 'native of Prague' – and his maternal grandfather, Robert Patterson, founded the Belfast Naturalists' Field Club, which Praeger joined as a boy. Despite an early interest in natural history, Praeger trained as an engineer, but soon swopped engineering for a job at the National Library* in Dublin, which he hoped would give him more scope to ramble and indulge his love of botany. Energetic and methodical, Praeger delighted in surveying the distributions of plant species. He divided Ireland into 40 botanical divisions, tramped the countryside identifying species and, in 1901, published a classic work *Irish Topographical Botany*. This methodical approach culminated in the Clare Island survey a decade later, and in two pioneering botanical studies in the 1920s and 1930s, when he untangled two complex botanical groups, the houseleeks (**Sempervivum**) and stone crops (**Sedum**). During World War I, his fieldwork was strictly curtailed when he and his German wife Hedwig were nearly interned. Praeger valued the contribution of amateurs and was good at motivating others. He wrote several popular books about Ireland's natural history, among them the much-loved *The Way That I Went,* which is seldom out of print. He was a founding president of the heritage organisation, An Taisce, and a president of the Royal Irish Academy.

These microscopic shells, from tiny single-celled marine animals called foraminifera, were among the 8,488 species found by the Clare Island survey.
Proc. RIA

British scientific institutions had agreed that it would be worth studying a larger Atlantic island. Several locations were considered, but Clare Island was chosen because of its substantial size, and because, being 5km offshore, it was at a reasonable remove from the mainland influence, yet not so far as to be difficult to reach; also, there was adequate accommodation for the visiting scientists.

Clare Island stands at the entrance to Clew Bay. In area, it measures about 16sq km, and at its highest point, Croaghmore, it rises to 465 metres. It is mostly treeless, windswept and cliff-bound, especially on its western Atlantic side. But its geology and topography are varied, and it has a great diversity of habitats – the western Atlantic cliffs, for example, are rich in snails, alpine plants and nesting seabirds. It has been inhabited for over 5,000 years and has a rich collection of archaeological remains. ⑧

For the island survey, Praeger was appointed the main organiser. Funding came from institutes such as the Royal Irish Academy,* Royal Dublin Society,* British Association for the Advancement of Science* and the Royal Society in London. Some of the fieldwork was done by professional scientists, but much of it was generously undertaken by volunteers and amateurs, many of them members of the Dublin and Belfast Naturalist Field Clubs. In all, well over 100 people took part, from, for example, Ireland, Britain,

Denmark, Germany and Switzerland. They came a dozen at a time – accommodation on the island being limited – and throughout the three years there was nearly always some scientist visiting. To complete the picture, in addition to the island, they surveyed the sea offshore to the 50 fathom line (100 metres), and the neighbouring coastline and islands, including parts of Achill.

The results of this mammoth undertaking filled 67 scientific papers that were published in 1915 in the *Proceedings of the RIA*. There were no earth-shattering discoveries – Clare Island is not the Galapagos – nevertheless the survey uncovered many new plants and animals, and increased the number of species known in Ireland by over 20 per cent. The species found on Clare Island were so similar to those found on the mainland, that the scientists concluded the island was once connected to the mainland by a land bridge – something which is at least possible, given how sea levels have fluctuated over the past 100,000 years.

In the 1990s, the RIA began a second Clare Island survey, using state-of-the-art equipment. In all, 24 project groups are taking part, with geologists from Glasgow University, botanists from the Natural History Museum in London, and lichen experts from Canada joining various Irish teams to study the island's geology, climate, botany, zoology, archaeology and history. Interim results are already accumulating and, by comparing their findings with the original survey, the scientists hope to investigate the changes have taken place in the intervening 80 years.

A pellucid green lake

Lough Carra, famous among anglers, is also noted for its unusual pale, pellucid green hue. The colour is due to the lake's chalky bottom, for this is Ireland's largest marl lake. Lough Carra sits in a basin of unusually porous limestone, and nearly all its water comes from springs bubbling up on the lake floor; the deepest spring, called Black Hole, is 18 metres deep. The spring water carries dissolved limestone, which settles on the bottom to form a chalky marl clay. This soapy chalk forms a limey crust that can be seen along the shore when the water level drops in summer. Few aquatic plants can grow on the chalk and, consequently, the water is clear and the lake bottom generally visible. The lake's limestone habitats, which have much in common with the Burren* of Co. Clare, are internationally important, and its flora and fauna have been studied over the past 200 years. Brilliant-blue Spring gentians flower there – the most northerly place in Ireland where you will see them – as well as 19 species of orchid. Lough Carra is also noted for its birdlife: an important colony of several hundred mallard, which have been studied continuously since 1968; also whooper and Bewick swans, snipe and woodcock, teal and goldeneye. Lough Carra is now connected to Lough Mask by a canal. There are nature and archaeological trails beside Lough Carra at the Doon Peninsula. ⑨

CONG

Lough Mask's underground connections

The limestone around Lough Mask is amazing: riddled with underground passages like a Swiss cheese, and as leaky as a sieve. If water there is not disappearing down a sinkhole, it is bubbling up from a spring. All the water that runs from Lough Mask into Lough Corrib, for instance, flows underground. Apocryphal tales are told of people digging foundations for buildings, only to find that they have broken in to a cave through which a subterranean river is rushing. The most spectacular vanishing act is that of the Aille river, which disappears down a large sink known as the **Aille Caves** close to the Croagh Patrick pilgrim path (10km southeast of Westport). Cave divers have mapped the passages and they contain deep water, which emerges 3km to the east at a resurgence, or rising, near Bellaburke. There are also numerous turloughs,* or seasonal lakes, in the Ballinrobe district, which fill and empty accordingly as the water table fluctuates. The unusual Lough Carra (see: **A**

pellucid green lake) is fed almost entirely by underwater springs, and its level can also drop significantly in summer.

The most remarkable part of this limestone landscape, however, is the isthmus between Lough Mask and Lough Corrib. The limestone at the southeast corner of Lough Mask is incredibly fissured and porous, and the water just seeps through it. East of the entrance to the notorious Cong Canal, there is a chaotic concentration of swallow holes and collapsed caves, and it is hard to tell where the lake ends and the shore begins. The underground rivers connecting Lough Mask and Lough Corrib are visible at a couple of places, notably the **Pigeon Hole**, 1km north of Cong, which provides a rare glimpse into this underground world. The subterranean rivers finally well up at Cong in a number of resurgences known locally as The Risings.

The **Cong Canal** has long been the butt of jokes. Dug through the notoriously porous limestone, the canal holds water like a sieve and, in summer, much of the channel is often dry. The jibes are unwarranted, however: the original intention was to seal the channel, but funds ran out before the project could be completed. The semi-arid canal was built in the late 1840s as a famine-relief project, to provide a navigational link between Lough Mask and Lough Corrib, and to relieve winter flooding in Lough Mask. Scottish engineer Alexander Nimmo* supervised the work, and the labourers were paid 4d a day. The canal could have been routed through harder, watertight rock a few kilometres to the east, but the limestone was easier to excavate. Even still, the 6-km channel took five years to dig. To bridge the 10-metre difference in height between the two lakes, three substantial canal locks were built in Cong village. When government funding ran out, the project was abandoned, and the lower part of the canal and one lock were sold to Lord Ardilaun of nearby Ashford Castle (the lock is now a boathouse). The canal water was diverted just north of Cong, so that the stretch of canal near the town's ball alley is permanently dry, giving visitors a rare opportunity to inspect the construction of a canal lock from the inside.

COUNTY GALWAY

Majestic Connemara, scenic Lough Corrib, the limestone wilderness of Aran, the dark stillness of Killary fjord – just some of the fine landscapes in this county which is blessed with riches. Remote Derrigimlagh bog was the unlikely setting for two important world firsts: the first transatlantic radio station and the first non-stop transatlantic flight. Noted Galwegians include Humanity Dick Martin, phlogistic Richard Kirwan and rebellious William MacNevin. Scientists who worked at University College Galway in the past included the man who named the Neanderthals and the first person to propose a black hole. Galway derives from *Gaillimhe* ('stoney river') a reference to the fast-flowing Corrib at Galway city.

Ireland's finest fjord

There are no fjords in the tropics, for the glacial grip of the ice ages never reached that far south. Norway's coastline, on the other hand, is all fjords. And in between lies Ireland, which, according to the old schoolbooks, has just one fjord, Killary Harbour. Strictly speaking, however, Ireland has several true fjords, but Killary is the most impressive. The classic fjord is a long, narrow sea lough, shallow at the mouth and deep at the head, and flanked by steep-sided mountains. It forms where a valley was excavated and deepened by glaciers, and later flooded when sea levels rose. Typically, at the head of the glen, the glacier was constricted by the valley walls and bit deep into the valley floor; downstream, however, as the valley broadened and the pressure eased, the glacier carved a shallower path. The result is a lough that has a deep head, and a shallow entrance that has a rocky sill and, usually, a mound or moraine of glacial debris.

Several Irish sea loughs meet these criteria, among them Carlingford Lough* and Lough Swilly, but Killary Harbour is the finest, with Mweelrea towering over it to the north and the Maumturks to the south. At the valley's head is Leenane village, a full 16km from the open sea; the lough's entrance is half-closed by islands, where seals regularly bask and dolphins can be seen. [10] In the 19th century, the British admiralty considered Killary's deep, sheltered waters as a haven for its North Atlantic fleet, with a possible rail link to the Galway–Clifden railway line, but these plans came to naught. Today, sheep farming is the main commercial activity on the hills around Leenane. [11]

Connemara – misty and mythical, bleak, barren and beautiful

Connemara's wild and wonderful scenery is justly famous, combining as it does two of Ireland's finest mountain ranges and a splendid Atlantic coastline. The region is also popular with naturalists on account of its rich flora and fauna, while its complex rock formations attract geologists from around the world. Connemara – the name derives from the territory of *Conmaicne Mara* (Conmaicne of the Sea) – is bounded to the north, west and south by the Atlantic Ocean, but has no official eastern boundary, prompting some to suggest that it is not so much a place as a state of mind. Oughterard proclaims itself the 'gateway to Connemara', but you could as easily argue that Connemara starts at Spiddal or Maam Cross. This wet, windy and mostly treeless corner of Ireland is exposed to the full onslaught of the Atlantic weather systems. Thanks to its location and mountainous terrain, it is one of the wettest places in Ireland, receiving 1,600mm of rainfall a year; proximity to the ocean means that the rain is among the saltiest in the country. The high rainfall favours the development of blanket bog* over much of the terrain, while the moist air creates soft lighting effects that are popular with artists.

Connemara is bisected by the Galway–Clifden Road (N59). The area also divides neatly into two **geological zones**:

In an octopus's garden... 560 different seaweeds grow in Irish waters. This "submarine garden on the coast of Yar-Connaught" appeared in an article about the commercial uses of Irish seaweeds, in the Quarterly Journal of Science *July 1869.*

a mountainous northern zone, encompassing the Twelve Bens and the Maumturk Mountains, that is predominantly quartzite, with some schist, gneiss and marble;* and a low-lying coastal southern zone of granite and, west of Carna, gritstone. Indeed, the GSI's map of the area reveals a very complex geology. Connemara's granite is unusual: due to a quirky chemical composition, it is easily dissolved and decomposed by sea and weather. This produces an intricate coastline that is a frilly lacework of water and rock, and a virtual archipelago of islands appropriately known in this Gaeltacht district as *na hOileáin* (the Islands). Some of the islands are accessible at low tide, others are linked to the mainland by sand and gravel spits. The lowlands behind the coastline are dotted with peaty lakes and bogs and equally wet. Given the high ratio of coastline to interior, it is not surprising that the traditional activities were maritime ones, notably fishing, making kelp* and harvesting seaweed,* all of which feature in a maritime heritage centre at Renvyle. (12)

Connemara's primary habitat is blanket bog: upland blanket bog in the mountains and lowland oceanic blanket bog by the coast. There is also some heathland and a few coastal patches of machair.* The main bog plants are purple moorgrass and heather, but you will also find lousewort, bog cotton, milkwort, bog asphodel and bog myrtle, and various orchids, lichens and mosses. Several arctic-alpine plants grow on the schisty slopes of Muckanaght, among them rose-root, starry saxifrage and mountain sorrel; and Connemara also has some Lusitanian* species, including four rare heathers. There is a rich birdlife, too, and the species to watch for are kestrel, sparrowhawk, merlin, fieldfare and brambling. Some 2,000 hectares have been set aside in a **national park**, which incorporates part of Kylemore Abbey Estate, Letterfrack's old industrial school and part of the former Martin* Estate. (13) (14)

Though most of Connemara is barren and treeless, two fragments of **ancient woodland** survive at Derryclare and Shannawoneen; both were protected from clearance and grazing by their rough and rocky ground. Shannawoneen oak wood, just north of Spiddal, is Connemara's only surviving semi-natural coastal woodland. Derryclare nature reserve in Glen Inagh, by the northeastern corner of Derryclare Lough, is a semi-natural remnant of ancient oak wood; the diversity of lichens and other species found there suggests that the wood is quite ancient. Oaks grow on the steep slopes and on some of the lough's islands, while hazel and ash grow on the more limey soils. There has been a salmon and trout fishery in the Inagh valley for 200 years; in 1986, the fishery was acquired by P J Carroll's tobacco company. Derryclare Lodge, now ruined, was built as a fishing lodge in the early 1800s. The Connemara Environmental Education Centre, based in Letterfrack, organises week-long summer schools, an annual sea week (October) and bog week (May), plus a varied schools programme. ⑮

CLIFDEN

1895 – the trans-Connemara railway

By the late 1880s, Ireland's mainline railways were in place, and the emphasis switched to building smaller local lines. To open up Connemara and revive the fortunes of Clifden, which was badly hit by the Great Famine, the government agreed to a Galway–Clifden railway. This it was hoped would aid the fishing industry by bringing the catch quickly to market, and might even lead to Clifden becoming a transatlantic shipping terminus. The initial plan envisaged a coastal route; a direct line through central Connemara would be shorter, but most of the population lived by the coast. Edward Townsend, UCG's professor of civil engineering, proposed an alternative route via Oughterard, with a branch line serving Killary* where the British navy was considering a deep-water harbour. A Royal Commission found in favour of Townsend's inland route, though not the Killary branch and, in so doing, sealed the railway's fate before it had ever opened.

The Midland & Great Western Railway was granted £250,000 to extend its Dublin–Galway line to Clifden. It used the standard Irish gauge (1.6 metres) and, to cross the Connemara bogs, employed the weight-spreading technique that George Hemans* had pioneered for the boggy Mullingar section of the Dublin–Galway line. The railway reached Clifden in 1895, the 65km of track having cost nearly £7,000 per kilometre. There were 30 bridges, plus a steel viaduct with a central lifting section over the Corrib river at Galway. But the line was not successful: Clifden's fishing, passenger and shipping industries never materialised and, for most of its route, the railway passed through uninhabited bog. Had the Royal Commission chosen a coastal route, the railway might still be running as a local and tourist line; instead it struggled for 40 years, before closing in 1935. Sections of the route survive and occasionally there is talk of rebuilding the line.

Clifden, 'capital' of Connemara and the railway's *raison d'être*, was a relatively new town, begun in 1810 by a Galway landlord, John D'Arcy. By 1815, there was still only one house, yet 20 years later, according to Lewis's *Topographical Dictionary of Ireland*, Clifden boasted 800 houses and 1,257 inhabitants. They were well served with chapels, a national school, a daily postal service to Galway, a dispensary and fever hospital, hotel, bridewell and constabulary police force. In the 1820s, Alexander Nimmo* built a substantial quay there, and a road to Galway through the centre of Connemara (now the N59), which the railway subsequently followed. Silting was a perennial problem and Clifden never became a successful port.

CLIFDEN

Prince of the prostate

Prior to 1900, a man with an enlarged prostate could suffer in silence, try one of several quack remedies or risk a painful operation that might make his problems worse. In 1900, however, an Irish surgeon changed all that when he pioneered a new technique to remove the enlarged gland. **Sir Peter Freyer** (1851–1921) was

Sir Peter Freyer, who developed the first successful surgical technique for removing an enlarged prostate. He is buried in the Church of Ireland graveyard at Clifden.
Institute of Urology, London

born near Clifden and studied at Galway, before moving to India, where he specialised in removing kidney and bladder stones. Freyer first came to prominence in 1888 when he successfully removed a bladder stone from the Rajah of Rampar, who gratefully paid him £10,000.

In 1896, Freyer moved to St Peter's Hospital in London (now the Institute of Urology at University College London) where, in 1900, he performed his first prostatectomy. The patient was a 71-year-old man who had had several unsuccessful operations to treat his enlarged prostate. Freyer's novel technique involved cutting through the bladder to access the prostate. Despite the problems caused by piercing the bladder – 5 per cent of Freyer's patients died of complications – the operation was a major improvement on previous treatments, and became the standard technique for 50 years. Freyer was a dextrous surgeon and showman: when performing before an audience of international surgeons, he would provide his own running commentary in French and Hindustani, with an assistant timing the work – having cut the bladder, it normally took him just three minutes to remove the prostate. Freyer's success earned him fame, fortune and a knighthood, and international urological conferences are still held in his honour.

His technique was superseded in the 1950s by a new approach, coincidentally developed by another Irish surgeon, **Terence Millin** (1903–1980) from Co. Down. Millin trained at the Royal College of Surgeons in Dublin* and played rugby for Ireland, but mostly worked in England, though he later returned to Ireland, working part-time as a consultant. Millin devised a way of removing the prostate without cutting the bladder, which was simpler, caused fewer complications and resulted in fewer deaths.

CLIFDEN

Connemara's green gold

Connemara's main mineral resource, aside from a few small abandoned lead mines, is a distinctive **green marble** that is probably Ireland's best-known rock. Several Irish limestones take a good polish and are called marble, but Connemara has the country's only true marble, formed from limestone that 600 million years ago was baked by a molten neighbouring rock. Connemara marble contains a green mineral, serpentine, but the stone's colour can vary from pale to dark green, and occasional sepia-brown and even white deposits also occur. The marble has been much battered and folded since it formed and many deposits exhibit pretty patterns, but the folding also makes the stone difficult to quarry. It occurs at various locations between Clifden and Recess, and people have quarried it for over 5,000 years, but large-scale extraction began only in the 19th century, when quarrying techniques improved, train services began and marketing started.

The best-known quarry was at Streamstown near Clifden; others were at Lissoughter (Recess); Derryclare and Ballynahinch. In 1895, the quarries were acquired by a New York firm and massive amounts were shipped to the USA to decorate cathedrals and other public buildings. Streamstown and Lissoughter are still producing, primarily building stone; fragments and off-cuts are used in tiles, jewellery and ornaments. Connemara marble can be seen *in situ* in the underground mine shafts at Glengowla,* and No 51 St Stephen's Green, Dublin, formerly the Museum of Economic Geology*, has a display of Irish marbles.

DERRIGIMLAGH

1907 – the first transatlantic radio station

Remote Derrigimlagh bog was once at the frontier of a telecommunications revolution. **Guglielmo Marconi** (1874–1937) chose it as the European terminus for his new transatlantic wireless telegraphy service, and established a large high-tech station there with 150 well-paid technical staff and nearly 300 local turf cutters (the boilers were fuelled by local turf, this being cheaper than imported coal). A massive hangar housed the enormous capacitors; there was also a blacksmith's forge, receiver room, and a staff compound with accommodation, social club, canteen and tennis court. Sparks and noise from the transmitter could reputedly be seen and heard for miles and, during World War I, the strategic installation was guarded by the British army.

Marconi is often credited as the inventor of radio, and indeed he was awarded a Nobel prize for physics in 1909. But, although he invented the earthed aerial, Marconi's contribution was as an entrepreneur, commercialising wireless telegraphy as a means of communication; mostly he exploited other people's ideas, conducting detailed experiments, improving their equipment, and building on the work of scientists such as Faraday, Maxwell, Hertz, Lodge and Popov.

Marconi was born in Bologna to a wealthy Italian father and an Irish mother, Anne Jameson of the whiskey family. He had a poor school record, but was always interested in the possibility of wireless communications, and at 21 sent his first wireless signal over several metres, using home-made equipment at the family home in Italy.

Two years later, he had improved his equipment sufficiently to send a signal across the English Channel, a distance of 15km, prompting him to establish the first wireless company. Despite several impressive demonstrations, Marconi failed to convince the British and Italian governments that his wireless technology was better than the existing wire-based telegraph.* So he turned to shipping applications and, in 1899, conducted a successful demonstration for Lloyd's of London at Rathlin* in Co. Antrim. As a result, the Royal Navy installed radio equipment on its fleet, and several coastal wireless stations were built, including one at Crookhaven in Co. Cork, specifically to communicate with passing ships.

In 1901, Marconi sent the first transatlantic wireless signals from a temporary station at Poldhu in Cornwall to a kite-born antenna at Newfoundland, proving that a signal could travel beyond the horizon. The event made him a household name, and ushered in the radio revolution. For a permanent transatlantic station, Marconi needed directional aerials, but these could not be used from Cornwall as the Irish terrain blocked the signal, so Marconi chose Derrigimlagh, and the station opened in 1907. Messages arrived at Derrigimlagh by wire – there being few other radio stations – from Dublin, London and elsewhere for onward transmission by Morse code. Transatlantic telegraph companies then charged a shilling a word, but Marconi charged half that, and half that again to newspaper reporters. By 1913, business was so brisk a second station was built at Letterfrack to act as a receiving station, while Clifden concentrated on transmission, an arrangement that doubled the system's capacity. Letterfrack's boom was short lived, however, as it was costly to run, and the development of shortwave radio equipment during World War I made it redundant. It closed in 1916.

Derrigimlagh operated until Republican forces burned it down during the War of Independence. By then, new powerful transmitters meant a westerly location was no longer needed, and the traffic was moved to Wales. Marconi pulled out of Ireland, to concentrate on building a radio network for the British government with stations across the empire. In later years, Marconi led a playboy, yacht-set lifestyle, divorcing his first wife, Beatrice O'Brien of Dromoland Castle, Co. Clare, for an Italian countess. He also supported Italy's Fascist movement and was buried with honours by Mussolini. But the radio revolution he started a century ago continues today with mobile phones and now wireless communications between computers. Remains of the world's only turf-powered radio station can be seen at Derrigimlagh, as can the line of the narrow-gauge railway used to shift the turf. Traces of Letterfrack's receiver station are on a hill above the village. Veldon's pub there has a photographic exhibition about the station.

1919 – the first non-stop transatlantic flight

The first non-stop flight across the Atlantic ended nose-down in Derrigimlagh bog near Clifden, at 8.40am on June 15th, 1919. The longest flight ever attempted then, it attracted considerable attention, not least because a £10,000 prize from the *Daily Mail* was at stake. **Capt John Alcock** (pilot) and **Lieut Arthur Brown** (navigator) had left Newfoundland the previous night, on course for Galway Bay. A *Daily Mail* reporter waited in Galway, an adjudicator waited in Dublin, and the staff at Marconi's wireless station in Derrigimlagh (see: **1907 – the first transatlantic radio station**) watched for the plane. Alcock and Brown's radio transmitter froze shortly after takeoff, so they could not send messages and, indeed, had trouble hearing themselves over the engine noise in the dark cockpit.

On spotting the tall Marconi aerials, the pair made for Derrigimlagh and landed their small bomber biplane beside the station. News of their arrival was immediately wired to London. The story was to have been a *Daily Mail* exclusive, but *Connaught Tribune* editor Tom Kenny scooped it and sold the story internationally. Alcock and Brown had intended flying to London, but their plane was stuck in the bog, so they went to Galway in the Marconi station motor car, and thence to Dublin and London, where they were wined and dined and greeted by crowds. With them was a small bag containing 800 letters, the first transatlantic airmail.

It had been a daring venture – the first transatlantic flight had been achieved just one month beforehand, by a convoy of US navy planes, which took days to island-hop across the ocean, accompanied by a flotilla of support ships. Alcock and Brown had left Newfoundland at 4am local time, and flew the 3,000-km journey in 16 hours. Commercial transatlantic flights began 10 years later with airships and, 10 years after that, a seaplane service started from Foynes*

in Co. Limerick. And Derrigimlagh might soon see more aeroplanes: Clifden Airport Company is planning an airfield on the site of the old Marconi station, serving the Galway/Mayo islands. (There is a memorial to Alcock and Brown at Ballinaboy near Clifden; a cairn marks their landing spot in Derrigimlagh bog.)

Humanity Dick and the RSPCA

The man responsible for founding the Royal Society for the Prevention of Cruelty to Animals was **Colonel Richard Martin** (1754–1834) from Ballynahinch. His family owned the biggest estate in Ireland – over 200,000 hectares in all, covering most of Connemara including the Twelve Bens and Maumturk Mountains – but it was mostly poor, unproductive, unimproved and inaccessible bog.

Richard Martin was educated in England, trained as a lawyer, and represented Galway first in the Irish Parliament and, after the Union, at Westminster. As a young man, he was a notorious dueller, earning the nickname Hairtrigger Dick, but he is remembered as Humanity Dick for his work on animal rights. Martin believed that animals had feelings and intelligence, though he distinguished between cruelty, of which he disapproved, and hunting, of which he did not. In 1822, he succeeded in having the first animal rights bill enacted, banning the ill-treatment of cattle. Two years later, he helped found the RSPCA. Martin reputedly fought duels on behalf of animals, arguing that "an ox cannot hold a pistol", and jailed tenants who mistreated animals. Richard Martin led an extravagant lifestyle – his stables were panelled with Connemara marble* – but died a debtor in France. Afterwards, the estate was so poor, it took 20 years to sell. The Martin family seat, Ballynahinch Castle, is now a hotel.

A Scotsman, a seals' stone and a new town

For thousands of years, there was little trade, traffic or communication across Connemara's boggy and mountainous interior. Communities developed along the coast instead, travelling and trading by boat. That pattern began changing in the early 1800s, thanks in part to a Scottish engineer, **Alexander Nimmo** (1783–1832). He designed a road (now the N59) that cut through the heart of Connemara, linking Galway to the new town of Clifden,* and establishing a route that would later be followed by the Galway–Clifden railway. Nimmo also designed the many quays that were built around Connemara's coast in the 1820s, and he developed a new town at Roundstone.

Alexander Nimmo came to Ireland in 1810 to work for the **Bog Commission**. The Westminster government established the commission shortly after the Act of Union to analyse the problems and commercial potential of Ireland's larger bogs; the survey, and similar assessments of canals, drainage, and later railways, was also a useful way for the new administration to learn about Ireland. Over two years, Nimmo surveyed 303,500 hectares of bog, equivalent to 6 per cent of the country, mostly in Kerry and Connemara; he recommended a system of canals to reclaim the bogs, but his plan was never implemented. Next, Nimmo designed numerous new roads (such as those around Kenmare), harbours (including Dunmore East) and bridges (among them the elegant castellated bridge over the Liffey at Pollaphuca*), all of which opened up remote and inaccessible areas. From 1820–22 he surveyed the entire Irish coast for the Fisheries Board, identifying landing places, designing new harbours, and specifying piloting and navigational instructions.

It was in Galway and Connemara, however, that Nimmo had most impact, working from Corrib Lodge (now Keane's Bar), his home and headquarters at Maum on Lough Corrib.

He designed and built a new quay at Galway city (still called Nimmo's Quay); harbours at Cleggan, Ros a'Mhíl, Clifden and Roundstone; and a bog road from Galway to Clifden across the Martin family's estate. Most of these works were famine-relief projects following a famine in 1822. Nimmo also developed Roundstone, which was only a small landing place before he built the new pier in 1822–25 (Roundstone's name derives from *Cloch na Rón*, meaning 'seals' stone'). He leased land, set about building a main street, sold plots and encouraged tenants to build two-storey houses. By 1837, according to Lewis's *Topographical Dictionary of Ireland*, Roundstone Harbour "could shelter the whole Navy of England".

Roundstone district is a botanist's heaven and home to some **unusual plants**, and a cosmopolitan mix that includes several Pyrenean and North American species. There are two rare pondweeds more usually associated with North America: pipewort (which resembles an upright knitting needle), and slender naiad, which was found for the first time in Ireland at Cregduff Lough, south of Roundstone, in 1852. **Dog's Bay** is noted for a brilliant-white sand composed of the beautiful shells of small marine creatures called foraminifera; shells from over 120 species have been found there. Behind Dog's Bay is Goirtín, a small rocky island joined to the mainland by a sandy tombolo; the lime-rich machair* there was once valued as a grazing ground for cattle crippled by the wet bog inland.

Rare heathers growing in the area include two that were discovered in the 1830s by Roundstone schoolmaster **William M'Calla*** (1814–49). M'Calla (sometimes McAlla) was an excellent, self-taught naturalist who knew Connemara well and made several important finds, notably the rare Irish heath (*Erica erigena*)* and Mackay's heath. The latter, which he discovered on a hill behind Roundstone in 1835, is one of Ireland's Lusitanians;* it grows in only the west of Ireland and in northwest Spain, where it was found subsequently. M'Calla, who was also the first person to realise the true nature of coral seaweed (see:

Connemara's coral

Connemara has several beaches that, instead of sand, have beautiful fragments of a white, rocky substance, commonly called coral. Until the mid-1800s, people thought this was a true coral, formed by a colony of tube-living animals, presumably living in a reef offshore. The first person to realise that this coral is made not by animals but by plants, was Roundstone schoolmaster William M'Calla. He identified the source as seaweeds, which produce a calcium-rich crust to protect their fronds from grazing shellfish. Two species, *Lithothamnium coralloides* and *Phymatolithon calcareum,* are found in Irish Atlantic waters up to 30 metres deep; they are slow-growing and can live for 50 years. Galway Bay has extensive beds, which are important spawning grounds for starfish and shellfish. While the plant is alive the coral is red; when it dies the crust washes ashore and bleaches white – the whiter the coral, the older it is. Coral beaches occur along Ireland's west coast, from west Cork to Donegal. Traditionally, the coral was used to lime land, especially in granite regions such as Connemara with little access to limestone. In recent decades, deposits were dredged from Bantry Bay,* and used instead of bonemeal in animal feed, as a soil additive, in water filtration systems and as a calcium diet supplement – unlike bonemeal, which could carry diseases such as BSE, the coral, or maërl as it is called on the continent, is thought to be relatively risk free. Maërl banks off Brittany are commercially harvested and the Irish Marine Institute has investigated the potential of the Irish deposits. The plants and the coral beaches are protected, and harvesting is regulated.

Connemara's coral) corresponded with naturalists in Ireland and Britain, and communicated his findings to institutions such as the RDS.* In 1845, he published *Algae Hibernicae*, a large two-volume *exsicatta* containing dried samples of Irish seaweeds. M'Calla died of cholera in 1849, and was buried at Roundstone. A seaweed, *Cladophora macallana*, is named after him.

ROUNDSTONE

A Connemara fractal

Question: How long is a coastline? Answer: As long as a piece of string. Roads are easy to measure, and have a finite length, but a coastline is infinitely long – start to include every rock pool, indentation and crevice, and there is literally no end to it. The bigger the scale of your ruler, the longer the coastline becomes. Roundstone is 15km from Ballyconneely by road, but the coastline in between could be 100km long, or 1,000km, even 100,000km. The Marine Institute claims Ireland has 6,500km of coastline, but if the institute chose a different scale, we could claim 650,000km. So a coastline is something more than a simple straight line, or as a mathematician might say, it has a dimensionality somewhere between one and two. And in the 1960s, a Polish-French mathematician Benoît Mandelbrot invented the concept of fractional dimensions, or fractals, to account for this kind of phenomenon.

The fractal nature of coastlines, and 3-dimensional landscapes in general, pose special problems for map-makers – because to attempt an adequate map is to attempt the impossible. **Tim Robinson,** a mathematician turned map-maker and writer, discovered this the hard way when he mapped Connemara. Robinson, whose Folding Landscapes studio is in Roundstone, is known for his elegant maps of the Burren, Aran Islands and Connemara. To map Connemara, he trekked its every road and track, its every stretch of coastline, a task which took five times longer than anticipated.[2] A map alone was insufficient for

Roundstone and its environs, from Tim Robinson's map of Connemara.
© Folding Landscapes

the information he collected, so Robinson produced an accompanying gazetteer with information on the many places featured on the map.[3] As an acknowledgement of the impossibility of producing the 'adequate map', Robinson's charts do not feature any broad strokes, or large areas of colour washes, which might suggest that he knew all there was to know about a district.

Atmospheric lookout station

Sniff the air at Mace Head and you inhale some of the world's cleanest air. Carried across the ocean on the prevailing southwesterly winds, the air admittedly contains detectable traces of industrial pollution from North America, but, by the time it hits Connemara, it has also been well washed by the Atlantic weather systems. Consequently, the low levels of air pollution there are a good baseline against which to compare the air in more industrial places. Since 1958, scientists have been measuring air pollution at Mace Head, in a small laboratory begun by Prof Tom O'Connor, a physicist at UCG. The laboratory is strategically located close to the ocean's edge, with no road or human habitation in front of it. In the 1980s, amid growing international concern about air pollution and especially ozone-destroying CFCs, Mace Head became increasingly important, and it is now a vital lookout station for the World Meteorological Organisation. In 1987, Mace Head took over from Adrigole* in Co. Cork, where English scientist James Lovelock had established the world's first CFC monitoring station. The Galway laboratory is part of several international networks monitoring atmospheric gases, and over 50 universities and research institutions from three continents use the facilities.

UCG also has a **shellfish research centre** at **Carna**, with laboratories, hatcheries, rearing tanks, and a seaweed farm that produces algae that are fed to the shellfish. Scientists there are researching ways of growing sea urchins, abalone and the crawfish, or 'spiny lobster'. Abalone, a gourmet shellfish that is relatively new to Ireland, is of interest for its flesh and its valuable mother-of-pearl shell. In the 1870s, the West of Ireland Oyster Company built oyster-rearing ponds around the Connemara coast, and exported oysters to England and Europe. The pond at Dawros near Renvyle was used in the early 1900s as a holding tank for lobsters, prior to exportation; the tidal pond, as big as a tennis court, can still be seen. The marine life of Galway Bay features in a new aquarium at Salthill. ⑯

Harvesting sea vegetables

Every day, without knowing it, you eat some seaweed. For **seaweeds** are an important source of gelling and stabilising agents, vital ingredients in a vast range of products, from beer, ice cream, ketchup and toothpaste, to paint and lipstick. Seaweed extracts are also used to glaze paper, as an absorbent gel in wound dressings, and in various cosmetics, creams and lotions. Over 560 types of seaweed grow in Irish waters, according to the Irish Seaweed Industry Organisation, but only a few have ever been put to any practical use. The main use was as a fertiliser and soil improver, and coastal communities traditionally divided land so that each family had access to a strip of shoreline where they could harvest weed; large boulders marked the boundaries on the lower shore, and extra boulders were sometimes added in an effort to farm the seaweed, by creating sheltered spaces where more wrack could grow. Kelp making,* a valuable source of seasonal income on Ireland's west and north coast for over 200 years, yielded alkali and iodine for the chemical industries. Edible seaweeds featured in traditional recipes – notably carrageen or sea moss (*carraigín, Chondrus crispus*); dillisk or dulce (*duileasc, Palmaria palmata*); sloke or laver (also called *nori, sleabhcán, Porphyra umbilicalis*); and kelp (*laminaria*). Seaweed was occasionally fed to animals as fodder or a salt lick.

Carrageen moss: this edible seaweed features in traditional Irish recipes and remedies. The illustration is from William Harvey's British Marine Algae *(1849 edition).*

UCG's shellfish research laboratory in Carna is pioneering a new use for seaweeds – farming them on ropes hung in the nutrient-rich effluent that comes from fish farms; this cleans the effluent, and produces a crop of seaweed for shellfish to eat. Cill Chiaráin is the headquarters of Ireland's seaweed industry: home to Arramara Teo, a State company founded in 1947 to harvest and process seaweed. Nearly 400 people are employed, mostly in seasonal work, harvesting the yellow-brown, knotted wrack (*Ascphyllum nodosum*); plants are cropped every five years, and left to regrow. The seaweed is processed at Cill Chiaráin to produce an alginate extract that is exported to Scotland.

Chorography – Ireland and Connaught described

Chorography – recounting the natural features of a district – was a popular activity in the 1600s, especially in Ireland where adventurers and Planters were keen to know more about the new territory. The first modern chorographical account by an Irish writer of at least part of Ireland, was a description of west Connaught written in 1684 by **Roderic O'Flaherty** (1629–1718), a scholarly historian from Moycullen. Ironically, the O'Flahertys had owned much of west Connaught until Cromwell dispossessed them of their lands but, despite his impoverished circumstances, O'Flaherty continued to study, write and publish.

The earliest-known description of Ireland's natural history was written 1,000 years before then, however, by an Irish monk **Augustin**. His *De Mirabilibus Sacrae Scripturae* (655) survived because, for centuries, it was attributed to his better-known namesake, St Augustin of Hippo. The manuscript is primarily a reflection on biblical events and miracles, but it includes a list of the mammals found in Ireland, of which Augustin counts nine: wolf,* hare,* badger, fox, deer,* wild boar, seal, otter and sesquivolos (squirrel).* Augustin concludes that Ireland was once joined to Britain and Europe by a land bridge,* for how else could these animals have recolonised Ireland after the biblical flood?

A more detailed, but occasionally fantastical account of Ireland's natural history was *Topographica Hiberniae* written in 1188 by a Welsh historian **Giraldus de Barri** (1146–1223), alias **Giraldus Cambrensis**. Grandson of Henry I, and related to the Geraldines who settled here, Giraldus spent several years in Ireland during the 1180s as secretary to Prince John, and as chronicler of the Anglo-Norman invasion of the country. Giraldus correctly noted the presence in Ireland of the capercaillie and crane, which subsequently became extinct here, and of three freshwater fish unique to Ireland – pollan, shad and char. He also records the absence of hedgehogs, frogs,* pike, roach and perch, which were not introduced until later. His account is, however, peppered with flights of fancy, although some of these were common beliefs then, such as that the barnacle goose* hatched from the goose barnacle.

The first modern scientific description of Ireland's natural history and resources, and the first one written in English, was compiled by Dutch physician **Gerard Boate** (1604–50). Published posthumously in 1652, its full title runs to over 100 words, but it is generally known by the first three – *Irelands Naturall History*. Based in London, Boate was physician to the king and friend to Robert Boyle,* with whom he started the Invisible College which later became the Royal Society. Boate subscribed £180 to a fund for "the reduction of the Irish" in return for which he was promised an estate in Tipperary. In the 1640s, he began compiling his guide to Ireland, aimed at promoting the country among developers and settlers. His sources were his brother Arnold, who had spent eight years in Ireland as physician to the army, and various Planters already settled in Ireland, among them William Parsons of Birr.* Gerard Boate himself spent little time in Ireland – he came in 1649 as physician to Cromwell's army, but died shortly afterwards. His text is a comprehensive account of Ireland's landscape, natural history, resources, climate and industry, with detailed descriptions of charcoal works and ironworks,* and bogs, which Boate believed could be drained and converted to agricultural land.

In the early 1680s, William Molyneux,* a philosopher and scientist at Trinity College Dublin, began supervising a major project aimed at compiling a detailed guide to Ireland's natural history that would accompany William Petty's* new map of Ireland. Sixteen correspondents, including Molyneux himself, Sir Henry Piers in Westmeath, and Roderic O'Flaherty in Galway, were commissioned to describe their region. O'Flaherty was responsible for the area that is bounded to the east by Lough Mask and Lough Corrib, to the north by Killary Harbour, to the west by the Atlantic, and to the south by Galway Bay. His *Chorographical Description of West or h-Iar Connaught*

CHOROGRAPHICAL DESCRIPTION
OF
WEST OR H-IAR CONNAUGHT,
WRITTEN A.D. 1684.
BY RODERIC O'FLAHERTY, ESQ.
AUTHOR OF THE "OGYGIA."

EDITED, FROM A MS. IN THE LIBRARY OF TRINITY COLLEGE, DUBLIN,
WITH NOTES AND ILLUSTRATIONS,
BY JAMES HARDIMAN, M.R.I.A.

DUBLIN:
FOR THE IRISH ARCHÆOLOGICAL SOCIETY.
MDCCCXLVI.

Title page from O'Flaherty's account of Iar Connaught, written in 1684 but not published until 1846. The entire work can be read online at galway.net/history.

is a factual account of the region, featuring mountains, mines, woodlands, soil, waterways, and ruins, as well as the fish, fowl and beasts, and the disposition of the natives. He itemises 24 species of fish, 16 mammals — several having been introduced since Giraldus Cambrensis's day — and 10 invertebrates, mostly edible shellfish. He also reports an early Galway experiment in salmon rearing: in 1684, the fins of 18 salmon* were notched to identify them; when the fish returned from the sea months later, the marks confirmed the salmon's amazing homing instinct. The only fantastical note is the "Irish crocodil" that reputedly lived in Lough Corrib and had black shiny skin and a bald tail; O'Flaherty admits, however, that he never saw one. Molyneux's grand project was never finished, but O'Flaherty's manuscript was finally published in 1846 by the Irish Archaeological Society. O'Flaherty also wrote *Ogygia*, a history of Ireland in Latin for the English reader. Impoverished by the Cromwellian land confiscations, he died in poverty.

Travel underground, and back in time

The only underground mine in Ireland currently open to the public is a 19th-century **lead and silver** mine at a farm in Glengowla. In the 1850s, there were half a dozen small lead mines around Oughterard, most of them operated by the local landlords, the O'Flahertys. They employed perhaps 300 people in all, and the ore was shipped via Lough Corrib to Galway. Most closed by 1870, but one mine near Maum was worked into the early 1900s. The mines produced primarily lead and some silver, though the district also has some zinc, copper, iron and barytes deposits. Glengowla was worked from 1850–65, initially for lead and silver – in the first two years it produced 4kg of silver – and later also for zinc. A mix of horse power and water power drove the pumps and the ore-crushing machines. Gunpowder for the blasting was initially carted from Oughterard barracks, but some of it regularly 'went astray', and so a small, secure powder house or magazine was built beside the mine. The magazine survives, as do several old mine buildings and a horse-powered whim engine or windlass, used to raise the ore from underground. Fine mineral formations can be seen in the mine shafts, as well as some Connemara marble. ⑰

Where east meets west

Lough Corrib, famed for its fishing, is one of Ireland's most scenic lakes. It divides Co. Galway in two, straddling the junction between the limestone plain of the midlands to the east, and the igneous and metamorphic rocks of the west. Consequently its eastern shore resembles the Burren,* while its western shore is an extension of mountainous Connemara. At 180sq km, Lough Corrib is also the second-largest lake in Ireland and Britain, after Lough Neagh.* Unlike Lough Neagh, however, which is relatively uniform, Lough Corrib is a complex lake with a great diversity of habitats.

Thanks to its size and strategic location, Lough Corrib has long been an important waterway and trade route, though the many islands and shallow shoals make navigation complex. **Ireland's first canal**, Friar's Cut, dug in 1178 by monks from Claregalway Abbey, is a 1-km channel cutting across an island near Galway city; it shortens the journey to Galway by 4km. In 1775, Lough Corrib acquired **Ireland's only inland lighthouse**: Ballycurrin lighthouse, in the northeastern corner, was built by a local landlord, Capt Lynch, who ran a sailing fleet of trading ships on the lake; his simple tower, the ruin of which survives, was topped with an open brazier where a turf fire was lit for the returning boats. The lighthouse ceased to function in the 1820s, after Lynch became bankrupt and scuttled his fleet. Lynch also established a navigation system with small pyramids marking a route through local shoals; some of these markers survive.

The 19th century saw heavy commercial traffic on the lake, and many piers were built around the 200-km shoreline: seaweed and fish were shipped inland from the coast at Galway; cattle and sheep were taken by boat to market; marble and lead ore were shipped to Galway port from Connemara's quarries and lead mines; and a passenger steamer service began in the 1850s. To facilitate all this traffic, a canal was built at the Claddagh (Galway city's fishing port) in 1830 and, in 1852, a shipping canal was added, with two navigational locks, connecting Lough Corrib to the sea (the canal closed in the 1950s when low road bridges were built blocking the navigation). The ill-fated Cong Canal,* dug in the late 1840s, was intended to extend the navigation to Lough Mask, but was never completed. Lough Corrib's modern navigation system of 160 numbered beacons and posts dates from the 1850s, and is maintained by the Corrib Navigation Trust.

Lough Corrib can be divided into four distinct sections, each with its own personality. The northern part is a large, triangular limestone basin, studded with numerous drowned drumlins, which make for difficult navigating. The northwestern corner is a long, narrow and deep glacial valley flanked by steep hills and leading into Maum Bridge. The rock there – quartzite, schist and gneiss – is at least 100 million years older than the limestone to the east, and the vegetation resembles that of Connemara; among the more unusual plants growing in this quarter are St Dabeoc's heath,* and pipewort,* a rare pondweed. The lake's midriff is long, narrow and irregular, with numerous jagged limestone reefs. In places, the lake there is fed from lime-rich underwater springs, which produce a pale marl deposit that gives the water a translucent appearance, as at Lough Carra.* The southern section, finally, is a shallow limestone basin where the water averages only 3 metres deep.

In all, Lough Corrib has over 200 islands. Many were inhabited or farmed in the past, but today only a few large islands close to the shore are used. Several islands date from the 1850s, when a major drainage programme lowered the lake's winter level by 1 metre. Trees growing on some of the more remote islands are fragments of the region's **aboriginal woodlands**. A small patch of ancient woodland also survives at the Curra peninsula on the western shore, thanks to the presence of a glacial field of enormous boulders, which meant that corner was never cleared.

Diverse habitats surround Lough Corrib, including callows* and turloughs,* marsh and fen,* raised bog* and limestone pavement, woodland and hazel scrub, and, consequently, the region has a rich flora and fauna. The limestone fern

Lough Corrib, its islands and shoreline are best explored by boat, bike and boot. The classic book about the area is *Loch Corrib* (1867) written by Oscar Wilde's father, Sir William Wilde,* whose holiday home was Moytura House near Cong (you can read Wilde's book online at galway.net/history). *Corrib Country: A Rambler's Guide & Map* (TírEolas, 1998), is a useful pocket guide; the Corrib Conservation Centre in Oughterard, a private research centre for exologiests, founded by the late Dr Tony Whilde, has produced several booklets on the region's natural history. Boat trips on the lough run in summer from Galway and Oughterard.

(*Gymnocarpium robertianum*) in Ireland grows only at Ballinduff Bay. Wonderful spring gentians and rare orchids, more usually associated with the Burren,* grow well on the limestone pavements along the eastern shore. The area is rich in birdlife, notably ducks, swans, waders and wildfowl and, in winter, Greenland white-fronted geese. Small populations of pearl mussel* survive in the Corrib waters. The lake's fish include native species such as arctic char, salmon and trout, alongside the introduced freshwater species of perch, bream and roach.

1866 – a new star in the Heavens

On the night of May 12th, 1866, **John Birmingham** (1816–84) was walking home near Tuam when he spotted a bright new star in the sky, where previously there had been only a faint star. It was a nova – a spectacular firework caused when the star shed some explosive material – and the brightest seen in 250 years. Birmingham, who had a small telescope, watched it for two hours, then wrote to the London *Times* to alert others. *The Times*, in its wisdom, ignored his letter, so Birmingham contacted noted English astronomer, Sir William Huggins,* who confirmed the discovery. Two years previously, Huggins had pioneered the use of a new instrument, the spectroscope, to study the chemical composition of starlight. Turning his spectroscope on Birmingham's nova, he showed that the bright display came from a shell of hot hydrogen gas. It was the first time a nova had been studied spectroscopically.

Birmingham, a Catholic landowner with a small estate at Millbrook near Tuam, was educated at St Jarlath's College. He studied engineering in Germany and was fluent in Irish and several European languages. A modest man, he campaigned against injustices and vivisection, and is said to have had a lifelong stutter.[4] His interest in astronomy was sparked by two spectacular comets, in 1858 and 1861; before that his great passion was geology, especially the

Donati's comet of 1858: Birmingham's interest in astronomy was awakened by this and another comet which appeared in 1861. Birmingham's telescope is now owned by his old school, St Jarlath's in Tuam. (Sketch by Irish astronomy writer, Mary Ward.)*

intriguing drifts of eskers* and erratic boulders* scattered around east Galway, which he thought had been deposited by a prehistoric sea. Emboldened by his nova discovery, Birmingham bought a decent astronomical telescope for £120 (with a 4.5-in refractor lens), and built a small wooden observatory in his garden. He went on to discover 49 red stars and, in 1876, compiled a catalogue of all the 658 then-known red stars. In 1881, he discovered a variable star in the constellation Cygnus which is named after him (variable stars brighten and dim periodically). Birmingham corresponded with many well-known astronomers of his day in Ireland, Britain, Europe and the US and a crater on the Moon is named in his honour.

Oats in, oatmeal out

In the 1830s, Tuam boasted a brewery, tannery, linen industry and several flour mills (two windmills, built *c*.1750 by an enterprising local archbishop, had ceased functioning by then). The remains of the town's brewery, and the stumps of the two windmill towers can still be seen. One of the flour mills was in use until 1964, driven by the River Nanny and grinding local oats to produce oatmeal. The raw oats were first winnowed to remove dirt and any grit that might damage the machinery; the grain was then dried in a kiln. Drying was important as Irish grain, especially in the west, was often damp (drying also made it easier to remove the husks and to grind the oats). The dried grain passed through a first set of millstones to remove the husks; the oats were winnowed again to separate out the husks, then ground into oatmeal by a second pair of stones, and finally bagged. After the mill closed, local schoolboys organised a plan to renovate the premises, which now houses a small exhibition on milling from the earliest times. ⑱ In recent years, a railway preservation group has run steam excursions on a restored part of the Tuam–Athenry line.

CORRANDULLA

The phlogistic 'Philosopher of Dublin'

Richard Kirwan: chemist, philosopher and eccentric. Cregg Castle is now a guesthouse; Kirwan's ruined laboratory is in the orchard. Of his other Galway homes, Cloughballymore is still a residence, Menlough Castle is in ruins.

Cregg Castle at Corrandulla was home for many years to an internationally renowned chemist, **Richard Kirwan** (1733–1812). Kirwan was friendly with some of the great scientists of his day, among them Priestley and Cavendish; and Catherine the Great invited him to head up the chemistry section of the Russian Academy of Sciences (he declined, recommending instead his friend and fellow countryman Bryan Higgins*). Today, chemists remember Kirwan for championing the theory of phlogiston, and for spurring the great French scientist Lavoisier into publishing his alternative theory of oxygen.

The Kirwans were one of Galway's founding tribes, related to the Martins* of Connemara and the Birminghams* of Tuam, and with seats at Cloughballymore near Kinvara (where Richard was born) and Cregg Castle (where he spent most of his youth). Richard Kirwan was educated at home, and then spent a year in France studying for the priesthood, before returning to manage the family estate after his brother was killed in a duel. In 1757, Kirwan married Anne Blake, and moved into her family home at Menlough Castle near Galway city, where he added a laboratory and library. Meanwhile, he studied law and, in order to practise at the Irish Bar, converted to Protestantism. After his wife died, Kirwan returned to Cregg, where he built another laboratory, before moving to London. There he pioneered the study of industrial resources, acquired a reputation as an analytical chemist and, within three years, had been elected a Fellow of the Royal Society.

Kirwan had a practical turn of mind, analysing everything from minerals to manures, and investigating dyeing and bleaching for the linen industry, and the chemistry of kelp.* His *Elements of Mineralogy* (1784) was the first systematic chemical analysis of minerals and was quickly translated into several languages, and his *Essay on Phlogiston* (1787) was hailed as a classic. The theory of phlogiston, then the dominant theory of combustion, held that everything combustible contained 'phlogiston', and that this element was given off when the substance was burned. A few chemists, among them Lavoisier and Priestley, believed instead that when a substance was burned it actually gained something from the atmosphere – a hypothetical gas that came to be called oxygen. When Kirwan's essay was published, Mme Lavoisier translated it into French for her husband, prompting him to publish a refutation of phlogiston, along with his own ideas concerning oxygen. Convinced by Lavoisier's arguments, Kirwan converted and began championing oxygen. He continued to believe in several other conservative ideas, though, preferring the Neptunist* theory for the origin of rocks, and a Biblical timescale for the history of the Earth.

In 1787, Kirwan moved to Dublin, where he was elected president of the newly established Royal Irish Academy,* and came to be called the 'Philosopher of Dublin'. He became interested in meteorology, collected weather data from dozens of places around the globe, and could reputedly predict seasonal weather patterns with some success. An eccentric in later life, he wore a hat and coat indoors, kept a pet eagle on his shoulder and paid his staff to kill flies. He died while "starving a cold" at 79.

GALWAY CITY

A French connection

Rue Darcy in Paris commemorates an Irish mathematician, **Patrick D'Arcy** (1725–79) who, though born in Galway, rose to become a distinguished member of the French Academy of Sciences, and a field marshal in the French army. His family was of French Catholic origin and Jacobite sympathies and, to avoid religious persecution at home, Patrick (later Patrice) was sent at 14 to live with an uncle in Paris. There he was befriended by a neighbour and noted mathematician Jean-Baptiste Clairant, under whose tutelage D'Arcy flourished.

At 17, he presented two treatises on dynamics to the Academy of Sciences and, at 24, was elected a member of the academy. His military career left little time for mathematics – he was, for example, involved in planning an invasion of England during the Seven Years War – and as a scientist he was something of a dilettante. Nevertheless, he found time to study the chemistry of gunpowder, electricity (he invented an early electrometer), and vision; he wrote about ballistics and mechanics and also wrote a book, *On the Duration of the Sensation of Sight* (*Sur la Durée de la Sensation de la Vue*).

Quick to anger, D'Arcy fought with many of the great French scientists of the time, among them d'Alembert and Condorcet, though when D'Arcy died Condorcet delivered an elegy in praise of the Irish mathematician.[5] D'Arcy never

returned to Ireland, on the grounds that the government remained unfriendly to Catholics, and is scarcely known in his homeland.

GALWAY CITY

Drinking the Corrib

Galway residents are never far from water. The city is by the sea, and straddles the Corrib estuary, where the river fragments into several channels, reflected in the city's original name – *Baile na Shruthán* (Streamstown). The Corrib has been a source of drinking and industrial water for centuries and, in the 1840s, powered nearly 30 industries, among them flour and woollen mills and distilleries. By then, however, there was a growing need for a pumped, or pressurised, water supply for the growing town and its fire-fighting service, as well as industry and the railways – steam locomotives in particular need to take on large volumes of water quickly. In 1867, Galway Corporation opened its first waterworks, at a site on the Corrib north of the city at Terryland. The head of water there was 2 metres, sufficient to turn two waterwheels; these drove the pumps that pushed the water around the city. A steam engine was added in 1895 to increase the water pressure, and meet the growing demand, but the cast-iron mains quickly rusted, and leaks meant supply was often low. In 1906, fresh mains were laid, and two water-powered turbines were installed; 20 years later a semi-diesel engine was added to supplement the turbines. In 1934, the pumps were electrified by the ESB, which was keen to sell the electricity being generated at Ardnacrusha.* A modern water treatment and pumping station was subsequently built alongside the original waterworks, but the old turbines and engines have been preserved and can be viewed by appointment with the city museum. (19)

GALWAY CITY

The Queen's "godless college"

Queen's College Galway (later University College and now National University of Ireland, Galway) was established in

1845 amid a blaze of controversy and a vitriolic Church–State debate. For 250 years, the only university in Ireland had been the predominantly Protestant Trinity College* in Dublin (St Patrick's College, Maynooth,* founded in 1795, was specifically for student Catholic priests). For years, Roman Catholics and Presbyterians, and those far from Dublin, had campaigned for greater access to third-level education. Caught between opposing camps, the government decided on a compromise: three provincial, non-denominational colleges, to be called the Queen's Colleges – a southern one in Cork; a northern one in either Belfast (which already had a noted academic institution), Derry or Armagh (an important astronomical centre); and a western one in either Limerick, Galway or Tuam (the latter the headquarters of an outspoken Catholic archbishop, Dr John MacHale). In the end, Galway and Belfast were selected for the western and northern colleges respectively.

All appointments, the government insisted, would be non-denominational, and there would be no State funding for theology, though the colleges could raise private funds for a theology chair. Sir Robert Inglis, a Protestant cleric, denounced the move as "a gigantic scheme of godless education". The Repeal Movement, campaigning for a repeal of the Act of Union, was split: Daniel O'Connell and 'old Ireland' opposed the colleges, while the Young Irelanders* favoured the liberal, non-denominational aspect. Northern Presbyterians, who had hoped for a Presbyterian College in Belfast, were disappointed, but the Catholic hierarchy was enraged. Historian Gearóid Ó Tuathaigh, in a history of Queen's College Galway, says the bishops insisted that Catholics could not attend lectures on history, logic, metaphysics, moral philosophy, geology or anatomy, without exposing themselves to danger, unless a Roman Catholic professor was appointed.[6] A massive State grant to Maynooth College failed to allay Catholic fears, the bishops continued to campaign for a separate college and, 10 years later, founded their own rival Catholic University in Dublin (later part of University College Dublin*).

Galway's new university was a welcome boost, bringing jobs and status to a relatively small and remote city. The college opened in 1849, but student numbers remained under 100 for many years, due partly to Catholic opposition, the aftermath of the Great Famine and a lack of second-level education in the region. Nonetheless, several noted academics joined the college, among them George Johnstone Stoney* (remembered as the man who named the electron) and geologist William King (see: **Naming the Neanderthals, debunking eozoon**). Galway's university is today noted as a marine biology centre; it runs the shellfish research laboratory at Carna,* the atmospheric research station at Mace Head,* and a field studies centre in the Burren.*

Naming the Neanderthals, debunking eozoon

In the 1860s, **William King** (1809–86), professor of geology at Queen's College Galway, tackled two major controversies, bringing considerable renown to the college: he was the first to recognise the significance of the Neanderthal fossils; and, with Thomas Rowney, professor of chemistry at Galway, King debunked a fossil that many thought was the world's earliest living thing.

In 1856, workers at a German quarry in the Neander valley found 16 bones in a cave. The bones were apparently human, yet oddly thick and curved, and the large skull had a prominent ridge above the eyebrows. Some thought they were the remains of a man with severe rickets and arthritis, others that it was a Cossack who, bow-legged from years in the saddle, his

A Neanderthal skull (redrawn by Sandra Minchin), based on a plaster cast from the James Mitchell Geological Museum in NUIG. The museum has 5,000 fossils (including Kiltorcan fossils and samples of Oldhamia*) and 3,000 minerals and rocks.*
James Mitchell Museum

face furrowed from riding into the wind, had crawled into the cave to die. No one considered that the fossils might be prehistoric, until three years later when Charles Darwin published his *Origin of Species*. William King pointed out that the cave sediments where the fossils were found were at least 30,000 years old and, in 1864, he argued that 'Neander man' was not a modern human, but another species entirely, a primitive human whom he called *Homo neanderthalensis*. Not everyone agreed, but as more Neanderthal remains were found around Europe, it became clear that the fossils were indeed prehistoric, and the original Neander bones have since been dated to 40,000 years old. The Neanderthals remain controversial: they were toolmakers, who cared for their sick and buried their dead, and had a larger brain than ours, but scientists can still not agree on whether they were a separate species, as King believed, or a subgroup of our own species. Some believe Neanderthals died out 30,000 years ago with the arrival of modern humans, while others think that the two groups interbred, and that modern Europeans are part Neanderthal.

With Rowney, King debunked a putative fossil called **eozoon** – "the creature from the dawn of time". Eozoon (*eos*, Greek for 'dawn') was found in Canada in 1859, and caused much controversy. The structures resembled a slice through a compact head of cabbage, with numerous concentric layers that were interpreted as growth rings; found in rocks that were probably 600 million years old, they were thought to be the world's oldest, most primitive life form. Charles Darwin even rewrote a later edition of *Origin of Species* to include this evidence of evolution from a simple life form. King and Rowney found eozoon fossils in Connemara marble,* but the more they studied them, the more they realised these were simple mineral formations and not the remains of living organisms. When they first published their ideas in 1865 they were dismissed as "wholly unworthy of consideration", but they spent four years compiling evidence from rocks around the world and, in 1869, comprehensively debunked eozoon.[7]

William King, who was born in England, had a varied career: he studied anatomy, ran a bookshop and, despite having no formal qualification, became a prominent figure in the geological circles of his day – geologists remember him as the person who sorted out the English fossils of the Permian era (280–250 million years ago). He once worked in a geology museum at Newcastle-upon-Tyne, but left after a dispute over ownership of the collection, and when he came to Galway, brought part of the collection with him. King established a geological museum at the Galway college (today called the James Mitchell Museum after a subsequent professor of geology there). [20]

Of black holes and gravity

The first person to suggest the existence of what we now call black holes was **Alexander Anderson** (1858–1936) who was born in Coleraine, Co. Derry and who, for 50 years, was professor of physics at Galway. Earlier astronomers had speculated about the existence of massive, dark and invisible stars. But these ideas fell out of favour in the 1800s as scientists came to believe that gravity would not affect a light beam. All that changed in 1919, when experiments conducted during a total solar eclipse dramatically confirmed Einstein's General Theory of Relativity, and proved that gravity could indeed bend a beam of light. Anderson, writing in response to that breakthrough, in an article published in February 1920, speculated on how the Sun's gravitational field might affect a ray of light. He suggested that, if the Sun was squeezed into a relatively small space, then its gravitational field would be so powerful no light would escape. Or, as Anderson put it: "If the mass of the sun were concentrated in a sphere of diameter 1.47 kilometres, the index of refraction near it would become infinitely great, and we should have a very powerful condensing lens, too powerful indeed, for the light emitted by the sun itself would have no velocity at its surface… [if] the sun should go on contracting there will come a time when it is shrouded in darkness, not because it has no light to emit, but because its gravitational field would become impermeable to light."[8]

Anderson was the first to speculate about what would happen if a star collapsed under its own gravity, and to postulate the existence of what physicists now call a black hole. Black holes and gravitational collapse were not widely accepted until much later, and Anderson's contribution became overlooked, though it is now acknowledged. Physicist Werner Israel, an international expert on black holes, in the book *300 Years of Gravitation* (1987), describes Anderson's "intriguing speculation" as "an extraordinary anticipation of the gravitational collapse scenario". During his long reign at Galway, Anderson ensured that the physics department had state-of-the-art equipment including, in 1902, some of the newly invented X-ray and radio apparatus, and cathode ray tubes. For his last 35 years there, Anderson was also president of the college.

The 'father of American chemistry'

Ireland's loss was New York's gain. **William MacNevin** (1763–1841), banished from Ireland for his involvement in the 1798 Rebellion, went on to become a professor of chemistry in New York, and earned the accolade 'father of American chemistry'. Born near Aughrim, MacNevin studied medicine at Vienna where his uncle, Baron William MacNevin (descended from one of the Wild Geese, who left Ireland in the 1600s to serve in the armies of Europe), was physician to the empress Maria Theresa. Returning to practise in Dublin, William joined the campaign for Catholic Emancipation and the United Irishmen, and when their Rebellion failed, he was arrested and exiled. He spent four years with the Irish Brigade in France and, in 1803, published *A ramble through Swisserland*, which included an account of the campaign for Swiss independence. On emigrating to the USA, he joined the staff of Columbia University, which had been founded 50 years previously and, by 1811, was professor of chemistry there. MacNevin introduced European scientific ideas to North America: he built a laboratory, then a novel idea, so that his students could conduct their own experiments, rather than merely watch demonstrations; used modern textbooks; and championed Dalton's new atomic theory. He analysed minerals and mineral waters, and believed that chemistry could help improve farming and industry in the New World. He founded the Duane Street medical school in 1826, worked as a hospital inspector during the cholera epidemic* in 1832 and started a society to aid Irish immigrants. A monument to him stands on New York's Lower Broadway.

Ireland in warmer times

Over the past 2 million years, the northern hemisphere's climate has oscillated between freezing ice ages and warmer intervals. Evidence for these events comes from the clay and boulder deposits dumped by the glaciers, and from the remains of plants that recolonised the land each time the ice retreated. One of the most famous Irish sites for these studies is a deposit of peat and mud beside **Boleyneendorrish river** in the Sliabh Aughty Mountains near Gort. The deposit, buried under boulder clay that was dumped by later glaciers, was discovered in 1865 and has been much studied since then. The mud and clay date from 420,000–300,000 years ago, during a mild interval, between two ice ages, which has come to be known as the **Gortian interglacial** (or warm) period. Pollen studies reveal that the first plants arriving after the end of the previous Ice Age were willow, juniper, birch and various herbs. As the climate warmed, these were followed by pine trees and eventually oak, elm, holly and hazel, much as happened in Ireland at the end of every other Ice Age.

Rhododendron grew at Boleyneendorrish during that warm time, though it is not native in Ireland today. In fact, 20 per cent of the plants found at Boleyneendorrish are no longer native here. Intriguingly, however, a few species, which today are rare in Ireland, grew at Boleyneendorrish over 300,000 years ago. They include several heathers – among them Mackay's heath,* St Dabeoc's heath* and Dorset heath, which are currently found in Ireland only in Connemara and may be glacial relicts* – and pipewort,* a tubular freshwater plant that grows at a few places in the west of Ireland but is now mostly found in North America. Towards the end of the Gortian interglacial period, the climate became colder and wetter, and the oak woods gave way to alder, yew and conifers. Eventually tundra conditions returned, followed by a severe Ice Age that gripped the whole country.

COOLE

Vanishing lakes and fairy shrimp

The limestone region of south Galway and the Burren is dotted with numerous lakes. Some are permanent, but many are **turloughs**, a type of seasonal or vanishing lake that is unique to Ireland. These are special places, with their own distinctive combination of plants and animals. When they drain in summer, they can provide lush, if short-lived, grazing; when they flood in winter they are an important wild-bird site. Turloughs form in depressions that have a sinkhole, and they occur only in areas with little or no soil, where the limestone is riddled with underground passages, and where there are no surface rivers. In winter, as the water table rises, the basin floods from below; during drier weather, the water will drain through the sinkhole. (Occasionally, the sinkhole is visible; more usually it is hidden under stones and debris.) Turloughs are most common in south Galway and the Burren, in Mayo around Lough Mask, and in Co. Roscommon, though some also occur in counties Sligo, Longford and Limerick. They were first described scientifically in 1684 when William King, Protestant Archbishop of Dublin, read an account to the Dublin Philosophical Society.*Many scientists have since studied them, but there is still much that is not known about them.

Some turloughs are mere ponds, filling and emptying in a few hours; others are substantial lakes that take months to drain. The largest turlough today is **Rahasane** near Craughwell, which, when fully flooded, covers 250 hectares and supports 40,000 wildfowl; the stone walls of the summer fields are often visible in the flooded lake. Many turloughs have been reclaimed for farmland, but the lush summer grazing of a turlough can happen only if the lake is allowed to flood in winter with the mineral-rich spring water; consequently, farmers must continually add fertiliser to a drained turlough. Sadly, draining and fertilising it destroys the unique ecosystem. Large numbers of turloughs were reclaimed during the 19th century, including the largest-known, Turlough More in east Galway, which, at 650 hectares, was almost three times larger than Rahasane. All that survives today are the Clare and Abbert rivers, two essentially artificial drainage channels that were created at the time. One unusual turlough is **Caherglassaun** near

Lime – quick, and slaked

Lime kilns were common in the 18th and early-19th centuries – every townland had at least one, as did most large estates. A fine example survives at Coole: a compact stone furnace, with an inlet at the top (now protected by a grill), and an outlet below. Alternating layers of firewood, turf and limestone were loaded into the kiln from above; a fire was lit and, after a few days, the smouldering pile was sealed and left to burn without oxygen for a week. This produced quick (dehydrated) lime, to which water was carefully added to produce slaked (hydrated) lime. Slaked lime was spread as fertiliser, and was especially valuable in reclaiming poor hill land; quicklime could be used in making mortar and whitewash, or as a raw ingredient in the chemical industry. The practice of liming land began to disappear in the mid-1800s with the arrival of imported guano from South America, and the development of artificial fertilisers.* Slaking lime generates heat (what chemists call an exothermic reaction) and, in seaside areas, this was sometimes exploited to evaporate sea water and make table salt. Today, two industrial lime factories, in Cork and Carlow, produce quicklime for the building industry; elsewhere, the remains of hundreds of small local lime kilns survive from an earlier time.

Coole which, despite being fresh water and 5km from the sea, shows a tidal effect. Thanks to a subterranean connection to the sea, water levels in Caherglassaun rise and fall twice daily, with a 3-hour lag between high tide on the coast, and high tide at Caherglassaun (the water in the turlough is not directly connected to the sea, but sucked and blown by pressure changes in the connecting channel).

The plants growing in and around a turlough are adapted to the harsh and unpredictable conditions, and form distinctive contours or tide marks. Scrubby vegetation grows above the winter high-water mark; shrubby cinquefoil (*Potentilla fruticosa*), with its pretty, yellow, rose-like flower, can tolerate having its roots submerged for a while and grows at the high-tide mark. Below that come four unusual violets: first, the early dog-violet, then the common dog-violet, followed by the heath dog-violet and at the bottom the fen violet. When a turlough drains, the boulders and rocky shore are covered in a distinctive black turlough moss (*Cinclidotus*) and carpeted with the white paper-like remains of the algae that lived in the water.

Birds to watch for on the larger turloughs include wigeon, pintail, golden plover, black-tailed godwit and mute and whooper swans. There are seldom fish in a turlough – perhaps the occasional stickleback, if the lake never fully drains – but there is usually a rich collection of aquatic creatures, which survive the drought by escaping to other pools or by forming drought-resistant spores. These include insects, flatworms and snails, frogs and newt, and the **fairy shrimp**. This transparent crustacean (*Tanymastix stagnalis*) was first discovered in Ireland at Rahasane in 1974, and has since been found at other turloughs. It is not known in Britain but does occur on the continent; birds may inadvertently have carried some shrimp eggs to Ireland. The adults are large – up to 20cm long – and are often seen swimming lazily on their backs.

There are turloughs at **Coole–Garryland nature reserve**, and Coole visitor centre has a model showing how they flood and drain.㉑ Coole House formerly drew its water from a turlough, using a horse-powered pump. Coole Estate was the home of writer Lady Gregory, and long associated with the poet W B Yeats who wrote about its woods and wild swans; the estate is now part of Coole–Garryland nature reserve. Fragments of aboriginal limestone woodland survive there, with guelder-rose, whitethorn, hazel, yew and oak, and several rarities, such as bird's-nest orchid and dark-red helleborine. The nature reserve is one of Ireland's most diverse, as it has forest, dwarf woodland, limestone reefs and turloughs. Most of the flowers of the Burren* occur on Coole's limestone pavements, while the fauna include badger, pine marten, bank vole* and seven species of bat.*

A limestone desert

Politically and administratively, the three Aran Islands are in Co. Galway, but geologically and botanically they belong with the Burren* in Co. Clare. And, like the Burren, they are predominantly a desert of naked, flat limestone pavement, with occasional oases of vegetation. The islands rise gently to the southwest, culminating in cliffs that, in places, drop nearly 100 metres to the sea. That cliff-bound Atlantic coast is constantly being eroded, resulting in spectacular blow holes, and stunning storm beaches with massive boulders weighing many tonnes tossed above the high-water mark. There is precious little soil, turf, trees or water on the islands. Rain quickly disappears into the porous limestone, so, wherever possible, rainwater was carefully caught and stored in tanks. Soil for the fields in this desert was, for centuries, made by crushing the local limestone, and mixing it with sand and seaweed. Turf was imported from Connemara, carried in a unique type of local boat, known as a hooker.㉒

The islands probably have the greatest concentration of **stone walls** in Ireland – 1,500km, according to some estimates. Aran's walls are built straight onto the limestone pavement, there being no soil for a foundation. The traditional Aran gate was a simple gap in the wall filled with something that was easy to roll – often a large fishing float, or a granite erratic* boulder. Granite erratics are common on the islands, where they were dumped during the Ice Age by glaciers flowing off the Connemara Mountains.

With no reeds or rushes on the islands, the traditional Aran roof was **rye grass thatch**, made from several varieties of local grass. This thatch typically lasted only two years, one side of the roof being replaced each alternate year. Reed thatch is now used increasingly, with reeds imported from the mainland or further afield, especially eastern Europe, where the low levels of fertiliser in the rivers produces a firmer, slower growing reed that is better for thatching. Dr Micheline Sheehy Skeffington from NUI Galway is studying the rye grasses that have been grown for centuries on Inis Meáin, the middle island, and has found several unique local varieties, and growing among them arable weeds such as cornflour and darnel – eradicated from fields elsewhere in Ireland, they survive on Aran thanks to the traditional farming ways.

Like the rest of Ireland, the islands were heavily wooded thousands of years ago, and some small copses of trees survived into mediaeval times. Botanists from NUI Galway led by Prof Michael O'Connell have found pollen from pine, oak, elm, hazel, birch and other trees, preserved in sediment at the bottom of a small lake on Inis Óirr, the smallest island. Some of the plant remains from about 2,000 years ago have unusually high lead levels, possibly contaminated by pollution from lead smelters in Roman Britain. The island's flora today resembles that of the Burren, though some of the species common on the Burren, such as mountain avens, are not known on Aran. The islands do have one rarity not found anywhere else in the country: purple milk-vetch (*Astralagus danicus*), which grows in a few patches on Inis Meáin and Árann, the big island (also known as Inis Mór).

The kelp makers

From about 1720 to 1940, the Aran islanders' main cash crop was kelp. It was made by burning kelp weed, and the resulting clinker and ash was a valuable source of soda (alkali) for use in glazing pottery, in making glass* and soap, in bleaching linen,* in dyeing fabrics and even making gunpowder.* Before kelp, wood-ash was the main source of soda but, by 1700, most of the trees had gone. French chemists discovered a way of making soda from seaweed, and the practice came to Ireland in the early 1700s, spreading from there to Scotland. Black weeds, red weeds and the sea rods of the lower shore were collected in spring, stacked and dried. This dried weed was burned in summer in makeshift kilns by the shore, a kiln load taking at least a day to burn. Ten tonnes of wet weed produced 5 tonnes of dried weed, which yielded 1 tonne of kelp. By the mid-1800s, Ireland was producing 3,000 tonnes of kelp a year, 20 per cent of which came from the Aran Islands. Writer and map-maker Tim Robinson,* who lived on Aran for several years, estimated that the islanders collected 120,000 large baskets full of weed every year.[9]

Kelp was an important cash crop on the Aran Islands for 200 years. Illustration from William Harvey's British Marine Algae *(1849 edition).*

The first quality-control test for kelp was devised by Galway chemist Richard Kirwan* – before that, the alkali content was assessed by rules of thumb based on taste and appearance. In the 1860s, there was an attempt to industrialise Aran's kelp production, when the Irish Iodine and Marine Salts Manufacturing Company opened a factory at Port Chorrúch on the big island; the company operated a monopoly, forcing the islanders to sell their kelp to the factory, often at a low price. By the mid-1800s, cheaper alkali sources were available, but kelp remained important as an iodine source for the chemical industry, medicine and photography.

Clare's riches include the Burren's amazing landscape and plantlife, an exhilarating Atlantic coastline, and some of the best whale-watching in Europe. Ardnacrusha power station was an engineering triumph when it opened in 1929, and the world's first successful submarine was designed by John Holland from Liscannor. Clare (*clár*, 'a plank', or 'level place') is named after a bridge at Clarecastle.

THE BURREN

A rich and rare place

The Burren is one of Ireland's great treasures. This relatively small area in Clare's northwest corner – covering in all perhaps 600sq km – is disproportionately blessed with riches: a strange, rocky limestone landscape; tremendous ecological diversity; amazing caves; several rare and vulnerable habitats, such as limestone pavements and turloughs, or seasonal lakes; and a rich plantlife that, in early summer, is a blaze of colour. Yet Cromwell's surveyor Capt Ludlow dismissed the Burren, as having "neither water enough to drown a man, nor a tree to hang him, nor soil enough to bury [him]". Had he looked closer, though, Ludlow would have found that there is plenty of water, it is just that most of it is underground; that there are trees, only they are growing stunted and prostrate to avoid the withering Atlantic wind; and that there is soil, but it lies concealed in crevices in the rock.

The name Burren (*boireann*) means bare, and some people, seeing the place for the first time, think of it as barren and lunar. The predominantly bald and rocky scenery certainly contrasts sharply with the boggy fields to the south and east, but the Burren is not barren: over half of the plant species that grow in Ireland can be found there, many of them rarities, and the region's rich and cosmopolitan assembly of plants is unique in the world. Moreover, if it was truly barren, no one would ever have lived there, yet the region is noted for its historic and prehistoric remains, from Neolithic stone walls to 19th-century big houses, and including no fewer than 350 ring-forts alone. There are still no towns or villages within the Burren, only small clusters of dwellings, and little in the way of industry, apart from farming and tourism, though these increasingly put pressure on the fragile terrain. Ireland has many other

A typical Burren view: rain-eroded limestone pavement, vegetation growing in crevices, and dry-stone walls and buildings.
© Enterprise Ireland

limestone landscapes, most of them also called Burren, but Clare's is the most extensive – indeed, it is the largest limestone, or karst, landscape in these islands.

To the north and west, the Burren is bounded by the Atlantic (the Aran Islands* in the distance are a western extension); the region's inland boundaries are the bog to the east, and the shale to the south along a line from Killinaboy to Kilfenora and Lisdoonvarna. **Limestone** dominates the Burren. The same limestone also covers most of central Ireland, but there it is hidden under a blanket of bog, whereas, at the Burren, it was gently uplifted and forms a plateau. This plateau slopes down towards the south until, at Lisdoonvarna, the limestone disappears under shale. Shale once covered the Burren also, but erosion removed it, apart from a small cap that survives on Slieve Elva, making it the Burren's highest point. Despite the relentless erosion, the limestone in places is nearly 1km thick. The rock is made from marine mud and the fossils of prehistoric animals, and it was laid down in a warm shallow sea 350 million years ago, during the Carboniferous era. Walk on the rock, and you are walking on the remains of long vanished creatures. Occasionally, the limestone beds are interspersed with layers of shale, which reflect a temporary change to shallower conditions in the prehistoric sea.

The Burren has been shaped over thousands of years by two powerful forces: water and people. Rain, which is slightly acidic, readily dissolves the limestone, carving cracks and fissures in the rock. The extent of the erosion can be seen under some surviving Neolithic stone walls, which were built 5,000 years ago, and under granite erratics, which glaciers deposited on the limestone nearly 20,000 years ago. The walls and boulders protect the underlying pavement from the rain and as the surrounding rock is eroded, the stone walls and boulders are left standing on a pedestal or platform of uneroded rock. From the thicknesses of these platforms, geologists calculate that rain removes 1mm of limestone every decade – equivalent to 1km every 10 million years.

For all the importance of water in creating the Burren, it is seldom seen there, as most of it runs underground. It percolates down through cracks in the limestone, until it hits an impermeable layer of shale, and then it begins to eat away at the limestone. As a result, the Burren is riddled with underground passages, making it porous and thirsty as a sponge. Occasionally, the roof over an underground cavern will collapse, to form a depression or *polje*. **Poulacarran** in the middle of the Burren, the finest *polje* in Ireland or Britain, is 2km long and 80 metres deep. Springs – the Burren's main source of drinking water – are common and usually occur at a junction between shale and limestone layers; some were venerated in the past as holy wells, and all are vulnerable to pollution from the surface, such as septic tank run-off. Frozen water also shaped the Burren: Ice Age glaciers removed rock and soil from the hills, then dumped the resulting boulder clay in the valleys and depressions so that, today, these valleys are fertile oases of rich, well-drained soil. At the end of the last Ice Age, torrents of meltwater excavated caves, sinkholes and deep spillways, as well as several valleys that today are dry.

After the ice retreated, the Burren was gradually recolonised. A thin soil developed and, by 6,000 years ago, the area was heavily wooded with pine and oak. But Neolithic farmers cleared the trees and, it is thought, profoundly altered the environment – once the trees were gone, the thin soil was quickly blown away, so that today's barren landscape is a relatively recent creation and, moreover, probably man-made. The Burren is still used for **farming**, especially cattle rearing, albeit on a relatively small scale. The lime-rich pastures offer good grazing, and the dry conditions mean liver fluke infection is unlikely. Because the limestone acts as a storage heater, the hillsides remain relatively warm in winter and there is year-round grass growth, enabling Burren farmers to practise an unusual form of transhumance: in conventional transhumance, cattle are taken in summer to the high meadows; but the Burren cattle are taken uphill in winter. Artificial fertiliser, which would encourage vigorous grasses, was seldom used in the past, and this favoured wild flowers. Grazing cattle also control the grasses, and help maintain the Burren's diverse plantlife. But farming in the region is changing and becoming more intensive, and this increasingly threatens the Burren's fragile ecosystems. If cattle are increasingly kept indoors in winter, for instance, then hazel scrub would spread uncontrolled.

It is for its **plantlife** that the Burren is world famous. For it is a strange, botanical melting pot: arctic and alpine plants, normally found only in tundra regions or high mountain pastures, grow there, and at sea level, and alongside Mediterranean species. Nowhere else in the world is there such an intriguing combination. In addition, the Burren has a rich limestone flora, with many rare plants in abundance. In all, over 600 plant species, including trees and ferns, have been recorded there. The arctic and alpine plants include the brilliant-blue Spring gentian, and the white-flowered Mountain avens; Mediterranean species include the rare Maidenhair fern, and the Dense-flowered orchid (one of 16 species of orchid found in the Burren). No one knows how this cosmopolitan collection came to be. The absence of frost certainly favours the Mediterranean species, while the arctic and alpine species may have survived there from the tundra conditions that existed shortly after the last Ice Age.

The Burren also has a tremendous **diversity of habitats**: dense hazel scrub and woodlands; dry grasslands and hay meadows; coastal dunes and salt marshes; upland plateau; turloughs,* or vanishing lakes, which are a special habitat; and, of course, the limestone pavements, which themselves contain a range of habitats, with microclimates in crevices and under boulders.

There are animals in the Burren too: dozens of different butterflies, including brown hairstreak and pearl-bordered fritillary; numerous dragonflies, hundreds of moths, and a tremendous range of snails (all that calcium in the limestone is good for shells). There are seabirds at the coastal cliffs, and waders and swans at the lakes and turloughs. Rarities spotted at Lough Atedaun include osprey, gadwall and Savi's warbler. Stoat, pine marten* and red squirrel* frequent the scrub and woodland, and herds of wild goat range over the hills. The Burren's most unusual creature is a recent arrival – the bright, bronze and snake-like **slowworm** (*Anguis fragilis*). Dr Ferdia Marnell, of the National Parks and Wildlife Service, describes it as a legless lizard that lives in burrows and, unlike a snake, can close its eyelids and shed its tail if caught. Slowworms grow to 40cm long, eat slugs and insects, have a strap-like tongue, and hibernate in winter. Native to Britain and Europe, they were introduced to Ireland a number of times, starting in the early 1900s. There have been occasional sightings around the country, but the only Irish breeding colony is in the Burren, where the slowworm was first seen in 1977.

The **Burren National Park** covers 1,200 hectares of the Burren's southeastern corner, including Mulloughmore Mountain; there is no visitor centre, after controversial plans to build one at Mulloughmore came to naught, but there are privately run information centres at Ballyvaughan and Kilfenora. (23) (24) Mid-May is considered the best time to visit the Burren for the wild flowers. Places to visit include Black Head for the limestone pavements, coastal views and plant spotting; Corcomroe Abbey for its 12th-century botanical carvings; Mulloughmore and nearby Lough Gealláin turlough and the Ballyeighter lakes; and Carran's *polje*.

Carran is also home to what may be Ireland's most attractive chemical laboratory, the **Burren Perfumery**. Founded in the early 1970s, this is Ireland's oldest perfumery, where plants are distilled in a traditional pot still to extract their essential oils. (25) The rocks and plants of the Burren are protected, and it is an offence to pick or remove plants, remove stone, or interfere with stone walls. The Burren Tourism & Environment Initiative campaigns for greater environmental awareness in the area (www.burren.ie).

An underground world

The hazel wood project

The Burren may look solid, but the rocky surface hides a hollow interior. For, below ground, the limestone is a veritable labyrinth, riddled like a Swiss cheese with underground passages and caverns. Many place names contain the element *poll* ('a hole'), and the region is popular with international cavers. Some Burren caves are substantial, many of them are still active and flood quickly after heavy rain, and most should be attempted only by experienced cavers. Pollnagollum on Slieve Elva is the longest cave in Ireland, with over 15km of mapped passages. Pol-an-Ionain, near Doolin, boasts the largest-known, free-hanging stalactite in the world, a massive structure 7 metres long and still growing, albeit slowly (it was at the centre of a controversy in the late 1990s over plans to open it as a show cave). Occasionally, a cave collapses, revealing itself on the surface. A glen at Clab, 3km northeast of Carran, is thought to be one such collapsed cave.

Currently, the Burren's only show cave is **Aillwee Cave**, which was discovered in 1940 by local man, Jack McGann, after his dog disappeared into the side of the mountain while chasing a rabbit. The cave was once an outlet for an underground river. The furthest reaches are still being explored, and already 1km of passageway has been mapped, though only part is open to the public. Excavations in the 1970s uncovered the bones of a brown bear,* and three bear hibernation pits. The only residents now, however, are a colony of lesser horseshoe bats.* Before the cave could be opened to the public, the entrance had to be widened, some sections made watertight and others cleared of rubble (the rubble was used in building the car park). The visitor centre draws its water from a deep lake far into the cave. ㉖ In the 19th century, lead and silver were mined in the hill above the cave.

The Burren is mostly treeless, but parts are covered by thick hazel scrub. This is an important habitat, rich in spring-flowering plants that can tolerate the shade, such as primrose, wood sanicle, bluebell and the early purple orchid. Hazel was one of the first trees to recolonise Ireland after the last Ice Age. The nuts were an important food for hunter–gatherers 9,000 years ago, while the flexible twigs and branches were used to make rods and stakes, scallops for thatching, and wattles for wattle-and-daub structures. It is a mark of how important hazel was to the ancient Irish that the generic word for a wood (*coill*) comes from *coll* (hazel). Hazel was still important in the 1700s when, in a bid to control the old Irish ways, the Penal Laws banned the cutting of hazel. Cutting hazel twigs was an informal type of **coppicing**, and almost the only kind practised in Ireland historically. In coppicing, the main trunk of a tree or shrub is removed close to the ground, and new stems are allowed to sprout from the stump; these can be harvested after several years, and a coppiced tree can continue producing wood for centuries. In managed hazel coppices, the trees are cut at intervals of six years for slender scallops, 15 years for producing firewood and charcoal. In 1994, Aillwee Cave Company established a hazel coppice, which is managed by the Coppice Association of Ireland (*Muintir na Coille*) as a demonstration project and for woodcraft courses.

The Spectacle Bridge and spa waters

Question: How do you build a relatively lightweight bridge across a deep and narrow gorge? Answer: With a pair of spectacles. In the early 1870s, **John Hill**, engineer for Co. Clare, designed the Lisdoonvarna–Ennistymon road (now the N67). Two kilometres outside Lisdoonvarna, the route crosses the Aille river gorge that, according to engineering historian Dr Ron Cox, is 25 metres deep in places.[10] Hill found a crossing point where the river was just 15 metres below the road level, but a solid bridge even that deep would still be exceedingly heavy. Hill's solution was to remove most of the bridge's filling, by running a large tunnel through the centre, to create a relatively lightweight structure. His 'Spectacle Bridge', still in use, was renovated in 2001. Beds of shale containing fossils of extinct molluscs known as goniatites, are exposed in the rocky sides of the Aille gorge. The shale is also the source of the minerals found in the spa waters at Lisdoonvarna.

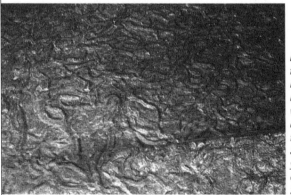

The ingenious Spectacle Bridge built in 1875 over the Aille river gorge near Lisdoonvarna. This photograph was taken before renovations began in 2001.
© Paddy Cusack/Clare County Council

They were drinking the waters at Lisdoonvarna in 1740, and they are drinking them still. Ireland had dozens of **spa resorts** in the 18th and 19th century, among them Mallow,* Lucan and Swanlinbar* – but only Lisdoonvarna survives. The town, which owes its existence to the spa, sits at the boundary between the karst limestone of the Burren to the north, and the waterlogged shale to the south. The shale is rich in iron-sulphide, otherwise known as iron pyrites or fool's gold. This chemical, which formed when the shale was a smelly, sulphurous black mud 320 million years ago, is the source of the iron and sulphur in Lisdoonvarna's wells.

The waters were first analysed in 1751, and the sulphur well, in particular, was recommended for skin problems, also for gout and "scrofulous swelling". The doses prescribed varied from tablespoonfuls to half-gallons, depending on the complaint. Local landowners, quick to spot the commercial potential, offered accommodation for "invalids and tourists" and, by 1837, there were several

holiday cottages in the vicinity. Lisdoonvarna does not appear on any map prior to 1845, yet by 1859 it had 60 houses, two hotels, a sub-post office and enough business to merit a Church of Ireland chapel of ease (most of the 1,500 visitors a year being well-to-do Protestant gentry).[11] The Midland & Great Western Railway even ran excursions from Dublin: by train to Galway, thence by steamer to Ballyvaughan, and finally by "well-appointed omnibus" to Lisdoonvarna. In 1895, when visitor numbers had swelled to 20,000 a year, thanks to improved train services, a local committee built a pump house and baths. These are still operating, run today by a local trust. ㉗

Fossils and flagstones

In the early 1900s, one of Ireland's most valuable non-agricultural exports was flagstone from Co. Clare. Thousands of tonnes of the stone were exported from the small harbour at Liscannor, whence the name **Liscannor flagstone**. Photographs of the day show the docks busy with workers and steam engines, and the harbour crowded with schooners. It was something of a 'flagstone fever': by 1890, English quarries could not meet the demand for paving stone for London, Liverpool, Manchester and elsewhere, so a dozen English companies opened quarries around Liscannor. The Clare rock was excellent, and being close to the sea was easily exported. Demand declined

Liscannor flagstone: the ripples were made by marine worms that crawled through the sand 325 million years ago.
© Liscannor Stone Ltd

during World War I, however, and all but disappeared afterwards with the switch to tarmac and concrete paving. Interest revived in 1980, and several quarries near the Cliffs of Moher and Doolin are again working. Flagstone (a type of sandstone) is widespread in Clare's shale and sandstone region. The best-known type is beautifully textured, with fossil patterns left by marine organisms that moved through the sediment 325 million years ago. An excellent building stone, it splits easily along its bedding planes to any thickness; is extremely hard, containing 85 per cent silica; and gives a good grip even when wet. The stone is typically silver-grey, though some rust-coloured beds occur. It was widely quarried for centuries, and used for walls, floors and even roofs – an old cottage on Lahinch's main street sags under a heavy flagstone roof. (28)

LISCANNOR

The world's first successful submarine

For centuries, military inventors dreamed of a submersible vessel that could attack enemy ships unseen. But a submarine is difficult to design: it must be watertight and reinforced to withstand the increased pressure; it must be able to dive and rise like a plane, and have some way to control its buoyancy; it should be able to travel both on the surface and underwater; it needs an air supply, some way of navigating underwater and, if it is to be useful as a military vessel, a weapons system.

Alexander the Great is said to have used a submersible vessel in 332 BC, to defend his ships at Tyre against divers attacking their anchors; if so, it was probably supplied with air from the surface. The first modern report of a submarine came 2,000 years later when, in 1620, Dutch inventor Cornelius van Drebel demonstrated a submersible in London. According to contemporary reports, it was rowed underwater by a crew of six, travelled 10km, carried 16 passengers and stayed submerged for three hours. It is not known how Drebel supplied air to the vessel, however, and

the secret died with him. Over the next 150 years, numerous other submarines were attempted. The most successful was the *Turtle*, built at New York in 1775 by David Bushnell, and deployed against the English fleet during the American War of Independence. The small, turtle-shaped craft had a hand-cranked propeller and a one-man crew. It was designed to clamp an explosive device onto the hull of a ship, but was defeated by the copper-covered hulls of the English warships.

The first Irish contribution to submarine development came in 1800, when Irish-American inventor Robert Fulton* built a submarine in Paris for Napoleon. His *Nautilus* successfully sank a warship in trials but, like Bushnell's *Turtle*, it was hand cranked and too slow to chase a ship under sail. The first steam-powered submarine was invented by an English cleric and inventor Rev George William Garrett in 1879. His *Resurgam* ('I will resurface'), failed to live up to its name, however, and sank on its maiden voyage (the wreck lies in shallow water off the Welsh coast). At the same time, an Irish man **John Philip Holland** (1841–1914) was building his first submarine across the Atlantic in New Jersey. Holland was born in Liscannor, where his father was a coastguard. He reputedly designed his first submarine when he was 17, motivated by nationalist ideals, and believing sub-marines were the only way Ireland could beat the might of the British fleet. Holland worked as a teacher in Ireland until 1872, when he emigrated to the USA, and began seriously to design submarines. He failed to interest the US navy, but the Irish Fenian Brotherhood funded his first three submarines out

John Philip Holland emerging from his 'torpedo boat'. There is a memorial to Holland in Liscannor Harbour. The first effective submarine periscope was designed by another Irish inventor, Howard Grubb.*

of its 'skirmishing fund'. The first sank on its maiden voyage in 1878, prompting one commentator to remark that "the professor has built a coffin for himself".

In 1881, Holland launched his cigar-shaped *Fenian Ram*. The three-man vessel was designed to ram ships and had a torpedo gun powered by compressed air. It cost $15,000, measured 10 metres long, by 2 metres wide and 2 metres high, had a 15hp engine and a toilet – but no privacy, periscope or navigational aid. In 1883, the Fenians, concerned at the escalating costs, stole the *Ram* and Holland severed his relationship with them. During the 1880s and 1890s, the US navy ran three competitions for submarine designs. Holland won all three, and this provided some funding for his research. Navy interference, however, meant the resulting submarines were often unstable. Eventually, Holland set up the Torpedo Boat Company (later the Electric Boat Company, which is still building submarines) and, in 1898, unveiled his seventh design, *Holland VII*. Previously, he used a petrol engine for both surface and underwater travel, although this was a complicated arrangement, and consumed valuable air while underwater. Now, he switched to a battery-powered electric motor for underwater travel, which was simpler and more effective. *Holland VII* cost $150,000 to build, was 20 metres long, with a 150hp engine and could dive to 20 metres. After successful trials, the US navy bought it in 1900 – its first submarine – and ordered six more, inaugurating the world's first submarine fleet. The submarine age had begun, and orders flooded in from Russia, Japan and the British Admiralty (though there are suggestions that Holland objected to the sale of his boats to Britain).

John Holland, whose many inventions included a device to help sailors escape from a sinking submarine, ended his days fighting over patent rights. In 1904, he resigned from the Electric Boat Company, but continued designing submarines, though the company tried to prevent him using the name 'Holland'. He died at the outbreak of World War I. A few weeks later a German submarine with a crew of 26 sank three British cruisers in under an hour, killing 1,400 men. Naval warfare had changed forever.

Holland's historic *Fenian Ram* was 'used' against the British after the 1916 Easter Rising – when it was exhibited in New York to raise funds for Sinn Féin; it is now in a museum at Paterson, New Jersey, where Holland last lived. Another Holland submarine, which sank off England in the early 1900s, is conserved at Gosport near Portsmouth. There is an exhibition about Holland at Clare County Museum. 29

A prehistoric Mississippi delta

Clare has a dramatic Atlantic coastline and, with clifftop walks along much of the coast, it is possible to explore a variety of features. The Cliffs of Moher are Clare's best-known coastal feature, but the most interesting stretch of coastline is between Loop Head and Kilkee. There the wild Atlantic has worked the rock into stunning cliffs, rocky headlands, teetering sea stacks and soaring arches, and you can see the fossilised remains of sand volcanoes and ancient landslides.

The northern coastline around the Burren is limestone but, south of there, the rock is a mixed geological sandwich, with layers of shale, flagstone and sandstone. These are the petrified remains of a vast prehistoric river delta, laid down 320 million years ago. 'Co. Clare' was underwater then, but for much of the time was close to the edge of a massive continent. And off that landmass flowed a river probably as big as today's Mississippi. Conditions in the prehistoric delta varied over time, perhaps as sea levels rose and fell, or there were changes in the type and amount of sediment washing in. Each change gave rise to a subtly different layer of rock – muddy layers, marine layers, layers that formed in delta swamps... and this layering can be seen in today's sea cliffs. Dark shale, for example, formed

The Atlantic river god: a wax model from the 1780s for one of the stone carvings that adorn the Custom House * in Dublin.
© Enterprise Ireland

from stinking, black sulphurous mud on the seabed. These shales are rich in an iron-sulphur ore known as pyrites, or fool's gold, and in places are so black they were mistaken in the past for coal. Inland, the shale weathers to form a heavy waterlogged soil, characterised by wet and rushy fields, and quite unlike the dry limestone to the north.

Shale forms the top 10 metres of the **Cliffs of Moher**. Below is a thick layer of yellow sandstone that formed where there were sandy channels in the ancient river delta; being harder than the shale, it resists erosion and forms a protruding ledge. Beneath the sandstone are nearly 200 metres of flagstone, of the kind quarried for Liscannor stone.* The cliffs plummet 200 metres to the sea, and nowhere else in Ireland will you find such a vertical drop (the cliffs at Slieve League* in Co. Donegal, though three times as high, are less sheer). The Cliffs of Moher are too steep for plants, but they are happily colonised by seabirds, and are home to Ireland's largest colonies of fulmar, razorbill and kittiwake. Samuel Lewis's *Topographical Dictionary of Ireland* (1837) reported that puffins were regularly collected from the cliffs by people "suspended over the lofty precipices".

The southern section of Clare's Atlantic coastline is more intricate, with sea stacks, explosive blow holes and numerous rocky headlands, each defended by a promontory fort. A mediaeval ruin perched on a sea stack at Intrinsic Bay reveals that the stack was once connected to the mainland. Two unusual features present along this coastline are fossilised sand volcanoes and turbidites, which can both be seen in place in rocks by the shore. **Sand volcanoes** form where sand is deposited so fast that water is trapped underneath; the sand accumulates until, eventually, its weight forces out the water, and it erupts like a miniature volcano. Occasionally these structures are preserved in the rock: some at Gowleen Bay, near the high-tide mark, are up to 10 metres across, and perhaps 3 metres high; smaller ones can be seen in rocks at Fisherstreet [Doolin].

Turbidites are the fossilised remains of underwater landslides that occurred in steep-sloping beds of sediment. The first evidence of submarine landslides came in 1929 when an earthquake triggered one off Newfoundland; it severed a number of transatlantic telecommunication cables, and from the time when each line went dead, geologists could calculate that the landslide moved at 130km an hour, and travelled for 800km. Submarine landslides like this can occasionally trigger catastrophic tsunami waves. Layers of turbidite can be seen around Clare's coast from Loop Head to Doolin; having formed from a jumble of mud and coarse-grained sand, their appearance is at odds with the surrounding fine-grained rock. They formed during landslides 270 million years ago, in water 1,000 metres deep, and contain occasional fossils of plants and animals that were caught up in the landslide, including even some tree trunks washed down from a river estuary. Dolphinwatch in Carrigaholt offers turbidite as well as whale-watching tours. (30)

The best whale-watching in Ireland

The waters off Loop Head are some of the richest in Europe, and regularly attract whales en route to their Arctic feeding grounds. The Shannon estuary has its own resident whales, however: a large group of about 115 **bottlenose dolphins** (*Tursiops truncatus*), one of only six such groups in Europe (the others are in Scotland, Wales, Brittany, Portugal and Croatia). There have probably been dolphins in the estuary for centuries, despite the fact that it is a busy waterway. Zoologists from UCC and the Irish Whale & Dolphin Group are studying the

The bottlenose dolphin: over 100 reside in the Shannon estuary.
© An Post

Shannon dolphins and, during 2001, eavesdropped on them with a hydrophone as part of an environmental impact study for a gas pipeline that will be laid across the estuary. Dolphin numbers make the estuary Ireland's most successful whale-watching location, and probably even the best in Europe, according to whale expert Dr Simon Berrow, who heads the Shannon Dolphin & Wildlife Foundation at Kilrush, which works to ensure that the whale-watching develops sustainably. The best whale-watching months are June–September (though some whale species can be seen all year round), ideally on a calm day (for visibility) and on a rising tide. Twenty-three species of whale are found in Irish waters. Rare visitors include the killer whale or orca (which is the largest of the dolphins) and the blue whale (at 26 metres long, the largest animal that has ever existed); the most common sightings are porpoises, dolphins and minke whales. The Irish were never noted as a whaling nation, yet the whaling gun harpoon was invented by a Donegal man, Thomas Nesbitt,* and in the early 1900s a Norwegian whaling company operated out of Inishkea and the Mullet* in Co. Mayo. In 1991, Irish waters as far as the 322-km limit, and including the Shannon estuary, were declared a whale and dolphin sanctuary.

SHANNON

Among the diamond makers

One-third of the world's synthetic diamonds are made in Ireland – half in Shannon by de Beers, and half in Dublin by General Electric. Using high temperatures and pressures, these companies produce 30 tonnes of diamonds every year, with individual stones measuring up to 10cm across.

People began trying to make diamonds in 1797, when English chemist Smithson Tennant proved that diamond, like graphite, was made entirely of carbon. At that time, the only source of diamonds were the gravel beds of certain rivers in southern India and Brazil. Some people thought these tough and brilliant gemstones were fragments of stars, others that they had been formed by lightning. It was only in 1870, when diamond-bearing volcanic rocks were discovered in South Africa, that people realised diamonds formed deep in the Earth's crust, at extreme pressures and temperatures.

For the next 75 years, people attempted to recreate these extreme conditions, and convert graphite to diamond. One would-be diamond maker was Charles Parsons* from Birr Castle, inventor of the steam turbine. Parsons conducted hundreds of diamond experiments – he subjected carbon to high currents, melted graphite at extreme temperatures and pressures, even fired high velocity bullets into carbon-rich material… all to no avail, though his experiments captured the public attention and he even featured in a novel, *Tom Swift among the Diamond Makers* (1911). One problem was generating the extreme pressures and temperatures needed, another was holding the carbon sample in place. These problems were solved in 1954 by scientists working for General Electric in New York, who finally achieved temperatures of 1,500°C, and pressures of 50,000 atmospheres, and started a billion-dollar business.[12] De Beers established a diamond factory at Shannon in 1962 and, 20 years later, General Electric opened its Dublin plant.

Diamonds have long been used industrially: the ancient Chinese used diamond-tipped engraving tools and, in the 19th century, microscope lenses were polished with natural diamond grit. Natural diamonds are still used in some applications, but now most industrial diamonds are synthetic. Prospecting for natural diamonds continues in Co. Donegal,* however, to date, only one diamond has been found in Ireland – the enigmatic Brookeborough diamond.* The crystal structure of natural diamond was first calculated by Irish scientist Dame Kathleen Lonsdale;* a type of diamond, lonsdaleite, is named in her honour.

Muck spreading and social experiments

In 1831, **John Scott Vandeleur**, a landowner at Ralahine near Bunratty, inspired by the ideas of the British socialist Robert Owen, handed his estate over to his tenants to run as Ireland's first commune. The **Ralahine Agricultural and Manufacturing Co-operative Association** elected a committee to run the commune of 53 men, women and children. This pioneering co-operative could buy large items of farm machinery, which would otherwise have been beyond the means of individual small farmers. It bought the first mowing machine in Ireland, also a horse-powered muck spreader imported from the USA, and now in the agricultural collection at Bunratty Folk Park. ㉝ Muck was traditionally spread off the back of a cart by men with pitchforks. Ralahine's horse-powered machine was state of the art: it had wooden slates and spikes and, as the horse walked forward, the whole device vibrated; this spread the muck, and did the work of four men. Ralahine commune was short-lived, however: in 1833, Vandeleur lost his money gambling, his estate was seized and the commune's rights were ignored.

Ralahine Commune's horse-powered muck spreader is now in the agricultural collection at Bunratty Folk Park.
© Shannon Development

The year after Ralahine commune started, a similar experiment began at **Ardfry** near Oranmore in Co. Galway. There Joseph Blake (Lord Wallscourt), described as "an eccentric aristocrat", ran his 700-hectare estate on a profit-sharing basis with his tenants. Technical advice was provided by Thomas Skilling, but the project closed in 1849 when Skilling was appointed professor of agriculture at Queen's College Galway.* It was another 44 years before Horace Plunkett* founded the Irish Agricultural Co-operative Society. Bunratty Folk Park's agricultural collection includes nearly 1,000 sizeable items. About half were donated by **Maurice Talbot** (1912–99), a dean of St Mary's in Limerick, who amassed an important collection of historic Irish farm implements.

Colza and canola

Brilliant yellow fields of **oilseed rape**, or colza, (*Brassica napus*) are often thought of as a late-20th century arrival into the Irish countryside, associated with EU subsidies. Yet rape was widely grown as an oil crop in Ireland in the 1600s and 1700s. The rapeseeds were crushed at oil mills and the oil, called canola, was used in lamps and as an ingredient in making soap and, later, margarine; a refined form of the oil was used as an industrial lubricant and the residue of the crushed seeds could be fed to animals. In 1696, George Pease built a large **oil mill and a soap factory** at Six-Mile-Bridge on the Owenagarney river. The mill, which functioned for 80 years, was of a Dutch design, according to current owner Hubert Roche Kelly, with an upstanding millstone that ran on its rim. It crushed locally grown rapeseeds, and produced oil for lamps and for the neighbouring soap factory.

The mill is in ruins, but the millstone survives, however, little remains of the soap factory. The fast-flowing Owenagarney was a noted milling river, with the added advantage of passing through good land and a well-populated region. It powered sawmills, cornmills and paper mills, and was navigable as far as Six-Mile-Bridge, which was a small but, for a time, locally important port.

The Earl of Cork's ironworks

Ireland's shyest mammal

Dromore Wood is a stronghold of the **pine marten** (*Martes martes*). This is Ireland's shyest mammal, and to see one is a rare treat. Chocolate-brown in colour, pine martens have a distinctive creamy bib and pale inner ears. Their body is long – up to 1 metre from tip to tip – with short legs and a bushy tail. They live in holes or crevices, but are good climbers, and their preferred habitat is scrub and woodland, especially the hazel scrub* of the Burren, where their numbers are increasing. Dromore Wood has the greatest concentration of pine martens in the country, and a study of their droppings there revealed they eat most things: from eggs, worms and berries, to fruit, bees and honey, and even a dead calf.[13]

The pine marten, probably Ireland's shyest mammal.
© An Post

The pine marten was once widespread in Ireland, but its numbers declined as woods were cleared and farmlands poisoned. It was also frequently hunted as vermin: Arthur Stringer, huntsman to Lord Conway in Co. Antrim and author of *The Experienc'd Huntsman* (1714), wrote that he hunted them relentlessly. Today, the pine marten is found mostly in the west and in the Slieve Blooms.* Dromore's other residents include stoat, otter and bank vole,* and Europe's largest breeding colony of lesser horseshoe bats.* The wood's diverse habitats include callows,* turloughs,* limestone pavement, fen,* reed beds and lakes.

In the 1600s and 1700s, there were extensive ironworks in east Clare, stretching from Feakle through Scarriff to Whitegate, where people smelted an ore that was dug from the local bogs. We tend to think of bogs as being poor in minerals, yet most contain **bog iron**,* a hard layer that forms below the surface where iron and other minerals are precipitated out. Bog iron was easy to smelt and ideal for cast-iron work, and was widely used in small foundries around the country. In 1870, some Irish entrepreneurs in London set up the Gas Purification & Chemical Company, successfully selling bog ore to British and European coal-gas companies, and huge amounts of ore were exported, especially from the Donegal ports of Ballyness and Buncrana – between 1860–80, some 80,000 tonnes were shipped from Ballyness alone. Bog iron was still mined in Co. Clare in the early 1900s.

In the early 1600s, east Clare's iron deposits attracted the interest of the wealthiest man in Ireland. **Richard Boyle** (1566–1643), first Earl of Cork, amassed a vast estate – some estimates suggest a quarter of the country – by buying out fellow Planters, defrauding the Crown of forfeited lands, and arranging judicious marriages for 10 of his 12 children (a son, Robert Boyle,* the noted chemist, did not marry; and a daughter, Mary, married against his wishes). His rent roll alone was £20,000 a year, and he reputedly made £100,000 profit from his ironworks. By 1632, Boyle had, according to local historian Gerard Madden, claimed 3,450 hectares of land in east Clare, including several ironworks. Where previously there had been small foundries, he organised a large industry, bringing in skilled Belgian workers for fear the native Irish would learn the secrets of modern smelting – likewise at Sliabh an Iarainn* in Co. Leitrim, where Boyle helped Sir Charles Coote to establish an extensive ironworks.

The east Clare iron was smelted at large blast furnaces fuelled by charcoal made from the local oak woods; the fires burned day and night, blown by big, water-powered bellows. This produced pig iron, which was then worked at a hammer works, usually located nearby – the Whitegate hammer works were at Meelick, for example, just 1km from the furnace. The only trace of these extensive ironworks today is a large furnace at Whitegate, a smaller one at Raheen and a large bog-iron mine hole at Ballymalone, near Tuamgraney. Several place names record this vanished industry, however, such as Furnacetown near Feakle, Curraghawillin (the moor of the mill) near Scarriff, and Woodford (in Irish Graig na Muilte Iarainn, village of the ironmills).

Scarriff is also home to the **Irish Seed Saver Association**, which 'curates' a wild range of living heirlooms. The ISSA rescues old Irish varieties of crops, grasses, potatoes* and apples,* many of which were previously thought to have been lost, if not extinct. Thanks to the association's husbandry, these old varieties are now on sale again. ㉞ The 4-hectare farm has a seed garden, orchard (with 38 Irish apple varieties), hay meadow and hardwood plantation. Twenty-three different potatoes are grown, including the Lumper – this piece of living history was one of the commonest varieties grown in Ireland at the time of the Great Famine.*

Construction under way at Ardnacrusha power station. The large horn-like structures are two of the four spiral casings that would deliver the water from the four intake pipes, or penstocks to the turbines.
© ESB Archives

ARDNACRUSHA
An eighth wonder of the world

In 1925, the new Irish Free State government began a most ambitious project. It was planning to divert the Shannon, build a massive hydroelectricity power station at Ardnacrusha and create the world's first national electricity grid system. The project was going to cost £5 million – 20 per cent of the national budget – and it would generate twice as much electricity as the country needed. Not surprisingly, many denounced it as a white elephant. Yet, when the Shannon scheme was finished four years later, on

time and more or less on budget, it was heralded internationally as a triumph, and one English commentator described it as "an eighth wonder of the world". It was the making of Siemens, and the birthplace of the Electricity Supply Board (ESB), the world's first semi-State body. It paved the way for the rural electrification programme, and was a model for similar large-scale hydro-schemes, such as the USA's Hoover Dam, and gave other small developing nations, especially in Africa, the confidence to attempt similar schemes. In 2002, Ardnacrusha won two major international honours: the Landmark Award ranks the power station on a par with the Eiffel Tower; and the Milestone Award, in recognition of the world standards Ardnacrusha set, puts it alongside the space shuttle and the Japanese bullet train.

The Shannon* is Ireland's largest river, yet over most of its course the gradient is so gentle that it was generally of little use as a power source. Between the Shannon Pot* and Killaloe, a distance of 200km, the river drops a mere 17 metres. Over the final 25km to Limerick, however, the Shannon drops 30 metres and, not surprisingly, there had long been proposals to harness that head of water. In 1844, Robert Kane* calculated, in his seminal book *The Industrial Resources of Ireland*, that the Shannon at Limerick was worth 34,000hp. Kane's calculation was forgotten in the darkness of the Famine years, and interest in the Shannon's potential revived only at the end of the 19th century with the development of hydroelectricity schemes. One early suggestion for a Shannon scheme envisaged four hydro-stations located at intervals between Killaloe and Limerick.

By the early 1920s, annual electricity consumption in Ireland was running at nearly 50MW, 75 per cent of which was used in Dublin. There were over 300 small power plants around the country, most of them hydro-powered, while Dublin's power was generated in a coal-burning power plant run by Dublin Corporation at the Pigeon House.* Demand for electricity was growing at an unprecedented rate, however, and the newly elected Free State government began to consider building a large hydroelectricity station. There were two competing proposals, one for the Liffey, the other for the Shannon. The Liffey scheme, masterminded by the 'grand old man of Irish engineering' Sir John Purser Griffith,* would entail creating a reservoir at Blessington,* but would otherwise have been relatively easy to construct. It had two added advantages: it was close to Dublin, the main market for the electricity; and the reservoir could double as a water supply for the city. The Shannon proposal, championed by a relatively young engineer from Drogheda, **Dr Thomas McLaughlin** (1896–1971), was a much larger scheme that would generate twice as much electricity as the country then consumed. It would entail major and costly building work and, moreover, was remote from Dublin, although as power distribution systems became more efficient, this became less of an issue. Many believed the Shannon scheme would bankrupt the new State, while naturalists worried that the works, and afterwards the varying water levels at Lough Derg,* would affect plants growing by the shore.

McLaughlin studied physics and engineering at UCD and lectured for a time at UCG, then went to Berlin in the early 1920s to work with Siemens-Schuckert. He persuaded the company to examine the potential of the Shannon scheme. Siemens had already worked on another pioneering Irish

The stamp issued in 1930 to mark the opening of Ardnacrusha power station.
© An Post

hydroelectricity project, the Bushmills tram* (opened in 1883) and, by the 1920s, there was growing interest in hydroelectricity in Germany, which had lost many of its coalfields after World War I. Siemens submitted a technical report and, despite ferocious opposition, McLaughlin convinced the government of the merits of the Shannon scheme. Work began in August 1925.

Locating a power station close to Limerick risked flooding the city with the discharge waters, so it was decided to build the station some distance away at Ardnacrusha, and channel the water to the station via a new canal. In all, the project called for: an intake weir to be built at Parteen; a 13km headrace canal to be excavated from there to Ardnacrusha; a dam, turbine hall and power station; two navigation locks; a 200-metre fish pass (then the world's largest); and a 2-km tailrace canal. Two rivers had to be diverted and four new bridges built where the intake canal cut across roads. It was a huge undertaking, and over 5,000 people were employed at its peak.[14] The chief engineer was Prof Frank Rishworth, a former colleague of McLaughlin's at UCG, who had worked in Egypt on the Aswan Dam and may have given McLaughlin the idea for Ardnacrusha.

A temporary power station was erected on the building site, to power the many workshops and a large electric crane. To move the massive amounts of clay and rock that were excavated, 100 locomotives and over 1,000 railway trucks were used. These ran on 100km of narrow-gauge railway, which could be lifted by crane and re-laid where needed. All this equipment was imported through Limerick port, which was extended to accommodate the extra traffic. Local roads were also improved, and every bridge on the route was load-tested. To improve freight delivery, a stretch of the mainline Limerick railway was electrified, the first electric mainline railway in these islands.

The intake weir diverted water from the Shannon to the headrace canal, but also raised the upstream water level by 8 metres, to match the level at Lough Derg and thus maximise the head of water. The resulting 40-metre drop between the head- and tailrace canals was breached by two navigation locks; these were equipped with diffusers, so that boats would not be rocked as the locks filled. (This approach was studied by a committee from North America and later adopted for the Great Lakes canals.) Large multi-bucket excavators were used to dig the headrace canal, while the tailrace canal was blasted through the rock at Ardnacrusha. Nothing was wasted – three large rock-crushing plants were erected on site, to crush the excavated rock so it could be re-used as hard core.

At a time of high unemployment, the construction jobs at Ardnacrusha were a boon, and men walked for days to join up. McLaughlin, who was in charge of the operation, ensured that Irish people were recruited whenever possible in preference to Germans. Many of the 5,000 workers were accommodated in temporary camps. Inevitably, there were accidents, some of them fatal, which led ultimately to new safety and training systems and industrial safety legislation.

In 1927, following a debate about how the electricity would be managed, the government established the ESB as an independent semi-State body, with headquarters at Ardnacrusha and with Tommy McLaughlin as chief executive – though some denounced the move as "creeping socialism". Throughout, the project attracted considerable media attention, both at home and abroad. Siemens took large numbers of publicity photographs, and one unexpected spin-off, according to engineering historian Paul Duffy, was a major boost for the Irish postcard industry. Noted Irish artist, Seán Keating, produced a series of paintings of the work in progress.

In July 1929, the sluice gates at Parteen weir were officially opened, and the intake canal began to fill; electricity generating started two months later. In the 1930s, Ardnacrusha produced nearly 90 per cent of the State's electricity needs. The new power supply facilitated a range of social, economic and industrial development programmes, not least the rural electrification programme and the Drumm battery train.* Ardnacrusha still generates electricity, although it now produces just 3 per cent of the national demand. The turbine hall was renovated in the 1990s, and capacity increased from the original 85MW to 110MW. The ESB has a fish hatchery at Parteen Weir, where it rears salmon and trout to restock rivers; eels*arriving in the Shannon are trapped below the power station and released upstream.

Chapter 8 sources

Co. Mayo

1. Dornan, Brian, *The Inishkeas: Mayo's Lost Islands* (Four Courts Press, 2000).

Co. Galway

2. Robinson, Tim, "A Connemara Fractal" *Technology Ireland* (June 1991).

3. Robinson, Tim, *Connemara: Gazetteer, Map* (Folding Landscapes, 1990).

4. Mohr, Paul, "John Birmingham of Tuam: a Most Unusual Landlord" *Journal Galway Arch. & Hist. Soc.* **46** 111–155 (1994).

5. *Eloge de M. le Comte D'Arcy, prononcé par le célèbre Marquis de Condorcet (1779)* (London, 1846).

6. O'Tuathaigh, Gearóid, "The Establishment of the Queen's Colleges: Ideological and Political Background" in Tadhg Foley (ed.), *From Queen's College to National University* (Four Courts Press, 1999).

7. King, William & Thomas Rowney, "On Eozoon Canadense" *Proc RIA* **10** 506–51 (1869).

8. Anderson, Alexander, "On the advance of the perihelion of a planet, and the path of a ray of light in the gravitational field of the Sun" *Phil. Mag.* **39** 626–628 (1920).

9. Robinson, Tim, *Stones of Aran* (Penguin, 1986).

Co. Clare

10. Cox, Ron & Michael Gould, *Civil Engineering Heritage: Ireland* (Thomas Telford, 1998).

11. 'P.D.', *A Handbook of Lisdoonvarna and Its Vicinity* (Dublin 1876; reprinted Clasp Press, 1998).

12. Hazen, Robert, *The Diamond Makers* (Cambridge, 1999).

13. Fairley, James, *An Irish Beast Book* (Blackstaff Press, 1984).

14. Duffy, Paul, *Ardnacrusha – Birthplace of the ESB* (ESB, 1990).

Chapter 9: *Tipperary • Kilkenny • Carlow*

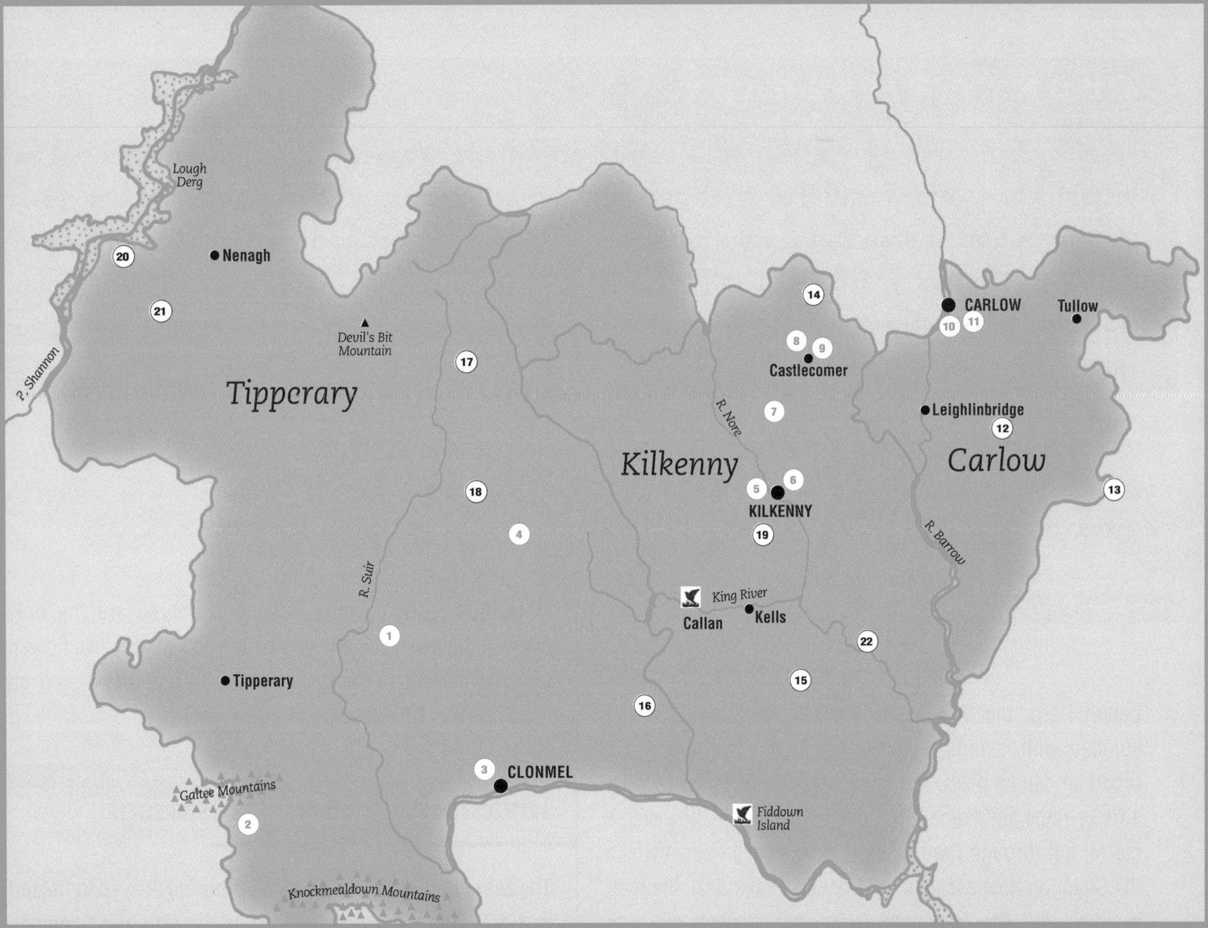

Lough Derg
Nenagh
R. Shannon
Devil's Bit Mountain
Tipperary
R. Suir
Tipperary
Galtee Mountains
Knockmealdown Mountains
CLONMEL
Kilkenny
R. Nore
Castlecomer
KILKENNY
King River
Callan
Kells
Fiddown Island
CARLOW
Tullow
Leighlinbridge
Carlow
R. Barrow

Visitor facilities

1. Rock of Cashel
2. Mitchelstown Cave
3. Richmond Mill Transport Museum
4. Slieve Ardagh Heritage Centre
5. Rothe House, Kilkenny
6. Patents Office, Kilkenny
7. Dunmore Cave, Ballyfoyle
8. Seamus Walsh Castlecomer Tours
9. Castlecomer Library
10. Carlow County Museum
11. Carlow Library

Other places of interest

12. Ballydarton
13. Bunclody
14. Clogh
15. Kiltorcan
16. Knockroe
17. Lisheen
18. Littleton
19. Millmount
20. Portroe
21. Silvermines
22. Thomastown

woodland

nature reserve or wildfowl sanctuary

Visitor facilities are numbered in order of appearance in the chapter, other places in alphabetical order. How to use this book: page 9; Directory: page 474.

COUNTY TIPPERARY

Fossils of the world's earliest land plants are among Tipperary's claims to fame. The brilliant and unconventional scientist, J D Bernal, was from Nenagh and, in the 19th century, Charles Bianconi revolutionised Ireland's transport and communications from Clonmel. Tipperary, Ireland's largest inland county, is named after Tipperary town (*Tiobraid Árann*, 'the well on the River Ara'). In turn, the river takes its name (*ára*, 'a ridge'), from high ground nearby.

The lower Shannon

Lough Derg, the Shannon's largest lake, is shallow and studded with islands, and lies where the river flooded a broad limestone plain. The lake is dammed in the south by a rocky ridge at Killaloe. During the last Ice Age, however, a glacier cut through the ridge, creating a gorge down which the Shannon now escapes to Limerick. Previously, the river probably left the lake near Dromineer, passed close to Nenagh, then took a low-lying corridor between the Arra Mountains and Silvermines Mountains. That old route may be millions of years old; today it is used not by the Shannon, but by the railway and N7 road. Lough Derg now forms part of the storage reservoir for Ardnacrusha* hydro-station, and the water level in the lake is controlled at the station's intake weir. The ESB also expanded the lake by flooding the valley between O'Brien's Bridge and Killaloe, creating a small appendix to Lough Derg. An early church on a small island that was submerged then has been rebuilt at Killaloe. Lough Derg's limestone shores are the only place in these islands where a large yellow daisy-like flower, the Irish fleabane (*Inula salicina*), is found. [Next stop on the Shannon: **The estuary at Limerick**.]

Invasion of the zebra mussel

The zebra mussel (*Dreissena polymorpha*) arrived in Ireland in 1997. This freshwater mollusc is the size of a fingernail and has distinctive zigzag bands of cream and brown. It is now found the length of the Shannon, from Limerick docks to Carrick-on-Shannon, and up the Grand Canal, but it is especially well settled in Lough Derg where it is upsetting the lake's natural balance. A prodigious breeder, it grows on anything and can reach densities of 100,000 animals per square metre. It can smother other species, clog spawning grounds, and block water intake pipes and filtration systems, boats and gear. It threatens our waterways and could become a costly nuisance.

Native to the Black and Caspian seas, the zebra mussel spread across Europe in the 1700s as canal navigation developed. The mussels travel as stowaways – as adults attached to boats and gear, or as microscopic larvae in bilge and ballast water – and they probably came here on barges imported to the Shannon. Alien species are not uncommon: most Irish freshwater fish were deliberately introduced centuries ago, and others, like the bank vole,* hitched here in recent years. New arrivals can upset the balance among existing species, and the zebra mussel is no exception.

The mussels are filter feeders and each animal filters 1 litre of water a day. This can produce a sparkling clear lake, but other species may starve, or find they cannot hide from predators in the clear water. Some duck will eat the zebra mussel, but, unfortunately, humans do not find them tasty. Scientists are studying the mussel's impact in Ireland, yet it is unlikely the visitor can be eliminated and the Marine Institute has begun a programme aimed at controlling its spread. People using inland waterways, whether boating or fishing, should comply with the regulations.

Slate quarries

The *Annals of the Four Masters* record that slate was quarried at Portroe when local man Brian Boru was king of Ireland in the 11th century. Some 850 years later, Robert Kane* noted in his *Industrial Resources of Ireland* (1844) that the quarry was still providing employment. The grey-green stone there is often called 'Killaloe slate' but the quarries are actually in Co. Tipperary, notably at Curraghbally and Gorrybeg in the Arra Mountains. Curraghbally quarry was nearly 100 metres deep, and horse-drawn windlasses were used to hoist the stone out of the pit, 5cwt at a time. In the 1840s, some 700 men and boys were

Slate, mud that was deposited in layers on an ocean floor and later compressed, cleaves easily along its original bedding layers. A versatile material, it was used as a flat surface (billiard tables, window sills); as an insulator (electrical switchboards); for roofing; and, instead of paper, for blackboards.
Source Library

employed in Portroe quarrying and finishing 10,000 tonnes of slate a year. These were prosperous times: traffic on the new canal, which bypassed the falls at O'Brien's Bridge, supported a hotel; skilled Welsh slate workers came to dress the stone (witness the Welsh names recorded in the local graveyard); and the quarry supported numerous businesses, including a boatyard. Portroe slate was ideal for roofing and was used throughout Ireland on buildings, such as Dublin Castle, St Canice's Cathedral* in Kilkenny and Coole Park House, and also in England, France and the USA. The quarries closed in 1956 after mixed fortunes and a revival in the early 1990s was short-lived.

From X-rays to D-Day

Brilliant, colourful, controversial, lifelong communist, decidedly unconventional, and with a reputation as a serial womaniser, **John Desmond Bernal** (1901-71) was a fascinating character. He laid the foundations of modern molecular biology, played a key role in planning the D-Day landings of Normandy, and helped found UNESCO and CND. His friends included Picasso (who painted a mural on the wall of his London flat, now preserved at London's Institute of Contemporary Art), and novelist C P Snow – the erudite Constantine in Snow's novel *The Search* is based on 'JD', who once joked that the only thing he knew nothing about was 6th-century Romanian churches.

Bernal was the son of prosperous Nenagh farmers who were Catholic converts (his father was a Sephardic Jew, his mother an American Presbyterian). Though educated in England – his mother felt Irish Catholic schools were "too rough" – JD was an ardent Sinn Féin supporter and spent his teens inventing weapons to advance the cause of Irish nationalism. He studied physics at Cambridge, where he earned the nickname 'Sage', joined the Communist Party and married his girlfriend, Eileen Sprague, in a civil ceremony. Appointed a lecturer at Cambridge, Bernal began working with crystallography, a new technique using X-rays to reveal the structure of small, simple crystals. But Bernal was interested in biological molecules, which are large and complex. In 1934, he succeeded in making crystals of pepsin, a digestive enzyme found in the stomach; from his X-ray photographs of the crystals, he could deduce the enzyme's structure. This pioneering work with complex molecules paved the way for further discoveries, including DNA's double helix and the structure of viruses. Many felt Bernal deserved a Nobel prize, but he never won one, though two of his students did: Max Perutz for the structure of haemoglobin (1962); and Dorothy Hodgkins for penicillin and vitamin B12 (1964).

JD's most influential book, *Social Function of Science* (1939), where he argued that scientific research must be well planned and funded, brought him to government attention, and he began working for the Ministry of Defence, though some objected to his communism. Among other things, Bernal played a crucial role in the 1944 **D-Day landings**: he designed the floating 'Mulberry' pre-fabricated harbours used by the troops, demonstrating his ideas to Churchill with paper models in a bathroom; and drew on his knowledge of mediaeval poetry to select the landing places in Normandy. (There were two other Irish contributions to D-Day: the weather forecast from Blacksod lighthouse in Co. Mayo was reputedly decisive in timing the invasion for June 6th, after it had been postponed some weeks before; and the tides were predicted using a calculator invented by William Thomson.*)

Despite his contribution, Bernal's politics meant he was refused entry to the USA after the war. He regularly visited the USSR instead, and was increasingly seen as a Soviet apologist. He became active internationally, helping to found UNESCO, the World Federation of Scientific Workers and the World Peace Council (at one time he was on 60 committees), and campaigning for better relations between scientists in the East and West. His many scientific and political honours included the Lenin Peace Prize. Public and establishment attitudes to Bernal's politics and numerous

love affairs overshadowed his brilliant research and his crucial role in encouraging a generation of young scientists, especially women in a traditionally male world. In his book *World Without War*, Bernal wrote: "I still resent the fact that… the only time I could get my ideas translated in any way into action in the real world was in the service of war."

Of salt and kidneys

Salts play vital roles in our bodies, from nerve function to mucus production, and our cells are bathed inside and out with saline solutions. One scientist who helped us understand how the body handles salts was Nenagh-born **Edward Conway** (1894–1968).

Our kidneys ensure that, regardless of what we eat and drink, the composition of our internal fluids remains constant. They do this by filtering the blood, reabsorbing essential chemicals, and excreting the waste as urine. Edward Conway investigated how tiny tubes, or tubules, in the kidney regulate levels of potassium and other salts. To extract and analyse the tiny volumes of fluid found in the tubules, Conway developed new instruments and a new

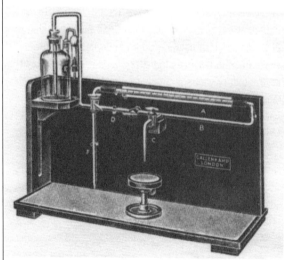

A Conway micro-burette: the instrument, made by Gallenkamp, was used to analyse tiny volumes of liquid. [From Conway's 1939 book, Microdiffusion Analysis and Volumetric Error.*]*

procedure. His practical technique was used by scientists worldwide in many other disciplines, and his book, *Microdiffusion Analysis and Volumetric Error* (1939), went to six editions and was translated into Japanese.

Conway next showed that the salt concentration inside cells is different to that outside, and that this is because the cell membrane lets some chemicals in but keeps others out. This concentration difference gives rise to a tension, or potential, across the cell membrane, a discovery that paved the way for research on nerve function by a team that later won a Nobel prize for its work. In 1948, Conway postulated, correctly as it turned out, that hydrogen ions, released when the body digests glucose in the presence of oxygen, are the source of stomach acid. Conway established a renowned research team at UCD, winning funding from the Rockefeller Foundation and the US National Institutes of Health, and many international honours, including membership of the New York Academy of Sciences. UCD's new Conway Institute of Biomolecular and Biomedical Research is named in his honour.

SILVERMINES

A mining tradition

In 1289, the sheriff of Tipperary permitted two Italians, John of Genoa and Scot de Vykes, to open a mine at Knockaunderrig near Silvermines. And, in 1375, the Anglo-Norman Butler family held the rights to all the lead and half the silver found there. These historic records make Silvermines the oldest documented Irish mine, though the area has probably been mined since the Bronze Age. The rich deposits supported a thriving local economy and, over the last 200 years, there have been mines at a dozen sites between Shallee White in the west and Ballygowan in the east (Béal Átha na nGabhann, 'the smith's or miner's ford'). The last mine closed in 1982: owned by Canadian company Mogul, it employed 500 people and, for 20 years, was one of Europe's largest lead-zinc producers.

One of the underground caverns at Shallee mine.
© Enterprise Ireland

Silvermines, despite the name, has little silver and some suggest the name was invented to attract miners. The main minerals were lead, zinc and, to a lesser extent copper, tin and iron. The minerals occur in a major fault line that runs from O'Brien's Bridge to Roscrea. This is part of the collision zone where the ancient continents of Laurentia and Avalonia* crashed together. The fault reaches deep into the Earth's crust, and provided a route through which mineral-bearing fluids could percolate upwards. The result is swarms of ore veins – most are 10–70 metres wide and perhaps 100–200 metres long – and miners chasing them created networks of underground tunnels and caverns. Despite the region's long mining history, it is still primarily an agricultural area. There are no large mining towns or settlements and most spoil heaps are now grown over, though some buildings and quarries remain, as do traces of the railway that transported the ore to Foynes port for shipping abroad.

THE DEVIL'S BIT

The world's oldest plants

The devil must have powerful teeth and jaws, for he is said to have bitten a chunk out of a mountain near Curraghbristy, leaving a distinctive gap known as the Devil's Bit. Later, however, flying over Cashel he caught sight of St Patrick and in shock dropped his mouthful, and so created the Rock of Cashel.* The devil must also be capable of powerful geological magic, for, in the 30km between Curraghbristy and Cashel, he changed the rock from its original 420-million-year-old gritstone into 320-million-year-old limestone.

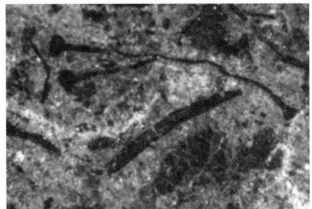

Geologists have an alternative theory: they believe the Devil's Bit was gouged out by a torrential river when the ice sheets melted at the end of the last Ice Age, releasing massive volumes of water. A deep lake of glacial meltwater was probably dammed in behind the mountain and the river escaped over the ridge onto the plain below. Sadly, because the Devil's Bit and Rock of Cashel are made of different rocks, geologists think it unlikely one came from the other.

The Devil's Bit has another claim to fame: in the 1980s, fossils of the world's oldest plants were found there and in the neighbouring Slieve Blooms* by a biologist, Dr John Feehan. Called **Cooksonia**, the fossils are 420 million years old and come from the time when life was starting to colonise dry land. This spectacular moment in the history of life began with simple plants like Cooksonia. The plants had to develop various features to survive in their harsh new environment: a skeleton or supporting structure to counter gravity now they were out of the water; a water circulation system, now they were living in the air; a waxy coating to prevent them drying out; and pores for respiration. Cooksonia was 5cm high, and probably grew in flat mats by the coast; some of the fossils have what look like spore pods.

LISHEEN

A new zinc mine

The largest, new zinc deposit found in Europe was discovered in 1990 at Lisheen and comprised 22 million tonnes of high-quality ore. Shortly before, a smaller deposit of 10 million tonnes was discovered in nearby Galmoy, Co. Kilkenny. Lisheen and Galmoy lie on a rich seam called the Rathdowney Trend, which stretches for 40km between Thurles and Crosspatrick. The ore occurs in 320-million-year-old limestone and the metals were spewed out by mineral-rich plumes erupting on the ancient sea floor.

Lisheen is jointly owned by an Irish company, Ivernia West, and international group, Anglo-American. By 2000, it was in full production with 300 employees, and has enough ore to last 14 years. The ore beds, with, on average, 13 per cent zinc and 2.2 per cent lead, are up to 30 metres thick and lie 200 metres below the surface. The ore is treated on-site to produce lead and zinc concentrates that are shipped to smelters around the world. Half the leftover crushed rock is mixed with cement and used to refill the underground caverns; the residue is deposited at a nearby tailings site. The Galmoy deposit is mined by Irish company, Arcon.

Twelve thousand years of history

In 1954, Hilda Parkes and Frank Mitchell took a core of peat 8 metres deep from a raised bog near Littleton. Parkes and Mitchell tediously separated out the pollen grains from each section of the core, identified which species of plant were present, and then counted the individual pollen grains. The result was a **pollen profile**, a comprehensive record – "a boggy archive" – showing how vegetation changed in the area over the previous 12,000 years since the end of the last Ice Age (the bottom of the 8 metres core), to the present day (at the core's top).

Each year, when plants come to reproduce, they release clouds of pollen grains. The tiny grains rain down on the earth, or settle at the bottom of pools, and are covered in sediment, before more pollen is deposited the following year. Over time, a sequence is laid down, trapping a record of the vegetation growing there each year. Each plant species produces a distinctive type of grain that can be identified using a microscope. And, because pollen grains are tough, they can survive being buried for thousands of years in acidic peat. In 1916, scientists learned how to analyse these pollen records and produce diagrams or profiles showing how the vegetation changed over time (a technique called palynology).

The pollen diagram that Parkes and Mitchell prepared for Littleton is a classic and, as a result, the period it covers in Ireland – the current warm interval after the last Ice Age – has been called the **Littletonian period**. Their diagram begins with plants recolonising the barren country after the ground had thawed. First came grasses, sedges and herbs of the tundra. These gave way, 10,000 years ago, to shrubby juniper and willow; next came birch, which gave way 9,000 years ago to hazel and pine. Oak, elm and alder, initially rare, gradually became more abundant and, 8,000 years ago, Littleton was thickly wooded with them.

The amount of pine dropped dramatically 7,000 years ago, while hazel increased. Frank Mitchell suggested this signalled the arrival of early-Neolithic farming, as hazel would have been an important crop. By then, the climate had reached its warmest and summer temperatures were about 3°C warmer than today. The next significant change in the pollen profile is the near total disappearance of elm: this elm decline, seen across Europe at the same time, was probably caused by an infection similar to Dutch elm disease. The later sections of Littleton's pollen profile reveal the eventual clearance of nearly all native woodland, and the arrival of cereals and weeds that signal a change to pasture and farming.

Years ago

Simplified version of the early part of the Littleton pollen profile (10,000–5,500 years ago), revealing the sequential arrival of new species and how the various species growing there changed over time. [After Mitchell & Ryan, Reading the Irish Landscape.]

Juniper   Willow   Birch   Hazel   Pine   Oak   Elm   Alder

Species

The sham shamrock?

While preaching at the Rock of Cashel and converting King Aenghus to Christianity, St Patrick is said to have used a small three-leaved plant, called shamrock, to explain the Holy Trinity. But, for centuries, botanists have argued about the plant's identity because, in truth, there is no such thing as a shamrock. The word shamrock derives from *seamróg*, meaning young clover. Three clovers grow in Ireland: yellow clover or the lesser trefoil (*seamair bhuí*, *Trifolium dubium*); white clover (*seamair bhán*, *T. repens*); and red clover (*seamair dhearg*, *T. pratense*). A fourth plant with a similar trefoil leaf, and known in Irish as *seamsóg*, is wood sorrel (*Oxalis acetosella*). In spring, and certainly on March 17th, the young leaves of all four species look remarkably alike.

The shamrock: a figment of our imagination? Wood sorrel (above) is one of the five species used as 'shamrock'.

In the 19th century, when shamrock was increasingly used as an unofficial Irish emblem, the plant's identity was hotly debated in Ireland and Britain by gardeners, amateur naturalists and professional botanists. In the 1890s, a Dublin naturalist, Nathaniel Colgan, conducted the first scientific study: he asked people around the country to send him samples of the "sweet little, dear little, shamrock of Ireland". These he planted and tended until they flowered in summer, when they could at last be identified. Colgan found five different species: most common was slender yellow clover, followed by coarser white clover, then red clover, black medic (*dumheidic*, *Medicago lupulina*) and finally wood sorrel.

Dr Charles Nelson recently repeated Colgan's experiment, planting 230 specimens sent to the National Botanic Gardens from around the country.[1] The most popular shamrock, and the one used by companies growing shamrock commercially, is still yellow clover. But Colgan's four other species are still identified as shamrock by many people. All this has implications for the timing of St Patrick's Day: in the 1960s the government considered holding Ireland's national holiday in summer to improve the chances of good weather. The decision was dropped and just as well: it would be difficult to find any shamrock in July.

The Rock of Cashel (*caiseal*, 'a stone fort') was the seat of Munster's kings and later an ecclesiastical centre. ① The Rock is a limestone knob that sits prominently above the surrounding plain; local limestone quarries are rich in fossils.

An erudite scholar

The **Rev John Jellett** (1817–88) from Cashel invented a device to measure the sugar content of liquids. He also wrote books about calculus, the Old Testament and *The Efficacy of Prayer* (1877), in which he proposed using science to investigate miracles and prayer.

Jellett's instrument, which he named a saccharimeter, exploits the fact that, when polarised light is shone through a liquid, any sugar present in the solution will rotate the polarisation; and the more sugar there is, the greater the rotation. By measuring the rotation caused by a mystery liquid, and comparing this with results for known sugar solutions, one could calculate the amount of sugar in the liquid.

Jellett's saccharimeter: prisms polarised the light, condensing lenses focused the light, containers held the liquids. The instruments were made by Spencer & Son, Dublin, and sold at the 1865 Great Exhibition*for £2.

Jellett was president of the RIA and professor of natural philosophy (science) at TCD, where he was succeeded by his son-in-law, the noted physicist George Francis FitzGerald.* A religious man, Jellett thought science could be used to investigate prayer, and suggested for example, a study to see if people who prayed for a long life lived longer. He would surely be interested in a recent US study which found that prayer helped heart patients recover even when they were not aware anyone was praying for them (*Archives of Internal Medicine* vol 159 p 2273).

An underground world

Mitchelstown Cave, discovered accidentally in 1833, lies in a limestone ridge between the Knockmealdown and Galtee Mountains, and in the long valley linking Mitchelstown and Cahir. The least commercialised of Ireland's show caves, it has numerous large galleries and a fine array of dripstone formations. Robert Lloyd Praeger* explored it, as did French caver Dr Edward Martel, who toured Ireland in 1895 when he also explored Fermanagh's Marble Arch Cave* in a folding boat. No boat was needed at Mitchelstown, which has not been active for centuries.

Despite the size of Mitchelstown Cave – there are over 3km of mapped passages – it was not much used historically, apart from a spell as a hiding place during the War of Independence, and few remains have been found. The cave was the first Irish cave to be electrified (in 1972); and the first to host an underground radio broadcast (1977), banquet (1972) and Mass – the latter attended by 300 people in 1980. Four species of animal apparently live there in the dark: a mite, two types of spring-tail which live on bacteria growing in damp patches, and a blind spider (*Porhomma myops*), which feeds on the spring-tails. In recent years, algae have also begun growing in the warm patches around the electric lights. ②

A long car at full gallop: a turntable at the front enabled it to turn corners.

A transport and communications revolution

Ireland's first mass transport system was set up by an enterprising Italian immigrant **Charles (Carlo) Bianconi** (1786–1875). His coaching service, which operated from Clonmel, carried people, goods and mail and revolutionised the country's transport, trade and communications.

Bianconi had shown no talent at school so his father apprenticed him to an itinerant art dealer who brought him to Dublin in 1802. Knowing little English, Bianconi was sent on foot to the provinces peddling prints, an experience which made him keenly aware of the need for cheap transport. In 1809, backed by his father, Bianconi established his own business and opened as a gilder in Clonmel. His business interests developed and he later dealt in bullion, then in demand as a hard currency, but he always felt the need for better communications, especially as the postal system was still limited and expensive.

After Napoleon's defeat at Waterloo, the export market for Irish horses collapsed and Bianconi bought a number of horses cheaply. In July 1815, he began ferrying passengers between Clonmel and Cahir at a $1\frac{1}{4}$d per mile and was an instant success. Soon, Bianconi cars were running to Wexford, Waterford, Cork and elsewhere, giving many towns a valuable new trade and communications link. Bianconi hired a large workforce and even opened a coach factory to make his 'long cars' (four-wheeled coaches with room for 15 passengers and luggage). Hearn's Hotel in Clonmel, Bianconi's former terminus, still has the clock used to time departures, also a replica long car. When the first railways were mooted, Bianconi saw them as an opportunity, not a rival, and bought shares in the Waterford–Limerick railway. He also changed his network to provide feeder services to railway stations.

Signed, sealed, stamped and delivered

From the outset in 1815, Bianconi's cars carried letters, but most of the country's post was carried by the rival mail coaches, run by the Post Office. Over the years, however, Bianconi won more mail contracts and, by 1853, had the major concession for the country. The same period saw phenomenal growth in the number of letters posted, as literacy rates improved, the mail network grew, and postage costs dropped.[2]

Before the advent of a postal service, those with money could hire a messenger to carry their letter, but poor people had to depend on travellers bringing news by word of mouth. In 1482, the English king, Edward IV, established a **Royal Mail** with relays of dedicated riders conveying the monarch's messages, and a special boat between Dublin and Chester to bring his letters to Ireland. The system gradually expanded: 'post boys', who walked 24km a day, were introduced; five post roads were established linking the major Irish cities and ports to Dublin; and the royal contractors were allowed to carry public mail for a fee – in the late 1500s, a letter from Waterford to London cost the then princely sum of £9.10s.0d The system was expensive, slow and unpredictable.

At the time, letters were also sent unpaid: the receiver paid when they collected the letter from either a post office or designated receiving house, and the charges depended on weight and distance. In 1657, Cromwell introduced a simplified system for all Ireland and Britain: a 1-page letter from Dublin to London cost 6d, and the volume of mail increased dramatically. By 1690, Ireland had 45 post offices, and a general post office located conveniently close to Dublin Castle. National mail was now carried on fast and secure mail coaches, often with an armed guard.

A separate **Dublin penny post** began in 1773. It provided a morning and evening delivery to designated receiving houses in the city, such as bookshops, and 4oz (113g) could be sent for 1d. Then John Lees, a Scot appointed as Irish postmaster in 1780, revised the national system: mail coach and road networks were extended and improved (often funded from tolls), and a new harbour was built at Howth,* Co. Dublin, for the mailboats. In 1831, the Dublin penny post and the national service merged, by which time over 400 towns were in the postal network. Remote areas were still badly served and, where neighbouring towns were not linked by a post road, the mail between them went via Dublin. The system was still complex, inefficient and unprofitable.

In 1840, the British and Irish postal system was modernised. Mail was now sent prepaid; a uniform national 1d rate was set;

For years, Bianconi cars carried Ireland's mail. An Post later paid tribute to Bianconi with commemorative stamps.
© An Post

and a gummed label, or stamp, had to be stuck to the envelope to indicate the charge had been paid. The labels ('penny blacks') bore the queen's head in profile, and were printed in black to indicate the letter was in credit. Public education programmes taught people how to use the stamps and to allay fears that the gum could spread disease. Pillar boxes were introduced, though at first people distrusted these inanimate letter collectors. And, as fees no longer had to be paid on receipt of a letter, delivery slots were cut in house doors. The first stamps had to be cut from a sheet. A few years later, **perforated stamps** were invented by a Dublin engineer, Henry Archer, who patented a machine to print and perforate 3,000 sheets of stamps an hour. He first used small wheels, or roulettes, to cut the holes; the system he perfected in 1854 punched the holes instead. By 1857, it was in use in North America.

In Ireland, the volume of mail trebled in just two years to 27 million items in 1842. Numbers employed in the postal service rocketed and sub-post offices opened around the country. In 1855, steam ships and trains made it possible to send a letter from Dublin to London in a record-breaking 12 hours. The trains and the new mailboats (by then operating from Kingstown [Dún Laoghaire]*) became mobile post offices, sorting letters en route. Services expanded: local post offices provided money orders, savings accounts, dog licences and, in 1909, the newly introduced old-age pension.

One relic from 1482 is free post for politicians: the original Royal Mail carried the monarch's messages for free, a concession eventually extended to MPs, but the system was much abused and MP's signatures were often forged (a crime punishable by seven years' deportation). In 1832, MPs could send and receive 10 letters free every day; now, TDs and senators can send 60 a day.

Bianconi lectured on transport management to the British Association for the Advancement of Science* when it met in Cork in 1843, and boasted that no vehicle of his was ever damaged or delayed even in "disturbed districts", evidence of the widespread appreciation of his service. He also claimed that his service resulted in cheaper goods: where previously calico cost 9d a yard, it now sold for 3d. In 1845, during his second term as mayor of Clonmel, Bianconi had 1,500 horses and 100 coaches covering 5,000km of road every day and serving 20 counties.

By all accounts Bianconi was a generous employer. During the Famine, he provided much needed employment at his Longfield Estate outside Clonmel, and when he retired in 1865, he distributed his business among his staff. He bought a house on Stephen's Green (No 86) in Dublin for Cardinal Newman's Catholic University (later UCD), and paid for the memorial erected in Rome over the heart of his friend, the politician Daniel O'Connell. Bianconi, however, is buried at Longfield in a vault he carved himself. The Transport Museum in Clonmel, housed in an old mill, has some original Bianconi timetables. In 1842, the 9am coach from Carlow would arrive in Dublin at 3.30pm, for instance, and the fare was 14s.6d. The museum's collection includes a unique 1914 tourer and early motorbikes, all in perfect running order, and hand-cranked petrol pumps from when petrol was 10d a gallon. ③

SLIEVE ARDAGH

Coal mines and rebellion

The small Slieve Ardagh coalfield between Killenaule and Commons has been mined on and off since the 1650s; the last mine only closed in the 1980s. The coal there is anthracite, mostly good quality, but some seams had a high sulphur content, which makes for a dirty fuel. The largest deposits lay in a depression and flooding was a perennial problem in the mines. Many of the seams were also fractured, or steeply angled, making them awkward and costly to work.

At first, local landowners opened pits on their estates: Baker in Kilcooley, Langley in Coalbrook, and Sir Vere Hunt in Glengoole. Much of the coal was used to fuel local lime kilns* to make fertiliser. In 1824, the Mining Company of Ireland leased most of the area from the local owners and began mining on an industrial scale. Steam engines were introduced to pump water from the lower mine levels and, in 1828, Ireland's first purpose-built mining town was founded at Mardyke. According to local historian Seán Watts, three streets were laid out, 33 houses were built with modern slate roofs, and a school, medical doctor and barracks (to guard the mine explosives) were installed. Great things were expected but the Mardyke pit closed in 1833 (ruins of the village and engine house survive).

In July 1848, the Commons colliery was at the centre of the short-lived **Young Irelander Rebellion**: William Smith O'Brien came to the mine to raise support for his revolution and demanding better pay and conditions for the miners. In the ensuing fracas, the police were outnumbered and took refuge in the Widow McCormick's farmhouse. Two rebels were killed in a shootout before police reinforcements arrived and arrested the Young Irelanders. The rebellion was all over in one weekend.

Mining continued there until the 1890s, employing 1,400 men at its peak, but demand declined in the face of competition from cleaner Welsh coal, and as new artificial fertilisers* replaced lime. The mines were worked sporadically during the 20th century, notably during the 1940s when the government-owned Míonraí Teo employed 100 men. In the 1950s, a new seam was discovered and a private company ran the mines, supplying coal to Thurles sugar company* for 15 years until the factory changed to oil. Most of the coal has now been exhausted. ④

Hellfire Jack and the mines

Hundreds of 19th-century miners owed their life and livelihood to a Tipperary engineer, **James Ryan** (c.1770–1847), who helped improve mine safety and efficiency. Ryan invented a drill that revealed accurate geological information about minerals present in the ground and also drilled holes to ventilate mines.[3]

Safety was a major problem, especially in coal mines: any methane in the air could be ignited by the miners' lamps and candles, and explosions regularly killed dozens of men. Prizes were offered for a safety lamp and, in 1813, an Irish doctor, William Clanny,* invented the first such lamp. (Humphrey Davy designed a less bulky one in 1815 that proved more popular.) But a decade before them, James Ryan had a more radical idea: ventilate the mines and remove the explosive methane altogether.

Ryan had invented a novel drilling machine, which he patented in 1804 following experiments near Castlecomer where he worked as a coal surveyor. His drill cut a round hole in the ground – various attachments could be used depending on the rock type – and a cylindrical core of material was then removed by tongs. Ryan intended his drill for prospecting: at the time, most surveyors simply dug a hole, but this provided little information about important factors, such as the slope and thickness of the seams and fractures in the rock. Ryan's core preserved the structures intact and the valuable geological information could be read straight off it. Ryan then realised that the drill holes themselves could be used to ventilate the mines.

Prominent figures – such as Richard Lovell Edgeworth* and Michael Faraday – championed Ryan's technique in England and, in 1808, his drill was used to clear Lord Ward's Staffordshire collieries of explosive gas. Though

popular with the miners, Ryan's approach was expensive – one early core cost the enormous price of £13 a yard – and few companies adopted it, as they would have had to bear the costs. When safety lamps became available, mining companies preferred to encourage these, since, traditionally, miners paid for their own lamps. There followed a long, controversial debate between those who favoured Ryan's approach, and those favouring safety lamps. (Although Davy's safety lamp was widely used, some say the number of accidents increased after its introduction.) Geologist and mining historian Hugh Torrens believes Ryan may also have been spurned by the British establishment because of his working-class Irish origins.

Core samples: Ryan's drill and core approach is still used by prospectors, and to extract sediment cores as at Littleton Bog. The GSI holds an archive of mineral cores in Dublin. © Enterprise Ireland.*

Nevertheless, in 1819, Ryan opened Britain's first purpose-built mining school and he worked tirelessly to improve mine safety. When inquests into mine disasters returned the usual verdict of 'accidental death', Ryan campaigned for one of 'wilful murder', earning himself the nickname Hellfire Jack. Some mines eventually adopted Ryan's approach and he later received a £50 pension from Lord Ward, whose collieries he first cleared in 1808.

Another ingenious Tipperary man who invented a range of mining equipment was **Richard Sutcliffe** (b.1849), whose inventions included a coal-cutting machine and a conveyor belt. In the 1870s and 1880s, Sutcliffe leased the Kilcoole coalfield in Slieve Ardagh, and some Castlecomer coalfields, before moving to England. His conveyor belt, launched in 1905, ran on rollers, the angle of which could be adjusted. It was designed for underground mines, and was made until recently at his factory at Wakefield in northern England.

COUNTY KILKENNY

Kilkenny can boast Ireland's last coal mine (it was still producing anthracite in 2000), two sets of world-famous fossils, the renowned scientific philosopher George Berkeley and the vast repository of Irish ingenuity that is the Patents Office. Kilkenny (*Cill Chainnigh*, 'St Cainneach's church'), takes its name from the church that stood on the site of St Canice's Cathedral in Kilkenny city.

KILTORCAN

Fabulous fossils – 1

A small, disused sandstone quarry near Ballyhale is world famous for its giant fossil ferns and primitive plants. When the fossils were discovered 150 years ago, nothing like them had been seen before and scientists came from abroad to visit this important site.

About 350 million years ago, Kiltorcan was a green oasis in a sandy, yellow desert. Enormous tree ferns grew around the shallow lake, also club mosses and horsetails. In the slow swampy streams, lived one of the first freshwater mussels, alongside primitive fish and crustaceans. There were no true trees then, no grasses or flowers, no land animals and not much oxygen in the atmosphere. We know all this because that desert became a yellow sandstone, and the oasis a green siltstone and, petrified within its layers, are the remains of the plants and animals that lived there. It is a perfect snapshot of what life was like then. **James Flanagan** (*d*.1858) discovered the site in 1851 when he was mapping Kilkenny for the Geological Survey of

Fronds of a giant tree fern that grew at a desert oasis in Kiltorcan 350 million years ago.
© Enterprise Ireland

Ireland.* The fossils include a giant fern (*Archaeopteris hibernia*), which grew to 20 metres tall; a mussel (*Archanadon*); primitive fish (among the earliest-known vertebrates); and the earliest seed-bearing plant in this part of the world, *Cyclostigma kiltorcense*, which is named after the area. Flanagan returned in 1858 looking for more specimens, but suddenly took ill and died while staying in Ballyhale. Irish museums with samples of Kiltorcan fossils include Rothe House* in Kilkenny, the Ulster Museum* in Belfast and TCD's Geology Museum.*

The King's river mills

There were once 16 mills on the King's river between Callan and Stonyford – one mill every kilometre on average, an indication of how important hydro-power was in the days before electricity. So many mills close together would have needed complicated agreements over water rights, but there may still have been disputes, especially during droughts. The monks from Kells Priory built many of these mills, including the substantial Hutchinson mill on the Callan side of Kells, which was still grinding flour in the 1970s. It was built in 1191 and the priory owned it until the monasteries were suppressed in 1536. Subsequent owners extended the building, and it had probably reached its present size when Richard Hutchinson bought it in 1872. Kells itself is an amazing village: a walled mediaeval town, it still boasts Ireland's largest monastic complex. The priory, founded in 1183 by Strongbow's lieutenant, Geoffrey fitz Robert, was a monastery-cum-barracks with substantial defences, much of which survive. The monks also had a mill and brewery on the site. (*The Book of Kells* comes from another monastic town, Kells in Co. Meath.)

A philosopher of vision

One of the great scientific philosophers of the 18th century, **George Berkeley** (1685–1753), was born at Castle Dysert near Thomastown. Berkeley's most important contribution was on the nature of perception and existence and he influenced philosophers such as Kant, Locke and Descartes. He also engaged in a celebrated argument about Newton's calculus and dividing by zero. Berkeley studied divinity at TCD* where, in an early experiment, he persuaded a friend to hang him until he was nearly senseless. In 1709, he published *A New Theory of Vision* that won him international praise. He left for England and the continent in 1713, returning for a while in 1724 as Dean of Derry, and, in 1734, as Anglican Bishop of Cloyne* in Cork.

Berkeley founded the philosophical school of **Immaterialism**. The mind was for him the ultimate reality and he viewed even intangible ideas as part of the real world. He saw time as a succession of ideas in a person's mind, and his great principle was *esse est percipi* ('to be is to be perceived'); put another way: things that cannot be perceived cannot exist. Berkeley believed his theories proved the existence of God, but his ideas were often controversial.

ÉIRE 44

Bishop George Berkeley 1685-1753

George Berkeley: in the 1720s he campaigned unsuccessfully to start a university in the new American colonies. The University of California, Berkeley, was later named in his honour. © An Post

A debate raged between his Immaterialists and the opposing Materialists, who believed that anything that exists, including ideas, must have some physical basis or cause. Berkeley, criticised by some as illogical, fought back in *A Discourse Addressed to an Infidel Mathematician* (1734), arguing that Newton's new calculus was also illogical: Newton's technique involved dividing by infinitesimal amounts, but Berkeley argued that, since infinitesimals could not be perceived, they could not exist.

In the 1720s, despairing of the Old World's decadence, Berkeley hoped for a new golden age in the Americas. He campaigned for a university there to educate colonist and native alike, and the government promised him £20,000. He moved his family to Rhode Island, but the money never arrived and, in 1731, they returned. Three years later, Berkeley moved to Cloyne, where for 20 years he proved a charitable bishop, donating rent from the demesne to help the poor. His last book, *Siris* (1744), advocates drinking tar water – which he had seen used in America – as a panacea and especially to treat dysentery. He died at Oxford in 1753.

Old woodlands

The elm-ash-oak woods near Callan at Ballykeefe, Garyricken and Kyleadohir are among the largest of their kind in the country. Protected now as national **nature**

reserves, together they cover 140 hectares. Ballykeefe is a semi-natural woodland with a rich diversity of species, including 12 mammals (from the pygmy-shrew to the badger) and an under-storey of orchids and wild raspberry and strawberry. Garyricken was originally a coniferous plantation, but the site is being returned to nature (the presence of brambles mean it is not suited to young children). Kilkenny's fourth nature reserve is Fiddown Island on the River Suir; a long narrow island covered with willow scrub and bordered by reed swamps, it is the only one of its kind in Ireland.

The man who invented the *Nautilus*

Jules Verne's fictional **submarine**, the *Nautilus*, was based on a real one built by an Irish-American inventor **Robert Fulton** (1765–1815). Fulton also invented the paddle-wheel steamboat, the double-ended ferry, a mechanical digger and a rope-making machine, and his submarines were reasonably successful for their time. Fulton's family came from Callan but emigrated to Philadelphia where Robert was raised. Though trained as a painter, he became interested in canals and boats, and a campaign for free trade that he hoped would free the seas of warships and "engines of oppression". To that end, perversely, Fulton designed a 'plunging boat' to plant bombs on the hulls of ships.[4] In 1800, Napoleon decided to try the idea against the English fleet and gave Fulton money to build his *Nautilus*.

The result was a sealed craft made from copper sheets fixed over iron ribs, with a glass-covered porthole and enough air for four men and two candles for three hours (a tank of compressed air was added later). It had a collapsible mast and sail, and a hand-cranked propeller for moving underwater. The craft dived by flooding its ballast tanks, and surfaced by expelling the water using compressed air. The plan was for the crew to drop underwater bombs either beside an enemy ship, or nearby, and let the bomb be carried alongside by the current. These bombs were copper cylinders stuffed with gunpowder, and equipped with a clockwork mechanism, flint and starting pin. Fulton named them **torpedoes** (a word now used for self-propelling bombs), after the fish that stuns its enemies with a bolt of electricity.

Fulton's Nautilus: sheathed in copper, it had a collapsing umbrella-like sail, a conning tower with glass portholes, and an auger to drill holes in enemy vessels; the ballast pumps and propellers were hand-cranked; a torpedo could be towed on a rope.

The *Nautilus* successfully sank a schooner in trials but, in a battle off Brittany, the English fleet set sail before the hand-cranked submarine could sneak close enough to plant any mines. Napoleon gave up on the idea. The English toyed with it for a while and, eventually, the US government funded Fulton's ambitious design for a steam-powered submarine to carry 100 men. The *Mute* was finished after Fulton died, but was never used and rotted at its mooring.

Meanwhile, Fulton invented an improved **steamboat**. Early steamboats used a jet of water ejected by a steam pump to work a simple paddle, but Fulton arranged several revolving paddles in a large wheel, and placed one of these paddle wheels on each side of his boat. In 1807, his first steamboat completed the 240km voyage from New York to Albany in a record 32 hours, compared with four days by sailing boat. Fulton started a shipping business in New Orleans, invented the double-ended ferry for short river crossings, and was instrumental in designing the Erie Canal on the Great Lakes, but he died of a chill before he could enjoy any success. Fulton's Philadelphia birthplace is now a museum and the USA issued a stamp commemorating him in 1965. In Paris, *Rue Fulton* is named after the man who first built the *Nautilus* there in 1800. The other great Irish submarine inventor was John Holland.*

The 'aerial chariot'

The history of aviation is littered with the ruins of fanciful flying machines. One such was the 'aerial chariot' invented in 1856 by Kilkenny man **Godwin Swifte** (Viscount Carlingford and a descendant of Dean Jonathan Swift*). He built his machine inside his home at Swifte's Heath, just outside Kilkenny city. When finished, a hole had to be knocked in the dining-room wall to let it out. Shaped like a boat on wheels, it had large wings and a screw propeller. Swifte persuaded his butler to pilot the machine on its maiden voyage but, when the chariot was launched from the top of the house, it came heavily to earth. The butler apparently escaped with some broken bones. The patent for Swifte's machine, and the remains of its rudder and a wheel, can be seen at Rothe House Museum. ⑤ The first successful Irish aeronaut was a balloonist, Richard Crosbie,* in 1785. In 1909, Harry Ferguson* (of tractor fame) achieved the first successful powered flight in Ireland with his home-made plane. Rothe House also has artefacts relating to the local coal mines and culm 'bomb' making.*

Wishful thinking? Godwin Swifte's 'aerial chariot', more like a boat with wings, made one (unsuccessful) voyage.
Courtesy Kilkenny Archaeological Society

Marble halls

The magnificent St Canice's Cathedral, noted for its carvings and stonework, is a perfect place for a **geological tour**: you can examine fossils crowded in the limestone, view local rocks in close-up, compare colourful marbles, and admire the work of mediaeval master stonemasons.[5]

The cathedral was built between 1215–32 on the site of an early Christian monastery (of which only the round tower survives), and it has been destroyed, restored and repaired at various times in its history. The earliest structures were built with local stone, as rock then was seldom hauled over land for more than 8km; more recent work, however, incorporated exotic stone from farther afield.

Limestone, dolomite and sandstone are the main local rock types and all were used in the cathedral. Kilkenny limestone is used throughout: dark grey, rich in fossils, and quarried at various sites near the city, it polishes beautifully to produce Kilkenny black 'marble'. Dolomite, a tough paler limestone, was used in the old round tower and as a decorative stone in the cathedral; it is still quarried near Bennettsbridge for road stone and fertiliser (lime). Local Old Red Sandstone – the remains of an ancient desert – was used in door and window arches. The roof is green Portroe slate.* The mediaeval stonemasons did use one imported stone: Dundry freestone, a soft, pale limestone from Bristol, was brought to Kilkenny by boat and barge and used for carved details. Dundry was popular with Norman stonemasons, who may have come from Britain to work on Kilkenny cathedral. Recent renovations incorporated Coolcullen sandstone from Castlecomer; black Welsh and Spanish slate; Caenstone from Brittany for the pulpit; white Italian Carrara marble behind the altar; and green Connemara marble,* along with Cork red marble (another polished limestone) in the sanctuary floor.

*Wax model for a carving of the Nore river god – Kilkenny city's river. The carving is on Dublin's Custom House. *
© Enterprise Ireland

An archive of ingenuity

The **Patents Office** is a veritable repository of Irish ingenuity. This is the national body that assesses and grants patents for Ireland, and it holds details of every invention, trademark and design submitted and granted since the office was established in 1927. All this information can now be examined at the office's library. ⑥

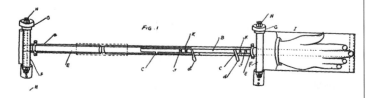

The first application received by the Irish Patents Office: a signalling device for motor cars, invented by Mrs Clara Boag of Dublin. Her telescopic tube was mounted on the windscreen; a flag at each end with a painted hand was activated by a spring. Courtesy Patents Office

With a patent, the State gives an inventor exclusive rights to make, use and sell their idea, or intellectual property, for a fixed term (10 years for a short-term patent; 20 years for a full one). In return, the inventor reveals the details of their design and pays an administration fee. Crucially, a patent prevents others in the jurisdiction exploiting the idea, unless licensed to do so by the inventor. To qualify, an invention must meet various criteria, so applications are assessed by skilled patent agents. The idea of a patent developed in 15th-century Venice and quickly spread. In Britain, 'letters patent' were granted by the Crown until, in 1852, the government set up a statutory Patents Office. The Irish office was established in 1927, and received its first application in January 1928 from Mrs Boag. The first Irish patent granted was in July 1929 to Hannah Mary Smith of Florida, USA, for a portable starting cage for racing dogs. (Mrs Boag's patent was granted the following month.) Notable Irish inventions, all featured in this book, include the industrial whiskey still, the first high-speed movie camera, the modern stethoscope, the ejector seat, artificial fertilisers, perforated postage stamps, the hypodermic syringe, cream crackers and bacon rashers.

MILLMOUNT

Colles' break

An unusual fracture of the wrist is named after **Abraham Colles** (1773–1843), a surgeon born at Millmount into "a Protestant English family… of good means, long settled in Kilkenny". Colles' passion for medicine was triggered when,

as a boy, he found an anatomy book in a field where it had been left by a flood. After studying in Dublin, Edinburgh and London, he worked in various Dublin hospitals including the Meath* and Dr Steevens'.* He was known as a dexterous operator, prepared to experiment, and was the first in Europe to tie the innominate artery. However, at one post mortem he is said to have admitted, "Gentlemen, it is no use mincing the matter; I caused the patient's death." In 1814, he was the first to describe a peculiar fracture of the arm just above the wrist that can happen when someone falls on their palm, still called Colles' fracture. Later, he specialised in treating the deformities caused by syphilis and recommended lowering the doses of mercury used to treat the infection, as he realised the mercury poisoning was often worse than the disease. Professor of surgery at the Royal College of Surgeons of Ireland,* he is buried at Mount Jerome, * Dublin.

DUNMORE CAVE

A Viking massacre

Dunmore, site of a Viking massacre, is the most historically interesting of Ireland's show caves. The *Annals of the Four Masters* relates that, in 928, Dutfrith leading "the foreigners of Áth Cliath [Dublin]… plundered Dearc Fearna [Dunmore] where more than 1,000 persons were killed". Excavations in the 1970s seemed to confirm this, when the remains of 44 people, mostly women and children, were found amid sediment on the cave floor. It is not known how they died, but they were probably from the nearby ring-forts and sheltering in the cave. The excavations also turned up silver pennies from 879–926, four were from York, from where Dutfrith was expelled in 927. The Irish did not use coins, so these were probably dropped by a Viking, perhaps as he lifted his arm in attack – in those pocket-less days, coins were often wrapped in cloth and attached by beeswax to underarm hair.

The cave's impressive entrance is a natural amphitheatre that formed 3,000 years ago when part of the roof collapsed. The cave is small, with under 400 metres of passages, but the fine formations include a 1-million-year-old stalagmite that is 6 metres high. The wildlife includes bats,* which inhabit the more remote recesses, and a fungus growing on candle grease dropped by earlier visitors. ⑦

Ireland's last coal mine?

Jim Saunders working a seam 45cm high in 1955. Anthracite was mined at Deerpark from 1928–69. Miners walked 5km underground to the coalface and spent their day in cramped conditions.
Courtesy Seamus Walsh

Six men were still mining coal near Castlecomer in January 2000. Their small operation – a cul-de-sac tunnel running 400 metres into a hillside – continued a 300-year-old tradition and produced anthracite for local use. The mine has since closed, however, because such old ways do not adapt easily to modern safety requirements.

Castlecomer plateau, home of the extensive **Leinster coalfield**, is a distinctive region that forms the watershed between the Barrow and Nore rivers. It has coal, iron ore, numerous mineral springs, extensive fire-clay deposits and famous fossils. In 1600, an iron mine opened at Ballinakill and, in 1630, the miners struck coal. Gerard Boate* recorded in his *Irelands Naturall History* (1652): "There be coals enough in this mine for to furnish a whole country… these coals are very heavy and burn with little flame, but lye like charcoal, and continue so for seven or eight hours, casting a very great heat." It was in fact good quality anthracite, low in sulphur, clean to burn and among the best in Europe.

The coal attracted Sir Christopher Wandesforde from Yorkshire, whose descendants developed Castlecomer demesne and ran mines there for 300 years. The first mines were small bell pits – a deep pit was dug down as an opening and a few men mined the coal at the bottom of this, forming a bell-shaped hole. The pits later caved in, leaving dozens of depressions that, in places, make for difficult terrain. The 19th-century mines were conventional underground networks; some more recent works were

opencast. The mid-1800s saw a massive influx of miners and, during the Famine, 13,000 people lived on the Castlecomer Estate. 'Assisted emigration' later cut this in half, with many of the miners settling at Minersville in Pennsylvania.

The largest modern mine was at Deerpark, near Castlecomer, which employed 1,000 men at its peak. Before they could start work, the miners walked 5km underground each day to the coalface, in a tunnel little bigger than the tubs used to cart out the coal.[6] Coal from neighbouring mines was processed at Deerpark: the anthracite arrived in large buckets via overhead ropes many kilometres long; it was graded and bagged according to size; then taken by rail to Castlecomer and Kilkenny.

Deerpark closed in 1969 when the government subsidy ran out. The mine openings were blocked, the pumps turned off and the levels allowed to flood. Few traces remain above ground of Castlecomer's industrial past: some old mine buildings are now silage pits; an occasional ruined chimney indicates a hidden site; even the spoil heaps are small as most of the rubble was taken in the 1970s to make cement. Author and former miner Seamus Walsh gives tours of the area. ⑧

Dancing the culm and building with clay

Dotted around this coal-mining area are some unusual stubby-looking grinding stones. These are **culm crushers**, used to grind culm, a poor-quality coal dust that, in the 19th and early-20th centuries, was the primary fuel for poor local farmers. Unlike conventional millstones which are laid flat, culm crushers are used standing upright on their rim. They were connected to a post at the centre of a small stone pavement and a horse walked the stone around in a tight circle, crushing the culm as it turned. The culm was ground to a putty with yellow marl clay, and the mix was shaped into ball-shaped lumps (also known as bums and bombs), which were left to dry before burning. Some estates provided private culm crushers for their tenants, while communal ones were set by the roadside. People without ready access to a stone sometimes crushed the culm under foot, a practice called 'dancing the culm'. Castlecomer is the only place in Ireland where you will find culm crushers; most of the grinding stones there date from the 1850s to the early 1900s and many were also used on occasion to grind corn, bones, chalk and mortar.[7] Culm crushers can be seen at Rothe House Museum in Kilkenny.

In the 1940s, Castlecomer was the scene of an unusual experiment, when a number of houses were built there with local clay. It was as part of a campaign during the Emergency (1940–47) by architect Frank Gibney to persuade people to use local building materials. As well as using cheap and readily available clay, Gibney believed that thick-walled, earthen houses could be superior to thin-walled modern ones. Castlecomer, with its long tradition of clay industries and brick and tile factories, was an obvious choice for Gibney's experiment.[8] First up was a traditional thatched cottage built in 1940 at nearby Railyard. Clay was packed into timber casings to form the walls, which were coated inside with a traditional layer of dung and lime. The only non-local material used was the glass in the windows.

In 1944, Gibney built a number of modern, semi-detached bungalows with tiled roofs on the Old and Yellow roads. Despite their success, people associated clay houses with mud cabins and poverty and Gibney's idea never took off.

The caterpillar tread

An early version of the caterpillar tread that enables vehicles to cross rough and soft ground, was invented by Castlecomer engineer **John Walker** (c.1841–1901). A man of many trades, it was said of him that, if he could not find the machine he needed, he invented it. Among his creations were a patent cuff link, a self-drying pen, and an overshot water turbine for his mill; he invented his caterpillar tread to help haul logs over rough ground to his saw mill. A century before, Longford-based engineer Richard Lovell Edgeworth* had suggested something similar to cope with boggy ground, a kind of 'portable railway' that was attached to and moved with a carriage, but his idea was never developed.

In Walker's design, a wide, chain-like belt was fitted around the pair of wheels on each side of the vehicle. The belt was made of flattened boxes bolted together, each box about 0.3 metres wide by 0.6 metres long, and the wheels, as they turned, moved the belt. A prototype was demonstrated in Castlecomer in 1896 and the *Railway Supplies Journal* described it as "a remarkable novelty".[8] Walker sent a version to London, but failed to interest the War Office in his invention. (During World War I, when the mud of Flanders forced the British army to develop a caterpillar tread, it used a similar design patented by the English company, Hornsby, in 1907.) Walker's design had slightly more success in the USA where his brother William helped commercialise it. Castlecomer Library has information about Walker and the local mines. [9]

Fabulous fossils – 2

Coal was mined on and off from 1797 at Clogh's Jarrow colliery (named after Jarrow in England). Fossil plants occasionally turned up in the coal but, in 1864, the miners found unusual lizard-like fossil animals. These were shown to the great English naturalist Thomas Huxley, who identified them as some of the world's earliest-known amphibians, including seven types that had never been seen before. Amphibians are the ancestors of all land animals, the first creatures that could survive for a time out of water, and today they are represented by frogs and newts. Jarrow is one of only a few sites where **early amphibian fossils** are found. Huxley, the man who famously championed Darwin's theory of natural selection, wrote about the Jarrow discovery in 1867 and brought it to international attention.

The Jarrow fossils date from the Carboniferous era 320 million years ago, when the northern hemisphere was blanketed by extensive swamps that later formed vast coal deposits (very different from the desert conditions 30 million years before then, which are preserved at Kiltorcan*). Complex plants had evolved by then and these gave off oxygen, enabling amphibians to emerge on land. The Earth's atmosphere was also now dense enough for enormous dragonflies to take to the air.

Keraterpeton galvani *was small and lizard-like with scaly armour. Nothing remains of the Clogh mine, but the Natural History Museum* (Dublin), the Ulster Museum* (Belfast), TCD's Geology Museum* and Rothe House Museum* have Jarrow fossils. [From Huxley's 1867 book on Jarrow's amphibians.]*
Courtesy TCD Geology Museum

The Jarrow fossils are small – *Keraterpeton galvani*, for example, named after Mr Galvan who found it, is less than 30cm long. Most have a small skull, and many have been flattened so that their four limbs are splayed out beneath them. The Jarrow fossils also include other creatures, such as fish that lived in the swampy lakes.

Megalithic astronomy

On the evening of the winter solstice, the rays of the setting sun illuminate the inner chamber of a passage grave at Knockroe. Smaller than Newgrange,* which is aligned with the rising sun on the same day, and less well known, Knockroe has, nonetheless, much in common with the Boyne valley tombs. Like the Boyne monuments, it is on raised ground near a river (in this case the Lingaun); consists of a large cairn with, in this case, two passage tombs (one opening east, the other west); and has numerous decorated megaliths (10 kerbstones and 20 more in the passages). The artwork here is simpler, however, consisting of cup marks and curved lines, but none of the ornate designs seen in the Boyne valley. Finally, like Newgrange, Knockroe was faced with white quartz quarried in Wicklow or Lambay and presumably transported to the site by boat. Knockroe was excavated in the 1990s by Prof Muiris O'Sullivan. Dúchas is preparing the site to facilitate access.

The brilliant John Tyndall, who first explained why the sky is blue, and Samuel Haughton, who invented the humane hangman's drop, were both from Co. Carlow. The heart of Ireland's sugar industry is there, along with a laboratory that creates new Irish potato cultivars. No trace remains, however, of the four lakes (*ceather-lough*) after which Carlow is named, but they may have been where the Burren and Barrow rivers meet.

CARLOW TOWN

The hangman's drop

The **Rev Samuel Haughton** (1821–97), a Carlow-born polymath with an interest in geology, medicine, mathematics and climate, invented the humane hangman's drop. Hanging was initially introduced as a prelude to drawing and quartering, and was not intended to kill the condemned person outright. After the practice of drawing and quartering was dropped, executioners would hang the condemned person from a short length of rope, usually about 1 metre long; but the drop was never enough to break their neck, and the unfortunate person slowly strangled, their agonies adding to the spectacle for the watching crowd. Samuel Haughton argued for a more humane approach and, in 1866, calculated that a 4.5-metre drop was needed to break the neck of someone weighing 160lb (72.5kg), thus minimising their pain and suffering.

Haughton was born into a well-known Quaker family and later ordained into the Church of Ireland. When he was professor of geology at TCD, he decided to study medicine, and then combined his knowledge of anatomy and geology to show that many animal fossils had been distorted by geological processes. From this he could measure the distortion in the rocks, a significant breakthrough at the time. Haughton also studied how muscles and bones are attached in fossils and used this to reveal how the extinct animals would have moved.

As well as the hangman's drop, Haughton produced the first reasonable estimate of blood pressure, based on how far blood spurted from a severed artery (the experiment was performed on a dog). He also calculated the age of the Earth,* using various approaches, including rates of sedimentation and cooling: at a time when most people were suggesting ages of at most a few hundred million years, Haughton calculated that the Earth was 2.5 billion years old. Being firmly anti-Darwin, however, and unable to conceive of such an age, he later revised this figure down to 100 million. This prolific author published numerous books on science, geology and mathematics that went to many editions. Tides and currents were another of his specialities and the police used his help on occasion to investigate the causes of shipwrecks. In 1862, Haughton published a manual on tides and currents in the Irish Sea that, he hoped, would reduce the number of shipwrecks in the channel.

Haughton's uncle, the anti-slavery campaigner **James Haughton** (1795–1873), was a vegetarian and teetotaller who supported the temperance movement and Catholic Emancipation. He was instrumental in having the rowdy Donnybrook Fair banned, and persuaded Dublin Zoo to open on Sundays as an alternative entertainment to pubs.

CARLOW TOWN

Sweet beet

Ireland's **sugar industry** probably owes its existence to the Napoleonic wars. After the Battle of Trafalgar, when Nelson blockaded the French ports and severed Europe's supply of sugar cane, Napoleon turned to a newly discovered process for extracting sugar from beet, a sweet vegetable. So began Europe's sugar-beet industry, helped in part by the fact that anti-slavery campaigners preferred locally grown beet to sugar cane imported from plantations. By the 1820s, sugar beet was being grown in the west of Ireland and processed in an Achill mill where the beet was ground and the juice extracted. (There are records of sugar mills in Belfast and Dublin in the 1500s that processed raw sugar cane; but most of Ireland's sugar then would have been imported.)

There were various attempts to establish an Irish sugar industry in the 19th century, notably the Royal Irish Beet-Root Sugar Factory, which opened at Mountmellick* in 1851 (heavy taxes imposed from London, however, made it uneconomic and it closed 10 years later). Ireland's modern sugar industry began in 1926 when a Belgian company opened a factory in Carlow with funding from the Free State. The plant was later bought by the government, and three further factories opened in Tuam, Thurles and Mallow.

In the 1980s, the Tuam and Thurles plants closed, but Mallow and Carlow continue to process 1.5 million tonnes of beet each year, producing over 200,000 tonnes of sugar. The beet campaign runs from September to January, when production is round the clock. Sugar is made by evaporating and crystallising the beet juice; by-products include molasses, and beet pulp, which is used as an animal feed. The company was privatised in the 1980s and is now part of Greencore plc. Though we in Ireland eat sugar, it can also be put to industrial use: replacing non-renewable petroleum, for example, in making plastics and chemicals; and fermenting it to produce ethanol (alcohol) for use as a fuel.

A sugar beet weighs 1kg and contains 14 teaspoons of sugar. Sugar is an Arabic word and similar words are used the world over: sucre, azucar, sokeri, zahar, siúcre…
© Irish Sugar

CARLOW TOWN

The potato breeders

The potato (*Solanum tuberosum*) is so closely associated with Ireland, it is easy to forget that this important staple came from the high Andes of South America. It is related to both the edible tomato and sweet pepper, and the poisonous nightshade (*S. dulcamara*). Note that while the potato plant's underground starch organs (the potatoes) are edible, its true fruit, a tomato-like 'potato apple', is toxic.

The potato arrived in Europe in the late 1500s and quickly became popular. The first Irish spuds were reputedly grown by Walter Raleigh* at his Youghal Estate, though others say the potato washed ashore from Armada shipwrecks and some early works referred to the potato as *an spáinneach* ('the Spaniard'). The potato is ideally suited to Ireland: it grows well even in our wet climate and on sour boggy land and, until Blight* arrived in the 1840s, no disease worried it. The potato is also useful, nutritious and versatile: it can be fed to people and animals, is used to make everything from potato cakes to *poitín*, can reclaim untilled ground, cooks quickly, and needs no processing (significantly, no miller need be paid to grind it).

There was one problem: in the cold Andes potatoes can be freeze-dried and stored for years, but in Ireland they do not keep beyond one season. Consequently, each spring there was usually a short famine between the end of one harvest and the arrival of the next. Despite this, the potato became the staple food of Ireland's poor: in the 1830s, 2 million hectares of potatoes were grown, usually in cultivation ridges* (or lazy beds), often fertilised by seaweed; and each hectare produced 7 tonnes of potatoes. This success may account for the rise in Ireland's population before the Great Famine: potatoes and milk provided a balanced diet and even the poorest family could usually grow enough for most of the year. Only in the 'oatmeal zone' of Ulster was the potato a minority crop.

The old varieties of the Famine era – the Black, Apple, Cup and the notorious Lumper potato – have been replaced by modern ones. Many were bred at **Teagasc's Oakpark research station,** where scientists have produced dozens of new varieties, including the internationally successful Cara, a high-yielding, blight-resistant variety that accounts for 20 per cent of Britain's potato crop. Teagasc's floury Rooster variety is popular in Ireland, where we still consume more potatoes per person than anywhere else in Europe. Ireland also sells seed potatoes to 40 countries, including South America, the plant's traditional home.

LEIGHLINBRIDGE

1999 – the day the sky fell

Can stones fall from the heavens? For centuries, tales of **meteorites** were dismissed as the stories of gullible country bumpkins. After all, as Aristotle had explained, each heavenly body occupied its own sphere in the sky, and nothing might move between these orbits. If a rock did fall, it must have been either an 'exhalation' from a volcano, or

formed by lightning. As late as 1807, the US president Thomas Jefferson preferred to believe that the professors were lying, than that a piece of Heaven could fall to Earth. Yet already, people studying meteorites realised that, regardless of where they fell, these rocks were similar – usually heavy, metallic and with a blackened exterior – and, in 1802, a chemist called Robert Howard concluded they came from outer space, and reported his finding to the Royal Society in London.

Leighlinbridge can claim its own piece of Heaven on Earth: a meteorite that fell shortly after 10pm on the night of November 28th, 1999. (Meteorite comes from *meteoros*, Greek for 'of the air', whence also meteorology.) The fireball was seen racing across the sky, and the impact, which some thought was an explosion, reportedly shook several houses. It was the first recorded fall in Ireland since the Bovedy meteorite of 1969, and the world's last recorded fall of the second millennium. Leighlinbridge's meteorite probably broke up as it entered the Earth's atmosphere, and fragments may have been widely scattered. Several were found on the Bagenalstown Road, weighing in all nearly 300g, and later sold to a Scottish meteorite dealer. The fragments are burned-looking, and rich in nickel and iron, and have since been identified as a 'rocky chondrite' meteorite (specifically, type L6, shock stage S3 and weathering grade W0), which is the most common type found. From its argon gas content, geologists conclude that the Leighlinbridge meteorite was thrown into space by a collision 500 million years ago, but the rock itself formed 4.5 billion years ago, and as such is a relic from the early days of our Solar System. Small slices of the Leighlinbridge meteorite are held by TCD's geology department and London's Natural History Museum.

In all, two dozen meteorites have been found in Ireland, including a hefty lump that fell at Brasky in Co. Limerick on September 10th, 1813 (now at the National Museum,* Collins Barracks, Dublin, apart from small pieces held by UCC's geology museum* and Schull planetarium*, Cork);

another fell at Dundrum in Co. Tipperary in 1865 (held at TCD's Geology Museum*); and the Bovedy meteorite, which fell in two pieces in counties Antrim and Derry in 1969, is on show at Armagh Planetarium*. These days, most meteorites hitting Ireland probably land unseen, as fewer people work outdoors, and a small meteorite would be hard to spot amid all the vegetation. (The best meteorite hunting grounds are Australia and Antarctica, where the blackened rocks are easily spotted against the desert or snowy background.)

LEIGHLINBRIDGE

Why is the sky blue?

The great, and often controversial, John Tyndall.

One of the great scientists of the 19th century, the often controversial **John Tyndall** (1820–93), was born in Leighlinbridge. Tyndall is best remembered as the first person to explain why the sky is blue, but he was also the first to suggest the greenhouse effect, he helped prove the germ theory of disease, and invented a precursor of the fibre-optic cable. Many of his discoveries stemmed from his interest in dust, which he called "the floating matter of the air".

Tyndall's family settled in Ireland in the 1600s. His father, an RIC constable, was not well off and Tyndall's early education went no further than Ballinabranagh National School. There, however, the noted Catholic schoolmaster John Conwill taught him science and surveying and prepared him for a job with the Ordnance Survey in Ireland.* Tyndall moved with the survey to England, where he later worked as a schoolteacher. At 28, he give up teaching to study physics at Marburg University under

John Tyndall explained that the **sky is blue** because dust in the air scatters much of the short-wavelength blue light (we now know clusters of molecules also contribute to this effect). This makes the Sun appear orange-red (if there was no scattering, the Sun would look white). The setting Sun, low in the horizon, is seen through a thicker slice of atmosphere that removes nearly all the blue light, leaving an even redder Sun.

Robert Bunsen (of burner fame) and, within two years, had completed a PhD. Back in England in 1853, he was invited to give a lecture at London's prestigious Royal Institution (still the venue for the annual Christmas science lectures for children). So successful was he that, within weeks, he was appointed a professor there, and he later succeeded his friend Michael Faraday as the institution's director. Tyndall was a talented lecturer and populariser and his books sold widely; some were even translated into Chinese, Indian and Japanese. A highly successful lecture tour of the USA in 1872 raised £7,000, which Tyndall donated to scholarships there. And when the *New York Tribune* published his lectures as a supplement, it sold an extra 50,000 copies in one day.

One of Tyndall's main interests was the atmosphere, and his experiments with dust particles in the air led him to discover why the sky is blue (see: panel above).[9] He demonstrated that our lungs filter particles from the air, that bacteria are present in air everywhere, and that food kept in sterile air does not putrefy. These findings confirmed Pasteur's germ theory of disease and delivered the final blow to the idea of spontaneous generation of life. The work led Tyndall to invent a sterilisation process (**tyndallisation**, still used in the food industry) in which food is heated briefly but repeatedly so that organisms not killed the first time will be killed the second time. (In pasteurisation, the food is heated once but for longer.)

Tyndall's early work on crystals led him to study ice crystals and glaciers. He visited the Alps frequently and discovered that glaciers move like ice-skaters: the pressure from the weight of the glacier melts the ice underneath, creating a layer of water that the glacier can 'skate' on. Meanwhile, Tyndall became a keen mountaineer and was the first person to climb the Weisshorn. He bought a summerhouse near the Matterhorn where a peak, Pic Tyndall, is named after him.

Another of his inventions was the **light pipe**: shine a light beam along a pipe filled with water and the light will be trapped within the water even when the water flows out the end of the pipe (a phenomenon called total internal reflection). Tyndall prophetically thought his pipe might one day prove useful – and modern fibre-optic cables now exploit the same principle to carry telephone conversations encoded in laser light beams.

Tyndall was a towering figure in Victorian science and friendly with Charles Darwin, Thomas Huxley and Michael Faraday. In 1876, aged 56, he married Louise Hamilton, daughter of Lord Claud Hamilton. Argumentative and some would say pugnacious, Tyndall was often embroiled in controversy, usually over science but sometimes religion and politics (he was a staunch Unionist and opposed Home Rule). The biggest controversy erupted after he addressed the 1874 annual meeting of the British Association for the Advancement of Science* which met that year in Belfast. Tyndall, the association's president, argued in favour of science and materialism, and against religion. It was a flashpoint in the science versus religion debate that also involved Darwin and Huxley and, in the ensuing storm, Tyndall was denounced from the pulpits.

His later years were plagued by ill health and insomnia. On the morning of December 4th, 1893, his devoted wife, intending to give him a large dose of indigestion remedy, mistook the bottles and gave him instead an overdose of sleeping draught. "My darling," he told her, "you have killed your John." Stomach pumps and emetics failed and Tyndall died that evening. His wife, who survived him by 47 years, erected a memorial to him at their alpine home. There are also commemorative plaques to Tyndall at Leighlinbridge

Main Street and at Ballinabranagh National School. The Tyndall Centre for Climate Change at the University of East Anglia, England, is also named after him. Carlow County Museum, [10] Carlow Library [11] and the Tyndall Bar in Carlow all have Tyndall memorabilia.

BALLYDARTON

The last Irish wolf

Ireland's last wolf was probably killed at Ballydarton in 1786 by a local man, Mr Watson. The animal was, ironically, hunted down by its domestic cousins – Watson's pack of hounds. The reindeer,* bear* and giant elk* that roamed Ireland with the wolf 10,000 years ago had long been extinct here, but the wolf survived, adapting to the changing climate and conditions and preying on sheep in the absence of reindeer. Its death marked the end of an era and the 'taming' of Ireland.

The wolf: hunted to extinction in Ireland in 1786.
© An Post

The wolf (*Canis lupus*, in Irish *mac tíre*, 'son of the country', also *fael*), was once common in Ireland. Its fossil remains are frequently found in caves alongside those of its prey. Early hunters probably used tame wolves and, according to the Brehon Laws, the Irish sometimes kept them as pets. As wilderness gave way to pasture, however, the wolf was seen as civilisation's enemy. By 1500, it had been hunted to extinction in England, after which it is said, sheep numbers dramatically increased and the price of wool plummeted. Ireland's wooded fastnesses gave the wolf sanctuary for a further 200 years, and something of a reputation – in Shakespeare's *As You Like It*, Rosalind refers to "a howling of Irish wolves against the moon". During the 17th century, a major campaign was waged against Irish wolves. The English declared that Ireland would be tamed only when the last wolf was gone and Cromwell offered a bounty for each one killed. Various places claim the dubious distinction of being the site where the last Irish wolf was killed – the Cork–Kerry border in 1710, for example – but the credit is generally given to Ballydarton.

TULLOW

Written in stone

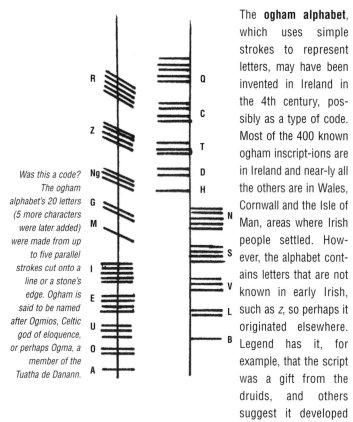

Was this a code? The ogham alphabet's 20 letters (5 more characters were later added) were made from up to five parallel strokes cut onto a line or a stone's edge. Ogham is said to be named after Ogmios, Celtic god of eloquence, or perhaps Ogma, a member of the Tuatha de Danann.

The **ogham alphabet**, which uses simple strokes to represent letters, may have been invented in Ireland in the 4th century, possibly as a type of code. Most of the 400 known ogham inscript-ions are in Ireland and near-ly all the others are in Wales, Cornwall and the Isle of Man, areas where Irish people settled. How-ever, the alphabet cont-ains letters that are not known in early Irish, such as *z*, so perhaps it originated elsewhere. Legend has it, for example, that the script was a gift from the druids, and others suggest it developed from a runic system or as a transliteration of the Roman alphabet. Ogham has none of the complexity of other writing systems, being more akin to Morse Code. But, if it was a code, why would it have been invented, as those who could read in AD 400 could surely read ogham? Perhaps, since most of the surviving examples are inscriptions on commemorative or burial stones, it was part of some druidic ritual.

Ogham is a clumsy and lengthy script, easy to misread and miswrite, and even a short inscription takes up considerable space. All surviving inscriptions are on stone, though wood and metal may also have been used. It was still employed in the Middle Ages and explanations from manuscripts, especially the 14th-century *Book of Ballymote*, have been used to decode it. The inscriptions provide us with the earliest-known examples of the Gaelic language. One stone found near Dingle, Co. Kerry, reads: *doveti maqqi cattini*, which has been decoded as: '[the stone] of Dovetos, son of Cattini'. In addition to standard ogham, other versions were later invented using words, such as place names and colours, but especially the names of trees, to denote letters. In tree-ogham, the old Irish names for trees stand for the letters of the alphabet, thus: *ailm* (Scots pine) for *alpha*, *beithe* (birch) for *beta*, and so forth. In 1998, the large woodland *Pairc Oghama* (Ogham Park) in Tullow was planted with the 20 native species of the tree-ogham alphabet to mark the 1798 Rebellion.

BUNCLODY

The flora of Carlow

Bunclody lies across the border in Co. Wexford but, for years, it was home to the botanist who conducted the most thorough survey of Co. Carlow's flora. **Evelyn Booth** (1897–1988) was born in Annamoe, Co. Wicklow, and her maternal grandmother, Caroline Hall-Dare, founded the Bunclody Lace School. Booth was a keen amateur botanist, angler and huntswoman who rode with the Carlow hounds, and knew the local territory well.[10] She specialised in the plants of Ireland's southeast in general, and Co. Carlow in particular, contributing information to numerous surveys and publications. By 1954, she had supplied information to surveys of Irish and British flora on 584 plant species for Co. Wexford, 579 for Co. Carlow and 584 for Co. Kilkenny. Her own book, *Flora of County Carlow*, was published in 1979 when she was 80 by the Royal Dublin Society (with help from another noted Irish botanist, Maura Scannell). The first Irish county flora by a woman, it includes chapters on Carlow's geology, soils and climate. The core 111 pages contain detailed data on over 700 plant species from the county, a testament to Evelyn Booth's dedicated work.

Chapter 9 sources

Co. Tipperary

1. Nelson, Charles, *Shamrock: botany and history of an Irish myth* (Boethius, 1991).
2. Reynolds, Mairead, *A History of the Irish Post Office* (Mac Donnell Whyte, 1983).
3. Torrens, Hugh, "James Ryan and the problems of introducing Irish new technology to British mines" in Bowler & Whyte (eds), *Science and Society in Ireland* (1997).

Co. Kilkenny

4. Colden, Cadwallader, *A Life of Robert Fulton* (New York, 1817).
5. Tietsczh-Tyler, Daniel, *The Building Stones of St Canice's Cathedral* (Geological Survey of Ireland).
6. Walsh, Seamus, *In the Shadow of the Mines* (Seamus Walsh, 1999).
7. Conry, Michael, *Culm Crushers and Graining Stones in the Barrow Valley and the Castlecomer Plateau* (Chapelstown Press, 1990).
8. Lyng, Tom, *Castlecomer Connections* (Castlecomer History Society, 1984).

Co. Carlow

9. Tyndall, John, *The Floating Matter of the Air* (London, 1881).
10. Scannell, Maura, "The power of the amateur tradition" in Mary Mulvihill (ed.), *Stars, Shells & Bluebells* (WITS, 1997).

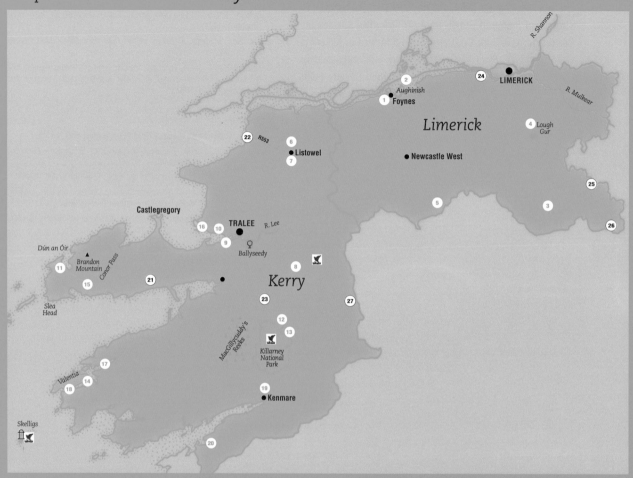

Visitor facilities

1 Flying Boat Museum
2 Aughinish Alumina
3 Kilfinane Outdoor Education Centre
4 Lough Gur Information Centre
5 Dairy Co-operative & Plunkett Museum, Drumcollogher
6 Listowel Library
7 Vintage Wireless Museum
8 Crag Cave, Castleisland
9 Blennerville Windmill
10 Tralee–Blennerville Steam Railway

11 Músaem Chorca Dhuibhne, Ballyferriter
12 Killarney National Park
13 Muckross Traditional Farm
14 Valentia Heritage Centre
15 Oceanworld, Dingle
16 Fenit Seaworld
17 Valentia Observatory, Caherciveen
18 Skellig Experience Visitor Centre
19 Kenmare Heritage Centre
20 Derreen Gardens, Lauragh

Other places of interest

21 Anascaul
22 Ballybunion
23 Ballydeanlea
24 Carrigogunnel
25 Galbally
26 Kilbeheny
27 Knocknageeha

♀ woodland
🦆 nature reserve or
    wildfowl sanctuary
🗼 lighthouse

*Visitor facilities are numbered in order of appearance
in the chapter, other places in alphabetical order.
How to use this book: page 9; Directory: page 474.*

COUNTY LIMERICK

Limerick highlights include romantic Lough Gur, volcanic Carrigogunnel (with basalt columns like those at the Giant's Causeway) and Foynes, which was once at the centre of transatlantic aviation – innovations used there included piggyback take-offs, in-flight refuelling and Irish coffee. The county also produced a pioneering aviator, Sophie Peirce. Limerick derives from *Luimneach* ('bare patch of ground') referring to the land around the lower River Shannon.

When boats could fly

Transatlantic flights began inauspiciously in May 1919 when four US Navy seaplanes left Newfoundland for Europe. With only primitive navigational instruments, they were guided by searchlights from ships strategically placed at 80km intervals. One plane completed the journey, taking 11 days to island-hop 6,000km across the Atlantic to Lisbon. Yet, just one month later, Alcock and Brown* daringly flew nonstop to Connemara in 16 hours, and not in a seaplane, but in an aeroplane.

Commercial transatlantic flights began in 1929 with Zeppelin airships, as the aeroplanes of the day were too small to carry commercial loads. The Hindenburg disaster in 1937, however, prompted the airlines to turn to seaplanes, or flying boats. These could land at sea if fuel ran low or the weather was rough and, moreover, they did not need a runway. Foynes was selected as the ideal eastern terminus for a transatlantic seaplane service: it had a deep

A Mayo Composite: a Short S20 seaplane locked onto an S21. On July 19th, 1938, this piggyback arrangement enabled the S20 to cross the Atlantic in a record-breaking 13 hours 29 minutes. Foynes Flying Boat Museum

and sheltered anchorage, a railway connection and a fuel depot. Critically, for pilots flying with simple instruments, the Shannon estuary was a large and recognisable landmark. With the addition of a weather station and radio equipment, Foynes became the first 'Shannon Airport'. The radio was used for air traffic control, communications, and weather forecasting, though the newfangled electric razors used by the US aircrews reputedly interfered with the signals.

Two test flights in 1937 with Short S23 craft proved the Foynes route was feasible. The journey time was 15 hours, and take-off was at night, so that the trickier job of landing could happen in daylight the following morning. Not being pressurised, the planes were forced to fly relatively low and through ice-clouds. This, coupled with the seaplane's limited range, prompted the airlines to investigate two fuel-saving techniques. First, the Mayo Composite, which had an ingenious piggyback system to get the seaplanes airborne: the departing plane was hoisted by crane onto a larger craft which then carried it aloft. It was tried for the first time in July 1938 and set a record of 13 hours 29 minutes to Newfoundland. Second, the hazardous technique of in-flight refuelling: a converted bomber aeroplane, equipped with a rocket harpoon to connect mid-air, acted as the petrol tank; it took off from flat land across the estuary at Rineanna in Co. Clare, later the site of the modern Shannon Airport. Both techniques were used for mail flights, but were not suited to passenger flights, which had to await the development of pressurised craft capable of flying above the ice-clouds.

The first scheduled passenger flight arrived at Foynes in July 1939. Soon, Pan Am, BOAC, Air France and Imperial (later British) Airways were flying from there. Journey times coming from North America were usually an hour or two faster than journeys heading west, thanks to westerly tailwinds. In summer, the route went via Newfoundland, but in winter a longer southerly route was taken via the Azores. These flights, which could carry 36 passengers, were the Concorde of their day: a return transatlantic ticket cost the equivalent of around €5,000, passengers dined on gourmet meals, some planes had beds – one even had a honeymoon suite – and there was a relief crew with separate sleeping quarters.

Politicians, military officials and celebrities all landed at Foynes and, thanks in part to Ireland's neutrality, services continued during World War II, though often under a veil of secrecy. The war, however, brought major improvements in aeroplane design, and the days of seaplanes were numbered. The last flights at Foynes were in 1947, by which time the modern Shannon Airport had opened at Rineanna. Shannon Airport's other claims to fame include the invention of duty-free shopping and of **Irish coffee**. The latter was created in 1942 by Foynes airport's chef, Tyrone-born **Joe Sheridan** (1902–1962), to warm miserable passengers whose flight had been forced back by bad weather. The Flying Boat Museum, located in the old airport building, has original radio equipment, flying boats and rare archive film. ①

The Shannon estuary

From Limerick city to Kerry Head and Loop Head, the Shannon estuary stretches over 100km. It is a varied stretch of water, with broad and narrow sections, deep water and sandy shallows, and home to an internationally important population of bottlenose dolphins.* Where the bedrock is soft limestone, the river spreads luxuriously; but where it crosses tougher shale, it is more constrained, notably at Foynes, where it squeezes through a deep gap in the shale that was cut by glaciers during a recent Ice Age. Shallow sandbanks occur at Scattery Island and Kilrush, where the same glaciers deposited thick beds of debris. The channel at Ballylongford, however, and from Foynes to Tarbert, is deep. As a result, there were once plans to develop Ballylongford as an oil refinery, and Foynes has long been an important port: steamers plying the Limerick–Kilrush route picked up passengers there in the 19th century, and ore from the Silvermines* was shipped from Foynes in the 20th century.

The estuary is naturally navigable, which was good for trade but made the area vulnerable to invasion, so a 24-gun battery was installed on Foynes Island during the Napoleonic wars. Access is more difficult at Limerick city, however, where the channel is narrow and shallow, and tidal conditions are tricky. Consequently, Limerick only began to develop as a port around 1750, though it now handles over 8 million tonnes of freight a year. The Shannon navigation survey in 1833 proposed various improvements to port facilities and suggestions to reduce flooding, but these were never fully implemented. An improved weir, lock and docks were completed in 2000, connecting the estuary with the navigable Shannon waterway upstream of Limerick. Meanwhile, new navigational aids at the mouth of the estuary guide the enormous coal tankers serving Moneypoint power station through a channel that, in places, is 300 metres wide.

Europe's largest **alumina plant** is on Aughinish Island in the estuary. It works nonstop, digesting 2 million tonnes of reddish bauxite ore each year, producing 1 million tonnes of alumina (Al_2O_3), the aluminium oxide that is the halfway stage in the process of extracting aluminium metal from bauxite ore. Aughinish Alumina, owned by US metals group Glencore, imports bauxite from Guinea. The ore is crushed and mixed with caustic soda, heated (to 1,100°C) and pressurised; the resulting slurry is filtered and dried to yield the white salt-like alumina. This is shipped to other European smelters for further processing, while the leftover red mud is stored at Aughinish. The huge Aughinish plant is a veritable small town, with its own services, and consuming an estimated €1 million worth of energy each month, more than any other Irish industry.

As an estuarine island, Aughinish is rich in birdlife. Its diverse habitats include salt marshes, scrubland, mud flats and freshwater lakes and the island is home to a colony of bank voles.* The alumina company established a bird sanctuary, nature trail and woodland area there, which are open to the public. ②

The basalt columns at Knockfeerina, sketched by George du Noyer* for the Geological Survey of Ireland in the 1850s. Geological Survey of Ireland

ago. Lava erupted from numerous vents, hot ash was spewed high in the air, and molten magma rose from the bowels of the Earth and cooked everything it touched. The legacy of those ancient events is a rich and complex volcanic landscape that stretches from Mungret south to Ballingarry and east to Pallas Green. There you will find all manner of volcanic and igneous rock, including basalt columns similar to those at the Giant's Causeway.* The most obvious feature is the numerous small hills, many of which are the remains of volcanic 'plugs', as at Derk and Kilteely. These formed where magma plugged a vent; erosion later removed the soft surrounding rock to reveal hills of hard volcanic material. Cromwell Hill, in contrast, is a basalt sill that formed where lava intruded between existing layers of rock. Limerick's basalt columns, though much smaller than the Giant's Causeway, are nearly 200 million years older. They can be seen at various places including Knockfeerina, Knockseefin near Linfield – where the pentagonal colonnades reach 60 metres high – and most notably at Carrigogunnel. This castellated crag overlooking the Shannon estuary has a high cliff face with layers of volcanic ash interspersed with basalt columns. Kilfinane Outdoor Education Centre runs field studies courses exploring the locale. ③

A volcanic landscape

1832 – the coming of cholera

This part of Ireland was rocked, scorched and smothered by lava during powerful volcanic eruptions 330 million years

The first person to realise that cholera victims need salt-and-water rehydration was an unusual Limerick-born physician, **William Brooke O'Shaughnessy** (1809–1889), who also introduced cannabis to Western medicine and the telegraph to India. Cholera is a virulent, often fatal disease that causes uncontrollable diarrhoea. A victim quickly dehydrates and can loose half their body weight in hours. Today it is a global problem, but, until 1817, it occurred only in India. Then, perhaps because a new virulent form had evolved, or because large numbers of people were on the move (1817 was a famine year), it began to spread. It moved along trade routes, reaching Britain in 1831, Dublin in March 1832 and Limerick two months later. A terrifying epidemic ensued. Towns and cities were worst affected, where up to 80 per cent of inhabitants died, and an estimated 50,000 fell to cholera that year in Ireland.[1] Panic-stricken survivors fled for the countryside. Worst of all, no one knew what caused the new disease, nor why some were struck and others spared. Some thought it was spread by contagion and that wearing poplin would keep it at bay. The physician Sir James Murray* believed it was caused by electricity and recommended insulating houses. Others prescribed bloodletting and turpentine enemas.

William O'Shaughnessy, who was studying medicine in Edinburgh, took an analytical approach. Examining urine, faeces and blood from cholera victims, he realised death was due to dehydration and loss of body salts, and recommended giving people a salt solution. This method was tried with some success by a few doctors, but most remained unconvinced and it was over 50 years before O'Shaughnessy's saline rehydration therapy was widely adopted. By then, people knew what caused cholera. The first breakthrough came during the 1854 epidemic when, in a classic study, a London doctor, John Snow, proved that water was the problem: people using contaminated water sources fell ill, while neighbours with clean wells were unaffected. In 1883, a Prussian scientist Robert Koch showed the cause was a bacterium, named *Vibrio cholerae*. Despite medical advances, cholera remains a problem, especially in crowded areas with poor sanitation.

After the 1832 cholera epidemic, O'Shaughnessy moved to Calcutta where he ran a successful medical school and became surgeon-major. He introduced hemp, or cannabis, a traditional Indian drug, to Western medicine – doctors in Ireland could prescribe it until the 1930s – and campaigned to build an electrical telegraph across India. In 1852, he was appointed director of the Indian Telegraph Company and, despite rough and remote terrain, built 5,000km of line over the next three years, creating an important communications tool for British rule in India. Thrice married, O'Shaughnessy retired to England where he died in 1889.

Woulfe's bottle

The Woulfe bottle is a simple yet important piece of scientific apparatus invented by an eccentric Limerick man, **Peter Woulfe** (1727–1803). Woulfe, whose family may have been among Limerick's Palatine settlers (who arrived from Germany in 1709), represents an era when medicine was still dominated by chemists concocting potions, and when alchemy was still in vogue: Woulfe himself was an excellent analytical chemist, but also an ardent alchemist and, for a time, surgeon-general at Guadalupe.

Peter Woulfe lived most of his professional life in London where his circle included the famous English chemist Joseph Priestley. The bottle Woulfe designed in 1767 merely had a second neck, yet this simple innovation

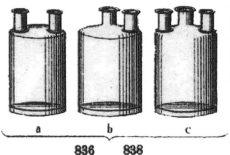

Woulfe bottles of various designs from a laboratory catalogue c.1900. TCD Chemistry Deptartment.

a   b   c

836   838

enabled him to bubble a gas through a liquid and thus purify the gas by removing "hurtful fumes". Water-soluble contaminants, for example, could be removed by bubbling a gas through water. The Woulfe bottle was a crucial piece of equipment in the development of modern practical chemistry and, in 1768, the Royal Society awarded Woulfe its highest prize, the Copley Medal.

Woulfe is also remembered as the person who discovered pure tin metal in Cornwall in 1766, and for being the first to make picric acid (from *pikros*, Greek for 'bitter'). This explosive yellow chemical, derived from indigo, was used as a dye and later in munitions; the Kynoch* factory produced it at Arklow, and derivatives of it are still used in armour-piercing shells. In 1782, Woulfe was asked to judge the claims of an alchemist, Dr Price of Guildford, who offered the king gold he claimed to have made from red and white powders; Price eventually killed himself rather than face an inquiry. Ironically, Woulfe was himself an alchemist who sought the 'Elixir of Life'. When he took ill in 1803, he refused medical attention, and prescribed himself a journey by mail coach to Edinburgh that, alas, proved fatal.

The colour of seaweed

Seaweeds grow on the shore according to their colour. Grass-green species hug the upper shore, where they are reached only by very high tides and by spray. Brown-green seaweeds occupy the mid and lower shore, where they are covered and exposed twice daily by the tides. The seaweeds of the lowest shore and the undersea are red however; they are seldom exposed to the air and have adapted to life in a dim underwater world were light is faint. The person who first realised that seaweeds could be classified in this way was a Limerick botanist, **William Henry Harvey** (1811–1866).

A Quaker, Harvey attended the noted Ballitore* school, where an accomplished botanist, James White, was the master. During school holidays, Harvey studied seaweeds on the beaches of Co. Clare and, when he was 20, he discovered a new moss, *Hookeria laete virens*, at Killarney. In 1835, Harvey went to botanise in South Africa, where his brother was a colonial treasurer. His brother died, however, and William took over as treasurer, but he preferred to collect plants and not taxes, and quickly became an authority on the South African flora. Despite his overseas posting, Harvey had enough information in 1841 to publish a comprehensive guide to the seaweeds of the British Isles, in which he outlined his classification of seaweeds according to colour and shore zone, a system still used today.

Alaria (a la, 'a wing'), from the 1849 edition of Harvey's guide to British seaweeds.

Thinking that sunlight could not penetrate water to any extent, Harvey wrongly thought that the red underwater seaweeds survived without light, perhaps by consuming some chemical form of energy. In fact, these seaweeds contain a red pigment, phycoerythrin, which is more efficient than green chlorophyll at utilising sunlight, and enables the seaweeds to survive in dark water. In 1844, Harvey was appointed curator of the botanic gardens at TCD,* and later the college's professor of botany. He continued collecting plants around the world, and various species are named after him, including a South African genus of parasitic plants, the *Harveya*.

Invasion of the giant hogweed

There is no mistaking giant hogweed (*Heracleum mantegazzianum*). Its hairy indented leaves are 2 metres wide, with dark reddish spots; the flowering stalks can be 4 metres tall and dwarf a grown person; and the white flowerheads are the size of an umbrella and shed tens of thousands of large seeds. No wonder that Victorian gardeners thought it was spectacular and imported it from the Caucasus. It was brought to Ireland in the 1880s, and spread slowly around the country at first, occasionally escaping into the wild. Since the 1970s, it has become widespread, however, and its burning, corrosive sap and vigorous growth make it both a health hazard and an ecological threat. Giant hogweed sap contains toxic furocoumarins that can cause a serious chemical burn and painful blisters, and leave the skin permanently sensitive to sunlight. The plant's vigorous growth results in dense canopies that shade out other riverbank plants. When this happens, the banks can destabilise and collapse, leading to erosion, murky waters, fish-kills and poor spawning grounds. The heavy seeds rely on streams for dispersal, so most hogweed plants are found on riverbanks, and nearly all Irish river systems are now infected.

Most invading alien species are tolerated – usually there is little that can be done, other than let nature find a new ecological balance. The Office of Public Works and the Central Fisheries Board (CFB), however, have decided to try to exterminate hogweed and, in 1997, began a pilot programme along the Mulkear, an important salmon and trout river. The plant was first recorded there in the 1930s by noted naturalist Robert Lloyd Praeger,* and it is one of the worst affected catchments. The eradication programme, led by Dr Joe Caffrey of the CFB, entails removing flower heads before they set seed, and spraying weed killer to control the hogweed's growth. If successful, the programme will be implemented elsewhere.

LOUGH GUR

A bear's cave

Lough Gur is an atmospheric place, almost encircled by hills. Lewis's *Topographical Dictionary* described it in 1837 as "romantic", with its lake, islands, rocky knolls and cave. The lake was partly drained later, so that two castles, formerly on islands, are now on the mainland. All that remains of the lake now is a crescent-shaped lake curled around Knockadoon Hill, and a smaller lake isolated behind

the hill. Knockadoon was also once an island, linked by a drawbridge to the mainland. Near Lough Gur is Red Cellar Cave, where several skeletons of prehistoric animals were found, including the remains of

The brown bear, extinct in Ireland for 8,000 years. Fossils have been found at various sites including Lough Gur and Aillwee Cave in the Burren.*
© An Post

a **brown bear** (*Ursus arctos*), which had probably been hibernating there, also arctic fox and lemming. Brown bears, the same species as North America's grizzly, were massive beasts that roamed Ireland until 8,000 years ago, and their fossil remains have been found at 23 sites. The bear disappeared from Ireland when early farmers began destroying forests, its preferred habitat.

Lough Gur is also a major neolithic site and, 4,000 years ago, had a thriving Stone Age community. When the lake level dropped in the 1850s, large numbers of artefacts were uncovered, some of which are on show at Limerick Museum and at museums worldwide. The wealth of monuments at Lough Gur includes Ireland's most impressive stone circle, with 120 standing stones. Lough Gur is a wildfowl sanctuary. There is open access to the sites; an information centre opens each summer. ④

NEWCASTLE WEST

A flying Irish woman

The Olympic Games admitted women athletes for the first time in 1928, and then only to three events and after much debate about whether athletics was a suitable activity for young ladies. One of those leading the campaign for change was a spirited young athlete from Newcastle West, **Sophie Peirce** (1896–1939). Later a pioneering aviator, Peirce was thrice married and thrice divorced, and led an adventurous if troubled life.[2] In 1919, she married Major William Elliott-Lynn and, to help run his coffee farm in Kenya, studied agriculture at UCD. Peirce was also a keen athlete and, in

1923, took the world high-jump record, clearing 4ft 11in (1.5 metres). Recognising the need to organise women athletes, she helped found the Women's Amateur Athletic Association in 1922 and, four years later, successfully argued the case for women athletes at a special meeting of the International Olympic Council.

Meanwhile, she discovered her real love – flying. She quickly gained a pilot's licence and broke numerous records, as the first woman to loop the loop and to parachute jump, for instance, and the first person to fly solo from Cape Town to London. She campaigned successfully for women to be allowed work as commercial pilots, and was the first woman in these islands to hold a commercial pilot's licence. Something of a celebrity, she toured the world working as an airline pilot, stunt flyer and instructor. In the 1930s, she bought Kildonan Aerodrome in Finglas, now a north Dublin suburb, where she ran Dublin Air Ferries. When that business failed, she moved to England and, despite a drink problem, sought flying work but, in 1939, she died after falling from a tram. At her request, her ashes were scattered over Newcastle West from a plane.

GALBALLY

Self-taught marine expert

Surprisingly little is known about **Annie Massy** (1867–1931), yet this modest, self-taught woman had an international reputation as a marine biologist.[3] Specimens dredged from around the world were sent to her for identification and, when she died, she was accorded an obituary in the prestigious science journal *Nature*. Massy, whose family estate was at Stagdale, near Galbally, seems to have been interested in wildlife from an early age and joined the Dublin Naturalists' Field Club as a young woman. In 1901, despite having no formal education, she was appointed to a temporary position in the fisheries section of the new Department of Agriculture and Technical Instruction (DATI), a post she held until her death.

Massy became an expert on molluscs, particularly the cephalopods, which include squid, octopus and cuttlefish. By 1907, she was skilled enough to realise that three cephalopod specimens recently trawled from Irish waters had never been seen before. Her reputation spread, and soon she was receiving specimens caught on voyages of discovery in the Antarctic Ocean. Other molluscs from India and Africa were later sent for her opinion. Many of those she analysed were new to science, and one Irish cephalopod, *Cirroteuthis massyae*, was named in honour of this woman who first identified it. In 1926, she almost single-handedly rescued Ireland's Society for the Protection of Birds, then on the point of being dissolved and, through her tireless efforts, rejuvenated the association which today, as BirdWatch Ireland,* is an important wildlife and conservation organisation. Her obituary in *Nature* lamented the untimely death of "a careful, critical, and efficient though retiring zoologist with no ambition but to do her work thoroughly". Annie Massy is buried at Malahide, Co. Dublin.

KILBEHENY

Circular (and triangular) reasoning

John Casey (1820–1891) came from a humble Catholic background, and did not attend university until he was 38, yet went on to become an internationally respected mathematician, wrote several classic textbooks and, unusually, given his background, was offered a professorship at

John Casey, a distinguished mathematician who came from Kilbeheny. He is buried at Glasnevin Cemetery in Dublin. Courtesy UCD

TCD,* which was then predominantly Protestant. He went to school at Kilbeheny, and afterwards took what was probably the only available opening, and trained as a national school teacher.[4] While teaching in Kilkenny, he was introduced to advanced mathematics, following a chance encounter with a student from TCD. On his own initiative,

Casey devised a proof of a complicated theorem of the French mathematician Jean-Victor Poncelet, which brought him to the attention of two senior mathematicians at TCD, George Salmon and Richard Townsend. They invited Casey to study mathematics at the college free of charge, and he graduated in 1862.

Casey spent the next 10 years engaged in complex mathematics research, especially the geometry of the circle and the triangle, while also working as a science teacher at Kingstown [Dún Laoghaire] School. In 1873, he was head-hunted for not one but two professorships – at TCD and at the Catholic University of Ireland (later University College Dublin).* Trinity College promised prestige and financial security; the much smaller Catholic University, which depended on donations from the Church, could offer neither of these. Yet Cardinal Cullen persuaded Casey to join the Catholic University, and to help advance the cause of Catholic education. To supplement his low salary, Casey gave grinds, yet he found time to develop his research interests, and wrote six major textbooks on geometry and trigonometry – his *A Sequel to the First Six Books of the Elements of Euclid* (1881) is still a classic. Despite his late start, Casey became a European authority on geometry and, when he died in 1891, he was accorded obituaries in some of the most prestigious mathematics journals of the day.

DRUMCOLLOGHER
Ireland's first dairy co-operative

Horace Plunkett, founder of the Irish co-operative movement. Irish Agriculture Wholesale Society

Drumcollogher is the cradle of Ireland's agricultural co-operative movement. There, at a small butter factory in 1889, **Horace Plunkett** (1854–1932) launched his national co-operative cam-paign and founded the first Irish creamery co-op. Local farmers agreed to pool, process and market their milk together; three people were employed to skim and churn the butter, which was sold through the Cork butter market;* and the residual skimmed milk was fed to calves. From this small start, Plunkett's co-operatives helped organise and modernise Irish farming and created a powerful self-help movement.

Plunkett was an agrarian reformer and politician from the Anglo-Irish Ascendancy. Irish farming was in disarray then, and the dairy industry in particular was close to collapse. Yet not long before, the Cork butter market had been a world leader. In 1880, the government, together with the RDS,* introduced measures to improve Irish farming and the quality of its produce. Cheese instructors came from Germany; poultry and various breed improvement programmes began; the Munster Institute and Model Farm in Cork bought modern mechanical cream separators to improve the quality of Irish butter; and the Albert Agricultural College in Dublin (now the site of Dublin City University) began training creamery managers. Finally, in 1899, the government established the Department of Agriculture and Technical Instruction (DATI), with Horace Plunkett as its vice-president, to develop these initiatives further.

Not everyone was pleased, though, and Nationalists suspected the DATI of undermining the Home Rule campaign. Moreover, farm women lost an important source of income when butter and egg production, formerly women's preserve, was taken over by the male-run co-operatives. Yet, by 1915, Plunkett's **Irish Agricultural Organisation Society** (IAOS) had 1,000 co-operatives and over 100,000 members. These later spawned successful businesses, such as Kerry Foods, which today are multi-million pound, multinational conglomerates. The Dairy Co-operative and Plunkett Museum in the restored Drumcollogher creamery has equipment dating from 1889, including steam engines and churns. ⑤

COUNTY KERRY

Kerry is blessed with an abundance of riches. A stunning landscape. An ancient Celtic cow and a colourful slug. An unusual railway that was inspired by a camel train. Ireland's most important woodlands. Historic transatlantic telegraph connections. Fascinating geology, including the oldest fossil footprints in the northern hemisphere. Numerous creatures and plants not found anywhere else in Ireland or Britain. The oldest copper mine in northwest Europe. And more nature reserves than any other Irish county, including the wonderful Skellig Rocks. No wonder they call it 'The Kindgom'. Kerry derives from *Ciarraí* (or *Ciar Ríacht*), 'Ciar's Kingdom'.

A camel-inspired railway

The world's first passenger-carrying monorail was the unique **Lartigue railway**, which ran between Ballybunion and Listowel. Inspired by a camel train, it ran on a single elevated rail that was held waist-high on A-shaped steel trestles. The engine, and passenger and freight carriages, straddled this construction, hanging either side like panniers, and concealing the central wheels and gears. Three locomotives specially made in England were used: each had two boilers and two cabs, one either side of the rail, the driver riding in one cab and the fireman in the other. Because the elevated railway crossed the country like a fence, bridges were needed to carry roads over the line.

The unusual system was invented by a French engineer **Charles Lartigue** (1834–1907) who, while visiting Algeria,

The unique Ballybunion–Listowel monorail, with its raised central track.
© *National Library of Ireland*

saw camels easily carry balanced heavy loads. In 1881, he built a 90km stretch of monorail in the Algerian desert where it was used to transport esparto grass, with mules pulling trains of panniers suspended on the rail.[5] A number of Lartigue railway companies then formed and a demonstration model was built in London. However, only two Lartigue railways were ever built: one near Lyon in France, which was never used, and the other linking Listowel and Ballybunion, which opened in 1888 and ran for 36 years. It followed the main Ballybunion–Listowel road (R553) and carried freight, cattle and passengers, brought tourists to Ballybunion and carted sand to Listowel from the beaches.

Loads had to be carefully balanced, a complex and time-consuming process especially where cattle were concerned. Even then, the Lartigue had a reputation for rolling sickeningly as it moved. It was also renowned for being noisy, unpunctual and slow, taking 40 minutes to complete the 15km journey. There was never enough traffic to support the route and, after part of the line was damaged during the Civil War, the railway closed in 1924. A few short sections of track were salvaged, but everything else was scrapped. Recently, however, work began at Ballybunion to rebuild part of this unusual line. Listowel Library has a model of the railway, also archive material and rare film footage of the train. ⑥

Ballybunion, at the mouth of the Shannon, was a popular sea-bathing resort in Victorian times. The picturesque rocky coastline has cliffs and caverns, and Lewis's *Topographical Dictionary of Ireland* (1837) relates a "curious occurrence" in 1753 when some minerals in the cliffs spontaneously ignited, it is said, and burned for some time. The headlands can be a good place to watch for the Shannon estuary's dolphins,* and Ballybunion hosts a dolphin festival each May bank-holiday weekend.

The bank vole discovered

Britain has four species of vole while Ireland, at least until 50 years ago, had none. Then, in 1964, a zoology student from UCC trapped Ireland's first bank vole at Listowel.

Dumpier than a fieldmouse, with a shorter tail and smaller ears, the bank vole (*Clethrionomys glareolus*) has rich brown upper fur, a pale belly and prominent chisel-like front teeth; it makes a squeaking noise, and prefers sheltered habitats, such as hedgerows and crevices. The bank vole presumably arrived in Ireland in the past century. Some suggest it came with equipment brought from Germany in the 1920s to build the Ardnacrusha* power station, though Dr Jim Fairley, who studied the distribution of Irish bank voles in the 1970s, thinks it arrived at Foynes port on the Shannon estuary *c.*1950. The Irish bank vole differs slightly from the British one, so a genetic comparison might clarify its origins. Meanwhile, it is spreading across Ireland at nearly 5km a year and has already reached counties Kildare and Galway. The main worry is its habit of stripping tree bark to reach the softer tissues underneath, a practice that, in extreme cases, can kill the tree. Therefore, in badly affected areas, forestry companies sometimes use spray repellent.

Vintage wireless museum

Listowel's Wireless Museum is an amazing private museum. The result of Eddie Moylan's fascination with the reproduction of sound, it contains over 1,000 items and reflects a century of changing technology. The earliest pieces of technology present are phonographs dating from the 1890s, and 200 accompanying wax cylinders, which are mostly music recordings. Next come gramophones (*c.*1901–1930) and old 78s. Then hundreds of wireless sets (1920–1950), many of them hand-built; plus old crystal sets, wet and dry batteries, valves, microphones, horn speakers and wire recorders. The last, a forerunner of the tape recorder, still provides the technology used in flight recorders, or black boxes. The museum's more unusual objects include 'Paris aerials' – receivers concealed in pictures and used in wartime France. ⑦

Peaty avalanches

Eight members of the Donnelly family perished when their cottage on Knocknageeha Hill was swept away by a **bog burst** during a storm in December 1896. It was probably the worst bog burst of the past 200 years: a stretch of waterlogged bog 1km across became dislodged and raced down the hillside, smothering 120 hectares under a thick layer of peat.[6] Some thought the catastrophe had been triggered by a recent earthquake in England, but an RDS investigation, led by Robert Lloyd Praeger,* concluded the peaty avalanche was triggered by the heavy rainstorm.

Bog bursts happen when the weight of waterlogged peat is more than the bog can hold, and they are most common in the west of Ireland, with its high rainfall and steep slopes. One of the worst on record killed 21 people at Poulevard in Co. Limerick in 1708. Such tragedies became less frequent, however, as the population living in mountainous areas dropped. Most bog bursts are small, quiet affairs in remote areas – in the past many occurred unseen and came to be attributed to 'the fairies'. They are usually chaotic avalanches but, occasionally, the bog will move as a raft, often complete with upright trees. In 1992, a bog burst on Cuilcagh Mountain in Co. Fermanagh flooded through nearby Marble Arch Cave.* Bog bursts have even featured in literature: Bram Stoker's early novel *The Snake's Pass* (1890) climaxes with a bog burst which conveniently buries the villain of the piece.

Crag Cave and climate change

A small stalagmite in Crag Cave has helped scientists piece together information about how Europe's climate has changed since the last Ice Age. Cave formations, such as stalagmites, are a valuable climate record stretching back thousands of years. The rate they grow at varies with the amount of rain, and their chemical composition also changes with temperature and rainfall. In the 1990s, climatologists studied stalagmites in several European caves, including one in Crag Cave that is 50cm high and 10,400 years old. Dr Frank McDermott from UCD examined its composition and its growth rings, or laminae, and concluded that over the past 10,000 years, Ireland's climate had oscillated between wet and dry. The stalagmite recorded four regular wet–dry phases, each lasting 2,500 years; a similar pattern was seen in the other European caves. The scientists are now combining this with information from other sources about how Europe's climate changed since the last Ice Age.

The caves around Castleisland were known for centuries but, in 1983, a Welsh cave diver, Martin Farr, explored a flooded passageway in one cave and, when he emerged beyond the short flooded section, found himself in a previously undiscovered cavern, which has now been opened as Crag Cave. Nearly 5km of passageway have since been mapped there, making it one of the 10 longest caves in Ireland. Crag Cave is perhaps 1 million years old, excavated by an underground river that exploited a weakness in the local limestone. The cave's lower level still contains an active river. [8]

Old ways restored

This tiny port boasts three 19th-century technologies: a ship canal, a steam railway and a towering 200-year-old windmill. The 5-storey mill, built *c.*1800 by local landowner **Sir Roland Blennerhassett** (1741–1821), is the most westerly windmill in Europe and the largest working windmill in Ireland or Britain. Though Ireland is a windy country, **windmills** were never popular. A windmill is inefficient, harnessing at best 10 per cent of the wind's energy; wind power is unreliable and, unlike water, which can be held in a pond and controlled using sluices, wind cannot be stored or controlled; windmills cannot operate on calm or very windy days – Blennerville can function only in

winds of Force 3–5 – and, lastly, the heavy and expensive sails were often damaged by gales. Of an estimated 6,000 mills here in the early-19th century, at most 500 were windmills, and, by 1880, most had closed or converted to steam power.[7]

Blennerville mill is unusual in being on the Atlantic coast, as most Irish windmills were on the less stormy east coast, but this could be because power was needed, and the local River Lee was not powerful enough to turn a waterwheel. The mill flourished until about 1820, producing 5 tonnes of stone-ground flour each week, yet by 1850 it was in ruins. It was restored in 1880, and a steam engine was installed, but the revival was short lived and, in 1891, the windmill was being used to store sleepers for the Tralee–Dingle railway. The building was restored again in the 1980s, complete with a new cap and sails, which are manually turned into the wind using a winding mechanism. (9)

When the mill was built, Blennerville served as Tralee's seaport: ships unloaded their cargo there onto lighter vessels that ferried the goods 3km up the shallow River Lee to Tralee. In 1846, the Tralee **ship canal** opened, effectively bypassing Blennerville. Three metres deep, and paved with limestone slabs, the canal could take ships of 250 tonnes displacement. There was a lock at the sea end, and a dock in Tralee, but that was filled in when the canal was abandoned in 1945. The canal was recently restored, along with a parallel stretch of the Dingle peninsula light railway (1891–1953), and in summer steam excursions run from Tralee to Blennerville. (10)

An ancient coppiced wood

The alder and ash woodland growing on the River Lee flood plain at Ballyseedy is probably unique in Ireland. Ecologists call it an "alder ash residual alluvial woodland", and it is a rare habitat in Europe. Ballyseedy is not virgin forest – in fact it was clearly managed in historic times – but it regenerated naturally and so it is a direct descendant of Ireland's ancient wild woods. Its boggy ground saved it: despite various attempts, the site was never successfully drained, and unsuspecting visitors should watch for old and treacherous drainage channels. This special place, rich in plant and animal species and wonderful trees, has been accorded some status as a special area of conservation, yet it was recently threatened by road improvement plans.

Intriguingly, some of Ballyseedy's oak and alder trees appear to have been coppiced in the past. **Coppicing** was much used in Britain as a way of farming timber: trees would be cut close to their base and the trunks allowed to regrow; after some time (16–24 years in the case of oak), this fresh growth was harvested and the cycle repeated. The result was a distinctive stocky tree, with a flat 'stool' base and several main shoots. Coppicing was not thought to have been much used in Ireland, where historically there were plenty of trees, however, coppiced oak and alder trees have been found at Ballyseedy by Ray Monahan, a lecturer at Tralee Institute of Technology who has studied the historic wood. They have that distinctive coppiced appearance, and some are 8 metres in girth and could be 400 years old.

Fool's Gold Fort

In September 1578, an English ship laden with gold ore ran aground in Smerwick Harbour. The *Emmanuel* was returning from a major gold-mining expedition to North America, led by a swashbuckling Elizabethan pirate,

Martin Frobisher (1535–1594). Most of the rock that landed at Smerwick was ferried to London, but some was left behind and, during the Desmond rebellion the following year, was used by Irish rebels building a fort there (*Dún an Óir,* 'Fort of the Gold'). The ore is still there but, alas, there is no gold: the rock was worthless and modern analyses show it contains even less gold than an average lump of stone.

Martin Frobisher: he left some fool's gold at Smerwick. In 1580, 700 armed Spaniards and Italians arrived at Smerwick to join the Desmond rebellion. Besieged at Dún an Óir by the English, the rebels surrendered but were put to the sword in a bloody massacre.

Frobisher fancied himself as an English 'Columbus' and, in 1576, he sailed in search of a Northwest Passage* to Asia around North America. Having found what he thought was Asia (actually the Canadian island of Baffin), he returned with a captured Inuit man, various plants, and a glistering black stone. The stone was treated like moon-rock, especially after an unscrupulous alchemist announced that it contained gold. In the ensuing gold fever, Frobisher raised £20,000 – about 10 per cent of the entire English exchequer then – for mining expeditions to his Cathay. He organised two major expeditions, in 1577 and 1578, with flotillas of ships and hundreds of Cornish miners. The miners quarried 1,200 tonnes of rock in harsh conditions from the frozen tundra, and ferried it home. Smelters were built near London to process the "Asian gold", and were England's largest industrial plants then. By the time the second expedition returned home, however, the first batch of ore had been properly analysed. The remaining rock was dumped, used to pave the streets of London with fool's gold, and to build a fort at Smerwick Harbour, a fort that, in Irish, might more accurately be called *Dún an Bhréagóir*. Músaem Chorca Dhuibhne in Ballyferriter features the local geology, flora, fauna and archaeology. ⑪

The running toad

There is no mistaking the natterjack (*Bufo calamita*). Ireland's only toad, it runs on its short back legs rather than hops, has a distinctive yellow stripe down its back, and a loud croaking call – during the April–June mating season males can be heard calling a mile away. The adults live in burrows in coastal sand dunes, emerging at night to dine on insects. When they mate, the male clasps the female from behind until she spawns, then he fertilises the eggs as they emerge. Up to 4,000 eggs can be shed at a time, but only a few will survive to adulthood. Strangely, it is still not known if the **natterjack toad** is native to Ireland or a recent introduction. (A similar mystery surrounds the frog,* another of our amphibians.) For, in Ireland, the natterjack is found only around Castlemaine and the Dingle peninsula, especially Castlegregory. Yet, in Europe, it is widespread, from Sweden to Russia, to the Mediterranean islands, to Iberia and Britain. Moreover, there are no confirmed findings of any fossil toads here, and the first official sighting in Ireland was not until 1805 at Callanfersy on the River Maine estuary.[8]

The enigmatic natterjack toad ('nether jack', Old English for 'lowly wee thing'). Sketch by the 19th-century naturalist Mary Ward. *

Some suggest the toad was introduced, possibly in the 1700s, by English boats dumping sandy ballast in Castlemaine, Dingle and Fenit harbours. This would at least explain why the toad is found on both sides of the mountainous Dingle peninsula. A recent genetic study found that the Irish toads are closely related to those on England's west coast, but clearly more research is needed. Despite a flourishing population at Castlegregory, the natterjack toad is endangered in Ireland and has already disappeared from some sites, notably Inch and Callanfersy, where it was first recorded. Drainage schemes remove the shallow pools where the tadpoles develop, while amenity developments and sand extraction intrude on dunes where the adults burrow. In a bid to conserve the Irish population, tadpoles were transplanted to the wildfowl sanctuary on Wexford's North Slobs,* where a small natterjack population is now successfully established.

Castlegregory is an interesting land formation: a tombolo of sand, 6km long, it links some of the Maharee Islands to the mainland, and impounds a shallow lagoon at Lough Gill. Most of the sand* around Ireland's coast was originally deposited offshore by glaciers, but the sand at Castlegregory comes from the ongoing erosion of the Kerry mountains, and is brought by rivers and sea currents to the northern side of the Dingle peninsula. This is part of the Earth's endless recycling of rock: Castlegregory's sand comes from the Old Red Sandstone mountains of MacGillycuddy's Reeks, which themselves formed 370 million years ago from sand eroded from even earlier mountains.

Which way is up?

The sea cliffs below Slea Head once made a small but significant contribution to geology. For it was there, in 1838, that **Patrick Ganly** (1809–1899) discovered a way of telling when rocks have been turned upside down. Geologists need to know the sequence in which rocks formed, a bit like having the pages of a novel in the correct order. If the rocks have not been disturbed this is simple as, in general, the youngest rocks lie on top of the older ones. But sometimes geological processes push the layers of rock up onto their side or even turn them over completely, making it difficult to know which rocks came first.

Patrick Ganly, born in Dublin to a professional Catholic family, was a field surveyor with the Ordnance Survey's valuation office. The office, run by Richard Griffith,* was responsible for calculating land valuations for the new rates system. Ganly joined when he was 18 and presumably learned his geology on the job – he did not attend university until his 30s and then only as a part-time science student at TCD – yet he was clearly a talented geologist. In 1836, while working at Carndonagh in Co. Donegal, a keen-eyed Ganly saw that the grains of sand in the ripples on a riverbed are graded from small to large, within the curve of each ripple, and in the direction of the current. As fresh ripple marks are laid on top of older ones, a pattern emerges reflecting the river's direction. Two years later, while studying vertical rock formations on the beach below Slea Head, Ganly saw this very pattern preserved as a fossil in the sandstone (the sand had been deposited in strong currents 370 million years before).

Further along the line of rocks, he found layers in which the fossil pattern was the mirror image of the original. Ganly realised this was because these rocks had been folded over and were now facing the wrong way, even though the fold was out of sight underground. He had found a way of telling which way was up. Twenty years later, he published his discovery in the Dublin Geological Society's journal, but this small report went largely unnoticed.[9] Only when US geologists rediscovered the technique in the early 1900s, was it finally accepted.

Ganly also did much of the detailed fieldwork for Richard Griffith's personal project, the first geological map of Ireland. Thanks to Ganly, this two-man map, published in

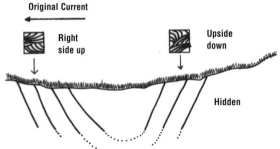

Patrick Ganly used fossil ripple patterns, in rocks near Slea Head, to tell which rocks were right side up and which were upside down, even though the fold was hidden.(Diagram based on Ganly's 1856 sketch.)

1839, is remarkably accurate. But Griffith never credited Ganly's contribution – perhaps because his employee was moonlighting – and his connection only emerged in the 1940s when hundreds of Ganly's letters to Griffith were discovered. Ganly's contribution was remembered in the 1970s, however, when the Geological Survey of Ireland* named its exhibition hall in his honour.

At the South Pole Inn

It is a long way from Anascaul to Antarctica. Yet **Tom Crean** (1877–1938), born on a small farm near Anascaul, took part in three Antarctic expeditions, including Robert Scott's ill-fated 1910–1912 expedition, and Ernest Shackleton's epic 1914–1916 one. Crean's indomitable good nature and great courage on these expeditions won him respect, honours and a place in exploration history.

Crean left school at 14 and enlisted in the Royal Navy. In 1901, he was stationed in New Zealand where Scott's *Discovery* expedition was being provisioned and, when a vacancy arose on the *Discovery*, Crean volunteered. Scott's team spent two years exploring Antarctica, then the world's last great wilderness. In 1910, Scott selected Crean for his *Terra Nova* expedition, which was racing to the South Pole. Crean took part in the gruelling 1,000km trek into Antarctica's interior, but he, Bill Lashly and Edward Evans were not chosen for the final, fatal 200km run. Instead, exhausted and with meagre rations, they had to man-haul themselves and their sledges back to base. At the end of the 2,000km round-trip, with Evans nearly dead from scurvy, Tom Crean set off alone to get help, leaving Lashly to keep Evans company. With no food, drink or shelter, Crean completed the last 50km in 18 hours, alerting base camp in time to save Evans' life, for which he was awarded the Albert Medal. Crean was later in the search party that found the bodies of Scott and his companions, and some say that, had Scott only chosen the Kerryman for the final dash to the Pole, they might all have returned safely.

In 1916, Crean took part in another epic Antarctic journey. He, Ernest Shackleton* and four others left their expedition companions on the remote Elephant Island, and set out to raise a rescue party. They sailed 1,100km in a small open boat across treacherous seas to South Georgia. Then, ill, cold, wet, exhausted and hungry, Shackleton, Crean and Frank Worsley spent a further two days crossing the island's unmapped mountainous interior nonstop to reach a whaling station on the other side. Their journey is still spoken of with awe. After returning to Europe, Crean served in World War I, then retired to Anascaul. There 'Tom the Pole', as he was known, married and opened the South Pole Inn.

In the grip of the ice

In 1837, a Swiss naturalist **Louis Agassiz** (1807–1873) proposed an astounding idea. He suggested that much of Europe had once been covered by ice sheets and glaciers, during what he called "*die Eiszeit*" or **Ice Age**. Agassiz believed this could account for many puzzling features in the landscape, such as mountain corries,* drumlins,* eskers,* erratic boulders* and gravel banks. Evidence slowly accumulated in favour of Agassiz's theory, but not everyone was convinced. Diluvialists believed the biblical flood had deposited the sand, gravel and erratics. Catastrophists believed there had been not one flood but a series of catastrophes, each bringing an epoch to a close. Yet others believed that Ireland and Europe had been submerged under an Esker Sea, and that eskers were the tide marks left by that sea.

In the 1830s, Agassiz visited Ireland and noted a number of features that indicated to him that the island had once been gripped by ice. However, it was another 10 years before **John Ball** (1818–1889) collected conclusive evidence of an Irish Ice Age from the landscape of the Connor Pass and particularly Lough Doon. Ball was born in Dublin, where his father was Attorney-General for Ireland, but he went to school in England and later to Cambridge University. After a brief career in law, he changed to politics, becoming a Poor Law Commissioner in Kerry and later an MP for Carlow. He was also a keen mountaineer, climbing in the Alps, Andes, Atlas mountains and elsewhere. Each place he visited, he studied the plants and local geology, being particularly interested in glaciers (moving rivers of ice) and how they flow.

Lough Doon, also known as Pedlar's Lake, sits in a basin in a rocky amphitheatre or corrie, a short walk from the northern car park at the Connor Pass. Ball saw similarities with the Swiss landscape, where glaciers are still present, and realised that Lough Doon's corrie and rock basin had been scooped out by ice. Glaciers form high up on mountains, often on the colder northern and eastern sides; as the ice accumulates, it begins digging into the mountainside, excavating a basin. As more ice accumulates, the glacier grows, eventually flowing up and over the basin's lip and down the mountainside. Rocky material caught in the underside of the ice will scratch the basin's rocky lip as it passes. Much of the debris entrained in the glacier will be scattered downstream in boulder fields, or deposited around the edge of the ice in banks known as **moraines**, some of which are big enough to dam a lake.

Evidence for all this can be seen at the Connor Pass: on Lough Doon's rocky lip are scratches, or *striae*, which the glacier cut; downstream in the valley are boulders the glacier dropped; across the road, a moraine impounds Loch Broin. The remains are as fresh as if the ice had melted yesterday, not 12,000 years ago. Ball reported his findings to the Dublin Geological Society in 1849, the first conclusive study that Ireland had been glaciated.[10] Today, we know there have been a number of ice ages at intervals over the past 2 million years, some more severe than others. During the most recent, which lasted from 35,000–13,000 years ago, the temperature is thought to have been on average only 5°C lower than today.

The impressive sandstone erratic of Carrigacapeen, evidence of the Ice Age, stands in thick scrub by a disused railway near Kenmare.
© Dr Peter Coxon

Kerry is particularly rich in **glacial features**. Northwest across the road from the Lough Doon car park, for example, is an impressive corrie on **Brandon Mountain** with an intriguing string of progressively bigger lakes, known locally as the Rosary Beads. The lakes sit in a series of basins scoured out by a glacier, each step separated by a bare rock wall over which the water cascades. This is a cold and dark, eerie and imposing place. It feels as if the Sun never shines there, as if the ice has just retreated. To see it close up, follow the pilgrim path from Cloghane grotto. The sign-posted route, a 75-minute walk, goes halfway up Brandon's northeastern ridge, then swings south and into the corrie (from the back of the corrie the path continues up the steep scree slope – note, this stretch is more testing – and after a further 40 minutes emerges at Brandon summit). Another impressive Kerry corrie is the Devil's Punchbowl on Mangerton mountain, so perfect it was once believed to have been a volcanic crater.

Kerry's scenic mountain gaps, notably the Gap of Dunloe, Moll's Gap and the Windy Gap, were all gouged out by the ice and are another glacial legacy. The Gap of Dunloe, for example, is a great U-shaped gash 500 metres deep that was cut when a glacier forced its way between Purple Mountain and MacGillycuddy's Reeks. Moraines have also shaped the Kerry landscape. Waterville, for instance, is built on a large end-moraine, which marks where a glacier stopped, and which today separates Lough Currane from the sea. Cromane is a long moraine – unlike Inch across the bay, which is a sandy spit – and contains debris that was dumped by glaciers flowing down off the reeks. (Cromane is connected now to the mainland by a stretch of bog.) Moraines also impound Lough Leane at Killarney and Caragh Lake.

Erratics are everywhere, plucked up by the ice and carried far from their original site. The beaches at Ballybunion and Ballinskelligs, for example, contain large granite boulders brought by ice 130,000 years ago from Galway; Castlegregory has cobbles of gneiss brought 200km from north Mayo around the same time. The most amazing Irish erratic, however, is surely **Carrigacapeen**. This massive sandstone boulder, measuring 2 metres by 1 metre and weighing 30 tonnes, was plucked from the Kerry mountains by a glacier 22,000 years ago, then dropped on limestone a few miles east of Kenmare (approx grid ref V937 725; ask locally, it is difficult to find). The sandstone boulder cap then protected the underlying limestone, so that, though the surrounding limestone dissolved away during thousands of years of rain, the boulder was left sitting on a limestone pedestal 1.5 metres high. As a result, Carrigacapeen has been used to estimate the rate at which limestone in the area is being eroded: 1.5 metres in 22,000 years, or about 1km every 12 million years.

MacGillycuddy's Reeks

Kerry alps and deserts

Kerry once had high alpine mountains, possibly 3,000 metres high. But that was 300 million years ago, and relentless erosion since has reduced the 'Kerry Alps' to a shadow of their former selves. Yet Kerry is still mountainous, and probably the highest county in Ireland. Two hundred peaks on the island are above 600 metres, and half are in Kerry, including three peaks above 1,000 metres, one of which is **Carrauntoohil** (1,039 metres), Ireland's highest mountain. Carrauntoohil (*Corrán Tuathail*, 'the

clumsy sickle'), is named for its sickle-shaped profile and because its serrated edge is on the outside of the blade and not the inside, as in a real sickle.

The Kerry Alps formed 300 million years ago as bits of Europe's geological jigsaw rearranged themselves. 'Ireland' then was 20° south of the Equator and was part of a vast new continent that was forming, called Pangaea, which included most of the world's landmasses. One geological collision at the time pushed up new mountain ranges in western Europe, particularly France's central massif. The force of this was also felt in the south of Ireland, where the surface was crumpled into parallel folds, like pleats in a piece of fabric, forming the mountains of the southwest. Because the force came from the south, these folds are aligned east–west, in a series of ridges and valleys that still persist. The force of this mountain building generated great heat and pressure, altering some surrounding rocks (at Valentia, for instance, siltstone was converted into slate*); it also opened faults and cracks in the rock, and melted minerals. Molten copper, lead and zinc then flowed through the cracks before solidifying to form deposits which people could later mine, as at Killarney* and Kenmare.*

The Kerry Alps were initially covered with a layer of limestone 2km thick (the limestone that still covers much of Ireland's central plain). This soft rock was quickly eroded from the high peaks, however, leaving the harder, older sandstone core that forms today's mountains. That sandstone is rich in iron, which gives the rock a deep red colour, even purple in places. It formed 370 million years ago and, to distinguish it from another, more recent red sandstone, it is called **Old Red Sandstone**, affectionately known to geologists as ORS.

There are various ORS layers in Kerry, reflecting the prevailing conditions when each was laid down. Some are true sandstones, laid down in fast-flowing streams and desert sand dunes; other fine-grained mudstone and siltstone layers were deposited in quiet estuaries and lakes;

conglomerates, which contain a jumble of sand and boulders cemented together, formed during catastrophic flash floods. From this evidence, geologists can piece together the ancient Kerry landscape, and from fossil ripple marks and layers in the rock they can even tell the direction river currents were flowing in, and which direction the winds blew across the dunes. So we now believe that, when the ORS formed 370 million years ago, Kerry was at the southern edge of a vast landmass. To the north was a massive mountain range, resembling today's Himalayas, which had been pushed up 30 million years earlier when the ancient continents of Laurentia and Avalonia* collided. As these great mountains were eroded, vast quantities of sand were carried down and deposited on Kerry – blown by the wind, swept by rivers, dumped by floods. The sand was turned to stone (the ORS) around 370 million years ago, and 70 million years later it was pushed up to form the Kerry Alps. Kerry's sandstone mountains are still being eroded, part of the endless recycling of geological material: the sand from the Reeks is carried down and deposited below, at Castlegregory, for instance, where it may one day again become sandstone.

The landscape of Kerry and southwest Cork is shaped by this geological history: the rugged mountain ridges; the parallel east–west river valleys of the Bandon, Lee and Blackwater;* the finger-like mountainous peninsulas; and the fine bays in the southwest which, being drowned valleys, or **rias**, are gentler than steep glacial fjords.*

At **Bennaunmore** there are basalt columns resembling those of the Giant's Causeway,* but 250 million years older. The Kerry columns date from 330 million years ago, the same time as Co. Limerick's volcanic landscape.* There is a self-guiding geology and archaeology trail around Loch a'Dúin (not to be confused with nearby Lough Doon), that explores the corrie, megalithic rock art and standing stones (the guidebook is sold locally).

A remarkable chalk quarry

Ireland was once covered with a thick layer of white chalk. It formed 100 million years ago when the land was submerged under a warm, shallow sea. The chalk, made of the shells and skeletons of dead marine organisms, was soft, and quickly removed by erosion, except in the northeast where a blanket of basalt protected it from the elements. Hence, there have been chalk mines and quarries in Antrim* for centuries, but none in the rest of Ireland. At least, that is, until 1966. Then, geologists discovered that a small limestone quarry at Ballydeenlea, which locals had been using for over a century, contained, not common limestone, but a tiny deposit of 100-million-year-old chalk. The Ballydeenlea chalk had been deposited in a sinkhole in the surrounding limestone, and this protected it from erosion. Similar small deposits could lie hidden under bog or gravel elsewhere around the country, but for the moment this is the only chalk known in Ireland outside of Co. Antrim.

Ireland's oldest mine

Killarney is an old mining town, and possibly Ireland's oldest-surviving settlement – you could even say that it is where Ireland's Bronze Age began. For **Ross Island**, just outside Killarney, is the site of a 4,500-year-old copper mine which is the oldest in northwest Europe and also the earliest known site in these islands where metal was made.[11] The mines were still worked in the early 1800s and, at one time, there were plans to drain Lough Leane, the main lake, to facilitate the works. But tourism won out and, when the mines closed in 1829, this former industrial site was landscaped.

Lough Leane is named after Lén, a giant mythical smith who guarded the treasures of Ross Island, and in the 8th century, a Welsh monk, Nennius, described those treasures as "a wonder of the known world". The rich mineral veins occur in a geological fault in the limestone, and contain mostly copper (which stains the rock blue in places), but also lead, zinc, silver and some cobalt. The richest deposits are at Ross Island, but neighbouring Muckross, Crow Island and Cahirnane were all mined at various times, mostly for copper and, to a lesser extent, lead and silver; the value of the cobalt was sadly not realised until after most of it had been thrown away as rubble.

Excavations in the 1990s, led by Dr William O'Brien from NUI Galway, revealed that the earliest mines at Ross Island are up to 4,500 years old. These are small, cave-like openings in the rock wall that were made by Neolithic or possibly Early Bronze Age miners. The openings were made by first setting a fire against the wall; then splashing cold water on the hot rock to crack it; and, after that, the cracked rock could be easily shattered using simple hammers made of cobblestones. At Ross Island, the archaeologists found thousands of discarded tools, charcoal (which they were able to use to carbon-date the workings), pottery, hut foundations, spoil heaps, pits (where metal was smelted) and even the remains of meals. Ross Island copper contains some arsenic and, thanks to this unusual chemical fingerprint, the archaeologists have identified Bronze Age artefacts in Britain that were made from the metal. This suggests Killarney was a major metal producer and had important trade links with Britain thousands of years ago.

The mines were worked again around AD 800 and at various times between the 1600s and early 1800s by Planters granted estates in Cork and Kerry. Indeed, Muckross Estate and the towns of Kenmare* and Killarney developed primarily because of the local mining and quarrying, and the works were fuelled by charcoal made from the great oak woodlands roundabout. The German writer Rudolf Raspe, author of the *Fabulous Adventures of the Baron von Munchausen*, was working in Killarney as a mining consultant in 1794 when he died of scarlet fever; he is buried at nearby Killegy.

Ross Island engine house c.1810; the engine worked a pump that drained the mine shafts. Courtesy William O'Brien

The largest mines were developed in the early 1800s by an English entrepreneur, Colonel Robert Hall, employing 500 miners brought in from Cornwall, Wales and Wicklow. The ore was carted to Tralee from where it was shipped to Swansea for smelting. Most of these 19th-century mines were underground tunnels dug beneath the lake level and accessed via vertical shafts (though some were open pits). Flooding was a continuous problem and, at one point, there were plans to drain the lake in order to dry the mines. In the end, small dams were built around the mine shafts and the impounded areas were pumped dry by a large steam engine installed in 1807 on Ross Island. In 1829, however, the noisy mining ceded to tourism. The pumps were switched off, the mine shafts were flooded, the engine house was demolished and the area was landscaped. There is a self-guiding trail at the site, where flooded mine shafts and dams can be seen.

The Kingdom's treasures

The richest and most amazing collection of plants and animals in Ireland is surely that found around Killarney. There is a fish there that is found nowhere else in the world; a Celtic cow; a herd of majestic red deer descended from those hunted by the legendary Fianna warriors; a yew tree that is probably Ireland's oldest living thing; beautiful ancient woodlands; and a motley collection of curious Lusitanians (among them a colourful slug and a strawberry tree) most of which are not found anywhere else in these islands. Colourful exotics, such as fuchsia and montbretia, thrive in Kerry's warm, wet and subtropical climate, as do luxuriant ferns, mosses and lichens.[12] (12)

First, those mysterious **Lusitanians**. This group includes the Kerry slug, a woodlouse, some insects and eight flowering plants – namely the strawberry tree; the insect-eating, large-flowered butterwort; the kidney saxifrage; St Patrick's cabbage; the Irish orchid; and three heathers which, in these islands, grow only in Connemara. The Lusitanians are so-called because they are usually found in northern Spain and Portugal, an area the Romans called Lusitania, though most are also found along the western Mediterranean and occasionally in southwest France and Brittany, all areas with mild winters. They are not found in Britain, and in Ireland they are mostly confined to the southwest of Ireland, making them something of a natural mystery. How did they get there? Did they survive the last Ice Age by sheltering in some refuge in the southwest? Did they arrive after the Ice Age, crossing from Lusitania by a land bridge before sea levels rose? Or were they brought by people?

The arbutus or strawberry tree – one of Kerry's mysterious Lusitanians.
© An Post

The lovely **strawberry tree** (*Arbutus unedo*, in Irish *caithne*) is particularly associated with Killarney. A member of the *Erica* family, and therefore related to heathers, it has a distinctive flaky bark and evergreen leaves and its twisting trunk grows from an underground tuber that can survive a forest fire. Masses of creamy flowers appear in October, followed by orange-red, strawberry-like fruits that take a year to ripen – hence, this year's flowers and last year's fruit appear together. The attractive berries are bitter, however, and *unedo* translates as 'I eat only one'. In Mediterranean regions, the arbutus is a small shrub, but in Kerry it can reach 12 metres. It likes light, and grows in

An ancient Celtic cow

The small, black **Kerry cow** is one of the oldest cattle breeds in Europe. This unique animal is related to the ancient black cattle of central Asia and was probably brought here by the Celts over 2,000 years ago. For centuries, it was the main cattle breed farmed in Ireland, but improved breeds were introduced during the Agricultural Revolution in the 1700s, and the native black cow became confined to poorer mountainous regions, particularly Co. Kerry, after which it came to be named. The Kerry is hardy, adaptable and agile, and can endure severe weather and meagre grazing. An adult cow is small (about 400kg), with a glossy black coat, white horns, and a pleasant character – it was once described as "gay and alert [and] light on her feet". They were renowned as efficient milkers, and the milk was said to be easily digested and recommended for infants and invalids. In 1887, the Kerry Breed Society was formed and a herd book was begun to register pedigree animals. Today, with just 500 animals in the herd book and only a handful of bulls available for breeding, the Kerry breed is rare and endangered. In the 1980s, conservation measures were introduced to preserve this living remnant of our Celtic heritage, and there are now a number of Kerry herds; the largest, with over 200 head, is in Muckross Traditional Farm at Killarney National Park. (13)

clearings and at woodland edges. Outside of Kerry, the strawberry tree in Ireland is found only at Lough Gill in Co. Sligo. It may be native to both counties: arbutus pollen was recently found in Sligo dating to 2,000 years ago and in Kerry to 3,000 years ago – or it could have been introduced to both sites by prehistoric people.

The colourful **Kerry slug** (*Geomalacus maculosus*) is a distinctive creature and especially obvious after rain. Strangely, it was not formally described until 1842, when it was first discovered near Lough Caragh; it has since been discovered in northern Spain (1868) and in Portugal (1873). Zoologists who studied specimens from the three countries recently concluded that the Spanish, Portuguese and Kerry slugs were indeed the same species. Two forms are known: the first, found in open country, is grey with numerous white spots and a white mucus; the second, found in woodland, is ginger with yellow spots and yellow mucus. In southwest Ireland, the slug frequents sandstone areas in particular and, as well as Killarney, is found in the woods at Uragh, and in Co. Cork at Glengarriff and on Cape Clear Island. It eats lichen and moss and can often be seen sliding over lichen-covered boulders. Reputedly the only slug that can curl into a ball to protect itself, it is also surprisingly long-lived and some have been kept in captivity for up to six years.

Kerry's other notable Lusitanian is the **large-flowered butterwort** (*Pinguicula grandiflora*), aka the bog violet. This insect-eating plant grows in great profusion on the bogs of Kerry and west Cork, but nowhere else in Ireland. In May, it produces large eye-catching violet flowers, up to 2.5cm wide, held on tall stalks above large rosettes of bright yellow-green leaves. (Its flower is larger than that of the similar common butterwort, which grows on bogs throughout Ireland.) Butterwort leaves produce a sticky substance that traps and digests insects, providing the plant with much-needed minerals. Traditionally, the leaves were used in cheese-making to curdle milk.

The **red deer** is our largest native mammal. It probably made its way to Ireland shortly after the end of the last Ice Age, but, unlike many other creatures of that era, such as the reindeer and bear,* the red deer survived climate change and the arrival of humans and still inhabits mountainous areas. Dark brown, with a yellowish rump, the deer can reach 1.2 metres at shoulder height. Stags have majestic antlers with up to seven points on each side. Red deer are found throughout Europe, mostly in forested regions, and were hunted since prehistoric times for their hide and venison. The Anglo-Normans kept herds in walled deer parks, along with introduced deer species, especially the Mediterranean fallow deer, with which the reds often interbred. As a result, most Irish red deer contain some exotic blood. The 700-strong herd at Killarney is thought to be the purest, though even it may have bred with red deer imported from Scotland

A red deer fossil: these magnificent animals still roam Killarney National Park.

in the 13th century. The best places to see them are the upland areas of the national park, especially Torc and Mangerton mountains. Killarney also has a herd of Japanese **sika deer**. These resemble the red deer but are smaller, with simpler antlers and a white rump bordered by black.

Ireland's rarest fish is possibly the goureen or **Killarney shad** (*Alosa fallax killarnensis*), which is found in Lough Leane and nowhere else in the world. Its ancestors originally migrated to the sea to breed, just as salmon do today. However, like a small number of other freshwater fish in these islands, such as the pollan and char, the Killarney shad dispensed with the sea and spends its life landlocked in a freshwater lake. This change probably happened during the last Ice Age when Lough Leane was cut off from the sea by ice. Since then, the Killarney shad has evolved in isolation. Rarely seen and seldom caught, it is related to the herring-like twaite shad, but is smaller and has more gill rakers, which may make it more efficient at capturing plankton.

Killarney is renowned for its **ancient woodlands** and deservedly so: oaks have been growing there for most of the past 8,000 years and the area still has Ireland's largest oak wood. The area was clear-felled in the 1600s, along with the rest of Munster's woodlands, to provide timber for barrels and buildings, charcoal for furnaces, and oak bark for tanning. In the rough mountainous areas, the woodlands regenerated naturally, so that today's trees are directly descended from Ireland's primeval forests, with a diverse and highly evolved ecosystem. Killarney's largest oak wood has over 1,000 hectares, mostly tall sessile oaks, but with groves of holly, and also birch, rowan and arbutus. Mosses, lichens and liverworts grow in thick profusion here, and there is a rich wildlife – an Irish oak tree supports over 300 species of insects and other wee beasties.

Reenadinna yew wood is a rare place, and probably prehistoric. There, across 25 hectares of bare limestone pavement, yew trees have taken root in crevices in the rock. No flowering plants can live in the dense shade under the trees; instead, the ground is carpeted with mosses that thrive in the cool, damp environment. The result is an eerie, green and very special woodland. It is the only pure yew wood in Ireland, and one of only three in Europe. A yew tree at Muckross Abbey, reputedly planted in 1344, is probably the **oldest living thing in Ireland**. Another unusual Killarney woodland is the carr, or alder swamp wood, growing on wetland by Muckross Lake.

Worryingly, much of Killarney's woodland is threatened by deer grazing and by the invasive **rhododendron** (*Rhododendron ponticum*). Rhododendron grew in Ireland 400,000 years ago during a warm interval between two ice ages, but disappeared during the subsequent cold period. The species that is now widespread around the west and southwest was introduced from the Black Sea area by Victorian gardeners in the 1800s. It thrives on acid soil and a wet climate, providing an effective evergreen windbreak and a profusion of pretty purple flowers in May. But nothing eats the rhododendron and, unlike the oak, it supports no other species. Worse, its dense shrubbery smothers seedlings, preventing regeneration of surrounding trees. Unfortunately, the shrub is now well established there, but a management programme is helping to limit its invasion of Killarney's woodlands.

Two less threatening aliens found in Kerry are **fuchsia** (*Fuchsia*) and **montbretia** (*Crocosmia*). Both were introduced in the 1850s, montbretia from South Africa and fuchsia (named after a German botanist, Leonard Fuchs) from Chile. They thrive in Kerry's subtropical climate where frost is rare, and provide a blaze of colour in summer hedgerows. In South America, fuchsia is pollinated by hummingbirds, but it is not known if anything in Ireland can penetrate the unusual pendant flowers. Some say bees do, but seedlings are seldom seen and, in general, if fuchsia is growing somewhere in Ireland, it is because someone planted a cutting there.

Killarney is one of the richest places in Europe for **ferns**. Victorian plant collectors flocked there in search of what some called "vegetable jewellery" or "emerald green pets", Torc waterfall being a popular spot. Over 50 fern species have been recorded in Kerry, among them the royal fern – at a stately 2 metres, this is Europe's tallest fern – and the famous **Killarney fern**, which is Kerry's rarest plant. A member of the filmy-fern group, it has dark-green, translucent leaves and grows near waterfalls and streams, in splash zones and under boulders. Such was the craze for it, that it was nearly collected to extinction, but it survives in some of Killarney's more remote spots. (There are Kerry ferns in the Botanic Gardens, Dublin.*)

Footprints in time

The **oldest footprints in the northern hemisphere** are preserved in rocks near Valentia radio station. They were made 385 million years ago, when a primitive, 4-legged creature walked along a mudflat while the tide was out. The footprints it made in the wet sand were later turned to stone, along with beautiful ripple marks left by the waves. In 1992, a Swiss geologist Dr Ivan Stossel discovered the meandering track in rocks that are now part of Valentia's sea cliffs. The creature probably resembled a salamander, lived partly on land and partly in water, and walked in a splayed way like a crocodile. From the size of its footprints, scientists estimate it was about 1 metre long. The footprints are fist-sized and originally were circular, but geological processes distorted them over time and they are now oval.

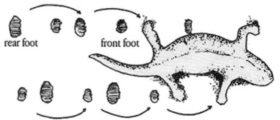

How the early amphibian made the footprints. The animal's tail, dragging in the sand, also left a groove in between the prints. Nigel Monaghan/Natural History Museum

The central groove in the track was made by the animal's tail. The site is thus internationally important and is now protected by legislation. The heritage centre in Knightstown and the geology museum in University College Cork* have plaster casts of the fossils, giving visitors some idea of what to look for.

Nearby **Valentia radio station**, part of the State's marine emergency service, was built by the British navy between 1912–14 during the early days of radio communications.* One of the first messages transmitted when it came into service was to the cruise ship *Lusitania*, warning of submarine activity in the area (shortly after, the liner was torpedoed and sank with the loss of 1,200 lives). After the war, Valentia, and a sister radio station at Malin Head, were run by the British Post Office, which held the stations until 1952, when they were handed over to the Irish Department of Posts & Telegraph. Despite mobile phones and satellite communications, the radio stations still play a vital role in marine communications, broadcasting weather forecasts, navigational warnings and commercial ship–shore links, and handling communications with lifeboats, helicopters and search vessels during emergencies at sea. Valentia Heritage Centre in Knightstown features the island's fossils, slate quarry and telegraph cable. ⑭ Valentia (sometimes 'Valencia') is not Spanish, but comes from the Irish *Béal Inse*, 'the mouth of the island'. In Irish the island is *Dairbhre* (Oak Island).

Linking the Old World and the New

The **first transatlantic telegraph cable** landed at Valentia amid much rejoicing in August 1858. It linked the Old World with Newfoundland in the New World, and Valentia was now as close to New York as to Dublin. Celebratory banquets were held on both sides of the ocean and small pieces of cable were sold as souvenirs. However, the celebrations were short-lived: the quality of the signal was poor and the high-voltage current soon burned through the cable's insulation. Still, at least it proved transatlantic telegraphy was possible.

Before the telegraph, information could travel only as fast as the messenger carrying it. In the 1760s, Richard Lovell Edgeworth* planned an optical telegraph, or 'semaphore', system, with signal towers erected at intervals to relay the coded messages. In one trial, Edgeworth succeeded in sending a message between Dublin and Galway in 8 minutes, but his system was never implemented. The French army did use a similar system in the 1790s, but on a limited basis. At the same time, scientists tried sending messages along electrical circuits. The instruments and codes they used were cumbersome and, initially, there were several different competing systems. Nonetheless, the new railroad companies found them useful for keeping in touch with train movements and installed telegraph lines along the railways.

Then, in 1840, a US artist, Samuel Morse, invented the simple dot–dash **Morse Code** and a tap-key for sending messages, and the telegraph revolution began. Information could be sent across long distances almost instantly and Morse's electric telegraph altered the world profoundly – stock exchange prices, international news, racing results, diplomatic messages, military information, business news,

Receiving messages at Valentia: skilled telegraphers sent and received the telegrams in Morse Code.

police work... all were affected. In 1845, the telegraph was used to apprehend a murderer when police in London, having been wired John Tawell's description, arrested him as he got off a train. Battalions of educated, middle-class women joined the workforce in the new profession of telegrapher. Telegrams – the Victorian e-mail – were sent in huge numbers, and call centres employing hundreds of telegraphers handled the burgeoning traffic.[13]

In 1843, gutta-percha gum was discovered in Asia. It was the perfect cable insulator and meant that telegraph cables could now be laid underwater. In 1851, England and France were connected and, by 1853, Ireland was connected to Britain, with cables coming ashore initially at Donaghadee* (Co. Down), and later at Howth* (Co. Dublin), Newcastle (Co. Wicklow) and Blackwater (Co. Wexford). The next stop was America.

A young US millionaire, Cyrus Field, financed the transatlantic project, and the ocean bed was surveyed to select the best route (marine samples dredged up on those expeditions shed new light on the deep sea bed). Field's consortium opted to use a high-voltage current and a thin wire core, just half an inch thick. Even then, the 3,300km of cable weighed 2,500 tonnes, so it was split in half, each coil carried on a separate boat. The mission was complex – the Victorian equivalent, perhaps, of the Moon shots of the 1960s – and the two crews practised paying out, laying and splicing the cable before attempting the real thing. The cable was eventually laid in 1858 from Valentia to Newfoundland, but burned out soon after. Subsequently, the star technical witness was Belfast-born **William Thomson**,* then professor of science at Glasgow University, who realised that the high voltage had destroyed the slim cable and its thin insulation.

For a second transatlantic attempt in 1865, Thomson designed a thicker cable, a heftier insulation, a low-voltage current, and a very sensitive instrument (a mirror galvanometer) to detect, or receive, the low-voltage signal.

The new cable was designed to sink slowly as it left the ship, had a breaking strain of 7 tonnes, so that 15km could be suspended underwater before it would break, and for simplicity was made in one continuous 3,300km length. Only one ship could carry all that weight: Brunel's 250-metre long *Great Eastern*, then the biggest ship in the world. Built, in 1859, as a passenger steamer, the *Great Eastern* was too big to be economic and the cable consortium bought it at a knockdown price. Captained by Wicklow-man Robert Halpin,* it left Ireland in July 1865 and began laying the new cable. In mid-ocean, however, the cable snapped and vanished into water 4km deep and the crew was unable to lift it. More cable was ordered, plus stronger lifting gear and, the following summer, *Great Eastern* again left Valentia; two weeks later she steamed successfully into Newfoundland. The cable was laid. Thomson's low-voltage system worked perfectly, and even at £1 per word, the company took £1,000 worth of business on its opening day. The following week, Halpin and his crew rescued the 1865 cable and, by September 8th, two lines connected Europe and North America.

Paying out the transatlantic cable from the deck of the Great Eastern: *a complicated system of gears and cogs insured the cable was paid out slowly and safely.*

William Thomson was made Baron Kelvin (he took the title from a Scottish river), and though today he is best known for his work on temperature (especially the absolute Kelvin scale), in 1866 he was famous as the man who made the transatlantic telegraph possible. Halpin and the *Great*

Eastern went on to lay 40,000km of cable around the world. By 1880, there were three more transatlantic lines – a fresh one at Valentia and two brought in by other companies at nearby Waterville and Ballinskelligs. By the 1890s, Kerry was at the hub of the 'Victorian internet', with connections to France, Britain and Germany as well as America. At each of the Kerry stations, skilled telegraphers, many of them initially recruited from abroad, sent and received the messages in Morse Code. World War I brought increased traffic and 300 people, including censors, worked round the clock at Valentia, which was then owned by Western Union. Competition from Marconi's wireless radio eventually closed Ballinskelligs, but Waterville remained open till 1962, and Valentia until 1966, a full century after it began. The cable landing sites and telegraph stations at Valentia, Waterville and Ballinskelligs are marked by commemorative plaques. QUB has one of Thomson's mirror galvanometers. Souvenir sections of transatlantic cable can be seen at various centres, including Valentia heritage centre and the National Museum, Collins Barracks.

Untangling the medusa

Maude Delap (1866–1953) helped unravel a puzzling biological phenomenon. Yet this indefatigable amateur marine biologist was self-taught, worked from a home-made laboratory on Valentia, and seldom ventured beyond Co. Kerry.[14]

Jellyfish are shape-shifters, and can occur in two forms. The first resembles the sea anemone, a hydra form that attaches to rocks; the second is the free-swimming jellyfish, or medusa. In many species, the animals alternate between the two, though some species have dispensed with one or other form. For a long time, it was thought the various hydra and medusa were different species, or it was not known which hydra went with which medusa. Maude Delap helped untangle the puzzle for a number of species, work that continues even today.

Born in Co. Donegal, Maude Delap spent most of her life on Valentia where her father was rector. The Rev Alexander Delap was a keen naturalist, regularly contributing observations to the *Irish Naturalist Journal*, and his daughter inherited his passion. Partly due to the Delap family's diligent work, English scientists chose Valentia for a detailed marine study in the 1890s. The expedit-ion team spent the summer of 1895 on the island collecting speci-mens, aided by Maude Delap and her sister Constance, who rowed around the coast trawling for samples with a net. After the scientists left, the sisters continued collecting for some years, sending the specimens to University College London. The survey results were eventually published in a 188-page report in the *Proceedings of the Royal Irish Academy* (1899), which acknowledges the role the Delap sisters played.

Three years later, Maude Delap began her complex experiments rearing jellyfish, to untangle which medusa went with which hydra. There was some trial and error to find the right conditions for the various species, but she persisted, using home-made aquarium tanks in a laboratory in the family house at Reenellen, replenishing the tanks with fresh sea water, and bubbling oxygen through the jars. Her many discoveries – mostly published in the Irish Fisheries scientific reports – meant that a number of species were reassigned. She discovered for example that two hydroids, *Laodicea undulata* and *Dipleurosoma typicum*, were species of *Cuspidella*. On the strength of her research she was offered a job at the Marine Biology Station in Plymouth, but her father ruled: "No daughter of mine will leave home, except as a married woman."

The white markings (in nature these appear on a colourless background), characteristic of Edwardsia delapiae, the jellyfish named after Maude Delap. A commemorative plaque on Knightstown main street marks the site of her former home.

Maude Delap never married, but remained on Valentia, continuing her research. She collected numerous specimens, sending most to the National Museum in Dublin, and corresponded with scientists there, in England and elsewhere. In the 1920s, she found the first complete skeleton of a True's beaked whale, which had been stranded on the Kerry coast. This rare and little-known whale (it was photographed for the first time only in 1993) was named, in 1913, from a partial skeleton, and only a few have ever been found, three of them in Ireland. The skeleton Maude Delap found is, therefore, an important specimen, and the fact that she could correctly identify the species is evidence of her expertise. In 1928, she was honoured by her colleagues when a sea anemone she had found near Valentia was named *Edwardsia delapiae*. There is more on Kerry's marine life at Oceanworld, Dingle and Fenit Seaworld. ⑮ ⑯

VALENTIA

Purple slate quarry

Sir Maurice FitzGerald, the 18th Knight of Kerry, opened a slate quarry at Dohilla in 1816 that produced a fine purple slate that was famous the world over. The slabs were renowned for their smoothness and immense size (up to 10 metres long and 2 metres wide) and were much prized for flooring, altars, billiard tables and shelving – London's Public Record Office has literally miles of Valentia shelving.

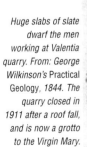

Huge slabs of slate dwarf the men working at Valentia quarry. From: George Wilkinson's Practical Geology, *1844. The quarry closed in 1911 after a roof fall, and is now a grotto to the Virgin Mary.*

It was even suggested that cottages could be built using four large slabs for the walls and a fifth for the roof. Thin slates of the stone were used in roofing, as at the Paris Opera House, London's Houses of Parliament, and even a railway station in El Salvador across the Atlantic. The slate was extracted from the northern side of Geokaun Hill in large blocks that were brought to Knightstown village by a steam-powered railway – possibly the earliest steam railway in Ireland. There they were cut using steam-powered saws, and cleaved by chisel. At its peak the quarry employed 400 men, many of them Welsh miners.

In addition to slate, Valentia exported corn and butter, and was known as "the granary of the southwest". For centuries, there were strong trading links with Spain, especially wine imports, and the sheltered harbour regularly provided storm refuge to transatlantic ships. Alongside Welsh miners and Spanish traders, this cosmopolitan community was home to battalions of elite English telegraphers (who had their own cricket pitch and tennis court), and the many scientists who came to study the island's marine life or visit its meteorological observatory.[*]

CAHERCIVEEN

1860 – the first shipping forecast

Weather forecasting is all about knowing what is happening upwind of you. In simpler times, people read clouds and other signs for an indication of what was coming, but the advent of the telegraph[*] meant it was possible to know nearly instantly what the weather was like many miles away. You just needed someone to take the measurements and wire you the information. In 1860, Britain's Royal Navy set up a network of 40 weather stations to do just that. Some 400 people had drowned off Anglesea during a recent storm, and it was hoped telegraphic storm warnings might prevent similar disasters.

The man in charge was **Vice-Admiral Robert Fitzroy** (1805–1865) who, in the 1830s, aboard his boat

HMS *Beagle*, had been the first to use Francis Beaufort's new wind scale.* (That voyage of the *Beagle* is now remembered for the young naturalist who travelled on it, Charles Darwin; ironically, Fitzroy was among those offended by Darwin's radical ideas.) One of Fitzroy's new weather stations was strategically located on Valentia, the first place in these islands to receive the weather arriving off the Atlantic and, crucially, also part of the national telegraphic grid. On October 8th, 1860, scientists working there from a rented house, later known as Valentia Observatory, were the first to telegraph weather data to Fitzroy's network.

It was the start of the shipping forecast. Gale warnings could now be telegraphed to ports, where semaphore signals were raised to pass the warning to ships – an upward pointing cone, for example, signalled a northerly gale. By 1861, Fitzroy was daringly predicting the weather, and the following year published the first weather forecasts in *The Times*. But with only limited information, the forecasts could never hope to be accurate and people soon ridiculed them. In 1865, after a run of poor predictions, but perhaps also guilty about his role in Charles Darwin's work, Robert Fitzroy took his own life.

Valentia Observatory, now part of Met Éireann,* continues to record strategic weather information, and data collected there are used by Irish and European meteorological services. In 1892, the observatory moved to the mainland at Caherciveen, though it is still referred to as Valentia in the weather reports. The staff now monitor solar radiation, the magnetic field over Ireland, earthquake tremors, rain pollution and ozone levels, as well as the usual weather variables, such as wind speed, sunshine levels, rainfall and temperature. Every day a radiosonde balloon is released to measure conditions in the upper atmosphere. The observatory has a collection of old scientific instruments and an unusual **phenological garden**: the dates when plants come into bud, leaf, flower and fruit, are noted, and the information is collated with data from similar European

gardens, and used to monitor climate change. [17] The JFK Arboretum* in Co. Wexford is part of the same network.

The edge of the known world

For over 500 years, these tiny rocky islands were an outpost at the edge of the known Western world, and home to a community of Christian ascetics. Now they are a nature reserve, providing sanctuary to tens of thousands of seabirds. Little Skellig is home to the world's second-largest **gannet colony**, an astounding 27,000 breeding pairs, nearly 12 per cent of the world's gannet population. So dense is the colony that the islet is white with guano, and the birds can only be counted in aerial photographs using special microscopes. Any remaining space on the island is taken by thousands of other seabirds, especially kittiwake, guillemot and razorbill. The underwater wildlife and landscape is equally rich: submerged sea cliffs and reefs teem with life and the area is popular with divers.

Neighbouring Great Skellig or Skellig Michael (*Sceilig Mhichíl*) is named after the saint of high places. Though famous for its vertiginous, early-Christian ruins, it too is an important bird site. Thousands of storm petrel, Manx shearwater, puffin and other seabirds spend the summer there, plus occasional choughs (a distinctive red-legged blackbird) and peregrine falcons. Great Skellig also claims the world's longest-lived storm petrel: No. 236 5699 was first ringed there in 1966 and returned every summer for over 26 years.

The Skelligs' remote location in clean, rich waters at the edge of the Atlantic make them attractive to seabirds, and the lighthouse beam is a positive invitation to passing migrants. Plantlife is limited by the salty spray and by the high concentration of nitrogen and phosphorous from the birds' droppings, yet sea pinks and sea campion are among those clinging to the rock. The presence of slugs and snails prompted suggestions that the Skelligs were once connected to the mainland, though the slugs could have come amid goods brought in by people.

The islets were once owned by the Butlers of Waterville, who harvested puffin feathers there. The Ballast Board bought the Skelligs, in 1821, to build a lighthouse on Skellig Michael, and a road was blasted out of the side of the island as part of the works. The lighthouse opened in 1826 but, despite the light, a US navy aeroplane searching for German submarines in February 1944 clipped the island's tip and crashed into the sea with the loss of all lives.

Great Skellig was home to lighthouse keepers until its light was automated in 1987. Children were born there – the last one in 1897 – and at least two are buried by the chapel. Life was harsh: there are no trees, no turf or other fuel; no fresh water, so rainwater must be collected; little flat ground and not much grass; and the mainland is hours away by currach. Before the telegraph and radio, communication was by semaphore using two large flags to send a signal which the lighthouse keeper at Bull Rock read by telescope, then relayed by semaphore to Dursey in west Cork. Yet monks lived on Great Skellig for nearly 600 years (AD 600–1200). They cultivated a small garden there, kept an occasional goat or cow, lived off fish and birds' eggs, and even had a small ironworks, the remains of which were found amid the ruins. ⑱

The Skelligs are national nature reserves and a visit is a must, for the boat trip, the birdlife and the monastic ruins. No one may land on Little Skellig, however, and only licensed boats may land at Great Skellig. Two other Kerry islands noted for their seabirds are the **nature reserves** of An Tiaracht (Tearaght Island) in the Blaskets, and Puffin Island south of Valentia, which has spectacular cliffs and is noted for puffin, shearwater and razorbill.

William Petty – scientist, inventor, coloniser

Kenmare was founded as a mining and industrial town in 1670 by the brilliant scientist, economist and inventor **Sir William Petty** (1623–1687), who also managed Oliver Cromwell's conquest of Ireland. Petty invented the modern sciences of political and statistical economy, and, perhaps more than any other, brought the Scientific Revolution to Ireland – or at least to TCD.* But he is best remembered here as Cromwell's secretary, and the man who supervised the Down Survey.*

William Petty made the first modern map of Ireland. Ardgillan Demesne, Co. Dublin, has a permanent exhibition on the Down Survey.*

Petty was born to a poor English family (his father was a tailor) and went to sea as a cabin boy at 15. The crew found him obnoxious, however and, after breaking his leg, left him stranded on a French beach. Undaunted, Petty learned French, raised money by teaching English, and eventually studied medicine at the famous Dutch college in Leyden. By 1650, he was studying medicine at Oxford where, in a notorious case, he successfully revived a convict, Anne Green, who had been only "half-hanged". Afterwards, Green was pardoned and Petty was made professor of anatomy at Oxford University (the following year he also became the college's music professor).

In 1652, at the age of 29, Petty came to Ireland as physician general with Cromwell's army, replacing Gerard Boate* who had just died. He reorganised the army's medical supplies, cut costs and waste, and revealed a talent for statistics and organisation. He was soon Cromwell's trusted private secretary and, in 1654, was given charge of the Down Survey* (so-called because everything was written down), which charted the confiscated lands that would be granted to settlers and used to pay soldiers. At the same time, Petty began his own personal project mapping all of Ireland, which resulted in the first modern atlas of the country, *Hiberniae Delineatio* (1685). The Down Survey was a massive undertaking, but William Petty completed it in time and under budget.

Controversially, Petty was also responsible for distributing the confiscated land – not always to everyone's satisfaction. Many soldiers sold their land to Petty, who amassed a vast estate in Kerry. He spent £10,000 settling an English colony at Kenmare, where he started a fishing industry, ironworks, and copper and lead mines. Traces of the mines and ironworks remain – the old copper mine on Coad Mountain above Caherdaniel, for example – but his lasting mark on the landscape was clearing the great ancient woodlands to make charcoal to smelt the iron. The many place names there containing the element Derr- (from *doire* 'oak wood') are evidence of how heavily wooded the area once was. [19]

Petty now divided his time between England and Ireland. He was a founder of the Royal Society in London, and first president of the Dublin Philosophical Society,* a similar Irish organisation. He invented political economy, applying the new scientific method to politics. He also pioneered the use of statistics and demographics to tackle problems of policy and administration, and was the first person to estimate national income. His two most influential books were *A Treatise on Taxes and Contribution* (1662), and *The Political Anatomy of Ireland* (1691). He analysed the growth of London, Dublin and a dozen other European cities, and was the first to tabulate the causes of deaths in cities, publishing his London study in 1662, and a Dublin one in 1682. Petty believed that information was essential in planning for emergencies, such as a plague, and he urged the government to collect data and establish a statistics office in Dublin. He also campaigned for the union of England and Ireland, believing it was the only way to protect the smaller island and ensure religious freedom there. A

union, he thought, would enable Roman Catholics to retain power in Ireland while protecting Protestant interests there, and also control extreme Protestant bigotry. Petty was also mechanically minded, and his inventions included the modern twin-hulled ship, or catamaran,* which he tested in Dublin Bay. Despite his friendship with Cromwell, and a talent for making enemies (once, when challenged to a duel, Petty chose an axe and dark cellar as his weapon and location), he also found favour with Charles I. When the monarchy was restored, he was allowed keep his Kerry lands that, under his descendants, became the Lansdowne Estate. Derreen Gardens, part of the Lansdowne Estate, were planted in the 1870s with exotic woodlands and tall New Zealand tree ferns. [20]

Chapter 10 sources

Co. Limerick

1. Robins, Joseph, *The Miasma: epidemic and panic in 19th-century Ireland* (IPA, 1996).
2. Scanlan, Mary, "The Flying Irish Woman" in Mary Mulvihill (ed.), *Stars, Shells & Bluebells* (WITS, 1997).
3. Byrne, Anne, "Bringing a shy biologist out of her shell" in Mary Mulvihill (ed.), *Stars, Shells and Bluebells* (WITS, 1997).
4. Gow, Rod, "John Casey" in Ken Houston (ed.), *Creators of Mathematics: The Irish Connection* (UCD, 2000).

Co. Kerry

5. Guerin, Michael, *The Lartigue Railway* (Listowel, 1988).
6. Feehan, John & Grace O'Donovan, *The Bogs of Ireland* (UCD, 1996).
7. Hogg, William, *The Millers and Mills of Ireland of about 1850* (Dublin, 1997).
8. Gleed-Owen, C P (*et al.*), "Origins of the Natterjack Toad (*Bufo calamita* Laurenti) in Ireland: reexamination of subfossil bones from Carrowmore, Co. Sligo" *Irish Biogeographical Bulletin* No 23 (1999).
9. *Journal Dublin Geological Society* **7** 164 (1856).
10. *Journal Dublin Geological Society* **4** 151 (1849).
11. O'Brien, William, *Ross Island and the Mining Heritage of Killarney* (NUI Galway, 2000).
12. Carruthers, Terry, *Kerry: A Natural History* (Collins Press, 1998).
13. Standage, Tom, *The Victorian Internet* (Phoenix, 1998).
14. Byrne, Anne, "Untangling the Medusa" in Mary Mulvihill (ed.), *Stars, Shells and Bluebells* (WITS, 1997).

Chapter 11: Cork • Waterford

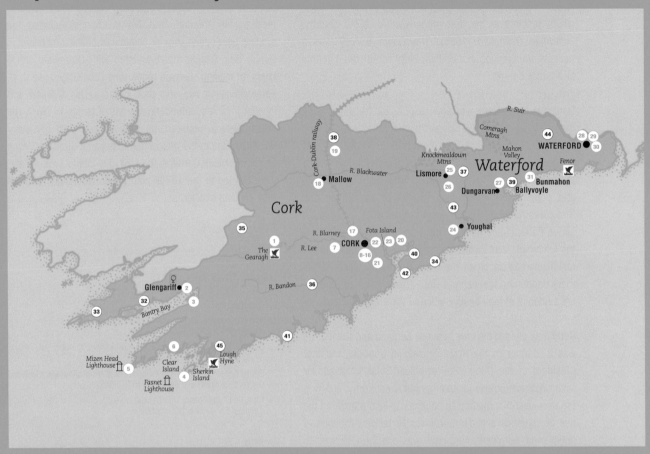

Visitor facilities

1. Bealick Mill, Macroom
2. Ilnacullin Garden
3. 1796 Armada Exhibition, Bantry
4. Cape Clear Bird Observatory
5. Mizen Vision
6. Schull Planetarium
7. Royal Gunpowder Mills, Ballincollig
8. Cork Public Museum
9. Cork City Archives
10. Cork Corporation Waterworks
11. Beamish & Crawford Brewery
12. Cork Heritage Centre
13. UCC Geology Museum
14. Radio Museum/Cork Gaol
15. Cork Butter Museum
16. Meitheal Mara
17. Monard Spade Mills & Wildlife Sanctuary
18. Spa House Energy Office
19. Doneraile Park
20. The Old Midleton Distillery
21. Cobh Museum
22. Fota Wildlife Park
23. Fota Arboretum & Garden
24. Youghal Heritage Centre
25. Lismore Heritage Centre
26. Lismore Castle & Garden
27. Dungarvan Museum
28. Jack Burtchaell Tours
29. Waterford Treasures Museum
30. Waterford Crystal
31. Bunmahon Heritage Centre

Other places of interest

32. Adrigole
33. Allihies
34. Ballycotton
35. Ballyvourney
36. Bandon
37. Cappoquin
38. Castlepook
39. Cloughlowrish
40. Cloyne
41. Duneen Bay
42. Inch
43. Molanna Abbey
44. Portlaw
45. Skibbereen

♀ woodland
☒ nature reserve or wildfowl sanctuary
⌂ lighthouse

Visitor facilities are numbered in order of appearance in the chapter, other places in alphabetical order. How to use this book: page 9; Directory: page 474.

COUNTY CORK

Substantial copper mines at Allihies, which featured in a Daphne du Maurier novel; a vast gunpowder plant at Ballincollig; and the largest butter market in the 19th-century world... just some of Co. Cork's rich industrial heritage. Other Cork highlights include mysterious Lough Hyne and the prehistoric flooded woodland of the Gearagh. Percy Ludgate, an accountant from Skibbereen, invented an early computer and George Boole, a professor at University College Cork, invented the mathematical logic that enables modern computers to think. Cork city was founded on marshy ground by the River Lee (Cork from *corcaigh*, 'a marsh'), and the marshes were a source of mosquitoes and malaria until the late-19th century.

BEARA PENINSULA
The copper mine of *Hungry Hill*

The southwest of Ireland was a major copper-producing region in the 19th century, with mines throughout the area, including Mount Gabriel, Crookhaven, West Carbery, the Bantry area, Kenmare* and Killarney.* A mineral-rich belt formed there during the powerful geological events that pushed up the Cork–Kerry mountains 300 million years ago, and deposited copper, barytes,* lead and silver there. In places, the copper is obvious at the surface, where it stains the rock green, and it has been mined on and off since Bronze Age times. Several large mines were worked during the late-17th and 18th centuries, especially by William Petty,* but, during the Napoleonic wars, when demand for metal rocketed, the mining began on a truly industrial scale.

The most substantial workings were at **Allihies**, where the hillside is riddled with shafts and opencast excavations. The shafts – now flooded and in a treacherous condition – drop 400 metres and frequently ended below sea level. To pump them dry, and to haul up the heavy ore, powerful steam engines were used – one came second-hand from Ross Island after the Killarney mine closed in 1829. Ruins of the substantial engine houses built for these leviathans can still be seen. Uniquely, one Allihies engine was a **man engine**: this was a primitive elevator, where the engine worked a series of wooden ladders to transport the men up and down the shaft. Only 19 such engines were built in the world. The Irish Mining Heritage Trust has begun work to conserve the building.

The Allihies mines were owned by an English family, the Puxleys, who built **Dunboy** mansion, near Castletownbere, with their copper profits. (The Puxleys appear as 'the Brodericks' in *Hungry Hill*, Daphne du Maurier's novel about the Allihies mines.) Dunboy Harbour was also the mine's ship-ping station. The ore was initially ferried there in small boats from Allihies; after 1824, when a road was cut over the back of the peninsula, the ore was brought by cart (the road is now part of the Beara Way). The mines used huge amounts of imported Norwegian timber, which was stored at Dunboy in a thick-walled storehouse that is now in ruins.

Miners stood on the man engine steps, and were lifted up or down with each stroke. This 1890s photograph was taken in a Cornish mine. Allihies has the sole-surviving purpose-built man engine house. Royal Institution of Cornwall

Allihies reached its peak around 1850, when 1,500 people were shifting 90,000 tonnes of ore a year. The ore was crushed on dressing floors beside the mines, and the copper picked out, usually by women and children. The waste quartz was dumped over the cliffs, forming an unusual white quartz beach. Much admired today, it is, strictly speaking, a mine spoil heap. Most workers were local, but the Puxleys brought in English supervisors and miners. These had favoured status: Cornish Village was built for them, and they had a chapel, school and free accommodation, which led to tensions with local people. In the 1880s, enormous copper deposits were found in the Americas, Africa and Australia, and the relatively small Irish copper mines could not compete. The mines of the Mizen and Beara peninsulas closed, and many of the men left to seek work in Montana, in mines run by Cavan man Marcus Daly.* Though some copper remains at Allihies, it is not economic. A leaflet on Allihies mines, published by the GSI, is sold locally. Bronze Age copper workings can be seen at Ross Island.* Ireland's only cable car connects the end of the Beara Peninsula to Dursey Island.

Where the CFC ban began

You could say that the international treaty to ban all ozone-destroying **CFCs** (chlorofluorocarbons) began at Adrigole, where English scientist **James Lovelock**, now famous for his Gaia theory about the Earth, had a holiday cottage. One summer's day in 1968, Adrigole was smothered in a smoggy haze, and Lovelock wondered if it might be air pollution from Europe wafting in on an easterly breeze. A decade earlier, Lovelock had invented a monitoring instrument, the electron capture device (ECD), that was extremely sensitive to CFCs and other pollutants in the atmosphere. (Lovelock has said that, if a few litres of a distinctive CFC were released over Japan, the ECD would be able to detect it in Brussels two weeks later; and a year later would detect its presence in the atmosphere anywhere around the world.) He returned to Adrigole the following year with his ECD and discovered that the haze was, indeed, a smog of industrial air pollutants, particularly CFCs.

CFCs are synthetic chemicals that were widely used in industry since the 1950s and, because they do not occur naturally, they are a useful indicator of industrial air pollution. After his Adrigole discovery, Lovelock wondered whether CFCs would be detected in other, relatively clean environments and, in 1971, brought his ECD on a 6-month voyage from England to Antarctica, sniffing the air all the while. Worryingly, he detected CFCs everywhere. Moreover, the amount he detected tallied with the amount of CFCs that had ever been released by industry. The CFCs were clearly accumulating in the atmosphere.

In separate research, other scientists had detected a thinning in the ozone layer high above Antarctica; subsequently it was realised that CFCs were the main culprit, and their use is now banned. In the 1970s, Lovelock installed equipment at his Adrigole cottage to monitor CFCs and other pollutants in the atmosphere, making Adrigole the first station in a global network to monitor the state of the world's atmosphere. In 1978, four other stations were added – Barbados, Oregon, Samoa and Tasmania – all in relatively clean, non-industrial environments making it easier to monitor background levels of pollution. In 1987, the Irish station moved from Adrigole to NUI Galway's atmospheric research station at Mace Head* in Connemara.

THE GEARAGH

An aboriginal forest in a lake

During the last Ice Age, glaciers flowing off the Cork–Kerry mountains deposited substantial mounds of sand and gravel around the upper Lee valley. These moraines impounded lakes along the river, most notably Lough Allua, but at the flat ground of the Gearagh, the sandy deposits braided the river into a thousand threads, creating a tortuous tangle of streams, soft mudbanks and small islands. A unique ecosystem evolved on this unusual landscape: willow and alder grew in the wetter swamps; oak woods, with birch, hazel and hawthorn, developed on the drier islands; and guelder rose and cherry trees colonised the edges.

This complex alluvial woodland was in place 7,000 years ago and, despite the arrival of humans, it remained untouched until 1952. Saved by its treacherous swampy nature, it was a true aboriginal forest. There was no other like it in western Europe. Locals navigated the maze using special flat-bottomed boats, outlaws and *poitín*-makers found sanctuary in the woods. However, in 1952, dams and hydro-generating stations were built across the River Lee at Carrigadrohid and Inniscarra, flooding the townland of Annahala East and the Gearagh's unique woodland. The people of Annahala had to move out – one man who refused, tragically drowned one night while returning home – and the trees were cut down and left to rot in the rising waters of the new reservoir.

The broad expanse of water is still studded with thousands of blackened tree stumps, and looks like some alien wasteland. The new wetland does provide a welcome home for **mudwort**. This rare, low-growing plant previously grew at only one site in Ireland (in Co. Clare); now it grows in profusion around the Gearagh when water levels are low in summer. At the upper end of the reservoir, where a section has been set aside as a nature reserve, plants are recolonising the swamp and the woodland is regenerating. When water levels are low you can cross by a causeway – the old Annahala quarry road – to explore the Gearagh nature reserve, its drowned quarry and abandoned buildings. (Note, the waters are treacherous on account of submerged structures and the tangled mass of felled trees that were dumped in the channels.)

Another local use of water power is at **Bealick Mill**, now renovated. The mill's 20-tonne wheel, built in the mid-1800s by McSwiney's foundry at MacCurtain Street, Cork, turned six grindstones. Oats were dried in a kiln prior to grinding and the mill's Macroom Oatmeal, which is still sold, had a distinctive roasted flavour. In 1899, a turbine was installed on the millrace to generate electricity for street lighting in Macroom, just one of many small electricity schemes in the days before the national grid. In the 20th century, Bealick functioned as a sawmill and foundry; now the renovated turbine generates electricity for local homes and the mill museum. ①

BALLYVOURNEY

A culinary experiment

Archaeologists conducted an unusual cooking experiment at Ballyvourney some years ago: they dug a pit in the ground, filled it with 450 litres of cold water, and added hot stones from a nearby fire; this brought the water to a boil in 30 minutes. By continually adding fresh hot stones, they managed to cook a leg of lamb in four hours, and prove that a *fulacht fiadh* ('ancient cooking site') could be used for cooking. *Fulachtaí fiadh* are found throughout northern

Europe and most date from about 3,500–4,000 years ago. Antiquarians in the 19th century referred to them as "old fireplaces of the primitive inhabitants". There are thousands in Ireland – at least 2,000 in Co. Cork alone – indicated in red on the 1:50,000 Ordnance Survey map. Most consist of an inconspicuous kidney-shaped mound of burned stones located beside a stream, along with a small depression, which is the remains of the pit or trough. Many are associated with traces of human settlement, such as trackways or hut sites. They may have been used for bathing, cooking or both: Geraldus Cambrensis wrote in *Topographica Hibernica* (1188) that a *fulacht fiadh* at the Magh Adhair clan inauguration site in east Clare, was used both to cook a sacrificial horse and bathe the newly elected chief.

BANTRY

Warm water, mussels, oil and ice

Thanks to the **Gulf Stream**, the sea off southwest Ireland is warmer in winter than it ought to be. That great ocean current acts like a vast conveyor belt-cum-storage heater, bringing warm Caribbean water across the Atlantic and up Ireland's west coast. As a result, Ireland has a mild climate and is warmer than anywhere else at the same latitude – Cork city for instance, is on the same latitude as London, but is warmer in winter by an average of 2°C. Sea water off southwest Munster can also be 3 degrees warmer than off Co. Antrim, which is the coldest point on the Irish coast; in August, when the sea has warmed to its fullest, the southwest averages a balmy 16°C, compared with 13°C off Antrim. The mild winters in the southwest mean subtropical plants can flourish in sheltered places – witness the many fine gardens in Cork and Kerry with lush exotic vegetation, such as Ilnacullin in Bantry Bay. ②

Sea temperature also influences the growth of algal tides, fish spawning and migration, and the spread of pollution. With mounting concern about climate change and rising sea levels, there is growing interest in sea temperatures and the direction of the Gulf Stream. Temperatures used to be measured by passing ships – simply sticking a thermometer into a bucket of sea water hauled on board; increasingly satellites are used to sense the temperature by monitoring infrared radiation reflected from the sea's surface. Since 1999, the Marine Institute has also measured sea temperature using buoys in Bantry Bay and at Fastnet. Powered by solar panels and a small wind turbine, these large, yellow buoys are equipped with instruments to measure sea temperature and current, and wind speed and direction, and beam the information to the institute's Dublin headquarters.

If the Gulf Stream did change direction, Ireland's climate would alter dramatically, becoming colder and probably more Scandinavian. Worryingly, this could be triggered by global warming, because, if the Greenland ice cap melts, this would dump a large volume of cold fresh water into the North Atlantic, which could push the Gulf Stream south and away from Ireland. Such a change happened about 10,500 years ago when temperatures plummeted by at least 7°C, throwing Ireland into a 500-year-long cold period known as the [Lough] Nahangan Phase,* when glaciers again formed in the Wicklow Mountains.

For the moment, though, Bantry's warm water is appreciated by **mussel growers**, and the bay is Ireland's largest mussel-processing centre. The mussels are grown on ropes suspended from floats and the surface of the bay is studded with these rafts. According to Dr John Joyce of the Marine Institute, this farming technique may have been invented in the early 1600s by Irishman Patrick Walton. The story goes that Walton was shipwrecked on the French coast where, starving and penniless, he slung rope nets in the water to catch birds and fish; mussel seeds colonised the ropes, however, and Walton later hauled in a fine crop of shellfish. Thus began the *bouchot* technique, now used by mussel farms worldwide.

The deep water around Whiddy Island made it ideal for an **oil terminal** and, in 1968, Gulf Oil opened its European trans-shipment terminal there. In 1979, however, the tanker *Betelgeuse* caught fire and 51 people were killed in Ireland's worst industrial accident. The wrecked terminal was later bought by the Irish National Petroleum Company (INPC) and reopened in 1998; it covers about 20 per cent of the island, and the national oil reserve is stored there. Huge tankers anchor offshore and unload their cargo via floating pipes; the INPC refines the oil at Whitegate, near Midleton. In 1981, while removing wreckage from the *Betelgeuse*, divers discovered the wreck of a French frigate, *La Surveillante*. The ship had been part of a French invasion fleet led, in 1796, by General Hoche and Wolfe Tone. Atrocious weather forced most of the expedition back, but *La Surveillante* was damaged by the storm and had to be scuttled in Bantry Bay. Archaeologists have since surveyed the site using various techniques, including side-scan sonar and seismic profiling deployed by the Irish marine research vessel *Celtic Voyager*. ③

The landscape around Bantry and Glengarriff bears the marks and scars of **ice** and **glaciers**. At the head of the bay is a small drumlin* swarm, the only significant swarm in Munster. Some of the drumlins drowned when sea levels rose, and these are now small islands. In the Shehy Mountains, glaciers gouged out the impressive Pass of Keimaneigh, a rock gorge 3km long and with walls nearly 100 metres deep. The beautiful wooded valley above Glengarriff was also deepened by glaciers flowing off the higher mountains behind.

Glengarriff's semi-natural **oak woodland**, now a national nature reserve, is internationally important. South-facing, sheltered from the wind by hills, and warmed by the Gulf Stream, the trees grow down to the shore. Alongside the oak are holly, strawberry trees (arbutus),* rowan, yew and willows, bilberry and hellebores, and filmy ferns and mosses. The wood is also noted for its rich insect life, including two rare species: the false click beetle, which lives under the bark of the strawberry tree, and the pseudo-scorpion.

The lady of the lichens

In 1815, a young woman died of consumption at Ardnagashel. Her passing was mourned by naturalists at home and abroad, when they learned the news some months later. For **Ellen Hutchins** (1785–1815) had an international reputation as an expert on mosses, ferns, lichens* and seaweeds.[1] Despite ill health, the conventions of the day regarding women, and the fact that she seldom travelled far from Bantry, Hutchins discovered numerous species new to science. Many were named after her, such as the lichen *Lecania hutchinsiae*, and a fern she discovered in Kerry, *Jubula hutchinsiae*, known as 'the dark companion of the Killarney fern'.

Ellen Hutchins was from a prosperous Bantry family – though her father was not above smuggling French wine – and attended school in Dublin. She was encouraged to take up an outdoor hobby for her health and a family friend, Dublin medic Dr Whitley Stokes,* suggested plant collecting. Stokes introduced her to some eminent Irish naturalists and, with their encouragement, Hutchins became skilled in identifying species before she was 20. She specialised in the difficult area of non-flowering plants, and had an unerring ability to spot rare species, especially plants that had never been scientifically described. Her name appeared frequently in botanical books of the day, she corresponded with top British naturalists, including Sir William Hooker of Kew Gardens, and once travelled to England to visit Dawson Turner, an algae expert. Ill health mostly confined her to Bantry, however, where naturalists often came to visit, keen to see the area made famous by her finds. Hutchins bequeathed her extensive collection of dried botanical samples and illustrations – she was a gifted artist and drew all her finds – to Dawson Turner, and most are now at Kew Gardens.

Watching birds pass

Keeping an eye on the sea

South of the Gaeltacht island of Oileán Chléire (Clear Island), at the edge of the continental shelf, cold oceanic water rises from the depths. This 'upwelling' brings with it minerals that nourish plankton (microscopic plants) at the surface, producing plankton swarms so dense they can sometimes be seen from the air. The plankton are an important food for fish, birds and whales, and feeding frenzies have occasionally been reported, when dolphins and flocks of birds gather to attack the shoals of fish grazing on the plankton. The coastline roundabout is thus noted for its bird- and whale-watching and, since 1959, Oileán Chléire has been home to an important bird observatory, where migrations are studied every spring and autumn. ⓐ A warden is there from March–November, and volunteers help at other times, so there is near-continuous monitoring of passing birds. Sooty shearwaters, Sabine's gulls, grey phalaropes are just some of the species seen at the island. Winter weather can blow rare strays onto the island, and late July produces an amazing phenomena, when up to 10,000 Manx shearwater can be seen each hour, as they pass to and from their nests on the Blaskets.

Cléire, Ireland's most southerly island, has a mild oceanic climate, a range of habitats – including freshwater lake, bogs, reed marsh, moorland, pasture and exhilarating sea cliffs – and a rich flora and fauna. Nearly 380 plant species grow there, including some Lusitanians* (species more usually found in northern Spain and Portugal). In mediaeval times, the island may have been a leper colony: the island's wind turbine is on Cnoc Carantáin ('quarantine hill'), and a nearby townland is called Gort na Lobhair ('the leper's field'). Apart from wind power, the island has photovoltaic street lighting and solar-powered water heating at the school.

Dotted along the shore between Cork Harbour and Bantry are 146 small monitoring sites where marine scientists regularly check conditions and count the plant and animal species present. This detailed work provides valuable baseline information and is used to spot changes in the environment. It is just one of the projects organised by **Sherkin Marine Station**, a private and self-funded research centre founded in 1975 by Matt Murphy, a man with a passion for the environment. Much of the scientific work is done by volunteers, aided by experts from international institutions, such as the British Museum and Amsterdam's Zoology Museum. The south coast is renowned for its sponges and the station has made a special study of these over the years, as well as of rock pools and local otter populations; divers from the centre also conduct regular underwater surveys. The station organises an annual research conference, and publishes books, reports and a newsletter. The 5-hectare complex has laboratories, an aquarium, a library, and a herbarium with hundreds of dried specimens of marine plants and seaweeds.

The 'tear drop of Ireland'

Fastnet Rock, Ireland's most southerly outpost, is a tiny jagged pinnacle 6km southwest of Clear Island. Its lighthouse was usually the last bit of the country seen by those emigrating to North America, and some called it 'the tear drop of Ireland'. Although there had been numerous shipwrecks in the area, work on a Fastnet light only began in 1848, and then only after a US liner, *Stephen Whitney*, was wrecked on neighbouring West Calf Island in 1847 with the loss of 91 lives.

Building a lighthouse on Fastnet was a major logistical undertaking because of its remote and exposed location,

and small size. George Halpin senior, a prominent Irish lighthouse engineer, designed a tower made from cast-iron flanged plates that could be assembled on site, and a colony of workmen was installed in wooden huts on the rock to build it. Frequent rough weather meant the tower was not finished until 1853. It stood 20 metres tall, and was secured to the rock with huge bolts 60cm across.[2] However, Halpin's tower vibrated when hit by the high seas that frequently washed over the rock and there were concerns for its stability, so, in 1866, it was strengthened. However, in 1881, a similar tower on nearby Calf Rock snapped in half during a storm – thankfully, while the keepers were on the ground floor – and the Commissioners of Irish Lights decided to replace Fastnet's metal tower. (Calf Rock light was abandoned, replaced by a new light on neighbouring Bull Rock; the stub of the original metal tower can still be seen.)

For Fastnet, William Douglass designed a new stone tower, to be made from 2,000 dove-tailed, inter-locking and precision-cut granite blocks. Building this 3-dimensional stone jigsaw was another logistical nightmare. To minimise the work done at Fastnet, the granite blocks were cut at a Cornish quarry, then delivered by a specially built steamer to a depot at Crookhaven. Frequent bad weather meant this tower took seven years to build, but it was completed in 1903 and the new lantern was lit in June 1904. The light was automated in 1989, and it still serves as a vital navigational aid to mariners.

Fastnet tower is just one of many lighthouses on this complex, rocky and often treacherous coastline. Another is the **fog signal station** on Mizen Head, which was built in 1909. The station is reached via an elegant footbridge that is an engineering gem: a prefabricated reinforced concrete suspension bridge, it was the largest of its kind at the time, and possibly the first to use precast elements. It spans nearly 100 metres across a chasm that drops 100 metres to the waves below. Everything was cast on-site using sand crushed from local rock. The largest elements are the two curved ribs that span the full width of the bridge; after being cast, they were swung into position; the load-bearing trestles and beams were then added. For years, all the station's supplies were ferried across this narrow bridge, often in wheelbarrows. The station was automated in 1993 and is now a visitor centre with an exhibition about the Fastnet lighthouse. ⑤ Mizen is also noted for bird- and whale-watching.

1909 – a novel Irish computer

An Irish accountant, **Percy Ludgate** (1883–1922), invented an early computer in 1909. His 'analytical machine' was designed to be programmable, portable, efficient and simple, yet capable of complex calculations. It was never made, but Ludgate's innovative design is accorded a place in the history of computing. Percy Ludgate was born in Skibbereen and studied at Rathmines College of Commerce, where he was a gold medal student, before joining a Dublin firm of accountants. From 1903, he spent his spare time designing an analytical machine, presenting the details to the RDS in 1909.[3] This brought him to international attention and, in 1914, the Royal Society invited him to lecture in Edinburgh at a special conference on mathematics and computing.

Since the invention of the abacus, people have been trying to build ever more complex calculating machines. The title 'grandfather of computing' is usually given to English inventor **Charles Babbage** (1791–1871), who spent years and a vast fortune attempting to build programmable calculating machines. His ideas were based on the Jacquard loom, where punched cards controlled the patterns. But Babbage's calculating machines proved impossible to build at the time and only small sections of them were attempted. Even these were enormous, ponderously slow – everything was reduced to addition or subtraction – and required thousands of precision-engineered metal parts. Percy Ludgate's design was also based on the Jacquard loom, but in most other respects it differed from Babbage's machine. Where Babbage used columns of toothed discs to store numbers, for example, Ludgate opted for a simpler shuttle mechanism.

Ludgate's design had all the elements of a modern computer: a mechanism for storing data, ways to input data and program the machine (using pre-punched formula tape and/or keyboards), a printer and even an operating system, all of which Ludgate designed himself. Uniquely, Ludgate's mechanical computer had two innovative features: it could be stopped at any stage in a calculation to add new variables and it could do subroutines. Ludgate's approach to calculations was also unusual: for multiplication he developed a technique using partial products; for division he used a table of reciprocals and a rapidly converging series using subroutines. He calculated that his engine would multiply two 20-digit numbers in under 10 seconds, and take two minutes to determine the logarithm of a number. It could also be set to solve algebraic equations and geometric problems. Ludgate envisaged that it would be powered by an electrical motor, that the calculations be automated, and that the complete device – measuring about 0.5 metres in every direction – would be portable.

The machine was probably never made. During World War I, Ludgate was diverted to work on a committee organising the supply of oats for the British cavalry, and he seems not to have returned to his computer. He died of pneumonia aged 39, after returning from a Swiss walking holiday. In 1991, the British Museum made the first complete Babbage difference engine. Sadly, Percy Ludgate's drawings have not survived, and we will probably never see a working model of this pioneering Irish calculating engine. In the 1960s, the Dundalk firm Queleq made innovative analog computers for the animal feed industry, but the company did not survive the transition to digital technology. Teagasc's Agricultural Museum at Johnstown Castle* collects early computers and has a Queleq machine.

Stars in her eyes

There is a crater on the moon named after Skibbereen woman **Agnes Clerke** (1842–1907), who was a renowned astronomical writer.[4] Clerke Crater, at the edge of the Sea of Serenity, was named in 1981 after someone who is still regarded as the "chief astronomical writer of the English-speaking world of her day". Clerke's father, who was a bank manager in Skibbereen, had a keen interest in science. He built a chemistry laboratory at home, taught his children astronomy, and used a telescope to observe the transit of stars – thus regulating the bank's clock and providing a local time service. Agnes Clerke, who suffered poor health most of her life, was schooled at home with her sister Ellen. In 1867, the sisters left to spend a decade in Italy and, by 1877, when the family settled in London, Agnes Clerke was fluent in several European languages.

Both sisters then began to work as writers. Ellen became a journalist with a Catholic paper, *The Tablet*, and Agnes wrote for several periodicals on art and Italian subjects, but increasingly on astronomy, and soon she was a regular contributor to the London-based science journal *Nature*. Her articles were thoroughly researched – she read European works in the original language – and carefully written, and won the respect of professional scientists.

Agnes Clerke wrote three major scholarly books that established her reputation internationally: *A Popular History of Astronomy during the 19th Century* (1885), still the definitive reference on 19th-century astronomy; *The System of the Stars* (1890), which explained current understanding about the evolution of stars and nebulae; and *Problems in Astrophysics* (1903), an ambitious 600-page survey of the new field of astrophysics. Despite her lack of a formal science education, the Royal Greenwich Observatory offered her a job, but the position was a lowly one – women being banned from the higher levels of the civil service – and she declined.

Agnes Clerke wrote several essays for *Encyclopaedia Britannica*, and 150 obituaries, mostly of scientists, for the *Dictionary of National Biography*. In 1903, she and the astrophysicist Margaret Huggins,* a friend and fellow Irishwoman, were granted honorary membership of the Royal Astronomy Society (the society did not open full membership to women until much later). A plaque to Agnes and Ellen Clerke marks their former home at Bridge Street, Skibbereen. Schull Community College, near Skibbereen, is home to the Republic's only planetarium. ⑥

Mysterious wonder of the natural world

Scenic Lough Hyne is a wonder, a natural curiosity, a national treasure – and something of a mystery. This small, landlocked, saltwater lake is probably the biggest, richest rock-pool in the world. It is certainly unique in Europe and scientists come from all over to study it. Despite its small size, it has a tremendous diversity of habitats and thousands of plant and animal species, including many that are rare or more usually found in warmer places. To date, scientists have clocked over 100 different types of sponge in the lough, 73 types of sea slug, 24 sorts of crabs, 18 species of sea anemone, numerous starfish and corals, various rare seaweeds, and unusual Mediterranean warm-water species, such as the trigger fish and the red-mouthed goby. This tremendous diversity is due to Lough Hyne's unpolluted water, sheltered site, unusual structure and location, and the fact that its water is a few degrees warmer than the sea outside.

The lough is roughly square, with sides 1km long, and surrounded on three sides by hills. On the fourth side, it is cut off from the Atlantic by a rocky sill, and its only connection to the sea is at the southeastern corner through a narrow opening known as 'the rapids'. Along its western side, a trough runs north–south for 500 metres, and there the water plunges to over 50 metres deep – an extreme depth for such a small lough.

This is arguably the world's most intensely studied square kilometre of sea, yet scientists still do not know how the rock basin formed. It was not made by waves, nor caused by subsidence – there is a drop of nearly 300 metres between the top of Knockomagh Hill to the north and the bottom of the lough's deep trough, but no evidence of any faulting in the rocks. It does not have the elongated shape of a fjord,* nor the circular shape of a glacial corrie* – in any case, being at sea level, it is too low-lying to be a corrie. The best current theory is that it was made during an Ice Age by 'glacial plucking': the local rock is a red sandstone, which breaks naturally in regular blocks, and scientists now think that the lough basin was formed by glaciers plucking blocks out of the sandstone.

Strangely, for much of its life, Lough Hyne was a freshwater lough. Fossils in its muddy sediment reveal a change to saltwater 4,000 years ago, when sea levels rose and the tide could at last breach the bar at the lough's mouth. The lough still has an unusual tidal system, thanks to the bottleneck at the rapids: generally, tides take 6½ hours to rise and the same time to fall; but Lough Hyne's shape is such that the tide only has four hours to rise, then spends the remaining 8½ hours falling, rushing through the rapids at a treacherous 15kmph. This asymmetric tide is one of Lough Hyne's special features.

At the entrance to the rapids, there is a dense forest of kelp. The fronds are 3 metres long and encrusted with sponges, anemones and corals; colourful fish and starfish hunt and graze among the shrubbery. The sea beyond is wild Atlantic – witness the many deep sea caves cut into the cliffs around – but Lough Hyne is sheltered by hills and sees very little wave action. The many habitats within the lough include rocky shoreline, underwater cliffs, the deep cold water of the western trough (which mixes with the rest of the lough water only in winter), and a bottom that varies from mud to silt to bare rock. Each habitat is home to a distinctive set of plants and animals. Much of the rocky shoreline, for instance, is covered with the white crust of a coralline seaweed, while just below water level are thick colonies of purple sea urchin.

This rich underwater treasure was discovered by accident in July 1886, when rough weather forced a steamer to shelter in the lough. As luck would have it, a Royal Irish Academy marine science expedition was on board. Since then, biologists from UCC, and the universities of East Anglia and Bristol, have studied Lough Hyne in detail; Dutch and Belgian scientists are regular visitors; small laboratories were built by the lough; and some 500 scientific papers have been published about it.[5] Fine views can be had from the top of Knockomagh (a woodland nature reserve) and Barloge Hill. The 8th-century church above the rapids is associated with St Brigid; the shore wall was a famine-relief project; and the castle on the island was reputedly home to Labhra Loingseach, the king with the ass's ears. The name Hyne (sometimes Ine) may derive from *iogheann* ('pot') or *dhoimhin* ('deep'). As the lough is a nature reserve, activities in and around it are controlled.

Barytes mines

The mountainous parts of Cork and Kerry are relatively rich in copper and barytes, minerals that were deposited 300 million years ago when the mountains there were formed. The copper was mined even in prehistoric times, but interest in the barytes did not develop until the industrial era, when mines were opened at various places, most notably Bantry and Duneen Bay.

Barytes (barium sulphate) is a soft, pink, clay-like mineral with various industrial uses – among other things, as a filler in paper and textile manufacture and, more recently, as a drilling mud on oil rigs, and in medicine for barium meals. Barytes was mined at several points around Bantry Bay, including Durrus, Mount Gabriel, Scart, Derryginagh and Letter. The first reference to the Duneen Bay barytes was its use in dressing calico at a Clonakilty factory in the early 1800s. In the 1870s, the mine, under local landowner William Beamish, was a major part of the Irish barytes industry. Duneen has been worked sporadically since, most recently in the early 1980s by a Texan company that used the barytes on its oil rigs.

1802 – the world's first suspension waterwheel?

A ruined cotton mill at Overton near Bandon may be where a suspension waterwheel was first used. The credit traditionally went to an English mill *c.*1811, but new detective work by Irish archaeologist, Dr Colin Rynne, reveals records of a suspension waterwheel at Overton in 1802. The waterwheel was invented over 2,000 years ago, and since then its design has been improved several times. Early wheels were laid horizontally – as at the 8th-century tidal mill at Nendrum,* Co. Down – and the first significant change came when new gearing meant wheels could be stood upright. Water was initially delivered to the base of a vertical waterwheel (under-shot design), but, in the 1700s, millwrights realised it was more efficient to deliver the water at the top (over-shot and breast-shot designs). Waterwheels were still all made of timber and it was not until the late-1700s that millwrights began experimenting with cast-iron components, and a Manchester mill engineer named Thomas Hewes invented the 'suspension waterwheel'. His wheel was braced internally by cast-iron rods that were held in place under suspension. Elegant and lightweight, it was more efficient than earlier designs, being easily turned by even a moderate water flow.

In 1802, Hewes was commissioned to build a 5-storey mill at Overton for **George Allman** (1750–1827), a major figure in the Irish cotton industry. Allman believed in using the latest technology and his new Overton mill was said to rival the best in England. As a precaution against fire, the stairs were of stone, and each floor was covered with iron sheeting. The mill held 10,000 spinning spindles, and was powered by a large wheel, 12 metres in diameter and 1.5 metres wide. Contemporary letters and a sketch unearthed by Dr Rynne reveal that it was a suspension wheel and that it was in place in 1802. Overton mill employed 300 people at its peak, before closing in 1830. This huge mill is now in ruins, but the axle of the historic waterwheel survives *in*

situ. A more modern example of hydro-power is at Bandon Weir. A water turbine there generated electricity for the town from 1919 until the 1960s. Now restored with a new 75kW turbine, the scheme is again generating power for the town.

A massive gunpowder plot

We do not usually associate Ireland with the international weapons industry, yet one of the biggest gunpowder mills in 19th-century Europe was by the banks of the Lee at Ballincollig.[6] It supplied gunpowder for the British Empire; ammunition for the Napoleonic, Crimean and Boer wars; and industrial explosives for mine and railway companies as far away as Africa and the West Indies. The mills were opened in 1794 by Charles Henry Leslie, scion of a Cork banking family, who saw commercial opportunities in the coming war with France. The Ballincollig site was perfect: the river could be used for water power and transport; there was good soil for growing the trees needed to supply timber and charcoal; and there was lime for mortar. The area was also relatively uninhabited – an important consideration in the event of an explosion – yet close to Cork port, making it easy both to import raw materials (notably sulphur from Mount Etna in Sicily and saltpetre from India), and to export the finished product.

The site was levelled, the Inniscarra Weir built on the river, embankments raised against flooding, trees planted, lime kilns built, and canals and mill races installed. Press houses, sawmills, corning houses, magazines, drying kilns and mills were erected. For safety reasons, these were widely spaced and separated by thick blast walls. The

The first gun to arrive in Ireland came in 1361 with Lionel, son of Edward III (he took it back to London afterwards). The first Irish gunpowder plant was probably at Dublin Castle; there were also two near Clondalkin, Co. Dublin, in the 1700s, and the Kynoch Company made explosives at Arklow* from 1895–1918.

resulting site stretched along the river for 2km, and had 7km of canals. In all, there were 30 water-powered processes, an unrivalled concentration of hydro-power for a single site.

Some of the heavy-duty grinding equipment at the gunpowder mills.
© *Michael Rice/Cork County Council*

To make the gunpowder, the various ingredients were first ground separately, then mixed in precise ratios, pressed, then broken and granulated. Finally, the grains were dried and coated with graphite to keep moisture out. No iron was allowed on site in case of stray sparks, so all barrels were bound with copper bands and shoes had to be wooden, and there was strictly no smoking. Despite these precautions, and the appointment in 1875 of safety inspectors and stricter standards, there were 16 serious explosions at Ballincollig over the years, killing 45 people.

In 1805, facing a renewed war with France under Napoleon, the British army bought the mills, partly for security and partly commercial reasons. The army built a barracks, developed Ballincollig village and created a massive 53-hectare industrial-military complex. When Napoleon surrendered 10 years later, however, the mills were mothballed. A Liverpool company, Tobin & Horsfall, makers of the Horsfall gun, reopened the mills in the 1830s. (The Tobins were originally Irish merchants who shipped slaves and, when slavery was abolished, switched to palm oil, candles and margarine, later becoming part of the Unilever Group.) The mill's biggest customers now were mine and

railway companies and, at its peak, Ballincollig employed 500 people.

By the late 1800s, newer explosives, such as nitrocellulose and cordite, were replacing gunpowder, but the Boer War brought a final reprieve: Curtis & Harvey (later part of ICI) bought the plant to make ammunition for rifles, and the mills began advertising for the first time, to the new gun-carrying public. In 1903, however, Ballincollig closed. ICI planned to convert the site into a staff resort but, in 1940, sold it to the Department of Defence. Fortunately, much of this historic industrial complex was intact – including an early French-style water turbine from *c.*1850 – and it has been restored by Cork County Council. ⑦

Gouldings 10-10-20

The first artificial fertiliser was a super-phosphate made, in 1817, by Sir James Murray* from Co. Derry. By the 1850s, **artificial fertilisers** were widely used by Irish farmers and, in 1856, William and Humphrey Goulding converted an old distillery in Blackpool to make Murray's super-phosphates. The process entailed crushing bones, then dissolving the powdered bones in strong sulphuric acid to extract the phosphates. Gouldings used bones collected around Ireland, along with phosphates imported from North America and Africa. They made the acid on site, using sulphur extracted from pyrites, a cheap sulphurous mineral, also known as fool's gold, which came to Cork as ballast on Spanish ships. The result was an extensive

A Gouldings advertisement from the 1880s.
Goulding Fertilisers/IAWS

Cork – an industrious city

Cork was founded in the 7th century as a monastic settlement on the banks of the River Lee, but, thanks to its ideal location and natural harbour, it quickly became an important market town and trading centre. Most of the settlement developed on reclaimed marshes. During the Viking era, Cork became part of an extensive Scandinavian trade network and, by the 1100s, it was a fortified port and market town. Numerous industries developed over the succeeding centuries; most relied on the Lee for water power, though there were also windmills on the high ground at Patrick's Hill.[7] By the mid-18th century, Cork had sizeable breweries* and distilleries,* a thriving butter market* and a strong textiles industry. The latter included Europe's largest **sailcloth factory,** a linen mill at Douglas begun in 1726 by the French-Huguenot Besnards family that, in 1750, employed 7,500 workers. Cork also had brickworks, which used mud from the estuary, and several limestone and 'red marble' quarries. The city did well out of the Napoleonic wars: it gained a major gunpowder factory at Ballincollig,* and benefited from the strong demand for sailcloth and rope, and for Irish produce. As a result, however, it also suffered during the economic depression that followed Napoleon's surrender in 1815 and, by 1850, was losing out to Belfast and the new industrial north. Cork Public Museum in Fitzgerald Park, the site of the 1903 international trade fair, has numerous local industrial items including butter branding irons and Monard* spades. ⑧ ⑨ Cork Archives record the city's commercial, manufacturing and maritime heritage.

chemical works and by 1917, according to industrial archaeologist Colin Rynne, Gouldings had 70 kilns working round-the-clock producing 500 tonnes of sulphuric acid every week, and powered by a variety of gas and steam engines.[7]

The extensive complex was demolished in the 1970s when Gouldings moved across the river to Marina. Now owned by the Irish Agriculture Wholesale Society (IAWS), Gouldings continues to make fertilisers, though it now blends imported ingredients, and no longer makes its own raw materials. Gouldings was just one of several chemical plants in Cork city in the 19th century: Shandon Chemical Works made paints and distempers, Brookefield Chemical Works produced various forms of magnesia, and there were numerous soap factories.

The Lee on tap

Cork's Victorian waterworks are a heritage jewel. This fine example of a 19th-century public health initiative, combined with elegant industrial technology, sits on the river's north bank beside the weir at the Lee Road. Two attractive cut-stone buildings house the old steam engines and turbines, while the tall rectangular brick-and-stone stack (1858), a prominent landmark, is surely Ireland's most elegant chimney. There are steam pumps and turbines, a pure-water holding tank or reservoir, and a sand filtration tunnel. In 1985, the Institute of Mechanical Engineers restored the steam plant and it is now open to groups by appointment. ⑩

Cork had a rudimentary water supply system as early as 1300, when water from the Lee was conveyed via rough wooden conduits to public fountains; there were also several artesian wells.[7] Cork's modern water supply system was started in 1761, by which time demand had increased significantly: millers and other industrial users wanted a regular power supply, and there was growing awareness of the need to supply clean and reliable drinking water to the populace; the mayor and sheriffs were empowered to collect water rates, the corporation established the Cork Pipe Water Company, and an Italian engineer Davis Ducart,* who was then working in Ireland, was commissioned to design a water supply system. By 1768, enough money had been collected for work to begin. The site by the Lee Road was chosen for several reasons: it was above the reach of the tide and the city's sewers; there was space for storage reservoirs; it was high enough for the water to flow to customers under gravity; and an existing salmon weir there made it easy to install waterwheels to power the pumps. A reservoir was built, wooden water pipes were laid, and the Pipe Water Company installed two large waterwheels and a pump house (since demolished). The pumps lifted the water to the reservoir, from where it gravitated to users.

One of the restored steam engines at the wonderful Cork waterworks. It was used to pump water from the river to the reservoir.
© Institution of Engineers of Ireland

Demand grew, particularly from Cork's many breweries and distilleries, and over time the network was upgraded and expanded. New reservoirs were built; steam-powered beam engines replaced the water-powered pumps, and were, in turn, replaced by steam expansion engines and later turbines; and the wooden conduits, which could not take the higher pressures, gave way to cast-iron mains. In 1875, the corporation installed a filter tunnel fitted with gravel and sand beds, to clean the river water before distributing it. International experts were regularly consulted on how to improve the system. In the 1870s, for example, there was concern that people were wasting water, and one English engineer recommended metering water use, but his suggestion was not taken up. In 1928, the water was sterilised for the first time, using chlorine, and, in 1938, the works switched to electric pumps. The steam plant was retained, however, and used during World War II when oil for generators was in short supply. Today, the Lee Road waterworks treats and pumps 11 million gallons of river water each day, supplying 70 per cent of the city's needs, and maintaining the high water pressure needed for fire-fighting supplies.

most of his working life in continental Europe. Bowles studied law in England, and natural history, metallurgy and chemistry in Paris. He spent 10 years touring French and German mines, possibly as a mining consultant, and must have established an international reputation, for, in 1752, he was invited to be Spain's superintendent of mines. That same year King Fernando VI established a natural history museum in Madrid and Bowles became its principal scientist.

William 'Guillermo' Bowles: a genus of Peruvian plants, the Bowlesia, is named after this Irish scientist, who is buried in Madrid.
Courtesy George Reynolds

Bowles spent the rest of his life in Spain, studying the country's mines, minerals, flora and fauna and, in 1775, published his *Introduccíon a la Historia Natural y a la Geografía Física de España*. It discussed everything from tides and extinct volcanoes, to Madrid's water supply and Spanish locusts. The book went to many editions and was translated into several European languages. Bowles was a modern scientific thinker: he realised the enormity of the geological timescale, and believed in evolution, long before these were acceptable concepts. As head of Spain's military laboratory, he studied platinum, and refuted the then-current idea that it was an alloy of iron and gold. He is also credited with introducing St Dabeoc's heath* to Britain (this Lusitanian heather is native only to northwest Spain and Connemara).

The man who described Spain

The first modern scientific description of Spain was written by **William 'Guillermo' Bowles** (*c*.1720–1780) who was born "near Cork".[8] Bowles is remembered in Spain, where he helped establish the country's Natural History Museum. Little is known about him in Ireland, however, as he spent

Cork's heady brew

In the 19th century, Cork boasted at least six breweries and eight distilleries, not including the large Jameson distillery at nearby Midleton.* This thriving industry created a strong demand for labourers, coopers, barrels, barley, grains, malt houses, mills, clean water and power sources. The city's

The substantial cooperage at the Beamish brewery c.1900. Beamish & Crawford

Museum. Lee Maltings on the Mardyke, an early Beamish & Crawford malt house, is now the high-tech National Microelectronics Research Centre (NMRC). The last major brewery to open in Cork was Murphy's Brewery on Leitrim Street, which also survives.

oldest-surviving brewery is **Beamish & Crawford**, which began in 1792 when two Cork businessmen, William Beamish and William Crawford, took over a brewery at Cramer's Lane, South Main Street (there has been a brewery on the site since at least 1640). By 1807, their brewery was producing 100,000 barrels of 'stout porter' each year, making it the largest in Ireland, a position it held until 1833 when it was superseded by Guinness's* of Dublin. The Cork brewery used horse-powered mills and pumps until 1818, when its first steam engine was installed. By the late 1800s, all the large Cork breweries had their own cooperage, barrel-washing plant, ammonia-cooling plant and malt mills. At Beamish & Crawford, one large roof space also held an enormous iron tank for storing water. Some of the firm's 19th-century equipment survives alongside the modern brewing plant, and a water-powered mill on the site has been renovated. The brewery, now owned by Scottish & Newcastle, still uses Lee water and a yeast strain developed 200 years ago by the first Beamish & Crawford brewers. ⑪

The Beamish & Crawford families contributed much to Cork city life: the Crawfords sponsored an art gallery and an observatory at UCC;* the Beamish family gave the city Fitzgerald Park and House, which now holds Cork Public

CORK CITY: BLACKROCK

Putting out fires

In the 18th century, insurance companies offered their paid-up customers an optional extra: a rudimentary **fire-brigade service*** that could be called upon should their premises go up in smoke. In time, the technologies improved, the local authorities took over responsibility, and they now provide a fire service for everyone. (Part of that service includes maintaining a high pressure in the local water supply.) Cork Heritage Centre recalls the early days of fire-fighting. The display includes primitive pumps and breathing apparatus, as well as models of fire tenders, and an exhibition about the burning of Cork during the War of Independence (1920). The centre, on the former Bessboro Estate, also explores Cork's maritime, transport and industrial history, and the ecology of the nearby estuary. Bessboro was once owned by the Pike family, Quakers with commercial interests in various local industries, such as shipbuilding, banking, market gardening, railways and woollen mills, all of which feature in the exhibition. ⑫

CORK CITY: EMMET PLACE

The Royal Cork Institution

One of the most prestigious scientific bodies in these islands is the Royal Institution (RI) in London. Its professors have included such noted scientists as Michael Faraday and John Tyndall,* and each Christmas its popular science lectures for children are televised. The RI was founded in 1799 and, three years later, a campaign began to establish a similar institution in Cork. Eminent citizens subscribed 400 guineas "for the purchase of various articles of apparatus... and for diffusion of scientific knowledge by means of lectures" and, in 1802, the Royal Cork Institution (RCI) was incorporated. Successful public lectures brought in a further £2,500, with sizeable donations from the earls of Cork and of Shannon (both were descended from the first Earl of Cork and related to Robert Boyle,* the great 17th-century chemist). The RCI soon had a library, a botanic garden and four professors – of chemistry, natural history, agriculture and natural philosophy – who conducted research and gave lectures.

Its best years began in 1813 with the arrival of **Edmund Davy** (1785–1857). Davy, born in Cornwall, was a cousin of Sir Humphrey Davy, inventor of a miner's safety lamp.* The cousins worked together at London's RI, until Edmund left to become professor of chemistry at Cork. Edmund Davy enjoyed tackling practical problems, especially in farming, and frequently organised lectures so the public could learn of new developments. He studied manures, looking for ways to improve the quality of natural fertilisers. He also devised tests for various poisons, and developed new explosives, a technique to detect adulterated milk, and industrial uses for turf, such as using peat charcoal as a deodorising filter. In one unusual experiment, he analysed the urine from a boa constrictor, and found it contained mostly uric acid. His greatest discovery was **acetylene**, which he made by accident in 1836. By then, Davy had left Cork to become professor of chemistry at the Royal Dublin Society.* It was in the RDS laboratories that he dropped water on some calcium carbide; this gave off a gas and Davy noted it burned with "a bright white flame of great splendour". (Acetylene is often credited to a French chemist, Bertholet, although his discovery came 23 years later.) After Davy left Cork, the RCI went into decline. In 1830, parliament withdrew its grant and, by 1838, the institution had stopped teaching. In the 1840s, its role was taken over by a new university, Queen's (University) College Cork (see: **Minerals and fossils of the world**). Edmund Davy was awarded a government pension, and died at his Dublin home in 1857. The institution's former premises are now home to the Crawford Art Gallery.

Minerals and fossils of the world

The Victorian passion for collecting things, and organising collections into museums, came to a head in the mid-19th century. Not surprising then, that, when **Queen's College Cork** was founded in 1845, one priority was to provide the college with a geological museum and, within 10 years, over 8,000 rocks, minerals and fossils had been acquired. The museum now boasts 20,000 items, with material from every continent, the oceans in between and even outer space. Among its fossils are: a 200-million-year-old ichthyosaur (marine reptile) collected by a renowned 19th-century English fossil hunter, Mary Anning; 350-million-year-old fossil plants from the famous Kiltorcan* quarry in Co. Kilkenny; and plaster casts of *Archaeopteryx* and of the fossil footprints recently discovered on Valentia Island* in Co. Kerry. There are also samples of Cork red marble (an iron-rich stone), various meteorites, basalt columns taken from the Giant's Causeway,* and cotterite, a rare quartz mineral discovered at Mallow in 1876 by Miss E Cotter and named after her by the college's then geology professor, Robert Harkness. ⑬

The Cork university was one of three Queen's colleges founded in 1845, the others being at Galway* and Belfast.* The colleges have had various identity changes since then, becoming the Royal University of Ireland in 1880, and the National University of Ireland (NUI) in 1908. The first president of Queen's College (now NUI) Cork was the renowned Irish chemist Sir Robert Kane.*

George Boole – AND, OR, NOT

Electronic computers and web search engines function thanks largely to **George Boole** (1815–1864). For this self-taught son of a shoemaker, who was the first professor of mathematics at Queen's (University) College Cork, invented a revolutionary way of translating abstract ideas into

George Boole invented the mathematical logic that modern computers use when 'thinking'. His appointment brought prestige to the new Queen's College Cork, and the college's modern library is named after him.

algebraic equations, which could be processed by a machine. Born in Lincoln, England, Boole was interested in mathematics and astronomy as a boy, but also had a special talent for languages. He learned Latin from a family friend and, by 12, had taught himself Greek, French and German. When he was 14, however, his father's shoemaking business collapsed – Boole senior preferred to make telescopes for his children, than shoes for customers – and George left school to support the family. He found work as a school assistant, but was considered a poor teacher as he spent more time reading French mathematics books borrowed from a local mechanics institute, than instructing his students. Yet, by 19, Boole had set up his own school, with the help of his brothers and sisters, and began to develop his mathematical ideas.

In 1840, George Boole published his first original mathematical research, mostly concerned with calculus. Over the next few years, he invented a powerful new branch of mathematics, known as invariant theory, which Einstein later used when developing his theory of relativity. Boole's novel ideas brought him to the attention of various mathematicians and, in 1844, despite his limited formal education, the establishment awarded him one of its top honours: a Royal Society gold medal.

Boole was also interested in logic, philosophy and the way the brain processes thoughts, and his greatest breakthrough came when he devised an algebraic way to represent logical statements. Boole realised that this could be done if ideas and objects could be assigned to classes, and statements could be formulated so that the answers were 'yes/no' or 'true/false'. [For example, 'all tall men' might be written as xy, being a class of objects drawn from a larger group containing all men (represented by x), but selecting only those who are also tall (the group y).] Such symbols and statements can be manipulated like algebraic equations. In 1847, Boole presented his ideas in a small book entitled *A Mathematical Analysis of Logic*. Others were working on similar ideas, among them an English logician Augustus de Morgan, with whom Boole was friendly. Boole's approach, however, proved to be the most powerful. Over the next five years, he refined his 'boolean algebra' and, in 1854, published *An Investigation of The Laws of Thought*, which philosopher Bertrand Russell later described as "the book where pure mathematics was discovered".

In 1849, thanks to his ground-breaking work, and though he had no university education, George Boole was appointed as the first professor of mathematics in the new Queen's College Cork. All his later work was done there, and he made many lasting contributions to mathematics. The full power of boolean algebra was not unleashed until 1939, however, when it was rediscovered by Claude Shannon, a young MIT graduate student. He realised that Boole's approach, with its 'yes/no' statements, could be used to translate logical statements into an 'on/off' format that could be used by electrical circuits. The modern electronic computer was born. Web search engines also rely on boolean algebra, especially operators, such as AND, OR and NOT, which are akin to addition and subtraction.

Boole has been described as a religious humanitarian, who viewed the human brain as God's greatest creation.[9] He was interested in literature, music and art, and wrote poetry and, despite his poor start in teaching, became a great educator. He believed in popularising scientific ideas, pioneered adult education programmes, and championed the mechanics institutes that had been so important in his own early education. In 1855, he married Mary Everest, niece of the man after whom the mountain is named. Their five daughters included **Alicia Boole Stott**, a self-taught mathematician like her father, who developed important concepts in 4-dimensional geometry; **Lucy Boole**, the first woman professor of chemistry in Ireland or Britain; and **Ethel Boole Voynich**, a bestselling novelist. In 1864, George Boole died of pneumonia, after he had walked to college in the rain and spent a day in wet clothes. He is buried at St Michael's Church, Blackrock.

CORK CITY: SUNDAY'S WELL

"Cork calling"

In 1927, Cork city gained its own **radio station, 6CK**. It broadcast from the unlikely setting of the former Women's Gaol, where the studio has been preserved and is now a radio museum. Cork was Ireland's third radio station, after Belfast (2BE), which the BBC started in 1924, and Dublin (2RN), which began on January 1st, 1926. Morse signals had been transmitted by radio since the 1890s, thanks mostly to Guglielmo Marconi,* and the world's first radio sound transmission took place in the USA on Christmas Eve 1906, when Reginald Fessenden read a Christmas greeting and poem, and some ships' radio operators, who previously received only Morse Code messages, heard his 'programme'. The first Irish sound broadcast was reputedly an appeal made during the 1916 Rising, but there is no record of anyone receiving that broadcast. Ten years later, when the Irish Free State first took to the airwaves, 2RN's opening programme included a speech by senator and Gaelic League founder Douglas Hyde, a selection of songs, a harp solo and a concert by the Army Band. To save money, that first programme was not recorded – though the BBC did record part of Douglas Hyde's speech – but 6CK's opening night was recorded, and that recording survives.

Music, songs and talks made up most of the early radio schedules. Well-known Irish entertainers, such as Jimmy O'Dea, starred in entertainment programmes; there were commentaries on theatre plays, movies, and art gallery openings; even carpentry and swimming lessons. Some talks were in Esperanto – a gesture, perhaps, to listeners abroad. Despite the predominantly urban audience, the Irish stations broadcast agricultural talks and, as the number of rural listeners grew, this service expanded to include livestock prices.

Stamp commemorating the 1995 centenary of Marconi's invention of radio. © An Post

There had been considerable debate before 2RN's launch about whether or not the new station should be run by a private company or by the State. Some argued that the Free State should not be engaged in broadcasting, and indeed could not afford to – the government, finding itself short of funds, had just cut 1 shilling from the old-age pension. Others felt that a radio station was just the vehicle for promoting Irish culture and a new sense of Irish identity. In the end, both 2RN and 6CK were run by the Department of Posts & Telegraphs, which erected low-powered transmitters in Dublin and Cork. Local listeners could use a crystal set (which cost about £3 and, moreover, needed no batteries), but listeners elsewhere in the country needed sensitive and expensive valve sets (which cost over £10) to pick up the weak signals. It was not until 1932, conveniently in time for the Eucharistic Congress, which was held that year in Dublin, that a high-powered transmitter was erected in Athlone.

About 5,000 radio licences were issued in the early years, but evasion was high and it is thought there were probably 25,000 radio sets in use, representing 5 per cent of households. In 1945, by which time 2RN had become Radio Éireann, 13 per cent of rural, and 45 per cent of urban households had a radio. In 1961, just as Irish television began, and Radio Éireann became Radio Telefís Éireann (RTÉ), there were 500,000 radio licences in Ireland, representing about half the population.

The radio museum at Cork has segments from some of the first transatlantic telegraph cables,* early radio sets and microphones – including the one Éamon de Valera used when replying to Churchill in 1945, and one used by John F Kennedy on his Irish visit in 1963. Visitors can see the Morse key used at Valentia to send a cryptic telegram to the USA, announcing the start of the 1916 Rising – "Mother was operated on today successfully signed Kathleen" – and send their own Morse message, or eavesdrop on shortwave broadcasts from around the world. Items from the RTÉ archive include books on radio pronunciation. ⑭

1928–1932 – the world's tractor factory

For several years, the largest tractor factory in the world was the Ford plant at Cork's City Park beside the River Lee. The plant had opened in 1919, after Henry Ford chose Cork as the location for his first manufacturing plant outside the USA. Colin Rynne suggests, in his *Industrial Archaeology of Cork City and Its Environs*, that the move was philanthropic: Ireland had neither the appropriate raw materials, nor any great need for tractors, but Ford's family originally came from Co. Cork.[7]

In July 1919, the first Ford tractor rolled off the Cork production line – a 22-hp, 4-cylinder model that ran on paraffin and kerosene – and over the next six months the factory produced 300 tractors. In the post-war slump, however, demand for tractors dropped, and the plant changed to making Model T cars. By 1927, it had made 10,000 cars, mostly for the Irish market. The following year, Ford moved all its tractor manufacturing from the USA to Cork, creating the world's largest tractor factory, an enormous industrial complex with purpose-built wharfs, workshops and warehouses. Four years later, however, tractor production moved again, this time to Ford's Dagenham plant in England. Cork continued making components until the 1980s, when Ford closed the plant. The original custom-built site survives, much of it now used for storage.

1769 – the world's largest butter market

In the late 1600s, an extensive dairy industry began to develop in the rich lowlands of Munster. Cattle numbers increased dramatically, new drover roads were built, and more and more markets and fairs took place. Even the landscape changed, as the ground was cleared of scrub, and sometimes people, to make way for new pastures. The multitudes of cattle and the plentiful grass produced large volumes of milk. Milk production was still seasonal, however – peaking in the summer – and, in the absence of refrigerators, the best way to preserve the milk was to make it into heavily salted butter. Cork city was already an important Atlantic port, so it was only natural that Munster's surplus butter should be exported from there, and, by 1727, Jonathan Swift* could write that Ireland had two industries: linen* in the north and butter in Cork.

A Cork-branded butter firkin from 1876. The markings indicate that it contains second or grade-2 butter, graded by Inspector 'S', approved by the Committee of Merchants (CM), with a total weight of 29lbs (13kg) and a tare of 14lbs (6.3kg).
Cork Butter Museum

Butter-making, which probably arrived from Roman Britain, was a radical innovation that meant milk could be 'saved' and make available year-round. The butter, an important source of dietary fat, still had to be preserved, either using salt (the salt was removed before use by soaking the butter in water) or by storing the butter in a bog pool where the acidic conditions and lack of oxygen prevented it from going off. Butter and other dairy products were such an important part of the traditional Irish diet that they were known as *bán bídh*, or 'white meats'. Sometimes they were flavoured with wild garlic or other herbs, or sweetened with honey. The 8th-century Brehon Laws and the ancient Gaelic tales contain numerous references to butter and, at times, butter was even used as a currency to pay taxes and fines.

Until the advent of modern creamery butter, every farmhouse had its own churn and butter-making equipment. Milking and butter-making were traditionally considered women's work, and a considerable body of superstition and folklore was associated with the work, especially churning.

Barrels of butter were first shipped from Cork in the 1630s, and, by the end of the 1700s, Cork had become the world's largest butter market, supplying British colonies as far away as South America and Australia, and with its prices quoted on the London commodity markets. The butter was packaged in distinctive small casks or firkins, and the volume of trade spawned a large coopering industry. In 1769, the city merchants established the **Cork Butter Exchange** to control the trade and chase the lucrative English market. Significantly, they introduced a quality control system, whereby trained inspectors graded the butter for taste and appearance, and producers were paid accordingly. (Inspectors had to pass a six-hour examination and be able to sample 150 firkins an hour – nearly three a minute.) Firkins were inspected at random to avoid corruption and bribery, and quayside patrols ensured counterfeit firkins could not be shipped surreptitiously.

All this helped establish Irish butter's worldwide reputation for over a century, a forerunner of the modern Kerrygold brand. For nearly 150 years, Cork led the world. By the late 1800s, however, the heavily salted Cork butter with its old-style wooden packaging was losing out to new products, such as margarine. New refrigeration technologies, which meant fresh butter could now be exported worldwide, were the final blow, and the Cork butter market closed in 1922. Refurbished in the 1990s, the exchange building is now home to the Shandon Craft Centre and the Cork Butter Museum, which celebrates the history of Cork's butter trade. ⑮ **Buttered eggs**, a Cork delicacy, are sold at the city market.

1855 – the first Dublin–Cork train arrives

The most difficult terrain on the Dublin–Cork railway line was the stretch between Mallow and Cork, where considerable engineering work was called for: a tunnel under Blackpool that took seven years to build; three substantial viaducts (at Mallow, Monard and Kilnap); and, at one point, a cutting 2km long to create a ledge for the line. Much of the route had to be supported by embankments, and millions of tonnes of earth were manually and laboriously moved to make these. William Dargan,* the redoubtable contractor who built most of Ireland's mainline railways, was in charge, and it is a tribute to the quality of the work that all of the original line is still in use. The last stretch to be built was the Blackpool tunnel, driven through a sandstone ridge just north of Cork city. Work began at both ends of the tunnel in 1847, but the two teams did not meet until 1855 (in the interim, the Dublin line terminated at a temporary station in Blackpool). At 2km, it is one of the longest railway tunnels in Ireland, and the inside is reputedly lined with 2 million bricks from the Youghal brickworks. The four ventilation shafts surface as stone circular towers at Barracktown, Assumption Road and Bellvue Park.

By 1915, Cork was served by seven railways, not counting tram services: the Great Southern & Western Railway (GS&WR) to Dublin; a cross-city train link (1912–76); a line to Blackrock and Passage (1850–1931); another to Bandon and the south coast (1849–1961); one to Macroom (1866–1953); a light railway to Muskerry (1887–1934); and lastly, the Cobh/Queenstown and Youghal line, which is still in use as far as Cobh.[7] The GS&WR loco No. 36 built in 1847 and last used in 1875, is on display at Cork's Kent railway station.

Currachs on the Lee

Currachs, naomhógs, prongs, even longboats – the craft workers of **Meitheal Mara** can build them all, from small-scale replica to the real thing. Meitheal Mara ('maritime working party') is a voluntary organisation with a passion for maritime heritage, from folklore and legends, to fishing and traditional boats. Begun in 1994 by Padraig Ó Duinnín, it is backed by Cork Corporation and FÁS among others. At its boatyard by the Lee, craft workers keep alive traditional skills. The team works with similar groups overseas, and many of its boats are crewed around the world at regattas for traditional craft. Over 50 types of traditional boat are still in use in Ireland. Most are associated with a particular river or lake, for example the Erne cot,* or an occupation, such as mussel raking in the Boyne estuary.* Boats from the north and east tend to be clinker-built (with overlapping boards), while those from the south and west are of a carvel type (boards butted edge to edge). The Meitheal Mara database has information on them all. ⑱

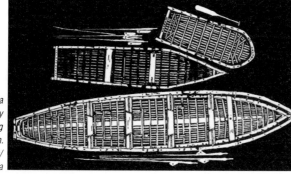

Three Irish currachs: a Kerry six-oar, Dunfanaghy four-oar, and Bunbeg paddling currach.
© Lawrence Hutson/ Meitheal Mara

1857 – malaria plague

Ireland is home to no fewer than 18 species of **mosquito** and in the past also had malaria. Many of the Irish mosquitoes are from the gnat family and, though most are harmless, some are especially bloodthirsty.[10] A number are capable of transmitting disease, and the 'ague', a malaria-like fever, is mentioned in the Irish Annals and as late as Cromwellian times, though some ague outbreaks may have been typhus. The female flies do the biting: they need blood as a high-protein food when laying eggs, and are equipped with a long syringe-like proboscis for sucking blood. Mosquitoes lay their eggs in still, fresh water and so are usually found in damp places. Bog pools are a favoured haunt, but urban areas near still water can be badly affected in summer.

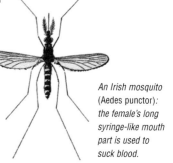

An Irish mosquito (Aedes punctor): the female's long syringe-like mouth part is used to suck blood.

The marshier parts of Cork city have long had a mosquito problem and, in the 1850s, suffered a major malaria outbreak. Infected soldiers returning from the Crimean War may have brought the disease home, while pools of water left by recent drainage works provided the ideal mosquito breeding ground. Ballintemple dispensary dealt with 175 cases of malaria in 1857, which the dispensary doctor described as "a perfect plague". The chances of another plague are slim, but with more people travelling to exotic areas, the risks increase.

The *King of Spades*

The Blarney river rushes headlong from its source down a narrow ravine between Coolowen and Monard. The drop from the top of the ravine to the bottom is about 20 metres, and for 200 years this head of water powered a series of mills and ironworks.[7] The mills closed in the 1960s, but were left untouched and this is now an important industrial heritage site, though some buildings are not in great repair. Good (albeit brief) views of the impressive complex can be had from the Cork–Dublin train, which crosses the ravine on a viaduct 8km out of Cork city.

The first works in the glen were two forges, built in the late 1700s by a Quaker industrialist, Abraham Beale. Eventually there were four large iron mills and 10 forges, plus dwellings, workshops and furnaces. One mill began as a logwood mill, crushing Brazilian hardwoods to make chips for a local dye-works, but, by 1850, it had converted to an ironworks. Because the glen is narrow, there was no room for a canal or millrace, and the mills themselves are on ledges cut out of the rock. So an ingenious water control and distribution system was built, most of it still in place, with millponds and weirs, cast-iron pipes, cisterns and sluices, and supplying water to each mill in turn. In winter, all four mills could have worked a full day but, when water levels fell in summer, they probably worked intermittently, stopping when the cistern above their waterwheel ran dry.

The waterwheels powered large forge bellows, and enormous industrial hammers, punches, guillotines, grindstones and sanders at the ironworks. The best known of the mills was Monard spade and shovel mill, and a 1930s film, *The King of Spades*, featured the mill and the men who worked it. Monard was one of many Irish spade mills,* each producing a vast range of spades for every conceivable job and condition, many of them following local designs – one Tyrone mill produced over 200 types of spade. Ireland's only working traditional spade mill is Patterson's* in Co. Antrim. The Monard mill site is now a private wildlife sanctuary, run by Tom O'Byrne. ⑰

"A second Sir Isaac Newton"

In 1817, Mallow produced a brilliant young mathematician, literally by accident, when 11-year-old **Robert Murphy** (1806–43), son of a poor, local shoemaker, was crushed by a cart.[11] He spent the next year in bed recovering and reading, and discovered a talent for mathematics. By 13, he had taught himself Euclid and, at 18, published his first work, a pamphlet criticising the work of an Irish mathematician, John Mackay, who claimed to have found a way of constructing the cube root of two using only a straight edge and compass. Murphy, however, showed how Mackay's principles "are proved erroneous". The pamphlet brought him to the attention of various mathematicians, one of whom said he was "a second Sir Isaac Newton… pray look after him". These benefactors raised funds to send him to college and persuaded Cambridge University to grant him a place, despite his lack of much formal education.

Murphy had a brilliant college career, becoming a Fellow, which entailed converting from Roman Catholicism to the Church of England. He was acclaimed as "a genius of mathematical invention", and published several books, notably on electricity and algebra. In 1832, however, he was forced to leave Cambridge on account of his "dissipated habits" and gambling debts, though he later found work as an examiner with the University of London. Murphy had much in common with George Boole,* another mathematician associated with Co. Cork: both were the sons of shoemakers, born around the same time, from a lowly background and largely self-taught. Both also died relatively young: Boole of pneumonia at 49, and Murphy of tuberculosis – probably exacerbated by drinking – at the age of 37, before he had fulfilled his early potential.

Warm springs and a spa resort

Ireland has plenty of mineral water springs, but those at Mallow are unusual in that the water is warm – a pleasant 20°C – and in the 18th and 19th centuries the town was a popular spa* resort. People no longer come to Mallow for "the cure", instead the warm springs are now used to heat a swimming pool. The springs are a product of local geology: Mallow sits on a layer of limestone that is over 1km thick, and riddled with cracks; water can percolate up through the cracks from the bottom of the limestone, where the temperature is about 40°C; by the time the water emerges at ground level it has cooled to 20°C.

Medical interest in Mallow's waters began in 1727 after Mrs Wellstead, a local woman who had been ill, drank nothing but spa water and made a remarkable recovery. Her physician invited a doctor from the famous Bristol spa to examine the Mallow waters and, thanks largely to his enthusiasm, the town soon became a fashionable resort and was known as 'the Bath of Ireland'. The waters were reputed to be good for lung and respiratory complaints, such as consumption and asthma, and each year from April to October people came to take the cure. The local landlord, Mr C D Jephson MP, built a fine spa house with a well in its front room, a pump-room, baths and medical consulting rooms. The fashion for such resorts declined, however, and Mallow's spa house closed in 1854. Some 130 years later, it was realised the thermal springs could be a geothermal energy source, and they now heat the local swimming pool. Mallow council restored the old spa house, and it is now an energy information office with a warm water spring in its front room. (18) Ireland's only surviving spa is at Lisdoonvarna.*

CASTLEPOOK

An Aladdin's fossil cave

The richest fossil cave in Ireland is at Castlepook, where an astounding 34,000 fossil bones have been found. Among them were bones from woolly mammoth,* giant deer,*

arctic lemming and the only spotted hyena so far found in Ireland. Reindeer and brown bear* fossils have been found in other caves nearby. The various remains have been carbon-dated to 26,000–35,000 years ago. The climate then was cold, but there was

Fossil reindeer antlers found in Ireland. Reindeer were among the species grazing by the Awbeg river 30,000 years ago.

clearly still enough grass for the mammoth, deer and lemming, and enough carcasses for scavenging hyenas. When the Castlepook fossils were found in the early 1900s, they were in a mess, having been jumbled by water flowing through the cave. The first excavations jumbled them further and one impatient naturalist allegedly blasted his way through the cave with dynamite. Castlepook is just one of many caves riddling the limestone region between Buttevant in Co. Cork and Dungarvan in Co. Waterford. There are other fossil caves along the Awbeg river, not to mention the Mitchelstown Show Cave* at Burncourt, and the Shandon* and Kilgreany* Caves near Dungarvan, which produced interesting finds in the 1800s.

DROMAGH

The Kanturk coalfield

Colliery Crossroads just east of Dromagh village is one of the few reminders that this was once a coal-mining area. When anthracite was discovered there in the early 1700s, there were high hopes for the mine, even plans to develop an export market, and, in 1755, work began on a canal that would link the colliery with the port of Youghal, via the River Blackwater. Only a short section of canal was ever built, for it was not to be: the coalfield was small, just 30sq km in all; and the seams were badly twisted and shattered, thanks to the powerful mountain-building forces that had crumpled this part of Munster 300 million years ago. This made the coal difficult and costly to mine and production was only ever sporadic. Despite being the first Irish coalfield with access to a railway – the Killarney–Mallow line of 1853 – Dromagh's coal never went further than Cork city.

DONERAILE

1887 – St Leger dies of rabies

Rabies, usually associated with mainland Europe, was once common in Ireland, and the disease was only eradicated here in 1903, following a major outbreak in 1884–96. Then, newspapers carried frequent reports of rabid dogs and foxes, and over 200 Irish bite victims travelled to the Pasteur Institute in France for treatment – in 1885 Pasteur made a vaccine that could be used to treat people after they had been bitten. Pasteur's treatment was usually successful, but a few patients still developed the disease and died. One of the more celebrated Irish cases was that of **Hayes St Leger** (Lord Doneraile). In January 1887 he and his coachman, Robert Barrer, were bitten by the lord's pet fox. Both travelled to Paris for treatment, but while Barrer recovered, St Leger developed clinical rabies and died.

In the 1890s, the government introduced regulations to eradicate rabies: all dogs had to be muzzled, strays were put down, and, from 1901, every animal entering Ireland had to spend time in quarantine. The disease was officially eradicated by 1903, though one case occurred in 1923 in Northern Ireland, after pet importation regulations were relaxed for soldiers returning from World War I. Rabies is still endemic across much of Europe and regulations remain in place to protect Ireland's disease-free status. Doneraile Park, former home of the St Legers, is now owned by the State. [19]

MIDLETON

From worsted wool to whiskey

The old distillery at Midleton began life as Ireland's first worsted woollen mill. Hand-spun wool had been a major cottage industry in the area, supplying yarn to the English market. In 1795, a local merchant, Marcus Lynch, thought the future lay in the new spinning machines which were then being tested at a couple of English mills. Machines for spinning cotton had been used for decades, but wool is a coarser fibre and it was not until the late 1780s that machines were developed which could produce a tight, or worsted, wool yarn. Lynch built a large factory, loom shops, waterwheels and ancillary buildings at Midleton, at a cost of £20,000, and provided employment for over 1,000 people. He was, however, ahead of his time: there were technical problems with the experimental machinery, and a shortage

The Midleton wheel is probably Ireland's largest working waterwheel.
© Irish Distillers

of skilled staff, and the business closed after eight years. Despite this early failure, Cork developed a sizeable woollen industry, with mills at Douglas, Donnybrook and, most notably, the Blarney Woollen Mills, which survive.

The Midleton mills found a new lease of life during the Napoleonic wars when they served as a military barracks, then, in 1825, the Jameson family converted it to a distillery. Whiskey is still made there by Irish Distillers 175 years later. When a new factory was installed in 1975, the original buildings and equipment were preserved. The site is a unique self-contained industrial complex, with barley stores, a grain-drying kiln, malting houses and mills. It includes the largest copper pot still in the world – capacity 14,387 litres – and one of the largest working waterwheels in Ireland, still in perfect condition. Built of cast iron in England in 1852, the wheel is 7 metres in diameter, and turns 2.5 times a minute. A steam-powered beam engine, which supplemented the wheel when water levels were low, can also still be seen. [20]

1838 – *Sirius* steams into history

A small Irish steam paddle-ship made nautical history when, in April 1838, it became the first to cross the Atlantic solely under steam power. Moreover, the **Sirius** beat one of the largest ships of the day, and made the journey in 18 days. The gauntlet had been thrown down in 1836 when Dionysius Lardner,* a famous Dublin scientist, lecturing to the British Association for the Advancement of Science,* said you might as well talk of travelling to the Moon as of crossing the Atlantic by steam power alone. Corkman **James Beale** (1798–1879), a director of the city's St George Steam Packet Co, was among those who rose to the challenge. For the voyage, he chartered a steamship newly built for his firm's Cork–London route. *Sirius* was 70 metres long with two paddle wheels, of the kind invented by Irish-American Robert Fulton,* plus two masts, for no one was yet ready to give up sail.

On April 4th, 1838, *Sirius* steamed out of Cobh bound for New York, with 35 crew, 40 passengers, no cargo, and as much coal as she could carry – 450 tonnes of it piled into every available space. She carried only 20 tonnes of fresh water, as she had condensers to recycle the boiler water, the first ship so equipped. The boilers consumed vast amounts of coal, however – 24 tonnes a day – and the ship nearly ran out of fuel. Refusing to hoist sail, Capt Richard Roberts resorted to burning some of the ship's timbers to keep the boilers going, and a mutiny was averted only when the New York coast was spotted.

Sirius reached New York Harbour on April 22nd – having averaged 300km a day – and just hours ahead of the much larger *Great Western*, which had also crossed wholly under steam. The *Great Western*, which was on its maiden voyage, completed the journey in a record 15 days, but *Sirius* arrived first and took the credit. The mayor of New York joined the celebrations onboard, before *Sirius* steamed home in triumph. She was returned to serve on Cork–British routes but, in 1847, sank near Ballycotton, with the loss of 50 lives. The main paddle shaft was salvaged and put to work at a spade mill, and is now on open display in Captain Roberts' hometown of Passage West; other souvenirs have been collected from the wreck over the years.

Though **Cobh** has a natural harbour, this was not exploited until the Napoleonic wars in the early 1800s. It quickly became an important military and naval base, a commercial and fishing port, a provisioning station and port of call for transatlantic voyages, and later a seaside resort. [21] From 1840–1940, nearly 3 million emigrants left from Cobh onboard convict and emigrant ships. Aside from *Sirius*, Cobh is associated with the ill-fated *Titanic** (1912) and *Lusitania* (1915). Spike Island in the harbour, now a prison, was once a livestock quarantine station.

The victorious Sirius *at New York Harbour, the first ship to cross the Atlantic solely under steam. The Maritime Museum* (Dún Laoghaire) has a model of the ship.*

Three nature lovers

During the first half of the 19th century, most of the studies of our natural heritage were completed by amateur naturalists. Among them were three members of a Cobh family: **Robert Ball** (1802–57), who helped found Dublin Zoo; and his younger sisters **Anne** (1808–72) and **Mary** (1812–98), who specialised in the biology of the seashore. They were from an old Devonshire family, the Stawell Balls, that came to Youghal in 1651. Robert and his sisters inherited their love of nature from their father, Robert Sr, who was a customs officer in Cobh. Robert Jr attended the renowned Quaker school at Ballitore,* but Anne and Mary were probably schooled at home. The family had no money to send Robert to university, so the Duke of Devonshire arranged for him to be employed as a civil servant at Dublin Castle. Robert Ball then became active in Dublin scientific circles: in 1830, he helped found the Royal Zoological Society of Ireland, which established Dublin Zoo;* he spent holidays with William Thompson,* a noted Belfast naturalist; became president of the Geological Society of Ireland; and amassed a sizeable collection of zoological specimens, such as stuffed animals, which he donated to TCD to start a zoological museum* (the college later awarded him an honorary degree). In 1851, he was appointed secretary to the Queen's colleges,* where he proved to be a competent administrator. He published papers on fossil Irish bears, had the RDS organise public lectures, and campaigned for a nominal one penny admission to Dublin Zoo. His son, Sir Robert Ball,* was a noted mathematician and Astronomer Royal for Ireland at Dunsink Observatory.

Anne and Mary Ball no doubt benefited from their brother's expertise and scientific contacts, yet they were skilled naturalists in their own right – Anne specialised in seaweeds, while Mary was an expert on insects and molluscs.[12] Given the conventions of the time, neither woman published under her own name; instead, they communicated their results to their brother or some other male naturalist, who then publicised the findings, sometimes ascribing them to an anonymous correspondent. Mary Ball, for example, is now known to have been the first person to observe stridulation (the noise made by rubbing body parts against each other) in water bugs, though the discovery was first ascribed to her brother Robert who reported it in 1846. Both sisters discovered several species new to science, some of which were named after them: *Rissoa balliae*, for example, a small spiral snail,

is named after Mary, who found the first one at Youghal; *Ballia callitricha* is a genus of marine plants named after Anne Ball. In the 1860s, the sisters lived in a Dublin house with no garden, so they made the basement into a fernery. Mary Ball's mollusc collection, reputedly the best in Ireland at the time, was sadly dispersed on her death.

On safari in Cork Harbour

One of the world's largest prides of captive cheetahs lives a few kilometres from Cork city. **Fota Wildlife Park** is helping to conserve the endangered cheetah, and is a major player in an international breeding programme. The ultimate aim is to reintroduce the cats to the wild, once appropriate reserves are ready; in the interim, the cheetahs bred at Fota wait in wildlife parks and zoos around the world. Central to the cheetah project is a managed breeding programme: detailed information on all captive cheetahs is available on a computer; mates are selected so as to avoid inbreeding, a major hazard when the population is small; and participating centres exchange breeding animals. Fota has been particularly successful, and at times has had over 40 cheetahs. Fota was set up in 1979 by the Zoological Society, which also runs Dublin Zoo,* as a modern wildlife park and to complement its conventional zoo. The 30-hectare site at Fota, provided by UCC, means many animals roam freely, though some, such as the cheetahs, are behind fences. Nearly 100 species from five continents are present, including zebra, giraffe, monkeys, lemurs and wallabies. [22]

Fota Island is also home to the renowned **Fota Arboretum**, which is now owned by the State. The gardens were laid out in the early-19th century by James Hugh Smith-Barry, who extended what had previously been a hunting lodge. The trees were planted at generous intervals, so that 150 years later they have been able to reach their full stature. Rare and tender plants also grow there, many from the southern hemisphere, such as tree ferns, palms and bananas. [23]

Nearby Slatty's Bridge, which links Fota with the mainland, is a good place to watch birds: ducks, waders and black-tailed godwits in winter, sand martin colonies in summer. The bridge is one of 18 points on the **Birds of East Cork Trail**. This route around the coast, from Fota to Youghal by way of Cobh, Midleton and Ballycotton, visits some of the area's best bird-watching sites, including mudflats, reed beds and freshwater lakes, and BirdWatch Ireland's nature reserve at Cuskinny.

INCH

Ireland's largest gas meter

The pipeline from the **Kinsale Head gasfield** comes ashore at Inch, where Marathon Petroleum meters the gas, before selling it to the State-owned Bord Gáis Éireann (BGÉ). The gasfield, which lies 50km off shore, was discovered by Marathon in 1971 and it has been supplying natural gas since 1978, though it is nearing the end of its life. The smaller Ballycotton gasfield, which lies 30km from shore, has been supplying gas since 1991. At both fields, the gas occurs in porous rocks that lie 1km below the seabed and are covered by a solid rock cap.

The first customers for Kinsale gas were local homes and industry. Irish Fertilizer Industries (IFI) built a large chemical plant nearby, and uses the gas in making ammonia and urea (the process generates carbon dioxide as a by-product, which is sold to Irish drinks companies for adding to carbonated drinks). Two gas-burning electricity-generating stations were also built, at nearby Whitegate (Aghada) and at Cork city (Marina) to exploit the new fuel. In 1983, a pipeline brought the Kinsale gas to the Dublin market and, via branch lines, to big industrial users en route, among them Carlow Sugar Company* and Waterford Crystal.* In 1987, when the pressure of the gas coming from the offshore wells had begun to drop, a compressor station was added to the pipeline at Midleton. The Kinsale and Ballycotton gasfields supplied nearly 15 per cent of Ireland's annual energy needs for over 20 years. As they wind down, attention is turning to the Corrib gasfield off Mayo, which will begin supplying gas in 2004. There are also inter-connector pipelines linking Ireland to British and international gas sources.

BALLYCOTTON

Scottish pebbles on the beach

Ailsa Craig is a distinctive dome-shaped islet 10km off Scotland's Ayrshire coast. It is made of an unmistakable cream-blue-green granite, containing small crystals of quartz, feldspar, mica and a blue mineral called riebeckite. The rock is traditionally quarried to make curling stones, or 'ailsas'. Strangely, pebbles of Ailsa granite can be found along Ireland's east coast, and as far around as Ballycotton beach where they were first noticed in the early 1900s. The pebbles are glacial erratics,* which were carried from Scotland by ice during the second-last Ice Age, 130,000 years ago. Most of the northern hemisphere was gripped by ice then, and a Scottish ice sheet flowed down the Irish Sea and round the Wexford corner, dropping Ailsa crumbs as it went. Geologists use these distinctive erratics to trace the route of the Scottish ice, and to reveal the extent of glaciation during that particularly deep Ice Age. Ballycotton's Ailsa pebbles are buried under rock and sand deposited 20,000 years ago, during the most recent, and less severe, Ice Age, by glaciers flowing off the Cork–Kerry Mountains. Ballycotton is on the East Cork Bird Trail; the cliff is noted for choughs, the beach for rare migrants in winter.

CLOYNE

Madame Dragonfly

Snake doctors, the devil's darning needles, horse stingers… they have been called many things, but we know them as **dragonflies**. There are 43 species in Ireland and Britain, and the person who contributed most to our knowledge of them was a self-taught naturalist and intrepid explorer, **Cynthia Longfield** (1896–1991) whose family home was at Castle Mary, near Cloyne.[13] Longfield's early interest in science was encouraged by her mother and by her mother's father, an English chemist who made a fortune from a technique he invented to extract copper from ore. Cynthia had a keen sense of adventure and, as a young woman, took part in expeditions to Latin America and Egypt. The family's wealth meant she could join these costly trips, and she never held a paying job – for most of her professional life she worked as a volunteer in the British Natural History Museum.

Cynthia Longfield's interest in dragonflies began with an expedition to the Pacific Islands in 1924, led by a noted entomologist, Cyril Collenette. On their return, Longfield worked at the British Museum cataloguing the dragonflies they had collected. She quickly became an international expert, and took part in several other expeditions to Latin America, southern Asia and Africa. Each time Longfield discovered numerous dragonfly species that were new to science – as many as 35 new species on one trip to Brazil. Many of her finds, such as *Castoraeschna longfieldae*, were named after her, the ultimate honour for a naturalist. In 1937, she published *The Dragonflies of the British Isles*, a definitive book that ran to many editions and was not bettered until the 1980s. She retired from the British Museum and returned to Castle Mary in 1956, but continued travelling, collecting specimens, lecturing and writing. This intrepid woman, once quoted as saying, "I find machetes so useful in the jungle", is buried at Cloyne Cathedral. The cathedral also has a monument to philosopher George Berkeley,* Bishop of Cloyne from 1734–52.

The first Irish potato?

When the potato* was introduced into Europe, people must not have appreciated its value, or they would surely have recorded the name of the person responsible. Long after the event, the credit was variously attributed to Spanish and English adventurers, including **Walter Raleigh** (*c*.1552–1618), who reputedly planted the first potato in Ireland at his Youghal estate in 1588. Another Irish story holds, however, that the potato washed ashore from the Spanish Armada that was wrecked around the coast that same year, and in Irish the potato was often called *an Spáinneach*, 'the Spaniard'.

The potato (*Solanum tuberosum*) is native to the Andes. The first Europeans to see it were the Spanish *conquistadores*, who brought the plant home, and there are records of potatoes growing in southern Spain in the 1570s. Sir Francis Drake bought and ate potatoes in Chile in 1578 while on a round-the-world voyage, and may have brought some back to England in 1586, although the first documentary evidence of potatoes in England comes 10 years later, when a London herbalist planted some in his garden. Water Raleigh's expeditions to the New World were to North America and the Caribbean, home of the unrelated sweet potato (*Ipomea batatas*). If Raleigh did plant potatoes at Youghal in the 1580s, he must have got them from Drake – it is unlikely he planted sweet potatoes, as these are not suited to the Irish climate. A more probable explanation is that the potatoes arrived from Spain, if not via the Armada, then by more conventional trade routes. Raleigh did introduce tobacco here, however: the plant was brought from Virginia by one of his expeditions, and several Planters tried to grow it on their Munster estates. Raleigh reputedly smoked the first tobacco in Ireland at his Youghal home of Myrtle Grove, where a fearful servant is said to have doused him and his pipe with cold water on seeing the smoke.

Raleigh had been granted 16,180 hectares of Munster as a reward for fighting the rebellious Irish. It included the monastic university town at Lismore,* and the walled mediaeval port of Youghal. Raleigh entertained his friend the mathematician Thomas Harriot* there, and the poet-soldier Edmund Spenser, who was then sheriff of Cork and living near Doneraile. When Raleigh's fortunes declined, he sold out for a paltry £1,500 to another Englishman, **Richard Boyle***(1566–1643). Boyle, the first Earl of Cork and father of the famous chemist Robert Boyle,* spent a fortune developing the estate, bringing in English settlers and starting a major ironworks and fishing industry. In Youghal, he rebuilt the church, added almshouses, and installed a monument that can still be seen – featuring himself, his two wives and 15 of his children – before moving his residence upriver to Lismore Castle where Robert was born.

Youghal (pronounced 'yawl', from *Eochaill*, 'a yew wood') retains many of the historic buildings associated with Boyle and Raleigh, including a college founded in 1464 as the 'university of the city of Youghal', which Boyle used as his residence. Strategically located at the mouth of the Blackwater, Youghal was an important port: granted a charter by King John, it had strong trade links with England, Europe and the Mediterranean. In the 19th century, it also had a large pottery, brickworks – which produced the 2 million bricks used to line Cork city's railway tunnel – and a lace industry. The latter was begun in 1847 by a nun, Sister Mary Ann Smyth, who unpicked a piece of Italian lace and, in an early example of reverse engineering, taught herself the stitches. Youghal lace declined in the early 20th century but was recently revived. The town's other major industry was Youghal Carpets: begun in 1954, it employed 3,500 at its peak across various locations in the 1960s and exported worldwide; the factory at Youghal closed in 1984, however. ㉔

COUNTY WATERFORD

Stunning Coumshingaun, the Blackwater's intriguing bend, and a bracing volcanic coastline with basalt columns like those at the Giant's Causeway, are some of this county's highlights. Robert Boyle, who gave his name to Boyle's Law, was born here, as was Ernest Walton, who helped split the atom and won a Nobel prize in physics. Waterford's ingenious industrialists invented the bacon rasher and the cream cracker and, thanks to the world's largest crystal factory, the county is an international brand name. That name comes from the Norse *vadrefjord* ('wether inlet', where wether, or castrated, rams were loaded onto boats). In Irish, the city and county are called *Port Láirge* ('port of the thigh'), because of the River Suir's shape at that point.

LISMORE

Robert Boyle's Law

Robert Boyle: the first modern chemist and the last alchemist.

Nature abhors a vacuum. At least, that is what philosophers and clerics believed in 1660. But **Robert Boyle** (1627–91), who was born in Lismore Castle and was a key figure in the Scientific Revolution, proved them wrong. Using an air pump, he demonstrated that Nature can tolerate a vacuum, and he made many discoveries about air and gases, including a fundamental law, relating pressure and volume, that is still called Boyle's Law. What Isaac Newton did for physics, Robert Boyle did for chemistry – and people are still debating his ground-breaking legacy.[14]

Robert was the youngest of 14 children born to Richard Boyle,* first Earl of Cork, and his second wife, Cathleen Fenton from Dublin. The earl was an Elizabethan adventurer who bought Walter Raleigh's vast Irish estates *c.*1600, including much of counties Waterford, Cork and Tipperary, for a bargain £1,500. He developed ironworks, planted his estates with English settlers, and became the richest man in Ireland. Though much of his fortune was lost in the 1641 Rebellion, his son Robert never had to take a job and could still pay for all his experiments and even employ an assistant.

At the age of eight, Robert Boyle left Lismore for Eton College, where he was a clever student but often sickly. When an apothecary's mistake made him ill, he developed a distrust of doctors and ever after prepared his own remedies. He spent his teens on a Grand Tour of Europe: in Italy he learned about Galileo's work, and in Germany he saw the air pump which Otto von Guericke had invented. Back in England and plagued by ill health, Boyle opted for a quiet life of reasoning and experiment, rather than a career in the army or the church. Ireland was "too troubled to have any hermetic thought", so he settled in Oxford where he established a celebrated circle of friends. Known as the Invisible College, this became, in 1660, the Royal Society, which continues today in London.

The world in 1660 was still enthralled by Aristotle, the occult and alchemy. Many believed that earth, air, fire and water were the fundamental elements. Experiments were frowned on as something done by lowly mechanics. And as yet there was no scientific method. Robert Boyle would change all that. He combined theory with practice and developed a rigorous experimental technique. He employed Robert Hooke as his assistant, an ingenious young mechanic later famous for Hooke's Law among other things. Hooke made Boyle an air pump that had a glass vacuum chamber where experiments with his pump could be observed. Boyle placed a candle in this chamber, pumped the air out and watched the flame die, and concluded that something in the air was essential for combustion. In other experiments, he discovered that a ringing bell could not be heard through a vacuum, that animals suffocated without air and that a feather falls faster in a vacuum. He also proved Galileo right: in a vacuum everything falls at the same velocity regardless of its weight.

In his most crucial experiment, Boyle investigated the elasticity of air. He put a partially inflated lamb's bladder into his vacuum chamber; as he pumped the air out of the chamber, the bladder expanded. When the air was let back in, the bladder shrank. Boyle discovered that the volume of the bladder depended on the pressure in the chamber: if the pressure increased, the volume dropped, and vice versa. This relationship, published in 1661, is Boyle's Law. In general it states that, at a given temperature, the volume of a gas multiplied by the pressure is a constant ($p.V =$ constant). It is a cornerstone of physics, and fundamental to our understanding of gases and the atmosphere.

In 1661, Boyle also published his most important book, *The Sceptical Chymist*. This classic text marks the end of alchemy and the start of modern chemistry and it paved the way for the chemical revolution of Priestley and Lavoisier a century later. In place of Aristotle's four elements, Boyle proposed a modern definition of a chemical element as something simple and unmingled, that could combine with other elements to form compounds. Conversely, compounds could be broken to release their constituent elements. He also dismissed magic potions in favour of well-defined and measured instructions, and described his own experiments in detail so others could replicate them, a key technique in the modern scientific method. Boyle developed some of the first analytical tests to identify substances and classify them according to their properties. Some – such as his flame tests – are still used. He introduced vegetable dyes as indicators of acidity or pH, and, following dissections on corpses, recommended hydrogen peroxide as a preservative.

But for all his radical ideas, Boyle was a man of his time. He retained a keen, often secret interest in alchemy, believed in angels, and quaffed various remedies depending on the direction of the wind. Though he scientifically challenged many beliefs held by the clergy, he was intensely religious, paying for the Bible to be translated into several languages including Arabic, Irish and Welsh, and leaving money for lectures to counter atheists and infidels. Robert Boyle was buried at St Martin's-in-the-Fields in London, but his grave was lost when the old church was destroyed in 1720. A statue of him stands at Government Buildings in Dublin (formerly the Royal College of Science*). His lasting legacy, however, is the science of modern chemistry, and the Royal Society in London, which he founded. The Heritage Centre, Lismore, has several books by Robert Boyle. ㉕

Lismore, his birthplace, is an historic town: St Carthagh (*d*.638) founded a celebrated seat of learning and a diocese there; Henry II met the chiefs of Munster there; and, in 1185, King John built a castle on a prominent site overlooking the River Blackwater. The land was granted to Walter Raleigh, who later sold it to Richard Boyle. The castle was destroyed in the turbulent 1640s, then restored a century later by the Duke of Devonshire who had acquired it by marriage. ㉖ He improved the castle and town, built fine bridges across the Blackwater, and ran a profitable salmon fishery, which exported frozen fish to Liverpool. The area has a number of mineral water springs, and there have been small lead, iron and copper mines at intervals in the past. North of Lismore is the scenic Vee, a stunning 200 metres gap gouged out of the hills by a glacier during a recent Ice Age.

CAPPOQUIN

The Blackwater's mysterious bend

The River Blackwater is internationally famous, both among anglers and those who study landscapes. For the river does something curious: instead of continuing its west–east course and taking the easy low-lying route to Dungarvan, it turns sharp right at Cappoquin and cuts through a series of sandstone ridges to reach Youghal. This is not normal behaviour – nothing stops it going to Dungarvan – and, in 1862, an English geologist **Joseph Beete Jukes** (1811–69), then director of the GSI,* realised it needed explaining.

This part of Munster has a definite east–west grain, thanks to the parallel ridges of Old Red Sandstone that were pushed up 300 million years ago. For the most part, the main rivers – the Blackwater, Lee and Bandon – follow these

east–west valleys. All three take an unexpected right-hand turn, however, and head south to the sea, the Blackwater's bend being the most significant of the three. Jukes's insight was to realise that these angled rivers are a mix of two river valleys: an old stretch, and a young 'captured' stretch. He thought the final north–south section of the Blackwater, from Cappoquin to Youghal, was the remains of an ancient Blackwater, a river that presumably originated further north in the distant geological past. The east–west section, ending at Cappoquin, was, he believed, a younger river that had grown by extending westwards up the valley, eventually reaching almost to Castleisland in Co. Kerry.

The Blackwater river god: crowned with fish and a basket of apples. © Enterprise Ireland

Millions of years ago, this region was buried under a thick blanket of limestone, much of which has since been removed by erosion, and it is nearly impossible to know now where rivers would have flowed in prehistoric times. Those who study how landscapes form agree, however, that today's Blackwater, Lee and Bandon are captured rivers with a young and an old section. But it is still not known why they bend, nor which section is the old river and which the new. Dr Pete Coxon, a geographer at TCD, suggests that, contrary to what Jukes thought, the west–east stretch of the Blackwater could be the old river, and that it did originally flow to Dungarvan. Ice sheets pushing in from the coast during the last Ice Age, however, could have blocked that route and forced the river to take a new route to Youghal which, because of the intervening rocky ridges, may have been protected from the coastal ice. The bend in the Blackwater remains a mystery 150 years after Joseph Beete Jukes first brought it to people's attention. His

The Blackwater is tidal at Cappoquin, 25km inland. This is because Ireland is tilted, dipping in the south, and the tide can reach high up the river's north–south valley. Likewise the Barrow, which is tidal to St Mullins in Co. Carlow, 50km inland.

technique of using rivers to examine how landscapes form, nevertheless, prompted others to study similar anomalies and won him international acclaim, and a hill in Utah, Jukes Butte, is named after him.

England's Galileo

Molanna Abbey was once owned by a brilliant English astronomer and mathematician, **Thomas Harriot** (c.1560–1621), whose discoveries rivalled those of Galileo. He also promoted the colonisation of North America, helped introduce tobacco to the Old World and, ironically, was probably the first European to die of a smoking-related cancer.

Thomas Harriot was the scientific adviser on Walter Raleigh's* 1585 expedition to Virginia and Carolina. During the voyage, Harriot devised several improvements to navigational techniques; on arrival, he made a detailed map of the coast, learned the Algonquin language and, like Adam in Eden, named and described everything he saw. He returned with tobacco, among other things, and a grateful Raleigh presented his friend with the former monastery of Molanna Abbey. There Harriot wrote his *Briefe and True Report of the New Found Land of Virginia* (1588). The first English description of the New World, it was soon widely translated and helped fuel European interest in North America.

Harriot was also a member of Raleigh's circle of powerful intellectuals. Known as the School of the Night, partly because of their melancholic black dress, partly because they were suspected of magic, they included Sir Francis Drake, the soldier-poet Edmund Spenser, Henry Percy (Earl of Northumberland), the writer Christopher Marlowe, the publisher Edward Blount and the 'wizard' John Dee. The notorious group is thought to have met at least once at the Waterford abbey, and a nymph in a Spenser poem is named Molanna. Later, there were accusations of atheism, dark

arts and treason, and Raleigh, Blount and Percy were eventually executed, while others, including Harriot, were imprisoned. But that was in the future, and, by then, Harriot had sold the abbey to Richard Boyle* of Lismore Castle, who bought most of Raleigh's Irish estate.

Thomas Harriot: astronomer, mathematician, "master of the dark arts" and one-time owner of Molanna Abbey.

Thomas Harriot spent the rest of his life engaged in scientific research, initially under Raleigh's patronage and later the Earl of Northumberland's. He was among the first to make a telescope, which he called a "perspective truncke" and which he reputedly showed in 1585 to the Native Americans. With the great European map-maker Gerard Mercator, he devised ways of making maps and navigation more accurate. Studying the trajectory of bullets, he came to understand the laws of motion and of falling bodies, and realised correctly that the flight of a projectile describes a parabola. He observed the comet of 1607 (later named after Halley), correctly calculated that its orbit was an ellipse, and predicted its return. Independently of Galileo, he used his telescope to observe other 'new worlds' – among them the craters of the Moon, the phases of Venus, and Jupiter's moons. He made the first telescopic map of the Moon and, in December 1610, simultaneously with Galileo, discovered sunspots.

Harriot's other great contribution was in mathematics, and he has been described as a founder of modern algebra. He devised the elegant concept of placing everything to one side of an equation and equating that to zero, and he also invented the useful concepts of, and symbols for, 'less than' (<) and 'greater than' (>). He discovered the fundamental sine law of refraction 20 years before Snell, and much more besides. Thanks to this mental magic, dabbling in the nothingness of zero, discovery of new worlds, and friendship with Raleigh, Harriot was imprisoned. He survived, however, and died of a tumour of the nose brought on by smoking tobacco. Yet, even after his death, friends feared to publish his work, condemning this brilliant mathematician and astronomer to relative obscurity.

DUNGARVAN

Shandon's mammoth and 'Kilgreany Man'

In March 1859, stonemasons at Shandon limestone quarry uncovered a cave containing a huge number of bones. A cartload was sold for grinding into bonemeal, but some of the bones were massive and these were paraded through nearby Dungarvan as belonging to "an antediluvian giant". The local postmaster, Edward Brenan, realised that the bones belonged to a mammoth* and, back in the cave, he identified remains from other species, including bear,* reindeer, horse and hare.* Brenan reported the discovery to the RDS that summer. In a pamphlet, *The occurrence of mammoth and other mammal remains*, he suggested the animals had died in the biblical flood. "Men fled to the tops of the highest mountains," he wrote, "and the wild animals… sought shelter in the caves… and there at once together died." Unable to believe that Ireland once had an arctic climate, he thought the flood had carried the mammoth's body from Siberia. He also believed the flood could explain why no human remains were found in the cave, and that the raging torrents had jumbled the bones.

Part of the Shandon mammoth's tusk; from Edward Brenan's 1859 report to the Royal Dublin Society.

Brenan was partly right about the flood. The area between Mallow and Dungarvan is low-lying limestone and riddled with caves. It escaped the worst of the last Ice Age (35,000–15,000 years ago, when the northern two-thirds of Ireland were covered by ice), and instead had a tundra climate. But torrents of glacial meltwater flooded down from the northern glaciers, excavating caves and depositing bones, as at Shandon. The Shandon mammoth has since been carbon-dated to 30,000 years old, and some of the other bones to 26,000 years ago. Similar fossil bones were found in other caves, notably Castletownroche and Castlepook* in Co. Cork. Like pollen profiles,* these discoveries shed light on how Ireland's climate and wildlife have changed over time.

As for early humans, if people were in Ireland over 12,000 years ago, then they must have been in the ice-free southern part. With this in mind, British geologists and cavers explored nearby Kilgreany Cave in 1928, where they found a human bone alongside bones from a giant deer. This '**Kilgreany Man**' was initially thought to be Ireland's earliest-known inhabitant, at least 10,000 years old and from the early Stone Age. But Kilgreany's fame was short-lived: later excavations revealed that, as at Shandon, the deer and human bones had been brought together by floods, and the human remains have since been dated to 4,500 years ago. (The earliest traces of people in Ireland so far date to 9,000 years ago at Mount Sandel on the River Bann.) (27)

DUNGARVAN

The man who split the atom

April 14th, 1932. A young Irish scientist was working in a Cambridge laboratory when he saw "a wonderful sight, lots of scintillations, looking just like stars". **Ernest Walton** (1903–95) had just split an atom and what he saw was the energy released as the atom broke apart. The brilliant experiment had transformed lithium into helium, a modern act of alchemy. It was a collaboration with John Cockcroft and it introduced the era of nuclear power. In 1951, the two men shared a Nobel Prize for their ingenious work.

Ernest Walton, son of a Methodist minister, was born in Dungarvan and schooled at Methody College in Belfast. There the emphasis was on science and engineering, and Walton learned woodwork and metalwork, which later proved useful when he built his own particle accelerator. After graduating from TCD, Walton went to work at Cambridge University under Ernest Rutherford, who had already discovered much about radioactivity and the structure of the atom. In natural radioactivity, large unstable atoms disintegrate, producing two smaller atoms and releasing nuclear energy in the form of energetic or radioactive particles. Walton and Cockcroft wanted to recreate this in a laboratory by artificially splitting an atom. Others were racing to do this, and two Swiss scientists were killed when they tried to harness lightning as a high-voltage power source for their experiment.

Cockcroft and Walton planned to fire high-energy particles (protons) at some lithium metal, hoping the collision would

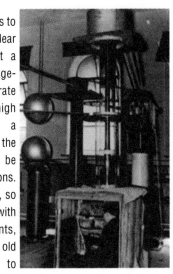

Ernest Walton in the home-made particle accelerator that he and Cockcroft used to split the atom in 1932. TCD Physics Department

force the lithium atoms to break and release nuclear energy. Walton built a device with voltage-doublers to accelerate protons to very high speeds; and also a detector to register the energy that would be released by the collisions. The budget was £500, so Walton improvised with recycled components, including parts from old petrol pumps and, to shield them from the radiation, lead from used car batteries. Even the screws were recycled.

On April 14th, 1932, sitting in a crate-like compartment underneath this home-made apparatus, Walton watched scintillations appear on a small screen. Cockcroft and Rutherford verified that the radiation came from the break-up of lithium atoms. The atom had been split. On May 3rd, newspapers around the world announced: "Atom split by British scientists – Alchemists dream realised – Three years of research leads to great discovery." Cockcroft and Walton had converted matter into energy (as predicted by Einstein's famous equation, $E=mc^2$). The nuclear power era had begun.

In the 17th century, Robert Boyle* had sought to transmute the elements, and 300 years later Walton and Cockcroft achieved it: in their experiment, after a proton collided with a lithium atom, the unstable atom-proton mix disintegrated into two helium atoms, releasing energy in the process. In 1934, Ernest Walton returned to Ireland as physics professor at TCD, where a plaque commemorates his contribution to science.

Nobel laureates

Ireland has just one Nobel science laureate (Ernest Walton in 1951), compared with four in literature (W B Yeats, George Bernard Shaw, Samuel Beckett and Seamus Heaney). It can claim, however, a number of other science laureates by association.

Guglielmo Marconi:* whose mother was Irish, won the Nobel Prize for physics in 1909.

Erwin Schrodinger:* Nobel laureate in 1933, came to Ireland in 1938 as an Austrian refugee at the invitation of Éamon de Valera.

J D Bernal:* many thought the Tipperary man deserved a Nobel, and though he never won one, two of his students did: Max Perutz (medicine, 1962), and Dorothy Hodgkins (chemistry, 1964).

Richard Millington Synge: from a Liverpool-Irish family, shared the 1952 Nobel Prize for chemistry.

Jim Watson and Maurice Wilkins: who shared the 1962 Nobel Prize for medicine with Francis Crick for their discovery of DNA's double helix, both have Irish family connections.

Gravity defying?

Rashers and cream crackers

There is a point on the scenic **Mahon Valley** drive, where it seems you can defy gravity. The spot is by a fairy thorn tree and near a cattle grid. There, if you put your car in neutral, it will seem to roll up hill. Similar anomalies, sometimes called 'magnetic hills' or 'electric braes', occur in other hilly parts of Ireland and elsewhere. There is a popular one at Ravensdale in Co. Louth, for example. All were discovered in the past hundred years with the arrival of the automobile – after all, you cannot put a horse and cart in neutral.

Far from defying gravity, these are powerful optical illusions created by the lie of the land. They generally occur in hilly regions, usually close to the top of a pass, where the true horizon is obscured by the surrounding hills and where the nearer hills appear higher than the distant ones. The result is that the brain perceives you are travelling uphill, when in fact you are rolling down. Like all optical illusions, the effect is best in broad daylight and disappears in the dark.

The next glen around from Mahon is **Coumshingaun**. The largest corrie* in Ireland, this gigantic rocky amphitheatre rises 200 metres vertically above a lake to the Comeragh summit plateau. It was carved by a glacier during the second-last Ice Age some 70,000 years ago. The corrie's vertical wall reveals successive layers of gritstone, sandstone and conglomerate, all layers of Old Red Sandstone rock. This part of Ireland was free of ice during the most recent Ice Age, which ended 15,000 years ago, and so the plateau still has a rough appearance, with rocky tors and spurs, which were not ironed smooth by any ice sheet. Coumshingaun can be seen from the road, but hill walkers get the best views.

Bacon rashers and cream cracker biscuits – world-famous food items most of us probably take for granted – were invented in Waterford in the 19th century. They are just two of Ireland's edible inventions, which also include cream liqueurs, Irish coffee,* fig rolls, flavoured crisps and iced caramel sweets.[15]

The rasher (a piece of bacon to be cooked quickly, or rashed), was invented in 1820 by **Henry Denny**, a Waterford butcher who patented several bacon-curing techniques. Until then, pork was cured by soaking large chunks of the meat in barrels of brine for weeks. Shelf-life was poor, as often the inside of the chunks did not cure properly, and the meat rotted from the inside out. Denny introduced two important innovations: he used long flat pieces of meat instead of chunks; and he dispensed with brine in favour of a dry or 'hard' cure, sandwiching the meat in layers of dry salt. This produced well-cured bacon with a good shelf-life and revolutionised Ireland's meat industry, and Irish bacon and hams were soon exported to Britain, Paris, the Americas and India. Denny's bought a Danish meat company in 1894, introducing Irish techniques to Denmark, while another Irish company, O'Mara's from Limerick, opened a factory in Russia. However, according to local historian Jack Burtchaell, not everyone was pleased: because the meat was now dry salted on open curing floors, barrels of brine were no longer needed and disgruntled coopers attacked some bacon companies.

The cream cracker was invented in 1885 by **W R Jacob**, who made dry biscuits for transatlantic voyages at his

In 1940, Dan Monaghan, of Clarnico Murray in Dublin, dipped toffee in icing and invented **iced caramels**. **Flavoured potato crisps** were invented in the 1950s by Dubliner Joseph Murphy, founder of Tayto, who introduced cheese 'n' onion crisps to the world. Gilbey's (Bailey's) invented **cream liqueurs** in the 1970s – stabilisers keep the homogenised cream-whiskey mix from separating.

Bridge Street bakery. The Jacobs were Quakers who came to Ireland in the 1670s, fleeing persecution in England. (Other Quakers who helped Co. Waterford's industrial development included the Penrose family who started Waterford Crystal (see: **Crystal city**), and the Malcolmsons, who founded Portlaw (see: **Cotton town**). The new biscuit was made from a yeast dough left to ferment for 24 hours, then flattened and folded several times to make a multi-layered biscuit. Jacob's cream crackers – "The original and best" – are today made by machines that can produce 1 million biscuits an hour, and are sold in 35 countries. Jacob's (now part of the French Danone group) were inventive bakers: they also created the Mikado biscuit (in 1888 to mark Gilbert & Sullivan's opera), the Kimberley (1893, when South Africa's diamond mine was in the news) and the Fig Roll (1903). This last ingenious biscuit, mentioned in James Joyce's *Ulysses*, was probably the world's first extruded product: unlike a sausage it has a definite structure and yet is not encased in a skin.

Waterford is also home to blaa, an unusual small batch loaf. This light breakfast bread, unique to Waterford, was reputedly introduced by Huguenot refugees who arrived in 1685. Blaa is one of many Irish **regional foods**; others include Cork's buttered eggs and drisheen (a blancmange-like pudding made from blood serum), Dingle's mutton pie, and potato boxty and soda farls in the northern counties.

For centuries, Waterford city was one of Ireland's busiest ports and our most important trade and communications centre with England. Among those who landed at Waterford were the Vikings, the Normans, Henry II, King John, Richard II and Cromwell; James II departed from Waterford for France the day after the Battle of the Boyne. The city has a long industrial tradition, attributed in part to a liberal policy, adopted in 1704, when the freedom of the city was granted to all traders, prompting European merchants to settle there. It also boasts Ireland's oldest civic building, Reginald's Tower (built 1003). ㉘

Crystal city

The first record of **glass-making** in Ireland was in the 1580s when an Englishman, George Longe, was granted permission to open a 'glasshouse' (glass manufactory) at Drumfenning near Dungarvan. Longe persuaded Elizabeth I that the industry would consume vast quantities of timber fuel, and that it was better to clear Ireland's rebel-infested forests, than denude England's fine woodlands. One glasshouse, he argued, was as good as 20 soldiers in a garrison, and moreover, the glass would have to be exported making it easy to tax. Among the objects Longe made were windowpanes of 'spread' glass, produced by splitting and spreading glass cylinders while they were still molten. A century later, several French Huguenot families were making windowpanes and glass drinking vessels at manufactories in Shinrone* and Portarlington.

The first Waterford Crystal – a 1790s Penrose lidded urn.
© Waterford Treasures

Glass has been made in Waterford city since at least 1720, but the crystal factory that made the county an international brand name was started in 1783 by two Quakers, **William Penrose** (1746–99) and his uncle **George Penrose** (1722–96). At that time, Irish glass was exempt from many of the taxes levied on British glass. This boosted the Irish industry and prompted William and George to open a "glass house of superior description" on Queen Street near Gibbet Hill.[16]

Their raw materials were silica sand, potash and litharge (a lead compound). From this they produced lead crystal "as fine a quality as any in Europe… in the most elegant style". Strong and hard-wearing, it rang sweetly when tapped. Penroses made clear, dark-green, blue and enamel glass, blown, moulded, cut and engraved glass. It was fashioned into smelling bottles, decanters, wine glasses and cruets, and aristocratic patrons such as the Countess of Westmoreland helped market the products. Within two years, the company was exporting to Britain, Europe, North America and the West Indies. In 1851, Waterford Crystal won gold medals and universal acclaim at the Great Exhibition in, appropriately, London's Crystal Palace, but that same year heavy taxes forced the business to close. ㉙ A century later, the Irish government revived the industry. Master European glass-makers were recruited to train local apprentices and, in 1951, Waterford Crystal reappeared on the world market. Now part of the international Waterford Wedgwood group, it is the world's largest producer of hand-crafted crystal, using materials, tools and techniques that have changed little since the Penrose days. ㉚

Cotton town

A number of purpose-built, industrial towns were created around Ireland in the late-18th and early-19th centuries, many of them by philanthropic Quaker industrialists. The most successful was Portlaw, where **David Malcolmson** opened a large cotton mill in 1825 at Curraghmore Park in the Marquis of Waterford's estate. The mill had three large waterwheels powered by the River Clodiagh, plus three steam engines. The flourishing business employed 1,600 people at its peak in the 1850s, and supported a further 3,000. It survived the American Civil War, which disrupted the raw cotton supply in the 1860s, and lasted into the early years of the 20th century. By contrast, a similar venture begun at Prosperous* in Co. Kildare in 1780 ended after just six years. At Portlaw, Malcolmson provided homes (including modern, slate-roofed cottages), a school, dispensary, surgeon and temperance society. The cotton was bleached on-site, and though most was sold in Ireland, some was also exported to America. When the mill closed in 1904, many workers found jobs at the family's other mills in Britain. Three Malcolmson brothers also played a part in the industrial development of Germany's Ruhr valley, when they financed William Mulvany* who opened the Ruhr's first deep coal mine in 1856.

A stone out of place

Beside Cloughlowrish Bridge is an enormous boulder, weighing tens of tonnes. It is of a lovely rock type, an Old Red Sandstone conglomerate that is 370 million years old. The rock probably formed in an ancient desert when a tremendous flash flood tossed stones and sand together, then cemented them in place. But Cloughlowrish stone lies on a very different, older igneous rock. So how did this anomaly or **erratic** come to be there?

Cloughlowrish stone, Ireland's largest glacial erratic; sketched in the 1850s for the Geological Survey of Ireland by George du Noyer.*

The surface of Ireland is littered with erratic stones of all types, and ranging in size from small pebbles to massive boulders. The one characteristic they all share is that they are conspicuously out of place and, for centuries, they were a curiosity and a puzzle. Until the mid-19th century, Catastrophists believed that a biblical flood was responsible and that erratics (and other glacial features such as eskers*) were evidence for the legend of Noah. Others thought Ireland had at some stage been submerged under what they called the Esker Sea. But, in 1837, the Swiss geologist Louis Agassiz* began propounding his theory of an Ice Age, and people eventually realised that the erratics had been deposited by glaciers and ice sheets.

Geologists can use erratics to trace the path taken by long-vanished glaciers and add to our understanding of the ice ages. Pebbles of a distinctive cream-blue granite found at Ballycotton,* Co. Cork, for example, are known to have come from the small Scottish islet of Ailsa Craig, and so must have been carried down the Irish Sea by marine ice sheets. Boulders of Connemara granite are often found in the midlands and on the Aran Islands, where they stand out from the surrounding limestone; these reveal that glaciers once flowed off the Connemara Mountains both eastward and westward. The Cloughlowrish stone was probably plucked from the Knockmealdown or Comeragh mountains to the north, its size demonstrating the awesome power of ice.

A volcanic coast

Volcanoes erupted. Hot ash was spewed high in the air. Molten lava rolled steaming into the ocean, broke up and cooled in pillow-shaped lumps. Anyone standing on the coast between Ballyvoyle and Tramore would have been killed instantly by the searing, suffocating conditions. But no one was there: this all happened 500 million years ago, at a time when life was still confined to the sea.

The fiery events were triggered by violent movements in the Earth's crust, as two ancient continents, Laurentia and Avalonia, lurched together and slowly closed off the Iapetus Ocean* (today North America and Asia squeeze the Pacific in the same way). Ireland did not yet exist: the southern half of the island was still attached to Avalonia, while the northern half was in Laurentia, along with North America and Greenland. Between these two continents was the Iapetus Ocean, dotted with volcanic islands, like the Hawaiian archipelago of today.

The remains of some of these volcanic events are preserved in Co. Waterford along this 30km roller-coaster rocky coastline. (There are other examples elsewhere in Ireland: Lambay,* for instance, an island off Co. Dublin, is the stump of a 500-million-year-old volcano.) This stretch of Waterford's coast has a wonderful variety of volcanic rocks: tuff that formed on dry land from hot volcanic ash, pillow-lavas that formed underwater, and numerous exotic

minerals. There are even basalt columns, like those of the Giant's Causeway, on sea cliffs east of Bunmahon, called the *Pipes of Baidhb*. Unlike soft Wexford next door, Waterford's hard coastline resists erosion, and makes for exhilarating cliff walks with caves, headlands and atmospheric coves.

Tramore seaside resort (*Trá Mhór*, 'big strand'), takes its name from the sandy spit that stretches 4km across the bay, almost to the far side, and impounds the broad tidal mudflat of Back Strand.

The Copper Coast

The quiet fishing village of Bunmahon was at the centre of a major copper-mining business 150 years ago. Traces of that industrial past can still be seen in the miners' cottages, the railway cuttings, the ruined engine house and chimney, and the mine shafts that bore into the cliffs between Ballydowane and Tankardstown, an area known as 'the copper coast'.

There may have been mines there thousands of years ago but, according to local historian Des Cowman, the first documented working belonged to Thomas Wyse, who mined lead and silver at Ballydowane and Danes Island from 1740–60. Mining began on an industrial scale in 1826 when the Mining Company of Ireland found a rich copper seam at Bunmahon. At its peak in 1838–45, there were 1,300 people working there, digging 8,000 tonnes of ore a year. Large waterwheels, and later steam engines, were used in crushing the ore and pumping the mines dry – many of the mine shafts and tunnels are below sea level, and the tall engine house at Tankardstown housed a steam engine that pumped water from tunnels 400 metres below. Bunmahon was profitable until 1865, but falling copper prices and the end of the lode brought the mine to a close in 1875. Many miners left to work in America, especially the large copper mines of Montana owned by Marcus Daly* from Cavan. At Bunmahon there is still just enough copper in places to stain the rock. Bunmahon's mine heritage trail and geological park are on open access. ㉛

A varied nature reserve

Great hairy willowherb, marsh St John's wort, water mint… some of the 118 plant species found on **Fenor Bog**, one of Co. Waterford's few peatland sites. Over 100 animal species have also been recorded there, from orange-tip butterflies and four-spotted chaser dragonflies, to fox and deer. Despite its name, the 13-hectare reserve is a fen,* that intermediate stage – half-water half-land – in the transition from freshwater lake through alkaline fen to acidic raised bog. Fenor Bog (*Fionnúr*, 'the sunny side of the hill'), began developing 10,000 years ago in a small lake basin at the foot of Ballymascanlon Hill. Vegetation gradually encroached and filled in the lake and, today, the area has a great variety of habitats, including swampy grassland, fen woodland and a small area of raised bog. The nature reserve is managed by the local Fenor Bog Development Group and the Irish Peatland Conservation Council. Walkways have been laid, non-native species are being removed and native wetland trees have been planted at the boundary with farmland.

Chapter 11 sources

Co. Cork

1. Chesney, Helena, "The young lady of the lichens" in Mary Mulvihill (ed.), *Stars, Shells & Bluebells* (WITS, 1997).
2. Long, Bill, *Bright Light, White Water – The Lighthouses of Ireland* (New Island Books, 1997).
3. Ludgate, Percy, "On a proposed analytical machine" *RDS Sci Proc* **12** 77–91 (1909).
4. Brück, Máire, "Agnes Mary Clerke, chronicler of astronomy" *Q.J.R. Astr Soc* **35**:59–79 (1994).
5. Norton, Trevor, *Reflections on a Summer Sea* (Century, 2001).
6. Kelleher, George, *From Gunpowder to Guided Missiles – Ireland's war industries* (Kelleher, 1993).
7. Rynne, Colin, *The Industrial Archaeology of Cork City and Environs* (Dúchas, 1999).
8. Reynolds, George, "William Bowles (1720–1780) Eurogeologist" *European Geologist* **5**: 67–70 (1997).
9. MacHale, Desmond, *George Boole: his Life and Work* (Boole Press, 1985).
10. O'Connor James P *et al.*, *Irish Indoor Insects* (TownHouse & CountryHouse Ltd, 2000).
11. Creedon, Leo, "Robert Murphy" in Ken Houston (ed.), *Creators of Mathematics: The Irish Connection* (UCD, 2000).
12. Hanly, Jane, "Stepping stones in science" in Mary Mulvihill (ed.), *Stars, Shells & Bluebells* (WITS, 1997).
13. Power, Monica, "Madame Dragonfly" in Mary Mulvihill (ed.), *Stars, Shells & Bluebells* (WITS, 1997).

Co. Waterford

14. Principe, Lawrence, *The Aspiring Adept: Robert Boyle and his alchemical quest* (Princeton University Press, 1998).
15. Cowan, Cathal & Regina Sexton, *Ireland's traditional foods* (Teagasc, 1997).
16. Dunlevy, Mairead, *Penrose Glass* (National Museum of Ireland, 1989).

Chapter 12: Wexford • Wicklow

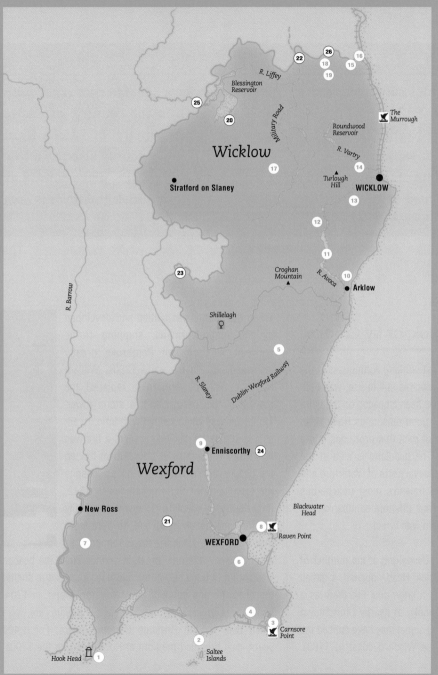

Visitor facilities

1. Hook Lighthouse
2. Kilmore Quay Maritime Museum
3. SWC Natural Heritage Tours
4. Tacumshane Windmill
5. Craanford Water Mill
6. Irish Agriculture Museum, Johnstown Castle
7. JFK Arboretum
8. Wexford Wildfowl Reserve
9. Carley's Bridge Pottery
10. Arklow Maritime Museum
11. Nick Coy Tours, Avoca
12. Avondale House
13. Kilmacurragh Arboretum
14. Mount Usher Gardens
15. Bray Heritage Centre
16. Sea Life Aquarium
17. Wicklow Mountains National Park, Glendalough
18. National Environmental Education Centre
19. Powerscourt Demesne, Enniskerry

Other places of interest

20. Ballyknockan
21. Camaross
22. Glencree
23. Liscolman
24. Oulart
25. Pollaphuca
26. The Scalp

⚲ woodland

☒ nature reserve or wildfowl sanctuary

🗼 lighthouse

Visitor facilities are numbered in order of appearance in the chapter, other places in alphabetical order. How to use this book: page 9; Directory: page 474.

Wexford has treacherous sandbanks strewn with shipwrecks; a 'great barrier coast' with shingle bars and coastal lagoons; the worst coastal erosion in the country; extensive reclaimed land at the North Slobs, now an important bird sanctuary; and Hook Head tower, one of the oldest-working lighthouses in the world. This flat county also has fertile soil, and a long tradition of farming and fruit growing. Wexford, a Viking place name, has been variously translated as West Harbour or Muddy Harbour; in Irish the town is *Loch Garman* ('the pool at the mouth of the Slaney').

HOOK HEAD

"By Hook, or by Crooke"

Hook Head has one of the oldest-working lighthouses in the world. The mediaeval tower, which is still in use, was built in the early 1200s, but there has been a beacon on the point since the 5th century, and for good reason, as Hook Head is at the end of a long finger of rock that protrudes into a busy Atlantic shipping lane. The headland is also at the entrance to the important mediaeval ports of Waterford and New Ross and this is one of Ireland's most treacherous coastlines – thanks to numerous offshore sandbanks, the sea off south Wexford is strewn with wrecks.

A 5th-century monk, Dubhán, despairing at the number of sailors who drowned off Hook Head, opened a small mortuary and hospital, and lit a large coal fire there as a beacon. For 600 years, the monks of nearby Churchtown tended the light (ironically, their signal may have enticed the Anglo-Normans to land on the Wexford coast in 1169).

When the county was settled and developed by the Anglo-Normans, shipping traffic between Wexford and Pembroke greatly increased and, in the early 1200s, William Marshall, Earl of Pembroke, replaced the monks' simple beacon with a stone tower. The tower was whitewashed with lime, to act as a landmark by day, and at night a fire was lit on top. The monks at Churchtown tended the beacon and continued to do so, despite the dissolution of the monasteries in 1542, until the Rebellion of 1641, when they finally left the area. Hook Head had no light from then until 1667, when the tower reopened as a lighthouse – the only break in service since the beacon was begun. A brick half-dome and flue were built to shelter the open fire and conduct away the smoke and, in 1704, lighthouses, which until then had been in private care, were taken over by the Revenue Commissioners. For 1,400 years, coal provided the light at Hook Head. In the 1790s,

The mediaeval tower at Hook Head, one of the oldest-working lighthouses in the world.
© Commissioners of Irish Lights

"For the Safety of All"

The 19th century was the golden age of Irish lighthouse building and engineering. There had been beacons at several coastal sites for centuries but, in the early 1800s, as the sea lanes became increasingly crowded, a major programme was begun to erect lighthouses at strategic points around the coast. This was a time when 'wreckers' patrolled the coastline, when navigation was an art and not a science, and when there was little or no lifeboat service.

Lighthouses call for special design and engineering requirements if they are to be effective – they need to be visible at a distance, and able to withstand the elements. Most were built with thick stone walls, but this meant massive amounts of building materials had to be carted to remote and inaccessible places. There were no cranes or hydraulic lifting devices, and horses could not be used if the site was offshore or near cliffs. Work was frequently halted by the weather, storms and tides, and some lighthouses took many years to complete (the Fastnet Rock tower* took seven years). Onshore towers were relatively easy to design and build, but offshore towers, especially those on sandbanks and remote rocky islets, posed special challenges. Not surprisingly, lighthouse work attracted some of the great engineers of the day, many of them Scottish. Among the engineers who worked on Irish lighthouses were Alexander Nimmo,* Alexander Mitchell,* John Rennie,* John Traill,* and the father and son team of George Halpin senior and junior. It was Mitchell, a blind Irish engineer, who invented screw piles for securing lighthouses on sandbanks.

Maintaining a beacon was another major technological challenge. The earliest lights were coal fires or candles, but these consumed vast amounts of fuel, and were usually extinguished by bad weather when, ironically, they were most needed. Glass panes to shield the lights from the elements were not readily available until at least the 1700s. Throughout the 19th and 20th centuries, lighthouse engineers experimented with various fuel options. John Wigham,* working at several Irish lighthouses, pioneered the use of gas lights and his innovations were later used worldwide. Several Irish scientists also served as consultants to the Irish and British lighthouse services, notably John Tyndall,* Robert Ball* and John Joly.*

Despite the development of sophisticated global positioning satellites, lighthouses remain vital navigational aids. Each has a unique flashing light sequence to identify it by night, and distinctive features to identify it by day, and they are used as position checks, to mark safe channels, and to warn against hazards. Ireland's modern coastal protection network also includes fog signals, and navigational buoys marking navigable channels and submerged hazards. Buoy lights were traditionally powered by acetylene gas, but today are solar-powered. All Irish lighthouses and lightships are now automated, maintained by the Commissioners of Irish Lights – whose motto is "**In Salutem Omnium**", "For the Safety of All" – and part-funded from levies on commercial shipping at Irish ports.

an oil lamp replaced the brazier and, in 1871, after successful experiments with coal-gas at the Baily* light in Howth, Co. Dublin, Hook Head switched to gas, and a coal-gas plant was built by the tower. In 1911, the fuel was again changed to paraffin vapour and finally, in 1972, to electricity.

The mediaeval tower, essentially three vaulted chambers stacked on top of each other, has been enlarged over the centuries. At one point, an outer wall was built around the original tower, and a winding stairs installed between the two walls. A fog signal was added in 1838: a simple bell at first, then a gun and later explosive charges, and now an electric horn which is triggered automatically by a fog detector. The light was automated in 1996, and the historic tower is now open to the public. ① Nearby is a lime kiln,* where local limestone was burned to make whitewash to paint the tower. The headland is good for bird-watching, and there are blow holes, and raised or fossil beaches* nearby. The long promontory of Hook Head is flat and relatively featureless and windswept and treeless, yet it is internationally noted for its geology and birdlife, and boasts several natural heritage areas. These include a salt marsh at Fethard, which is home to several rare plants, and the sea cliffs at Baginbun Head. Inland at Tintern Abbey is an important bat colony. Two townlands on the coast named Saltmills are where the monks from Tintern Abbey had tidal, or salt, mills.* The phrase "by Hook, or by Crooke", reputedly coined by Oliver Cromwell, refers to the approaches to Waterford Harbour, via Hook Head or the village of Crooke across the estuary.

Lightships and compasses

Sandbanks are a major shipping hazard, but it can be difficult to build a lighthouse on them. So, for much of the 19th and 20th centuries, lightships were used, tethered to the spot by three strong anchors. Among them was *Guillemot II*, which marked the sandbanks off Co. Wicklow from 1922–1968. A crew of seven lived on board in relatively cramped conditions, and maintained the paraffin lantern. Their place has been taken by an illuminated buoy, and the retired lightship – now moored in concrete beside Kilmore Harbour – has been preserved with its original fittings as a maritime museum. ② Five kilometres offshore are the **Saltee Islands**. Now an important bird sanctuary, they were linked to Forlorn Point when sea levels were lower.

Kehoe's pub in Kilmore village also has a collection of marine memorabilia, including a fine example of a **dry-card compass** invented in 1880 by William Thomson (Lord Kelvin).* Thomson, who made a fortune from instruments he invented for the transatlantic tele-graph cable,* bought a yacht in 1870 and became interested in navigation. He invented several nautical and nav-igational devices, including an improved dry-card compass for use on the new iron-clad ships. Instead of a single needle, it had eight thin wires fixed with fine thread to a lightweight card; the card rested on an aluminium point, and was not affected by the iron ship's magnetic field. It was adopted by the British navy, and used until 1900 and the development of liquid compasses that did not leak.

William Thomson's ingenious compass could be used even on iron ships.

The great barrier coast

Between Carnsore Point and Cullenstown lies Ireland's 'great barrier coast'. The barrier is a substantial bank of sand and gravel, stretching for 22km, and protecting three coastal lagoons that are internationally important for their rare plants and birdlife. The barrier is bisected by Forlorn Point, creating two separate stretches: Ballyteige Burrow which runs for 9km west of the point, and Grogan Burrow, which runs east of the point for 13km. Of the three lagoons or lakes, only Lady's Island Lake is a true lagoon; Tacumshin Lake was tidal until 1972 when the barrier there

closed naturally, but a one-way overflow pipeline was installed, so it still has an outlet to the sea; the Cull behind Ballyteige Burrow is open to the sea at Cullenstown, and is still a tidal marsh, rather than a classic lagoon. The barriers consist of sand and gravel that was deposited offshore by glaciers during the last Ice Age, when this part of Ireland was a vast meltwater delta. The barriers themselves probably formed offshore much later, perhaps 5,000 years ago, and have since been gradually rolled over by storms, shifting landwards at rates that averaged, depending on conditions, between 5 and 50cm a year.

Lady's Island Lake is brackish, a mix of fresh and salt water: a dozen small freshwater streams flow into the lake, while the salt water comes as wind-blown aerosol, or when high waves wash over the barrier.[1] The lake's only outlets are evaporation and natural seepage across the gravel barrier. Occasionally, to relieve flooding, a local lake management committee will arrange for a trench to be dug through the barrier; usually this 'cut' heals naturally after a few days of tidal action. There are three islands in the lake: Our Lady's Island, a pilgrimage site for 900 years, is connected to the mainland by a causeway; and Inish and Sgarbheen islands are home to an internationally important tern colony, including roseate terns, which are Europe's rarest seabird. By 1989, the number of terns there had dropped to 160, mostly due to predation from local cats and rats. A management and conservation programme was begun, with 24-hour warden coverage during the breeding season, and numbers have risen to 4,000. Terns are ground nesting, so the islands are off limits, but the colony can be watched from the shore. In winter, the lake and the surrounding marshland and reed swamp are noted for wildfowl. Peat beneath the sand at nearby St Margaret's beach was dug for fuel in 1941 during the Emergency (1940–47). Coastal peat deposits like this are evidence of lower sea levels in prehistoric times. Europe's largest stand of cottonweed grows on top of the shingle barrier there, the only place in Ireland and Britain where this plant grows.

Ballyteige Burrow, the stretch of barrier on the other side of Forlorn Point, is a substantial shingle spit topped by a fine sand-dune system. As its name suggests, it was once a rabbit* farm: the monks from nearby Tintern Abbey kept a warren there that reputedly produced 4,000 rabbits a year, an important meat and fur source. Ballyteige Burrow is a nature reserve, noted for rare plants including wild asparagus (*Asparagus officinalis*) – it is Europe's main site for this plant – and henbane. It is also known for invertebrates, especially beetles, wasps and ants, particularly the turf ant, which is otherwise uncommon in Ireland. The site's diversity is due, in part, to cattle grazing, which helps control the plant cover. Behind the barrier is **the Cull**, the remains of a large estuary that was reclaimed in the late 1800s. It is still tidal, with extensive salt marshes and areas that are partially a lagoon. ③

TACUMSHANE

A smuggler's mill?

Wexford's climate and soil are perfect for tillage and fruit growing. This is truly 'the sunny southeast' – the only part of Ireland that receives over 4.5 hours of sunshine on average every day – and, for centuries, the county was at the heart of the Irish grain belt. Not surprisingly, there were numerous cornmills, both water-powered and wind-driven.

The windmills of the south Wexford coast, however, unusually for agricultural buildings, were closely connected with the sea. Much of the wood used to build them was driftwood, or salvaged from shipwrecks, and the sail arms were often used to signal to smugglers (if customs officials were in the vicinity, the arms would be stopped in the upright position; otherwise they were stopped in the normal X position). According to Dr Austin O'Sullivan, curator of the agricultural museum at Johnstown Castle,* the contraband – goods such as brandy, wine and tobacco – might also be stashed in the mills, until it could be sent on in sacks of corn.

The thatched windmill preserved at Tacumshane is one such mill. Built in 1846 by millwright Nicholas Moran, who trained in Rotterdam, it operated until 1936, and has been preserved relatively intact. ④ In 1890, there were a dozen similar windmills in the area, but only Tacumshane survives. The thatch cap can revolve, turning the arms into the wind, and each sail was set with linen sailcloth, according to the strength of the wind. A unique, 17th-century, domestic watermill is preserved at **Craanford**, where it has been restored to full working order. An overshot timber wheel, 4 metres in diameter and with 48 buckets, is built into the side of a farm building, where it drives three grinding stones. ⑤

CARNSORE POINT

When Ireland and Britain were one

Fifteen thousand years ago you could probably have walked from Carnsore Point to Cornwall. Sea levels were at their lowest then (over 60 metres lower than today according to some estimates), and **land bridges** are thought to have connected Ireland and Britain at several points – indeed, most of the Irish Sea may have been dry land. Similarly, Britain was linked to northern Europe, and Alaska to Siberia. The land bridge between Carnsore Point and Cornwall may have been due in part to a rise in land level, as to a fall in sea level: the late Prof Frank Mitchell has suggested that, during the last Ice Age, the weight of the ice in the north pushed up a "fore bulge" of land in the south; as the ice melted and retreated, the bulge would have migrated north after it, before gradually disappearing.[2]

Many of the plants and animals that recolonised Ireland after the Ice Age probably came across these land bridges, spreading out of Europe at about 0.5km a year. The very south of Ireland had remained free of ice, however, so some of the recolonisers were probably here to begin with. By 7,000 years ago, the land bridges between Ireland and Britain had disappeared under the rising sea levels, Ireland became an island, and the main immigration route into the country closed. The bridge connecting Britain and Europe remained open for a further 2,500 years, however, and, consequently, Britain has more plant and animal species than Ireland – 1,172 native plant species, as against 815 in Ireland, and 32 native land mammals, as against 14 in Ireland. New species still arrive in Ireland from time to time, though most are introduced by people, whether accidentally or deliberately, such as the bank vole* and the rabbit;* others reach Ireland independently, such as the magpie,* which was blown onto the Wexford coast during a storm around 1680.

Carnsore Point is strewn with large glacial erratics* and, for years, this boulder field was a handy source of ready-made boulders, some of which were used in building nearby piers. The boulders create their own ecosystem: the dry micro-climate around each boulder sustains a complex community of ants, and the warmth from the ant hills sustains heat-loving plants. Boulders of Carnsore pink granite were also dumped by the glaciers on the Saltee Islands. In the 1970s, the ESB proposed building a nuclear power station at Carnsore Point. Today it has plans for a wind farm at the site.

JOHNSTOWN CASTLE

1169 – invasion of the haymakers

The **history of farming** in Ireland is partly a history of invasions. Cattle, for example, in the shape of the Kerry cow,* were probably brought in by the Celts 2,000 years ago. Butter-making* is thought to have come from Roman-Britain. The Anglo-Normans introduced haymaking, rabbits* and dovecotes,* all intended to provide meat, especially in winter. And the potato* arrived in the late 1500s, possibly with English planters or Spanish traders.

The major challenge throughout the history of farming, in Ireland and elsewhere in northern Europe, is providing food

A wheelwright's workshop, recreated at the agricultural museum in Johnstown Castle. © Irish Agricultural Museum

during the lean winter months. By the 8th century, European farmers had learned to make hay, which enabled them to keep most of their cattle over winter, thus ensuring a continued supply of fresh meat and dairy produce. Some historians argue that haymaking was one of the great achievements of the last 2,000 years, facilitating a tremendous growth in Europe's population, which was cut short only by the Black Death* in the 14th century. The Irish, however, had little need of hay: the mild winters meant that most years cattle could be grazed on reserved pastures (in summer, rough hill pastures were used); and, in any case, the summers were often too wet for haymaking. But the Irish system broke down in severe winters, and then people were forced to slaughter their milch cows, a drastic act that left them with neither cow nor calf, neither milk nor meat. The Anglo-Norman technique of haymaking helped prevent a return to those hard times. (The final breakthrough in the challenge to keep animals fed over winter, was the 18th-century development of the Swede turnip, which could be stored and used as winter fodder.) The haymaking Anglo-Normans turned Co. Wexford into an important tillage centre. Their legacy can be seen in the county's regular field pattern, its many windmills, and its long tradition of agricultural engineering, examples of which can be seen at the Agricultural Museum in Johnstown Castle.

The FitzGerald family gave the castle to the State in 1946, on condition that it be used as an agricultural facility. The national soil-testing laboratory was installed there initially. **Soil** is of fundamental importance to farming – an iodine deficiency in soil, for instance, can have a major effect on animal health – and, in the mid-1800s, Richard Griffith's* land survey, used to determine rateable valuations, took account of soil type. In the 1940s, when the Department of Agriculture began a major farm improvement drive, soil testing was a key part of the programme. In 1959, with the

1845 – potato blight arrives in Ireland

On June 24th, 1845, a new disease was noticed on potatoes growing in Flanders. Within two weeks, the blight was in northern France. Its rapid spread and destructive nature caused anxiety, and its progress was monitored across Europe. By mid-July, the blight had crossed the English Channel, and on August 20th, David Moore, curator of Dublin's Botanic Gardens,* ominously noted its arrival in Ireland. There was mounting concern about its impact here and, in September, the **London Gardeners' Chronicle**, in a special editorial, asked: "Where will Ireland be in the event of a universal potato rot?" By mid-October the potato crop across most of Ireland was blighted, and the disease had spread to Scotland, Scandinavia, Poland and even northern Spain.

The new disease was surprising because, until then, Europe's potato* crop had been relatively trouble-free. In the 300 years since the plant's introduction from South America, there had been occasional poor harvests caused by bad weather, but little disease to report. Then, in the 1830s, a new disease called wet rot affected potatoes in Europe for the first time. It was not a major problem, but Belgian potato growers decided to breed resistant varieties. To this end, in 1843, they imported several varieties from North America, but they got more than they paid for: potato blight had been noticed on the east coast of North America in 1842, probably having come from Mexico, and scientists now believe it reached Europe as part of that Flanders shipment. Faster crossings, thanks to the new steamships, made it easier for a disease to survive the transatlantic journey.

In 1845, as the spread of blight caused public consternation, a heated, often vitriolic debate was waged about its cause. All manner of explanations were invoked, mostly the wet summer, but also lightning, volcanoes, fumes from the newly invented sulphur matches, guano fertiliser, newfangled electricity, fairies, degeneration caused by inbreeding, even divine retribution on the lazy Irish peasants. Some scientists suggested that a fungus, which could be seen growing on blighted plants, caused the disease. But the 'germ theory of disease' had not yet been established, and most scientists believed that the fungus was merely taking advantage of an already ailing plant, and that it was a consequence of disease and not a cause. As the disease took its toll on Europe's potato crop, the scientific debate raged on in learned journals, newspapers, gardening magazines and public inquiries. By the end of 1845, it was generally agreed that bad weather was the problem and those who believed in a fungal cause were silenced, at least for a while.

In Ireland, the British government appointed a scientific commission of three wise men to investigate the disease, and recommend how best to preserve the crop and procure seed for 1846. All three – two English experts, Dr John Lindley and Dr John Playfair, and a noted Irish chemist Sir Robert Kane* –

believed the problem was caused by the unusually cool, wet summer. Writing in 1945, for the 100th anniversary of the Great Famine, the late historian T P O'Neill suggested that, if even one advocate of the fungal theory had been appointed to the commission, it might have been more effective. As it was, the commission recommended storing harvested potatoes in ventilated pits, which may have made things worse by allowing wind to disperse the fungal spores.

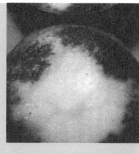

Blighted potato tubers; the fungal infection also damages the leaves.
© Teagasc

In a way, those who said the problem was the weather were right, if for the wrong reasons. The late Dr Austin Bourke, a former director of Ireland's Meteorological Office,* studied weather records for 1845. He found that there were frequent easterly winds that summer, which could have carried the fungal spores from the continent; moreover, the weather was wet, which would have favoured the blight's spread. It was 1863, however, before the fungal cause was firmly established, and the 1880s before an effective treatment for blight was discovered – the 'Bordeaux spray' made from copper sulphate or bluestone. Ironically, in the 1840s some people came tantalisingly close to finding an effective treatment: they tried soaking seed potatoes in bluestone solution, which was already being used successfully with cereals to prevent rust, another fungal disease; unfortunately, it is not enough to soak the seed potatoes – the mature plants must also be sprayed.

Potato blight is caused by a fungus (*Phytophthora infestans*) and, in Ireland, this microscopic organism triggered the Great Famine. The worst affected areas were those most dependent on the potato, especially the susceptible Lumper variety. The Famine changed the course of Irish history, and altered forever the face of Irish farming, society and landscape. There was a move to a more modern form of agrarianism, with associated land clearances. Reform programmes sought to develop the country's infrastructure and fishing industry, to introduce new plant and animal varieties, and to encourage industries such as lace- and carpet-making. Potato blight remains one of the most important crop diseases in the world, but despite over a century's research, the versatility of the fungus is such that no fully resistant potato variety has yet been developed. Hence most potato growers spray their crop if the weather is likely to favour the spread of the fungus. Weather forecasters around the world use a system devised in Ireland by the late Dr Bourke to predict when to spray. A major exhibition about the history of the potato and blight, prepared in 1995 for the 150th anniversary of the Great Famine, is at the Agricultural Museum in Johnstown Castle.

establishment of the Agricultural Research Institute (An Foras Talúntais, now Teagasc), Johnstown Castle became an agricultural research centre. The headquarters of the Environmental Protection Agency (EPA) are now also located at Johnstown.

The agricultural museum in the castle courtyard has a major exhibition devoted to potato and blight (see: **Potato blight arrives in Ireland**) and an extensive farming collection of tools and machinery, much of it made by the renowned Wexford foundries; also domestic items, and country furniture. ⑥ The oldest item – a horse-powered butter churn – dates from 1860, and the exhibition traces the increasing mechanisation of Irish farming. The curator, Dr Austin O'Sullivan, also collects early laboratory and computing equipment, and has one of the first computers used in Ireland. An analog computer made in the 1960s by an innovative Dundalk company, Queleq Corporation, it was used by an animal feed firm to compute the optimum ration mix. The machine, "as big as an office and looking like a space station", was in use until the 1980s. Queleq computers were world class, but the pioneering Irish company did not survive the transition to digital technology.

Pingos in the fields

The fields around Camaross village are strewn with fossil pingos. These small doughnut-shaped features in the landscape, named from an Eskimo word, date from a time during the last Ice Age when the ground around Camaross was permafrost. Pingos form when water just below the surface freezes to form a lens of ice in the soil. As the lens grows it can bulge up, to produce a large soil-covered pimple or mound, perhaps 10 metres high, with an icy core. When the climate starts to warm, the surface of the mound thaws first, and the soil slumps down to form a rim; when the core melts it leaves a depression in the centre, and the result is a small crater perhaps 30–50 metres across. Today, pingos occur in northern Canada, Scandinavia and

Russia, but 15,000 years ago, thousands of them formed around Ireland, and their fossil remains can still be seen in places. The late Prof Frank Mitchell mapped several hundred in the fields around Camaross.[3] Some are circular, and a few resemble dumbbells, but many are crescent-shaped, either because they formed like that, or because they were subsequently damaged by land reclamation. They occur on poorly drained soil, and often contain a small pond; some were used as wells or watering holes. Fossil pingos are of scientific interest, a clue to conditions in the area during the Ice Age. (The fossil pingos are on private farmland; groups wanting to see them should seek permission from the landowner.) Other relics in Co. Wexford from that time include kettle moraines at Curracloe* and kettle holes at Blackwater.*

When the first bud forms

Each year the gardeners at John F Kennedy Arboretum help to monitor climate change, by recording the dates when plants in their **phenological garden** first produce buds, flowers and fruit, and shed their leaves. It is one of four such gardens in Ireland – the others are at Valentia,* Johnstown Castle* and the National Botanic Gardens, Dublin.* All their records are collected by Met Éireann,* the meteorological service, and sent to Berlin University, which has been collating similar records from across Europe since 1990. JFK Arboretum opened in 1968 in honour of the former US president, whose family came from nearby Dunganstown. It spans 250 hectares and is both a botanic garden and research station. The garden has 4,500 species of tree and shrub, both broadleaf and conifer. There are also 200 forestry plots, organised geographically by continent, and a native woodland section. The garden is growing elm trees as part of an EU comparative trial, and conducting 'provenance trials' of commercial conifers, such as Sitka spruce, with seed sourced across the species' range from California to Alaska. ⑦

Salmon cots and snap nets – working heritage

The salmon fishermen who work on the Slaney and Barrow estuaries use boats and techniques that date back thousands of years. The two main fishing techniques on these rivers are snap netting and draft-net fishing, and both are done from salmon cots, traditional boats that are unique to the area. In draft-net fishing, one person stands on shore holding an end of the net; the rest of the net is paid out from the back of the boat as it moves away to bring the net into an arc. After a while, the net is brought ashore upriver of the starting point, to encircle any catch. In snap netting, two boats drift together on the tide with a net stretched between them, fishing out on the ebb tide and back on the flood. This demanding technique calls for skill in maintaining the correct distance and speed. Unknown outside of Ireland, it was once common on the Barrow, Nore and Suir rivers, but is now little used, and the current generation could be the last to practise this ancient technique. Boat historian Darina Tully describes the Slaney and Barrow cots as "one step up from a dugout canoe". They have low sides, a flat bottom, are at most 4 metres long, with a pointed bow and a narrow transom, and ideal for estuary conditions – at low tide they can be launched like a toboggan from the mudflats, and easily recovered by winch. Some are punted with a pole, others are rowed with long-bladed oars; increasingly, the cots carry an outboard motor.

Speed the plough

Ulster's linen mills, with their enormous steam engines and massive machines, required heavy engineering backup. This gave rise to a strong engineering sector there, and generated numerous shipbuilding yards and world-class companies such as Harland & Wolff.* By contrast, at the other end of the country in Wexford, the mills were mostly cornmills, which required a light-engineering support industry. But, by the 1850s, this had spawned a thriving farm machinery industry, and at least one world-class player, **Pierce & Company**. For over 100 years, Pierce's was Ireland's largest farm machinery manufacturer, with subsidiaries in Europe and Argentina, and exports to every continent. Most of the farm machinery firms were located around Enniscorthy and Wexford towns, where there was a long tradition of ironworks and foundries. For centuries, these had smelted iron ore using the oak woods of Shillelagh* for charcoal, and the Slaney river for water power.

A Pierce horse-drawn mower and reaper: one person drove the horses, another operated a foot pedal to control the rake. When a sheaf's worth of corn was cut, it was pushed off the rake, to be bound by people following the reaper (such 'mower and reapers' predate 'reaper and binders'). This illustration is from an early-20th century catalogue; note the cast-iron wheels, and absence of tyres. Irish Agriculture Museum

Pierce & Company was begun by a Wexford millwright and inventor, **James Pierce** (1813–68), who was famous for his hearth fan. That invention replaced the bellows and, by the late 1800s, most Irish houses had a Pierce Fire Machine. Pierce had considered emigrating to Argentina, but then he is said to have seen imported farm machinery waiting at Wexford docks, and realised that he could make it locally. By 1847, he had set up a company in Wexford town, making a wide range of machines and implements, including threshers, root cutters, ploughs and churns, all to his own designs. By 1914, the company, still owned by the Pierce family, was employing 1,000 people, and exporting to Africa, North and South America, Europe and Australia, with subsidiary offices in Paris and Buenos Aires. The company lost most of its international business during World War I, though it remained an important player in the Irish market, developing a range of accessories for use with tractors, until it closed in 1980.

One of Pierce's main competitors was **Wexford Engineering Company**, known for its Star and Shamrock brands. It was started in 1881 at Cappoquin, Co. Waterford, by Sir Richard Keane. By 1890, it had a substantial export business and moved to Wexford Harbour. The company exported to Europe, Africa and North America and survived until 1964. The factory was bought by the Smith Group from Cavan, which bought Pierce & Company at the same time, and, for a while, served as a car-assembly plant. Another well-known Wexford farm company was **Selskar Ironworks**, begun in 1870 by William and Andy Doyle, who had trained with Pierce's. It made a range of equipment including hay rakes, ploughs and diggers, until it closed in the 1930s. Examples of Wexford-made farm machinery can be seen at the Agricultural Museum, Johnstown Castle* and most Irish local museums.

WEXFORD TOWN

1853 – the Northwest Passage at last

Ever since Europeans 'discovered' America in 1492, they have been trying to get around it. In 1497, Henry VII of England sent what would be the first of many expeditions seeking a Northwest Passage around America to the riches of the Orient. Navigators also sought a Northeast Passage above Russia, but it was the Northwest Passage that captured most attention – and lives. Even today, with the best of ice-breaking technology, the route is a severe challenge, though, if global warming melts the sea ice sufficiently, it might at last become more commercially attractive. The route lies 800km inside the Arctic Circle, and stretches for 1,500km through a tangled archipelago. Just reaching the start of the passage entails a hazardous voyage through water strewn with giant icebergs. Numerous expeditions set out in search of a passage, but all failed, many of them ending in disaster. The most famous was the Franklin expedition,* which left London in 1845 with 129 men, never to return. It was while searching for news of the Franklin expedition, that Wexford man

Southeast passage: the Dublin–Wexford rail line is Ireland's most scenic, taking in the Wicklow coast, Avondale* and the Vale of Avoca,* before emerging from the Wicklow Mountains to follow the Slaney into Wexford town.

Robert McClure (1807–73) finally discovered the Northwest Passage.

McClure had joined the British navy in 1824 and, 12 years later, served on his first Arctic expedition. In 1850, he was given command of *Investigator*, one of two ships sent to search for the Franklin expedition. McClure's party started north of Alaska at the Bering Strait, where it was hoped Franklin's team would be found. Over the next three years, *Investigator* sailed up Prince of Wales Strait and Melville Sound, and finally discovered McClure Sound, the last stage in the much sought northwest route. After three winters, McClure's ship was trapped in ice north of Banks Island, and had to be abandoned. The party sledged east, before being rescued. McClure returned to London in 1854 and, despite being censured for returning without his ship, was knighted. He later served as an admiral in 'Chinese waters'.

Wexford-born Robert McClure, who discovered the Northwest Passage between the Atlantic and Pacific oceans.

ARDCAVAN

Ireland's nether lands

The **North** and **South Slobs** are polder lands (slob from *slab*, 'muddy' or 'soft'). Flat and relatively featureless, this artificial landscape is more usually associated with the Netherlands, or Norfolk in England. A high embankment keeps the sea at bay, and the fields are separated by drainage channels rather than walls or hedgerows, and maintained only by continuous pumping. Ireland has several reclaimed polder lands – at Lough Foyle,* for example, which is used for farming, and the North and South Lotts beside Dublin's docklands, which are residential – but the most extensive are the Wexford Slobs, where 2,000 hectares were reclaimed from the sea.

The project was begun in 1846 by the Wexford Harbour Improvement Scheme, and promoted by the ubiquitous William Dargan.* The aim was to enclose the slobs (tidal mudflats) on the north and south side of Wexford Harbour, improve the deepwater approach to the port, and corset the Slaney river. The embankments were begun in 1849 and took 10 years to complete, though, by 1851, farmers were growing crops on part of the North Slobs. The reclaimed ground was 3 metres below sea level, so pumping stations with steam-powered pumps were built at Drinagh in the south and Ardcavan in the north. Initially, the new land was used for rough summer grazing and a little tillage, but the farming intensified when more efficient pumps were installed in the 1960s. The steam pumps have long since been replaced by electric ones, but Ardcavan's old steam equipment has been preserved.

The North Slobs are Ireland's premier **wildfowl site**, and over 240 species of bird have been recorded there. A century ago, the area was popular with fowlers, who hunted from flat-bottomed punts and with long-barrelled punt guns. When a hunting party there failed to hit any golden plover in 1951, it prompted an after-hunt discussion about what was Europe's fastest game bird, a discussion that led soon after to the founding of the *Guinness Book of Records*. (The fastest game bird, according to the *GBR*, is the Canadian long-tailed duck, which can fly at 115kmph.) In 1981, part of the North Slobs was set aside as a national nature reserve, and hunting is now prohibited there. The reserve's location and nature make it particularly attractive to migrating birds, and geese first arrived there in the winter of 1898. Now, the slobs provide sanctuary and winter grazing to 10,000 Greenland white-fronted geese, one-third of the world's population of this species, making the reserve internationally important. Tens of thousands of other birds also congregate there, including six other species of goose, as well as swans, ducks and waders. The best time to visit is winter, when the air is thick with wheeling flocks of geese. ⑧

One for sorrow, two for joy?

Many uninvited guests have landed at Wexford over the centuries, but one of the more unusual arrivals was a party of **magpies** reputedly blown ashore during a storm around 1680. Colonel Solomon Richards of Wexford town recalled in 1682 how: "There came with a black easterly wind, a flight of magpies, under a dozen as I remember, out of England or Wales, none having been seen in Ireland before. They lighted in the Barony of Forth, where they have bred and are so increased that they are in every wood and village in the County… The natural Irish much detest them, saying 'they shall never be rid of the English while these magpies remain'." Magpies are now common across the country. Two later arrivals are the **fulmar** and **collared dove**, both now also common. The fulmar, a seabird with a characteristic stiff-winged, soaring flight, was first spotted in 1911 at Portnacloy, Co. Mayo; the birds may have come by following trawlers from the Faroes. The collared dove, first spotted at Galway and Dublin in 1959, originated in Asia, reaching Turkey in the 1500s, and then crossing Europe at a rate of 10km a year.

The North Slobs and the nature reserve at the nearby Raven are also a sanctuary for the **Irish hare** (*Lepus timidus hibernicus*). This russet-coloured animal has short ears and a white tail, and is unique to Ireland. It is a subspecies of the white arctic hare (*L. timidus*) but, though the Irish hare

The Irish hare finds sanctuary at the Wexford Slobs and the Raven.
© An Post

moults twice a year, its coat seldom turns white in winter. There have been hares in Ireland since tundra conditions prevailed here at the end of the last Ice Age. Hare bones found at Shandon Cave* in Co. Waterford have been dated to 28,000 years ago, and our forebears dined on hare 9,000 years ago at Mount Sandel in Co. Derry. In the 1870s, a second species, the European brown hare (*L. europaeus*), was introduced as game into some northern counties, where small numbers of them are still found. It has a speckled coat and very long ears. Ironically, it was probably the long-eared brown hare that featured on the Irish pre-decimal threepenny bit, and not the native species.

The Irish hare is found throughout Ireland, on farmland, on sand dunes and on bog, and from mountain summit to sea level. Numbers have been declining since 1980, however, probably because of habitat loss and because, like the corncrake,* hares make their nest, or form, on the ground, where they are vulnerable to silage-making equipment. Unusually, hares are both a protected wildlife species, and a designated game species that can be hunted under special licence. Hunting is prohibited, however, in the Wexford Wildfowl Reserve and the Raven, where their numbers reach 40 per hectare, compared with six per hectare on farmland.

The Raven nature reserve, at the eastern end of the North Slobs, has some of the best sand dunes on the east coast. Originally a long sand spit, the Raven was incorporated into the reclaimed slobs when the embankment was built in the 1850s. It formed from sand and sediment generated by coastal erosion to the north, notably at Blackwater Head,* and deposited at the Raven by the currents. The southern tip is still active, and the sandbanks and lagoons there are constantly shifting and reforming. Much of the Raven was afforested in the 1930s, but it retains a rich flora and fauna. Several rare plants grow there, notably round-leaved wintergreen, lesser centaury and green-flowered helleborine. The area is also noted for caterpillars, moths and butterflies, the seals which frequent the southern tip, and birdlife – in summer little terns nest on the sandbanks; in winter the banks are used as a night-roost by the thousands of geese grazing on the slobs.

The natterjack toad* was successfully introduced from Kerry in 1991 as part of a conservation programme, and is now found at the Raven alongside the common frog, common newt and **common**, or **viviparous**, **lizard**. The lizard (*Lacerta vivipara*) is Ireland's only native reptile. Shy and seldom seen, It is yellow-brown in colour, and at most 20cm long. When it is to be seen, it is usually basking in the sun: the lizard is cold-blooded and, before it can hunt insects, needs to warm its body to 30°C (in winter it hibernates). It lives in burrows in bogs and sand dunes, and the young are born live (hence viviparous), in litters of up to a dozen.

The viviparous lizard, Ireland's only native reptile. The Raven is one of only two places in Ireland where you can find the common frog, common newt, natterjack toad and lizard.
© An Post

ÉIRE 32
Common Lizard *Lacerta vivipara*

Macamore and redmore and kettle moraines

Wexford has some of the best soils in Ireland – fertile, and neither too wet nor too acidic, they are perfect for farming and, in places, for pottery. Most Irish **clays** are a legacy of the last Ice Age: glaciers swept the soil off the hillsides and deposited it in lowland areas, along with massive amounts of sand and gravel. Our soils formed from that debris, or from rock that has weathered in the interim. Wexford clays are unusual, however, as they also contain shelly material and marine mud taken from the sea bed. During the last Ice Age, glaciers flowing down the channel now occupied by the Irish Sea, were kept at bay by the Wicklow Mountains; but Co. Wexford is flat and low-lying and the ice could sweep inland there. As the ice came, it brought shelly mud dredged from the then-dry seabed. This lime-rich, marine mud yielded an unusual soil, **macamore**, which is ideal for farming. Where the top layer weathers it forms a lime-free clay known as **redmore**, which is perfect for pottery and brickwork, and is currently the only available Irish pottery clay. Ireland's oldest pottery was founded near Enniscorthy in 1654 by the Carley family who came from Cornwall in search of pottery clay. The company is still run by their descendants, making handmade pots from the local redmore. ⑨

The glaciers were also responsible for the unusual, hummocky terrain at Curracloe. This undulating landscape, with small sandy mounds and numerous water-filled depressions, is a **kettle moraine** and it has a distinctive appearance on the 1:50,000 OSI map. It is a thick layer of sand and gravel that was dumped by a retreating glacier. The depressions are kettle holes that formed where a huge block of ice, which had been trapped in the gravel, melted. (Not to be confused with a fossil pingo,* which is a depression surrounded by an earthen rim.)

The thick blanket of glacial debris covering Co. Wexford is very amenable to being reworked by the forces of nature, and makes Wexford vulnerable to **coastal erosion**. The most dramatic coastal erosion in Ireland is at **Blackwater Head**, where the cliff is crumbling into the sea at a rate of about 5 metres a year. Abandoned houses teeter on the edge, and the remains of others can be seen below. As part of the give and take of coastal erosion, the material that is removed is deposited further south at the Raven.* A different type of coastal erosion can be seen on Wexford's southern coast at **Bannow Island**: encroaching dunes smothered Bannow village in mediaeval times, and all that remains is a ruined church on the hill. As a result, Bannow was sometimes called the 'Herculaneum of Ireland', a reference to the Roman town that, with Pompeii, was buried under volcanic debris in AD 79. Boulders were dumped at the base of Blackwater Head in 1988; these absorb some of the sea's energy and have slowed the rate of erosion, but have not stopped it. Marram grass and hay bales are used to stabilise the dunes at Ballyteige Burrow. If climate change brings rising sea levels and stormy weather, however, the coastal erosion could get worse.

The Wexford coast is dynamic, constantly being sculpted and shifted by wind and waves, creating majestic dune systems, sand and shingle spits, lagoons and treacherous offshore sandbanks. Co. Wexford is also particularly flat and low-lying – the underlying rock is visible at just a few places on the coast, notably the rocky headlands of Hook Head,

The Slaney river god: crowned with crabs, oysters and ears of corn, this is Enniscorthy's river.
© Enterprise Ireland

Forlorn Point and Carnsore Point – and, but for the blanket of glacial debris, much of the county would be underwater.

1851 – the modern stethoscope

For centuries doctors struggled to elicit information about their patient's inner workings. In 1761, Leopold Auenbrugger, son of a German innkeeper, devised a technique of tapping his patient's back, and listening to the sounds made – just as his father tapped casks to ascertain the volume of wine inside. In 1819, a Frenchman, René Laënnec, invented the monaural stethoscope (*stethos*, Greek for chest), a wooden cylinder for listening to sounds from his patients' chest. Laënnec had spent years noting the sounds, then correlating these with the conditions he found in autopsies after the patients died. The modern stethoscope with its two earpieces was invented in 1851 by a Wexford doctor, **Arthur Leared** (1822–79).[4]

Leared studied medicine at Trinity College Dublin, and worked at Dublin's renowned Meath Hospital,* before becoming the dispensary doctor at Oulart. Leared realised Laënnec's instrument would be more effective and easier to use if it had two earpieces connected to the listening cylinder by rubber tubes. His binaural design was shown in 1851 at the Great Exhibition in London's Crystal Palace, and was quickly and widely copied. Leared's other varied interests included travel, tuberculosis, and the digestive system and, in 1854, he discovered the important role pancreatic juices play in digesting fats. He travelled around India, Iceland, Morocco and elsewhere, and wrote a classic travel book, *Morocco and the Moors* (1876), which is still in print, "being an account of travels, and a general description of the country and its people". In 1879, he returned to North Africa, intending to open a TB sanatorium, but contracted typhoid fever and died.

COUNTY WICKLOW

Wicklow is a county of mountains and minerals: historic Glendalough was once a thriving mining community; Ireland's only gold rush was at Croghan in 1795; and Avoca mine, which closed in 1982, is the longest worked of any Irish mine. Wicklow also has Ireland's oldest fossils, its biggest electrical battery, and its most radioactive hot spot, as well as a fine barrier coast and one of the largest artificial lakes in Europe. Wicklow (*Vikingr Ló*, 'Viking meadow') is a Norse name; in Irish the town and county are *Cill Mhantáin* ('St Mantán's church').

Walling in the Avoca

Arklow is at the mouth of the Avoca river, and in Irish the town is *an tInbhear Mór*, 'the great estuary'. For thousands of years, however, the estuary was a wandering, shifty affair. The coastline there is sandy and, every time a storm shifted the sandbanks, it changed its route. Two hundred years ago, for instance, the river entered the sea 3km north of Arklow. In the early 1800s, however, major engineering works were undertaken to improve Arklow port, and the Avoca was finally tied down, corseted between substantial sea walls, and forced to exit at the harbour. The sea walls changed the currents, and caused extensive sandbanks to be deposited north of the town; dunes and salt marshes developed there, and this new waterfront was eventually reclaimed. In 1872, a fertiliser factory was built there, later used by the Kynoch explosives company.* Extensive coastal protection works had to be built in the 1990s, however, to stabilise the beach and the reclaimed land.

The port improvement scheme was prompted by the growth of the Avoca mines* upstream, and from the 1830s, massive amounts of ore were shipped through Arklow, spawning a thriving **shipbuilding business** there. Arklow had long been a fishing port, but it also had a substantial merchant fleet, partly because the road network in the county's mountainous interior was poor – at least until the Military Road* was built in 1803 – and, consequently, much of the trade with Dublin went by sea. There have been shipyards at Arklow since at least 1715. The best-known was a yacht and trawler yard begun in 1864 by John Tyrrell, from a local seafaring family, which still builds boats at Arklow. One of the last timber boats built at Arklow was the Irish sail-training ship, *Asgard II*. [10] There are plans for a large offshore wind farm on the Arklow Bank, a sandbank that lies 10km off Arklow.

From fool's gold to fertilisers

There is plenty of fool's gold at Avoca.* Fool's gold, or iron pyrites, contains sulphur, but any unearthed at Avoca was initially dumped. Around the 1850s, however, sulphur became a key ingredient in several chemical processes, especially to make sulphuric acid, which was crucial to the manufacture of the new artificial fertilisers.* In 1872, to exploit the Avoca pyrites, John Morrison built a chemical factory on the newly reclaimed land just north of Arklow town, making fertilisers, sulphuric acid, acetone and soap until it closed in 1887; from 1895–1918 the plant was home to Kynoch's explosives factory.*

In the 1960s, a new fertiliser factory opened at Arklow, again using Avoca pyrites. The factory was run by a State company, **Nitrigin Éireann Teo** (NÉT), which merged, in 1987, with ICI's fertiliser operation to form Irish Fertilizer Industries. IFI still makes nitrogen fertiliser for Irish farmers under the NÉT Nitrate brand. Ammonia and limestone are combined to make the nitrogen fertiliser: the ammonia is made at IFI's Cork plant from Kinsale natural gas;* the ammonia is converted into nitric acid at Arklow, then mixed with more ammonia and limestone to produce fertiliser. By-products include carbon black, an anti-static substance used in electrical cabling; carbon dioxide, which is added to soft drinks; and urea, a raw material for the plastics industry.

Kynoch's cordite

In 1895, a big new **explosives factory** opened at Arklow. It was built by Kynoch's, a British company, and, at its peak during World War I, it employed 5,000 people. The company had been started by a Scotsman, George Kynoch, who in 1886 famously offered to arm Ulster, if Gladstone persisted in introducing Home Rule. In 1887, however, the company was taken over by Arthur Chamberlain, uncle of the future British prime minister, and George Kynoch was squeezed

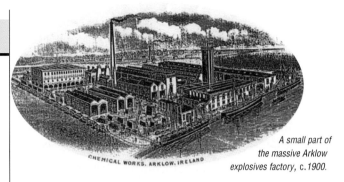

A small part of the massive Arklow explosives factory, c.1900.

out. The company's main ammunition factory was at Birmingham but, in 1894, it won the contract to supply the British army with ammunition, and needed a new factory.[5]

To this end, Kynoch's inspected the Ferrybank site at Arklow in 1894. The location, previously a fertiliser factory, was perfect: it was by a train line and beside a port; the old chemical factory was still usable; and it was big – an explosives factory needs lots of space. Within months, a coal-gas plant and electricity system had been installed, and the factory was soon in production. The Boer War was good for business and, in 1900, several chemical facilities were added, along with a large nitroglycerine plant. In all, the factory stretched for 3km along the shore, with hundreds of buildings – mostly of light wooden construction, in case of explosions – and making a range of explosives, among them cordite, dynamite and gelignite, also sulphuric and nitric acids, and acetone. The company had its own fire brigade and steamship fleet, and bought the Clondalkin paper mills in Dublin to print its labels and catalogues. There were also company houses, a canteen, social hall and hospital (which was subsequently a TB sanatorium and later the Arklow Bay Hotel).

In 1903, Chamberlain calculated that the minimum weekly sum needed to keep a worker, his wife and three children was 21s.8d. This was paid to employees in England but, until 1914, Arklow workers received little over half that. Production at Arklow peaked during World War I: the plant was extended, cargo-handling facilities at the harbour were increased, and 5,000 people were employed, many commuting by train each week from Wexford, Wicklow and Dublin, and staying at company dormitories. The average wage rose to £2 a week, and Arklow prospered.

The company's safety record was generally good, but there were three fatal explosions, the biggest in 1917 killing 27 people. It was the worst industrial accident in Ireland, until the 1979 *Betelgeuse* explosion that killed 51 people at Bantry Bay.* Theories as to its cause ranged from a German submarine shelling the factory, to sabotage by Sinn Féin, and an accidental spark. In 1918, several British explosives firms merged to form Nobel Industries, later ICI (which by then also owned the Ballincollig gunpowder mill*). In the rationalisation, Arklow was closed and, within weeks of Armistice Day, the last workers left. The site was bought by the Hammond Lane Metal Company of Dublin, which systematically scrapped the entire works.

Ireland's most radioactive 'hot spot'

In the early 1970s, nuclear power was at its most popular, and the **uranium*** market was strong. Mining companies turned their attention to Ireland, where uranium prospecting began in earnest. Uranium occurs naturally in granite, and interest concentrated on the granite mountains of Donegal and Wicklow. The Geological Survey of Ireland* conducted its own reconnaissance survey, as part of which geologists toured the country with monitoring devices. In September 1976, they found Ireland's most radioactive spot, a boggy field in Liscolman townland, which has come to be known as the **Liscolman anomaly**. The peaty soil in the field appears to be naturally rich in uranium, possibly because the local rock has a high concentration of uranium, or because the acidic conditions in the bog enable the uranium to accumulate. The soil contains up to 2 per cent uranium, and its radioactivity has been variously estimated

at 10,000–60,000 becquerels per kilogram (Bq/kg) – soil radioactivity levels naturally vary from place to place depending on local geology, but the usual range is between 100–1,000Bq/kg. The amount of uranium in the field was thought to be between 0.25–1,000 tonnes. The Liscolman anomaly is not a health threat, however, and the radioactivity is localised and vented to the open air; no one lives close to it, and the land is not farmed. The uranium price fell in the late 1970s, and Ireland's uranium was never mined, but there is renewed interest today because of **radon**, a radioactive gas emitted when uranium naturally decays. This can pose a health threat if it collects in a building, where people can inhale it and, in the 1990s, the Radiological Protection Institute of Ireland surveyed radon levels in Irish buildings.

Remnant of the ancient Irish wild woods

The Shillelagh woods were famous and vast and, even in the 1600s, covered thousands of hectares. The logs were rafted down the River Slaney to Wexford Harbour, and Cambridge University, Westminster Abbey in England, and Trinity College in Dublin were all built with Shillelagh timber. Shillelagh was also unusual in Ireland, in that the wood was coppiced. **Coppicing** is a form of timber farming: the tree is cut at the stump, leaving a 'stool' from which new shoots grow; these are later harvested in turn – in the case of oak, usually every 25 years. Coppicing was common in England, where there was a strong forestry tradition but, in Ireland, trees were so plentiful few people saw the need to manage them. At Shillelagh, however, estate records show that the trees were coppiced, and there was a strong local tradition of foresters skilled in handling trees and timber. For the Arklow shipbuilding industry, they could produce 'crooks' of timber, by training trees into just the right shape. The timber was also burned to make charcoal for iron foundries downstream at Enniscorthy, while Wicklow tanneries used the bark to cure leather. During the 18th century, demand for timber increased, and the Shillelagh woods were increasingly clear-felled. A remnant of these ancient coppiced woods survives at Tomnafinnoge. Wicklow County Council granted a felling licence for the trees in 1991, but following a public outcry, bought the woods.

the discovery of creosote (produced by distilling coal-gas). John Kyan later emigrated to the USA, and he was developing a filter system for New York city's waterworks when he died. The word 'kyanisation' is still listed in the Oxford English Dictionary.

Kyanising wood

The greatest of Irish mines

An early wood preservative was invented in the 1820s by Wicklow man **John Kyan** (1774–1850). It was used to treat building timbers, railway sleepers and even rope and sailcloth. It won the British Admiralty's seal of approval, and even featured in a song: "Wood will never wear out, thanks to Kyan, thanks to Kyan/He dips in a tank any rafter or plank, and makes it immortal as Diane, as Diane." Kyan's father once owned the Avoca mines, so he was familiar with the problems of timber supports rotting in the damp mine shafts. The mine's fortunes declined, however, and, when Kyan senior died in 1804, the family was penniless, so John Kyan went to work at an English vinegar factory.

In 1812, he started experimenting with ways of preserving wood, and found that impregnating timbers with a solution of 'corrosive sublimate' (bichloride of mercury) gave the best result. Without revealing his process, he sent a block of treated oak to the British Admiralty. They put the wood in a 'fungus pit' for three years, and it emerged in perfect condition in 1831. Kyan patented his technique in 1832, and it quickly attracted public attention, helped by noted English scientist Michael Faraday, who made it the subject of his inaugural lecture at the Royal Institution. The Anti-dry Rot Company bought the rights, and built huge treatment tanks at various locations around London. Many large buildings, including the British Museum, used kyanised timber, as did the railway companies, when they changed from stone sleepers to timber ones. Although the bichloride of mercury was expensive, and it corroded iron fittings, Kyan's process was widely used for 20 years, until

The beautiful Vale of Avoca sits on the richest suite of minerals in Ireland. There is gold and silver there, lead and zinc, and copper, sulphur and iron. The iron and sulphur stain everything in glorious red and yellow, and the mountain streams are rich in minerals – the waters, traditionally recommended for skin disorders, can dissolve an iron nail in just a few months. Not surprisingly, there have been mines at Avoca since prehistoric times. Ptolemy's map of Ireland in AD 150 marks Oboka (Avoca), suggesting that the Romans knew the Wicklow ore; and the Early Irish, the Anglo-Normans and the Elizabethans all had mines there. Mining continued on and off until 1982, making this the longest worked of all Irish mines. The Avoca river divides the valley in two, and there have been mines at various places along both the east and west banks. The ore belt runs through the mountains like ham in a sandwich. The 'ham' is thin – at most 30 metres wide – but it stretches for 15km. The minerals were laid down 450 million years ago, during violent volcanic events when the

County Wicklow's first railway: this arch was part of a steam tramway (built 1846) connecting the Avoca mines with Arklow port.
© Nick Coy

ancient continents of Laurentia and Avalonia collided and closed the Iapetus Ocean.*

When steam pumps developed in the 1720s, deep underground mining below the water table became possible. Initially, Avoca was mined for its copper and several companies operated in the valley, many of them Welsh or Cornish, notably the Williams Brothers of Perranarwortal. By 1810, according to mining historian Nick Coy, some 1,000 people worked at Avoca; boys and men underground, girls and women above ground, picking over the crushed ore. Most of the work was manual, apart from the crushing and pumping which was water-driven. In those labour-intensive days, miners chased only seams with at least 3 per cent copper, as anything less was uneconomic to process. Gunpowder was sometimes used to blast the tunnels: a hole was made with hammer and chisel, filled with a charge of powder, then lit with a straw or quill fuse. The work was hard and dangerous, and miners were noted for being thin and anxious looking, and old and infirm before they were 30. There were 140 vertical shafts at Avoca, and 50km of tunnel. All the supports were timber, mostly larch, which could survive in the corrosive underground streams for over a century, and the mining companies had their own forestry plantations nearby; iron nails quickly corroded, so all the fittings were wooden.

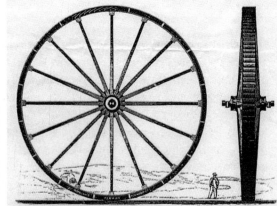

A 50-ft (15-metre) waterwheel. One was used at Avoca to provide power for the mines in the 19th century.
© Nick Coy

Boom-time for Avoca began in the 1830s, when the king of Sicily imposed a levy on sulphur exported from the slopes of Mount Etna. Sulphur was an important raw material for the Industrial Revolution, used in making sulphuric acid and artificial fertiliser,* among other things. Although Avoca had been known for its copper, the most plentiful mineral there is sulphur. Previously, the sulphur was dumped as waste, now it was mined in massive amounts and carted to Arklow, for use in new chemical and fertiliser factories, or shipped abroad. In 1846, a steam tramway, Co. Wicklow's first railway, was built linking the mine to Arklow, where the harbour had been improved to cope with the traffic. Later, the main Dublin–Wexford railway line was diverted from the coast via Avondale to serve the mines (it still runs through Avoca today). Some 2,000 miners worked at Avoca during this time. Steam engines were installed in engine houses on the hillsides, to drain the mines, haul ore and crush rock. Enormous waterwheels were also used, one of which was 15 metres in diameter; power was transmitted from the wheel by a series of rods and cranks, and the spent water was channelled around the mountain and reused to wash ore.

This intensive mining ended in 1880. By then, all the high-grade copper ore had been exhausted, and with large mines starting in America and Australia, the price of copper had fallen. Small-scale works continued on and off, and an ochre factory opened at Ballygahan in 1911, making pigments from the yellow clay. There was large-scale opencast mining at east Avoca in the 1950s, 1960s and again in the 1970s. These modern, highly mechanised operations processed 11 million tonnes of ore in 20 years – compared with 17 million tonnes in the previous 200 years – before closing in 1982. An estimated 100 million tonnes of ore remains, but at 0.5 per cent copper, it is uneconomic. The 20th-century workings destroyed most of the older shafts, though some 18th- and 19th-century galleries survive, but these are flooded. Occasionally, an old tunnel collapses, and the ground above subsides. There is still much to be seen at Avoca, including the remains of seven engine houses, the 1846 tramway arch, colourful spoil heaps, precipitation ponds and a vast 20th-century excavation pit. Part of west Avoca is now a landfill site; the eastern side (Cronebane) is owned by the Department of Public Enterprise, and public access there has been tolerated in recent years. ⑪

Gold mining in Wicklow c.1800. Gravel and sand dug from the mountainside are washed and sieved, to extract even the smallest gold particle.

1795 – the gold rush

Ireland's only 'gold rush' happened in 1795 and lasted six weeks. It began in early September, when word went around that there was gold in the Aughatinavough river, on the northeastern slope of Croghan Mountain.[6] All work stopped as hundreds flocked to pan the Croghan streams. Despite being inexperienced – in the rush, they missed much, and their spoil heaps were later scavenged profitably – they collected 28kg of gold, worth £5,000, then a small fortune. On October 15th, however, the Wicklow militia arrived and took charge of operations. By then, the Aughatinavough had been renamed Gold Mines River.

The government worked the Croghan sites on and off until 1803 – work stopped during the 1798 Rebellion, for example – and since then the area has been worked at various times by locals, and by commercial and State companies, most recently in 1963. In all, about 250kg of gold was found at 19 sites, mostly during the 1800s (returns in the 1900s were generally poor). The most profitable site was Red Hole near Ballinagore bridge. All the finds were placer gold – small particles and occasionally nuggets found in gravel beds. Despite extensive searching, no one has yet found the mother lode. Indeed, some geologists believe the lode no longer exists, removed millions of years ago by geological events or simple erosion. Traces of gold had previously been found at several places around the country, but Croghan was the only commercial Irish gold mine for over 2,000 years. Ireland's first modern gold mine is a small opencast mine near Omagh,* Co. Tyrone, which opened in 2000.

Ireland is rich in **prehistoric gold** artefacts – the National Museum in Dublin has 47kg of Bronze Age gold alone – so, where did this gold come from? Gold-bearing rocks were found in the late-20th century at Croagh Patrick,* the Sperrins* and Clontibret,* but it is unlikely these were known in prehistoric times. Yet in the Late Bronze Age (1500–600 BC) it was as if a rich lode had been discovered: most of the fabulous gold hoards which have been found date from this time, and many of the objects are heavy. Interestingly, gold was scarce in the Early Bronze Age (2500–1500 BC), and only a few lightweight artefacts are known from this period, and again after the Late Bronze Age, when it was as if the gold supply evaporated. Significantly, perhaps, this same pattern is seen in Denmark at the same time.

It was long presumed that Wicklow was the primary source of Ireland's prehistoric gold, partly because the *Annals of the Four Masters* record that, in 1600 BC, King Tighearnmas Mac Follaigh was smelting gold at Cualann (between Bray and Wicklow town). The masters do not reveal where the gold was mined, however, and the Bronze Age miners probably panned in rivers, a type of mining that leaves no trace. Fortunately, each gold ore has a unique chemical fingerprint (with traces of other elements, such as silver and bismuth, and reflecting its geological history). In a bid to identify the source of Ireland's prehistoric gold, scientists are now analysing samples of Irish, British and European gold, and using sophisticated instruments to compare their chemical fingerprints. The gold detectives include Mary Cahill (National Museum of Ireland), Richard Warner (Ulster Museum), and Norman Moles (Queen's University Belfast). Surprisingly, their initial results suggest that Wicklow is not the source; rather, Ireland's prehistoric gold may have come from the north of Ireland, Scotland or possibly further afield.

Sweet Avondale's arboretum

The wooded estate of Avondale is the spiritual and physical home of Irish forestry. It was planted in the 1770s by **Samuel Hayes** (1743–95), the first person to win a preservation order for a tree and, in 1904, it became the home of the new State afforestation programme, though it is more usually associated with another of its owners, the politician **Charles Stewart Parnell** (1846–91).

Avondale's Samuel Hayes, the first person to win a preservation order for a tree. Avondale Museum

By the end of the 18th century, the great woodlands of Ireland were gone. Having been a net timber exporter for centuries, the country began importing timber, and one visitor was prompted to remark that "the most striking thing… is the total absence of trees of any kind". Samuel Hayes, a barrister and an MP for Wicklow, and an active member of the Royal Dublin Society's agriculture committee, believed the future lay in managed woodlands and reforestation. He planted his Avondale Estate – then 3,000 hectares in extent, which was relatively modest for Wicklow – and wrote the first book about trees in Ireland, *A Practical Treatise on Planting; and the Management of Woods and Coppices* (1794). The book is historically important as it contains information about Irish estates at the time, and even individual trees. Hayes died the following year, however, and was buried at Rathdrum, and Avondale passed to his cousin John Parnell.

Parnell's son, Charles Stewart, was born at Avondale in 1846. Had he not gone into politics, he might have been a successful industrialist. As it was, he opened lead mines on his estate, helped make Arklow into a thriving port, developed a water-powered sawmill to cut timber from the estate, and even devised a calipers mounted on a long pole, so he could measure the diameter of trees halfway up the trunk, to get a better indication of their size. He opened a quarry at Ballesse, employing 200 stonemasons and producing setts to pave much of Dublin city. These regular, rectangular stones – often erroneously called 'cobbles' – can still be seen at Dublin's Smithfield, the docklands and elsewhere. Parnell also prospected for gold: he panned the Avonmore river and, reputedly, collected enough to make a wedding ring for Mrs Kitty O'Shea.

In 1904, the Department of Agriculture and Technical Instruction (DATI) acquired 200 hectares of Avondale, and began experiments there that would dramatically alter the Irish landscape. The man behind the experiments was **Augustine Henry** (1857–1930), a world-famous plant collector from Portglenone, Co. Derry. Henry spent 20 years working as a doctor in China for the Imperial Customs Service. Intrigued by Chinese herbal remedies, he began collecting plants there and, despite having no formal botanical training, started sending specimens of Chinese plants to Kew Gardens in London. In all, Henry discovered 25 plant genera that were new to science, and over 1,000 new species and varieties – many of which were named after him – and he sent nearly 250,000 plants to Kew Gardens. He was also responsible for introducing hundreds of Chinese and oriental plants into European gardens, including *Buddleia*, which is now a common escape in cities. Henry was always on the lookout for plants of commercial or medicinal value, and among those he promoted were the Chinese gooseberry, now better known as the kiwi fruit. He later studied forestry in France and, in 1913, was appointed professor of forestry at the Royal College of Science* (later UCD).

When the DATI began planning a reforestation programme, the trees commonly planted were Scandinavian conifers. Augustine Henry realised that Ireland's climate is closer to that of northwest USA and Canada, and that species from there, such as Sitka spruce and Douglas fir, might suit Ireland better than Scandinavian ones. To test his hypothesis, Henry and the forestry director, A C Forbes, planted several types of tree in 0.4 hectares experimental plots at Avondale. As predicted, the Sitka spruce grew well, and it came to be the tree of choice for the Irish State forestry programme. Their experimental plots survive, alongside trees Samuel Hayes planted in the 1770s. Avondale is now owned by Coillte, the Irish forestry board. ⑫ Wicklow is Ireland's most heavily forested county, with nearly 20 per cent under trees, mostly conifers. Ireland as a whole has just 9 per cent forest cover, compared to 30 per cent in Europe.

RATHDRUM

An acidic garden

The arboretum at **Kilmacurragh** is the 'acidic wing' of the National Botanic Gardens.* For the soil at Kilmacurragh is granite-based and acidic, unlike the basic, limestone soil of the Botanic Gardens and, in the 19th century, Kilmacurragh provided a welcome refuge for acid-loving plants that could not be grown in Dublin. Kilmacurragh Estate was developed in the mid-19th century by Janet Acton and her brother Thomas. They planted Himalayan rhododendrons and exotic conifers collected from around the world, and laid the seeds of a fine arboretum.

This was a golden age for plant collecting, and the botanic gardens at Dublin were regularly offered seeds and plants from around the world. The Actons were friendly with David Moore, curator of the Dublin gardens, and provided space

at Kilmacurragh for conifers and acid-loving plants. After Thomas Acton died in 1908, the house and grounds fell into disrepair. They were acquired in 1996 by the State, and formally became an outstation of the Botanic Gardens with which, for 75 years, they had been informally twinned. ⑬

The Actons often donated plants from Kilmacurragh to **Mount Usher Gardens**, which were begun in 1860 beside the Vartry river at nearby Ashford. Unlike many show gardens, Mount Usher is of botanic importance, for it features over 5,000 types of plant and, according to the garden's plantsman John Anderson, has three nationally important collections: of southern hemisphere beech (*Nothofagus*, 14 species); *Eucalyptus* (50 species); and *Eucryphia*, another genus of southern hemisphere plant (25 species). The garden still receives plants from plant-collecting expeditions and botanic gardens in Ireland and overseas. ⑭

WICKLOW TOWN

The 'Cable King'

In the 1860s and 1870s, submarine telegraph cables were strung across the oceans, continents were connected and instant international communications became possible. The man who connected the continents, and laid over 30,000km of cable – enough to girdle the Earth – was a

The massive Great Eastern, entering port. Wicklow's Robert Halpin captained the vessel for most of its illustrious cable-laying career.

master seaman from Wicklow, **Robert Halpin** (1836–94). Halpin went to sea at 10, survived a shipwreck at 16 and, at 22, was given command of his first steamship. He worked the South American routes first, shipping guano fertiliser to Europe, then the Galway–Newfoundland emigrant route, and, during the American Civil War, broke the Yankee blockades, bringing supplies to the Confederates and returning to Europe with cotton. In 1865, he was appointed master navigator on the world's biggest ship, the steamship *Great Eastern*. Five times bigger than any other ship, it was being used to lay a transatlantic submarine telegraph cable between Valentia Island* and Newfoundland. Thanks to Halpin, the crew successfully laid one cable in 1866, and also located and raised a cable which had broken and been lost in mid-ocean during a previous attempt, so that, by September 1866, there were two transatlantic telegraph cables. In 1868, Halpin was given full command of the *Great Eastern*, and he went on to lay cables from France to Newfoundland, from Bombay across the Indian Ocean to Aden and Suez, from Madras to Singapore, followed by Australia to Indonesia. At 40, he retired and built Tinakilly House overlooking Wicklow Bay. No expense was spared, and the bill was reputedly paid by a grateful British government. When local MP Charles Stewart Parnell* died in 1891, Halpin contested the election but failed to win the seat. This master Irish mariner died of gangrene, seemingly contracted when he cut himself trimming his toe nails. A memorial to Robert Halpin stands in Wicklow town. The National Maritime Museum* in Dún Laoghaire has his collection of maritime memorabilia. Tinakilly House is now a hotel.

THE MURROUGH

Wicklow's barrier coast

The coastline from Greystones to Wicklow town is dominated by a long and substantial shingle bank. Salt marshes, lagoons, fens and swamps shelter behind it, and the area is called the Murrough (from *Murbhach*, 'salt marsh'). Though smaller than Wexford's barrier coast, the Murrough is internationally important for its varied habitats and rich birdlife. Erosion is a problem, though, and boulders are in place to protect the coast, especially the railway line that runs along the shingle bank. The two most interesting sites are the **Kilcoole Marshes** and **Broad Lough**. Much of the Kilcoole Marshes has been reclaimed for farming, but there are still substantial lagoons, salt marshes, reed beds and tidal channels. These quiet waters attract Brent and greylag geese in winter, and the reserve is managed by BirdWatch Ireland. Deep drainage channels prevent access to the inner wetlands, but you can walk along the beach side. At Broad Lough, the shingle barrier blocks the Vartry's access to the sea, and forces the river south for 3km, forming a sheltered tidal estuary that finally exits at Wicklow Harbour. The area has been colonised by salt-tolerant plants, such as scurvy grass, sea beet and sea aster. The lough is an important wildfowl sanctuary but, because of its proximity to Wicklow town, is increasingly under pressure from development.

BRAY

The 'Brighton of Ireland'

The coming of the railways changed many things, not least the small coastal town of Bray, which was deliberately developed as a resort in tandem with the railway.[7] The train line to Kingstown [Dún Laoghaire] – Ireland's first railway* – opened in 1834, but it was 20 years before the line was extended to Bray and later to Wicklow town and Wexford. William Dargan,* the man responsible for most of Ireland's mainline railways, was behind both the new coastal railway and the development of Bray. ⑮

The natural and cheapest route through Bray was inland of Bray Head and across the Earl of Meath's estate at Kilruddery. The 10th earl was not greatly enamoured of the idea, however, and by way of an alternative, donated land for a coastal route around Bray Head to Greystones. This posed an engineering challenge, and Dargan used the great English engineer Isambard Kingdom Brunel as a consultant.

Dargan's ornate Turkish baths at Bray were demolished in 1980. From: Heffernan's Illustrated Plan of Bray *(1870).*

1844 – the oldest fossils in the world

In 1840 a Dublin geologist **Thomas Oldham** (1816–78) found some puzzling marks in rocks at the foot of Bray Head, which were then among the oldest rocks known, at 550 million years old. No fossils of that age had yet been found, so if Oldham's marks were fossils, they were the oldest in the world. The marks were small – a few centimetres across – but numerous. There were two types: one looked like a wheel, with spokes radiating from a centre, the other resembled fronds on a stem. It was not clear whether they were the remains of plants or animals, or if they were fossils at all, and Oldham delayed publicising his discovery until 1844. Geologists quickly agreed they were fossils, however, and Bray Head became famous as the site of the world's oldest fossils. **Oldhamia**, as the fossils are called, have since been found in several countries, but they are no longer the oldest in the world – in the 20th century, fossils of simple algae were found which are over 1 billion years old. They remain the oldest Irish fossils, though it is still not known whether they were made by plants or animals. Examples can be seen at TCD's Geology Museum.*

Several tunnels had to be driven through the rocky headland, bridges had to be built over river valleys, and the line had to cope with the relentless coastal erosion. The erosion continues to force the railway inland: the original tunnels were bypassed in 1876, 1889 and again in 1917. Rockfalls remain frequent, and the tracks are still inspected every day.

From the early 1800s, Bray was a starting point for tourists visiting Powerscourt and the Wicklow Mountains; in the 1830s, with the advent of sea-bathing, it also became a resort. Dargan and a local landowner, John Quin, set about developing Bray, modelling it directly on Brighton. They built a regular new town; demolished several properties to enable Quinsborough Road to run through to the seafront, and strategically placed the train station at the end of Quin's Walk. Ironically, because the railway line has to climb towards Bray Head, it blocks the sea view from most of the town. Dargan built and gave to the town the Carlisle grounds, an ornate Turkish baths and an esplanade – all designed to attract visitors from Dublin by train. The grounds, now home to a football club, had an indoor skating rink in the 1870s, and there were plans once for an aquarium there (one finally opened on the esplanade in the 1990s). ⑯ The Turkish baths operated for five years, before being converted to an assembly hall.

In 1846, Thomas Oldham became director of the newly established Geological Survey of Ireland* and subsequently director of the Geological Survey of India. His son **Richard Dixon Oldham** (1858–1936), was born in England and also joined the Geological Survey of India; he made important contributions to the developing science of seismology* and, from a study of seismic shockwaves, was the first to propose that the Earth had a core. Dixon Oldham thought the core was liquid, but it is now known to be a solid nickel-iron core surrounded by a liquid layer.

The biggest battery in Ireland

Demand for electricity drops at night, leaving a surplus of energy in the National Grid. It would be costly to turn off a big fuel-burning power station, so the surplus energy is used to re-charge a battery at the Wicklow Gap; the stored electricity can then be released when it is needed. The battery is an ingenious system made from two reservoirs, one on top of a mountain, the other lower down, and connected by an underground pipe that contains four reversible pump-turbines. At night, the turbines are reversed and surplus power from the national grid is used to pump water to the upper reservoir; when the energy is needed, the pipeline is opened and water flows down to turn the turbines to generate electricity. Pumping the water consumes more energy than the system generates, but it means fuel-burning stations can operate continuously, and that there is stored power that can be tapped instantly to meet demand.

There are similar pumped-storage facilities in Britain and elsewhere, but this is Ireland's only one. Built by the Electricity Supply Board (ESB), with funding from the World Bank, it opened in 1973. The lower reservoir is a natural corrie lake, **Lough Nahanagan**; but an artificial upper reservoir had to be built on top of the mountain. A vast cavern was excavated inside the mountain to house the generating plant and controls, and the pressure tunnel that carries the water and holds the turbines. From outside, only the access road can be seen and the regular outline of the summit water tank. The power station is called Turlough Hill because, like a true turlough,* the lakes regularly flood and empty. Water from Lough Nahanagan also generated power in the 1830s, when it turned the waterwheels for Glendalough's lead mines.*

Lough Nahanagan sits in a corrie on the north side of Camaderry Mountain, dammed by a moraine of debris that was deposited in the last Ice Age, over 15,000 years ago.

The artificial lake at the top of Turlough Hill, where the ESB stores surplus power.
© *Electricity Supply Board*

During building work for the station, the lake was temporarily drained and this uncovered a small glacial moraine inside the lake margin. This inner moraine must have been deposited by a glacier that formed *after* the last Ice Age. Confirmation of this came from plant material found in the inner moraine, which was carbon-dated to about 10,000 years ago. At that time, there was clearly a cold spell, perhaps lasting 500 years, when temperatures dropped by at least 7°C, and it was cold enough for a small glacier to form again in Lough Nahanagan. This cold period, known to geologists and climatologists as the **Nahanagan phase**, may have been caused by cold meltwater from the Arctic temporarily switching off the warm Gulf Stream current, and is an indication of the kind of climate Ireland might expect if this were to happen again, because of global warming.

Lead mines and granite glens

Had you visited Glendalough in 1850, you would have seen a busy mining town, not a quiet monastic ruin. For Glendalough, and the neighbouring valleys of Glendasan and Glenmalure, were once home to **Ireland's biggest lead mine**. At their peak, they employed 500 miners, plus at least as many more labourers and tradesmen, and a mining community of several thousand people lived in the vicinity.

Looking west into Glendalough, which has the classic U-shape of a glaciated valley; sketched for the GSI by George du Noyer.

Plenty of evidence of this mining activity remains, including spoil heaps, miners' tracks, ruined buildings, old settling ponds, and ore-crushing floors, notably at the upper end of Glendalough Valley, at the Wicklow Gap (Luganure) and in Glendasan. Wild goats in the area are said to be descended from animals the miners kept.

The ore in this part of Wicklow is 85 per cent lead, with some iron, zinc and silver. It was deposited 400 million years ago, when the ancient continents of Laurentia and Avalonia* collided. Iron was probably mined at Glendalough in prehistoric times, and certainly in the Middle Ages: there are charcoal-making platforms around Glendalough's upper lake, which are possibly mediaeval, and where charcoal was burned to make fuel for the smelters.

Lead mining probably began in the late 1700s. By 1800, there was a lead mine in Glenmalure, and, a few years later, another mine was opened at Luganure. In 1824, the Mining Company of Ireland began developing the area's lead mines, building mine roads, and installing water-powered machinery and a railway to cart the ore from the mine shafts to the crushing floors below (the wagons were pulled by mules and ponies). The mines were well located, being close to the Military Road* which gave ready access to the lead smelter at Ballycorus* in South Co. Dublin. The hammers that crushed the ore and the pumps that drained

the shafts were driven by waterwheels, but these were vulnerable to drought and, during dry summers, the mines closed for want of power. Many of the mine tunnels run horizontally into the hillside and some go all the way through Camaderry Hill, linking Glendalough and Glendasan. Most Irish mines employed women and children to pick over the crushed ore, but it is said that no women or children worked at the Glendalough mines, as easier tourism work was available at the monastic site at the other end of the valley.

The most profitable years were 1848–1875. The international lead price was high, some 40,000 tonnes of ore were produced, and the mines regularly reported annual profits of nearly £8,000. Optimistic about the prospects, the firm planted a million trees, including Scots pine trees that are still growing on Camaderry. By 1876, however, the world lead price had plummeted and the venture went into decline. The mines were worked sporadically since then, notably during World War I when there was a lead shortage, and, on a small scale, even into the 1950s and 1960s.

Glendalough (*Gleann dá Locha*, 'glen of the two lakes') is an appropriate name for a valley that, for the time being, has two lakes. But a few thousand years ago, there was only one lake there, and in a few thousand years' time there will again be just one lake. In fact, if the valley floor continues to silt up, the lakes will disappear altogether, leaving just a narrow river. At the end of the last Ice Age, Glendalough was flooded by a lake that was nearly 7 metres higher than today. As the water level receded, it left sand and gravel banks, and a fossil shoreline – Glendalough's 6th-century monastic site is built on a beached gravel bank, and the road on the valley's north side runs along the prehistoric shore. Two rivers fed into the lake, then as now: Glenealo river at the head of the valley, and Lugduff Brook, which tumbles down Pollanass waterfall halfway along the southern side of the glen. Both rivers deposit substantial amounts of sediment where they enter the lake. At the foot of Pollanass waterfall, an estimated 20 million tonnes accumulated over the past 10,000 years, and eventually cut the original lake in two. The upper lake is also shrinking, thanks to debris from the old mine spoil heaps, which Glenealo river has been washing into the lake for 200 years.

Glendalough is a classic glaciated valley – a U-shaped trough, with a flat bottom and steep sides. It is one of several excavated by glaciers in the Wicklow Mountains, among them Glenmalure which, at over 35km, is Ireland's longest valley. Other glacial features in the mountains are corries,* as at Lough Nahanagan* on Turlough Hill; glacial moraines,* such as the mound impounding Kelly's Lough above Glenmalure; the smoothly rounded mountain summits; and hanging valleys – these carry tributaries which are now left high above the main valley floor, and they are marked by waterfalls as at Powerscourt and Pollanass.

The Wicklow Mountains, part of which are now in **Wicklow National Park**, are a substantial mass of granite running from South Co. Dublin almost to New Ross in Co. Wexford, and which is the largest granite intrusion in Britain or Ireland. The granite formed 400 million years ago, when five enormous balloons of molten granite rose up from deep within the Earth; they lodged several hundred metres below the surface, where they cooled slowly. Heat from the molten granite baked the surrounding rocks up to a distance of 1km, converting shale to schist, and sandstone to quartzite. The rocks were later uplifted, and erosion removed most of the schist, to unroof the granite (patches of schist remain on some summits, notably Djouce and Scarr). ⑰

Wicklow's distinctive, conical **Sugar Loaf** is not a volcano, as is sometimes thought, but the remains of a prehistoric sandy beach. It is made of quartzite, which was originally sandstone and formed at the edge of the ancient Iapetus Ocean.* Quartzite shatters easily, to form pointed peaks surrounded by slopes of broken scree; other familiar quartzite peaks are Errigal and Croagh Patrick. Apart from a few quartzite peaks, Wicklow is dominated by its granite. Unusually, it is the only Irish county with no limestone,

unless you count boulders carried from the midlands by glaciers. The granite influences Wicklow's landscape and farming – most of the soil is poor and peaty – and even social development, which was shaped by the uplands and valleys. The mountains seem wild, but this is no wilderness, and it has been greatly shaped by human intervention. Most of the trees were planted, many of them by mining companies in the 19th century, and the herds of deer are descended from reds and sikas introduced at Powerscourt* around the same time. Turf cutting within the national park has ceased, grazing is controlled in a bid to enable natural vegetation to regenerate and, once the existing coniferous plantations have been harvested, the area will be allowed to revert to its natural state.[8]

A calico town

At the end of the 18th century, several purpose-built manufacturing towns were established around the country, often by well-intentioned and entrepreneurial landowners. They included Prosperous* in Co. Kildare, Portlaw* in Co. Waterford and Stratford-on-Slaney in Co. Wicklow, all of which were cotton towns. They met with varying success: the factory at Prosperous closed after just six years, Stratford-on-Slaney lasted 70 years, and Portlaw continued into the 20th century. Stratford-on-Slaney was founded in 1780 by Edward Stratford, second Earl of Aldborough, who owned estates in Wicklow, and elsewhere in Ireland and England. He built a factory, bleach greens, church and chapel, and 36 workers' houses with slate roofs and sash windows. He leased his new town to John and William Orr, who invested £20,000 on printing and other equipment, and opened a calico factory, all water-powered. At its peak, the factory employed 1,000 workers (the town's population today is just 200). They wove cotton yarn imported from Scotland, and produced an estimated 40,000 metres of fabric a week. Designs were printed onto the fabric with printing blocks and later high-speed rollers. By 1850, however, the business had collapsed, and all that remains are the bleach greens and a ruined dye house.

Drinking the Vartry

By the 1850s, Dublin city had outgrown its water supply.* Having tapped the rivers Liffey, Dodder and Poddle, and the Grand and Royal canals, Dublin Corporation turned its sights to the clean mountain waters of the Vartry river. Plans were drawn up for a reservoir at Roundwood, and an underground pipe to convey the water to the metropolis. The scheme was championed by **Sir John Gray** (1815–75), chairman of the corporation's waterworks committee and proprietor of the *Freeman's Journal*. He bought land at Roundwood to prevent it being acquired by property speculators, then sold it without profit to the corporation. Construction began in 1862 and took five years: a reservoir, dam and embankment were built at Roundwood, and a smaller reservoir in the Dublin suburb of Stillorgan; the two were connected by a 50km aqueduct, 5km of which was tunnelled through rock. The pipe crosses the Dargle and Cookstown rivers on iron bridges, and delivers water to Bray en route. Much of the detail of the construction features in James Joyce's novel *Ulysses*: Leopold Bloom, running his kitchen tap, remarks on how the water comes from Roundwood, via a subterranean pipe, which cost £5 per linear yard to build. In the 1920s, the scheme was extended, and there are now two reservoirs at Roundwood. Marshes and woodland surround the shore, the reservoirs are popular with winter fowl, and there is public access to the southern reservoir. Water levels vary more than in a natural lake, and this creates problems for plants and animals inhabiting the shore. One common plant is shoreweed (*Littorella uniflora*), a white-flowered plantain that can tolerate the sudden changes from drought to flood. A statue to Gray, erected by public subscription, stands in Dublin's O'Connell Street.

1940 – the coming of the waters

In the early 1920s, demand for electricity grew dramatically. The new Free State government, acutely aware of problems caused by coal shortages during World War I, opted to develop hydro-power. But it faced two competing proposals: to harness the river Shannon at Ardnacrusha* or dam the Liffey at **Pollaphuca**, and create a reservoir at Blessington. The Shannon scheme was proposed by a young Irish engineer, Thomas McLaughlin, and the Liffey scheme by the 'grand old man of Irish engineering', **Sir John Purser Griffith** (see panel). Irish engineers were divided: some argued that Ardnacrusha was too far from Dublin, others that electricity distribution was now so efficient distance was not an issue; some thought that buying land for a Liffey reservoir would prove costly, others that the reservoir could supply drinking water to Dublin. In the end, the Shannon scheme won. The Liffey, it was decided, could be dammed another time if demand for electricity continued to grow.

As it was, Ardnacrusha was in operation for only six years, when Dublin Corporation and the Electricity Supply Board

'Grand old man of Irish engineering'

Sir John Purser Griffith (1848–1938), a Welshman, came to study engineering at TCD, and stayed to make Ireland his adopted home. As senior engineer with the Dublin Port & Docks Board,* he made Dublin into a modern port and distribution centre – until the late 1800s, shipping companies preferred Belfast. He initiated extensive land reclamation projects and dredging schemes, and electrified the port in the early 1900s with its own power station. A visionary, Griffith championed the Pollaphuca scheme, and believed in exploiting turf as a fuel, especially to generate electricity in peat-burning power stations built beside large bogs. In the 1920s, to prove his idea, he invested £70,000 of his personal fortune buying Turraun Bog* in Co. Offaly, where he developed a big, mechanised, turf-cutting business and a power station. It was a prototype for the Turf Board (Bord na Móna), which was established in the 1930s. Purser Griffith, a major figure in Irish and British engineering circles, was known as the 'grand old man of Irish engineering'.[9]

(ESB) agreed to build the Liffey scheme, to supply both power and water. Work began in 1937, and 300 people were moved from their homes in the valley; burials at Burgage graveyard were reinterred at Blessington; roads were diverted; and new bridges built. Three dams and generating stations were constructed: the main one on a dramatic gorge at Pollaphuca, smaller ones at Golden Falls by Ballymore Eustace, and at Leixlip, Co. Kildare. A water treatment plant was built at Ballymore Eustace, supplied by an underground pipe from the reservoir.

In March 1940, the sluice gates were closed, and the reservoir began to fill. According to local historian Seamus Ó Maitiú, the day is still known locally as 'the coming of the waters'. The reservoir took 18 months to fill, and is one of the largest artificial lakes in Europe. Locals resented it, however, as it flooded homes and severed the community – Ballyknockan* was cut off from its parish church at Valleymount – and all for the benefit of distant Dubliners. On the plus side, however, the reservoir provides good fishing – the ESB stocks it annually with 12,000 fish – and it has virtually eliminated flooding on the Liffey. Previously, winter flooding was common at Straffan and Kilcullen; now, flood waters can be retained and slowly released.

The flooding of Blessington valley recreated a lake that had existed 15,000 years ago. **Glacial Lake Blessington** was a vast meltwater lake that, towards the end of the last Ice Age, filled the valley to a depth of at least 150 metres. It was much larger than the modern reservoir, which is 25 metres at its deepest. The glacial lake left massive deposits of sand and gravel in the valley. These were mapped in the 1930s before the reservoir was created, by a Cork geologist **Anthony Farrington** (1893–1973), and revealed much about the prehistoric Liffey.* The Liffey, it seems, has changed course several times, so that there are a number of dry valleys in the region (one is crossed by Dry Bridge near Pollaphuca), and at one point during the last Ice Age, the Liffey's route north was blocked by ice, and it was forced to drain south into the Slaney. Blessington's glacial

Pollaphuca power station is built in a gorge on the River Liffey. Pollaphuca Reservoir (aka the Blessington Lakes) is owned by the ESB.
© ESB

sand and gravel has been quarried for centuries. Commercial quarrying continues, and the heavy traffic of gravel trucks makes the N81 among Ireland's dustiest roads. The ornate castellated bridge at Pollaphuca was designed and built in the 1820s by Scottish engineer Alexander Nimmo.*

BALLYKNOCKAN

A stonecutters' village

Ballyknockan village is steeped in the ancient craft of stone working.[10] Even small cottages have elegant stone lintels and porches, ornate gateways and chimney pots, and many houses have a granite fireplace, each unique and intricately carved. The village dates to the 1820s, when a granite quarry opened near Valleymount, on the northern slope of Silsean. Nearly 200 years later, the quarry is still producing stone and still employing master stonemasons.

There had been a quarry about 12km north of Ballyknockan at Kilbride until the early 1800s. It produced granite for the great Georgian buildings of the day, among them Dublin's General Post Office. That building era came to an end with the Act of Union, however, and the Kilbride quarry closed. A new wave of building began in the 1820s, mostly banks, large Catholic churches – as Catholic Emancipation was introduced – and, from the 1840s on, railway stations. A new quarry opened at Ballyknockan to service this trade, employing the former Kilbride quarrymen and stonemasons, and their first order was for a Jesuit church on Dublin's Gardiner Street. All of the work then was done by hand: quarrymen cut the stone out of the mountainside with sledges and wedges; the elite stonemasons – their apprenticeship was seven years – dressed the stone; and finally, the carters brought the finished stone to the building sites, leaving about midnight, so as to arrive in Dublin for the morning shift. The other group of quarry workers were the blacksmiths, who made and sharpened the various saws and chisels. At its peak then, Ballyknockan employed 500 men. Today, most work is done by machine – the sledges and wedges replaced by air drills, the simple saws by diamond-tipped cutting tools – though some stonecutters still execute individual pieces by hand.

Dublin Corporation's new Civic Offices by the Liffey is among the modern buildings clad with Ballyknockan granite. It is prized as a high-quality building stone, and has a distinctive creamy colour. There are currently four working quarries, two run by the local families and two by building contractors.

SALLY GAP

1798 and the Military Road

Wicklow's mountainous interior was for centuries a barrier to communication, and conversely a boon to Irish rebels. Prior to 1800, the only roads were four mountain tracks following the main valleys and running east–west. They were not connected, and there was no direct road to Dublin. This greatly frustrated the English army in its pursuit of the 1798 insurgents, who had taken to the hills under Michael Dwyer. In February 1800, with the rebels still at large, local estate owners successfully petitioned the government for a new road. They were concerned about security, but probably also hoped to benefit commercially from improved access to their estates.

A Scottish engineer, **Alexander Taylor** (1745–1827), was appointed to design a road and supervise the work. Some 200 infantrymen of the Scottish Fencibles were issued with 200 shovels, 30 pickaxes, 40 hand barrows and 20 crow irons and, in August 1800, work began on what would become the Military Road.[11] Taylor's route minimised the amount of climbing: starting in Rathfarnham, it went south to Rockbrook and Killakee, then across the Feather Beds to Glencree, the Sally Gap, Glenmacnass, Laragh, Glenmalure and Aughavanagh. Water pavements and stone culverts were built across the numerous mountain streams; many of these small bridges survive. Michael Dwyer surrendered in 1803, before the road had reached Glenmacnass; nevertheless, army barracks were built along the route at Glencree, Laragh, Drumgoff and Aughavanagh. They were fully occupied by 1805 though, and after the Napoleonic threat had passed, only a small guard remained. Two are still in use: Aughavanagh as a youth hostel, and Glencree as a reconciliation centre. Plans to settle a colony of Scottish Fencibles at Glenmacnass and establish "a sturdy race of loyal mountaineers", came to nought.

Taylor's road ran for 34 Irish miles (70km; 1 Irish mile equals 2,048 metres) and cost the then enormous sum of £40,000, twice the original estimate. The Military Road was the making of Alexander Taylor, who had come to Ireland as a surveyor in 1778 with the Aberdeen Highlanders. After the road was finished in 1805, he became the engineer for the Post Office's mail coach survey and later a paving commissioner for Dublin. He is buried at Naas.

ENNISKERRY

Dry valleys and a deer park

The Scalp is a dramatic rocky cleft that leads north from Enniskerry to Kilternan in Co. Dublin. One account from the 1830s described it as "a deep natural chasm in the

mountain, forming a defile with lofty and shelving ramparts on each side, from which large detached masses of granite of many tons weight have fallen [and] are heaped together in wild confusion... threatening at every moment to crush the traveller". Strangely, this deep rocky gorge is dry: the river that flows through it is of traffic, not water. Wicklow has many such **dry valleys**, among them the Glen of the Downs, Hollywood Glen and Piperstown Glen. It was the early 1900s before geologists realised these formed during the Ice Age, and that Enniskerry was the site of a long vanished glacial lake. At the end of the last Ice Age, meltwater collected in a vast lake there, called **Glacial Lake Enniskerry**. Torrents of meltwater flowing into the lake from the north carved out the Scalp, and water flowing out of the lake to the south carved the **Glen of the Downs**. The water has long since drained away, leaving several dry valleys and thick gravel deposits on the former lake bed around Enniskerry.

The narrow road and speeding traffic make it difficult to stop in the Scalp, and the best views are from the car park of the Kilternan Hotel and from the Ballycorus* road. The granite of the Scalp is cracked, and freeze-thaw action easily fractures it into boulders; the bare rock walls reveal the valley's V-shape. The Glen of the Downs, in contrast, is thickly wooded, mostly with oak, ash and hazel; Ireland's longest glacial meltwater channel, it is now a nature reserve. **Knocksink Wood** nature reserve by Enniskerry is noted for its rich plantlife, insect life and petrifying springs. An environmental education centre there is run jointly by Dúchas, the Dublin Institute of Technology and the Irish Wildlife Federation. [18]

In the 13th century, neighbouring Glencree Valley was a **deer park** and royal forest. The Anglo-Normans, who introduced deer parks to Ireland, hunted the great herds of native red deer,* as well as fallow deer, which they introduced from the continent. The first record of fallow deer in Ireland is at Glencree where, in 1296, a dozen were presented to Eustace le Poer. By the mid-1300s, the great oak woods at Glencree had been felled and the deer park was abandoned. The area was still known as Deerpark and, in 1860, by which time it was Powerscourt Demesne, it again lived up to its name. The native red deer had disappeared from Wicklow by then, so **Viscount Powerscourt** introduced red deer from Germany and Britain, along with exotic deer, including eland, wapiti, sambar and Japanese sika. He had, he said "a fancy to try to acclimatise various kinds of deer and other animals", but only the red deer and sika survived. The sika is smaller than the red deer, with a white rump bordered with black stripes; the fawn has white spots and a black tail. Sika and red deer readily interbreed, so that now there are no pure red deer in Wicklow. [19]

Powerscourt waterfall, the tallest in these islands (120 metres), from: Heffernan's Illustrated Plan of Bray (1870).

Chapter 12 sources

Co. Wexford

1. Hurley, Jim, *Water Level at Lady's Island Lake 1984–96* (SWC Promotions, 1997).

2. Mitchell, Frank & Michael Ryan, *Reading the Irish Landscape* (TownHouse & CountryHouse Ltd, 1997).

3. Mitchell, G F, "Fossil pingos in Camaross townland, County Wexford" *Proc RIA* **73** (B): 269–284 (1973).

4. Coakley, Davis, *Irish Masters of Medicine* (TownHouse & CountryHouse Ltd, 1992).

Co. Wicklow

5. Kelleher, George, *From Gunpowder to Guided Missiles – Ireland's war industries* (Kelleher, 1993).

6. Reeves, T J, "Gold in Ireland" *Geol. Surv. Ireland Bull.* **1** 73–85 (1971).

7. Davies, K M, "Bray" *Irish Historic Towns Atlas* No.9, (RIA, 1998).

8. Nairn, Richard, & Miriam Crowley, *Wild Wicklow: Nature in the Garden of Ireland* (TownHouse & CountryHouse Ltd, 1998).

9. Cox, Ronald, *Grand old man of Irish engineering* (Institution of Engineers of Ireland, 1998).

10. Ó Maitiú, Seamus, & Barry O'Reilly, *Ballyknockan: A Wicklow Stonecutters' Village* (Woodfield Press, 1997).

11. O'Keeffe, Peter, "Building the Military Road" *Technology Ireland* **30** No.2 22–25 (May 1988).

Directory of centres

Chapter 1: *Dublin, city & county*
Liffey, port and bay
(1) North Bull Island Interpretive Centre · (tel) 01 833 8341 daily (2) Dublin Port Company, Alexandra Road · (tel) 01 855 0888 by appointment (3) Custom House Visitor Centre · (tel) 1890 202021 half-time (4) Waterways Visitor Centre · (tel) 01 677 7510 daily (5) ENFO St Andrew Street · (tel) 1890 200191 Mon–Sat

Dublin City
(6) TCD Library · (tel) 01 608 1171 daily (7) TCD Geology Museum· (tel) 01 608 1477 Mon–Fri (8) Centre for Civil Engineering Heritage, TCD · (tel) 01 608 2544 by appointment (9) TCD Zoology Museum · (tel) 01 608 1679 by appointment (10) TCD Chemistry Department · (tel) 01 608 1726 by appointment (11) TCD Botanic Gardens, Dartry · (tel) 01 497 2070 by appointment (12) Royal Irish Academy, Dawson Street · (tel) 01 676 2570 Mon–Fri (13) Royal College of Physicians of Ireland, Kildare Street · (tel) 01 661 6677 Mon–Fri (14) National Library of Ireland, Kildare Street · (tel) 01 603 0200 Mon–Sat (15) National Photographic Archive, Temple Bar · (tel) 01 603 0200 Mon–Sat (16) National Gallery of Ireland, Merrion Square · (tel) 01 661 5133 daily (17) Natural History Museum, Merrion Street · (tel) 01 677 7444 Tues–Sun (18) Government Buildings, Merrion Street · (tel) 01 662 4422/661 5133 Sat (19) Irish Architectural Archive, Merrion Square · (tel) 01 676 3430 Tues–Fri (20) Apothecaries Hall, Merrion Square · (tel) 01 676 2147 by appointment (21) Marsh's Library, St Patrick's Close · (tel) 01 454 3511 Mon–Sat (22) St Patrick's Cathedral, Patrick Street · (tel) 01 475 4817 daily (23) Guinness Storehouse, James's Street · (tel) 01 408 4800 daily (24) Dublin Brewing Company, Smithfield · (tel) 01 872 8622 daily (25) The Old Jameson Distillery, Bow Street · (tel) 01 807 2355 daily (26) Dr Steevens' Hospital, Steevens' Lane · (tel) 01 679 0700 by appointment (27) Phoenix Park Visitor Centre · (tel) 01 677 0095 daily (28) Ordnance Survey of Ireland, Phoenix Park ·

(tel) 01 802 5300 Mon–Fri (29) Dublin Zoological Gardens · (tel) 01 677 1425 daily (30) St Michan's, Church Street · (tel) 01 872 4154 Mon–Sat (31) City Hall, Dame Street · (tel) 01 672 2204 daily (32) National Museum, Collins Barracks · (tel) 01 677 7444 Tues–Sun (33) Chester Beatty Library, Dublin Castle · (tel) 01 407 0750 daily

Co. Dublin
(34) Geological Survey of Ireland, Beggar's Bush · (tel) 01 604 1420 Mon–Fri (35) National Print Museum, Beggar's Bush · (tel) 01 660 3770 daily (36) Institution of Engineers of Ireland, Clyde Road · (tel) 01 668 4341 by appointment (37) National Maritime Museum, Dún Laoghaire · (tel) 01 280 0969 May–Sept (38) Rathfarnham Castle · (tel) 01 493 9462 May–Oct (39) Mount Jerome Cemetery, Harold's Cross · (tel) 01 497 1269 daily (40) Dunsink Observatory · (tel) 01 838 7911 (open nights by ticket; tours for groups only) (41) National Botanic Gardens · (tel) 01 837 4388 daily (42) Met Éireann, Glasnevin · (tel) 01 842 4411 by appointment (43) National Metrology Laboratory · (tel) 01 808 2000 third-level groups only (44) Dublin Fire Brigade Museum, Marino · (tel) 01 833 8313 groups only (45) Howth Transport Museum · (tel) 01 832 0427/848 0831 summer (46) Baily Lighthouse · (tel) 01 833 9594 adult groups only (47) Hedgerow Society, c/o A Lynch, Richardstown, Ballyboughil (48) Ardgillan House, Balbriggan · (tel) 01 849 2212 daily (49) Malahide Demesne · (tel) 01 846 2184 daily (50) Skerries Mills · (tel) 01 849 5208 daily

Chapter 2: *Westmeath, Meath & Louth*
Co. Westmeath
(1) Locke's Distillery, Kilbeggan · (tel) 0506 32134 daily
(2) Tullynally Castle, Castlepollard · (tel) 044 61159 summer
(3) Belvedere House, Mullingar · (tel) 044 49061 daily

Co. Meath
(4) Hill of Tara · (tel) 046 902 5903 centre: May–Oct
(5) Brú na Boinne (Newgrange) · (tel) 041 988 0300 daily

Co. Louth

⑥ Millmount Museum · (tel) 041 983 3097 daily ⑦ White River Mills, Dunleer · (tel) 041 685 1141 May–Sept ⑧ Louth County Museum, Dundalk · (tel) 042 932 7057 daily ⑨ Táin Holiday Village, Omeath · (tel) 042 937 5385 daily

*Chapter 3: **Armagh, Down, Belfast & Antrim***

Co. Armagh

① Slieve Gullion Visitor Centre · (tel) 048 4173 8284 Easter–Sept ② Palace Stables, Armagh · (tel) 048 3752 9629 daily ③ Armagh (Robinson) Library · (tel) 048 3752 3142 Mon–Fri ④ Armagh County Museum · (tel) 048 3752 3070 Mon–Sat ⑤ Armagh Planetarium · (tel) 048 3752 3689 Mon–Fri ⑥ Navan Fort/Emain Macha · (tel) 048 3752 1801 by appointment ⑦ The Argory, Derrycaw · (tel) 048 8778 4753 April–Sept ⑧ Ardress Estate, Annaghmore · (tel) 048 3885 1236 April–Sept ⑨ Lough Neagh Discovery Centre, Oxford Island · (tel) 048 3832 2205 daily

Co. Down

⑩ Fergusons Irish Linen, Banbridge · (tel) 048 4062 3491 Mon–Thurs ⑪ Irish Linen Tour, Banbridge · (tel) 048 4062 3322 by appointment ⑫ Canal Visitor Centre, Scarva · (tel) 048 3883 2163 April–Oct ⑬ Moneypenny's Lockhouse, Brackagh · (tel) 048 3832 2205 Sat–Sun summer ⑭ Mourne Countryside Centre, Newcastle · (tel) 048 4372 4059 Mon–Fri ⑮ Silent Valley Reservoir · (tel) 048 9074 6581 daily ⑯ Annalong Cornmill · (tel) 048 4376 8736 Feb–Nov ⑰ St Patrick Centre, Downpatrick · (tel) 048 4461 9000 daily ⑱ Down County Museum, Downpatrick · (tel) 048 4461 5218 daily ⑲ Downpatrick Railway Museum · (tel) 048 4461 5779 daily ⑳ Quoile Countryside Centre, Saul · (tel) 048 4461 5520 daily ㉑ Castle Ward, Strangford · (tel) 048 4488 1204 Centre: May–Oct ㉒ Exploris Aquarium, Portaferry · (tel) 048 4272 8062 daily ㉓ Castle Espie, Comber · (tel) 048 9187 4146 daily ㉔ Nendrum Monastery, Mahee Island · (tel) 048 9754 2547 Centre: April–Sept ㉕ Scrabo Country Park, Newtownards · (tel) 048 9181 1491 Centre: Easter–Sept ㉖ Grey Abbey · (tel) 048 4278 8585 Centre: April–Sept ㉗ RSPB Belfast · (tel) 048 9049 1547 ㉘ North Down Heritage Centre, Bangor · (tel) 048 9127 1200 Tues–Sun ㉙ Ballycopeland Windmill · (tel) 048 9181 1491 July–Aug ㉚ Ulster Folk & Transport Museum, Cultra · (tel) 048 9042 8428 daily

Belfast

㉛ Lagan Lookout, Donegall Quay · (tel) 048 9031 5444 daily ㉜ Belfast Harbour Commissioners, Corporation Square · (tel) 048 9055 4422 by appointment ㉝ Linen Hall Library, Donegall Square · (tel) 048 9032 1707 Mon–Sat ㉞ Botanic Gardens · (tel) 048 9032 4902 daily ㉟ Ulster Museum · (tel) 048 9038 3000 daily ㊱ W5, Queen's Quay · (tel) 048 9046 7700 daily ㊲ Cave Hill Heritage Centre, Belfast Castle · (tel) 048 9077 6925 daily ㊳ Belfast Zoo, Belfast Castle · (tel) 048 9077 6277 daily

Co. Antrim

㊴ Irish Linen Centre, Lisburn · (tel) 048 9266 3377 Mon–Sat ㊵ Ulster Aviation Centre, Crumlin · (tel) 048 9445 4444 Sat ㊶ Patterson's Spade Mill, Templepatrick · (tel) 048 9443 3619 March–Sept ㊷ ECOS, Ballymena · (tel) 048 2566 4400 daily ㊸ Flame, Carrickfergus · (tel) 048 9336 9575 March–Oct ㊹ Railway Preservation Society of Ireland, Whitehead · (tel) 048 2826 0803 by appointment ㊺ Larne Interpretive Centre · (tel) 048 2826 0088 Mon–Sat ㊻ Larne Museum · (tel) 048 2827 9482 varies ㊼ Giant's Causeway Centre, Bushmills · (tel) 048 2073 1855 daily ㊽ Ballycastle Museum · (tel) 048 2076 2942 June–Sept ㊾ RSPB Rathlin · (tel) 048 2076 3948 by appointment ㊿ Causeway Railway, Bushmills · (tel) 048 2073 2594 April–Oct (51) Bushmills Distillery · (tel) 048 2073 1521 May–Aug (52) Portrush Countryside Centre · (tel) 048 7082 3600 July–Aug

Chapter 4: Derry, Donegal & Tyrone

Co. Derry

① Tower Museum, Derry · (tel) 048 7137 2411 Mon–Sat ② Roe Valley Country Park, Limavady · (tel) 048 7772 2074 Centre: June–Aug ③ Wm Clarke & Sons Ltd, Upperlands · (tel) 048 7944 7200 groups only ④ Harbour Museum, Derry · (tel) 048 7137 7331 Mon–Fri ⑤ Amelia Earhart Cottage, Ballyarnet · (tel) 048 7137 7331 by appointment

Co. Donegal

⑥ Maritime Museum, Greencastle · (tel) 074 938 1363 May–Sept ⑦ Newmills, Letterkenny · (tel) 074 912 5115/913 7090 June–Sept ⑧ Tullyarvan Museum, Buncrana · (tel) 074 936 1613 April–Sept ⑨ Glenveagh National Park · (tel) 074 913 7090 centre: April–Oct ⑩ *An Mhuc Dhúbh* Railway, Fintown · (tel) 074 954 6280 June–Sept ⑪ CDR Railway Heritage Centre, Donegal · (tel) 074 972 2655 daily ⑫ Donegal County Museum, Letterkenny · (tel) 074 912 4613 Mon–Sat ⑬ St Connell's Museum, Glenties · (tel) 086 394 5399 April–Sept ⑭ Dolmen Centre, Kilclooney · (tel) 074 954 5010 Mon–Sat

Co. Tyrone

⑮ Gray's Printing Press, Strabane · (tel) 048 9751 0721 April–Sept ⑯ Sperrin Heritage Centre, Cranagh · (tel) 048 8164 8142 April–Oct ⑰ An Creagán/Creggan · (tel) 048 8076 1112 daily ⑱ Fivemiletown Library · (tel) 048 8952 1409 Tues–Sat ⑲ Wellbrook Beetling Mill, Corkhill · (tel) 048 8675 1735 April–Sept ⑳ Benburb Valley Linen Museum · (tel) 048 3754 9752 Easter–Sept ㉑ Kinturk Cultural Centre · (tel) 048 8673 6512 daily ㉒ Cornmill Heritage Centre, Coalisland · (tel) 048 8774 8532 Mon–Fri ㉓ Tyrone Crystal, Dungannon · (tel) 048 8772 5335 daily

Chapter 5: Fermanagh, Cavan & Monaghan

Co. Fermanagh

① Castle Archdale Country Park · (tel) 048 6862 1588 centre: May–Aug ② Enniskillen Castle · (tel) 048 6632 5000 Mon–Fri ③ Explore Erne, Belleek · (tel) 048 6865 8866/ 6632 3110 June–Sept ④ Crom Estate, Newtownbutler · (tel) 048 6773 8118 April–Sept ⑤ Belleek Pottery · (tel) 048 6865 8501 daily ⑥ Castle Caldwell, Belleek · (tel) 048 6863 1253 ⑦ Florence Court · (tel) 048 6634 8249/8497 daily ⑧ Marble Arch Cave, Marlbank · (tel) 048 6634 8855 Apr–Sept

Co. Cavan

⑨ Cavan County Museum, Ballyjamesduff · (tel) 049 854 4070 Tues–Sat

Co. Monaghan

⑩ Monaghan County Museum · (tel) 047 82928 Tues–Sat ⑪ Vintage Museum, Ballinode · (tel) 047 89840

Chapter 6: Sligo, Leitrim, Roscommon & Longford

Co. Sligo

① Stokes Summer School, c/o Maths Dept, DCU, Dublin 9 · (tel) 01 700 5000 ② Carrowmore visitor centre · (tel) 071 916 1534 March–Oct ③ Lissadell House, Drumcliff · (tel) 071 916 3150 June–Sept

Co. Leitrim

④ Cavan & Leitrim Railway Museum, Dromod · (tel) 071 963 8599 daily ⑤ Shannon–Erne Promotions, Ballinamore · (tel) 071 964 4855 ⑥ Sliabh an Iarainn Visitor Centre, Drumshanbo · (tel) 071 964 1522 April–Sept ⑦ County Library, Ballinamore · (tel) 071 964 4012 Mon–Fri

Co. Roscommon

⑧ Arigna Mining Museum · (tel) 071 964 6185/964 6186 varies ⑨ Lough Ree Environmental Summer School, Lough Ree Co-op, Lanesborough, Co. Longford · (tel) 043 27070 ⑩ Clonalis House, Castlerea · (tel) 094 962 0014 June–Sept ⑪ Hell's Kitchen Railway Museum, Castlerea · (tel) 094 962 0181 daily ⑫ Clay-pipe Museum, Knockcroghery · (tel) 090 666 1923 May–Sept ⑬ Strokestown House · (tel) 071 963 3013 Apr–Oct

(14) Elphin Windmill · (tel) 071 963 5627 April–Sept

Co. Longford
(15) Longford County Library · (tel) 043 41124 Mon–Fri
(16) Corlea Trackway Centre, Kenagh · (tel) 043 22386 April–Sept

Chapter 7: *Laois, Offaly & Kildare*
Co. Laois
(1) Steam Museum, Stradbally · (tel) 0502 25154 by appointment (2) Abbeyleix Heritage Centre · (tel) 0502 31653 daily (3) Mountmellick Enterprise Centre · (tel) 0502 24525 Mon–Fri

Co. Offaly
(4) Birr Castle · (tel) 0509 20336 daily (5) Slieve Bloom Centre, Kinnitty · (tel) 0509 37299 Tues–Fri (6) Bog Rail Tour, Shannonbridge · (tel) 090 967 4114 April–Oct (7) Tullamore Heritage Centre · (tel) 0506 25015 daily

Co. Kildare
(8) Peatland Museum, Lullymore · (tel) 045 860133 Mon–Fri (9) Quaker Museum, Ballitore · (tel) 059 862 3344 Tues–Sat (10) Crookstown Mill, Ballitore · (tel) 059 862 3222 Apr–Sept (11) Athy Heritage Centre · (tel) 059 863 3075 daily (12) County Library, Newbridge · (tel) 045 431 109 Tues–Fri (13) National Stud, Kildare · (tel) 045 521617 Feb–Nov (14) Straffan Steam Museum · (tel) 01 627 3155 Easter–Sept (15) Science Museum, St Patrick's College, Maynooth (tel) 01 628 5222 Tues, Thur, Sun

Chapter 8: *Mayo, Galway & Clare*
Co. Mayo
(1) Céide Fields, Ballycastle · (tel) 096 43325 Apr–Oct (2) Mayo North Heritage Centre, Crossmolina · (tel) 096 31809 Mon–Fri (3) Belderrig Research Centre/Ionad Taighde agus Staidéir Bhéal Deirg · (tel) 096 43987 by appointment (4) Basking Shark Survey, Merchant's Quay, Kilrush, Co. Clare (5) Foxford Woollen Mills · (tel) 094 925 6756 daily (6) Knock Folk Museum · (tel) 094 938 8100

April–Oct (7) Clew Bay Heritage Centre, Westport · (tel) 098 26852 April–Oct (8) Centre for Island Studies, Clare Island · (tel) 098 25412 (9) Doon Nature Peninsula, Burriscarra · (tel) 094 936 0287 Aug–Sept

Co. Galway
(10) Killary Cruises, Leenane · (tel) 1800 415151 April–Oct (11) Leenane Cultural Centre · (tel) 095 42323 April–Oct (12) Ocean's Alive, Renvyle · (tel) 095 43473 daily (13) Connemara National Park, Letterfrack · (tel) 095 41054 centre, April–Oct (14) Walking Ireland Tours, Clifden · (tel) 1850 266636 (15) Connemara Environmental Education Centre, Letterfrack National School · (tel) 095 41034 (16) Atlantaquaria, Salthill · (tel) 091 585100 daily (17) Glengowla Mine, Oughterard · (tel) 091 552360 March–Nov (18) Tuam Mill Museum · (tel) 093 25486 June–Sept (19) Galway City Museum · (tel) 091 567641 daily (20) James Mitchell Museum, NUIG · (tel) 091 524411 Mon–Fri (21) Coole Park, Gort · (tel) 091 631804 centre, Easter–Sept (22) Ionad Árann, Árann (Inishmore, Aran Islands) · (tel) 099 61355 April–Oct

Co. Clare
(23) Burren Exposure, Ballyvaughan · (tel) 065 707 7277 March–Oct (24) Burren Centre, Kilfenora · (tel) 065 708 8030 March–Oct (25) Burren Perfumery, Carran · (tel) 065 708 9102 daily (26) Aillwee Cave, Ballyvaughan · (tel) 065 707 7036 daily (27) Spa Wells, Lisdoonvarna · (tel) 065 707 4023 June–Oct (28) Liscannor Stone Story · (tel) 065 708 1930 daily (29) Clare County Museum, Ennis · (tel) 065 682 3382 Mon–Fri (30) Dolphinwatch, Carrigaholt · (tel) 065 905 8156 April–Oct (31) Scattery Island Ferries, Kilrush · (tel) 065 905 1327 April–Oct (32) Shannon Dolphin Foundation, Scattery Island Visitor Centre, Kilrush · (tel) 065 905 2139 June–Sept (33) Bunratty Castle & Folk Park · (tel) 1800 269 811 daily (34) Irish Seed Saver Association, Capparoe · (tel) 065 921 866 (35) Ardnacrusha Power Station · (tel) 061 345588 groups only

Chapter 9: Tipperary, Kilkenny & Carlow

Co. Tipperary

① Rock of Cashel · (tel) 062 61437 daily ② Mitchelstown Cave, Burncourt · (tel) 052 67246 daily ③ Transport Museum, Clonmel · (tel) 052 29727 daily ④ Slieve Ardagh Heritage Centre, Killenaule · (tel) 052 56165 Mon–Fri

Co. Kilkenny

⑤ Rothe House, Kilkenny · (tel) 056 772 2893 Mon–Sat ⑥ Patents Office, Kilkenny · (tel) 1890 220 223 Mon–Fri ⑦ Dunmore Cave, Ballyfoyle · (tel) 056 776 7726 March–Oct ⑧ Seamus Walsh Castlecomer Tours · (tel) 056 444 1504 ⑨ Castlecomer Library · (tel) 056 444 0055 Tues–Sat

Co. Carlow

⑩ County Museum, Carlow · (tel) 059 923 1759 Tues–Sun ⑪ Carlow Library · (tel) 059 917 0094 Mon–Fri

Chapter 10: Limerick & Kerry

Co. Limerick

① Flying Boat Museum, Foynes · (tel) 069 65416 April–Oct ② Aughinish Alumina, Askeaton · (tel) 061 604000 refinery tours: groups only ③ Kilfinane Outdoor Education Centre · (tel) 063 91161 ④ Lough Gur Centre · (tel) 061 385186 May–Sept ⑤ Dairy Co-operative & Plunkett Museum, Drumcollogher · (tel) 063 83433 May–Sept

Co. Kerry

⑥ Listowel Library · (tel) 068 23044 Tues–Sat ⑦ Vintage Wireless Museum, Listowel · (tel) 068 50346 by appointment ⑧ Crag Cave, Castleisland · (tel) 066 714 1244 March–Oct ⑨ Blennerville Windmill · (tel) 066 712 1064 Easter–Oct ⑩ Tralee–Blennerville Steam Railway, Tralee · (tel) 066 712 1064 May–Sept ⑪ Músaem Chorca Dhuibhne, Ballyferriter · (tel) 066 915 6333 Easter–Sept ⑫ Killarney National Park · (tel) 064 31440 centre: March–Oct ⑬ Muckross Traditional Farm, Killarney National Park · (tel) 064 31440 March–Oct ⑭ Heritage Centre, Knightstown, Valentia · (tel) 066 947 6411 May–Aug ⑮ Oceanworld, Dingle · (tel) 066 915 2111 daily ⑯ Fenit Seaworld · (tel) 066 713 6544 March–Oct ⑰ Valentia Observatory, Caherciveen · (tel) 066 947 2939 Thur ⑱ Skellig Experience, Valentia · (tel) 066 947 6306 daily ⑲ Kenmare Heritage Centre · (tel) 064 41233 Apr–Oct ⑳ Derreen Gardens, Lauragh · (tel) 064 83588 daily

Chapter 11: Cork & Waterford

Co. Cork

① Bealick Mill, Macroom · (tel) 026 42811 June–Oct ② Ilnacullin, Bantry Bay · (tel) 027 633040 Mar–Oct ③ 1796 Armada Exhibition, Bantry House · (tel) 027 50047 Apr–Oct ④ Cape Clear Bird Observatory, Clear Island (details from BirdWatch Ireland · (tel) 01 280 4322) ⑤ Mizen Vision, Mizen Head · (tel) 028 35115 March–Oct ⑥ Planetarium, Schull · (tel) 028 28552 March–Aug ⑦ Royal Gunpowder Mills, Ballincollig · (tel) 021 487 4430 April–Sept ⑧ Cork Public Museum, Fitzgerald Park · (tel) 021 427 6222 Sun–Fri ⑨ Cork Archives, South Main Street · (tel) 021 427 7809 by appointment ⑩ Cork Corporation Waterworks, Lee Road · (tel) 021 454 1761/496 6222 groups only ⑪ Beamish & Crawford, South Main Street · (tel) 021 427 6841 Tues, Thur ⑫ Cork Heritage Centre, Blackrock · (tel) 021 435 8854 May–Sept ⑬ Geology Museum, UCC · (tel) 021 490 2533 Mon–Fri ⑭ Radio Museum, Cork City Gaol · (tel) 021 430 5022 daily ⑮ Cork Butter Museum, Shandon · (tel) 021 430 0600 May–Sept ⑯ Meitheal Mara, Crosses Green House · (tel) 021 431 6813 by appointment ⑰ Monard Mills & Wildlife Sanctuary · (tel) 021 438 5564 groups only ⑱ Energy Agency Office, Mallow · (tel) 022 43610 Mon–Fri ⑲ Doneraile Park · (tel) 022 24244 daily ⑳ The Old Midleton Distillery · (tel) 021 461 3594 daily ㉑ Cobh Museum, Cobh · (tel) 021 481 4240 June–Sept ㉒ Fota Wildlife Park · (tel) 021 481 2678 March–Oct ㉓ Fota Arboretum & Garden · (tel) 021 481 2728 daily ㉔ Youghal Heritage Centre · (tel) 024 20170 Mon–Fri

Co. Waterford

(25) Heritage Centre, Lismore · (tel) 058 54975 Mon–Sat
(26) Lismore Castle · (tel) 058 54424 gardens: Easter–Sept
(27) Dungarvan Museum · (tel) 058 45960 Mon–Fri (28) Jack Burtchaell, Historic Waterford Tours · (tel) 051 873711
(29) Waterford Treasures, Waterford · (tel) 051 304500 daily
(30) Waterford Crystal · (tel) 051 373311 daily
(31) Bunmahon Heritage Centre · (tel) 051 292249 June–Aug

Chapter 12: *Wexford & Wicklow*

Co. Wexford

(1) Hook Lighthouse, Hook Head · (tel) 051 397055 daily
(2) Maritime Museum, Kilmore Quay · (tel) 051 561144 May–Sept (3) SWC/Jim Hurley, Natural Heritage Tours · (tel) 053 29671 (4) Windmill, Tacumshane (key held locally) (5) Water Mill, Craanford · (tel) 055 28392 Easter–Aug (6) Irish Agriculture Museum, Johnstown Castle · (tel) 053 42888 daily (7) JFK Arboretum, New Ross · (tel) 051 388 171 daily (8) Wexford Wildfowl Reserve, Ardcavan · (tel) 053 23129 daily (9) Carley's Bridge Pottery, Enniscorthy · (tel) 054 33512 Mon–Fri

Co. Wicklow

(10) Maritime Museum, Arklow · (tel) 0402 32868 daily
(11) Nick Coy, Avoca Tours · (tel) 045 866400 (12) Avondale House, Rathdrum · (tel) 0404 46111 house: March–Oct
(13) Kilmacurragh Arboretum, Rathdrum (tel) 01 837 4388 daily (14) Mount Usher Gardens, Ashford · (tel) 0404 40205 March–Oct (15) Heritage Centre, Bray · (tel) 01 286 7128 Mon–Sat (16) Sea Life Aquarium, Bray · (tel) 01 286 6939 May–Sept (17) Wicklow Mountains National Park, Glendalough · (tel) 0404 45425 May–Sept (18) Knocksink Environmental Education Centre, Enniskerry · (tel) 01 286 6609 by appointment (19) Powerscourt Demesne, Enniskerry · (tel) 01 204 6000 daily

Contact details for some associations, institutes and non-governmental organisations

Astronomy

– Astronomy Ireland, PO Box 2888, Dublin 1 · (tel) 01 459 8883 · www.astronomy.ie
– Irish Astronomical Society, PO Box 2547, Dublin 14
– Irish Astronomy Association, c/o R McLaughlin, 31 Manse Road, Ballygowan, Co. Down BT23 6HE

Geography and geology

– Geographical Society of Ireland, c/o Department of Geography, UCD Belfield, Dublin 4
– Institute of Geologists of Ireland, c/o Dr Julian Menuge, Geology Department UCD, Belfield, Dublin 4 · www.igi.ie
– Irish Geological Association, c/o Geology Department, TCD, Dublin 2
– Mining Heritage Trust of Ireland, 36 Dame Street, Dublin 2 www.mhti.com

Natural history and environment

– Belfast Naturalists' Field Club, c/o Richard Clarke, 78 Kings Road, Belfast BT5 6JN · (tel) 048 9079 7155
– BirdWatch Ireland, Rutledge House, 8 Longford Place, Monkstown, Co. Dublin · (tel) 01 280 4322 · www.birdwatch.ie
– Coppice Association of Ireland, c/o Aillwee Cave, Ballyvaughan, Co. Clare · (tel) 065 707 7036
– Dublin Naturalists' Field Club, c/o 35 Nutley Park, Dublin 4
– Irish Wildlife Trust, 1 07 Lower Baggot Street, Dublin 2 · (tel) 01 676 8588 · www.iwt.ie
– Irish Genetic Resources Conservation Trust, c/o TCD Botanic Gardens, Palmerston Park, Dublin 6 · (tel) 01 497 2070
– *Irish Naturalists' Journal*, c/o Ms Catherine Tyrie, Ulster Museum, Botanic Gardens, Belfast BT9 5AB
– Irish Peatland Conservation Council, 119 Capel Street, Dublin 1 · (tel) 01 872 2397· www.ipcc.ie

- Irish Seed Savers' Association, Capparoe, Scariff, Co. Clare · (tel) 061 921866 · www.irishseedsavers.ie
- Irish Whale & Dolphin Group, c/o Zoology Department, University College Cork · iwdg.ucc.ie
- Orchard Trust, c/o Department of Agriculture & Rural Development, Manor House, Loughall, Co. Armagh · (tel) 048 3889 2300
- Royal Society for the Protection of Birds (RSPB), Belvoir Park Forest, Belfast, BT8 4QT · (tel) 048 9049 1547 · www.rspb.org.uk
- An Taisce (National Trust for Ireland), Tailors' Hall, Back Lane, Dublin 8 · (tel) 01 454 1786 · www.antaisce.org
- Tree Council of Ireland, Cabinteely House, The Park, Cabinteely, Dublin 18 · (tel) 01 284 9211 · www.treecouncil.ie

Industry & Engineering:
- Industrial Heritage Association of Ireland, The Tailors' Hall, Back Lane, Dublin 8 · (tel) 01 454 1786 · www.steam-museum.ie/ihai
- Inland Waterways Association, c/o Stone Cottage, Claremont Rd, Killiney, Co. Dublin · www.iwai.ie

- Institution of Engineers of Ireland, 22 Clyde Road, Dublin 4 · (tel) 01 668 5508 · www.iei.ie
- Irish Railway Records Society, Box 9, Heuston Station, Dublin 8 · www.irrs.ie
- Railway Preservation Society of Ireland, PO Box 171 Larne, Co. Antrim BT40 1UU · (tel) 048 2826 0803 · www.rpsi-online.org

Other:
- Institute of Biology of Ireland, c/o UCD, Belfield, Dublin 4 · www.may.ie/ibi
- Institute of Chemistry of Ireland, c/o Science Section, RDS, Ballsbridge, Dublin 4 · www.instituteofchemistry.org
- Institute of Physics, c/o Alison Hackett, School of Science & Technology, IADT, Kill Avenue, Dún Laoghaire, Co. Dublin · www.iop.org
- Irish Mountaineering & Exploration History Society, c/o K Higgins, 6 Park View, Kilkenny · (tel) 056 61044
- Maritime Institute (and museum), Haigh Terrace, Dún Laoghaire, Co. Dublin · (tel) 01 280 0969 · www.mii.connect.ie

Bibliography

The research for this book drew on numerous sources; some of the more specific ones are cited at the end of the relevant chapter, while the more general titles are listed below. This list is by no means exhaustive, but it would be a starting point for anyone interested in reading further, although sadly a few titles are out of print. Particularly useful are the *Dictionary of National Biography* – published at intervals over the past century, it contains obituaries of many of the historical scientists featured here – and Lewis's *Topographical Dictionary of Ireland*, which records the Ireland of the early 1830s (occasionally reprinted, 'Lewis' is now also available on CD-ROM).

*indicates a core source

Biography

Dictionary of National Biography (Oxford University Press).

The Cambridge Dictionary of Scientists (Cambridge University Press, 1996).

Boylan, Henry (ed.), *A Dictionary of Irish Biography* (Gill & McMillan, 1998).

Coakley, Davis, *The Irish School of Medicine: Outstanding Practitioners of the 19th Century* (TownHouse & CountryHouse Ltd, 1988).

*Coakley, Davis, *Irish Masters of Medicine* (TownHouse & CountryHouse Ltd, 1992).

Garvin, Wilbert & Des O'Rawe, *Northern Ireland Scientists & Inventors* (Blackstaff Press, 1993).

Hart-Davis, Adam & Paul Bader, *100 Local Heroes* (Sutton Publishing, 1999).

Houston, Ken (ed.), *Creators of Mathematics: The Irish Connection* (University College Dublin, 2000).

Praeger, Robert Lloyd, *Some Irish Naturalists* (Dundalgan Press, 1949).

Mac Cnáimhín, Séamus, *Éireannaigh San Eolaíocht* (Oifig an tSoláthair, 1966).

Mollan, Davis & Finucane (eds), *Some People and Places in Irish Science & Technology* (Royal Irish Academy, 1985).

Mollan, Davis & Finucane (eds), *More People and Places in Irish Science & Technology* (Royal Irish Academy, 1990).

Mulvihill, Mary (ed.), *Stars, Shells & Bluebells: women scientists and pioneers* (WITS, 1997).

Newman, Kate, *Dictionary of Ulster Biography* (Institute of Irish Studies, 1993).

Nudds, McMillan, Weaire, McKenna & Lawlor (eds), *Science in Ireland: 1800–1930: Tradition and Reform* (D L Weaire, 1988).

Geology, landscape and maps

Field Guide to the Geology of Some Localities in County Dublin (TCD/ENFO, 1993).

The Ordnance Survey in Ireland: An Illustrated Guide (Ordnance Survey Ireland/Ordnance Survey Northern Ireland, 1991).

*Aalen, Whelan & Stout (eds), *Atlas of the Irish Rural Landscape* (Cork University Press, 1997).

Andrews, J H, *Shapes of Ireland: Maps and their Makers 1564–1839* (Geography Publications, 1997).

*Cole, Grenville, *Memoir of Localities of Minerals of Economic Importance and Metalliferous Mines in Ireland* (Geological Survey of Ireland, 1922) (Mining Heritage Society of Ireland facsimile edn, 1998).

Herries Davies, Gordon L, *North from the Hook* (Geological Survey of Ireland, 1995).

Kennan, Pádhraig, *Written in Stone* (Geological Survey of Ireland, 1995).

*Mitchell Frank & Michael Ryan, *Reading the Irish Landscape* 2nd edn (TownHouse & CountryHouse Ltd, 1997).

*Whittow, J B, *Geology and Scenery in Ireland* (Penguin, 1974).

Williams, Michael & David Harper, *The Making of Ireland: Landscapes and Geology* (Immel Publishing, 1999).

History

*Bowler, Peter & Nicholas Whyte (eds), *Science & Society in Ireland 1800–1950* (Institute of Irish Studies (QUB), 1997).

Burnett J E & A D Morrison-Low, *Vulgar & Mechanick: The Scientific Instrument Trade in Ireland 1650–1921* (Royal Dublin Society, 1989).

Byrne, Liam, *History of Aviation in Ireland* (Blackwater Press, 1980).

Connolly, S J (ed.), *The Oxford Companion to Irish History* (Oxford University Press, 1998).

*Foster, John Wilson & Helena Chesney (eds), *Nature in Ireland: A Scientific & Cultural History* (The Lilliput Press, 1997).

Hogg, William, *The Millers and Mills of Ireland of about 1850* (W E Hogg, 1997).

Kearney, R, *The Irish Mind: exploring intellectual traditions* (Wolfhound, 1985).

Kelleher, George, *From Gunpowder to Guided Missiles – Ireland's War Industries* (Kelleher, 1993).

*McMillan, Norman (ed.), *Prometheus's Fire* (Tyndall Publications, 2000).

Mollan, Charles, *The Mind and the Hand: Instruments of Science 1685–1932* (Trinity College Dublin, 1985).

Mollan, Charles, *Irish National Inventory of Historic Scientific Instruments (interim report)* (Royal Dublin Society/Eolas, 1990).

Mollan, Charles & John Upton, *The Scientific Apparatus of Nicholas Callan & Other Historic Instruments* (St Patrick's College Maynooth/SAMTON, 1994).

Shields, Lisa (ed.), *The Irish Meteorological Service 1936–1986* (Stationery Office Dublin, 1986).

Williams, Trevor, *Our Scientific Heritage: An A–Z of Great Britain & Ireland* (Sutton Publishing, 1996).

Natural history

The Flora of Dublin (Dublin Naturalists' Field Club, 1999).

*Cabot, David, *Ireland: A Natural History* (HarperCollins, 1999).

Carruthers, Terry, *Kerry: A Natural History* (The Collins Press, 1998).

Fairley, James, *An Irish Beast Book* (Blackstaff Press, 1984).

*Feehan, John & Grace O'Donovan, *The Bogs of Ireland* (University College Dublin, 1996).

Foss, Peter & Catherine O'Connell, *Irish Peatland Conservation Plan 2000* (Irish Peatland Conservation Council, 1996).

Giller, P (ed.), *Studies in Irish Limnology* (Marine Institute, 1998).

Hall, Valerie & Jonathan Pilcher, *Flora Hibernica* (The Collins Press, 2001).

Hayden, Tom & Rory Harrington, *Exploring Irish Mammals* (TownHouse & CountryHouse Ltd, 2000).

Mitchell, Frank, *The Way That I Followed: a naturalist's journey around Ireland* (TownHouse & Country House Ltd, 1990).

Moriarty, C (ed.), *Studies of Irish Rivers & Lakes* (Marine Institute, 1998).

Nairn, Richard & Miriam Crowley, *Wild Wicklow: Nature in 'the Garden of Ireland'* (TownHouse & CountryHouse Ltd, 1998).

Neeson, Eoin, *A History of Irish Forestry* (The Lilliput Press, 1991).

Nelson, Charles & Wendy Walsh, *Trees of Ireland: Native and Naturalised* (The Lilliput Press, 1993).

*Praeger, Robert Lloyd, *The Way That I Went* (1937) (The Collins Press, 1997).

Reynolds, Julian, *Ireland's Fresh Waters* (Marine Institute, 1998).

Engineering and railways

*Cox, Ron & Michael Gould, *Civil Engineering Heritage: Ireland* (Thomas Telford, 1998).

Hughes, Noel, *Irish Engineering 1760–1960* (Institution of Engineers in Ireland, 1982).

Long, Bill, *Bright Light, White Water: The Lighthouses of Ireland* (New Island Books, 1997).

O'Connor, Kevin, *Ironing the Land* (Gill & McMillan, 1999).

Wilson, T G, *The Irish Lighthouse Service* (Allen Figgis, 1968).

Miscellaneous

Cowan, Cathal & Regina Sexton, *Ireland's Traditional Foods* (Teagasc, 1997).

Cronin, Kevin, *Off the Beaten Track – Irish railway walks* (Appletree Press, 1996).

Deane, Ciarán, *Guinness Book of Irish Facts & Feats* (Gullane Publishing, 1994).

*Lewis, Samuel, *Topographical Dictionary of Ireland* (1837).

McAfee, Patrick, *Irish Stone Walls – history, building, conservation* (O'Brien Press, 1997).

Rothery, Seán, *A Field Guide to the Buildings of Ireland* (The Lilliput Press, 1997).

Permission notices

The lines from Patrick Kavanagh's poem "Monaghan Hills", reprinted on *page 229* are from his *Selected Poems,* edited by Antoinette Quinn (Penguin Books), reprinted with the permission of the trustees of the estate of the late Katherine B Kavanagh, through the Jonathan Williams Literary Agency.

The photographs of the Lartigue train (*page 381* ref R10591), the Boyne Viaduct (*page 120* SP 2107, the Lawrence Collection), the Fastnet lighthouse and Mizen Bridge (both *page 409*, the Commisisoners of Irish Lights Collection) and an early telephone (*page 43*, the Eircom Collection) are reproduced courtesy of the National Library of Ireland. The image of construction work on the Silent Valley Reservoir (*page 140*) is reproduced by kind permission of the Public Record Office of Northern Ireland (ref WAT/1/3G/1/8). The image of the mining 'man engine' (*page 404*) appears with the permission of the Royal Institution of Cornwall (ref MiDol 01a).

Every effort has been made to trace all copyright owners of the images used in this book. If any images have inadvertently been reproduced without permission, we would like to rectify this in future editions, and encourage owners of copyright, not acknowledged here, to contact us.

Index

Page references in bold indicate an image.Page references in **bold** indicate an image.

THE MUSEUM
INTERIOR

THE MUSEUM INTERIOR

TEMPORARY +
PERMANENT DISPLAY
TECHNIQUES

MICHAEL BRAWNE

with 330 illustrations, drawings and plans

THAMES AND HUDSON

First published in Great Britain in 1982
by Thames and Hudson Ltd, London

Copyright © by Verlag Gerd Hatje, Stuttgart

Printed and bound in West Germany

Encyclopedia of the

ANCIENT

GREEK

WORLD

Encyclopedia of the
ANCIENT GREEK WORLD

David Sacks

Historical Consultant:
Oswyn Murray,
*Fellow of Balliol College and
Lecturer in Ancient History at Oxford University*

Original drawings by Margaret Bunson

Constable • London

First published in Great Britain 1995
by Constable and Company Limited
3 The Lanchesters, 162 Fulham Palace Road
London W6 9ER
Copyright © 1995 David Sacks
All Drawings © 1995 Margaret Bunson
ISBN 0 09 475270 2

Published under license from Facts On File, Inc., New York

A CIP catalogue record for this book
is available from the British Library

Title page illustration: The great god Zeus and his emblematic bird, the eagle,
in a black-figure scene painted on the bottom of a cup from Laconia (the region of Sparta), circa 560 B.C.
The god is portrayed wearing his hair long, in the Spartan fashion.

Printed in the United States of America

This book is for
Rebecca and Katie Sacks.

CONTENTS

AUTHOR'S ACKNOWLEDGMENTS

There are several people to whom I am most grateful for help. My former teacher Oswyn Murray, fellow of Balliol College and lecturer in ancient history at Oxford University, vetted the manuscript with the same patience and receptiveness that distinguish his tutorial sessions. I owe him a great deal, not only for the book's preparation but also for my wider fascination with the ancient world. Two other scholars kindly donated their time to read sections and make comments: Gilbert Rose, professor of classics at Swarthmore College, and Christopher Simon, currently a visiting assistant professor of classics at the University of California, Berkeley. (Any factual errors here remain my own, however.)

I wish to thank my parents, Louis and Emmy Lou Sacks, for their unstinting encouragement. Ditto my good friend Jeffrey Scheuer. My thanks to Facts On File editor Gary Krebs, who kept the door open while the manuscript came together. Especially I must thank my wife, Joan Monahan, who brought home the family's bacon, day after day, during the four years consumed by this project.

Mainland Greece, the Aegean Sea, and Western Asia Minor

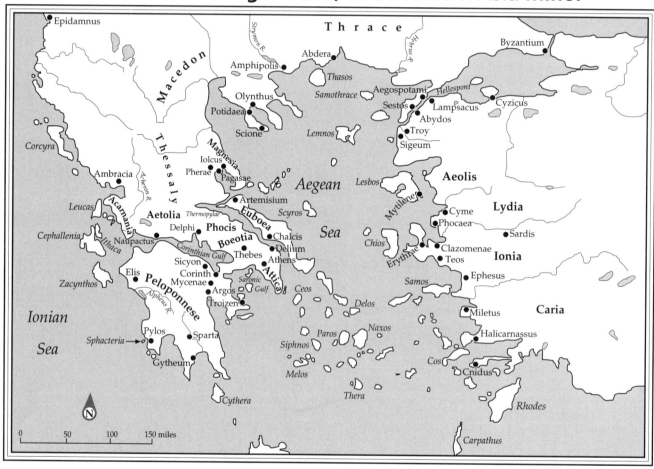

INTRODUCTION

About 2100 B.C. a migrant, cattle-herding, pony-riding people made their way into the Mediterranean land mass that today is called Greece. They entered overland from the north, probably the Danube Basin, but their origins may have been farther northeast, for they spoke a language of the Indo-European linguistic family. Modern philologists believe that the ancestral Indo-European language—whose modern descendants include English, German, Gaelic, French, Farsi, Hindi, and modern Greek—evolved in the fourth millennium B.C. on the plains of southern Russia. This mother tongue then branched into different forms, carried in all directions by nomadic tribes. The group that reached Greece circa 2100 B.C. brought with it an early form of the Greek language. These people can be called the first Greeks.

The land that they invaded was held by farmers who had probably immigrated centuries earlier from Asia Minor, a place with which they perhaps remained linked via an eastward trade network that included the Aegean island of Crete. They apparently knew seafaring and stone masonry—two skills that the nomadic Greeks did not yet have.

But the Greeks were the stronger warriors. They took over the country, probably by violence in the most desirable locales, but elsewhere perhaps by intermarriage (as may be reflected in the many Greek myths in which the hero marries the foreign princess). One apparent sign of conquest is the wrecked remnant of a pre-Greek palace that modern archaeologists call the House of the Tiles, at Lerna on the plain of Argos. Destroyed by fire circa 2100 B.C., this may have been the home of a native ruler who led an unsuccessful defense of the fertile heartland of southern Greece. Yet at certain other sites, archaeologists have found no clear signs of violence—only continued habitation and the abrupt emergence of a new style of pottery, betokening the Greeks' arrival.

The region that the Greeks now took over—and that would henceforth be their homeland—is a huge, jagged, southward-pointing peninsula, with a coastline stretching nearly 2,000 miles. Beyond its shores, particularly to the southeast, are islands that beckon to sea travelers and traders. Through the peninsula's center, from north to south, runs an irregular line of mountain ranges, whose slopes in ancient times held forests of oak, beech, and fir—timber for generations of house builders and shipwrights. In a later era, the limestone formations in these mountains would yield marble for sculptors and architects. But the mountains also occupied most of the mainland's total area, leaving only 20 percent as arable land.

Aside from scattered pockets, the farmland lay mainly in three regions: the plains of Argos, Boeotia, and Thessaly, in southern, central, and northern Greece, respectively. These territories were destined to become early Greek centers of power, especially the region of Argos, with its capital at Mycenae.

The soil of much of Greece is red or orange from clay deposits, which served centuries of potters and sculptors. In ancient times the farmed plains and foothills produced wheat, barley, olives, grapes, figs, and pomegranates—crops that could survive the ferociously hot, dry Greek summer. Summer, not winter, is the barren season in Greece, as in other parts of the Mediterranean. Winters are relatively mild—cool and rainy, but far rainier on the mainland's western side. The eastern regions, although traditionally densely populated, are blocked by the central mountains from receiving the westerly rainy weather. Athens gets only about 15 inches of rainfall a year; Corfu, on the west coast, has three times that much.

In such a country, where farmland and water supplies were precious, the Greek invaders of circa 2100 B.C. found most of the best locales already settled. The Greeks took over such settlements but kept their pre-Greek names. For that reason, the names of most ancient Greek cities do not come from the Greek language. Names such as Athens, Corinth, and Mycenae are not etymologically Greek; their original meanings are lost in prehistory. Relatively few ancient mainland sites have recognizably Greek names, among them Pylos ("the gate"), Megara ("the great hall"), Chalcis ("bronze city"), and Marathon ("fennel").

Eventually the Greeks acquired the civilizing arts of the people they had conquered. The Greeks learned shipbuilding, seamanship, and stoneworking—skills at which they excelled. More significantly, they borrowed from the non-Greeks' agrarian religion, which perhaps involved the worship of a mother goddess and a family of fertility deities. Non-Greek goddesses and beliefs, imported into Greek religion, served to complement and refine the warrior Greeks' Indo-European–type worship of a sky father and male gods. A new spirituality was born.

Thus in the centuries after 2100 B.C. came the creative fusion of two cultures—one primitive Greek, one non-Greek. To these two elements was added a third: the example and influence of the dynamic, non-Greek, Minoan civilization of Crete. By 1600 B.C. such factors had produced the first blossoming of the Greeks, in the Bronze Age urban society called the Mycenaean civilization.

For reasons never adequately explained, the Greeks of the next 15 centuries showed a spiritual and intellectual genius that expressed itself in religious awe, storytelling, poetry, sports, the material arts, trade, scientific studies, military organization, and in the governments of their self-contained city-states, particularly Athens. Their legacy to modern global society is immense. The Greeks invented democracy, narrative history writing, stage tragedy and comedy, philosophy, biological study, and political theory. They introduced the alphabet to European languages. They developed monumental styles of architecture that in the United States are used for museums, courthouses, and other public buildings. They created a system of sports competitions and a cult of physical fitness, both of which we have inherited. In sculpture, they perfected the representation of the human body. In geometry, they developed theorems and terminology that are still taught in schools. They created the idea of a national literature, with its recognized great writers and the libraries to preserve their work. And (perhaps what most people would think of first) the Greeks bequeathed to us their treasure trove of myths, including a hero who remains a favorite today—Heracles, or Hercules.

The early Greeks learned much about art and technology from Near Eastern peoples such as the Egyptians and Phoenicians. But more usually the Greeks became the teachers of others. They were an enterprising, often friendly people, and—as sea traders, colonists, mercenary soldiers, or conquerors—they traveled the world from southern Spain to Pakistan. Everywhere they went, they cast a spell through the magnetic appeal of their culture and style of life.

Their most fateful protégés were the Romans, a non-Greek people of Italy. Influenced by imported Greek goods and ideas from the 700s or 600s B.C. onward, the Romans modeled their religion largely on the Greeks', using Greek deities to shape their native Roman gods. This early stage was followed by a more elaborate copying—of Greek coinage, architecture, and other arts—starting in the 300s B.C. When the Romans sought to create their own national literature, they naturally turned to Greek models in epic and lyric poetry, history writing, rhetoric, tragedy, and comedy. They also became important patrons of Greek artists and craftsmen.

But meanwhile Roman armies were capturing Greek cities and kingdoms—first in Italy and Sicily (300s–200s B.C.), then in mainland Greece, Macedon, and Asia Minor (100s B.C.), and finally in Syria and Egypt (first century B.C.). Roman generals and governors plundered centuries' worth of Greek sculptures and other art treasures, removing them from temples and public squares and shipping them to Rome. In most locales, the inhabitants became tax-paying subjects of the Roman Empire. The Romans more or less put an end to the Greek achievement, even as they inherited it. The Roman poet Horace found a more hopeful phrasing for this when he wrote, about 19 B.C., "Captive Greece took mighty Rome captive, forcing culture onto rustic folk."

The Romans went on to conquer a domain that, at its greatest extent, stretched from Scotland to Mesopotamia. Their borrowed Greek culture became part of the permanent legacy of Europe and the eastern Mediterranean. Today we speak automatically of our "Greco-Roman" heri-

tage. But there was no necessary reason for the Romans to imitate the Greeks (the two did not even speak the same language), except that the ambitious Romans saw these people as superior to them in the civilizing arts.

The Romans were by no means the only ones to fall under the Greek spell. Another such people were the Celts. Extant Celtic pottery and metalwork clearly show that the "La Tène" culture, emerging circa 500 B.C. in Gaul (modern France, Switzerland, and Belgium), was inspired by Greek goods and influences, undoubtedly introduced up the Rhone River by Greek traders from Massalia (modern Marseilles, founded by Greeks circa 600 B.C.). By the first century B.C. the Celts of Gaul were writing in the Greek alphabet and had learned from the Greeks how to grow olive trees and grape vines (the latter mainly for wine-making). The creation of the French wine industry is a legacy of the ancient Greeks.

Similarly, from the 200s B.C. onward, the powerful African nation of Nubia, in what is now northern Sudan, traded with Greek merchants from Ptolemaic Egypt. In time the Nubian upper class adopted certain Greek styles of life: for instance, queens of Nubia were using the Greek name Candace down to the 300s A.D.

Nor were the Jews immune to Greek influence, especially after the conquests of Alexander the Great (334–323 B.C.) created a Greco-Macedonian ruling class in the Near East. In religion, Jewish monotheism was not much affected by Greek paganism. But in society and business, many Jews of Near Eastern cities adapted enthusiastically to the Greek world. They attended Greek theater, exercised publicly in Greek gymnasiums, and used the Greek language for commerce and public life. In Egyptian Alexandria (although not everywhere else), Greek-speaking Jews forgot their traditional languages of Hebrew and Aramaic. For the benefit of such people, a Greek translation of the Hebrew Bible began being produced in Alexandria during the 200s B.C. Thus, for many assimilated Jews of this era, Judaism was preserved in Greek form. Today a Jewish house of worship is known by a Greek word—synagogue (from *sunagogē*, "gathering place")—which is but one reminder of the Jews' fascination with the Greeks.

This encyclopedia attempts to give all the essential information about the ancient Greek world. Aimed at high-school and college students and general readers, the book tries to convey the achievements of the Greek world, while also showing its warts. (And warts there were, including slavery, the subordination of women, brutal imperialism, and the insanely debilitating wars of Greek against Greek.)

The encyclopedia's entries, from "Abdera" to "Zeus," range in length from about 100 to 3,000 words. The entries embrace political history, social conditions, warfare, religion, mythology, literature, art, philosophy, science, and daily life. Short biographies are given for important leaders, thinkers, and artists. Particular care is taken, by way of several entries, to explain the emergence and the workings of Athenian democracy.

The book's headwords include the names of real-life people (for example, Socrates), mythical figures (Helen of Troy), cities (Sparta), regions (Asia Minor), and institutions (Olympic Games), as well as many English-language common nouns (archaeology, cavalry, epic poetry, marriage, wine). Supplementing the text are more than 70 ink drawings, based mainly on photographs of extant Greek sculpture, vase paintings, architecture, and metalwork.

My research has involved English-language scholarly books and articles, ancient Greek works in translation, and many of the ancient Greek texts themselves. (I have used my own translations for quotations from Greek authors.) In writing this encyclopedia, I have tried to be aware of recent archaeological finds and other scholarly developments. My manuscript has been vetted by an eminent scholar. However, I have chosen and shaped the material for the general reader, not the scholarly one.

I have assumed that the reader knows nothing about the ancient Greeks and that he or she wants only the "best" information—that is, for any given topic, only the main points, including an explanation of why the topic might be considered important in the first place. I have tried to keep my language simple but lively and to organize each entry into a brisk train of thought. Although facts and dates abound in this book, I hope they only clarify the bigger picture, not obscure it.

In choosing the entries, I have had to abbreviate or omit much. Names or topics that might have made perfectly good short entries—Antaeus, grain supply, or Smyrna—have been reduced to mere cross-references in the text or to listings in the index. The reader is therefore urged to consult the index for any subject not found as an entry.

In time frame, the encyclopedia covers over 2,000 years, opening in the third millennium B.C. with the beginnings of Minoan civilization and ending with the Roman annexation of mainland Greece in 146 B.C. Occasionally an entry will trace an ongoing tradition, such as astronomy, beyond the cutoff date. And short entries are given for a few Roman-era Greek authors, such as Plutarch (circa A.D. 100) and the travel-writer Pausanias (circa A.D. 150), because their work sheds important light on earlier centuries. But most Greek personages and events of the Roman Empire, including the spread of Christianity, are omitted here as being more relevant to the Roman story than the Greek.

Within its 2,000-year span, the encyclopedia gives most attention to the classical era—that is, roughly the 400s and 300s B.C., which produced the Greeks' greatest intellectual and artistic achievements and most dramatic military conflicts. The 400s B.C. saw the Greeks' triumphant defense of their homeland in the Persian Wars, followed by Athens' rise as an imperial power. This was the wealthy, democratic Athens of the great names—the statesman Pericles, the tragedians Aeschylus, Sophocles, and Euripides, the historian Thucydides, the sculptor Phidias, and the philosopher Socrates. In these years the Parthenon arose and the fateful Peloponnesian War was fought, ending in Athens' defeat. The 300s B.C. brought the rise of Macedon and the conquest of the Persian Empire by the Macedonian king Alexander the Great. This was the time of the philosophers Plato and Aristotle, the historian Xenophon, the orator Demosthenes, and the swashbuckling Macedonian prince Demetrius Poliorcetes. Many of the topics that will bring readers to a book about ancient Greece fall within these two centuries.

In a book of this scope written by one person, certain preferences are bound to sneak in. I have tried always to be thorough and concise. But I have allowed slightly more space to a few aspects that I consider more likely than others to satisfy the general reader's curiosity. When I studied Greek and Latin at graduate school, my happiest hours were spent reading Herodotus. He was an Ionian Greek who, in the mid 400s B.C., became the world's first historian, writing a long prose work of incomparable richness about the conflict between the Greeks and Persians. And I find, with all humility, that I have favored the same aspects that Herodotus tends to favor in his treatment—namely, politics, personalities, legends, geography, sex, and war.

David Sacks

Mainland Greece and Neighboring Regions

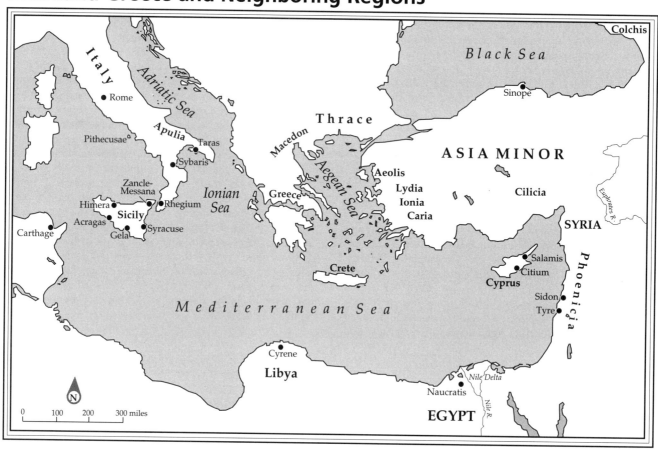

Abdera Important Greek city on the north Aegean coast in the non-Greek region known as THRACE. Located on a coastal plain near the mouth of the River Nestos, Abdera served as a depot for TRADE with local Thracian tribesmen and as an anchorage on the shipping route between mainland Greece and the HELLESPONT. After a failed beginning in the 600s B.C. due to Thracian attacks, Abdera was reestablished by Greek colonists circa 544 B.C. These settlers came from Teos—a city in the Greek region of western ASIA MINOR called IONIA—which they had abandoned to the conquering Persians under King CYRUS (1). Among the Tean settlers was a young man, ANACREON, destined to become the most famous lyric poet of his day.

Like other Greek colonies of the northern Aegean, Abdera prospered from Thracian trade, which brought GOLD and SILVER ore, TIMBER, and SLAVES (available as war captives taken in Thracian tribal wars). These goods in turn became valuable Abderan exports to mainland Greece and other markets. Local wheatfields and fishing contributed to prosperity. The disadvantages were periodic Thracian hostility and the northern climate (cold and wet by Greek standards).

Lying directly in the path of the Persian invasion of the spring of 480 B.C., Abdera submitted to the Persian king XERXES and hosted him at legendary expense. After the Persian defeat (479 B.C.), Abdera became an important member of the Athenian-controlled DELIAN LEAGUE (478 or 477 B.C.). In 457 B.C. wealthy Abdera was paying an annual Delian tribute of 15 TALENTS (as much as BYZANTIUM and more than any other state except AEGINA).

Although other Greeks considered the Abderans to be stupid, the city produced at least two important thinkers of the middle and late 400s B.C. the sophist PROTAGORAS and the atomist philosopher DEMOCRITUS. In these years Abdera, like other cities of the silver-mining north Aegean, was famous for the beauty of its COINAGE. The city's symbol on coins was an ear of wheat.

Abdera passed briefly to Spartan influence after Athens' defeat in the PELOPONNESIAN WAR (404 B.C.), but by about 377 B.C., Abdera was a member of the SECOND ATHENIAN LEAGUE. Seized by the Macedonian king PHILIP II circa 354 B.C., Abdera remained within the Macedonian kingdom over the next 180 years. Sacked by Roman troops in 170 B.C. during the Third Macedonian War, Abdera recovered to become a privileged subject city in the Roman empire.

The ancient site, excavated since the 1950s, has yielded the outline of the city wall and the admirably precise grid pattern of the city's foundations.

(See also COLONIZATION; PERSIAN WARS; ROME.)

Abydos See SESTOS.

Academy The Akademeia was a GYMNASIUM and park about a mile outside ATHENS, sacred to the local hero Akademos. In around 386 B.C., PLATO bought land and buildings there and set up a school of PHILOSOPHY, which can be counted as the Western world's first university.

Plato's aim was to train future leaders of Athens and other Greek states. Students at the early Academy did not pay fees, and lessons probably took place in seminars similar to the disputations portrayed in Plato's written *Dialogues*. Teachings emphasized MATHEMATICS and the Platonic reasoning method known as dialectic. In its breadth of inquiry, the Academy of 386 B.C. was distinct from all prior Greek schools of advanced study, which taught only RHETORIC, poetry, or the argumentative techniques of the SOPHISTS.

Two great students of the early Academy were the mathematician-astronomer Eudoxus of CNIDUS and the philosopher ARISTOTLE. Aristotle was considered Plato's possible successor as president, but after the master's death Academy members voted Plato's nephew Speusippus as head (347 B.C.). Aristotle evetually set up an Athenian philosophical school of his own, called the LYCEUM.

Under Speusippus and his successors, the Academy's curriculum became more mathematical and abstract, until Arcesilaus of Pitane (president circa 265–242 B.C.) redirected it toward philosophical SKEPTICISM. Arcesilaus and his distant successor Carneades (circa 160–129 B.C.) both were known for their criticisms of the rival school of STOICISM.

After the Romans annexed Greece (146 B.C.), the Academy attracted students from all over the Roman—and later the Byzantine—Empire. The Academy survived more than 900 years from its founding, until the Christian Byzantine emperor Justinian closed it and the other pagan philosophical schools in A.D. 529.

The school's name has produced the English common noun *academy*, meaning a place of rigorous advanced study. (See also EDUCATION.)

Acarnania A region of northwest Greece between the Gulf of Patras (to the south) and the Gulf of AMBRACIA (to the north). Although largely mountainous, the region contains a fertile alluvial plain along the lower Acheloüs River. Acarnania was inhabited by rough Greek "highlanders" who in the 400s B.C. were still known for carrying weapons in public. Their main town was named Stratos, and their political structure was a loose-knit union of rural cantons (later, of towns). Acarnania was bordered west and east by hostile neighbors—the Corinthian colonies of the seaboard and the inland people of AETOLIA. Because of these threats, the Acarnanaians sought alliances with several great states of the Greek world.

As allies of ATHENS in the PELOPONNESIAN WAR, Acarnanian troops under the Athenian general DEMOSTHENES (2) wiped out most of the army of the Corinthian colony of Ambracia in three days (426 B.C.). In 338 B.C. Acarnania (with the rest of Greece) passed to the control of the Macedonian king PHILIP II. The Acarnanians were staunch allies of King PHILIP V in his wars against Aetolia and shared his defeat in the disastrous Second Macedonian War against ROME (200–196 B.C.). Thereafter, Acarnania passed into Roman hands.

(See also ALCMAEON (1)).

Achaea For most of ancient Greek history, the place-name Achaea was applied to two different regions of Greece: (1) the hilly northwest corner of the PELOPONNESE, and (2) a small area in THESSALY. The Peloponnesian Achaea (the more important of the two) was at some early date organized into a 12-town Achaean League, with shared government and citizenship. In the late 700s B.C., Achaean colonists founded or cofounded important Greek cities in south ITALY, including CROTON and SYBARIS.

Under the commander Aratus of SICYON (active circa 250–213 B.C.), the Achaean League emerged as the strongest power of mainland Greece. By tapping the Greeks' hatred of Macedonian overlordship, Aratus united the northern Peloponnese against MACEDON (and in defiance of the rival Peloponnesian state of SPARTA). For a few years the democratic league was the last hope for that unfulfilled dream of Greek history: the creation of an independent, federal state of Greece.

But the maneuverings of two greater powers, Macedon and ROME, made it impossible for the league to survive alone. As a Roman ally (198 B.C. and after), the Achaean League encompassed most of the Peloponnese, including the important city of CORINTH. However, resistance to Roman interference led to the disastrous Achaean War of 146 B.C., in which the Romans sacked Corinth, dissolved the league, and made Achaea part of a Roman province.

(See also ACHAEANS.)

Achaeans The word Achaioi (Achaeans) is one of the terms used by the poet HOMER (circa 750 B.C.) as a general name for the Greeks. In this, Homer probably preserves a usage of the Mycenaean Age (circa 1600–1200 B.C.), when Achaioi would have been the Greeks' name for themselves. As a result, modern scholars sometimes use the name *Achaeans* to mean either the Mycenaeans or their ancestors, the first invading Greek tribesmen of about 2100 B.C.

Intriguingly, a place-name pronounced *Ahhiyawa* has been deciphered in the cuneiform annals of the Hittite people of ASIA MINOR (1300s–1200 B.C.). In the documents, the name indicates a strong foreign nation, a sea power, with which the Hittite kings were on polite terms. Possibly this foreign nation was the mainland Greek kingdom ruled from the city of MYCENAE. The Hittite rendering *Ahhiyawa*

may reflect a Greek place-name, Achaiwia or "Achaea," meaning the kingdom of Mycenae.

In later centuries, the Greek place-name ACHAEA came to denote a region of the northwestern PELOPONNESE, far from Mycenae.

(See also MYCENAEAN CIVILIZATION.) Probably that name arose because surviving Mycenaeans took refuge there after their kingdom's downfall.

Achilles Preeminent Greek hero in the legend of the TROJAN WAR. Achilles (Greek: Achilleus, perhaps meaning "grief") was son of the hero PELEUS and the sea goddess Thetis. He figured in many tales, but received his everlasting portrait as the protagonist of HOMER's epic poem the *Iliad* (written down circa 750 B.C.). At the story's climax, Achilles slays the Trojan champion HECTOR in single combat, fully aware that his own preordained death will follow soon.

To the Greek mind, Achilles embodied the old-time heroic code, having specifically chosen a brief and glorious life over one that would be safe and obscure. Achilles recounts the terms of this choice in a well-known passage in the *Iliad*:

> My goddess mother says that two possible destinies bear me toward the end of life. If I remain to fight at Troy I lose my homecoming, but my fame will be eternal. Or if I return to my dear home, I lose that glorious fame, but a long life awaits me [book 9, lines 410–416].

The *Iliad*'s announced theme is "the anger of Achilles" (book 1, lines 1–2). Opening in the war's 10th year, the poem portrays Achilles as a glorious individualist, noble and aloof to the point of excessive pride. Still a young

Achilles—the most formidable of all Greek warriors who sailed against Troy—waits at a temple of Apollo to ambush the Trojan prince Troilus, in a black-figure scene on a cup from Laconia, circa 550 B.C. Achilles is shown driving away the temple's guardian serpents.

man, he has come to the siege of TROY from his native THESSALY at the head of a contingent of troops, his Myrmidons ("ants"). After quarreling justifiably with the commander-in-chief, King AGAMEMNON, over possession of a captive woman named Briseis, Achilles withholds himself and his men from the battlefield (book 1). Consequently, the Greeks suffer a series of bloody reversals (books 8–15). Achilles rebuffs Agamemnon's offered reconciliation (book 9) but relents somewhat and allows his friend PATROCLUS to lead the Myrmidons to battle (book 16). Wearing Achilles' armor, Patroclus is killed by Hector, who strips the corpse.

Mad with grief, Achilles rushes to battle the next day wearing wondrous new armor, forged for him by the smith god HEPHAESTUS at Thetis' request (books 18–20). After slaying Hector, he hitches the Trojan's corpse to his chariot and drags it in the dust to the Greek camp (book 22).

His anger thus assuaged, Achilles shows his more gracious nature in allowing Hector's father, the Trojan king PRIAM, to ransom the body back (book 24). At the Iliad's end Achilles is still alive, but his death has been foretold (for example, in book 19, lines 408–417). He will be killed by the combined effort of the Trojan prince PARIS and the god APOLLO, patron of the Trojans.

Greek writers later than Homer provide details of Achilles' life before and after the Iliad's events. At Achilles' birth his mother tried to make him immortal by dipping him into the River Styx (or into fire or boiling water, in other versions). But she was interrupted or otherwise forgot to immerse the baby's right heel, and this later proved to be the hero's vulnerable "Achilles' heel."

Knowing at the Trojan War's outset that her son would never return if he departed, Thetis arranged with Lycomedes, king of the island of Scyros, to hide Achilles, disguised as a girl, in the WOMEN's quarters of the king's palace. There Achilles fathered a son with Lycomedes' daughter Deidameia; the boy was named NEOPTOLEMUS. The Greeks, having heard a prophecy that they could never take Troy without Achilles' help, sent ODYSSEUS and other commanders to find Achilles, which they did.

At Troy, Achilles showed himself the greatest of warriors, Greek or Trojan. Among the enemy champions he slew were, in sequence: Cycnus, TROILUS, Hector, Queen Penthesilea of the AMAZONS, and the Ethiopian king MEMNON. At last Achilles himself died, after his vulnerable heel was hit by an arrow shot by Paris and guided by Apollo. (Either the arrow was poisoned or the wound turned septic.) In Homer's Odyssey (book 11), Odysseus meets Achilles among the unhappy ghosts in the Underworld. But later writers assigned to Achilles a more blissful AFTERLIFE, in the Elysian Fields.

(See also FATE; PROPHECY AND DIVINATION.)

Acragas (Greek: Akragas, modern Agrigento) The second most important Greek city of SICILY, after SYRACUSE. Today the site of ancient Acragas contains some of the best-preserved examples of Doric-style monumental Greek ARCHITECTURE.

Located inland, midway along the island's southern coast, Acragas is enclosed defensively within a three-sided, right-angled mountain ridge. The city was founded in about 580 B.C. by Dorian-Greek colonists from the nearby city of GELA and the distant island of RHODES. Acragas lay close to the west Sicilian territory of the hostile Carthaginians, and the city soon fell under the sway of a Greek military tyrant, Phalaris, who enlarged the city's domain at the expense of the neighboring native Sicans, circa 570–550 B.C. (Notoriously cruel, Phalaris supposedly roasted his enemies alive inside a hollow, metal bull set over a fire.)

Acragas thrived as an export center for grain to the hungry cities of mainland Greece. Local WINE, olives, and livestock added to the city's prosperity. Under the tyrant Theron (reigned 488–472 B.C.), Acragas became the capital of a west Sicilian empire. Theron helped defeat the Carthaginians at the Battle of HIMERA (480 B.C.) and used Carthaginian war captives as labor for a grand construction program at Acragas. Among Theron's works was a temple of ZEUS, never finished but intended to be the largest building in the Greek world.

After ousting Theron's son and successor, Thrasydaeus, the Acragantines set up a limited DEMOCRACY (circa 472 B.C.). Associated with this government was Acragas' most illustrious citizen—the statesman, philosopher, and physician EMPEDOCLES (circa 450 B.C.). In Empedocles' time, Acragas underwent a second building program, whose remnants include the temples that stand today along the city's perimeter ridge, as if guarding the site. The most admired of these is the beautifully preserved Temple of Concord (so-called today, perhaps really a temple of CASTOR AND POLYDEUCES).

Captured and depopulated by the Carthaginians in 406 B.C., Acragas was resettled by the Corinthian commander TIMOLEON (circa 338 B.C.) but never recovered its former greatness. By about 270 B.C. the city was again a Carthaginian possession. As a strategic site in the First Punic War between CARTHAGE and ROME (264–241 B.C.), Acragas was twice besieged and captured by Roman troops. In 210 B.C. it was again captured by the Romans and soon thereafter repopulated with Roman colonists.

(See also DORIAN GREEKS; ORPHISM; TYRANTS.)

acropolis The "upper city," or hilltop citadel, was a vital feature of most ancient Greek cities, providing both a refuge from attack and an elevated area of religious sanctity. The best-known acropolis is at ATHENS, where a magnificent collection of temples and monuments, built in the second half of the 400s B.C., remains partially standing today; the most famous of these buildings is the PARTHENON, the Temple of the Virgin ATHENA. In terms of natural setting, the highest and most dramatic Greek acropolis was on the 1,800-foot mountain overlooking ancient CORINTH.

Most primitive societies naturally concentrate their settlements on hilltops. In Greece, the hilltop citadels of pre-Greek inhabitants were attacked and captured by Greek-speaking tribesmen in around 2100 B.C. On the choicest of these hills arose the royal palaces of MYCENAEAN CIVILIZATION (1600–1200 B.C.). The Mycenaean Greeks favored hilltops close to farm plains and not too close to the sea, for fear of pirate raids. Typical Mycenaean sites include MYCENAE, Athens, and Colophon (meaning "hilltop"), a Greek city in ASIA MINOR. Of the great classical Greek

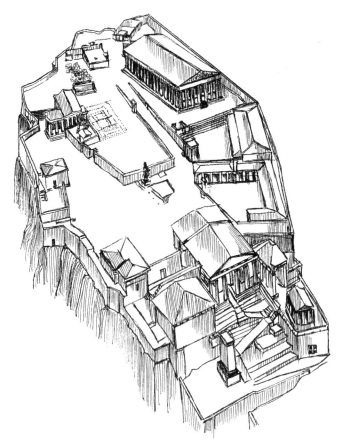

The Athenian acropolis, in a bird's-eye view from the northwest, circa 400 B.C. On the highest part of the summit stood Athena's most glorious temple, the Parthenon, *rear right*. The Erectheum, *rear left*, was a temple housing several patriotic cults. The acropolis entrance was in the west, through the colonnades and roofed gateway of the Propylaea, *front center*. The little temple of Athena Nike stood perched outside the Propylaea, *front right*.

cities, only SPARTA—a post-Mycenaean settlement—had a puny, unfortified acropolis. Rather than rely on a defensive citadel, Sparta relied on its invincible army and on the mountain ranges enclosing the region.

The Athenian acropolis is a limestone-and-schist formation, rising about 300 feet above the lower town. The hill's association with the goddess Athena probably dates from Mycenaean times (circa 1200 B.C.), when the king's palace stood there. ARCHAEOLOGY reveals that the acropolis' upper sides were first enclosed in a man-made wall circa 1200 B.C.; a later wall from ancient times still encloses the upper rock face today.

Like other Greek citadels, the Athenian acropolis played a role in its city's turbulent politics. CYLON (circa 620 B.C.) and PISISTRATUS (circa 560 B.C.) each began an attempted coup by seizing the acropolis; later, as dictator (546–527 B.C.), Pisistratus beautified the site with marble temples to the gods. But these were burned down by the occupying Persians of 480–479 B.C.

The remarkable monuments now standing on the Athenian acropolis—a focus of international tourism in the ancient world, as today—derive from the building program

prompted in around 448 B.C. by the Athenian statesman PERICLES and supervised by the sculptor PHIDIAS. To celebrate a peace treaty with PERSIA and to glorify the city, the Athenians voted to erect temples to replace those destroyed by the Persians 30 years before. The resulting group includes the small, Ionic-style temple of Athena Nikē (Athena of Victory), the Ionic-style Erectheum (which housed the cults of Athena, POSEIDON, and the legendary king Erectheus), and the monumental Propylaea (gateway). But the pride of the group is the Parthenon.

Financing for the acropolis building program came from tribute paid by Athens' allies within the DELIAN LEAGUE. The fact that other Greek states paid for Athens' beautification caused angry debate at that time, both among the allies and within Athens itself.

(See also ARCHITECTURE; CALLIAS; PERSIAN WARS; THUCYDIDES (2).)

Adonis   A beautiful mortal youth who was a lover of the goddess APHRODITE. According to the usual version of the MYTH, Adonis was the son of a Cypriot or Syrian princess who had fallen in love with her own father and became impregnated by him. Growing up, Adonis became the beloved of both PERSEPHONE and Aphrodite. When the two rival goddesses appealed to ZEUS, he decreed that Adonis should spend part of the year with each. (This myth resembles the similar tale of DEMETER and Persephone).

Out hunting one day in the mountains of what is now Lebanon, Adonis was gored to death by a wild boar—the disguised form of the jealous god ARES, Aphrodite's occasional lover. Roses or anemones sprang from the dying youth's blood; these scarlet flowers recall Adonis' beauty and mortality.

At ATHENS and other cities of classical Greece, the death of Adonis was commemorated each summer in a WOMEN's festival lasting about eight days. At the culmination, women of all social classes would stream out of the city in a mourning procession, wailing for the slain Adonis and carrying effigies of him to be thrown into the sea. For this occasion, women would cultivate "gardens of Adonis"—shallow baskets of earth in which seeds of wheat, fennel, and flowers were planted, to sprout quickly and then die and be thrown into the sea. While probably symbolizing the scorched bleakness of the eastern Mediterranean summer, this strange rite also invites a psychological interpretation—as a socially permitted emotional release for Greek women, amid their repressed and cloistered lives.

The worship of Adonis is a prime example of Greek cultural borrowing from non-Greek peoples of the Near East. According to modern scholars, the Greek cult of Adonis derived from a Phoenician festival of the mother goddess Astarte and her dying-and-reborn lover Baalat or Tammuz. The center of this worship was the Phoenician city of BYBLOS. Around the 700s B.C., Greek or Phoenician merchants brought this worship from Byblos to Greece perhaps by way of the Greco-Phoenician island of CYPRUS. In the Greek version of the myth, the sex-and-fertility goddess Astarte becomes the love goddess Aphrodite.

The name Adonis is not Greek, but rather reflects the Phoenician worshippers' ritual cry of *Adon*, meaning "lord." (Compare Hebrew *Adonai*, "the Lord.")

Adrastus See SEVEN AGAINST THEBES.

adultery See MARRIAGE.

Aegean Sea The approximately 80,000-square-mile section of the eastern Mediterranean stretching between Greece and ASIA MINOR, bounded on the north by the coast of ancient THRACE and on the south by the island of CRETE. The Aegean contained or bordered upon most of the important ancient Greek states. The sea was supposedly named for the mythical King Aegeus, father of the Athenian hero THESEUS; Aegeus was said to have drowned himself in this sea. But its name may in fact come from the Greek word *aigis*, "storm."

(See also CHIOS; CYCLADES; GREECE, GEOGRAPHY OF; LESBOS; RHODES; SAMOS.)

Aegina Small island state in the Saronic Gulf, in southeast-central Greece. Only 33 miles square, the triangular island lies 12 miles southwest of the Athenian coast and five miles northeast of the nearest point on the Argolid. Aegina's capital city, also called Aegina, stood in the northwest part of the island, facing the Argolid and Isthmus.

In prehistoric times, Aegina was inhabited by pre-Greek peoples and then by Mycenaean Greeks before falling to the invading DORIAN GREEKS in circa 1100–1000 B.C. The unfertile island gave rise to merchant seamen who claimed descent from the mythical hero Aeacus (son of ZEUS and the river nymph Aegina). In the late 600s and the 500s B.C., Aegina was a foremost Greek sea power, with a Mediterranean TRADE network rivaling that of CORINTH. In the 400s B.C., however, the island became a bitter enemy of nearby ATHENS. The Athenian statesman PERICLES (mid-400s B.C.) called Aegina "the eyesore of Piraeus"—a hostile presence on the sea horizon, as viewed from Athens' main harbor.

The Aeginetans' trade routes have been difficult to trace, because they were simply the middlemen in the selling of most wares. Specifically, they manufactured no POTTERY of their own for modern archaeologists to find in far-off locales. But we know that Aeginetan trade reached EGYPT and other non-Greek Near Eastern empires. Circa 595 B.C. Aegina became the first Greek state to mint coins—an invention probably learned from the kingdom of LYDIA, in ASIA MINOR. Made of SILVER and stamped with the image of a sea turtle, Aegina's COINAGE inspired other Greek states to start minting.

The 500s B.C. were Aegina's heyday. Relations with Athens had not yet soured. Aeginetan shippers brought Athenian black- and red-figure pottery to the ETRUSCANS of western ITALY; they probably also brought WINE, metalwork, and textiles. In exchange, the Etruscans gave raw metals such as silver and tin.

Aegina's prosperity is reflected in the grand temple of the goddess Aphaea (a local equivalent of ATHENA or ARTEMIS), built of local limestone soon after 500 B.C. Located near the island's northeast coast and still partly standing today, this Doric-style structure is the best-preserved early temple in mainland Greece. The building's pediments contained marble figures of mythical Greek heroes fighting at TROY; these important archaic SCULPTURES were carted off in A.D. 1811 and now are housed in a Munich museum.

Hostility with Athens flared in the late 500s B.C. The two states had become trade rivals, and Athens feared Aegina's navy, the largest in Greece at that time (about 70 ships). Hatred worsened when Aegina submitted to envoys of the Persian king DARIUS (1), the enemy of Athens (491 B.C.).

By about 488 B.C., Aegina and Athens were at war. The Athenians, urged by their statesman THEMISTOCLES, built 100 new warships, doubling their navy's size. But this Athenian navy fought alongside the Aeginetans in defending Greece against the invasion of the Persian king XERXES (480 B.C.).

After the Persians' retreat, Aegina joined SPARTA's alliance for protection against Athens. Nevertheless, in 459 B.C. the Athenians defeated an Aeginetan fleet, landed on the island, and besieged the capital. Defeated, Aegina was brought into the Athenian-controlled DELIAN LEAGUE and made to pay the highest tribute of any member, 30 TALENTS. Probably at this time Athens settled a garrison colony on the island. Aegina's anger and defiance in these years are suggested in certain verses by the Theban poet PINDAR, who had friends there.

According to the Thirty Years' Peace, agreed to by Athens and Sparta in 446 B.C., the Athenians were supposed to grant Aegina a degree of self-determination (*autonomia*). This promise was never kept. The resentful Aeginetans continually urged the Spartans against Athens until, at the outbreak of the PELOPONNESIAN WAR (431 B.C.), Athens evicted the Aeginetans and repopulated the whole island with Athenian colonists. The Aeginetans, resettled by Sparta, eventually were reinstalled on Aegina by the triumphant Spartan general LYSANDER (405 B.C.).

After some renewed hostility toward Athens in the CORINTHIAN WAR (395–386 B.C.), Aegina fades from history.

(See also NAUCRATIS; PERSIAN WARS; PIRAEUS; WARFARE, NAVAL.)

Aegospotami Aigospotamoi, "goat's rivers," was a shoreline on the European side of the HELLESPONT, opposite the city of Lampsacus, where the strait is about two miles wide. There in September 405 B.C., the final battle of the PELOPONNESIAN WAR was fought. At one swoop, the Spartan commander LYSANDER eliminated the Athenian fleet and left the city of ATHENS open to blockade and siege. Within eight months of the battle, Athens had surrendered.

The battle was waged over possession of the Hellespont. In the summer of 405 B.C., Lysander slipped into the Hellespont with a fleet of about 150 warships. There he captured the Athenian ally city Lampsacus and occupied its fortified harbor. In pursuit came 180 Athenian warships—almost the entire Athenian navy—led by six generals drawn from a depleted Athenian high command.

The Athenians encamped opposite Lampsacus, on the open shore at Aegospotami. The next morning they rowed out toward Lampsacus to offer battle. But Lysander kept his fleet inside the harbor's defenses, which the Athenians were unwilling to attack. Returning late in the day to Aegospotami, the Athenians beached their ships and went ashore for firewood and food. Lysander sent out a few fast ships to spy on them.

This procedure continued for several days. The Athenian generals did not withdraw to the nearby port city of SESTOS, where a fortified harbor could offer defense; apparently they thought a withdrawal would allow Lysander to escape.

On the fifth evening the Athenian crews beached their ships as usual at Aegospotami and went ashore. This time the Spartan scout ships signaled back to Lysander's fleet—which immediately rowed out from Lampsacus and attacked. The Athenians were completely unprepared: many of their ships still lay empty as the Spartans reached them. Only one Athenian leader, CONON, got his squadron away; the other 170 or so Athenian ships were captured, with most of their crewmen. The Spartans collected their prisoners—perhaps 5,000 Athenians and allies—and put to death the 3,000 or so Athenians among them.

(See also WARFARE, NAVAL.)

Aeneas In Greek MYTH, Aeneas (Aineias) was a Trojan hero of royal blood, the son of the goddess APHRODITE and the mortal man Anchises. Aeneas' earliest appearance is as a minor character in HOMER's epic poem the *Iliad* (written down circa 750 B.C.). He is shown as a respected figure, pious to the gods (who protect him in his overambitious combats with the Greek champions DIOMEDES and ACHILLES). The god POSEIDON prophesies that Aeneas will escape Troy's doom and that his descendants will rule future generations of Trojans (book 20).

Over later centuries, partly in response to Greek exploration and COLONIZATION in the western Mediterranean, there arose various non-Homeric legends describing how, after the fall of Troy, Aeneas voyaged westward, establishing cities in SICILY, ITALY, and elsewhere. In the first century B.C. the Roman poet Vergil amalgamated these tales in his patriotic Latin epic poem, the *Aeneid*. Vergil's Aeneas endures hardships and war in order to found the city of Lavinium and initiate a blood line that will eventually build the city of ROME. Aeneas was thus one of the very few Greek mythological figures who was more important in the Roman world than in the Greek.

(See also TROJAN WAR.)

Aenus Rich and important Greek trading city on the northeastern Aegean coast, in the principally non-Greek region known as THRACE. Aenus (Greek: Ainos) was founded circa 600–575 B.C. by colonists from CYME and other AEOLIAN GREEK cities of ASIA MINOR. The city lay advantageously at the mouth of the Hebrus River, in the territory of the powerful Odrysian Thracians. Like its distant Greek neighbor ABDERA, Aenus prospered from TRADE with the Thracians, who brought TIMBER, SLAVES, SILVER ore, and other precious resources for overseas export to the major markets of Greece.

Circa 477 B.C. Aenus became an important member of the Athenian-controlled DELIAN LEAGUE. Around this time Aenus was minting one of the most admired silver coinages in the Greek world; the Aenian coins showed the head of HERMES, god of commerce. The city remained an Athenian ally during the PELOPONNESIAN WAR (431–404 B.C.), and it came under Spartan rule after Athens' defeat. Later Aenus passed to the region's dominant powers: MACEDON, PERGA-MUM, and, in the 100s B.C., ROME. A late tradition connected the founding of Aenus with the mythical Trojan-Roman hero AENEAS.

(See also AEOLIS; COINAGE.)

Aeolian Greeks Ethnic branch of the ancient Greeks, distinct from the two other main groups, the IONIAN GREEKS and DORIAN GREEKS. The Aeolians spoke a dialect called Aeolic and claimed a mythical ancestor, Aeolus (not the ruler of the winds in the *Odyssey*, but another Aeolus, son of the first Greek man, HELLEN).

During the epoch of MYCENAEAN CIVILIZATION (circa 1600–1200 B.C.), the Aeolians seem to have been centered in central and northeastern Greece. But amid the Mycenaeans' violent end (circa 1100–1000 B.C.), displaced Aeolians migrated eastward across the AEGEAN SEA. First occupying the large eastern island of LESBOS, these people eventually spread along the northwest coast of ASIA MINOR, in the region that came to be called AEOLIS.

By the 600s B.C. Aeolian Greeks inhabited Lesbos and Aeolis (in the eastern Aegean) and BOEOTIA and THESSALY (in mainland Greece). The strong poetic traditions of Aeolian culture reached their peak in the poetry of SAPPHO and ALCAEUS, written at Lesbos in the early 500s B.C.

(See also GREEK LANGUAGE; LYRIC POETRY.)

Aeolic dialect See GREEK LANGUAGE.

Aeolis Region inhabited by AEOLIAN GREEKS on the northwest coast of ASIA MINOR. Extending from the Hermus River northward to the HELLESPONT, Aeolis was colonized by Aeolians in eastward migrations between about 1000 and 600 B.C.; the nearby island of LESBOS apparently served as an operational base for these invasions. The major city of Aeolis was CYMĒ.

Aeolis prospered from east-west TRADE. However, the loose confederation of Aeolis' cities never achieved international power in the Greek world, and Aeolis was dwarfed in importance by its southern Greek neighbor, the region called IONIA.

Aeolus See ODYSSEUS.

Aeschines Athenian orator who lived circa 400–320 B.C. and who is remembered mainly as a political enemy of the famous orator DEMOSTHENES (1). In 346 B.C., when the Macedonian king PHILIP II was extending his power by war and intimidation throughout Greece, Aeschines and his mentor, Eubulus, belonged to an Athenian party that sought a negotiated peace with Philip; Aeschines served on two Athenian embassies to Philip that year. Aeschines' conciliatory speeches in the Athenian ASSEMBLY brought him into conflict with Demosthenes, who staunchly advocated war.

Soon Philip's flagrant expansionism had borne out Demosthenes' warnings, and Demosthenes brought Aeschines to court twice (346 and 343 B.C.) on charges that he had advised the Athenians irresponsibly, acting as Philip's paid agent. Although the bribery charge was probably false, Demosthenes' second prosecution nearly succeeded, with Aeschines winning the jury's acquittal by merely one vote.

Thirteen years later Aeschines struck back with a charge against an associate of Demosthenes named Ctesiphon, who had earlier persuaded the Athenians to present Demosthenes with a golden crown, in gratitude for his statesmanship. By a procedure known as *graphē paranomon*, Aeschines accused Ctesiphon of having attempted to propose illegal legislation in the assembly. Demosthenes spoke in Ctesiphon's defense; his speech, *On the Crown*, which survives today, is considered to be Demosthenes' masterpiece of courtroom oratory. Defeated and humiliated, Aeschines retired to the island of RHODES.

Three of Aeschines' speeches are extant, each relating to one of his three court cases against Demosthenes. In the speech *Against Timarchus* (346 B.C.), Aeschines successfully defended himself by attacking Demosthenes' associate Timarchus, who was coprosecuting. Invoking an Athenian law that forbade anyone of bad moral character from addressing the court, Aeschines argued persuasively that Timarchus had at one time been a male prostitute. The speech is a valuable source of information for us regarding the classical Greeks' complex attitudes toward male HOMOSEXUALITY. Aeschines' speech *Against Ctesiphon* (330 B.C.) also is interesting, for it gives a negative assessment of Demosthenes' career.

(See also LAWS AND LAWCOURTS; RHETORIC.)

Aeschylus Earliest of the three classical playwrights of fifth-century-B.C. ATHENS. (The other two were SOPHOCLES and EURIPIDES.) He wrote 90 plays, of which only seven survive under his name; and of these, *Prometheus Bound* may not really have been written by him. Like other Athenian tragedians, Aeschylus wrote mainly for competition at the annual Athenian drama festival known as the City Dionysia, where three playwrights would each present three tragedies and a satyr play. Among Aeschylus' extant plays is the only complete Greek tragic trilogy to come down to us, the *Oresteia*, or *Oresteian Trilogy*—one of the greatest works of Greek literature.

Aeschylus' place in Western culture is due to his solemn vision of divine justice, which orders events on earth. He drew largely on MYTHS for his stories, and described his plays as morsels from the banquet of HOMER. He was also a pioneer of stage technique at a time when Greek drama was still crude and was a spokesman for the big, patriotic emotions that had been aroused by Athens' victory in the PERSIAN WARS (490–479 B.C.). Aeschylus won first prize at the Dionysian competition 13 times; after his death, his plays came to be seen as old-fashioned in theme and language.

Aeschylus was born in 525 B.C. into an aristocratic family of Eleusis, a city in Athenian territory. His father's name was Euphorion. Little is known of Aeschylus' life, but as a teenager he would have witnessed two great public events: the expulsion of the dictator HIPPIAS (1) (510 B.C.) and the institution of Athenian DEMOCRACY as fashioned by the reformer CLEISTHENES (1) (508 B.C.). In 490 B.C. Aeschylus took part in the single most important moment in Athenian history, fighting as a soldier in the Battle of MARATHON, which repulsed a Persian invasion (and in which his brother Cynegeirus was killed). Aeschylus also may have fought 10 years later at the sea battle of SALAMIS

(1), where a much larger Persian invasion was defeated. His participation in these great events shaped his patriotism and his faith in an ordering divinity—themes that echo throughout his plays. These were beliefs shared by his audiences in the 480s–460s B.C.

One anecdote from Aeschylus' early years mentions a competition circa 489 B.C. to choose the official epitaph for the Athenian dead at Marathon. Aeschylus' submitted poem was not selected, although he was an Athenian who had fought at the battle; the judges, finding that his poem lacked sympathy of expression, preferred the poem submitted by the poet SIMONIDES of Ceos. A modern scholarly reconstruction of the two poems has shown that Simonides' poem characterized the dead men as saviors of Greece, but Aeschylus' as saviors of Athens. The episode is significant in showing Aeschylus' pro-Athenian outlook and his inclination toward the grand vision rather than the human details. Both of these traits tend to contrast Aeschylus with the younger tragedian Euripides, and it is no coincidence that the comic playwright ARISTOPHANES fictionally showed Aeschylus and Euripides competing in his comedy *Frogs* (405 B.C.).

Aeschylus presented his first tragedies in around 499 B.C. and won his first festival victory in 484 B.C., with a trilogy whose name we do not know. His tragedy *The Persians* was presented in 472 B.C. as part of a trilogy that won first prize, and its *chorēgos* (paying sponsor) was the rising young politician PERICLES. *The Persians* apparently is modeled somewhat on *The Phoenician Women*, by the tragedian PHRYNICHUS. It is unusual in that its subject matter is drawn not from myths but rather from a recent events—namely, the Persian disaster at Salamis, as seen from the Persian viewpoint. The play's title describes the chorus (a group of Persian councillors), and the protagonist is Atossa, mother of the Persian king XERXES. In the simple plot, arrival of news of the calamity is followed by an invocation of the ghost of the great Persian king DARIUS (1), Xerxes' father. The Persians are presented theatrically, but with pathos and dignity, as victims of Xerxes' insane HUBRIS.

Shortly afterward, Aeschylus traveled to SICILY, to the wealthy court of the Syracusan tyrant HIERON (1) (patron also of such poets as Simonides and PINDAR). It was probably at this time that Aeschylus wrote a new play, *Women of Aetna* (now lost), to commemorate Hieron's founding of a city of that name, near Mt. Etna. Aeschylus returned to Athens to compete at the City Dionysia of 468 B.C., but he lost first place to a 28-year-old first-time contestant named Sophocles.

Of Aeschylus' other extant work, the *Seven Against Thebes*—a pageant centering on the Theban king Eteocles' decision to meet his brother, Polynices, in combat to defend his city—was presented in 467 B.C. Another play, *The Suppliants*—about the Danaid maidens' flight from their suitors, the sons of Aegyptus—dates from around 463 B.C. The three plays of the *Oresteia*—*Agamemnon*, the *Libation Bearers*, and the *Eumenides*—were performed in 458 B.C.

Perhaps alarmed by growing class tensions at Athens, Aeschylus traveled again to Sicily, where he died at the age of 69 (456 B.C.). *Prometheus Bound*, if it is in fact by Aeschylus, may have been presented in Sicily in his final years. He was buried at GELA. Aeschylus' brief verse epitaph, which he supposedly prepared himself, ignored his many literary honors and mentioned only that he had fought at Marathon.

The Oresteian trilogy, Aeschylus' greatest work, describes the triumph of divine justice working through a series of horrific events on earth. In the first play, *Agamemnon*, the vainglorious Agamemnon, fresh from his victory in the TROJAN WAR, is so misled by pride that he cannot see that his wife, the adulterous CLYTAEMNESTRA, plans to murder him. After the killing, their son, ORESTES, must avenge his father by slaying his mother in the trilogy's second play, the *Libation Bearers* (*Chōephoroi*). But this act in turn incites the wrath of supernatural fiends, the FURIES (Erinues), whose divine function is to avenge a parent's blood. In the third play, the *Eumenides*, Orestes is pursued by the chorus of Furies to Athens, where he is cleansed of his curse with the help of ATHENA and APOLLO. Tried for his murder before the Athenian law court of the AREOPAGUS, Orestes is acquitted, and his persecutors are invited to stay on at Athens as protective spirits—the "Kindly Ones" of the play's title. The *Eumenides* is simultaneously a bit of Athenian nationalism and a profound vision of civilized society as a place where the old, violent code of blood vengeance has been replaced by law.

Aeschylus was responsible for many innovations that soon became standard on the Athenian stage. He developed the use of lavish costumes and introduced a second speaking actor, thereby greatly increasing the number of possible speaking roles (since each actor could "double" or "triple" on roles). Aeschylus had a fondness for visual affects and wild, demonstrative choral parts, which his successors found crude. Yet in places his language has a spellbinding solemnity, and in the scenes leading up to the murders in the *Agamemnon* and *Libation Bearers*, he is a master of suspense.

His later life saw a period of serious political strife at Athens, between radical democrats and more right-wing elements. The brilliant left-wing statesman THEMISTOCLES was ostracized (circa 471 B.C.) and forced to flee to PERSIA to avoid an Athenian death sentence, but his policies eventually were taken up by the young Pericles and his comrades. It is evident that Aeschylus was a member of this democratic party, not only from his 472 B.C. association with Pericles, but also from the plays he wrote. The Athenian navy is indirectly glorified in *The Persians*; there are muted, approving references to Themistocles in *The Persians* and *The Suppliants*; and a major aim in the *Eumenides* (458 B.C.) is to dignify the Areopagus, which in real life had recently been stripped of certain powers by left-wing legislation.

But Aeschylus' work was never partisan in a petty way: plays such as *The Suppliants* and the *Eumenides* end with hopeful reconciliation between opposing forces, and in this we can see the lofty, generous spirit of an artist who sought out the divine purpose in human affairs.

(See also DIONYSUS; ELECTRA; EPHIALTES; PROMETHEUS; SEVEN AGAINST THEBES; THEATER.)

Aesop Supposed author of a number of moralizing fables, many involving animals as characters. According to legend, Aesop was a slave on the island of SAMOS in the 500s B.C. In fact, Aesop may have been no more than a name around which certain folktales gravitated.

One of the best known of Aesop's fables tells of the race betwen the tortoise and the hare. The overconfident hare, stopping to nap in midrace, loses to his slower but steadier opponent.

Animal parables also occur in extant verses by ARCHILOCHUS (circa 650 B.C.), whose writings may have inspired some of the tales that we know as Aesop's.

Aetolia Mountainous region of central Greece, north of the Corinthian Gulf, bordered on the west by ACARNANIA and on the east by the Mt. Parnassus massif. Interior Aetolia contained good farmland, but the southern mountains blocked Aetolia from the gulf and from outside influences. Through the 400s B.C. the Aetolians remained rugged Greek "highlanders," divided by tribal feuds and known for carrying weapons in public for self-defense.

During the PELOPONNESIAN WAR, Aetolia was invaded by the Athenian general DEMOSTHENES (2), who hoped to seize the eastward mountain route into enemy BOEOTIA. But the Aetolians, arrayed as javelin-throwing light infantry, defeated the cumbersome Athenian HOPLITES in the hills (426 B.C.).

In the late 300s B.C. Aetolia emerged as a force in the Greek resistance to the overlordship of MACEDON. By now the Aetolians had united into a single federal state—the Aetolian League. Aetolian towns shared a common citizenship, representative ASSEMBLY, and a war captain, elected annually; the capital city was Thermon. By the late 200s B.C. the aggressive league dominated most of central Greece, with alliances extending to the PELOPONNESE. Aetolia fell into conflict with its southern rival, the Achaean League, as well as with Macedon. In 218 B.C. the dynamic Macedonian king PHILIP V invaded Aetolia and sacked Thermon.

Aetolia was a natural ally for Philip's enemy, the imperialistic Italian city of ROME. Allied to Rome in the Second Macedonian War, the Aetolians helped defeat Philip at the Battle of Cynoscephalae (197 B.C.). However, disappointed by the mild Roman peace with Macedon, the Aetolians allied with the Seleucid king ANTIOCHUS (2) III against Rome (192 B.C.). After the Romans had defeated Antiochus, they broke the Aetolian League's power and made it a Roman subject ally (189 B.C.).

(See also ACHAEA; WARFARE, LAND.)

afterlife Throughout ancient Greek history, nearly all Greeks believed in some form of life after death. Only the philosophy called EPICUREANISM (after 300 B.C.) maintained unequivocally that the human soul died with the body. Because Greek RELIGION had no specific doctrine on the subject, beliefs in the afterlife varied greatly, from crude superstition to the philosopher PLATO's lofty vision (circa 370 B.C.) of an immortal soul freed of its imperfect flesh and at one with absolute reality in another world.

The primitive concept that the dead somehow live on in their tombs never disappeared from Greek religion. The shaft graves at MYCENAE—datable to 1600–1550 B.C., at the dawn of MYCENAEAN CIVILIZATION—were filled with armor, utensils, and even pets and SLAVES, killed in sacrifice, to comfort the deceased in the afterlife. This practice may have been inspired by the burial rites of Egyptian pharaohs, but the general idea seems to have survived in Greece for over 1,000 years. Greeks of the 400s and 300s B.C. were still offering food and drink at graveside, as nourishment for the dead.

Another belief was that the souls of the dead traveled to an Underworld, the realm of the god HADES and his wife, PERSEPHONE. Unlike the modern concept of Hell, this "House of Hades" (as the Greeks called it) was not primarily a place of punishment. It was, however, a cold and gloomy setting, where the souls—after being led from the living world by the messenger god HERMES—endured a bleak eternity.

The earliest extant description of Hades' kingdom comes in book 11 of HOMER's epic poem the *Odyssey* (written down circa 750 B.C.), when the living hero ODYSSEUS journeys there by ship to seek prophecy. The site is vaguely described as a grim shoreline of OCEANUS, at the edge of the living world. (Later writers tended to situate it underground.) There Odysseus recognizes the ghosts of some of his family and former comrades. He also sees the torments of three sinners—Tityus, TANTALUS, and SISYPHUS—who had betrayed the friendship of the gods. The only people to be excused from Hades' realm were those who had been granted divinity and who now resided with the other gods on Mt. OLYMPUS. These lucky few included HERACLES and the twins CASTOR AND POLYDEUCES.

Gradually, concepts of reward and punishment were enlarged. Poets wrote of a place called Elysium (*Elusion*) where certain souls, chosen by the gods, enjoyed a happy afterlife. Also known as the Islands of the Blessed, this locale is described by the Theban poet PINDAR (476 B.C.) in terms of shady parklands and athletic and musical pastimes—in other words, the ideal life of the living Greek aristocrat. Post-Homeric sources placed ACHILLES there with other heroes, including the Athenian tyrannicides of 510 B.C., HARMODIUS AND ARISTOGITON.

Similarly, legend began to specify a lowermost abyss in Hades' realm, a place called Tartarus. This was the scene of punishment for the evil TITANS and for the worst human sinners. (Post-Homeric sources add the DANAIDS and IXION to the group.) Typical punishments require the prisoner to endure eternal frustration of effort (Sisyphus, the Danaids) or desire (Tantalus).

Greek writers such as Plato began to describe the mythical judges who assigned each soul to Elysium, Tartarus, or the netherworld. These judges were MINOS and Rhadamanthys (who were brothers) and Aeacus, all of whom had once been mortal men. The concept of eternal judgment contains an obvious ethical message—a warning to act justly in this life—that resembles the later Christian view.

The well-known rivers of the Underworld are best described in Vergil's Latin epic poem, the *Aeneid* (circa 20 B.C.). But the idea of rivers or lakes in Hades' kingdom goes back at least to the Greek poet HESIOD's *Theogony* (circa 700 B.C.). The Greeks associated these Underworld waters with actual rivers of mainland Greece, apparently believing that the waters continued their course underground. The Styx ("hated") was an actual river in ARCADIA.

The Acheron ("woeful") flowed in EPIRUS, near an oracle of the Dead. The Underworld's other rivers were Lethe ("forgetting"), Cocytus ("wailing"), and Pyriphlegethon or Phlegethon (burning). These dire names probably referred to Greek FUNERAL CUSTOMS rather than to any punishment for the souls.

Legend usually described the Styx as the Underworld's boundary. New arrivals were brought across by the old ferryman Charon, and Greek burial rites often included placing a coin in the corpse's mouth, to pay for this final passage. The monstrous many-headed dog CERBERUS stood watch on the Styx's inner bank, preventing the souls from leaving.

This grim Greek picture of the common man's afterlife eventually inspired a reaction: A number of fringe religious movements arose, assuring their followers of a happy afterlife. These were called mystery cults or mysteries (*mustēria*, from *mustēs*, "an initiate"). While centering on a traditional deity such as DIONYSUS, DEMETER, or Persephone, the mysteries claimed to offer the correct beliefs and procedures for admittance into Elysium. In Greek tombs of southern ITALY from about 400 B.C., archaeologists have discovered golden tablets inscribed with precise directions for the soul entering the Underworld: The soul is warned not to drink from the attractive spring of forgetfulness—"seen on the right, where the white cypress grows"—but from the lake of remembrance, beyond.

One mystery faith, ORPHISM, emphasized reincarnation (also known as transmigration). According to this belief, each person's soul passed, at death, into a newborn body, whether human or not. The new assignment was based on the person's conduct and belief in the prior life; bad souls descended through criminals, slaves, and animals, but a right-living soul ascended to kings and heroes, eventually gaining admittance to Elysium. This concept was adapted by the philosophers PYTHAGORAS (circa 530 B.C.) and EMPEDOCLES (circa 450 B.C.), who influenced Plato.

(See also ELEUSINIAN MYSTERIES; HELLENISTIC AGE; PROPHECY AND DIVINATION.)

Agamemnon   In MYTH, the king of MYCENAE and ARGOS, son of ATREUS, husband of CLYTAEMNESTRA, and commander of the allied Greek army in the TROJAN WAR. Agamemnon's earliest appearance in literature is in HOMER's epic poem the *Iliad* (written down circa 750 B.C.), where he is portrayed negatively. Contrary to his name, which means "very steadfast," Agamemnon is shown to be an irresolute, arrogant, and divisive leader. His quarrel with the Greek champion ACHILLES over possession of a female war captive provokes Achilles to withdraw from the fighting and sets in motion the *Iliad*'s tragic plot.

The events leading up to Agamemnon's command are told by Homer and later writers. The Greek-Trojan conflict began when Helen, wife of MENELAUS (Agamemnon's brother) was seduced by the Trojan prince PARIS and eloped with him. Agamemnon organized an expedition against TROY to recover Helen, but incurred Clytaemnestra's hatred by sacrificing their daughter IPHIGENIA as a blood offering to the hostile goddess ARTEMIS, who was sending contrary winds to prevent the Greek ships' departure.

After the Greeks sacked Troy, Agamemnon sailed for home with his war booty, which included the captured Trojan princess CASSANDRA. But on the very day that they stepped ashore, Agamemnon and Cassandra were murdered by henchmen of Agamemnon's treacherous cousin Aegisthus, Clytaemnestra's illicit lover. (This is Homer's version in the *Odyssey*; in later tales the king dies while emerging from his bath, stabbed by Aegisthus or axed by Clytaemnestra.) It was left to Agamemnon's son ORESTES and daughter ELECTRA to avenge his murder.

Agamemnon's downfall is the subject of Athenian playwright AESCHYLUS' tragedy *Agamemnon* (458 B.C.), the first play in the Oresteian Trilogy.

(See also HELEN OF TROY)

Agathocles   Ruthless and flamboyant ruler of the Sicilian Greek city of SYRACUSE from 316 to 289 B.C. Agathocles was the last of the grandiose Syracusan TYRANTS. He challenged the mighty African-Phoenician city of CARTHAGE and captured most of SICILY from Carthaginians and fellow Greeks. His imperial reign in the Greek West was inspired partly by the example of ALEXANDER THE GREAT's successors in the East.

Agathocles did not come from the ruling class. Born in 361 B.C. in Thermae, in the Carthaginian-controlled western half of Sicily, he was the son of a Greek manufacturer of POTTERY; enemies later derided Agathocles as a mere potter. Emigrating to Syracuse, he came to prominence as an officer in the Syracusan army. In 316 B.C. he overthrew the ruling Syracusan OLIGARCHY and installed himself as *turannos*, or dictator, with the common people's support.

Many adventures followed. Suffering a major defeat in battle against the Carthaginians, Agathocles was besieged by land and sea inside Syracuse (summer 311 B.C.). But he solved this predicament with an amazingly bold action: In August 310 B.C., when the Carthaginians briefly relaxed their naval blockade, Agathocles sailed from Syracuse harbor with 60 ships and a mercenary army of about 13,000 to invade Carthage itself.

His was the first European army to land in Carthaginian North Africa. But despite his victories over Carthaginian armies in the field, Agathocles failed to capture the city. Meanwhile, in Sicily, Syracuse held out against the Carthaginians; but a Sicilian-Greek revolt against Agathocles induced the tyrant to abandon his African army under his son Archagatus and return to Sicily. Eventually Archagatus and his brother were murdered by the army, which evacuated North Africa.

Agathocles made peace with the Carthaginians, giving up territories in west Sicily (306 B.C.). But he soon became the sole ruler of Greek-held eastern Sicily. In 304 B.C., patterning himself on Alexander the Great's heirs who were reigning as supreme monarchs in the East, Agathocles adopted the absolute title of king (*basileus*).

He then extended his power to Greek south ITALY and western mainland Greece. Circa 300 B.C. he drove off the Macedonian king CASSANDER, who was besieging CORCYRA. Agathocles took over Corcyra and gave it, twice, as a dowry for his daughter Lanassa's two influential MARRIAGES, first to the Epirote king PYRRHUS (295 B.C.) and

then to the new Macedonian king, DEMETRIUS POLIORCETES (circa 291 B.C.). The aging Agathocles himself married a third wife, a daughter of the Greek Egyptian king PTOLEMY (1). But his hope of founding a grand dynasty faded when his son Agathocles was murdered by a jealous relative.

The elder Agathocles died at age 72, probably from jaw cancer (289 B.C.). Although he had thwarted the Carthaginian menace, he left no legacy of good government for Sicily. However, his military exploits were influential in demonstrating that mighty Carthage was susceptible to invasion. The Romans would invade Carthage more effectively during their Second Punic War (202 B.C.).

Agathon   Athenian tragic playwright of the late 400s B.C., considered by the Athenians to be their fourth greatest tragedian after AESCHYLUS, SOPHOCLES, and EURIPIDES. Less than 40 lines of his work survive; these show a clever, polished style, influenced by the contemporary rhetorician GORGIAS and by the SOPHISTS. Agathon won his first drama competition at the annual festival known as the City Dionysia, in 416 B.C., with a tragic trilogy whose titles have not survived.

We know little of Agathon's life. But we do know that in 407 B.C. he left Athens—as Euripides had done—for the court of the Macedonian king Archelaus, and there (like Euripides), Agathon died, circa 401 B.C.

From references in ARISTOTLE's *Poetics* we know that Agathon was an innovator. He often removed the chorus from the story's action, reducing the choral odes to mere interludes. In his day he was noteworthy for his tragedy *Antheus*, of which he invented the entire plot himself, rather than drawing on MYTH or history. His plots were overinvolved; Aristotle once criticized him for having crammed the entire tale of the TROJAN WAR into one play.

Agathon had personal beauty and apparently an effete manner. He appears as a fictionalized character in ARISTOPHANES' comedy *Thesmophoriazusae* (411 B.C.) and is burlesqued for his effeminacy—but not for his writing, which Aristophanes calls "good" (*agathos*). Agathon also appears as a character in the philosopher PLATO's dialogue the *Symposium* (circa 370 B.C.), which is set at a drinking party at Agathon's house to celebrate his 416 B.C. competition victory.

(See also THEATER.)

Agesilaus   Spartan king who reigned 399–360 B.C., leading SPARTA's brief phase of supremacy after the PELOPONNESIAN WAR. Although a capable battlefield commander, Agesilaus steered Sparta into a shortsighted policy of military domination in Greece. Eventually this policy provoked the rise of a challenger state, THEBES.

The Athenian historian XENOPHON was a friend of Agesilaus' and wrote an admiring biography of him, as well as including his exploits in the general history titled *Hellenica*. These two extant writings provide much of our information about Agesilaus.

Born in 444 B.C., a son of King ARCHIDAMUS of Sparta's Eurypontid royal house, Agesilaus was dynamic, pious, and lame in one leg. He became king after the death of his half brother, King Agis II. In 396 B.C. he took 8,000 troops to ASIA MINOR to protect the Spartan-allied Greek cities there from Persian attack. Marching inland through Persian-held west-central Asia Minor, Agesilaus defeated the Persians in battle before being recalled to Sparta (394 B.C.). His raid probably helped inspire the future conquests of ALEXANDER THE GREAT.

Agesilaus had been summoned home to help his beleaguered city in the CORINTHIAN WAR. Bringing his army overland through THRACE and THESSALY, he descended southward into hostile BOEOTIA, where he narrowly defeated a coalition army of Thebans, Argives, Athenians, and Corinthians at the Battle of Coronea (394 B.C.). Wounded, and with his army now too weak to occupy Boeotia, Agesilaus withdrew to Sparta. He commanded subsequent Spartan actions against ARGOS (391 B.C.), CORINTH (390 B.C.), ACARNANIA (389 B.C.), and defiant Thebes (378–377 B.C.).

His hostility toward the Theban leader EPAMINONDAS at a peace conference resulted in renewed war and a disastrous Spartan defeat at the Battle of LEUCTRA (371 B.C.). In the disarray that followed, Agesilaus helped lead the defense of the Spartan homeland, culminating in the stalemate Battle of MANTINEA (362 B.C.).

After peace was made, the 82-year-old king sailed to EGYPT with 1,000 Spartan mercenary troops to assist an Egyptian prince's revolt against the Persians. (The expedition's purpose was to replenish Sparta's depleted revenues.) The revolt went awry, and Agesilaus died on the voyage home (360 B.C.).

(See also EURYPONTID CLAN.)

Agiad clan The Agiads (Agiadai) were the senior royal family at SPARTA, which had an unusual government in that it was ruled simultaneously by two kings. The Agiads, "descendants of Agis," traced their ancestry back to a legendary figure who was one of the sons of HERACLES. As the senior house, the Agiads enjoyed certain ceremonial privileges over their partners, the EURYPONTID CLAN. Notable Agiad kings include the brilliant CLEOMENES (1) and Leonidas, the commander at the Battle of THERMOPYLAE.

agora The open "place of assembly" in an ancient Greek city-state. Early in Greek history (900s–700s B.C.), free-born males would gather in the agora for military duty or to hear proclamations of the ruling king or COUNCIL. In the more settled centuries that followed, the agora served as a marketplace where merchants kept open-air stalls or shops under colonnades.

Classical ATHENS boasted a grand agora—the civic heart of the city that dominated Greece. Under the Athenian dictators PISISTRATUS and HIPPIAS (1) (second half of the 500s B.C.), the agora was cleared to a rectangular open area of about 600 by 750 yards, bordered with grand public buildings. Devastated by the occupying Persians in 480–479 B.C., the agora was rebuilt in the later 400s to include temples, government buildings, and several colonnades, of which the best known was the Stoa Poikilē (painted colonnade).

Today the ancient Athenian agora has been partly restored. The pride of the reconstructed buildings is the Doric-style temple of HEPHAESTUS (previously misidentified as a temple of THESEUS), built circa 449 B.C.

(See also ARCHITECTURE; ASSEMBLY; DEMOCRACY; OSTRACISM; PAINTING; STOICISM.)

agriculture See FARMING.

Ajax (1) In the legend of the TROJAN WAR, Ajax (Greek Aias) was king of SALAMIS (1) and son of Telamon. After ACHILLES, he was the bravest Greek warrior at Troy. In HOMER's epic poem the *Iliad*, Ajax engages in many combats—for example, dueling the Trojan hero HECTOR to a standoff (book 7) and leading the Greeks in defense of their beached ships (book 13). Giant in size, stolid, and slow-spoken, Ajax embodies the virtue of steadfastness. He carries a huge oxhide shield and is often called by the poetic epithet "bulwark of the Achaeans." Homer implicitly contrasts him with his chief rival among the Greeks, the wily ODYSSEUS. Although the stronger of the two, Ajax loses a WRESTLING match to Odysseus' skill (*Iliad* book 23).

As described by Homer in the *Odyssey* (book 11), Ajax's death came from his broken pride over this rivalry. After Achilles was killed, Ajax and Odysseus both claimed the honor of acquiring his wondrous armor. The dispute was arbitrated by Trojan prisoners of war, who agreed that Odysseus had done more to harm the Trojan cause. Maddened with shame, Ajax eventually killed himself with his sword. This tale is the subject of the extant tragedy *Ajax*, written circa 450–445 B.C. by the Athenian playwright SOPHOCLES. Sophocles' Ajax is brought down by his flaws of anger and pride—and by a deep nobility that prevents him from accepting a world of intrigue and compromise, personified by Odysseus.

(See also ACHAEANS; HUBRIS.)

Ajax (2) Greek warrior—son of Oileus and leader of troops from LOCRIS—in the legend of the TROJAN WAR. Often known as the Lesser Ajax, he was brave, swift-footed, arrogant, and violent. His savage behavior at the sack of TROY, unmentioned by the poet HOMER, is described by later writers. Finding the Trojan princess CASSANDRA in sanctuary at the goddess ATHENA's altar, Ajax pulled her away and raped her. He was hated by Athena, but his death, as described in Homer's *Odyssey* (book 4), occurred at POSEIDON's hands. On the homeward voyage from Troy, Ajax's ship was wrecked; he reached shore safely but sat atop a cliff declaring hubristically that he had beaten the gods. Poseidon, enraged, blasted him back into the sea.

(See also HUBRIS.)

Alcaeus Lyric poet of the city of MYTILENE on the island of LESBOS. Born in about 620 B.C. into an aristocratic family, Alcaeus was a contemporary and fellow islander of the poet SAPPHO. Like her he wrote love poems, but unlike her, he also wrote of his involvement in great events, such as the civil strife between Mytilene's traditional ARISTOCRACY and ascendant TYRANTS. Better as a poet than as a political analyst, Alcaeus was a spokesman for the old-fashioned aristocratic supremacy, which in his day was being dismantled throughout the Greek world.

Although no one complete poem by him has come down to us, the surviving fragments show his talent and give a dramatic biographical sketch. When Alcaeus was a boy in around 610 B.C., his elder brothers and another noble, PITTACUS, expelled the local tyrant. Soon Alcaeus was fighting under Pittacus' command in Mytilene's war against Athenian settlers in northwest ASIA MINOR. Alcaeus threw away his shield while retreating and (like ARCHILOCHUS, an earlier Greek poet) wrote verses about it.

When another tyrant arose in Mytilene, Alcaeus went into exile until the tyrant died. Alcaeus may have gone home—but only for a brief time—because soon his former comrade Pittacus was ruling singly in Mytilene, and Alcaeus and many other nobles were expelled. In his poetry Alcaeus raved against Pittacus as a "low-born" traitor and expressed despair at being excluded from the political life that was his birthright.

Apparently Alcaeus went to EGYPT, perhaps as a mercenary soldier. (Meanwhile, his brother joined the army of the Babylonian king Nebuchadnezzer and took part in the campaign that captured Jerusalem in 597 B.C.) At some point Alcaeus and his friends planned to attack Mytilene and depose Pittacus, but the common people stood by their ruler. Supposedly Pittacus at last allowed Alcaeus to come home.

Like Sappho, Alcaeus wrote in his native Aeolic dialect and used a variety of meters. A number of his extant fragments are drinking songs, written for solo presentation at a SYMPOSIUM. Even by ancient Greek standards, Alcaeus seems to have been particularly fond of WINE. He wrote hymns to the gods, including one to APOLLO that was much admired in the ancient world. In accordance with the upper-class sexual tastes of his day, he wrote love poems to young men.

More than 550 years after Alcaeus' death, his poetry served as a model for the work of the Roman poet Horace.

(See also GREEK LANGUAGE; HOMOSEXUALITY; HOPLITE.)

Alcestis In MYTH, the wife of King Admetus of Pherae in THESSALY. She became Admetus' wife after he was able, with the god APOLLO's help, to fulfill her father Pelias' onerous precondition of yoking a lion and wild boar to a chariot and driving it around a racecourse.

Alcestis is best known for the story of how she voluntarily died in her husband's place. Apollo, discovering from the Fates that his mortal friend Admetus had only one day to live, arranged that Admetus' life be spared if a willing substitute could be found; the only one to consent was Alcestis. In EURIPIDES' tragedy *Alcestis* (438 B.C.), this old tale of wifely duty is recast as a disturbing account of female courage and male equivocation.

(See also FATE.)

Alcibiades Athenian general, politician, and social figure who lived circa 450–404 B.C. and strongly influenced the last 15 years of the PELOPONNESIAN WAR between ATHENS and SPARTA. The mercurial Alcibiades embodied the confident Athenian spirit of the day. Although brilliant as a leader in battle, he was prone to dangerously grandiose schemes in war strategy and politics. His fellow citizens repeatedly voted him into high command, yet they mistrusted him for his private debaucheries and for his ambition, which seemed aimed at seizing absolute power at Athens. After his political enemies organized the people

against him (415 B.C.), Alcibiades spent three years as a refugee turncoat, working for the Spartans (414–412 B.C.). Pardoned by Athens in its hour of need, he led the Athenians through a string of victories on land and sea (411–407 B.C.) that could have saved the city from defeat. But the Athenians turned against him once more, and he died in exile, murdered at Spartan request, soon after Athens surrendered.

As a flawed genius of tragic dimensions, Alcibiades is vividly portrayed in extant Greek literature. His Athenian contemporaries, the historians THUCYDIDES (1) and XENOPHON, describe him in their accounts of the Peloponnesian War. A biography of Alcibiades comprises one of PLUTARCH's *Parallel Lives*, written circa A.D. 100–110. A fictionalized Alcibiades appears in dialogues written by the philosopher PLATO in around 380 B.C.

Alcibiades was born into a rich and powerful Athenian family during the Athenian heyday. His mother, Deinomache, belonged to the aristocratic Alcmaeonid clan. After his father, Cleinias, was killed in battle against the Boeotians (447 B.C.), Alcibiades was raised as a ward of Deinomache's kinsman PERICLES, the preeminent Athenian statesman. Breeding and privilege produced a youth who was confident, handsome, and spoiled, and he became a rowdy and glamorous figure in the homosexual milieu of upper-class Athens; Plutarch's account is full of gossip about men's infatuated pursuit of the teenage Alcibiades. Later he also showed a taste for WOMEN, especially for elegant courtesans. He married an Athenian noblewoman, Hipparete, and they had two children, but Alcibiades' conduct remained notoriously licentious. We are told that he had a golden shield made, emblazoned with a figure of the love god EROS armed with a thunderbolt.

In his teens Alcibiades became a follower of the Athenian philosopher SOCRATES, who habitually tried to prompt innovative thought in young men bound for public life. This is the background for the scene in Plato's *Symposium* where a drunken Alcibides praises Socrates to the assembled drinkers: According to Plato's version, the middle-age Socrates was in love with Alcibiades but never flattered the younger man or had sexual relations with him, despite Alcibiades' seductive advances.

Alcibiades reached manhood at the start of the Peloponnesian War. At about age 18 he was wounded in the Battle of POTIDAEA (432 B.C.) while serving as a HOPLITE alongside Socrates. (Supposedly Socrates then stood guard over him during the combat.) Alcibiades repaid the favor years later at the Battle of Delium (424 B.C.): On horseback, he found the foot soldier Socrates amid the Athenian retreat and rode beside him to guard against the pursuing enemy.

Although an aristocrat, Alcibiades rose in politics as leader of the radical democrats, as his kinsman Pericles had done. He was only about 30 years old when he was first elected as one of Athens' 10 generals. Meanwhile, he pursued fame with scandalous extravagance, sponsoring no less than seven CHARIOTS at the OLYMPIC GAMES of 416 B.C.—the most ever entered by an individual in an Olympic contest. His chariots took first, second, and fourth places, and inspired a short poem by EURIPIDES. Many right-wing Athenians were alarmed by this flamboyance, so reminiscent of the grandiose TYRANTS of a prior epoch.

Alcibiades' political leadership was similarly reckless. In 420 B.C. he helped to sabotage the recent Peace of NICIAS (which had been meant to end Spartan-Athenian hostilities), by convincing the Athenians to ally themselves with Sparta's enemy, the city of ARGOS. The outcome was a Spartan field victory over an army of Argives, Athenians, and others at the Battle of MANTINEA (418 B.C.).

In 415 B.C. Alcibiades led the Athenian ASSEMBLY into voting for the most fateful undertaking of Athenian history—the expedition against the Greek city of SYRACUSE. (This huge invasion—by which Alcibiades hoped eventually to conquer SICILY and CARTHAGE—would later end in catastrophe.) Not trusting Alcibiades as sole commander, the Athenians voted to split the expedition's leadership between him and two other generals, including the cautious Nicias. But the force, with 134 warships, had barely reached Sicily when Athenian envoys arrived, summoning Alcibiades home to face criminal charges of impiety.

One accusation (possibly true) claimed that on a prior occasion Alcibiades and his friends had performed a drunken parody of the holy ELEUSINIAN MYSTERIES. A second accusation concerned a strange incident that had occurred just before the Sicilian expedition's departure: An unknown group had gone around overnight smashing the herms (*hermai*, marble figures of the god HERMES that stood outside houses throughout Athens), perhaps to create a bad omen against the invasion. Alcibiades was charged with this mutilation—although this charge was certainly false.

Knowing that these accusations had been orchestrated by his enemies to destroy him, Alcibiades accompanied the Athenian envoys by ship from Sicily but escaped at a landfall in southern ITALY. Crossing on a merchant ship to the PELOPONNESE, he sought refuge at Sparta, where his family had ancestral ties. The Athenians condemned him to death in absentia and confiscated his property.

In Alcibiades the Spartans found a most helpful traitor. At his urging, they sent one of their generals to Syracuse to organize that city's defense; within two years the Athenian invasion force was totally destroyed. Also on Alcibiades' advice, the Spartans occupied Decelea, a site about 13 miles north of Athens, to serve as their permanent base in enemy territory (413 B.C.). By now Athens had begun to lose the war.

In 412 B.C. Alcibiades went on a Spartan mission to the eastern Aegean to foment revolt among Athens' DELIAN LEAGUE allies and to help bring PERSIA into the war on Sparta's side. Yet the Spartans soon condemned Alcibiades to death—they mistrusted him, partly because he was known to have seduced the wife of the Spartan king Agis. With Sparta and Athens both against him, Alcibiades fled to the Persian governor of western ASIA MINOR. From there he began complex intrigues with commanders at the Athenian naval base on the nearby island of SAMOS in hopes of getting himself recalled to Athenian service.

His chance came in June 411 B.C., after the government at Athens fell to the oligarchic coup of the FOUR HUNDRED, and the Athenian sailors and soldiers at Samos defiantly proclaimed themselves to be the democratic government-in-exile. Alcibiades was invited to Samos and elected general. After the Four Hundred's downfall (September 411 B.C.), he was officially reinstated by the restored DEMOC-RACY at Athens, although he stayed on active duty around Samos.

Then about 40 years old, Alcibiades began a more admirable phase of his life. The theater of war shifted to Asia Minor's west coast and to the HELLESPONT seaway, where Spartan fleets, financed by Persia, sought to destroy Athens' critical supply line of imported grain. Alcibiades managed to keep the sea-lanes open. His former ambition and recklessness now shone through as bold strategy and magnetic leadership. (For example, he once told the crews of his undersupplied ships that they would have to win every battle, otherwise there would be no money to pay them.) His best victory came in 410 B.C. at CYZICUS, where he surprised a Spartan fleet of 60 ships, destroying or capturing every one. In 408 B.C. he recaptured the strategic but rebellious ally city of BYZANTIUM. In 407 B.C., at the height of his popularity, he returned ceremoniously to Athens to receive special powers of command. Then he sailed back to war, destined never to see home again.

In 406 B.C. a subordinate of Alcibiades was defeated in a sea battle off Notium, near EPHESUS, on Asia Minor's west coast. The fickle Athenian populace blamed Alcibiades and voted him out of office. Alarmed, he fled from his fellow citizens a second time—only now he could not go to Sparta. He eventually settled in a private fortress on the European shore of the Hellespont. But, with Athens' surrender to Sparta in 404 B.C., Alcibiades had to flee from the vengeful Spartans, who now controlled all of Greece.

He took refuge with Pharnabazus, the Persian governor of central Asia Minor. But too many people desired Alcibiades' death. The Spartan general LYSANDER and the Athenian quisling CRITIAS feared that Alcibiades would lead the defeated Athenians to new resistance. At Spartan request, Pharnabazus sent men to kill him. Legend says that Alcibiades was abed with a courtesan when he awoke to find the house on fire. Wrapping a cloak around his left arm as a shield, he dashed out naked, sword in hand, but fell to arrows and javelins. The woman escaped and later had him buried.

So died the foremost Athenian soldier of his day. The historian Thucydides sums up the dual tragedy of Athens and Alcibiades: "He had a quality beyond the normal, which frightened people. . . . As a result his fellow citizens entrusted their great affairs to men of lesser ability, and so brought the city down." The Athenians' puzzlement over him is suggested in ARISTOPHANES' comedy *Frogs*, staged in 405 B.C., the year before Alcibiades' death. The play involves a poetry contest in the Underworld; the final question to the contestants is: "What do you think of Alcibiades?"

(See also ALCMAEONIDS; HOMOSEXUALITY; PROSTITUTES.)

Alcmaeon (1)   In MYTH, a hero of the city of ARGOS. After his father, Amphiaraus, was killed in the exploit of the SEVEN AGAINST THEBES, Alcmaeon led the expedition of the Epigoni (Descendants) and captured THEBES. Then, in accordance with a prior vow to Amphiaraus, Alcmaeon murdered his own mother, Eriphyle, for her treacherous role in convincing Amphiaraus to join the doomed expedition. The dying Eriphyle cursed her son, wishing that no land on earth might welcome him.

Tormented by the FURIES for his crime, Alcmaeon fled from home. On advice he journeyed to ACARNANIA, where a recent strip of alluvial shore from the River Acheloüs supplied a "new" land, unaffected by Eriphyle's curse. There Alcmaeon received absolution.

The story is similar to the myth of ORESTES, who was also compelled by duty to kill his mother. The horror and legalistic dilemma of this situation appealed to the classical Greek mind. Alcmaeon was the subject of at least two tragedies by the Athenian playwright EURIPIDES (late 400s B.C.).

Alcmaeon (2) See CROTON.

Alcmaeonids The Alkmaionidai (descendants of Alcmaeon) were a noble Athenian clan, active in politics in the 600s–400s B.C.. The family claimed descent from a certain Alcmaeon (not the same as the Argive hero ALCMAEON or the physician Alcmaeon of CROTON). Although the Alcmaeonids were aristocrats, a few of them played major roles in the Athenian DEMOCRACY of the late 500s–mid-400s B.C. These included CLEISTHENES (1) and (by maternal blood) PERICLES and ALCIBIADES.

As a group, the Alcmaeonids were not greatly trusted by other Athenians. They were suspected of plotting to seize supreme power, and they were considered to be living under a hereditary curse from the days when an Alcmaeonid commander had impiously slaughtered the conspirators of CYLON (632 B.C.).

The Alcmaeonids were thought to be responsible for the treasonous heliographic signal—the showing of the shield—that accompanied the Battle of MARATHON (490 B.C.). In the rash of OSTRACISMS of the 480s B.C.—aimed against suspected friends of the fallen tyrant HIPPIAS—two Alcmaeonid figures were expelled: Cleisthenes' nephew Megacles and XANTHIPPUS, who had married into the clan and was Pericles' father.

Pericles gained the people's trust by disassociating himself from the family; he avoided Alcmaeonid company and aristocratic gatherings in general. In 431 B.C., on the eve of the PELOPONNESIAN WAR, the hostile Spartans demanded that the Athenian people "drive out the curse," that is, by expelling Pericles. But the demand was ignored.

(See also ARISTOCRACY; HARMODIUS AND ARISTOGITON; TYRANTS.)

Alcman Famous poet of SPARTA, circa 630 B.C. A plausible later tradition says Alcman was an Ionian Greek who had immigrated from ASIA MINOR; he was notorious for his supposed gluttony. Only a few fragments of his work have come down to us, but they include the earliest surviving example of a choral ode—a type of poem sung by a chorus of girls or men, to musical accompaniment, at a religious festival or other event. Choral poetry was in those years a distinctly Spartan art form, and the Doric Greek dialect of Sparta was the language of the genre.

Alcman's 101-line fragment presents the final two thirds of a *partheneion*, a "maiden song." The fragment begins by recounting one of the adventures of the hero HERACLES, then abruptly switches topic and starts praising by name the individual teenage girls who are singing the words.

Certain passages seem intended for delivery by half choruses in playful rivalry. The poem obviously was composed for a specific occasion, perhaps a rite of female adolescence connected with the Spartan goddess ARTEMIS Orthia. Alcman's technique of layering mythology and personal references seems to anticipate the work of the greatest Greek choral poet, PINDAR (born 508 B.C.).

Intriguing for a modern reader are the emotionally charged statements that Alcman wrote for public recitation by these aristocratic girls. ("It is Hagesichora who torments me," the chorus says, referring to the beauty of a girl who may have been the chorus leader.) The nuances are sexual, and presumably the verses commemorate genuine emotions within this exclusive girls' group. In the surviving fragments of another *partheneion*, Alcman seems to be addressing the same topic. The situation resembles the female-homosexual style of life later described by the poet SAPPHO (circa 600 B.C.).

In other extant verse, Alcman celebrates simple aspects of the natural world: birds, flowers, food. His buoyant, sophisticated poetry reflect a golden phase of Spartan history—the period of the city's triumph in the hard-fought Second Messenian War, before Sparta had completely become a militaristic society.

(See also GREEK LANGUAGE; HOMOSEXUALITY; IONIAN GREEKS; LYRIC POETRY; MESSENIA.)

Alexander the Great Macedonian king and conqueror who lived 356–323 B.C. Alexander was the finest battlefield commander of the ancient world, and when he died of fever just before his 33rd birthday he had carved out the largest empire the world had ever seen, stretching 3,000 miles from the Adriatic Sea to the Indus River.

His principal achievement was the conquest of the empire of PERSIA, an event that remade the map of the ancient world. For 200 years previously, the Persian kingdom had been a menacing behemoth on the Greeks' eastern frontier. With Alexander's conquests, Persia ceased to exist as a sovereign power. The Persians' former territory—including their subjugated regions such as EGYPT and Mesopotamia—became Alexander's domain, garrisoned by Macedonian and Greek troops.

Alexander's sprawling realm quickly fell apart after his death, and there arose instead several Greco-Macedonian kingdoms of the East, including Ptolemaic Egypt, the SELEUCID EMPIRE, and Greek BACTRIA. These rich and powerful kingdoms carried Greek culture halfway across Asia and overshadowed old mainland Greece, with its patchwork of relatively humble city-states. Historians refer to this enlarged Greek society as the Hellenistic world.

At the start of his reign, the 20-year-old Alexander was the crowned king only of MACEDON—a crude Greek nation northeast of mainland Greece—and some of the credit for his triumphs must go to his father, King PHILIP II. When the tough, hard-drinking Philip fell to an assassin's knife in 336 B.C., he himself was preparing to invade Persian territory. Philip had devoted his reign to building a new Macedonian army, invincible in its CAVALRY and its heavy-infantry formation known as the PHALANX. He bequeathed to Alexander troops, home-base organization, and propaganda program needed for the Persian campaign.

Alexander the man. The conqueror's yearning and determination, as well as a certain premature aging, are conveyed in this marble bust ascribed to the workshops at Pergamum, circa 180 B.C. The likeness may be based on a bronze portrait statue, apparently sculpted not long before Alexander's death (323 B.C.).

Alexander's campaigns, written in Greek, is believed to derive from the now-vanished campaign memoirs of Alexander's friend and general PTOLEMY (1). Plutarch's biography of Alexander, written circa A.D. 110, is one of his *Parallel Lives* (paired short biographies comparing noble Greeks with noble Romans); here, Alexander's life story is paired with that of Julius Caesar.

Alexander (Alexandros, "defender") was King Philip's eldest legitimate child. His mother, Olympias, came from the ruling clan of the northwestern Greek region called EPIRUS. The tempestuous Olympias remained for 20 years the foremost of the polygamous Philip's wives. But the royal MARRIAGE was unhappy, and mother and son sided together against Philip.

As a mystical follower of the god DIONYSUS, Olympias was said to sleep with a giant snake in her bed as a pet and spiritual familiar. Apparently she convinced the young Alexander that in conceiving him she had been impregnated not by Philip but by ZEUS, king of the gods. This divine parentage would have put Alexander on a par with the noblest heroes of MYTH.

Such childhood influences gave Alexander a belief in his preordained greatness, a need to surpass his father and all other men, and an imperviousness to danger, pain, and fatigue. Ancient writers describe Alexander's yearning for adventure and exploration. He modeled his behavior on two legendary heroes—the world-civilizing HERACLES and the great soldier ACHILLES.

Alexander was fair-skinned, fair-haired, and not tall. Although a dedicated soldier, he disliked all SPORT except hunting. He was sexually abstemious, once remarking that sleep and sexual intercourse both made him sad since they reminded him that he was mortal. When he did pursue

Alexander conquered to rule, not to plunder. Whereas most Greeks despised non-Greeks as barbarians (*barbaroi*, meaning "those who speak gibberish"), Alexander planned to introduce the Persian ruling classes into his army and government. This plan is sometimes referred to as Alexander's fusion policy.

Still, the man had serious flaws. He neglected his kingdom's future by exhausting himself in warfare while he delayed in fathering a royal successor. He was capable of dire cruelty when opposed. His heavy drinking led to disastrous incidents and hastened his death. His lack of long-range planning is shown by his conquest of the far-off Indus Valley (327–325 B.C.)—how could he have hoped to manage such an immense domain? It has been said that Alexander died just in time, before he could see his empire collapse.

Knowledge of Alexander comes mainly from the surviving works of four ancient authors who lived centuries after him: Arrian, PLUTARCH, DIODORUS SICULUS, and the first-century A.D. Roman writer Curtius Rufus. Arrian, the most reliable of the four, was an ethnic Greek Roman citizen who served in the Roman government in the 120s–130s and devoted his retirement to writing. His thorough account of

Alexander the god. A generation after his death, Alexander is shown with the ram's horns of the Greek-Egyptian deity Zeus-Ammon, on this silver tetradrachm, or four-drachma coin, minted circa 300 B.C. The coin comes from the north Aegean region of Thrace, where one of Alexander's followers, Lysimachus, ruled after the conqueror's death.

love, he tended in his youth to prefer males: His lifelong intimate friend was Hephaestion, a Macedonian noble. Later in life, Alexander had sexual relations with WOMEN; when he died he left behind two wives, one of whom was pregnant with his son. Alexander's bisexual development was in keeping with Greek upper-class custom.

Legend claims that at about age nine Alexander tamed the stallion Bucephalas (*Boukephalus*, "ox head"), after its trainer failed to do so. Bucephalus became Alexander's war-horse, carrying him into each of his major battles over the next 20 years.

To tutor the teenage Alexander, King Philip brought the Greek philosopher ARISTOTLE to Macedon. Between about 343 and 340 B.C. Aristotle taught Alexander political science and literature, among other subjects. For the rest of his life, Alexander is said to have kept with him Aristotle's edited version of HOMER's *Iliad*. Also, in future years King Alexander supposedly sent specimens of unfamiliar plants and animals from Asia to his old tutor, to assist Aristotle's biological studies at the LYCEUM.

At age 16 Alexander held his first battlefield command, defeating a Thracian tribe on Macedon's northern frontier. Two years later, in 338 B.C., he commanded the Macedonian cavalry at the Battle of CHAERONEA, in which King Philip won control over all of Greece. (In all his battles, even as king, Alexander remained a cavalry commander, personally leading the charge of his 2,000 elite mounted assault troops known as the Companions.)

With Greece subjugated, King Philip next planned to "liberate" the Persian-ruled Greek cities of ASIA MINOR— and to seize the fabled wealth of Persian treasuries. For propaganda purposes, Philip arranged the creation of a federation called the Corinthian League, representing all the major mainland Greek states, except resistant SPARTA. League delegates dutifully elected Philip as war leader. But before Philip could invade Persian territory, he was assassinated in Macedon (summer 336 B.C.).

The killer, an aggrieved Macedonian nobleman, was slain as he tried to escape. The official verdict claimed that he had been bribed by the Persian king, Darius III. Yet suspicion also lights on Alexander and his mother, who had both recently fallen from royal favor. After Philip's death, Alexander was immediately saluted as the new Macedonian king.

The invasion was delayed for two more years. First Alexander was elected as the Corinthian League's new war captain, empowered to raise troops from mainland Greece and to make war against Persia, in revenge for the Persian king XERXES' invasion of Greece over 140 years before. But despite this pretence of alliance, the Greek city of THEBES revolted (335 B.C.). Alexander—who feared a Greek rebellion as the worst threat to his plans—angrily captured the city and destroyed it. Six thousand Thebans were killed; 30,000, mostly women and children, were sold as SLAVES.

Leaving behind a regent, ANTIPATER, to guard Greece and Macedon and organize reinforcements, Alexander invaded Persian territory in the spring of 334 B.C. He sailed across the HELLESPONT to northwestern Asia Minor with a small Macedonian-Greek force—about 32,000 infantry, 5,100 cavalry, and a siege train. His second-in-command was the 60-year-old Parmenion, who had been Philip's favorite general.

The events of the next 11 years, culminating in Alexander's conquest of Persia and the Indus Valley, can only be summarized here. Alexander's first battle victory came within a few days, at the River Granicus, where he defeated a smaller Persian army commanded by local Persian governors. The central action was a cavalry battle on the riverbank, after Alexander had led a charge across the river.

The Granicus victory opened Asia Minor to Alexander. There he spent the next year and a half, moving methodically south and east. Local Persian troops fell back to a few strongholds. The fortress of Sardis surrendered without a fight. The Persian garrisons at MILETUS and HALICARNASSUS resisted but succumbed to siege.

As local Greek cities opened their gates to Alexander, he set up democratic governments and abolished Persian taxes. But in Asia Minor's non-Greek territories, he merely replaced Persian overlordship with a similar system of obligation toward himself. Most non-Greek cities continued to pay tribute, now to Alexander. His governors were chosen either from his staff or from cooperative local gentry.

By the spring of 333 B.C. he had reached the province of Phrygia, in central Asia Minor. At the Phrygian capital of Gordium stood an ancient wagon supposedly driven by the mythical king MIDAS. The wagon's yoke carried a thong tied in an intricate knot, and legend claimed that whoever could untie the Gordian Knot was destined to rule Asia. According to the most familiar version, Alexander sliced the knot apart with his sword.

In November 333 B.C. Alexander defeated a second Persian army, this one commanded by King Darius himself. The battle took place in a seaside valley on the Gulf of Issus—the geographic "corner" where Asia Minor joins the Levantine coast—about 15 miles north of the modern Turkish seaport of Iskendrun. The mountainous terrain presents a string of narrow passes; it was a natural place for the Persians to try to bottle up Alexander.

At the Battle of Issus, Alexander had about 40,000 troops; Darius had about 70,000. Although half of Darius' army was inferior light infantry, they could have won the battle had not Darius fled in his chariot when Alexander and his Companion Cavalry charged into the Persian left wing. Darius' retreat caused most of his army to follow.

Darius had lost more than a battle. The captured Persian camp at Issus contained the king's wife (who was also his sister) and other family, who now became Alexander's hostages. Ancient writers emphasize that Alexander not only refrained from raping Darius' beautiful wife, as was his due, but he also became friends with Darius' captive mother!

Oddly, rather than pursue the beaten Darius, Alexander chose to let him go. (Upon reaching Mesopotamia, Darius began raising another army.) Alexander turned south to the Levant, to capture the Persians' remaining Mediterranean seaboard. Most of PHOENICIA submitted. But the defiant island city of Tyre provoked an immense siege, which lasted eight months (332 B.C.). The siege's turning point came when the Greeks of CYPRUS rebelled from Persian rule and declared for Alexander, sending him 120 badly needed warships. Lashing these ships into pairs as needed, Alexander equipped some with catapults and others with siege ladders. When the ships' catapults had battered down a section of Tyre's wall, he led his shipborne troops inside. With exemplary cruelty, Alexander sold most of the 30,000 inhabitants into slavery.

Envoys from Darius offered peace: Darius would give his daughter in marriage to Alexander and cede all territory west of the Euphrates River. When the Macedonian general Parmenion commented that he would accept such terms if he were Alexander, the young king replied, "So would I, too, if I were Parmenion." Alexander dismissed the envoys.

In Egypt in the fall of 332 B.C., he received the Persian governor's surrender and was enthroned by the Egyptians as their new pharaoh (as Persian kings had customarily been). In the spring of 331 B.C., west of the Nile's mouth, he founded a city destined to be one of the greatest of the ancient world, ALEXANDRIA (1).

The final campaign against Darius came in the fall of 331 B.C. Alexander marched northeast from Egypt. King Darius, with a huge army levied from all remaining parts of the empire, awaited him east of the northern Tigris, near Gaugamela village, not far from the city of Arbela (modern Erbil, in Iraq).

The Battle of Gaugamela, also called the Battle of Arbela, was a huge, clumsy affair that has defied modern analysis. Alexander was greatly outnumbered. Against his 7,000 cavalry and 40,000 foot, the Persians had perhaps 33,000 cavalry and 90,000 foot. The Persians remained weak in the quality of their infantry, most of whom were Asiatic light-armed troops. But Darius intended a cavalry battle.

Darius launched a massive cavalry attack against both wings of Alexander's army. Somehow the cavalry's departure left Darius' own center-front infantry open to attack; Alexander led his Companion Cavalry charging across the open ground between the two armies' center fronts and struck the Persian infantry there. The melee brought the Macedonians close to Darius—who turned and fled in his chariot, just as at Issus. Soon all Persian troops were in retreat, despite having inflicted heavy losses on Alexander's men.

With his kinsman Bessus, Darius fled into the mountains toward Ecbatana (modern Hamadan, in Iran). For the second time, Alexander turned away from a defeated Darius and marched south, into Mesopotamia.

Alexander now declared himself king of Asia. At Babylon he received the submission of the Persian governor and appointed him as *his* governor there. This move is the first sign of Alexander's fusion policy. Soon he would reappoint other such governors and organize well-born Persian boys into units of cadets, to train for a Macedonian-Greek-Persian army.

From Babylon, Alexander headed southeast, overcoming some fierce resistance, into the Persian heartland. The royal cities of Susa and Persepolis surrendered, opening to him the fantastic wealth of the Persian kings. In April 330 B.C., at Persepolis, Alexander burned down the palace complex designed by King Xerxes. (Its impressive ruins are visible today.) According to one story, Alexander—usually so respectful of Persian royalty—set the fire during a drunken revel.

Alexander's claim to the Persian throne was confirmed by Darius' death (July 330 B.C.). Stabbed by Bessus' men in the countryside near what is now Tehran, Darius died just as Alexander's pursuing cavalry arrived. Alexander is said to have wrapped his cloak around the corpse. Bessus fled farther east and declared himself king, but Alexander had him hunted down and executed.

Still Alexander did not rest. The years 330–327 B.C. saw him campaigning in Bactria (northern Afghanistan) and Sogdiana (Uzbekistan and Tajikistan), where rugged mountain dwellers and horsemen had lived semi-independent of any Persian king. Subduing these people, Alexander took one of them as his bride—Roxane, the beautiful daughter of a Sogdian baron. No doubt the marriage helped to pacify the defiant region. But it is puzzling that the 29-year-old conqueror did not choose a marriage of wider political advantage.

By then he had won the entire domain over which his adversary Darius had ruled. Alexander's northeast frontier became dotted with garrison towns named ALEXANDRIA (2). But these years brought worsening relations between Alexander and certain Macedonian nobles who resented his solicitude toward the defeated Persians and his adoption of Persian customs. Between 330 and 327 B.C. several of Alexander's associates were executed for suspected treason. These included the 70-year-old Parmenion and the army's official historian, Callisthenes, who was Aristotle's nephew. History does not record Aristotle's reaction.

In the summer of 327 B.C., Alexander led his army across the Hindu Kush Mountains and down into the plain of the Indus River (called India by the Greeks but today contained inside Pakistan). This region had at one time been part of the Persian Empire. Alexander's conquest required three arduous years, during which he encountered a fearsome new war machine—the Indian elephant, employed as "tank corps" by local rulers.

A battle fought in monsoon rains at the River Hydaspes (an Indus tributary, now called the Jhelum) was Alexander's military masterpiece. There he defeated his most capable adversary, the local king Porus. The captured Porus was confirmed as Alexander's governor of the region (May 326 B.C.).

In fall 326 B.C., at the Beas River, Alexander's men mutinied, refusing to continue east to the Ganges River. Angrily and reluctantly, Alexander turned west and south. The army had reached its easternmost point and was now on a roundabout route home.

At the resistant fortress of the Malloi (perhaps modern Multan, in Pakistan), Alexander's siege ladder collapsed behind him as he went over the enemy wall. Trapped inside, he was hit in the lung by an arrow. Although rescued, he nearly died. The damaged lung surely hastened his death, now less than two years away.

After a disastrous march west through the southern Iranian desert (325 B.C.), Alexander returned to the Persian royal cities. At Susa in 324 B.C. he held his famous marriage of East and West. Although he already had a wife, he now also married Darius' daughter Barsine, and 90 other Macedonian and Greek officers married high-born Persian women.

That year Alexander sent messages to the mainland Greeks requesting that they honor him as a living god. This request was granted but was met with derision at ATHENS and elsewhere—the Greeks of that era did not generally deify living people. If Alexander had a political purpose in this, it failed. By now he may have been losing his grip on reality.

At Ecbatana in the late summer of 324 B.C., Alexander's close friend Hephaestion died from fever and drinking. Alexander, frantic with grief, ordered a stupendously extravagant monument and funeral. He himself was drinking heavily. Reportedly, in these last days, the brooding Alexander would go to dinner dressed in the costumes of certain gods, such as HERMES or ARTEMIS.

Planning a naval expedition to Arabia, Alexander traveled in the spring of 323 B.C. to Babylon (which is humid and unhealthy in the warmer months). After several nights of drinking, he fell ill. He died 10 days later, on June 10, 323 B.C., in the palace of Nebuchadnezzar. The story that he was poisoned by Antipater's sons CASSANDER and Iolas—to preserve their father's power—is probably false. Legend claims that the dying Alexander, when asked to name his successor, replied, "The strongest."

Of his two widowed wives, Barsine was murdered by order of the pregnant Roxane, who gave birth to Alexander's only legitimate son, also named Alexander. Cassander, as ruler of Macedon, later murdered Roxane, her son, and Olympias. The empire broke into warring contingents under Alexander's various officers, known as the DIADOCHI, or Successors.

(See also ANTIGONUS (1); HELLENISTIC AGE; HOMOSEXUALITY; RELIGION; WARFARE, LAND; WARFARE, NAVAL; WARFARE, SIEGE.)

Alexandria (1) Major Mediterranean port of EGYPT, in ancient times and still today. Alexandria was founded in 331 B.C. by ALEXANDER THE GREAT, one of the many Eastern cities that he established. Located 20 miles west of the Nile's westernmost mouth, the city was immune to the silt deposits that persistently choked harbors along the river. Alexandria became the capital of the hellenized Egypt of King PTOLEMY (1) I (reigned 323–283 B.C.). Under the wealthy Ptolemy dynasty, the city soon surpassed ATHENS as the cultural center of the Greek world.

Laid out on a grid pattern, Alexandria occupied a stretch of land between the sea to the north and Lake Mareotis to the south; a man-made causeway, over three-quarters of a mile long, extended north to the sheltering island of Pharos, thus forming a double harbor, east and west. On the east was the main harbor, called the Great Harbor; it faced the city's chief buildings, including the royal palace and the famous Library and Museum. At the Great Harbor's mouth, on an outcropping of Pharos, stood the lighthouse, built circa 280 B.C. Now vanished, the lighthouse was reckoned as one of the SEVEN WONDERS OF THE WORLD for its unsurpassed height (perhaps 460 feet); it was a square, fenestrated tower, topped with a metal fire basket and a statue of ZEUS the Savior.

The Library, at that time the largest in the world, contained several hundred thousand volumes and housed and employed scholars and poets. A similar scholarly complex was the Museum (Mouseion, "hall of the MUSES"). During Alexandria's brief literary golden period, circa 280–240 B.C., the Library subsidized three poets—CALLIMACHUS, APOLLONIUS, and THEOCRITUS—whose work now represents the best of Hellenistic literature. Among other thinkers associated with the Library or other Alexandrian patronage were the mathematician Euclid (circa 300 B.C.), the inventor Ctesibius (circa 270 B.C.), and the polymath Eratosthenes (circa 225 B.C.).

Cosmopolitan and flourishing, Alexandria possessed a varied population of Greeks and Orientals, including a sizable minority of JEWS, who had their own city quarter. Periodic conflicts occurred between Jews and ethnic Greeks.

The city enjoyed a calm political history under the Ptolemies. It passed, with the rest of Egypt, into Roman hands in 30 B.C., and became the second city of the Roman Empire.

(See also ASTRONOMY; HELLENISTIC AGE; MATHEMATICS; SCIENCE.)

Alexandria (2) Name of several cities founded by ALEXANDER THE GREAT on his conquests eastward (334–323 B.C.). Among these were:

1. Modern-day Iskenderun (previously known as Alexandretta), in southeastern Turkey. Founded after Alexander's nearby victory at Issus (333 B.C.), the city guarded the mountain passes linking ASIA MINOR with the Levantine coast.

2. Modern-day Herat, in northwest Afghanistan.

3. A city at or near modern-day Kandahar, in southeast Afghanistan.

The grandiose Greek city recently excavated at the site called Aï Khanoum, in northern Afghanistan, may have been the Alexandria-in-Sogdiana mentioned by ancient writers.

(See also COLONIZATION.)

Al Mina Modern name for an ancient seaport at the mouth of the Orontes River, on the Levantine coast in what is now southern Turkey. In around 800 B.C. the town was settled as an overseas Greek trading depot—the earliest such post-Mycenaean venture that we know of—and it seems to have been the major site for Greek TRADE with the East for several centuries. Al Mina was surely the main source for Eastern goods that reached Greece and the islands during the "Orientalizing" period, roughly 750–625 B.C. This seaport is not mentioned in ancient Greek literature and was discovered purely by ARCHAEOLOGY in A.D. 1936.

The first Greeks must have arrived with permission of the Armenian-based kingdom of Urartu, which then controlled the Al Mina region. From types of POTTERY found at the site, we can guess that Greeks from EUBOEA predominated, with Cypriot Greeks and Phoenicians also present. Being a place where East and West mingled, Al Mina is the most probable site for the transmission of the Phoenician ALPHABET to the Greeks (circa 775 B.C.).

Al Mina apparently remained important until about 300 B.C., when it was displaced by nearby ANTIOCH and its seaport, Seleuceia.

(See also CHALCIS; ERETRIA; PHOENICIA.)

alphabet The Greek alphabet, containing 24 to 26 letters (depending on locale and era), originated by being adapted from the 22-letter alphabet of the ancient Phoenicians, sometime between 800 and 750 B.C. Prior Greek societies had relied on cumbersome syllabic scripts, in which each character represented a whole syllable: e.g. in modern English, one symbol for "pen," two for "pencil." (Although simple in concept, a syllabic system needs several dozens or even hundreds of symbols to accommodate the various sounds in a language.) The genius of the alphabet is that it reduces the number of symbols by assigning each symbol a precise sound, not an entire syllable. These alphabetic symbols (we call them letters) can be used flexibly in innumerable combinations, to fit different languages.

Presumably, the Greeks first learned about the Phoenician alphabet from observing the record keeping of Phoenician traders. This observation may have taken place at the north Levantine seaport of AL MINA, where Greeks and Phoenicians mingled from about 800 B.C. on. Alternatively, CYPRUS, CRETE, or mainland Greece may have supplied the point of contact. We know that one of the earliest forms of the Greek alphabet was written by the Greeks of CHALCIS (who were also prominent at Al Mina); by 700 B.C. many regional versions of the Greek alphabet had emerged. The Ionic version, as later adopted and modified at ATHENS, is the ancient form most familiar today.

The Greeks imitated the general shapes, names, and sequence of the Phoenician letters. Phoenician ✝ (aleph, "an ox") became Greek A and was renamed *alpha*. Phoenician ⏀ (*bayt*, "a house") became Greek *beta*, B, and so on. All of the Phoenician letters represented consonantal sounds, and most of these were retained by the Greeks—Greek B imitates the "b" sound of Phoenician ⏀, for instance. But the Greeks changed the meaning of seven of the letters, so as to supply vowels. Thus Greek A represented the vowel sound "a," replacing the *aleph*'s glottal stop.

The Phoenician alphabet is loosely preserved in the modern Arabic and Hebrew alphabets. The ancient Greek letters live on in modern Greek, but were also adapted by the ETRUSCANS and the Romans to produce a Roman alphabet (circa 600 B.C.) that is the direct ancestor of our English alphabet, among others.

(See also PHOENICIA; ROME; WRITING.)

Amazons In Greek MYTH, a tribe of female, horse-riding warriors imagined as dwelling in northeast ASIA MINOR, or along the east coast of the BLACK SEA, or at other locales on the northeast fringe of the known world. Beneath layers of poetic elaboration, the Amazon myth may owe something to travelers' tales of real-life Scythian male shamans, who dressed as women but worked and fought as men. Alternatively, the myth may recall Hittite armed priestesses in Asia Minor in the second millennium B.C.

Whatever its origin, the Amazon story exerted a strong influence on the Greek imagination. Amazon society was thought of as savage and exclusively female. To breed, the Amazons periodically mated with foreign males, and they discarded or crippled any resulting male babies. The Amazons wore clothes of animal skin and hunted with bow and arrow; to facilitate use of the bow, they would sear off their young girls' right breasts—hence their name, "breastless" (Greek: *amazoi*).

Appropriately, the Amazons worshipped ARTEMIS, virgin goddess of the hunt, and ARES, the war god. In Greek art, Amazons usually are shown wearing Scythian-style trousers, with tunics that reveal one breast; they are armed with bow, sword, or ax and carry distinctive crescent-shaped shields.

In HOMER's epic poem the *Iliad* (written down circa 750 B.C.) the Amazons are mentioned as a distant people, previously warred upon by the Trojan king PRIAM and the Greek hero BELLEROPHON.

Later writers give the Amazons a role in the TROJAN WAR. The beautiful Amazon queen Penthesilea led a contingent of her tribeswomen to Troy to aid the beleaguered city after HECTOR's death; she was slain by the Greek champion ACHILLES, who then grieved over her death. Other tales develop the sexual overtones; for his ninth Labor, HERACLES journeyed to the Amazons' land and fought them to acquire the belt (often called the "girdle") of their queen, Hippolyta. Similarly, the Athenian hero THESEUS abducted the Amazon queen Antiope; when her outraged subjects pursued them back to Athens and besieged the city, Theseus defeated them and (Antiope having been killed) married their leader. This Amazon bride, also named Hippolyta, bore Theseus' son, HIPPOLYTUS.

Certain tales associated the Amazons with the founding of EPHESUS and other Greek cities of Asia Minor. A later legend claimed that an Amazon queen met the real-life Macedonian king ALEXANDER THE GREAT on his Eastern campaign and dallied with him, hoping to conceive his child (circa 330 B.C.).

Surely the Amazons were in part a reverse projection of the dowdy, housebound lives of actual Greek WOMEN, most of whom were excluded from the men's world, bereft of both political power and sexual freedom. Imaginary "male women" apparently were both fascinating and frightening to Greek men. On the one hand, the Greeks found tall, athletic women generally attractive, and the legend of hard-riding, overtly sexual Amazons seems designed in part to provide an enjoyable male fantasy.

On the other hand, the Amazons represented the kind of foreign, irrational power that was felt to threaten life in the ordered Greek city-state. A favorite subject in Greek art was the Amazonomachy, the battle between Greeks and Amazons; by the mid-400s B.C. the Amazonomachy had come to symbolize the Greeks' defeat of Asian invaders in the PERSIAN WARS. An Amazonomachy is portrayed among the architectural marble carvings of the Athenian PARTHENON.

Ambracia Corinthian colony of northwestern Greece, located north of the modern Gulf of Ambracia. The city was founded in around 625 B.C. as part of a string of northwestern colonies along CORINTH's trade route to ITALY.

Ambracia soon came into conflict with its non-Corinthian Greek neighbors in ACARNANIA and Amphilochia. In the PELOPONNESIAN WAR (431–404 B.C.), Ambracia fought as a Corinthian ally against ATHENS, but Ambracia was effectively neutralized in 426 B.C., when most of its army was wiped out by the brilliant Athenian general DEMOSTHENES (2).

In 338 B.C. Ambracia was occupied by troops of the Macedonian king PHILIP II. In 294 B.C. the city passed into the hands of King PYRRUS of EPIRUS, who made it his capital. As a member of the Aetolian League, Ambracia was besieged and captured by the Romans in 189 B.C. It later became a free city of the Roman empire.

(See also AETOLIA; PERIANDER.)

Amphiaraus See SEVEN AGAINST THEBES.

Amphictyonic League Confederation of different peoples in central Greece, organized originally around the temple of DEMETER at Anthela (near THERMOPYLAE) and later around the important sanctuary of APOLLO at DELPHI. The league's name derives from the Greek *amphictiones*, "dwellers around." The 12 member states included THESSALY, BOEOTIA, LOCRIS, and PHOCIS. The league maintained its two sanctuaries, holding regular meetings of members' delegates and raising and administering funds. It was the league, for instance, that managed Delphi's PYTHIAN GAMES.

Amphipolis Athenian colony near the north Aegean coast of the non-Greek region known as THRACE. Located about three miles inland, Amphipolis ("surrounded city") stood on a peninsula jutting into the Strymon River. Originally a Thracian town had occupied the site. The Athenian colony was established in 437 B.C., after a failed attempt in 462 B.C., when native Thracians massacred the settlers.

Amphipolis controlled the local bridge across the Strymon and hence the east-west route along the Thracian coast, as well as the north-south riverine route to the interior. As a local TRADE depot, Amphipolis was an important source of certain raw materials exported to ATHENS. Among these were GOLD and SILVER ore (mined from Thrace's Mt. Pangaeus district) and probably SLAVES, purchased as war prisoners from the feuding Thracian tribes. Shipbuilding TIMBER was another valuable local product, and it seems that Athenian warships were constructed right at Amphipolis.

In 424 B.C., during the PELOPONNESIAN WAR, the city was captured without a fight by the brilliant Spartan commander BRASIDAS. In 422 B.C., at the Battle of Amphipolis, Brasidas and the Athenian leader CLEON were both killed as the Athenians tried unsuccessfully to retake the city. Thereafter Amphipolis remained beyond Athenian control. Captured by King PHILIP II of MACEDON in 357 B.C., it became a Macedonian city and coin-minting center. After ROME's final defeat of Macedon in the Third Macedonian War (167 B.C.), Amphipolis passed into Roman hands.

(See also THUCYDIDES (1).)

amphora See POTTERY.

Anacreon Celebrated Greek lyric poet of the late 500s B.C., active at SAMOS and ATHENS but born circa 565 B.C. at Teos, a Greek city of western ASIA MINOR. Witty, decadent, and evidently bisexual, Anacreon was the poet of pleasure. His sophisticated verses celebrate WINE, WOMEN, boys, and song—the vital ingredients at the SYMPOSIUM, or upper-class drinking party. Anacreon exemplified the wealth and sophistication of his native region of IONIA, which during the poet's own lifetime fell disastrously to Persian invasion.

Anacreon was in his teens or 20s when he joined the evacuation of Teos to escape the attacking Persians, circa 545 B.C. Sailing north to the coast of THRACE, these refugees founded the city of ABDERA; one poem by Anacreon, presumably from this period, is an epitaph for an Abderan soldier slain in local fighting.

Before long, however, Anacreon had emerged at one of the most magnificent settings in the Greek world—the court of the tyrant POLYCRATES of Samos. There (where another great poet, IBYCUS, was installed) Anacreon won fame and fortune and became a favorite of the tyrant. In keeping with upper-class taste (and with Polycrates' own preference), Anacreon wrote many poems on homosexual themes, celebrating the charms of boys or young men.

In around 522 B.C., when the Persians killed Polycrates and captured Samos, Anacreon went to Athens. According to one tale, he escaped the Persian onslaught aboard a warship sent expressly by Hipparchus, the brother and cultural minister of the Athenian tyrant HIPPIAS (1). Anacreon thrived at Hippias' court (where the poet SIMONIDES was another guest). By the time of Hippias' downfall (510 B.C.), Anacreon had found new patrons among the aristocrats of THESSALY. Soon, however, he was back in Athens, where he seems to have been welcome despite his prior association with the tyrant.

His Athenian friends of these years include XANTHIPPUS (later the father of PERICLES) and a young man named Critias (the future grandfather of the oligarch CRITIAS and an ancestor of PLATO), to whom Anacreon wrote love poems. Anacreon seems to have created a cultural sensation at Athens; a red-figure vase circa 500 B.C. shows a symposium scene with a figure labeled "Anacreon" wearing an Asian-style turban and playing an Ionian-style lyre. Surely Anacreon's long stay in Athens helped to introduce Ionian literary tastes, thus contributing to the city's grand cultural achievements in the following decades. He probably died at Athens, perhaps in around 490 B.C.; legend assigns to him an appropriate death, from choking on a grape seed. In later years the Athenians set up a statue of him on the ACROPOLIS.

Most of Anacreon's surviving poems are short solo pieces for lyre accompaniment, written to be sung at a symposium. In simple meters and simple Ionic language, these verses combine yearning with merriment. Whereas a poet like SAPPHO (circa 600 B.C.) might earnestly describe love as a fire under the skin, Anacreon writes of love as a game of dice or a boxing match. "Boy with the virginal face," he writes, "I pursue you but you heed me not. You do not know you are the charioteer of my heart."

One poem (later imitated by the Roman poet Horace) compares a young woman to a frisky colt who needs the right fellow to mount her gently and break her in. Another poem clearly presents the poet's sophisticated world: "Now golden-haired Love hits me again with a purple ball and tells me to play with the girl in colored sandals. But she comes from Lesbos, a cosmopolitan place, and finds fault with my gray hair. And she gapes at someone else— another girl!"

Among his other surviving work are hymns to ARTEMIS, EROS, and DIONYSUS. The corpus also includes 60 anonymous poems, not written by Anacreon but penned centuries after his death, in imitation of his style. These verses, called Anacreonta, had great literary influence in the Roman era.

(See also GREEK LANGUAGE; HOMOSEXUALITY; LYRIC POETRY; MUSIC.)

Anaxagoras Greek philosopher (circa 500–428 B.C.), born in Clazomenae but active at ATHENS. Coming from the intellectually advanced region called IONIA, in Greek ASIA MINOR, Anaxagoras played a vital role in introducing the study of PHILOSOPHY at Athens. As such, he was an important forerunner of the philosopher SOCRATES.

A teacher and friend of the Athenian statesman PERICLES, Anaxagoras is said to have lived at Athens for 30 years, probably circa 480–450 B.C. Supposedly he taught philosophy to the tragic playwright EURIPIDES. Eventually, however, Anaxagoras was charged with the criminal offense of impiety (asebeia, the same charge that would destroy Socrates 50 years later). The supposed offense was Anaxagoras' theory that the sun is really a huge, red-hot stone—an idea that would logically deny the existence of the sun god, HELIOS—but probably the accusation was meant to harm Pericles. Anaxagoras fled Athens, apparently before the case went to trial, and was condemned to death in absentia. He settled at Lampsacus, in Asia Minor, near the HELLESPONT, where he founded a philosophical school and lived as an honored citizen.

Anaxagoras carried forth the Ionian tradition of natural philosophy—that is, of theorizing about the natural world. He is said to have written only one book (now lost). Our knowledge of his work comes mainly from sometimes contradictory references by later writers. He seems to have accepted PARMENIDES' doctrine that reality is unchanging and eternal, but he also was influenced by the atomist theories of DEMOCRITUS and LEUCIPPUS to the extent that he pictured a system of various "seeds" that bunch together in different combinations to constitute different material. Behind the movements of these seeds is "Mind" (nous), the universe's animating force, which is infinite and aloof, but which is somehow reflected in human intelligence and other phenomena. These concepts help to explain Anaxagoras' best-remembered (although enigmatic) statement: "In everything there is a portion of everything except Mind. And there are some things in which there is Mind also."

An important figure in early Greek ASTRONOMY, Anaxagoras followed ANAXIMENES' theory that the earth is flat and suspended in air, with the heavenly bodies rotating around it. Anaxagoras guessed that the moon is closer to us and smaller than the sun, and that its light is reflected from the sun. His belief that the sun, moon, and stars are really huge stones was probably influenced by the fall of a large meteorite in the Hellespont district in 467 B.C.

(See also ANAXIMANDER; PERSIAN WARS; THALES; THUCYDIDES (2).)

Anaximander Early brilliant scientist and philosopher of the 500s B.C. A pupil of THALES, Anaximander lived between about 610 and 545 B.C. in MILETUS, in the flourishing Greek region called IONIA, in western ASIA MINOR. In modern opinion, Anaximander is the most distinguished of the three thinkers who comprised the Milesian School of natural philosophers. (The third is ANAXIMENES.) Anaximander also can be called the West's first astronomer and geographer.

Anaximander wrote the earliest known Greek prose work, a theoretical description and history of the natural world. This treatise has not survived, but a number of later ancient writers refer to it. It contained the first Greek map of the heaven and the earth, and described the movements of the constellations. It was probably the first written attempt in the West to substitute SCIENCE for MYTH in explaining the universe.

Anaximander pictured the earth as a cylinder suspended upright, with the flat ends at top and bottom; humans live on the top surface, surrounded by the heavens. While Thales had theorized that the earth floats on water, Anaximander believed that the earth is suspended in air, equidistant from all other heavenly bodies. This idea looks remarkably like a guess at the celestial law of gravity.

According to Anaximander, the primal element in the universe is not water, as Thales believed, but a more mysterious substance that Anaximander refers to as the apeiron, the "boundless" or "indefinite." He apparently imagined this apeiron as partaking of characteristics more usually ascribed to the gods. The apeiron is immortal, indestructible, the source of creation for the heavens and the earth, and also the receptacle that receives and recirculates destroyed matter. Although abstruse, Anaximander's theory seems to point toward a pantheistic or monotheistic notion of a life force animating the universe; it also anticipates modern chemistry's discovery that basic matter is never really destroyed but only undergoes change.

Anaximander seems to have guessed at the biological process of evolution. He is recorded as having believed that humankind originally emerged from fishes to step forth onto land. He is also credited with constructing the first sundial in the Greek world, probably based on Babylonian examples.

Anaximander is said to have died around 545 B.C., the same year in which the proud, accomplished Ionian Greek cities were conquered by the Persian armies of King CYRUS (1).

(See also ASTRONOMY; PHILOSOPHY.)

Anaximenes Early Greek philosopher and scientist. He lived between about 585 and 525 B.C. at MILETUS, a prosperous Greek city in the region called IONIA, in western ASIA MINOR. Following his two greater predecessors, THALES and ANAXIMANDER, Anaximenes was the last important member of what is called the Milesian School of natural philosophers. Nothing is known of his life except that he wrote a book in "simple and unpretentious Ionic language," as one later writer described it. Although the book is lost to us, enough of it is paraphrased by other authors to give an idea of its message.

Like Thales and Anaximander, Anaximenes sought to identify a single, primal substance that is the basic element in the universe. Thales had said that this primal substance is water. Anaximenes, apparently observing that water is itself part of a larger process of condensation and evaporation, identified the universal element as *aër*—"air" or "mist."

The air, he said, is infinite and eternally moving. As it moves, it can condense into different forms: into wind, which produces cloud, which creates rain, which can freeze into ice. Contrarily, air can rarify itself to form fire. Cold is a result of condensation; heat, of rarefaction. This is why a person puffs through compressed lips to cool down hot food, but puffs through open lips to heat cold hands.

Anaximenes believed that the earth was flat and floated on air in the cosmos. He considered air to be the divine, ordering force in the universe; it is said that he did not deny the existence of the gods, but claimed that the gods arose from air. Although much of Anaximenes' theory seems to derive from his two forerunners, his commonsensical attempt to explain material change was to have great influence on later philosophers, particularly on the atomist DEMOCRITUS.

(See also PHILOSOPHY; SCIENCE.)

Andocides   Athenian political figure and businessman who lived circa 440–390 B.C. and whose adventures were associated with the downfall of his city in the PELOPONNESIAN WAR (431–404 B.C.). Four speeches are preserved under his name, although one of these "Against Alcibiades," is considered a later forgery.

Andocides was born into a prominent aristocratic family at ATHENS. As a member of a right-wing club, he was named as one of the conspirators in the Mutilation of the Herms in 415 B.C. To gain immunity from prosecution, Andocides confessed and named his co-conspirators—whether truthfully or not is unclear. Departing from Athens, he prospered elsewhere as a merchant, supplying needed oars at cost to the Athenian fleet at SAMOS.

Unluckily, Andocides' return to Athens in 411 B.C. coincided with the oligarchic coup of the FOUR HUNDRED, which brought to power men from the right-wing circles that had been harmed by his confession of four years before. He was then thrown into prison. Released, he left Athens again and resumed his trading. Eventually he was reinstated at Athens, after defending himself against certain charges in his speech "On the Mysteries" (circa 400 B.C.).

As an Athenian ambassador to SPARTA in 392 B.C., during the CORINTHIAN WAR, Andocides helped negotiate a proposed Athenian-Spartan peace treaty, the terms of which are preserved in his speech "On the Peace."

Unfortunately, the Athenians rejected the terms and prosecuted the ambassadors, whereupon Andocides left Athens yet again. He died soon after.

(See also OLIGARCHY; RHETORIC.)

Andromache   In Greek MYTH, a princess from southeastern ASIA MINOR who became the wife of the Trojan prince HECTOR. During the 10-year TROJAN WAR, her father, brothers, and husband were slain by the Greek hero ACHILLES. Her son, Astyanax, was executed by the Greeks after their capture of TROY, and Andromache became the slave of Achilles' son, NEOPTOLEMUS. She accompanied him home to EPIRUS and bore him a son, Molossus, ancestor of the royal Molossian tribe. She ended her days in Epirus, as wife of the refugee Trojan prince Helenus.

In HOMER's epic poem the *Iliad*, she appears as a gracious and stalwart lady whose future misery is clearly foreshadowed. As an embodiment of female suffering at the hands of conquerors, Andromache was a natural subject for the intellectual Athenian playwright EURIPIDES; his tragedy *Andromache* (presented circa 426 B.C.) survives today, and Andromache also is prominent among the characters in his *Trojan Women* (415 B.C.).

Andromeda   See PERSEUS (1).

Antaeus   See HERACLES.

Antalcidas   See KING'S PEACE, THE.

Antigone   Mythical princess of the central Greek city of THEBES. A daughter of the incestuous union of OEDIPUS and Jocasta, Antigone is the heroine of an extant tragedy by the Athenian playwright SOPHOCLES. His *Antigone* (performed circa 442 B.C.) examines the conflict between law and moral obligation. Although Sophocles was drawing on an existing MYTH, most of the information available to us about Antigone comes from his play.

Antigone had two brothers, Eteocles and Polynices. In the disastrous expedition of the SEVEN AGAINST THEBES, Polynices led a foreign army to Thebes in an attempt to depose Eteocles, who, contrary to prior agreement, was monopolizing the kingship. The two brothers killed each other in single combat. The new Theban ruler, Creon, decreed that the invader Polynices' corpse go unburied, thereby—according to Greek belief—denying Polynices' ghost a resting place in the Underworld. But Antigone chose her obligations of KINSHIP and RELIGION over her obligations as a citizen, and she covered the body with dust and did honors at the graveside. As punishment, Creon sentenced her to be sealed alive in a vault. After being warned prophetically that he was offending the gods, Creon relented: too late, for Antigone had killed herself. This brought calamity to Creon's house, in the suicides of Creon's son Haemon (who had been betrothed to Antigone) and of his wife, Eurydice.

Antigone was the subject of a lost play by EURIPIDES that seems to have followed a familiar folktale pattern. In this version, the condemned Antigone is handed over to Haemon for execution; instead he hides her in the countryside, and they have a son. Unaware of his royal lineage, the boy eventually makes his way to Thebes, where adventures and recognition follow.

(See also AFTERLIFE; FUNERAL CUSTOMS.)

Antigonus (1)   Macedonian general and dynast who lived circa 382–301 B.C. and ruled ASIA MINOR and other parts of the Greek world in the tumultuous years after the death of ALEXANDER THE GREAT (323 B.C.). Antigonus is counted as one of the DIADOCHI, who carved up—and fought over—Alexander's vast domain. Antigonus and his son, the dynamic soldier DEMETRIUS POLIORCETES, came close to reconquering and reknitting Alexander's fragmenting empire. But Antigonus died in battle at the hands of a coalition of his enemies (301 B.C.). Much of our information about him comes from the later writer PLUTARCH's short biography of Demetrius.

Born in about 382 B.C., Antigonus was nicknamed One-eyed (Monophthalmos), possibly from a war injury. Serving as one of Alexander's generals in the East, Antigonus was appointed governor of Phyrgia (central Asia Minor), circa 333 B.C. His ascent truly began in 321 B.C., when the Macedonian regent ANTIPATER appointed him chief commander in Asia. In the next two decades, Antigonus' ambition of reuniting the empire brought him and his son into wars on land and sea against the four secessionist Diadochi—PTOLEMY (1) (who claimed EGYPT as his domain); CASSANDER (who claimed MACEDON and Greece); LYSIMACHUS (who claimed THRACE); and SELEUCUS (1) (who, having deserted from Antigonus' command, claimed vast tracts in Mesopotamia and the Iranian plateau).

Based at Celaenae, in southern Phrygia, Antigonus and Demetrius fought against the allied Diadochi over two periods, 315–312 B.C. and 307–301 B.C., in Syria, Asia Minor, Greece, and the Mediterranean. After Demetrius' spectacular naval victory over Ptolemy near Cyprian Salamis (306 B.C.), Antigonus adopted the title king (*basileus*)—that is, king of Alexander's empire. Soon after, each of the other Diadochi took the title king as well.

The downfall of Antigonus, aged about 80, came from a concerted campaign against him. With Ptolemy and Cassander helping elsewhere, Seleucus marched an army west out of Asia and joined Lysimachus in northern Phrygia. The Battle of the Kings was fought at Ipsus, in central Phrygia, in 301 B.C. Antigonus and Demetrius, with an army allegedly of 70,000 infantry, 10,000 cavalry, and 75 elephants, opposed Seleucus and Lysimachus' force of 64,000 foot, 10,500 horse, and 480 elephants. Demetrius, leading his cavalry, was drawn too far forward in the field and was cut off by the enemy's elephants. Antigonus was surrounded and killed; his last recorded words were: "Demetrius will come and save me." Demetrius survived the battle and fled to EPHESUS, to fight another day. Antigonus' kingdom was divided between Seleucus and Lysimachus.

Antigonus (2)   See MACEDON.

Antioch   Rich and important Greek city on the northern Levantine coast, situated beside the Orontes River, about 15 miles inland from the Mediterranean Sea. King SELEUCUS, (1) creator of the SELEUCID EMPIRE, founded Antioch in 300 B.C. to be his Syrian provincial capital. He named it for his son, Prince Antiochus. The city thrived, eventually replacing Seleuceia-on-the-Tigris as the empire's capital. Antioch's port was another city named Seleuceia, at the Orontes' mouth.

Antioch profited from its fertile plain and especially from its position on the age-old TRADE route between Mesopotamia and the Mediterranean. Militarily, it gave access north to ASIA MINOR, west to the Mediterranean, and east to the Asian continent. Antioch was the second city of the eastern Mediterranean, after Egyptian ALEXANDRIA (1); like Alexandria, it had an international population, including a large minority of JEWS.

Along with the remaining Seleucid domain, Antioch became a Roman possession in 63 B.C.

(See also HELLENISTIC AGE.)

Antiochus (1)   See ANTIOCH; SELEUCID EMPIRE.

Antiochus (2) III   Capable and ambitious king of the SELEUCID EMPIRE (reigned 223–187 B.C.). Surnamed "the Great," Antiochus reconquered prior Seleucid holdings in the Iranian plateau and eastward, but came to grief on his western frontier against the expansionism of ROME. The Romans were alarmed by Antiochus' conquests in the Levant at the expense of Ptolemaic EGYPT (202–198 B.C.), by his invasion of THRACE (196 B.C.), and finally by his invasion of Greece at the invitation of the Aetolian League (192 B.C.). In the Romans' Syrian War (192–188 B.C.), Antiochus was defeated in Greece and western ASIA MINOR and at sea. His peace treaty with Rome prohibited any further Seleucid military activity on the Mediterranean seaboard. This Treaty of Apamaea (188 B.C.) was a major step toward Rome's absorption of the Greek East. Antiochus died the following year.

(See also AETOLIA.)

Antipater   Macedonian general who lived 397–319 B.C. During King ALEXANDER THE GREAT's eastern campaigns (334–323 B.C.), Antipater served as Alexander's regent over MACEDON and the conquered land of Greece. He destroyed two Greek rebel uprisings: one in 331 B.C., led by the Spartan king Agis III, and the second in 323–322 B.C., when ATHENS and other states arose at news of Alexander's death.

With the Macedonian royal house in disarray, Antipater became nominal regent of Alexander's whole empire in 321 B.C. In his last years' struggle to keep the vast domain together, Antipater was aided by his friend ANTIGONUS (1), whom he made chief commander in Asia. But on Antipater's death, his own son CASSANDER seized Macedon and Greece, in defiance of Antigonus.

(See also DIADOCHI.)

Antiphon   Athenian orator who lived circa 480–411 B.C. and who masterminded the abortive right-wing coup of the FOUR HUNDRED in 411 B.C. at ATHENS. After the coup's failure, Antiphon chose to remain behind when most of his co-conspirators fled. Arrested, he was condemned for treason and executed. The historian THUCYDIDES (1) knew Antiphon and describes him as one of the most capable

Athenians of the day. Antiphon rarely spoke in public, before the ASSEMBLY or law courts, preferring instead to advise or to compose speeches for clients. His speech in his own defense at his treason trial was, according to Thucydides, probably the best courtroom speech ever made.

Antiphon's work survives in three rhetorical exercises for courtroom-speaking practice, plus two or three speeches written for actual court cases. The exercises are known as tetralogies, from their four-part structure—two speeches each by both prosecution and defense. The best known of the tetrologies concerns an imagined criminal case, apparently popular among contemporary thinkers, in which a boy at the GYMNASIUM has been killed accidentally by a thrown javelin. Who is guilty, the speech inquires: the one who hurled the weapon, or the weapon itself?

(See also OLIGARCHY; PROTAGORAS, RHETORIC.)

Aphrodite Goddess of love, sex, regeneration, and bodily beauty. She is one of several deities (along with ZEUS, ATHENA, and APOLLO) whose earthly influence is most celebrated in Greek art and poetry.

At some time between about 1200 and 900 B.C. Aphrodite's cult arrived in mainland Greece, probably imported from the island of CYPRUS, which had attracted Greeks in the copper TRADE. In MYTH, Cyprus was said to be Aphrodite's birthplace, and in historical times the Cyprian city of Paphos had an important temple of the goddess.

The early Aphrodite of Cyprus may originally have been a fertility deity similar to the Semitic goddess Ishtar-As-tarte. Among the Greeks she came to personify sexual urge and pleasure, and was devoutly worshipped as a universal force. Aphrodite oversaw the mating and reproduction of animals. She was a protector of seafarers (a trait perhaps derived from mercantile Cyprus). In some locales

The goddess Aphrodite teaches her son Eros how to use his bow and arrow, in this scene incised onto a bronze mirror cover, 300s B.C. The goddess wears only a pair of slippers.

she was a war goddess (which may be the background of her mythical association with the war god ARES). Not surprisingly, Aphrodite was also the patron deity of PROSTITUTES. Her temple at CORINTH was famous for its official harlots, whose fees helped enrich the goddess. (This feature, unique in a Greek cult, may have owed something to tradeborne Syrian-Phoenician influence in the 700s B.C.) Among Aphrodite's other cult centers was the island of Cythera, off the southern Peloponnesian coast.

In myth, Aphrodite was married to the smith god HEPHAESTUS but had Ares as a frequent lover. Her human paramours included the Trojan prince Anchises (their union produced the hero AENEAS) and the Cypriot youth ADONIS, whom jealous Ares eventually killed.

The origin of the name Aphrodite is unknown; the Greeks fancifully explained it as meaning "foam-born." A passage in HESIOD's epic poem the *Theogony* (circa 700 B.C.) tells the best-known version of her birth: how the primeval god CRONUS cut off the genitals of his father, Uranus, and threw them into the sea, where they generated a white foam (Greek: *aphros*). From this the goddess arose and stepped ashore at Cyprus. But in HOMER's *Iliad* and *Odyssey*, written down perhaps 50 years before Hesiod's time, Aphrodite is described as the daughter of Zeus and Dione, an Oceanid.

Homer's Aphrodite is an oddly undignified character. In the *Iliad*, she tries to protect the Trojans and her Trojan son, Aeneas, but she ignominiously flees the battlefield after being wounded by the Greek hero DIOMEDES, whereupon Zeus reminds her that her province is love, not war (book 5). An episode in the *Odyssey*'s book 8 tells how Aphrodite's cuckolded husband, Hephaestus, used a chain-link net to trap her and Ares together in bed, then dragged the ensnared pair before the assembled gods on MT. OLYMPUS.

Later legends made Aphrodite the mother (by Ares) of the boy-god EROS and also the mother (by the god HERMES) of Hermaphroditus, a creature with the sex organs of both genders.

Aphrodite's attributes included the dove, the myrtle leaf, and the woman's hand mirror; in later centuries these traits, like the rest of Aphrodite's worship, were borrowed by the Roman goddess Venus. Aphrodite's titles included Pandemos (of the whole people), Ourania (celestial), and Philommeides (laughter-loving or genital-loving).

Naturally enough, Aphrodite was the subject of some of the most ambitious and inspired ancient artwork. The marble statue known today as the Venus de Milo, displayed in the Louvre in Paris, France, is a second-century-B.C. Aphrodite found on the Greek island of MELOS; apparently it is a copy of a lost original from the 300s B.C. The finest statue of the ancient world was said to be the Aphrodite of CNIDUS, carved by PRAXITELES in around 365 B.C., which shows the goddess standing naked, having disrobed for her bath.

(See also OCEANUS.)

Apollo Important Greek god whose attributes typify the classical Greek ideal. Apollo was lord of MUSIC, POETRY, dance, intellectual inquiry, shepherding, COLONIZATION, and MEDICINE; he was also the god of religious healing, offering ritual purification to those guilty of murder and other impieties. Through his various human oracles, Apollo could reveal the future and the will of his father, ZEUS. Apollo's temple and oracle at DELPHI, in central Greece, comprised the most influential religious center in the Greek world, and every four years the important PYTHIAN GAMES were held there in the god's honor.

The two famous precepts carved on the Delphi temple wall—"Know thyself" (meaning "Know your human limitations") and "Nothing in excess"—expressed the Apolline ideals of moderation, harmony, and sanity. These were also the ideals of the Greek city-state.

This most Greek of gods seems to have been a relative latecomer to Greek RELIGION, introduced perhaps circa 1200–1100 B.C., at the end of MYCENAEAN CIVILIZATION. Apollo's worship may have been brought south by the invading DORIAN GREEKS, or it may have been introduced from the East, from Hittite ASIA MINOR. Apparently Apollo was originally a god of flocks and herds, of shepherds, and of the shepherd's enemy, the wolf; two of his prominent cult titles were Lukeios (of the wolf) and Nomios (of pastures). Certain of the god's attributes, such as archery, medicine, and music-making, probably began as aspects of shepherd life.

During the Greek colonizing era, circa 750–550 B.C., Apollo's influence was enlarged by his priests at Delphi, who sanctified many colonizing expeditions. One of Apollo's titles was Archigetes (leader of colonists).

Greek art portrayed Apollo as a young man of idealized face and body, often carrying an archer's bow or a lyre. In MYTH, he was the twin brother of the goddess ARTEMIS; their parents were Zeus and the demigoddess Leto. The divine twins were born on the island of DELOS, where Leto had hidden to escape the malice of Zeus' jealous wife, HERA. From the smith god HEPHAESTUS, Apollo received an archer's bow of SILVER for use against his enemies; sometimes the arrows that he shot were intangible and "gentle," bringing disease.

Apollo's first adventure was to slay the primeval serpent Python, which inhabited the sanctuary Pytho, beside Mt. Parnassus. Apollo then took on the title Pythian and occupied the sanctuary, which became known as Delphi.

Other myths emphasized Apollo's relation to the arts, particularly to the ordering principle. Persuading the MUSES on Mt. Helicon to give up their frenzied style of dancing, Apollo led them in stately measures, and so earned the title Mousagetes (leader of the Muses). When the rude satyr Marsyas challenged the god to a music contest (Marsyas' flute against Apollo's lyre), Apollo defeated him, as judged by the Muses, and then flayed Marsyas for his arrogance.

Apollo, like his father, had many love affairs. He seduced Creusa, daughter of the Athenian king Erechtheus, to produce the hero ION (1), ancestor of the IONIAN GREEKS. Falling in love with the athletic nymph CYRENE (2), Apollo carried her to Libya, to the site of the future city that bore her name. In another story he fathered ASCLEPIUS with a Thessalian woman named Coronis. Less successful was his courtship of the Trojan princess CASSANDRA; to woo her he offered the gift of prophecy. She took the offer but still refused his advances. Although he could not recall his gift, he vengefully decreed that no one would believe her soothsaying. Apollo also loved the Spartan lad HYACINTHUS, who was eventually slain by the jealous Zephyrus, god of the west wind.

According to myth, it was Apollo, lord of colonists, who helped Cretan or Arcadian colonists establish the city of TROY. In HOMER's epic poem the *Iliad*, the god is portrayed as ardently pro-Trojan, dealing harm whenever possible to the Trojans' enemies, the Greeks; at the poem's beginning; Apollo shoots arrows of plague into the Greek camp.

But even dearer to Apollo than the Trojans were the Hyperboreans, a virtuous race that lived (as their name indicates) beyond the north wind. Apollo would spend the winter months among them, and in historical times it was claimed that gifts from the Hyperboreans reached Delphi every year, wrapped in straw packing.

Among Apollo's traditional titles was Phoebus (Phoibos), meaning "bright" or "golden," and in the 400s B.C. there arose a philosophical theory that said Apollo was the sun. Although this idea caught on as an intellectual fancy, it seems never to have been a sincere religious belief. In Homer (circa 750 B.C.) and other canonical poets, the sun has its own mythological character, named HELIOS.

(See also CYRENE (1); POLIS; AND PROPHECY AND DIVINATION.)

Apollonia Several Greek cities bore this name, in honor of APOLLO, god of colonists. The most important Apollonia was located inland of the eastern Adriatic coast, in a non-Greek territory that is now Albania. The city was founded circa 600 B.C. by colonists from CORINTH and perhaps also from CORCYRA, as part of a Corinthian TRADE network extending westward to ITALY.

Other Greek cities named Apollonia included one on the Aegean island of Naxos, one on the western BLACK SEA coast, and one that was the port city of CYRENE (1).

(See also COLONIZATION; EPIDAMNUS; ILLYRIS.)

Apollonius Greek poet and scholar of ALEXANDRIA (1), in EGYPT, who lived circa 295–230 B.C. His surviving epic poem, the *Argonautica*, presents the legend of JASON (1) and the Argonauts in their quest for the Golden Fleece. Impressive in its verbal beauty and presentation of character, the *Argonautica* is the only Greek epic to be preserved from the HELLENISTIC AGE (300–150 B.C.). The poem's scholarly subject matter, its often playful tone, and its concern with male and female sexual feelings all reveal the values of the Alexandrian literary movement. Apollonius—along with his rival, CALLIMACHUS—epitomizes this sophisticated movement.

Although probably born in Alexandria, Apollonius is often known by the surname Rhodios (of RHODES), referring to the Greek island city where he spent the last part of his life. According to various sources, he began as the pupil of the established poet Callimachus, but the two men became antagonists in a famous literary feud. This bitter quarrel—which the poets also pursued in their verses—was perhaps based partly on literary tastes. (The experimental Callimachus objected to the writing of Homeric-style EPIC POETRY, with its familiar subject matter and long plot.) Another cause of enmity may have been King PTOLEMY (2)

II's appointment of Apollonius as director of the great Library at Alexandria (circa 265 B.C.). This prestigious job made Apollonius the most influential person in the Greek literary world and incidentally placed him in authority over Callimachus, who also was employed at the Library (and who may have been passed over for the directorship).

Whether the disruptive feud played a part or not, Apollonius resigned from the Library post and withdrew to Rhodes, in order to write, or rewrite, his *Argonautica*. This relatively short epic—5,834 dactylic hexameter lines, in four books—skillfully combines traditional MYTH, scholarly erudition, and romance. The first two books describe the outward voyage from THESSALY to Colchis, at the eastern shore of the BLACK SEA; the admirable third book describes the Colchian princess MEDEA's self-destructive love for the hero Jason and the exploits relating to the fleece's capture; the last book recounts the Argonauts' escape homeward by way of a fantastical route that calls forth much geographical lore from the poet. The poem's flaw is a lack of cohesion in theme and tone, but it is noteworthy for being the first Greek epic to include psychological descriptions of a woman in love (possibly inspired from the stage tragedies of EURIPIDES). In this and other aspects, the *Argonautica* had a great affect on subsequent poetry, particularly on the Roman poet Vergil's epic work, the *Aeneid* (circa 20 B.C.).

Apollonius also wrote prose treatises, epigrams, and scholarly poems on the foundations of certain cities, according to the literary taste of the day. Other than one extant epigram attacking Callimachus, these writings survive only in fragments.

(See also LYRIC POETRY; WOMEN.)

Aratus (1)   See ACHAEA.

Aratus (2)   Greek poet and philosopher of the mid-200s B.C., best known for his long poem about ASTRONOMY, titled the *Phaenomena*. Born circa 315 B.C. in the city of Soli in southeastern ASIA MINOR, Aratus studied at ATHENS under the Stoic founder Zeno. There he met the future Macedonian king Antigonus II (reigned circa 276–239 B.C.). Aratus became a court poet to Antigonus in MACEDON.

The *Phaenomena* survives today. It combines religious and philosophical lore with the astronomical theories of Eudoxus of CNIDUS (active circa 350 B.C.) in explaining the heavenly bodies' movements. The poem was immensely popular in the ancient world, and later generations regarded Aratus as one of the four great Hellenistic poets, alongside CALLIMACHUS, APOLLONIUS, and THEOCRITUS. But the *Phaenomena*'s abstruse subject matter leaves it virtually unread today.

(See also HELLENISTIC AGE; STOICISM.)

Arcadia   The mountainous, landlocked, central portion of the PELOPONNESE, bordered on the south by the territory of SPARTA and on the northeast by that of ARGOS. Arcadia was inhabited by a rugged breed of highlanders who claimed to have inhabited their mountain glens since before the moon was born. The poverty and hardiness of the Arcadians is shown in their ancient reputation for eating acorns. Serving as HOPLITES, they were formidable warriors, and by the 400s B.C. they were producing mercenary soldiers for wars abroad.

Arcadia had few cities aside from the important group of TEGEA, MANTINEA, and Orchomenus in the eastern plains, and (later) Megalopolis in the west. The area's history in the 500s–300s B.C. mostly involves feuding between Tegea and Mantinea and periodic resistance to Spartan domination. The Theban statesman EPAMINONDAS liberated Arcadia from Sparta and founded Megalopolis (circa 365 B.C.) after his victory over the Spartans at LEUCTRA (371 B.C.).

The Arcadians' dialect bore resemblance to that of the distant island of CYPRUS; scholars believe that this shared, unique dialect goes back to the language of the MYCENAEAN CIVILIZATION (circa 1600–1200 B.C.). Both Arcadia and Cyprus seem to have been points of refuge for the Mycenaeans, whose civilization was destroyed by internal wars and by the invading DORIAN GREEKS of 1100–1000 B.C. Corroboratively, archaeological evidence reveals little settlement in Arcadia before about 1000 B.C.

In the cosmopolitan circles of the Hellenistic and Roman worlds (and later, in the European Renaissance), Arcadia was romanticized as the home of rustic virtues amid a mythical Golden Age. The Sicilian-Greek poet THEOCRITUS (circa 265 B.C.) imagined Arcadia as an idyllic haunt of lovelorn shepherds and shepherdesses.

(See also HELLENISTIC AGE.)

archaeology   The systematic study of the past through recovery and interpretation of material remains, including building debris, metal weapons and utensils, clay utensils, human and animal skeletons, and inscriptions in stone or fired clay. Typically these items are preserved by being buried protectively underground or by lying undisturbed for centuries on the sea floor.

POTTERY, widespread in ancient use and surviving up to 10,000 years in the ground, provides by far the single most common source of archaeological data. Likewise stone, in building and SCULPTURE, is nearly indestructible underground, but many ancient structures have disappeared because they were quarried for materials in later antiquity or the Middle Ages. Moisture plays a major role in the deterioration of wood, textiles, and papyrus. Only in the dry sands of EGYPT and other southern and eastern Mediterranean sites have a few such materials been preserved from ancient Greek times.

The specific study of Greek and Roman antiquities is traditionally called classical archaeology. It had its origins in art collecting, going back at least to the London-based Society of Dilettanti (founded A.D. 1733), which financed a series of expeditions to Italy for the sketching, written description, and purchase of visible remains. The practice of excavation had begun by 1738, when Queen Maria of Naples sponsored the dig that discovered the ancient Roman city of Herculaneum. Many archaeological treasures reached Western Europe in these early years, as French, British, and German enthusiasts purchased excavation rights at sites in the Eastern Mediterranean, then stole whatever they could find. Today, increasingly sophisticated host countries such as Greece and Turkey monitor digs to ensure that all discovered items remain government property.

The most spectacular archaeological discoveries regarding the ancient Greek world took place in the later 19th century. Using intuition and reliance on descriptions in HOMER's *Iliad*, the wealthy German businessman and amateur archaeologist Heinrich Schliemann discovered and excavated the site of ancient TROY (1871–73). Schliemann next turned his attention to the mainland Greek city of MYCENAE (whose location was already known). Digging just inside the famous Lion's Gate in summer of 1876, Schliemann discovered the group of treasure-filled, second-millennium tombs now known as Grave Circle A. Schliemann also excavated at Boeotian ORCHOMENUS and at TIRYNS. His work proved the historical basis of the Homeric poems and established the existence of a previously unsuspected Greek prehistory, circa 1600–1200 B.C.—the MYCENAEAN CIVILIZATION. Similarly, starting in A.D. 1900, excavations by the British scholar Sir Arthur Evans at CNOSSUS and other sites in CRETE brought to light the earliest great Aegean culture, the MINOAN CIVILIZATION.

In the century since Evans, archaeology has benefited from advanced technology, including aerial photography and electromagnetic search. Aerial viewing of terrain from aircraft at several thousand feet can reveal variations in the color of topsoil or of ripening crops that indicate the presence of buildings underneath. Electromagnetic search—typically conducted on foot, with hand-held equipment—can reveal the presence of buried material by indicating an obstruction to the soil's natural conductivity of electrical flow. Similarly, buried items of IRON or burned clay create a perceptible distortion in the area's natural magnetic field.

Modern underwater equipment such as submersible vessels, scuba gear, and the vacuum cleaner–like suction dredger have opened the sea floor to archaeology since the mid-20th century. The most dramatic Aegean find of recent years is the sunken ancient trading ship lying off the island of Dokos, near the Argolid, in southern Greece. Possibly a remnant of the pre-Hellenic inhabitants of Greece, circa 2200 B.C., this very early find testifies to those people's shipbuilding skill and overseas TRADE routes—assets later taken over by the conquering Greeks.

Despite modern equipment, archaeology still relies much on "low-tech" tools as the trowel, brush, sieve, and icepick-like piolet. After removing the area's topsoil and sifting it for displaced remains, the archaeological team might divide the site into numbered quadrants, indicated by a gridwork of strings held aloft on poles. As digging proceeds with trowel, piolet, and brush, the team keeps records regarding the depth at which each discovered item was found. With some variation, earlier items tend to be located deeper underground, later items nearer the surface.

Perhaps the greatest modern advance has been scholars' ability to assign a time frame to recovered items. Until recently, dates were assigned largely on the basis of associated information, such as excavation level, or how the item's shape compared with similar ones of known date. But recent technology allows for dating on the basis of some items' molecular structure. For example, carbon-14 dating measures the radioactive type of carbon that occurs in all living matter. Thermoluminescence reveals the number of loose electrons in certain material. Both methods can give approximate dates to some archaeological remains.

The most fruitful archaeological sites are ones that are no longer inhabited. For instance, much of our information about housing and town planning in the 300s B.C. comes from OLYNTHUS, a north Aegean Greek city abruptly destroyed by war in 357 B.C. and never reoccupied. Conversely, archaeology at ATHENS has been greatly restricted by the problem of how to requisition excavation sites in the middle of the modern Greek capital city. THEBES and PIRAEUS are two examples of modern Greek cities sitting atop ancient layers that remain largely untouched and inaccessible.

Among the more successful classical archaeological projects of recent years has been the excavation of the original city of Smyrna, near the modern Turkish seaport of Izmir. The site—containing remnants of houses, a temple of the goddess ATHENA, and an encircling wall—provides the best surviving example of a Greek city-state of the early 500s B.C.

(See also DELOS; DELPHI; EPHESUS; EPIDAURUS; LEFKANDI; LINEAR B.)

Archidamus Dynastic name among the Eurypontid kings of SPARTA. The best known was King Archidamus II, who reigned from about 469 to 427 B.C. His reign was clouded by Spartan-Athenian hostility, culminating in the PELOPONNESIAN WAR. Impressed by Athenian sea power and wealth, Archidamus unsuccessfully urged his fellow Spartans to vote against war in 431 B.C. He predicted (correctly) that they would be bequeathing the war to their children.

After hostilities began, the aged Archidamus led the Spartan invasion of Athenian territory in 431, 430, and 428 B.C. and the attack on PLATAEA in 429 B.C. The war's first decade, from 431 to the Peace of NICIAS in 421 B.C., is often known as the Archidamian War.

(See also EURYPONTID CLAN.)

Archilochus One of the earliest and greatest Greek lyric poets. Archilochus, who lived from about 680–640 B.C., was the bastard son of Telesicles, an aristocrat of the island of Paros, and a slave woman. Archilochus emigrated with a colonizing expedition that his father led to the northern Aegean island of THASOS (in a GOLD- and SILVER-mining region). There Archilochus served as a soldier, defending the colony against native Thracians. At some point he may have taken work as a mercenary to fight elsewhere. He is said to have died in battle.

As a poet, Archilochus was an innovative genius, the first person in Western culture to write movingly about his own experiences and emotions. There had been lyric poets before him, but he gave to his verses the kind of strong personal content that is considered to be the identifying feature of more modern poetry. Later generations of Greeks revered him.

Archilochus' work survives in some 100 items, most of which are fragments of once-longer poems, quoted by later writers. His favorite verse forms are iambic meters (which he pioneered as a form for satire) and the elegy (generally composed to be sung or recited to flute accompaniment). The personality that emerges in these verses is cynical, angry, proud, and vigorously heterosexual. Archilochus' illegitimate birth and adverse life seem to have given him an outsider's sardonic view of the world, yet he could also feel intense passions.

Archilochus writes of his love for and anger with the woman Neoboule and of his tender seduction of her younger sister. Other subjects include shipwreck, war, WINE, and the male organ. One famous fragment describes with jovial regret how he had to abandon his (expensive but heavy) shield in order to run away from the Thracians: "Let the shield go,' the fragment ends, "I'll get another just as good.' The antiheroic tone sounds almost modern and sets Archilochus apart from other voices of his age. Greek society frowned on a soldier's retreat and loss of shield, but that did not stop Archilochus from writing about it. Among later Greeks he had a reputation for being abusive in verse. According to legend, when Neoboule's father, Lycambes, reneged on his promise to give her in MARRIAGE to Archilochus, the infuriated poet circulated such withering satirical verses that Lycambes, Neoboule, and the rest of the family hanged themselves out of shame.

See also COLONIZATION; HOPLITE; LYRIC POETRY.

Archimedes   Inventor and mathematician from the Sicilian Greek city of SYRACUSE, circa 287–211 B.C. Archimedes' discoveries in geometry and hydrostatics (the study of the properties of standing water) were monumental, and represent a high point of Greek achievement.

A friend and advisor of King Hieron II, Archimedes was famous for his inventions, which included the Archimedes screw, still used today for drawing a continuous flow of water upward. Yet he dismissed engineering feats as pandering to a vulgar public and was prouder of his work in geometry, particularly his discovery that a sphere contained inside a cylinder will always have an area two-thirds that of the cylinder.

Regarding the weight-displacing abilities of the lever, Archimedes made the famous statement, "Give me a place to stand and I will move the world." But his best-remembered utterance concerns his discovery of the principle of specific gravity, after he was asked by King Hieron to determine whether a golden crown had been adulterated by baser metal. Pondering the problem in his bath at the GYMNASIUM (the story goes), Archimedes suddenly realized that he could compare the amount of water displaced by the crown with that displaced by an equal weight of pure GOLD. Delighted with his discovery, he ran naked through the street shouting "I have found it!—*heurēka!*" (or, as we render it today, "eureka").

During the Roman siege of Syracuse (213–211 B.C.), he constructed elaborate devices of defense, including a giant glass lens that focused sunlight on Roman warships in the harbor and set them afire. When the Romans captured the city, Archimedes was killed, supposedly because he enraged a Roman solider by commanding "Don't disturb my circles" as the man found him pondering diagrams in the sand.

Of his written work, nine treatises survive in Greek and two others in later Arabic translation. Most of these deal with geometrical inquiries, particularly regarding circles, spheres, and spirals.

(See also MATHEMATICS; SCIENCE.)

architecture   Perhaps the most visible legacy from ancient Greece is the three famous architectural "orders," or styles, that the Greeks developed in stone in the 600s–300s B.C. These orders—whose most distinctive feature was the use of stone columns to hold up a solid entablature (upper structure)—were known as the Doric, the Ionic, and the Corinthian. These architectural styles can be seen today on the exteriors of such modern neoclassical public buildings as banks, museums, libraries, and city halls.

The Doric and Ionic orders (both older than the Corinthian) were developed specifically for the construction of temples of the gods. Of course, temples were being built long before these orders appeared, and the orders themselves seem to have emerged, in the 600s B.C., as imitations in stone of existing woodworking techniques. Unfortunately, because wood decays, no wooden structures survive from ancient Greece for us to compare with their stone successors. But certain purely ornamental details of the

The Doric order of architecture emerged with the limestone temple of Artemis at Corcyra, built circa 600–580 B.C. Little of the building remains today. Pictured here is an artist's conception of the western facade. The pediment's center shows the running figure of the demon Medusa with her son Chrysaor, flanked by panthers. A few panels of the pediment—including the carved limestone Medusa, over 9 feet tall—are preserved today in the Corfu Museum. Medusa supplied a favorite Greek artistic motif, used to scare away evil-wishers.

The Ionic order reached its zenith in the temple of the goddess Artemis at Ephesus, shown here in a side view through the entrance porch. Built between the mid-500s and mid-400s B.C., this marble temple was the biggest in the Greek world at that time—180 × 380 feet—and was counted as one of the Seven Wonders of the World. The drawing is conjectural: Little of the temple remains today. The scene conveys the lushness of Ionic architecture, with its scroll-like column capitals and other decorative patterns, borrowed (probably) from Near Eastern architecture.

Doric style are best understood as preserving (in stone) the shapes of wooden beams and pegs that had been necessary elements of earlier, wooden temples. Also, the fluting of Doric and Ionic columns surely commemorates the grooving done by adze in the tree-trunk pillars of earlier temples.

The Doric order emerged in mainland Greece. The name (Dorikē archē in Greek) refers to the style having originated at such prosperous Dorian-Greek cities as CORINTH and the Corinthian colony of CORCYRA. The earliest all-stone building that we hear of was the goddess ARTEMIS' temple at Corcyra, completed circa 590 B.C.; it was built of carved limestone (cheaper and softer than marble, which later became the stone of choice). Although only the temple's foundations and parts of its western pediment survive today, modern scholars have re-created the building's probable appearance. The heart of the temple was a walled, rectangular, roofed structure (the *naos,* or cella) that housed

the cult statue of the deity. The cella's single doorway was typically in the east, perhaps with two columns in the entranceway. The roof extended on all four sides around the cella and was supported on each side by at least one row of columns. At Corcyra—as at the most famous Doric temple, the Athenian PARTHENON (completed by 438 B.C.)—the proportions of these outer columns are eight across front and back, with 17 along each side (counting the corner columns twice).

A temple's sloped roof had a triangular gable, or pediment, at front and back. These pediments were often adorned with SCULPTURES that were fastened to the wall behind. The figures often showed a scene from MYTH, and Greek sculptors used great ingenuity in devising arrangements that would accommodate the pediment's narrowing height at the outside edges. The sculptures were painted, to project the flesh, hair, clothing, and weaponry, and the pediment's background was also painted, usually a solid blue or red. Similarly, painting would enliven the carved reliefs of the panels known as metopes—or "intervals"—that ringed the outside of the entablature, alternating with unadorned, corrugated panels known as triglyphs.

Meanwhile the Ionic order emerged at such wealthy Ionian-Greek cities as EPHESUS and MILETUS, on the west coast of ASIA MINOR. Lighter but more ornate than the Doric order, the Ionic employed certain distinctive details, the most obvious of which were the scroll-like volutes (or curls) at four corners of the capital (or head) of each column; this lovely design may derive from the Tree of Life motif on Near Eastern architecture, known to the Ionians through their trading contact with Oriental kingdoms of Asia Minor and the Levant. A forerunner of the

A Corinthian-style capital, from inside the rotunda at Asclepius' shrine at Epidaurus, circa 330 B.C. The Corinthian column, basically an elaboration of the Ionic type, was invented in the later 400s B.C. The Corinthian capital shows acanthus leaves and small volutes (curls).

Ionic capital may be the treetoplike shape of the Aeolic capital, which survives in a carved stone form of the 600s B.C. from the Greek island of LESBOS (near the Asia Minor coast).

Ionic columns were more slender than their Doric counterparts. Other distinctive Ionic features included the use of a column base and the absence of Doric-style metopes and triglyphs along the frieze beneath the roof and pediments; unlike the Doric structure, the Ionic entablature could show continuous carvings around the frieze. The most glorious building in the Ionic order was the huge temple of Artemis at Ephesus, constructed circa 550–450 B.C. and considered to be one of the SEVEN WONDERS OF THE WORLD. Unfortunately, it and most other Ionic temples have disappeared over the centuries; among the few Ionic structures standing today are the Erechtheum and the temple of Athena Nike, both on the Athenian ACROPOLIS.

The Corinthian order arose in the 400s B.C. as an ornate variant on the Ionic form, using stylized acanthus leaves around the column capital. Early use seems to have been confined to the interiors of certain structures in southern Greece. Beginning as a single capital inside the Doric temple of APOLLO at BASSAE (circa 430 B.C.), the Corinthian design was employed in the 300s B.C. at the rotunda of ASCLEPIUS' temple at EPIDAURUS and at the temple of ATHENA at TEGEA. Later the design was thought to communicate imperial splendor, and became a favorite of the Romans.

(See also DORIAN GREEKS; IONIAN GREEKS; RELIGION.)

archon Meaning "leader" or "ruler," the archon was a political executive in numerous ancient Greek states. In the democratic ATHENS of the mid-400s B.C. and later, the archonship was a prestigious but relatively narrow job, with executive and courtroom duties. Nine archons were selected annually by lot, from the Athenian upper and middle classes.

The three senior Athenian archons were the archon *basileus* (or king), the *polemarchos* (war leader), and the archon *eponumos* (eponymous). The *basileus* oversaw state religious functions and any related lawsuits. Religious and judicial duties also were assigned to the *polemarchos* (whose role as a military commander was discontinued soon after 490 B.C.). The *eponumos* had jurisdiction over cases of inheritance and other property rights. The man who served as archon *eponumos* also gave his name to the calendar year— that is, the year was henceforth known as that in which so-and-so had been archon.

A man (WOMEN were ineligible) might be an archon only once at Athens. After the end of his office, barring any disqualifying offense, he was enrolled for life in the judiciary council known as the AREOPAGUS.

(See also DEMOCRACY; LAWS AND LAW COURTS.)

Areopagus In the democratic ATHENS of the mid-400s B.C. and later, the Areopagus was a special law court of 200–300 members, comprised of former ARCHONS. With regard to its origin, the Areopagus (Areiopagos means "hill of ARES," indicating the site where the court's building stood) was a remnant of Athens' old-time aristocratic government.

In the days of aristocratic rule, circa 900–600 B.C., the Areopagus probably ran the city, acting as a legislative body and high court. As Athens developed in stages toward DEMOCRACY, however, the Areopagus gradually was shorn of power. Under SOLON (circa 594 B.C.), a new COUNCIL preempted the Areopagus' legislative-executive duties, and a newly created court of appeals made the Areopagus' legal decisions no longer final.

As the job of Athenian archon became less exclusive and demanding (500s–400s B.C.), so did the Areopagus cease to function as a right-wing bastion. In 462 B.C. the radical reforms of EPHIALTES deprived the Areopagus of most of its important legal jurisdictions and distributed these among the citizens' ASSEMBLY, the council, and the other law courts. The Areopagus henceforth heard only cases of deliberate homicide, wounding, and arson.

In its capacity as a homicide court, the new, diminished Areopagus is celebrated in AESCHYLUS' tragedy the *Eumenides* (458 B.C.). In the second half of this play, actors portray the ancient Areopagus sitting in judgment over the mythical hero ORESTES for the murder of his mother. Aeschylus wrote the *Eumenides* during the period of civil turmoil following Ephialtes' reforms, and one of his intentions was to soothe the class strife of his fellow citizens.

(See also ARISTOCRACY; LAWS AND LAW COURTS.)

Ares The Greek god of war. Ares was one of the 12 principal gods, but he never stood in the Greek mind as a benevolent guardian figure, as did his Roman counterpart, Mars. Ares was a god with little moral aspect—he seems to have been mainly a personification of war. His cult was never neglected, but it was of second rank. Only in THEBES was Ares' worship important.

Ares was associated with the land of THRACE and may have been Thracian in origin, although his name seems to be Greek. In MYTH he was the son of ZEUS and HERA, and the adulterous lover of the goddess APHRODITE. A picturesque passage in HOMER's epic poem the *Odyssey* (book 8) describes how the two lovers were surprised and trapped in a chain-link net sprung by Aphrodite's husband, the smith god HEPHAESTUS. This irreverent tone is typical of much of Ares' treatment in myth and poetry; one passage in the *Iliad* (book 5) mentions that Ares was subdued by the demigods Otus and Ephialtes, and imprisoned in a metal casket.

Fierce and impetuous, Ares had unhappy associations. Disguised as a wild boar, he jealously killed Aphrodite's lover, the beautiful youth ADONIS. Ares' children by various mortal women included such violent characters as the Thracian Diomedes and the outlaw Cycnus, both slain by HERACLES. Ares and Aphrodite were the parents of EROS and of Harmonia, who became CADMUS' wife.

(See also ATHENA.)

Argonauts See JASON (1).

Argos Major city of the eastern PELOPONNESE, located three miles inland, on the western rim of the Argive plain, at the neck of the large peninsula known as the Argolid. The city's patron deity was the goddess HERA, whose Argive cult was very ancient, going back to the pre-Greek mother goddess worshipped there. In its early days, Argos was one of the foremost cities of Greece. But after about 600 B.C. SPARTA displaced Argos in the control of the Peloponnese, and thereafter Argos' story was one of decline and subordination, despite flashes of ambition.

With its farmland and twin citadels, Argos was a fortress and cult center well before the first Greek invaders arrived (circa 2100 B.C.). The Greeks' conquest of the region may perhaps be shown by local ARCHAEOLOGY in the burned remnants of the House of the Tiles at Lerna, near Argos. During the Greek MYCENAEAN CIVILIZATION (circa 1600–1200 B.C.), Argos shared preeminence with the overlord city of MYCENAE, just across the plain, and with nearby TIRYNS.

Argos' importance in this era is suggested by the city's prominent role in Greek MYTH, as the home of PERSEUS (1), of ATREUS and his royal family, and of many other heroes. In general, Argos seems to have supplanted Mycenae in later folk memory (or perhaps Mycenae was once called Argos). In the epic poems of HOMER, the term *Argive*—along with *Achaean* and *Danaan*—simply means "Greek."

During the Dorian-Greek invasion of about 1100–1000 B.C., Argos became a Dorian city and the Dorians' base for their conquest of southern Greece. Thereafter Argos remained supreme in the Peloponnese for over 400 years until the rise of Sparta, farther south. In the mid-600s B.C. Argos enjoyed a brief peak of power under its dynamic King PHEIDON, who defeated the Spartans at the Battle of Hysiae (669 B.C.) and extended his rule across the Peloponnese. But soon Sparta and CORINTH had become the great southern Greek powers, while Argos withdrew into isolation, occasionally emerging to fight (and lose to) Sparta. Argos' worst defeat came at the Battle of Sepeia (494 B.C.), at the hands of the Spartan king CLEOMENES (1). This disaster brought the Spartans right up to the walls of Argos (where, according to legend, the Argive poet Telesilla rallied her fellow citizens and led a counterattack of armed Argive WOMEN).

During the Persian king XERXES' invasion of Greece (480–79 B.C.), Argos alone of the Peloponnesian states remained neutral, in effect siding with the Persians. Internal strife, possibly caused by the trauma of Sepeia, now resulted in the OLIGARCHY's overthrow and the rise of a DEMOCRACY on the Athenian model. Argos allied itself to ATHENS in 461, 420, and 395 B.C., but did so without realizing its potential as a rival to Sparta.

(See also DORIAN GREEKS; PELOPONNESIAN WAR; THEMISTOCLES.)

Ariadne See THESEUS.

Arion Greek lyric poet active at CORINTH, at the wealthy court of the tyrant PERIANDER, circa 620 B.C. Arion was born on the Aegean island of LESBOS—where strong artistic traditions produced other poets of this era, including SAPPHO, ALCAEUS, and TERPANDER—and he became famous as a performer who sang his own verses, as well as other poets', while accompanying himself on the cithara (a type of lyre). The Greek historian HERODOTUS described Arion as "the best singer in the world."

Arion is credited with imposing artistic order on the dithyramb, a song sung by a chorus in honor of the god DIONYSUS. Apparently he systematized the performance so that the singers stood stationary, grouped in a circle; he may have assigned a specific subject to each song. These developments would later contribute to the emergence of Greek stage tragedy. None of Arion's compositions survives today.

But Arion is best remembered for the charming legend that he rode on a dolphin's back through the sea. According to the tale told by Herodotus, Arion—returning from a lucrative performance tour of the Greek cities of SICILY and ITALY—took passage on a ship from Italy to Corinth. But he was forced to jump overboard when the crew decided to steal his money. As the ship sailed on, Arion was picked up at sea by a dolphin—an animal sacred to the god APOLLO, who was also the patron of poets. Carried to Cape Taenarum, on the southern coast of the PELOPONNESE, Arion made his way back to Corinth, arriving ahead of the ship. When the ship's crew reached Corinth, they were confronted by Arion and the ruler, Periander.

The legend may have a kernel of truth. According to modern scientific studies, the sea-mammals known as dolphins and porpoises are the only wild animals to be attracted to humans. It is not unusual for dolphins to flock around a human swimmer at sea.

(See also MUSIC; LYRIC POETRY; THEATER).

Aristides Athenian soldier and statesman (circa 530–465 B.C.) who was one of the founders of the Athenian empire. Aristides (or Aristeides) was instrumental in creating the Athenian-controlled alliance of Greek states known as the DELIAN LEAGUE. He was surnamed "the Just" for his reputedly fair dealings in politics and in diplomacy with other Greek states.

According to the later Greek writer PLUTARCH, Aristides was one of the Athenian generals at the Battle of MARATHON (490 B.C.). Yet in 482 B.C. the Athenians voted to ostracize him—amid a rash of OSTRACISMS in the 480s B.C. This might have been an outcome of Aristides' political conflict with THEMISTOCLES.

As required by the ostracism law, Aristides withdrew from ATHENS, expecting to be in exile for 10 years. However, two years later he was recalled to help against the Persian king XERXES' invasion of Greece. Elected as a general by the Athenians, he served alongside his former enemy Themistocles. Aristides fought at the Battle of SALAMIS (1) (480 B.C.) and, reelected as general for the following year, commanded the Athenian contingent at the land BATTLE OF PLATAEA (479 B.C.).

As a general again in 478 B.C., Aristides led the Athenian squadron in the allied Greek naval liberation of BYZANTIUM and much of CYPRUS. When the arrogance of the Spartan commander Pausanias alienated the other Greeks, Aristides began his diplomatic efforts to secure a mutual-defense alliance between Athens and the eastern Greek states. When the resulting Delian League was formed, in the summer of 478 or 477 B.C., Aristides won admiration for his fair assessment of the annual contribution due from each member state.

Aristides seems to have been widely contrasted with the wiliness and rapacity of Themistocles. At the performance of AESCHYLUS' tragedy SEVEN AGAINST THEBES in Athens in 467 B.C., when certain lines were spoken concerning the hero Amphiaraus' wisdom and righteousness, the entire audience turned to look at Aristides in his seat.

Aristides is said to have died poor, having refused to enrich himself dishonestly through office.

aristocracy (*aristokratia*) "Rule by the best," an early form of government in some Greek city-states whereby power was shared by a small circle whose membership was defined by privilege of noble birth. This ruling circle—often confined to one clan—tended to monopolize wealth, land, and military and religious office as well as government. The typical instrument of aristocratic rule was the COUNCIL (*boulē*), a kind of omnipotent senate in which laws and state policies were decided.

Aristocracies arose to supplant the rule of kings, after the fall of MYCENAEAN CIVILIZATION circa 1200–1100 B.C. (Some Greek states, such as ARGOS and SPARTA, seem to have retained kings who were merely the first among aristocratic equals.) By the 600s B.C., however, growth in TRADE and revolutionary developments in warfare had resulted in the breakup of the old aristocratic monopoly, as a new class of citizen—the middle class—acquired wealth and military importance. In some cities, such as CORINTH, conflict between aristocrats and commoners produced revolutions that placed popular TYRANTS in power; elsewhere, the new middle class gradually was granted political power. In these cases, the aristocracy became an OLIGARCHY (rule by the few—the mass of poorer citizens still were excluded from power). Although oligarchies resembled aristocracies in certain ways, such as the council, they notably lacked the old requirement of noble blood. (See also HOPLITE; POLIS.)

Aristogiton See HARMODIUS AND ARISTOGITON.

Aristophanes Athenian comic playwright who lived circa 450–385 B.C. In antiquity, Aristophanes was recognized as the greatest classical Athenian writer of comedy. Eleven of his 30 or so plays have survived in their entirety. These 11 plays supply our only complete examples of fifth-century-B.C. Athenian "Old Comedy" and show the highly political nature of that art, with plots and jokes devoted to current events. Aristophanes' work provides valuable information about Athenian public opinion during the PELOPONNESIAN WAR (431–404 B.C.).

As suggested by his name (meaning "noblest showing"), Aristophanes probably came from an aristocratic family. On the basis of a reference in Aristophanes' play *Acharnians*, some scholars believe that the playwright owned land on the Athenian-occupied island of AEGINA. The satirical spirit of his comedies provides no clear proof of Aristophanes' own political beliefs, but he probably identified with the intellectual, anti–left-wing, and basically antiwar Athenian noble class (the class that provided the military's CAVALRY and that is sympathetically portrayed in Aristophanes' play *Knights*). Without being fully antidemocratic, Aristophanes' comedies mock the excesses of Athenian DEMOCRACY: namely, the people's fickle abuse of power and the political pandering and vulgarity of the demagogues.

But any aspect of Athenian life was fair game for Aristophanes, whose weapons were parody, burlesque, and comic exaggeration. Against members of his own leisured class, he derides the dishonest quibbling of the SOPHISTS, and the abstruse brainwork of the philosopher SOCRATES and the tragedian EURIPIDES. These men and many of his other targets were Aristophanes' contemporaries, who might be found in the audience at a play's performance.

Athenian comedy was performed at two state-sponsored festivals: the Lenaea, in midwinter, and the grander holiday called the City Dionysia, in early spring. Each play was part of a three-way competition.

By a convention of the era, most of Aristophanes' plays are titled according to the group character of the onstage chorus. His earliest play, *Banqueters* (427 B.C.), was presented when he was probably in his early 20s. Neither that work nor his comedy of the next year, *Babylonians*, has survived, but we know that *Babylonians* daringly portrayed the Greek subject allies of the Athenian empire as SLAVES, grinding at the mill. Coming at a moment when ATHENS was relying heavily on allied loyalty against the Spartan enemy in the Peloponnesian War, the play provoked the powerful left-wing politician CLEON to prosecute Aristophanes. (We do not know the exact charge.) The prosecution failed, and Aristophanes' mocking pen was eventually turned against Cleon.

Acharnians, performed at the Lenaea of 425 B.C., is Aristophanes' earliest surviving work. The plot involves an Athenian farmer who makes a private peace with the Spartans while the Athenian military fights on. The play's pro-peace message foreshadows the more urgent antiwar themes of Aristophanes' *Peace* (421 B.C.) and *Lysistrata* (411 B.C.). Without being defeatist, *Acharnians* spoke to a city disillusioned in its hopes of an early victory, and the play received first prize.

Knights, performed at the 424 B.C. Lenaea, is Aristophanes' most political play: It amounts to a vicious attack on the politician Cleon, who at that time was standing for election to Athens' board of generals. In *Knights*, Cleon clearly appears in the character of a flattering, scheming slave named Philodemos ("lover of the people"). Philodemos' foolish and elderly master is Demos ("the people"), and the plot involves Philodemos' vulgar competition with a sausage vendor for Demos' affection. The Athenians approved of Aristophanes' mockery of Cleon, and *Knights* received first prize. However, the real-life Cleon still won the election.

Clouds ridicules the real-life Socrates as a crackpot scientist and corrupt teacher of sophistry and RHETORIC. The play won only third prize at the 423 B.C. City Dionysia, being defeated by CRATINUS' masterpiece, *The Bottle*. The next year saw the *Wasps*, which parodies Athenian litigiousness and the elderly Athenians' enthusiasm for jury duty. Like *Knights*, this play contains liberal mockery of Cleon (who had sponsored a law increasing jury pay). Cleon's death in battle in 422 B.C. deprived Aristophanes of his favorite comic butt.

Peace was presented at the City Dionysia of 421 B.C., just a few days before or after the Peace of NICIAS was concluded, which supposedly ended the Peloponnesian War. The plot involves the rescue of the goddess Peace from her prison pit. Despite its topical subject, *Peace* received only second prize.

Aristophanes' comedies of the years 420–415 B.C. have not come down to us. By the time of his next extant work, *Birds* (414 B.C.), the playwright had begun to move away from political topics in favor of themes more generally social. In *Birds*, two Athenians, in despair over the city's litigiousness, fly off and create a bird city in the sky, Cloud-cuckoo-land (Nephelokokkugia). *Birds* contains some of Aristophanes' finest theatrical effects, particularly in its onstage chorus of birds. The play won second prize at the City Dionysia.

The year 411 B.C. brought two comedies about WOMEN's issues. *Thesmophoriazusae* (women celebrating the Thesmophoria festival) amounts to a satire on the tragedies of Euripides; the story involves a scheme by the Athenian women to destroy Euripides on account of his plays' revealing portrayal of women. *Lysistrata* imagines the Athenian women organizing a sex strike to compel the men to make peace in the Peloponnesian War (which by now was turning disastrously against Athens). The comedy combines bawdiness with a sincere, conciliatory, antiwar message.

Frogs, which won first prize at the Lenaea in 405 B.C., is considered Aristophanes' masterpiece. The comic protagonist is the god DIONYSUS, who journeys to the Underworld to fetch back the tragedian Euripides (who in fact had died the previous year). But, once in the Underworld, Dionysus is compelled to judge a contest between Euripides and AESCHYLUS (died 456 B.C.), as to who was the greater tragedian. Deciding in favor of Aeschylus, the god brings *him* back, instead, to save Athens. Part of *Frogs'* appeal is its pathos: Performed barely a year before Athens' final defeat in the war, the play conveys a giddy sense of desperation.

Aristophanes' career continued after the war's end for perhaps 20 years, but we have only two plays from this final period. *Ecclesiazusae* (women in the assembly) was presented in 392 or 391 B.C.. In this play, the Athenian women, tired of the men's mismanagement, take over the government and proclaim a communist state.

Wealth (388 B.C.) is Aristophanes' latest surviving comedy. Plutus, the god of wealth, is blind, which explains why riches are inappropriately distributed in the world. But when Wealth regains his sight in the story, attempts at redistribution create social chaos. *Wealth* clearly belongs to the less political, more cosmopolitan "comedy of manners" of the 300s B.C.

(See also ASSEMBLY; EUPOLIS; MENANDER; THEATER.)

Aristotle   Greek philosopher and scientist who lived 384–322 B.C. A man of immense learning and curiosity, Aristotle (Aristoteles) can be seen as the most influential Western thinker prior to the 19th century A.D. After spending 20 years as a student of the great Athenian philosopher PLATO, Aristotle rejected Plato's otherworldly doctrines, partly because he was fascinated by the multifaceted material world around him. Aristotle was as much interested in defining problems as in finding answers, and he more or less established the scholarly tradition of systematic research, creating categories—including "logic," "biology," and "physics"—that are still in use today. The Athenian scholarly community that he founded, the LYCEUM, was

Aristotle's combination of mental power and compassion is apparent in this likeness, from a Roman marble bust. The bust is thought to be copied from a bronze portrait statue made by the Greek sculptor Lysippus, circa 330 B.C.

one of the greatest centers for advanced study in the ancient world.

His chief accomplishment in PHILOSOPHY was in devising the first system of logic—the system now known as Aristotelian syllogistic, which provided a cornerstone of logic studies for centuries. In SCIENCE his work was epoch-making. He pioneered the studies of biology and zoology (among others), and he divorced science from philosophy, setting scientific method on its future course of empirical observation.

Of his many writings, less than one-fifth have survived—about 30 treatises. Yet these few works came to dominate European higher learning during the Middle Ages and Renaissance. Among his other accomplishments, Aristotle was one of the first political scientists, cataloguing and analyzing the various forms of government in the Greek world and beyond. His extant treatise *Politics* is now probably his most widely read work, having become a virtual textbook for college political science courses.

Although he spent more than half his adult life in ATHENS, Aristotle was never an Athenian citizen. He was born in a minor town named Stagirus (later Stagira), in the Greek colonial region called CHALCIDICĒ, on the northwest Aegean coast near the Greek kingdom of MACEDON. Aristotle's father, Nicomachus, was the court physician to the Macedonian king Amyntas III. Aristotle's career was destined to be shaped by both the Macedonian connection and by the medical tradition's emphasis on observation and diagnosis. The family was probably rich; hence Aristotle's ability to travel and to devote himself to study.

After Nicomachus' death, 17-year-old Aristotle was sent to Athens to join Plato's philosophical school, the ACADEMY (367 B.C.). Eventually recognized as Plato's foremost pupil and possible successor at the Academy, Aristotle remained there for 20 years until Plato's death (347 B.C.). In the election of the Academy's new president, Aristotle was passed over—which may be why he then left Athens with a number of friends and followers. They traveled to northwest ASIA MINOR, where they set up a school at the town of Assos at the invitation of Hermias, a local Greek ruler who was a Persian vassal. Evidently Aristotle became a close friend of Hermias, for he married Hermias' adopted daughter, Pythias. But when Hermias was arrested and executed for some unknown reason by the Persians, Aristotle and his followers sailed to the nearby Greek island of LESBOS, and stayed in the city of MYTILENE. By then Aristotle's group included a well-born man of Lesbos named Theophrastus, who became his protégé and later his successor as head of his philosophical school.

At the landlocked lagoon of Pyrrha, in central Lesbos, Aristotle conducted many of the observations of marine animals recorded in his zoological writings. His method of field observation and documentation—an innovation in his own day—was destined to set the pattern for all future biological study.

In around 343 B.C. the 41-year-old Aristotle accepted an invitation from King Philip, by then the most powerful man in the Greek world, to go to Macedon to tutor Philip's 13-year-old son, Alexander (who was destined to conquer the Persian Empire and be known to history as ALEXANDER THE GREAT). Aristotle spent about three years providing a higher education for the prince and some of his retinue. The curriculum probably included political science—Aristotle wrote two (lost) treatises for Alexander, *On Kingship* and *On Colonists*—as well as studies in literature and biology. Aristotle gave Alexander an edited version of HOMER's *Iliad* that the young warrior supposedly carried with him for the rest of his life. According to legend, King Alexander would later send Aristotle specimens of unfamiliar Eastern plants and animals to study. Beyond this, however, Aristotle's influence on Alexander was not profound, and the teacher probably went home to Stagirus after Alexander began serving as regent for his father, in 340 B.C.

In 335 B.C., after Alexander had inherited the throne, Aristotle returned to Athens to establish his own philosophical school there. He opened the school in rented buildings outside the city, on the grounds of a GYMNASIUM known as the Lyceum ([Lukeion], named for its grove sacred to the god APOLLO in his cult title Lukeios). Aristotle's Lyceum period (335–323 B.C.) is the third and most important phase of his life.

The Lyceum enjoyed special resources and status—Alexander supposedly donated the immense sum of 800 TALENTS—and was protected by the local Macedonian authorities. Without doubt, the Lyceum provided Aristotle with the means to engage in the encyclopedic array of inquiries for which he is remembered. The school resembled a modern university in some ways, with general courses alongside "graduate" research projects under the master's guidance. Among such delegated projects was a description of the individual governments (or "constitu-

tions") of the 158 most important Greek cities; of these 158 analyses, only one survives today—the *Constitution of Athens* (*Athenaion Politeia*), our major source of information for the workings of the Athēnian DEMOCRACY.

Supposedly, Aristotle would lecture to general audiences in the morning and to the advanced scholarly circle in the afternoon. It is said that he had a habit of walking around (*peripatein*) while lecturing and that his listeners walked with him—which gave to the community its title of the Peripatetic School. (However, the name may come from the roofed courtyard, *peripatos*, that was a physical feature at the Lyceum or at a later site.) In personal demeanor, Aristotle is said to have dressed elegantly or even foppishly and to have spoken with an upper-class lisp.

During these years at Athens, Aristotle's wife, Pythias, died, and he lived with a common-law wife, Herpylla. Aristotle had two children: a daughter named Pythias, after her mother, and a son, Nicomachus, by Herpylla. Nicomachus is remembered in the title of Aristotle's *Nicomachean Ethics,* so called because the son supposedly edited the work after his father's death.

With Alexander's sudden death in the East (323 B.C.), anti-Macedonian feeling erupted at Athens, and a criminal charge of impiety was lodged against Aristotle. The 61-year-old scholar fled, supposedly remarking that he was saving the Athenians from committing a second sin against philosophy (a reference to their execution of SOCRATES for impiety in 399 B.C.). Aristotle took refuge at the nearby city of CHALCIS, where his mother had been born and where a Macedonian garrison was in control, and there he died in the following year (322 B.C.).

As previously mentioned, less than one-fifth of Aristotle's writings are represented in the 30 works that have come down to us. The vanished work—bits of which survive as quotations cited by later ancient authors—included poems, letters, essays, and Platonic-style dialogues. Many of these were polished literary pieces, intended for a general readership: In a later century, the Roman thinker Cicero described Aristotle's writing style as "a river of gold." This praise would not be appropriate for the 30 surviving items, which often make for difficult reading.

Ironically, most of these extant writings probably were never meant for publication. Many seem to be Aristotle's lecture notes for his more advanced courses of study. These "treatises" or "esoteric writings," as they are sometimes called, contain passages that are notoriously difficult—either overly condensed or repetitive—with apparent cross-references to works now lost. The titles and sequence of these preserved treatises are not Aristotle's choice, but rather are the work of ancient editors after Aristotle. However, the 30 treatises do provide a good sampling of Aristotle's mature thought, covering logic, metaphysics, ethics, scientific inquiries, literary criticism, and political science.

Aristotle's six treatises on logic (*logikē,* meaning "the art of reasoning") are sometimes mentioned under the collective title Organon, or "tool"—that is, of thought. Of these works, the best known is the one titled *Prior Analytics,* which contains Aristotle's system of syllogistic. (A syllogism is a form of reasoning consisting of two premises and a conclusion. The classic form of syllogism is: [1] Socrates is a man. [2] All men are mortal. Therefore, [3] Socrates is mortal.) The *Prior Analytics* examines the various forms of syllogistic thought.

Other extant treatises include studies of the natural world: *Physics, On the Heavens, Meteorology, Generation of Animals*, and the admired *History of Animals* (so called traditionally, although *Zoological Researches* would better translate the Greek title), which presents many careful descriptions of different species. The very notions of genus and species—general type, specific type—are among the categories devised by Aristotle.

Aristotle's theory of reality is found in his *Metaphysics*, or "Beyond Physics" (a title created by an ancient editor, originally signifying only that the book followed the *Physics* in sequence). Here—in a criticism and radical adaption of Plato's theory of Forms—Aristotle explains how universal qualities are rooted in particulars. For example, a specific dog partakes of universal "dogness" along with all other canines, yet this dogness has no existence apart from the world's flesh-and-blood dogs. Book 12 of the *Metaphysics* presents Aristotle's well-known picture of God as the unmoved mover—pure intelligence, uninvolved in the world's day-to-day occurrences.

In book 1 of the *Nicomachean Ethics*, Aristotle identifies the attainment of happiness as the highest good, and he defines happiness as an activity of the soul in conformity with virtue (*aretē*). Book 2 is famous for defining virtue as the mean (*mesotēs*) between two extremes of behavior: Courage, for example, falls between cowardice and recklessness; generosity falls between stinginess and extravagance.

In the *Rhetoric*, Aristotle examines the art of public speaking in terms of its desired goal, persuasion. The *Poetics* analyzes the nature of literature and offers the famous Aristotelian observation that stage tragedy provides a cleansing (*katharsis*) of the audience's emotions of pity and fear.

(See also ASTRONOMY; EDUCATION; MEDICINE.)

Aristoxenus See MUSIC.

armies See ALEXANDER THE GREAT; CAVALRY; HOPLITE; PELOPONNESIAN WAR; PERSIAN WARS; PHALANX; WARFARE, LAND.

arms and armor See CAVALRY; HOPLITE; WARFARE, LAND.

art See ARCHITECTURE; PAINTING; POTTERY; SCULPTURE.

Artemis Greek goddess of wilderness, wild animals, and the hunt, also associated with childbirth and the moon. Although her worship was of secondary rank in most cities of mainland Greece, she was a principal deity for many Greeks of ASIA MINOR. At EPHESUS, in Asia Minor, she was honored with a huge temple, begun in the mid-500s B.C. and counted as one of the SEVEN WONDERS OF THE WORLD.

In MYTH, Artemis was the twin sister of the god APOLLO; their parents were the great god ZEUS and the demigoddess Leto. Impregnated by Zeus, Leto fled from Zeus' jealous wife, HERA, and found refuge on the island of DELOS, in the AEGEAN SEA. There she gave birth to the divine twins. Artemis and Apollo grew to be skilled archers. With their

arrows, they slew a number of Leto's enemies, including the children of the arrogant NIOBĒ. Artemis also shot the enormous hunter ORION, whose body now hangs as a constellation in the sky.

Like ATHENA and HESTIA, Artemis was a virgin goddess. Imagined as a lithe young woman, she was said to roam mountain forests and uncultivated lands, hunting the beasts and, contrarily, overseeing their safety and reproduction. Her most famous title, mentioned by the poet HOMER (circa 750 B.C.), was Potnia Theron, "lady of wild animals." In art she often was shown accompanied by deer, bears, or similar beasts, and myths told of her retinue of NYMPHS—female creatures embodying the spirit of wilderness places. Among her human devotees described in myth were the Athenian prince HIPPOLYTUS and the female warriors known as AMAZONS. Two regions dear to Artemis were the mountains of CRETE and ARCADIA.

In the classical Greek mind, Artemis' virginity probably suggested the sanctity of wilderness places. A legend tells how the Theban prince Actaeon, while out hunting with his hounds, accidentally found Artemis bathing naked in a stream. Enraged to have her modesty compromised, the goddess changed Actaeon into a stag, whereupon his own hounds mauled him to death.

Although she had no children of her own, Artemis was concerned with birth and offspring among both animals and humans. One of her titles was *Kourotrophos* ("nurse of youths"). Like the goddess HERA, Artemis watched over WOMEN in labor: They would call on her in their distress, remembering the birthing pains of Leto. Women who died in childbirth, or who died suddenly from other natural causes, were said to have been killed by Artemis' arrows.

The goddess Artemis, with her emblematic bow and arrows, commands the hounds of the Theban prince Actaeon to maul their master to death. This red-figure scene was painted circa 475 B.C. on an Athenian vase of the type known as a bell krater.

The puzzle of why the Greeks would have a virgin fertility goddess can perhaps be explained. Artemis may combine two different heritages, of which the fertility concept is the older. The name Artemis seems not to come from the Greek language, and scholars believe that this goddess—like Athena and Hera—was not originally Greek. Rather, Artemis represents a religious survival from the non-Greek peoples who inhabited the Aegean region before the first Greek invaders arrived circa 2100 B.C. A goddess of mountains and beasts is portrayed on surviving gemstones and other artwork of the second millennium B.C. from Minoan Crete. This unnamed goddess, or series of goddesses, is shown walking with a lion and standing between rampant lions. She appears to be a deity of wilderness abundance. Presumably the conquering Greeks appropriated this regal Lady of Animals into their own RELIGION, during the second millennium B.C., changing her into Zeus' daughter. This origin would explain the Greek Artemis' ties to Crete. Artemis' virginity could have been an aspect created by the Greeks, perhaps because the original fertility-Artemis had no husband in the mythology.

Even in historical times, certain cults of Artemis emphasized fertility over virginity. The Artemis worshipped at Ephesus was a mother figure, influenced by the contemporary cult of the goddess Cybele, from interior Asia Minor. The Ephesian Artemis' cult statue, of which copies survive today, portrayed a crowned goddess with at least 20 breasts—very unlike the boyish huntress of mainland Greek religion.

Another peculiar cult existed at SPARTA. At the annual festival of Artemis Orthia (the surname's meaning is lost to us), Spartan boys endured a public ordeal of whipping as they tried to steal cheeses from the goddess' altar. This brutal rite, dating at least from the 400s B.C., had deteriorated into a tourist attraction by the days of the Roman Empire.

(See also MINOAN CIVILIZATION.)

Artemisia See HALICARNASSUS.

Artemisium Harbor and ship-beaching site on the north shore of EUBOEA, off the eastern coast of Greece, named for a local temple of the goddess ARTEMIS. The site overlooks a six-mile-wide channel whose opposite shore lies along the Magnesian peninsula of THESSALY. In this channel, during the PERSIAN WARS, the allied Greek navy first opposed the invading navy of King XERXES (summer 480 B.C.). The Battle of Artemisium was fought nearly simultaneously with the land battle of THERMOPYLAE, 40 miles away, and was part of a Greek strategy to block the Persian southward advance at two neighboring bottlenecks, on land and sea, north of central Greece. Like Thermopylae, Artemisium was a marginal Persian victory that neverthless helped to boost Greek morale.

The allied Greek fleet had about 380 warships, with the largest contingent supplied by ATHENS (180 ships). The Persian ships—which in fact were manned by subject peoples, such as Phoenicians, Egyptians, and IONIAN GREEKS—may have numbered 450 or more. This fleet recently had been reduced, from perhaps 600 ships, by storms off Thessaly.

The battle began when the Persian fleet left its base on the channel's Thessalian shore and rowed out in an enveloping crescent formation against the Greeks. In its Phoenician contingent, the Persian side had the better sailors and faster vessels, but these advantages were reduced in the chaotic press of battle.

The fight was something of an infantry battle on the water. The Persians preferred boarding tactics to ramming: They fought by bringing their ships alongside the Greeks' and sending over the thirty Persian foot soldiers who rode aboard each ship. The Greeks fought back with their own ships' soldiers, about 40 per vessel. After substantial mutual damage, the two fleets returned to their harbors. But upon receiving news of the Greek defeat at Thermopylae, the Greeks withdrew southward.

Artemisium contributed to the strategic Greek naval victory at SALAMIS (1), about three weeks later. The Artemisium losses weakened the Persian fleet and narrowed Xerxes' options for using it; he decided against sending a squadron to raid Spartan territory (which might have been successful in breaking up the Greek alliance). For the Greeks, Artemisium supplied a lesson. It probably helped convince the Greek commanders that, to offset the enemy fleet's better maneuverability, they would have to offer battle in a narrower channel than the one at Artemisium—a channel such as the one at Salamis.

In modern times, the Artemisium waters have yielded up one of the finest artworks of the ancient world: a larger-than-life BRONZE statue of the god ZEUS, from about 455 B.C. Lost with an ancient shipwreck and recovered by underwater ARCHAEOLOGY in the 1920s, the statue now stands in Athens' national museum.

(See also SCULPTURE; WARFARE, NAVAL.)

Asclepius Mythical Greek physician-hero who was eventually worshipped as a god of MEDICINE. The cult of Asclepius blossomed in the 300s B.C., partly in response to a new spirit of individualism, which sought a more personal RELIGION. Asclepius' cult was centered on his sanctaury at EPIDAURUS, in the northeastern PELOPONNESE. Sick people flocked there to acquire, for a fee, cures supposedly provided by the god.

The myth of Asclepius is told in a choral ode by the poet PINDAR, performed circa 470 B.C. At this time Asclepius was still regarded as a mortal hero who had lived and died long before. He was the son of APOLLO and the Thessalian woman Coronis, but was brought up by the centaur Chiron after Apollo destroyed Coronis for her infidelity. Medicine was one of Apollo's arts, and his son, taught by Chiron, came to excel as a physician. Unfortunately, Asclepius overstepped the boundary of human knowledge when, at the plea of the goddess ARTEMIS, he resurrected a previously dead man (HIPPOLYTUS, a favorite of Artemis'). The high god ZEUS, recognizing Asclepius' skill as a threat to natural order, immediately killed both men with a thunderbolt.

In the latter 400s B.C., Asclepius became associated with divine healing. It is unclear why he should have grown to rival his mythical father, Apollo, as patron of physicians, but in any case Asclepius began to be worshipped as a god. His cult center at Epidaurus became a shrine for invalid pilgrims, comparable to that of Lourdes, in France.

The Epidaurus cult was based on a process known as incubation, whereby a worshipper who sought a cure would spend a night in a dormitory associated with the temple, and be visited by the god in a dream. The next morning, perhaps, the god's priests would interpret the dream and dispense specific medical advice. Many existing inscriptions attest to this procedure's success, whether the explanation lies in the worshippers' autosuggestion or in pious fraud on the part of priests who might impersonate the nocturnal god. But Epidaurus and other Asclepian shrines contained genuine health-promoting facilities, such as baths and GYMNASIUMS. Major Asclepian sanctuaries, established through the Epidaurus priesthood, arose at ATHENS (420 B.C.), PERGAMUM (300s B.C.), and the Italian, non-Greek city of ROME (291 B.C.). At Rome, the god's name was latinized to Aesculapius.

Asclepius' totem was a sacred snake; a snake normally lived at each of his shrines. Greek art often portrayed the Asclepian snake as wrapped around another of the god's emblems, the physician's staff. Asclepius' snake and staff survive today as the symbol of the modern medical profession.

(See also CENTAURS; HIPPOCRATES.)

Asia See EUROPE and ASIA.

Asia Minor Peninsular landmass, 292,260 miles square, forming a western subcontinent of Asia. Also known as Anatolia, this territory now comprises much of Turkey. In Greek and Roman times, Asia Minor contained important Greek cities, particularly in the central west coast region called IONIA.

The name Asia Minor (lesser Asia) came into use among the Romans not long before the birth of Jesus; for the Greeks of prior centuries, the region was simply called Asia. At its northwest corner, Asia Minor is separated from the European continent by the narrow BOSPORUS and HELLESPONT waterways and the SEA OF MARMARA. The hospitable west coast—fertile, temperate, with fine harbors—opens onto the AEGEAN SEA; large inshore islands include (from north to south) LESBOS, CHIOS, SAMOS, and RHODES. Asia Minor's southern coast faces the greater Mediterranean; its north coast faces the BLACK SEA. North and south coasts give rise to steep mountains. The Pontic Range, alongside the north coast, offered TIMBER and raw metals in ancient times. The Taurus Range, in the south, intrudes on the shoreline and limits the number of harbors and habitable sites. Stretching eastward, the long Taurus chain also restricts land travel between southeast Asia Minor and northern Syria, narrowing the route to a series of mountain passes, such as the Cilician Gates, north of Tarsus. Inland of the mountains, Asia Minor is an often-arid plateau.

By the early second millennium B.C., before the first Greeks arrived, Asia Minor was controlled by the Hittites, who ruled out of the north-central interior. Probably with the Hittites' permission, Mycenaean Greeks set up trading posts along the west coast in around 1300 B.C. (as we can guess from the troves of Mycenaean POTTERY found at

MILETUS and other sites). The Greeks were looking for raw metals—copper, tin, GOLD, and SILVER, all mined in the Asia Minor interior—and for luxury goods, such as carved ivory, available through TRADE routes from Syria and Mesopotamia. The legend of the TROJAN WAR surely reflects (distortedly) the Mycenaean Greeks' destruction of an actual, non-Greek fortress town in northwest Asia Minor, a town we call TROY, which evidently was interfering with inbound Greek shipping in the Hellespont (circa 1200 B.C.).

After the Hittite Empire collapsed (circa 1200 B.C.), central Asia Minor came under control of an Asian people called the Phrygians. Their name survived in the central region, known as Phrygia. In these years the new technology of IRON-working, previously monopolized by the Hittites, spread from Asia Minor to mainland Greece (before 1050 B.C.). Greek legend claimed that the world's first ironworkers were the Chalybes, a mountain people of northeastern Asia Minor.

Asia Minor's west coast then came under steady Greek invasion, amid the migrations accompanying the downfall of Myceneaen society in mainland Greece. By 1000 B.C. Greeks of the Ionian ethnic group had settled in the region thereafter called Ionia; the foremost states here were Miletus, Samos, Chios, and EPHESUS. Greeks of the Aeolian type, using Lesbos as their base, colonized Asia Minor's northern west coast, thereafter called AEOLIS (circa 1000–600 B.C.). Greeks of Dorian ethnicity occupied sites around the southern west coast (circa 900 B.C.), including HALICARNASSUS, CNIDUS, and the islands of Cos and Rhodes. Seizing the best harbors and farmlands, the arriving Greeks ejected the non-Greek peoples. One such people were the Carians, who survived in the southwestern region called Caria.

Greek settlement spread to Asia Minor's northern and southern coasts during the great age of COLONIZATION (circa 750–550 B.C.). In the north, CYZICUS, on the Sea of Marmara, and SINOPE, midway along the Black Sea coast, were among the colonies founded by Miletus in an attempt to control the Black Sea trade route. On the south coast, the principal Greek cities were (from west to east) Phaselis, Perge, Aspendus, and Side.

The Greek settlements prospered along their sea routes and caravan routes. The cities of Ionia produced the poet HOMER (circa 750 B.C.) and the earliest Greek achievements in SCIENCE, PHILOSOPHY, and monumental ARCHITECTURE (600s–500s B.C.). But there were also non-Greek nations of the interior to contend with. The marauding Cimmerians overran Phrygia and besieged Ephesus (circa 600 B.C.). The kingdom of LYDIA dominated the west coast (by 550 B.C.) and then fell to the Persian king CYRUS (1) (546 B.C.).

By the late 500s B.C. all of Asia Minor was under the jurisdiction of the Persian king, who ruled and collected tribute through governors called satraps. The two chief satraps of western Asia Minor were based at the former Lydian capital of Sardis and at Dascylium, on the Sea of Marmara. At the end of the PERSIAN WARS, the Greek naval counteroffensive liberated most of Asia Minor's western Greek cities (479–477 B.C.). Persian puppet governments were expelled in favor of Athenian-style DEMOCRACY, and the Greek cities became tribute-paying members of the Athenian-dominated DELIAN LEAGUE. Passing to Spartan control by the end of the PELOPONNESIAN WAR (404 B.C.), these cities were ignominiously handed back to PERSIA by the terms of the KING'S PEACE (386 B.C.).

ALEXANDER THE GREAT's invasion of Asia Minor (334 B.C.) reliberated the Greek cities and brought them into a suddenly enlarged Greek world. Alexander restored local democracies and abolished tribute. After his death (323 B.C.), Asia Minor became a battleground for his warring successors. ANTIGONUS (1) ruled from his Phrygian citadel of Celanae, but upon his death at the Battle of Ipsus (in Phrygia, in 301 B.C.), Asia Minor was parceled out between LYSIMACHUS and SELEUCUS (1). Seleucus destroyed Lysimachus at the Battle of Corupedium (in Lydia, in 281 B.C.) and brought Asia Minor into the sprawling SELEUCID EMPIRE. But the emergence of the city of PERGAMUM, inland on the middle west coast, created a local rival for power (mid-200s B.C.). By the early 100s B.C., Pergamum was one of the most beautiful and prosperous Greek cities.

The Pergamene kings cooperated with the encroaching nation of ROME, and—after helping a Roman army defeat the Seleucids at the Battle of Magnesia (189 B.C.)—Pergamum obtained control of most of Asia Minor. The entire region passed into Roman hands with the death of the last Pergamene king (133 B.C.). Later the non-Greek kingdom of Pontus arose against Roman authority.

(See also AEOLIAN GREEKS; BRONZE; DIADOCHI; DORIAN GREEKS; HELLENISTIC AGE; IONIAN GREEKS; MYCENAEAN CIVILIZATION.)

Aspasia

Common-law wife of the Athenian statesman PERICLES. After divorcing his legal wife, Pericles lived with Aspasia from about 450 B.C. until his death in 429 B.C.

Aspasia was an immigrant from MILETUS and thus was a resident alien at ATHENS. Pericles, being an Athenian citizen, was prohibited from marrying her (according to a law of 451 B.C., which he himself had sponsored). Historians traditionally have theorized that Aspasia was either herself a prostitute (*hetaira*) or that she at least managed a "house" of *hetairai,* but some modern scholars dismiss this information as being a daring joke of Athenian stage comedy. Contemporary comic playwrights such as CRATINUS, EUPOLIS, and ARISTOPHANES made vicious public fun of Pericles' liaison with Aspasia.

Witty and well educated, Aspasia seems to have had a captivating personality. She is said to have given lessons in RHETORIC, and the Athenian philosopher SOCRATES supposedly enjoyed conversing with her.

The 430s B.C. at Athens saw a rash of politically motivated prosecutions indirectly aimed at Pericles; the immediate targets included Aspasia and Pericles' friends PHIDIAS and ANAXAGORAS. Aspasia was accused of impiety (*asebeia*)—a common charge against intellectuals—and she also may have been accused of procuring freeborn Athenian ladies for Pericles' pleasure. At trial she was acquitted after Pericles made a personal appeal to the jurors, during which he uncharacteristically burst into tears.

She and Pericles had a son, also named Pericles. Although ineligible for Athenian citizenship because his mother was alien, the younger Pericles was enrolled as a citizen at his dying father's request, circa 429 B.C. In 406 B.C. the younger Pericles was one of six Athenian generals executed by public order for failing to rescue Athenian survivors after the sea battle of Arginusae, during the PELOPONNESIAN WAR.

Aspasia provides an example of the kind of success that an ambitious woman might find in a society that excluded WOMEN from power and wealth. Although further disadvantaged by being alien, she achieved a degree of influence and security by attaching herself to a powerful man.

(See also EDUCATION; MARRIAGE; METICS; PROSTITUTES.)

assembly

The word used to translate the Greek word *ekklēsia*—the official gathering of citizens in a Greek DEMOCRACY for the purpose of public debate and vote. At democratic ATHENS in the 400s and 300s B.C., the assembly was the sovereign body of government.

Admission to the Athenian assembly was open to all male citizens over age 18 (in theory about 30,000–40,000 men; in practice about 5,000). Under the radical democracy there were no property requirements for admission, and the 300s B.C. saw the introduction of a small payment for attendance, comparable to our modern jury pay.

The Athenian assembly met at least 40 times per year, with extra meetings as called for by the COUNCIL or by the board of generals. In the 400s B.C. the usual place of meeting was the Pnyx ("Packing Place"), a smoothened hillside west of the ACROPOLIS. There the people might vote on issues by show of hands; if written balloting was required—such as in an OSTRACISM vote—then the AGORA would be used. In debate, any Athenian had the right to address the assembly; a chairman of the day presided; and rules of order were maintained. Foreign ambassadors and other noncitizens might be allowed to address the assembly on issues of state.

Usually the assembly could debate and vote only on those topics placed on the agenda by the council; however, the assembly could (by vote) require the council to list a certain topic for the next meeting. Like other instruments of Athenian government, the assembly enjoyed courtroom powers. For example, it had the final verdict in certain serious criminal cases. By its vote the assembly passed laws, declared war, made peace, inflicted individual sentences of death or exile, and elected the army's generals and other important executives. The power of the assembly during Athens' imperial heyday can be seen in the Mytilenean Debate—described in THUCYDIDES' (1) history of the PELOPONNESIAN WAR (book 3)—where the fate of every man, woman, and child of the rebellious city of MYTILENE was decided in public debate at Athens in 427 B.C.

astronomy

The ancient Greeks pioneered the study of astronomy in the Western world, cataloguing the stars and identifying five of the planets (besides the earth). Most important, they developed several geometric models that tried to explain the movements of the heavenly bodies in terms of concentric spheres and other orbital paths. Unfortunately, the Greeks incorrectly placed earth at the center of the universe. This geocentric theory was brought to a false perfection by Ptolemy of ALEXANDRIA (1), circa A.D. 135; with Ptolemy's revisions, the geocentric model seemed to account for all heavenly motion observable to the unaided eye. (The Greeks possessed no telescopic lenses, only crude sighting devices.)

Ptolemy's system was so persuasive that it remained canonical in late antiquity and the Middle Ages, before it was finally ousted in A.D. 1543 by Copernicus' heliocentric model, which correctly placed the sun at the center of the solar system. Ironically, the ancient Greek astronomer Aristarchus of SAMOS (circa 275 B.C.) had produced a simple heliocentric model, but this had been bypassed in favor of the more apparently promising geocentric one.

Like other ancient peoples lacking precise calendars, the earliest Greeks relied on the rising and setting of the constellations in order to gauge the FARMING year. The poet HESIOD's versified farming almanac, *Works and Days* (circa 700 B.C.) mentions certain duties as signaled by the stars—for instance, grapes are to be picked when the constellation ORION has risen to a position overhead, in early autumn.

In the verses of HOMER (circa 750 B.C.) and other early poets, the earth is imagined as a disk surrounded by the stream of Ocean (*Okeanos*). The early philosopher-scientist THALES of MILETUS (circa 585 B.C.) elaborated on this concept, picturing a world that floats like a log on a cosmic lake of water. Thales' successor ANAXIMADER (circa 560 B.C.) devised the first, crude astronomical theory; he saw the world as a cylinder or disk suspended in space, with the cylinder's flat top constituting the inhabited world and the visible sky comprised of a series of fitted rings forming a hemisphere. The sun, moon, and stars are holes in these rings, through which we glimpse a distant celestial fire.

An obvious weakness of Anaximander's theory was its inability to explain the puzzling movements of the planets. The five planets known to the Greeks were Mercury, Venus, Mars, Jupiter, and Saturn (to use modern names). To the naked eye they look like bright stars, but their individual, looping progress through the night skies earned them the Greek name *planētai,* or "wanderers."

The planets' movements were first plausibly explained by the brilliant mathematician Eudoxus of CNIDUS, who was a student of PLATO at the ACADEMY at ATHENS (circa 350 B.C.). Eudoxus held that all heavenly movements are caused by the circular rotations of 27 concentric spheres, with the earth at their center. All of the proper stars exist on the single, outermost sphere (according to Eudoxus), but each planet employs no fewer than four concentric spheres, which spin along different axes to produce the planet's irregular course. Similarly, sun and moon are governed by three spheres each. Greater precision seemed to be achieved when more spheres were added by Eudoxus' younger contemporary Callipus of CYZICUS and by the great ARISTOTLE (both circa 330 B.C.).

Another astronomer of this Athenian intellectual heyday was Heraclides Ponticus (like Eudoxus, a younger associate of Plato at the Academy). Using the geocentric model, Heraclides suggested that the earth itself rotates on its axis. All of these thinkers approached astronomy as a challenge in geometry, not physics; there seem to have been no theories regarding what mechanical forces would cause the spheres to move. Understandably, the Greeks had no knowledge of astrophysics; topics such as gravity or the chemical compositions of stars were largely unexplored.

Aristarchus of Samos was a follower of the Aristotelian Peripatetic School at Athens, circa 275 B.C. Rejecting the Eudoxan model, Aristarchus invented the heliocentric theory, which correctly posited that the earth and planets revolve around the sun and that the earth spins on an axis. Like prior astronomers, Aristarchus imagined the stars as spread along a vast outer sphere—but thought that this sphere was stationary. He also correctly believed that the stars' apparent movement is an illusion created by the earth's orbit. Few of Aristarchus' writings survive, and we know of his theory mainly from later authors.

This heliocentric theory was not accepted in antiquity for two reasons. First, it ran counter to prevailing religious-philosophical beliefs by removing humankind from the center of creation. Second, it failed to account satisfactorily for the absence of stellar parallax—in other words, if the earth moves through space, then the stars should be seen to slide sideways in their progress overhead, which they do not seem to do. (In fact, the stars do shift their angles minutely as the earth orbits past, but their immense distance from the earth makes this variation invisible to the naked eye.)

The next great name in Greek astronomy is Apollonius of Pergē (a Greek town in southern ASIA MINOR), who was active at Alexandria circa 200 B.C. A mathematical genius, Apollonius revolutionized the geocentric theory by abandoning Eudoxus' cumbersome spheres in favor of two different models explaining celestial movement in terms of irregular orbits called eccentrics and epicycles. For example, the epicyclical theory imagined a planet shooting around in a small circle while at the same time orbiting in a much larger circle around the earth.

One of antiquity's few practical astronomers was Hipparchus of Nicaea (a Greek city in western Asia Minor). Active at RHODES circa 135 B.C., Hipparchus invented or improved the sighting device known as the dioptra, and was the first to map and catalogue approximately 850 stars.

The work of Apollonius and Hipparchus laid the groundwork for the greatest astronomer of the ancient world, Claudius Ptolemaeus, usually known as Ptolemy (but not a member of the Macedonian-Egyptian royal dynasty that used that name). A Greek-blooded Roman citizen of Alexandria, Ptolemy was active circa A.D. 127–140, under the Roman Empire. His written masterpiece was the *Almagest,* which has survived in medieval Arabic translation. (That title is the Arabic form for the unofficial Greek title *Megistē,* "the greatest," i.e. textbook.) In 13 books, the *Almagest* presented the sum total of astronomical knowledge of the day, as enlarged by Ptolemy's own work on the geocentric model. Aided by his pioneering use of the sighting device known as the armillary astrolabe (the forerunner of the medieval astrolabe), Ptolemy also enlarged Hipparchus' star map, among his many other achievements.

(See also ARATUS (2); MATHEMATICS; OCEANUS; PYTHAGORAS.)

Astyanax   See ANDROMACHE; HECTOR.

Atalanta   In MYTH, an Arcadian or Boeotian heroine of the athletic, forest-ranging, virginal type (in the pattern of the goddess ARTEMIS and of the AMAZONS). Beloved by the hero Mcleager, Atalanta joined him in the CALYDONIAN BOAR HUNT. But the best-known legend about Atalanta concerns her MARRIAGE. Unwilling to be wed, she declared that she would submit only to the man who could beat her in a footrace. According to one version, she would follow the racing suitor with a spear and stab him as she overtook him. She was finally won by a suitor named Melanion (or Hippomenes) who, with the goddess APHRODITE's help, had acquired three golden apples. During the race he dropped these treasures one by one in Atalanta's way; pausing to acquire them, she lost the contest. Their son was named Parthenopaeus ("born of the virgin").

Athena (often Athene) Important Greek goddess, guardian of the city of ATHENS—whose name she shared—and patron of wisdom, handicraft, and the disciplined aspect of war. Athena was worshipped on the ACROPOLIS at several Greek cities, including ARGOS and SPARTA. But her most significant shrine was on the Athenian acropolis, where a succession of temples culminated, in 438 B.C., in the building that is the gem of ancient ARCHITECTURE: the PARTHENON, or Temple of Athena the Virgin.

The name Athena, like the name Athens, is not actually Greek; the ending -na belongs to the language of the non-Greek people who preceded the Greeks as the inhabitants of mainland Greece. In her pre-Greek form, Athena was probably a patron deity of kings, whose hilltop palaces she

The head of Athena, from a Roman marble bust copied from a classical Greek work. Scholars dispute whether this head portrays the famous Lemnian Athena—a bronze statue by the great Athenian sculptor Phidias, circa 440 B.C. Regardless, the face conveys the sternness and intelligence of the goddess of wisdom, handicraft, and military strategy.

guarded as an idol shown clad in armor; in her ultimate origins she may be related to armed goddesses of the ancient Near East, such as the Mesopotamian Ishtar. After about 2100 B.C., when ethnic Greek invaders began taking over the land of Greece, Athena became incorporated into the developing Greek RELIGION (as did other pre-Greek goddesses, including HERA and ARTEMIS). Keeping her armor and her association with the hilltop, Athena entered Greek MYTH as the daughter of the great god ZEUS. This transition was under way by the Mycenaean era (1600–1200 B.C.), as is suggested by the mention of a "Lady of Atana" on a surviving Mycenaean clay tablet, inscribed in LINEAR B script and dated by archaeologists to 1400 B.C.

The charming myth of Athena's birth from the head of Zeus is told by the poets HESIOD (circa 700 B.C.) and PINDAR (467 B.C.). It seems that Zeus swallowed his first wife, Metis, when she was pregnant. He did this because a prophecy had warned him that Metis would otherwise bear a son destined to depose his father. But Zeus soon felt a terrible headache. And when the smith god HEPHAESTUS helpfully split open Zeus' skull, the goddess Athena sprang out—adult, in full armor, giving a war shout. This tale surely contains a primitive religious-political motive: namely, it shows absolutely that the goddess is Zeus' daughter and hence subordinate to him. The story also symbolizes how Zeus and Athena both partake of cleverness (Greek: *mētis*). The birth scene is a favorite subject on surviving Greek vase paintings from the 600s B.C. onward.

As goddess of organizational wisdom, Athena was thought to guide the typical Greek city-state (not only Athens). Her cult titles included Boulaia, "goddess of the COUNCIL," and Polias, "goddess of the city." In these functions she resembled Zeus, who was god of cosmic order and whose titles included Poleios, "lord of the city." In later centuries she naturally became associated with the academic wisdom of PHILOSOPHY.

As a protector of civilized Greek life, Athena was prominent in myths of order versus chaos. She was imagined as fighting gallantly in the gods' primeval war against the rebellious GIANTS: A famous section of the Great Altar of PERGAMUM, sculpted circa 180 B.C. and now in Berlin, shows Athena hauling the subdued Giant leader, Alcyoneus, by his hair. In other myths she repeatedly appears as a counselor to heroes struggling against monsters and villains: She helps HERACLES with his Twelve Labors, ODYSSEUS with his homecoming, and PERSEUS (1) with the killing of MEDUSA.

As a war goddess, Athena overlapped with the war god ARES. But where the brutal Ares embodied war's madness and waste, Athena tended to represent the more glorious aspects—strategy, discipline, national defense. She was said to have introduced such military inventions as the ship, the horse bridle, and the war chariot. In HOMER's epic poem the *Iliad* (written down circa 750 B.C.), she is the Greeks' staunch ally against the Trojans: She accompanies her Greeks into battle; intervenes to aid her favorite, DIOMEDES; and, in the poem's climactic section, helps ACHILLES to slay the Trojan HECTOR. Yet early in the poem she restrains Achilles from an act of violence against the Greek commander, AGAMEMNON, for she is also a deity of reason and control.

In poetry and art, Athena's war gear includes a breast-plate, helmet, spear, and shield of some sort. Sometimes her shield is the supernatural aegis (aigis), associated also with Zeus. Homer describes the aegis as a fearsome storm cloud, but by the mid-600s B.C. vase painters were picturing it as a goatskin mantle on Athena's left arm, tassled with live snakes and showing the vanquished Medusa's face. Later the aegis became styled as a round metal shield, embossed with Medusa's image.

Among peaceful duties, Athena was said to have invented weaving, spinning, and other domestic crafts performed by WOMEN in the ancient Greek world. She was also the patron of carpenters, potters, and (like the god Hephaestus) of metalworkers. Her relevant title was Erganē, "worker woman."

The Athenians credited her with introducing the cultivation of the olive, a staple of the ancient Greek diet and economy. Legend said that Athena and the god POSEIDON had competed publicly over who would become the Athenians' patron deity. Poseidon stabbed his trident into the acropolis summit, bringing forth a saltwater well; but alongside it Athena planted the first olive tree, and so was judged the winner by the people. The saltwater well and sacred olive tree were features of the Athenian acropolis in historical times.

Athena was pictured as stately and beautiful, although stern. In Homer's verses, Zeus calls her by the fond nickname glaukopis, "gray eyes" or "bright eyes." Her virginity—a trait she had in common with the goddesses Artemis and HESTIA—may perhaps harken back to her primeval form as a defender of the citadel: She is inviolate, like the fortress that she oversees. Her famous Athenian cult title was Parthenos, "the virgin," and her very old title Pallas was likewise understood as meaning "virgin." In myth, the Palladium (Palladion) was an image of Athena, sent down from heaven and worshipped by the Trojans.

By the late 400s B.C. Athena had two sanctuaries atop the Athenian acropolis. One was the Parthenon, housing a 35-foot-tall, GOLD-and-ivory statue of Athena Parthenos; the other was the Erectheum, housing a smaller, immemorably old, olivewood statue of Athena Polias. This simple wooden idol was revered at the midsummer festival of the Panathenaea ("all Athens")—considered to be Athena's birthday—when the people ceremoniously brought a new woolen gown (peplos) up to the goddess. An especially large celebration, held every fourth year, was called a Great Panathenaea. At another festival, in early summer, the Polias statue was carried down to the sea to be washed.

The Greeks identified their gods with certain animals; Athena was associated with snakes and, particularly, owls. Folklore claimed that owls were wise, like their goddess. Modern scholars believe that Athena's links with the owl, snake, and olive tree were all survivals of her primitive, pre-Greek cult, dating back to the third millennium B.C.

(See also FARMING; MINOAN CIVILIZATION; PARIS; PHIDIAS.)

Athenian democracy See DEMOCRACY.

Athens Foremost city of ancient Greece, and one of the three most important ancient cities that shaped Western civilization, along with ROME and Jerusalem. The culture produced by the rich and confident Athens of the mid-400s B.C.—in THEATER, ARCHITECTURE, SCULPTURE, and PHILOSOPHY—provides a cornerstone of our modern society. Most important, Athens was the birthplace of DEMOCRACY (dēmokratia, "power by the people").

The physical city, beautified by the wealth of empire, was a marvel to the ancient world—as it is today for the tourists and scholars who visit Athens (now the modern Greek capital). With some justification the Athenian statesman PERICLES described his city as "the greatest name in history . . . a power to be remembered forever." Defeat in the PELOPONNESIAN WAR (431–404 B.C.) broke Athens' imperial might, but the city remained politically important for the next century and culturally influential for the rest of antiquity.

An advantageous site helped make the city great. Located in southeast-central Greece, in the peninsular landmass called ATTICA, Athens lies amid the region's largest plain, where grain and olives were grown in historical times. The site's specific attraction was a 300-foot-tall, rock-formed hill (later known as the ACROPOLIS). This defensible position dominated the plain and, being four miles inland of the Saronic Gulf, gave access to the sea without inviting naval attack.

The locale was inhabited long before the first Greek-speaking tribesmen arrived circa 2000 B.C.: Archaeological excavations at Athens have yielded artifacts datable prior to 3000 B.C. The name Athens (Athenai) is pre-Greek and is associated with the city's patron goddess, ATHENA, who is likewise pre-Greek. Evidently the early Greek invaders appropriated these two related names (whose original meanings are lost to us). During the flourishing Greek MYCENAEAN CIVILIZATION (1600–1200 B.C.), Athens was a second-rank power—overshadowed by MYCENAE, THEBES, and other centers.

Greek MYTH claims that the unification of Attica under Athenian control was brought about by the Athenian king THESEUS. This legend probably recalls a real-life event of the late Mycenaean era. In later centuries this unification (sunoikismos, "synoecism") was celebrated at an annual Athenian festival called the Sunoikia. Among the resources of the Attic countryside that then passed to Athenian control were the fertile plain of Eleusis, the marble of Mt. Pentelicus, and, most important, the raw SILVER of Mt. Laurium. In a later era, Laurium silver would provide the Athenian COINAGE and help finance the social and military programs of the democracy.

A united Attica was able to resist the invading DORIAN GREEKS who swept southward through Greece after the collapse of Mycenaean society (circa 1100 B.C.). Legend tells how the last Athenian king, Codrus, sacrificed his life to save Athens from Dorian capture. Other legends describe Athens as a rallying point for Greek refugees whose homes in the PELOPONNESE had been overrun by the Dorians. These refugees, of the Ionian ethnic group, migrated eastward to establish cities on the ASIA MINOR coast that in later centuries retained cultural and political ties to Athens.

Following the depressed years of the DARK AGE (1100–900 B.C.), Athens emerged as a mainland commercial power, alongside CORINTH, CHALCIS, and a few other cities. In the

800s–700s B.C., Athenian workshops produced a widely admired Geometric-style POTTERY eventually imitated throughout Greece. At Athens, as at other prosperous Greek cities, there arose a monied middle class—mainly manufacturers and farmers—who resented being excluded from political power.

Like other Greek states, Athens at this time was governed as an ARISTOCRACY. A small circle of noble families monopolized the government and judiciary and owned most of the land. The general population was unrepresented, its political function being mainly to pay taxes, serve under arms, and obey the leaders' decrees. The emergence of democracy involved a drastic revamping of Athens' government in the 500s and 400s B.C., to the point where the citizens' ASSEMBLY comprised the sovereign, decision-making body.

This process required improvization, compromise, and bloodshed. In about 632 B.C. an adventuring nobleman named CYLON tried to exploit middle-class unrest to make himself tyrant of Athens, but the coup failed. Popular demand for a permanent, written law code was left unsatisfied by the aristocratically biased legislation of DRACO (circa 624 B.C.), but SOLON's radical reforms (circa 594 B.C.) averted revolution and laid the foundations for Athenian democracy. In foreign affairs, Athens defeated its nearby rival MEGARA (1) for control of the island of SALAMIS (1). Athenian COLONIZATION in the HELLESPONT district—at Sigeum and the CHERSONESE—foreshadowed Athens' imperial control of that region in future years.

Despite Solon's legislation, class tensions remained keen enough for PISISTRATUS to seize power as tyrant (mid-500s B.C.). Under his enlightened rule, Athens advanced commercially and artistically. Supplanting Corinth as the foremost mainland mercantile power, Athens monopolized the pottery market, exporting Athenian black-figure ware throughout the Mediterranean. Athens became a cultural center, as Pisistratus and his son and successor, HIPPIAS (1), attracted poets, sculptors, and other craftsmen from throughout the Greek world.

The expulsion of Hippias in 510 B.C. brought on the birth of democracy. The reforms of CLEISTHENES (1) (508 B.C.) broke the aristocracy's residual influence by means of administrative changes based on the redivision of the population into 10 new tribes. A citizens' COUNCIL of 500 became the guide and executive for the popular assembly, which in practice numbered about 5,000 men. The assembly held regular meetings in the Pnyx ("packing place"), a hollowed-out hillside about 500 yards west of the acropolis. In around 507 B.C. the Athenians repulsed an attack by the Spartan king CLEOMENES (1) that was aimed at dismantling the fledgling democracy. The optimism of the following years helped the Athenians to repel a Persian invasion force at the Battle of MARATHON, outside Athens (490 B.C.).

By 500 B.C., the physical city had begun to take on its classical layout. Probably by then a surrounding wall had been built (although only later walls have survived in remnants). Atop the acropolis, which was sacred to Athena and the other gods, old temples of wood or limestone were being replaced by marble structures. At the acropolis' southeast base stood an open-air THEATER dedicated to the god DIONYSUS, where performances of the newborn

The goddess Athena mourns, in an Athenian marble bas-relief, probably a funerary stone, circa 460 B.C. The short column at the lower right may represent a list of Athenian war dead. Showing all the grace and simplicity of the Early Classical style, this 21-inch-tall carving dates from the height of the Athenian Empire, when an overconfident Athens was leading its Delian League allies against the Persians in Cyprus and Egypt, and—closer to home—against the Spartans, Corinthians, Boeotians, and other Greeks. Eventually Athenian aggression would bring on the start of the Peloponnesian War (431 B.C.).

Athenian art-forms of tragedy and comedy were held. Northwest of the acropolis lay the AGORA—the city's commercial center. Some distance northwest of the agora, beyond the Dipylon Gate, stood an industrial suburb, the Ceramicus (*Keiramikos,* or "potters' quarter"), at the outer edge of which lay the aristocratic families' elaborate graveyards, still visible today.

Under leadership of the left-wing statesman THEMIS-TOCLES, the Athenians developed PIRAEUS as a seaport and amassed the biggest navy in Greece, 200 warships. This navy played a paramount role in defending Greece against the invasion of the Persian king XERXES (480 B.C.). Athens was evacuated before the advancing Persians—who burned and sacked the empty city twice, in 480 and 479 B.C.—but the Athenians fought on in their warships and helped destroy the Persian navy at the sea battle of Salamis.

The defeat of Xerxes' invasion created an outpouring of confidence—and an influx of money—that made Athens into the cultural capital of the Greek world. First the city wall, torn down by the Persians, was rebuilt and enlarged, despite Spartan objections (479 B.C.); remnants of this four-mile-long "Themistoclean Wall" survive today. In the Greek naval counteroffensive against Persian territories, Athens now took the lead, organizing the Greek city-states of the AEGEAN SEA and the Asia Minor coast into an Athenian-controlled coalition, the DELIAN LEAGUE (478 or 477 B.C.). The league's most prominent officer was the conservative Athenian statesman CIMON. Delian member states paid dues—in the form of silver or military service—which in time came to be used strictly for Athens' advantage.

Athenian aggressions in mainland Greece—against AE-GINA, BOEOTIA, PHOCIS, and Corinth—brought Athens into intermittent conflict with those states' powerful ally, Sparta (459–446 B.C.). Meanwhile the left-wing statesman Pericles became heir to Themistocles' policies of sea empire and resistance to Sparta. Pericles tightened Athens' control over the Delian allies, and he built the LONG WALLS, connecting Athens to its naval base at Piraeus in a continuous fortification (461–456 B.C.). After arranging peace with PERSIA, he initiated a building program (448 B.C.) to replace the temples destroyed in 480–479 B.C. This program, supervised by the sculptor PHIDIAS, turned Athens into the grandest city of the Greek world. Among the famous constructions still standing today are the PARTHENON, Propylaea, Erectheum, and Temple of Athena Nikē (all on the acropolis), and the admirably preserved temple of the god HEPH-AESTUS in the agora.

The middle and late 400s B.C. saw the Athenian high noon. The tragic playwrights AESCHYLUS, SOPHOCLES, and EURIPIDES, the painter Polygnotus, the intellectual movement of the SOPHISTS, the comic playwright ARISTOPHANES, the historian THUCYDIDES (1), and the West's first great ethical philosopher, SOCRATES, all lived in Athens during these years. From other Greek cities craftsmen, poets, and businessmen flocked to Athens, swelling Attica's population to over 200,000 (including SLAVES) and making the city fully dependent on imports of Ukrainian grain, supplied by Pericles' far-reaching naval program.

Fear of Athenian expansionism led Sparta, Corinth, and Boeotia to start the Peloponnesian War (as we call it), in 431 B.C. Despite the outbreak of plague (430 B.C. and later), the Athenian navy and maritime network could have brought the city safely through the conflict, had the citizens not wasted lives and resources in a vainglorious attempt to conquer SYRACUSE, in Greek SICILY (415–413 B.C.). This calamitous defeat threw Athens on the defensive, while the Spartans, with the help of Persian money, were able

to develop a navy, subvert Athens' Delian allies, and attack the Athenian lifeline in the Hellespont. With their fleet destroyed (405 B.C.), the Athenians surrendered after an eight-month siege (404 B.C.). The victorious Spartans pulled down the Long Walls. Despite urgings from the Thebans and Corinthians to destroy Athens, the Spartans spared the city, installing the short-lived dictatorship of the THIRTY TYRANTS .

Athens quickly recovered, joining another war against Sparta (395–387 B.C.) and organizing another maritime coalition, the SECOND ATHENIAN LEAGUE (377 B.C.). In spite of the eloquent harangues of the statesman DEMOSTHENES (1), the Athenians proved ineffectual against the military-diplomatic campaigns of King PHILIP II of MACEDON. After sharing in the defeat at CHAERONEA (338 B.C.), Athens received lenient treatment as Philip's subject city; the league, however, was disbanded. Athens revolted against Macedonian rule after the death of ALEXANDER THE GREAT (323 B.C.), but defeat brought an end to Athens' foreign policy ambitions, and a Macedonian garrison was then installed at Piraeus (322 B.C.). Under the Macedonian king CASSANDER, Athens was governed by the enlightened De-metrius of Phalerum (317–307 B.C.). Liberated by the warrior prince DEMETRIUS POLIORCETES, Athens reverted periodically to Macedonian control, until the coming of ROME broke Macedon's power (167 B.C.).

The 300s B.C. marked the greatest days of Athenian philosophy, starting with the establishment of PLATO's school of higher learning, the ACADEMY (circa 385 B.C.). In 335 B.C. Plato's former pupil ARISTOTLE opened a rival school, the LYCEUM. Among other influential philosophical movements to emerge at that time at Athens were EPICURE-ANISM (circa 307 B.C.) and STOICISM (circa 300 B.C.). Athens still retained a foremost position in the arts, producing, among other talents, the sculptor PRAXITELES (circa 350 B.C.) and the playwright MENANDER (circa 300 B.C.).

But Athens' creative output had begun to diminish by the early 200s B.C., as the HELLENISTIC AGE produced new centers of wealth and patronage, such as ALEXANDRIA (1). Under Roman rule, Athens remained a prestigious center of EDUCATION.

(See also CORINTHIAN WAR; PAINTING; PERSIAN WARS; POLIS; TRADE; TYRANTS.)

Atlantis The legend of Atlantis ("island in the Atlantic") first appears in two dialogues, the *Timaeus* and the *Critias*, written circa 355 B.C. by the Athenian philosopher PLATO. Plato describes Atlantis as a vast, wealthy island or continent that used to exist, 9,000 years earlier, in the Atlantic Ocean. This domain was an ideal society, inhabited by an advanced and virtuous people who conquered eastward to Europe and Africa. But their greed for power grew, until the angry gods sent a deluge to submerge the island.

Despite our modern-day fascination with this legend, it seems clear that Atlantis never existed as described. Plato probably made up the tale as a parable of self-destructive pride. Or, Plato's story may possibly record some 1,100-year-old folktale telling of the decline of MINOAN CIVILIZA-TION in the Aegean region. ARCHAEOLOGY indicates that in around 1480 B.C. the mighty island of CRETE was devastated by earthquakes and tidal waves. Plato's Atlantis may be a distorted memory of Crete's downfall, projected beyond the Mediterranean world.

(See also HUBRIS; THERA.)

Atlas   In MYTH, one of the TITANS (the race of demigods that ruled the universe before the emergence of the Olympian gods). After the Titans' overthrow by the gods, Atlas was condemned by ZEUS to hold up the sky on his shoulders for eternity. The Greeks associated Atlas with the far West. By the mid-400s B.C. he was identified with the Atlas mountain range in North Africa, and his name had been given to the Atlantic Ocean (Greek Atlantikos, "of Atlas").

Atlas was said to be the father of the HESPERIDES (daughters of Evening), who tended the fabled garden in the West where golden apples grew. When the hero HERACLES arrived to fetch a few such apples for his twelfth Labor, Atlas offered to get them for him if he would kindly support the sky in the meantime. Returning with three golden apples, Atlas intended to leave Heracles holding up the celestial canopy, but Heracles was able to trick Atlas into resuming his assigned burden. The incident is portrayed on a famous marble carving from the temple of ZEUS at OLYMPIA.

(See also OLYMPUS, MT.)

Atreus   Mythical king of MYCENAE. Atreus was the father of the heroes AGAMEMNON and MENELAUS, who in poetry are often referred to as the Atridae (Atreidai), the sons of Atreus.

Atreus and his brother Thyestes labored under a hereditary curse received by their father, PELOPS. The events of their own lives perpetuated this curse. Thyestes offended Atreus by seducing his wife and stealing a golden ram, a token of the kingship. Atreus banished his brother, but then, pretending to be reconciled, invited him to a banquet. At the feast, Atreus treacherously served the cooked flesh of Thyestes' own children, which the unwitting guest ate. Upon realizing the horror, Thyestes invoked a curse on his brother and departed. This curse was fulfilled when Thyestes' surviving son Aegisthus helped murder Agamemnon after the latter's return from the TROJAN WAR. The banquet of Atreus and Thyestes was a favorite reference in later Greek and Roman poetry. The Roman statesman-philosopher Seneca (circa A.D. 60) wrote a tragedy *Thyestes*, presenting the monstrous events onstage.

(See also CLYTAEMNESTRA; ORESTES.)

Attalus   See PERGAMUM.

Attica   Territory of ATHENS. Named perhaps from the Greek word *aktē*, "promontory," Attica is a triangular peninsula in southeast-central Greece, extending southward into the AEGEAN SEA. The peninsula's neck is bordered by BOEOTIA (to the north and northwest) and by the territory of MEGARA (1) (to the northwest). Attica's western sea is the Saronic Gulf, containing the islands SALAMIS (1) and AEGINA, among others.

With 1,000 square miles, Attica is about half the size of Massachusetts. The mountainous terrain reaches four separate peaks: Mt. Hymettus in the southwest (near Athens); Mt. Pentelicus halfway up the east coast; Mt. Parnes in the north-central area; and Mt. Laurium at the southern tip.

Attica's natural resources played vital roles in making Athens a major commercial power, starting in the 500s B.C.; particularly significant was the SILVER mined at Laurium. An admired marble was quarried at Pentelicus and Hymettus, and the superior clay in the Attic soil was obviously important for Athens' domination of the Greek POTTERY market in the 800s–700s and 500s–400s B.C..

The mountains and hills of Attica enclose four discrete farming plains, which from prehistoric times determined the groupings of settlement. There is (1) the large plain of Athens, (2) the southern plain of the Mesogeia ("midland"), (3) the small northeastern flatlands of MARATHON and Aphidna, and (4) the one truly fertile plain, at Eleusis in the west. The prominence of Eleusis as a food source is reflected in the ancient, state-run ELEUSINIAN MYSTERIES, which celebrated the goddess DEMETER's gift of grain to humankind. Other crops grown in the rather thin Attic soil included vines, figs, and olives.

Among other important locales was Sunium, at Attica's southern tip, where a famous temple to Poseidon was built in the 440s B.C..

(See also ATHENA; CLEISTHENES (1); DEME; DEMOCRACY; PELOPONNESIAN WAR; PERSIAN WARS; PIRAEUS; THESEUS.)

B

Bacchae   See EURIPIDES; MAENADS.

Bacchus   See DIONYSUS.

Bacchylides   Greek lyric poet from the island of Ceos, circa 520–450 B.C. His work survives mainly in the form of 20 choral poems—14 victory odes and six dithyrambs—discovered in an Egyptian papyrus in A.D. 1896. He is known to have written many other verses for choruses, including hymns that survive in a few fragments. Despite his lack of genius, Bacchylides showed genuine artistic virtues, including clarity of expression and a talent for narrative.

Bacchylides was the nephew (sister's son) and protégé of the famous poet SIMONIDES, and it was to Simonides' sponsorship that Bacchylides owed much of his worldly success. He seems to have followed his uncle's footsteps from early on and to have been employed by some of the same patrons. Like Simonides, Bacchylides wrote dithyrambs for poetry competitions at ATHENS. (The dithyramb, a precursor of Athenian stage tragedy, was a narrative poem on a mythological subject; it was sung in public performance by a chorus, one of whose members would take on a solo role in character.) Two of Bacchylides' extant dithyrambs present imaginative episodes from the life of the Athenian hero THESEUS. In around 476 B.C., Bacchylides accompanied the 80-year-old Simonides to the court of HIERON (1), ruler of the Sicilian-Greek city of SYRACUSE. There Bacchylides and Simonides are said to have been engaged in an unfriendly rivalry with the Theban poet PINDAR, who was visiting Sicily in those years and was probably close to Bacchylides' age.

Bacchylides and Pindar both composed victory odes for Hieron. Hieron's victory in the horserace at the OLYMPIC GAMES (476 B.C.) is commemorated in Bacchylides' Ode 5 as well as in Pindar's Olympian Ode 1; similarly, Hieron's victory in the chariot race at the PYTHIAN GAMES (470 B.C.) produced Bacchylides' Ode 4 and Pindar's Pythian 1. But when Hieron won his coveted Olympic chariot victory in 468 B.C., it was Bacchylides, not Pindar, who was assigned to write the celebratory ode.

This poem, Ode 3, is Bacchylides' most successful work. It addresses a rich dictator who, although at his peak of prestige, was in failing health, with only a year to live. The poem is full of sad reminders of human transience, couched in terms that are lovely despite being wholly conventional: *Life is short and troubled*, the poet says, in essence, *and hope is treacherous. Be ready for death tomorrow or 50 years hence. The best a mortal man can do is worship the gods and live a good life*. The poem is spun around the central legend of the Lydian king CROESUS, who fell from wealth and power to defeat, before being saved by the god APOLLO. The ode's consolation in the face of death and its muted warning against the sin of pride may have proved poignant to Hieron. In closing, the poet refers to himself as "the nightingale of Ceos"—a charmingly modest and apt self-description.

(See also HUBRIS; LYRIC POETRY; THEATER.)

Bactria   Corresponding roughly to what is now northern Afghanistan, Bactria was an important province of the Persian Empire in the mid-500s to mid-300s B.C. In 330 B.C. the region was captured by the Macedonian king ALEXANDER THE GREAT, during his conquest of the Persian domain. Bactria's significance for Greek history is that, after being assigned heavy settlements of Alexander's veteran soldiers, the region developed into a far-eastern enclave of Greek culture.

Following Alexander's death (323 B.C.), Bactria passed into the east Greek SELEUCID EMPIRE, but by 255 B.C. it had revolted under a leader named Diodotus, who became its first king. Prospering from its central Asian TRADE routes, this Greek kingdom lasted for over a century and enveloped what is now southern Afghanistan, western Pakistan, southern Uzbekistan, and southeastern Turkmenistan. The Bactrian kings—of which there seem eventually to have been two rival dynasties, reigning out of different capitals—minted a superb COINAGE, stamped with royal portraits. Today surviving Bactrian coins provide some of the most realistic portraiture from the ancient Greek world.

Among the principal Bactrian cities were two named ALEXANDRIA (2)—modern Kandahr and Herat, in Afghanistan. The remarkable remnants of a Greek-style GYMNASIUM and temple discovered at Aï Khanoum, in northern Afghanistan, mark the site of another city—probably also named Alexandria—that thrived under Bactrian kings. But by 100 B.C. Bactria had fallen, overrun by Asian nomads.

(See also HELLENISTIC AGE.)

Bassae   Site, in southwest ARCADIA, of a remarkable temple to APOLLO, now reconstructed. Bassae (meaning "ravines") is located on the remote slopes of Mt. Cotilion. The austere Doric-style temple was begun in about 450 B.C. and completed perhaps 30 years later. It was designed by Ictinus, the architect of the Athenian PARTHENON, and it almost certainly predates the Parthenon (begun 447 B.C.).

Built atop a narrow mountain ridge, the temple necessarily has a north-south orientation (as opposed to the typical east-west). Its most famous feature is the single Corinthian-style column that stands in one part of the interior. The column's original capital disappeared in the 19th century A.D. but was previously noted and sketched as having acanthus-leaf decorations on four sides. This makes it the earliest known example of a Corinthian column.

The temple's architectural SCULPTURES, stripped off and carted away in the 19th century, now stand in the British Museum in London. They portray battles between Greeks and AMAZONS and between Lapiths and CENTAURS—two favorite subjects of Greek temple carving, conveying the theme of (Greek) order versus (barbarian) chaos, and appropriate to a Greek god of culture such as Apollo. Similar scenes existed among the sculptures of the Parthenon.

(See also ARCHITECTURE.)

Bellerophon Mythical Corinthian hero. Bellerophon came to be associated with a flying horse named Pegasus, but Pegasus is not mentioned in the earliest surviving references to Bellerophon, in HOMER's *Iliad* (written down circa 750 B.C.).

Like HERACLES and other Greek heroes, Bellerophon was assigned a series of impossible-seeming tasks, which he accomplished. His adventures began when he resisted the advances of Anteia (or Stheneboea), the wife of King Proteus of ARGOS (or TIRYNS). Humiliated, she claimed he had tried to seduce her, and so Proteus sent him away with an encoded letter for the king of Lycia, in ASIA MINOR. The letter requested that the king kill the bearer. Obligingly, the king sent Bellerophon to fight the Chimera and the AMAZONS, among other hazards. But the hero triumphed and eventually married the king's daughter; Philonoë.

Later versions of the MYTH—such as in the poet PINDAR's 13th Olympian ode (466 B.C.)—had Bellerophon assisted by the winged Pegasus, which the goddess ATHENA helped him to capture. Pegasus appeared as a symbol of Corinth on the city's COINAGE.

Bion See CYNICS.

birth control See PROSTITUTES.

Black peoples The ancient Greeks were acquainted with people of Negro race. During the second and first millennia B.C. the Greeks had periodic contact with EGYPT, culminating in ALEXANDER THE GREAT's conquest of the land in 322 B.C. Although most Egyptians were of Semitic blood, some were of Negro blood, with also an intermingling of the two races. As is the case in Egypt today, many inhabitants may have had dark skin, with Semitic or Negro physical features in varying degrees. People of black African descent played important roles as soldiers, administrators, priests, and sometimes pharaohs.

Moreover, the Greeks knew about Egypt's southern neighbor—the powerful nation called Nubia or Kush (also spelled Cush). Located along the Nile in what is now northern Sudan, Nubia was inhabited by black Africans, with possibly an admixture of Semitic blood. Nubia was a trading partner and periodic enemy of the pharaohs' Egypt; during the third to first millennia B.C. the two nations warred intermittently over a shared, shifting frontier that lay south of the first Nile cataract (modern-day Aswan). Nubia was an important supplier of GOLD ore to Egypt, and Nubian troops often served as mercenaries in Egyptian service.

By the mid 700s B.C. the Nubians had reorganized themselves and had begun a large-scale invasion of Egypt, the conquest of which was completed in around 715 B.C. by the Nubian ruler Shabako (or Shabaka). The period from about 730 to 656 B.C., when Nubian pharaohs ruled part or all of Egypt, is called by modern historians the 25th Egyptian Dynasty. But Assyrian invaders toppled the dynasty, and Egypt passed to other hands.

In the early 500s B.C. Nubia came under military pressure from a resurgent Egypt. The Nubians removed their capital southward to Meroë, situated between the fifth and sixth Nile cataracts (about 120 miles north of modern-day Khartoum). This "Meroitic" Nubian kingdom flourished for 900 years as an African society independent of Egypt and of Egypt's successive conquerors—the Persians, the Greco-Macedonian Ptolemies, and the Romans.

Among the Greeks, such facts became the stuff of legend. HOMER's epic poems, the *Iliad* and *Odyssey*, written down circa 750 B.C., make reference to a people known as *Aithiopes* (Greek: "burnt-face ones"). Beloved of the gods, they dwelt far to the south, in the land called *Aithiopia*. It was the force of the southern sun, the early Greeks believed, that crisped these people's faces. Greek MYTH told of an Aithiopian king, MEMNON, who led an army northward to help defend TROY during the TROJAN WAR but was killed in single combat with the Greek hero ACHILLES. This episode (not described by Homer) was recounted in a now-vanished Greek epic poem usually called the *Aethiopis*. It is not known whether this legend contains any kernel of historical truth.

Homer and other early writers give no clear location for Aithiopia. But by the early 400s B.C. the Greeks had come to equate it specifically with the nation of Nubia, south of Egypt. The Greek historian HERODOTUS, visiting Egypt circa 450 B.C., heard tales about Aithiopia-Nubia. According to Herodotus, the Aithiopians were said to be "the tallest and most attractive people in the world" (book 3). Herodotus also mentions that Aithiopian troops, armed with spear and bow, served in the Persian king XERXES' invasion of Greece in 480 B.C. (Although not part of the Persian Empire, Nubia seems to have been enrolled as a diplomatic friend of PERSIA.)

In modern times, the place-name Aithiopia (latinized to Ethiopia) has been applied to an African nation located far southeast of ancient Nubia. Confusingly, however, modern historians of the ancient world sometimes use the terms Ethiopia and Ethiopian in the old, Greek sense, to denote ancient Nubia. Egypt's 25th Dynasty is often referred to as the Ethiopian Dynasty.

As a result of contact with black people of Egypt and Nubia, ancient Greek artists began portraying blacks in statuary, metalwork, vase paintings, and the like; the earliest surviving examples date from the 500s B.C. One of the finest pieces is an Athenian clay drinking cup, realistically shaped and painted to represent the head of a black youth (circa 525 B.C.). The hero Memnon appears on several extant vase paintings; usually he is shown with distinctly Negro facial features.

Another black man in Greek art was the evil Busiris. According to Greek legend, Busiris was an Egyptian king, a son of the god POSEIDON; capturing any foreigners who entered Egypt, he would sacrifice them on the altar of the god ZEUS. But at last the Greek hero HERACLES arrived and—in a scene popular in vase paintings—killed Busiris and his followers. This odd tale may distortedly recall Greek relations with Egypt amid the downfall of the 25th (Nubian) Dynasty in the mid 600s B.C.

Alexander the Great's conquest of the Persian Empire brought Egypt into the Greek world, thus opening Nubia and other parts of West Africa to Greek exploration and TRADE. Circa 280–260 B.C. Egypt's king PTOLEMY (2) II sent explorers and merchants sailing along the Red Sea coast as far as modern-day Somalia, and up the Nile to Nubian Meroë and beyond. Ptolemy's goal was the acquisition of gold and other goods, particularly live African elephants for military use. Subsequent Ptolemies maintained these African trading and diplomatic ties, which incidently brought Greek influences to the ruling class in Nubia. We know, for example, that the Greek name Candace was used by Nubian queens from this era until the 300s A.D.

(See also NAUCRATIS; POTTERY; XENOPHANES.)

Black Sea Modern name for the 168,500-square-mile oblong sea bordering ASIA MINOR on the south and Ukraine and southwestern Russia on the north. Attractive for its access to Asia Minor's metal TRADE and Ukraine's vast wheatfields, the Black Sea became ringed by Greek colonies, from circa 700 B.C. onward. The Greeks called it Pontos Euxeinos, "the hospitable sea." But this name was intentionally euphemistic, insofar as fierce native inhabitants and a cold and stormy climate might make the region distinctly inhospitable.

The Black or Euxine Sea was opened up by explorers from the Greek city of MILETUS, on the west coast of Asia Minor. The Milesians' first goal was to acquire copper, tin, GOLD, and other raw metals of interior Asia Minor, and one of their early colonies was SINOPE, located halfway along the southern Black Sea coast. In the next 100 years, colonies from Miletus (and a few other east Greek cities) arose around the entire Black Sea.

Along the northern coastlines, the principal Greek cities were Olbia (meaning "prosperous"), at the mouth of the River Bug, and PANTICAPAEUM, on the east coast of the Crimean peninsula. These sites offered the very valuable resource of grain, grown as surplus by farmers in the interior and purchased by the Greeks for export to the hungry cities of mainland Greece. By 500 B.C. the north Black Sea coast was a major grain supplier, especially for ATHENS. This shipping route placed strategic importance on two narrow waterways—the BOSPORUS and the HELLESPONT—that help connect the Black Sea to the AEGEAN SEA.

(See also AMAZONS; COLONIZATION; HECATAEUS; HERODOTUS; JASON (1); SHIPS AND SEAFARING.)

Boeotia Northwest-southeast–elongated region of central Greece, bordered by PHOCIS to the northwest, by the Straits of EUBOEA to the northeast, and by the Corinthian Gulf to the southwest. Central Boeotia consists of a mountain-girt plain that provided rich FARMING and the raising of horses. In the plain's southeast corner lay the major Boeotian city, THEBES. The second city was ORCHOMENUS, in the plain's northwest corner, opposite Thebes.

South of Thebes, Boeotia shared an ill-defined border with the Athenian territory, along the east-west line of Mt. Parnes and Mt. Cithaeron. The historically important Boeotian town of PLATAEA lay just north of Cithaeron. Farther northwest stood Boeotia's tallest mountain, Helicon (about 5,800 feet).

The Boeotian heartland comprised one of the centers of MYCENAEAN CIVILIZATION (circa 1600–1200 B.C.)—as suggested by Thebes' prominent place in Greek MYTH. The name Boeotia refers to the raising of cattle (*boes*) and specifically commemorates the Boiotoi, a Greek people who invaded the region, migrating south from THESSALY circa 1100 B.C. The Boeotians of historical times spoke a form of the Aeolic dialect, related to the dialect of Thessaly and that of the east Aegean island of LESBOS. Boeotians were reputed to be boorish and ignorant, although prosperous. "Boeotian pig" was a Greek epithet. But Boeotia's literary tradition produced the poets HESIOD, PINDAR, and Corinna.

With the exception of Plataea, the Boeotian towns followed Thebes as enemies of ATHENS, starting in the late 500s B.C. Boeotia was a target of Athenian expansionism, and the Athenians actually occupied Boeotia for a decade, 457–447 B.C. Soon thereafter the towns formed a Boeotian League, under Theban dominance, for mutual defense and a jointly decided foreign policy. The Boeotian army then emerged as one of the best in Greece, distinguished by a large CAVALRY force in addition to strong infantry. At the Battle of Delium in 424 B.C., during the PELOPONNESIAN WAR, the Boeotians defeated the invading Athenians and ended their hope of reconquering the region. Later, under the Theban commander EPAMINONDAS, the Boeotian League broke the might of SPARTA (371 B.C.) and became the foremost power in Greece, before falling to Macedonian conquest (338 B.C.).

Boeotia's central location and flat interior often made it a battleground—"the dancing floor of War," Epaminondas called it. Famous Boeotian battlefields included Plataea (479 B.C.), LEUCTRA (371 B.C.), and CHAERONEA (338 B.C.).

(See also GREECE, GEOGRAPHY OF; GREEK LANGUAGE; MUSES; PERSIAN WARS; PLATAEA, BATTLE OF; WARFARE; LAND.)

Bosporus (or Bosphorus) Narrow, zigzagging, 18-mile-long channel flowing southwestward from the BLACK SEA to the SEA OF MARMARA. Beyond the Marmara, the current continues south and west through the HELLESPONT channel to the AEGEAN SEA. Like the Hellespont, the Bosporus borders part of northwestern ASIA MINOR; it was considered to be a dividing line between EUROPE AND ASIA.

Shorter and generally narrower and swifter-flowing than the Hellespont, the Bosporus ranges in width from 2.5 miles to 400 yards. Its name, "cow ford" or "ox ford," was in ancient times said to refer to the mythical wanderings of IO, a woman loved by the god ZEUS and transformed into a cow by Zeus' jealous wife, HERA. But the name may refer to a more mundane cattle crossing.

Around 513 B.C., the Persians under King DARIUS (1)—preparing to cross from Asia to Europe for their invasion of Scythia—spanned the Bosporus with a pontoon bridge consisting of about 200 ships anchored in a row. This was a remarkable engineering feat in the ancient world, although not as amazing as the Persians' bridging of the Hellespont, a wider channel, 30 years later. In modern times the Bosporus, now a part of Turkey, was not bridged until 1973.

The Bosporus and Hellespont were the two bottlenecks along the shipping route between the Black Sea and the Aegean. This route had become crucial by about 500 B.C., when ATHENS and other cities of mainland Greece were becoming dependent on grain imported from the northern Black Sea coast. As a natural site where shipping could be raided or tolled, the Bosporus, like the Hellespont, offered wealth and power to any state that could control it. This, combined with the excellent commercial fishing in the strait and its value as a ferry point, helps to explain the prosperity of the Bosporus' most famous city, BYZANTIUM, located at the southern mouth. Athens controlled the Bosporus in the 400s B.C. by holding Byzantium as a subject ally.

(See also THRACE.)

boulē See COUNCIL.

boxing Important SPORT among the ancient Greeks, although less popular than the two other combat sports, WRESTLING and PANKRATION.

Our knowledge of Greek boxing comes mainly from extant literature, artwork, and inscriptions (on tombs and religious offerings). Because HOMER's epic poem the *Iliad* describes a boxing match in book 23, we know that the Greeks were practicing boxing by at least the mid-700s B.C., when the Homeric poems were written down. Possibly the early Greeks learned the sport from MINOAN CIVILIZATION of the ancient Aegean region (2200–1400 B.C.). A surviving fresco from Minoan THERA, painted circa 1550 B.C., shows two boys engaging in what appears to be a stylized form of boxing.

The sport's patron god was APOLLO, a deity of the civilizing arts. During historical times boxing was the sort of discipline that a wealthy young man in a Greek city-state might practice at a local GYMNASIUM. Like other Greek sports of the pre-Roman era, boxing was purely an amateur pastime. The best boxers could hope to compete in the men's or boys' category at the OLYMPIC GAMES or at one of the other great sports festivals.

The Greeks had no boxing rings; official contests might take place on an unfenced sand floor in an outdoor stadium, where the referee would keep the two opponents in a fighting proximity. More brutal than today's sport, Greek boxing did not recognize different weight classes; the advantage tended to go to the heavier man. The match had no rounds, but continued until one man either lost consciousness or held up a finger, signaling defeat. Boxers were allowed to gouge with the thumb but were forbidden to clinch or grab. Certain vase paintings show a referee using his long stick to beat a clinching boxer.

Down through the 400s B.C. boxers often wore protective rawhide thongs wrapped around their hands. During the 300s B.C. the thongs developed into a heavier, more damaging form, with a hard leather knuckle pad. For practice only, boxers might use soft gloves similar to our modern boxing gloves. Boxers tended to attack the face—ancient artwork and inscriptions commemorate broken noses and damaged eyeballs. It was not unusual for a boxer to die from injuries.

Although champions came from all over the Greek world, boxing was particularly associated with the grim discipline of SPARTA. The Athenian philosopher PLATO's dialogue *Gorgias* (circa 386 B.C.) mentions "the boys with the cauliflower ears," referring to the antidemocratic, upper-class Athenian youth who practiced boxing in imitation of Spartan training.

(See also EDUCATION; OLIGARCHY.)

Brasidas Dynamic Spartan general of the PELOPONNESIAN WAR. His successes against Athenian holdings on the north Aegean coast in 424–422 B.C. helped to offset Athenian victories elsewhere and bring about a mood of stalemate, resulting in the Peace of NICIAS. Brasidas' great triumph was his capture of the vital Athenian colony of AMPHIPOLIS (early 423 B.C.). He died at the Battle of Amphipolis, successfully defending the city from an Athenian army under CLEON (422 B.C.).

Brasidas saw that the way to attack an impregnable ATHENS was to destroy its northeastern supply lines. His northern campaign was an early, crude, land-bound version of the strategy that would later win the war for SPARTA, in the naval campaigns of 413–404 B.C.

(See also THUCYDIDES (1).)

bronze Alloy of copper and tin, usually in a nine-to-one ratio. Bronze supplied the most useful metal known during the third and second millennia B.C., replacing prior uses of copper in weaponry, tools, and artwork throughout the Near East and Mediterranean.

Fortified by its measure of tin, bronze is harder than copper, but it melts at the same relatively low temperature—2,000 degrees Fahrenheit. Molten bronze can be intricately shaped by casting—that is, by being poured into a mold. Once cooled, the bronze item can be sharpened or shaped further.

This technology was invented in the Near East before 3000 B.C. and had spread to Minoan CRETE by around 2500 B.C. Whether the first arriving Greek tribesmen of about 2100 B.C. already had their own bronze weapons is unknown, but in the centuries after they occupied mainland Greece they steadily increased their bronzeworking skill, especially for warfare. As shown by ARCHAEOLOGY, bronze was providing the Greek MYCENAEAN CIVILIZATION with swords and spear-points before 1500 B.C., and by 1400 B.C. Mycenaean metalsmiths had mastered the technique of casting bronze plates, then hammering them out to make helmets and body armor. The Mycenaeans used bronze plowshares, sickles, ornaments, and vessels for drinking and cooking.

The potential disadvantage of bronze was that one of its two component metals, tin, could be difficult for the Greeks to acquire. Whereas the Greeks mined copper on the islands of EUBOEA and CYPRUS, tin had to be purchased through expensive, long-range TRADE with ASIA MINOR and western Europe. Tin was mined intensively by native peoples in what is now Cornwall, in southwest England; it then traveled overland—brought by non-Greek middlemen—to trade outlets on the Mediterranean.

The collapse of Mycenaen society and the invasion of mainland Greece by the DORIAN GREEKS (circa 1100–1000 B.C.) temporarily destroyed the tin routes. Compelled to find a replacement for bronze, the Greeks began mining and working IRON. Just as bronze had supplanted copper, so iron now replaced bronze in many objects, such as plowshares and sword blades. But bronze-working returned, alongside ironworking, once the tin supplies were renewed. In ancient technology, bronze was far more malleable than iron, and bronze remained essential wherever shaping or thinness was required.

During the great trading expansion of the 800s–500s B.C., the mainland Greeks imported and copied the artful bronzework of cauldrons, hand mirrors, and other artifacts from the Near East. Bronze helmets, breastplates, and shield facings were standard equipment for Greek HOPLITE armies (circa 700–300 B.C.). The foremost Greek city of about 700 B.C. was CHALCIS, whose name probably refers to bronze (Greek: *chalkos*). Now and later, bronze provided a favorite material for SCULPTURE; one technique was to pour the molten metal into a wax mold that had a clay or plaster core, to produce a hollow statue.

Tin supplies were always a concern. The lure of tin brought Greek traders into the western Mediterranean by 600 B.C., when the Greek colony of MASSALIA was founded. A major tin supplier was the Celtic kingdom of Tartessus, in southern Spain. (Perhaps none of this tin was mined locally; it may all have come from Cornwall.) Competition over metal supplies brought the western Greeks into conflict with similarly aggressive traders—the Carthaginians. Circa 500 B.C. the Carthaginians probably destroyed Tartessus, but by then the Greeks had found new tin suppliers, in ITALY and what is now France.

(See also BRONZE AGE; CARTHAGE; SINOPE; WARFARE, LAND.)

Bronze Age   Term used by modern archaeologists and historians to describe the phase of Asian and European human prehistory that falls roughly between 3500 and 1000 B.C. Coming after the Neolithic or New Stone Age, the Bronze Age is considered to have spread from the Near East to various other regions over several centuries. The era was marked by improved metallurgy that produced the alloy BRONZE as the prime substance for tools of war, agriculture, and industry. For the kings and lords who could produce or buy it, bronze replaced copper, obsidian, and flint. IRON—destined to replace bronze for many uses—was not yet in use.

The Bronze Age saw the birth of the earliest great civilizations of the Near East and the eastern Mediterranean. These include the Sumerian kingdom in Mesopotamia (which arose circa 2800 B.C.), the Egyptian Old Kingdom (circa 2660 B.C.), the MINOAN CIVILIZATION in CRETE (circa 2000 B.C.), the Hittite kingdom in ASIA MINOR (circa 1650 B.C.), and the MYCENAEAN CIVILIZATION of mainland Greece (circa 1600 B.C.).

By scholarly convention, the Bronze Age in mainland Greece is divided into three phases, called Early Helladic (circa 2900–1950 B.C.), Middle Helladic (circa 1950–1550 B.C.), and Late Helladic (circa 1550–1100 B.C.), with each phase subdivided into stages I, II, and III. Much of Early Helladic covers an epoch prior to the Greeks' appearance on the scene, when mainland Greece was still inhabited by a non-Greek people (who used bronze, although not extensively). The violent arrival of the first primitive Greeks is usually placed between Early Helladic II and III, circa 2100 B.C. The flowering of Greek Mycenaean culture corresponds to the Late Helladic era. The Mycenaeans' downfall marks the end of the Bronze Age and the advent of the DARK AGE in Greece.

Bucephalus   See ALEXANDER THE GREAT.

building   See ARCHITECTURE.

burial   See FUNERAL CUSTOMS.

Byblos   The northernmost of ancient PHOENICIA's three major seaports. (The other two were Sidon and Tyre.) Today the site of Byblos is about 26 miles north of Beirut, in Lebanon.

Called Gebal ("citadel") by the Phoenicians, Byblos was one of the earliest cities of the Near East. In the Phoenician heyday of the 900s–700s B.C., Byblos was the nation's capital, with a powerful navy and TRADE routes extending to Greece and EGYPT.

Contact between Greek and Bybline traders, at such ports as AL MINA, in Syria, and Citium, in CYPRUS, was a crucial factor in transmitting certain Near Eastern advantages to the emerging Greek culture, including the ALPHABET (circa 775 B.C.). Byblos at this time was the major export center for papyrus (primitive "paper" made from the pith of Egyptian water plants and used as a cheap substitute for animal skins in receiving WRITING). Byblos' monopoly in this trade is commemorated in the early Greek word for papyrus, *biblos*. That word, in turn, yielded Greek *biblion*, "book," and *bibliotheka*, "library," which survive in such familiar words as Bible, bibliophile, and French *bibliothèque* ("library").

Byblos' political fortunes in the era 700–300 B.C. followed those of greater Phoenicia, and the city quickly faded in importance after the founding of the nearby Seleucid-Greek city of ANTIOCH (300 B.C.).

(See also ADONIS.)

Byzantium   Celebrated Greek city on the European side of the southern mouth of the BOSPORUS channel. Byzantium is now the Turkish metropolis of Istanbul.

Founded in the mid-600s B.C. by Greeks from MEGARA (1), the city was called Buzantion in Greek. Supposedly it took its name from the colonists' leader, Buzas; more probably, it was the name of a preexisting settlement of native Thracians. The Greek city thrived amid Thracian tribes hungry for Greek goods.

Byzantium enjoyed a superb location on a peninsula jutting between the Bosporus mouth and the SEA OF

MARMARA. Alongside the peninsula's landward base lay the mouth of the Golden Horn River, providing rich fishing and access to the interior. More important, Byzantium controlled the Bosporus. Greek merchant ships—full of precious grain from the BLACK SEA coast, bound for cities of mainland Greece—would sail south through the channel. Byzantium's navy was able to impose tolls on this passing traffic.

Byzantium was held as a strategic point by successive imperial powers: PERSIA (circa 513–478 B.C.), ATHENS (478–404 B.C.), and ROME (mid-100s B.C. onward). As a member of the Athenian-dominated DELIAN LEAGUE in the 400s B.C., Byzantium played a vital role in Athens' control of the Black Sea grain route. We know that in 457 B.C. wealthy Byzantium was paying a relatively high yearly Delian tribute—15 TALENTS. The city revolted unsuccessfully against Athens twice (440 and 411 B.C.).

Passing to Spartan influence at the end of the PELOPON-NESIAN WAR (404 B.C.), the Byzantines soon became disenchanted with their new masters. By about 377 B.C. they were again Athenian allies. The Macedonian king PHILIP II unsuccessfully besieged Byzantium in the winter of 340–339 B.C.

Amid the wars of the DIADOCHI in the late 300s B.C., Byzantium was able to keep its independence. The city suffered attacks from the CELTS, who invaded much of the Greek world in the early 200s B.C., and eventually Byzantium seems to have paid ransom to keep them away. The city went to war in 220 B.C. against the emerging naval power of RHODES over the issue of Byzantium's Bosporus tolls.

In A.D. 330 Byzantium became a foremost city of the world when the Roman emperor Constantine made it the eastern capital of his empire and renamed it Constantinople.

(See also THRACE.)

C

Cadmus In Greek MYTH, Cadmus (Greek: Kadmos) was a prince of the Phoenician city of Tyre and founder of the Greek city of THEBES. Young Cadmus' sister at Tyre was EUROPA, whom the god ZEUS abducted in the shape of a bull. Cadmus was assigned by his father, King Agenor, to find the vanished Europa. Leading a band of men to central Greece, he consulted the god APOLLO through the oracle at DELPHI. Apollo advised Cadmus to abandon the search and instead follow a cow that he would find outside the temple; he should establish a city wherever the cow lay down to rest. Accordingly, the cow led Cadmus to the future site of Thebes, about 50 miles away. There Cadmus built the Cadmea, which became the citadel of the later city.

Several adventures accompanied this foundation. To gain access to the local water supply, Cadmus had to slay a ferocious serpent, the offspring of the war god ARES. When Cadmus consecrated the dead monster to the goddess ATHENA, she appeared and told him to sow the serpent's teeth in the soil. Immediately there sprang up a harvest of armed men to oppose him, but Cadmus cleverly threw a stone into their midst, thereby setting them to fight one another. The surviving five of these warriors joined Cadmus' service and became the ancestors of the Theban nobility, known as the Sown Men.

After enduring eight years' servitude to Ares in expiation for having killed his serpent son, Cadmus was allowed to marry Harmonia, daughter of Ares and APHRODITE. As a wedding gift, Harmonia received a wondrous necklace that bestowed irresistible attraction on its wearer; this necklace would play a role in the next generations' misfortunes, in the adventure of the SEVEN AGAINST THEBES.

Reigning as king of Thebes, Cadmus civilized the crude local Greeks by teaching them how to write. He and Harmonia had four daughters, all of whom suffered unhappy fates: Semele (later the mother of the god DIONYSUS); Ino (who was driven mad by the goddess HERA); Agave (whose son PENTHEUS would be destroyed by Dionysus); and Autonoë (whose son ACTAEON would be destroyed by the goddess ARTEMIS). In old age, Cadmus and Harmonia emigrated to the northwest, where they ruled over the Illyrians and eventually were changed into serpents.

The Cadmus legend—a Phoenician prince civilizing and ruling part of Greece—presents an odd combination. Although the Greeks had extensive contact with the Phoenicians in the 900s–700s B.C., this myth seems to commemorate an earlier period of Greek contact with the Near East, possibly during the Mycenaean era (circa 1200 B.C.).

(See also ALPHABET; ILLYRIS; MYCENAEAN CIVILIZATION; PHOENICIA.)

Callias Athenian nobleman and diplomat, active in the early and mid-400s B.C. Callias' family was the richest in ATHENS, renting out slave labor to the state SILVER mine at Laurium. Despite his upper-class background, Callias became a political follower of the radical democrat PERICLES. During Pericles' preeminence, Callias was the foremost diplomat for Athens.

Callias made at least one embassy to King Artaxerxes I of PERSIA, circa 461 B.C. Most modern scholars accept the theory, previously disputed, that circa 449 B.C. Callias negotiated an end to the Greek-Persian hostilities known as the PERSIAN WARS. Apparently the Peace of Callias in part protected the Athenian-allied Greek cities of western ASIA MINOR, prohibiting the Persians from sailing or marching west past certain set boundaries. For their part, the Athenians may have agreed to dismantle the fortifications of the Asia Minor Greek cities.

(See also DELIAN LEAGUE; IONIA; SLAVES.)

Callimachus The most admired lyric poet of the HELLENISTIC AGE. Born in the Greek city of CYRENE (1) in North Africa, Callimachus (Greek: Kallimachos) lived circa 310–240 B.C. and wrote at Egyptian ALEXANDRIA (1) during the brief literary golden age under King Ptolemy II (reigned 285–247 B.C.). Of Callimachus' voluminous output—a reported 800 scrolls' worth—only six hymns and some 60 epigrams survive whole. Several dozen fragments of other poems exist, and more of his verses are being discovered in the sands of Egypt. (One sizable fragment was published from papyrus in 1977.) Callimachus' best-known work, now lost, was the *Aetia* (Origins), a narrative elegy of about 7,000 lines; this erudite and digressive poem presented MYTHS and descriptions explaining the origins of places, rites, and names throughout the Greek world.

Callimachus was the Alexandrian poet par excellence: witty, scholarly, and favoring brief forms and cerebral topics. He was an "in" poet, composing for an "in" group of sophisticated readers or, more accurately, *listeners*, to whom he would read aloud. His work remained immensely popular in later literary circles; 200 years after his death, his style and values were influencing such Roman poets as Catullus, Propertius, and Ovid. Callimachus also is remembered for his bitter rivalry with his former pupil and intimate, the poet APOLLONIUS. Their dispute, based partly on differing poetic tastes, was the most famous literary quarrel of the ancient world.

51

Emigrating from Cyrene to EGYPT, the young Callimachus began his career as a schoolteacher in a suburb of Alexandria. Eventually attracting royal patronage, he was installed at Alexandria's famous Library, with the huge job of cataloguing the several hundred thousand books there. His resultant catalogue supposedly ran to 120 volumes and must have taken 10 years; it would have included lists of titles, biographical sketches of authors, and literary criticism. Apparently it was an admirable piece of scholarship, and it testifies to the scholarly, cataloguing urge that also infused his poetry.

Callimachus never reattained the prestigious post of director of the Library, but rather was passed over in favor of Apollonius after the retirement of the director Zenodotus. This development sparked or fueled the two men's quarrel. The enmity was part of a larger literary dispute between the writers of lengthy, narrative EPIC POETRY, based on HOMER's *Iliad* and *Odyssey*, and poets such as Callimachus, who considered epic irrelevant to modern society. "A big book is a big evil," Callimachus wrote. Although he did compose certain lengthy poems toward the end of his life, he seems to have objected specifically to the continuous plot and familiar subject matter of the Homeric-style epic; rather, Callimachus' longer poems were innovative and episodic.

Callimachus' epigrams have been praised for their sincere emotion and their charming word use; they include epitaphs and expressions of sexual desire. The most admired epigram poignantly describes the writer's feelings on learning of the death of his fellow poet Heraclitus. Of Callimachus' five surviving hymns, Hymn 1 describes the mythical birth and rearing of the god ZEUS. Famous in antiquity, this Hymn to Zeus set a standard of court poetry by drawing subtle parallels between the king of the gods and King Ptolemy II.

No information exists regarding a wife or children for Callimachus. He seems to have been one of those strictly homosexual Greek men who shunned MARRIAGE. Of his surviving epigrams, every one of the erotic poems celebrates the charms of boys. He lived to about age 70, highly esteemed and enjoying the patronage of Ptolemies; certain of his verses, such as his poem on Queen Arsinoë's death (270 B.C.), sound like the public presentations of a court poet laureate.

(See also HOMOSEXUALITY; LYRIC POETRY.)

Callinus One of the earliest Greek lyric poets whose work (in part) has come down to us. Callinus lived in the mid-600s B.C. in the Greek city of EPHESUS in ASIA MINOR, and wrote patriotic verses encouraging his countrymen in their defense against the Cimmerians, a nomadic people from southwestern Russia who were ravaging Asia Minor in those years. In the single substantial fragment by him that survives, Callinus reminds his audience how honorable it is to fight for city and family and how death finds everyone eventually.

Callinus is the first known writer of the verse form known as the elegy, which was intended for recital to flute accompaniment.

(See also LYRIC POETRY, MUSIC.)

Calydonian Boar Hunt One of many Greek MYTHS recounting the destruction of a local monster. King Oeneus of Calydon (a region in AETOLIA) offended the goddess ARTEMIS by forgetting to sacrifice to her. In retaliation, she sent a monstrous boar to ravage the countryside. To hunt the beast, Oeneus' son Meleager collected a band of heroes, including the virgin huntress ATALANTA, with whom Meleager was in love. After much effort, the heroes succeeded in killing the creature, the honor of the first spear thrust going to Atalanta. At Meleager's insistence, she then received the edible portions of the boar. But this insulted Meleager's maternal uncles, who were part of the hunt, and in the ensuing fight, Meleager slew them.

Upon hearing the news, Meleager's mother, Althaea, took vengeance on her son. She had in her possession a half-burned log, with the following significance: years before, just after Meleager's birth, Althaea had been visited by the three goddesses of FATE, the Moirai, who informed her that her baby son would live only until the log then on the fire should be burned away. Subverting the prophecy, Althaea had quenched the fire and preserved the half-burned log in a chest. Now, in anger at her brothers' murder, she threw the log on the fire, and Meleager died.

The boar hunt was a favorite subject in vase painting and other artwork. Among surviving representations is a panel on the Athenian black-figure François vase (circa 570 B.C.).

(See also POTTERY.)

Calypso A beautiful nymph who played a role in the MYTH of ODYSSEUS, as told in HOMER's epic poem the *Odyssey* (book 5). The name Kalupso means "concealer." A daughter of ATLAS and the sea goddess Thetis, she lived alone on a remote island, where she received the shipwrecked Odysseus. As Odysseus' lover, she detained him for seven years, promising to make him immortal if he stayed; but he insisted on returning to his wife and kingdom. At ZEUS' command, Calypso helped him build a boat on which to put to sea.

In the structure of the *Odyssey*, Calypso supplies a benevolent doublet to the sinister CIRCE (who also detains Odysseus seductively, for one year). Odysseus' seven years with Calypso can be interpreted as representing the distractions of pleasure versus the duties of leadership.

(See also NYMPHS; WOMEN.)

Cappadocia See ASIA MINOR.

Caria See ASIA MINOR.

Carthage Major non-Greek city of North Africa, located about 10 miles from the modern city of Tunis. Carthage was founded by Phoenician settlers from Tyre circa 800 B.C. to be an anchorage and trading post for Phoenician merchant ships in the western Mediterranean. The Phoenician name Kart Hadasht means "New City"; the Greeks called it Karchedon and the Romans, Carthago.

Governed as an OLIGARCHY under two presiding officials called shophets, Carthage thrived by commerce. It became a foremost power of the ancient world, with a feared navy to protect its trading monopolies in the western

Mediterranean. The Carthaginians maintained trading bases such as Gades and Malaca (modern Cadiz and Malaga, both in southern Spain), from which they acquired valuable raw metals. The Carthaginians traded with the Greeks, as indicated by the troves of Corinthian POTTERY discovered at Carthage by modern archaeologists. But Carthage is important in Greek history mainly as an enemy and occasional overlord of the western Greeks, particularly in SICILY.

Greeks and Carthaginians first came into conflict when the Greeks intruded into Carthaginian trading waters. Circa 600 B.C. a Carthaginian fleet was defeated by Phocaean Greeks in the vicinity of MASSALIA (modern-day Marseille, in southern France). About 60 years later a joint fleet of Carthaginians and ETRUSCANS successfully drove the Greeks away from Corsica. Carthage also dominated Sardinia and most of western Sicily.

For the next 300 years, Greeks and Carthaginians fought for control of Sicily, whose coast lies about 130 miles from Carthage. Carthage's greed for Sicily eventually brought on conflict with the rising Italian city of ROME. The resulting three Punic Wars (264–241 B.C., 218–201 B.C., and 149–146 B.C.) ended in the complete destruction of Carthage by the Romans.

Prosperous Carthage won the admiration of the philosopher ARISTOTLE (mid-300s B.C.), whose treatise *Politics* contains a remarkable section that favorably compares the Carthaginian government with that of certain Greek cities. But another Greek writer describes—and ARCHAEOLOGY confirms—the gruesome Carthaginian custom of sacrificing upper-class children in time of public crisis, to ensure the god Baal-Ammon's care of the city.

(See also AGATHOCLES; BRONZE; DIONYSIUS (1); GELON; HIMERA; PHOCAEA; PHOENICIA; SILVER; TRADE.)

caryatid The Greeks' name for their type of decorative pillar represented as a clothed woman, holding up a ceiling structure with her head. The most famous caryatids are the six in marble from the south porch of the Athenian Erectheum (built 421–406 B.C.); the best preserved one of these is now in the British Museum in London.

Supposedly the name caryatid referred to the region of Caryae in LACONIA, where WOMEN traditionally danced with baskets on their heads. In any case, the sight of women carrying water pitchers or laundry baskets in this way was a familiar one in the ancient Greek world and certainly contributed to the design. The male equivalents of caryatids were called Atlantes, from the mythological figure ATLAS, who held up the heavens with his head.

(See also ARCHITECTURE.)

Cassander Macedonian general and ruler, one of the several DIADOCHI (Successors) who carved up the empire of ALEXANDER THE GREAT after the latter's death in 323 B.C. Born circa 360 B.C., Cassander (Greek: Kassandros) was son of the Macedonian general ANTIPATER, who served as regent in MACEDON during Alexander's eastward campaigns (334–323 B.C.). Cassander joined Alexander's army in Asia in 324 B.C. He and the king seem to have disliked each other bitterly, and Cassander is mentioned by ancient writers as a suspect in theories that Alexander died from poisoning. After Antipater's death (319 B.C.), Cassander seized control of Macedon and most of Greece. He executed Alexander's mother, Olympias, Alexander's widow, Roxane, and Alexander's young son, Alexander III.

In the following years Cassander joined the other secessionist Diadochi—SELEUCUS (1), PTOLEMY (1), and LYSIMACHUS—in resisting the efforts of ANTIGONUS (1) to reunite Alexander's empire. Cassander died circa 297 B.C., and his dynasty in Macedon lasted only a few years before falling to the descendants of Antigonus.

Cassandra In MYTH, a daughter of King PRIAM of TROY. She is mentioned in HOMER's *Iliad* simply as the loveliest of Priam's daughters, but later writers—particularly the Athenian playwright AESCHYLUS, in his *Agamemnon* (458 B.C.)—present her in pathetic terms. Cassandra was beloved by the god APOLLO, who wooed her by offering her the power of prophecy. She accepted the gift but still refused his advances. Apollo could not revoke his gift, but he decreed vindictively that no one would ever believe her predictions.

In the TROJAN WAR, the Trojans ignored Cassandra's frantic plea not to take the Greeks' giant wooden horse into the city. When the Greeks emerged from the horse to sack Troy, Cassandra was dragged away from the altar of the goddess ATHENA and raped by AJAX (2) the Locrian. Afterward, Cassandra was allotted as concubine to the Greek grand marshal, King AGAMEMNON. On her arrival with Agamemnon at his kingdom of MYCENAE, Cassandra and Agamemnon were murdered by Agamemnon's wife, CLYTAEMNESTRA. Naturally, Cassandra foresaw her own and Agamemnon's death.

Our modern expression "to be a Cassandra" or "to play Cassandra" is applied to someone who habitually predicts that bad things will happen.

(See also PROPHECY AND DIVINATION.)

Castor and Polydeuces Mythical twin Spartan heroes, worshipped as gods. The cult of Castor and Polydeuces was important at SPARTA and other Dorian-Greek cities, and was an early cultural export to the Romans, who latinized the youths' names to Castor and Pollux.

The distinction between mortal hero and immortal god played a central role in the MYTH. According to the story, the Spartan king Tyndareus had a wife, LEDA, whom the god ZEUS raped or seduced. Leda bore two pairs of twins: the girls CLYTAEMNESTRA and Helen (later known as HELEN OF TROY) and the boys Castor and Polydeuces. One version says that both boys were the immortal sons of Zeus—hence they were called Dioscuri, or Dios kouroi, "the youths of Zeus." But a different tradition says that only Polydeuces was Zeus' child and thus immortal, while Castor, fathered by Tyndareus, was doomed to die like other humans.

The boys grew up inseparable and devoted to each other. Polydeuces became a champion boxer, Castor a famous horseman. They had three major adventures: they raided Athenian territory to rescue their sister Helen after she was kidnapped by the Athenian king THESEUS, sailed with the Argonauts under the Thessalian hero JASON (1) to capture the Golden Fleece, and abducted the two daughters of the Messenian nobleman Leucippus, who sent in pursuit his nephews Idas and Lynceus (to whom the young women had been betrothed). In the ensuing fight, Idas, Lynceus, and Castor were all killed. But Polydeuces prayed to his father, Zeus, who resurrected Castor on the condition that the twins thereafter divide their lot, living together one day among the gods and the next day in the Underworld. HOMER's epic poem the *Odyssey* mentions that "the grain-giving earth holds them, yet they live . . . One day they are alive, one day dead" (book 11).

Like HERACLES (another hero turned god), the Dioscuri were thought to be sympathetic to human needs and were worshipped widely by the common people. As patron gods of seafarers, they averted shipwreck, and their benevolent presence supposedly was indicated by the electrical phenomenon that we call St. Elmo's fire (whereby the mast and rigging of a sailing ship would seem to sparkle during a thunderstorm). They also were identified with the constellation known as the Twins (Gemini).

Often pictured as riding on white horses, the Dioscuri embodied the spirit of military youth. At the Battle of the Sagras River (late 500s B.C.), the divine twins were believed to have appeared in person to aid the army of LOCRI (a Dorian-Greek city in south ITALY) against the army of CROTON (a nearby Achaean-Greek city). This miracle was copied by the Romans, who claimed that the Dioscuri helped them to win the Battle of Lake Regillus, against the Latins (circa 496 B.C.).

(See also AFTERLIFE; BOXING; DORIAN GREEKS; ROME; SHIPS AND SEAFARING.)

Catana Greek city on the east coast of SICILY, at the foot of volcanic Mt. Etna. Set at the northeastern rim of a large and fertile plain, this site is now the Sicilian city of Catania. The locale was seized circa 729 B.C. by Greek colonists from the nearby city of NAXOS (2), who drove away the region's native Sicels. This attack was part of the Greeks' two-pronged capture of the plain; the other captured site was Leontini.

Catana (Greek: Katanē) possessed one of the Greek world's first law codes, drawn up by a certain Charondas, probably in the early 500s B.C. As an Ionian-Greek city, Catana took part in ethnic feuding between Ionians and DORIAN GREEKS in Sicily. By 490 B.C. the city had fallen under the sway of its powerful Dorian neighbor to the south, SYRACUSE. In 476 B.C. Syracusan tyrant HIERON (1) emptied Catana, exiling its inhabitants to Leontini, and repopulated the site with his Greek mercenary troops. The "new" city, renamed Aetna, was celebrated by the visiting Athenian playwright AESCHYLUS in a play (now lost), titled *Women of Aetna*. But after Hieron's death, the former Catanans recovered their city by force and gained independence (461 B.C.).

Catana provided the Athenians with a base against Syracuse in 415 B.C., during the PELOPONNESIAN WAR. In 403 B.C. Catana was recaptured by the Syracusans under their tyrant DIONYSIUS (1). For the next 150 years, despite moments of independence provided by such saviors as the Corinthian commander TIMOLEON (339 B.C.) and the Epirote king PYRRHUS (278 B.C.), Catana remained a Syracusan possession. Seized by the Roman in 263 B.C., during the First Punic War, Catana become an important city of the Roman Empire.

(See also COLONIZATION; IONIAN GREEKS; LAWS AND LAW COURTS; ROME.)

cavalry Ancient Greek warfare emphasized the foot soldier over the horse soldier. It took the tactical genius of the Macedonian kings PHILIP II and his son ALEXANDER THE GREAT (mid-300s B.C.) to raise cavalry to even a prominent secondary position. One reason for cavalry's inferior status lay in the mountainous terrain of mainland Greece, which was resistant to the strategic movement of horsemen and to horse-breeding itself; in most regions, only the very rich could afford to raise horses.

During the 700s to 300s B.C., when Greek citizen-soldiers supplied their own equipment, a city's cavalry typically consisted of rich men and their sons. Cavalry contingents were therefore small in most Greek armies. Only on the horse-breeding plains of THESSALY, BOEOTIA, and Greek SICILY did large cavalry corps develop.

In those days, cavalry was not so effective a "shock" troop as it would become in later centuries. The Greeks knew nothing of the stirrup—a vital military invention that enables a rider to "stand up" in the saddle and lean forward strenuously without falling off. (The stirrup probably came out of Siberia circa A.D. 550.) Nor had the horseshoe or the jointed bit yet been invented. The Greeks knew only small breeds of horses prior to the late 300s B.C. (when Alexander's conquests introduced larger breeds from the Iranian plateau). Therefore, in battle, cavalry was not usually strong enough to ride directly against formations of infantry. The juggernaut charges of medieval Europe's armored knights were still 1,500 years in the future.

Ancient Greek artwork and certain writings—such as the historian XENOPHON's treatise *On Horsemanship* (circa 380 B.C.)—suggest what a horse soldier looked like. He might wear a corset of linen or leather, with a BRONZE helmet (open-faced, to leave his vision clear). He probably carried no shield, or at most a small wooden targe attached to his left forearm. The Greek cavalry weapon was a spear for jabbing (not usually for throwing); unlike the long lance of a medieval knight, this spear was only about six feet long.

Cavalry in action against infantry, in a painted scene on a vase from the Greek city of Gela, in Sicily, circa 490 B.C. The horsemen are perhaps overtaking an enemy in retreat.

The Athenian cavalry received its basic training in organizations of cadets—wealthy young men who supplied their own horses and who drilled at local riding tracks and in countryside maneuvers. This vivid marble carving from the Parthenon frieze, circa 432 B.C., shows Athenian cadets riding in procession to the acropolis, at a sacred festival of the goddess Athena.

We know from battle scenes in Greek art that the horseman used his spear for a downward thrust, often overhand from the shoulder: the spear was not couched in the armpit for the charge, since the impact probably would have knocked the stirrupless rider off his horse. If the spear was lost, the horseman would rely on the IRON sword tied into a scabbard at his waist. As in other eras of military history, the preferred cavalry sword was a saber—that is, it had a curved cutting edge, designed to slash downward rather than to stab.

Horsemen of the 500s and 400s B.C. were needed for scouting and supply escort, and in combat they had the job of guarding the vulnerable infantry flanks. When one side's infantry formation began to dissolve into retreat, the cavalry of both sides might have crucial roles to play, either in running down the fleeing enemy, or (on the other side) in protecting the foot soldiers' retreat. Descriptions of ancient battles make clear that a disorderly retreat could become a catastrophe once the withdrawing infantry was overtaken by enemy horsemen. Among such examples is the plight of Athenian HOPLITES pursued by Boeotian cavalry after the Battle of Delium (424 B.C.), during the PELOPONNESIAN WAR.

The kingdom of MACEDON had a strong cavalry in its horse-breeding barons, whom King Alexander I (circa 480 B.C.) organized into a prestigious corps of King's Companions. The subsequent innovations of Philip and Alexander brought cavalry into the heart of battle. Cavalry became the offensive arm, to complement the more defensive role of the heavy-infantry PHALANX. The Macedonian phalanx would stop the enemy infantry attack and rip gaps in its battle order, and the cavalry would then attack these vulnerable gaps before the enemy could reorganize. In this case cavalry could charge against massed infantry, since the charge was directed not against the enemy's waiting spear-points but against open ground. Into such a gap Alexander led the Macedonian cavalry at the battles of CHAERONEA (338 B.C.) and Gaugamela (331 B.C.). Prior to Alexander's conquests, the Persians possessed the most numerous and formidable cavalry known to the Greeks. All noble-born Persian boys were taught to ride and shoot with the bow.

The HELLENISTIC AGE (300–150 B.C.) saw Greek kings in the eastern Mediterranean experimenting with such Persian-influenced cavalry as mounted archers and javelinmen, and horse and riders protected by chain-mail armor and known as *kataphraktai*, "enclosed ones." These cataphracts pointed the way toward the Parthian heavy cavalry of the Roman era and the armored knights of the Middle Ages.

(See also PERSIA; WARFARE, LAND.)

Celts   Race of people speaking an Indo-European language who emerged from central Europe in a series of invasions after about 750 B.C. Among the places they occupied were France, Spain, and Britain. Today Celtic languages and culture still exist in Ireland, Scotland, Wales, and Britanny (in France), and the Celts are commemorated in place names such as Galicia (in modern Spain) and Gaul (Gallia, the Roman name for ancient France). The Celts (Greek: Keltoi) were known for their physical size and beauty, their natural spirituality, and their undisciplined fierceness in battle. The distinctive Celtic armament was the long, oblong shield.

Celtic society was tribal and agricultural, with walled towns for commerce and defense, but few cities. Although not a literary people (at that time), they developed an admirable material culture, especially in metallurgy. Early Celtic culture reached maturity circa 400–100 B.C., in the La Tène style of art and metal design (named for an archaeological site in Switzerland).

By 600 B.C. Greek merchants in the western Mediterranean were dealing with Celtic peoples in Spain and France, to acquire metals such as tin and SILVER. The Spanish Celtic kingdom of Tartessus had friendly dealing with Greek merchants from PHOCAEA. From the Phocaean colony at MASSALIA (modern Marseille), Greek goods and culture slowly spread inland among the Celts of Gaul, and prompted the subsequent emergence of the La Tène civilization.

But the Greeks and Celts collided in the 200s B.C., when Celtic tribes descended the Danube, invaded mainland Greece from the north, and overran the sanctuary at DELPHI (279 B.C.). Another column of invaders crossed the HELLESPONT to attack ASIA MINOR, where they were contained after being defeated in battle by the Seleucid king Antiochus I (circa 273 B.C.). The valor and brutishness of these invaders is portrayed in Greek statues like the Dying Gaul, commissioned by the people of PERGAMUM and known to us from later Roman copies.

Settling in north-central Asia Minor, immigrant Celts formed the kingdom of Galatia. Amid the neighboring Greeks and Phrygians, the Galatians maintained Celtic customs and language for centuries. In the first century

A.D. Galatia contained an early Christian community, and these Galatians are commemorated in the New Testament, as recipients of a letter from St. Paul.

(See also BRONZE; CARTHAGE; TRADE; WINE.)

centaurs Legendary tribe of creatures who were part human, part horse. A centaur was usually imagined as having a male horse's body with a male human torso and head emerging above the horse's chest. This shape is portrayed on the earliest surviving likeness of a centaur—a baked-clay figurine from the 900s B.C., decorated in Geometric style, discovered at LEFKANDI, in central Greece. However, certain other early images in art show human forelegs, with a horse's body and rear legs. The original meaning of the Greek word *kentauros* is not clear to us.

In Greek MYTH, centaurs had occupied the wild regions of THESSALY and ARCADIA in the old days; they represented the uncivilized life, before the general establishment of Greek laws and city-states. Although capable of wisdom and nobility, they were fierce, oversexed, and prone to drunkenness. The best-known myth about the centaurs, mentioned in HOMER's *Iliad* and *Odyssey*, is their battle with the Lapiths, a human tribe of Thessaly. The gathering began as a friendly banquet to celebrate the wedding of the Lapith king Pirithous, but the centaurs got drunk on WINE and tried to rape the Lapith women. In the ensuing brawl the centaurs were defeated. As a symbol of savagery versus civilization, this battle was a favorite subject of Greek art. It appeared, for instance, in the architectural SCULPTURES on the Athenian PARTHENON and on the Temple of ZEUS at OLYMPIA (mid-400s B.C.).

Another myth tells how a centaur named Nessus offered to carry HERACLES' second wife, Deianira, across a river but then tried to rape her. Heracles shot the creature with poisoned arrows, but before Nessus died he gave to Deianira the poisoned blood-soaked garment that would later cause Heracles' death. A more benevolent centaur was Chiron, or Cheiron, the wise mountain-dweller who served as tutor to heroes, including JASON, ASCLEPIUS, and ACHILLES.

(See also AMAZONS; GIANTS.)

Cerberus In MYTH, a monstrous dog that guarded the inner bank of the River Styx, at the entrance to the Underworld. Cerberus (Greek: Kerberos) had 50 heads, according to HESIOD's epic poem the *Theogony*. A later, more familiar version gave him three heads and outgrowths of snakes. Cerberus would fawn on the ghosts arriving at the infernal kingdom, but became vicious toward anyone who tried to leave.

Cerberus was the target of HERACLES' eleventh Labor. The hero visited the Underworld and won permission from the god HADES to bring the dog temporarily to the upper world. Dragging Cerberus to the city of TIRYNS, Heracles mischievously frightened his taskmaster, King Eurystheus. This was a favorite scene in ancient Greek artwork. One famous Athenian black-figure vase (530s B.C.), now in the Louvre, shows the timid king hiding inside a storage jar as the hound of Hell is led in.

(See also AFTERLIFE.)

Chaeronea Northernmost town of BOEOTIA, located in the Cephissus River valley, on the main route between central and northern Greece. There in the summer of 338 B.C. the Macedonian king PHILIP II defeated an allied Greek army, to make Greece a subject state of MACEDON.

The decisive Battle of Chaeronea was one of the most consequential fights of ancient history—not only for politics, but for the development of battlefield tactics. The Greek army, comprised mainly of contingents from Boeotia and ATHENS, had a slight advantage in numbers: 35,000 HOPLITES (heavily armed infantry) against Philip's 30,000, with about 2,000 CAVALRY on each side. The allied Greeks comfortably guarded the route southward—their battle line filled the valley side to side, from the town's citadel on their left, to the river on their right. The Athenian hoplites, 10,000 strong, occupied the army's left wing; in the center were various levies and a company of 5,000 mercenaries; and on the right wing were 12,000 Boeotians, including the men of THEBES, the best soldiers in Greece. The position of honor, on the extreme right wing, was given to the Theban Sacred Band, an elite battalion of 300, consisting entirely of paired male lovers.

Philip, on horseback behind his army's right wing, brought his men southward through the valley in an unusual, slantwise formation, which was destined to become a model Macedonian tactic. The Macedonian PHALANX was angled so that the right wing advanced ahead of the center and the center ahead of the left wing. Holding back on the Macedonian left wing was the cavalry, commanded by Philip's 18-year-son, Prince Alexander (later known to history as ALEXANDER THE GREAT).

Philip's battlefield plan was a refinement of tactics used by the Thebans themselves at LEUCTRA more than 30 years before. Like the Thebans at Leuctra, Philip planned to hit his enemies at their strongest point: their right wing. But Philip's inspired innovation was to precede this blow with a disruptive feint against the enemy's left wing in the hope of creating a gap along the long line of massed Greek soldiery.

According to a statement by a later ancient writer, it is possible that Philip staged a retreat of his right wing during the battle. Presumably the Macedonians withdrew by stepping backward in good order, with their 13-foot-long pikes still facing forward. The disorganized Athenians followed, in deluded triumph. But in fact, the overexcited Athenians were drawing their army's left wing forward, with the Greek troops in the center following suit. Eventually gaps appeared in the Greek battle line, as various contingents lost contact with one another. On the Greek far right, the Theban Sacred Band stood isolated from the rest of the army. It was then that Prince Alexander led his cavalry charging down the valley, followed by the reserve Macedonian infantry.

The Sacred Band was surrounded and overrun by Alexander's cavalry. Meanwhile Philip, off on the Macedonian right, ordered an end to his false retreat. His men pressed forward against the Athenians, who scattered and fled. The Macedonians pursued, killing 1,000 Athenians and taking 2,000 prisoners. Among the fugitives was DEMOSTHENES (1), the great Athenian orator and enemy of Philip's.

After the battle, the corpses of the Sacred Band lay in the serried ranks of their disciplined formation. Of their 300 men, only 46 had survived. The rest were buried on the battlefield, where their 254 skeletons were discovered by archaeologists in the 20th century. Today the site is marked by a marble lion, sculpted in ancient times, overlooking the Sacred Band's graves and the burial mound of the Macedonian dead.

(See also HOMOSEXUALITY; WARFARE, LAND.)

Chalcedon   See BYZANTIUM.

Chalcidicē   Northwestern coast of the AEGEAN SEA, located in what eventually became Macedonian territory, north of Greece proper. Chalcidicē is distinguished by three peninsulas, each about 30 miles long, jutting into the Aegean and providing harbors and natural defense. These three, west to east, were called Pallene, Sithonia, and Acte (meaning "promontory" in Greek).

Originally inhabited by Thracians, the region was colonized in the latter 700s B.C. by Greeks from the city of CHALCIS; they established about 30 settlements and gave the region its Greek name, "Chalcidian land." Among later Greek arrivals were Corinthians who, circa 600 B.C., founded the important city of POTIDAEA on the narrow neck of the Acte peninsula. Another prominent city was OLYNTHUS, founded in the 400s B.C., north of Potidaea.

Chalcidicē offered precious TIMBER, SILVER ore, and SLAVES, and controlled the coastal shipping route to the HELLESPONT. The prosperous Chalcidic towns had become tribute-paying members of the Athenian-dominated DELIAN LEAGUE by the mid-400s B.C., but in the spring of 432 B.C. Potidaea revolted unsuccessfully against the Athenians. More revolts followed the arrival of the Spartan general BRASIDAS in 432 B.C., during the PELOPONNESIAN WAR. The town of Scione, recaptured by the Athenians in 421 B.C., was treated with exemplary cruelty: All men of military age were killed and the WOMEN and children were sold as slaves.

Chalcidicē endured a generation of Spartan rule after SPARTA's victory in the Peloponnesian War (404 B.C.), yet meanwhile the Chalcidic towns organized themselves into a federation, with a shared citizenship and government and with Potidaea as the capital (circa 400 B.C.). In the 370s B.C. this Chalcidic League joined the SECOND ATHENIAN LEAGUE. But renewed Athenian interference with Potidaea drove the Chalcidic Greeks to make an alliance with the ambitious Macedonian king PHILIP II (356 B.C.). Resistance to Philip led to war, in which Philip captured and destroyed Olynthus and Potidaea (348 B.C.). Henceforth the region was ruled by MACEDON.

Chalcidicē revived under the Macedonian king CASSANDER (reigned 316–298 B.C.). He built a grand new city, Cassandreia, on the site of the ruined Potidaea.

(See also ARISTOTLE.)

Chalcis   Important city midway along the west coast of the large inshore island called EUBOEA, in east-central Greece. Chalcis was strategically located on the narrow Euripus channel, so as to control all shipping through the Euboean Straits. Inland and southward, the city enjoyed the fertile plain of Lelanton.

Inhabited by Greeks of the Ionian ethnic group, Chalcis and its neighbor ERETRIA had emerged by about 850 B.C. as the most powerful cities of early Greece. Chalcis—its name refers to local copper deposits or to worked BRONZE (both: *chalkos*)—thrived as a manufacturing center, and its drive for raw metals and other goods placed it at the forefront of Greek overseas TRADE and COLONIZATION circa 800–650 B.C. Seafarers from Chalcis and Eretria established the first Greek trading depot that we know of, at AL MINA on the north Levantine coast (circa 800 B.C.). A generation later, in western ITALY, they founded the early Greek colonies PITHECUSAE and CUMAE.

Around 735 B.C. the partnership of Chalcis and Eretria ended in conflict over possession of the Lelantine plain. Chalcis apparently won this LELANTINE WAR (by 680 B.C.), and the city and its new ally CORINTH then dominated all Greek westward colonization. Among the westward colonies founded by Chalcis in this era were RHEGIUM, in southern Italy, and NAXOS (2) and ZANCLĒ, in SICILY. In the late 700s B.C., Chalcis also sent colonists to the north Aegean region eventually known as CHALCIDICĒ (Chalcidian land). But by the mid-600s B.C. Chalcis was being eclipsed in commerce by Corinth.

According to tradition, Chalcis' last king was killed in the Lelantine War, after which the city was governed as an ARISTOCRACY led by the *hippobotai* (horse-owners), Chalcidian nobles. In 506 B.C. these aristocrats, with help from nearby BOEOTIA, made war on ATHENS and its newly democratic government. Chalcis was completely defeated, and a portion of the *hippobotai*'s land was confiscated for an Athenian garrison colony.

Chalcis contributed 20 warships to the defense of Greece in the PERSIAN WARS (480–479 B.C.). Soon after, Chalcis and all the other Euboean cities were compelled to join the Athenian-led DELIAN LEAGUE. In 446 B.C. Chalcis led a Euboean revolt from the league, but Athens crushed this harshly, exiling the *hippobotai*. A successful revolt followed later, amid the Delian uprisings of the later PELOPONNESIAN WAR (411 B.C.). However, after a bitter taste of Spartan overlordship, Chalcis joined the SECOND ATHENIAN LEAGUE (378 B.C.).

The Macedonian king PHILIP II conquered the Greeks in 338 B.C. and placed a garrison in Chalcis. For the next 140 years or more, Chalcis served as a Macedonian stronghold—one of the Macedonians' four "fetters" of Greece.

In 194 B.C. the Romans ejected the Macedonian garrison; however, as a member of the anti-Roman Achaean League, Chalcis was besieged and captured by the Romans in 146 B.C. The city recovered, to play a role in the empire of ROME.

(See also ACHAEA; ALPHABET; GREECE, GEOGRAPHY OF; IONIAN GREEKS; MACEDON.)

chariots   The chariot—an axled, two-wheeled vehicle typically pulled by horses—played a small but picturesque role in Greek history. Circa 1600–1200 B.C. the kings of the MYCENAEAN CIVILZATION kept fleets of war chariots, as indicated by archaeological evidence (including an extant Mycenaean tomb-carving and texts of LINEAR B tablets listing inventories of chariots and chariot parts). The Mycenaean chariot probably was made of wood plated with BRONZE, with upright sides and front. It was drawn by two horses and carried two soldiers—the driver and a passenger armed with spears or arrows.

Although the chariot would have served well for display and for sportsmanlike Mycenaean battles on the plains of ARGOS, BOEOTIA, and THESSALY, it must have had very limited use elsewhere, in the hilly Greek terrain. Modern scholars believe the Mycenaeans copied the chariot's use from Near Eastern armies—either from the Hittites of ASIA MINOR or the New Kingdom Egyptians—without themselves having a clear tactical need for such a vehicle. Evidently the Mycenaeans were attracted to the machine's pure glamour.

Mycenaean chariots are commemorated in HOMER's *Iliad* (written down circa 750 B.C. but purporting to describe events of circa 1200 B.C.). Homer describes chariots as two-horse, two-man vehicles that bring aristocratic heroes such as ACHILLES and HECTOR to and from the battle. For actual fighting, the warrior leaps to the ground while the driver whisks the chariot to the sidelines. Modern scholars doubt that this poetic picture is accurate. Homer, presumably without realizing his mistake, has represented Mycenaean chariots as being used exactly as horses were used in his own day—to carry noble champions to battle. In fact (says the modern theory), like their Hittite and Egyptian counterparts, Mycenaean chariots seem to have taken part directly in combat—in charges, sweeps, and similar tactics—with an archer or spearman stationed inside each vehicle.

After the destruction of Mycenaean society and the ensuing DARK AGE (circa 1100–900 B.C.), the Greek chariot became used purely for SPORT, in races held at the great religious-athletic festivals. Because of this new function, the chariot evolved a new design. They were usually a one-man vehicles drawn by four horses or, sometimes, by four mules. During the 700s–100s B.C., the most prestigious competition in the entire Greek world was the race of horse chariots at the OLYMPIC GAMES. Coming last in the sequence of Olympic events, this race involved dozens of chariots in 12 laps around the stadium's elongated track, for a total distance of almost nine miles. Like the later chariot races of imperial ROME, the Greek sport offered considerable danger of collisions and other mishaps. For instance, at the PYTHIAN GAMES of 482 B.C., the winning chariot finished alone in a starting field of 41 vehicles.

The official competitors in a chariot race were not usually the drivers (who were professionals), but the people sponsoring the individual chariots. It was a sport for the rich, requiring large-scale breeding and training of horses. For Greek aristocrats and rulers, an Olympic chariot victory was the crowning achievement of public life. To help celebrate such a triumph, a winner might commission a lyric poet such as PINDAR (circa 476 B.C.) to write a victory song.

Among the more famous chariot owners was the flamboyant Athenian politican ALCIBIADES. For the Olympic Games of 416 B.C., he sponsored no less than seven chariots, which finished first, second, and fourth.

(See also ARISTOCRACY; PELOPS; WARFARE, LAND.)

Charon   See AFTERLIFE; FUNERAL CUSTOMS.

Charybdis   See SCYLLA.

Chersonese   The Chersonēsos ("peninsula") was a 50-mile-long arm of the northeastern Aegean coast, alongside northwestern ASIA MINOR. The peninsula's eastern shore forms the western side of the HELLESPONT—the 33-mile-long strait that was a crucial part of the ancient TRADE route to the BLACK SEA. To control the Hellespont, great powers such as ATHENS and MACEDON sought to hold the Chersonese. In modern times this region is known as Gallipoli, the scene of bloody fighting in World War I.

Although part of the non-Greek land of THRACE, the Chersonese began receiving Greek colonists from LESBOS and MILETUS by the 600s B.C. Athenian settlers arrived in the early 500s B.C. The elder Miltiades, an Athenian, ruled most of the Chersonese as his private fief (mid-500s B.C.). His nephew and successor, the famous MILTIADES, abandoned the area to Persian invasion.

After the PERSIAN WARS, the Chersonese settlements were dominated by Athens through the DELIAN LEAGUE. In 338 B.C. the region passed to the Macedonian king PHILIP II. Thereafter it was held by various Hellenistic kingdoms, until the Romans conquered Greece and Macedon in the mid-100s B.C..

The Chersonese had two chief cities, both Greek: SESTOS, which was the commanding fortress of the Hellespont; and Cardia, on the peninsula's western side.

Chilon   Spartan EPHOR who lived in the mid-500s B.C. and who is credited with certain developments in SPARTA's government and foreign policy. Apparently, Chilon increased the ephors' power to counterbalance the two Spartan kings, and he probably launched Sparta's policy of hostility to TYRANTS throughout Greece. He also organized Sparta's individual alliances with other states—including CORINTH and SICYON—into a permanent network of mutual defense, which modern scholars call the Peloponnesian League.

Revered at Sparta after his death, Chilon was counted as one of the SEVEN SAGES of Greece. His political heir was the Spartan king CLEOMENES (1), who reigned circa 520–489 B.C..

Chios   Large Aegean island, 30 miles long and 8–15 miles wide, lying close to the coast of ASIA MINOR. The main city, also named Chios, was situated on the island's east coast (five miles across from the Asian mainland) and was one of the foremost Greek city-states.

Occupied circa 1000 B.C. by IONIAN GREEKS from mainland Greece, Chios thrived as a maritime power whose exports included a renowned WINE as well as textiles, grain, figs, and mastic (a tree resin used for varnish). Chios figured prominently in the cultural achievements of the east Greek region known as IONIA. The island is said to have been the birthplace of the poet HOMER (born probably circa 800 B.C.); in later centuries a guild of bards, the Homeridae, or sons of Homer, were active there. Chios was also known as an international slave emporium, supplying the markets of Asia Minor and itself employing many SLAVES for FARMING and manufacturing.

During the Greek wars of the 700s–600s B.C., Chios tended to ally itself with MILETUS, against such nearby rivals as Erythrae and SAMOS. In government, the emergence of certain democratic forms at Chios is shown in an inscription (circa 560 B.C.) that mentions a "People's Council"—possibly a democratically elected COUNCIL and court of appeals. This democratic apparatus may have been modeled on the recent reforms of SOLON at ATHENS.

Like the rest of Ionia and Asia Minor, Chios fell to the conquering Persian king CYRUS (1) in around 545 B.C. In the ill-fated IONIAN REVOLT against Persian rule, Chios contributed 100 warships, which fought gallantly at the disastrous Battle of Lade (494 B.C.). Later the avenging Persians devastated Chios.

Following the liberation of Ionia during the PERSIAN WARS (479 B.C.), Chios became a prominent member of the Athenian-dominated DELIAN LEAGUE. Chios chose to make its Delian annual contributions in warships rather than in SILVER—one of the few league members to do so. For 65 years Chios proved the most steadfast of any Athenian ally. The Athenians paid Chios the compliment of coupling its name with that of Athens itself in public prayer at each Athenian ASSEMBLY. But in 412 B.C., with Athens' power dissolving during the PELOPONNESIAN WAR, Chios initiated the revolt of the Delian subject cities. Chios staved off the vengeful Athenians, and by 411 B.C. the island was firmly controlled by Athens' enemy, SPARTA.

After 30 years of Spartan domination, the Chians joined the SECOND ATHENIAN LEAGUE, but in 354 B.C. revolted from Athens again. During the 200s B.C., in resistance to the SELEUCID Empire's designs, Chios joined the Aetolian League. The island passed to Roman control in the next century.

(See also AETOLIA; DEMOCRACY.)

choral poetry   See LYRIC POETRY.

chorus   See LYRIC POETRY; THEATER.

Cilicia   See ASIA MINOR.

Cimon   Athenian soldier and conservative statesman in the 460s B.C. who lived circa 505–450 B.C. Cimon briefly dominated Athenian politics but succumbed to the failure of his pro-Spartan policy. He was the political enemy of the radical democrats, and his decline after 462 B.C. marked the rise of the young left-wing politician PERICLES. In foreign policy, Cimon was the last great Athenian enemy of PERSIA, and his death ushered in a Greek-Persian peace treaty that officially concluded the PERSIAN WARS.

Cimon (Greek: Kimon) was born into a rich and eminent family, the Philaïds. His father was the Athenian soldier MILTIADES; his mother, Hegesipyle, was daughter of a Thracian king. By the 470s B.C., Cimon was regularly being elected to the office of general. He assisted ARISTIDES in the organization of the DELIAN LEAGUE (circa 478 B.C.), and from 476 to 462 B.C. he was the premier Athenian soldier, leading the league's expeditions against the Persians. He became known particularly as a sea commander.

Cimon's height of success came in 469 or 466 B.C. (the exact date is unknown) when, with 200 league warships, he totally destroyed a Persian fleet and army at the River Eurymedon, midway along the south coast of ASIA MINOR. In purely military terms, the Eurymedon was the greatest Greek victory over Persia prior to the campaigns of ALEXANDER THE GREAT (334–323 B.C.).

At home, this success established Cimon in Athenian politics. Gracious and well-connected (his first or second wife, Isodice, was of the powerful Alcmaeonid clan), Cimon now emerged as leader of the conservative opposition. He blocked left-wing reforms and advocated an old-fashioned policy of hostility toward Persia and friendship with SPARTA. He even gave one of his sons the striking name Lacedaemonius, "Spartan."

Cimon's downfall came after he persuaded the ASSEMBLY to send him with an Athenian infantry force to assist the Spartans against their rebellious subjects, the Messenians (462 B.C.). This expedition ended in fiasco; the Spartans—fighting a serf rebellion—apparently found the Athenian soldiers' pro-democratic sentiments alarming and sent the Athenians home. Humiliated, Cimon now saw his conservative party swept out of power by democratic reforms sponsored by EPHIALTES and Pericles. In the following year, 461 B.C., the angry Athenians voted to ostracize Cimon.

Although the OSTRACISM law allowed a victim to return home after 10 years, this event ended Cimon's power and policies. Soon Athens began a full-fledged war against Sparta and its allies (460 B.C.). In 457 B.C., when the Athenian army was about to battle the Spartans near Tanagra, in BOEOTIA, the exiled Cimon arrived, asking permission to fight alongside his countrymen. Permission was refused but, according to one story, Cimon was specially recalled to Athens soon after.

In 450 B.C. he died while leading Athenian troops against the Persians in CYPRUS. He was remembered for his nobility and bravery, but his policies were at odds with Athens' destiny as a radical democracy that would dominate the rest of Greece.

(See also ALCMAEONIDS; CALLIAS; MESSENIA; WARFARE, NAVAL.)

Circe   In MYTH, a beautiful goddess and witch, daughter of the sun god HELIOS and his wife, Perse. Circe dwelt on a magical island in the West, in a stone house with enchanted wolves and lions. Her name (Greek: Kirkē) seems derived from kirkos, "hawk."

Circe plays a sinister but exciting role in Homer's epic poem the *Odyssey* (book 10). When the homeward-bound Greek hero ODYSSEUS brings his ship to her island and sends half his crew ashore to scout, she welcomes them with a magic drink that turns them into pigs. Odysseus goes to their rescue, armed with a magical herb called moly (*molu*)—the gift of the god HERMES—that makes him immune to Circe's spells. Odysseus forces her to restore his men's human shapes. Then, at her invitation, he lives with her for a year as her lover, but finally demands that she give him directions for his continued voyage home. Circe's advice (book 12) enables Odysseus to resist the deadly SIRENS and avoid other dangers. Circe is one of two supernatural women who become Odysseus' lovers during his wondrous voyage; the other is CALYPSO.

Generations of Greeks after Homer elaborated Circe's story. Her home was sometimes identified with Monte Circeo, a promontory midway along the west coast of ITALY. According to one story, she bore Odysseus' son Telegonus, who later unwittingly killed his father. In APOL-

LONIUS' epic poem, the *Argonautica*, Circe welcomes JASON (1) and MEDEA after their escape from the kingdom of Colchis.

city-state See POLIS.

Cleisthenes (1) Athenian statesman of the late 500s B.C., usually considered to be the father of Athenian DEMOCRACY. Cleisthenes began his career as a privileged aristocrat in a political arena of TYRANTS and aristocrats; but, whether through pure ambition or genuine convictions, he used his influence to reorganize the government to enlarge the common people's rights.

Born circa 560 B.C. into the noble Athenian clan of the ALCMAEONIDS, Cleisthenes (Greek: Kleisthenes) was son of the politician Megacles and of Agariste, daughter of the Sicyonian tyrant CLEISTHENES (2). The younger Cleisthenes served as ARCHON (525 B.C.) under the Athenian tyrant HIPPIAS (1), but was later banished with the rest of the Alcmaeonids, on Hippias' order.

After Hippias' ouster (510 B.C.), Cleisthenes returned to Athens and became leader of one of two rival political parties. When his opponent Isagoras was elected archon (508 B.C.), Cleisthenes struck back. According to the historian HERODOTUS, writing some 70 years later, Cleisthenes "took the common people (*dēmos*) into partnership." Cleisthenes proposed, in the Athenian ASSEMBLY, certain radical reforms to increase the common people's rights at the expense of the aristocrats. This program made Cleisthenes the most powerful individual at Athens, with all the common people behind him. His reforms continued a process begun by the lawmaker SOLON nearly 90 years prior and transformed Athens into a full democracy, the first in world history.

Cleisthenes' changes were extensive and complicated. The enabling first step was to improve the rights of the mass of poorer citizens (the laborers and peasants, called *thetes*.) The *thētes* were disadvantaged by the traditional system of four Athenian tribes (*phulai* or phylae). These tribes, which supplied the basis for public life in the city, were traditionally dominated by aristocratic families. Cleisthenes overhauled the tribal system, replacing the four old phylae with 10 new ones. Each new tribe was designed to include a thorough mix of Athenians—farmers with city dwellers, aristocrats and their followers with middle-class people and *thētes*. The effect was to reduce greatly the influence of the nobles within each tribe.

To create his new, "mixed-up" tribes, Cleisthenes reorganized the political map of ATTICA, the 1,000-square-mile territory of Athens that included all Athenian citizens. It was now, if not earlier, that Attica became administratively divided into about 139 DEMES—*dēmoi*, "villages" or local wards. By means of Cleisthenes' complicated gerrymandering, each new tribe was made to consist of several demes (an average of about 14, but the actual numbers varied between six and 21). Typically these tribal-constituent demes were unconnected by geography, traditional allegiances, and the like. The new tribes thus were relatively free from aristocratic domination and from the localism and feuding associated with aristocratic domination. Across the map of Attica, the traditional pockets of local-family influence were, in effect, broken up.

With his 10 tribes as a basis, Cleisthenes democratized other aspects of the government. The people's COUNCIL was enlarged from 400 to 500 members, now consisting of 50 citizens from each tribe, chosen by lot from a pool of upper- and middle-class candidates. The Athenian citizens' assembly received new powers, such as the judicial right to try or review certain court cases. But these radical changes did not go unchallenged. Cleisthenes' rival Isagoras appealed to the Spartan king CLEOMENES (1), who marched on Athens with a small Spartan force. Cleisthenes and his followers fled (507 B.C.). But when Cleomenes attempted to replace the new democracy with an OLIGARCHY consisting of Isagoras and his followers, the Athenian populace rose in resistance. Cleomenes and his army were besieged atop the Athenian ACROPOLIS, then were allowed to withdraw, taking Isagoras with them.

Cleisthenes returned to Athens, but soon his prominence was over. He may have involved himself in diplomatic overtures to the Persian king DARIUS (1), and if so he would have been disgraced in the ensuing anti-Persian sentiment. But when Cleisthenes died (circa 500 B.C.), he received a public tomb in the honorific Ceramicus cemetery, just outside Athens. His democratization of Athens would be taken further by the radical reforms of EPHIALTES and PERICLES, in the mid-400s B.C..

(See also KINSHIP; OSTRACISM.)

Cleisthenes (2) Tyrant of the Peloponnesian city of SICYON and maternal grandfather of the Athenian statesman CLEISTHENES (1). Cleisthenes ruled Sicyon from about 600 to 570 B.C., when the city was one of the foremost commercial and military powers in Greece. He extended his influence to DELPHI by leading a coalition against PHOCIS in the First Sacred War (circa 590 B.C.) and crowned his achievements with a chariot victory at the OLYMPIC GAMES (circa 572 B.C.). His wealth and prestige are apparent in the historian HERODOTUS' tale of how Cleisthenes hosted his daughter Agariste's 13 suitors at his palace for a year, observing them in SPORT, discourse, and so on, and gauging their aristocratic qualifications for marrying his daughter. The suitors came from various parts of the Greek world, but Cleisthenes favored two Athenians: Megacles, son of Alcmaeon of the clan of the ALCMAEONIDS, and Hippocleides, son of Teisander of the Philaïd clan.

On the day appointed to announce his choice, Cleisthenes held a great feast, at which the suitors competed in two final contests, lyre-playing and public speaking. Hippocleides, who outshone the others, was the one whom Cleisthenes had by now secretly chosen. But as more WINE was drunk, Hippocleides requested a tune from the flute-player and boldly began to dance. Then he called for a table and danced atop it, while Cleisthenes watched with distaste. And when Hippocleides started doing handstands on the table, beating time with his legs in the air, Cleisthenes cried out, "O, son of Teisander, you have danced away your wedding!" To which the young man replied, "Hippocleides doesn't care" (*ou phrontis Hippokleidei*).

The story became proverbial as an example of aristocratic detachment and *joie de vivre*. Agariste's hand in MARRIAGE went to Megacles, and their son was Cleisthenes, the Athenian statesman (born circa 560 B.C.).

(See also MUSIC; TYRANTS.)

Cleomenes (1) Dynamic and ambitious king of SPARTA who reigned circa 520–489 B.C.. Cleomenes' efforts to expand Spartan power beyond the PELOPONNESE, along with his resistance to Persian encroachment, mark him as one of the dominant personalities of the late 500s B.C. Unfortunately, our major source for his reign, written by the historian HERODOTUS (circa 435 B.C.), is tainted by Spartan official revisionism that tries to diminish Cleomenes' importance.

The son of King Anaxandridas of the Agiad royal house, Cleomenes (Greek: Kleomenēs) was awarded the kingship in victorious rivalry against his half brother Doreius. Soon Cleomenes was taking aim against the Athenian dictator HIPPIAS (1). Cleomenes wanted to end Hippias' reign in order to bring ATHENS into the Spartan alliance. In addition, Hippias' diplomatic overtures to the Persian king DARIUS (1) had made Cleomenes fear that Hippias was maneuvering to assist a Persian invasion of mainland Greece. In 510 B.C. Cleomenes entered Athens with an army, ejected Hippias, and withdrew. But Cleomenes was mistaken in expecting that this would result in an Athenian OLIGARCHY friendly to Sparta. Instead, there occurred peaceful revolution, producing the Athenian DEMOCRACY (508 B.C.).

Cleomenes returned with an army in 508 or 507 B.C., intending to overthrow the new government. He captured the Athenian ACROPOLIS but found himself besieged there by the Athenian populace and withdrew under truce. His later attempt to organize a full Spartan-allied attack on Athens was blocked by the other Spartan king, Demaratus, and by Sparta's ally CORINTH (which at that time was still friendly with Athens).

Despite a request from the Greek city of MILETUS, Cleomenes wisely declined to send Spartan troops overseas to aid the IONIAN REVOLT against the Persians (499 B.C.). He had an enemy nearer home to attend to—Sparta's Peloponnesian rival, ARGOS. In 494 B.C., at the Battle of Sepeia, on the Argive plain, Cleomenes obliterated an Argive army and marched his Spartans to the walls of Argos. But the city withstood Cleomenes' siege, a failure that later was to be used against him by his enemies at Sparta.

In around 491 B.C. Persian envoys went to Sparta asking for earth and water, the tokens of submission to PERSIA. Probably at Cleomenes' prompting, the Spartans threw the Persians down a well, telling them they would find plenty of earth and water there. But other states, notably AEGINA, did submit. When Demaratus blocked Cleomenes' desired retaliation against Aegina, Cleomenes decided to rid himself of this uncooperative partner. He persuaded Demaratus' kinsman Leotychides to claim the kingship on grounds that Demaratus has been an illegitimate child, and he bribed the oracle at DELPHI to support this claim. Demaratus was deposed and succeeded by Leotychides.

Circa 490 B.C. Cleomenes was in THESSALY. According to Herodotus, the king was in hiding because his disgraceful intrigues against Demaratus had been exposed. However, it seems more likely that Cleomenes was still at the height of power and working to organize Greek resistance to Persia. He then visited ARCADIA, a Peloponnesian region traditionally under Spartan control. By then he may have been aiming at an ambitious goal—to resist the Persians by creating a unified Peloponnesian state, under his personal dictatorship. Possibly he even tried to gain support from the Spartan serfs known as HELOTS, promising them freedom in exchange.

After returning to Sparta, Cleomenes died violently. According to the official version, he took his life in a fit of insanity. But it looks as if he were assassinated by conservative Spartan elements. His successor was his half brother, Leonidas, who was fated to die at the Battle of THERMOPYLAE (480 B.C.), resisting the Persian invasion that Cleomenes had tried to prevent.

(See also AGIAD CLAN; CLEISTHENES (1); PERSIAN WARS.)

Cleomenes (2) See SPARTA.

Cleon Athenian demagogue (rabble-rouser) of the early PELOPONNESIAN WAR. For seven years after the death of the statesman PERICLES (429–422 B.C.), the left-wing Cleon was the foremost politician in ATHENS. Although not really a soldier, he was the one person at the time who—by his bold planning and determination—could have won the war for Athens. But he was killed in battle in 422 B.C., and the Athenian war leadership passed temporarily to the cautious NICIAS.

Most information about Cleon comes from the written work of his contemporary, the Athenian historian THUCYDIDES (1). The normally objective Thucydides viewed Cleon with distaste and underestimated his importance. Similarly, Cleon was despised by the Athenian comic playwright ARISTOPHANES, who mocked him in the Knights (424 B.C.) and other comedies as a crowd-pleasing opportunist. Yet despite the disapproval of such intellectuals, Cleon probably was widely viewed as Pericles' legitimate successor. Although Cleon came to prominence by attacking the aging Pericles (431 B.C.), he also imitated Pericles in his obsessive loyalty to the Athenian common people (dēmos), who were his power base. That is why, in the Knights, Aristophanes uses the name Philodemos (lover of the people) for his character who is a caricature of Cleon.

The son of a rich tanner of hides, Cleon (Greek: Kleon) was an accomplished orator who could whip up public opinion in the Athenian ASSEMBLY; he is said to have introduced a more vulgar and demonstrative mode of public speaking. He reached his peak of power in 425 B.C., after Athens had gained the advantage in the fighting against the Spartans at PYLOS. Accusing the Athenian generals of incompetence, Cleon won command of the entire Pylos campaign by acclamation of the assembly. He journeyed to Pylos, where—helped by the Athenian general DEMOSTHENES (2)—he won a total victory over the supposedly invincible Spartans.

Then preeminent in Athenian politics, Cleon presumably is the one who prompted the notorious "Thoudippos Decree" (425 B.C.), which authorized a reassessment of the annual tribute to be paid to Athens by its DELIAN LEAGUE allies. The decree resulted in a doubling or tripling of the amounts due from individual allied states. Around the same time, Cleon sponsored a law increasing jury pay, to the advantage of lower-income citizens.

Elected as a general for 422 B.C., he led an Athenian army to the north Aegean coast, to recapture the area from the Spartan commander BRASIDAS. But at the Battle of AMPHIPOLIS, Cleon was defeated and killed. Thucydides mentions that he was running away when an enemy skirmisher cut him down.

(See also DEMOCRACY; LAWS AND LAW COURTS; MYTILENE.)

Cleopatra Dynastic female name of the Macedonian royal family in the 300s B.C., later used by the Macedonian-descended Ptolemies of EGYPT. The name Kleopatra means "glory of her father" in Greek. The famous Cleopatra—Cleopatra VII, daughter of Ptolemy XII—was the last Ptolemaic ruler of Egypt and also the last Hellenistic ruler outside Roman control. Upon her death in 30 B.C., Egypt was annexed by ROME.

(See also HELLENISTIC AGE; PTOLEMY (1).)

clothing The ancient Greeks had no fashion industry; most clothing, typically made of wool or linen, was woven at home. Greek female servants and wives might spend much of their day indoors, spinning yarn and weaving on the loom. Clothing was simple and loose-fitting, with only a few basic forms and much similarity between garments for men and WOMEN.

Information about Greek clothing comes from written sources and from scenes in extant artwork, particularly vase paintings. Males and females wore the tunic, which looked like a sleeveless nightgown or dress. The woman's tunic, called a *peplos*, would typically fall to the feet, with extra material pinned around the shoulders—it could be worn with or without a belt. In addition, there might be an undergarment and a cloak around the shoulders.

Equipped for the gentlemanly pastime of hunting, a youth wears a woolen riding cloak and leather boots, with his broad-rimmed Greek traveler's hat slung back. This red-figure scene was painted on the inside bottom of an Athenian cup, circa 475 B.C.

The man's tunic was the *chiton*. Simpler and lighter than the Roman toga, the Greek *chiton* might fall to the feet (in the Ionian fashion) or merely to the knees or above (in the Dorian fashion). The shorter style was favored by workmen, farmers, and soldiers, who wore tunics underneath their body armor. Like women, men might wear a cloak (*himation*). For horseback riding, younger men often wore a distinctive wide cloak (the *chlamus*, or chlamys).

Travelers, workmen, and sailors might wear a conical cap known as a *pilos*; travelers, hunters, and others sometimes wore the low, broad-rimmed hat (*petasos*), to protect against the hot Greek sun. The footwear of choice for heavy walking was sandals or leather boots; otherwise clogs or slippers were worn.

(See also POTTERY).

Clytaemnestra In MYTH, the wife and murderer of King AGAMEMNON of MYCENAE. Clytaemnestra was the daughter of the Spartan queen Leda. According to different versions of the tale, Clytaemnestra's father was either the god ZEUS or Leda's husband, the Spartan king Tyndareus. Clytaemnestra had the famous HELEN OF TROY for a twin sister and CASTOR AND POLYDEUCES for her brothers. Her name, Klutaimnēstra, probably means "famous wooing" or "famous intent."

As Agamemnon's wife, Clytaemnestra bore a son, ORESTES, and two daughters, ELECTRA and IPHIGENIA. But she grew to hate her husband when he chose to sacrifice Iphigenia to appease the goddess ARTEMIS, who was sending contrary winds to block the Greeks' sailing fleet at the start of the TROJAN WAR. During Agamemnon's 10-year absence at TROY, Clytaemnestra became the lover of Agamemnon's cousin Aegisthus, and the two plotted to kill the king on his return. In the act, they also killed the Trojan princess CASSANDRA, whom Agamemnon had brought home as his war prize.

Afterward, Clytaemnestra and Aegisthus ruled as queen and king at Mycenae. They were eventually killed in vengeance by Orestes, with Electra assisting.

Clytaemnestra's character shows a growth in strength and evil during the centuries of Greek literature. Where she is first mentioned in HOMER's epic poem the *Odyssey* (written down circa 750 B.C.), she is overshadowed by the vigorous Aegisthus, who seduces her and plots the murder. But in AESCHYLUS' tragedy *Agamemnon* (performed 458 B.C.), Clytaemnestra is the proud and malevolent prime mover, and Aegisthus is just her effete, subordinate lover. In the play, she persuades the newly arrived Agamemnon to tread atop a priceless tapestry; he thereby unwittingly calls down heaven's anger on himself. Then she leads him to his bath, where she murders him and Cassandra.

Agamemnon is the first play of Aeschylus' Oresteian Trilogy. In the following play, the *Choëphoroi* (or *Libation Bearers*), Clytaemnestra and Aegisthus are slain by Orestes. As she dies, she calls down the avenging FURIES upon her son. A hateful Clytaemnestra is portrayed also in SOPHOCLES' tragedy *Electra* and EURIPIDES' tragedy *Electra* (both circa 417 B.C.). In vase painting, Clytaemnestra sometimes is shown as wielding a double ax against Agamemnon or Cassandra or (unsuccessfully) against Orestes.

(See also HUBRIS; THEATER.)

Cnidus Prosperous Greek city of the southwestern coast of ASIA MINOR. Founded circa 900 B.C. by DORIAN GREEKS from mainland Greece, Cnidus and its neighbor HALICARNASSUS were the two important Dorian cities of Asia Minor, and were part of a larger eastern Mediterranean Dorian federation including the islands of Cos and RHODES.

Cnidus (Greek: Knidos) stood on the side of a lofty promontory jutting into the sea, and it thrived as a seaport and maritime power. In search of raw metals such as tin, the Cnidians founded trade depots on the Lipari Islands and elsewhere in the western Mediterranean. When the Persians attacked Asia Minor under King CYRUS (1) in 546 B.C., the Cnidians tried to convert their city into an inshore island by digging a canal across the peninsula. Finding the work slow and injurious, they consulted the Oracle of DELPHI and were told to abandon resistance and make submission.

After the liberation of Greek Asia Minor in 479 B.C., Cnidus became a tribute-paying member of the DELIAN LEAGUE, under Athenian domination. In 413 B.C., with ATHENS losing ground in the PELOPONNESIAN WAR, Cnidus joined the general Delian revolt and became a Spartan ally. In 394 B.C., during the CORINTHIAN WAR, the Athenian admiral CONON destroyed a Spartan fleet in a battle off Cnidus.

Returned to Persian overlordship by the terms of the KING'S PEACE (386 B.C.), Cnidus was reliberated in 334 B.C. by the Macedonian king ALEXANDER THE GREAT. Circa 330 B.C. the city's public and commercial buildings were rebuilt near the tip of its peninsula to take advantage of the superior harbor there. By then the city had reached its height of prosperity, being renowned for its exported WINE and its school of MEDICINE. Cnidus housed a statue that was considered to be the most beautiful in the world: the naked APHRODITE carved in marble by PRAXITILES circa 364 B.C. According to one ancient writer, people would voyage to Cnidus expressly to view this work.

During the HELLENISTIC AGE, Cnidus became subject to Ptolemaic EGYPT (200s B.C.). By the later 100s B.C. the city was part of the empire of ROME.

(See also BRONZE; PERSIAN WARS.)

Cnossus Chief city of the island of CRETE for most of the 2,000 years before Christ. Located in the north-central part of the island, about three miles inland, Cnossus had a good harbor—the site of the modern seaport of Heraklion—and also guarded a major land route southward across the middle of Crete. Cnossus was the capital city of the MINOAN CIVILIZATION, which thrived on Crete circa 2200–1450 B.C. and controlled a naval empire in the AEGEAN SEA. Although the Minoans were not themselves Greek, they stand at the threshold of Greek history insofar as they deeply influenced the emerging Greek MYCENAEAN CIVILIZATION (circa 1600–1200 B.C.). The name Knossos is in origin a pre-Greek word, probably from the lost Minoan language.

Cnossus today is one of the world's most important archaeological sites, containing remnants of the largest Minoan palace, the seat of the Minoan rulers. First erected in around 1950 B.C. and then rebuilt and enlarged after earthquakes in around 1700 and 1570 B.C., the palace was first revealed to modern eyes by the British archaeologist

Sir Arthur Evans in A.D. 1900. The building and environs, excavated almost continuously since then, have yielded such well-known Minoan artwork as the Bull's Head Rhyton and the Toreador Fresco, now in the Heraklion archaeological museum.

Built mainly of large blocks of Cretan limestone, the palace survives today mostly as a network of foundations and wall remnants covering five and a half acres. Around a central, square courtyard measuring 82 by 180 feet, the building contained pillared hallways, staircases, and hundreds of rooms and storage chambers in a mazelike configuration. There were two or even three upper floors, now partly reconstructed. Among the palace's splendors was running water, carried in clay pipes. The building's size and complexity—unequaled in 1700 B.C. outside the urban centers of EGYPT and Mesopotamia—probably inspired the later Greek legend of the Cretan LABYRINTH.

Remarkably, the palace had no enclosing wall; the Minoans evidently trusted in their navy for defense. Nevertheless, the palace was completely destroyed by fire in around 1400 B.C. or soon after, and was never rebuilt. As shown by ARCHAEOLOGY, every other known Minoan site on Crete also was destroyed at this time. The cause for such widespread ruin may have been an invasion of Crete by warlike Mycenaeans from the Greek mainland. Intriguingly, it seems that the Cnossus palace already had been occupied by certain Mycenaeans prior to its destruction. This conclusion is reached because the palace debris of this era has yielded up nearly 4,000 clay tablets, inscribed with inventory notes written in the Greek LINEAR B script, identical to the script later used at Mycenaean sites on the mainland. But whoever it was who destroyed the palace, evidently Cnossus and all of Crete were abandoned thereafter by the Mycenaeans.

In around 1000 B.C. Crete was occupied by new conquerors—the DORIAN GREEKS, invading from the PELOPONNESE. Cnossus, with its superior location, became Dorian Crete's foremost city-state. Like the other Cretan cities, it was governed as a military ARISTOCRACY, with social and political institutions similar to those at SPARTA. Dorian-ethnic nobles ruled over a rural population of non-Dorian serfs.

Cnossus shared in Crete's general decline after the 600s B.C. By the late 200s B.C. it had fallen into a debilitating conflict with its Cretan rival city, GORTYN. When the Romans annexed Crete in 67 B.C., they chose Gortyn over Cnossus as their provincial capital.

For the classical Greeks, Cnossus kept its associations with a dimly remembered, fictionalized Minoan past. The myths of DAEDALUS, MINOS, and THESEUS describe Cnossus as a city of grandeurs and horrors.

(See also GREEK LANGUAGE; HELOTS.)

coinage The first coins came into use in the 600s B.C. But long before then the ancient Greeks traded (by barter) and understood the concepts of value and profit. Among the Greeks, the forerunners of the coin were certain implements used for representing high value. One such was the bar of SILVER, usually called an ingot. Several ingots might comprise the monetary unit known as the TALENT (Greek *talanton*, "weight"), equivalent to just under 58 pounds of silver. Another precious metal of those days was IRON.

A silver Athenian tetradrachm, or four-drachma piece, minted circa 440 B.C. One side shows the city's patron goddess, Athena, garlanded with olive leaves; the other side shows Athena's bird, the owl. The Greek lettering reads: *ATHE*. This drawing presents the coin at three times its actual size.

The Greeks developed a high-level currency based on the iron rods (*oboloi*), traded by the handful (*drachmē*). These rods remained in circulation down through the 500s B.C. and even later in the reclusive city of SPARTA.

The invention of coins—easily stored, transported, and counted—marked an improvement over ingots or rods. The first nation on earth to mint coins was most likely the kingdom of LYDIA, in ASIA MINOR, circa 635 B.C. The Lydians were not Greek, but the Greeks seem to have imported and studied this new Lydian invention. Modern archaeologists have uncovered Lydian coins composed of electrum (an alloy of silver and GOLD) at the ancient Greek city of EPHESUS, in Asia Minor. Stamped with a lion figure "seal of approval" as a royal guarantee of weight and purity, these Lydian coins may have been given to pay large groups of artisans or mercenary soldiers, who surely included Greeks. Before long, Greeks were minting their own coins, copied from the Lydian ones.

According to ancient writers, the first Greek minting state was AEGINA (probably circa 595 B.C.). The Aeginetan coins were silver, in three sizes that cleverly employed the denominations already existing for iron rods. The smallest coin was called the obol, the next the drachm, and the largest the didrachm (two drachmas), or statēr. The Aeginetan statēr showed a sea turtle, the city symbol. The other two early Greek coining states were ATHENS (circa 575 B.C.) and CORINTH (circa 570 B.C.)—both, like Aegina, trading and maritime powers.

By the 400s B.C. every important Greek state was minting its own coins, as a sign of independence and an aid to TRADE. There was not yet a uniform system of denominations; rather, each city tended to follow one of two available systems. Silver remained the prime coining metal (gold being a scarce resource for the Greeks). But coinages of BRONZE, to cover lower denominations, began in the mid-400s B.C. And by the mid-300s B.C. gold coins were being minted by certain states such as MACEDON, which had supplies of gold ore.

Unlike some modern currency, a Greek coin was intended to contain an unadulterated amount of precious metal equal to the face value: The stamp on the coin was meant to guarantee this. The earliest Greek coins were stamped on only one side. To prepare such a coin, the smith would place a blank, heated disk of metal on an anvil atop a shallow die bearing the engraved shape of the intended imprint. The disk was then punched into the die by a rod (Greek: *charaktēr*), hammered at one end. The resulting coin would bear a relief created by the die on one side (the "obverse") and a mere indentation on the other (the "reverse"). By about 530 B.C. the Greeks had learned to make two-sided coins by using a rod whose lower end also bore an engraved die.

To judge from the hundreds of specimens surviving today, Greek coins were intended as objects of beauty and as means of civic propaganda. Extant coins bear such charming or stirring emblems as the Athenian owl, the Corinthian winged horse Pegasus, and the man-faced, bull-bodied river god of GELA. Coins also became a medium for accurate profile portraiture; it is from coins that we know what certain ancient personages looked like. In the late 300s B.C., for instance, the successors of ALEXANDER THE GREAT paid homage to his memory—and laid claim to his empire—by issuing coins showing his portrait.

Greek coinage changed at this time, in the aftermath of Alexander's conquest of the Persian Empire. Under Hellenistic kings ruling vast territories in an enlarged Greek world, there arose uniform coinages used over large areas and not necessarily differing from city to city. Important coining states of the 200s and 100s B.C. included Macedon, the SELEUCID EMPIRE, Hellenistic EGYPT, PERGAMUM, and the Greek leagues of ACHAEA and AETOLIA. Also, the Greco-Macedonian kings of BACTRIA commemorated themselves in a series of remarkably vivid personal portraits on coins. Of the older Greek states, only Athens and Rhodes continued minting.

Most Greek coinage ceased during the Roman conquests in the two centuries before Jesus. Henceforth the Greeks relied mainly on coins minted by the imperial city of ROME.

(See also ATHENA; BELLEROPHON; DIADOCHI; HELLENISTIC AGE; HIMERA.)

colonization Even prior to the conquests of ALEXANDER THE GREAT (circa 334–323 B.C.), the ancient Greek world stretched from Libya and southern France to CYPRUS, ASIA MINOR, and the Crimean peninsula. This Greek proliferation around the Mediterranean and BLACK SEA took place during the great colonizing era of the 700s through 500s B.C., when Greek cities extended their TRADE routes (and alleviated domestic food shortages and other population problems) by sending out shipborne expeditions of young male colonists. Sailing along sea routes already scouted by Greek traders, these colonists might travel considerable distances to descend on a land typically occupied by a vulnerable, non-Greek people (such as in eastern SICILY). But coasts defended by powerful kingdoms such as EGYPT or Assyria were never colonized by early Greeks.

The colonists' new city would retain the laws, traditions, and religious cults of the mother city, or *mētropolis*. Metropolis and colony would typically enjoy cordial relations and trade agreements; for example, SYRACUSE, in Sicily, was surely bound to export surplus grain to its hungry mother city, CORINTH.

The earliest Greek colonizing took place during the MYCENAEAN CIVILIZATION, circa 1600–1200 B.C. The Mycenaean Greeks founded settlements on Cyprus and on the west coast of Asia Minor. But the great age of colonization began in the mid-700s B.C., after trade and improved seamanship had begun the expansion of the Greek world. Parts of HOMER's epic poem the *Odyssey* (written down circa 750 B.C.) seem colored by a contemporary interest in trade and exploration around Sicily and the Adriatic Sea. Most modern knowledge of this period comes from archaeological evidence—building sites, POTTERY types, occasional inscriptions—and from literary sources such as the fifth-century-B.C. historians HERODOTUS and THUCYDIDES (1) and the Roman geographer Strabo.

The earliest datable colony was established circa 775 B.C. by Greeks from the commercial cities of CHALCIS and ERETRIA, in central Greece. It was a trading station on the island of PITHECUSAE, six miles off the Bay of Naples, on the central west coast of ITALY. The location reveals a purely mercantile motive; the settlement was intended as a safe haven for trade with the ETRUSCANS of mainland Italy. Later the Pithecusans crossed to the mainland, to found CUMAE.

In the 730s B.C. expeditions from mainland Greece ventured to the fertile east coast of Sicily; NAXOS (2) and ZANCLĒ were founded by Chalcis, and Syracuse by Corinth. By then the colonists were setting sail primarily for farmland. The later 700s B.C. saw colonies planted in southern coastal Italy: RHEGIUM (from Chalcis), TARAS (from SPARTA), and SYBARIS and CROTON (from ACHAEA). The colonists' hunger for Italian grainfields is symbolized in the ear-of-wheat emblem on coins minted by the Achaean colony of Metapontum, on the Italian "instep." In time, the entire southern Italian coastline, from "heel" to "toe," became dotted with Greek cities.

The north Aegean coast of THRACE received colonists by the late 700s B.C. Ousting the native Thracians, Greeks mainly from Chalcis occupied the region subsequently called the Chalcidian land—CHALCIDICĒ. Not far away, the island of THASOS was occupied by Greeks from the island of Paros. Later colonies on the Thracian coast included ABDERA, AENUS, and AMPHIPOLIS. The region offered farmland and valuable resources for export: grain, TIMBER, SLAVES, SILVER, and GOLD.

Farther east, the Black Sea and its approaches were colonized almost single-handedly by MILETUS. Of perhaps two dozen Milesian colonies here, the most important included SINOPE (founded circa 700 B.C.), CYZICUS (circa 675 B.C.), and PANTICAPAEUM (late 600s B.C.). But the city with the grandest destiny was BYZANTIUM, founded circa 667 B.C. by colonists from MEGARA (1). Among the assets of the Black Sea region were the boundless wheatfields of Ukraine and the metals trade of Asia Minor.

Circa 730 B.C. the Corinthians established CORCYRA (modern Corfu) on the northwestern coast of GREECE, 80 miles across the Adriatic from the "heel" of Italy. Later the Corinthians compensated for a Corcyrean rebellion by creating new northwestern colonies, including AMBRACIA, APOLLONIA, and Leucas (630–600 B.C.).

The North African Greek city of CYRENE (1), destined for commercial greatness, was founded by colonists from the humble island of THERA, circa 630 B.C.. To the northwest, Greeks from PHOCAEA endured Carthaginian hostility in order to establish MASSALIA (modern Marseille, in southern

France) and other far-western Greek colonies, circa 600 B.C..

Like other great endeavors in the ancient world, colonization was a mixture of the utilitarian and the religious. The mother city would appoint an expedition leader, or "oikist" (*oikistēs*), who would organize the departure, lead the conquest of new land, and rule the new city as king or governor. Before departure, the oikist would seek approval for the project from the god APOLLO's oracle at DELPHI. Apollo was the patron of colonists—one of his titles was Archagetēs, "leader of expeditions"—and several colonies were named for him, including Apollonia on the Adriatic and Apollonia on the Black Sea. He was the only god who could sanction a colonizing expedition, and it was probably by careful politicking that the priesthood at Delphi was able to make its oracle of Apollo more important than any other. In the late 500s B.C. a Spartan prince Dorieus set out to found a colony without gaining Delphi's approval; eventually he and his followers were massacred by non-Greeks in western Sicily.

By the 500s B.C. colonizing had become regulated as a tool of imperialism, with great powers planting garrison colonies in militarily desirable areas. Athenian designs on the HELLESPONT in the late 500s B.C. brought Athenian colonists to Sigeum, in northwest Asia Minor, and to the Thracian CHERSONESE. The imperial ATHENS of the 400s B.C. punished its rebellious subject allies by establishing cleruchies (*klērouchiai*), which were land-grabbing Athenian colonies that acted as garrisons. Among the states to receive these onerous colonies were Chalcis, AEGINA, and LESBOS.

The use of military colonies was developed in the 330s–320s B.C. by Alexander the Great. The many cities that he established in the wake of his conquests, including several named ALEXANDRIA (1 and 2), were intended to settle veterans while also guarding lines of supply. These far-flung Greco-Macedonian settlements—such as the ancient city represented by Greek-style ruins at Aï Kanoum, in northern Afghanistan—played a crucial role in hellenizing the East during the HELLENISTIC AGE (300–150 B.C.).

Alexander's successors in EGYPT and the SELEUCID EMPIRE continued his practice of founding garrison settlements, either within the kingdom or along a frontier. The most successful of such foundations was the Seleucid city of ANTIOCH (330 B.C.). A major purpose of Hellenistic rulers' colonization was to allow the valuable Greco-Macedonian soldier class to breed and create a new generation to become heavy infantry.

(See also BOSPORUS; BRONZE; ILLYRIS; MARMARA, SEA OF; SHIPS AND SEAFARING.)

comedy See THEATER.

commerce See TRADE.

Conon Athenian admiral active from 414–392 B.C. Born circa 444 B.C. into a noble family at ATHENS, Conon (Greek: Konon) commanded fleets in the final years of the PELOPONNESIAN WAR. At the disastrous Athenian naval defeat at AEGOSPOTAMI (405 B.C.), Conon alone of all the commanders got his squadron safely away from the Spartan ambush.

In the CORINTHIAN WAR (395–386 B.C.) he was active in developing the navy of Athens' ally PERSIA. Leading an Athenian squadron fighting alongside the Persians, Conon destroyed a Spartan fleet at the Battle of CNIDUS (394 B.C.). With funds from the Persian king Artaxerxes II, Conon oversaw the rebuilding of Athens' LONG WALLS. But his dreams of re-creating the Athenian empire died when the Persians switched sides in the war (392 B.C.). Conon died soon thereafter.

Corcyra Major Greek city of the Adriatic coast, on the lush and attractive inshore island now known as Corfu. The city was situated on a peninsula on the island's east coast, opposite the Greek mainland region called EPIRUS. Corcyra (Greek: Kerkura) was one of the earliest Greek colonies, founded circa 734 B.C. by settlers from CORINTH.

Located only 80 miles across the Adriatic from the Italian "heel," Corcyra provided a vital anchorage on the coastal TRADE route from Greece to ITALY and SICILY. Corcyra also enjoyed local trade with the Illyrians, a non-Greek people who supplied SILVER ore, TIMBER, SLAVES, tin (for BRONZE-making), and wildflowers (for perfume-making). Agriculturally, the island was (and is) known for dense growths of olive trees.

Unlike most Greek colonies, Corcyra rebelled violently against its mother city. In 664 B.C., as mentioned by the Athenian historian THUCYDIDES (1), Corcyra and Corinth fought the first sea battle on record; Thucydides neglects to state who won. Circa 610 B.C. the Corinthian tyrant PERIANDER brought Corcyra temporarily to heel. At some point the two cities cooperated in founding the Adriatic colony of EPIDAMNUS, but Corinth continued to establish its own colonies in the area, partly to guard against Corcyrean hostility.

It was at prosperous Corcyra circa 580 B.C. that the first large, all-stone Greek building was erected—a temple of ARTEMIS, designed in the newly emerging Doric style. This building has been called the Gorgon temple, on account of its pedimental sculptures showing the Gorgon MEDUSA flanked by beasts. Today the temple survives only in its west pediment, preserved in the Corfu Museum.

Despite having one of the most powerful navies in Greece, Corcyra remained neutral in the PERSIAN WARS. Later its bitter relations with Corinth played a role in igniting the PELOPONNESIAN WAR (431–404 B.C.). These events began in 435 B.C., when Corcyra was drawn into conflict with Corinth over relations with Epidamnus. A Corcyrean fleet of 80 ships defeated 75 invading Corinthian ships at the Battle of Leukimme, off Corcyra's southeast coast. To defend against retaliation, the Corcyreans allied with the powerful city of ATHENS. The next episode saw 10 Athenian warships ranged alongside 110 Corcyrean ships against a Corinthian fleet of 150. This was the important Battle of Sybota (433 B.C.), fought off the mainland near Corcyra and ending in a Corinthian victory. As a consequence of this battle, Corinth urged SPARTA to declare war against Athens.

Corcyra fought in the Peloponnesian War and suffered greatly. The city was being governed as a pro-Athenian DEMOCRACY when a group of Corcyrean right-wingers launched a coup, hoping to swing Corcyra over to the Peloponnesian side. The coup failed, bringing on a gruesome civil war (427–425 B.C.). Thucydides' sympathetic but objective description of the Corcyrean *stasis* (civil strife) is one of the great set pieces in his history of the Peloponnesian War (book 3).

Later in the war, Corcyra was a staging base for Athens' disastrous invasion of SYRACUSE (415 B.C.). In 410 B.C. Corcyra ended its alliance with Athens. But in 375 B.C. Corcyra joined the SECOND ATHENIAN LEAGUE as protection against Spartan domination.

With its wealth and strategic position, Corcyra became a bone of contention among the rival dynasts after the death of the Macedonian king ALEXANDER THE GREAT (323 B.C.). The city was occupied variously by Macedonians, Syracusans, Epirotes, and Illyrians, before passing to the empire of ROME (229 B.C.). It served for many years as a Roman naval base.

(See also AGATHOCLES; ARCHITECTURE; COLONIZATION; ILLYRIS; WARFARE, NAVAL.)

Corinth Major city of the northeastern PELOPONNESE, known for its manufacturing and seaborne TRADE. Located on the narrow isthmus that connects southern and central Greece, Corinth became great through its geography: It guarded the landroute along the isthmus, and it controlled harbors on both shores, eastward on the Saronic Gulf and westward on the aptly named Corinthian Gulf. Also, the city's lofty ACROPOLIS—the "Acrocorinth," atop a limestone mountain standing outside the lower town—gave Corinth a nearly impregnable citadel.

Corinth's heyday came in the 600s and early 500s B.C., when its shipping network dwarfed that of the other mainland Greek cities. Although it eventually lost this preeminence to ATHENS, Corinth remained a center for commerce and luxury throughout ancient history. Among its most lucrative tourist attractions were the sacred PROSTITUTES at the temple of APHRODITE.

The name Korinthos is not Greek, containing as it does the *nth* sound that identifies certain words surviving from the language of the pre-Greek people of Greece. The invading Greek-speaking tribesmen of circa 2100 B.C. took over an existing, pre-Greek settlement and kept the non-Greek name. After the fall of MYCENAEAN CIVILIZATION the region was overrun by the invading DORIAN GREEKS, circa 1100–1000 B.C. Later MYTHS connect the Dorian conquest of Corinth with that of ARGOS, farther south. The Dorian city eventually became governed as an ARISTOCRACY, dominated by an endogamous clan called the Bacchiads. By the 700s B.C. Corinth was the foremost Greek port and manufacturing center, known to HOMER and other poets by the epithet *aphneios*, "wealthy." Corinthian shipbuilding was renowned. Imports included textiles, worked metal, and carved ivory from the non-Greek kingdoms of western Asia. Exports included POTTERY in the beautiful painted styles now known as Protocorinthian and Corinthian. This pottery dominated all markets from about 700 to 550 B.C., until it was superceded by Athenian black-figure ware.

Corinth's exports traveled beyond the Greek world. One avid market was the powerful, non-Greek people called the ETRUSCANS, in northern and central ITALY. Archaeological excavations have revealed immense troves of Corinthian pottery at Etruscan sites; such remnants surely indicate other export items, now vanished, such as perfumes, textiles, and metalwork.

To provide anchorages and local depots along the western trade route, Corinth (circa 734 B.C.) founded two colonies destined to become great cities in their own right: SYRACUSE, on the southeast coast of SICILY, and CORCYRA, on an Adriatic island. But Corcyra rebelled against its mother city, and Corinthian-Corcyrean hostilities would later be an important cause of the PELOPONNESIAN WAR (431–404 B.C.).

Meanwhile, prosperity brought violent political change to Corinth, as the middle class chafed under the Bacchiads' monopoly of power. Circa 657 B.C. a revolution toppled the aristocrats and installed a popular leader, CYPSELUS, as dictator (turannos). Cypselus usually is considered to be the earliest of the Greek TYRANTS; soon this pattern of revolution was sweeping the other major cities of Greece.

Under Cypselus and his son PERIANDER (reigned circa 625–585 B.C.), Corinth reached new commercial heights. New northwestern colonies—AMBRACIA, APOLLONIA, and others—were founded to develop the western trade route and guard the approaches to the Corinthian Gulf. With Periander's paving of a five-mile-long dragway across the isthmus' narrowest section, merchant ships could be trundled on trolleys between the eastern and western seas, thus eliminating the voyage around the Peloponnese and bringing to Corinth a rich revenue in tolls from non-Corinthian shipping. In artwork, Corinth became the birthplace of the monumental ARCHITECTURE that we know as the Doric order. (The more ornate Corinthian order would emerge later.) The influential ISTHMIAN GAMES were instituted around this time.

After Periander's death, his successor was quickly deposed in favor of a constitutional OLIGARCHY that remained Corinth's typical form of government down to Roman times. During the late 500s B.C. Athens arose as Corinth's trade rival. However, the two cities remained friendly, and 40 Corinthian warships fought alongside the Athenian fleet against the invasion of the Persian King XERXES (480 B.C.).

By the mid-400s B.C., however, Corinth—along with its ally SPARTA—was feeling the threat of an expansionist, democratic Athens. Alarmed by an Athenian alliance with Corcyra (433 B.C.), Corinth urged Sparta and the other Peloponnesian states to declare war on Athens. The resulting Peloponnesian War enveloped the Greek world for nearly 30 years and saw Corinthian war fleets and troops in many battles. The fighting ended in defeat for Athens and exhaustion for Corinth.

Rebelling from the onerous rule then imposed by Sparta, Corinth fought against its former ally (and alongside Athens, Argos, and other states) in the CORINTHIAN WAR of 395–386 B.C.. In these troubled years the city also underwent a brief democratic coup. In 338 B.C. Corinth shared in the failed defense of Greece against King PHILIP II of MACEDON. Corinth received a Macedonian garrison and remained a Macedonian holding until 243 B.C., when it was liberated by the statesman Aratus for the Achaean League. But in 222 B.C. Corinth returned to Macedonian control.

When the Romans freed Greece from Macedonian rule in 196 B.C., Corinth again became the foremost city of the Achaean League. But the Greeks' resistance to Roman interference led to the disastrous Achaean War of 146 B.C., in which Corinth was captured and sacked by Roman troops. The Roman senate decreed that the city should be burned down and its art treasures either be sold or carried off to Rome. This pillaged art is said to have increased the Greek influence on the emerging Roman imperial culture.

Historians usually consider the destruction of Corinth in 146 B.C.—and the associated annexing of ACHAEA to become part of a Roman province—as the end of the ancient Greek world. Although other Greek cities thrived under Roman rule and Corinth itself eventually revived, mainland Greece as a political entity had ceased to exist. Greece had become part of ROME.

(See also BELLEROPHON; COINAGE; COLONIZATION; GREECE, GEOGRAPHY OF; HOPLITE; LELANTINE WAR.)

Corinthian order   See ARCHITECTURE.

Corinthian War   Name given by modern historians to the conflict of 395–386 B.C., fought between SPARTA (with its allies) and an alliance of CORINTH, ATHENS, BOEOTIA, ARGOS, EUBOEA, and the kingdom of PERSIA. The grand alliance was remarkable for combining traditional enemies in a united campaign against Spartan supremacy. This anti-Spartan axis introduces the politics of the next generation, when the Boeotian city of THEBES would emerge to challenge and finally defeat Sparta (371 B.C.).

The background of the Corinthian War is Sparta's victory over Athens in the huge PELOPONNESIAN WAR (404 B.C.), which marked the beginning of Sparta's oppressive rule over all the Greek states, former friend and foe alike. Claiming also to protect and rule the Greek cities of ASIA MINOR, Sparta came into conflict with the Persian king Artaxerxes II (399–395 B.C.). As a result, Persian funds became available for the anti-Spartan alliance in Greece.

The course of the war is described in the account titled *Hellenica*, by the Athenian historian XENOPHON (who was present at some of the events). The Spartan general LYSANDER invaded Boeotia at the head of a Spartan-allied army, but was defeated and killed by the Thebans at the Battle of Haliartus (395 B.C.). The next summer the Spartans defeated an allied army in battle outside Corinth, but this was counterbalanced when a Spartan fleet was destroyed by the Athenian admiral CONON, leading a Persian-Athenian fleet at the Battle of CNIDUS, off the coast of Asia Minor (394 B.C.).

Meanwhile, the Spartan king AGESILAUS, summoned home from campaigning against Persian land forces in Asia Minor, led his army along the north Aegean coastline into northeastern Greece. Descending southward, he invaded Boeotia. At the desperate Battle of Coronea (394 B.C.), Agesilaus narrowly defeated an allied army but then withdrew southward toward Sparta without attacking any Boeotian cities.

Thereafter the war became bogged down with maneuverings around Corinth and Argos. In around 392 B.C. Corinth, traditionally an OLIGARCHY, underwent a short-lived democratic coup. But the most important event of the war was its ending: The fighting stopped in 386 B.C., when Persia withdrew its support after King Artaxerxes II had negotiated a separate peace with Sparta. According to this treaty, known as the KING'S PEACE, Sparta ceded the Greek cities of Asia Minor and CYPRUS back to Persian control. This notorious treaty unmasked Sparta once and for all as an oppressive power, bent on dominating mainland Greece even at the cost of selling out the eastern Greek cities. The rise of Thebes in the following years was buoyed partly by widespread anger at this treaty.

(See also EPAMINONDAS; IONIA.)

council   Evidence in HOMER's epic poems and in the LINEAR B tablets indicates that a council (*boulē*) of king's advisers was an important facet of government in primitive Greek societies (circa 1600–1000 B.C.). As the age of kings gave way to the age of ARISTOCRACY (circa 1000–600 B.C.), a Greek city-state's council more or less became the government, combining legislative, executive, and judiciary powers. In this era ATHENS was governed by an aristocratic council called the AREOPAGUS, whose several hundred members, probably serving for life, were drawn exclusively from the city's narrow circle of noble families.

The democratic reforms of CLEISTHENES (1) (circa 508 B.C.) removed the Areopagus from the heart of government in Athens and created instead a council of 500 members selected annually from upper- and middle-income citizens. (By the late 400s B.C., lower-income citizens apparently were admitted.) Under the Athenian DEMOCRACY of the 400s and 300s B.C., the council had important legislative and executive duties but was always subordinate to the sovereign ASSEMBLY of citizens.

The most important job of the Athenian council was to prepare the agenda for the 40 assembly meetings each year. Specifically, the council drafted the proposals to be debated at the next assembly—apparently the assembly could vote only on proposals prepared beforehand by the council. Council meetings might include debate and voting, the hearing of citizens' petitions, the summoning of witnesses, and other information-gathering, for which the council had high authority. Like the assembly, the council had judicial duties. The councillors could act as judge and jury in trying cases of certain state offenses, such as misconduct by officials.

The council also served as an executive body, responsible for the enactment and enforcement of the assembly's decisions and for much of the day-to-day running of the state. The council received foreign ambassadors, had responsibility for the care of public buildings, and oversaw the construction of new warships and the maintenance of the fleets and dockyards for Athens' all-important navy. The council seems to have been especially important in administering finances: Among other duties, it audited the books of all outgoing officials who had handled public funds.

Yet the citizens who served on the council were not career politicians or public servants, but amateurs fulfilling a public duty not so different from our modern jury duty. The 500 council members for the year were selected by lot, in procedures at the 139 local city wards, or DEMES (*dēmoi*). Only male citizens over the age of 30 were eligible—there, as everywhere else in the Athenian government, WOMEN

were excluded. By the late 400s B.C. councillors were receiving state pay for their service. (This important provision, enabling less-wealthy citizens to serve, probably began circa 457 B.C. amid the left-wing reforms of EPHIALTES and PERICLES.)

Each new council began its term in midsummer, at the start of the Athenian year. Aside from festivals and days of ill omen, the members met daily, usually in the council chamber in the AGORA. By nature of its selection, the council always consisted of 10 50-man contingents from the 10 Athenian tribes. Each of these 50-man groups took one turn, through the year, as the council's presiding committee, or *prutaneis*.

(See also LAW AND LAW COURTS; POLIS.)

courtesans   See PROSTITUTES.

Cratinus   One of the three great comic playwrights of classical ATHENS, alongside ARISTOPHANES and EUPOLIS. Cratinus (Greek: Kratinos) was Aristophanes' older contemporary and rival, and greatly influenced him. Apparently Cratinus' plays relied even more on obscene farce and personal derision of public figures than those of Aristophanes.

Cratinus' comedies, of which 27 titles and some 460 fragments survive, were presented in the mid-400s down to 423 B.C. We know he won first prize nine times in competitions at the two major Athenian drama festivals, the City Dionysia (in early spring) and the Lenaea (in midwinter). His titles, often styled as plural nouns, include *The Thracian Women, The Soft Fellows, The Dionysuses,* and *The Odysseuses* (which apparently used the chorus as ODYSSEUS' crew and included a large model of his ship).

No complete play by Cratinus survives, but there exists a synopsis of his *Dionysalexandros* ("Dionysus-Paris"), performed circa 430 B.C. There the god DIONYSUS substitutes for the Trojan prince PARIS in judging the famous beauty contest of the three goddesses. Deciding in 'Aphrodite's favor, the god acquires the beautiful HELEN OF TROY as his reward but is terrified by the approach of Paris himself; farcical action follows, including the start of the TROJAN WAR. The play—clearly a burlesque on the recent outbreak of the very real and serious PELOPONNESIAN WAR—contained much indirect ridicule of the Athenian statesman PERICLES, who was cast as the meddling Dionysus, while his common-law wife, ASPASIA, doubtlessly received a few hits as Helen.

Cratinus' final success came in 423 B.C., with his comedy *The Bottle*. The year before, Aristophanes' *Knights* had publicly mocked Cratinus as a drunken has-been; this year Cratinus responded with a comic self-portrait featuring: a character named Cratinus; his estranged wife, Komoidia (comedy); his sluttish girlfriend, Methē (drunkenness); and the tempting pretty-boy Oiniskos (little wine). The play contained the proverb that an artist who drinks only water can never create anything worthwhile. *The Bottle* won first place at the City Dionysia, well over Aristophanes' *Clouds*, which took third. Cratinus died shortly thereafter.

(See also THEATER.)

cremation   See FUNERAL CUSTOMS.

Crete Largest and southernmost island of the AEGEAN SEA. Crete is long and thin in shape—about 160 miles long and between circa 30–40 miles wide—and on a map it extends horizontally east-west. The island's western shore lies only about 65 miles southeast of the mainland Greek PELOPONNESE and about 200 miles northeast of the Libyan coast. Crete's eastern shore is 130 miles from the coast of ASIA MINOR. Due partly to this dominant position on the sea routes linking the Aegean with the Egyptian and Near Eastern worlds, Crete in the late third millennium B.C. gave birth to a brilliant BRONZE AGE culture, the MINOAN CIVILIZATION. The Minoans were not Greeks, but they are considered as marking the start of Greek prehistory, since they deeply influenced the emerging Greek MYCENAEAN CIVILIZATION of the mainland.

The traditional Greek name Krētē is not a genuine Greek word and may reflect the vanished language of the Minoans. Crete's terrain rises in mountainous humps, peaking at 8,000 feet midisland at Mt. Ida and in the west at the White Mountains. Limestone formations include caves, some of which housed important religious cults in antiquity. Arable land is found in small lowland pockets, where olives, grapes, and grain were farmed, and the many upland plateaus offered grazing for livestock. Mountain TIMBER provided shipbuilding material and a prized export.

Ancient populations tended to favor the warmer and drier eastern half of the island, especially the flatter north coast. ARCHAEOLOGY reveals that the humans first came to Crete in the Neolithic era, around 5000 B.C. These seaborne pioneers, coming perhaps from the Levant, brought the island's first pigs, sheep, and cattle. Newcomers arrived around 2900 B.C., possibly from Asia Minor, bringing BRONZE weapons to Crete and intermarrying with the existing people. From this fusion there arose, circa 2200 B.C., the Minoan civilization, the earliest great nation on European soil. The seafaring Minoans traded with EGYPT and the Levant, and dominated the Aegean.

The Minoan kings' capital was the north-central city of CNOSSUS, inland of the modern seaport of Heraklion. At Cnossus in A.D. 1900 the British archaeologist Sir Arthur Evans discovered the remnants of an elaborate palace complex, begun in around 1950 B.C. and reaching its existing form in around 1700 B.C. Other surviving Minoan monuments on Crete include the palace at Phaestus (mid south coast, across the island from Cnossus), the villas at Hagia Triada (Phaestus' harbor town), the palace at Mallia (eastern north coast), and the remnants of a Minoan town at Gournia (eastern north coast). Current scant knowledge of the Minoans comes largely from the archaeology of these sites. Minoan artworks found in excavation are collected in the Heraklion archaeological museum.

The Minoan sites on Crete tell a tale of vigorous construction after 1950 B.C. and of fiery ruin in around 1400 B.C. or soon after. The cause of this simultaneous destruction was probably an invasion of Crete by warlike Mycenaean Greeks from the mainland. Oddly, the Mycenaeans seem to have abandoned Crete after the palaces' destruction.

In around 1000 B.C. a new people occupied the island. These were the DORIAN GREEKS, who had previously invaded southward through mainland Greece, overrunning the Peloponnese and continuing their conquests by sea.

For the rest of antiquity, Crete remained a Dorian-Greek island, with governmental and social institutions that resembled those at SPARTA.

Dorian Crete was divided into city-states governed as military aristocracies. The chief of these were Cnossus, GORTYN (in the south-central island), and, later, Cydonia (modern Khania, on the western north coast). As was the case at Sparta, Dorian-Greek nobles ruled over a population of rural serfs—probably the descendants of the subjugated non-Dorians.

The island was active in the seaborne expansion of the Greek world in the later 700s and 600s B.C. Cretan colonists helped to found the city of GELA, in southeastern SICILY, and Cretan workshops exported an admired Geometric-style POTTERY and contributed to the style of statuary now known as Daedalic. But gradually the Cretan cities withdrew into isolation, and Crete declined amid internal conflicts, mainly between Cnossus and Gortyn. By the late 200s B.C. the island had become notorious as a haunt of pirates. Order was restored by the Romans, who annexed Crete in 67 B.C. and made it part of a Roman province.

The Cretans were the best archers in the Greek world (where archery was generally not practiced), and many Greek armies from the 400s B.C. onward employed Cretan bowmen as mercenaries. Cretans also had the reputation of being liars.

The RELIGION of Dorian Crete was distinguished by certain cults and beliefs that probably contained pre-Greek, Minoan elements. It was said that the great god ZEUS, as a baby, had been hidden in a cave on Crete's Mt. Dicte to save him from his malevolent father, CRONUS. More amazing to classical Greeks was the Cretans' claim that the immortal Zeus was born and died annually on Crete and that his tomb could be seen at Cnossus. This "Zeus" may have been the surviving form of a mythical son or consort of the prehistoric Minoan mother goddess.

(See also ARISTOCRACY; ARTEMIS; ATLANTIS; COLONIZATION; GREEK LANGUAGE; HELOTS; ROME; SCULPTURE; SHIPS AND SEAFARING.)

Critias Athenian oligarch who lived circa 460–403 B.C. As leader of the THIRTY TYRANTS, who ruled ATHENS by terror in 404–403 B.C., Critias was one of the true villains of Athenian history.

Born into an aristocratic clan, Critias (Greek: Kritias) was pupil of certain SOPHISTS and of the philosopher SOCRATES. Critias was an elder kinsman of the philosopher PLATO (specifically, cousin to Plato's mother) and was also the uncle and guardian of Plato's uncle Charmides. Both Critias and Charmides appear as glamorous, youthful figures in several of Plato's (fictional) dialogues, written in the first quarter of the 300s B.C.

As a suspected enemy of the DEMOCRACY, Critias was arrested but then released after the incident of the Mutilation of the Herms (415 B.C.), during the PELOPONNESIAN WAR. Later Critias proposed to the Athenian ASSEMBLY the decree recalling the exiled Alcibiades (probably 411 B.C.), but when the Athenians turned against Alcibiades again (406 B.C.), Critias apparently suffered incidental blame and was banished. He went in exile to THESSALY.

Like other exiles, Critias was recalled as part of the peace terms imposed on Athens by the victorious Spartans at the end of the Peloponnesian War, in 404 B.C. He was then elected to the dictatorial, pro-Spartan government commonly known as the Thirty Tyrants. The Athenian historian XENOPHON, who lived through these events, portrays Critias as leader of the Thirty's extremist faction, against the more moderate THERAMENES. Critias directed the Thirty's reign of terror, executing anyone who was likely to organize resistance or whose personal wealth was attractive. He eventually denounced his colleague Theramenes and had him executed. Critias' crowning outrage was to arrange the mass execution of 300 men (probably the entire male citizen population) of the nearby town of Eleusis.

In early 403 B.C. the Thirty were toppled. Critias was killed, with Charmides and others, in street-fighting at PIRAEUS. The hatred of Critias' memory played a role in the prosecution of his former teacher Socrates under the restored democracy (399 B.C.).

A man of intelligence and talent, Critias wrote LYRIC POETRY, tragedies, and prose. His prose style was admired, and one of his works, now lost, was titled "Conversations" (Homilia). It is possible that this was a model for Plato's literary dialogues. As a thinker, Plato seems to have inherited some of his uncle's authoritarian nature.

Critius See SCULPTURE.

Croesus Last king of LYDIA, a powerful non-Greek nation of west-central ASIA MINOR. Croesus (Greek: Kroisos) inherited the throne at age 35 from his father, Alyattes, and reigned circa 560–546 B.C., until his country was conquered by the Persian king CYRUS (1).

Croesus' wealth was proverbial. Friendly toward the Greek world, he was a patron of the god APOLLO's shrine at DELPHI and was the first foreign ruler to form an alliance with a mainland Greek state, SPARTA. Closer to home, he subdued the Greek cities of western Asia Minor, but dealt benevolently with them as subjects.

On his eastern frontier he met disaster. Seeking to conquer east-central Asia Minor from the Persians, Croesus led an army across the River Halys, the border between Lydia and the Persian domain. After an inconsequential battle against King Cyrus, Croesus returned to Lydia, where Cyrus, following quickly, defeated and captured him. Croesus' subsequent fate is unknown. One legend says he was carried off by his divine protector Apollo to safety in the magical land of the Hyperboreans.

Like other Eastern despots, Croesus inspired the Greek imagination. The tales told about him by the historian HERODOTUS (circa 435 B.C.) are a fascinating mix of fact and storytelling. With his grand style and sudden fall, Croesus represented for the Greeks a real-life example of HUBRIS—excessive pride that leads to a divinely prompted blunder in judgment, which leads to disaster.

One legend claims that, at the height of Croesus' reign, he was visited by the Athenian sage SOLON. Croesus asked Solon to name the happiest man he had ever seen, expecting to hear himself named, but Solon explained to Croesus that no man may be called "happy" until he is dead; before then, he is merely lucky. This Greek proverb later reappears at the end of SOPHOCLES' tragedy *Oedipus the King.*

The best-known story tells how, before his campaign against PERSIA, Croesus consulted the Delphic Oracle. He was advised that if he crossed the Halys River he would destroy a mighty empire. Heartened by this prophecy, he decided to march. Unfortunately, the mighty empire turned out to be his own.

(See also MIDAS; PROPHECY AND DIVINATION.)

Cronus In MYTH, the primeval king of the TITANS and father of ZEUS and other gods. According to HESIOD's epic poem the *Theogony,* Cronus was born to GAEA (Mother Earth) and Uranus (Ouranos, "Sky"). On his mother's advice, Cronus castrated his father with a sickle and ruled in his place.

Cronus married his own sister, Rhea, who bore him the gods HESTIA, DEMETER, HERA, HADES, POSEIDON, and, last, Zeus. But Cronus had been warned that his offspring would subdue him, and so he swallowed each newborn child. Finally, at Zeus' birth, Rhea tricked her husband—presenting him with a stone wrapped in swaddling clothes—and spirited the infant off to CRETE. Eventually Cronus was made to vomit up all his swallowed children, who (still alive) followed Zeus in revolt against their father. A different vein of legend, at odds with this grim picture, described Cronus as supervising a time of innocence and blessing—if not in heaven, at least on earth. Cronus' reign was said to have marked the Golden Age of human history, when people lived without greed, violence, toil, or need for laws.

The name Cronus (Greek: Kronos) seems to have no meaning in the Greek language. Like other elements of the Greek Creation myths, the Cronus story may have derived from non-Greek, Near Eastern sources. Specifically, it resembles the tale of the Mesopotamian god Kumarbi, which was current in the second millennium B.C. Modern scholars believe that the Cronus myth may have come to Greece via the Phoenicians, with whom the Greeks had extensive trading contacts in 900–700 B.C.

At classical ATHENS there was a feast of Cronus, the Kronia, celebrated in midsummer, just after the Greek New Year. The Kronia was probably a harvest festival, and Cronus was often portrayed in art as carrying a sickle, the harvester's tool. (The sickle also was the weapon he used to castrate his father.)

The primitive figure of Cronus was the object of attempted rationalization during the Athenian Enlightenment of the 400s–300s B.C. A theory claimed that his name was not really Kronos but rather Chronos, "time," and that his myth of impious violence to father and children was simply an allegory for the ravages of time. This confused interpretation created an image that, surviving 2,500 years, can be seen every New Year's—old Father Time, holding his emblematic sickle.

(See also PHOENICIA; RELIGION.)

Croton Important Greek city on the southern rim of the Gulf of Taranto, on the "sole" of south ITALY. Located high above a small double harbor, near fertile farmland, Croton (Greek: Kroton) was founded by colonists from ACHAEA, probably around 700 B.C. The city carved out a sizable domain at the expense of the local Italian inhabitants, the Brutii.

Circa 530 B.C. the Samian-born philosopher PYTHAGORAS founded his mystical school of study at Croton and apparently helped to run the city government, as an OLIGARCHY. By the late 500s B.C., prosperous Croton had a famous school of MEDICINE, producing, among others, the philosopher-physician Alcmaeon. Another distinguished citizen of the day was Milon the Strongman, a champion in WRESTLING at the OLYMPIC and PYTHIAN GAMES.

The city's zenith came in 510 B.C., when it destroyed its archrival, the Italian Greek city SYBARIS. But Croton soon was weakened by conflicts with two other Italian Greek cities, RHEGIUM and LOCRI, and with the Brutii. Croton was captured and plundered in 379 B.C. by the Syracusan ruler DIONYSIUS (1). Exhausted by the wars that accompanied ROME's expansion through Italy in the 300s and 200s B.C., Croton eventually became a Roman subject state.

Cumae Ancient Greek city of Campania, on the west coast of ITALY, located just north of the Bay of Naples, 10 miles west of modern-day Naples.

Cumae (Greek: Kumē) was one of the earliest datable Greek colonies, established by Euboean Greeks who moved ashore from their nearby island holding of PITHECUSAE circa 750 B.C. One story claims that the new city included eastern Greek settlers who named it for their native CYMĒ, in ASIA MINOR.

The founding of Italian Cumae marks an early milestone in Greek COLONIZATION, namely the decision by a band of Greeks to make a first landing on the Italian coast. Possibly they went ashore at the invitation of the ETRUSCANS, a powerful Italian people who were avid consumers of Greek goods and who had a stronghold at nearby Capua.

Cumae thrived, supplying the Etruscans with Corinthian painted POTTERY, Euboean worked BRONZE, and other wares. It sent out colonists of its own, founding ZANCLĒ (modern Messina, in SICILY, circa 725 B.C.) and Neapolis ("new city," modern Naples, circa 600 B.C.), among other settlements. As a northern outpost of Greek culture in Italy, Cumae played a crucial role in the hellenization of the Etruscans.

But Cumae's relations with the Etruscans eventually worsened, and in the late 500s B.C. the Cumaean leader Aristodemus twice defeated Etruscan armies (and then made himself tyrant of the city). In 474 B.C., in alliance with the Syracusan tyrant HIERON (1), Cumae won a great sea battle against the Etruscans and crushed their power. But in 421 B.C. the city fell to another Italian people, the Samnites, and later passed to the Romans. Throughout antiquity Cumae was known for its priestess called the Sibyl, who was a Greek oracle of the god APOLLO similar to that at DELPHI.

(See also PROPHECY AND DIVINATION; ROME; TYRANTS.)

Cyclades Group of islands in the AEGEAN SEA roughly forming a circle, *kuklos,* around the holy isle of DELOS. Aside from Delos, the major Cyclades were Naxos (1), Paros, Andros, Ceos, THERA, and MELOS. Paros was famous for its marble; Naxos, the largest and most fertile of the

Cyclades, was known for its WINE and its associated worship of the god DIONYSUS.

The Cyclades enjoyed a flourishing civilization in the third and second millennia B.C., well before the Greeks arrived. Archaeological evidence such as surviving POTTERY gives a picture of two-way TRADE with Minoan CRETE, including the export of obsidian (volcanic glass, mined on Melos and used as a cutting edge for axes, adzes, and the like, circa 2000 B.C.). By 1600 B.C. these pre-Greek Cyclades had come under direct control of Crete. After 1400 B.C. the Cyclades seem to have been ruled by Mycenaean Greeks from the mainland. Before 1000 B.C. most of the Cyclades were occupied by migrating IONIAN GREEKS. For the main centuries of Greek history, the Cyclades were inhabited by ethnic Ionian Greeks; of the principal islands, only Melos and Thera were settled by DORIAN GREEKS.

The Cyclades fell to the invading Persians in 490 B.C.; Naxos, resisting, was ravaged. At the end of the PERSIAN WARS, most of the islands joined the DELIAN LEAGUE, headquartered at Delos but led by ATHENS (478 B.C.). The league died with Athens' defeat in the PELOPONNESIAN WAR (404 B.C.), but by 377 B.C. most of the islands had joined the SECOND ATHENIAN LEAGUE. With the rest of Greece, they passed into the hands of the Macedonian king PHILIP II in 338 B.C.. During the HELLENISTIC AGE (300–150 B.C.), the Cyclades were a bone of contention mainly between the Greek dynasties of MACEDON and EGYPT. In the mid-100s B.C. they passed to the domain of ROME.

Cyclops (plural: Cyclopes) A type of mythical, gigantic, semihuman monster, first described in HOMER's epic poem the *Odyssey* (written down circa 750 B.C.). The name Kuklops, "round eye," refers to the one large eye that such creatures had in midforehead. Descended from the earth goddess GAEA and her husband, Uranus (Ouranos, "Sky"), the Cyclopes dwelt in a distant land (sometimes identified as eastern SICILY), with no cities or laws. The *Odyssey* (book 9) describes the hero ODYSSEUS' violent encounter with the Cyclops POLYPHEMUS.

A slightly different tradition, presented in HESIOD's epic poem the *Theogony* (circa 700 B.C.), made the Cyclopes a guild of supernatural, one-eyed blacksmiths, who forged thunderbolts for the great god ZEUS. These Cyclopes were associated with the smith god, HEPHAESTUS. The classical Greeks ascribed to the Cyclopes the building of certain ancient walls made of huge blocks without mortar, such as the walls at TIRYNS. In reality, these had been built by Mycenaean Greeks circa 1300 B.C., but to later generations they looked like the work of giants. The term cyclopean was used to describe such masonry.

(See MYCENAEAN CIVILIZATION.)

Cylon Athenian aristocrat who attempted to seize supreme power at ATHENS, circa 632 B.C., at a time when the city was still governed as an ARISTOCRACY. Cylon (Greek: Kulon) and his followers captured the Athenian ACROPOLIS; yet the common people did not rise in support, and government forces besieged the conspirators. Cylon escaped but his men were massacred, despite having claimed sanctuary at religious altars. This mishap was the origin of the "Alcmaeonid curse"—the taint of pollution considered to lie perpetually on the entire clan of the ALCMAEONIDS, because one of the clan, Megacles, had overseen the massacre.

(See also FURIES; TYRANTS.)

Cymē Southernmost and main city of AEOLIS, a Greek region on the northwestern coast of ASIA MINOR. Like other cities of Aeolis, Cymē was settled circa 900 B.C. by AEOLIAN GREEKS who had migrated eastward, via LESBOS, from mainland Greece. As indicated by the name Kumē (wave or sea), the city was a port, advantageously located alongside the mouth of the Hermus River (which allowed trading contact with interior Asia Minor).

Cymē's fortunes followed those of the Aeolis region. Captured by the Persians circa 545 B.C., Cymē took part in the failed IONIAN REVOLT in 499–493 B.C. Liberated by the Greeks at the end of the PERSIAN WARS (479 B.C.), Cymē became part of the Athenian-controlled DELIAN LEAGUE. After ATHENS' defeat in the PELOPONNESIAN WAR (404 B.C.), Cymē reverted to Persian control by the terms of the KING'S PEACE (386 B.C.). By about 377 B.C. Cymē had joined the SECOND ATHENIAN LEAGUE. After ALEXANDER THE GREAT'S death (323 B.C.), Cymē passed through the SELEUCID EMPIRE and the kingdom of PERGAMUM, before becoming part of the Roman Empire in the 100s B.C.

Cymē's best-known citizen was the historian Ephorus (circa 400–330 B.C.), who wrote a world history as well as a history of Cymē, neither of which has survived. An inferior scholar, Ephorus was notorious for including in his world history such comments as "In this year the citizens of Cymē were at peace."

(See also CUMAE.)

Cynics The English words *cynic* and *cynicism* describe an informal, embittered frame of mind. But the original Cynics comprised a major Greek philosophical movement, which arose in the early 300s B.C. The name Kunikoi, "doglike," derived from one of the movement's pioneers, DIOGENES of SINOPE (400–325 B.C.), whose austere and immodest way of life won him the nickname the Dog (Kuon). Unlike other philosophies, Cynicism was never organized around a formal place of study. Although overshadowed in the 200s B.C. by two new and partly derivative schools of thought, STOICISM and EPICUREANISM, Cynicism nevertheless continued to attract a following in response to the spiritual crisis in the HELLENISTIC ERA (300–150 B.C.). In the first century A.D. the cities of the Roman Empire teemed with traveling Cynic beggars.

The Cynics taught a radical doctrine of moral self-sufficiency. They renounced wealth and social convention, preferring to go homeless, do no work, wear simple clothes or rags, and avoid such niceties as finding privacy before relieving themselves. They sought knowledge and harmony through a liberation from materialism. But, far from withdrawing to live as hermits, they chose to dwell in cities and (often) to travel from city to city, preaching. They would sermonize and hold philosophical discussions on streetcorners and were notorious for their outspokenness. Their lampoons against vice and pride were often directed personally against passersby.

The movement's roots lay with the Athenian philosopher SOCRATES (469–399 B.C.). Although not himself a Cynic, Socrates was known for his austerity and his habit of questioning people on the street. After Socrates' death, one of his followers, Antisthenes, an Athenian, lived in exaggerated imitation of Socrates' plain living. In a modified Socratic belief, Antisthenes taught that happiness comes from personal virtue (*aretē*), not from pleasure, and that the best virtue comes from triumph over hardship.

Antisthenes was the first Cynic, but the movement's most famous master was Diogenes, who began as Antisthenes' pupil. Other important Cynics included Bion of Borysthenes and Menippus of Gadara (early 200s B.C.); Menippus adapted the Cynics' haranguing speaking style to a written form that was a precursor of the literary satire later produced by the Romans.

With their doctrine of self-improvement through painful effort, the Cynics revered the mythical hero HERACLES, whose twelve Labors were seen as an ideal example of moral victory and self-liberation.

(See also PHILOSOPHY; POLIS.)

Cyprus Large eastern Mediterranean island, 140 miles long, which was an eastern frontier of the Greek world. Located 60 miles off the coast of northern Syria and 50 miles south of ASIA MINOR's southern coast, Cyprus played a vital role in TRADE and warfare between East and West. The island's Oriental contacts and remoteness from mainland Greece gave rise to a distinctive, sometimes peculiar, Greek culture.

As shown by ARCHAEOLOGY, the island was host to a flourishing non-Greek civilization, derived from Asia Minor, during the first half of the second millennium B.C.. After about 1400 B.C. Cyprus was intensively colonized by Mycenaean Greeks from mainland Greece. What lured them to Cyprus was copper—the major component of the alloy BRONZE, upon which the Mycenaeans depended for war and FARMING. Cyprus had the richest copper deposits in the eastern Mediterranean; the English word *copper* comes from the Latin term *Cyprium aes*, "the Cypriot metal." A copper ingot eventually became the island's emblem on COINAGE.

Greek Cyprus was largely undisturbed by the collapse of MYCENAEAN CIVILIZATION in mainland Greece (circa 1200–1000 B.C.). The island's spoken Greek dialect, unaffected by any change or immigration, retained certain characteristics of old Mycenaean Greek. Other Mycenaean survivals included the Cypriot Syllabary (as it is now called), a primitive form of writing using pictograms, employed on Cyprus circa 700–200 B.C.

In the 800s B.C. the seafaring Phoenicians, seeking copper and shipping stations, established a stronghold in southeast Cyprus: Kart Hadasht, "new city," called Kition (Citium) by the Greeks and Kittim in the Bible. The Phoenicians, in steady conflict with the Greeks, remained a powerful presence and later served as puppets of the Persian overlords. Meanwhile Cyprus became a place where Greeks could observe and imitate Phoenician shipbuilding, trade techniques, and religious beliefs. The Phoenician cult of ADONIS may have reached Greece by way of Cyprus, as probably did a more important religious borrowing from the Near East—the cult of the goddess APHRODITE.

The Cypriot Greeks became expert seafarers. By 700 B.C. their population was centered in eight cities, roughly ringing the island's east-south-west perimeter. Each city had its own king; foremost of these Greek cities was Salamis (often called Cypriot Salamis, to distinguish it from the island near ATHENS). Located on a hospitable bay on Cyprus' east coast, site of modern Famagusta, Salamis commanded the shipping run to north Syria. The ninth city of Cyprus was Phoenician Citium; in later centuries it became partially Greek.

In 525 B.C. Cyprus submitted to the conquering Persians. As Persian subjects, the Cypriots—both Greeks and Phoenicians—supplied warships and crews for their overlords' campaigns. The Cypriot Greeks joined the IONIAN REVOLT against Persian rule but were defeated (circa 497 B.C.). Modern archaeology at the southwest city of Paphos has uncovered the lower levels of an ancient Persian siege mound, littered with hundreds of bronze arrowheads—the remnants of Greek defensive arrow volleys from the city's walls.

In the mid-400s B.C. Cyprus was a bone of contention between the Persians and imperial Athens. The Athenian commander CIMON had begun a large-scale attempt to liberate Cyprus when he died there (450 B.C.), and the island relapsed into Persian-Phoenician control, despite the hellenic cultural program of the dynamic king Evagoras of Salamis (circa 411 B.C.). By the terms of the KING'S PEACE (386 B.C.), the mainland Greeks recognized Persian mastery of Cyprus.

In 332 B.C., after the Macedonian king ALEXANDER THE GREAT had invaded the Persian Empire, Greek Cyprus dramatically joined his side and contributed 120 warships for the conqueror's massive siege of Tyre. After Alexander's death, PTOLEMY (1), the Macedonian-born ruler of EGYPT, captured Cyprus despite suffering a naval defeat by his enemy DEMETRIUS POLIORCETES off Salamis (306 B.C.). Ptolemy suppressed Cyprus' traditional nine kingships, replaced them with democratic forms, and established an Egyptian overlordship that lasted until 58 B.C., when the island was taken over by the Romans.

(See also GREEK LANGUAGE; PHOENICIA; SHIPS AND SEAFARING; WARFARE, NAVAL.)

Cypselus   Corinthian dictator who reigned circa 657–625 B.C., after seizing the city from the oppressive, aristocratic Bacchiad clan. CORINTH was at that time the foremost commercial power in Greece, and Cypselus gained control by leading a revolution of the affluent middle class that had previously been denied political power. Cypselus (Greek: Kupselos) reigned mildly and is said to have kept no bodyguard. But as usurper and dictator, he usually is counted as one of the first Greek TYRANTS, who began arising in the mid-600s B.C. to wrest power violently from the aristocratic ruling class throughout the Greek world.

According to a folktale recorded by the Greek historian HERODOTUS (circa 435 B.C.), Cypselus was born to a Bacchiad mother and a non-Bacchiad father. Members of the Bacchiad clan, alerted by a prophecy that the baby was destined to destroy them, arrived to kill him. But Cypselus' mother hid him in a wooden chest (kupselē) and outwitted the assassins.

This concept of the blessed but hunted infant is a Near Eastern mythological theme that recurs in the stories of Moses, Sargon the Assyrian, CYRUS (1) the Persian, and Jesus. The Greeks—having learned this type of legend from the Phoenicians, circa 900–700 B.C.—applied it to certain Greek mythical heroes, such as OEDIPUS. Herodotus' tale evidently derives from progaganda that Cypselus himself issued during his reign; a cedarwood chest, supposedly the one used to hide the infant Cypselus, was eventually kept on display in the Temple of HERA at OLYMPIA.

Cypselus and his son and successor, PERIANDER, increased Corinth's power through TRADE and manufacturing. Export markets were developed in Etruscan ITALY and other western locales, and Corinthian vase painting blossomed into the perfection of style that modern scholars call Early Corinthian.

(See also ETRUSCANS; POTTERY; PROPHECY AND DIVINATION.)

Cyrene (1)   Major Greek city of North Africa, near modern-day Benghazi in eastern Libya, founded by DORIAN GREEKS from the island of THERA circa 630 B.C. Located inland, on a hill surrounded by fertile plains, Cyrene (Greek: Kurēnē) thrived in TRADE with mainland Greece, which lay 300 miles due north across the Mediterranean. (CRETE was 200 miles northeast.) Through its harbor town of Apollonia, Cyrene exported grain, woolens, oxhides, and silphium—a local plant, prized as a laxative and digestive. Far surpassing Thera in wealth and power, Cyrene established local colonies of its own and was an important early outpost of the Greek world.

The circumstances of Cyrene's foundation are well known, thanks to an account by the historian HERODOTUS (circa 435 B.C.) and a surviving inscription, dating from the 300s B.C., claiming to record the Therans' decision to send out ships of colonists 300 years earlier. The expedition was organized to relieve overpopulation and water shortage on Thera. Although volunteers were welcome, one son from each family was required to join, the penalty for default being death. The sea route to the intended landfall had already been explored by Greek traders from Crete. The Theran colonists, landing on the then-fertile Libyan coast, were at first welcomed by the native Libyans.

Cyrene's foundation offers one of the clearest examples of the role played by APOLLO, patron god of colonists, and by Apollo's oracle at DELPHI, in central Greece. The Theran expedition and its leader were approved beforehand by Delphic Apollo, and the new city was named for the god. (The name Kurēnē is related to Kouros, "young man," a cult title of Apollo.)

Early Cyrene was governed by kings of a family named the Battiads, or "sons of Battus." (Battos, "stammerer," was supposedly the name of the original expedition's leader.) The dynasty customarily used the royal names Battus and Arcesilaus (meaning "leader of the people"). A prominent king was Battus II, surnamed the Lucky, under whom the city attracted a new influx of Greek settlers and fought off a Libyan-Egyptian attack, circa 570 B.C. In this era Cyrene founded its own nearby colonies: Barca and Euhesperides (modern Benghazi). The territory of these Greek cities, in the "hump" of eastern Libya, was named Cyrenaica, as it is still called today.

Cyrene submitted to the Persians in around 525 B.C., but the Battiads remained in power, amid some dynastic violence, until about 440 B.C., when Arcesilaus IV was deposed in favor of an Athenian-style DEMOCRACY. Remote from the political centers of the Greek world, Cyrene was unaffected by the PELOPONNESIAN WAR (431–404 B.C.) but was absorbed into the domain of ALEXANDER THE GREAT after he reached EGYPT in his invasion of the Persian empire (332 B.C.). On Alexander's death in 323 B.C. the city passed to PTOLEMY (1), the Macedonian-born ruler of Egypt. Ptolemy removed the city's democracy and installed a liberal OLIGARCHY, which he could better control.

Under lenient Egyptian rule, the city reached a peak of prosperity in the late 300s and 200s B.C., with a public building program and such cultural ornaments as a native school of PHILOSOPHY—the Cyrenaics, founded by Aristippus, a follower of SOCRATES. In the mid-200s B.C. the Cyrenaic cities banded into a federation, which remained part of Egypt until the 100s B.C., when the Romans began to interfere in Egyptian administration. In 74 B.C. Cyrenaica became a province of the Roman domain.

(See also COLONIZATION; CYRENE (2); ROME.)

Cyrene (2) In MYTH, a nymph who was beloved by the god APOLLO. Cyrene (Greek: Kurēnē) lived in THESSALY, where she delighted in the woodlands and in hunting wild beasts.

Apollo spied Cyrene one day as she wrestled with a lion on Mt. Pelion. He immediately fell in love with her for her beauty and "manly" strength. The god carried her off in his golden chariot to Libya in North Africa. There, on a hill surrounded by fertile plains, Apollo founded the Greek city that he named CYRENE (1), for her to rule. She bore him a hero son, Aristaeus, who introduced to humankind such civilizing arts as bee-keeping and cheese-making.

The legend of Cyrene reveals at least two layers. The earlier layer—describing Apollo and the nymph in Thessaly—appears in a fragment by the epic poet HESIOD, who wrote circa 700 B.C. (before the Greeks founded the city of Cyrene, in circa 630 B.C.). In the original version, Apollo probably did not carry his beloved off to North Africa—a detail not needed for the story. But at some point, perhaps circa 600 B.C., the older tale was developed into a political "foundation myth," explaining the origin of the city of Cyrene. This later legend is told in the poet PINDAR's ninth Pythian ode, written circa 474 B.C. for an athlete of Cyrene.

(See also NYMPHS.)

Cyrus (1) Persian king and conqueror who ruled from 559–530 B.C. and built PERSIA into a vast empire. Cyrus (Greek: Kuros) also brought Persia to its first, hostile contact with the Greek world. Despite this, Cyrus was viewed by later Greek thinkers as a model of the wise and righteous ruler; the historians HERODOTUS (circa 435 B.C.) and XENOPHON (circa 380 B.C.) wrote about his life in legendary terms.

Cyrus—sometimes called Cyrus the Great or Cyrus II—was the grandson of a prior King Cyrus and son of King Cambyses. He ruled after the death of his father. The Persians at that time were a simple, seminomadic people of the southwestern Iranian plateau; they and their king were subordinate to a kindred people, the Medes. Cyrus revolted against the Medes, captured their capital city of Ecbatana, and made himself supreme king. Henceforth the Persians were the dominant people. The Medes became their inferiors, and there arose a new imperial dynasty—Cyrus' family, called the Achaemenids (descendants of Achaemenes).

The domain that Cyrus had seized was one of the three great powers of western Asia; the others were the kingdoms of LYDIA and Babylonia. Over his 20-year reign, Cyrus eliminated these two rival kingdoms and carved out an empire stretching from the AEGEAN SEA to Afghanistan.

First, provoked by the (undoubtedly welcome) aggression of the Lydian king CROESUS, Cyrus marched to ASIA MINOR in 546 B.C. and conquered Lydia. It was then that he encountered the Greek cities of IONIA, in western Asia Minor, which had been privileged subject states under Croesus. After some resistance, the Greeks of Asia Minor submitted to the Persians (circa 545 B.C.). These events could be described as the starting point of the PERSIAN WARS.

Later Cyrus seized the city of Babylon and its empire—in effect avenging the Babylonian subjugation of the JEWS—a triumph for which he is praised in the Old Testament book of Isaiah. He seems to have governed benevolently, allowing nations to keep their laws and religious customs once they had offered submission. He died in battle on his northeastern frontier (perhaps modern-day Uzbekistan) and was succeeded by his son Cambyses (530 B.C.).

Cyrus (2) See XENOPHON.

Cyzicus Greek seaport of northwestern ASIA MINOR, on the south coast of the SEA OF MARMARA, about 60 miles east of the HELLESPONT. Lying in the territory of the non-Greek Mysians, Cyzicus (Greek: Kuzikos) was founded by Greeks from MILETUS, circa 675 B.C. According to legend, the city was named for an ancient native King Cyzicus, who had been killed in a mishap involving the Greek hero JASON (1) and his Argonauts.

Cyzicus was located on the southern tip of an inshore island called Arctonnessus ("bear island," now a peninsula of the Turkish mainland). The site commanded a double harbor, formed by the narrow channel between island and shore; this superior anchorage became an important station along the Greek shipping route to the BLACK SEA. Cyzicus thrived from TRADE by sea and by the land routes from interior Asia Minor; it also enjoyed rich local fisheries.

From the late 500s until the late 300s B.C., Cyzicus minted coins of electrum (an alloy of GOLD and SILVER) in the high denomination known as the statēr. Stamped with the city's emblem, a tuna, Cyzicene statērs were renowned for their beauty and reliable measure.

By about 493 B.C. the city had submitted to the conquering Persian king DARIUS (1). Liberated by the Greek counteroffensive after 479 B.C., Cyzicus joined the Athenian-led

DELIAN LEAGUE and (a sign of its prosperity) contributed a sizable annual tribute of nine silver TALENTS.

Revolting from Athenian rule in the Delian uprising of 411 B.C., during the PELOPONNESIAN WAR, Cyzicus welcomed a Spartan-Peloponnesian fleet under the Spartan commander Mindarus. The following spring, 410 B.C., saw one of the war's great sea battles, off Cyzicus. The Athenian commander ALCIBIADES, arriving with 86 warships, surprised the 60 Spartan ships at their naval exercises and cut them off from Cyzicus harbor. The Spartans fled inshore and moored. When Alcibiades brought some of his ships ashore, the Spartan commander, Mindarus, also led some of his men ashore, but was killed in the fighting.

The Athenians captured the entire enemy fleet with the exception of one squadron, set afire by its fleeing crew. The victorious Alcibiades extorted large sums of money from Cyzicus to pay his crews, but sailed off without doing other damage to the recaptured rebel city.

After passing into the empire of ALEXANDER THE GREAT (334 B.C.), Cyzicus was absorbed into the east Greek SELEUCID EMPIRE circa 300 B.C. The city continued to prosper, enjoying good trade relations with the Seleucids' rivals, the kings of nearby PERGAMUM. Like other portions of western Seleucid territory, Cyzicus came under Roman control circa 188 B.C.

(See also COINAGE.)

D

Daedalus Mythical Athenian inventor, regarded as a patron hero of craftsmen. According to legend, Daedalus (Greek: Daidalos) invented carpentry and woodworking, among other skills. But he fled from ATHENS after murdering his nephew-apprentice, who showed signs of surpassing him in talent. Traveling to CRETE, Daedalus constructed a number of fabulous works, including the mechanical cow in which Queen Pasiphaë was able to quench her unnatural lust for a bull. Imprisoned by the enraged king MINOS on account of the Pasiphaë episode, Daedalus and his son, Icarus, escaped by flying away on mechanical wings attached to their arms. But Icarus, despite his father's warning, soared too close to the sun, which melted his wings' adhesive wax, and plummeted to his death in the AEGEAN SEA. Icarus' fatal flight has often been interpreted as a metaphor for HUBRIS and overreaching ambition.

Landing in SICILY, Daedalus was protected by the native king Cocalus. The mighty Minos, arriving in pursuit, was murdered at Cocalus' order. Daedalus stayed in Sicily and built a number of works.

In real life, the classical Greeks—especially the Sicilian Greeks—ascribed to Daedalus many existing ancient monuments whose origin they could not otherwise explain. Supposedly he was the first sculptor to free the statue's arms from its sides and show the eyes open.

The Athenians commemorated Daedalus in a DEME (city ward) named Daidalidai, "the descendants of Daedalus." The Athenian philosopher SOCRATES (469–399 B.C.), a sculptor or stonecutter by profession, claimed personal descent from Daedalus—perhaps jokingly or symbolically.

(See also LABYRINTH; SCULPTURE.)

Damocles Syracusan courtier during the reign of the tyrant DIONYSIUS (1), in the early 300s B.C. Damocles praised the luxury and power of a ruler's life, until one day Dionysius ordered Damocles to be feasted at a banquet, underneath a downward-pointing sword, suspended by a single hair. The object lesson was that a ruler's life, for all its splendors, is fraught with worry and fear.

Danaids See DANAUS.

Danaus In Greek MYTH, Danaus and Aegyptus were brothers who lived in EGYPT and had 50 children each; Aegyptus had all boys, Danaus, all girls (the Danaidai, "daughters of Danaus.") When the children reached maturity, Aegyptus insisted that his sons marry their cousins, the Danaids. Danaus refused and fled with his daughters

to the Greek city of ARGOS, where they were received as suppliants. When the sons of Aegyptus arrived in pursuit, Danaus—to spare the Argives harm—consented to the 50-fold marriage. But he secretly instructed his daughters to murder their husbands on their wedding night. All obeyed except for Hypermnestra (whose name is translatable as "excessive wooing" or "special intent"); she spared her husband, Lynceus, and helped him to escape.

Danaus was enraged to learn of Hypermnestra's disobedience, but she was acquitted by an Argive law court following a plea by the love goddess APHRODITE. Lynceus later returned, killed Danaus, ruled Argos with Hypermnestra, and founded a dynasty of Argive kings. The other Danaids remarried among the Argives; their descendants were known as the Danaans (Danai), a name that in the epic poems of HOMER simply means "Greeks." The tale of the Danaids was the subject of a trilogy by the Athenian tragedian AESCHYLUS; one play from this trilogy survives, the *Suppliants* (463 B.C.).

According to one story, the 49 guilty Danaids were punished after death for their outrage to the MARRIAGE bed. In the Underworld, they are forced forever to fetch water in leaky jars or sieves.

The odd myth of Danaus possibly presents a distorted reflection of some contact between Egypt and the Mycenaean Greeks, circa 1600–1200 B.C.

(See also AFTERLIFE; MYCENAEAN CIVILIZATION.)

Dardanelles See HELLESPONT.

Darius (1) King of PERSIA from 522 to 486 B.C., known for his soldiering and administrative abilities. His reign saw hostilities intensify between the Persian Empire and its relatively weak western neighbors, the Greeks.

Although not the son of a king, Darius was of Persian royal blood. Coming to the throne at around age 30, following the death of the childless king Cambyses, Darius soon advanced the empire's boundaries. On the northwestern frontier, he subdued several eastern Greek states, including SAMOS, CHIOS, LESBOS, and BYZANTIUM. He was the first Persian king to cross into Europe, bridging the BOSPORUS and the Danube and leading an expedition against the nomads of western Scythia (circa 512 B.C.). This expedition failed, but Darius did gain the submission of THRACE and MACEDON, which brought his domain right up to the northeastern border of mainland Greece. The Persian Empire had reached its greatest extent, stretching (in modern terms) from Pakistan to Bulgaria and southern EGYPT. It was the largest empire the world had yet seen.

Darius ruled strictly but wisely. He was the first Persian ruler to mint coins, and he reorganized the empire into 20 provinces, or "satrapies," each governed by a satrap, answerable to the king. (In a future century, this apparatus would be taken over en bloc by the Macedonian conqueror ALEXANDER THE GREAT.) Darius built a road system with relay stations for mounted messengers: This Persian "pony express" became a marvel of the ancient world. He aggrandized the city of Susa as his capital and began work on a summer capital at Persepolis. Darius practiced the Zoroastrian faith, but he enforced Persian tolerance toward the religions of subject peoples, such as JEWS and Greeks. One ancient Greek inscription records Darius' show of respect for the god APOLLO.

Darius employed Greek subjects as soldiers and craftsmen; his personal physician was a Greek, Democedes of CROTON. But in 499 B.C. the empire's Greek cities erupted in rebellion. This IONIAN REVOLT was finally crushed by Darius (in 493 B.C.), but meanwhile it had drawn in two free cities of mainland Greece to fight against Persia: ATHENS and ERETRIA. Angry at these two cities, Darius sent a seaborne expedition to capture them. At the ensuing Battle of MARATHON (summer 490 B.C.), the Persians were unexpectedly defeated by the Athenians.

This distant defeat must have mattered little to the mighty Darius. He died in 486 B.C. It was his son and successor, XERXES, who mobilized the full strength of Persia against mainland Greece.

(See also PERSIAN WARS.)

Darius (2) III See ALEXANDER THE GREAT.

Dark Age Term sometimes used by modern historians to describe the approximately 200 years of barbarism that descended on mainland Greece after about 1100 B.C., when the invasion of the DORIAN GREEKS finally eradicated the remnants of MYCENAEAN CIVILIZATION. The beginning of the Dark Age corresponds to the start of the Iron Age in Greece.

This era is "dark" because of modern ignorance of the events of these years and because of its grimness—archaeological evidence gives a picture of widespread destruction and impoverishment. The Greek Dark Age was named in imitation of the Dark Ages of medieval Europe (circa A.D. 476–1000), which similarly saw chaos after the fall of a dominant civilization, the Roman Empire.

(See also IRON; GREEK LANGUAGE; LEFKANDI.)

death See AFTERLIFE; FUNERAL CUSTOMS; HADES; PHILOSOPHY; RELIGION.

Delian League Modern name for the Athenian-controlled alliance of Greek states, mostly in and around the AEGEAN SEA, that arose at the end of the PERSIAN WARS (circa 478 B.C.) and lasted, with much erosion, until ATHENS' defeat in the PELOPONNESIAN WAR (404 B.C.). The league—based for its first 25 years at the sacred island of DELOS—was formed as a mutual-defense pact against PERSIA, but eventually became the basis for an Athenian naval empire in the Aegean. The number of league member states swelled to about 200 by the mid-400s B.C. In exchange for Athenian protection, these member states paid an annual tribute (although a few powerful allies, such as SAMOS, CHIOS, and LESBOS, chose to supply warships instead). The allies' tribute, paid in TALENTS of SILVER, was originally used to finance league naval operations, but later came to be used by the Athenians to aggrandize and fortify their own city. By the start of the Peloponnesian War (431 B.C.), the Delian League had become an Athenian weapon against not Persia but SPARTA.

The history of the league is one of Athens' increasing arrogance and authoritarianism toward its subject allies. The milestones in this process are: (1) the removal of the league treasury from Delos to Athens (454 B.C.); (2) the Peace of CALLIAS, which formally ended hostilities with Persia but also ended the league's logical reason for existence (449 B.C.); (3) the revolt of Samos, which, although not the first rebellion within the league, came close to succeeding (440 B.C.); and (4) the reassessment of 425 B.C., by which league members had their annual dues doubled or trebled, as Athens strove to raise funds for the Peloponnesian War. After the Athenian disaster at SYRACUSE (413 B.C.), many league members revolted and went over to the Spartan side in the war; among these rebellious states were Chios, MILETUS, BYZANTIUM, MYTILENE, EPHESUS, THASOS, and the cities of EUBOEA. But after the war, Spartan domination proved so loathsome that many former allies returned to Athens to form a SECOND ATHENIAN LEAGUE.

Partial lists exist of the Delian League members and their annual tributes for certain years of the mid-400s B.C. These lists were preserved as inscriptions cut into marble and set up on the Athenian ACROPOLIS. The state that paid by far the most tribute was Athens' old enemy AEGINA—a crushing 30 talents a year. For other states, the tribute may have been demanding but not punitive: ABDERA and Byzantium paid 15 talents each in 457 B.C. (more later, after the reassessment); Paros was paying 16 talents by the mid-400s; AENUS, CYMĒ, and Lampsacus each paid 12 talents; Miletus and Perinthus, 10. League membership offered many attractions—and not just access to the cultural glories of Athens. The league was based partly on the notion that Athenian-style DEMOCRACY was available to the member states, and lower- and middle-class citizens usually supported league membership. (Sparta, on the other hand, always appealed to the upper class; "liberation" by Sparta meant local rule by an OLIGARCHY.)

(See also ARISTIDES; CIMON; CLEON; EGYPT; THUCYDIDES [2].)

Delium See PELOPONNESIAN WAR; WARFARE, LAND.

Delos Island in the AEGEAN SEA, in the center of the CYCLADES group. Inhabited by pre-Greek peoples in the third millennium B.C., Delos was occupied circa 1050 B.C. by IONIAN GREEKS from the mainland. As the mythical birthplace of the divine APOLLO and ARTEMIS, Delos eventually became the second-greatest sanctuary of Apollo, after DELPHI. By the 700s B.C. Delos was the scene of a yearly Aegean festival in the god's honor.

The island's scant two-square-mile area contained temples and monuments, including a famous artificial lake with swans and geese sacred to Apollo. Much of ancient

Delos can be seen today, thanks to more than a century of French ARCHAEOLOGY on the site. Among the preserved monuments is the famous Lion Terrace, with its five surviving marble lions from the 600s B.C.

Delos' sanctity made it politically attractive to various Greek rulers. Both the Athenian tyrant PISISTRATUS (circa 543 B.C.) and the Samian tyrant POLYCRATES (circa 525 B.C.) held public ceremonies there. In 478 or 477 B.C. Delos became the headquarters and treasury of the DELIAN LEAGUE, the Aegean-Greek alliance organized by ATHENS against PERSIA. But in 454 B.C., amid increased authoritarianism, the Athenians removed the league treasury from Delos to Athens.

However, Athenian stewardship of Delos continued. In 426 B.C., during the PELOPONNESIAN WAR, the Athenian general NICIAS conducted ceremonies to resanctify the island in gratitude for Apollo's ending of the plague at Athens. The Athenians built a new temple to the god and inaugurated a new festival, called the Delia, to be celebrated every four years.

After Athens' defeat in the war (404 B.C.) and a generation of hated Spartan rule, Delos in 387 B.C. became the capital of a new Athenian-led alliance, the so-called SECOND ATHENIAN LEAGUE. In the late 300s B.C. Delos became the center of another Aegean alliance, the League of Islanders, most likely established by the Macedonian commander ANTIGONUS (1). In the 200s B.C. Macedonian kings were patrons of the island; one of their gifts was the Stoa of PHILIP V, a colonnade that still is partly standing today. This friendship with MACEDON brought Delos into hostility with ROME. After their victory in the Third Macedonian War (167 B.C.), the Romans handed over Delos to their allies the Athenians, who colonized the island.

Under Roman-Athenian control, Delos became a thriving free port. In the 100s B.C. it was the notorious center of the Greek slave trade. Later it dwindled in importance and was abandoned.

(See also SLAVES.)

Delphi   Sanctuary of the god APOLLO in central Greece. Delphi was the most influential of all ancient Greek shrines and contained the most famous of the Greek world's oracles, or priestly soothsayers. The Delphic oracle was a priestess, the Pythia, who would go into a trance or seizure to speak Apollo's answers to suppliants' questions. During Delphi's busy heyday, in the 500s B.C., as many as three Pythias held the office at once.

Located about 2,000 feet above sea level, Delphi sits on a southern spur of Mt. Parnassus, north of the central Corinthian Gulf, in the region once known as PHOCIS. The name Delphi may commemorate Apollo's cult title *Delphinios* (meaning dolphin or porpoise). The site, terraced into the mountainside, was thought to be the center of the world, and one of Delphi's relics was a carved "navel stone" (*omphalos*) symbolizing this geographic position. Delphi served as a meeting place for the AMPHICTYONIC LEAGUE, a powerful coalition of central Greek states. Because of its oracle and its sports competition held every four years, the PYTHIAN GAMES, Delphi attracted religious pilgrims and other visitors from all over the Greek world.

Delphi's buildings were ruined by an earthquake and scavengers in late antiquity but have been excavated and partly reconstructed by French archaeologists since the end of the 19th century A.D. They now comprise the single most spectacular collection of surviving Greek monuments.

Long before the first Greek invaders arrived circa 2100 B.C., Delphi was a holy place of the pre-Greek inhabitants. Modern ARCHAEOLOGY at Delphi has uncovered traces of these people's religious sacrifices but do not reveal what kind of deity was worshipped. Perhaps in those early days Delphi was the seat of a non-Greek oracular priestess. According to Greek tradition, the shrine's original name was Pytho. Greek MYTH tells how the young Apollo took over Pytho after slaying its protector, the serpent Python. In the Greek historical era (700s B.C.), a sacred serpent was part of Apollo's cult there. Also, a second god, DIONYSUS, was said to inhabit Delphi during the winter months, when Apollo was absent.

It is not known when the earliest temple to Apollo was built at Delphi but it is known that the building was replaced twice during historical times. A stone temple, dating from the 600s B.C., burned down in 548 B.C. and was replaced by an elaborate Doric-style building, completed circa 510 B.C. with the help of the powerful Alcmaeonid clan of ATHENS. That structure collapsed in an earthquake in 373 B.C. and was replaced by the Doric-style, limestone temple that partly survives today.

Amid a clutter of smaller buildings, the grand temple stood atop its terraced platform in the center of a walled, rectangular, hillside enclosure. The site's secondary structures eventually included colonnades, meeting halls, a theater where the Pythian Games' musical and dramatic contests were held, and a dozen or so "treasuries" (small, elegant stone buildings erected by various Greek states to hold statuary, relics, and other precious offerings to the god).

Today the Doric-style Treasury of the Athenians is the only treasury standing, reconstructed by modern archaeologists to look as it did in 490 B.C., when the Athenians built it to commemorate their victory over the Persians at MARATHON. Another important treasury was built circa 525 B.C. by the island of Siphnos; financed by the island's GOLD mines, this treasury was meant to surpass all others in size and beauty. The Delphi Museum now holds part of the Siphnian treasury's frieze; its marble carvings, showing the mythical combat between gods and GIANTS, supply one of the best surviving examples of early Greek architectural SCULPTURE.

Within the temple was the sanctum where the Pythia gave Apollo's prophecies. After a ritual purification, a suppliant (males only) could enter the Pythia's presence and hand over his written question for the god. Some suppliants might seek guidance on personal matters—MARRIAGE, business, and the like—while others would be representing city governments on questions of public importance. The Pythia would go into a trance or fit (despite a common belief, it is not certain that she inhaled some sort of narcotic vapor) and would then deliver the god's answer in sometimes-unintelligible exclamations, which an attending priest would render into hexameter verse.

The facts behind this soothsaying cannot be fully explained. Clearly some pious fraud was involved. Like the medieval Vatican, Delphi was a rich, opportunistic organization, with widespread eyes and ears. The suppliants tended to be wealthy men, whose dealings and affiliations might be well known. The priesthood could learn about upcoming questions while individual suppliants or their messengers made the long and public pilgrimage to Delphi. Nor were the god's answers often clear: Of the 75 Delphic responses that are reliably recorded by ancient historians or inscriptions, most are either vague, commonsensical, or nonsensical.

The most notorious answers, as reported by the historian HERODOTUS, were gloriously ambiguous. When the Lydian king CROESUS inquired circa 546 B.C. if he should invade Persian territory, the oracle replied that if he did so a mighty empire would be destroyed. So Croesus invaded— but it was his own empire that was destroyed.

Delphi reached its peak of power early in Greek history. The shrine became important amid the great movement of overseas Greek COLONIZATION, beginning in the mid-700s B.C. Apollo was the god of colonists, and by skillful politicking, the priests at Delphi were able to position "their" Apollo as the official sanction for all Greek colonizing expeditions. Delphi's prestige was further enhanced circa 582 B.C., when a reorganization of the Pythian Games made them into a major event of the Greek world.

By endorsing TYRANTS such as the Corinthian CYPSELUS (mid-600s B.C.), the Delphic oracle attracted gifts and support from this new class of Greek rulers. Delphi also won the patronage of non-Greek kings in ASIA MINOR—including Croesus and the Phrygian MIDAS—who sought to establish contacts and allies in mainland Greece.

But Delphi's Eastern contacts brought about its undoing, for when the Persians prepared to invade Greece shortly before 480 B.C. the Delphic oracle advised the Greeks to surrender. The Delphi priesthood may have hoped to see their shrine become the capital of Greece under Persian occupation. Several major Delphic prophecies in these years warned (with only slight equivocation) that Greek resistance would be futile. But in fact the Greeks triumphed over the invaders in the PERSIAN WARS of 490–479 B.C. The Delphic priesthood was henceforth stigmatized for its defeatism as well as for its reputed susceptibility to bribery.

Nevertheless, Delphi remained politically influential through the 300s B.C. and remained a central religious shrine for more than 600 years after that, down to the late Roman Empire. The oracle was shut down circa A.D. 390 by order of the Christian Roman emperor Theodosius, in his campaign to eliminate pagan rites throughout his domain.

(See also ALCMAEONIDS; CELTS; PHILIP II; PROPHECY AND DIVINATION; RELIGION.)

deme (*dēmos*) Village or city ward constituting part of a larger territory. The term usually is used to describe the political wards of ATTICA (the 1,000-square-mile territory of ATHENS), as organized by the democratic reformer CLEISTHENES (1) in 508 B.C. The Attic demes numbered 139 and ranged in type from city neighborhoods to townships to patches of rural area.

The demes were the foundation blocks of the Athenian DEMOCRACY: for example, the 500-man Athenian COUNCIL drew its members from each deme, in proportion to population. The demes' headquarters maintained local census figures, with each male citizen formally enrolling on his 19th birthday. There were kept deeds of property and other legal documents and there "town meetings" were held.

Demeter Greek goddess of grain and fertility. Unlike many other major Greek goddesses, Demeter seems to have been purely Greek in origin. Her name is Greek: *dēmētēr* probably means "spelt mother." ('Spelt' is a form of grain.) In MYTH, Demeter and ZEUS had a daughter, Korē ("virgin"). However, at an early date Korē seems to have been assimilated into a more complex, pre-Greek diety, PERSEPHONE, who was goddess of the dead.

Demeter's main cult was at the town of Eleusis, located 15 miles northwest of ATHENS in the fertile Thriasian plain. There the prestigious and stately ELEUSINIAN MYSTERIES were held in honor of Demeter and her daughter. This cult, celebrating the goddess' gift of grain to mankind, surely arose from Eleusis' function as a barley and wheat-growing center.

According to myth, Korē-Persephone was carried to the Underworld by the amorous HADES, god of death. Frantically seeking her daughter, Demeter neglected her duties and let the earth go barren and the grain wither. Her search eventually brought her to Eleusis, where, disguised as an old woman, she was hospitably received at the home of the local king. Later Zeus decreed that Korē might return to the upper world. But because Korē had eaten several pomegranate seeds while in Hades' realm, she was partly obligated to him, and Zeus decided that she would henceforth spend four months of every year belowground as Hades' wife and eight months on the earth with her mother. The myth offers an explanation or allegory for the soil's unproductivity in the fierce Greek summer, while at the same time "explaining" why Eleusis is Demeter's cult center.

In some versions of the myth, Demeter's search is localized in SICILY—for instance, to light her way by night she lit two torches at volcanic Mt. Etna. Such details evidently emerged from the important cult of Persephone observed by the Sicilian Greeks.

(See also FARMING; ORPHISM.)

Demetrius Poliorcetes Macedonian-born soldier and ruler who lived 336–283 B.C. Brilliant but unstable, he received the surname Poliorkētēs (the besieger) on account of his spectacular yet unsuccessful siege of the city of RHODES in 305 B.C.

Demetrius' father, ANTIGONUS (1), served as a general under the Macedonian king ALEXANDER THE GREAT and was his governor in ASIA MINOR. After Alexander's death (323 B.C.), Antigonus sought to maintain Alexander's empire against the secessionist claims of the other DIADOCHI (the successors to Alexander), including CASSANDER, PTOLEMY (1), LYSIMACHUS and SELEUCUS (1). Demetrius was positioned for power at age 15, when he married Phila, the daughter of the Macedonian regent ANTIPATER. By age 25, Demetrius had become his father's field commander and admiral, in this doomed cause of reuniting the empire.

From Antigonus' base in central Asia Minor, Demetrius led his fleets to and fro across the eastern Mediterranean. In 307 B.C. he captured ATHENS, ejected Cassander's governor, and proclaimed that Athens was free. After nearly 17 years of Cassander's overlordship, the Athenians welcomed the conquering Demetrius with adulation.

Demetrius' most famous victory came the following spring, 306 B.C., in the sea battle off the Cyprian city of Salamis. With 108 warships, he totally defeated Ptolemy's fleet of 140 (albeit smaller) warships. Ptolemy lost 120 ships and thousands of shipboard mercenaries; he also lost CYPRUS and command of the sea for the next 20 years. On the basis of this victory, Antigonus proclaimed himself and Demetrius to be joint kings of Alexander's empire. However, in the following year (spring 304–spring 303 B.C.) Demetrius failed in his efforts to capture Rhodes. Rhodes was an important trading city and potential naval base for Ptolemy, but Demetrius' willful decision to waste a year on the siege when he had already captured the more important island of CYPRUS demonstrates his flaws as commander.

Meanwhile Demetrius' enemy Cassander had recaptured central Greece and was besieging Athens. With a fleet of 330 warships and transports, Demetrius sailed from Rhodes and rescued Athens a second time, then defeated Cassander's army at THERMOPYLAE (304 B.C.). Demetrius spent that winter at Athens, carousing with his courtesans, whom he installed in the PARTHENON. The following year he liberated central and southern Greece from Cassander's troops. Although he still had Phila as his wife, he also married Deidameia, sister of the Epirote king PYRRHUS. These years mark Demetrius and his father's peak of power.

Soon, however, their enemies had allied against them. At the huge Battle of Ipsus, in central Asia Minor (301 B.C.), Antigonus and Demetrius were defeated by the combined forces of Seleucus and Lysimachus. Demetrius, leading 10,000 CAVALRY, pressed his attacks too far forward and was cut off by the enemy's elephant brigade, containing 480 beasts. Meanwhile Antigonus was surrounded and killed. Surviving, Demetrius fled to EPHESUS and rallied a fleet, but the disastrous defeat had left him a mere pirate, without a kingdom.

Cassander's death (297 B.C.) and the subsequent turmoil in MACEDON provided an opportunity to acquire that country's throne (to which Demetrius had a claim through his marriage with Phila). By 294 B.C. he was king of Macedon and by 289 had reestablished his old control over Greece. Yet the following year he lost this kingdom also, when his foolhardy scheme to attack Seleucus in Asia Minor resulted in his army's desertion. In 285 B.C. the 51-year-old Demetrius was captured in Asia Minor by Seleucus. Comfortably imprisoned, Demetrius idly drank himself to death with WINE (283 B.C.). Antigonus II, who was Demetrius' son by Phila, eventually left a stable dynasty on the Macedonian throne.

(See also WARFARE, NAVAL; WARFARE, SIEGE.)

democracy The form of government that we call democracy (Greek: *dēmokratia*, "power by the people") was invented at ATHENS in the late 500s and early 400s B.C. "It is called a democracy because it is directed, not by a privileged few, but by the populace," explained the Athenian statesman PERICLES in his Funeral Oration of 431 B.C. (as recounted by the historian THUCYDIDES (1)). The Athenian democracy was a marvel and inspiration to the Greek world, as it remains today; one reason for Athens' successful imperialism in the 400s B.C. was the city's appeal to the underprivileged classes in Greek cities governed as old-fashioned oligarchies. It was not the Athenian army and navy that SPARTA feared so much as the menace of Athenian-sponsored democracies taking over the rest of Greece.

Current knowledge of the Athenian democracy comes partly from literary sources, such as the political speeches of DEMOSTHENES (1) (mid-300s B.C.) and other orators and the comedies of ARISTOPHANES (late 400s B.C.). The most important extant literary source is a treatise preserved among the works of ARISTOTLE (mid-300s B.C.) titled the *Constitution of Athens* (*Athēnaion Politeia*), which briefly describes the different branches of Athenian government. Filling out this meager picture is the archaeological evidence, such as several dozen surviving decrees of the Athenians of this era, preserved as inscriptions in stone.

Democracy at Athens was forged by the political reforms of SOLON (circa 594 B.C.) and CLEISTHENES (1) (508 B.C.). The immediate purpose was to assuage the Athenian middle class, who supplied the backbone of the HOPLITE army and whose discontent was creating the threat of civil war and of usurpation by TYRANTS. Similarly, lower-class Athenians benefited politically when Athens' need to maintain a large navy made the urban poor important as naval oarsmen (400s B.C.). The losers in these reforms were members of the traditional Athenian ARISTOCRACY, who had previously monopolized political office and decision making. Yet by the early 400s B.C. individual aristocrats such as THEMISTOCLES and Pericles had adapted to the new reality and were holding power as left-wing champions of the common people.

The sovereign governing body of the Athenian democracy was the citizens' ASSEMBLY (*ekklēsia*). This was open to all 30,000 to 40,000 adult male Athenian citizens but was usually attended by only about 5,000. (Although individual WOMEN were classed as citizens or noncitizens, they had no political voice either way.) The assembly convened 40 times a year, normally in a natural hillside auditorium called the Pnyx. There, by public debate and vote, the people directed foreign policy, revised the laws, approved or condemned the conduct of public officials, and made many other state decisions. The assembly's agenda was prepared by the 500-member COUNCIL (*boulē*), which also had important executive duties in enacting the assembly's decrees. Councillors were ordinary Athenian citizens of the upper or middle classes, chosen by lot to serve for one year.

It was essential to the Athenian democracy that its public officials not have much individual power. There was no such office as president of Athens. Pericles enjoyed 20 years of nearly unrivaled influence because he was the leader of a political party (the left wing) and because he was able to win the people's trust. But, in terms of elected office, Pericles was always just a *stratēgos*—a military general, one of 10 such elected annually at Athens. Having endured the tyrannies of PISISTRATUS and HIPPIAS (1) (late 500s B.C.), the Athenians feared that other politicians might try to take over the government, and the odd procedure known as OSTRACISM existed as a safeguard against anyone even suspected of thirsting for supreme power.

Aside from the generalships, Athenian state offices tended to be narrowly defined. The top civilian jobs were the nine archonships, which by the mid-400s B.C. were filled by lot, annually, from candidates among the upper and middle classes. An ARCHON might supervise various judicial or religious procedures; the most prestigious post was the *archon eponumos,* where the officeholder's personal name was used to identify the calendar year. Beneath the archonships were the 400-odd lesser executive jobs for the day-to-day running of the Athenian state, such as the commissioners of weights and measures and the commissioners of highway repair. These posts were filled by Athenian citizens, normally chosen by lot to serve for one year with pay. The prominent use of lottery was part of a radical-democratic theory of government which held that an honest lottery is more democratic than an election, because lotteries cannot be unduly influenced by a candidate's wealth or personality. However, elections were used to award certain posts, such as naval architects, army officers, and liturgists.

A liturgy (*leitourgia,* "public duty") was a prestigious service that a rich Athenian might undertake at his own expense. The most prominent liturgy was the *triērarchia,* whereby a citizen would finance, out of pocket, the maintenance of a naval warship for a year. In the mid-300s B.C. 1,200 citizens were recognized as rich enough to be nominated as trierarchs.

Jury duty was a vital aspect of the democracy. Athenian jury panels had considerable courtroom authority and discretion (more akin to modern judges than modern juries), and jurymen were paid reasonably well (although less than an able-bodied man could earn in a day). Consequently, jury duty was known as a resort of old men wishing to pass the time profitably. This social pattern is mocked by Aristophanes in his comedy *Wasps* (422 B.C.). Juries were filled by volunteers—male citizens over age 30—who were enrolled at the start of the Athenian year.

Jury pay, like the salaries for other government jobs, presented an important political card for Athenian statesmen in the 400s and 300s B.C. By increasing such payments, left-wing politicians could please the crowd and genuinely broaden the democracy's base by making such jobs more accessible to lower-income citizens. Pericles introduced jury pay (circa 457 B.C.) and his imitator CLEON increased it (circa 423 B.C.).

Democracy at Athens was an expensive proposition. Classical Athens was an imperial power, enjoying monetary tribute from its DELIAN LEAGUE subject cities as well as a publicly owned SILVER mine. Such revenues financed the democracy's programs and helped prevent power from being monopolized by the wealthiest citizens.

(See also DEME; LAWS AND LAW COURTS; OLIGARCHY; POLIS.)

Democritus   Philosopher of the Greek city of ABDERA, circa 460–390 B.C. Democritus shares with his teacher LEUCIPPUS the credit for having developed the concept of material PHILOSOPHY that is now called the atomic theory. In a remarkable anticipation of 20th-century physics, the atomists decided that the basic components of matter are tiny particles, which they called *atomoi,* "indivisible." These indivisible particles, infinite in shape and number, move in the void and combine variously; their movements and changes produce the compounds of the visible world.

Modern scholars are unsure how to apportion credit between Leucippus and Democritus for the atomic theory. Democritus' contribution probably lay in blending the idea into a theory of ethics. He believed, for example, that moderation and knowledge can produce happiness because they protect the soul's atoms from turmoil.

Democritus was a prolific author, said to have written 70 treatises on subjects such as MATHEMATICS and MUSIC as well as physics and ethics. His atomist theory was presented in a book called the *Little World-system,* probably written to complement Leucippus' *Big World-system.* (Alternatively, both works may have been by Democritus.) None of Democritus' books survives, and, although there are many fragments quoted in works by later authors, these come mostly from the ethical works, not from the atomist writings.

Within a century of Democritus' death, his ideas greatly influenced the philosopher Epicurus (circa 300 B.C.), founder of EPICUREANISM. In later centuries Democritus was sometimes remembered as the "laughing philosopher," possibly because of his ethical emphasis on happiness or cheerfulness (*euthumiē*).

(See also SCIENCE.)

Demosthenes (1)   Greatest of the Athenian orators, best known for opposing the Macedonian king PHILIP II's ambitions in Greece in 351–338 B.C. Although Demosthenes' rallying of his countrymen against Philip brought defeat for ATHENS and could not save Greece from subjugation, it also produced the finest surviving political speeches of the ancient Greek world. Demosthenes was not a great statesman; his defiance of Philip led him to urge foolhardy extremes, and he was badly distracted by Athenian political infighting. But he was a true patriot, for which, under Macedonian rule, he eventually paid with his life.

Demosthenes lived 384–322 B.C. He was born into the well-off middle class—his name means "strength of the people"—but his father, who owned an arms-manufacturing business, died when Demosthenes was seven, and the boy's brothers dissipated the estate. The young Demosthenes' first public speech was directed against his brother Aphobus and an associate, in a lawsuit over Demosthenes' withheld inheritance. Although skinny, awkward, and dour, Demosthenes soon gained prominence as a paid attorney in private cases, and then began serving as an assistant prosecutor in public trials. He also began speaking in the Athenian political ASSEMBLY on questions of public importance.

Alarmed by Philip's aggressions in the northeast, Demosthenes in 351 B.C. delivered his *First Philippic,* urging the Athenians to recapture their old colony of AMPHIPOLIS from the Macedonians. Demosthenes' military advice was not heeded. When Philip besieged the Greek city of OLYNTHUS, in the CHALCIDICĒ region (349 B.C.), Demosthenes, in his three speeches called the *Olynthiacs*, again urged an Athenian military expedition to save the town; again the Athenians voted down such an action.

Meanwhile, Demosthenes was embroiled in partisan conflicts with his Athenian enemies, notably Eubulus and the orator AESCHINES. Demosthenes and Aeschines had been amicable partners on an embassy to MACEDON to make peace with Philip in 346 B.C. But after Philip made a mockery of the peace by immediately marching into central Greece, occupying THERMOPLYLAE and taking control of DELPHI, Demosthenes began his long and bitter feud with Aeschines, denouncing him as Philip's paid lackey.

In 340 B.C. Demosthenes engineered an Athenian alliance with the vital northeastern city of BYZANTIUM; this spurred Philip to declare war on Athens and besiege Byzantium (unsuccessfully). As Philip led his army south into Greece in 338 B.C., Demosthenes proposed and established an alliance with THEBES, resulting in an Athenian-Theban army that confronted the Macedonian advance at CHAERONEA, in northern BOEOTIA. The Battle of Chaeronea, fought in the summer of 338 B.C., was a disastrous Greek defeat. Demosthenes fought on the field as an Athenian HOPLITE and joined the Athenian retreat; according to one story, when the fleeing Demosthenes' cloak got tangled in a bush, he shouted, "Don't kill me! I'm Demosthenes!"

After the battle, however, he became the leader of the moment, organizing Athens' defense and arranging for grain imports. Demosthenes was not persecuted by the Macedonians when they came to terms with Athens, and he was chosen by the city to deliver the funeral oration for the Athenians slain at Chaeronea. The Athenians later awarded him an honorific crown or garland for his state services, and when Aeschines accused him of improper conduct in this (330 B.C.), Demosthenes won such a resounding acquittal with his extant defense speech *On the Crown* that Aeschines emigrated from Athens in humiliation. Later, however, Demosthenes was condemned for embezzlement and was exiled from the city.

Meanwhile, Philip's son and successor ALEXANDER THE GREAT had died at Babylon after conquering the Persian Empire (323 B.C.), and the time looked ripe for a Greek rebellion against Macedonian rule. Recalled to Athens, Demosthenes helped organize this revolt, which was crushed by the Macedonian governor ANTIPATER. Demosthenes fled from Athens and committed suicide as Antipater's men were closing in. He is said to have died by drinking the poison that he always carried in a pen (322 B.C.).

(See also FUNERAL CUSTOMS; LAWS AND LAW COURTS; RHETORIC.)

Demosthenes (2) Accomplished Athenian general of the PELOPONNESIAN WAR, active 426–413 B.C. By developing the use of light-armed javelin men in hilly terrain, Demosthenes made a lasting contribution not only to the Athenian war effort but to the progress of Greek military science. He was the first commander to compensate for the limited mobility of the heavy-infantry HOPLITE.

Defeated by light-armed skirmishers in the hills of AETOLIA (426 B.C.), Demosthenes took the lesson to heart in his next campaign that same season, against the northwestern Greek city of AMBRACIA. With a force of Athenian hoplites and light-armed local allies, Demosthenes destroyed nearly the entire Ambracian army.

In 425 B.C. came his greatest triumph: the Battle of PYLOS and the capture of nearly 200 Spartan hoplites on the inshore island of Sphacteria. Archers and javelin men played a crucial role in the final assault. The operation was commanded by the Athenian politician CLEON who undoubtedly used Demosthenes' troops and battle plan. By then Demosthenes was acting as Cleon's unofficial military adviser; after Cleon's death (422 B.C.), Demosthenes seems to have fallen into disfavor. He was not elected to another generalship until 413 B.C., when he was sent to reinforce the Athenian general NICIAS at the disastrously stalled Athenian siege of SYRACUSE.

There Demosthenes commanded the rear guard in the calamitous Athenian retreat. Surrounded, he eventually surrendered, and was later put to death by the Syracusans.

(See also WARFARE, LAND.)

Diadochi The "successors" (Greek: diadochoi) to the Macedonian king ALEXANDER THE GREAT. The term applies to those of his generals who emerged after his death (323 B.C.) to lay claim to—and fight over—parts of Alexander's vast empire, which stretched from Greece to the Indus River. The main Diadochi were: ANTIGONUS (1) and his son DEMETRIUS POLIORCETES (who, based in ASIA MINOR, sought to preserve and rule the empire); PTOLEMY (1) (who seized EGYPT and founded his dynasty there); SELEUCUS (1) (who did likewise in the Near East and other parts of the old Persian domain); CASSANDER (who claimed MACEDON and Greece); and LYSIMACHUS (who took THRACE). The Macedonian regent ANTIPATER was active in the early years, as the overlord to Antigonus. In addition, an able commander named Eumenes posed an early threat but was driven eastward out of Asia Minor by Antigonus, then was killed.

The chaotic campaigns and maneuverings of the Diadochi lasted for over 25 years. Antigonus' death at the Battle of Ipsus (301 B.C.), in Asia Minor, destroyed the possibility that Alexander's domain could be reknit. The boundaries of the Hellenistic kingdoms were more or less settled by circa 280 B.C.

(See also HELLENISTIC AGE.)

dialects See GREEK LANGUAGE.

Diodorus Siculus Greek historian born at Agyrium, in central SICILY. Diodorus the Sicilian (as his surname means) lived in the first century B.C. and wrote a world history, from earliest times to the end of Julius Caesar's Gallic Wars (54 B.C.). Much of the work is lost, but books 11–20 survive fully and provide much information (and some misinformation) about events in Greek history of the 400s and 300s B.C. While Diodorus sometimes relies on inferior source material, he provides the best available account for many events of the 300s, such as the wars of the DIADOCHI.

Diogenes Best known of the Cynic philosphers (circa 400–325 B.C.). As an extreme devotee of Cynic simplicity and rejection of social convention, Diogenes voluntarily lived in a brutish and immodest manner that supposedly won him the nickname "the Dog" (Kuon). From this, his followers became known as Kunikoi, or CYNICS. Diogenes did not, however, start the Cynic movement; it had been founded by an Athenian, Antisthenes.

Diogenes was born in the Greek city of SINOPE, on the southern coast of the BLACK SEA. He learned the Cynic's creed upon moving to ATHENS in middle age. According to one story, he was taught by Antisthenes himself, after pursuing him relentlessly for instruction.

Living homeless and impoverished, sleeping in public colonnades or in an overturned clay storage jar (not exactly the "bathtub" of legend), Diogenes traveled to CORINTH and probably to other cities, preaching the Cynic doctrine of spiritual self-sufficiency. Despite his avowed apathy to social matters, he seems to have been a skilled exhibitionist, attracting the attention of passersby while preaching his sermons. He is said to have walked the streets with a lit lamp in daylight, remarking that he was searching for an honest man. Another story claims that he threw away his only possession, a cup, after watching a boy scoop up cistern water with his hands. Supposedly Diogenes also had the scandalous habit of masturbating in the public places where he lived—a practice he defended by saying he wished he could satisfy hunger as easily, by just rubbing his stomach.

According to legend, when the young Macedonian king ALEXANDER THE GREAT visited Corinth in 335 B.C. he found Diogenes at some public haunt and asked if he could do anything for Diogenes. "Yes," replied the sage. "Move a little, out of my sunlight." Later the king declared, "If I were not Alexander, I would be Diogenes." (That is, If I did not have my conquests to pursue, I would emulate Diogenes' asceticism.)

Diomedes   Mythical Greek hero of ARGOS who played a prominent role in the TROJAN WAR. The son of the Calydonian hero Tydeus and the Argive princess Deipyle, Diomedes as a young man avenged his father's death at THEBES by helping to capture the city in the expedition of the Epigoni (Descendants).

Diomedes appears in HOMER's epic poem the *Iliad* as a superb fighter and good councillor, somewhat resembling ACHILLES in prowess but without the latter's complexity. Helped by the goddess ATHENA, Diomedes has the odd distinction of wounding the goddess APHRODITE and the war god ARES himself in two separate combats (book 5). Later he and ODYSSEUS make a night raid on the Trojan allies' camp and kill the Thracian champion Rhesus (book 10).

In tales told by later writers, Diomedes was sometimes paired with Odysseus in bringing the wounded hero PHILOCTETES from the island of Lemnos and in stealing the sacred image of Athena from inside Troy, thus fulfilling one of the fated preconditions for the city's capture.

Diomedes survived the war and returned to Argos. According to one version of the tale, he found that his wife, Aegialeia, had been unfaithful, and so departed again, this time for ITALY.

(See also SEVEN AGAINST THEBES.)

Dionysius (1)   Ruthless and dynamic dictator of the Sicilian Greek city of SYRACUSE (reigned 405–367 B.C.). Like other Sicilian TYRANTS, Dionysius was able to seize supreme power largely due to the external threat of Carthaginian domination. A brilliant battlefield commander—he introduced the first Greek use of siege artillery, in the form of arrow-shooting giant crossbows—Dionysius confined the expansionist Carthaginians to the western third of SICILY. Ruling also over Greek southern ITALY, he was, for a time, master of the Greek West. He is best remembered for briefly hosting the Athenian philosopher PLATO at his court (387 B.C.).

Born circa 430 B.C., Dionysius became a promising young soldier-politician in the democratic Syracuse of the late 400s, and he married the daughter of the statesman Hermocrates. Amid renewed Carthaginian attacks on the Sicilian Greeks, Dionysius was elected to the Syracusan board of generals (406 B.C.). Supplanting his colleagues, he easily convinced the Syracusan people to vote him supreme powers, for they were seeking a military savior.

Over the next 15 years Dionysius stymied the Carthaginian advance in Sicily, while quelling uprisings at home and subduing many Sicilian Greeks and native Sicels. He destroyed the Sicilian Greek city of NAXOS (2) and captured and sacked the Carthaginian stronghold of Moyta, at Sicily's western tip, after a famous siege (circa 398 B.C.). Twice he beat back the Carthaginians from the walls of Syracuse. By 390 B.C. peace had been made with the island's two domains—Carthaginian (west) and Syracusan (east)—now officially recognized.

Dionysius next sought conquest over southern Italy. Allied with the Italian-Greek city of LOCRIS and the native Lucanii, he destroyed the Italian-Greek city of RHEGIUM (circa 387 B.C.). But a renewed Carthaginian war in Sicily brought battlefield defeat and loss of territory.

Dionysius' cruelty and greed were notorious: He constantly needed money to finance his campaigns. Yet he was also a cultured man who wrote tragedies and owned relics (including a desk and writing tablets) that had belonged to the great Athenian playwrights AESCHYLUS and EURIPIDES. Dionysius' death, at about age 64 in 367 B.C., supposedly came from the effects of his carousing after receiving the news that his play *Hector's Ransom* had won first prize at an Athenian tragedy competition.

He was succeeded by his son Dionysius II, a far less capable figure.

(See also CARTHAGE; THEATER; WARFARE, SIEGE.)

Dionysius (2) II   See SYRACUSE.

Dionysus   Greek god of WINE, vegetation, and religious possession. Like the goddess DEMETER, Dionysus was an agrarian deity whose cult was more popular among the Greek common people than among the aristocrats. He had probably entered Greek RELIGION by about 1200 B.C., since his name evidently appears on LINEAR B tablets from Mycenaean-era PYLOS.

The distinguishing feature of Dionysus was that he was a god with whom humans could commune, either through intoxication or through spiritual ecstasy. Worshippers achieved a divinely inspired frenzy or "ecstasy" (*ekstasis*,

The god Dionysus, pictured as a youth garlanded with fruit and flowers, in a mosaic from the Syrian-Greek city of Antioch, circa A.D. 175. Originally worshipped in the second millennium B.C. as a rustic deity of wine and vegetation, Dionysus had, by the later Greek era, become a favorite god of urban mystery cults. In this role he was associated with rebirth and the promise of a happy afterlife.

"standing outside oneself"), in which the human personality briefly vanished, supposedly replaced by the identity of the god.

In MYTH, Dionysus was the son of ZEUS and the Theban princess Semele. Semele was pregnant with Zeus' child when Zeus' jealous wife, HERA, approached her in disguise and convinced her to ask Zeus this favor: that he should visit her in his true shape. When Semele made this request, Zeus came to her as a thunderbolt, which killed Semele but made her child immortal. Zeus wrapped the unborn baby in his thigh until it was born. Hera persecuted the child by sending monsters, but the boy evaded or survived their torments, and was at last welcomed by the other deities as the 12th god on Mt. OLYMPUS.

Some ancient Greeks believed that the worship of Dionysus had been introduced to Greece from the uncivilized, non-Greek land of THRACE, northeast of Greece. Other evidence points to origins in the non-Greek region of Phrygia, in central ASIA MINOR. (The Thracians and Phrygians might have been kindred peoples.) The name Dionusos may mean "son of Zeus." Alternatively it may mean "god of Nysa" (referring to Mt. Nysa in Phrygia). It is probable that by the 400s B.C. the Greek Dionysus had come to include religious elements from both lands—the drunken frenzies of the Thracians and the vegetation cult and child-god tradition of Asia Minor.

The Greeks sometimes called him Bacchos (the name later adopted as Bacchus when the Romans borrowed the god for their religion). Greek legend told of Dionysus' human female followers called Bacchae or MAENADS, who would roam the mountains in frenzied bands, dancing and singing the god's honor. EURIPIDES' tragedy The Bacchae (405 B.C.) recounts how the god, on returning to his mother's city of THEBES to claim divine worship, discovered that his followers had been persecuted and his worship prohibited by the young Theban king Pentheus (lamenting), who was also his uncle. In retaliation, Dionysus inspired the Theban women to wander the countryside as Bacchae; then, in human disguise, he convinced Pentheus to spy on them. Outraged at this intrusion, the WOMEN hunted down their king—whom, in their divine frenzy, they did not recognize—and tore him limb from limb. Pentheus' mother, Agavë, led these horrible proceedings.

But in actual Greek ritual, Dionysian worship and possession took less grotesque forms. No doubt Dionysus' worshippers got agreeably drunk at his festivals, as people in more recent centuries have done at county fairs and on St. Patrick's Day. There was also symbolic possession, in the wearing of masks. Dionysus had several annual festivals at ATHENS, the most important one being the City Dionysia or Greater Dionysia, held in early spring. At this festival in primitive times, masked worshippers would impersonate the god, speaking words in his character. By the late 500s B.C. these rites had developed into the beginnings of Athenian stage comedy.

Athens, BOEOTIA, and the Aegean island of Naxos were the main sites of Dionysus' worship in Greece; all were important vine-growing regions. As a vegetation god, Dionysus was associated with all fruit-bearing trees and with the pine and plane trees. His fertility aspect was suggested in the unquenchable lust of his attendant creatures, the SATYRS and Silenoi. Sexual relations did not play a major role in his myths (that was the province of the goddess APHRODITE), but Dionysus did have one well-known lover, the Cretan princess Ariadne, whom he carried off from Naxos after she had been abandoned by the Athenian hero THESEUS.

As a favorite subject of vase painting, Dionysus was at first (500s and 400s B.C.) always portrayed as a bearded man. Later representations tended to show him as a beardless youth or child.

For the classical Greeks, Dionysus represented the irrational aspect of the human soul. The frenzies wrought by Dionysus were the natural complement to the virtues of reason and restraint embodied in the god APOLLO. Symbolically, Dionysus was said to inhabit Apollo's sanctuary at DELPHI during the latter's annual winter absence. Later, amid the hunger for personal religion in the HELLENISTIC AGE (300–150 B.C.), Dionysus was one of the few Greek deities whose worship produced a mystery cult, offering the hope of a happy AFTERLIFE to its initiates.

(See also LYRIC POETRY; MUSIC; NYMPHS; ORPHISM; THEATER; PRAXITELES.)

Dioscuri   See CASTOR AND POLYDEUCES.

dithyramb   See LYRIC POETRY; THEATER.

divination   See PROPHECY AND DIVINATION.

divorce   See MARRIAGE.

Dodona Famous sanctuary and oracle of the god ZEUS, in a region of northwest Greece called EPIRUS. Although purely Greek in origin, the sanctuary was very ancient, and it retained primitive aspects in historical times. It lay mainly out in the open, and its central feature was an ancient oak tree, sacred to Zeus. In response to an applicant's questions, the god spoke in the oak leaves' rustlings, which were then interpreted by the priests; a sacred dove may have been part of the cult. HOMER's epic poem the *Iliad* (written down circa 750 B.C.) describes Zeus' priests at Dodona "sleeping on the ground, with unwashed feet." By the time of the historian HERODOTUS (circa 435 B.C.), three priestesses had somehow replaced the male priests. These priestesses called themselves "doves" (*peleiades*).

Although never as influential as the oracle of the god APOLLO at DELPHI, Dodona was the religious-political center of northwest Greece. The Epirote king PYRRHUS made it the religious capital of his domain and beautified it with a building program (circa 290 B.C.). Later Dodona was destroyed by the Aetolians but rebuilt by the Macedonian king PHILIP V (late 200s B.C.). However, during the Third War, Dodona was wrecked again, this time by the Romans (167 B.C.).

Excavations at the site since the 1950s A.D. have restored the beautiful amphitheater, originally constructed under Pyrrhus. There are also remnants of the temple of Zeus and the council house, both built or enlarged by Pyrrhus.

(See also AETOLIA; PROPHECY AND DIVINATION.)

Dorian Greeks One of the three main ethnic branches of the ancient Greek people; the other two were the IONIAN and AEOLIAN GREEKS. Archaeological evidence, combined with a cautious reading of Greek MYTH, indicates that Dorian invaders overran mainland GREECE circa 1100–1000 B.C., in the last wave of violent, prehistoric Greek immigration. Emerging (probably) from the northwestern Greek region called EPIRUS, the Dorians descended southward, battling their fellow Greeks for possession of desired sites. They bypassed central Greece but occupied much of the PELOPONNESE and the isthmus and Megarid. Taking to ships, the Dorians then conquered eastward across the southern AEGEAN SEA and won a small area of southwestern ASIA MINOR. The Dorian invasion obviously was associated with the collapse of MYCENAEAN CIVILIZATION, but recent scholarship considers the invasion to be an effect, rather than a cause, of the Mycenaeans' downfall. Similarly, some historians have believed that the Dorians' success was due to their possession of IRON weapons—superior to the defenders' BRONZE—but it seems equally likely that all Greeks acquired iron-forging only after the Dorian conquest.

The word Dorian (Greek: *Dorieus*) may be related to the Greek word *doru*, "spear." Ancient legend also connected

the name to the hero Dorus, son of HELLĒN. A small Dorian region called Doris, in the mountains of central Greece, was erroneously thought to be the people's original homeland.

By the classical era (400s B.C.), the important Dorian states included the Peloponnesian cities of CORINTH and ARGOS, the Aegean islands of CRETE and RHODES, and the Asian Greek cities of HALICARNASSUS and CNIDUS. Farther afield, prosperous Dorian colonies existed in SICILY (particularly SYRACUSE), in southern ITALY, and in Libya, at CYRENE (1). But the most important Dorian site, and the one that other Dorians looked to as their protector, was the militaristic city of SPARTA, in the southeastern Peloponnese.

Dorian states were distinguished by their dialect (called Doric) and by their peculiar social institutions, including a tripartite tribal division and the brutal practice—at Sparta, Crete, Syracuse and elsewhere—of maintaining an underclass of serfs, or HELOTS. Due largely to the superior armies of Sparta, the Dorians were considered the best soldiers in Greece, until the 300s B.C. saw the emergence of non-Dorian BOEOTIA.

Although Dorian cities such as Corinth were at the forefront of Greek TRADE and culture in the 700s and 600s B.C., by the following centuries the Dorians had acquired the reputation for being crude and violent (at least in the eyes of the Athenians and other Ionian Greeks). Ethnic tensions between Dorians and Ionians in mainland Greece and Sicily reached a bloody crisis in the PELOPONNESIAN WAR (431–404 B.C.). During the HELLENISTIC AGE (300–150 B.C.), Syracuse, Rhodes, and Halicarnassus were among the Dorian states important in commerce and art.

(See also ACRAGAS; AEGINA; CORCYRA; DARK AGE; GELA; GREEK LANGUAGE; LOCRI; MEGARA (1); MELOS; MESSENIA; SICYON; TARAS.)

Doric dialect See GREEK LANGUAGE.

Doric order See ARCHITECTURE.

Doris See DORIAN GREEKS.

Draco Athenian statesman who supposedly gave the city its first written code of law, circa 621 B.C. Draco's laws were aristocratically biased and egregiously harsh, involving wide use of the death penalty. One later thinker remarked that the laws had been written in blood, not ink. Draco's severity is commemorated in the English word "Draconian," used to describe excessively harsh law or administration. Within 30 years, Draco's code was replaced by that of the great lawgiver SOLON (circa 594 B.C.).

(See also ARISTOCRACY; LAWS AND LAW COURTS.)

drama See THEATER.

E

education Literacy became widespread at a surprisingly early date in the ancient Greek world. Surviving public inscriptions and historical anecdotes suggest that, in the more advanced Greek states, a majority of at least the male citizens could read and write by circa 600 B.C.—barely 175 years after the Greeks had first adapted the Phoenician ALPHABET. The impetus for this learning was probably not love of literature but the necessity of TRADE—a need that included the middle and lower-middle classes.

As today, literacy and numeracy were taught to boys at school. The earliest surviving reference to a school in the Western world occurs in the history of HERODOTUS (circa 435 B.C.), in reference to the year 494 B.C. on the Greek island of CHIOS. Similarly, school scenes first appear on Athenian vase paintings soon after 500 B.C. Schools were private, fee-paid institutions: There were no state-funded schools at this time and no laws requiring children to receive education. There were separate schools for girls (coeducation did not exist), but girls' schooling was generally not as widespread or thorough as boys'.

In classical ATHENS (400s–300s B.C.), a boy's schooling usually began at age seven, and many of the lower-income students probably left after the three or four years needed to learn the basic skills. For the rest, there might be as much as 10 years' elementary school, under three types of teachers. The *grammatistēs* gave lessons in reading, WRITING, arithmetic, and literature. Literary studies emphasized the rote memorization of passages from revered poets, particularly HOMER and (in later centuries) the Athenian tragic playwrights of the 400s B.C. SPORT comprised the second branch of Greek education, under a coach (*paidotribēs*). WRESTLING and gymnastics were among the preferred disciplines. Militaristic states such as SPARTA greatly emphasized sports as a preparation for soldiering; the rugged art of BOXING was considered a typically Spartan boy's sport. Sparta was unusual in encouraging gymnastics training for girls.

The third teacher, the *kitharistēs*, gave instruction in MUSIC—specifically, in singing and playing the lyre for the recitation of LYRIC POETRY. This branch of Athenian-style education may have been less esteemed outside of Athens. Supposedly, the Macedonian king PHILIP II once rebuked his young son Alexander after the boy's musical performance at a banquet. "Are you not ashamed, my son, to play the lyre so well?" Philip asked the future king ALEXANDER THE GREAT (340s B.C.).

At age 18 an Athenian male became known as an *ephebos* (youth) and began two years of military training; similar programs existed at other Greek states. That a rich young man might then resume his studies was a practice that evolved at Athens in the mid- and late 400s B.C. The pioneers in this practice were the SOPHISTS and teachers of RHETORIC who came flocking to Athens. Charging high fees, tutors such as PROTAGORAS of ABDERA and GORGIAS of Leontini gave lessons in disputation and public speaking to young men planning to enter public life. The Athenian philosopher SOCRATES (469–399 B.C.), while not himself a sophist, occupies an important place in this evolution, as does the Athenian orator ISOCRATES (436–338 B.C.). Such

This image from an Athenian red-figure cup, circa 480 B.C., may represent a *paidotribēs*, or a boys' sports instructor. His power to punish is embodied in the stick he carries. In classical Athens, sports provided a major facet of boys' primary education, along with reading, writing, arithmetic, and music.

tutors answered a growing need at Athens for higher education.

Socrates and the sophists paved the way for the Western world's first university—the ACADEMY, established by the Athenian philosopher PLATO in parkland buildings just outside the city wall, circa 385 B.C. The Academy was known as a school of PHILOSOPHY, but offered lectures and advanced study in many areas that might not be associated with philosophy, such as MATHEMATICS. An even broader range of study was offered at the LYCEUM, founded at Athens by ARISTOTLE (circa 335 B.C.). Two other major philosophical schools, of EPICUREANISM and STOICISM, were established at Athens by about 300 B.C. During the HELLENISTIC AGE (300–150 B.C.), a new center of higher learning arose at the royal court of ALEXANDRIA (1).

(See also ASTRONOMY; HOMOSEXUALITY; SCIENCE; WOMEN.)

Egypt Located in northeast Africa, roughly 600 miles southeast of mainland Greece, Egypt was (and is) a non-Greek land lying about 700 miles along the lower Nile River, with coastlines along the Mediterranean and the Red seas. At the beginning of the Greek era (in the second millennium B.C.), Egypt was a magnificent kingdom that awed the humble Greeks. By 300 B.C. the conquests of ALEXANDER THE GREAT had brought Egypt into the Greek world as a land ruled by a Macedonian-Greek court, where some of the finest Greek poetry of the ancient world was being written.

The Egyptian Middle and New Kingdoms probably influenced the emerging MYCENAEAN CIVILIZATION of mainland Greece (circa 1600 B.C.). The royal tombs at MYCENAE may have been inspired by the pharaohs' pyramids and graves. The Greek MYTH of DANAUS and Aegyptus—among other tales—seems to commemorate some early Greek-Egyptian contact. The Mycenaeans would have visited Egypt for TRADE, for service as hired mercenaries, or for pirate raids. Almost certainly the "Sea Peoples" who ravaged Egypt's Mediterranean coast around 1100 B.C. included groups of displaced Mycenaeans, fleeing from social collapse in Greece.

For several centuries in the first millennium B.C., the pharaohs banned Greek merchants from Egypt. (As a result, Egypt never influenced the ripening Greek culture of the 700s B.C. as strongly as did another Near Eastern civilization, PHOENICIA). But in around 650 B.C. Greek traders were allowed to set up an emporium in the Nile delta. This was NAUCRATIS, where the Greeks offered SILVER ore and SLAVES in exchange for Egyptian grain and luxury goods, such as carved ivory. The Greek poet BACCHYLIDES (circa 470 B.C.) mentions grain ships bringing a fat profit home from Egypt. It was the Naucratis trade that brought Egyptian artwork in quantity to Greece, to be imitated by Greek artists: The famous Greek *kouroi* statues of the early and mid-500s B.C. preserve the postures of Egyptian statuary.

Greek soldiers were more welcome than traders. By 600 B.C. pharaohs were hiring and bringing to Egypt Greek HOPLITE mercenaries to serve in their wars. Among the oldest surviving Greek inscription is a Greek soldier's graffito, scratched onto a colossal statue at Abu Simbel, 700 miles up the Nile (591 B.C.).

Egypt was conquered by the Persian king Cambyses in 525 B.C. As a Persian subject state, Egypt provided warships and crews in the PERSIAN WARS against the Greeks. In 460 B.C. a Greek invasion force—120 ships of the Athenian-led DELIAN LEAGUE—sailed into the Nile to attack the Persian-garrisoned city of Memphis. The invasion was destroyed by the Persians in 455 B.C., and Egypt remained securely within the Persian Empire for another 120 years. But in 332 B.C. the Macedonian king Alexander the Great captured Egypt, in his conquest of the Persian Empire. On Egypt's Mediterranean coast Alexander founded a city destined to be one of the greatest of the ancient world, ALEXANDRIA (1).

At Alexander's death, Egypt fell into the hands of a Macedonian general, PTOLEMY (1), who founded a dynasty. With its brilliant capital at Alexandria, Ptolemaic Egypt became the wealthiest and most important kingdom in the HELLENISTIC AGE (300–150 B.C.). The Ptolemies took over the apparatus of the pharaohs and modernized it, creating an immense, efficient civil service for drawing taxes from the Egyptian peasantry who worked the fertile Nile valley.

In foreign affairs, the Syrian-based SELEUCID EMPIRE soon emerged as the enemy of Ptolemaic Egypt. Between 274 and 168 B.C., the two kingdoms fought the six Syrian Wars, disputing their common boundary in the Levant. The Ptolemies ruled Egypt until the Roman conquest of 30 B.C. The last Ptolemaic ruler was the famous Cleopatra VII.

(See also BLACK PEOPLES; GOLD; HERODOTUS; PTOLEMY (2) II; SCULPTURE; WARFARE, LAND; WARFARE, NAVAL.)

ekklesia See ASSEMBLY.

Elea See PARMENIDES.

Electra In MYTH, a daughter of King AGAMEMNON of MYCENAE and his queen, CLYTAEMNESTRA. Not mentioned in HOMER, Electra made her first known literary appearance in STESICHORUS' poem the *Oresteia* (circa 590 B.C.). In Athenian tragedy, she appears as a main character in the story—handled by all three of the great tragedians—of how her brother ORESTES avenged Agamemnon's murder by slaying Clytaemnestra and her lover, Aegisthus.

In AESCHYLUS' play the *Libation Bearers* (458 B.C.), Electra welcomes Orestes and supports his scheme to kill her hated mother and Aegisthus. In SOPHOCLES' *Electra* (circa 418–410 B.C.) she acts similarly, and is the focus of the play. But in EURIPIDES' *Electra* (circa 417 B.C.) she appears obsessively hateful and jealous of her mother, who has denied Electra the chance to marry, thereby punning on her name: *alektron* ("without marriage"). In this play, Electra wields an ax and actually helps Orestes to kill Clytaemnestra (offstage), then goes wild with guilt. Euripides' treatment of Electra provides one of the best examples of that playwright's innovative interest in psychology and the plight of WOMEN in Greek society.

elegy See LYRIC POETRY.

Eleusinian Mysteries Important Athenian religious cult in honor of the grain goddess DEMETER, observed at the town of Eleusis (which lies 15 miles northwest of ATHENS, in the region's main wheat- and barley-growing plain). The cult was given the title "Mysteries" (*mustēriai*, from *mustēs*, "an initiate") on account of its secretive nature. Only those who had been formally initiated could participate, and details of the rites (which seem to have been harmless enough) were forbidden to be revealed publicly. The Eleusinian Mysteries were a rare form of worship in classical Greece (mid-400s B.C.), where city-states emphasized public cults; later, however, in the HELLENISTIC AGE (300–150 B.C.), mystery religions began to proliferate.

The cult was run by the Athenian state and was officiated by two noble families of Eleusis: the Eumolpidae and Kerykes. Despite the emphasis on secrecy, the requirements for initiation were lenient: Anyone of Greek speech and without blood guilt was eligible to join. This inclusion of SLAVES and WOMEN is remarkable. The initiate's oath of secrecy was taken seriously; no knowledgeable ancient writer has left us a description of the rites, and in the 400s B.C. such prominent men as the playwright AESCHYLUS and the politician ALCIBIADES were investigated or prosecuted for supposedly revealing the mysteries. Clearly the initiates' conduct was thought to affect the goddess Demeter's goodwill and the all-important fertility of the grainfields.

The "bible" of the mysteries was the central myth of Demeter. According to legend, the death god HADES stole away Demeter's daughter Korē (also called PERSEPHONE). Demeter searched the world for her, letting the fields go barren in her grief; and the god ZEUS restored order by allowing Korē-Persephone to remain with her mother for eight months out of every year.

The mysteries probably reenacted this story every spring, as a kind of pageant, with dance and incantation. At the climactic, secret, nocturnal ceremony in Demeter's temple, the priest would hold up an ear of grain amid reverent silence. The "doctrine" of the mysteries was probably very simple: thanksgiving for Demeter's gift to the living and the hope that she and her daughter Persephone would take further care of the initiates' souls in the Underworld. This hope of a happy AFTERLIFE became a cornerstone of later mystery cults.

(See also DEMETER; ORPHISM.)

Eleusis   See ATTICA; ELEUSINIAN MYSTERIES.

Elis   Plain in the western PELOPONNESE. In about 471 B.C., the inhabitants established a city, also called Elis, as their political center. The Eleans had charge of the highly important OLYMPIC GAMES, the sports-and-religious festival held every four years in honor of the god ZEUS. The actual site of the games was the sanctuary and sports complex known as OLYMPIA. Since Olympia lay closer to the city of Pisa than to the city of Elis, there was intermittent strife between Elis and Pisa for control.

Formidable soldiers, the Eleans remained staunch allies of SPARTA until 420 B.C., when events led them to make alliances with Sparta's enemies ATHENS and ARGOS, in the PELOPONNESIAN WAR. Later, still an enemy of Sparta, Elis made an alliance with ARCADIA (369 B.C.). In the 200s B.C. it was an enemy of Arcadia and a member of the Aetolian League.

(See also AETOLIA; PHEIDON.)

Empedocles   Influential early philosopher who lived circa 495–430 B.C. Empedocles was born to an aristocratic family of the Sicilian Greek city of ACRAGAS. As a thinker, poet, statesman, and physician Empedocles became a semi-legendary figure among his contemporaries. He is plausibly said to have helped establish DEMOCRACY at Acragas after the expulsion of the reigning tyrant, circa 472 B.C., and he supposedly declined a public offer of kingship. He is said to have later been exiled from Acragas. He was an admired orator, who reportedly tutored the greatest Greek orator of the next generation, GORGIAS of Leontini.

His philosophy, partly inspired by the Italian-based Greek philosophical schools of PARMENIDES and PYTHAGORAS, was presented in two epic poems in hexameter verse, *On Nature* and *Purifications*; he is said to have recited the latter at the OLYMPIC GAMES. About 450 verses of these poems (approximately one-tenth of their combined total) survive today as quotations in works by later writers. The poems' ideas had a profound effect on subsequent thinkers, including ARISTOTLE.

On Nature introduced the concept, later fundamental to Aristotle's physics, that all matter derives from four elements: air, water, fire, and earth. These elements are eternal and unchanging; apparent creation and destruction in the world merely indicates the ceaseless reorganizations of these four elements into new ratios. Love (Philia) and Strife (Neikos) are cosmic forces seeking to unify and divide, respectively. Empedocles' theory of elements has been interpreted as a correction of Parmenides' belief that ultimate reality is unified and immobile. Here Empedocles probably helped inspire the elemental theories of the early "atomists," DEMOCRITUS and LEUCIPPUS.

The *Purifications* was concerned with humankind's Original Sin and restoration—ideas that had already been developed in the mystic ORPHISM of Acragantine aristocratic circles. Empedocles apparently identified the primal sin as the first shedding of blood and eating of meat. Tainted by this ancestral pollution, the individual human soul must be purified through a series of incarnations, bringing the soul through the round of elements to a renewed state of bliss. In the *Purifications'* most famous verse, Empedocles declares, "Already I have been a boy and a girl, a bush and a bird, and speechless fish in the sea." The Orphic-derived notion of the soul's transmigration was also a feature of Pythagorean belief, taught at certain Greek cities of nearby southern ITALY. While not a Pythagorean himself, Empedocles was clearly influenced by such teachings.

(See also AFTERLIFE; MEDICINE; SCIENCE.)

Epaminondas   Brilliant Theban statesman and general who lived circa 410–362 B.C. Epaminondas (Greek: Epameinondas) engineered the rise of THEBES as the foremost Greek city, in defiance of SPARTA. He destroyed the legend of Spartan military invincibility and ended Sparta's domination of Greece that had lasted for 35 years since Sparta's victory in the PELOPONNESIAN WAR.

Thebes and Sparta were already at war when Epaminondas first came to prominence, in 371 B.C. As an elected commander for that year, he came to quarrel with the Spartan king AGESILAUS at a peace conference. But the resulting Spartan invasion of Theban territory ended in a Theban triumph, when Epaminondas destroyed a Spartan army at the Battle of LEUCTRA.

After the battle, Epaminondas simply dismantled the Spartan empire. Marching into the PELOPONNESE in winter 370–369 B.C., he liberated ARCADIA from Spartan overlordship and (then or later) established his "big city," Megalopolis, to be the center of an Arcadian league. Soon afterward Epaminondas entered the Spartan-ruled region called MESSENIA, where he founded another city, Messene, to be a political center against Sparta. Epaminondas' liberation of Messenia had a devastating affect on Sparta, which had traditionally relied on Messenian grain, grown by the Messenian serfs known as HELOTS.

Epaminondas' later exploits included further invasions of the Peloponnese and a naval expedition against the Athenians. In 362 B.C. he again led an army into the Peloponnese, to oppose a Spartan threat against Arcadia. Although the Battle of MANTINEA was another Theban victory, Epaminondas died there from wounds. He was the greatest leader to emerge in the tumultuous half century between the end of the Peloponnesian War (404 B.C.) and the rise of MACEDON (350s B.C.).

(See also WARFARE, LAND.)

Ephesus Greek city of IONIA, on the central west coast of ASIA MINOR, known for its TRADE and its elaborate cult of the goddess ARTEMIS. Founded circa 1050 B.C. by IONIAN GREEKS from mainland Greece, Ephesus lay at the mouth of the Cayster River, at a site that commanded the coastal farming plain, the riverine route inland, and the sea passage to the nearby Greek island of SAMOS.

Ephesus was one of the Greek world's foremost cities during the Ionian heyday of the 600s and 500s B.C. Although never a great seagoing power like its neighbor and trade rival MILETUS, Ephesus thrived as a terminus for caravans from the Asian interior and as an artisan center; its ivory carving was famous. In the mid-600s B.C. Ephesus withstood attack by a nomadic people, the Cimmerians, who had swept westward through Asia Minor. Patriotic resistance was urged in verses by the Ephesian poet CALLINUS (whose surviving fragments are among the earliest extant Greek LYRIC POETRY). But Ephesus' most famous resident was the early Greek philosopher HERACLITUS (circa 500 B.C.).

Around 600 B.C. the city's oligarchic government gave way to a line of TYRANTS. One of these rulers, in the mid-500s B.C., began construction of a magnificent temple of Artemis, later reckoned as one of the SEVEN WONDERS OF THE WORLD. This temple, known as the Artemisium, was the largest Greek building of its day, measuring 358 feet in length and 171 feet in width. Apparently not completed until about 430 B.C., it burned down in 356 B.C. at the hands of an arsonist who wanted his deed to be remembered forever.

Like the rest of Ionia, Ephesus fell to the Lydian king CROESUS in the mid-500s B.C. and soon thereafter to the Persian king CYRUS (1). The city fared better than other Ionian cities under the Persians. Although it joined in the ill-fated IONIAN REVOLT of 499 B.C., it apparently made a timely surrender and avoided the worst retaliations, while the Persians' sack of Miletus in 494 B.C. removed Ephesus' main rival. In 479 B.C. Ephesus was liberated, along with the rest of Ionia, by the mainland Greeks.

Ephesus then became part of the Athenian-controlled DELIAN LEAGUE (circa 478 B.C.), but joined the general Delian revolt against ATHENS toward the end of the PELOPONNESIAN WAR (in 411 B.C.). As an important Spartan naval base, Ephesus served the commander LYSANDER in his defeat of an Athenian fleet at the nearby Battle of Notium (406 B.C.), and after the war Ephesus continued as a Spartan base for sea operations against PERSIA. But, with the KING'S PEACE of 386 B.C., Ephesus and all other Asian Greek cities passed back to Persian rule, until they were again liberated in 334 B.C., by ALEXANDER THE GREAT.

The city then embarked on its second era of greatness. Rebuilt at a new site by the Macedonian dynast LYSIMACHUS around 294 B.C., Ephesus became an emporium of the HELLENISTIC AGE, rivaled only by ALEXANDRIA (1) and ANTIOCH. Ephesus passed into the influence of the Seleucid kings in the mid-200s B.C., then to the kingdom of PERGAMUM and, finally, in 133 B.C., to ROME. In Roman times it continued to be a great city of the East.

Ephialtes Left-wing Athenian statesman and mentor of PERICLES. Born circa 500 B.C. into a humble family, Ephialtes arose in the 460s B.C. as the democratic leader against the conservative party of CIMON; in this, he was the political heir of THEMISTOCLES.

In 462 B.C., taking advantage of Cimon's recent disgrace, Ephialtes and Pericles proposed stripping the powers and jurisdictions from the conservative law court known as the AREOPAGUS. The proposals were passed, but within a year Ephialtes was dead—probably murdered by political reactionaries. The mantle of leadership of the left wing passed to Pericles.

(See also DEMOCRACY.)

ephor Title of an annually elected chief official at SPARTA and other Dorian Greek states; the name suggests "one who watches over." The Spartan ephors, who numbered five by the 400s B.C., served as an important counterweight to the two Spartan kings. The ephors oversaw the kings' administration and personal conduct; they could summon the kings to their presence or prosecute them. Two ephors always accompanied a king on campaign (a sign of the Spartan fear of the corrupting outside world). Every month ephors and kings exchanged oaths to observe each other's authority.

(See also CHILON; LYCURGUS (1).)

epic poetry The earliest and greatest works of extant Greek literature are the two long narrative poems titled the *Iliad* and *Odyssey*, ascribed to the poet HOMER. Employing dactylic hexameter verse and a distinctive, elevated vocabulary, the *Iliad* and *Odyssey* typify the Greek literary genre called *epikē* (from *epos*, "word").

Today most scholars recognize that the *Iliad* and *Odyssey*—written down circa 750 B.C., soon after the invention of the Greek ALPHABET—represent the final stage in a prior traditon of unwritten verse composition, stretching back some 500 years to the second millennium B.C. This unwritten poetry-making is today known as oral composition. The foremost geographic region for this poetic tradition was the eastern Greek area called IONIA.

In preliterate cultures, poetry serves a mnemonic purpose: The rhyme and rhythm of a poem or song make the words easier to remember. Many preliterate societies have used poetry (in unwritten forms) for the memorization and recital of legends and folktales. In the early Greek world, oral poetry typically recounted tales from Greek MYTH—tales of olden-day heroes and their interractions with the gods.

These stories were preserved and retold by a professional class of bards or minstrels known as *aoidoi* or *rhapsoidoi* (rhapsodes, "song-stitchers"). The skill of these bards lay in their knowledge of the vast mythological material and in their ability to select and shape episodes for public recitation. This oral poetry was sung or recited in the flowing rhythm of dactylic hexameter. (Greek poetry did not employ rhyme: The GREEK LANGUAGE had too many natural rhymes for this to be considered beautiful.) An idea of the bard's function is communicated in book 8 of the *Odyssey*, where the blind minstrel Demodocus recites to his own accompaniment on the lyre at a royal feast. The technique by which a bard might draw on familiar subject matter while composing individual lines spontaneously from a mental trove of formulaic expressions has been illuminated in modern times by Milman Parry's studies of oral composition in Serbia and Croatia in the 1930s A.D.

The second well-known Greek epic poet is HESIOD (circa 700 B.C.). The rural, middle-class Hesiod was an individualist in the art: In his *Works and Days* he fitted the epic verse form to a most unusual content, a moralizing farming calendar.

But the main epic tradition is exemplified by Homer's poems of aristocratic war and voyaging. We know that, by about 550 B.C., there existed about 10 other epic poems—written down after the *Iliad* and *Odyssey*, shorter than they, and probably inspired by them. Known collectively as the epic cycle (*epikos kuklos*), these poems taken together recounted a loose mythical world history from the Creation to the aftermath of the TROJAN WAR. Among the non-Homeric epics were the *Oedipodia*, the *Thebaïd*, and the *Epigoni*, recounting the tragic history of the ruling house of the city of THEBES (including the tales of OEDIPUS and of the SEVEN AGAINST THEBES). But most of the epic cycle described episodes of the Trojan War not recounted in Homer; these poems included the *Cypria, Little Iliad, Destruction of Troy,* and *Homecomings.*

Today the non-Homeric epics exist only in fragments quoted by later writers. But many of the legends that they described have been preserved for us—in Athenian stage tragedy (400s B.C.), in the prose works of ancient scholars, and in such later epics as the Roman Ovid's *Metamorphoses.*

By the mid-500s B.C., epic composition had died out in favor of newer forms, such as LYRIC POETRY and then THEATER. Epic was revived in a more self-conscious, literary form in the mid-200s B.C. by the Alexandrian poet APOLLONIUS. The Alexandrian epic tradition strongly affected Roman literature, culminating in the Roman poet Vergil's patriotic masterpiece, the *Aeneid* (circa 20 B.C.).

(See also ACHILLES; JASON (1); MUSIC; ODYSSEUS.)

Epicureanism Influential Greek philosophical school of the HELLENISTIC AGE (300–150 B.C.) and after, founded by Epicurus. Born at SAMOS to an Athenian father circa 341 B.C., Epicurus (Greek: Epikouros) resided at ATHENS from the late 300s B.C. until his death in 270 B.C. There he and his followers lived in privacy and simplicity on his property (a house and garden), where he taught pupils. The fact that these pupils included SLAVES and WOMEN is a sign of Epicurus' innovative and liberal outlook, by ancient Greek standards.

His doctrine of Epicureanism was a daring intellectual breakthrough. For the Epicureans, the purpose of life was pleasure (*hēdonē*), as derived from a simple, even ascetic, mode of existence. Pursuits that brought pain, ill health, frustration, anxiety, or unending desire were not considered appropriate pleasures. Since the chief cause of pain is unsatisfied desire, Epicureans were taught to limit their desires rather than seeking to satisfy each one. The ideal state of the soul is freedom from agitation (*ataraxia*); this can be achieved through temperance and study. Teaching his followers to renounce political ambition, with its cares and corruptions, Epicurus went boldly against the traditional grain of the Greek city-state and anticipated the more personal, individualistic values of Hellenistic society.

On the metaphysical side, Epicureanism relied on the atomist theory of DEMOCRITUS and LEUCIPPUS. Epicurus concluded that, because the human soul is composed of atoms like the body, these atoms must disperse at the person's death. In other words, the soul dies with the body. And thus there is no such thing as an AFTERLIFE. It was this revolutionary idea that led to much criticism of Epicureanism by religious circles in the ancient world.

Also offensive to religious thinkers was the Epicurean idea that the gods, although they exist, are completely detached from human events. Like good Epicureans themselves, the gods live in contentment and self-restraint. Humans should revere them, but not hope for favors or fear their anger.

Although it remained overshadowed by the rival school of STOICISM, Epicureanism did influence later thinkers, such as the Roman poets Lucretius and Horace (who lived more than 200 years after Epicurus). Ironically, because Epicurus' own voluminous writings have been lost to us, scholars have relied largely on Lucretius' philosophical epic poem *De Rerum Natura (On the Nature of Things)* in order to reconstruct the tenets of Epicureanism.

Epicureanism suffered from "bad press" through the centuries, insofar as its beliefs flatly contradicted both ARISTOTLE and Christianity. The Epicurean doctrine of pleasure has been misunderstood as vulgar hedonism, and this is reflected in our modern English word *epicure*, meaning someone with well developed tastes in food, wine, and the like.

Epicurus See EPICUREANISM.

Epidamnus   A Greek city located on the eastern Adriatic shore, at the site of modern-day Durazzo in Albania. Founded in the late 600s B.C. by colonists from CORCYRA and CORINTH, Epidamnus occupied an isthmus beside a harbor in the non-Greek region known as ILLYRIS. The city was ideally located as an anchorage on the coastal route to ITALY and as a depot for TRADE with the Illyrians. Prosperity eventually brought class warfare, however, as Epidamnus' middle class rose up against their aristocratic rulers.

By 435 B.C. the citizens had established a DEMOCRACY and expelled the aristocrats, who in turn besieged the city. The democrats appealed to Corinth for help. But nearby Corcyra sent out its powerful navy, which captured Epidamnus for the aristocrats and also defeated the Corinthian fleet at the sea battle of Leukimme (435 B.C.), fought off of Corcyra island. This conflict between Corinth and Corcyra was an important cause in igniting the PELOPONNESIAN WAR (431–404 B.C.).

Epidamnus itself remained remote from the great events of the 400s and 300s B.C. Eventually known by a new name, Dyrrhachium, the city passed to King PYRRHUS of EPIRUS (circa 280 B.C.) and thereafter into the expanding empire of ROME.

(See also ARISTOCRACY.)

Epidaurus   Region of the northeastern PELOPONNESE, facing the Saronic Gulf, known principally for its sanctuary of the physician hero god ASCLEPIUS. A great temple of Asclepius, built in the 300s B.C., was the center of a large complex, devoted to healing and worship. The site included a GYMNASIUM, baths, hostels, lesser shrines, and a THEATER for dramatic presentations. The temple boasted a huge GOLD-and-ivory statue of Asclepius, shown seated with two of his emblems: a staff in one hand and a serpent under the other. Excavated since the 1880s A.D., the site is noteworthy today for its beautifully restored theater.

As the focus of pilgrimages by invalids seeking cures from the hero god, Epidaurus combined aspects of a modern spa with those of medieval healing shrines. Suppliants would spend the night in a dormitory associated with the temple, and there (supposedly) they would be visited by the god in a dream. This process, known as incubation, is attested to in many surviving inscriptions of the ancient Greek world. The next morning or so, Asclepius' priests would give the worshipper specific medical advice, including regimens to be followed in the nearby facilities.

(See also MEDICINE; PELOPONNESIAN WAR.)

Epirus   Northwest region of mainland Greece, bordered on the west by the Adriatic Sea and on the east by the Pindus mountain range. Epirus (Greek: *epeiros*, "the mainland") was a humid and forested region, inhabited by 14 tribes or clans, some of Dorian-Greek descent and others of non-Greek, Illyrian blood. The ruling clan, the Molossians, claimed to be descended from the hero NEOPTOLEMUS.

Epirus contained two primitive but important religious sanctuaries: the ancient oracle of ZEUS at DODONA; and an oracle of the dead, situated along the River Acheron, where an entrance to the Underworld was believed to be.

A political backwater for most of Greek history, Epirus came into prominence briefly under its dynamic Molossian king PYRRHUS (319–272 B.C.), who made AMBRACIA his capital. Later Epirus sided with MACEDON against the encroaching Romans. After the Roman victory in the Third Macedonian War (167 B.C.), Epirus passed to Roman rule.

(See also AFTERLIFE; DORIAN GREEKS; ROME.)

Eratosthenes   See MATHEMATICS.

Erectheum   See ACROPOLIS; ARCHITECTURE; ATHENA.

Eretria   City of the large island of EUBOEA, in east-central Greece. Located midway along the island's west coast, Eretria was established circa 800 B.C., apparently by Ionian-Greek refugees from the nearby site now called LEFKANDI. Eretria allied itself with its important neighbor, CHALCIS, and together these cities were among the most prominent in the Greek world during the early era of TRADE and COLONIZATION (circa mid-800s to mid-600s B.C.). The Eretrians and Chalcidians set up trading depots on the coasts of Syria and ITALY, and founded the first Greek colonies in Italy and SICILY. Later these two cities warred over the fertile Lelantine plain, which lay between them; Eretria seems to have fared somewhat the worse in this LELANTINE WAR (circa 720–680 B.C.), and soon both Euboean cities had been surpassed by CORINTH and ATHENS.

In 499 B.C. the Eretrians sent five warships to aid the Greek IONIAN REVOLT against Persian rule in ASIA MINOR. This tiny expedition bore disastrous fruit in 490 B.C., when a Persian seaborne army captured the city and burned it in retaliation. After the PERSIAN WARS (490–479 B.C.), Eretria was forced to join the Athenian-run DELIAN LEAGUE. With the rest of Euboea, Eretria revolted from Athens unsuccessfully in 446 B.C. and successfully in 411 B.C. Chafing under Spartan domination, Eretria joined the SECOND ATHENIAN LEAGUE (377 B.C.) but again revolted (349 B.C.). Along with the rest of Euboea, Eretria came under control of the Macedonian king PHILIP II in 338 B.C.. With the rest of Greece, Eretria passed to Roman control in the mid-100s B.C.

(See also AL MINA; CATANA; CHALCIDICĒ; CUMAE; IONIAN GREEKS; MARATHON; NAXOS (2); PITHECUSAE; ZANCLĒ.)

Erinna   Female poet of the Greek island of Telos, near southwestern ASIA MINOR. Despite one story that makes her a contemporary of SAPPHO (circa 590 B.C.), Erinna probably lived in the 300s B.C. She was known for her 300-line poem, *The Distaff*, composed in memory of a young woman or girl who had died unwed. The poem's surviving fragments, written in Erinna's native Doric dialect, present a touching picture of the dead girl's sweet personality and unfulfilled life.

(See also GREEK LANGUAGE; WOMEN.)

Erinyes   See FURIES.

Eros   Greek god of love, often described as the son of the love goddess APHRODITE. While Aphrodite personified a universal sexual principle, Eros represented more the romantic feelings that one has for a specific person.

Eros was a relative latecomer to Greek religion. In the earliest Greek literature—the epic poems of HOMER (written down circa 750 B.C.)—the word *eros* is a common noun denoting sexual desire. Later HESIOD's epic poem the *Theogony* mentions a personified Eros as an attendant, not son, of Aphrodite (circa 700 B.C.). Lyric poets such as ANACREON (circa 520 B.C.) imagined him as the cruel, mischievous boy who is familiar to us as Cupid; his bow and arrow were first mentioned, as far as we know, by the Athenian playwright EURIPIDES (circa 430 B.C.). As a god of male beauty or of fertility, Eros had cults in BOEOTIA and at ATHENS. The love story of Eros and Psyche (soul) was a philosophical allegory that arose in the 300s B.C.

Borrowed by the Romans, Eros became the Roman god Cupid (Latin *cupido*, meaning "desire").

Eteocles   See SEVEN AGAINST THEBES.

Ethiopia   See BLACK PEOPLES.

Etruscans   Powerful non-Greek people who occupied the region of northwest ITALY now known as Tuscany. By the 700s B.C. they controlled a domain extending south to Campania, on the Bay of Naples. When the first Greek traders began arriving at the Campanian coast after 775 B.C., it was probably the Etruscans, as overlords of the region, who invited the newcomers to make a permanent colony at CUMAE.

The Etruscans' significance for Greek history is twofold: (1) as avid consumers of Greek goods, they provided an important overseas market for the Greeks in the 700s–500s B.C.; and (2) having imbibed Greek material culture and RELIGION, they then transmitted aspects of this to their non-Etruscan subject cities, including the Latin town of ROME. The Etruscans were a crucial early link between the Greek and Roman worlds.

The Greeks called the Etruscans Tursenoi and believed, perhaps rightly, that they had emigrated from ASIA MINOR. The Etruscans were able seafarers, with a reputation for piracy. Their craving for Greek goods has been proven by ARCHAEOLOGY: Excavations in Tuscany since the early 1800s A.D. have unearthed vast troves of Corinthian POTTERY. This surviving pottery is a clear sign of a much-larger TRADE in goods that have not survived, such as textiles, metalwork, and WINE. The importance of this foreign market for CORINTH can be seen in the systematic founding of Corinthian western colonies—CORCYRA, AMBRACIA, and others (mid-700s to 600 B.C.)—to serve as anchorages on the coastal route to Italy. Later (500s B.C.) the Athenians dominated this market—as evidenced from archaeological finds of Athenian black- and red-figure pottery in Italy. What the Etruscans may have supplied in exchange was raw metal: tin (for BRONZE-making), lead, and SILVER.

Eventually the Etruscans came into conflict with their proliferating Greek guests. In alliance with the Carthaginians, the Etruscans defeated the Phocaean Greeks in a naval campaign near Corsica (circa 545 B.C.). Soon, however, they suffered defeats on land from the Greek leader Aristodemus of Cumae (c. 524–505 B.C.). In 474 B.C. the Etruscans lost their claim to Campania forever, when they were defeated in a sea battle off Cumae, at the hands of the Syracusan tyrant HIERON (1). The Etruscans continued as a power in their northern home region for another 200

years but were finally absorbed by their former subjects, the Romans.

(See also CARTHAGE.)

Euboea   Large inshore Greek island, nearly 100 miles long, running parallel to the east coast of central Greece, alongside the regions of LOCRIS, BOEOTIA, and ATTICA. Euboea lies closest to the mainland at a narrow channel called the Euripus. There lay the most important Euboean city, CHALCIS. Close to Chalcis in locale and importance was ERETRIA; the other significant town was Carystus, on the island's southern coast.

The island's name means something like "rich in cattle." It emerged from the DARK AGE circa 900 B.C. as an Ionian-Greek region. Chalcis and Eretria soon became the two foremost cities of early Greece, both in TRADE and in COLONIZATION. Eventually, however, the two cities fought for possession of the fertile Lelantine plain, which stretched between them; this LELANTINE WAR (circa 720–680 B.C.) was remembered as the first major conflict in Greek history. Later the island was a target of Athenian expansionism, as ATHENS and Chalcis became enemies. The Athenians defeated the Chalcidians and seized part of the Lelantine plain for Athenian colonists (506 B.C.).

During the PERSIAN WARS, the Persians landed an expedition on Euboea and destroyed Eretria (490 B.C.); in 480 B.C. the Euboeans supplied crews and warships against the invaders, and the major sea battle of ARTEMISIUM was fought off the northwestern tip of Euboea. After the war the Euboean cities were coerced into joining the Athenian-controlled DELIAN LEAGUE, from which the whole island unsuccessfully revolted in 446 B.C. The Athenian reconquest of Euboea was personally led by PERICLES, who installed Athenian garrison colonies. However, a successful revolt followed amid Athens' decline during the PELOPONNESIAN WAR (411 B.C.).

In about 377 B.C., after chafing under Spartan domination, the Euboean cities joined the SECOND ATHENIAN LEAGUE, but revolted again, in 349 B.C. After 338 B.C. the island was under control of the Macedonian king PHILIP II, who garrisoned Chalcis as one of his strategic holds on the Greek mainland. Thereafter Euboea remained in Macedonian hands until 196 B.C. Eventually it became part of the Roman province of Macedonia (146 B.C.)

(See also GREECE, GEOGRAPHY OF.)

Eubulus   See AESCHINES.

Euclid (Eukleides)   See MATHEMATICS.

Eumenes (1)   See DIADOCHI.

Eumenes (2)   See PERGAMUM.

Eumenides   See EURIPIDES; FURIES.

Euphorion   Poet and scholar of the HELLENISTIC AGE, born circa 260 B.C. at CHALCIS, in CENTRAL GREECE. After enriching himself by marrying a wealthy widow, Euphorion was appointed head of the library at ANTIOCH by the Seleucid king Antiochus II (223–187 B.C.). His poems—

typically short epics on mythological subjects—were influential in the Greco-Roman world but have not survived, and now exist only as a few fragments quoted by later writers. His work apparently provided models for Roman poets such as Catullus and his circle (circa 60 B.C.) and Vergil (circa 30 B.C.).

(See also EPIC POETRY.)

Eupolis One of the three masters of classical Athenian comedy, along with ARISTOPHANES and CRATINUS. Titles of 19 plays by Eupolis are known. None of these has survived, but many fragments of his work exist, quoted by later writers. In his relatively short career (429–411 B.C.), Eupolis won first prize in annual Athenian drama-competitions at least four times—three times at the midwinter festival known as the Lenaea and once at the City Dionysia, a grand event held in early spring. Eupolis died circa 411 B.C. in the HELLESPONT region, on duty in one of the Athenian naval campaigns of the later PELOPONNESIAN WAR.

In the tradition of comedy under the Athenian DEMOCRACY, Eupolis' plays seem to have been political, obscene, and insulting to various public figures. *The Golden Age* (424 B.C.) ridiculed the powerful left-wing politician CLEON. *The Demes* (412 B.C.) showed great Athenians of bygone years arising from the dead to counsel a city in turmoil.

(See also THEATER.)

Euripides Youngest and most controversial of the three great Athenian tragedians of the 400s B.C. Euripides lived approximately 485–406 B.C. His place in literature alongside AESCHYLUS and SOPHOCLES is due to his insight into human psychology (especially abnormal psychology, such as madness or obsessive love), his frequent and sympathetic use of female protagonists (unusual for a classical Greek male), and his bold questioning of the traditional religious ideas in the Greek MYTHS. His most well-known plays include *Medea* (431 B.C.), *Hippolytus* (428 B.C.), *Trojan Women* (415 B.C.), and *Bacchae* and *Iphigenia in Aulis* (both presented posthumously, perhaps in 405 B.C.). These and other works show protagonists (often female) caught in the grip of obsession or disaster, in tales that combine suspense and human detail with brooding questions about the nature of the universe and of the supposed gods.

On the other hand, as critics ancient and modern have observed, Euripides' distortion of certain myths can approach mere caricature, with certain gods shown as repulsive (e.g., HERA in the *Heracles Insane* and APOLLO in the *Ion*). One of Euripides' faults is the intrusion of his love of ideas. This has the effect of turning his characters into mere mouthpieces for clever arguments that are unnecessary to the action. Euripides' contemporaries clearly found his religious irreverence to be disturbing: Although he competed 22 times at Athenian festival drama competitions, he was awarded first place only five times (as compared with Aeschylus' 13 and Sophocles' 24 in their lifetimes). Yet he was always considered thought-provoking enough to be allowed a hearing. The Athenian officials never rejected his application to compete. The ambivalent Athenian attitude toward Euripides is shown in ARISTOPHANES' comedy *Frogs* (405 B.C.), presented soon after Euripides' death. In this play, Euripides and Aeschylus hold a contest in the Underworld for the privilege of returning to ATHENS, and Euripides loses.

Euripides was said to have written 92 plays, of which 80 titles are known. His large number of surviving works make him better represented than any other Athenian playwright: 19 of his tragedies have come down complete, as opposed to only six or seven by Aeschylus and seven by Sophocles. This generous legacy has perhaps hurt Euripides' reputation among modern thinkers, since, unlike the other two tragedians, some of the plays that he is represented by include work that is not his best.

Little is known about his life. He came from a family of Phyle outside Athens that might have been wealthy. Tradition holds (probably falsely) that he was born on the same day as the Greek naval victory over the Persians at SALAMIS (1) (480 B.C.). Unlike Sophocles, Euripides was not politically active, but he did compose an epitaph for the Athenians slain in the disastrous expedition to SYRACUSE during the PELOPONNESIAN WAR (415–413 B.C.).

He was associated with the intellectual movement of the SOPHISTS, who often presented disturbing ideas to their contemporaries. It was supposedly at Euripides' home that the sophist PROTAGORAS recited a treatise that cast doubt on the existence of the gods. Euripides also was said to be friends with the philosophers ANAXAGORAS and SOCRATES as well as with the politician and general ALCIBIADES. Twice married, Euripides was supposedly surly and reclusive; he is said to have done his writing in a cave on the island of Salamis. Like many intellectuals, he was thought to be hostile to the radical Athenian DEMOCRACY.

Medea, a play about the mythical queen who eventually kills her children out of obsessive hatred for her husband, is now considered one of the greatest Greek tragedies. But at the City Dionysia drama competition in 431 B.C. the Athenian judges awarded *Medea*'s three-play group only third prize. However, in 428 B.C. they voted first prize to Euripides' three-play group, which included *Hippolytus*, an insightful play that features a psychological portrait of the mythical queen Phaedra and her hopeless love for her stepson, HIPPOLYTUS.

Trojan Women—one of a trilogy of plays about the TROJAN WAR—was performed in 415 B.C., at the height of the Peloponnesian War. Opening after TROY's capture by the Greeks, the play catalogues the woes that befall the royal Trojan female captives (HECUBA, ANDROMACHE, and CASSANDRA) at the hands of the bullying conquerors—who are themselves unknowingly doomed to meet disaster on their voyage home. The play is a clear protest against Athenian imperialism, including the Athenians' contemporary expedition against Syracuse and their destruction of the resistant, helpless Aegean island of MELOS. The Athenians voted the trilogy second prize.

In 408 B.C. Euripides, then nearly 80 years old, departed Athens, having accepted an invitation from the Macedonian king Archelaus, an enthusiastic literary patron. The reason why Euripides decided to leave home remains a mystery. Possibly his right-wing views were making him unpopular in the increasingly desperate atmosphere at Athens, after the Sicilian disaster and the aborted right-wing coup of the FOUR HUNDRED. At any rate, Euripides died in MACEDON, probably in 406 B.C. It is said that Sophocles, at the next Dionysia, dressed his chorus in mourning for his dead rival.

The *Bacchae*, produced after Euripides' death, presents the story of the god DIONYSUS' destruction of his kinsman, the Theban king Pentheus. The play is admired as a study of human delusion and a disturbing inquiry into the cruelty of the gods.

(See also IPHIGENIA; MAENADS; MEDEA.)

Europa In Greek MYTH, Europa (Greek Europē, "wide eyes") was a Phoenician princess of the city of Tyre. Seeing her by the Mediterranean shore, the great god ZEUS, became inflamed with desire. He changed himself into a handsome bull and enticed her to climb atop his back, then swam quickly out to sea to the island of CRETE. There Europa bore the god two sons, the Cretan princes MINOS and Rhadamanthys. (Some versions add the hero Sarpedon.) Meanwhile, Europa's brother CADMUS began a futile search for her, during which he crossed to Greece and eventually founded the city of THEBES.

The myth may reflect in some dark way Mycenaean-Greek relations with the Near East and Crete circa 1400–1200 B.C. The modern English word Europe, meaning the continent, seems to derive from this mythical character's name.

(See also EUROPE AND ASIA; MYCENAEAN CIVILIZATION; PHOENICIA.)

Europe and Asia The English words *Europe* and *Asia* come from the ancient Greeks. By the 700s B.C. the Greeks had developed the notion that these were two different continents, separated by the AEGEAN SEA. The Greek word Europē seems to have been derived from the name of the mythical princess EUROPA. As used as a place-name, the word originally referred to central Greece. *Asia* originally denoted interior ASIA MINOR, east of the Greek-settled region called IONIA.

As Greek TRADE and COLONIZATION opened up the Mediterranean and other waterways (800–500 B.C.), the name *Europe* was applied to the coastlines of SICILY, ITALY, France, Spain, and the western and northern BLACK SEA. Similarly, the region indicated by the name *Asia* grew with time. After the Eastern conquests of ALEXANDER THE GREAT (334–323 B.C.), Asia came to include the Near East, PERSIA, and other lands, east to the Indus and Ganges river valleys. But EGYPT was usually considered to be outside both Asia and Europe.

In Greek MYTH, a character named Asia is a nymph, sometimes described as the wife of PROMETHEUS. In fact, the Greeks may have taken that name circa 1200 B.C. from the Hittites, whose word *Assuiuva* denoted western Asia Minor.

(See also HECATEAUS; HERODOTUS; NYMPHS.)

Eurydice See ORPHEUS.

Eurypontid clan The Eurypontids (eurupontidai) were the junior royal family at SPARTA, a city that had an unusual government insofar as it was ruled simultaneously by two kings. The Eurypontids traced their ancestry back to a legendary King Eurypontis, one of the sons of HERACLES.

Eurypontid kings took second place in protocol to their partners, the Agiad kings, but they tended to share power equally. Among the best-known Eurypontids was the King ARCHIDAMUS who reigned circa 469–426 B.C.

(See also AGIAD CLAN; DORIAN GREEKS.)

Euxine See BLACK SEA.

family See KINSHIP.

farming Mainland Greece is a mountainous country, and ancient Greek society was shaped by the scarcity of farmland. Individuals owning land in the plains tended to be much richer and more elevated socially than those who farmed the foothills. Hilly terrain dictated cultivation of olive trees, grape vines, and other fruit-bearing plants that thrive on rough ground.

MYCENAEAN CIVILIZATION (1600–1200 B.C.) was a feudal society based on sheep- and cattle-ranching, which involved baronial estates worked by serfs. Cattle rustling and meat feasts are two favorite activities of the aristocrats in HOMER's *Iliad* and *Odyssey* (written circa 750 B.C. but reflecting the Mycenaen world of some 500 years prior). The classical Greek religious custom of sacrificing cattle and sheep to the gods is clearly a Mycenaean vestige. Preoccupation with grazing is reflected in such regional names as EUBOEA and BOEOTIA, containing the Greek root *boes*, "cattle."

The social upheavals of the DARK AGE and years following (circa 1100–700 B.C.) saw the emergence of smaller holdings, owned by nobles, yeomen, or poor subsistence farmers. Developments in metallurgy brought an improved plowshare, tipped with IRON rather than BRONZE (circa 1050 B.C.). By the start of the historical era (750 B.C.), Mycenaean-style ranches survived mainly on the plains of THESSALY; elsewhere, a more efficient use of the land arose in the raising of crops.

Information on Greek farming in this epoch comes from the Boeotian poet HESIOD's verse calendar *Works and Days* (circa 700 B.C.) and from agricultural writers of later years. Details are added by archaeological evidence, ranging from farming scenes on vase paintings to recovered ancient pollen spores. The primary Greek farm crops were grain, olives, grapes (mainly for WINE-making); other fruits such as apples, pears, figs, and pomegranates; and beans and other greens. Barley was the most common grain grown in mainland Greece, but by the 400s B.C. imports of Ukrainian and Sicilian wheat were displacing barley on the market at many cities. Millet was grown for fodder. Flax was grown for weaving into linen. The ancient Greeks knew about cotton, which was cultivated in EGYPT but could not be grown easily on Greek terrain. However, North American–type corn (maize) was unknown to the Greeks, as were tomatoes and potatoes.

Crucial to the ancient Greek economy was olive cultivation, which in the Aegean dates back at least to the Minoan era (circa 2000 B.C.). Eventually the Greeks had more than 25 varieties of olive; they used olive oil for cooking, washing, lamp fuel, religious devotions, and as a treatment for athletes. Olive oil was a principal export of the city of ATHENS, where the olive tree was honored in MYTH as the goddess ATHENA's gift to the city.

Early Greek farmers had to let a field lie fallow every other year (or sometimes two years for each one cultivated),

Harvesters gather the olive crop, in this black-figure painting on an Athenian amphora of the late 500s B.C. Olives were a staple of the Athenian economy, supplying the city's principal export, olive oil, which the Greeks used in cooking, washing, and religious ceremonies.

so the soil could replenish its nutrients. But by circa 400 B.C. the Greeks had discovered the much more productive method of raising different crops in annual rotation on the same land. The Greeks did their plowing and grain-sowing in October, at the start of the rainy season. In order to plow a field, a farmer would guide a wooden rig behind a pair of yoked oxen. (Ancient Greek horse breeds were too small for draft, aside from pulling CHARIOTS.) Plowing was hard work: Unlike the much-improved plowshare of medieval northern Europe, the ancient Greek tool did not turn the soil but only scratched the surface, and the plowman had to keep pushing the blade into the earth as he proceeded.

Grain grew through the relatively mild Greek winter and was harvested in May. During the parching Greek summer the crop was winnowed and stored. September brought a second harvest—of grapes, olives, and other fruit.

(See also ASTRONOMY; ELEUSINIAN MYSTERIES; GREECE, GEOGRAPHY OF; MINOAN CIVILIZATION; SPORT.)

fate   A supernatural power imagined as preordaining certain or all human events. In Greek MYTH and RELIGION, a belief in fate paradoxically existed alongside the belief in the divine guidance and decision making of ZEUS and his subordinate gods.

The main words for fate were *moira* and *aisa*, also translatable as "lot" or "share"—a person's fate is what has been apportioned to him or her. Obviously the final and most dramatic item of human fate is death. Greek notions of fate are best understood as elaborating on the simple knowledge that every person will die and that the gods are unwilling or unable to prevent this.

The god Zeus was imagined as working in harmony with fate or somehow causing fate. One of his titles was Moiragetēs, "leader of fate." HOMER's epic poem the *Iliad* (written down circa 750 B.C.) repeatedly shows Zeus holding a pair of golden scales and weighing the respective dooms (*kēres*) of two human antagonists on the battlefield; the warrior indicated by the heavier pan is doomed. Zeus will not avert a doomed warrior's fate—even his beloved human son Sarpedon is fated to die, slain by the Greek warrior PATROCLUS (book 16). Conversely, the angry god POSEIDON in the *Odyssey* cannot completely destroy the hero ODYSSEUS, because Odysseus is fated to survive his voyage and return home (book five). But the hero ACHILLES in the *Iliad* is aware of a choice in his fate: He can either stay and die gloriously, fighting at TROY, or can return home and live to a safe, undistinguished old age (book 9).

The personification of fate as three goddesses, the Moirai, is first clearly described in HESIOD's epic poem the *Theogony* (circa 700 B.C.). Hesiod presents the Moirai as the daughters of Zeus and the goddess Themis. They are imagined as working at the womanly task of spinning—drawing out a thread of yarn that determines or represents each person's life. Into the thread may be woven sorrow, wealth, travel, and the like. Hesiod names the three fates Clotho (spinner), Lachesis (disposer of lots), and Atropos (unavoidable). Later writers distinguished the goddesses' duties, with Clotho spinning the thread, Lachesis measuring it out, and Atropos snipping it. Either the cutting produced the person's death or, in another version, the

thread was entirely spun out and cut at the baby's birth, to contain the person's future.

The pretty conceit of divine spinners was probably more important to poets than to the Greek religious public. But the Moirai were widely worshipped as goddesses of childbirth and as promoters of good harvest. Athenian brides, for instance, brought cuttings of their hair as offerings to the fates.

(See also PROPHECY AND DIVINATION.)

Four Hundred, The   Oligarchic committee that briefly seized power at ATHENS in 411 B.C., during the PELOPONNESIAN WAR. The Four Hundred were led by the politician Pisander and by the orator ANTIPHON. Taking advantage of the absence from Athens of large numbers of working-class, pro-democratic citizens (who were manning the large fleet in operation at SAMOS), the conspirators began intimidating the populace in the spring of 411 B.C. with a series of political murders and demonstrations. The coup d'etat was performed in June in the Athenian ASSEMBLY. Led by Pisander, the plotters pushed through a number of revolutionary decrees, amounting to a dismantling of the DEMOCRACY; all existing executive posts were abolished and a new COUNCIL, of 400 men (replacing the democratic council of 500), was appointed.

The conspirators ruled Athens for the duration of the summer. But the Athenian troops on distant Samos remained staunchly pro-democratic, perpetuating the Athenian democracy by holding their own assemblies and elections. After a home-defense war fleet dispatched by the Four Hundred was defeated by the Spartans in the Straits of EUBOEA, the oligarchs were discredited and overthrown (in September 411 B.C.). Pisander fled to SPARTA; Antiphon, disdaining to flee, was tried and executed under the restored democracy.

(See also OLIGARCHY.)

funeral customs   Despite the cremation of slain heroes described in HOMER's epic poems, the Greeks outside of the DARK AGE (circa 1100–900 B.C.) generally buried their dead. The elaborate tombs of Mycenaen kings have supplied archaeologists with most of their information about Greek prehistory for the era 1600–1200 B.C. The Greeks of historical times, after about 750 B.C., had their own burial customs. Their need to cover a corpse with earth was not simply a matter of hygiene or decorum; it was believed that the sight of a dead body would offend the Olympian gods, and that the dead person's ghost could not enter the Underworld until the body had been covered. This helps explain the insistence of SOPHOCLES' tragic heroine ANTIGONE on sprinkling dirt over her slain brother's corpse, although forbidden by a ruler's decree from doing so. As Antigone's action suggests, such a "burial" could be merely ceremonial, performed with a few handfuls of dirt.

For a regular funeral (Greek: *taphē*), the body was bathed and clothed. At the home, it was mourned over by relatives and household—but excessive lamentation actually was forbidden by law at ATHENS; there seems to have been a deep-seated fear that WOMEN's lamenting might bring the dead back to life. The body would then be carried on a stretcher to the place of burial, where it would be placed

might be left—to "refresh" the dead in the grave or for the journey to the Underworld.

After the burial, the grave would receive some kind of marker. Again, for the wealthy, there was marble—specifically, the type of monument known as a *stēlē*, with

An Athenian marble gravestone, or *stēlē*, circa 535 B.C. Standing nearly 14 feet tall, the monument has a carved bas-relief of the dead person and is topped with a statuette of a guardian sphinx. The sphinx—a mythical winged monster, with a woman's head and lion's body—was adapted by the Greeks from a Near Eastern design originating in Egypt. The palmette pattern, appearing below the sphinx, was a familiar Greek funerary motif, originally inspired by the Tree of Life of Near Eastern art.

in a coffin or—for the rich of a certain era—a stone sarcophagus ("flesh-eater"). Certain marble sarcophagi, carved with beautiful reliefs showing scenes of hunting and war, are among the finest Greek SCULPTURE of the late 300s–200s B.C.. At the graveside, offerings of food and drink

This broken marble gravestone dates from about 510 B.C. and commemorates an Athenian named Aristion. (The name was carved into the stone's base.) The life-size bas-relief shows Aristion in soldier's armor: The top parts of his spear and helmet have broken off, as has the extension of his beard. Gravestone carvings conventionally portrayed mature men as bearded soldiers and younger men as clean-shaven athletes. However, it is possible that Aristion actually died in battle, living up to the aristocratic creed expressed in his name, "one of the noblest."

an idealized portrait of the dead. At Athens the Ceramicus district, outside the northwest city wall, contained monumental graveyards owned by aristocratic families.

For public funerals, such as the mass observances by which the classical Athenians honored their war dead, a prominent citizen might be chosen to deliver a speech (or, alternatively, compose a poem) honoring the departed. This kind of funerary speech was known by the adjective *epitaphios*, from which comes our word "epitaph." The most famous funeral oration of the ancient world was delivered by the Athenian statesman PERICLES in the first year of the PELOPONNESIAN WAR (431 B.C.). As recounted by the historian THUCYDIDES (1), Pericles' speech presents a glorification of the Athenian DEMOCRACY, for which the men had died.

(See also AFTERLIFE; MYCENAEAN CIVILIZATION.)

Furies In Greek RELIGION and MYTH, the Furies (Greek: Erinues) were horrible female spirits of divine retribution.

Tormenting their victims by nonphysical means inciting fear and madness—they especially punished the crime of murdering a family member. But they might pursue oathbreakers or anyone else who had broken a bond of society. Their best-known appearance in literature is in AESCHYLUS' tragedy *Eumenides* (458 B.C.), where they constitute the play's 12-member chorus. In the story, they torment the hero ORESTES, who has slain his mother.

In HESIOD's Creation poem, the *Theogony* (circa 700 B.C.), they are described as being the daughters of GAEA (Mother Earth), born from the spattered blood of the castration wound that CRONUS inflicted on his father, Uranus (Ouranos, "Sky").

The Furies were usually worshipped under some euphemistic name; Aeschylus' title *Eumenides* means "kindly ones." An altar to them as the Dreaded Goddesses, at the foot of the Athenian ACROPOLIS, became associated with the curse of the Alcmaeonid family.

(See also ALCMAEON (1); ALCMAEONIDS; CYLON; ELECTRA.)

G

Gaea Greek earth goddess. Unlike the grain goddess DEMETER, Gaea (Greek: Gaia) was not a central deity of the Greek RELIGION. Rather she was a personification of the element of land (as opposed to sky or sea). Her name reflects the Greek common noun for "land," *gē* or *gā*. Gaea had a cult at several locales, particularly in Greek SICILY.

According to the creation myth in HESIOD's epic poem *Theogony* (circa 700 B.C.), Gaea emerged from primeval Chaos, then gave birth to Uranus (*Ouranos*, "Sky"), whom she married. Their monstrous children included the TITANS, the Cyclopes, and the "Hundred-handed Ones" (Hekatoncheires). When their son CRONUS castrated his father, fertile Gaea was impregnated by the spattered blood and gave birth to the FURIES and the GIANTS. Later she produced the monster TYPHON. In turn, the Giants and Typhon impiously attacked the immortal gods atop Mt. OLYMPUS.

(See also CYCLOPS; ZEUS.)

Galatea See POLYPHEMUS; PYGMALION.

Galatia See ASIA MINOR; CELTS.

Games See ISTHMIAN GAMES; NEMEAN GAMES; OLYMPIC GAMES; PANATHENAEA; PYTHIAN GAMES; SPORT.

Ganymede In Greek MYTH, Ganymede was a handsome Trojan prince, son of King Tros. The earliest surviving mention of him, in HOMER's epic poem the *Iliad* (written down circa 750 B.C.) says that Ganymede was abducted by the gods to be cupbearer for the great god ZEUS on Mt. OLYMPUS; in exchange, Ganymede's father received a herd of wondrous horses.

Later versions add that the boy was stolen by Zeus himself in the shape of an eagle, or that the eagle was sent by Zeus. Unlike Homer, poets such as THEOGNIS (circa 542 B.C.) portrayed the abduction as a rape, assuming that Zeus desired Ganymede sexually. Modern scholars believe that this sexual nuance is not intended in the original tale told by Homer, the added homosexual coloration would seem to reflect new upper-class values that arose after Homer's time, perhaps in the late 600s B.C. The version adopted by the ETRUSCANS of ITALY included the sexual aspect. The Etruscan rendition of the boy's name was Catamitus, a word that eventually went into Latin and subsequently into our own language as the pejorative noun *catamite,* meaning "a male who receives sexual penetration."

In Greek vase painting of the 500s and 400s B.C., Ganymede is portrayed as an idealized youth or boy with long hair. Sometimes he holds a playing hoop (suggesting boyhood) and a rooster (Zeus' courting gift).

(See also HOMOSEXUALITY.)

Gaugmela, battle of See ALEXANDER THE GREAT.

Gela Greek city on the southeastern coast of SICILY, near the mouth of the River Gelas. Founded circa 688 B.C. by DORIAN GREEKS from CRETE and RHODES, Gela was a small city on a fertile plain that enjoyed prosperity. In around 580 B.C. it established its important daughter city, ACRAGAS, farther west on the south coast. Gela was ruled by a horse-breeding ARISTOCRACY until the late 500s B.C., when (like other cities of the Greek world) it passed into the hands of TYRANTS.

Gela became the most powerful Sicilian city circa 490 B.C., after the Geloan tyrant Hippocrates conquered much of eastern Sicily. But Gela's preeminence ended when Hippocrates' successor, GELON, captured the great city of SYRACUSE in 485 B.C. and made it his capital, going so far as to transplant the richer half of Gela's population to Syracuse.

By the mid-400s B.C. the tyrants had fallen from power, but Dorian Gela remained an ally of Dorian Syracuse against the region's Ionian-Greek cities. Gela assisted Syracuse's triumphant defense against the Athenian invasion of 415–413 B.C. In the 300s B.C., with Sicily a battleground between the Greeks and the Carthaginians, Gela suffered from the exploits of Greek tyrants such as AGATHOCLES, who executed 4,000 Geloans for suspected treason. In 311 B.C. Gela's population was entirely transplanted to a new city by the tyrant Phintias of Acragas.

Gelon Dictator or tyrant (*turannos*) of the Greek cities GELA and SYRACUSE, in SICILY. Gelon lived circa 540–478 B.C. An able military commander, he is remembered for his defeat of the Carthaginians at the Battle of HIMERA (480 B.C.)—an event that made him the most powerful individual in the Greek world.

As his name suggests, Gelon was descended from the first Greek colonists of Gela. He was an aristocrat, and before seizing power he served as a leader of CAVALRY for the Geloan tyrant Hippocrates (who carved out a small empire in eastern Sicily). After Hippocrates' death, Gelon dispossessed the tyrant's sons and assumed full power, at about age 50 (circa 490 B.C.).

In 485 B.C. Gelon captured Syracuse, the foremost city of Sicily. Making Syracuse his new capital, Gelon grandiosely transferred the richer half of Gela's population there, as well as the richer citizens of other captured Greek towns. (Other Sicilian tyrants soon copied this ruthless new practice of transplantation.) Gelon installed his brother HIERON (1) as tyrant of Gela and allied himself with Theron, tyrant of ACRAGAS, taking Theron's daughter Demareta as one of his wives.

Meanwhile, the Greek world was being threatened by two foreign empires, the Persians in the East and the Carthaginians in the West. Around 481 B.C. a delegation of mainland Greeks went to Syracuse to appeal for Gelon's aid against the coming invasion of Greece by the Persian king XERXES. Gelon offered to contribute a large force on the condition that he himself receive chief command of either the Greek army or navy. When these two options were rejected, Gelon sent the ambassadors away, remarking that they seemed better equipped with generals than with troops.

In 480 B.C.—coinciding with the Persian invasion of Greece—the Carthaginian leader Hamilcar sailed from North Africa with an army and landed on Sicily's north coast. But a Greek army under Gelon destroyed Hamilcar and his force at the Battle of Himera.

After making peace with CARTHAGE and imposing on them a huge indemnity of 1,000 TALENTS, Gelon began a program of propagandistic cultural display. He dedicated lavish offerings at the sanctuaries of DELPHI and OLYMPIA, in Greece; he enlarged and adorned Syracuse with public building; and he minted a celebrated victory-issue coin, called the Demareteion (after his wife). Gelon died in 478 B.C., and was succeeded by his brother Hieron, who proved to be a less popular ruler.

(See also COINAGE; PERSIAN WARS; TYRANTS.)

Geometric pottery   See POTTERY.

geometry   See MATHEMATICS.

Geryon   See HERACLES.

Giants   In MYTH, the Gigantes were a primeval race of monsters who unsuccessfully rebelled against ZEUS and the other gods, piling Mt. Ossa atop Mt. Pelion in order to assault Mt. OLYMPUS. In the Creation myth in HESIOD's epic poem *Theogony* (circa 700 B.C.), the Giants are described as children of Mother Earth, GAEA. (Her name is reflected in theirs.) She gave birth to them after being impregnated by the blood of the god Uranus (Ouranos, "Sky"), which fell to earth when Uranus was castrated by his son CRONUS. The Giants were sometimes portrayed in art as large, snake-legged humanoids.

The Gigantomachy (battle with the giants) was one of the most popular mythological subjects in Greek art and poetry, symbolizing the victory of civilization over savagery. On grand artistic vistas, such as the architectural sculptures of the PARTHENON, the battle might be shown in conjunction with the fight between CENTAURS and Lapiths or the battle between Greeks and AMAZONS. Later writers embellished the tale, adding that the gods enlisted

the hero HERACLES after being warned that they must include a mortal in order to win.

(See also ATHENA; THESSALY; ZEUS.)

gold   Ancient Greeks prized gold for the casting of precious SCULPTURE and other artifacts. However, gold as a mineral deposit was scarce in the Aegean region. The source for the gold used for the royal tomb offerings at MYCENAE circa 1550 B.C. is unknown. The ore may have come from EGYPT or ASIA MINOR, traveling by way of TRADE or plunder of war.

By the 600s B.C. Greek pioneers were prospecting for gold at Mt. Pangaeus, on the Aegean coast of THRACE. Despite the hostility of the native Thracians, the Greeks, based in the nearby island of THASOS, panned the streams of Pangaeus and perhaps dug mines; SILVER seems to have been more abundant there than gold. Eventually Pangaeus was seized by imperial ATHENS (mid-400s B.C.) and then by the Macedonian king PHILIP II (mid-300s B.C.). The nearby Athenian colony at AMPHIPOLIS was strategic to these interests. Another Aegean gold source, closer to mainland Greece, was the Greek island of Siphnos, in the western CYCLADES. The Siphnian mines financed the construction of an elegant treasury building at DELPHI, circa 525 B.C.; later the mines were ruined by natural flooding.

Most gold probably reached early Greece through trade with non-Greek nations beyond the Aegean basin. Trading partners in the 600s and 500s B.C. included the Asia Minor kingdoms of Phrygia and LYDIA, where gold could be panned from certain rivers, and where such kings as MIDAS and CROESUS, fabled for their wealth, ruled. In Lydia, a gold-silver alloy called electrum—panned from the rivers where it formed naturally—supplied the metal for the world's first coins, circa 635 B.C. The Greeks, however, rarely used gold for their COINAGE; since gold was so difficult for them to acquire, they minted with silver and, eventually, with BRONZE.

In the mid-500s B.C. the gold fields of Asia Minor and Mesopotamia passed into the hands of the conquering Persians, who continued the gold trade. Like other precious goods, gold became more plentiful for the Greeks of the HELLENISTIC AGE (circa 300–150 B.C.), after the conquests of the Macedonian king ALEXANDER THE GREAT had opened up the mines and treasure troves of the Persian empire.

(See also PHIDIAS.)

Gordian Knot   See ALEXANDER THE GREAT.

Gorgias   Greek orator of the east-coast Sicilian Greek city of Leontini (circa 483–376 B.C.). Visiting ATHENS on a political embassy in 427 B.C., he created a cultural sensation with his highly wrought style of public speaking. His display pieces, using rhyme, alliteration, and rhythmical parallel clauses, were soon widely imitated at Athens; his visit marked a revolution in ancient Greek RHETORIC. Gorgias periodically returned to Athens to teach; his greatest pupil was the Athenian orator ISOCRATES. Gorgias' influence has been noted in the works of the historian THUCYDIDES (1) and the playwright AGATHON.

Because of his love of the ingenious and shameless argument, Gorgias was associated with the intellectual movement of the SOPHISTS. One of his surviving speeches, the *Encomium of Helen,* is a defense of the mythical HELEN OF TROY and her adulterous, disastrous elopement with the Trojan prince PARIS. The speech shows Gorgias' soothing, jingling style, which apparently seeks to charm the listener by incantation.

A dignified, amiable Gorgias appears as a character in PLATO's fictional dialogue *Gorgias* (written circa 380 B.C.).

Gorgons See MEDUSA.

Gortyn City of central CRETE, settled by the DORIAN GREEKS after about 1000 B.C. By 500 B.C. it was a flourishing city, the largest on the island. In the first century B.C. it became the capital of the Roman province of Crete and Cyrenaica.

Archaeological excavations of the ancient town in the 1880s uncovered the now-famous marble inscription called the Law Code of Gortyn, carved circa 500 B.C. The code, written in the Doric dialect, addresses such civil-law issues as land tenure, mortgages, and the status of SLAVES. Although the surviving inscription contains no sections on criminal law, the Gortyn code remains the most important, single, extant source for Greek law prior to 300 B.C.

(See also GREEK LANGUAGE; LAWS AND LAW COURTS; WRITING.)

grain supply See BLACK SEA; EGYPT; FARMING; HELLESPONT; SICILY; THRACE.

Greece, geography of The key to Greece's history lies in its geography. The mountainous terrain compartmentalized the country into separate population centers, with distinctive dialects, cultures, and politics. This is the background of the Greek "city-state," or POLIS. The rough terrain also placed great economic importance on the few rich FARMING plains. The earliest Greek culture, the MYCENAEAN CIVILIZATION (circa 1600–1200 B.C.) was centered mainly on the two major farmlands of southern and central Greece: the Argive plain, containing the cities of MYCENAE, ARGOS, and TIRYNS; and the twin Boeotian plains, containing THEBES and ORCHOMENUS.

Threatened by famine, Greek cities fought desperate wars over farmland. The first fully remembered conflict in Greek history was the LELANTINE WAR, fought for control of the plain lying between the central Greek cities of CHALCIS and ERETRIA (circa 720–680 B.C.). In the south, the rising state of SPARTA waged a series of wars to annex the plain of MESSENIA (mid-700s–600 B.C.).

Cities unable to retain or capture farmland might take to shipping—for Greece comprises a large, ragged peninsula that juts into the northeastern Mediterranean, and many Greek cities lie near the sea. After acquiring seafaring skills from the prehellenic inhabitants of the land (second millennium B.C.), and after copying the technologies and trade routes of the brilliant Phoenicians (circa 900–800 B.C.), the Greeks emerged as a great seagoing people. The vast enlargement of the Greek world through overseas TRADE and COLONIZATION (circa 800–500 B.C.) was in part due to

individual states using seafaring to solve their problems of lack of food supply and employment. The tiny Greek state of AEGINA, perched on a rocky island, became rich as a middleman in overseas trade (500s–early 400s B.C.). The city of ATHENS became a superpower through its strong navy (400s–300s B.C.).

Cities could prosper by controlling critical points on a trade route—a sea channel, an anchorage, or a mountain road. The best example is CORINTH, superby located on the narrow isthmus stretching northeastward from the PELOPONNESE. By land, Corinth commanded the route north to central Greece. By sea, Corinth possessed two harbors, facing east and west across the midpoint of Greece, which enabled the city to develop trade routes in both directions, to the Near East and to ITALY and SICILY.

Of strategic importance were the mountains of north-central Greece. Any invading army marching southward from the northeast was forced to face a defensible bottleneck at the mountain pass of THERMOPYLAE, in southern THESSALY. Themopylae played a repeated role in Greek military history, most famously as the site of a failed Greek defense in the PERSIAN WARS (480 B.C.), but also in the machinations of the Macedonian king PHILIP II (in 346–338 B.C.).

The Macedonian kings who kept Greece subjugated in the 200s B.C. recognized four "fetters" or "keys" of Greece—four critical geographic points, possession of which were vital for control. The fetters were Corinth, Chalcis (guarding the narrows of the Euboean Straits), the Athenian port city of PIRAEUS, and the coastal stronghold of Demetrias, beside Thessaly's Gulf of Pagasae (near modern Volos).

(See also ACARNANIA; AEGEAN SEA; AETOLIA; ATTICA; BOEOTIA; CHALCIDICĒ; EPIRUS; EUROPE AND ASIA; EUBOEA; IONIAN SEA; LOCRIS; MACEDON; PHOCIS; SHIPS AND SEAFARING; THRACE.)

Greece, history of See ALEXANDER THE GREAT; ATHENS; COLONIZATION; CORINTH; DARK AGE; DIADOCHI; HELLENISTIC AGE; MACEDON; MINOAN CIVILIZATION; MYCENAEAN CIVILIZATION; PELOPONNESIAN WAR; PERSIAN WARS; ROME; SPARTA; THEBES; TRADE; TYRANTS.

Greek language The ancient Greek language belonged to the Indo-European family of languages that now includes English, German, Russian, Persian, and such Latin-derived tongues as French and Spanish. These languages share certain resemblances in grammar and vocabulary (e.g., English "father," ancient Greek *patēr,* ancient Latin *pater*). According to modern linguistic theory, the prototype of all these languages was spoken by a people living in the Caucasus region (now in southern Russia) in the third millennium B.C. At some later date, this people—or, at least, their language—began "radiating" outward, presumably by migration and conquest.

The decipherment in A.D. 1952 of LINEAR B inscriptions from the MYCENAEAN CIVILIZATION confirmed that the Mycenaeans of 1400–1200 B.C. spoke an early form of Greek. It seems that probably in about 2100 B.C. the developing Greek language entered the land of Greece with invaders from the Danube basin. These invaders—who can be called the first Greeks—overran the land, subduing the prior inhabitants by war and assimilation. In later centuries the Greek language contained many words of non-Greek etymology, presumably derived from the pre-Greek inhabitants' language. Often distinguished by the endings -*nthos,* -*sos,* or -*ēnē,* these words included names of local plants, certain gods' names, and most of the place-names in mainland Greece. Examples are CORINTH (Korinthos), HYACINTHUS (Huakinthos), acanthus (akanthos), NARCISSUS (Narkissos), and ATHENA (Athēnē)—also, words such as *thalassa,* "sea," and *nēsos,* "island."

The final forging of the Greek language came with the end of the Mycenaeans, soon after 1200 B.C. As Mycenaean society toppled, Greece became subject to a new immigration of Greeks—the DORIAN GREEKS, speaking their own dialect. The story of what happened in the centuries following has been pieced together by ARCHAEOLOGY, philology, MYTH, and a few precious references by later Greek writers.

Entering central Greece from the northwest (circa 1100 B.C.), the Dorians failed to capture the territory of ATHENS. But farther south they overran almost all of the PELOPONNESE (although mountainous ARCADIA held out), and they continued conquering by sea across a band of the southern AEGEAN SEA, to CRETE and southwestern ASIA MINOR. In historical times, this swath of territory was distinctive for the Dorian culture and Doric dialect shared by its various inhabitants. The best-known Dorian city was SPARTA. Modern linguists tend to group the Doric dialect into a larger category known as West Greek, which includes the language of the ancient Greeks around the western Corinthian Gulf.

Meanwhile (circa 1000 B.C.), other migrations from mainland Greece were taking place. The AEOLIAN GREEKS of BOEOTIA and THESSALY, buffeted by the invading Dorians, sent refugees fleeing eastward across the Aegean to the island of LESBOS and beyond, to northwestern Asia Minor. These disparate areas became known by their shared Aeolic dialect.

But the most fateful migration was of those Ionian-ethnic Greeks whom the Dorians had chased from the Peloponnese. These Ionians emigrated across the Aegean to the central west coast of Asia Minor, where they eventually flourished as the rich and advanced society of IONIA. Athens and the CYCLADES islands remained Ionian territory, all distinguished by an Ionic dialect. The Athenian language developed as a subcategory of Ionic, called Attic (from the Athenian home territory of ATTICA).

Such were the three main divisions of Greek language and culture, although other categories existed. The highlanders of Arcadia spoke a very old form of Greek that retained aspects of ancient Mycenaean speech and that resembled the speech of another, but very distant, Greek enclave, the island of CYPRUS. Modern scholars call this shared dialect Arcado-Cyprian.

(See also ALPHABET; EPIC POETRY; IONIAN GREEKS; LABYRINTH; LYRIC POETRY; MYCENAE; RELIGION; SHIPS AND SEAFARING; TIRYNS; WRITING.)

gymnasium Sports complex. The Greek word *gumnasion* comes from the verb *gumnazo,* "exercise" (from the adjective *gumnos,* "naked" or "loinclothed," which was how Greek athletes usually trained and competed). Typically located outside a city's walls, the gymnasium featured a running track, a WRESTLING court, and fields for throwing the javelin and discus; also, rooms for changing, oiling down, and so on. A more elaborate type might also include parklands, colonnaded walks, and a horse-riding track.

Gymnasiums first appeared in the 500s B.C. They were usually state-run institutions open to men and boys of the citizen classes. Within this group, the majority of regular customers would be aristocrats and other leisured rich, who did not need to work for a living. The gymnasium played a central role in the Greek city-state, fostering male competition and camaraderie. Schoolboys would receive training in SPORT, and men would pass their time in exercise and conversation. The gymnasium was also central to the male-homosexual climate of the 500s–300s B.C., being the place where men could observe, and try to meet, boys and youths.

With its gathering of educated people, the gymnasium was a natural scene for philosophical discussion—at least in ATHENS during its age of intellect (400s and 300s B.C.). The Athenian philosopher SOCRATES often could be found in one or another of the city's gymnasiums, conversing with his following of aristocratic men. Two Athenian gymnasiums, the ACADEMY and the LYCEUM, gave their names to nearby philosophical schools, set up by PLATO and ARISTOTLE, respectively (300s B.C.). The intellectual and social-sexual energy of the Athenian gymnasium is conveyed in certain of Plato's fictional dialogues, such as the *Charmides.*

(See also ARISTOCRACY; EDUCATION; HOMOSEXUALITY; POLIS.)

Hades In MYTH, the god of the Underworld, and brother of ZEUS and POSEIDON. The name Hades ("the unseen") properly refers to the god and not his kingdom. Unlike the Judeo-Christian Satan, Hades is not evil; as king of the dead, he rules a domain in nature that complements the happier realm of his brother Zeus. Hades is, however, rather colorless. He has no MYTH, except for his abduction of PERSEPHONE. By Zeus' decree, she lives with Hades as his queen for four months of every year.

Like other gods of death or ill omen, Hades often was worshipped under euphemistic titles. One of the most common of these was Pluton, "the Rich One," probably referring to the rich metal ore in the ground. The early Romans, in adapting Greek RELIGION to their own use, adapted this title as the name for their god of the dead, Pluto.

(See also AFTERLIFE.)

Halicarnassus Greek seaport of southwestern ASIA MINOR, in the non-Greek territory known as Caria. Founded circa 900 B.C. by DORIAN GREEKS from Troezen (in the eastern PELOPONNESE), Halicarnassus lay at the northern shore of what is now called the Bay of Gökova, in Turkey. As a gateway to and from the southeastern AEGEAN SEA, the city had a lively commercial culture, combining Carian and Dorian-Greek elements. Halicarnassus was part of the local Dorian federation, centered at nearby CNIDUS.

Like other east Greek cities, Halicarnassus flourished in the 600s and 500s B.C., was captured by the Persians circa 545 B.C., and endured a series of Persian-run Greek TYRANTS. Remarkably, one such tyrant was a woman, named Artemisia, who accompanied the Persian king XERXES on his invasion of Greece and led a naval squadron against the mainland Greeks at the Battle of SALAMIS (1) (480 B.C.).

Halicarnassus' most famous native—the first writer of history, HERODOTUS—fled the city circa 460 B.C. after taking part in a failed coup against another tyrant, Lygdamis. By 450 B.C., however, Halicarnassus was a DEMOCRACY and a member of the Athenian-led DELIAN LEAGUE. In 412 B.C., during the PELOPONNESIAN WAR, the city became a major Athenian naval base, after other sites had been lost to widespread revolt by the Delian allies. Later Halicarnassus reverted to Persian control, by the terms of the KING'S PEACE (386 B.C.).

Under Persian rule, Halicarnassus became one of the most beautiful and dynamic cities of the Greek world. The ruler Mausolus—Carian by birth but reigning under Persian suzerainty—glorified the city as his capital, build-ing a circuit wall, public monuments, dockyards, and a citadel (circa 370 B.C.). The crown of the building program was his own monumental tomb, the famous Mausoleum, erected after his death by his adoring widow (who was also his sister, Artemisia). Now vanished, but described by ancient writers, the Mausoleum was a squared, pyramidlike structure of white marble, adorned with a colonnade at its base and ascending in layers to a sculpted chariot group at the top. Numerous exterior sculptures (some now housed in the British Museum) showed a battle between Greeks and AMAZONS and other mythological scenes. Because of its tremendous size and splendor, the Mausoleum was counted as one of the SEVEN WONDERS OF THE WORLD.

Defended by a Persian garrison, Halicarnassus was besieged and captured by the Macedonian king ALEXANDER THE GREAT in 334 B.C., and passed into his empire. His successors warred over it, with the Ptolemies of EGYPT eventually winning possession (200s B.C.). The city came under Roman rule in the 100s B.C.

Harmodius and Aristogiton Known as the tyrannicides (tyrant slayers), these two young male Athenian aristocrats became semilegendary figures after they died while assassinating Hipparchus, younger brother of the Athenian dictator HIPPIAS (1), in 514 B.C. Far from ending Hippias' reign at ATHENS, the assassination brought on repression from the ruler; but Harmodius and Aristogiton's bravery and sacrifice—and the fact that the two men were linked by romantic love—struck a sentimental chord in their fellow Athenians.

Harmodius and Aristogiton were kinsmen in an ancient noble clan. In accordance with the usual pattern of upper-class Greek homosexual pairings, Aristogiton was the protective older lover, perhaps in his late 20s; Harmodius was probably in his late teens. When Harmodius ignored the unwelcome advances of Hipparchus, the thwarted Hipparchus insulted Harmodius' sister in public.

In revenge, Aristogiton plotted with Harmodius and a few sympathizers to murder the dictator Hippias and overthrow the regime. But the plot went awry. At the summer festival of the PANATHENAEA, Harmodius and Aristogiton, attacking with daggers, had the chance to kill only Hipparchus. The young Harmodius was slain by bodyguards. Aristogiton, escaping into the crowd, was later captured and tortured to death.

After Hippias was expelled from Athens by other forces in 510 B.C., the dead tyrannicides came to be seen as forerunners of the newly installed DEMOCRACY and were commemorated with yearly religious sacrifices and a famous public statue. At the same time, they were viewed by the upper class as models of aristocratic conduct. One symposiastic song praised them in mythical terms, including the stanza "Dearest Harmodius, they say surely you are not dead but live forever in the Islands of the Blessed, with swift-footed Achilles and Diomedes."

The Athenian historian THUCYDIDES (1), in his history of the PELOPONNESIAN WAR, devotes an unusual digression to the subject of the tyrannicides, trying to correct the exaggerated, unhistorical stories about them.

(See also ACHILLES; AFTERLIFE; ARISTOCRACY; DIOMEDES; HOMOSEXUALITY; SYMPOSIUM; TYRANTS.)

harpies In MYTH, the harpies (Greek: harpuiai) were winged female demons who would fly down to steal food, people, and so on; their name means "snatchers-away." They are mentioned in HOMER's epic poem the *Odyssey* (written down circa 750 B.C.) as carrying off the daughters of the hero Pandareus. HESIOD's *Theogony* (circa 700 B.C.) describes them as three in number, named Celaeno ("dark"), Ocypete ("swift-wing"), and Aello ("storm"). The harpies are most familiar from APOLLONIUS' epic poem, the *Argonautica* (circa 245 B.C.). There they are said to torment the Thracian king Phineus on a regular basis by flying down, snatching away his food, and departing, leaving their feces on everything.

Possibly the harpies originated as spirits or personifications of the wind. In their general role as winged, malevolent females, they resemble the FURIES and the SIRENS.

Hecataeus Early geographer and "logographer" (travel writer) of the Greek city of MILETUS, in ASIA MINOR. He lived circa 550–490 B.C. On the basis of travels in EGYPT, Scythia, the Persian interior, and elsewhere, Hecataeus wrote a prose treatise called the *Periodos Gēs*, or "Trip Around the World" (which today survives only in fragments quoted by later writers). The book accompanied a map Hecataeus had made—perhaps painted onto textile or incised on a copper plate—showing the world as he had encountered it; although this was not the world's first map, the map and book together were revolutionary for Greek learning. We know that the book gave accounts of local histories, customs, and so on. There were two volumes: *Europe* and *Asia*. Hecataeus is an important forerunner of the world's first historian, HERODOTUS (born circa 485 B.C.).

We know almost nothing about Hecataeus' life. His father was said to have been a landowner at Miletus, and Hecataeus must have been wealthy to undertake his travels. He surely had contacts among Miletus' merchant community, for his journeys included sites along the Milesian shipping route northeastward to the BLACK SEA. Herodotus mentions Hecataeus amid events of the IONIAN REVOLT against Persian rule (499–493 B.C.): Hecataeus was present at the first Milesian war council and advised against the revolt. Herodotus tells us no more about Hecataeus; perhaps he was killed in the Persian capture of Miletus in 493 B.C.

In addition to his travel book, Hecataeus wrote a treatise called *Geneologies*, about the legendary pedigrees of noble Greek families and mythical heroes. The few surviving fragments of this work suggest a strong rationalizing purpose: For example, DANAUS could not have had 50 daughters as the MYTH claims, but perhaps 20. The book's opening words announce the writer's logical approach: "I write what I believe to be the truth. For the stories of the Greeks are many and laughable."

(See also EUROPE AND ASIA; IONIA.)

Hecatē Mysterious and sinister goddess, associated with night, witchcraft, ghosts, the moon, and the supernatural danger of the crossroads. Hecatē was very much a chthonian deity, whom the Greeks attempted to propitiate by leaving out monthly "Hecatē's suppers." She had shrines at crossroads, apparently intended to keep her away. To our knowledge, she had no MYTH.

Her name means "One Hundred" in Greek and may have originated as a euphemism for some unspoken name. Hecatē was probably a survival of a pre-Greek goddess of black magic.

(See also RELIGION.)

Hector In myth, a Trojan prince, eldest son of King PRIAM and Queen HECUBA, and commander of TROY's forces in the TROJAN WAR. His lasting portrait occurs in HOMER's epic poem the *Iliad*, which reaches its climax in Hector's death at the hands of the Greek hero ACHILLES.

Although Achilles is the *Iliad*'s protagonist, Hector is by far the poem's most sympathetic character—brave, generous, a devoted husband, father, and son. His poetic epithets include *hippodamos*, "tamer of horses." He is a civilized figure, remarkably unselfish as compared with his Greek counterparts. Also unlike them, he is shown as having a domestic role: In the *Iliad* (book 6) he visits his wife, ANDROMACHE, and their child, Astyanax, for what will be the last time. This poignant scene reveals his certainty that doom will overtake him and his family.

Hector's deeds in the story include his single combat with the Greek hero AJAX (1) (who gets somewhat the upper hand before it ends in a draw) and his attack with his men against the Greeks' camp and the beached Greek ships (which he almost succeeds in burning up). In the course of this action, Hector slays the Greek warrior PATROCLUS, who has entered the fray dressed in the armor of his friend Achilles. This fateful act sets the scene for Hector's single combat with the enraged Achilles the next day.

In the *Iliad*'s climactic scene (book 22), Hector waits for Achilles on the plain in front of Troy, despite the appeals of his parents from the city walls. But he turns and flees as Achilles approaches. After a long chase on foot, the heroes duel. Achilles kills Hector, ties the corpse to his chariot, and drags it back to camp. Later relenting, he allows Priam to ransom the body, and the *Iliad* ends with Hector's funeral.

Unlike many of Homer's Trojans, Hector has a Greek name ("warder-off"). We know that this name was used by at least one Greek prince in ASIA MINOR in the 700s B.C.; Hector's name and portrayal might possibly be based on some Greek leader living in the era when the *Iliad* was written down (circa 750 B.C.).

Hecuba   In MYTH, the wife and queen of the Trojan king PRIAM. She was the mother of the princes HECTOR and PARIS and 18 of Priam's other sons. Later writers also assign to her a daughter, Polyxena.

In HOMER's epic poem the *Iliad* (written down circa 750 B.C.), Hecuba (Greek: Hekabē) is a secondary character, regal and gracious. Later writers, describing her relentless sorrows and decline after TROY's fall, treat her as a symbol of ruined happiness. She is portrayed as a Greek captive in two of EURIPIDES' surviving tragedies, *Trojan Women* (415 B.C.) and *Hecuba* (424 B.C.). *Hecuba* presents her vengeance on the Greek champion Polymnestor, who had killed Polyxena as a sacrifice to ACHILLES' ghost. Lured into Hecuba's tent, Polymnestor is blinded, and his two young sons murdered, by her waiting women. Later authors, such as the Roman Ovid, describe Hecuba's death; forgiven for her violence against Polymnestor, she was handed over as a prize to ODYSSEUS. But her hatred of him was such that she was supernaturally transformed into a snarling, barking dog. In this shape, she leapt into the sea. Her grave became a landmark, Cynos Sema (dog's tomb), along the HELLESPONT.

Helen of Troy   In MYTH, a Spartan princess of great physical beauty, daughter of the god ZEUS and the Spartan queen LEDA.

Helen was married to the Spartan king MENELAUS. But at the prompting of the love-goddess APHRODITE, Helen eloped with the handsome Trojan prince PARIS. This infatuation was the cause of the disastrous TROJAN WAR, which was to end, after a 10-year siege, in Troy's destruction and the death of Paris and many others, Trojan and Greek. HOMER's epic poem the *Iliad* shows Helen installed at Troy, protected and admired by the people to whom she has brought so much trouble. She is portrayed as a glamorous and gracious lady, well aware of her reprehensible position. In one famous scene (book 5), the Trojan elders, admiring her beauty, agree once again that she is worth fighting for. In Homer's *Odyssey*, Helen is shown living contentedly again with Menelaus back in Sparta, after the war.

Later writers elaborated on Helen's story. In one version, Menelaus intends to kill her after the Greeks take Troy, but, on seeing her, forgives her. The most famous reworking of Helen's story was the *Palinodia* of the Sicilian Greek poet STESICHORUS (circa 590 B.C.), which claimed that Helen had never left Sparta at all; rather, the gods had sent a phantom-Helen with Paris to Troy, so that Troy's doom might be fulfilled.

At SPARTA Helen was worshipped as a goddess associated with TREES and nature. This divine Helen was probably a survival of a pre-Greek goddess—one of many who infiltrated Greek RELIGION in various guises.

(See also EURIPIDES; GORGIAS.)

Helios   In MYTH, a god personifying the sun. Helios was not important in Greek RELIGION, except for his nationalistic cult on the island of RHODES. Elsewhere he was more of a poetic fancy—a charioteer who drives his horses across the sky each day and returns by night, sailing in a giant cup around the stream of ocean, to start again next morn-

ing. His best-known myth told how he had a human son, Phaëthon, who won Helios' permission to drive the solar chariot in Helios' place for one day. But Phaëthon, unable to control the reins, careened too close to the earth, until the god ZEUS was forced to kill him with a thunderbolt, to save the world from fire.

As the god who sees everything from above, Helios was sometimes invoked as the guarantor of oaths. Some myths give him the minor function of bringing news to the gods of certain events that he has witnessed.

In the philosophical climate of the 400s B.C., there arose a theory identifying Helios with the major god APOLLO. Later writers of the ancient world sometimes referred to Apollo as a sun god, and this idea persists in popular belief today; but in Greek myth the sun generally had its own god.

As the patron deity of Rhodes, Helios was honored with a grand annual festival. This cult was clearly shaped by non-Greek, Oriental beliefs. The famous Colossus of Rhodes, built circa 275 B.C. outside the harbor of Rhodes city, was a giant statue of the sun god.

(See also OCEANUS.)

Hellas   See Hellēn

Hellēn   In MYTH, the father of the Greek people. Hellēn—who should not be confused with the princess HELEN OF TROY—had three sons: Dorus, Aeolus, and Xuthus. From these came the main ancient Greek ethnic branches—the DORIAN GREEKS, the AEOLIAN GREEKS, the IONIAN GREEKS, and the Greeks of ACHAEA (the latter two categories are named for Xuthus' sons, Ion and Achaeus). According to the myth, this is why the Greek people are known collectively as the Hellēnes, "the sons of Hellēn."

The legend's purpose was simply to explain the origins of these various ethnic groups. In true fact the name *Hellēnes* seems to be derived from the word *Hellas*, which was the Greeks' name for mainland Greece.

Hellenistic Age   Term used by scholars to describe the era of the enlarged, cosmopolitan Greek world of 300–150 B.C. Unlike the word *Hellenic*, which refers straightforwardly to the Greeks, the word *Hellenistic* comes from a verb *Hellazein*, "to speak Greek or identify with the Greeks," and refers to the Greek-influenced societies that arose in the wake of ALEXANDER THE GREAT's conquests (334–323 B.C.). This Hellenistic world extended from southern France to northern Afghanistan. Its characteristic nature was a mingling of Greek and Eastern cultures, particularly in the Near East and in ASIA MINOR. The political units of this world were the rich and large kingdoms of Alexander's successors—MACEDON, Ptolemaic EGYPT, the SELEUCID EMPIRE, and, eventually, PERGAMUM—and the Syracusan monarchy. Gone was the society built around the traditional Greek POLIS, or city-state, where citizens debated public policy in political assemblies and served as soldiers in time of war. In the Hellenistic world, kings did the governing, war was the business of professionals, and citizens turned to more individualistic pursuits: mystery religions and new, more personal philosophies.

Corinthian-style columns and other Greek architectural remnants at Palmyra, in the Syrian desert. Palmyra, an oasis town on the caravan route between Damascus and Mesopotamia, flourished in the Hellenistic world that arose in the wake of Alexander's conquests. From circa 300 B.C. to 64 B.C. Palmyra owed allegiance to the Seleucid kings; thereafter, it was a half-Greek, half-Syrian frontier city of the Roman Empire. These columns were part of a grandiose sanctuary of the local god Bel, built in A.D. 32.

Mainland Greece lay under Macedonian rule for much of this era. ATHENS, still a revered "university town," was no longer the cultural center of the Greek world. That honor had passed to Egyptian ALEXANDRIA (1) and secondarily, to ANTIOCH, Pergamum, and SYRACUSE. At these places, rich kings sponsored courts full of scientists and poets.

The Hellenistic world was absorbed by ROME in several stages. Syracuse fell to Roman siege in 211 B.C. The Romans dismantled the Macedonian kingdom in 167 B.C., after the Third Macedonian War. Mainland Greece was occupied by the Romans in 146 B.C., after the Achaean War. In 133 B.C. the Greek cities of Asia Minor were bequeathed to the Romans by the last king of Pergamum (effective 129 B.C.), and the remnants of the Seleucid kingdom and Ptolemaic Egypt were annexed by Rome in the first century B.C.

(See also AFTERLIFE; APOLLONIUS; ARCHIMEDES; ASTRONOMY; BACTRIA; CALLIMACHUS; DIONYSUS; EPICUREANISM; MATHEMATICS; MEDICINE; PHALANX; RELIGION; RHODES; SCIENCE; SCULPTURE; SKEPTICISM; STOICISM; THEOCRITUS; WARFARE, LAND; WARFARE, NAVAL; WARFARE, SIEGE.)

Hellespont Thirty-three-mile-long sea strait that separates the Gallipoli peninsula from the northwestern coast of ASIA MINOR. Through this channel, which ranges between one and five miles wide, a strong current flows southwestward from the SEA OF MARMARA to the AEGEAN SEA. Also known as the Dardanelles, the Hellespont is the more western of two bottlenecks—the other being the BOSPORUS—along the shipping route to the BLACK SEA. Like the Bosporus, the Hellespont traditionally has been considered a boundary between the continents of EUROPE AND ASIA.

The ancient Greek name Hellēspontos, "Hellē's Sea," was said to commemorate the young daughter of Athamas, a mythical king of EUBOEA. According to legend, Hellē fell to her death in the strait when she slipped off the back of the flying ram that was carrying her and her brother to safety from their stepmother. (This was the first episode in the story of JASON (1) and the Golden Fleece.)

As a bottleneck where shipping could be attacked or tolled systematically, the Hellespont was a critical link in the eastern TRADE route. The opportunities of this locale, combined with the channel's excellent commercial fishing, raised several local cities to wealth and prominence. The earliest and most important of these was TROY, situated on the Asian side, just outside the strait's western mouth. Archaeological evidence suggests that a wealthy Troy, inhabited by non-Greeks, was destroyed abruptly around 1220 B.C.—probably by Mycenaean Greeks driven to remove Troy's disruption of metal imports to Greece.

As the Black Sea region began supplying grain for the cities of mainland Greece, control of the Hellespont became crucial to the ambitious, food-importing city of ATHENS. By 600 B.C. Athenian colonists were warring with settlers from LESBOS over possession of Sigeum, near the old site of Troy. By the mid-400s B.C. Athens had a naval base at the Hellespontine fortress of SESTOS, on the European shore. There all westbound, non-Athenian, merchant grain ships were subject to a 10 percent tax, which would be reimbursed if the ship brought its cargo to Athens' port.

During the later PELOPONNESIAN WAR, Athenian and Spartan navies fought no less than three battles in the Hellespont (411 and 405 B.C.), as the Spartans sought to destroy Athens' lifeline of imported grain. The third such battle, the Spartan victory at AEGOSPOTAMI, won the war for SPARTA.

Besides Sestos, prominent Greek cities of the Hellespont included Abydos and Lampsacus, both on the Asian shore. The fact that Abydos faced Sestos across the channel's narrowest part (one mile wide) helped to inspire the charming legend of Hero and Leander. This tale described how a man of Abydos (Leander) swam the Hellespont every night to visit his mistress in Sestos.

(See also ALCAEUS; CHERSONESE; PERSIAN WARS.)

helots Publicly owned serfs, or non-citizens, who farmed the land in virtual captivity, providing food for the citizen population of a given state. Unlike SLAVES, these serfs kept their own language, customs, and communities on the land (which of course they did not own). They were typically the descendants of a once-free people, either Greek or non-Greek, who had been conquered in a prior era.

The public use of serfs may have been a tradition specifically of the DORIAN GREEKS, because most of the attested locales were Dorian (including CRETE and SYRACUSE). The serfs were known by various names: At ARGOS, for instance, they were called *gumnētes,* or "naked ones." But the most notorious use of such people was at SPARTA, where they were called *helotai,* or helots. This name supposedly came from the town of Helos ("marsh"), annexed and subjugated by Sparta in the late 700s B.C. By 600 B.C. further Spartan conquests had created helots in two large geographic areas: LACONIA (the territory around Sparta) and MESSENIA (the large plain west of Laconia's mountains). These people were Dorian Greeks, just like their conquerors, the Spartans.

Although helot farm labor freed Sparta's citizens to concentrate on war, this subjugated population had a warping affect on the Spartan mentality. Between Messenia and Laconia, the helots outnumbered their Spartan masters, and the fear of a helot revolt—such as that of 464 B.C.—kept the Spartan army close to home and drove Sparta to repressive measures. The Krypteia ("secret society") was an official Spartan group dedicated to eliminating subversive helots.

Helots might accompany Spartan armies to war, serving as soldiers' servants or as skirmishers. But such loyalty did not win Spartan trust. In 424 B.C., during the PELOPONNESIAN WAR, the Spartans decreed that they would honor those helots who claimed to have done the best battlefield service for Sparta. Two thousand helots came forward in the belief that they would be made Spartan citizens. But the treacherous Spartans, having thus identified the most spirited of the helots, eventually murdered them.

(See also PYLOS.)

Hephaestus   Lame smith god, patron of craftsmen, worshipped mainly at ATHENS and other manufacturing centers. Hephaestus was also the god of fire and volcanoes, and this more primitive aspect was probably his original one. He had a cult on the volcanic Greek island of Lemnos, and he was associated with volcanic Mt. Etna, in SICILY. To the imaginative Greeks, these volcanoes must have suggested a smithy furnace.

Hephaestus' MYTHS are few but picturesque. According to the poet HOMER, Hepaestus was the son of ZEUS and HERA. When he sided with his mother in a quarrel, Zeus threw him off of Mt. OLYMPUS. He fell for nine days and nights and hit the island of Lemnos. His subsequent return to heaven was a favorite subject in art.

Hephaestus' wife was the love-goddess APHRODITE. Tired of her adulteries, he fashioned a marvelous chain-link net, with which he captured his wife and the war god ARES together in bed. Then he hauled them, netted, before the assembled gods. Hephaestus also had a minor role in many other tales, providing such marvelous handiwork as the invincible armor of ACHILLES. The Romans adopted Hephaestus' myth and cult, attaching these to their fire god Vulcan (*Volcanus*).

(See also ATHENA; SEVEN AGAINST THEBES.)

Hera   One of the most important Olympian deities, both wife and sister of the great god ZEUS, and patron of WOMEN and MARRIAGE. Hera had major cult centers at the city of ARGOS (a very ancient cult) and on the island of SAMOS, but

This head of the goddess Hera, carved in limestone 20 inches high, is all that remains of the colossal statue that stood in her temple at Olympia, circa 600 B.C. As Zeus' jealous wife, Hera often plays malevolent roles in Greek myth; but in daily religious life she was honored as a goddess of marriage and childbearing, and the guardian deity of Argos, Samos, and other Greek states.

she was worshipped throughout the Greek world. Her name means "lady" in Greek.

In MYTH she cuts a strong figure—an independent wife, but jealous of Zeus' many extramarital amours. In one famous scene in HOMER's epic poem the *Iliad,* she seduces her husband as part of a scheme to distract Zeus from observing the TROJAN WAR and aiding the Trojans (book 14). Other stories show her as oppressively cruel toward her rivals—that is, toward the female humans and demigods who involuntarily attracted Zeus' lust. Among such persecuted rivals were Leto (who gave birth to Zeus' children APOLLO and ARTEMIS) and IO. But the most important example is Hera's enmity toward the hero HERACLES, son of Zeus and the mortal woman Alcmene. Hera is Heracles' constant antagonist and is reconciled with him only at the end of his labors, at Zeus' command.

Her children by Zeus are ARES, HEPHAESTUS, Eileithyia (goddess of birth), and Hebe (goddess of youth). In art, she appears as a mature, physically attractive woman. She has a definite sexual nuance and is sometimes paired with APHRODITE. There was reportedly an ancient statue of her at SPARTA called Hera Aphrodite. The Romans identified Hera with their goddess Juno.

Despite her Greek name and her place in Greek RELIGION, Hera was probably not Greek in origin. It has been plausibly suggested that she is a vestige of the mother goddess who was the chief deity of the non-Greek occupants of Greece, before the first Greeks arrived circa 2100 B.C. The immigrating Greeks conquered the land and assimilated this goddess' cult into their own religion, to a position subordinate to the Greek father god, Zeus. Whether this was done as a political expedient or because it was religiously attractive to the early Greeks we will never know, nor can we know this goddess' original name.

(See also CRONUS.)

Heracles   The most popular of all Greek mythical heroes, famous for his strength, courage, and generosity of spirit. By virtue of his 12 Labors performed for the good of mankind, Heracles (Greek: Herakles) was worshipped as both god and man throughout antiquity. Philosophical schools such as STOICISM saw him as an ideal of human fortitude. The early Christians called him a forerunner of Christ. Better known today by his Roman name, Hercules,

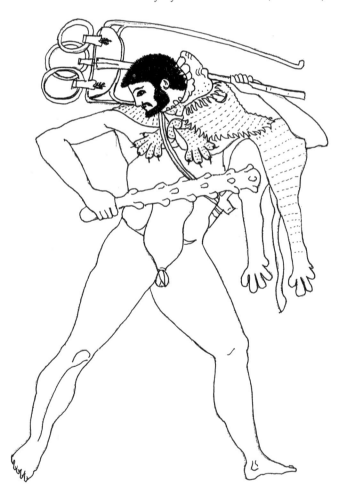

An angry Heracles carries off the holy tripod from the god Apollo's sanctuary at Delphi, after the god has refused to grant advice. From an Athenian red-figure vase, circa 480 B.C. The hero wears the invincibility-bestowing skin of the Nemean Lion, his opponent in his First Labor.

the hero continues to have a life of his own, although in debased form, in such media as comic books, films, and advertising. Heracles is one of our single most vital legacies from ancient Greece.

His name means "glory of HERA." In the legends that have come down to us, the goddess Hera is his implacable enemy. But, given the hero's name, it seems plausible that at some early stage of Greek mythology (circa 1200 B.C.), Hera was his divine patron and he her servant, performing his helpful labors at her command. Possibly the Greek Heracles derived from a pre-Greek god or hero—a servant or consort of the mother goddess who was worshipped on the Argive plain long before the first Greeks arrived, circa 2100 B.C.

In MYTH, Heracles was associated with both of the two centers of Mycenaean Greece—the Argive plain and the Theban plain. Born at THEBES, he was the son of Alcmēnē, a princess of MYCENAE or TIRYNS. Alcmēnē was married to Amphitryon, but Heracles' true father was the great god ZEUS, who had visited Alcmēnē disguised as her husband. (Heracles had a mortal twin brother, Iphicles, begotten by Amphitryon.)

The goddess Hera, chronically jealous of her husband's infidelities, was Heracles' enemy from the first. She sent two snakes to kill the twins in their cradle, but the baby Heracles strangled them. Hera continued to plague him throughout his life, and was reconciled with him only after his death and transformation into a god.

As a child at Thebes, Heracles showed strength but wildness. When his MUSIC teacher tried to beat him for misbehavior, Heracles brained the man with a lyre. Later his weapons were the club and the archer's bow. He defended Thebes from an attack from the city of ORCHOMENUS and married the Theban princess Megara. But when Hera blighted him with a fit of madness, he killed Megara and their children. (This episode is described in EURIPIDES' surviving tragedy *Heracles*, circa 417 B.C.) Seeking expiation at the god APOLLO's shrine at DELPHI, Heracles was instructed by the priestess to return to his parents' home region in the PELOPONNESE and place himself in servitude for 12 years to his kinsman Eurystheus, king of Tiryns.

The tasks set by Eurystheus comprise the famous 12 Labors (*athloi* or *ponoi*), which are the heart of the Heracles myth. In each case, Heracles had either to destroy a noxious monster or to retrieve a prize. The Labors' objectives begin with the elimination of certain local monsters in the Peloponnese, but gradually the goals become more distant and fabular until, by the last two Labors—capturing the monstrous dog CERBERUS from the Underworld and fetching apples from the supernatural sisters called the Hesperides—Heracles was symbolically conquering death.

The 12 Labors were: (1) kill the lion of Nemea, (2) kill the hydra (water snake) of Lerna, (3) capture alive the boar of Erymanthus, (4) capture the hind (female deer) of Ceryneia, (5) destroy the birds of Stymphalia, (6) cleanse the stables of King Augeas of ELIS, (7) capture the bull of CRETE, (8) capture the horses of the Thracian king Diomedes, (9) fetch the belt ("girdle") of the queen of the AMAZONS, (10) steal the cattle of the monster Geryon, in the far West, (11) fetch Cerberus from the realm of HADES, and (12) bring back some of the golden apples from the

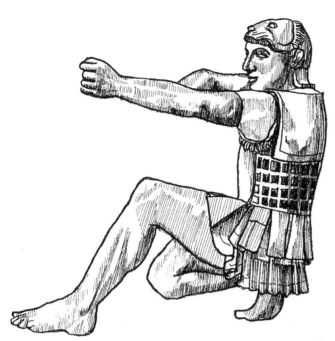

A beardless, soldierly Heracles, as portrayed in a marble sculpture from the east pediment of the goddess Aphaea's temple at Aegina, circa 480 B.C. Wearing his lion skin, Heracles bends his bow (since lost from the sculpture)—this and the club were his favorite weapons. The figure was part of a scene showing the "first" Trojan War, in which Heracles led his fellow Greeks in capturing the city of Troy.

Hesperides' garden, in the far West. The hero accomplished these tasks, albeit with the goddess ATHENA's occasional help, and so he won purification for his crime.

Besides the Labors, Heracles had many mythical exploits, in part because so many Greek cities wanted to claim an association with him. In a dispute with the Delphic oracle, he tried to steal Apollo's holy tripod from the sanctuary. He sailed with the Thessalian hero JASON (1) in the quest for the Golden Fleece. He vanquished the wrestler Antaeus, a Giant who was son of POSEIDON. And, after a quarrel with the Trojan king Laomedon, Heracles raised an army and captured TROY, in the generation before the TROJAN WAR.

When Nessus the centaur tried to rape Heracles' second wife, Deianira, the hero killed him with arrows dipped in venom of the Lernean hydra. But the dying centaur gave Deianira his bloody shirt, convincing her that it could be used as a love charm on Heracles, if needed. Later, seeking to retain the love of her unfaithful husband, Deianira laid the shirt on his shoulders. Heracles died in rage and excruciating pain from the venom-soaked blood (as described in SOPHOCLES' tragedy *The Women of Trachis*, circa 429–420 B.C.). But on the funeral pyre Heracles' mortal part was burned away, and he ascended as a god to Mt. OLYMPUS. There he was welcomed by his divine father, reconciled with Hera, and married to Zeus and Hera's daughter, the goddess Hebe.

In a politically angled legend, the DORIAN GREEKS claimed descent from Heracles. Supposedly the Heraclidae (the "sons of Heracles"), reclaiming their lost birthright, had led the Dorians in their invasion of central and southern Greece, circa 1100–1000 B.C.

(See also AFTERLIFE; AGIAD CLAN; CENTAURS; CYNICS; EURYPONTID CLAN; GIANTS.)

HERACLITUS Early philosopher, of the Greek city of EPHESUS, in western ASIA MINOR. Almost nothing is known of his life, but he apparently he lived circa 500 B.C. He is said to have been of high aristocratic lineage and of an arrogant and reclusive nature. Some of the writings ascribed to him are derisive of the democratic politics at Ephesus.

Heraclitus (Greek: Herakleitos) was the last of the great Ionian thinkers who, in the 500s B.C., pioneered the study that we call PHILOSOPHY. His writings consisted of a collection of prose proverbs, fashioned in an oracular style. A number of individual proverbs have come down to us in quotations by later writers. Although often obscure in wording, these aphorisms seem to show remarkable originality in their search for universal order amid worldly flux.

Where his Ionian predecessors—THALES, ANAXIMANDER, and ANAXIMENES—had looked for cosmic unity in some elemental substance such as water or mist, Heraclitus sought it in the universe's defining arrangement (Greek: *logos*). He saw everything in the world as part of single, continuous process of change—"All things flow" and "You cannot step into the same river twice" are two of his more famous sayings. This change, he said, is regulated by a balance or measure typically involving the conflict of opposing forces. Heraclitus found deep significance in the opposing tensions employed in the stringing of the musician's lyre and the archer's bow (both of which were inventions associated with the god APOLLO, lord of harmony and order). Heraclitus evidently saw the lyre and bow's "backward-stretched unity" (Greek: *palintonos harmoniē*) as a key to understanding the cosmos.

Similarly, many of Heraclitus' proverbs seek to show that opposites such as hot and cold are actually related or connected. "The road up and the road down are one and the same," he wrote, apparently referring to a footpath that leads both ways (so to speak), up and down a mountain. Other surviving proverbs suggest an innovative belief in one God, reflecting this notion of cosmic unity—for example, "The god is day, night, winter, summer, war, peace, satiety, hunger."

The cosmic order was somehow maintained by fire, according to Heraclitus. He believed that a person's soul was a kind of fire, which could be harmfully "dampened" by bad behavor such as sexual excess or drunkenness. Spiritually healthy souls remained dry; after death they would be able to reach higher places in the heavens.

A brilliant but eccentric thinker, Heraclitus did not inspire an immediate following. However, in the 200s B.C. the philosophical movement called STOICISM honored his memory and adapted some of his ideas.

(See also IONIA.)

Hercules See HERACLES.

Hermes In MYTH, Hermes was the messenger of the gods, and son of the great god ZEUS and the demigoddess Maia (daughter of ATLAS). In human society, Hermes was the patron of land travel, heralds, commerce, weights and

The affable god Hermes holds a ram, in a bronze figurine of the late 500s B.C. As the patron deity of travel, Hermes wears his familiar winged shoes; his peaked cap and short, Dorian-style tunic are both of a type favored by travelers, workmen, and shepherds. Like his half brother Apollo and son Pan, Hermes was worshipped as a guardian of shepherds and flocks.

measures, RHETORIC, guile, thieves, WRESTLING, and other SPORT. Often portrayed as a young man in traveler's or herald's garb, Hermes was an attractive and picturesque deity, somewhat resembling his half brother, the god APOLLO. As the mythical inventor of the lyre, Hermes was a minor patron of poetry, which was normally Apollo's province.

In origin he may have been a protector of wayfarers, commemorated at roadside stone piles or carved images. Hermes' name most probably comes from the Greek word *herma* (plural: *hermai*), meaning "pile of marker stones." In an era when travel was uncomfortable and dangerous, Hermes safeguarded travelers.

The legend of his birth is told in the charming *Homeric Hymn to Hermes*, written circa 675 B.C. Hermes was conceived when his mother was ravished by Zeus. At dawn he was born on Mt. Cyllene in ARCADIA; by noon he was playing the lyre (which he had invented, using a tortoise shell as a sounding board); and that evening he ventured out and stole a herd of cattle from Apollo. The story shows the mixture of creativity and dishonesty that characterized the mythical Hermes—and the Greeks themselves. Hermes was believed to guide the spirts of the dead to the Underworld. Other legends name him as a lover of APHRODITE, who bore him a child, an androgynous creature named Hermaphroditus.

By the 400s B.C., cities such as ATHENS contained stylized *hermai*, erected at streetcorners and in the AGORA. These were bronze or marble pillars sacred to the god, with sculpting showing only the god's face and genitals. Shortly before the Athenian sea expedition against SYRACUSE during the PELOPONNESIAN WAR (415 B.C.), the hermai throughout the city were mutilated overnight—probably by gangs of defeatist right-wingers, seeking to cast bad omens over the expedition by doing outrage to the god of travel.

(See also AFTERLIFE; ALCIBIADES; ANDOCIDES; MUSIC.)

Herodas (or Herondas) Writer of "mimes" (Greek: *mimiamboi*, "satirical sketches") of the 200s B.C. We are unsure when and where he lived, but an ancient papyrus discovered in the 1800s A.D. has left us with seven of his detailed urban scenes in verse. The best of these give delightful portraits of matrons, pimps, and vendors.

Herodotus Greek historian from HALICARNASSUS in southwest ASIA MINOR. He lived circa 485–420 B.C. and wrote a detailed, surviving account of the PERSIAN WARS, which had culminated when he was a child. Herodotus' lengthy history describes how the Persian expansion westward after the mid-500s B.C. was eventually defeated by the Greeks' defense of their homeland in 480–479 B.C. Herodotus is considered the world's first historian, the first writer ever to make systematic factual inquiries into the past. He has been called the father of history.

Prior to him, there had been only "logographers"—writers of travelogue, recounting local sights and local histories. Although Herodotus himself traveled widely to acquire local lore, his major accomplishment was that he arranged his voluminous material around a central theme and that he was the first to try to explain historical cause and effect. This approach is summed up in his opening words: "Here is the account of the inquiry (Greek: *historiē*) of Herodotus of Halicarnassus, in order that the deeds of men not be erased by time, and that the great and miraculous works—both of the Greeks and the barbarians—not go unrecorded, and, not least, in order to show what caused them to fight one another."

Herodotus was inspired by a predecessor. Forty or 50 years before his time a Milesian Greek logographer named HECATAEUS had written a travel account of EGYPT, Scythia, and other locales. Herodotus mentions Hecataeus' work with disdain, while apparently borrowing from it liberally. Herodotus began by imitating Hecataeus, visiting Egypt and the Scythian BLACK SEA coast early on. He probably intended to write Hecataean-style travelogue. But at some point Herodotus warmed to a second inspiration—the mythical TROJAN WAR, as described in HOMER's *Iliad* and other Greek EPIC POETRY. If the Trojan War was seen as the original clash between East and West, then the Persian Wars could be explained in the same light, as part of an ongoing, destructive pattern.

Another idea of Herodotus', that pride goeth before a fall, was perhaps borrowed from contemporary Athenian stage tragedy. In Herodotus, the grandeur of monarchs often leads to foolhardy decisions resulting in disaster. This is conveyed clearly in his literary portrait of the Persian king XERXES, whose vanity and arrogance resulted in the Persians' failed invasion of Greece and the deaths of so many good men on both sides. Herodotus himself never observed Xerxes, and his treatment of the king probably owes much to literary imagination. Herodotus' lecherous, violent, cowardly but aesthetic-minded and occasionally gracious Xerxes is one of the most enjoyable villains in Western literature.

We know little of Herodotus' life. He was born in Halicarnassus, a Greek city of international commerce that at the time was still ruled by Persian overlords and included a second non-Greek people, called Carians. Herodotus' father had a Carian name, Lyxes; the family was probably an affluent merchant clan of Greek-Carian blood.

Perhaps around age 25, Herodotus fled into exile after his involvement in a failed coup against the city's reigning Greek tyrant. By 454 B.C. the tyrant had fallen and Halicarnassus was part of the Athenian-controlled DELIAN LEAGUE. Despite this, Herodotus never returned home, but rather spent the next decade traveling and writing. Beside Egypt and Scythia, he evidently visited mainland Greece and the Levantine seaboard. His travels were probably helped by a Persian-Greek peace treaty that ended hostilities at that time (circa 449 B.C.).

Sometime in the 440s–420s B.C. Herodotus became known in the Greek world for giving paid readings aloud of his work-in-progress. He is said to have sought quick notoriety by reading at OLYMPIA during the Games. The Athenians supposedly voted to pay him the astounding sum of 10 TALENTS out of gratitude for his favorable portrayal of their city. He certainly had ties to ATHENS (which by then had become the cultural center of Greece). Supposedly, the Athenian playwright SOPHOCLES wrote verses to

Herodotus, and the noble Athenian clan of the ALCMAEONIDS seems to have been a major source of information for him. Although Herodotus' history is generally respectful of Athens' rival, SPARTA, he does in one passage credit Athenian naval resistance as the single factor that saved Greece in the Persian Wars (book 7).

In 443 B.C. Herodotus joined an Athenian-sponsored project to colonize a city, Thurii, in southern ITALY. He died there in around 420 B.C. These last 20 years probably saw him traveling, giving readings, and revising his work. His history was being published in (or near) its present form by 425 B.C., when the Athenian comic playwright ARISTOPHANES parodied its opening passages in his play *Acharnians*.

Herodotus' native tongue was probably the Doric Greek dialect of Halicarnassus, but he wrote his history in Ionic—the dialect generally used for prose explication—and used a clear, pleasant, storytelling style. On its completion, this history was the longest Greek prose work ever written. Later editors divided it into nine "books." Not only is it far and away our major written source for the Persian Wars and prior events in Greece and PERSIA in the 700s–500s B.C., but it also makes for delightful reading. In telling his story, Herodotus shows himself interested in war, politics, and the gods, also in personality, gossip, and sex. Unusual for a classical Greek writer, Herodotus is fascinated by WOMEN and their influence in a man's world both East and West.

Remarkably, a large part of his story is told from the Persian viewpoint. In this Herodotus was imitating such Athenian stage tragedy as AESCHYLUS' play *The Persians*, which describes the Battle of SALAMIS (1) from the Persian side. Some of Herodotus' Persian scenes were surely fabricated by the author; yet it is equally clear that he conducted original research, interviewing individual Persians as well as Greeks. Despite the fact that Persia was the Greeks' enemy, Herodotus portrays many Persians as brave and noble, and he shows a deep respect for their culture. The later Greek writer PLUTARCH accused Herodotus of being *philobarbaros*, "overfond of foreigners."

Literary pioneer though he was, Herodotus shows certain endearing flaws. He often strays from his main narrative with digressions. Although dedicated to finding out historical causes, he sometimes fastens on trivial or fairy-tale causes. For example, he claims (in book 3) that the attack of the great Persian king DARIUS (1) on Greece was prompted by the bedroom persuasion of his wife. Further, Herodotus is not above telling the occasional tall tale—claiming, for instance, that he visited Babylon, even though his work lacks any description of that majestic Mesopotamian city. Besides being called the father of history, Herodotus has been called the father of lies.

Nevertheless, his history was a monumental achievement: the first rational inquiry into the past. In this, he paved the way for his greater successor, the Athenian historian THUCYDIDES (1).

(See also CALLIAS; EUROPE AND ASIA; GREEK LANGUAGE; HUBRIS.)

heroes See MYTH.

Herondas   See HERODAS.

Hesiod   Greek epic poet, one of the earliest whose verses have survived. Hesiod lived circa 700 B.C., perhaps 50 years after HOMER. Together with Homer, he is often considered a pioneer of early Greek literature. Two major works by him are extant: the *Theogony*, or *Birth of the Gods*, which describes the world's creation and the origins of the Olympian gods and lesser deities; and *Works and Days*, a kind of farmer's almanac laced with advice on how to live a good life through honest work. For modern readers, the *Theogony* is a treasury of information for Greek MYTH, while the *Works and Days* supplies a unique picture of early Greek rural society.

The two poems give some autobiographical details. Hesiod lived in the central Greek region of BOEOTIA, in a town called Ascra ("bad in winter, worse in summer, not good anytime," as he describes it in *Works and Days*.). His father, having abandoned a seafaring life, had come from CYMĒ, in ASIA MINOR. According to the *Theogony*, Hesiod was tending sheep on Mt. Helicon as a boy when the MUSES appeared and gave him the gift of song. Later he won a prize at nearby CHALCIS, in a song contest at certain funeral games. Hesiod quarreled with his brother Perses, who had apparently stolen part of Hesiod's inheritance, and the *Works and Days* was written partly as an instructional rebuke to Perses. The poet's personality, as conveyed in the poems, is surly, practical, and conservative—an old-fashioned Greek yeoman farmer, who happened to have the poetic gift.

In addition to providing facts and genealogies about the gods, the *Theogony* supplies evidence of Near Eastern influence on the formative Greek culture of the 700s and 600s B.C., for a number of Hesiod's myths closely resemble certain older legends from Mesopotamia. This influence came from Greek TRADE with the Near East, via Phoenician middlemen.

(See also CRONUS; EPIC POETRY; PHOENICIA.)

Hesperides   See HERACLES.

Hestia   Goddess of the hearth and of domestic fire. The least important of the 12 Olympian deities, Hestia ("hearth") was the sister of ZEUS. Little MYTH was attached to her name; having refused to marry, she remained a virgin and lived as a kind of respected spinster aunt on Mt. OLYMPUS.

Her cult, however, was important. In an era before manufactured matches, the hearth was the crucial place where cooking fire was maintained perpetually, winter and summer. The hearth helped to preserve order and civilization, and many governmental buildings had public hearths that symbolized the public good. Accordingly, Hestia was the patron goddess of town halls and similar; one of her epithets was *boulaia*, "she of the council house."

(See also COUNCIL.)

hetairai   See PROSTITUTES.

Hieron (1)   Dictator of the Greek cities GELA and SYRACUSE in SICILY, and the most powerful individual in the Greek world circa 470 B.C. Hieron ("holy one") came to power under his brother GELON, who ruled in Syracuse and appointed him as his governor in Gela. On Gelon's death, Hieron succeeded as lord (*turannos*, "tyrant") of Syracuse, the foremost city of the Greek West (478 B.C.). Although less popular than his brother, Hieron ruled in grand style. He formed an alliance with the other great Sicilian-Greek tyrant, Theron of ACRAGAS. When the Greeks of CUMAE, in western ITALY, appealed for help against the encroaching ETRUSCANS, Hieron achieved his greatest triumph—his sea victory at Cumae, in the Bay of Naples, which broke the Etruscans' sea power forever and removed them as a threat to the Italian Greeks (474 B.C.).

At home he consolidated his power through social engineering and cultural display. He continued the Sicilian rulers' ruthless practice of transplantation, removing 10,000 citizens of CATANA in order to reestablish the site as a new, Dorian-Greek city, Aetna. The "founding" of Aetna was commemorated in the tragedy *Women of Aetna* by the Athenian playwright AESCHYLUS, who visited Hieron's court at the ruler's invitation around 470 B.C.

Other luminaries of Hieron's court were the great Theban choral poet PINDAR and the poets SIMONIDES and BACCHYLIDES of Ceos. Pindar and Bacchylides were commissioned to write poems celebrating Hieron's prestigious victories in horse- and chariot-racing at the OLYMPIC and PYTHIAN GAMES (476–468 B.C.). Hieron is the addressee of Pindar's famous First Olympian ode and First Pythian ode.

Hieron died, probably of cancer, in 467 B.C. He was succeeded by his son Deinomenes, but by 466 B.C. Syracuse had overthrown the tyrant and installed a DEMOCRACY in its place.

(See also CHARIOTS; TYRANTS.)

Hieron (2) II   See SYRACUSE.

Himera   Greek city located midway along the northern coast of SICILY, founded circa 649 B.C. by colonists from ZANCLĒ in northeastern Sicily. Himera's most famous citizen was the poet STESICHORUS (circa 590 B.C.). Himera was a Greek frontier town, close to the Carthaginian-controlled western part of Sicily. In the late summer of 480 B.C., the vicinity of Himera was the site of a great battle in which a Greek army totally defeated an invading force from CARTHAGE.

In the years prior to the battle, the town was ruled by a Greek tyrant, Terillus, who eventually was ejected by Theron, tyrant of ACRAGAS. Terillus appealed to his ally, the Carthaginian leader Hamilcar, and in 480 B.C. Hamilcar sailed from North Africa with a large army to reinstate Terillus and conquer Greek Sicily. Landing in Sicily, the Carthaginians marched on Himera, but outside the town Hamilcar and his army were destroyed by an allied Greek force under GELON, the tyrant of SYRACUSE. According to legend, the battle was fought on the very same day as another great victory of liberation, the sea battle of SALAMIS (1), against the invading Persians in mainland Greece.

To commemorate the victory at Himera, Gelon minted what is perhaps the most beautiful coin of the 400s B.C.—the SILVER Syracusan decadrachm known as the Demareteion.

The Carthaginians' defeat stymied their ambitions in Sicily for three generations. But in 409 B.C. they captured Himera and razed it in vengeance.

(See also COINAGE; TYRANTS.)

Hipparchus See HIPPIAS (1).

Hippias (1) Athenian dictator who reigned 527–510 B.C., the son and successor of PISISTRATUS. Succeeding his father as ruler at about age 40, Hippias aggrandized ATHENS with public works, ambitious diplomacy, and economic projects. Under him, the city continued its emergence as the future cultural capital of Greece. Hippias' younger brother Hipparchus brought to Athens two of the greatest poets of.the era: ANACREON and SIMONIDES. But the days of Greek TYRANTS like Hippias were numbered by the late 500s B.C., and Hippias eventually was ousted by Athenian opposition and Spartan intervention.

In 510 B.C. the Spartan king CLEOMENES led an army to Athens, against Hippias. The Spartans had been urged to do so by the oracle at DELPHI, but Cleomenes himself surely wanted to stop Hippias' diplomatic overtures to the Persian king DARIUS (1). Cleomenes probably feared that Hippias was planning to submit to Darius in order to retain personal power.

Aided by Athenians who were hostile to the tyranny, the Spartans surrounded Hippias and his followers on the ACROPOLIS. Hippias abdicated in exchange for a safe-conduct out. His departure brought to an end more than 35 years of Pisistratid rule at Athens, setting the stage for the political reforms of CLEISTHENES (1) and the full-fledged Athenian DEMOCRACY.

Traveling eastward to PERSIA, Hippias became an adviser at Darius' court. In 490 B.C., when the Persians launched their seaborne expedition against Athens, Hippias (by now nearly 80) accompanied the Persian army. Apparently he was intended as the puppet ruler.

He guided the Persian fleet to the sheltered bay at MARATHON, about 26 miles from Athens. It was there, more than 50 years before, that Hippias had helped his father, Pisistratus, bring a different invading army ashore. But this time fortune favored the defenders, and the Persian force was totally defeated by the Athenians at the famous Battle of Marathon. Hippias withdrew with the Persian fleet and, according to one story, died on the return voyage.

(See also ALCMAEONIDS; HARMODIUS AND ARISTOGITON; PERSIAN WARS; POTTERY; THEATER.)

Hippias (2) See SOPHISIS.

Hippocrates (1) Greek physician and medical writer, usually considered to be the founder of scientific medical practice. Hippocrates lived circa 460–390 B.C. and was a native of the Dorian-Greek island of Cos, near southwestern ASIA MINOR.

At Cos, Hippocrates established a school of MEDICINE that became renowned in the ancient world. While not the first Greek doctor, Hippocrates was apparently the first to systematize the existing knowledge and procedures and to ground the medical practice in solid observation rather

than theory. The later writer Celsus (circa A.D. 30) remarked that Hippocrates separated medicine from PHILOSOPHY.

Of the 60 medical treatises that have survived from the Hippocratic school, possibly none was written by Hippocrates himself. But certain shared traits of these writings, such as their emphasis on observation and diagnosis, convey the spirit of the school. The treatises include *Airs, Waters, and Places*, which describes the effects of different climates on health and psychology, and *The Sacred Disease*, a discussion of epilepsy (concluding that there is nothing sacred about it).

The "Hippocratic Oath" taken by graduates of that school is still administered to new doctors today, in modified form, 2,400 years later. In the ancient oath, the swearer promised to honor the brotherhood of the school, never to treat a patient with any purpose other than healing, never to give poison or induce abortion, and never as a doctor to enter a house with any ulterior motive, such as seduction of SLAVES.

Hippocrates (2) See GELA; GELON.

Hippodamus Town planner of the mid-400s B.C., born at MILETUS but active in the service of Periclean ATHENS. Circa 450 B.C. he designed the grid pattern for the Athenian port city of PIRAEUS (his design is still in use there today), and he probably did likewise for the Athenian-sponsored colony of Thurii, in southern ITALY. He was one of the colonists who emigrated to Thurii circa 443 B.C. ARISTOTLE's *Politics* (circa 340 B.C.) mentions Hippodamus' affected physical appearance—long hair and adorned robes in the Ionian manner.

(See also IONIAN GREEKS.)

Hippolyta See AMAZONS; HERACLES; THESEUS.

Hippolytus In MYTH, the son of the Athenian king THESEUS and the Amazon queen Hippolyta. After his mother's death, Hippolytus grew to manhood at ATHENS as a hunter and male virgin, devoted to the goddess ARTEMIS. The love goddess APHRODITE, irked by Hippolytus' celibacy, caused Theseus' young wife, the Cretan princess Phaedra, to fall in love with Hippolytus, her stepson. Rebuffed by him, she hanged herself, but left behind a letter accusing him of rape. Theseus, not believing his son's declarations of innocence, banished him and then cursed him. The curse was effective (as being one of three wishes that Theseus had been granted by his guardian, the god POSEIDON), and a monstrous bull emerged from the sea while Hippolytus was driving his chariot on the road. Hippolytus' terrified horses threw him from the chariot, and so his death was fulfilled in the manner suggested by his name: "loosed horse," or stampede. Theseus learned the truth from Artemis after it was too late.

Our main source for the legend is the admirable, surviving tragedy *Hippolytus* (431 B.C.) by the Athenian playwright EURIPIDES. In the play, Hippolytus appears as priggish and lacking in compassion, while Phaedra is convincingly imagined as an unhappy woman in the unwelcome grip of an obsession.

A later legend claimed that Hippolytus was restored to life by the physician-hero ASCLEPIUS. But then the god ZEUS, fearing a disruption of natural order, killed both men with a thunderbolt.

(See also AMAZONS.)

Hipponax Greek lyric poet of EPHESUS who lived in the mid-500s B.C. Banished by one of the city's Persian-controlled TYRANTS around 540 B.C., he supposedly lived as a beggar in nearby Clazomenae. His wrote satirical poems in various meters, with the flavor of the gutter. He was said to have been the inventor of parody; one of his surviving fragments is a mock-Homeric description of a glutton.

Hipponax is credited with inventing the *skazon,* or "lame" iambic meter—which, to the Greek ear, had a halting, comic affect, appropriate to satire.

(See also IONIA; LYRIC POETRY.)

Homer According to tradition, Homer was the earliest and greatest Greek poet. Two epic poems were ascribed to him: the *Iliad* and *Odyssey,* which present certain events of the mythical TROJAN WAR and its aftermath. These two works, totaling about 27,800 lines in dactylic hexameter verse, supplied the "bible" of ancient Greece. Greeks of subsequent centuries looked to them for insight into the

"Blind is the man, and he lives in rocky Chios." This portrait of Homer comes from a marble bust of the Roman era, probably copied from a Greek original from circa 150 B.C. Although plausible-looking, the likeness is imaginary: Homer himself—if he was an individual person—lived during or before the mid-700s B.C., and later generations had no authentic portrait of him.

gods' nature, for answers to moral questions, and for inspiration for new literature. The Athenian playwright AESCHYLUS (circa 460 B.C.) described his own tragedies as "morsels from the banquet of Homer." It is a token of the Greeks' reverence for the *Iliad* and *Odyssey* that both these long poems survived antiquity. Each poem's division into 24 "books," still used in modern editions, was made long after Homer's time, by ancient editors at ALEXANDRIA (1) in the early 100s B.C.

Despite a few legends, nothing is known about the life of Homer (Greek: Homeros, possibly meaning "hostage"). It is not even known that "he" was a man, except that the Greek world's social structure makes it likely. Similarly, it is not impossible that the *Iliad* and *Odyssey* were each composed by a different person. The two poems differ in tone and narrative style: The *Odyssey,* which relies far more than the *Iliad* on fable and folktale, may have been produced much later than the *Iliad.* Furthermore, given the collaborative nature of preliterate Greek EPIC POETRY, either poem could have been created by a group of poets rather than by one alone.

If there was a single Homer, he may have lived sometime between 850 and 750 B.C. The *Iliad* and *Odyssey* were probably first written down around 750 B.C., after the invention of the Greek ALPHABET, but evidently both poems were fully composed before that date. Apparently they were created by the oral techniques of preliterate Greek epic poetry—that is, by a centuries-old method of using memorized verses, stock phrases, and spontaneous elaboration, without WRITING. It is not known if Homer wrote the poems down (or dictated them to scribes) in the mid-700s B.C., or whether Homer was long dead when others wrote down his verses that had been preserved by oral retelling.

The *Iliad* and *Odyssey* reveal no autobiographical information. A single supposed autobiographical item is found in the *Hymn to Delian Apollo,* one of 33 choral songs that the ancient Greeks ascribed to Homer. In the hymn, the unnamed poet describes himself with these words: "If anyone should ask you whose song is sweetest, say: 'Blind is the man, and he lives in rocky Chios.' " Many ancient Greeks believed this to be a true description of the mysterious Homer, although modern scholars are wary of accepting it.

Homer could have been a bard who performed at the court of some Greek baron or king—much like the character Demodocus in book 8 of the *Odyssey,* who earns his livelihood singing traditional tales to entertain noblemen and ladies. Homer may have lived on the east Greek island of CHIOS, as the hymn claims, or elsewhere in the Greek region called IONIA, on the west coast of ASIA MINOR. That Homer was Ionian is suggested by the mainly Ionic dialect of the poems, and that Ionia was at that time the most culturally advanced part of the Greek world. In later centuries Chios was home to a guild of bards who called themselves the Homeridae, the sons of Homer. But other Ionian locales, such as Smyrna, also claimed to have been the poet's home.

More important than the poet's identity are the values and traditions that his poetry is based on. Homer and other bards of his day sang about an idealized ancestral society that was 500 years in the past. Homer's audience imagined this bygone era as an Age of Heroes, when men of superior strength, courage, and wealth lived in communion with the gods. Homer's aristocratic protagonists—ACHILLES in the *Iliad* and ODYSSEUS in the *Odyssey*—live by a code of honor that shapes all their actions. For them, disgrace was the worst thing of all—far worse than death. If a man's honor was slighted, then the man was obliged to seek extreme or violent redress.

Greek MYTH contains hundreds of stories from the Age of Heroes. Homer's genius lay in selecting certain tales and imposing order on the material, to fashion (in each poem) a cohesive, suspenseful tale, with vivid character portraits. Contrary to some popular belief, the *Iliad* does not describe the entire Trojan War and fall of TROY; rather, it presents a tightly wrapped narrative, focusing on certain episodes from war's 10th and last year, involving the Greek champion Achilles. The poem tells of the "anger of Achilles"—that is, his quarrel with the Greek commander AGAMEMNON, after Agamemnon has needlessly slighted Achilles' honor. Achilles' withdrawal from the battlefield brings reversals for the Greeks, culminating in the death of Achilles' friend PATROCLUS at the hands of the Trojan prince HECTOR. At the story's climax (book 22), Achilles slays Hector in single combat, even though he knows that this act is ordained by FATE to seal his own doom. The *Iliad* ends not with Achilles' death or Troy's fall—those are still in the future—but with Hector's funeral, after Achilles has magnanimously allowed the Trojans to ransom back the corpse.

The *Odyssey*, although it stands on its own merits, presents a loose sequel to the *Iliad*. There the war is over, Troy having fallen to the Greeks. The wily and resourceful Greek hero Odysseus, king of ITHACA, is making his way home to his kingdom amid supernatural adventures, both violent and sexual. Like the *Iliad*, the *Odyssey* displays narrative skill in maintaining suspense. Daringly not introducing the hero until book 4, the poem opens with scenes at the disrupted kingdom of Ithaca, where more than 100 arrogant suitors (assuming Odysseus to be dead) have taken over the palace and are individually wooing his intelligent and gracious wife, PENELOPE, while threatening his adolescent son, TELEMACHUS. The reader or listener thus observes the consequences of Odysseus' absence and shares in the longing for his return. We then meet Odysseus, who is nearing the end of his journey. Aided by the goddess ATHENA, he returns to Ithaca in disguise (book 14) and scouts out the dangerous situation at the palace, then reveals his identity, slays the suitors, and reclaims his wife and kingdom (books 22–23). The suspense that precedes this violent climax is remarkably modern in tone.

(See also GREEK LANGUAGE; MYCENAEAN CIVILIZATION; WARFARE, LAND.)

homosexuality   Ancient Greek literature and art clearly show that certain types of homosexual relationships were considered natural and even admirable in many Greek cities during the epoch between about 600 B.C. and the spread of Christianity. Especially, male homosexuality was encouraged in some (not all) forms. Love between males was seen as harmonious with other Greek social values,

The hero Achilles, at right, bandages a wound for his friend Patroclus, in a red-figure scene from an Athenian cup, circa 500 B.C. The painting conveys Patroclus' distress and Achilles' sympathetic concentration; the sexuality of the two men is suggested by the gratuitous peek at Patroclus' genitals. Although the *Iliad* never portrays Achilles and Patroclus as lovers, the Greeks after Homer came to view the pair as models for aristocratic, military, male homosexuality.

such as athletic skill, military courage, and the idealization of male youth and beauty (reflected also in surviving Greek SCULPTURE). Such relationships provided males with a romance not usually found in MARRIAGE, since Greek society viewed WOMEN as morally and intellectually inferior.

Relatedly, female homosexuality was an approved practice in some locales, at least in the 600s–500s B.C. Extant verses by the poets SAPPHO and ALCMAN document sexual feelings and acts between aristocratic young women on the Aegean island of LESBOS and at SPARTA, both circa 600 B.C. Also, a single sentence in the work of the later writer PLUTARCH suggests that at Sparta it was usual for mature women to have affairs with unmarried girls. But little other information survives regarding love between women. The silence is due partly to scarcity of extant writings by Greek women and partly to the fact that female homosexuality was not encouraged as widely as its male counterpart. Relations between females may even have been forbidden in certain cities that had male-homosexual traditions, such as ATHENS in the 400s B.C.

The ancient literary sources for information on Greek homosexuality include LYRIC POETRY composed between about 600 and 100 B.C., Athenian stage comedy of the 400s B.C., the works of the philosopher PLATO and the historian XENOPHON (both of them Athenians writing in the early 300s B.C.), and Athenian courtroom speeches of the 300s B.C. The visual evidence consists mostly of vase paintings from the 500s and 400s B.C., some showing courtship or sex between males. As is usual for any aspect of ancient

Greek society, much of the extant source-material is from Athens. But other Greek states—including Sparta, ELIS, CHALCIS, and especially THEBES—had important male-homosexual cultures, linked to the training and esprit of citizen armies.

Homosexuality as a social norm evidently arose in Greece at a specific time, the late 600s B.C. Modern scholars have found no homosexual content in the poetry of HOMER (circa 750 B.C.), HESIOD (circa 700 B.C.), or ARCHILOCHUS (circa 660 B.C.). Expressions of homosexual desire first appear in the extant verses of Sappho (circa 600 B.C.) and SOLON (circa 590 B.C.).

This new social custom probably derived in part from the military reorganizations that swept Greek cities after the arrival of HOPLITE tactics in the 600s B.C. Other, related social changes at this time included the glorification of masculinity and (at Sparta) the elimination of family life by the mass military training of boys. The Greeks tended to associate homosexuality with manliness and soldiering. Significantly, the IONIAN GREEKS of ASIA MINOR were reputed to be the softest, the least military, and the least interested in homosexual pursuits, of all Greek peoples.

The ancient Greeks did not classify a person as strictly homosexual or heterosexual, as modern society tends to do. The Greeks assumed that an attractive, young individual of either gender could inspire sexual desire in either gender. Adult male citizens—the one class of people who had sexual freedom—often led private lives that were bisexual. (Yet not always: The Greeks did recognize that some men preferred one or the other gender exclusively.)

Male citizens were expected to marry female citizens and beget children, but evidently most men were not in love with their wives, as the society did not encourage this. Instead, a husband was legally and morally free to seek partners outside of marriage. (Wives enjoyed no such privilege.) Possible partners for a married or unmarried man included male or female PROSTITUTES and SLAVES, who were of the lower social ranks and who received payment or sustenance in exchange for giving sexual favors. However, if a male citizen wished for romance with someone who was his social equal—that is, if he wished to conduct courtship and seduction, with the possibility of mutual love and admiration—then his choices were limited. In most cities, the wives and daughters of citizens were often kept away from public places, and their chastity was protected by severe laws. There was only one kind of publicly approved romance available for people of the citizen classes—namely, the romance that might arise between a mature man and younger male.

This pairing—older male citizen/younger male citizen—was the classic pattern of ancient Greek homosexual love, as idealized in legend and art. This was the love pursued especially by wealthier and aristocratic citizens. Perhaps the best-known couple in this tradition were the Athenian tyrannicides of 514 B.C., HARMODIUS AND ARISTOGITON.

The younger male was typically a well-bred boy between about 12 and 20 years old—that is, between early puberty and full maturity. Youths around ages 16 and 17 were considered especially desirable as being in the prime of beauty. The young man or boy would be the passive partner, the recipient of the older man's courtship and gifts. The most handsome and accomplished youths became glamorous social figures, over whom men would conduct fierce rivalries. The teenage aristocrat ALCIBIADES was one such figure, and Plato's dialogue *Charmides* includes a vivid scene where the teenage Charmides enters a GYMNASIUM followed by a boisterous crowd of quarreling admirers. But only citizen males were allowed to woo such love objects; any male slave who pursued a citizen boy in this way was liable to dire punishment.

Among the qualities the classical Greeks admired in their boys were masculinity and bodily strength. Such attributes are clearly indicated in several hundred surviving Greek vase paintings showing images of boys or young men, often labeled with the inscribed word *kalos,* "beautiful." (These are the famous kalos vases, the homosexual "pin-ups" of ancient Greece.) Boys with more feminine bodies or mannerisms were apparently not much sought after in the 500s and 400s B.C., although they seem to have come into vogue by the late 300s.

Painted scenes of Greek homosexual couples nearly always show the older male as bearded, indicating adulthood. He might be in his 20s, 30s, or possibly 40s—anywhere from about five to 25 years older than his partner. The younger male always appears as beardless. Written evidence reveals that—in fifth-century B.C. Athens, but not always elsewhere—young men were considered no longer desirable once they began sprouting facial hair. Probably at around age 20 a young Athenian would feel social pressure to relinquish his junior sex role. He might maintain a close friendship with his former lover(s), but he would now be ready to take on the adult role, as the active pursuer of a younger male. This change by him was part of his larger transition to adulthood and to his full identity of a citizen.

Men could meet boys and youths at public places and upper-class venues such as gymnasiums, riding tracks, religious sacrifices and processions, and city streets where boys traveled to and from school. The sons of wealthy homes went out accompanied by a *paidagogos,* a slave whose duties included keeping would-be suitors away. Part of the pursuer's challenge might lie in intimidating or evading the paidagogos, in order to make the boy's acquaintance.

Once a suitor had won the approval of the boy's father, the courtship progressed through stages that included the suitor giving gifts; one gift often appearing on vase paintings is a live rooster. Together the two males would partake of upper-class recreations such as SPORT, hunting, and the drinking party known as the SYMPOSIUM, where politics or intellectual topics might be discussed.

In a society that did not foster close ties between father and son, the suitor served as a role model. He played a vital part in the boy's EDUCATION, helping to improve the boy's athletic skills, military aptitude, and general readiness for manhood. At Sparta, for example, legend claimed that after a boy once cried out in pain during a fistfight, the boy's lover was punished for failing to teach manliness.

Beyond mere instruction, a lover might provide a boy with financial help and career contacts that supplemented, perhaps vastly, what the boy's family could provide. At Thebes the lover customarily supplied the younger man's first suit of armor (no small expense). At Athens many a politician, lawyer, and poet seems to have gotten his career-start as a handsome boy, meeting older men who would become his benefactors and allies.

The relationship between man and boy was thought to be mutually inspiring: The man strove to be admirable in his public conduct; the boy strove to be worthy of the man. In Elis and Thebes, where love relationships often continued after the younger male reached adulthood, it was customary to station lovers side-by-side in battle, on the theory that each would fight more fiercely if observed by his partner. At Thebes this theory led to the creation, in around 378 B.C., of an elite, 300-soldier unit called the Sacred Band, comprised entirely of paired lovers.

Exactly what sexual activity was involved in such relationships is not clear to modern scholars; sexual customs evidently varied from region to region. In general the Greeks valued sexual restraint, much as they valued the ability to endure hunger or fatigue. Apparently one school of thought believed that lovers should practice abstinence. Plato, in his dialogue *The Symposium*, glorifies male homosexual love as a search for beauty and truth, yet he argues that love in its most exalted form involves no sexual contact—the famous "Platonic love." Similarly Xenophon, while describing homosexual pairings at Sparta, makes the surprising statement that Spartan law severely punished any man who had sexual relations with a boy. On the other hand, extant vase paintings (mostly from Athens) show more than one form of genital activity between males.

The question is: Was the younger male typically subjected to sexual penetration by the older? The answer seems to be that at Thebes and certain other Greek cities this was socially permitted, while at Athens and at many other cities it was officially discouraged but it sometimes occurred anyway. At Athens the "better sort" of older partner did not try to suduce his beloved, as revealed by an anecdote in Xenophon's memoirs where the Athenian philosopher SOCRATES rebukes the young man CRITIAS over his unseemly lust for the boy Euthydemus. Socrates says that Critias wants "to rub himself against Eurthydemus the way itchy pigs want to rub against stones." For sexual outlet, Greek men had recourse to their wives, concubines, and male and female prostitutes and slaves. Citizen boys—at Athens and Sparta, at least—were supposed to be kept pure (although that rule might be disobeyed).

This complicated outlook was part of the Greeks' attitude toward sex in general. Sex was seen as a form of power: One partner was considered dominant and one subordinate in any relationship. Typically the dominant partner was the male whose penis entered a bodily orifice of the other person. The recipient's submission was proof of inferior status. Although there was an awareness that women could enjoy sex—as suggested in the legend of the seer TIRESIAS—the Greeks were uncomfortable with the idea that any subordinate partner could feel pleasure. Sex was for the dominant person's benefit.

In all sexual relations, an adult male citizen was expected to dominate. Males who willingly received sexual penetration were supposed to be either slaves or prostitutes—noncitizens, unlucky in their servitude or poverty. Any male citizen who wanted to be penetrated sexually was considered bizarre and morally debased. A citizen who gave his body for money was deemed a prostitute and was liable to lose most of his citizenship rights. This was the background of the speech *Against Timarchus* (346 B.C.), in which the Athenian orator AESCHINES convinced an Athenian court that his enemy Timarchus had prostituted himself in his younger days.

The contradiction in Greek homosexual love was that it placed young male citizens in danger of being sexually subordinated and thus dishonored. Any boy might receive expensive gifts from an admirer; many a boy probably succumbed to a lover's seduction. What distinguished this behavior from a prostitute's? The answer lay partly in monogamy. A boy who gave sexual favors might avoid disrepute by not being promiscuous and by choosing a worthy, discreet lover. A second way to avoid dishonor may have been by limiting sexual contact to an activity known today as intercrural intercourse (shown on vase paintings). This involved the two males standing or lying face to face, with the older man moving his penis between the younger's clamped thighs. Because the younger male was not actually being penetrated, this submission was probably thought of as being less degrading.

Modern scholars believe that the homosexual themes in Greek MYTH all represent a relatively late layer, added after 600 B.C. In other words, legends about friendships between males had existed without homosexual nuance for centuries prior; these included the tales of GANYMEDE and the god ZEUS, of PELOPS and the god POSEIDON, and of the heroes ACHILLES and PATROCLUS. Eventually these myths received a sexual coloring, reflecting real-life social norms of the 500s B.C. and later. Legend claimed that Greek homosexual practices had been initiated by the Theban king Laius (father of OEDIPUS), when he carried off the boy Chrysippus.

(See also ANACREON; CALLIMACHUS; CHAERONEA; HYACINTHUS; IBYCUS; PINDAR; POTTERY; THEOGNIS.)

hoplite Heavy infantryman of the Greek world, from about 700–300 B.C. The famous land campaigns of the 400s B.C.—in the PERSIAN WARS and the PELOPONNESIAN WAR—were fought by hoplite armies. In the late 300s B.C., hoplite, tactics were superceded by the tactics of the Macedonian PHALANX.

The early hoplite was named for his innovative shield, the *hoplon,* which was round, wide (three feet in diameter), heavy (about 16 pounds), and deeply concave on the inside; it was made of wood reinforced with BRONZE, often with a bronze facing. The soldier held the shield by passing his left forearm through a loop on the inside center and then gripping a handle at the far inside edge. This arrangement helped with the necessary task of keeping the shield rigidly away from the man's chest. The shield was notoriously difficult to hold up for a long period of time; a hoplite fleeing from battle always threw away his cumbersome shield, and even victorious soldiers could lose their shields in the melee. In militaristic societies such as SPARTA, keeping your shield meant keeping your honor, as indicated by the Spartan mother's proverbial command to her son: "Return with your shield or upon it."

The standard hoplite helmet was the "Corinthian" type, developed in the early 600s B.C. and modified in the 400s B.C. to the shape shown here; the higher crown of the new design better protected the top of the head. For display and further protection, a horsehair crest might be attached to the helmet's top. Beaten skillfully from a single sheet of bronze, the Corinthian helmet had no ear-holes—which must have made the wearer practically deaf. Away from combat, the helmet could be pushed up to rest above the face.

The rest of the hoplite's armor, or panoply, included a helmet—typically beaten from a single sheet of bronze and topped with a crest of bronze or horsehair—and a bronze breastplate and greaves (metal shin guards); under the breastplate the man would wear a cloth tunic. The offensive weapons were a six- to eight-foot-long spear for thrusting (not throwing) and a sword of forged IRON, carried in a scabbard at the waist. By various modern estimates, the whole panoply weighed 50–70 pounds, and it seems that, on the march and up until the last moments before battle, much of a hoplite's equipment was carried for him by a servant or slave.

A hoplite did not normally fight alone; he was trained and equipped to stand, charge, and fight side by side with his comrades, in an orderly, multiranked formation. The hoplite relied foremost on his spear, thrusting overhand at the enemy while trying to shield himself from their spear-points. The sword was used if the spear was broken or lost.

The armor's weight, plus the need to keep in formation, meant that hoplites could not charge at full speed for any distance. Two hundred yards would seem to be the farthest that hoplites actually could run and still be in a condition to fight.

Yet this heavy armor did not make the hoplite invulnerable. It was not practical for armor to cover a man's neck, groin, or thighs, and these were left exposed. References in ancient poetry and art make it clear that deadly wounds to the neck and groin were common, as were fatal blows to the head (possibly received from an inward denting of the helmet). Sometimes the bronze breastplate could be pierced—as demonstrated by evidence that includes the recently recovered remains of a Spartan hoplite, buried at a battle-site with the fatal, iron spear-point lodged inside his chest.

Hoplites could fight effectively only on level ground; hilly terrain scattered their formation and left the individual soldiers open to attacks from lighter-armed skirmishers. Similarly, hoplites who broke ranks and became isolated—in retreat, for example—were easy prey for enemy CAVALRY.

On warships, hoplites served as "marines." There they were armed mainly with javelins (for throwing) and were employed in grapple-and-board tactics. Soldiers unlucky enough to fall overboard would be dragged to the bottom by their heavy armor.

Hoplite armies began their history as citizen armies. In most Greek cities, each man up through middle age who could afford the cost of a panoply was required to serve as a hoplite if his city went to war. (Alternatively, those rich enough to maintain horses might serve in the cavalry.) In states governed as oligarchies, a man had to be of hoplite status or better in order to be admitted as a citizen. What distinguished a DEMOCRACY such as ATHENS was that the Athenian citizenry included men whose income level was below the hoplite level.

There also existed professional hoplites, recruited for service in the pay of some other power, whether Greek or foreign. The best-known mercenary from Greek history is the Athenian XENOPHON who, with 10,000 Greek hoplites, marched deep into the Persian Empire in 401 B.C., in the service of a rebel Persian prince.

(See also OLIGARCHY; PHEIDON; ARCHILOCHUS; WARFARE, LAND; WARFARE, NAVAL.)

hubris (sometimes written as *hybris*) Human arrogance or excessive pride, which usually leads to disaster. Implicit in much of Greek MYTH, the concept of hubris received its full expression in Greek tragedy. A prime example of hubris is AGAMEMNON's decision to tread on the purple tapestry in AESCHYLUS' tragedy *Agamemnon*—a wanton desecration of expensive finery that called down upon Agamemnon the wrath of the gods. Extreme examples of hubris include mythical villains such as IXION or SISYPHUS, who betrayed the friendship of the gods.

In Athenian legal parlance, hubris had a second meaning: assault against an Athenian citizen. The term was used in the sense of statutory rape or any kind of violence. As in the primary definition of hubris, the concept involved an unacceptable flouting of boundaries.

(See also AJAX (2); TANTALUS; THEATER; XERXES.)

Hyacinthus In Greek MYTH, Hyacinthus (Greek: Huakinthos) was a handsome Spartan youth loved by the god APOLLO. While Apollo was teaching him how to throw the discus one day, the jealous West Wind god, Zephyrus, sent the discus flying back into the young man's skull. Hyacinthus lay dying, and from his blood there sprang the type of scarlet flower that the Greeks called the hyacinth (perhaps our iris or anemone, but not what is called the hyacinth now).

Hyacinthus was honored in a three-day early-summer festival throughout the Spartan countryside. His tomb was displayed at Apollo's shrine at Amyclae, near SPARTA. One ancient writer noted that Hyacinthus' statue at Amyclae showed a bearded, mature man, not a youth.

The name Hyacinthus is pre-Greek in origin, as indicated by its distinctive *nth* sound. Modern scholars believe that Hyacinthus' cult dates back to pre-Greek times and that he was originally a local non-Greek god, associated with a local flower but not imagined as a young man and not yet associated with the Greek god Apollo. Sometime in the second millennium B.C., Hyacinthus was adapted to the RELIGION of the conquering Greeks and was made into a human follower of Apollo. The element of romantic love between them may have been added later, after 600 B.C.

(See also GREEK LANGUAGE; HOMOSEXUALITY.)

I

Ibycus Lyric poet of circa 535 B.C., known particularly for his love poetry. Coming from the Greek city of RHEGIUM, located in the "toe" of ITALY, he supposedly refused an offered dictatorship and traveled instead to the east Greek island of SAMOS, where he flourished at the wealthy court of the tyrant POLYCRATES.

Only a few fragments of Ibycus' verses survive, either quoted by later ancient writers or discovered recently in ancient papyri from EGYPT. It is known that Ibycus' work featured two very different genres of Greek LYRIC POETRY: mythological storytelling, as developed previously by the poet STESICHORUS, another western Greek; and short, personal love poems.

The love-poem tradition comes from the eastern Greeks, a product of the sophisticated cities of IONIA and LESBOS. In keeping with the upper-class tastes of the time, the feeling expressed in these poems was homosexual. Such poems usually were written by a man of aristocratic blood to proclaim his infatuation for some teenager or young man of equal social status. Ibycus was considered one of the great poets of this genre, and he had the personal reputation of being crazy for the love of boys.

Ibycus' verses included choral poetry—poems to be sung or chanted by choruses at religious festivals or other great occasions. Recently discovered fragments suggest that Ibycus pioneered the choral form known as the victory ode (*epinikion*), years in advance of the poets SIMONIDES and PINDAR.

Ibycus is perhaps best known for the dubious story of his death. He was supposedly attacked by robbers in a deserted place and died saying that his murder would be avenged by the cranes that were flying overhead. Later, in the city, one of the murderers, seeing some cranes, declared, "Look! The avengers of Ibycus." The statement drew an inquiring crowd, and the killers were apprehended. Appropriately, Ibycus' extant verses show a love of the natural world, especially of birdlife.

(See also HOMOSEXUALITY.)

Icarus See DAEDALUS.

Iliad See ACHILLES; HOMER; TROJAN WAR.

Ilium See TROY.

Illyris Non-Greek territory of the Adriatic coast, northwest of Greece, corresponding roughly to modern Albania. Organized into warlike tribes that were hungry for Greek goods, the Illyrians played a role in the development of northern Greece. The Corinthian colonies EPIDAMNUS (founded circa 625 B.C.) and APOLLONIA (founded circa 600 B.C.) conducted TRADE with the Illyrians, acquiring TIMBER, SLAVES, raw metals, and wildflowers (for perfume-making).

In the 300s and early 200s B.C., the Illyrians were enemies of the kings of MACEDON and EPIRUS. The Illyrians raided these kingdoms and, in turn, suffered annexation of their own territories. By the late 200s B.C. the Illyrians were fighting against Roman armies, until both Illyris and Macedon were defeated and occupied by the Romans in the Third Macedonian War (171–167 B.C.). Illyrian territory supplied part of the Roman province of Illyricum.

(See also PHILIP II; PYRRHUS; ROME.)

Io In MYTH, a woman of the city of ARGOS. The great god ZEUS loved her, but to conceal her from his jealous wife, HERA, he changed Io into a young cow. After several misadventures she was restored by Zeus to human shape in EGYPT, and there bore him a son, whose descendants were the DANAIDS.

This odd myth might possibly reflect Mycenaean-Greek attempts to connect the Egyptian cow-goddess Hathor with the Argive worship of Hera in the shape of a cow.

(See also DANAUS; MYCENAEAN CIVILIZATION; RELIGION.)

Ion (1) Mythical ancestor of the IONIAN GREEKS. According to legend, Creusa, daughter of the Athenian king Erechtheus, was seduced by the god APOLLO and bore Ion in secret. Apollo took Ion away. Later Ion discovered the secret of his birth and subsequently became king of ATHENS. The legend was the subject of an extant tragedy by EURIPIDES (circa 410 B.C.).

(See also HELLĒN.)

Ion (2) Tragic playwright and social figure of the island of CHIOS, circa 490–421 B.C. A wealthy aristocrat, Ion wrote tragedies (all now lost) on mythological subjects for competition in the theater festivals at ATHENS. After winning first prize in two categories (tragedy and dithyramb) at the major annual festival known as the City Dionysia, he made a gift of Chian WINE to every Athenian citizen.

Ion is better remembered for his lively memoirs, which survive as fragments quoted by later writers. Equally at home in Athens or Chios, he hobnobbed with some of the greatest Athenians of the mid-400s B.C., including the statesman CIMON (whose affability Ion contrasts with the coldness of PERICLES) and the tragedian SOPHOCLES (whose charm and wit Ion conveys).

Ion embodied the spirit of the Athenian empire, which saw cultured and talented people flocking to Athens from all over the Greek world.

(See also LYRIC POETRY; THEATER.)

Ionia Greek-occupied central part of the west coast of ASIA MINOR. The area was named for Greeks of the Ionian ethnic group who invaded circa 1050–950 B.C., displacing prior, non-Greek inhabitants. In later centuries the Asian IONIAN GREEKS retained links with the most important Ionian-Greek city of old Greece, ATHENS.

By the 700s B.C. an Ionian League had been formed from 12 states of the Asia Minor coast. These were: the islands of CHIOS and SAMOS, and the mainland cities of PHOCAEA, Clazomenae, Erythrae, Teos, Colophon, Lebedus, EPHESUS, Priēnē, Myus, and MILETUS. At least one other local city, Smyrna, could claim Ionian kinship but was not a league member. The league held meetings at the Panionium, a sanctuary of the god POSEIDON at Cape MYCALĒ. There were internal rivalries and even wars, especially between the two preeminent powers, Miletus and Samos.

In TRADE, naval power, and culture, Ionia was at the forefront of the early Greek world. The poet HOMER lived and composed in Ionia, possibly in Chios (circa 750 B.C.). In the 600s and 500s B.C., Miletus and Phocaea planted colonies from the BLACK SEA to southern Spain, and traded with EGYPT and other non-Greek empires. The JEWS of the Levant knew Ionia as "Javan"—mentioned in the biblical book of Ezekiel (early 500s B.C.) for its exports of SLAVES and worked BRONZE. Woolen textiles were another prized Ionian export.

The term *Ionian Enlightenment* sometimes is used to describe the intellectual explosion that occurred, chiefly at Miletus, in the 500s B.C. There Western SCIENCE and PHILOSOPHY were born together, when THALES, ANAXIMANDER, and ANAXIMENES first tried to explain the world in rational, nonreligious terms. In Ionia there arose the grandest Greek temples of the 500s B.C., at Ephesus, Samos, and Didyma (near Miletus). The decorative schemes developed for such buildings are still employed today, in the Ionic order of ARCHITECTURE.

But this confident culture existed precariously alongside the restless Asiatic empires of the interior. The non-Greek kingdom of LYDIA warred constantly against the Ionians and finally conquered them under King CROESUS (mid-500s B.C.). Croesus dealt favorably with Ionia, but in 546 B.C. the Persian king CYRUS (1) rode out of the East to defeat Croesus and seize his kingdom.

Ionia, after a brief resistance, became a tribute-paying part of the Persian Empire. Many Ionians left home; the populations of Phocaea and Teos sailed away en masse. Throughout the cities, the Persians installed unpopular Greek puppet rulers. After the doomed IONIAN REVOLT (499–493 B.C.), Persian rule was harshly reaffirmed.

In the Persian invasions of mainland Greece (490 and 480 B.C.), Ionian troops and ships' crews were made to fight as Persian levies against their fellow Greeks. In 479 B.C. a seaborne force of mainland Greeks landed in Ionia, beat a Persian army at the Battle of Mycale, and liberated Ionia.

The leadership of an exhausted Ionia now passed to Athens, which established the DELIAN LEAGUE as an Ionian kinship–based mutual alliance against PERSIA (478 B.C.). Eventually, however, the onus of tribute and the fading of the Persian threat served to disenchant the allies. Samos revolted spectacularly (but unsuccessfully) against Athens in 440 B.C., and once the tide had turned against Athens in the PELOPONNESIAN WAR (after 413 B.C.), Chios, Miletus, Ephesus, and other Ionian states went over to Athens' enemy, SPARTA.

After Athens' defeat by Sparta (404 B.C.), Ionian freedom was short-lived. Spartan overlordship was worse than Athenian, and in 386 B.C. Sparta cynically handed Ionia back to the Persians, by the terms of the KING'S PEACE. Ionia was liberated again in 334 B.C., by ALEXANDER THE GREAT's invasion of the Persian Empire. Under Alexander, the Ionian cities enjoyed DEMOCRACY and freedom from paying tribute.

Several Ionian cities, notably Ephesus and Smyrna, went on to thrive in the HELLENISTIC AGE (300–150 B.C.) and Roman eras. But Ionia as a distinctive culture was finished.

(See also ANACREON; ANAXAGORAS; CALLIAS; HECATAEUS; HERACLITUS; PERSIAN WARS; PYTHAGORAS; XENOPHANES.)

Ionian Greeks Greek cultural and linguistic group, distinct from other ethnic groups, such as the DORIAN GREEKS and AEOLIAN GREEKS. The adjective *Ionic* usually refers to dialect of the Ionians or to the distinctive architectural style—the Ionic order—that was developed in Ionia for monumental buildings of stone.

After the collapse of MYCENAEAN CIVILIZATION and the subsequent Greek migrations (circa 1150–950 B.C.), the Ionians were left occupying various sites in and around the AEGEAN SEA—namely, ATHENS, EUBOEA, most of the CYCLADES islands, and IONIA (as it came to be called), on the west coast of ASIA MINOR. Starting in the mid-700s B.C., Ionian-Greek seafarers colonized parts of ITALY, SICILY, the BLACK SEA coasts, and the northwestern Aegean region known as CHALCIDICĒ, among other regions.

Claiming descent from a common ancestor, ION (1), the far-flung Ionian Greeks retained a sense of kinship. In Sicily and southern Italy, Ionian cities banded together for protection against hostile Dorian settlements. During the 400s B.C., imperial Athens made the propagandistic claim to be the protector of all the Ionian Greeks.

In contrast to the stolid Dorians, the Ionians had the reputation for being intellectual, artistic, unsoldierly, elaborate in dress, and luxury-loving. Intriguingly, Ionian societies—other than Athens—were thought to be the least conducive to HOMOSEXUALITY.

(See also ARCHITECTURE; DELIAN LEAGUE; GREEK LANGUAGE.)

Ionian Revolt Failed rebellion of the Greek cities of ASIA MINOR against the Persian king DARIUS (1), marking the beginning of the PERSIAN WARS. Emanating from the prosperous Greek region called IONIA, the revolt lasted more than five years (499–493 B.C.) and might have succeeded but for internal rifts and weak leadership. The war demonstrated to the Persians that, in order to secure their western frontier, they would have to invade and subjugate mainland Greece itself.

In the mid-500s B.C. the Greeks of Asia Minor and the eastern Aegean islands had been engulfed by the Persian Empire. Persian rule was moderate (in the absence of resistance), but the Greeks had to pay tribute and contribute ships, soldiers, and craftsmen to Persian wars and other projects. Seagoing states such as MILETUS, which had previously built up networks of TRADE, found commerce dampened by the Persians (who at this stage in their history were antimercantile). Greek discontent was rife—particularly against the Greek TYRANTS whom the Persians had set up as their puppet rulers in the cities.

Information about the revolt comes from the Greek historian HERODOTUS (circa 435 B.C.). The rebellion began in an Ionian-Greek fleet that was returning home from naval duty for the Persians. The ships' crews and officers rose up and arrested those tyrants who were serving aboard as squadron commanders. The ringleader of this mutiny was Aristogoras of Miletus, himself one of the tyrants. Advised by his father-in-law, Histiaeus, Aristagoras became leader of the revolt.

The home cities followed. At SAMOS, CHIOS, LESBOS, and EPHESUS, tyrants were deposed in favor of DEMOCRACY. Two cities of mainland Greece, ATHENS and ERETRIA, sent warships as aid—an anti-Persian act that would later have dire consequences for both Greek cities.

The Athenians and Eretrians sailed home again after an allied land raid against Sardis, the Persians' main base in Asia Minor (498 B.C.). Meanwhile the revolt spread to other Greek regions under Persian rule: BYZANTIUM, the HELLESPONT, and the Greek cities of CYPRUS.

There followed the methodical Persian counteroffensive. With a navy supplied by their subject state PHOENICIA, the Persians landed in Cyprus and besieged the rebel cities. As Persian armies campaigned through western Asia Minor, the erratic Aristagoras found himself unpopular with his fellow Greeks. He relinquished command and sailed to the north Aegean coast to prepare a refuge in case of defeat, but was killed by native Thracians.

The climactic battle came in 494 B.C. Seeking to destroy the Persians at sea, the Greeks assembled a large fleet (353 warships) from the nine major states still combatant. But the Greek side was divided by jealousies and mistrust, while the Persians offered preferential treatment for quick surrender. When the Greek fleet rowed out to fight the Battle of Lade, off Samos, 49 Samian warships hoisted sail and fled. Most of the other Greek ships followed, leaving only the Chians and Milesians to fight and lose.

The Persians took fierce vengeance on the defeated Greeks. Samos was spared, but Miletus was besieged, sacked, and depopulated, its people transported to interior PERSIA or sold as SLAVES. Returning eventually to their more usual leniency, the Persians reduced the cities' tribute and replaced the old systems of tyrant puppets with democratic governments. But Ionia—the birthplace of Western SCIENCE and PHILOSOPHY—had ceased to exist as a culture or a mercantile power.

(See also MARATHON; PHRYNICHUS; WARFARE, NAVAL.)

Ionian Sea   The name, both modern and traditional, for the southward extension of the Adriatic Sea, which

separates western Greece from southern ITALY and eastern SICILY. Principal islands in this sea are ITHACA and Corfu (ancient CORCYRA).

Confusingly, the Ionian Sea is nowhere near the ancient Greek territory of IONIA (located some 300 miles to the east, on the ASIA MINOR coast). Nor was the Ionian Sea region inhabited by Greeks of the Ionian ethnic group. The name probably derives from the Greek term *Ionian Gulf* (Ionios Kolpos), used by ancient writers to denote the Adriatic. That term possibly dates back to the 800s–700s B.C., when the Adriatic was being explored by Ionian-ethnic Greek seafarers from EUBOEA.

(See also COLONIZATION; IONIAN GREEKS.)

Ionic dialect   See GREEK LANGUAGE.

Ionic order   See ARCHITECTURE.

Iphigenia   In MYTH, the daughter of CLYTAEMNESTRA and AGAMEMNON, king of MYCENAE and grand marshal of the Greek army in the TROJAN WAR. At the war's outset the Greek fleet assembled at Aulis—a port in BOEOTIA, in the Euripus channel—but departure for TROY was perpetually delayed by contrary winds, sent by the hostile goddess ARTEMIS. Agamemnon, learning through a seer the divine cause of his troubles, agreed to sacrifice his daughter Iphigenia to propitiate the goddess' anger. Iphigenia was summoned to Aulis on the pretense that she was to marry the Greek champion ACHILLES; on arrival, she was either killed or (in some versions) carried away by the goddess Artemis to safety among the distant Tauroi, a tribe on the Crimean peninsula of the northern BLACK SEA. The Athenian playwright EURIPIDES follows the latter version in his tragedy *Iphigenia at Tauris* (circa 413 B.C.). In any case, this affair was the cause of Clytaemnestra's hatred toward her husband, and she then schemed to murder him after his return from Troy.

iron   Cheaper and easier to acquire than BRONZE, iron began replacing bronze in mainland Greece circa 1050 B.C. as the metal of choice for swords, spear-points, ax heads, hammerheads, and other cutting or striking tools.

As produced by ancient foundries, iron and bronze were about equal in their toughness; iron's advantage lay in the fact that it was far more plentiful than bronze, both for the Greeks and for other ancient peoples. Bronze, on the other hand, is an alloy of copper and tin, and tin is very scarce in the eastern Mediterranean. Ancient Greek bronze production depended on long-range routes of TRADE, to provide tin. But iron ore—which needs only refining, not mixing—could be found in parts of mainland Greece and ASIA MINOR, among other Mediterranean locales.

Ancient iron forging involved repeated heating and hammering of the metal in order to refine it, weld it into workable quantities, and finally shape it. This process—far different from the casting of molten bronze—apparently was invented circa 1500 B.C. by non-Greek peoples in the region now known as Armenia. At first the technology was kept secret and monopolized by the Hittite overlords of Armenia and Asia Minor. But after the Hittite Empire's collapse circa 1200 B.C., ironworking quickly spread through eastern Mediterranean regions, both Greek and non-Greek. This was the era of social upheaval and migration that historians refer to as the early Iron Age. One reason why ironworking spread amid violence was that iron democratized warfare: A warrior no longer needed to be rich, or a rich man's follower, in order to have a superior weapon.

The arrival of ironworking in Greece can be connected to the final disappearance of MYCENAEAN CIVILIZATION and the invasion of the DORIAN GREEKS, circa 1100–1000 B.C. Either the Dorians used iron weapons in conquering Greece or (more likely) they arrived with bronze weapons, but, having destroyed the old Mycenaean trade routes, found themselves without bronze supplies and immediately embraced the new metal. The Dorians possibly learned iron forging via maritime contact with the eastern Mediterranean island of CYPRUS, which was a meeting place of East and West.

The foremost early Greek ironworking cities included ATHENS, CORINTH, CHALCIS, and ERETRIA. Iron's importance in this economy can be seen in the Greeks' use of iron rods as a primitive form of currency. Only gradually were these replaced by the use of coins, in the 500s B.C.

By the 600s B.C. the Greeks had developed seaborne trade routes to major iron sources outside of Greece, in the western Mediterranean. These locales included western ITALY (where the ETRUSCANS traded with iron ore mined on the island of Elba) and southern Spain.

Ancient foundries never developed the technique, mastered during the European Middle Ages, of casting iron—that is, of heating iron hot enough to pour into a mold. For this application, bronze remained the premier metal. Throughout antiquity, bronze continued to supply such items as military breastplates, helmets, decorative tripods, and SCULPTURE, where casting or intricate shaping was required.

(See also COINAGE; HEPHAESTUS; PITHECUSAE; WARFARE, LAND.)

Iron Age   See IRON.

Isaeus   Greek orator, active at ATHENS, who lived circa 420–350 B.C. He may have been an Athenian. He is said to have been a pupil of the orator ISOCRATES and a teacher of the great DEMOSTHENES (1). Of Isaeus' speeches, 11 survive: They are all courtroom speeches, dealing with disputed inheritance and other civil matters. Isaeus provides us with much of our information about Athenian laws of inheritance in the 300s B.C.

(See also LAWS AND LAW COURTS; RHETORIC.)

Ischia   See PITHECUSAE.

Isocrates   Influential Athenian orator, educator, and pamphleteer who lived 436–338 B.C. Isocrates' importance for Greek history is in his attempts to make the Greeks unite in a military crusade against the Persian empire. His most significant written work was his *Philippus,* or *Address to Philip* (346 B.C.). This work, which survives today, was an "open letter" addressed to King PHILIP II of MACEDON, calling on the king to lead such a campaign. The letter surely came as a propaganda blessing to the opportunistic Philip, who was preparing to subdue Greece. Perhaps it

was Isocrates who first inspired Philip's further ambition to conquer PERSIA (an ambition eventually fulfilled by Philip's son ALEXANDER THE GREAT).

Born into a wealthy family, Isocrates studied RHETORIC under the famous GORGIAS. Like certain other Greek orators, Isocrates wrote speeches for his clients to deliver themselves. He rarely, if ever, argued in court or spoke in the Athenian political ASSEMBLY. But his indirect influence was great. His students of rhetoric included the future historian Androtion and the future orator ISAEUS. At some point Isocrates had a school on the island of CHIOS.

It was the news of the humiliating terms of the KING'S PEACE (386 B.C.) that fired Isocrates' vision of the Greeks uniting to liberate the Greek cities of ASIA MINOR. Isocrates' first published pamphlet on the subject was the *Panegyricus* (380 B.C.), an idealistic tract that pictured ATHENS and SPARTA leading the crusade. The *Panegyricus* strongly anticipates the letter to Philip, written 24 years later. In the intervening years Isocrates somewhat shamelessly addressed similar pleas to other rulers of the Greek world, including the Spartan king AGESILAUS and the Syracusan tyrant DIONYSIUS (1).

But Isocrates' idealism could not stem Philip's ambition of conquering Greece. After the failure of a final plea following the Battle of CHAERONEA (338 B.C.), the 98-year-old Isocrates starved himself to death.

Ancient writers knew of 60 works by him; we possess 21. Six of these are court speeches; the rest are political pamphlets.

(See also EDUCATION; LAWS AND LAW COURTS.)

Isthmian Games   One of the four Greek-international sports-and-religious festivals; the other three were the OLYMPIC, PYTHIAN, and NEMEAN GAMES. The Isthmian Games, or Isthmia, took place every two years at CORINTH (located on the isthmus of central Greece). The festival honored the god POSEIDON, and the ritual prize for the victors was a garland of wild celery. The games were first organized during the early 500s B.C.

isthmus   See CORINTH; PELOPONNESE.

Italy   Like SICILY, Italy contained a number of powerful Greek cities in ancient times. Two separate regions of the Italian peninsula saw intensive COLONIZATION by the land- and commerce-hungry Greeks, circa 750–550 B.C. One region is now called Campania, on the Bay of Naples on the Italian west coast. There the Euboean settlement of CUMAE became the first Greek landfall in Italy and one of the very earliest Greek colonies anywhere (circa 750 B.C.). The attraction was TRADE with the powerful ETRUSCANS.

The other major focus of COLONIZATION was in the south, along the south coast of the Italian "toe" and "instep." There the Spartan colony of TARAS (founded circa 700 B.C.) occupied the best harbor in Italy, on what is now called the Gulf of Taranto. Other Greek cities of south Italy included SYBARIS, CROTON, RHEGIUM, LOCRI, Metapontum, and Siris. The main attraction at these sites was farmland, which the Greeks seized from the native Brutii and Lucanii.

The Greek name Italia—applying originally only to the south coast of the Italian "toe"—was probably a Greek rendering of a local Italian place-name: Vitelia, "calf land."

The "heel" of Italy originally had a different Greek name, Iapygia. Later the name Italia, passing into the Latin language of the conquering Romans, came to be applied to the entire peninsula, north to the Alps.

To the Greeks of southern Italy, coming from mountainous and overpopulated Greece, the land of Italy seemed to extend forever. The Greeks gave the region a nickname—Great Greece, Megalē Hellas—or (as it has been more traditionally known, in its Latin name) Magna Graecia.

By the 400s B.C. the Italian Greeks were subject to influence and control by the successive rulers of the Sicilian-Greek city of SYRACUSE. In the 200s B.C., after centuries of war with each other and with their Italian neighbors, the Greek cities of Italy became absorbed by the expansionist power of ROME.

(See also AGATHOCLES; DIONYSIUS (1); HIERON (1); PARMENIDES; PITHECUSAE; POSEIDONIA; PYTHAGORAS.)

Ithaca Small (44 square miles) island of the northwest coast of Greece, located outside the Gulf of Patras. Beautiful but unfertile, its coastline fretted with inlets of sea, Ithaca was inhabited throughout ancient times yet played almost no role in Greek history. It is best remembered as the domain of the Greek king ODYSSEUS. Archaeological excavations have uncovered POTTERY and house foundations from the Mycenaean era, confirming the island's importance circa 1200 B.C., when Odysseus would have lived.

(See also IONIAN SEA; MYCENAEAN CIVILIZATION.)

Ixion Villain of Greek MYTH. Ixion was a prince of the Lapith tribe of THESSALY who murdered his father-in-law so as to avoid paying the bride price for the MARRIAGE. Although pardoned by the great god ZEUS, Ixion next schemed to seduce Zeus' wife, the goddess HERA. But Zeus, aware of Ixion's plan, deceived Ixion with a facsimile of Hera, shaped from a cloud. (Supposedly Ixion's semen impregnated the cloud-Hera, which gave birth to the race of CENTAURS.)

At Zeus' order, Ixion was whipped until he repeated the words "Benefactors deserve honor." Then he was tied to a fiery wheel and sent spinning through the sky—or through the Underworld—for eternity. Ixion was one of several great sinners of Greek legend—others include SISYPHUS and TANTALUS—whose crime involved betraying the friendship of the gods.

(See also AFTERLIFE; HUBRIS.)

J

Jason (1) In MYTH, a Thessalian hero who led the Argonauts (sailors of the ship *Argo*) to the distant land of Colchis, to acquire the fabulous Golden Fleece. The tale of the Argonauts is mentioned in HOMER's *Odyssey* (written down circa 750 B.C.) and elaborated in PINDAR's fourth Pythian ode (462 B.C.), but the main source is APOLLONIUS' clever Alexandrian epic, the *Argonautica* (circa 245 B.C.).

The Argonaut tale seems to combine two different layers: (1) a very ancient legend going back to Mycenaean times (second millennium B.C.), of significance for Mycenaean centers in THESSALY and BOEOTIA; and (2) later traders' tales of the Sea of MARMARA and the BLACK SEA, developed during the Greek exploration of that region in the 700s and 600s B.C. In the earlier version (now lost), the Argonauts may have sailed to a fabled land called Aea, somewhere at the edge of the world. But the surviving version has the Argonauts voyaging to Colchis, a real (although distant), non-Greek region on the eastern Black Sea. (Colchis was an emporium for caravans from north-central Asia, and by the 500s B.C. the Milesian Greeks had a TRADE depot there.)

In the myth, Jason (Greek: Iason, "healer") was the son of Aeson and heir to the kingship of the Thessalian city of Iolcus (modern Volos, the region's only seaport). As a child, he was smuggled away by his mother after Aeson's stepbrother Pelias had seized the throne. Raised in the Thessalian wilds by the wise Chiron—the centaur who regularly tutors mythical young heroes—Jason returned as a young man to Iolcus, to claim his birthright. He entered the city with one sandal missing, having lost it crossing a river. Pelias—knowing by that foretold sign to beware of this man—immediately persuaded the newcomer to go out in search of the Golden Fleece.

This fleece was the skin of a winged, golden-fleeced ram, which in prior times the gods had supplied to carry the young Phrixus and Hellē (the children of the Boeotian hero Athamas) away from their evil stepmother. Carried eastward through the sky, Hellē had lost her grip and fallen off, drowning in the waterway thenceforth called the HELLESPONT (in her honor). Her brother had safely reached the far-off land of Colchis, where he sacrificed the ram to ZEUS the Savior.

Jason, aided by his divine patron, HERA, then assembled an expedition consisting of the noblest heroes in Greece. The roster varies, but the most familiar names include: the Calydonian hero MELEAGER; the female warrior ATALANTA; the brothers PELEUS (father of ACHILLES) and Telamon (father of AJAX [1]); the Thracian musician ORPHEUS; the Spartan twins CASTOR AND POLYDEUCES; the twins Calais and Zetes (sons of the North Wind); and the greatest of heroes,

HERACLES (who soon gets "written out" of the story, so as not to monopolize the adventures).

Their ship was the *Argo*, built by a shipwright named Argos, who went along on the expedition. The Argonauts were sometimes called by the name MINYANS (*Minuai*), which was also the name of the ruling family of the Boeotian city of ORCHOMENUS. These Boeotian Minyans may have figured in the early (lost) version of the myth.

Sailing northeastward from Greece to the Black Sea, the heroes had several adventures. They dawdled for a year on the Aegean island of Lemnos, busily impregnating the man-hungry Lemnian WOMEN, who had all murdered their husbands out of resentment of the latter's Thracian slave girls. Finally continuing to the Sea of Marmara, the Argonauts were welcomed by the native king Cyzicus, but he was killed by them in a mishap. Soon Heracles was separated from the expedition, having gone ashore to search for his page, Hylas.

Still in the Marmara, the Argonauts rescued the old, blind Thracian king Phineus who, for some offense, was living in eternal torment. The HARPIES, hideous winged female demons, would relentlessly fly down to snatch away his food and leave behind their feces. The Argonauts Calais and Zetes, who could fly, chased away the harpies forever. In gratitude, Phineus gave Jason advice on how to slip through the Symplegades, the Clashing Rocks (which are a mythical rendering of the narrow BOSPORUS, channel, leading into the Black Sea).

At the eastward end of the Black Sea, the adventurers reached Colchis, where the evil king Aeëtes set his conditions for surrendering the fleece. One of the Argonauts must plow a field with a pair of fire-breathing BRONZE bulls and sow the magical dragon's teeth (remnants of the hero CADMUS' adventure at THEBES), which Aeëtes had in his possession. From the seeds, there would arise armed men, who must be conquered.

Aeëtes was sure that these conditions could never be met. But his sorceress daughter MEDEA, having fallen in love with Jason (at APHRODITE's hand, from Hera's bidding), provided the hero with a magic ointment to make him invulnerable for a day. Thus Jason was able to fulfill Aeëtes' terms. Then the Argonauts fled Colchis with the fleece and Medea (who had stolen the fleece for them after charming to sleep the dragon that guarded it).

Aeëtes led his ships in pursuit. But Medea, who had brought her young brother Apsyrtus aboard the *Argo*, now murdered the child and cut him up, throwing the body sections into the sea. Aeëtes and his followers halted, collecting the royal corpse for burial.

The Argonauts returned to Thessaly, either the way they had come or via a fantastical route up the Danube and into the Mediterranean. Apollonius' *Argonautica* ends with the return to Iolcus. Other writers describe how Jason and Medea took vengeance on Pelias by fatally tricking him into climbing into a cauldron of boiling water that was supposed to rejuvenate him.

Driven from Iolcus by Pelias' son, Jason and Medea hung up the Golden Fleece in the temple of Zeus at Orchomenus. They settled at CORINTH, where the final chapter of Jason's drama was played out, as described in EURIPIDES' tragedy *Medea* (431 B.C.). Jason divorced Medea, intending to marry Glauce, daughter of the Corinthian king Creon. Medea, insane with rage, avenged herself by killing Glauce, Creon, and her own two children by Jason. She then fled to ATHENS. Jason, meanwhile, had set up the *Argo* on land and dedicated it to the god POSEIDON. As he slept beneath the ship's stern one day, a section fell off and killed him.

(See also CENTAURS; CYZICUS; EPIC POETRY; SHIPS AND SEAFARING.)

Jason (2) Tyrant of Pherae in THESSALY, reigning circa 385–370 B.C. In a period of political turmoil throughout Greece, he united Thessaly under himself and tried to forge it into a major power. An ally of THEBES, he negotiated a treaty between Thebes and SPARTA after the Spartan defeat at LEUCTRA (371 B.C.) but was assassinated the following year. In history he stands as a precursor to the Macedonian king PHILIP II (reigned 359–336 B.C.), who may have modeled himself on Jason to some extent. In warfare, Jason developed the use of primitive siege artillery—a military science later taken up by Philip. (See also TYRANTS; WARFARE, SIEGE.)

Jews The Jews and the Greeks, despite their respective importance in ancient history, had little to do with each other prior to the 300s B.C. An independent Jewish kingdom, with its capital at Jerusalem, ceased to exist in 586 B.C., when the Babylonians captured Jerusalem and removed the population to Babylon. This Babylonian Exile ended in 539 B.C., when the Persian king CYRUS (1) conquered much of the Near East and permitted his new Jewish subjects to return to their homeland in the Levant. (Not all did so, however.)

It was now that the Jews came into contact with another subject-people of the Persians—Greeks from IONIA, in western ASIA MINOR. The biblical prophet Ezekiel (late 500s B.C.) wrote disdainfully of the profit-minded Greek traders of "Javan" (Ionia), trading in SLAVES and worked BRONZE. The Greek historian HERODOTUS (circa 435 B.C.) knew of the Jews—he called them "Palestinian Syrians"—and listed them among the naval levies serving in the Persian king XERXES' invasion of Greece in 480 B.C.

The destruction of the Persian Empire by the Macedonian king ALEXANDER THE GREAT between 334 and 323 B.C. left Macedonian and Greek governors over the various Jewish pockets of the Near East. By about 300 B.C. Alexander's domain had fragmented into the large Greco-Macedonian kingdoms of the HELLENISTIC AGE. Ptolemaic EGYPT seized Jerusalem and the southern Levant, while the SELEUCID EMPIRE ruled Babylon and other Jewish-inhabited regions. Consequently Jewish immigrants flooded to such newly founded Hellenistic cities as Egyptian ALEXANDRIA (1) and Seleucid ANTIOCH. Alexandria in particular developed an important Jewish minority that occasionally endured ethnic violence from the Greek population.

Jewish monotheism was not deeply affected by the polytheistic RELIGION of the Greeks. However, the glamorous Greek style of life attracted many of the wealthier Jews, creating an assimilated, pro-Greek class. This process occurred not only in the Hellenistic cities but also in Jerusalem itself (where at least one Greek-style GYMNASIUM and THEATER were each built in the 100s B.C.). Greek nomenclature infiltrated Judaism: The word synagogue, for example, is Greek ("assembly place"). Among surviving customs, Greek influence can be seen in the ceremonial Jewish Passover meal: The ritual drinking of cups of WINE and the prayerbook references to dining in a reclining position are best understood as elements borrowed from the Greek drinking party known as the SYMPOSIUM.

With Greek the language of commerce, administration, and secular law in the Hellenistic kingdoms, the emigrant Jewish communities began to forget the Hebrew tongue. In Alexandria this process had taken hold by around 260 B.C., when certain books of the Jewish Bible began appearing in Greek translation. The complete Greek translation of the Jewish Bible—a work supposedly ordered by the Macedonian-Egyptian king PTOLEMY (2) II and conducted by 70 scholars—became known as the Septuagint (from the Latin *septuaginta*, "seventy"). In the Greco-Roman world of later centuries, the Septuagint contributed greatly to the survival of Judaism and the spread of Christianity.

In 198 B.C. the Seleucid king ANTIOCHUS (2) III conquered much of the Levant, including the Jewish heartland. There the next generation saw the best-known, tragic encounter between Jews and Greeks—the Maccabean Revolt of 167–164 B.C., which is commemorated in the Jewish festival of Hanukkah. The biblical First Book of Maccabees describes how certain rural Jews rebelled after the Seleucid king Antiochus IV tried to impose Greek religious customs and convert the temple at Jerusalem into a temple of the Greek god ZEUS. The spreading revolt was led by Judas Maccabee ("the hammer"), of the priestly Hasmonaean family. Judas defeated Seleucid armies in battle and recaptured the Jerusalem temple, but the rebellion decayed into a Jewish civil war of anti-Greek versus pro-Greek factions. Finally, in 142 B.C., the rebels ejected the Seleucid garrison from the citadel at Jerusalem. For the next 80 years the Jews of the Levant comprised a sovereign nation ruled by a Hasmonaean dynasty. But in 63 B.C. the country fell to the Roman legions of Pompey the Great.

Jocasta See OEDIPUS.

K

King's Peace   The treaty drawn up in 386 or 387 B.C. between SPARTA and the Persian king Artaxerxes II, sometimes known as the Peace of Antalcidas (from the name of a Spartan ambassador). This treaty ended recent Spartan-Persian hostilities in the CORINTHIAN WAR and severed PERSIA's alliance with Sparta's enemies, ATHENS, CORINTH, BOEOTIA, and ARGOS. But the notorious aspect of the treaty was Sparta's renunciation of its former claim to protect the Greek cities of IONIA and of other parts of ASIA MINOR. The King's Peace ceded the Greek cities of Asia Minor and CYPRUS back to the Persian king, even though these cities had been liberated by the Greeks after the PERSIAN WARS.

For the Persians, the King's Peace marked a high point in their designs against Greece, a return to the western conquests of DARIUS (1) (circa 500 B.C.). For the Greeks, the peace unmasked the "real" Sparta from its pretense of being a liberator. Amid the resulting anti-Spartan anger, Athens was able to attract allies for its new SECOND ATHENIAN LEAGUE (circa 377 B.C.), and the city of THEBES began to emerge as a rival to Sparta. It was then (380 B.C.) that the Athenian orator ISOCRATES began to publish pamphlets urging the Greeks to unite and liberate Ionia by invading the Persian Empire—a plea that would later bear fruit in the conquests of the Macedonian king ALEXANDER THE GREAT (334–323 B.C.).

kinship   From earliest times, Greek society was organized along multiple levels of kinship. The largest and simplest divisions were the different Greek ethnic groups, chiefly the IONIAN, AEOLIAN, and DORIAN GREEKS. All Dorian and many Ionian settlements were originally subdivided into citizens' groupings called *phulai* (or phylae, usually translated as "tribes"). Dorian cities such as SPARTA or SYRACUSE had three phylae. Ionian cities such as ATHENS or MILETUS might have had four or more. Undoubtedly the phylae predate the Greek cities and reflect the tribal organizations of a nomadic era in the BRONZE AGE.

Appropriate to this primitive origin, the phylae were religious-military societies with inherited membership, centered on aristocratic families. Phylae were themselves divided into groups called *phratriai* (brotherhoods), whose members were *phrateres* (brothers). This Greek word—connected in origin to other Indo-European words such as Latin *frater* and English *brother*—described men who were

not literally brothers but who fought together in the retinue of aristocratic war leaders. In HOMER's epic poem the *Iliad* (written down circa 750 B.C.), the Greek leader AGAMEMNON is advised to marshal the army into battalions consisting of men from the same tribe and phratry. Early Sparta (circa 700 B.C.) had 27 phratries, probably nine from each tribe.

Like other aspects of Greek citizenship and military life, the phratry was mainly the province of adult, citizen males. During the era of Greek aristocratic rule (circa 1000–600 B.C.), a man did not need to be of noble blood to belong to a phratry, but he did have to be associated with one of the aristocratic families—for example, as a spear-carrier in war. For the individual citizen, the phratry provided aristocratic patronage and an extended family, giving assurance of support in legal proceedings, blood feuds, and similar events.

At the head of each phratry was a *genos* or (plural) *genē*. The *genos* was an aristocratic clan—a group of kinsmen claiming a single noble ancestor through male descent. A *genos* typically had a name formed with a Greek suffix meaning "the sons of"—for example, the Athenian Alcmaeonidae (sons of the ancestor Alcmaeon) and the Corinthian Bacchiadae (sons of Bacchis). As a center of political power, religious authority, and military might, the *genos* dominated the (larger and less elite) phratry, but presumably a phratry might be headed by more than one *genos*. We know the names of about 60 Athenian *genē*; they were also called by the blanket term *Eupatridai*, "sons of noble fathers."

With the arising middle-class challenge to aristocratic rule in the 600s–400s B.C., the old-fashioned, aristocratic-based phyle and phratry in many Greek cities were completely reorganized by political decree. At Sparta this occurred probably under the reforms of LYCURGUS (1), circa 650 B.C. At Athens such revamping was at the heart of the democratic reforms of CLEISTHENES (1) (circa 594 B.C.). Abolishing the four traditional Athenian phylae, Cleisthenes created 10 new phylae and reconstituted the phratries so as to include newly enfranchised, lower-income citizens.

In the democratic Athens of the 400s and 300s B.C., the phratry remained a vital social-political entity, connecting the individual citizen with the political life of the state. As a young child and again as a teenager, the young male citizen was presented by his father and near kinsmen to his *phrateres* at the altar of ZEUS Phratrios, the Zeus of the

Phratry. The adolescent presentation—at which the youth dedicated his newly shorn childhood hair to the god—signified the young man's entrance into the community. Later in the man's life, his *phrateres* witnessed his betrothal ceremony and feasted at his MARRIAGE. Less elaborately, it was at the local phratry office that a female Athenian's name would be enrolled, thus assuring her of the rights (regarding marriage, public assistance, etc.) available to citizen WOMEN.

(See also ARISTOCRACY; DEMOCRACY; POLIS.)

Knossus See CNOSSUS.

Korē See DEMETER; PERSEPHONE.

Labyrinth In MYTH, the Labyrinth was the mazelike palace designed by the Athenian craftsman DAEDALUS at CNOSSUS, in CRETE, to house the monstrous Minotaur. The Athenian hero THESEUS, making his way through the Labyrinth's corridors with the help of the Cretan princess Ariadne, slew the Minotaur. The classical Greeks described any kind of architectural maze as a labyrinth (*laburinthos*)— much as we use the word today.

Like other myths, this one contains a kernel of fact: The name *Labyrinth* was probably applied to the Minoan palace at Cnossus, sometime in the second millennium B.C. *Laburinthos* was not originally a Greek word, as shown by its distinctive *nth* sound (which distinguishes such other pre-Greek names as CORINTH and HYANCINTHUS). It was probably a Minoan word, meaning "house of the double ax"—the two-headed ax (*labrus*) being a Minoan royal symbol.

(See also GREEK LANGUAGE; MINOAN CIVILIZATION; MINOS.)

Laconia Local territory of SPARTA, in the southeastern PELOPONNESE. The word apparently was a shortened form of the region's alternate name, Lacedaemon.

Of the Peloponnese's three southern peninsulas, Laconia included the eastern (Cape Malea) and middle one (Cape Taenarum). The two capes and much of the interior consist of limestone mountains; two ranges—Mt. Parnon in the east and Mt. Taygetus in the west—run north to south. Between these mountain ranges lay the fertile Eurotas River valley, which widens southward to the Laconian Gulf. In the north, Laconia shared a mountainous (and often-disputed) border with ARCADIA and the territory of ARGOS. To the west, beyond the Taygetus range, lay the region of MESSENIA.

In the second millennium B.C. Laconia was the site of a thriving Mycenaean-Greek kingdom; the semimythical king MENELAUS would have ruled there. Overrun by DORIAN GREEKS circa 1100–1000 B.C., Laconia eventually produced a number of Dorian settlements, including Sparta, in the upper Eurotas valley. By about 700 B.C. Sparta had conquered the Eurotas down to the sea. The town of Gytheum, on the western Laconian Gulf, became Sparta's port. The Eurotas farmland became the home of the Spartan elite, the "true" Spartans (Spartiatai). Other parts of Laconia were inhabited by Spartan citizens of lesser social rank—the PERIOECI (Greek: *peroikoi*, "dwellers about"). Together these two groups comprised the Spartan free population, the Lacedaemonians (*Lakedaimonioi*).

The name *Laconia* has provided the modern English word *laconic*, referring originally to the Spartan brevity of speech. (See also HELOTS.)

Lampsacus See HELLESPONT.

Laocoön Trojan priest of APOLLO, in the MYTH of the TROJAN WAR. Laocoön was the brother of Anchises and uncle of the hero AENEAS. After the Greeks had apparently abandoned the siege of TROY, leaving behind the wooden Trojan Horse, Laocoön objected vehemently to the Trojans' plan to bring the horse within the city's walls. (In fact, the horse was full of Greek soldiers.) To stifle Laocoön's objections, and thus fulfill the ordained doom of Troy, the goddess ATHENA sent two huge serpents from the sea.

The horrible death of Laocoön and his two sons is shown in an original marble statue group, variously dated between about 150 B.C. and A.D. 30 and ascribed to three sculptors of Rhodes. Notice the serpent's head at center, biting Laocoön's hip. Now in the Vatican Museum, this ambitious work adorned the Roman emperor Tiberius' grotto at Sperlonga, south of Rome, circa A.D. 30.

These enwrapped Laocoön and his two small sons and crushed them to death.

lapiths See CENTAURS.

laws and law courts By the 700s B.C.—and probably long before—Greek states had developed official and public procedures for administering justice. No information is available on laws or law courts during the Mycenaean era (circa 1600–1200 B.C.), but HOMER's epic poem the *Iliad* (written down circa 750 B.C.) mentions a public trial as one of the scenes embossed on the shield of ACHILLES (book 18). In the scene, elders seated in a city's AGORA arbitrate a dispute between two men over payment of blood money for a murder done by one of the two. A crowd stands around, and heralds call for order.

The judges at such court cases would have been the individuals who ran the city government—that is, the aristocratic COUNCIL or a committee thereof. The ancient Greeks never distinguished between the judiciary and the legislative or executive branches, as modern American society does; there was no class of professional judges. Rather, in early Greek history (circa 900–500 B.C.), the judges might be the same men who decreed the laws, commanded the army, oversaw the state RELIGION, and owned most of the land. Such law courts obviously were biased in favor of aristocratic plaintiffs and defendants, and this unfairness contributed to the popular anger that produced the Greek TYRANTS (600s B.C.) and, later, the Athenian DEMOCRACY (500s B.C.).

Systems of law differed from one city-state to another. Before the invention of the Greek ALPHABET, the law would have been handed down by oral tradition, with officials and their underlings memorizing whole legal codes. But in the 600s B.C., with the spread of WRITING in the Greek world, laws came to be written down for permanence and easy reference. (By contrast, written law codes had existed in the Near East since at least 1800 B.C.)

The writing-down of Greek legal codes came in response to the tense political climate of the 600s B.C., as middle-class citizens demanded fair, permanent, and publicly accessible laws. Naturally this was also an occasion for legal reforms and revisions. The earliest written Greek law code was supposedly composed by a certain Zaleucus at LOCRI, in Greek southern ITALY, circa 662 B.C. At ATHENS, new laws were drafted and written down under DRACO (circa 625 B.C.) and SOLON (circa 594 B.C.). Solon's laws, carved into wooden blocks, were displayed for centuries afterward in the Athenian agora. Today the law code of the city of GORTYN partly survives in a lengthy inscription, carved in stone circa 450 B.C.

Law codes covered civil cases such as disputes over inheritance and land boundaries, as well as criminal cases such as homicide and forgery. A comparison with American laws shows some surprising differences. For instance, in Athens in the 400s–300s B.C., raping a nonslave woman was merely a finable offense, but a man who seduced such a woman could be punished much more severely. (The legal theory here was that rape was a spontaneous crime while seduction was not only premeditated but corruptive to the woman's morals and to the Athenian household.)

The most serious crime at Athens was homicide; conviction brought the death penalty for an intentional killing or exile for an unintentional one (aside from an excusable accident or self-defense). Penalties for other crimes included loss of citizenship or confiscation of property.

Most information on the day-to-day workings of Greek law courts comes from the democratic Athens of the 400s–300s B.C. Sources include inscriptions, courtroom speeches, and factual references in the writings of PLATO, ARISTOTLE, and other thinkers. Cases were heard by large groups of jurors, often numbering 501 but otherwise ranging between about 201 and 2,501. Odd numbers were employed to avoid a tie jury vote—unlike modern American juries, ancient Greek juries did not need to reach a unanimous verdict.

The jurors were ordinary Athenian citizens, chosen by lottery. But their responsibilities far surpassed those of modern American jurors. Although an Athenian courtroom had an officiating magistrate to maintain procedure, there was no learned judge to interpret the law, enforce the rules of evidence, or pass sentence. These decisions were made by the jury itself, on the basis of the speeches and examination of witnesses by prosecutor and defendant (or, in a civil case, by the two disputants).

Similarly, there was no state-employed district attorney, whose job was to prosecute in court. State prosecutions were usually brought by volunteers; any adult male citizen could do so. A private individual might decide to prosecute out of civic duty or for public attention toward a political career. In case of conviction, the prosecutor might be entitled to a portion of the defendant's paid fine or confiscated property. Frivolous prosecution was discouraged by a law requiring the prosecutor to pay a fine himself if his case won less than one-fifth of the jury vote. Nevertheless, the courts undoubtedly became the scene of personal and political vendettas, and it is no coincidence that the Athenians were known to be a litigious people.

(See also AESCHINES; ANDOCIDES; ANTIPHON; AREOPAGUS; ARISTOCRACY; ARISTOPHANES; CATANA; CLEON; DEMOSTHENES (1); HOMOSEXUALITY; ISAEUS; ISOCRATES; LYCURGUS (1); LYCURGUS (2); LYSIAS; MARRIAGE; PERICLES; PITTACUS; PROSTITUTES; PROTAGORAS; RHETORIC; SLAVES; SOCRATES; WOMEN; ZEUS.)

Leda In MYTH, the beautiful wife of the Spartan king Tyndareus. The god ZEUS became infatuated with her and raped her, having approached her in the shape of a swan. Leda bore two sets of twins, all of whom had important destinies: HELEN OF TROY and CLYTAEMNESTRA, and CASTOR AND POLYDEUCES. According to a familiar pattern of Greek storytelling, it was sometimes claimed that only one member of each pair of twins was Zeus' child, the other being Tyndareus'. Helen and Polydeuces were Zeus' children, and hence eligible for immortality. In some versions, the two pairs of twins were hatched from two giant bird eggs.

Lefkandi Modern village and archaeological site on the Lelantine plain, on the west coast of the island of EUBOEA. In the 1960s A.D. British excavations there found traces of an ancient settlement whose earliest level predated the Greeks' arrival (circa 2100 B.C.). As a Greek town, the site apparently prospered during the Mycenaean era and later, down to about 825 B.C., before being abandoned circa 700 B.C.

Lefkandi has shed rare light on Greek life during the DARK AGE (circa 1100–900 B.C.). The pride of Lefkandi's archaeological yieldings is the so-called Hero's Tomb, a surprisingly rich grave of an unknown baron who died circa 900 B.C. and was buried amid his finery, including ornaments of GOLD, imported from the East. The tomb's evidence of wealth and TRADE has forced historians to revise their otherwise grim picture of Dark Age Greece.

Lefkandi's apparent abrupt decline after 825 B.C. seems related to the emergence of the nearby ancient city of ERETRIA. ARCHAEOLOGY suggests that Eretria arose abruptly, in prosperity, around 825 B.C., and modern historians have guessed that Lefkandi was the original city of the Eretrians. Lefkandi sits halfway along the Lelantine plain between Eretria (southeast) and CHALCIS (northwest). Probably under pressure from nearby Chalcis, Lefkandi was abandoned and a new settlement, Eretria, was founded farther away from Chalcis.

(See also LELANTINE WAR.)

Lelantine War Earliest Greek conflict for which any reliable historical record exists. The war was fought circa 720–680 B.C. between the neighboring cities of CHALCIS and ERETRIA, on the west coast of the large island of EUBOEA, in central Greece. Chalcis and Eretria were the most powerful Greek cities of the day; they had previously cooperated in overseas TRADE and COLONIZATION ventures. But now they fought over possession of the fertile plain of Lelanton, which stretched between them.

According to the Athenian historian THUCYDIDES (1) (circa 410 B.C.), the Lelantine War marked the first time that the Greek world divided itself into alliances on one side or the other. It seems to have been a primitive world war, in which the most powerful Greeks states squared off with each other according to traditional local enmities: MILETUS (Eretria's ally) versus its old rival and neighbor SAMOS (Chalcis' ally); CHIOS (Eretria's side) versus its mainland neighbor Erythrae (Chalcis' side); and SPARTA (Chalcis') against its bitter enemy, the neighboring region of MESSENIA (Eretria's).

The fighting between Chalcis and Eretria on the Lelantine plain may have been an old-fashioned, gentlemen's affair. One later writer reports seeing an old inscription recording the belligerents' agreement not to use "long-distance missiles"—that is, arrows, javelins, and sling-stones. Another inscription mentions CHARIOTS, a very old-fashioned military device. These battles were probably fought by dueling aristocrats rather than by the massed concentrations of HOPLITE citizen-soldiers who would come to define warfare in the mid-600s B.C.

The war ended circa 680 B.C., apparently with Chalcis as the marginal winner. In any case, both cities soon were overtaken as commercial and military powers by Corinth, Sparta, and ATHENS. But the war's alliances had lasting effects. With the help of Corinth and Samos, Sparta later conquered Messenia. And seafaring Corinth, shut out of the BLACK SEA by its "Lelantine" enemy Miletus, instead enlarged its grain supplies and foreign markets in SICILY and ITALY.

(See also WARFARE, LAND.)

Leonidas   See THERMOPYLAE.

Leontini   See CATANA; SICILY.

lesbianism   See HOMOSEXUALITY; SAPPHO.

Lesbos   Largest island (630 square miles) of the eastern AEGEAN SEA, close to the northwestern coast of ASIA MINOR. The first inhabitants of Lesbos may have been descendants of the civilization of TROY. In around the 900s B.C., Lesbos was conquered by AEOLIAN GREEKS who had migrated east from the Greek-mainland regions of THESSALY and BOEOTIA. In the following centuries, Lesbos was the departure point for further Aeolian expeditions that colonized the northwest Asia Minor coast.

By the 600s B.C. Lesbos was the homeland of a thriving Aeolian culture, with MYTILENE its foremost city. The island prospered by seaborne TRADE, with mainland Greece and with the Eastern kingdoms of LYDIA and EGYPT. Lesbos' tradition of poetry-making found its culmination in the verses of SAPPHO (circa 600 B.C.). Lesbos' heyday also saw the political reforms of the Mytilenian statesman PITTACUS (circa 600 B.C.) and the short-lived appearance of Aeolic-style temple architecture at Mytilene and Nape (early 500s B.C.). But the golden age had ended by the time the Persians captured the island in 527 B.C.

(See also PERSIAN WARS.)

Leto   See APOLLO; ARTEMIS.

Leucippus   Greek philosopher of the mid-400s B.C., usually credited—alongside his follower DEMOCRITUS—as the inventor of the "atomic" theory of natural philosophy. This theory held that the basic components of matter and reality are tiny particles, which Leucippus called *atomoi*, an adjective meaning "indivisible". All creation, destruction, and change is accomplished by the perpetual reorganization of these atoms into new combinations.

Basic facts of Leucippus' life and works are unknown. He was probably older than Democritus (born 460 B.C.) and was probably his teacher. Leucippus may have been a native of ABDERA (Democritus' home) or MILETUS or Elea, in Greek southern ITALY. No written works survive under Leucippus' name, but two treatises preserved under Democritus' name—*The Great World-system* and *On Mind*—may in fact have been written by Leucippus.

Leuctra   Ancient village in northern BOEOTIA, site of one of the most eventful battles in Greek history. There, in the summer of 371 B.C., the legend of Spartan invincibility was destroyed when an army of Spartans and allies under King Cleombrotus met defeat by a force of Thebans and allies commanded by the Theban leader EPAMINONDAS. In the battle's aftermath, Epaminondas simply removed SPARTA as a first-class power, mainly by liberating the Spartan-ruled HELOTS of MESSENIA (362 B.C.). Sparta, previously master of Greece, never recovered its former power, and the mantle of Greek leadership passed (briefly) to THEBES.

At the heart of the Theban victory was Epaminondas' decision to deepen his battle line's left wing to 50 rows, against the Spartans' more usual 12-man depth. Epaminondas' intent was to destroy the Spartans along their strongest front—their right wing, where the elite "Spartiates" (the Spartan upper class) were arrayed and where the king commanded personally. Traditionally, Greek armies had won their battles on their right wings. The better troops customarily were stationed there, and, due to the battle rows' natural drift rightward, the advancing right wing tended to overlap (advantageously) the enemy left—for both armies. Victory, prior to Leuctra, lay in winning on the right while withstanding the enemy's advantage on its right; traditionally, the left wing performed more poorly. Epaminondas' tactic was to make his left wing the better-performing, by deepening it.

The success was devastating. The Spartan right was pushed back and crushed by the heavy Theban left, and the Spartan left wing scattered in retreat. Nearly 1,000 from the Spartan-led army were killed, including King Cleombrotus and 400 other Spartiates. The Theban side lost 47 men.

Epaminondas' innovation—striking at the enemy's strongest point—changed the nature of Greek warfare. Within a few decades, the technique had been elaborated by the Macedonian king PHILIP II.

(See also CHAERONEA; WARFARE, LAND.)

Linear B Modern name for a pre-alphabetic form of WRITING used by the Mycenaean Greeks, circa 1400–1200 B.C. This was a syllabary script: It basically employed about 90 symbols, each representing a vowel-consonant combination, with pictograms sometimes added to help identify certain words. Linear B was revealed to modern eyes in 1900 A.D. by archaeologist Sir Arthur Evans in his excavations at Mycenaean-era CNOSSUS, on the island of CRETE. The writing, incised into clay tablets, had been incidentally preserved when the tablets were fired during the violent burning of the palace, circa 1400 B.C. Evans named the then-undeciphered script Linear B to distinguish it from an earlier, Minoan script found at Cnossus, called Linear A.

Soon similar tablets were found at PYLOS, THEBES, and MYCENAE, at archaeological levels corresponding to about 1200 B.C. In each case the inscriptions had been preserved by unintended firing of the clay. To date, over 5,000 tablets have been discovered.

Linear B was finally deciphered in 1952 by British architect Michael Ventris. Ventris' breakthrough showed that the writing's language was an early form of Greek—that is, when the symbols are correctly sounded out, they yield words that are Greek. This discovery confirmed the assumption (not previously proven) that the Mycenaeans had spoken Greek. The deciphered tablets were revealed to be palace records—inventories, lists of employees and administrative notes—that have shed valuable light on the organization and material culture of late Mycenaean society.

Evidently the Mycenaeans acquired Linear B writing by adapting Minoan Crete's Linear A script to the Greek language. Linear A—which records a vanished, non-Greek language—remains largely undeciphered to this day.

(See also ALPHABET; MINOAN CIVILIZATION; MYCENAEAN CIVILIZATION.)

literacy See WRITING.

literature See EPIC POETRY; LYRIC POETRY; THEATER; WRITING.

Locri City on the east coast of the "toe" of ITALY, founded circa 700 B.C. by Dorian-Greek colonists from LOCRIS, in central Greece. The city's official name was Lokroi Epizephurioi, "West Wind Locrians." After ejecting the native Sicels from the area, the Greek settlers established a well-run OLIGARCHY, called the Hundred Houses. Locri boasted the Greek world's earliest written law code, attributed to the statesman Zaleucus (circa 622 B.C.).

A sprawling community defended by a wall more than four miles long, Locri feuded with neighboring Italian peoples and with CROTON and RHEGIUM, two nearby Greek cities inhabited by non-Dorians. In around 500 B.C. the Locrians—supposedly aided by the divine twins CASTOR AND POLYDEUCES—won a famous victory over the Crotonians at the Battle of the Sagras River.

Locri's allies included the powerful Dorian-Greek city of SYRACUSE, in eastern SICILY. Circa 387 B.C. Locri helped the Syracusan tyrant DIONYSIUS (1) to destroy Rhegium. But Locri was later weakened by the wars that accompanied the expansion of ROME through the Italian peninsula, and the city surrendered to the Romans in 205 B.C. It remained an important town of Roman Italy.

(See also DORIAN GREEKS; LAWS AND LAW COURTS.)

Locris Bipartite region in the mountains of central Greece, divided by the states of PHOCIS and Doris. West Locris—or Ozolian Locris, named for a local tribe—bordered the northern Corinthian Gulf from NAUPACTUS eastward to Crisa, the harbor of DELPHI. The heart of West Locris was the fertile coastal valley of Amphissa. East Locris—or Opuntian Locris, named for its center at Opus—lay along the Euboean Straits on the mainland's east coast, from THERMOPYLAE southward to a border with BOEOTIA.

East Locris, governed as a 1,000-man OLIGARCHY, was the more advanced of the two regions. It sent colonists to found LOCRI, in southern ITALY, circa 700 B.C., and was minting coins in the 300s B.C. The Locrians were DORIAN GREEKS, but their dialect (called Northwest Greek by modern scholars) was one shared by many of the peoples dwelling around the western Corinthian Gulf.

(See also GREEK LANGUAGE.)

Long Walls, the Walls built between 461 and 456 B.C. at the urging of the Athenian statesman PERICLES, to connect ATHENS with its port city of PIRAEUS, four miles away. The completed Long Walls consisted of two parallel walls, about 200 yards apart. The result was to make Athens and Piraeus into a single, linked fortress, suppliable by sea and easily in contact with its navy. By building the Long Walls, Periclean Athens became the Greek naval power par excellence.

Greek poetry was originally meant to be sung to the music of a lyre or flute. The singer here accompanies himself on a cithara, an elaborate form of lyre, in a scene from an Athenian red-figure vase, circa 500 B.C. The cithara was played with a plectrum, which this singer holds in his right hand. He wears a long, Ionian-style tunic and a short cloak over his shoulder.

But the walls affected civilian concerns as well. The city was henceforth connected with the "party of the Piraeus"—the pro-democratic, lower-income citizens who supplied the bulk of the navy's crews. These people began exercising more power in the Athenian ASSEMBLY, resulting in the left-leaning Periclean legislation of these years, such as the introduction of jury pay (457 B.C.). This political association was so odious to certain Athenian right-wingers that in 457 B.C. a band of Athenian extremists hatched a treasonous plot to deliver the Long Walls to the Spartans. The scheme was discovered and averted.

After the outbreak of the PELOPONNESIAN WAR (431 B.C.), the Long Walls permitted Pericles to employ his tortoiselike land strategy of abandoning the countryside to Spartan invasion and collecting the rural Athenian populace within the city's fortifications. The area between the Long Walls became crowded with refugees, resulting in the disastrous plague of 430–427 B.C.

After Athens' defeat in the war, the Long Walls were pulled down by Spartan order, to the MUSIC of flutes (404 B.C.). The Athenian leader CONON rebuilt the walls in 393 B.C.; but after 322 B.C., under Macedonian overlordship, they gradually fell into disuse. Hardly any trace of them remains today.

(See also DEMOCRACY; OLIGARCHY; WARFARE, NAVAL; WARFARE, SIEGE.)

Lotus Eaters   See ODYSSEUS.

love   See APHRODITE; EROS; HOMOSEXUALITY; LYRIC POETRY; MARRIAGE.

Lyceum   Grove and GYMNASIUM outside ATHENS, sacred to the god APOLLO under his title *Lukeios*, "wolf god." It was at the Lukeion or Lyceum that ARISTOTLE opened his philosophical school, in 335 B.C. The Lyceum operated as a rival to the ACADEMY, which was the Athenian philosophical school founded by PLATO.

In around 286 B.C., after the death of Aristotle's successor, THEOPHRASTUS, the Aristotelian school relocated to new buildings at Athens, donated by Theophrastus in his will. According to one version, the new site included a colonnaded walk (Greek: *peripatos*), which gave the institution its famous name, the Peripatetic School. Theophrastus' successor, Straton of Lampsacus, was the last great thinker to preside over the school (circa 286–269 B.C.).

The Peripatetics suffered a decline in influence in the later 200s B.C. Whereas Aristotle and Theophrastus had overseen inquiries into every known branch of SCIENCE and PHILOSOPHY, later scholars focused mainly on literary criticism and biography writing; the movement acquired a reputation for pedantry. The decline may have resulted partly from the gradual disappearance of many of Aristotle's original writings, and when these were rediscovered and published in the first century B.C., the school did enjoy a partial revival. But by then the Peripatetics were being permanently overshadowed by other Greek philosophical movements, primarily STOICISM and EPICUREANISM, which were flurishing among the Romans.

Lycia   See ASIA MINOR.

Lycurgus (1)   Semilegendary, early Spartan lawgiver who founded the government and social organization of classical SPARTA. The Spartans revered the memory of Lycurgus (Greek: Lukourgos), yet today little is known of his life or even when he lived.

By the most plausible modern theory, Lycurgus arose as a political savior circa 665 B.C., after Sparta had been disastrously defeated in a war with ARGOS. In the decades following this defeat, Sparta completely reorganized itself as a militaristic state, devoted to the HOPLITE style of warfare and governed as a moderate OLIGARCHY built around the hoplite class of citizens. The name given to this new form of government was *eunomia*: "good law" or "good discipline." It is reasonable to see this swift, comprehensive change as the work of individual political genius.

In later centuries the Spartans preserved the text of a brief "commandment" (*rhētra*), supposedly written by Lycurgus as the basis for his reforms.

Lycurgus (2) Athenian orator and statesman of the mid-300s B.C. Known for his fierce public prosecutions of various enemies, Lycurgus was an able administrator, who received a special commission to oversee the city's finances. He was active in opposition to the Macedonian king PHILIP II, before Philip's conquest of Greece (338 B.C.). One speech by Lycurgus survives, *Against Leocrates* (331 B.C.).

(See also LAWS AND LAW COURTS.)

Lydia Wealthy and advanced non-Greek kingdom of western ASIA MINOR, centered at the city of Sardis, on the Hermus River. As a military and trading power—with riverborne GOLD and SILVER deposits, and access to eastern caravan routes—Lydia dominated Asia Minor in the 600s–500s B.C., before falling to the Persian conqueror CYRUS (1), in 546 B.C. Lydia's significance for Greek history lies in its relations, both hostile and friendly, with the Greeks of IONIA, on the coast of Asia Minor.

Lydia was the first non-Greek nation to take an active part in Greek affairs. King Gyges (circa 670 B.C.) sent gifts to the god APOLLO's shrine at DELPHI, in mainland Greece. King CROESUS, another patron of Delphi, also made a treaty of friendship with SPARTA (circa 550 B.C.). Yet in that same era the Lydian kings repeatedly attacked the Ionian cities, particularly MILETUS, coveting them as outlets to the sea. Finally submitting to Croesus, these Greek cities became privileged subject states under Lydian rule.

Lydia was probably the first nation on earth to mint coins (circa 635 B.C.). This invention was quickly copied by certain Greek cities.

Like certain other Asian peoples, the Lydians had a formidable CAVALRY; they also pioneered early techniques of siege warfare. But after Croesus succumbed to the Persian blitzkrieg, Lydia became a western province of the Persian Empire.

Although not independent thereafter, Lydia retained a cultural identity. After the conquests and death of ALEXANDER THE GREAT (334–323 B.C.), Lydia became a province of the Hellenistic SELEUCID EMPIRE. In the 100s B.C. Lydia passed briefly into the kingdom of PERGAMUM, before being absorbed into the empire of ROME.

(See also COINAGE; PERSIA; WARFARE, SIEGE.)

lyric poetry Like Greek EPIC POETRY, lyric had its origins in MUSIC. *Lurika mele* (lyric verse) meant a solo song in which the singer accompanied him- or herself on the stringed instrument known as the lyre. These brief songs were distinct from epic verse, which told traditional tales of the deeds of heroes, using an on-flowing rhythm (the hexameter). Lyric used other meters, to convey more personal or immediate messages. Of the earliest lyric poets whose verses have survived, SAPPHO (circa 600 B.C.) is the art's greatest practitioner.

Early on, the term *lyric* was enlarged to include different forms of brief song. One such was the elegy, which had its own distinctive meter and was sung to the music of a flute (*elegos*); obviously, for this type of song, the singer used an accompanist. One natural setting for the elegy was the aristocratic drinking party known as the SYMPOSIUM, where female SLAVES would provide flute music as well as other pleasures. The Ephesan poet CALLINUS (circa

640 B.C.) produced what are probably our earliest surviving elegiac verses, urging his countrymen to defend themselves against the marauding Cimmerians. The Spartan TYRTAEUS (circa 630 B.C.) wrote elegies in a similar vein. The elegies of ARCHILOCHUS (circa 640 B.C.) are more angry and personal, and those of ANACREON (circa 520 B.C.) reflect a life that was fun-loving and decadent. One subcategory of elegy was the epigram, which was typically a brief poem, composed for inscription in stone; the epitaph, or gravestone inscription, was one type of epigram.

Another specialized lyric meter was the *iambos*, used by Archilochus, HIPPONAX (circa 540 B.C.), and others for conveying satire and derision. Eventually iambic meter was adopted by Athenian playwrights as sounding natural for individual speaking parts in stage comedy and tragedy.

The term *lyric* also encompassed the category of choral poetry. These verses, in various meters, were written for public performance by a chorus—of men, boys, or girls—usually at a religious festival. Typically accompanied by lyre music, the chorus (usually 30 in number) would dance and present interpretive movements.

The many categories of choral lyric include the *paian*, (religious hymn of praise), the maiden song (*parthenaion*), the dirge (*threnos*), and the victory ode (*epinikion*). The greatest Greek choral poet, PINDAR (circa 470 B.C.), is best remembered for his victory odes, which celebrated various patrons' triumphs at such panhellenic festivals as the OLYMPIC GAMES. Another form of choral was the dithyramb, which was a song narrating a mythological story. At ATHENS, dithyrambs were performed at festivals of the god DIONYSUS and included the wearing of masks and the impersonation of characters in the story. By the mid-500s B.C., this practice had given birth to a crude form of stage tragedy.

(See also ALCMAN; BACCHYLIDES; CALLIMACHUS; SIMONIDES; STESICHORUS; THEATER.)

Lysander Leading Spartan commander of the later PELOPONNESIAN WAR and aftermath. He lived circa 445–395 B.C. First appointed as an admiral in 408 B.C., he pursued a strategy to capture the HELLESPONT waterway from the Athenians, and it was in the Hellespont that he won the sea battle of AEGOSPOTAMI (405 B.C.), which virtually won the war for SPARTA. After besieging ATHENS, Lysander received the Athenian surrender, occupied the city, and installed the puppet government of the THIRTY TYRANTS (404 B.C.). Similarly, he set up oligarchic governments in former Athenian ally cities.

However, the Spartan king Pausanias reversed Lysander's policies, abandoning the Thirty and permitting the restoration of DEMOCRACY at Athens and elsewhere (403 B.C.). Lysander became shut out of power by Pausanias and his colleague AGESILAUS. As a field commander at the outbreak of the CORINTHIAN WAR (395 B.C.), Lysander invaded BOEOTIA but was killed in battle against the Thebans.

Lysias Athenian orator, circa 459–380 B.C.. Of his reportedly 200 speeches, 32 survive today. These provide an important source for our knowledge of the Athenian law courts.

Like other orators, Lysias wrote speeches for clients but

did not personally plead in court. (He was forbidden to do so, since he was a resident alien, or metic). His father, Cephalus, a Syracusan by birth, had a lucrative shield manufactory in PIRAEUS. (It is this Cephalus who, with Lysias' brother Polemarchus, is featured in the opening episode of PLATO's fictional dialogue *The Republic*.) Evidently the family knew SOCRATES and moved in intellectual circles.

Under the THIRTY TYRANTS (404 B.C.), Polemarchus was put to death and the family's wealth confiscated. Lysias escaped to MEGARA (1), returning under the restored DEMOCRACY of 403 B.C. His major speeches, such as his *Against Eratosthenes*—in which Lysias prosecuted a surviving member of the Thirty—were delivered in the period after this return.

(See also LAWS AND LAW COURTS; METICS; RHETORIC.)

Lysimachus Macedonian general, and one of the DIADOCHI (successors) who carved up the empire of ALEXANDER THE GREAT. Born circa 360 B.C., Lysimachus was a friend and bodyguard of Alexander's. After Alexander's conquest of the Persian Empire and sudden death (323 B.C.), Lysimachus claimed a domain encompassing THRACE and northwest ASIA MINOR; his main enemy was ANTIGONUS (1), who sought to reconquer Alexander's empire from a base in Asia Minor. At the Battle of Ipsus (301 B.C.), in Asia Minor, Lysimachus and his ally SELEUCUS (1) destroyed Antigonus and divided his kingdom between themselves. By then Lysimachus was calling himself king and had built a royal capital, named Lysimacheia, on the west coast of the CHERSONESE.

In 285 B.C. he briefly captured MACEDON and THESSALY from Antigonus' son, DEMETRIUS POLIORCETES. But in 281 B.C. Lysimachus was defeated and killed by his former ally Seleucus at the Battle of Corupedium, in Asia Minor. Lysimachus' patchwork kingdom did not survive him, but was divided between the Seleucids and the Antigonids of Macedon.

Lysippus   See SCULPTURE.

M

Macedon Outlying Greek kingdom north of THESSALY, inland from the Thermaic Gulf, on the northwest Aegean coast. In modern reference, the name *Macedon* usually refers to the political entity, as opposed to the general territory called Macedonia. Macedon's heartland was the wide Thermaic plain, west of the modern Greek city of Thessaloniki, where the rivers Haliacmon and Axius flowed close together to the sea. The widely separated upper valleys of these rivers supplied two more regions of political and economic importance. Elsewhere the country was mountainous and forested. Its name came from an ancient Greek word meaning highlanders.

Macedon was inhabited by various peoples of Dorian-Greek, Illyrian, and Thracian descent, who spoke a Greek dialect and worshipped Greek gods. Prior to the mid-400s B.C. Macedon was a mere backwater, beleaguered by hostile Illyrians to the west and Thracians to the east, and significant mainly as an exporter of TIMBER and SILVER to the main Greek world.

Unification and modernization came gradually, at the hands of kings of Dorian descent. Alexander I "the Philhellene" (reigned circa 485–440 B.C.) began a hellenizing cultural program and minted Macedon's first coins, of native silver. The ruthless Archelaus I (413–399 B.C.) built forts and roads, improved military organization, chose PELLA as his capital city, and glorified his court by hosting the Athenian tragedians EURIPIDES and AGATHON (both of whom died in Macedon).

Macedon emerged as a major power in the next century. The brilliant Macedonian king PHILIP II (359–336 B.C.) created the best army in the Greek world and annexed territory in THRACE, ILLYRIS, and Greek CHALCIDICĒ, then subjugated Greece itself (338 B.C.). His son and successor, ALEXANDER THE GREAT (336–323 B.C.), conquered the Persian Empire and made Macedon, briefly, the largest kingdom on earth.

In the turmoil after Alexander's death, Macedon was seized by a series of rulers until the admirable king Antigonus II anchored a new, stable dynasty there (circa 272 B.C.). This century saw Macedon as one of the three great Hellenistic powers, alongside Ptolemaic EGYPT and the SELEUCID EMPIRE. Macedonian kings controlled Greece, with garrisons at CORINTH, at the Athenian port of PIRAEUS, and elsewhere. In 222 B.C. the Macedonian king Antigonus III captured the once-indomitable city of SPARTA.

To counter Macedonian domination, there arose two new Greek federal states, the Achaean and Aetolian leagues (mid-200s B.C.). The Macedonian king PHILIP V (221–179 B.C.) punished the Aetolians but came to grief against a new European power—the Italian city of ROME. Philip's two wars against Roman-Aetolian armies in Greece—the First and Second Macedonian wars (214–205 and 200–197 B.C.)—ended with Macedon's defeat and the Roman liberation of Greece (196 B.C.).

Macedon's last king was Philip's son Perseus (179–167 B.C.). After defeating Perseus in the Third Macedonian War (171–167 B.C.), the Romans imprisoned him and dismantled his kingdom. In 146 B.C. the region was annexed as a Roman province, called Macedonia.

(See also ACHAEA; AETOLIA; CASSANDER; DORIAN GREEKS; HELLENISTIC AGE; PHALANX; PYRRHUS.)

maenads In MYTH, the maenads (Greek: *mainadai*, "mad-women") were the frenzied female devotees of the god DIONYSUS. Usually imagined as Thracian WOMEN, the maenads symbolized the obliteration of personal identity and the liberation from conventional life that came with Dionysus' ecstatic worship. Clothed in fawn- or panther-skins, crowned with garlands of ivy, the maenads would rove across mountains and woods, to worship the god with

A figure of divine madness, this dancing maenad appears on an Athenian cup, painted in white-ground technique, circa 490 B.C. She grasps a leopard cub and the type of ivy-bound staff known as a thyrsus. Her headband is a live serpent.

dancing and song. In their abandon, they would capture wild animals and tear them limb from limb, even eating the beasts' raw flesh. Also known as the Bacchae, the maenads were female counterparts of the male SATYRS, but, unlike the satyrs, they were never comical figures. The maenads' rites are unforgettably portrayed in the Athenian tragedian EURIPIDES' masterpiece, *The Bacchae* (405 B.C.).

Although certain women's religious groups of central Greece did practice a midwinter ritual of "mountain dancing" (Greek: *oreibasia*) in Dionysus' honor, the maenads as presented in art and literature were not real. Rather, they were mythical projections of the self-abandonment found in Dionysus' conventional cult. Perhaps the idea of the maenads originated in savage, real-life Dionysian festivals in the non-Greek land of THRACE.

Magna Graecia See ITALY.

Mantinea Important city of ARCADIA. Located in the central Arcadian plain, Mantinea emerged circa 500 B.C. from an amalgamation of villages: the name *Mantineia*, suggesting holiness, may refer to a religious dedication at the city's founding. Mantinea was the implacable rival of its 10-mile-distant neighbor, TEGEA. The two cities fought over their boundary and the routing of the destructive water courses through the plain. Also, Mantinea tended to be the enemy of the mighty Spartans, who coveted Arcadia.

Mantinea became a moderate DEMOCRACY circa 470 B.C.—probably due to Athenian involvement—and in 420 B.C., during the PELOPONNESIAN WAR, Mantinea joined an anti-Spartan alliance of ATHENS, ARGOS, and ELIS. This led to a Spartan invasion and a Mantinean defeat at the first Battle of Mantinea (418 B.C.). In 385 B.C. the Spartan king AGESILAUS captured the city and pulled down its walls; but after the Spartans' defeat at LEUCTRA (371 B.C.), the Mantineans built the majestic fortifications whose remnants still stand today. In 362 B.C. Mantinean troops fought on the Spartan side and shared in the Spartan defeat at the second Battle of Mantinea, where the Theban commander EPAMINONDAS was killed in his hour of victory. This battle marked the final Theban invasion of the Peloponnese.

The city was captured and destroyed by the Macedonian king Antigonus III in 223 B.C., during his campaign that captured SPARTA. Mantinea was later reestablished under the name Antigoneia.

Marathon Town in the Athenian territory of ATTICA, three miles from the sea and about 26 miles northeast of ATHENS, along a main road. The coastal plain at Marathon saw one of history's most famous battles, when an Athenian army defeated a seaborne invasion force sent by the Persian king DARIUS (1) (summer 490 B.C.). The Persians, masters of western Asia, were thought to be invincible. The victory at Marathon saved Athens from subjugation and came to be viewed as a nearly mythical event—the gods were said to have smiled on the city and its fledgling DEMOCRACY. Among many later commemorations of the battle was a famous wall PAINTING in Athens' AGORA.

The name *Marathon* may have referred to wild fennel (Greek: *marathon*) in the area. More than 50 years before the battle, Marathon town played a role in the rise to power of the Athenian tyrant PISISTRATUS. Like other towns of eastern Attica, Marathon held traditional allegiance to Pisistratus' family, and it was at Marathon's hospitable coast that Pisistratus landed an army by sea, to capture Athens (546 B.C.). His son HIPPIAS (1) rode with the army that day. Hippias later succeeded his father as tyrant of Athens, but was ousted in 510 B.C. Making his way to PERSIA, he became one of King Darius' advisers. Then, in 490 B.C., Hippias returned with a Persian invasion fleet, which he guided to the shore near Marathon.

Darius had sent the Persian force to punish two mainland Greek cities, ERETRIA and Athens, for their role in the recent IONIAN REVOLT against Persia. Hippias, then nearly 80 years old, was intended as the Persians' puppet ruler in Athens. Apparently he had been in secret contact with partisans inside Athens who had promised to hand the city over to the invaders.

The ancient Greek historian HERODOTUS, who is the main extant information source for the Marathon campaign, says that the Persian fleet numbered 600 ships. A more accurate figure may have been 200. The fleet carried a land army of perhaps 20,000: armored spearmen, archers, and the formidable Persian CAVALRY, well suited to fighting on the Marathon plain.

After destroying Eretria, the Persian troops disembarked at Marathon to attack Athens. Against them came marching a HOPLITE army of about 9,000 Athenians, plus perhaps 600 from Athens' faithful ally PLATAEA. The Spartans too had promised to send aid, but their troops were delayed at home by certain religious obligations—a typical case of excessive Spartan piety. By the time they arrived, the fight was over.

The opposing armies faced each other for several days, retiring to their respective camps at night. The Greeks, who had no horsemen or archers, were unwilling to attack: They were outnumbered, probably two to one, and they feared the Persian cavalry. The Greeks' advantage was their BRONZE armor (heavier and more extensive than the Persians') and their organized hoplite tactics; man for man, they were to prove the better foot soldiers.

The Athenian command, at this point in history, consisted of 10 generals overseen by a *polemarchos*, or "war leader." Five of the generals wanted to march back to defend the walls of Athens; but the other five, led by the brilliant MILTIADES, convinced the polemarch, Callimachus, to attack. Apparently the timing involved a chance to catch the Persians without their cavalry; the Persian horsemen may have been reembarking to sail elsewhere, when the Athenian army charged.

Advancing quickly across the mile that separated the armies, the Greeks crashed into the Persian infantry ranks and sent both wings fleeing. In the center, the Persian spearmen stood firm but were destroyed when the two Greek wings—Athenians on the right and Plataeans on the left—wheeled inward in an unusual pincer movement.

The retreating Persians were slaughtered as they ran for their ships waiting inshore. The Greeks pursued them into the sea, right up to the ships. There Callimachus was killed. All but seven Persian ships got away, leaving behind 6,400 dead soldiers. The Athenians had lost 192. Later the

Athenian and Plataean dead were buried on the battlefield, under a mound still visible today.

A runner was sent to Athens to report the victory. (Twenty-five centuries later this run would be commemorated in the marathon footrace, first organized for the A.D. 1900 Paris Olympics.) Meanwhile, on distant Mt. Pentelicus, behind the battlefield, a shield was seen flashing in the morning sun—probably a heliographic signal to the Persian fleet. Presumably Hippias' Athenian partisans were reporting themselves ready to launch their coup d'etat. But no coup was ever attempted: The traitors were frightened off by news of the Athenian victory.

The Persian fleet sailed south and west around Cape Sunium, intending to land closer to Athens. But when they found the Athenian army waiting there, having marched from the battlefield, the Persians set sail for Asia. For the Persian command, the battle had demonstrated that future military action against mainland Greece would require a large army of occupation; 10 years later they would return with one.

The victory was a source of Greek religious awe. Legend claimed that the mythical Athenian hero THESEUS had been seen fighting alongside the Greeks and that an Athenian long-distance messenger had met the god PAN while running through the Arcadian wilds en route to SPARTA. For centuries afterward the battlefield was said to be haunted at night by the noise of combat.

Marathon inspired the confident outlook of a young generation of Athenians, who would defeat the Persians again (480–479 B.C.) and go on to forge an Athenian Empire. When the Athenian playwright AESCHYLUS died in 456 B.C., after a life full of honors, he recorded only one achievement in his epitaph: that he had fought at Marathon.

(See also OSTRACISM; PERSIAN WARS.)

marble See ARCHITECTURE; SCULPTURE.

Marmara, Sea of Modern name for the roughly diamond-shape saltwater body, 140 miles long and 40 miles wide at most, that connects the BOSPORUS and HELLESPONT seaways, along the northwestern coast of ASIA MINOR. The Greeks called this sea the Propontis—the "foresea" (i.e., in front of the BLACK SEA). The Propontis supplied a link in the crucial Greek TRADE route that brought metals and grain westward from the Black Sea.

Although not originally part of the Greek world, the Propontis had begun attracting Greek COLONIZATION by the early 600s B.C. Greek exploration is reflected (distortedly) in certain episodes in the legend of JASON (1) and the Argonauts.

The major Greek cities on the shores of the Propontis were CYZICUS and BYZANTIUM. The sea's abundant fish runs were commemorated in Cyzicus' COINAGE, which used a tuna as the city's emblem.

marriage Ancient Greeks were monogamous; only the Macedonian royal house and certain other marginal traditions allowed for multiple wives. Brides could be as young as 14 in certain eras and locales, but more likely 18. The groom might be as much as 20 years older than the bride.

The earliest form of marriage was a purchase: The groom acquired the bride for a price, paid to her father or male guardian. The bride's consent was not necessary. At ATHENS in the 400s–300s B.C. (the one ancient Greek society about which there exists considerable information), the exchange of money was usually a dowry, paid to the groom, from the bride's family. The marriage was considered complete upon the groom's ritualized acceptance of the dowry or promise thereof.

The goddess of marriage was HERA, but the wedding did not require any clergy. The principal ceremony, for families with means, was a wedding feast, typically held at the bride's father's house. Afterward the veiled bride left her father's home and accompanied her husband to his house. A procession of revelers followed, and when bride and groom entered the wedding chamber the well-wishers stood outside the closed door and sang the epithalamion (outside-the-bedroom song).

The purpose of marriage was to make children. In the Athenian marriage ceremony the bride's guardian announced, "I give this woman for the begetting of legitimate offspring." The question of legitimacy bore on citizenship: By laws prompted by the statesman PERICLES in 451 B.C., an Athenian citizen could legally marry only another citizen, and citizenship for children was confined to the legal offspring of such marriages.

The exclusive emphasis on procreation rather than love, in a society that otherwise devalued WOMEN, tended to make Greek marriages unromantic. Husbands might find adventure outside marriage—with SLAVES, PROSTITUTES, or in the homosexual pursuits of the GYMNASIUM and other male gathering places. The husband's businesslike view of marriage is conveyed in the words of a male Athenian courtroom speaker of the mid-300s B.C.: "Prostitutes we have for our pleasure, concubines for our daily refreshment, but wives to give us legitimate children and be faithful guardians of our homes" (*Against Neaera*, 122; ascribed to DEMOSTHENES (1)).

At Athens, a husband's extramarital liaisons were not punishable as adultery unless they happened to involve another man's wife. Adultery was defined as the liaison of a wife with a man not her husband—the offense was that it defied the husband's authority and created the possibility of illegitimate children. In cases of adultery, both the erring wife and her male lover could be severely punished. (Due to a lack of evidence, it is not known what the classical Athenians' attitude might have been toward wives taking female lovers, but it too was probably very disapproving. In any case, a wife's secluded, housebound life offered few chances to find lovers of any type.)

Greek divorce was simple. A husband could just dismiss his wife from the house, or the wife could decide to walk away. But to petition for return of her dowry, a woman had to present herself in a court of law, which was biased in favor of her husband. For example, when Hipparete, the estranged wife of the Athenian statesman ALCIBIADES (circa 420 B.C.), arrived at court for this purpose, Alcibiades appeared, lifted her up, and carried her home unchallenged.

(See also ADONIS; ASPASIA; HOMOSEXUALITY; LAWS AND LAW COURTS.)

Massalia Greek seaport in southern France, located just east of the mouth of the Rhone River; in modern times the major French city of Marseille. Massalia was established circa 600 B.C. by long-range Greek seafarers from PHOCAEA, in ASIA MINOR; the founding was associated with a sea battle in which the Phocaeans defeated rival Carthaginians in the area.

The Ligurian tribes of the region apparently welcomed the Greeks. Massalia prospered through TRADE with the Ligurians as well as with the CELTS of interior Gaul and the Celtic kingdom of Tartessus in southern Spain. The Massaliote Greeks soon established their own Mediterranean colonies, including Nicaea (modern Nice) and Emporion (meaning "seaport"—modern Ampurias, in southern Spain). These contacts supplied the raw metals—SILVER, IRON, tin, and lead—that the Massaliotes craved for manufacturing or lucrative export to Greece. In exchange the Greeks offered luxury goods, such as worked BRONZE. Greek goods made their way into the continent: Modern ARCHAEOLOGY has discovered Greek bronzes of the 500s B.C. (cauldrons, a helmet, vasehandles) in southern Spain, southern France, and at a site 100 miles southeast of Paris.

Another important Greek export was WINE, the key that unlocked the wealth of native peoples. Like the French fur traders of the 1600s A.D. who took brandy into the Canadian wilderness to sell to Indians for pelts, so did the ancient Greeks take Greek wine into interior Gaul, introducing it among the Celtic tribes. A later Greek writer (circa 20 B.C.) described how cheaply the Celts sold their SLAVES for wine, "trading a servant for a drink." In time, the peoples of southern France learned how to cultivate their own grapes. The creation of the French wine industry is not the least cultural legacy of the ancient Greeks.

This infusion of Greek influences up the Rhone in the 500s B.C. played a vital role in shaping the Celtic "La Tène" culture that was to emerge in Gaul in the next century. From the Greeks, the Gauls learned improved methods of FARMING, fortification, and local administration. They copied the Greek ALPHABET, which Julius Caesar would find still in use throughout Gaul in 59 B.C. Aspects of the Gauls' religious beliefs may have been shaped by images of the Greek gods. One ancient writer observed that it seemed as if Gaul had become part of Greece.

Circa 310–306 B.C. a Massaliote seafarer named Pytheas sailed through the Straits of Gibraltar and around the British Isles, perhaps reaching as far as Norway, in search of the sources of traded tin and other goods. Later ancient writers plausibly describe his exploration.

Massalia was known for the stability of its aristocratic government. Remote from the agitations of mainland Greece, the city thrived without social conflict, while keeping contact with the Greek homeland through such means as a Massaliote treasury house at DELPHI. For centuries, Massalia's only principal enemy was CARTHAGE, and Massalia finally allied itself with the Italian city of ROME against Carthage in the Second Punic War (218–201 B.C.). In 125 B.C. hostility from local tribes compelled Massalia to appeal to Rome, resulting in the Roman occupation of the region. The Romans called this territory "the province," Provincia—modern Provence.

(See also ARISTOCRACY; SHIPS AND SEAFARING.)

mathematics The study of mathematics represents the most permanent achievement of ancient Greek thought. Many aspects of modern mathematics, particularly the rules and nomenclature of geometry, derive from the Greeks.

The Greek term *mathematikē* simply means "learning." From the 500s B.C. the Greeks saw mathematics as an ideal study and exercise of pure intellect, relevant to PHILOSOPHY. The mystic philosopher PYTHAGORAS (circa 530 B.C.) sought the secrets of the universe in numbers, while PLATO (427–347 B.C.) undoubtedly found inspiration for his theory of Forms in the analogy of perfect mathematical concepts. The Greeks were not the only ancient mathematicians—the Babylonians (especially) and Egyptians were also masters in the field—but the Greeks' abiding contribution was their practice of devising rigorous proofs for their theorems.

Like modern peoples, the Greeks counted by 10s, but their numerical symbols were awkward. (Arabic numerals had not been invented yet.) By the 400s B.C. Greek numbers were represented mainly by modified letters of the Greek ALPHABET in accumulative combinations. For example, 6,000 was written as ⌐X; 6,852 was ⌐X ⌐HHH⌐II. While arithmetic calculations could be difficult, plane and solid geometry came more naturally to the Greeks, partly from the age-old inspiration of FARMING patterns and boundary demarcation—the Greek word *geometria* means "land survey." The study of geometry was at the heart of early Greek ASTRONOMY, as brilliant mathematicians such as Eudoxus of CNIDUS (circa 350 B.C.) and Apollonius of Pergē (circa 200 B.C.) devised geometric models to explain the heavenly bodies' movements.

As for the accumulation of mathematical knowledge, a brief sketch must suffice here. By about 500 B.C. Pythagoras had discovered the geometric theorem that still bears his name, which states that the squared value of the hypotenuse of a right triangle is equal to the sums of the squares of the two adjacent sides. The very concept of square numbers was a Pythagorean discovery; the category was used for numbers that could be shown as dots forming a square, such as:

These and other principles probably had been uncovered by the Babylonians centuries before, but the Greeks most likely made independent discoveries rather than somehow copying Babylonian technique.

Most likely, the earliest Greek mathematical textbook was the now-lost *Elements of Geometry* by Hippocrates of CHIOS, circa 430 B.C., (He is not be confused with his contemporary, the physician HIPPOCRATES (1) of Cos.) Mathematics were emphasized at Plato's Athenian school of higher learning, the ACADEMY (founded circa 385 B.C.). The most important mathematician of the day was Plato's pupil Eudoxus, who lived circa 390–340 B.C. None of his writing survives, but he apparently developed the general laws of geometric proportion that now comprise Euclid's book 5. Eudoxus is probably the one who discovered that a cone will always cover one-third of the area covered by a cylinder with the same base and height.

By 300 B.C. the Greek-Egyptian city of ALEXANDRIA (1), with its royally endowed Library and Museum, had emerged as a new Greek center of learning. Among the city's mathematicians was Euclid (Eukleidēs), who wrote a textbook titled *Elements* (*Stoicheia*), which survives today. Divided into 13 "books," the *Elements* sets forth the total mathematical knowledge of the day. Euclid was not an original thinker, but he did provide a systematic statement of definitions and problems, featuring clear and methodical proofs. His terminology and methods became canonical, and not just for the ancient world. The *Elements'* first six books, on topics in plane geometry, were used in barely revised form as a textbook for European and North American schools up to the early 20th century.

The next century saw the acme of Greek mathematics in the work of two geniuses, ARCHIMEDES of SYRACUSE (circa 287–212 B.C.) and Apollonius of Pergē (active circa 200 B.C.). Although best remembered today as an inventor and physicist, Archimedes did groundbreaking mathematical work; his extant writings include part of *On the Sphere and the Cylinder, On Conoids and Spheroids*, and the *Sand Reckoner* (in which he invented a system for denoting very large numbers). Of all his life's achievements, he is said to have been most proud of a certain geometric proof, namely, that a cylinder circumscribing a sphere will always have an area in ratio 3–2 to the sphere.

Apollonius was born at a Greek city of southern ASIA MINOR and studied mathematics at Alexandria with the followers of Euclid. Later he sought the patronage of King Attalus I of PERGAMUM. Apollonius' major treatise, *Conics*, has survived in most part and provides a milestone in post-Euclidean solid geometry. From Apollonius, for example, come the modern names by which the three types of conic sections are known: ellipse, parabola, and hyperbola.

The most famous practical application of mathematics in the ancient world was that of Eratosthenes of CYRENE (1), who was head of the Alexandrian Library circa 245–194 B.C. and who used simple geometric principles to calculate the earth's polar circumference. Eratosthenes discovered that an upright stick would cast no shadow at noon on the summer solstice at Syene (modern Aswan, in southern EGYPT), while at the same moment a stick at Alexandria would cast a small shadow at a 7.2-degree angle. Believing Alexandria to be due north of Syene at a distance of 5,000 stades, Eratosthenes multiplied 5,000 by 50, in ratio to the calculation 50 × 7.2 = 360. To the resultant figure of 250,000 stades he added 2,000 more, to offset possible error. Assuming Eratosthenes' stade measure to equal 200 yards, then his figure for the earth's north-south circumference equals about 28,636 miles; assuming a stade measure of 175 yards (as seems more likely, from the Alexandria–Syene figure), then Eratosthenes' figure equals about 25,057 miles. This is remarkably close to the modern measurement of 24,805 miles.

(See also SCIENCE.)

Mausolus See HALICARNASSUS.

Medea In MYTH, the daughter of King Aeëtes of Colchis and the niece of the witch CIRCE. Medea herself was a sorceress; her name, Medeia, means "cunning." In the story of JASON (1) and the Argonauts, as told by the epic poet APOLLONIUS (circa 245 B.C.), the young Medea fell in love with the handsome hero when he arrived at Colchis in search of the Golden Fleece. Against her father's interest, she helped Jason with magic ointments and instructions, and escaped with him and the fleece aboard the ship *Argo*.

Later, on arriving at the city of Ioclus, in THESSALY, Medea helped Jason avenge himself on his enemy, King Pelias. First, she boiled a cauldron of water and magic herbs for Jason's old father, Aeson. After submerging himself in the boiling water, Aeson emerged as a young man. Medea then suggested that Pelias' daughters do the same for him. But this time she supplied useless herbs, and when the king entered his daughters' bath he was boiled to death.

Ousted from Ioclus by Pelias' son, Jason and Medea settled at CORINTH, where they had a son and daughter. What followed is the subject of EURIPIDES' tragedy *Medea* (431 B.C.). Learning that Jason planned to desert her and marry Glauce—daughter of the Corinthian king—Medea went insane with anger. She sent Glauce a robe and tiara smeared in poison, resulting in the death of both the bride and her father. Next, to avenge herself fully on Jason, Medea stabbed both of their young children to death, then flew off in a magic chariot, taking refuge with the Athenian king Aegeus (father of THESEUS). Euripides' *Medea* provides an almost modern study of pathological jealousy and hatred, emanating from a woman wronged.

Medes See PERSIA.

medicine Today knowledge of Greek medicine comes largely from the surviving works of medical writers of antiquity, from references in the works of nonmedical writers, and from artwork depicting medical scenes. The greatest Greek medical writer was Galen of PERGAMUM, who lived in the 100s A.D. In the centuries before Galen, existing medical knowledge was organized and written down by HIPPOCRATES (1) of Cos and his followers (late 400s B.C. and after) and by various physicians based at ALEXANDRIA (1).

To a surprising degree, the Greeks distinguished medicine from RELIGION or superstition. Although APOLLO, ASCLEPIUS, and other gods were believed to have healing powers—and although crude healing magic existed—such beliefs did not prevent the development of scientific methods. Since the Greeks lacked such modern marvels as antibiotics and anesthesia, surgery was a drastic and dangerous recourse. There was also no knowledge of microorganisms as the cause of disease. But a knowledge of dietetics, healing herbs, primitive orthopedics, and techniques of bandaging provided the beginnings of Greek medicine in the centuries prior to 400 B.C.

There was no medical licensing in ancient Greece. Physicians were considered to be on par with skilled craftsmen such as poets or architects. The historian HERODOTUS mentions a famous Greek doctor, Democedes of CROTON, who in the late 500s B.C. became the court physician to the tyrant POLYCRATES of SAMOS and then to the Persian king DARIUS (1). But it is Hippocrates (circa 460–377 B.C.) who provides the first milestone.

On his native island of Cos, Hippocrates in circa 430 B.C. established a school to create standards of medical and ethical procedure: The modern "Hippocratic Oath" is one legacy of this school. The Hippocratics emphasized careful observation and diagnosis. Among the writings known as the Hippocratic Collection (produced by Hippocrates' followers in the late 400s and early 300s B.C.), the two books entitled *Epidemics*—presenting 42 case histories of severely ill patients—are unequaled in clarity by any extant European medical writing prior to the 16th century A.D. Modern readers of the *Epidemics* are able to identify cases of diphtheria and typhoid fever from the symptoms·described. It is a testament to the medical ignorance of the day, and to the honesty of the writer, that about 60 percent of the *Epidemics'* cases end with the patient's death.

Another Hippocratic writing is a treatise whose title is traditionally translated as *On the Nature of Man*. Written circa 400 B.C., this work presents the famous (but incorrect) theory of the Four Humors, destined to dominate medical thought for the next 2,000 years. According to the theory, the human body consists of four elementary fluids—blood, phlegm (*pituita*), yellow bile (*cholē*), and black bile (*melancholia*)—whose correct proportion maintains health and whose imbalance causes illness. One crude procedure of the day was to apply heated metal cups to a patient's skin, in an attempt to draw off any excessive humors; gravestones and other surviving artwork sometimes show the physician with two metal cups, as a badge of profession. The practice of bleeding a patient, to purge "excessive" blood, was another misguided procedure inspired by the Four Humors belief.

The 300s B.C. saw a growing interest in anatomy, due partly to the biological studies of ARISTOTLE and his school, the LYCEUM. Anatomical knowledge was greatly advanced by two Alexandrian physicians of the middle and late 200s B.C.—Herophilus of Chalcedon and Eristratus of Ceos, whose respective writings survive only in fragments but whose work is described by the Roman writer Celsus (first century A.D.). Working separately, Herophilus and Eristratus were among the first to perform dissection on the human body—and, according to rumor, they also practiced human vivisection in the name of medical knowledge, using condemned criminals handed over by royal permission. These doctors and their colleagues added greatly to the knowledge of human circulation and respiration, and several of the anatomical terms coined by them survive in use today.

(See also SCIENCE.)

Medusa In MYTH, a winged female monster with hair of snakes and a face that caused any humans who looked at it to turn to stone. Medusa ("cunning") was mortal but had two immortal sisters, Sthenno ("strong") and Euryale ("wide jumping"). Known collectively as the Gorgons ("grim ones"), they lived together in the far West. With the aid of the goddess ATHENA, the Greek hero PERSEUS cut off Medusa's head and put it in a bag. The head—with its abiding ability to petrify an onlooker—aided Perseus in his subsequent adventures. The head was later given to Athena, who set it in the center of her cloak known as the aegis.

Clearly, in early Greek or pre-Greek RELIGION, Medusa and her sisters had an apotropaic function; that is, they were worshipped or represented as protective demons who could scare away evil from the community. The protective head of Medusa was said to be buried under the AGORA at Perseus' city, ARGOS. By the 600s B.C., Medusa (or an unspecified Gorgon) had became a favorite subject in PAINTING and SCULPTURE, particularly at CORINTH and at Corinthian colonies such as CORCYRA. Portrayed as a winged running demon grimacing at the viewer, the monster would be shown in vivid colors. Probably the best-known surviving example is the Gorgon pediment sculpture from the temple of ARTEMIS at Corcyra (circa 600–580 B.C.).

Megara (1) City of south-central Greece, located in the widening, northeastern isthmus region known as the Megarid. Lying in a fertile valley near the Saronic Gulf and the island of SALAMIS (1), Megara is surrounded on three sides by mountains, which separate the city from the regions of ATHENS (to the north east), CORINTH (to the southwest), and THEBES (to the north). In particular, the southwestern mountain range, the Geranea, formed a natural barrier between central Greece and the PELOPONNESE, and was of strategic importance in Greek history.

After a promising start, the city slipped into a largely unhappy pattern of resistance to its powerful neighbors. Megara was founded by DORIAN GREEKS who had invaded the Peloponnese (circa 1100 B.C.). Unlike older Greek cities founded by pre-Greek inhabitants, Megara has a name that actually means something in Greek: "the great hall." In the age of TRADE and COLONIZATION (circa 800–550 B.C.), Megara was a foremost power, exporting its prized woolens and establishing the colonies Megara Hyblaea, on the east coast of SICILY (circa 750 B.C.), and Heraclea Pontica, on the south shore of the BLACK SEA (circa 560 B.C.). In the LELANTINE WAR (circa 720–680 B.C.), Megara sided with ERETRIA, the enemy of Megara's neighbor and rival, Corinth.

Like other commercial cities of this era, Megara came under the sway of TYRANTS. A certain Theagenes (circa 630 B.C.) seized power after leading the common people in slaughtering the nobles' cattle. Theagenes was the father-in-law of the Athenian aristocrat CYLON and seems to have played a role in Cylon's failed coup at Athens (circa 625 B.C.). Later the tyrant fell.

Engulfed by class war between nobles and commoners, the city declined, losing border land to Corinth and the coveted island of Salamis to Athens (mid-500s B.C.). The poems of the Megarian aristocrat THEOGNIS (circa 540 B.C.) are filled with hatred toward the common people, signifying the discord that drained Megarian power in these years.

As part of the Spartan alliance, the Megarians fought on sea and land against the Persian king XERXES' invasion of Greece (480–79 B.C.). Later, after seeking help against Corinth in a border dispute, Megara allied itself with Athens (460 B.C.). The Athenians occupied the region and helped set up a DEMOCRACY at Megara. They also built "long walls," extending the mile's distance between Megara and its port of Nisaea, on the Saronic Gulf. The walls turned Megara and Nisaea into a linked fortress, suppliable directly by sea and impregnable by land.

Ironically, Megara's defenses came into use against Athens. In the 430s B.C. the Megarians withdrew from the Athenian alliance, encroached on Athenian territory, and murdered an Athenian herald. The Athenian statesman PERICLES responded with the notorious Megarian Decree, which attempted to starve the Megarians into submission by placing embargoes on their food imports and their export trade. This hostile act alarmed the Spartans and other Greeks, and played a role in causing the PELOPONNE-SIAN WAR (431–404 B.C.).

As a Spartan ally near Athenian territory in the war, Megara was a constant target for the Athenians, who invaded the Megarid with a huge force in 431 B.C. and kept invading—without capturing the city—every year until 421 B.C. In 424 B.C. a faction within Megara attempted to betray the city to the Athenian general DEMOSTHENES (2); he seized Nisaea and the Long Walls, but Megara itself was saved by the Spartan general BRASIDAS.

A brief phase of prosperity came in the 300s B.C., when the city was host to a school of SCULPTURE and to the Megarian School of PHILOSOPHY (founded by Eukleides, a disciple of the Athenian SOCRATES). Thereafter, although still inhabited, Megara fades from history.

Megara (2)   See HERACLES.

Meleager   See CALYDONIAN BOAR HUNT.

Melian Dialogue   See THUCYDIDES (1).

Melos   Small island of the western AEGEAN SEA, in the island group known as the CYCLADES. ARCHAEOLOGY reveals that in the pre-Greek "Cycladic" culture of the third millennium B.C. Melos was the exclusive source of obsidian—a volcanic glass that when quarried and honed, supplied axblades, plowshares, and the like, in the days before general use of metals. By the 1600s B.C. Melos was an ally or subject of Minoan CRETE and by 1400 B.C. was part of the Aegean empire of the Mycenaean Greeks.

Circa 1000 B.C. Melos was occupied by DORIAN GREEKS from LACONIA. With its capital city, also called Melos, in the north of the island, Melos existed as a Dorian state amid the predominantly Ionian-Greek Cyclades. By about 475 B.C. the Cyclades were the heart of the Athenian-dominated DELIAN LEAGUE, of which Melos was not a member.

Melos remained neutral at the outbreak of the PELOPON-NESIAN WAR between ATHENS and SPARTA (431 B.C.). The Athenians attacked Melos in 426 B.C., and in 416 B.C. they again landed there, demanding that Melos join the Delian League. The Melians refused, preferring to look for help from their kinsmen the Spartans. This "Melian Dialogue" provides one of the set pieces of THUCYDIDES' (1) history of the Peloponnesian War (book 5). Although the dialogue often has been interpreted as an argument between Might and Right, it seems clear that Thucydides' heart is with the Athenians, who deliver their argument in terms of expedience and common sense. In any case, Melos city was besieged and captured by the Athenians, who enslaved and deported the islanders and planted an Athenian garrison colony there.

The celebrated statue called the Venus de Milo, now in the Louvre, is a marble APHRODITE discovered at Melos in A.D. 1820. ("Milo" being the Italian form of Melos.) Dated to the 100s B.C., it is probably a copy of a lost original of the 300s B.C..

(See also BRONZE AGE; MINOAN CIVILIZATION; MYCEN-AEAN CIVILIZATION.)

Memnon   In Greek MYTH, Memnon was an Ethiopian king, son of the dawn goddess Eos and the mortal man Tithonus. Memnon, whose name means "resolute," led a contingent of his countrymen to TROY, to help defend the city against the Greeks in the TROJAN WAR. In battle Memnon slew NESTOR's son Antilochus, but was himself killed by the Greek champion, ACHILLES. The dawn goddess carried away her son's corpse, and the great god ZEUS revived him as an immortal in heaven. The story of Memnon was told in a now-lost epic poem, the *Aithiopis*.

Memnon was a favorite subject of Greek vase painting in the 500s and 400s B.C. Like other mythological figures from Ethiopia, he was sometimes shown with distinctly negroid facial features. During the HELLENISTIC AGE (300–150 B.C.), the Greeks in EGYPT associated Memnon with certain preexisting monuments of the ancient pharaohs, including the so-called Colossi of Memnon—two huge sandstone statues of Pharaoh Amenophis III, seen at Luxor today.

(See also BLACK PEOPLES.)

Menander   Athenian comic playwright of the late 300s–early 200s B.C. Menander (Greek: Menandros) is the last great Athenian literary artist whose writings have survived from antiquity. He wrote over 100 plays, yet only one complete work exists today: *The Misanthrope* (Greek: *Duskolos*). Portions of other plays are known from quotations by later ancient authors and by modern archaeological discoveries of papyrus remnants in EGYPT, where Menander was apparently a favorite author of the Greco-Roman educated classes during the Roman Empire.

Little is known of his life. Born in 342 or 341 B.C., he grew up in an ATHENS that was well past its imperial prime and was dominated (like all of Greece) by the powerful kingdom of MACEDON. Menander apparently moved in the highest circles of Athenian society and politics. He studied under the Aristotelian philosopher THEOPHRASTUS and was a friend of Demetrius of Phalerum, an Athenian who served as governor for the Macedonian king CASSANDER. Demetrius' overthrow in 307 B.C. is said to have put Menander in temporary danger.

In 321 B.C., at about age 20, Menander won his first drama-victory, when his comedy *Anger* (not extant today) took first prize at an Athenian festival. His surviving masterpiece, *The Misanthrope*, won first prize in 316 B.C., at the midwinter festival called the Lenaea. Among his plays that survive in part today are *The Woman from Samos*, *The Hated Lover*, and *The Girl with the Short Haircut*.

Menander died in around 291 B.C., having won first prize eight times—a respectable but not dazzling record. However, his reputation grew soon after his death, and he came to be considered one of the classic Athenian authors. His plays served as models for the work of the Roman comic playwrights Plautus (circa 200 B.C.) and Terence (circa 150 B.C.).

For modern readers, *The Misanthrope* was discovered on an ancient papyrus only in A.D. 1958. (The French comic playwright Molière may have known the title but did not have the play as a model for his own *Misanthrope*, performed in A.D. 1666.) Set in the Athenian countryside, Menander's fanciful tale involves an old farmer, Knemon, who loathes humanity and lives, secluded, with his daughter. A rich young man, falling in love with the daughter, poses as a rustic laborer in order to win her hand in MARRIAGE. He succeeds after pulling Knemon out of a well where the old man has fallen.

Ancient and modern scholars have seen Menander's work as the high point of the "New Comedy" (circa 321–264 B.C.), which marks the last phase of evolution for Athenian stage comedy. "Old Comedy," as developed by ARISTOPHANES and others in the 400s B.C., was the product of the powerful Athenian city-state; Menander and the New Comedy are products of the more cosmopolitan and uncertain HELLENISTIC AGE, where most citizens lived cut off from major political events. Whereas Old Comedy presented fantasy, political satire, and obscene farce, New Comedy offered realistic settings and domestic plots, often "boy meets girl." In keeping with other conventions of New Comedy, Menander's plays show a great reduction in the onstage role of the comic chorus, which is used only for song-and-dance interludes.

(See also THEATER.)

Menelaus In MYTH, king of Lacedaemon and younger brother of King AGAMEMNON of MYCENAE. Menelaus was married to the Spartan princess Helen, later known as HELEN OF TROY. At the prompting of the love goddess APHRODITE, Helen abandoned Menelaus to elope with the Trojan prince PARIS. It was to punish this outrage that Agamemnon organized the Greek expedition against TROY, in the TROJAN WAR.

In the *Iliad* (book 3) Menelaus agrees to settle the entire war by single combat with Paris. Menelaus overwhelms his rival, but is deprived, by Aphrodite, of the chance to kill him. In the *Odyssey* Menelaus is shown briefly, back home with Helen and reconciled with her.

Messana See ZANCLE.

Messenia Southwest region of the PELOPONNESE. Messenia's eastern frontier is separated from the Spartan region of LACONIA by the lofty Mt. Taygetus range; on the west and south, Messenia is bounded by the sea; and in the north, it borders ELIS and ARCADIA. Largely mountainous, Messenia includes the westernmost of the Peloponnese's three southern peninsulas, Cape Acritas.

The heart of the region is the fertile Messenian plain, opening southward around the River Pamisus to the Mes-

senian Gulf. On the west coast, a second, narrower plain was home to a thriving kingdom in Mycenaean times, circa 1400–1200 B.C. Later Greeks remembered this kingdom by the name PYLOS and associated it with the mythical king NESTOR. This domain vanished prior to or during the invasion of the DORIAN GREEKS, circa 1100–1000 B.C., and consequently, Messenia became a Dorian region.

Messenia had a tragic history, being the most complete victim of Spartan domination. In the 700s B.C. the Spartans, crossing the Taygetus range, began their long campaign of conquest. The First and Second Messenian wars (circa 730–710 B.C. and circa 650–620 B.C.) were bitter conflicts; the Messenian leaders Aristodemus and Aristomenes were later remembered for their failed, heroic defense. By 620 B.C. the Messenian plain was in Spartan hands, and those Messenians who had not fled were virtually enslaved as Spartan-owned HELOTS.

Under the helot system, the Messenians were serfs, owned in common by the Spartan state. They were not deported to Sparta, but were left on the Messenian land (which they no longer owned), to farm it and produce food for their overlords. Half of all their produce went to Sparta.

The Messenians were allowed to maintain their local cults, customs, and family structures, but they were brutalized and terrorized. The Spartan poet TYRTAEUS writes of a Messenian's hopeless drudgery and his dutiful mourning when his Spartan master dies. Discipline was maintained by Spartan garrisons and by the Krypteia—the Spartan "secret society" whose job was to identify and do away with subversive helots. The Spartans' ferocity toward the Messenians—who were, after all, their fellow Dorian Greeks—comprised the single most warped aspect of the Spartan mentality.

The Third Messenian War was a rebellion sparked by news of a devastating earthquake at Sparta (464 B.C.). The revolt died with the surrender of the Messenian hilltop fortress of Ithome ("the step"), circa 460 B.C. At least some of the Messenian defenders were permitted to depart under safe-conduct, and the Athenians later relocated them to NAUPACTUS, a city on the northwest shore of the Corinthian Gulf.

During the PELOPONNESIAN WAR (431–404 B.C.) Messenian helots served their Spartan masters loyally while, on the other side, refugee Messenians fought as valuable Athenian allies. After the Athenian victory at the site called Pylos, on Navarino Bay (425 B.C.), the Athenians built a fortified naval base there. But the hoped-for Messenian revolt against Sparta never materialized.

Messenia was finally liberated in 369 B.C. by the Theban statesman and general EPAMINONDAS, after his destruction of a Spartan army in 371 B.C. Under Epaminondas' guidance, the Messenians founded a city, Messene (modern Messini), to be the capital of the new, free territory. Messene and Messenia thrived for a century, but suffered in the 200s and 100s B.C. from the intervention of MACEDON and the Achaean League, before passing into the hands of ROME after 146 B.C..

(See also ACHAEA; DEMOSTHENES (2); ZANCLE.)

meter See EPIC POETRY; LYRIC POETRY.

metics Resident aliens of legal status. A metic (Greek: *metoikos*, "dweller among") was usually a Greek who had immigrated to a Greek city other than his birthplace. Metics abounded in the wealthier and more populous Greek cities, but their existence is best attested at ATHENS.

The rules for acquiring metic status and the legal restrictions applying to metics somewhat resembled those surrounding resident aliens in the United States today. The applicant had to be sponsored by a citizen, had to register in an Athenian DEME (city ward), and had to pay a special annual tax. Metics owed the state certain public duties and military service, usually as crewmen in the Athenian navy. They could not set foot inside (much less vote at) the Athenian political ASSEMBLY, usually could not own land, and could not marry an Athenian citizen. In exchange, metics enjoyed the courts' protection, the right to engage in business, and a recognized position in the community. At Athens, much of the import-export and manufacturing was in the hands of metics, some of whom became very rich. One example is the Syracusan-born Cephalus, father of the orator LYSIAS. The most famous person who was a metic at Athens was ARISTOTLE.

(See also POLIS.)

Midas Last king of the wealthy, non-Greek nation of Phrygia, in ASIA MINOR. Midas reigned through the late 700s B.C. and was said to be the first non-Greek to send gifts to the god APOLLO's shrine at DELPHI, that is, the first foreign ruler to open diplomatic relations with mainland Greece. He also was said to have married the daughter of the king of the east Greek city of CYMĒ, in Asia Minor. Midas died when his kingdom was overrun by the nomadic Cimmerians, circa 696 B.C.

On account of his fabulous wealth, Midas—like the later king CROESUS of LYDIA—became a figure in Greek legend. Best known is the tale of the Midas Touch. Offered any wish by the god DIONYSUS, Midas asked that whatever he touched might turn to GOLD. But, finding that this ruined the food that he touched, he prayed to lose the gift. In another version, it was his daughter whom he accidentally turned to gold.

Miletus Greek city on the central west coast of ASIA MINOR, preeminent in TRADE, COLONIZATION, and cultural achievements in the 600s–500s B.C. Situated at the mouth of the River Maeander, Miletus enjoyed communication seaward and inland. According to archaeological evidence, this advantageous site was occupied in the second millennium B.C. first by Minoans, then by Mycenaean Greeks, and last by non-Greek Carians. Around 1000 B.C. IONIAN GREEKS arrived and founded the historical Miletus, which became the southernmost city of the Greek region known as IONIA. The city's patron god was APOLLO.

Expert seafarers, the Milesians led the way in the expansion of the Greek world in the 600s B.C. They exported prized woolens and metalwork, and founded a remarkable string of colonies along the trade route to the BLACK SEA, as far as distant Crimea.

By the late 600s B.C. Miletus was ruled by TYRANTS. In the mid-500s B.C. Miletus and the rest of Ionia were conquered by King CROESUS of LYDIA, only to be conquered shortly thereafter by the Persian king CYRUS (1). From about 546 to 499 B.C. Miletus enjoyed privileged status under the Persians, who maintained control through a series of Greek puppet rulers.

In the 500s B.C. the wealthy city witnessed the twin birth of PHILOSOPHY and SCIENCE, as the Milesian School of natural philosophers—THALES, ANAXIMANDER, and ANAXIMENES—took the first, revolutionary steps toward explaining the universe in nonreligious terms. Another innovative Milesian thinker of the day was the geographer HECATEUS.

But the city's heyday was ending. After leading the doomed IONIAN REVOLT against Persian overlordship, Miletus was captured, sacked, and depopulated by the vengeful Persians (probably in 493 B.C.). The survivors were resettled inside PERSIA. Miletus recovered, but not as a world power. After Ionia was liberated by the Greeks in 479 B.C., Miletus joined the Athenian-dominated DELIAN LEAGUE. Then governed as a DEMOCRACY, the city was an Athenian ally for much of the PELOPONNESIAN WAR. But in 412 B.C., Miletus revolted from Athens and became a naval base for the Spartan side. In 386 B.C. the Spartans returned Miletus and all Ionia to Persian rule, by the terms of the KING'S PEACE.

In 334 B.C. Ionia was again freed of Persian rule, this time by the Macedonian king ALEXANDER THE GREAT, who stormed Miletus to capture it from a Persian garrison. In the HELLENISTIC AGE (circa 300–150 B.C.), Miletus enjoyed self-rule, prosperity, and an admirable building program, but the gradual silting up of the harbor brought economic decline. In Roman times Miletus was eclipsed by its neighbors EPHESUS and Smyrna.

(See also CYZICUS; MINOAN CIVILIZATION; MYCENAEAN CIVILIZATION; PANTICAPAEUM; PHRYNICHUS; SINOPE.)

Miltiades Athenian general and politician who lived circa 550–489 B.C. Miltiades masterminded the Athenian victory over the invading Persians at the Battle of MARATHON (490 B.C.) and afterward enjoyed a brief preeminence at ATHENS before succumbing to political enemies. He was the first of a series of dynamic Athenian leaders in the 400s B.C.

Born into the rich and aristocratic Philaïd family, he began his career under the Athenian tyrant HIPPIAS (1). In around 522 B.C. Miltiades left Athens for the Thracian CHERSONESE—the 50-mile-long peninsula that forms the European side of the HELLESPONT—to rule the region, inherited from his maternal uncle, the elder Miltiades.

There the young Miltiades reigned as *turannos* (dictator) over native Thracians and Athenian colonists, but served as a vassal of the Persian king DARIUS (1). Miltiades married a Thracian king's daughter, who bore his son CIMON. Like other eastern Greek rulers, Miltiades took part in Darius' expedition across the Danube into Scythia (circa 513 B.C.). However, after participating in the doomed IONIAN REVOLT against Darius, Miltiades fled back to Athens (493 B.C.).

Athens, by then a full-fledged DEMOCRACY, was entirely different from the place that Miltiades had left 30 years before. In the turbulent political climate, Miltiades had enemies who resented his prior dictatorship in the Chersonese and his association with Hippias. But before long

Miltiades moved to the political fore, being elected as one of Athens' generals for the year 490 B.C.

That summer, when a seaborne Persian army landed at Marathon with the aim of capturing Athens, it was Miltiades who convinced the Athenian commander-in-chief to attack, rather than just defend the city walls. The Battle of Marathon—a complete victory by an outnumbered force over a supposedly invincible foe—was one of the crucial moments in Greek history. The unique battle plan, involving an enveloping tactic by the army's wings, probably came from Miltiades.

His glory lasted barely a year. Ambitious, he led a 70-ship Athenian fleet to seize Paros, a wealthy Greek island that had submitted to Darius. But the attack failed, and the 60-year-old Miltiades, with a badly injured knee, returned to face public anger at Athens. The left-wing leader XANTHIPPUS accused Miltiades of deceiving the Athenian people. The hero of Marathon was tried and convicted in the ASSEMBLY and fined a ruinous 50 TALENTS. Soon Miltiades died from his gangrenous injury, and the fine was paid by the young Cimon (himself destined to become the foremost Athenian soldier and statesman of his day).

Among the ancient artifacts now displayed at the OLYMPIA museum is a Greek HOPLITE helmet discovered by archaeologists in A.D. 1961 and inscribed with the words in Greek, "Miltiades dedicated this to Zeus." Presumably it is Miltiades' own helmet, worn at the Battle of Marathon and afterward given as an offering to the king of the gods and the lord of victory.

(See also PERSIAN WARS; ZEUS.)

Mimnermus Early Greek lyric poet of Colophon, in western ASIA MINOR (latter 600s B.C.). His work survives only in fragments. He was a writer of the elegy—a form meant to be sung to flute accompaniment—and many of his verses were love poems addressed to a flute girl named Nanno. Mimnermus was a forerunner of the later love elegists of ALEXANDRIA (1) and imperial ROME. His best-known fragment deals with a topic destined to become a favorite of Greek and Latin poets: the transience of youth and the implacable approach of old age and death.

(See also LYRIC POETRY.)

Minoan Civilization A name invented in A.D. 1900 by the British archaeologist Sir Arthur Evans, to describe the civilization of ancient CRETE, of roughly 2200–1400 B.C. Evans was the first to discover remnants of this accomplished society, the earliest imperial power in Europe. Evans' adjective "Minoan" refers to a hero in Greek mythology, the powerful Cretan king MINOS.

The Minoans were not Greeks, and their language, RELIGION, and social structures were not Greek. Most of what is known or can be guessed about the Minoans comes from modern ARCHAEOLOGY on Crete. (The little island of THERA also has yielded an important Minoan site.) Evidence suggests that the Minoans emerged from a fusion between existing Cretan inhabitants and invaders from ASIA MINOR during the era 2900–2200 B.C. These people became master seafarers and built a society inspired partly by contact with the Egyptian Old Kingdom (circa 2650–2250 B.C.). By about 1900 B.C. the Minoans were acquiring an AEGEAN SEA em-

The famous Toreador Fresco from the palace at Cnossus in Crete, circa 1500 B.C. The scene shows the mysterious Minoan practice of bull-leaping: A male dancer, painted as red-skinned, vaults over the bull's back, with two female dancers nearby. The speed of the bull's charge is denoted by the elongated body and outstretched legs. This sport may have been a religious rite, intended to capture the strength and sexual power of the bull.

pire and were constructing palaces on Crete—at CNOSSUS, Phaestus, Mallia, and Khania—that were bigger and more elaborate than any buildings outside the Near East. So confident were the Minoans in their naval power that they declined to encircle their palaces with defensive walls.

Wealth came from Cretan FARMING and fishing, from taxes paid by subject peoples in the CYCLADES and other Aegean locales, and from long-distance TRADE. Minoan objects discovered by archaeologists outside Crete indicate two-way commerce with EGYPT, Asia Minor, and the Levant as well as with western ITALY (a region that offered raw tin and copper, the components of BRONZE). But much Minoan trade, especially after 1600 B.C., was with the northwestern Aegean mainland now called Greece, where Greek-speaking tribes had been settling since about 2100 B.C.

The Minoans' importance for Greek history is that they supplied the model for the Greeks' MYCENAEAN CIVILIZATION, which arose on the mainland circa 1600 B.C. The Mycenaean fortress palaces at MYCENAE, TIRYNS, and elsewhere were warlike imitations of Minoan palaces on Crete. Mycenaean skills in metalworking, POTTERY-making, and other handicrafts were improved by copying Cretan models. The Mycenaean form of WRITING—a syllabary script that modern scholars call LINEAR B, invented soon before 1400 B.C.—was copied from the Minoan system (a yet-undeciphered script called Linear A). Eventually the Mycenaeans were ready to challenge Minoan supremacy in the Aegean.

Daily scenes of the Minoans' life are preserved on some of their beautiful art objects, which include cut gems, worked GOLD, terra-cotta figurines, vase paintings, and frescoes. Sensuous and modern-seeming in design, Minoan pictorial art favors sea animals and other subjects from nature. Religious scenes often show a goddess (or priestess) with a subordinate male figure or with wild beasts, such as lions, in tame postures. Evidence of this kind leads many scholars to conclude that Minoan religion was centered on a mother goddess or a group of goddesses overseeing nature and bounty. Aspects of Minoan worship apparently infiltrated Greek religion in the cult of certain goddesses, such as ARTEMIS and HERA.

The Minoans ascribed religious or magical power to dancing and to the remarkable athletic performance now known as bull leaping. Minoan reverence for the bull is probably reflected in Greek MYTHS of later days, such as the interrelated tales of Minos and of THESEUS and the Minotaur, or the tale of HERACLES and the Cretan bull.

Minoan high society probably revolved around a priest-king or priest-queen whose capital city was Cnossus and whose royal emblem was the *labrus*, a double-headed ax. Scenes in art suggest a confident, vivacious life at court. Upper-class WOMEN—portrayed as wearing flounced skirts and open-breasted tunics—apparently played prominent roles in court life (as opposed to the secluded existence of women in Greece in later centuries).

The material level enjoyed by the Minoan ruling class was probably unsurpassed anywhere before the late 19th century A.D. The Cnossus palace, reaching three stories in parts, boasted clay-piped plumbing and a clever system of air wells to bring light and ventilation to interior rooms. COINAGE had not yet been invented, but Minoan wealth was measured in luxury items and in farm surplus such as sheep, pigs, and olive oil (great quantities of which were stored at Cnossus).

The Minoan golden age on Crete, circa 1900–1450 B.C., was a time of peace but was troubled by natural disasters. Archaeology at Cnossus shows that the palace was destroyed twice by earthquake, circa 1730 and 1570 B.C. Circa 1480 B.C. Cretan coastal regions suffered damage and depopulation, possibly caused by tidal waves from the volcanic explosion of Thera, 70 miles away.

The Cnossus palace, on high ground, survived, but new archaeological signs of distress in the mid-1400s B.C. include proliferation of war equipment and the first appearance on Crete of the horse (presumably imported as a tool of war). Overseas, Minoan pottery from this time is absent from certain sites—a sign of disrupted trade routes. Presumably a foreign enemy or number of enemies, taking advantage of Cretan natural disaster, had begun to cut into the Minoan Empire. These enemies surely included groups of Mycenaean Greeks.

In about 1400 B.C. or soon after, all the Cretan palaces were destroyed by fire, presumably in war. The most obvious explanation for this simultaneous destruction is a Mycenaean invasion of Crete. Intriguingly, archaeological evidence suggests that, prior to this invasion, Mycenaean Greeks had already taken over the Cnossus palace and that it was they who were destroyed in the palace's ruin. There may have been rival Mycenaean armies, battling each other for control of Crete.

Although the Mycenaean victors seem to have abandoned Crete soon after 1400 B.C., the Minoan culture was finished.

(See also ATLANTIS; BRONZE AGE; DAEDALUS; LABYRINTH; SHIPS AND SEAFARING.)

Minos   According to Greek MYTH, Minos was a Cretan king who ruled the AEGEAN SEA with fleets of warships in olden times. A son of the god ZEUS and the Phoenician princess EUROPA, Minos was born in CRETE with his two brothers, Rhadamanthys and Sarpedon. As a young man at the royal city of CNOSSUS, Minos married Pasiphaë, daughter of the sun god, HELIOS. During a dispute over who should become the next king of Crete, Minos prayed to the god POSEIDON to send a bull from the sea as a sign of divine favor. The bull emerged, thus assuring Minos of the kingship; but the pure-white animal was so beautiful that Minos neglected to sacrifice it to Poseidon, as he had promised to do.

In retaliation, the god inspired Pasiphaë with an unnatural lust for the animal, and she acquired—from the immigrant Athenian craftsman DAEDALUS—a wooden device that disguised her as a cow. In this costume she approached the bull and was mounted by it. She conceived its child, which proved to be a grotesque creature, half human, half bull—the Minotaur (Greek: Minotauros, "the bull of Minos").

To hide the Minotaur, Minos angrily ordered the meddling Daedalus to build the palace known as the LABYRINTH. Then Minos imprisoned Daedalus and his son, Icarus, inside a tower. But Daedalus escaped from Crete on mechanical wings. (Icarus fell to his death en route.)

By ship, Minos pursued Daedalus to SICILY, where he was treacherously murdered in his bath by King Cocalus' daughters, who did not want to relinquish the miraculous inventor Daedalus. Zeus then installed Minos as one of the judges of the dead in the Underworld.

The Minos legend probably represents the Greeks' distorted memory of the great days of Cretan wealth and naval power, circa 1950–1450 B.C. For this reason, in A.D. 1900 the pioneering British archaeologist Sir Arthur Evans coined the adjective "Minoan" to describe that real-life civilization on Crete. The name Minos is not Greek in origin and may possibly have been a hereditary royal title of the Minoan rulers.

(See also AFTERLIFE; MINOAN CIVILIZATION.)

Minotaur   See LABYRINTH; MINOS; THESEUS.

Minyans   In MYTH, the Minyans (Greek: Minuai) were a powerful Greek clan that had controlled parts of BOEOTIA and THESSALY in the old days. This legend probably commemorates an actual northern Greek dynasty of the Mycenaean era, circa 1200 B.C. Relatedly, in the Thessalian-based legend of JASON (1) and the Argonauts, the Argonauts traditionally are referred to as Minyans.

The modern term *Minyan Ware* was coined by archaeologist Heinrich Schliemann in A.D. 1880 to describe a kind of POTTERY he had discovered at the Boeotian city of ORCHOMENUS. The pottery, now dated to circa 1900 B.C., usually is considered to have been made by early Greeks—that is, by early descendants of the Greek-speaking invaders who occupied mainland Greece after about 2100 B.C.

(See also MYCENAEAN CIVILIZATION.)

Mitylene   See MYTILENE.

Muses   Greek goddesses of poetry, MUSIC, dance, and the arts in general. The word *Muse* (Greek: Mousa) is related to *mousikē*, "music." In MYTH, they were the daughters of ZEUS and Mnemosyne—that is, the god of universal order and the goddess Memory—and this is surely a metaphor for the reliance on verse rhythm and the singer's memory in the traditional technique of oral composition, prior to the spread of literacy.

The Muses, together or individually, were imagined as inspiring human song and poetry. HOMER's *Odyssey* (written circa 750 B.C.) opens with an appeal to an unspecified Muse to help the poet sing about his subject. In his epic poem *Theogony* (circa 700 B.C.), HESIOD describes how the Muses approached him on Mt. Helicon, in BOEOTIA, and breathed the gift of song into him. Hesiod named nine goddesses, but it was only a later elaboration that assigned to each Muse a separate function. The nine were Calliope, Clio, Euterpe, Terpsichore, Erato, Melpomene, Thalia, Polyhymnia, and Urania.

(See also APOLLO; EPIC POETRY; WRITING.)

music Greek music—*mousikē*, "the art of the MUSE"—was closely related to the recitation of Greek poetry. Greek poetic meter was a form of rhythm, and verses were sung or chanted to instrument accompaniment. The two principal instruments were the lyre (*lura*), a stringed instrument played by plucking, and the *aulos,* a wind instrument often described as a flute but really more akin to our clarinet or oboe. The lyre's sound was considered dignified and soothing, while flute music was more exciting.

The lyre was associated with the god APOLLO, lord of order and harmony. The ennobling and civilizing power of music was emphasized in legends of the great lyre musicians, such as ORPHEUS, who could charm wild beasts with his song, or Amphion, whose music brought stones trooping of their own accord to build the perimeter wall of THEBES. By contrast, the flute was associated with the riotous god DIONYSUS.

Different forms of poetic verse were considered appropriate to each instrument. The lyre had very ancient associations with the singing or chanting of EPIC POETRY. LYRIC POETRY (*lurikē melē*) too was developed primarily for lyre accompaniment, but by the mid-600s B.C. lyric had come to include verse forms such as the elegy, intended for flute music. Early surviving verses of elegy—such as those by the poets CALLINUS (mid-600s B.C.) and TYRTAEUS (circa 630 B.C.)—convey military-patriotic themes, intended to rouse and encourage an audience. Other forms of flute poetry included the dithyramb, which was a choral song sacred to Dionysus. Descended from the dithyramb (by the late 500s B.C.) was the choral ode of Athenian stage tragedy, also accompanied by flutes.

Flute and lyre had important functions in other walks of life. Instructions in lyre-playing formed an important part in the EDUCATION of upper- and middle-class boys at ATHENS. Flute-playing was more the resort of professional musicians, including SLAVES and PROSTITUTES. Flutes supplied lively dance music, also background music for SPORT competitions and for the drinking party known as the SYMPOSIUM. From about the mid-600s B.C. onward, Spartan armies always marched into battle with flute players, to keep the soldiers in step and rouse their courage. Stately flute music was used as background at sacrifices and other religious ceremonies.

Greek MYTH ascribed the lyre's invention to the god HERMES. In fact, the Greeks probably adopted the instrument in the second millennium B.C. from the Cretan MINOAN CIVILIZATION. The simplest form of lyre had a sound box of tortoiseshell (or wooden facsimile), with strings of gut stretched down from a crossbar between two extended horns. Seven strings was the conventional number, although archaeological evidence suggests some lyres had as many as 12. Unlike modern harp strings, lyre strings were uniform in length, but a skilled musician might vary the sound by stopping a string partway. A bigger, more elaborate form of lyre was the cithara (*kithara*), used mostly for public performance. With the instrument held in place by a strap, the cithara player typically used both hands—the bare fingers of his left hand and an ivory plectrum in his right.

The flute was made of wood, ivory, or bone, with a double reed in the mouthpiece. Artwork often shows a player blowing a pair of flutes, sometimes strapped around his or her head.

The question of Greek musical notation is obscure. Surviving artwork never shows musicians reading musical notes, but it is believed that some form of written musical notation did exist by the late 400s B.C.

Being a form of measurement susceptible to mathematical laws, yet with an emotional appeal, music was studied reverently by certain philosophers. PYTHAGORAS (circa 530 B.C.) sought in music the secrets of the universe. Similarly, PLATO (427–347 B.C.) saw music as a powerful force for either good or ill, with the ability to mold human character permanently.

The best-known musical theorist of ancient Greece was Aristoxenus (active circa 330 B.C.), who was born at TARAS and studied under ARISTOTLE at the LYCEUM. Of his works there currently exists, in part, *Principles and Elements of Harmonics* and *Elements of Rhythm.*

(See also APOLLO; PYTHIAN GAMES; SYMPOSIUM.)

Mycalē Mountain peninsula on the central west coast of ASIA MINOR, opposite the island of SAMOS, in the Greek-inhabited region known as IONIA. On Mycalē's north side was a sanctuary area called the Panionium, where delegates from the 12 Ionian cities met and where an annual Ionian festival was held. The site was sacred to POSEIDON.

In late summer of 479 B.C., at the end of the PERSIAN WARS, Cape Mycalē was the scene of a small but significant land battle, resulting in the liberation of Ionia after about 75 years of Persian domination. This battle supposedly occurred on the same day as the decisive Battle of PLATAEA, in mainland Greece.

At the Battle of Mycalē, a Greek force of about 4,400 HOPLITES landed from ships and destroyed a Persian army twice as large. More significantly, the Greeks then were able to burn the Persians' beached warships—numbering perhaps 100—which had comprised the last remnant of the Persian navy. (The rest had been destroyed at the Greek sea victory at SALAMIS the year before.) Without a navy, PERSIA became temporarily helpless to defend its other east Greek possessions, such as the HELLESPONT district or the island of CYPRUS, and these regions fell to the Greek counteroffensive in the following months.

The Greek commander-in-chief at Mycalē was the Spartan king Leotychides, at the head of a Spartan contingent. But—according to the version told by the historian HERODOTUS—the brunt of the fighting was borne by the Athenian contingent, under command of the soldier-politician XANTHIPPUS.

In the battle's aftermath, Xanthippus and other Athenian commanders enrolled the nearby Aegean island states in a common alliance against Persia. This union led quickly to the creation of the DELIAN LEAGUE, the power base for the Athenian Empire during the next 75 years.

Mycenae Greek city in the northeast PELOPONNESE, of great significance in early Greek history. Situated in the hills at the northeastern edge of the Argive plain, Mycenae was the capital of a rich and accomplished early Greek culture (circa 1600–1200 B.C.). The name MYCENAEAN CIVILIZATION was coined by the pioneering German archaeologist Heinrich Schliemann on the basis of his excavations at Mycenae in A.D. 1876. The ARCHAEOLOGY of the site has provided the single most important source of information about the Mycenaean Greeks, who left behind no written history.

Mycenae was one of the first places occupied by the invading Greek tribesmen of about 2100 B.C., but it predates their arrival. Archaeology shows that the site—commanding the fertile plain to the south and the land route northward to the isthmus—was first inhabited circa 3000 B.C. by Neolithic settlers. The Greek takeover of the region may be indicated in the destruction of a pre-Greek palace (often called the House of the Tiles) at Lerna, at the opposite edge of the Argive plain, circa 2100 B.C. Greek presence is almost certainly indicated in changes in style of POTTERY found at Mycenae, datable to about 1900 B.C. The Greek name Mukēnai has no apparent meaning in the GREEK LANGUAGE, and surely preserves a pre-Greek name.

Greek Mycenae's preeminence by the mid-millennium is clearly shown by the 31 lavish royal tombs now called Grave Circle A and Grave Circle B and dated to about 1550 B.C. and 1650 B.C., respectively. Circle A, located atop Mycenae's ACROPOLIS, was discovered by Schliemann; the tombs' treasure of GOLD and SILVER gave the first archaeological proof of the existence of an early-Greek high civilization.

Mycenae's heyday came in 1400–1200 B.C., when the huge limestone walls and hilltop citadel were built. Circa 1260 B.C. the city received its most distinctive surviving feature: the Lion Gate, surmounted by rampant carved-limestone lions (now headless).

To this era belongs the supreme Mycenae later recalled in the MYTHS of the classical Greeks—the city "rich in gold" (as the poet HOMER called it), with a high king who was overlord of many lesser Greek rulers. The legendary AGAMEMNON, king of Mycenae, raises an army against TROY, by summoning his various vassal kings. This is probably an accurate reflection of Mycenae's feudal dominance in the Mycenaean age.

Archaeology also tells the tale of the city's decline in the 1200s B.C. A series of burnings culminated circa 1200 B.C. in major destruction, which probably indicates internal strife or several defeats at the hands of other Mycenaean-

The monumental Lion Gate at Mycenae, built circa 1250 B.C., WHEN MYCENAE WAS THE GREATEST CITY IN GREECE SEEN FROM OUTside the wall, the 10-foot-tall, carved-limestone slab shows two lions (now minus their heads) flanking a pillar. The lions' heraldic stance may derive from the adoring postures of animals in certain Minoan religious art. Architecturally, the slab relieves the weight on the massive lintel, which stretches over a 10.5-foot-high gateway.

Greek cities. There followed perhaps three generations of depopulation, culminating in Mycenae's final ruin, possibly at the hands of the invading DORIAN GREEKS (circa 1120 B.C.).

Later Mycenae existed as a Dorian town whose inhabitants probably lived amid the ruins of vanished grandeur. Men of Mycenae fought as allies of SPARTA against the Persians at THERMOPYLAE (480 B.C.) and PLATAEA (479 B.C.). This Mycenae was destroyed by its powerful neighbor ARGOS circa 468 B.C., but reemerged in the 200s B.C.

(See also ATREUS; TROJAN WAR.)

Mycenaean civilization The term used by modern scholars to describe the earliest flowering of mainland Greek culture, circa 1600–1200 B.C. The Mycenaeans were Greeks whose warlike society rose and fell long before the era of classical Greece. The classical Greeks of circa 400 B.C. half remembered their Mycenaean forebears as a race of heroes, celebrated in MYTH and EPIC POETRY.

In world prehistory, the Mycenaeans comprised the last of several great civilizations to emerge in the eastern Mediterranean during the BRONZE AGE. The Mycenaeans' urban building, military organization, and TRADE seem to have been partly copied from a few preexisting, non-Greek, Bronze Age cultures—namely, the Middle and New Kingdoms of EGYPT, the Hittite kingdom of ASIA MINOR, and especially, the MINOAN CIVILIZATION of CRETE.

The adjective "Mycenaean" was coined in A.D. 1876 by the pioneering German archaeologist Heinrich Schliemann, on the basis of his spectacular discoveries at the site of MYCENAE, in the northeastern PELOPONNESE, in southern Greece. The Mycenaeans lived before the era of history-writing, and thus most details of their story—such as their rulers' names or the reasons why their entire society collapsed in fiery ruin around 1200 B.C.—remain unknown. Modern knowledge relies mostly on artifacts uncovered by ARCHAEOLOGY at a few sites, such as Mycenae, TIRYNS, and PYLOS (in the Peloponnese) and THEBES, ORCHOMENUS, and ATHENS (in central Greece). The artifacts include POTTERY, stone carvings, jewelry, and armor—most of it found in the tombs of rulers—as well as the remnants of Mycenaean stone palaces and defenses. Particularly, the sites of Mycenae and Tiryns still show huge fortifications built by Mycenaean inhabitants in the 1300s and 1200s B.C.

In addition, a few sites have yielded primitive Mycenaean written records, inscribed on clay tablets that seem

The face of early Western civilization. This beaten-gold mask, probably the best-known artifact from Mycenaean times, was among the treasures discovered at Mycenae in A.D. 1876 by German archaeologist Heinrich Schliemann. The mask had been placed over the face of a male corpse, presumably a king, who had lived circa 1550 B.C. His name is unknown to us, but he sometimes is misidentified as the mythical king Agamemnon (who belonged to a later century). The mask—shaped by being hammered over a hard form, such as carved wood—is probably the man's portrait, made during his lifetime.

to date from about 1400 B.C. or 1200 B.C., depending on the site. Written in a script that modern scholars call LINEAR B, the records have been deciphered mainly as lists of inventory—produce, livestock, military equipment—and accounts of goods-distribution, religious rites, and similar daily events. The tablets provide precious information on the social structure, economy, and RELIGION of the Mycenaeans, as well as on the early-stage GREEK LANGUAGE that they spoke.

Aside from archaeology, some insight into the Mycenaeans has been gained from a cautious reading of HOMER's epic poems, the *Illiad* and *Odyssey*. Although written circa 750 B.C., more than 400 years after the Mycenaeans' disappearance, these poems derive from oral tradition that stretches back to the Mycenaeans. It is believed that the poems faithfully record certain aspects of Mycenaean upper-class life—such as the warrior code and the network of local kings—amid distortions and overlays.

The first Greek-speaking tribes arrived in mainland Greece circa 2100 B.C., from the Danube region. But 500 years went by before the emergence of the culture that we call Mycenaean: The remarkable social and technological changes of these intervening centuries can only be guessed at. No doubt the Greeks were deeply influenced by the non-Greek people they had conquered, and from them the Greeks probably learned skills such as stone masonry, ship-building, navigation, the cultivation of the olive and certain other crops, and the worship of certain female deities (with associated, new spiritual concepts). Similarly, the Greeks were inspired by the palace society of Minoan Crete.

The Mycenaean era began around 1600 B.C., as archaeology reveals. Several sites in Greece came under control of powerful rulers who were buried in elaborate tombs, unlike the simple graves of prior centuries. And within a few generations the tomb designs altered again, suggesting further dynastic changes and evolving organization. The six treasure-filled tombs at Mycenae known as Grave Circle A—built in the era 1550–1500 B.C. and discovered intact by Schliemann—provide clear proof of the rulers' wealth and overseas contacts. For example, the tombs contain items of GOLD that were shaped by Greek smiths, but the raw metal probably came from Asia Minor or Egypt. The warlike nature of these leaders is suggested by the many weapons left as offerings in the tombs.

In Greece's terrain, where mountain ranges separate the flatlands, the Mycenaeans apparently emerged as four or so major kingdoms, each based at a large farming plain. Two of these domains were in the Peloponnese: the plain of ARGOS (with its capital at Mycenae) and the plain of MESSENIA (capital at Pylos). One was in central Greece: the plain of BOEOTIA (with the cities Thebes and Orchomenus vying for supremacy). And one was in the north, on the great plain of THESSALY (capital at Iolcus). Lesser kingdoms probably existed as well. But the greatest domain was Mycenae, as indicated by its signs of superior wealth and by the testimony of Greek myth. In Homer's *Illiad*, the Mycenaean king AGAMEMNON is the supreme commander, to whom all other kings, such as ODYSSEUS and NESTOR, owe obedience.

One event of the Mycenaean era that modern scholars are sure of is that by around 1450 B.C. Mycenaeans had

A lion hunt scene, inlaid on a Mycenaean dagger found at Mycenae. The dagger was deposited as an offering in a royal tomb circa 1575 B.C., not long after the dawn of Greek civilization. Two of the hunters carry the distinctively Mycenaean hourglass-shape ox-hide shields. In later centuries, wild lions did not inhabit mainland Greece: Either they were exterminated in the Mycenaean era, or this scene is imaginary, based on lion hunts shown in Near Eastern art.

taken over the Cretan palace at CNOSSUS—probably as the result of a Mycenaean naval invasion of Crete. Mysteriously, the Mycenaeans seem to have abandoned Crete soon thereafter, circa 1400 B.C. But the years of occupation there taught Mycenaean rulers certain organizational skills—such as improved architectural techniques and the use of Cretan WRITING (adapted at this time, as the Linear B script)—that ushered in 200 years of the Mycenaean heyday in mainland Greece, circa 1400–1200 B.C.

It was now that the Mycenaeans built their own palaces, adapted from the Minoan palaces on Crete. Mycenae and Tiryns were turned into elaborate, high-walled castles; other palaces, such as at Pylos, arose without huge defenses. The social and economic structure of these centers is partly revealed by the Linear B tablets. The palace was the seat of the king (*wanax* in Mycenaean Greek); beyond the capital city, a network of outlying villages paid taxes, obeyed the king's laws, and relied on him for defense against other rulers. Tha palace was also a center of industry, where metalworkers, weavers, perfumers, and many other crafts people turned out finished goods, to enrich the king or to be distributed by him. Raw materials came from local taxes (sheep's wool, for example) and from overseas trade.

The premier metal for war and industry was BRONZE (the use of IRON being introduced to the Greek world only later). The search for bronze's two components—copper and tin—led Mycenaean sea traders far and wide. Large remains of Mycenaean pottery in CYPRUS show that parts of that copper-rich island were colonized by Mycenaeans. On the western Asia Minor coast, the site of MILETUS probably became a Mycenaean trading colony, mainly for the acquisition of raw metals. Toward the other end of the Mediterranean, extant pottery suggests a Mycenaean presence in western ITALY, where tin could be found.

The Mycenaean rulers commanded armies of heavy infantry. The soldiers' standardized equipment, including bronze breastplates and helmets, is recorded on Linear B tablets. Various evidence paints a picture of Mycenaean kings or princes leading Viking-like raids overseas, of which the biggest were the (presumed) invasions of Crete and Cyprus. On certain Linear B tablets, SLAVES are men-

tioned by names that suggest they came from Asia Minor; probably they were captured in Mycenaean raids there. The Greek myth of JASON (1) and the Argonauts may distortedly commemorate such an overseas expedition. But the Mycenaean kingdoms fought also against each other: the legend of the SEVEN AGAINST THEBES seems clearly based on actual warfare between Mycenae and Thebes.

By about 1250 B.C. the Mycenaean world had come under pressure, due partly to upheavals in the Near East. The decline of the Hittite kingdom in Asia Minor probably brought a gradual closing of the Mycenaeans' eastern trade routes. Deprived of raw metals for industry and conquest, Mycenaean society began to whither. The Greek legend of the TROJAN WAR may recall the Mycenaeans' attempt to keep trade routes open by removing the interfering, non-Greek, Hellespontine city TROY, circa 1220 B.C.

Finally, it seems, the Mycenaean kingdoms turned against each other and destroyed each other, in a desperate bid for survival. Archaeology clearly reveals the fiery ruin of Thebes, Mycenae, and other centers in the 50 years leading down to 1200 B.C. At Pylos, the final days are dramatically indicated in emergency troop movements and religious sacrifices recorded on Linear B tablets.

Modern historians used to believe that this wholesale destruction was the work of outsiders—specifically, DORIAN GREEKS invading from the northwest. But more recent scholarship concludes that the Dorian invasion, circa 1100 B.C., was merely opportunistic: The Mycenaeans had already exhausted themselves through internal war.

In the villages outside of the wrecked palaces, Mycenaean society survived on an improverished scale during the 1100s B.C. Social change in these rural areas can be glimpsed in the development of a certain Greek word: The official title *quasireu*, which during the Mycenaean heyday had indicated a local sheriff (a relatively low position), gradually changed to *basileus* and took on a new meaning, "king." These men became the new local rulers within the disintegrated Mycenaean kingdoms.

(See also ACHAEA; ACHAEANS; ARCADIA; ATREUS; CHARIOTS; DARK AGE; FUNERAL CUSTOMS; PERSEUS (1); SHIPS AND SEAFARING; WARFARE, LAND.)

mystery cults See AFTERLIFE; DIONYSUS; ELEUSINIAN MYSTERIES; ORPHISM; SAMOTHRACE.

myth The Greek word *muthos* means simply "a tale." In modern use, that word has come to mean a popular tale, elaborated by generations of storytelling, that may contain a kernel of historical fact and that is significant for understanding a people's mass mentality.

Greek myths may reflect events in the distant past or popular unfulfilled aspirations. Some myths have a moral, such as the need to be hospitable to strangers. Other myths are "aetiological"; that is, they attempt to explain local geographical features, religious rites, or other phenomena not fully understood by other means. Related to this type is the propagandistic "charter myth," which seeks to sanctify a custom or institution—examples include the myth connecting the DORIAN GREEKS with the prestigious sons of HERACLES, or the myth claiming that the city of CYRENE (1) was founded by the god APOLLO for his mistress, named CYRENE (2).

Modern scholarship has shown that the most important Greek myths tend to present distorted memories of the Greek MYCENAEAN CIVILIZATION (circa 1600–1200 B.C.), which later generations of Greeks remembered as an Age of Heroes. Among many examples of such historical myths are the SEVEN AGAINST THEBES and the TROJAN WAR. Because the Mycenaeans kept no written histories, certain real-life events were commemorated by heroic songs, handed down through the years and elaborated in a tradition of oral EPIC POETRY.

A remarkable number of Greek myths have survived antiquity. Their quantity and diversity are due partly to the fact that Greek society passed from a preliterate stage to a literate one in (probably) a single generation, circa 775–750 B.C., after the invention of the Greek ALPHABET. Many legends previously maintained by oral versifying were written down then or soon thereafter. HOMER's *Iliad* and *Odyssey* are the earliest and most important of these writings with mythological content. Among the many other ancient authors whose work recounts Greek myth are: the Boeotian epic poet HESIOD; the Theban choral poet PINDAR; the Athenian tragedians AESCHYLUS, SOPHOCLES, and EURIPIDES; the Alexandrian poets CALLIMACHUS and APOLLONIUS; and the Roman poet Ovid.

(See also CALYDONIAN BOAR HUNT; JASON (1); ODYSSEUS; OEDIPUS; ORESTES; PERSEUS; THESEUS; ZEUS.)

Mytilene The principal city of the eastern Greek island of LESBOS, in ancient times and still today (under the name Mitilini). Located on the island's southeastern shore, 12 miles from the northern west coast of ASIA MINOR, Mytilene was a prosperous Greek seaport, connecting East and West.

Archaeological study of the ancient site has been hampered by the modern city, but we know that this advantageous locale was occupied before the arrival of the first Greeks. We also know that the name Mytilene—sometimes rendered as Mitylene—is not Greek, for the ending -*ene* is similar to the endings of other pre-Greek place-names. Mycenaean-Greek POTTERY from about 1200 B.C. has been discovered on Lesbos; very possibly the seafaring Mycenaeans captured Mytilene at this time. Later, circa 1000 B.C., the city became a refuge for eastward-migrating AEOLIAN GREEKS. During 1000–900 B.C., Mytilene was an important departure point for further Aeolian colonizing, along the northwest Asia Minor coast.

By 625 B.C. Mytilene was a leading city of the eastern Greek world. Lesbos' traditions of LYRIC POETRY reached their zenith with the Mytileneans SAPPHO (circa 600 B.C.) and ALCAEUS (circa 590 B.C.). In this era the city was governed as an ARISTOCRACY. The chief clan was the Penthilidae, who had the oppressive habit of arbitrarily beating people with clubs in the street. Class tensions were enflamed by an overseas military failure—Mytilene's loss of the Hellespontine colony of Sigeum to Athenian settlers (circa 600 B.C.).

TYRANTS arose in Mytilene to lead the common people against the nobles. But civil war was averted by the statesman PITTACUS, who served as an elected 10-year dictator (circa 590–580 B.C.) and was later enshrined as one of the SEVEN SAGES of Greece. The turmoil of this era is conveyed in the poems of Pittacus' enemy Alcaeus.

Circa 522 B.C. Mytilene and the rest of the island fell to the advancing Persians under King DARIUS (1). The Mytileneans joined the doomed IONIAN REVOLT against Persian rule and distinguished themselves by their cowardice at the sea battle of Lade (494 B.C.). Liberated by the Greek counteroffensive at the end of the PERSIAN WARS (479 B.C.), Mytilene became a leading member of the Athenian-controlled DELIAN LEAGUE. For the next 50 years, Mytilene was one of the few league states to pay its annual obligation in the form of warships and crews rather than in SILVER.

During the PELOPONNESIAN WAR (431–404 B.C.), Mytilene twice revolted unsuccessfully against Athenian rule. The first revolt (428–427 B.C.) ended with Athenian troops occupying the island. Afterward, there occurred a famous debate in the ASSEMBLY at ATHENS, where a decision to destroy Mytilene and execute the adult male population was rescinded by a revote. This Mytilenean Debate forms a set piece in book 3 of the war history written by THUCYDIDES (1). After the second revolt was crushed (412 B.C.), the Mytileneans lost their fleet, their city walls, and much of their farmland to the vengeful Athenians. With the defeat of Athens in 404 B.C., a reduced Mytilene came under Spartan control.

After a generation of Spartan oppression, Mytilene joined the SECOND ATHENIAN LEAGUE, circa 377 B.C. In 333 B.C. the city fell briefly to the Persian navy but was liberated by the conquering Macedonian king ALEXANDER THE GREAT. After Alexander's death (323 B.C.), Mytilene and Lesbos passed to various Hellenistic rulers. Acquired by the kings of nearby PERGAMUM in 188 B.C., Mytilene was part of the domain bequeathed to the Romans by the Pergamene king Attalus III in 133 B.C. The city soon became part of the Roman province of Asia.

(See also GREEK LANGUAGE.)

N

names Unlike modern Americans, ancient Greek males and females typically carried only one personal name—for example, Socrates. In formal address, a man or boy might be specified by a patronymic (his father's name): "Themistocles, son of Neocles." However, at some Greek cities, specifically ATHENS, the use of the patronymic was discouraged due to social-leveling legislation aimed at removing distinctions and stigmas of lineage. Hence "Themistocles of the DEME of Phrearrus" was the preferred form. The masculine or feminine ending of a name clearly distinguished the person's gender—Diotima and Elpinice are female; Pericles and Diodorus, male.

Greek names, like those in German and certain other Indo-European tongues, usually contained common words in compound form. The Greek female name Cleopatra means "glory of her father." One of the elements might typically be a deity's name, as in Herodotus, "given by HERA." Two male names popular after the 500s B.C. were Demetrius and Dionysius, which were adjectival forms of the names of well-loved agrarian gods, DEMETER and DIONYSUS.

Because the keeping of horses was a sign of wealth, many aristocratic names included the proud element *hippo*—: Philip (Philippos), "horse-lover," Hippocrates, "horse power," Xanthippe, "yellow horse." For some reason, an aristocratic boy usually was not named after his father, but often after a grandfather—for example, CLEISTHENES (1) and (2), and THUCYDIDES (1) and (2). Collectively, members of a noble clan might be known by a family name—such as the ALCMAEONIDS (descendants of Alcmaeon).

Not only bluebloods had identifying names. Children of left-wing families might receive names with such politically charged elements as Demo—("the people") or Iso—("equality")—for instance, Demosthenes, "the people's strength," or Isodice, "equal justice."

Acquired nicknames came into use, mainly for royalty, starting in the late 300s B.C. The Macedonian soldier-prince Demetrius was honored with the surname Poliorketes, "the city-besieger"; his father, ANTIGONUS (1), was Monophthalmos, "the one-eyed." And the conquering Macedonian king Alexander was surnamed Megas, "the Great."

(See ALEXANDER THE GREAT; DEMETRIUS POLIORCETES.)

Narcissus In MYTH, a handsome Boeotian youth, son of the river god Cephisus and the nymph Liriope, who fell in love with his own reflection in a pond. In one version, he pined away and died of hopeless longing, in another, he stabbed himself with a dagger in frustration. From his body or blood there arose the white flower that the ancient Greeks called the *narkissos* (possibly a type of iris or lily, but not the same as our modern narcissus flower). This tale has produced the English words *narcissism* and *narcissistic*.

Narcissus' connection with a local flower, along with the pre-Greek *issos* ending of his name, suggest that he was originally a god or demigod of the pre-Greek peoples, absorbed into Greek mythology during the second millennium B.C.

(See also GREEK LANGUAGE; HYACINTHUS; NYMPHS; RELIGION.)

Naucratis Ancient port city of EGYPT, about 50 miles inland, on the westernmost branch of the Nile River. In the later 600s B.C. the pharaohs assigned Naucratis as the one emporium for all Greek TRADE in Egypt. The city then became the site of Greek temples and offices where various Greek states were represented. As listed by the historian HERODOTUS, these states included the great seagoing powers of the day: AEGINA, SAMOS, CHIOS, MILETUS, PHOCAEA, MYTILENE, and RHODES. The name *Naukratis* seems to be Greek, meaning "ship power"—that is, shipping place.

Archaeological excavations at Naucratis, combined with written references, give some idea of the commerce between Greece and Egypt in the 600s–500s B.C. Much of this trade apparently consisted of Greek SILVER ore and SLAVES (both acquired in the northern Aegean) exchanged for Egyptian grain, which was shipped by Greek merchants at a large profit to the hungry cities of the Greek world. Egyptian luxury goods, such as carved ivory, were also exported.

Like other port cities throughout history, Naucratis offered its share of men's entertainment. Herodotus describes Naucratis as "a place for lovely courtesans" (*hetairai*). Certain verses by the Greek poet SAPPHO (circa 600 B.C.) lament the predicament of Sappho's brother, ensnared at Naucratis by a fascinating *hetaira* on whom he has squandered his fortune.

Naucratis' fortunes declined after 525 B.C., when the Persian occupation of Egypt disrupted the Greek trade. After ALEXANDER THE GREAT's conquest of Egypt (332 B.C.), Naucratis was completely eclipsed by the founding of nearby ALEXANDRIA (1).

(See also PROSTITUTES; THRACE.)

Naupactus Seaport of West LOCRIS, situated at the mouth of the Corinthian Gulf. Naupactus' position controlling the gulf's narrow outlet made the town a natural naval base—its name in Greek means "shipbuilding." In the mid-400s B.C. Naupactus was captured by ATHENS and repopulated with fugitives from MESSENIA who had unsuccessfully revolted from Spartan rule (464–460 B.C.). During the PELOPONNESIAN WAR (431–404 B.C.), Naupactus was the main Athenian naval base in western Greece; it was off Naupactus that the Athenian admiral Phormion won his two brilliant sea victories over Peloponnesian fleets in 429 B.C.

After Athens' defeat in the war, Naupactus passed to the state of ACHAEA. In 338 B.C. the Macedonian king PHILIP II gave Naupactus to his ally, the state of AETOLIA. Naupactus had lost its importance by the 100s B.C., when the Aetolian League was defeated by ROME.

Nausicaa Princess of the virtuous Phaeacians in HOMER's *Odyssey*. In one of the poem's most charming episodes (book 6), she leads her waiting women to wash clothes at a stream and there encounters the shipwrecked hero ODYSSEUS. Approaching naked and shielded only by a leafy branch, Odysseus unintentionally scares away the servants, but not Nausicaa. Inspired by the goddess ATHENA, she supplies food and clothing to the hero, then shows him the way to her father's palace, where he is suitably received.

navies See WARFARE, NAVAL.

navigation See SHIPS AND SEAFARING.

Naxos (1) See CYCLADES.

Naxos (2) First Greek colony in SICILY, and one of the earliest Greek colonies anywhere. Situated on a promontory on the island's east coast, at the foot of Mt. Etna, Naxos was a natural landfall for westbound ships rounding the "toe" of ITALY from the northeast. Naxos was founded circa 734 B.C. by Greeks from CHALCIS under a leader named Thucles. Supposedly, the expedition included Greeks from the Aegean island of Naxos, who gave the new city its name.

Naxos itself seems to have been intended only as a Greek foothold in Sicily: Six years after its founding, Thucles and his followers drove the native Sicels from the fertile plain of Catania and founded the cities of CATANA and Leontini. These cities and others, such as nearby SYRACUSE, soon exceeded Naxos in importance.

By the early 400s B.C. Naxos was ruled by Syracusan TYRANTS. As an Ionian-Greek city in a realm dominated by DORIAN GREEKS, Naxos became a target of ruthless social engineering when the Syracusan tyrant HIERON (1) depopulated it and moved the people to Leontini (476 B.C.).

Reconstituted, Naxos made an alliance with ATHENS against Syracuse in 415 B.C., during the PELOPONNESIAN WAR, and served as an Athenian base for the disastrous expedition against Syracuse. In 403 B.C. Naxos was captured and razed by the Syracusan tyrant DIONYSIUS (1).

(See also COLONIZATION.)

Nemean Games One of the four great sports-and-religious festivals of ancient Greece, along with the OLYMPIC, PYTHIAN, and ISTHMIAN GAMES. Sacred to the god ZEUS, the Nemean festival was held every other year at the valley and sanctuary called Nemea in the northeastern PELOPONNESE, in the region of ARGOS but close to the town of Cleonae. There, according to myth, the hero HERACLES instituted the games after slaying the Nemean Lion—the first of his 12 Labors. In fact, the Nemean festival first became important in 573 B.C., when Argos took over its administration from Cleonae and enlarged it on the model of the Olympic Games. The prize for victors at Nemea was a garland of wild celery (as at the Isthmian Games).

The ruins of a Doric-style temple of Zeus, built circa 340–320 B.C., have been excavated in Nemea. The temple was destroyed by earthquakes in late antiquity.

Neoptolemus Greek mythical hero of the island of Scyros, the son of the Thessalian hero ACHILLES and the princess Deidameia. Neoptolemus ("new warrior") was begotten while Achilles was hiding among the WOMEN of Scyros in an attempt to avoid serving in the TROJAN WAR.

HOMER's epic poem the *Odyssey* mentions that Neoptolemus himself went to TROY after his father's death. He was summoned by the Greeks, who had learned in a prophecy that Neoptolemus' presence was a fated precondition of the city's fall. Although Neoptolemus would have been only about 10 years old at that time, this detail was overlooked in the legend, and he was said to have fought fiercely at Troy, winning the nickname Pyrrhus ("fiery" or "red").

Neoptolemus was one of the select Greek commandoes who hid inside the Trojan Horse. During the capture of the city, he slew the Trojan king PRIAM—despite the fact that Priam had taken sanctuary at the altar of the great god ZEUS. This brutal act brought the hatred of the Trojans' patron god APOLLO against Neoptolemus.

According to one version, Neoptolemus was killed after the war in a dispute at Apollo's shrine at DELPHI. Another version says he sailed to EPIRUS, in northwestern Greece, where he fathered the ruling clan, the Molossians. In historical times the Molossians used the hero's nickname, Pyrrhus, as a given name. The most famous such person was the Epirote king PYRRHUS (reigned 297–272 B.C.).

(See also ANDROMACHE.)

Nereids See NEREUS.

Nereus In MYTH, Nereus was a minor sea god, a kind of old man of the sea. He and his wife, Doris (the daughter of OCEANUS), had 50 daughters, known as the Nereids (Greek: Nereidai "daughters of Nereus"). These sea-dwelling young goddesses, often imagined as fish-tailed, have been favorite subjects of art and poetry since ancient times. The best known of the Nereids were Amphitritē, who married the Olympian sea god POSEIDON, and Thetis, who married the mortal PELEUS and gave birth to the hero ACHILLES.

The name Amphitritē is not Greek, and surely derives from the language of the prehistoric people who occupied the land of Greece before the first Greek-speaking tribes arrived, circa 2100 B.C. The Nereids are probably survivals of the pre-Greek people's RELIGION.

(See also ARTEMIS; NYMPHS.)

The Nereids, elusive sea goddesses, were among several types of Greek female deities embodying the spirit of wilderness places. This figure, sometimes identified as a Nereid, appears on a marble carving of about 500 B.C. from a temple of Hera near the ancient Greek city of Poseidonia, in southeastern Italy.

Nestor   In MYTH, the king of PYLOS, in the southwestern PELOPONNESE, and the most elderly of the Greek commanders in the TROJAN WAR. Nestor—the son and successor of King Neleus—was probably imagined as being over the age of 50 when the war began and over the age of 60 at its conclusion. HOMER's *Iliad* and *Odyssey* give a charming portrait of an admirable but garrulous old man, often ineffectual in combat and councils of war. In both poems he delivers meandering but lively speeches recalling his youthful achievements.

The *Iliad* presents Nestor's fond relationship with his soldier-son Antilochus but omits the tale of Antilochus' death. We know from other ancient sources that Antilochus was killed while defending his father from the Ethiopian champion MEMNON. In the *Odyssey*, Nestor is shown safely back in Pylos, where he welcomes the prince TELEMACHUS in the latter's search for his father, ODYSSEUS.

Nicias   Athenian general and politician of the PELOPONNESIAN WAR, who lived circa 470–413 B.C. Rich and devoutly religious, Nicias served his city loyally. But his hesitancy and befuddlement in the Athenian campaign against SYRACUSE (415–413 B.C.) produced an epic disaster that destroyed an Athenian force of perhaps 50,000 men, including Nicias himself.

Nicias (Greek: Nikias, "victorious") was the son of Niceratus, of a distinguished Athenian family whose income came from the leasing of large numbers of SLAVES to work the Athenian SILVER mines at Laurium. Upon the death of PERICLES (429 B.C.), the soldierly Nicias became the heir to Pericles' defensive strategy in the Peloponnesian War.

As a politician, Nicias headed the conservative peace party, composed of the upper class and the smallholding farmers. Nicias' opponent and personal enemy was the

pro-war radical democrat CLEON. After Cleon's death, Nicias arranged the short-lived peace with SPARTA that bears his name—the Peace of Nicias (421 B.C.).

In 415 B.C. Nicias was appointed alongside the generals ALCIBIADES and Lamachus to command a 134-ship armada to besiege and capture Syracuse, in SICILY. Unfortunately, the 55-year-old Nicias disapproved of the ambitious venture and, after Alcibiades was recalled, Nicias proved to be a dangerously indecisive leader. The death of Lamachus, combined with Nicias' kidney ailment, contributed to the deteriorating situation, despite Nicias' field victory at the River Anapus (late 415 B.C.). By summer of 413 B.C. the Athenians—although reinforced by fresh troops—were surrounded by the enemy on land and sea, after Nicias had hesitated too long to abandon the siege and sail away. Leading a hopeless attempt to escape overland, Nicias was captured by the Syracusans amid the slaughter of his men at the River Assinarus. He was later executed by the Syracusans.

(See also DELOS; DEMOSTHENES [2].)

Nikē   Goddess of victory, in war or SPORT. Like many other minor deities, Nikē was more a symbol than an important character in Greek MYTH. The name Nikē was sometimes reduced to an epithet of the goddess ATHENA—Athena Nikē, the patron of victory through strategy.

From the 500s B.C. onward, Nikē was picturesquely shown in art as having two feathery wings. She became a favorite subject after the Greek triumph in the PERSIAN WARS, and was often associated with the god ZEUS—most famously in the colossal statue of Zeus sculpted by PHIDIAS for the temple at OLYMPIA. In this monument (completed circa 430 B.C.; now lost, but represented on coins), Nikē appeared standing on the god's upturned palm.

The best-known surviving Nikē in art is the marble statue called the Winged Victory of Samothrace, sculpted circa 190 B.C. and now in the Louvre. In the Greco-Roman world, Nikē often was shown on coins and medallions, sometimes elevated, garlanding a victorious general's head.

(See also PARTHENON; SAMOTHRACE.)

Niobē   In MYTH, the wife of the Theban hero Amphion. As the mother of six youths and six maidens, Niobē boasted arrogantly that she was superior to the demigoddess Leto, who had borne only two children—the deities APOLLO and ARTEMIS. Angered at this affront to their mother, Apollo and Artemis hunted down, with bow and arrow, all 12 of Niobē's children. Niobē wept ceaselessly until, on the 10th day of her lamenting, the god ZEUS turned her to stone. This tale is first told in HOMER's *Iliad*, as background to the grief of the parents of the slain Trojan hero HECTOR. As a symbol of maternal grief, Niobē became a popular subject in Greco-Roman art and poetry.

(See also HUBRIS.)

numbers   See MATHEMATICS.

nymphs   The numphai were mythical female spirits of the wilderness, representing the beauty and fertility of nature. They were daughters of ZEUS or of other gods, but they themselves were not usually immortal. Rather, they were like leprechauns, living for centuries in the wild and avoiding contact with humans.

Usually imagined as young and amorous, nymphs were associated with the god DIONYSUS and his coterie of lusty male SATYRS. (This association is behind the pseudopsychiatric term *nymphomaniac*, meaning a woman with obsessive sexual desire.) Individual nymphs of MYTH include CALYPSO, lover of the hero ODYSSEUS, and CYRENE (2), lover of the god APOLLO. In the ancient GREEK LANGUAGE, *numphē* could also mean a young marriageable woman.

Eventually these pretty wilderness creatures became the subject of poets' elaborations, especially in the Hellenistic and Roman eras, after 300 B.C. There we find specialized categories of nymphs, such as the Naiads (stream nymphs) and the Dryads or Hamadryads (tree nymphs).

Like the NEREIDS and other demigoddesses, the nymphs may date back to the RELIGION of the pre-Greek people who inhabited the land of Greece before 2100 B.C.

O

Oceanids See OCEANUS.

Oceanus In MYTH, Okeanos was a river that encircles the world and serves as the underground source for all earthly rivers. The poet HESIOD described the river god Oceanus as one of the primeval offspring of Uranus (Greek: Ouranos, "Sky") and GAEA (Mother Earth). Oceanus and his wife, Tethys, produced the 3,000 Oceanids, the "daughters of Oceanus." These Oceanids were NYMPHS inhabiting bodies of water and other wilderness sites. In HOMER's *Iliad* and *Odyssey*, Oceanus was associated with the far West; there he flowed past shores inhabited by such fabulous creatures as MEDUSA, Geryon, and the Hesperides.

The notion of an encircling world stream is common to many mythologies. The Greeks may have borrowed it from the Near East, since the word Okeanos does not seem to be Greek. But the Greeks soon added their own layer, when their concept of *Okeanos* became colored by rumors of the Atlantic sea, lying west of the Mediterranean. The Greeks chose the word Okeanos to identify the Atlantic, once they began to venture beyond the straits of Gibraltar in the wake of the pioneering Phoenicians (600s B.C.). The English word *ocean* of course derives from Okeanos.

(See also ATLANTIS; ATLAS; HERACLES; PHOENICIA.)

Odysseus In MYTH, the wily king of ITHACA (a small island on the northwest coast of Greece) and a captain of the Greeks in the TROJAN WAR. Odysseus' 10-year-long journey home from TROY, and his strategy to rid his kingdom of troubles brewed by his absence, constitute the story of the 12,000-line epic poem the *Odyssey*. Written down circa 750 B.C. and ascribed in ancient times to the poet HOMER, the *Odyssey*—with its associated poem, the *Iliad*—stands at the beginning of Western literature, as one of the greatest works of Western literature.

The name Odusseus means "angry." The Romans, in retelling the myth, latinized the name to Ulixes, from which we get the form Ulysses. In some versions, Ulysses or Odysseus was said to be secretly the son of the cunning hero SISYPHUS, begotten on Anticleia, wife of the Ithacan king Laertes. In any case, Odysseus was raised as Laertes' son and succeeded him as king after Laertes abdicated in old age. Soon, however, Odysseus was summoned, along with other Greek vassal kings, to bring troops and ships to the expedition against Troy organized by King AGAMEMNON of MYCENAE. Odysseus left behind a dedicated and highly intelligent wife, PENELOPE, and their infant son, TELEMACHUS. The Trojan War itself was to last for 10 years.

In the *Iliad*, which describes events of the war's 10th year, Odysseus appears as one of the foremost Greek commanders, valiant in combat and wise in counsel. He often is associated with the Greek hero DIOMEDES. Together they make a night raid against the Trojan allies (book 10) and later, amid fierce fighting, they stop a Greek retreat (book 11). Being the swiftest-running Greek aside from ACHILLES, Odysseus wins the footrace at the Funeral Games of PATROCLUS (book 23). He also wrestles the hero AJAX (1) to a draw, using skill to offset his opponent's greater strength. Later Greek writers, such as SOPHOCLES (400s B.C.), enlarged on the rivalry between these two heroes, contrasting the cleverness or deviousness of Odysseus with the simplicity or honesty of Ajax.

The chronological sequence of Odysseus' homeward progress—not presented in this simple order in the *Odyssey*—is as follows: Leaving the ruins of Troy, he and his 12 ships raid the coast of THRACE but are beaten off by the warlike Cicones (book 9). Thereafter the voyage enters the realm of fable and is no longer geographically recognizable. (However, certain of its supernatural landmarks seem to be associated with the shores of southern ITALY and eastern SICILY, which real-life Greek mariners were exploring in the 800s and 700s B.C.)

Lashed to the mast, Odysseus endures the Sirens' hypnotic song while his men row on, their ears plugged with beeswax. This famous incident from Homer's *Odyssey* is shown in a red-figure scene from an Athenian vase, circa 480 B.C. The Sirens, although never precisely described by Homer, were pictured by Greek artists as being part woman, part bird.

The weathered, wily face of Odysseus, as imagined by an unknown sculptor, probably from the early first century A.D. This carving, discovered in 1957, was part of a marble statue group in the Roman emperor Tiberius' grotto at Sperlonga, south of Rome.

The voyagers' next landfall is among the friendly Lotus-eaters, who are perpetually narcotized by their magical *lotos*-fruit food. Fleeing from this seductive danger, the squadron puts ashore in the territory of the one-eyed monsters known as Cyclopes (book 9). Odysseus is captured by one of them, named POLYPHEMUS, but escapes after blinding the creature. Polyphemus prays to his father, the god POSEIDON, for revenge.

Next Odysseus and his ships reach the island of the hospitable wind king, Aeolus (book 10). Aeolus gives Odysseus a tied-up bag containing all the winds except the favorable one needed for Odysseus' voyage home. But out at sea, while Odysseus sleeps, his men greedily open the mysterious bag: The released winds blow the ships back to the isle of Aeolus, who refuses to help again. Later, Odysseus loses all ships and crews except his own to the cannabalistic Laestrygonians.

The hero's ship arrives at the island of the beautiful witch CIRCE (book 10). After saving his men from Circe's evil magic, Odysseus becomes her lover. She tells him that in order to reach home he must journey down to the Underworld and consult the ghost of the seer TIRESIAS. In the Underworld (book 11) Odysseus observes the ghosts

of several prominent people, including his own mother and such former comrades as Achilles and Ajax.

After lingering with Circe for a year, Odysseus reembarks, having received Circe's directions for reaching home. His ship bypasses the SIRENS, survives the channel of SCYLLA and the whirlpool Charybdis, and reaches the island where the cattle of the sun god, HELIOS, graze. There, despite his warnings, his men slaughter the cattle for food. Consequently, the ship is destroyed at sea by the thunderbolt of the great god ZEUS. Everyone aboard perishes except Odysseus, who drifts to the isle of the amorous CALYPSO (book 12).

Seven years later, at Zeus' command, Calypso allows Odysseus to depart in a boat of his own construction (book 5). But he is shipwrecked again, this time by Poseidon, in vengeance for Polyphemus' distress. With the help of the goddess ATHENA, Odysseus reaches the land of the virtuous Phaeacians and encounters the princess NAUSICAA, who directs him to the palace of her father, King Alcinous (book 6). There the hero, welcomed royally (books 7–8), tells the tale of his wanderings since Troy (books 9–12).

By then nearly 10 years have passed since Troy's destruction and Odysseus' departure for home and nearly 20 years have gone by since he first left Ithaca. Unknown to Odysseus, Ithaca has meanwhile fallen prey to more than 100 visiting lords and petty kings, who crowd the palace, competing to marry the presumably widowed Penelope; she remains elusive, in the hope that Odysseus might yet return. The prince Telemachus, now a young man, has gone abroad in search of his father, hoping to free his home from the loathsome suitors.

The *Odyssey*'s entire second half (books 13–24) describes Odysseus' return to Ithaca, his espionage there in disguise as a beggar, and how—with the help of Athena, Telemachus, and two trusted retainers—he destroys the suitors, reclaims his wife and throne, and restores harmony to the island.

The story of Odysseus' death, not told in the *Odyssey*, was described in a later epic poem, now lost, called the *Telegonia*. Tiresias in the Underworld had predicted that Odysseus' death would come from the sea, but Odysseus reigned for many happy years on Ithaca. Meanwhile his three illegitimate sons by Circe grew to manhood. The youngest, Telegonus ("distant born"), set out to meet his father and somehow encountered Odysseus en route, not knowing who he was. A fight ensued, and Telegonus slew Odysseus. (This episode bears comparison with the similar tale of OEDIPUS.) The weapon that Telegonus used was a sharp fishbone, thereby fulfilling the terms of Tiresias' prophecy.

Odysseus is the best-known example of the "trickster" type of Greek mythical figure. Other tricksters from Greek myth include the god HERMES, the demigod PROMETHEUS, and Odysseus' putative father, Sisyphus.

(See also AFTERLIFE; EPIC POETRY; PROPHECY AND DIVINATION.)

Odyssey See HOMER; ODYSSEUS.

Oedipus In MYTH, a Theban king who could not escape his FATE. Oedipus married a woman who, unbeknownst to him, was his own mother, Jocasta (Iocastē). The story is mentioned in HOMER's *Odyssey*, but the classic telling of the Oedipus tale is the Athenian playwright SOPHOCLES' extant tragedy *Oedipus the King* (presented circa 429–420 B.C.).

Oedipus was the son of Laius, king of THEBES, and Jocasta, his wife. Warned by prophecy that his son would kill him, Laius abandoned the infant on Mt. Cithaeron, after first running a spike through the baby's feet—from which the child later got his name, Oidipous, "swollen foot." As in all such folktales, the infant did not die but was rescued, in this case by a Corinthian shepherd, servant of the Corinthian king. This king and his wife, being childless, were happy to adopt the boy and pass him off as their own.

Nevertheless, the young Oedipus was taunted by others for being adopted, and eventually he went to DELPHI to ask the god APOLLO who his real parents might be. The oracle withheld this information but told Oedipus that he would kill his father and marry his mother. Disgusted, Oedipus decided never to return to CORINTH. Journeying by chance toward Thebes, he fell into a dispute at a crossroads with a stranger. Not knowing that this was Laius, his own father, Oedipus killed him. Then he reached Thebes, which was at that time being terrorized by a supernatural female monster, the Sphinx.

When the Sphinx encountered people, it would ask them a riddle, and eat them when they failed to guess the answer. The riddle asked: "What goes on four legs at morning, two at noon, and three at evening?" Oedipus deduced the correct answer: a human being. (The "legs" represent, respectively, the baby's hands and knees, the adult's upright legs, and the elderly person's legs and cane.) At this point the Sphinx killed itself or was killed by Oedipus. Acclaimed by the Thebans, Oedipus now married the newly widowed queen, Jocasta, whom he did not know to be his mother. In most versions they had four children: the girls ANTIGONE and Ismenē and the boys Eteocles and Polynices. These children were also doomed to unhappy ends.

In Sophocles' play, these events are revealed through Oedipus' own careful investigations (prompted by a plague and famine in Thebes that can be resolved only by discovering the murderer of Laius). Jocasta hanged herself in grief, and Oedipus blinded himself and abdicated his throne, going into exile with Antigone. In Sophocles' play *Oedipus at Colonus* (performed 401 B.C.), the aged hero is shown as having wandered to Colonus, an outlying Athenian village (and Sophocles' own home). There Oedipus disappeared from earth, having been taken up by the gods.

The subject of Oedipus and the Sphinx was a popular one of vase painting in the 400s B.C.

(See also PROPHECY AND DIVINATION; SEVEN AGAINST THEBES.)

oligarchy   "Government by the few." In ancient Greece, an oligarchy was a city-state in which only a small minority of citizens could be admitted to political power. While the mass of citizenry might enjoy certain protections, they lacked any important say in government.

ARISTOTLE and other political thinkers recognized the distinction between an oligarchy and an ARISTOCRACY (where familial lineage was the sole means of deciding who could hold power). In oligarchies, the exclusive ruling circle consisted of rich men, not only men of noble birth. Of course, there was some overlap between these two groups, and oligarchy usually represented a development from aristocracy, due to urbanization and commercialization in the 700s–400s B.C. The inroads made by wealth against privileges of birth are well documented in the indignant verses of aristocratic poets such as THEOGNIS (circa 550 B.C.).

Oligarchy and DEMOCRACY were the two opposing government forms of classical Greece (400s–300s B.C.). The champion of oligarchies everywhere was SPARTA, and its enemy was ATHENS, the beacon of democracy (whose citizens, nevertheless, included an angry pro-oligarchic minority). The PELOPONNESIAN WAR (431–404 B.C.) was in part a conflict between these two forms of government.

Prominent oligarchies of the 400s B.C. included CORINTH, THEBES, and LOCRI (in Greek ITALY).

(See also HOPLITE; POLIS.)

olives   See FARMING.

Olympia   Sanctuary and sports complex, sacred to the great god ZEUS, in the region known as ELIS, in the western PELOPONNESE. This was the site of the most important panhellenic festival of ancient Greece, the OLYMPIC GAMES. The name Olympia refers to Olympian Zeus—that is, Zeus, king of the gods on Mt. OLYMPUS. (However, Mt. Olympus itself stands in THESSALY, in northeastern Greece, hundreds of miles from Olympia.)

The elaborate complex, which has been partly restored by archaeologists since 1829, lay in a pleasant valley, near the confluence of the rivers Alphaeus and Cladeus. ARCHAEOLOGY has revealed traces of pre-Greek settlement, and it may be that Olympia—like DELPHI and certain other Greek sanctuaries—was a holy place for the inhabitants of the land long before the first Greek invaders arrived circa 2100 B.C. The site was inhabited by Greeks of the Mycenaean era (circa 1600–1200 B.C.), but Olympia truly emerged after 776 B.C., the traditional date for the first Olympic Games.

In its heyday, Olympia contained hostels, restaurants, a meeting hall, and many other amenities, but it was never a city in the political sense, for it had neither citizens nor a government. The heart of Olympia was the sacred precinct, or Altis—a walled enclosure whose name apparently comes from *alsos*, "grove." From early times the Altis contained a temple of the goddess HERA and a shrine to the hero PELOPS (mythical founder of the Games). In 457 B.C. the magnificent Doric-style temple of Zeus was completed inside the Altis. This building was wrecked by an earthquake at the end of antiquity, but the foundations and platform are visible today, as are the temple's huge, collapsed column drums (made of a limestone visibly comprised of fossilized shellfish). The temple's pedimental sculptures, now partly preserved in the Olympia Museum, showed two mythical scenes: the battle between Lapiths and CENTAURS at King Pirithous' wedding, and the chariot race of Pelops and Oenomaus. The temple housed a colossal cult statue of Zeus, fashioned by the Athenian sculptor PHIDIAS (circa 430 B.C.) and reckoned as one of the SEVEN WONDERS OF THE WORLD for its size and solemn majesty.

The statue, about 40 feet high, showed the god seated, enthroned. One ancient viewer objected that Zeus' head was so close to the roofbeams that he could not have stood up without taking off the roof, but in the temple's darkened interior this disproportion was probably not obvious. The god's flesh was made of ivory; his cloak, sandals, and accoutrements were GOLD. A human-size figure of the goddess NIKĒ, in ivory and gold, stood atop his right-hand palm, and his left hand held a scepter of various precious metals, tipped with a golden eagle. His cloak was inlaid with images of animals and lilies, and on his head was a golden wreath of facsimile olive leaves.

This statue had the power to awe those who stood before it. The Roman orator Quintilian (circa A.D. 70) stated that it added something to human religion. The great work survived through the Roman era until the 400s A.D. when, having been removed to the eastern Roman capital at Constantinople, it was accidentally destroyed in a fire. Our scanty knowledge of the statue's appearance comes from its image on ancient Elean coins and its description by the travel writer PAUSANIAS (2) (circa A.D. 150).

Modern archaeological excavations at the site of Phidias' workshop at Olympia have uncovered traces of ivory and gold, and terra-cotta molds (apparently for the casting of Zeus' golden cloak). Also, rather spectacularly, a red-figure ceramic jug has been found, inscribed with Phidias' name—his "office coffee mug."

Other ancient splendors of Olympia included many statues of prior victors and a row of 12 treasury houses, built between about 600 and 480 B.C. by various rich states of the Greek world, to hold offerings to the god. Of these, the Treasury of the Sicyonians has been reconstructed by modern archaeologists. Along the valley were the structures for the sporting events: a huge GYMNASIUM; a hippodrome for horse and chariot races; and a stadium (now fully restored) for footraces, where 40,000 spectators could be seated on grass embankments that had no need of benches.

(See also ARCHITECTURE; MILTIADES; SCULPTURE.)

Olympic Games   Oldest and most important of the ancient Greek sports-and-religious festivals. Open to all male, free-born Greeks, the Olympic Games were held every four years in honor of the god ZEUS, at OLYMPIA, in the western PELOPONNESE. Traditionally the games were said to have been established by the hero HERACLES, or alternatively by the hero PELOPS, but in practical terms the Greeks recognized 776 B.C. as the inaugural date, the date at which a record of victors was first begun.

The games were administered by the people of ELIS (the general region of Olympia). However, after King PHEIDON of ARGOS marched his troops into Olympia and took over the games in (probably) 668 B.C., he handed Olympia over to be run by the people of the nearby town of Pisa. The Pisans held Olympia for about 90 years, until the Eleans, with Spartan support, won it back (circa 580 B.C.).

In the earliest years, the contests took place all on one day and consisted of only two events: WRESTLING and the

footrace (one lap of the stadium). Later, perhaps after King Pheidon's intervention, the games were enlarged to include races for saddle horses and for horse- and mule-drawn CHARIOTS. By 471 B.C. the games spanned five days—involving competitions, religious sacrifices, and feasting—and included most of the famous sports events of classical Greece: BOXING; PANKRATION; the PENTATHLON; boys' categories in footrace, wrestling, and boxing; and, the final event of the sequence, the footrace for men in armor. A single event for girls, a footrace, also may have been part of the games. For every contest, the prize for victory was nothing more than a garland of olive leaves.

The games were of immense political and social importance. To Olympia every four years crowded the most influential people (almost exclusively men) of the Greek world, as spectators and competitors. Statesmen such as the Syracusan tyrant HIERON (1) (circa 468 B.C.) and the Athenian ALCIBIADES (circa 416 B.C.) spent fortunes developing chariot teams to enroll at the games. Famous writers such as HERODOTUS and EMPEDOCLES gave readings at Olympia, probably at privately organized side events. Drinking parties of the rich and powerful took place. The spirit of the Olympic Games—religious fervor and aristocratic pride—can be found in the victory odes of the Theban poet PINDAR (circa 470 B.C.).

The four-year cycle between Olympic festivals was known as an Olympiad and was used as a chronological device in the ancient world. For instance, "the 67th Olympiad" signified the period that we would call 512–508 B.C., which began 264 years (66 by 4 years) after the first Olympiad (776 B.C.). Sometimes the ancients remembered a specific year by noting that that was the year when so-and-so won the footrace at Olympia.

The most striking aspect of the games was the Olympic Truce, which was announced by heralds in all major cities of mainland Greece months before the start of each Olympic festival. Whatever wars might be in progress elsewhere, the region of Elis and the site of Olympia were sacrosanct for the truce. During the PELOPONNESIAN WAR, for example, visitors from enemy cities coexisted peaceably at Olympia during the games. In 420 B.C. the Spartans, having attacked Elean territory after the truce was declared, were barred from competing that year. To enforce the ban, several thousand troops from Elis, ARGOS, and MANTINEA (all enemies of Sparta) guarded Olympia during the competition. A more notorious truce-breaking occurred in 364 B.C., when the Pisans and Arcadians invaded Olympia and fought a battle against the Eleans that raged within the sanctuary itself.

The games remained prestigious—although less central—in the Roman era (after about 150 B.C.), and survived nearly to the end of antiquity. In A.D. 393 they were abolished by an edict of the Byzantine emperor Theodosius I that prohibited all pagan festivals.

(See also SPORT.)

Olympias   See ALEXANDER THE GREAT.

Olympus, Mt.   Tallest mountain in mainland Greece and, in MYTH, the celestial home of ZEUS and most of the other important gods. The Olympus massif, whose height reaches 9,570 feet, rises in the region of THESSALY, in northeastern Greece.

The religious awe that Mt. Olympus inspired probably had to do with the mystery of its summit, often cloaked in stormclouds appropriate to the weather god Zeus. The massif also served as a barrier to any invasion force moving southward from MACEDON, and it may have represented an important milestone for the first immigrating Greeks of circa 2100 B.C.

Throughout Greek literature, Mt. Olympus is identified as the royal court of Zeus—the "Olympian"—and of his foremost subordinate gods. These 12 Olympian gods, aside from Zeus, were: HERA, POSEIDON, APOLLO, ARTEMIS, ATHENA, APHRODITE, ARES, DEMETER, DIONYSUS, HERMES, and HESTIA. They comprised the important deities for most spheres of life, with the exception of the Underworld.

(See also AFTERLIFE; HADES; RELIGION.)

Olynthus Major Greek city of the northwestern Aegean region known as CHALCIDICE. Located on the coastal plain of the Chalcidic mainland, at the head the Gulf of Torone, Olynthus was originally a settlement of the non-Greek, native Bottiaeans. It was occupied by local Chalcidic Greeks after the non-Greek settlement was destroyed by the army of the Persian king XERXES, in 479 B.C.

With its southern neighbor POTIDAEA, Olynthus became one of the two premier cities of Chalcidicē. After Potidaea revolted from the Athenian-controlled DELIAN LEAGUE (432 B.C.), Olynthus served as a regional fortress against the retaliatory Athenian invasion. Soon thereafter, Olynthus became the capital of the newly formed local federation known as the Chalcidic League.

In 349 B.C. war broke out between the Chalcidic League and the aggressive Macedonian king PHILIP II. Philip besieged Olynthus. Despite the efforts at ATHENS of the statesman DEMOSTHENES (1) to organize military aid for Olynthus (in his three extant speeches called the *Olynthiacs*), Philip captured the city with the help of traitors within (348 B.C.). He razed Olynthus to the ground and sold the inhabitants as SLAVES. Reports of the pitiful condition of these captives provided Demosthenes with material for more of his speeches, the famous *Philippics*. Demosthenes' *Third Philippic* (341 B.C.) contains the comment that a visitor to the Olynthus site would never know that a city had stood there.

Olynthus was excavated by American archaeologists in A.D. 1928–1934. The site, free from any later buildings, has provided one of the clearest pictures of the floor plans of ordinary ancient Greek houses. The city was laid out on a grid pattern, an admirable example of the Greek town planning that developed in the mid-400s B.C.

(See also ARCHAEOLOGY.)

orators See RHETORIC.

Orchomenus Name of three ancient Greek cities, the most important being "Minyan" Orchomenus in BOEOTIA, in central Greece. The other two cities called Orchomenus were in ARCADIA and THESSALY. The name seems to be Greek, meaning "place of the strong battle rank."

The site of Boeotian Orchomenus was first excavated by archaeologist Heinrich Schliemann in A.D. 1880. Buildings uncovered from the Mycenaean era (1600–1200 B.C.) include a palace and an elaborate tomb. Such remnants—along with certain hints in extant Greek MYTH—indicate that Orchomenus was one of the wealthiest Mycenaean-Greek cities, surpassing its neighbor and enemy, THEBES.

This Mycenaean Orchomenus was the seat of a rich and powerful clan named the MINYANS (Minuai), who may originally have come from the Thessalian Orchomenus. The Boeotian city thrived from its location—controlling the land route through central Greece—and from its fertile farmland, claimed by the engineered draining of nearby Lake Copaïs.

But by the main centuries of ancient Greek history, Orchomenus had declined greatly, due to Thebes' dominance and the refilling of Lake Copaïs. Orchomenus fought as a Theban ally against ATHENS in the PELOPONNESIAN WAR (431–404 B.C.) but sided with SPARTA against Thebes during the CORINTHIAN WAR (395–386 B.C.). After Thebes had replaced Sparta as the foremost Greek power (371 B.C.), revenge was swift. Orchomenus was destroyed by the Theban-led Boeotian League in 364 B.C. It was rebuilt, then again destroyed in 349 B.C.

(See also ARCHAEOLOGY; MYCENAEAN CIVILIZATION; POTTERY.)

Orestes In MYTH, the son of King AGAMEMNON and CLYTAEMNESTRA, of the city of MYCENAE (alternatively, of the city of ARGOS). In the best-known version of the tale, told by the Athenian playwright AESCHYLUS in his Oresteian Trilogy (presented 458 B.C.), Orestes murders his mother and her lover, Aegisthus, in vengeance for their murder of his father. But, having broken the bonds of nature by committing matricide, Orestes is driven mad by the demonic FURIES, the agents of divine punishment. Orestes flees to DELPHI—and then, in Aeschylus' version, to ATHENS—and is eventually cleansed of his sin. The Greeks were intrigued by the horror and legalism of Orestes' dilemma—compelled by filial duty to avenge his father (even against his mother), but doomed to be punished for doing so. A similar myth developed around the Argive hero ALCMAEON (1).

The earliest surviving version of the story is a mention in HOMER's epic poem the *Odyssey* (written down circa 750 B.C.). There Orestes has killed Aegisthus, who (in this version) is the principal culprit in Agamemnon's murder. It is not stated specifically whether Orestes has also killed Clytaemnestra, but it is implied. Homer says nothing of the Furies.

The Furies may have been introduced into the story by STESICHORUS (circa 580 B.C.), the sophisticated Sicilian-Greek poet whose work had a strong influence on Aeschylus and other later writers. Stesichorus wrote a narrative poem called the *Oresteia*, which survives only in fragments. In this version the Furies pursued Orestes and Apollo gave him a special bow, designed to ward them off.

The Orestes story also is told in surviving plays by SOPHOCLES and EURIPIDES. A side-by-side comparison of the three great Athenian tragedians' versions sheds light on differing values and messages. Aeschylus is concerned (primarily) with settling Orestes' controversial position in the world order and (secondarily) with glorifying the playwright's beloved Athens. Orestes is introduced in the second play of Aeschylus' trilogy, the *Libation Bearers* (*Choēphoroi*), as having been ordered by Apollo to take vengeance on Clytaemnestra and Aegisthus. Assisted by his friend Pylades and sister ELECTRA, Orestes can hardly bring himself to kill his mother but, having done so, he is immediately assailed by the Furies. The trilogy's third play, the *Eumenides* (kindly ones), features Orestes' legal acquittal at the hands of the Athenian homicide court, the AREOPAGUS, with the help of Apollo and ATHENA.

The version presented in Sophocles' *Electra* (circa 418–410 B.C.) is closer to the simple Homeric story. Orestes shows only a little natural reluctance to do the deed, and afterward there are no Furies. Euripides in his *Orestes* (408 B.C.) and *Electra* (probably 417 B.C.), seeks to bring out the myth's most disturbing implications. His Orestes emerges, not as a heroic figure, but as an enthusiastic killer who has sexual feelings for his own sister. After the crime, the Furies (not shown onstage) are made to be simply phantoms of Orestes' troubled mind. In Euripides' *Iphigenia in Tauris* (circa 413 B.C.), Orestes must expiate his guilt by journeying to the Crimean peninsula, at the edge of the known world.

Orestes' slaying of Aegisthus was a popular subject of vase painting and SCULPTURE in the 500s and 400s B.C.

Orion   In MYTH, a hunter of enormous size, skilled with bow and arrow. He is variously described as the son of the earth goddess GAEA or as the son of the god POSEIDON and Euryalē the Gorgon. Orion was killed by the goddess ARTEMIS. His offense may have been his arrogant challenge of Artemis to a discus-throwing contest or his attempted rape of one of her attendants. Or his great hunting prowess may have proved to be an affront to the jealous goddess of the hunt. In any case, his huge corpse was set in the sky as the constellation Orion.

Orpheus   In Greek MYTH, Orpheus was a Thracian lyre-player and singer of supernatural ability. Although THRACE was a savage, non-Greek land, Orpheus was a civilizing influence; his song could charm not only humans and beasts, but trees and stones as well. He served as one of the Argonauts who sailed with the Thessalian hero JASON (1) to win the Golden Fleece. But the best-known story about Orpheus tells how when his wife, Eurydicē, died from snakebite, he ventured down into the Underworld to seek the god HADES' permission to bring her up from the dead. Permission was granted, on condition that Orpheus not look back as Eurydicē followed, until they should reach the upper world. But as they approached the upper world Orpheus, not hearing her footsteps behind him, did look back, and with a cry Eurydicē disappeared forever.

Despondent, Orpheus neglected all other WOMEN, thereby infuriating the Thracian women—or, more specifically in some versions, those female followers of the god DIONYSUS known as MAENADS. These women tore Orpheus limb from limb in the course of a Dionysian frenzy and pitched his head into the Thracian River Hebrus. The head floated, singing, out to sea, to the Aegean island of LESBOS.

The background to these legends is uncertain. Orpheus may have been an actual poet and musician, perhaps of royal blood, in Thrace circa 700 B.C. He was said to be the founder of the mystic religious cult known as ORPHISM, which arose in the Greek world in the 600s B.C. and revered certain religious poems supposedly written by him. He also was said to have introduced male HOMOSEXUALITY among the Thracians, in consequence of his later aversion to women. Alternatively, he was said to have lived celebate after Eurydicē's death.

For the classical Greeks, Orpheus symbolized the civilizing or soothing power of MUSIC. One well-known Athenian vase painting of circa 500 B.C. shows him playing his lyre and singing to a group of charmed (but fierce-looking) Thracian warriors. The Athenian playwright AESCHYLUS wrote a tragic trilogy (now lost) about Orpheus.

(See also LYRIC POETRY.)

Orphism   Religious cult, arising probably in the 600s B.C., with followers eventually at ATHENS and the Greek cities of SICILY and southern ITALY. Orphism's beliefs were set down in religious poems supposedly composed by the Thracian musician ORPHEUS. These poems, now lost, are known from mentions by later writers.

Orphism emphasized a belief in the transmigration of souls. This concept, later adapted by the Greek philosophers PYTHAGORAS (circa 530 B.C.) and EMPEDOCLES (circa 450 B.C.), claimed that a dead person's soul generally goes on to inhabit another life form, whether human, animal, or vegetable. Like the ELEUSINIAN MYSTERIES and other mystery cults, Orphism promised its followers a happy existence in the AFTERLIFE.

The Orphic MYTH centered on a holy infant named Zagreus, who was seen as an alter ego of the god DIONYSUS. In this peculiar tale, perhaps derived from Thracian legend, the great god ZEUS seduced or raped his own daughter, the goddess PERSEPHONE. She bore the baby Zagreus, whom Zeus intended to make supreme among the gods. However, the jealous TITANS, after luring the child to their midst, tore him limb from limb and ate him. Zeus angrily blasted the Titans with a thunderbolt, incinerating them. From their ashes emerged the first humans. Zagreus' living heart, saved from the butchery, was now swallowed by Zeus, who passed it on supernaturally in begetting his next divine son, Dionysus.

In Orphic belief, Zagreus' ghoulish murder was a kind of Original Sin, in which all humans shared. Consequently, atonement had to be made to Zagreus' bereaved mother, Persephone, who, as queen of the Underworld, judges people after death. Apparently Orphism viewed the cycle of transmigration as a kind of purgatory: Only by living successive lives (on earth and also in the Underworld) could a person's soul be cleansed of its Original Sin. Reincarnation might mean reward or punishment, depending on Persephone's judgment of one's conduct in a prior life. The soul might descend to inhabit beasts or paupers, or ascend to statesmen and heroes. The Orphic devotee hoped eventually for release from this round of reincarnations. "I have escaped from the wheel of pain," one Orphic verse announces, referring to the achievement of bliss in the other world. The golden leaves discovered by archaeologists in Greek tombs of southern Italy from circa 400 B.C., containing precise instructions for the soul's procedure on reaching the Underworld, are thought to be Orphic documents.

Related to this spiritual preparation, Orphism also involved strict behavioral taboos. Orphics were notorious for abstaining from hunting and meat-eating, and even from wearing wool. They also were known for sexual abstinence—or at least for discouraging heterosexual intercourse—in order to thwart pregnancy; pregnancy was lamentable because it imprisoned another soul in a human body. The Orphics said that the body (*soma*) was the tomb (*sēma*) of the soul. The higher life was the purified soul's existence in Persephone's realm.

Orphism's otherworldliness and contempt for bodily existence recommended it to the Athenian philosopher PLATO (circa 370 B.C.). But its strict code of behavior struck many Greeks as crankish and kept Orphism on the mystic fringe.

ostracism Political practice unique to Athenian DEMOCRACY in the 400s B.C., whereby the people could vote to banish any citizen for 10 years. Created in reaction to the tyrannies of PISISTRATUS and HIPPIAS (1), ostracism was intended for use against wealthy politicians who, while not guilty of wrongdoing, might still be suspected of hoping to seize supreme power.

Once a year, at an appointed time in winter, the citizens in ASSEMBLY voted on whether an ostracism should be held that spring—no candidates were named. If the majority voted yes, then the ostracism vote itself took place a few months later. There each citizen had the chance to write down the name of one person for exile. Because clay potsherds were the ancient world's equivalent of scrap paper, each voter used a sherd (Greek: *ostrakon*, plural: *ostraka*), on which to scratch the intended victim's name. The potsherds gave this unique practice its name, *ostrakismos*. The vote was secret, with officials making sure no one handed in more than one ballot. If a quorum of 6,000 votes was reached, then the man with the most votes had to remove himself from the city within 10 days.

Unlike more punitive forms of banishment, ostracism allowed the victim to retain his Athenian citizenship and property while absent and to return after the allotted 10 years. During the state emergency of the PERSIAN WARS, two Athenians under ostracism, ARISTIDES and XANTHIPPUS, were allowed to return immediately (480 B.C.).

According to ARISTOTLE's treatise the *Constitution of Athens*, the ostracism law was created by the reformer CLEISTHENES (1) circa 508 B.C., soon after the expulsion of Hippias. Clearly, the law was aimed against Hippias' friends who might dream of reinstating a dictator. But what puzzles modern scholars is that the people did not vote their first ostracism until 487 B.C. (The victim was Hippias' kinsman Hipparchus.) This 20-year delay has led some historians to suggest that the ostracism law really was created soon after the Battle of MARATHON (490 B.C.), but was later falsely ascribed to Cleisthenes.

After Hipparchus, four more men were soon ostracized amid the political turmoil of the 480s B.C., and it is thought that about eight others fell victim during the rest of the 400s B.C. These included the left-wing statesman-soldier THEMISTOCLES (circa 471 B.C.) and the conservative statesman-soldier CIMON (462 B.C.). This clearly shows that ostracism had quickly become a tool by which the two political parties could attack each other. Many other Athenians were named as ostracism candidates but did not receive majority votes—as indicated by the 64 names compiled from the several thousand *ostraka* discovered in modern archaeological excavations of the Athenaian AGORA and other sites.

The last successful ostracism was of the demogogue Hyperbolus (417 B.C.), after which the law fell out of use.

(See also THUCYDIDES [2].)

Votes for banishment: four *ostraka* (potsherds), incised with the names of four Athenian politicians of the 400s B.C., found in archaeological excavations of the Athenian agora. The ballots read, clockwise from top left: "Aristides, son of Lysimachus," "Themistocles, son of Neocles, of the deme Phrearrus," "Pericles, son of Xanthippus," and "Cimon, son of Miltiades." Each of these prominent men earned suspicion that he was aiming to make himself dictator, and of the four only Pericles avoided being voted into ostracism.

P

paean See LYRIC POETRY.

Paestum See POSEIDONIA.

painting For the painting of scenes and portraits, the Greeks used various surface materials, including POTTERY, wooden panels, stone walls, leather, and textiles. Of these, only pottery has preserved Greek painting in any quantity, due to the ceramic's durability and the paint being baked into the clay during firing. Secondarily, there exist some examples of tomb paintings—from walls of graves or the insides of coffins—that were preserved underground. But most other Greek paintings have vanished over time, as their surfaces crumbled to dust.

The paintings on pottery can reveal only a certain amount about general Greek painting, because pottery was a miniaturist discipline, with colors limited to those that would survive firing. Modern scholars must therefore rely partly on indirect evidence, such as ancient writers' descriptions of specific paintings and techniques and certain surviving mosaics and wall paintings of the Roman era, believed to be copies of earlier Greek paintings.

Aegean painting traditions predate the Greeks. By the third millennium B.C. the Minoan peoples were making pigments by processing and grinding down various vegetation, and were painting lively murals at their palaces in CRETE. Among several existing examples is the famous Toreador Fresco (circa 1500 B.C.). Minoan whorl designs, sea-life subjects, and human figures helped inspire Mycenaean-Greek paintings on walls and pottery, from the 1500s B.C. onward.

Examples of Greek painting in the centuries after the Mycenaeans' fall are represented now mainly by works on pottery: the whorls and other designs of the Protogeometric style (circa 1050–900 B.C.), the stick-figure human and animal silhouettes of the Geometric style (circa 900–700 B.C.), and the colors, detail, and primitive depth perception of the ripe Corinthian style (circa 625–550 B.C.).

CORINTH was a center for painting in the 600s B.C., and one rare surviving example of nonvase painting of this era is a wooden plaque of circa 500 B.C., found at a site near Corinth. The painting shows a sacrificial procession, in a style reminiscent of the Corinthian vase painting that had peaked a century before. With its vivid red, blue, and flesh tones, the work suggests the color range available to the ancient Greek painter—well beyond the restricted colors of the vase painter. The painting's background is white, which would remain the chosen background color for decades to come.

ATHENS arose to replace Corinth as the artistic center of Greece in the mid-500s B.C. By about 460 B.C. the splendors of the Athenian AGORA included a Painted Colonnade (Stoa Poikilē) where large wall paintings on mythological and patriotic subjects were displayed. This era produced the first Old Master of Greek painting—the artist Polygnotus, born at THASOS but active at Athens in the 470s–450s B.C. He is best remembered for two mythological wall paintings, neither of which survives: a scene of the fall of TROY, at the Painted Colonnade, and a scene of the Underworld, painted at the Meeting House at DELPHI. Polygnotus apparently used no foreshortening but did innovatively convey depth by showing his human figures standing at different levels—a device that subsequently appears in Athenian vase painting. He also pioneered the technique called encaustic, which used heated wax for applying colors. Micon of Athens is another painter remembered for public wall paintings in these years.

Use of perspective is said to have developed at Athens in the mid-400s B.C. partly through the production of painted backdrops for Athenian stage performances. The second half of the 400s B.C.—the Athenian cultural heyday—saw greater use of shading to indicate depth: The great name of the day was Apollodorus of Athens, surnamed *skiagraphos*, "shade painter." His younger contemporary and imitator Zeuxis—from Greek ITALY but active at Athens circa 430–400 B.C.—was known for his illusionary realism: Supposedly Zeuxis' outdoor paintings of grapes attracted hungry birds. Apollodorus, Zeuxis, and the painter-sculptor Parrhasius (active at Athens in these years) were all known for their mythological subjects. But scenes from everyday life—a woman spinning, young WOMEN playing at dice—are indicated by later Roman copies of lost originals.

With Athens' decline after the PELOPONNESIAN WAR (431–404 B.C.), the Peloponnesian city of SICYON emerged briefly as a painting center. But it was the Macedonian king ALEXANDER THE GREAT (reigned 336–323 B.C.) who inspired a revival of Greek painting, at his wealthy court. The best-known surviving painting of this school comes from the Macedonian city of Aegae (modern Vergina, in northern Greece), from a wall of the royal tombs excavated in A.D. 1977. The scene shows HADES, the god of death, abducting the maiden goddess PERSEPHONE. Factual subjects are indicated by the admirable Roman mosaic from the "House of the Faun" at Pompeii (first century A.D.), showing the Persian king Darius III fleeing from Alexander at the Battle of the Issus; this is believed to be a copy of a Greek painting of the late 300s B.C., presumably commissioned by Alexander or one of his followers. The crowded battle scene shows intricate use of detail and foreshortening.

The HELLENISTIC AGE (300–150 B.C.) saw changes that anticipate the surviving examples of Roman painting (first centuries B.C. and A.D., mainly). These include development of landscape and other backgrounds to the extent of dwarfing the foreground figures and a new value placed on *trompe l'oeil* affects, such as mural-painted false windows or corridors, opening to realistic background scenes. Subjects became more whimsical while expert techniques were forgotten, and by about A.D. 50 the Roman writer Pliny the Elder could describe painting as a dying art.

(See also ARCHITECTURE.)

Palladium   See ATHENA.

Pallas   See ATHENA.

Pan   Minor but picturesque Greek god of flocks, herds, and mountain pasturage, worshipped mainly in the rural area of ARCADIA. Pan's name probably meant "feeder" or "pasturer." He was described as a son of HERMES and was pictured as having a human face, torso, and arms, with a goat's legs, hooves, ears, and horns. He played tunes on a reed pipe, traditionally called a panpipe. As a god of animal fertility, he was thought of as highly sexed.

At ATHENS, Pan had a shrine on the ACROPOLIS and yearly festivities involving torch races. The Athenians had adopted his worship after an incident associated with the Battle of MARATHON (490 B.C.), in which the god supposedly appeared to an Athenian dispatch runner in Arcadia.

Panathenaea   Athenian festival celebrated annually in midsummer, and especially elaborately every fourth year. Although its roots were in prehistory, the Panathenaea ("all-Athens") was instituted by the Athenian tyrant PISISTRATUS in the mid-500s B.C., to serve as a nationalist-religious holiday. Considered to be the birthday of the city's patron goddess, ATHENA, the festival included sacrifices, sports competitions, and a procession of citizens (illustrated on the famous PARTHENON frieze). At the fourth-yearly celebration—called the Great Panathenaea—the procession would carry a new *pelops*, or lady's gown, up to the Athenian ACROPOLIS. There the *pelops* would be presented to the ancient olivewood statue of Athena Polias ("of the city") that was housed, after about 410 B.C., in the Erectheum.

Pandora   See PROMETHEUS.

pankration   One-on-one combative SPORT combining BOXING and WRESTLING. The rules allowed for punching, kicking, strangleholds, and twisting of arms and fingers; only biting and gouging were forbidden. Opponents fought naked, sometimes with light boxing thongs to protect their hands. Victory occurred when a contestant either signaled surrender or was unable to continue. The bout was overseen by a referee, but pankration ("total strength" or "complete victory") was a brutal sport: Contestants sometimes died from injuries.

Pankration was an important event at the OLYMPIC GAMES and other major athletic festivals. Successful pankratiasts were celebrities throughout the Greek world. Ancient vase paintings and stone carvings show pankration scenes, and further details are given by ancient writers and athletes' tombstone inscriptions.

Pankration was a relative latecomer to Greek sport. It is not mentioned in HOMER's poems or in any writing before the 400s B.C.

Panticapaeum   Important Greek colony on the northern BLACK SEA coast, founded circa 625 B.C. by settlers from MILETUS. Panticapaeum (Pantikapaion, "ditch town") guarded the Straits of Kerch in the eastern Crimean Peninsula, in what is now southern Ukraine. With access to the interior steppes along the River Don, the city thrived from the export of Ukrainian grain to the distant cities of the central Greek world. In the era circa 438–404 B.C., under the powerful Thracian-Greek dynasty of the Spartocids, Panticapaeum became the premier supplier of grain to imperial ATHENS. Threatened in later centuries by local Scythian and Sarmatian peoples, Panticapaeum passed into the hands of King Mithridates VI of Pontus, who made it his capital (110 B.C.). After Mithridates' defeat by the Romans, the city came under control of ROME (63 B.C.).

papyrus   See WRITING.

Paris   In MYTH, a handsome Trojan prince, son of King PRIAM and Queen HECUBA. The name Paris (which has nothing to do with the capital city of France) means "wallet" in Greek. It may have been a nickname, since Paris had another name, Alexander (Greek: Alexandros, "defender").

With the help of the goddess APHRODITE, Paris seduced Helen, wife of King MENELAUS of Lacedaemon, and stole her away to TROY. This provoked the TROJAN WAR, in which the Greeks besieged Troy for 10 years to get Helen back. At the war's end, Paris was dead, Troy destroyed, and Helen was contentedly reunited with Menelaus.

In HOMER's *Iliad*, which presents events of the war's 10th year, Paris is portrayed as affable but irresponsible, in contrast to his elder brother, the champion HECTOR. In the *Iliad* (book 3) Paris challenges Menelaus to single combat but is nearly killed, and is saved only by Aphrodite's intervention. The goddess carries him back to his bedchamber within the walls of Troy and summons Helen to join him there.

The other tales of Paris' life were told in various post-Homeric works, including two epic poems now lost, the *Cypria* ("Tales of Cyprus") and *Little Iliad*. The narrative is as follows. Hecuba, while pregnant with Paris, dreamed that she gave birth to a flaming torch (a reference to the fated destruction of Troy). On advice of soothsayers the infant Paris was handed over to be killed, but—as in all similar Greek legends, such as that of OEDIPUS—the entrusted servant merely abandoned the baby in the wild. Rescued by shepherds, Paris grew to manhood and reclaimed his royal title after an incident of recognition at Troy.

The Judgment of Paris, the hero's most famous adventure, was recounted in the *Cypria* and was a favorite subject in Greek vase painting and other art from the 600s B.C. onward. After the goddess Eris ("strife") had set off a female rivalry in heaven (by offering a golden apple inscribed "For the fairest"), the goddesses HERA, ATHENA, and Aphrodite invited Paris to judge a beauty contest among them. They presented themselves naked to the young man. He chose the love goddess Aphrodite, whose bribe had been to promise him the world's most beautiful woman. Paris thus won Helen and the favor of Aphrodite. But henceforth he and all the Trojans suffered the hatred of Hera and Athena.

In the war's 10th year, sometime after the Greek champion ACHILLES had slain Hector, it was Paris, with the god APOLLO's help, who killed Achilles. Paris shot an arrow that struck Achilles in his only vulnerable spot, his right heel, and the wound proved fatal. Later Paris was slain by the Greek hero PHILOCTETES. Helen was briefly married to Paris' brother or half brother Deiphobus, before rejoining with Menelaus after the war.

(See also EPIC POETRY; HELEN OF TROY.)

Parmenides   The most important Greek philosopher before SOCRATES. Parmenides lived circa 515–440 B.C. in the Greek city of Elea, in southwest ITALY. A member of a rich and influential family, he supposedly served Elea as a statesman and lawgiver. Supposedly one of his teachers was the immigrant philosopher XENOPHANES. Parmenides founded a school of philosophy at Elea, the Eleatic School. He is said to have visited ATHENS at least once, circa 448 B.C.; this occasion is the setting for the Athenian philosopher PLATO's fictional dialogue *Parmenides.*

Parmenides' importance to PHILOSOPHY lies in his breakthrough idea that the material world as we observe it is not real. Prior to Parmenides, the foremost philosophers sought out essential reality by searching for the material world's fundamental ingredient (hence the use of the name *material philosophers* to describe such thinkers). THALES (circa 570 B.C.) decided that the basic element is water; ANAXIMENES (circa 546 B.C.) called it *aer,* or mist. For Parmenides, however, the real world is not what we see and feel around us; rather, it can be understood only by mental contemplation. Reality is an ideal world, a timeless unity, free of change, variation, generation, or destruction. Parmenides' notion of an ideal world, inherited in part from the Pythagoreans, would later contribute significantly to Plato's Theory of Forms (mid-300s B.C.).

Parmenides' best-known written work was a didactic poem in hexameters, couched in the oracular language of religious revelation and perhaps titled *On Nature.* The poem was comprised of two parts, "The Way of Truth" and "The Way of Seeming." Of these, the former seems to have been the more important part, and, luckily, we possess about 150 lines of it. "The Way of Truth" describes how an unnamed goddess instructs the poet that only the statement "It is"—or "It exists"—can be valid. Nothing, the goddess says, does not exist.

Parnassus, Mt.   See DELPHI.

Parthenon   Temple of the goddess ATHENA the Virgin (*Parthenos*), built atop the highest part of the Athenian ACROPOLIS in 447–432 B.C. Still partly standing today, the Parthenon is considered the most glorious building of the ancient world—the epitome of the Doric order of ARCHITECTURE and the embodiment of the classical Greek values of harmony and proportion. The Parthenon crowned a rich and powerful ATHENS that was the cultural capital of Greece. The temple was a focus of international tourism in ancient times, as it still is today; its survival for nearly 2,500 years is due to its superior design and workmanship. In fact, the Parthenon would be in far better shape today had it not been blown half apart in A.D. 1687 amid fighting between Turks and Venetians, when a Venetian cannonball ignited a Turkish powder magazine inside the building.

The Parthenon was constructed in the statesman PERICLES' building program, overseen by the sculptor PHIDIAS and meant to celebrate the conclusion of hostilities with PERSIA. Designed by the architects Ictinus and Callicrates, the temple was intended as the building program's masterpiece. The Parthenon stood close to the site of a prior temple of Athena, which had been burned down by the occupying Persians of 480 B.C..

Intriguingly, the Parthenon never housed the ancient, olivewood statue of Athena Polias (which was eventually placed in the nearby Erectheum). The Parthenon was devoted to a grander, more modern vision of goddess and city. The Parthenon's cult statue was a 35-foot-high, GOLD-and-ivory figure of Athena, sculpted by Phidias. Although this statue is lost, marble copies and descriptions by ancient writers give some idea of its appearance. The goddess wore a warrior's helmet, sleeveless robe, and biblike aegis. On the palm of her right hand stood a human-size statue of the goddess NIKĒ (Victory). Athena's left hand grasped the top rim of an upright shield, which stood grounded by her left leg; a giant snake coiled within the shield, and a long spear stood upright in the crook of her left arm. The

The Parthenon's western facade, circa 430 B.C. The steps held dedicatory plaques and larger-than-life statues commemorating various citizens. On the building's western pediment, the sculpture group showed the mythical contest between Athena and Poseidon for possession of Athens.

The northwestern view of the Parthenon today. Many of the decaying sculptures of the upper structure were purchased and removed in 1800–1812 by the British ambassador to Greece, Thomas Bruce, Lord Elgin. These Elgin Marbles, as they are called, remain today in London's British Museum.

flesh of Athena and Nikē was sculpted in ivory; the rest was gold.

The temple, measuring 228 feet by 101 feet at the top of its entrance stairs, is built almost entirely of marble quarried at Mt. Pentelicus, near Athens. It has 17 columns at each side and eight columns across the front and back, rather than the more usual six.

Like other elaborate ancient Greek temples, the Parthenon had its upper structures adorned by painted, marble SCULPTURE. There were carved figures at the two pediments (the triangular gables at front and back), at the metopes (plaques along the sides), and along the frieze (the continuous paneling that ran around the outside top of the inner block, or cella). The pedimental sculptures were carved entirely in the round. The east pediment showed the mythical birth of Athena from the head of ZEUS; the west showed the contest between Athena and POSEIDON for mastery of Athens.

The metopes' figures were carved nearly in the round. They showed various mythical combats, each symbolizing the triumph of civilization over barbarism or West over East—a victory that, to the classical Greek mind, had been actually reenacted in the recent PERSIAN WARS. The combats in sculpture were: on the east, gods versus GIANTS; on the south, Lapiths versus CENTAURS; on the west, Greeks versus AMAZONS; and on the north, the TROJAN WAR.

The frieze carvings were in bas-relief. They were unique in Greek temple sculpture for showing not a mythical scene but a scene from real-life Athens—namely, the procession at the festival of the PANATHENAEA. This daring glorification of contemporary Athens must have struck some ancient observers as crass or sacriligious.

Through decay and vandalism over the centuries, many of the sculptures fell or were torn from their high places. In A.D. 1800–1812, the British ambassador to Greece—Thomas Bruce, Lord Elgin—bought and removed many of

the sculptures; these "Elgin Marbles" are displayed in the British Museum. At the Parthenon today, modern casts of the Panathenaic frieze replace those carted off by Lord Elgin.

(See also CAVALRY.)

Patroclus In HOMER's epic poem the *Iliad*, Patroclus is the close friend of the Greek hero ACHILLES in the TROJAN WAR. Patroclus is the older of the two. In the *Iliad* (book 22) he charges into battle wearing Achilles' armor (with Achilles' permission) to rally the retreating Greeks. Slain by the Trojan champion HECTOR, Patroclus is avenged ferociously by Achilles, who kills numerous Trojans, including Hector. Book 23 describes Patroclus' funeral games.

The Greeks of the 500s B.C. and later, judging Patroclus and Achilles by the social standards of those days, assumed that the two men were sexual partners, in the familiar aristocratic pattern of an older man educating and protecting a younger. But the *Iliad* never states this to be, and modern scholars believe that such socially sanctioned male couplings were not part of the poet Homer's world (circa 750 B.C.) nor part of the Mycenaean world (circa 1200 B.C.) that inspired the *Iliad*. The modern consensus is that Homer imagined Patroclus strictly as Achilles' comrade and confidant.

(See also HOMOSEXUALITY.)

Pausanias (1) See PERSIAN WARS; PLATAEA, BATTLE OF.

Pausanias (2) Greek travel writer, circa A.D. 150 said to have been a physician from ASIA MINOR. His surviving prose work, *A Description of Greece*, is a valuable source of information on the physical monuments and local histories of all the important mainland Greek cities and sanctuaries, including ATHENS, SPARTA, DELPHI, and OLYMPIA.

Pegasus   See BELLEROPHON.

Peirithous   See PIRITHOUS.

Peisistratus   See PISISTRATUS.

Peleus   Mythical hero, father of ACHILLES. Peleus' father was Aeacus, king of AEGINA and a son of the god ZEUS. As an adult, Peleus had various misadventures—including murdering his half brother and accidentally killing a man during the CALYDONIAN BOAR HUNT—before arriving in THESSALY. There he married the sea goddess Thetis, a daughter of NEREUS. The match was arranged by Zeus, who (knowing that Thetis' son was preordained to be greater than his father) was anxious that she marry a mortal, rather than a god. But in order to win the resentful Thetis' hand in MARRIAGE, Peleus had to defeat her in a WRESTLING bout, during which she took on various harmful shapes: fire, water, a lion, a serpent, and an ink-squirting cuttlefish.

Later, when Thetis abandoned him and the baby Achilles, Peleus took the boy to be raised by the centaur Chiron in the glens of Thessaly's Mt. Pelion. Eventually Peleus and Thetis were reconciled, and Peleus was made immortal in repayment for his hardships.

(See also CENTAURS.)

Pelias   See JASON (1).

Pella   Capital city of the kingdom of MACEDON, from about 410 B.C. until the Roman conquest of 167 B.C. Located in the Thermaic coastal plain, about 25 miles northwest of the modern Greek seaport of Thessaloniki, Pella was in existence by the mid-400s B.C. and was chosen by the Macedonian king Archelaus I as his capital. In this, Pella replaced the old royal capital at nearby Aegae (modern Vergina). Archelaus, who reigned 413–399 B.C., made Pella a showplace of Greek building and culture.

At Pella ruled such mighty Macedonian kings as PHILIP II, ALEXANDER THE GREAT, and PHILIP V, in the 300s and 200s B.C. But under Roman rule Pella was totally overshadowed by the development of Thessalonica (Thessaloniki).

Archaeological excavations at Pella since A.D. 1957 have turned up the remnants of grand homes containing superb mosaics of circa 300 B.C. These works include the famous Lion Hunt of Pella and a scene of the god DIONYSUS riding on a leopard's back.

Peloponnese (Peloponnēsos)   "Island of PELOPS," the wide, jagged-shape, mountainous peninsula of southern Greece, measuring 132 miles north to south and 134 east to west (maximum) and connected with central Greece by the narrow (seven-mile-wide) Isthmus of CORINTH. Other major sites in the Peloponnese include MYCENAE, SICYON, EPIDAURUS, OLYMPIA, and SPARTA. In classical times the Peloponnese was the heartland of Dorian-Greek culture—the "other" Greece, in conflict with democratic, Ionian-Greek ATHENS.

Despite a coastline hundreds of miles long, the Peloponnese has few natural harbors; only Corinth and the Argos region became great from seaborne TRADE. Coastal geography includes four mountainous peninsulas jutting southward into the AEGEAN SEA; from west to east, these four capes were named Acritas, Taenarum, Malea, and (far to the northeast) Scyllaeum. Between these peninsulas, the sea forms three wide gulfs, whose shoreline typically features a fertile alluvial plain. Important events of Greek history centered on these three coastal flats named, from west to east, the plains of MESSENIA, LACONIA, and ARGOS.

It was on the plain of Argos that Greek civilization began, circa 1600 B.C. There stood the chief group of Mycenaean-Greek cities: Mycenae, Argos, and TIRYNS. After the MYCENAEAN CIVILIZATION's downfall, the Peloponnese was largely overrun by DORIAN GREEKS, circa 1100–1000 B.C. But one Mycenaean contingent fled inland to the mountains of the central Peloponnese, to reemerge in historical times as the hard-bitten "highlanders" of ARCADIA. Of the Dorian settlements, one in particular grew to power—Sparta ("the sown land"), located in interior Laconia. By 600 B.C. the Spartans had conquered Messenia as well as Laconia, thus controlling the entire southern Peloponnese.

Two other principal Peloponnesian regions were ELIS, in the west, and ACHAEA, in the northwest. The inhabitants of these two regions were not Dorians but had affinities in race and language with the Greeks of AETOLIA and other areas north of the Corinthian Gulf.

(See also FARMING; GREECE, GEOGRAPHY OF; GREEK LANGUAGE.)

Peloponnesian War   The world war of ancient Greece, in which SPARTA and its allies defeated ATHENS and its allies after 27 years of intermittent fighting, 431–404 B.C. The defeat ended Athens' 75-year reign as the richest and most powerful Greek city, and left Sparta in control of an exhausted Greece.

The Greeks called this conflict "the war between the Peloponnesians and the Athenians." One side consisted of Sparta and its core allies throughout the PELOPONNESE, grouped into a Spartan-led alliance known as the Peloponnesian League. The Athenian side included most of the islands and cities of the AEGEAN SEA and western ASIA MINOR—perhaps 200 states in all—organized into the Athenian-controlled DELIAN LEAGUE. Most of the Delian League states were peopled by IONIAN GREEKS, while the Spartan side consisted largely of DORIAN GREEKS. With some important exceptions, the war was fought along these ethnic lines, Ionians versus Dorians.

The fighting ranged across the Greek world, from eastern SICILY to BYZANTIUM. Yet the war often was mired in stalemate: Sparta remained invincible on land and Athens supreme at sea. Against the mainly defensive Athenian strategy, the cautious Spartans frequently proved ineffectual.

Low military technology contributed to the stalemate: Greek armies of the 400s B.C. had not yet developed the tactics and machinery for storming fortified cities. Catapults and elaborate siege-towers were still unknown. Siege warfare was a slow and expensive game of encircling the enemy city and waiting for its defenders to succumb to starvation, fear, or betrayal. A fortified city suppliable by sea, such as Athens, was impossible to capture unless the besieger also could block the harbor.

Sparta finally gained the upper hand in the war after two critical changes occurred. First, the Athenian disaster at SYRACUSE (413 B.C.) inspired a widespread revolt of Athens' unhappy Delian League allies. Second, Sparta allied with the non-Greek empire of PERSIA (412 B.C.), which brought Persian funds to finance Spartan war fleets. Gradually Sparta was able to challenge Athenian naval supremacy, and the Spartan naval victory at AEGOSPOTAMI (405 B.C.) made Athens' surrender inevitable (404 B.C.). The victorious Spartans—over the objections of some of their allies—spared the defeated city from destruction.

Modern knowledge of the war comes partly from archaeological evidence (such as inscribed Athenian legislative decrees) but mostly from the works of ancient writers—principally the Athenian historian THUCYDIDES (1), who served briefly as a general in the war and who produced a highly reliable history of it. Thucydides' account, unfinished, breaks off in describing the events of 411 B.C. The war's final years, 411–404 B.C., are recorded in the first part of the *Hellenica,* by the Athenian XENOPHON. Xenophon, like Thucydides, witnessed some of the events he describes. A third contemporary source was the Athenian comic playwright ARISTOPHANES, who addressed war politics in such plays as *Acharnians* (performed 425 B.C.) and *Peace* (421 B.C.). Later sources include the biographer PLUTARCH (circa A.D. 100).

In a famous passage, Thucydides states that the ultimate cause of the war was Spartan fear of the growth of Athenian power. For decades prior to 431 B.C., Athens showed military aggression toward its Delian League allies and toward the supposedly sovereign states of central Greece. But, more than Athens' navy and army, Sparta feared Athens' magnetic cultural appeal and the appeal of Athenian DEMOCRACY. Thus, the Peloponnesian War was partly a conflict between two political ideologies—radical democracy, embodied by Athens and most of its allies, and the old-fashioned government of OLIGARCHY, championed by Sparta. Sparta's control of the Peloponnese depended on the oligarchic ruling class in the ally cities. The politically excluded underclass in these cities posed a tremendous potential threat, liable to be exploited by the agents of democratic Athens.

Similarly, in almost every democratic city allied with Athens, a few rich right-wingers secretly wished for oligarchy. Athens itself contained such people; they seized power briefly in 411 B.C. The defections and attempted defections of cities on each side in the war came partly from this internal struggle between rich and poor, one side naturally pro-Spartan, the other naturally pro-Athenian. The most devastating civil strife occurred at the northwestern Greek city of CORCYRA, which at the war's start was an important Athenian ally but which soon (427 B.C.) succumbed to riots between democrats and oligarchs.

An immediate cause of the Peloponnesian War was Athens' cavalier attitude toward the Thirty Years' Peace of 446 B.C., which had ended prior hostilities between Athens and Sparta. Although this religiously sanctioned treaty had guaranteed that Athens would not interfere with other mainland Greek cities, the treaty was flouted by the Athenian statesman PERICLES, who apparently saw a future war as inevitable. The most provocative Athenian violations were domination of AEGINA and designs against MEGARA (1). The notorious Megarian Decree, passed at Athens circa 433 B.C., effectively prohibited the Megarians from conducting TRADE throughout the Athenian empire.

Next, Athens slipped into an undeclared war against CORINTH, an important Spartan ally. First, the Corinthian navy defeated a joint Corcyrean-Athenian fleet at Sybota, near Corcyra island (433 B.C.), then Corinthian and Athenian land troops fought outside POTIDAEA (432 B.C.). Incited by the Corinthians, the Peloponnesian League voted for war against Athens (432 B.C.). The Athenian ASSEMBLY, on Pericles' urging, rejected Spartan ultimatums (fall and winter 432–431 B.C.). Both sides prepared for war.

The Spartan bloc was joined by THEBES and other cities of BOEOTIA (a central Greek region traditionally hostile to Athens). Because Boeotia bordered Athenian territory to the northwest and the Peloponnese bordered it to the west, the peninsular Athenian home region of ATTICA now faced enemies along its entire land frontier.

But Athens was ready, with its navy of 200 warships, its impregnable fortifications (including the LONG WALLS, linking the city to its harbor at PIRAEUS), and its huge financial reserve, supplied by the tribute-paying Delian League allies. Foremost among the Delian states were the eastern Aegean islands of LESBOS and CHIOS, which contributed warships and crews to Athens' navy. Outside of the league, Athens had several allies on a non-tribute-paying basis, including Corcyra, the Boeotian city of PLATAEA (a traditional enemy of Thebes), and the northwestern regions known as ACARNANIA and Amphilochia.

The Athenian war strategy was innovative and odd. Faced with a numerically superior enemy land army that boasted the best soldiers in Greece (the Spartans, Thebans, Eleans, and Arcadians), Pericles planned to avoid land battles. Instead, Athens would rely on static defense by land and amphibious attacks by sea. If the Spartans attacked Athens, as expected, the Athenians would wait out the assault, relying on the city's fortifications and sea imports of food and supplies. Meanwhile, Athenian war fleets could sail from Piraeus even if a Spartan army was outside the walls. Carrying land troops aboard, Athenian fleets could attack enemy coastal sites. These attacks might be intended as either harassments or as wholesale attempts to capture enemy cities.

The drastic aspect of Pericles' strategy was that it would abandon the Athenian countryside to the Spartans. The Athenian rural population would be housed in camps within the Long Walls and the walls of the city and Piraeus—at least for the summer campaigns if not for the whole war. The farmers' livestock would be evacuated to the Athenian-controlled island of EUBOEA.

Pericles apparently believed that his one-sided strategy would not be needed for long. He seems to have expected the war to end in about three years, with Sparta's allies demoralized. But events were destined to take a far different turn.

The war's first military action was a failed Theban night attack against Plataea (spring 431 B.C.). Then, in midsummer, the Spartan king ARCHIDAMUS led a Peloponnesian army of 24,000 HOPLITE infantry into Athenian territory. Events followed much as Pericles had forseen: Archidamus devastated crops and farm buildings in the evacuated countryside but was powerless against Athens' walls. After perhaps a month the Peloponnesians withdrew. This pattern of ineffectual Spartan invasion would recur in several following summers. The war's first 10 years (431–421 B.C.) are sometimes known as the Archidamian War, after the Spartan king who led these initial attacks.

After Archidamus had marched away that first year, Pericles himself led 13,000 hoplites and 100 warships against nearby Megara. But, like so many assaults against fortified cities in the Peloponnesian War, this one failed—as did Pericles' similar attack on EPIDAURUS the following year.

In the summer of 430 B.C., Athens suffered its first catastrophe of the war. When the Spartan army invaded for the second time, and the Athenian rural population again gathered within the walls, a deadly plague broke out at Athens (the disease's identity is unknown to us.) The cramped, unsanitary conditions of the Athenian refugees provided a breeding ground for the contagion, which raged for the next year and a half, and recurred briefly two years after that (427 B.C.). Eventually one-fourth of the population, perhaps 50,000 people, had died.

The plague removed Athens' initial advantage in the war, creating problems in manpower and leadership. Pericles himself died of the sickness in 429 B.C., at about age 65. Most significantly, the plague encouraged the Spartans by seeming to show the will of the gods.

In 429 B.C. the Spartans began besieging the defiant little city of Plataea. Two years later, the starving inhabitants surrendered (after half the garrison had escaped in a nighttime breakout through the Spartan siege lines). To please their Theban allies, the Spartans executed the male prisoners—200 Plataeans and 25 Athenians—and sold the WOMEN as SLAVES. Later, the empty town was leveled. Many Greeks felt that Plataea's destruction was an impious act, because during the PERSIAN WARS Plataea had been the site of a famous oath sworn by the Spartans and other Greeks, promising forever to honor the city.

The Athenians, now deprived of Pericles, sought more aggressively to drive Sparta's allies out of the war. Based at the West Locrian seaport of NAUPACTUS in northwestern Greece, the Athenian admiral Phormion won two spectacular naval victories against larger, Corinthian-led fleets (429 B.C.). In 426 B.C. the Athenian general DEMOSTHENES (2) brought an army of light-armed Acarnanians and Amphilochians to victory against their traditional enemy, the Corinthian colony of AMBRACIA. Demosthenes failed to capture Ambracia, but his experiences in the hilly, wooded terrain taught him the possible advantage of javelin men and slingers over heavier-armed hoplites. The subsequent Athenian development of light-armed projectile troops was one of the few tactical advances of the Peloponnesian War.

Meanwhile the important Delian League city of MYTILENE, capital of Lesbos, revolted against Athens (428 B.C.) but fell to a besieging Athenian force (427 B.C.). Back in Athens, the citizens' assembly debated how to punish the captured Mytileneans. Rescinding a vote led by the Athenian politician CLEON for massacre of the entire adult male population and enslavement of the women and children, the assembly voted a lesser penalty—execution of the 1,000 Mytileneans judged responsible for the rebellion. Certain speeches of this Mytilenean Debate provide a famous scene in Thucydides' history (book 3). Later, amid the war's increasing brutality, the Athenians would begin treating rebellion more cruelly.

Cleon, now emerging as Pericles' political heir, was a leader who could have won the war for Athens. Although not a military man by training, he embodied an aggressive, can-do spirit that was at odds with the Pericles-derived passivity of much of the Athenian officer class. In partnership with Demosthenes, Cleon would soon give Athens the upper hand in the war.

This critical campaign began when Demosthenes established an Athenian naval base at PYLOS, in the southwestern Peloponnese, only about 50 miles west of Sparta (425 B.C.). Pylos had a superb harbor, sheltered by the forested, inshore island of Sphacteria ("wasp place"). The site was especially apt because it lay in the region of MESSENIA, whose subjugated population might be incited to rebel against their Spartan overlords.

The threat of an Athenian-inspired Messenian revolt brought Spartan naval and land forces rushing against Pylos. In the ensuing sea battle, the victorious Athenians trapped 420 Spartan hoplites on Sphacteria island. The loss of these soldiers—then confined by Athenian naval patrols around the island—was disastrous for Sparta; all of the young men belonged to the dwindling Spartan warrior class, and perhaps a third of their number were "Spartiates" (Spartiatai), members of the military elite.

At the summer's end, 425 B.C., Sphacteria was overrun by an Athenian force of 800 hoplites and several thousand light-armed troops, under Demosthenes. Of the Spartans, 292 surrendered and survived. The political impetus for this Athenian assault had come from Cleon, who was present at Pylos as an elected Athenian general.

Taken as prisoners to Athens, the 292 Spartans served as diplomatic pawns until 421 B.C., when they were returned home. That their government welcomed them back rather than punishing them for their surrender—as a prior generation of Spartan leaders might have done—is a sign of how valuable their bloodstock was thought to be, amid Sparta's long-term problem of falling birthrate.

Sparta then faced its low point in the war. Its supposedly invincibile army had been humiliated, while the Athenian fort at Pylos became a rallying point for rebellious Messenians. At Athens, the Pylos victory made Cleon politically supreme, ushering in a period of Athenian military boldness in pursuit of total victory (425–421 B.C.). Spartan peace overtures were rebuffed. An imperialistic tone is evident in the surviving Athenian inscription called the Thoudippos Decree, which effectively doubled the tribute owed to Athens by its Delian League subject allies (autumn 425 B.C.). But Cleon's aggressive policy brought some Athenian setbacks. Particularly, a foolhardy Athenian invasion of Boeotia was defeated at the land battle of Delium, fought near a temple of Delian Apollo alongside the Straits of Euboea (autumn 424 B.C.).

More bad news followed for Athens: The brilliant Spartan general BRASIDAS, having marched a small Peloponnesian army to the north Aegean coast, convinced several Delian League subject towns to defect to the Spartan side. Then, in a night attack, he captured the strategic Athenian colony of AMPHIPOLIS (December 424 B.C.). The city's loss caused an outrage at Athens: Without possessing a single ship, Brasidas had dealt a blow to Athens' northeastern shipping route and had opened the threat of a wider north Aegean revolt. The Athenian general inside Amphipolis apparently had been killed or captured, while the other Athenian general in the vicinity had arrived with a naval squadron too late to save the city. This general was Thucydides, the same man whose written history of the war is our major source. Blamed by his countrymen for the loss, Thucydides was banished from Athens.

However, Brasidas was not supported by a new Spartan king, Pleistoanax, who viewed these successes mainly as bargaining chips to end the war. Yet a year's truce for peace talks passed inconsequentially (spring 423–422 B.C.), while in the north, Brasidas—ignoring the truce—continued to bring local towns into revolt against Athens.

Promising to recapture Amphipolis and eliminate Brasidas, Cleon sailed from Athens in the summer of 422 B.C. with 30 warships and a small army. But for all his pugnacity, Cleon was not an experienced tactician; while scouting the terrain around Amphipolis, he and his force were surprised by a sortie from the city, led by Brasidas. In the fight, the Athenians retreated with 600 dead, including Cleon—but Brasidas too had been killed. The Battle of Amphipolis removed at once the two most warlike leaders of the day.

In March 421 B.C. the two sides agreed to peace, which was to last 50 years. (In fact, it lasted barely three.) Modern scholars call this treaty the Peace of NICIAS, named for the cautious soldier-politician who was then the leading Athenian. The peace's terms amounted to a triumph for Athens, which was left free to maintain its empire and to recapture rebellious subject cities. It was all that Pericles could have hoped for. The peace was followed quickly by an alliance between Athens and Sparta, designed to give them dual hegemony over the rest of Greece.

The losers in this process were the Spartan ally states, which had endured hardship only to see their interests ignored. Certain states then began looking for a new leader in the Peloponnese. This they found in the previously neutral, east Peloponnesian city of ARGOS.

By 420 B.C. an anti-Spartan axis had arisen, consisting of Argos, MANTINEA, ELIS, and—surprisingly—Athens, which flouted the Peace of Nicias. The driving force behind this Athenian policy was ALCIBIADES, a volatile, left-wing aristocrat, perhaps 30 years old, who had emerged as Nicias' political opponent. Alcibiades embodied the reckless mood of an Athenian citizenry that—far from being exhausted by 11 years of war—was overconfident from the advantageous peace.

Provoked by the four-city alliance, the Spartans marched. At the Battle of Mantinea, in the central Peloponnese, a Spartan-led army of 3,600 hoplites crushed a roughly equal force of Argive, Mantinean, Elean, and Athenian troops (418 B.C.). The alliance dissolved, bringing Elis and Mantinea back into the Spartan fold. Hostilities between Athens and Sparta did not recommence outright, but the angry Spartans awaited their chance.

In 416 B.C., the Athenians—tightening their control over their Aegean sea routes—landed a force on the Cycladic island of MELOS. The Melians were Dorian Greeks, descended from Spartan colonists. They had shown prior hostility to Athens and were now commanded to submit to Athenian rule. Under truce, Melian leaders refused the Athenian demands: Political-philosophical arguments purporting to be the positions of both sides are presented in the famous Melian Dialogue in Thucydides' history (book 5). Besieged, the Melians were starved into surrender. The Athenians killed all adult male prisoners and sold the women and children into slavery. The island was later repopulated with Athenian colonists.

The year 415 B.C. brought the most fateful event of the war—the Athenian expedition against the city of Syracuse, in southeastern Sicily. A wealthy, seafaring, Dorian-Greek democracy, Syracuse was an ally of Sparta and the enemy of the Ionian Greeks of Sicily and southern Italy.

At Alcibiades' urging, the Athenian assembly voted to send out a major force. The intended capture of Syracuse would give Athens control of the wheatfields that supplied grain to Corinth and other Spartan-allied cities. Although Alcibiades was the natural choice to lead the expedition, the Athenians mistrusted him because of his flamboyant and licentious private life. So they split the command among three generals—Alcibiades, Nicias (who had opposed the whole idea of the venture), and the able Lamachus.

Just before the departure, an act of religious terrorism occurred: Under cover of darkness, a group (probably pro-peace, right-wing conspirators) smashed the carved faces and genitals of the stone effigies of the god HERMES, called herms (*hermai*,) that stood in front of houses throughout Athens. This sacrilegious event is known as the Mutilation of the Herms. Undoubtedly intended to create a bad omen against the expedition, the act succeeded well, because the Athenian people irrationally turned their suspicion on the commander Alcibiades.

The armada sailed in late spring. It consisted of 134 warships, 130 supply ships, 5,100 hoplites, and 1,300 archers, javelin men, and slingers (mostly allies and mercenaries), plus about 20,000 crewmen. But Alcibiades was soon summoned back to Athens to face impiety charges. He instead fled to Sparta, where he gave strategic advice against his fellow Athenians. Among other things, he advised that a Spartan officer be sent to Syracuse to organize the city's defense.

The Athenian attack on Syracuse stalled under the indecisive, ailing, 55-year-old Nicias, who took nearly a year to settle down to a siege (the spring of 414 B.C.). Outside the city wall, Athenian troops began constructing a siege wall to cut off the city's landward side; on the seaward side, Athenian naval patrols kept the Syracusan navy (about 80 warships) bottled inside the harbor. When Lamachus was killed in a skirmish at the siege works, Nicias took sole command. Meanwhile, the pessimistic Syracusans were urged to resistance by their statesman-general Hermocrates.

The tide turned when a Spartan general, Gylippus, led a small army of Peloponnesians and Sicilian Greeks into Syracuse on the landward side, around the incomplete Athenian siege wall. Gylippus blocked completion of the Athenians' wall and captured their naval base at the mouth of the bay known as the Great Harbor. This loss forced the Athenians henceforth to beach their ships on an exposed shore at the head of the bay, where they could be bottled in by the Syracusan navy. As troops and supplies trickled into Syracuse by land and sea, the besiegers became the besieged.

In the spring of 413 B.C., in response to Nicias' appeal for help, a second Athenian armada sailed for Syracuse: 73 warships carrying 5,000 Athenian and allied hoplites and 3,000 light-armed troops. The force brought two generals, Eurymedon and (Athens' finest battlefield commander of the war) Demosthenes.

By then—with Nicias' warships having been defeated in a sea battle in the Great Harbor—the Athenian position at Syracuse was so vulnerable that the reinforcements were simply entering a death trap. Demosthenes and Eurymedon urged an evacuation by sea, but Nicias (still the senior general) hesitated, believing he could yet take the city. An unlucky eclipse of the moon (August 27, 413 B.C.) led the superstitious Nicias to delay further. Another sea battle in the Great Harbor brought a second Athenian defeat and the death of Eurymedon. When Nicias finally gave the order to evacuate by sea (September 413 B.C.), the Syracusan navy blocked the Athenians from leaving the harbor mouth. Driven back to shore, the Athenians abandoned their dead and wounded and marched inland, hoping to escape to the neutral territory of the native Sicels. The bedraggled refugees numbered perhaps 40,000, largely crewmen from the ships.

"No prior Greek army had ever endured such a reversal. They had come to enslave others, but were now fleeing in fear of slavery themselves." So writes Thucydides, whose solemn account of the Athenians' final destruction (book 7) is the most moving section in his history. On the seventh day, the Athenian rear column, under Demosthenes, was surrounded and captured. The Athenian main column, under Nicias, was surrounded the next day in a riverbed and massacred with arrows and javelins. Of the approximately 7,000 captured, most died in captivity or were sold as slaves. Nicias and Demosthenes were captured and put to death.

It was the worst disaster in Athenian history: Losses included perhaps 50,000 men, 173 warships, and immense expense. Yet such were Athens' financial and spiritual resources that the city fought on for nearly another nine years (413–404 B.C.). In mainland Greece, the war had recommenced. In the spring of 413 B.C., the Spartans had invaded Athenian territory for the first time since 425 B.C. However, they did not withdraw entirely at the summer's end, but fortified the site called Decelea, about 13 miles north of Athens, to be a permanent base in Athenian territory.

Under personal command of the Spartan king Agis II, Decelea served Sparta much as Pylos had served Athens—as a front-line fort for raids and for rallying of the enemy's disaffected underclass. Thucydides says that 20,000 skilled Athenian slaves escaped to Decelea during the remaining war years.

The Spartan presence at Decelea threw Athens finally onto the defensive: Overland communication outside the walls became a hardship. The Athenian CAVALRY was now obliged to make daily patrols against Spartan war parties in the Athenian countryside. Local resources, such as food and SILVER ore, became less available. The Athenians grew increasingly reliant on purchased supplies imported by ship.

The Peloponnesian War's final nine years are sometimes known as the Ionian War, because the major campaigns shifted to the eastern Aegean and the shorelines of Asia Minor (particularly to the Greek region of western Asia Minor called IONIA). There lay the most important Athenian allies and the sea routes bringing resources westward to Athens; the most essential of these imports was grain, brought by merchant ships from the ports of the BLACK SEA. The vulnerable links in the sea route were the narrow waterways of the HELLESPONT and BOSPORUS, which Athenian forces controlled out of the naval base at SESTOS. It was against these regions that the Spartans directed their operations, aiming first to dislodge the unhappy and heavily taxed Athenian allies, and then to block Athens' supply route and bring the city to starvation.

Sparta's new role as a naval power was made possible because Athens had lost nearly three-fourths of its navy at Syracuse, and because Sparta now found a source of funds for its own shipbuilding and crews' wages, in an alliance with the Persian king Darius II (412 B.C.). In exchange, Darius received the right to reclaim the Greek cities of Asia Minor after the war.

The year after the Sicilian disaster saw the Greek world turn against Athens. Warships from the Peloponnese and Dorian Sicily began to invade the eastern Aegean. The Delian League unraveled amid revolts. The first rebel was Chios, with its considerable navy, followed by MILETUS, EPHESUS, PHOCAEA, CYMĒ, RHODES, Mytilene, and others. Of Athens' greatest east Aegean possessions, only the island state of SAMOS remained loyal. The Athenians responded with new war fleets, built with emergency funds. Samos became Athens' eastern naval base, as amphibious operations began against the rebels. Miletus and Dorian-Greek CNIDUS became Sparta's eastern naval bases.

By winter of 412–411 B.C., the scene was set for Alcibiades' return. After earning the mistrust of his Spartan masters, Alcibiades had fled to Persian protection in western Asia Minor, and from there conducted secret intrigues with the Athenians at Samos to get himself recalled to Athens' service. His chance came when the oligarchic FOUR HUNDRED seized power at Athens, and the Athenian sailors and soldiers at Samos formed a democratic government-in-exile (June 411 B.C.). Alcibiades was brought to Samos and elected as a general.

At Athens, the government of the Four Hundred, using untrained emergency crews, suffered a naval defeat against Peloponnesian ships in the Straits of Euboea (summer 411 B.C.). With all of Euboea now in revolt against Athens, the disgraced Four Hundred soon fell from power.

The next four years marked the final Athenian high point of the war, under Alcibiades' leadership. The Spartans suffered a disastrous naval defeat at Alcibiades' hands at CYZICUS (410 B.C.) and soon offered peace on generous terms. The overconfident Athenians, however, rejected the offer as insufficient. In 408 B.C. Alcibiades captured the strategically vital, rebellious Delian city of Byzantium. In 407 B.C. Alcibiades was recalled to Athens to receive extraordinary powers of command.

But when Alcibiades returned to the east Aegean in the autumn of 407 B.C., he faced a ruthless new opponent—the Spartan LYSANDER, commanding a fleet at Ephesus. After Lysander defeated a lieutenant of Alcibiades in a sea battle (spring 406 B.C.), Alcibiades was blamed at Athens and was not reelected as general. Knowing that his life was in danger from his ungrateful fellow citizens, the greatest naval commander of the day retired to an estate beside the Hellespont.

Athens—foolishly trying to re-create its lost empire—was running out of resources. To man the newest fleet off the blocks, the Athenians resorted to enlisting slaves and resident aliens as oarsmen (a duty usually taken by lower-income Athenian citizens). In August of 406 B.C. these motley crewmen won the huge sea battle of Arginusae ("the white islands") off Mytilene. But the triumph was soured when the crews of the Athenian disabled vessels—about 5,000 men—perished in a storm immediately afterward. At Athens, hysteria reigned over the failure to rescue the crews: Six of the responsible naval commanders, including Pericles' son, were executed. Worse, on the urging of the demagogue Cleophon, the assembly again rejected Spartan peace proposals.

Now a weak and inexperienced Athenian naval command led an irreplaceable force of about 200 ships. In the summer of 405 B.C., the wily Lysander sailed from Ephesus with 150 ships and slipped into the Hellespont, where he captured Lampsacus. When an 180-ship Athenian fleet arrived in pursuit, Lysander ambushed and destroyed it at the sea battle of Aegospotami ("the goat's rivers"). The Athenian navy had ceased to exist.

Spartan forces converged on the fortress of Athens, which endured six months of siege, without food imports. At last, the starving inhabitants voted to accept Sparta's terms—peace in exchange for the destruction of certain city fortifications, the surrender of overseas holdings and of most of the remaining warships, the recall of political exiles, and a military alliance with Sparta. The city capitulated in April 404 B.C. Lysander's forces occupied Athens and Piraeus. The Long Walls were torn down to the MUSIC of flutes. It was the end of Athens as the dominant power in Greece.

(See also CONON; CORINTHIAN WAR; EURIPIDES; SOCRATES; SOPHOCLES; THERAMENES; THIRTY TYRANTS, THE; WARFARE, LAND; WARFARE, NAVAL; WARFARE, SIEGE.)

Pelops   In Greek MYTH, Pelops was a prince of a non-Greek kingdom in ASIA MINOR—either LYDIA or Paphlagonia. His father was TANTALUS, who was a son of the god ZEUS and who offended the gods in an infamous episode. Pelops eventually emigrated to southern Greece, where he supposedly founded the OLYMPIC GAMES and fathered the line of kings at MYCENAE. He is commemorated in the name of the southern Greek peninsular region, the PELOPONNESE (Greek: Peloponnēsos, "island of Pelops"). The name Pelops (dark face) may refer to his Asian origin.

The best-known version of the Pelops story is told in the poet PINDAR's First Olympian ode, written in 476 B.C. As a youth in Asia Minor, Pelops became the favorite of POSEIDON (god of horses as well of the sea), who taught Pelops to be a great horseman and charioteer. Arriving in Greece, Pelops went to Pisa (the future site of OLYMPIA). There he asked for the hand in MARRIAGE of Hippodameia, the daughter of King Oenomaus.

Oenomaus secretly intended that his daughter never marry, either because he harbored an incestuous desire for her or because he knew a prophecy that his son-in-law would kill him. His terms of courtship were that anyone might carry Hippodameia off by chariot, but Oenomaus would follow and, upon overtaking, could spear the suitor. Twelve or 13 men had already died in this manner; their heads were nailed above the gates of Oenomaus' palace.

But Pelops cleverly bribed Oenomaus' groom, Myrtilus, to remove the linchpins from his master's chariot axle and replace them with wax. In the contest, Oenomaus' chariot collapsed and the evil king was dragged to death behind his horses. Pelops then married Hippodameia and became king of Pisa. Later he deceitfully killed Myrtilus, either to avoid acknowledging his help or to avoid sharing Hippodameia's favors with him, as promised. But the dying Myrtilus cursed Pelops and his family, and the curse proved effective. Pelops' son ATREUS and grandson AGAMEMNON ruled households riven by bloodshed.

The myth of Pelops and Oenomaus was in part aetiological—that is, it explained the origin of a custom, namely the Olympic chariot race. In historical times Pelops was worshipped as a hero at Olympia, where his purported tomb was located. At Olympia's Temple of Zeus, the sculptures on the east pediment showed Pelops and Oenomaus preparing for the fateful chariot race, with Zeus standing between them and looking toward his grandson Pelops.

(See also CHARIOTS.)

Penelope   In MYTH, the Spartan-born wife of the Ithacan king ODYSSEUS; their son was TELEMACHUS. Penelope faithfully awaited her husband's return during his 20-year absence—10 years fighting the TROJAN WAR and another 10 years for his homeward voyage.

In HOMER's epic poem the *Odyssey*, Penelope appears as a gracious and clever lady. Since about the seventh year of Odysseus' disappearance during his voyage from TROY, she has been pressured to assume him dead and to marry one of the 108 suitors—local nobles and nearby kings—who have flocked to the palace at ITHACA. She puts them off by claiming she cannot remarry until she finishes weaving a shroud for Odysseus' elderly (but still living) father, Laertes. Every night she secretly unravels what she has woven that day, so that the work never progresses. At length the deceit is discovered, and she is compelled to make known the day on which she will choose a new husband. Finally she announces that she will marry whoever can bend and string Odysseus' mighty bow, then shoot an arrow through the empty sockets of 12 axheads sunk in a line. In fact, it is Odysseus himself who performs this feat, having returned to Ithaca disguised as a beggar. Then he slaughters the suitors and reclaims his throne (*Odyssey* books 21–22).

Afterward, Penelope and Odysseus lived happily for years. After his death—according to one post-Homeric version—Penelope married Telegonus, who was Odysseus' son by CIRCE.

pentathlon The "fivefold contest," one of the most prestigious events at such Greek sports festivals as the OLYMPIC GAMES. Contestants competed in the footrace, long jump, throwing the discus, and throwing the javelin. The two men scoring highest in these events were selected for the final contest, a WRESTLING match. The winner of this was declared to be the pentathlete.

(See also SPORT.)

Penthesilea See AMAZONS.

Pentheus See DIONYSUS.

Pergamum One of the wealthiest and most important Greek cities of the later HELLENISTIC AGE (circa 250–133 B.C.). Located on the north-central west coast of ASIA MINOR, about 15 miles inland in the hills along the River Caïcus, Pergamum (Greek: Pergamon) thrived from local SILVER mines and exports of grain, textiles, and parchment (Greek: *pergamēnē*).

We first hear of Pergamum after about 400 B.C., when it was one of many Greek cities of the region called AEOLIS. Soon after 300 B.C. Pergamum was a subject city of the sprawling SELEUCID EMPIRE. But by 262 B.C. the Pergamenes, under their leaders Philetaerus and his adopted son Eumenes, had revolted from Seleucid rule. The royal dynasty of the Attalids—"the sons of Attalus" (Philetaerus' father)—brought the city to its zenith. Pergamum dominated Asia Minor and became a center of Greek culture, with a royal library that was second in size only to the one at Egyptian ALEXANDRIA (1).

Eumenes' successor was King Attalus I (reigned 269–197 B.C.). He defeated the invading CELTS, resisted the ambitions of the Macedonian king PHILIP V, and wisely sided with the Romans against Philip in the Second Macedonian War (200–197 B.C.). Attalus' friendship with ROME helped assure Pergamum's survival amid Roman conquest of the eastern Mediterranean. When Attalus III died in 133 B.C., he bequeathed his kingdom to the Romans.

The grandeur of ancient Pergamum is suggested in the Great Altar of ZEUS, erected by King Eumenes II (reigned 197–159 B.C.) and now housed in a Berlin museum. Carved with marble bas-reliefs showing the gods' mythical combat with the GIANTS, the altar is one of the most admired surviving Hellenistic artworks.

The site of Pergamum, terraced cleverly into high hills, has been excavated since the late 19th century A.D. The partly restored city is one of the most dramatic ancient Greek sites in Turkey. Particularly grand is the ancient THEATER, nestled into a hill near the ACROPOLIS.

(See also WRITING.)

Periander Dictator of CORINTH from about 625 to 585 B.C. The son and successor of CYPSELUS, Periander reigned energetically over a city that was the most prosperous in Greece. Combining total power with wise statesmanship, he was the most famous example of the breed of Greek rulers known as TYRANTS. Later generations counted him among the SEVEN SAGES of Greece.

Periander extended his power through the Greek world by war and diplomacy. He brought Corinth's rebellious colony CORCYRA to heel and captured the Peloponnesian city of EPIDAURUS. Looking to western ASIA MINOR, he forged ties with the Greek city of MILETUS and the non-Greek kingdom of LYDIA. Asked to arbitrate in a war between ATHENS and MYTILENE, he in effect judged in favor of Athens. This act inaugurated a period of Corinthian-Athenian friendship. (The two cities had a mutual trading rival, AEGINA.)

In TRADE, Periander brought Corinth to new heights. Responding to the great demand for Corinthian goods among the ETRUSCANS of western ITALY, Periander enlarged the westward shipping route, establishing Adriatic colonies such as APOLLONIA and EPIDAMNUS to serve as anchorages and local depots. In the distant northeastern region called CHALCIDICE, Periander founded the colony of POTIDAEA. Apparently he also pursued trade with EGYPT, as suggested by the Egyptian royal name Psammetichus that he gave to his son.

Exploiting Corinth's geography, Periander constructed a ships' dragway (Greek: *diolkos*) across the seven-mile-wide isthmus that separates the Saronic Gulf from the Corinthian Gulf. A stretch of this stone roadbed, with grooves for trolley wheels, still can be seen today. The *diolkos* provided Greek merchant ships with an east–west overland shortcut across central Greece—and provided Corinth with revenue from tolls.

Periander's city was the cultural capital of mainland Greece. Poets such as the famous ARION came to Periander's court. The tyrant's public-building program stimulated the development of the form of monumental ARCHITECTURE known as the Doric order. In painted POTTERY, Corinthian artists achieved the style that we call Corinthian and that monopolized the Mediterranean market in that day.

Periander could be ruthless. He used surveillance and intimidation against Corinthians who might pose a threat to his power, and he executed the most important of them. After one of Periander's sons was killed in Corcyra (probably as governor), Periander took 300 boys from Corcyra's foremost families and shipped them as a gift to the Lydians, to be turned into eunuchs. (As it happened, they were rescued.) Periander's passions were uncontrolled: Gossip claimed that he murdered his beloved wife, Melissa, in an irrational rage of jealousy and then, in remorse, had sexual intercourse with her corpse.

Dying at an old age in 585 B.C., Periander left his dynasty beleaguered. His son Psammetichus reigned for three years before being ousted in favor of a liberal OLIGARCHY.

(See also COLONIZATION.)

Pericles The greatest Athenian statesman, living circa 495–429 B.C. Despite his aristocratic blood, he entered politics as a left-wing radical and by the mid-440s B.C. was preeminent in the government, steering ATHENS through its heyday of power and cultural achievement. More than any other individual, Pericles shaped classical Athens as a radical DEMOCRACY and naval power, as a center of learning, and as a city of architectural splendors, ascending to the PARTHENON. Pericles' ambitious vision created this Athens that (to use his own phrase) was an EDUCATION to the rest of Greece. Yet he ended his days in disfavor, after his belligerence toward other Greek states had provoked the PELOPONNESIAN WAR and his defensive military strategy had brought plague and demoralization to his city.

Pericles never held any such post as "president" of Athens, since no such job existed under the democracy. His highest title was *stratēgos*, "military general," a post to which the Athenians elected 10 different men annually. Unlike most other generals, however, Pericles was elected at least 20 times; we know that he served without interruption from 443 to 430 B.C..

Pericles' source of power was the *dēmos*—the common people, the mass of lower- and middle-income citizens. He was a politician they trusted to protect them against the

Fleet commander, imperialist, left-wing politician: Pericles was the statesman who made Athens into the greatest city of its day. This likeness comes from a surviving Roman marble bust, which was probably copied from an Athenian bronze portrait statue (now lost) sculpted by Cresilas, a contemporary of Pericles.

rich and to guide the city. A rigorously intellectual man, Pericles was an accomplished orator, whose speeches in the public ASSEMBLY could sway his listeners to vote for his proposals.

Most of our knowledge of him comes from the Athenian historian THUCYDIDES (1), who wrote circa 430–400 B.C., and from the biographer PLUTARCH, who wrote circa A.D. 100. Thucydides is especially helpful because he knew Pericles personally and understood his politics. Some details are added by other writers and by inscriptions recording Periclean-inspired legislation.

Pericles' father was XANTHIPPUS, a distinguished Athenian soldier-politician. Pericles' mother was Agariste, a niece of the Athenian statesman CLEISTHENES (1) and a member of the noble clan of the ALCMAEONIDS. The family's wealth and influence no doubt helped Pericles enter politics, but he soon disassociated himself from the Alcmaeonids and from other right-wing connections.

At about age 23 he served as *chorēgos* (paying sponsor) for AESCHYLUS' stage drama *The Persians* (472 B.C.). Pericles' involvement in this politically charged play identifies him as a follower of the beleaguered left-wing Athenian statesman THEMISTOCLES. In the 480s B.C. Themistocles had pushed two interrelated programs—radical democracy and a buildup of the navy. The lower-income citizens who followed Themistocles were the same men who served as crews in the labor-intensive warships. Gradually Pericles emerged as Themistocles' political heir.

In 463 B.C., at about age 32, Pericles (unsuccessfully) prosecuted the conservative leader CIMON for bribery. The following year Pericles assisted the radical reformer EPHIALTES in proposing legislation that dismantled the power of the AREOPAGUS. Ephialtes' death and Cimon's OSTRACISM (both in 461 B.C.) left Pericles as the foremost Athenian politician. His influence can be seen in the next decade, in Athens' left-leaning domestic policies and in its bellicose stance toward SPARTA and certain other Greek states, such as AEGINA.

In 461 B.C. Pericles urged the creation of the LONG WALLS, which were to stretch four miles from Athens down to its harbor at PIRAEUS. Completed circa 457 B.C., the walls turned Athens into an impregnable naval fortress. They also brought Piraeus' laborers and citizen-sailors more directly into Athenian politics.

Circa 457 B.C. Pericles sponsored a bill creating jury pay. This important democratic measure effectively opened jury duty to lower-income citizens. It was probably also Pericles who, sometime in the mid-400s B.C., instituted public assistance for the poorest citizens.

By 460 B.C. Athens had slipped into an undeclared war against Sparta and its allies. Repeatedly elected as general, Pericles led a seaborne raid against the enemy city of SICYON (circa 454 B.C.) and led an Athenian fleet to the BLACK SEA to make alliance with Greek cities there that could supply precious grain to Athens (circa 452 B.C.).

Circa 449 B.C. Pericles convinced his fellow citizens to accept an offered peace with PERSIA, officially ending the PERSIAN WARS. To mark the event and to glorify the city, Pericles initiated a public building program, directed by the sculptor PHIDIAS. This program—which created most of the famous buildings still standing today on the Athenian

ACROPOLIS—amounted to a show of dominance over Athens' Greek ally states in the DELIAN LEAGUE. The league had been formed (circa 478 B.C.) as a mutual-defense alliance against Persia, but now Pericles was demonstrating that, rather than reduce or forgive the allies' war dues on account of the new peace, Athens would continue collecting dues and would use them as it pleased.

Pericles' imperialism is also evident in the use of Athenian garrison colonies (known as cleruchies) to punish and guard resistant Delian League states. Among the Greek regions that received these hated, land-grabbing, Athenian settler communities were Aegina, EUBOEA, and the Cycladic island of Naxos.

As reported by Thucydides, Pericles' grim imperialism is summarized in a speech to the Athenian assembly in 430 B.C., after the outbreak of the Peloponnesian War. "It is no longer possible for you to give up your empire," Pericles told the people. "You hold the empire as a tyrant holds power: you may have been wrong to take it, but you cannot safely let it go." He went on in the speech to glorify Athens as "the greatest name in history . . . the greatest power that has ever yet arisen, a power to be remembered forever."

In 446 B.C. the Spartans invaded Athenian territory. Although Pericles was able to appease them with bribery and diplomatic concessions, the resulting Thirty Years' Peace was unpopular at Athens. Pericles was attacked by an Athenian right-winger named THUCYDIDES (2)—not the historian, but his maternal grandfather. By 443 B.C. Pericles and his followers had rallied, convincing the Athenian people to ostracize Thucydides.

Having become preeminent, Pericles became inaccessible. The biographer Plutarch records a change in Pericles' personality at this point; no longer the straightforward man of the people, Pericles adopted a formal style that Plutach calls "aristocratic or even kingly." More than just indulging in arrogance, Pericles apparently was seeking to keep control of the people and to make himself impervious to bribery or undue influence. He became known for never accepting an invitation to dinner or to the drinking parties where rich men discussed politics. His cold, rational aloofness earned him the nickname "the Olympian" (as though he were a god of Mt. OLYMPUS).

His supremacy was severely tested by the revolt of the Delian League allies SAMOS and BYZANTIUM (440 B.C.). The defeat of Samos required a huge Athenian naval effort, and Pericles' generalship seems to have come in for some criticism on the Athenian comic stage. In 440 or 439 B.C. Pericles sponsored a law suspending the performance of all comedy for three years.

He endured an indirect political attack in around 438 B.C., when his friends Phidias and ANAXAGORAS were charged with the crime of impiety. ASPASIA, Pericles' common-law wife, might have been a third defendant. Phidias and Anaxagoras each departed from Athens at this time.

In the late 430s B.C., when hostilities with Sparta and CORINTH loomed, Pericles apparently saw war as inevitable, and refused to appease. It was undoubtedly he who advised the Athenians to make their fateful defense pact with CORCYRA (433 B.C.), which Corinth found so alarming, and he was likewise the author of the bellicose Megarian Decree

(circa 432 B.C.), directed against the nearby city of MEGARA.

When the Peloponnesian War broke out (431 B.C.), Pericles advised defense by land and offensive strikes by sea. With his faith in Athens' immense resources and fortifications, he convinced the rural citizens to evacuate their homes and contain themselves within the city walls, while the invading Spartans ravaged the countryside. This strategy backfired, however, when plague broke out among the Athenian refugees in their unsanitary encampments and swept the city (430 B.C.). At the same time, large Athenian expeditions failed to capture either Megara or (in a siege directed by Pericles himself) EPIDAURUS.

The angry Athenians voted to depose the 65-year-old Pericles from office and fine him a crushing 10 TALENTS. Relenting, they elected him as general for the following year. But in that year, 429 B.C., he died, probably of the plague.

Pericles had two sons by an Athenian wife, whom he divorced for Aspasia circa 450 B.C. By a law of his own sponsoring, Pericles, being an Athenian citizen, could not legally marry Aspasia, who was an immigrant from MILETUS. Still, they lived together for 20 years until his death. Their son, named Pericles, was made an Athenian citizen by a special vote of the assembly, but he was one of the six Athenian generals executed by the people after the sea battle of Arginusae in 406 B.C.

(See also CALLIAS; CLEON; CRATINUS; FUNERAL CUSTOMS; LAWS AND LAW COURTS; MARRIAGE; PROTAGORAS; RHETORIC; SOPHOCLES; THEATER; WARFARE, NAVAL; WOMEN.)

perioeci The Greek term *perioikoi* (dwellers around, often latinized to perioeci) describes certain second-class citizens in some Greek states. Perioeci were typically free-born and native; they had basic legal protections and obligations, such as military service, but were excluded by birth from the taking part in government. By occupation, many were smallholding farmers.

This citizen class existed at ARGOS, CRETE, THESSALY, and elsewhere. But the perioeci usually are spoken of in connection with SPARTA, where they were literally "dwellers around," inhabiting the foothills and secondary sites around the Eurotas River valley. The heart of this fertile plain was reserved for true Spartans—the Spartiatai. Together, Spartiates and perioeci constituted the united people known as Lakedaimonioi.

(See also LACONIA; POLIS.)

Peripatetic School See LYCEUM.

Persephone Greek goddess who was queen of the Underworld and wife of the god HADES. In MYTH, Persephone was identified with Korē, the daughter of the grain goddess, DEMETER. Modern writers, for the sake of thoroughness, sometimes refer to her as Korē-Persephone.

When the invading Greeks first entered the Greek mainland circa 2100 B.C., they probably encountered a pre-Greek, native goddess whom they called Persephone. This name, translatable as "destroyer" in Greek, presumably imitates the sound of some non-Greek name now lost to us. The pre-Greek goddess may have been a judge and ruler among the spirits of the dead. The Greeks amalgamated her into their own RELIGION, identifying her with their minor goddess Korē ("maiden").

The Persephone or Korē of Greek myth was the daughter of Demeter and the great god ZEUS. While gathering flowers one day, she was abducted by her uncle, the god Hades, to become his queen in the Underworld. After searching frantically for her, Demeter appealed to Zeus. He ruled that Persephone should spend four months of every year with her husband and eight months in the upper world with her mother. The classical Greeks interpreted the goddess' double name as a sign of her dual nature: Korē the Maiden aboveground and the sterner Queen Persephone below. (But neither Persephone nor Hades was thought of as evil; death was seen as a natural part of the world.)

By the 600s B.C. Persephone also had become part of the fringe cult called ORPHISM. Orphism involved a complicated belief in reincarnation after death: Only by living successive lives (on the earth and also in the Underworld) could a person's soul ascend to eternal bliss. Persephone, as the judge of the dead, directed each soul's sequence of transmigrations, according to how piously the person had lived each prior life.

For some reason, Orphism became quite popular among the Greeks of SICILY and southern ITALY. Consequently Persephone became a major goddess of the Sicilian Greeks, and her myth became localized in Sicily. Supposedly it was near the city of Henna (modern Enna) that Hades stole Persephone away, and supposedly volcanic Mt. Etna served as the hearth where Demeter lit two torches for night-searching. Obviously, such details represent a later layer of the story.

(See also AFTERLIFE; ELEUSINIAN MYSTERIES.)

Perseus (1) In MYTH, a hero of ARGOS and TIRYNS, and the slayer of the monster MEDUSA. The story of Perseus' birth is one of the most picturesque Greek legends: King Acrisius of Argos had a beautiful daughter, Danaë, but he was warned by prophecy that he was destined to be killed by any son born to her. So he locked her in a tower or a bronze chamber, to keep anyone from approaching her. But the great god ZEUS came to her in the form of a shower of GOLD, and in due time she bore a son, whom she named Perseus (destroyer).

Acrisius, surprised to learn of these developments, put Danaë and Perseus to sea in a chest, which drifted south through the Aegean. Coming ashore at Seriphus, an island of the CYCLADES, mother and child were welcomed by the local king, Polydectes. There Perseus grew to manhood. Polydectes—intending to force Danaë to marry him, and wanting Perseus out of the way—sent the young man on a seemingly deadly mission: to go to the far West (or far South), find the three demon sisters known as the Gorgons, and bring back the head of the one called Medusa. The Gorgons were so hideous that merely looking at any one of them would cause a person to turn to stone. Nevertheless, Perseus set off, with sword and shield.

On advice of the goddess ATHENA, Perseus consulted the Graiae—three old sisters who had one eye and one tooth among them. Stealing their one eye until they would help him, Perseus received directions for finding the Gorgons' home. He also received from them a pair of winged flying shoes and a cap that would turn him invisible, as well as a pouch for carrying Medusa's severed head. Flying invisible to the land of the Gorgons, he sneaked up on Medusa with his face averted, guiding himself by watching her harmless reflection in his shield. Shearing off her head, he immediately hid it in his satchel, since the head itself could turn people to stone. Perseus flew off, still invisible, evading Medusa's two sisters.

Soaring over the sea toward Seriphus, he then saved the Ethiopian princess Andromeda from a sea monster. (Andromeda's parents had planned to sacrifice her, to save their kingdom from the god POSEIDON's wrath.) Perseus killed the monster, married Andromeda, and resumed his journey by ship with her. At Seriphus, Perseus rescued his mother, using Medusa's head to turn Polydectes and his followers to stone. Then, leaving the island in the hands of Polydectes' righteous brother Dictys, Perseus returned with Andromeda to his native Argos.

The prophecy that Perseus would kill his grandfather Acrisius was fulfilled when the hero, throwing the discus in competition, inadvertently struck and killed Acrisius, who was sitting among the spectators. Leaving the throne of Argos to a kinsman, Perseus became king of Tiryns and, according to one tradition, founded the nearby city of MYCENAE. Supposedly one of his sons, Perses, became the father of the Asian people known as the Persians.

The Perseus tale may distortedly recall the political unification of the Argive plain in early Mycenaean times (circa 1600 B.C.), as well as Mycenaean naval conquests in the Cyclades and beyond.

(See also MYCENAEAN CIVILIZATION.)

Perseus (2) See MACEDON.

Persia The vast, rich, non-Greek kingdom of Persia played a crucial role in Greek history—largely as an antagonist, but also as an employer, overlord, or ally for many Greeks, especially those in ASIA MINOR. At its height, circa 500 B.C., the Persian realm stretched westward from the Indus valley (modern-day Pakistan) to THRACE and MACEDON, on the northeastern border of Greece, with EGYPT included. Encompassing about 1 million square miles, unified by a network of roads, and ruled by an absolute monarch, Persia was the greatest empire the world had yet seen. Persian plans to conquer mainland Greece in 490 and 480 B.C. were miraculously defeated by the relatively puny Greek states. The empire's size eventually proved to be its undoing, as the delegation of power to local governors gradually weakened the king's authority. When the Macedonian king ALEXANDER THE GREAT invaded the Persian Empire with a mere 37,000 soldiers in 334 B.C., he found a disunified realm under a weak monarch, ripe for destruction. By the time of Alexander's death (323 B.C.), Persia was ruled as a Greek-Macedonian kingdom, eventually known as the SELEUCID EMPIRE. But the Iranian cultures of the central plateau continued to thrive, reclaiming power by the second century B.C.

The Persians enter recorded history circa 850 B.C. as a nomadic, pastoral people in the western Iranian plateau. They called themselves the Parsa; their territory (which shifted steadily in these years) was known as Parsua. By the 600s B.C. the Persians had settled in the southwest Iranian plateau, bounded on the west by the River Tigris and on the south by the Persian Gulf; this region would henceforth be their heartland. In these years the Persians were subjects of the kindred Medes, another Iranian people. As vassals of the Medes, the Persians kings reigned out of the provincial capital at Susa, near the head of the Persian Gulf.

In 549 B.C. the Persian king CYRUS (1) led a revolt against the Median overlords. The Persians captured the Median capital and imposed a new dynasty—a Persian one. Henceforth the Persians were the ruling class; the Medes became their subordinates, although still partners in the empire.

It was Cyrus' conquest of western Asia Minor (546 B.C.) that brought the Persians into contact with the Greeks, in the cities of IONIA.

(See also AGESILAUS; CALLIAS; CAVALRY; CYPRUS (1); DARIUS (1); DELIAN LEAGUE; HERODOTUS; KING'S PEACE; PERSIAN WARS; XENOPHON.)

Persian Wars Although the Greeks and Persians were in intermittent conflict for over 200 years, the term *Persian Wars* refers mainly to the campaigns of 490 and 480–79 B.C., in which the Greeks successfully defended mainland Greece against two Persian invasions. These were the famous campaigns of MARATHON, THERMOPYLAE, and SALAMIS (1).

The Persian Wars were immensely important for Western civilization. The Greeks, a relatively small and disunified nation, unexpectedly defeated the greatest empire on earth. Afterward, this experience compelled the Greeks to identify themselves culturally. The world's first real historian, the Greek HERODOTUS, writing in the mid-400s B.C., chose the Persian Wars as his story to tell; it was the greatest understandable event in human memory, and it distinguished the Greek from the "barbarian." Similarly, the amazing Athenian cultural achievements of the 400s and 300s B.C.—in THEATER, PHILOSOPHY, SCULPTURE, ARCHITECTURE, and the development of DEMOCRACY—were products of a confidence or arrogance adopted because of the Persian Wars. Had the invading Persians won and had Greece become just another province of their empire, there would have been no brilliant Athenian century to serve as the foundation of modern culture.

The first stage of the conflict was the Persians' westward push into ASIA MINOR in the mid-500s B.C., when the Persian king CYRUS (1) conquered the east Greek region known as IONIA. Cyrus and his successors ruled moderately but demanded tribute, labor, and military service from their Greek subjects, governing them by means of Greek puppet rulers.

In 513 B.C. the Persian king DARIUS (1) led an army across the BOSPORUS and received the submission of THRACE and MACEDON. Southwest of Macedon, the land of Greece itself was still free, but for how long? It seemed obvious that the Persians would attack eventually, and equally obvious—to many—that Greece would fall. Certain Greeks began making friendly overtures to their expected overlords. A new verb appeared in the Greek language: Mēdizein, "to Medize," or to collaborate with the Medes (i.e., with the Persians). Among the most egregious Medizers were those local Greek despots known as TYRANTS (*turannoi*), who hoped to retain power and become the Persians' puppets after the conquest. Another Medizing force was the priesthood of the god APOLLO at DELPHI. The Persian kings had traditionally sent gifts to Delphi, to develop contacts inside Greece; apparently the priesthood now decided that a Persian conquest would be to their god's advantage. Therefore, during the Persian Wars, the Delphic oracle consistently gave advice that was defeatist in tone.

The IONIAN REVOLT of 499–493 B.C. saw the Greek cities of Asia Minor rise up, ill-fatedly, against their Persian masters. Before being crushed, the revolt attracted help from two cities of mainland Greece: ATHENS and ERETRIA. This intervention backfired, convincing the Persians to subdue Greece immediately. Soon Persian heralds appeared in the cities and islands of Greece, demanding the tokens of earth and water that were the formal symbols of submission to the Persian king. Many frightened Greek states obeyed, but the Athenians, in a violation of diplomatic sacrosanctity, threw the Persian envoys into the condemned criminals' pit. At SPARTA the envoys were thrown down a well, where (they were told) there was plenty of earth and water. Thus did the militaristic state of Sparta, with the best army in Greece, signal its intention to stand alongside Athens, against Persia.

Darius had at his court an Athenian adviser, HIPPIAS (1), the former tyrant of Athens who had been expelled in 510 B.C. Darius resolved to capture Athens and install the aged Hippias as his puppet. In the summer of 490 B.C. a Persian seaborne expedition crossed the AEGEAN SEA, landed at Eretria, destroyed that city, and sailed south to the coastal town of Marathon, about 26 miles from Athens. There a Persian army of perhaps 20,000 troops disembarked; with them was Hippias. But at the Battle of Marathon the attackers were totally defeated by 9,000 Athenians and a few hundred Plataeans, in a plan devised by the general MILTIADES.

After Darius' death, his son and successor, XERXES, decided to invade Greece in full force. He gathered a massive army, perhaps 300,000 land troops, including subject peoples as well as Persian regulars. His navy, numbering perhaps 600 warships, was drawn from seafaring provinces such as PHOENICIA, EGYPT, and Greek Ionia. Other preparations included spanning of the HELLESPONT with twin bridges of boats, a marvel of Persian engineering skill.

Preparations were under way on the Greek side as well. At Athens the brilliant stateman and general THEMISTOCLES convinced the democratic ASSEMBLY to vote to build 100 new trireme warships, thereby giving Athens about 180 ships in all—the biggest navy of any Greek city. Themistocles foresaw that the defense of Greece would depend on fighting at sea. The Persians' land army was too vast to destroy; but cripple their navy, and the army would founder—from lack of transport, supply, and communication. With a large fleet, Athens could be protected or, if need be, evacuated.

The uneasy alliance of Greek states, dominated by the land power Sparta, mapped out a grand strategy. The terrain and shoreline of Greece naturally compel a southward-moving invader through a series of bottlenecks on land and sea, and the Greek plan was basically to fight at certain defensible bottlenecks—a mountain pass, a sea channel—where the invader's advantage of superior numbers would be canceled. On land, the Greeks would avoid battle in the open plains, where the mighty Persian CAV-ALRY would rule. Similarly, for a sea fight, a narrow channel would tend to offset the superior seamanship of the Persians' elite contingent, the Phoenicians.

One advantage for the Greeks, of which they were not yet fully aware, was that their HOPLITE heavy infantryman was better than his Persian counterpart—not in courage, but in equipment and training. The hoplite wore a BRONZE breastplate and helmet, and carried a bronze-plated shield and a thrusting spear up to eight feet long; the Persian trooper wore less armor and carried a shorter spear and a shield made only of wickerwork. In the crush of battle, the hoplites usually could push through. But hoplites were not invincible, and the Persians had the great advantage in numbers.

Led by Xerxes, the Persians crossed the Hellespont and descended through Thrace and Macedon (spring 480 B.C.). A Greek army marched to the mountain gorge of Tempe, in northern THESSALY, but then abandoned the site as being too far forward, in hostile territory. Thessaly submitted to the Persians, and the Greeks drew up a line of defense on Thessaly's southern frontier, at two interrelated sites: the mountain pass of Thermopylae and the six-mile-wide sea channel at ARTEMISIUM, on the northern shore of EUBOEA.

Simultaneous battles were fought on land and sea. The Battle of Thermopylae was a heroic Greek blunder; an army under the Spartan king Leonidas held the pass for three days but never received reinforcements, and was overrun. More creditable was the sea fight at Artemisium, where about 370 Greek warships met an enemy fleet that had been badly reduced—perhaps to 450 ships, from 600—by recent storms. The battle was a marginal Persian victory, but it showed the Greeks what they could do against the much-vaunted Phoenician crews, and it apparently encouraged the Greek admirals in their strategy of relying on narrow waterways to offset the enemy's advantages.

The defenders retreated south through central Greece. BOEOTIA submitted to the invaders and, like Thessaly, was forced to supply troops to fight against their fellow Greeks. In a dramatic decision, the Athenians voted to evacuate their city to the enemy: They would not stay and surrender. The noncombatants were transported to safe locales nearby, and the Athenian troops and warships remained on duty in the Greek forces. Athens was occupied and sacked by the Persians.

The summer was ending. Certain Greek commanders wanted to retreat farther south, to the naturally defensible Isthmus of CORINTH, but Themistocles and the Spartan commanding admiral, Eurybiades, pressed to offer battle at sea, in the one-mile-wide channel east of the island of Salamis, in the Saronic Gulf. There the Greek fleet stationed itself. Although Xerxes could have chosen to bypass, he overconfidently ordered his navy to attack the Greeks inside the channel, and the result was a complete Greek victory at the Battle of Salamis.

Suddenly the tables had turned. With his armada now in tatters, Xerxes was in danger of being trapped in Europe. Taking much of his army, he hurried back to the Hellespont

before the Greek navy could arrive to sever the bridges there. Meanwhile, his able general Mardonius stayed in Greece with a force of perhaps 60,000.

Mardonius wintered in Thessaly and marched south again in the spring of 479 B.C.; his army was supplemented by collaborationist Greek troops from Thessaly and Boeotia. Athens, still evacuated, was reoccupied and burned. Mardonius reached the northeastern outskirts of MEGARA (1), the Persian high-water mark in Europe, but on news of an approaching Greek force he withdrew toward THEBES, the main Boeotian city, which was friendly to the Persians. The Greek army—nearly 39,000 Spartan, Peloponnesian, and Athenian hoplites, commanded by the Spartan general Pausanias—approached the town of PLATAEA in the late summer. Pausanias declined to attack, probably fearing what Mardonius' cavalry could do on the Boeotian plain. After almost two weeks of waiting—while both armies were plagued by a shortage of food and water—Mardonius launched a surprise attack. He was killed and his army destroyed at the BATTLE OF PLATAEA.

Historians debate whether it was Salamis or Plataea that saved Greece from Persian conquest. Probably the Persians' invasion was doomed once they had lost most of their fleet at Salamis. Greece, a mountainous country, does not lend itself to being conquered by a strictly land-bound army. Had the Persians won at Plataea, they would probably have lost the war anyway. The Greeks would have retreated south to the isthmus, to beat them there.

That same year, 479 B.C., saw the Greek counteroffensive begin. Supposedly on the same day as the Battle of Plataea, an amphibious Greek force under the Spartan king Leotychides landed at Cape MYCALE in western Asia Minor; the Greeks beat a Persian army and burned the remaining Persian fleet. Ionia had been liberated; over 90 years would pass before the Persians could reclaim it. By winter an Athenian fleet under the general XANTHIPPUS had captured the Hellespont after destroying the Persian garrison at SESTOS. In 478 B.C. the Spartan Pausanias took BYZANTIUM and the Bosporus. There the Spartans dropped out of the counteroffensive, but the Athenians—calculatedly building an empire—carried on their liberation of the east Greeks in Asia Minor and CYPRUS.

Around 449 B.C., Persia negotiated a peace treaty with Athens, and hostilities ceased. But the Persian kings continued to scheme for the recovery of their lost territory of Ionia. As the patriotic Greek alliances of 480 B.C. dissolved into the PELOPONNESIAN WAR (431–404 B.C.) and CORIN-THIAN WAR (395–386 B.C.), Persia eventually began to help Sparta in exchange for a free hand in Ionia. It took ALEXAN-DER THE GREAT, with his dismantling of the Persian Empire in 334–323 B.C., finally to make peace between Persia and the Greeks.

(See also CALLIAS; CIMON; DELIAN LEAGUE; KING'S PEACE; WARFARE, LAND; WARFARE; NAVAL.)

Phaedra See HIPPOLYTUS; THESEUS.

Phaethon See HELIOS.

phalanx Battle formation of heavy infantry. The noun *phalanx* is used by ancient Greek writers from HOMER (circa

750 B.C.) onward, but modern scholars usually reserve the word to describe the distinctive battle order of the Macedonian and Hellenistic heavy infantry (mid-300s to mid-100s B.C.). These soldiers employed equipment and tactics invented by the Macedonian king PHILIP II (circa 357 B.C.) and perfected by his son ALEXANDER THE GREAT. In several battles of the latter 300s B.C., the Macedonian phalanx proved itself superior to the classic HOPLITE armies of the Greek city-states.

Designed as an improvement on hoplite tactics, the Macedonian phalanx consisted of about 9,000 men arranged in orderly rows, one behind the other, up to 16 in number, with several hundred men in each row. Every man carried a 13- to 14-foot-long pike, called a *sarissa*. When the first five rows presented their sarissas forward, a hedge of metal pike-points was formed, extending in serried rows to about 10 feet ahead. As the phalanx moved forward, these points pushed toward the enemy with great force. Behind the first five rows, the soldiers kept their sarissas upright, waiting to move forward as their comrades in front fell.

Because hoplites and other heavy infantrymen of the day were armed only with six- to eight-foot-long jabbing spears, the phalanx enjoyed an advantage in reach and in density of offered weapons; the enemy formation could present its spears only from a depth of two or three rows. Also, the rows of the phalanx tended not to crush together as tightly as did rows of hoplites or other troops, and this made combat somewhat less exhausting for men in the phalanx.

Because the sarissa needed to be held with two hands, the man's shield was strapped to his left forearm or shoulder, without engaging the hand. This shield was necessarily smaller than the kind carried by Greek hoplites. In general, Macedonian soldiers seem to have worn less armor than their hoplite counterparts, using leather or cloth for helmets and corsets, in place of BRONZE. While this allowed the phalanx men greater maneuverability and endurance (and less expense for equipment), it also left them vulnerable to archers and javelin men, who might easily stay out of reach. Another weaknesses of the phalanx (one shared by the hoplite formation) was its unshielded right flank and rear.

The proper use of the phalanx involved coordination with CAVALRY and light-armed infantry to guard the flanks and chase away enemy projectile troops. In combat, the phalanx's natural function was defensive—to hold the enemy's charge and damage his formation, while the cavalry looked for a weak point to attack.

The Hellenistic kingdoms of the 200s and 100s B.C. continued the legacy of Macedonian-style warfare, but the phalanx became enlarged to such unwieldy sizes as 20,000 men. The wars of ROME against MACEDON and the SELEUCID EMPIRE saw the phalanx beaten repeatedly—at Cynocephalae (197 B.C.), at Magnesia (186 B.C.), and at Pydna (167 B.C.)—by the more maneuverable Roman legions.

(See also CHAERONEA; HELLENISTIC AGE; PHILIP V; WARFARE, LAND.)

Pheidias See PHIDIAS.

Pheidon Powerful king of ARGOS who probably reigned circa 675–655 B.C. Pheidon brought his city to a brief preeminence at the expense of its perennial enemy, SPARTA. With his army, Pheidon seized the sacred site of OLYMPIA and personally took over management of the OLYMPIC GAMES—"the most arrogant thing ever done by a Greek," according to the ancient Greek historian HERODOTUS. Probably after this coup, Pheidon introduced a uniform system of weights and measures throughout the PELOPONNESE (a step meant to extend Argive control over TRADE).

The philosopher ARISTOTLE, in his treatise *Politics*, written circa 340 B.C., states that Pheidon began as a king and ended as a tyrant (Greek: *turannos*). This means—not that Pheidon's rule grew more harsh—but that, after coming to the Argive throne by legal succession, he seized absolute power, probably at the expense of the city's aristocrats and with the support of the middle class. As such he anticipated the first wave of Greek TYRANTS, who began taking power violently at CORINTH, SICYON, and other cities in the mid-600s B.C.

It was almost certainly under Pheidon that the Argives reorganized their army for HOPLITE tactics and became the best soldiers in Greece. In 669 B.C. an Argive army soundly defeated a Spartan army at the Battle of Hysiae, in the eastern Peloponnese. According to the most plausible modern explanation, this battle marked the triumph of Argive hoplites over an enemy still using the older, disorganized tactics. It may have been in the following year that Pheidon marched his army across the Peloponnese to Olympia.

(See also WARFARE, LAND.)

Phidias Athenian sculptor who lived circa 490–425 B.C. His masterpiece was the colossal GOLD-and-ivory cult statue of the god ZEUS, constructed circa 430 B.C. for Zeus' temple at OLYMPIA and counted as one of the SEVEN WONDERS OF THE WORLD for its size and solemn majesty. Today, however, Phidias is better remembered for his work on the Athenian PARTHENON.

Phidias (Greek: Pheidias) was the son of a man called Charmides (an Athenian aristocratic name that also crops up in the family of PLATO). The family was probably well off, which allowed the young Phidias to pursue his art. His genius encompassed not only SCULPTURE but also PAINTING, engraving, and metalworking. He eventually achieved great prestige as an associate of the Athenian statesman PERICLES.

One of Phidias' earlier works (circa 456 B.C.) was a 30-foot-high BRONZE statue of the armored Athena Promachos (the Defender), which stood among the outdoor statues atop the Athenian ACROPOLIS. The gleam from the point of the goddess' upright spear was said to be visible to ships as far as 15 miles away. Another early work was Phidias' bronze statue known as the Lemnian Athena, commissioned by inhabitants of the island of Lemnos (probably Athenian colonists who had settled there recently). This work's grace and dignity were much admired in ancient times.

Circa 448 B.C., Phidias was chosen as the artistic director of the public-building program organized by Pericles to commemorate peace with the Persians. It was in this job that Phidias oversaw the building of the Parthenon and the execution of the temple's famous architectural sculptures. We do not know whether Phidias himself carved any of these. But he did sculpt one central work, the 35-foot-high gold-and-ivory cult statue of the goddess ATHENA that stood inside the Parthenon.

The statue was completed by about 438 B.C., but soon Phidias had to pay for his success, for criminal charges were brought against him as part of a political scheme to discredit his friend Pericles. Phidias was accused of having stolen some of the gold entrusted to him for the statue's construction. This charge was disproved when the gold plates—which Phidias had constructed as detachable—were taken off the statue and weighed.

But a second accusation, of impiety, was more damaging. Phidias had incurred this charge by rashly including likenesses of himself and his patron Pericles amid the repoussé figures on the shield of Athena. There, in a battle scene of Greeks versus AMAZONS, Phidias had portrayed himself as a bald old man hurling a rock and Pericles as a warrior about to spear an Amazon.

Perhaps as a result of his legal troubles, Phidias left Athens soon after the Parthenon Athena was in place. (A later legend, claiming that he died in an Athenian prison, is untrue.) Around this time he won the prestigious commission to sculpt his second giant gold-and-ivory statue, at Olympia. The Olympian cult statue of Zeus was arguably the most important religious statue of the Greek world.

Philemon   See THEATER.

Philip II   Macedonian king who reigned 359–336 B.C. and fell to an assassin's knife at about age 46. A brilliant soldier and diplomat, Philip took his backward Greek kingdom of MACEDON and turned it into the mightiest nation in the Greek world. His creation of a new-style army and his subjugation of mainland Greece both proved essential for the subsequent career of his son, ALEXANDER THE GREAT. Although he did not live to see Alexander's conquest of the Persian Empire (334–323 B.C.), Philip made it possible.

In the absence of better surviving sources, our knowledge of Philip comes mostly from book 16 of the *World History* of DIODORUS SICULUS, a later Greek writer (circa 60–30 B.C.) prone to distortion. However, the following narrative seems plausible. Philip (Philippos, "horse lover") was the youngest son of the Macedonian king Amyntas, circa 382 B.C. As a hostage for three years at the powerful city of THEBES, in central Greece, the teenage Philip probably had contact with the great Theban leader EPAMINONDAS. Another model that may have inspired the young Philip was the dynasty of rulers at Pherae, in the northern Greek region called THESSALY. In 359 B.C. Philip's brother Perdiccas, king of Macedon, was killed in battle. Philip, then about 23, became king.

Soon he began his military reforms, which were destined to change ancient warfare. Previously, the Macedonian army had consisted of a capable, aristocratic CAVALRY alongside a ragamuffin light infantry of peasant levies. With training and new equipment, Philip created an expert heavy infantry, armed with 13- to 14-foot-long pikes, fight-

King Philip II of Macedon revolutionized military tactics, subdued mainland Greece, and became the most powerful man in the Greek world; he then bequeathed these advantages to his son Alexander. The shrewd, hard-drinking Philip is probably portrayed on the carved ivory head shown here. This damaged artifact, just over an inch tall, was discovered in 1977 in northern Greece, at excavations of royal tombs near the ancient Macedonian city of Aegae.

ing in the formation known as the PHALANX. Philip also pioneered new tactics of siege warfare, daring to storm fortified sites in direct attacks using siege-towers and arrow-shooting catapults. Among his other accomplishments, Philip was the most successful besieger of his day.

His early imperialism involved two strategies: the subjugation of Macedon's warlike, non-Greek neighbors, the Thracians and Illyrians; and the seizure of Greek cities of the north Aegean coast, to provide revenue and shipping outlets. Philip's rival for possession of this seacoast was the distant city of ATHENS, with its mighty navy.

In 357 B.C., the second year of his reign, Philip captured the north Aegean city of AMPHIPOLIS, a former Athenian colony located near the GOLD- and SILVER-mining region of Mt. Pangaeus. The Athenians declared war, but hostilities trailed off amid Athenian reluctance to commit land troops so far from home (a recurring factor that worked to Philip's advantage during the next two decades). Meanwhile the

Pangaeus mines provided Philip with an enormous yearly sum of 1,000 TALENTS.

Another foreign policy success of 357 B.C. was Philip's MARRIAGE to Olympias, a young noblewoman from the northwestern Greek region called EPIRUS. For 20 years Olympias was the foremost of Philip's multiple wives. (Eventually he had seven.) Although the royal marriage proved unhappy, it produced a son and a daughter; the boy, born in 356 B.C., was Alexander.

By involving himself in the Third Sacred War against the central Greek state of PHOCIS (355–346 B.C.), Philip began to influence affairs in Greece. Soon he dominated the leadership of Thessaly and made an alliance with formidable Thebes. Amid these events, Philip lost an eye when he was hit by an arrow during his siege of the rebellious Macedonian city of Methone (354 B.C.).

In 349 B.C. Philip moved to devour the Greek cities of the north Aegean region known as CHALCIDICĒ. He besieged the Chalcidic capital, OLYNTHUS, whose inhabitants appealed for help to Athens. The Athenians again declared war on Philip but again declined to send troops, despite the fiery oratory of the Athenian statesman DEMOSTHENES (1). Captured by Philip, Olynthus was leveled to the ground.

The year 346 B.C. saw Philip readying to conquer mainland Greece. As was typical of his style, he first made peace with Athens. When the Third Sacred War finally ended in Phocis' defeat, Philip marched his Macedonians unchallenged through THERMOPYLAE and into Phocis. At meetings of the influential AMPHICTYONIC LEAGUE, at DELPHI, Philip was admitted in place of the Phocian delegates and was personally awarded the two votes previously controlled by the Phocian people. Through influence, he also controlled the votes of the Thessalian delegates and others. King of Macedon, overlord of Thessaly, master of Phocis, and boss of Delphi—he was by now the most powerful man in Greece.

It was in this year that the Athenian orator ISOCRATES published his pamphlet *Philip*, urging the Macedonian king to liberate the Greek cities of ASIA MINOR from Persian rule. No doubt Philip welcomed the propaganda value of this romantic plea, but he also took the idea to heart. In his way, Philip was a lover of Greek culture. He may have sincerely wished to lead a united Greece against its traditional enemy, PERSIA.

The idea of conquest also held a strong financial appeal for Philip, who was running out of money, despite his Mt. Pangaeus revenues. His ceaseless military campaigning and frequent resort to bribery had taken a heavy financial toll. Only the booty of a rich conquest could save him from bankruptcy and downfall.

Meanwhile his Greek enemies organized against him, with funding from the alarmed Persian king, Artaxerxes III. Supported by the Athenian navy, the northeastern Greek colonies of BYZANTIUM and Perinthus held out against Philip's siege (340 B.C.). Then his erstwhile Greek ally Thebes turned against him, ejecting his garrison from Thermopylae (338 B.C.).

Philip invaded central Greece in the spring of 338 B.C. Bypassing the Theban garrison at Thermopylae, he marched through Phocis toward Thebes. In response, Athens and Thebes, although traditional enemies, patched together a hasty alliance and faced Philip's advancing army at the Battle of CHAERONEA, northwest of Thebes.

The battle was a complete Macedonian victory, and Thebes and Athens surrendered. Philip treated Athens leniently but Thebes harshly, staging executions and selling war prisoners as SLAVES. After 20 years of scheming, he had conquered Greece in one campaign.

Almost immediately, Philip began preparations for an invasion of Persian-held Asia Minor, for which his trusted general Parmenion secured a crossing of the HELLESPONT (337 B.C.). But in 336 B.C. Philip was assassinated at the old Macedonian royal city of Aegae (modern Vergina), on the morning of his daughter's wedding. The killer, an aggrieved Macedonian noble, probably acted at the instigation of Olympias (whom Philip had recently divorced) and the 20-year-old Alexander. Philip's new number-one wife had recently borne Philip a son, and no doubt Olympias feared this threat to Alexander's succession to the throne. The young wife and child were later murdered on Olympias' orders. Alexander, acclaimed as king, took up Philip's invasion plan and crossed to Asia Minor in 334 B.C.

In A.D. 1977–1980 archaeological excavations at Vergina uncovered two tombs dating from the 300s B.C. that were filled with treasure. The cremated remains in the tombs are thought to be those of Alexander's father and son (Alexander IV, who was murdered at age 13, in 310 B.C.). The skull believed to be Philip's apparently shows signs of an injured right eye socket—the wound from an arrow at the siege of Methone.

(See also ARCHAEOLOGY; ILLYRIS; THRACE; WARFARE, LAND; WARFARE, SIEGE.)

Philip V   Last great king of MACEDON (reigned 221–179 B.C.). Seeking to enlarge Macedon's traditional control over mainland Greece, Philip came to grief against a rival contender—the Italian city of ROME.

Philip ascended the Macedonian throne at age 17, on the death of his grandfather, Antigonus III. During the so-called Social War against the powerful Aetolian League in Greece, Philip led an army into the heart of enemy AETOLIA and sacked its capital, Thermon (218 B.C.). Faced with Roman domination of the nearby Illyrian coast, Philip in 215 B.C. made an anti-Roman alliance with the North African city of CARTHAGE (whose brilliant general Hannibal seemed on the verge of capturing Rome in the Second Punic War).

The Carthaginian alliance proved to be a disastrous decision. The Romans, despite their troubles with Hannibal, declared war on Philip and dispatched a fleet and land troops. This First Macedonian War (214–205 B.C.) saw Philip campaigning in Greece against an alliance of Romans and Aetolians. Hostilities ended in a treaty. But after defeating Carthage (201 B.C.), the Romans lent a ready ear to reports from their Greek allies PERGAMUM and RHODES, complaining of Philip's naval aggression in the eastern Mediterranean.

Again the Roman senate declared war on Philip. The Second Macedonian War (200–197 B.C.) culminated in Philip's defeat by a Roman-Aetolian army at the Battle of Cynocephalae ("dog heads," a hill in THESSALY). At the battle, the Roman general Titus Quinctius Flamininus managed to send his troops around the flank of the Macedonian PHALANX, to destroy it from behind. Cynocephalae marked the end of an independent Macedon; more significantly, it marked the end of the military era of the phalanx.

Philip was compelled to pay the Romans a huge cash penalty, hand over his son Demetrius as a hostage, and withdraw his troops forever from mainland Greece, which was declared by the Romans to be free of foreign taxation and interference. Philip became Rome's ally; his life's last decade was spent repairing Macedon's finances and warring against non-Greek peoples in the Balkans. Philip died on a military campaign and was succeeded by his son Perseus, Macedon's last king.

(See also ILLYRIS; WARFARE, LAND.)

Philoctetes Mythical Greek hero of the TROJAN WAR. Philoctetes (his name means "lover of possessions") was an expert archer from central Greece who led five ships in the allied Greek expedition against TROY. En route he went ashore on Lemnos or another Aegean island and was bitten on the foot by a serpent. The wound was agonizing, foul-smelling, and incurable, and so the Greeks—on ODYSSEUS' advice—abandoned Philoctetes on Lemnos. There he languished, neither dying nor recovering. He lived by hunting, using the unerring bow and arrows that had once belonged to the hero HERACLES (and that had been bequeathed by Heracles to Philoctetes' father, Poeas).

Nearly 10 years went by. As the Greek siege of Troy became stalled, the Greeks learned of a prophecy that said the city could be captured only with the aid of Heracles' bow and arrows. Consequently two Greek leaders—in most versions, Odysseus and DIOMEDES—sailed to Lemnos and brought Philoctetes and the charmed weapons to the siege of Troy. There, at the Greek camp, Philoctetes was healed by the physician sons of ASCLEPIUS, Machaon and Podalirius. Taking to the field with Heracles' bow, Philoctetes killed the Trojan prince PARIS in an archery duel. After Troy's fall, Philoctetes survived the homeward voyage that claimed the lives of many other Greeks. He was said to have traveled to the West to establish Greek cities in eastern SICILY and southern ITALY.

The Athenian tragedian SOPHOCLES wrote a *Philoctetes* (performed in 409 B.C.), which survives today. In presenting the tale of the wounded hero on Lemnos, the play investigates the question of private wishes versus public duty. In Sophocles' version, the two arriving Greek captains are Odysseus and NEOPTOLEMUS (ACHILLES' son). Philoctetes loathes Odysseus as the one responsible for his being abandoned; but finally the honest, young Neoptolemus, with help from the gods, persuades Philoctetes to accompany them to the war to fulfill his destiny.

(See also FATE.)

Philosophy *Philosophia* is a Greek word, literally meaning "love of wisdom," but, in effect, translating to something like "love of arcane knowledge" or "a desire to find out the truth." The Greeks invented the concept of philosophy; for them, it meant a way of looking for reality and truth without strict reference to the traditional gods of RELIGION and mythology.

Philosophy began as a form of SCIENCE. The first philosophers, such as THALES of MILETUS (circa 560 B.C.), sought to discover the essential element in the material world and to explain the causes of physical change. The Italian-Greek philosopher PARMENIDES (circa 515–445 B.C.) revolutionized Western thought by theorizing that ultimate reality lies outside of the material world. Parmenides' younger contemporary SOCRATES of ATHENS (469–399 B.C.) redirected philosophy toward the pursuit of ethical truths and such moral questions as "What is virtue?" Both of these thinkers helped pave the way for PLATO (427–347 B.C.), who produced the first surviving philosophical system, in which questions about reality might shed light on ethical and epistomological questions. Plato's onetime pupil ARISTOTLE (384–322 B.C.) developed a system of logic and forever divorced science from philosophy by reorganizing them as separate disciplines. During the HELLENISTIC AGE (circa 300–150 B.C.) new Greek philosophies such as STOICISM and EPICUREANISM sought to assure people about their place in the universal order.

(See also ACADEMY; ANAXAGORAS; ANAXIMANDER; ANAXIMENES; CYNICS; DEMOCRITUS; EMPEDOCLES; HERACLITUS; LEUCIPPUS; LYCEUM; PYTHAGORAS; SKEPTICISM; SOCRATES; XENOPHANES.)

Phocaea Greek city of IONIA, on the central west coast of ASIA MINOR. During the 600s and 500s B.C. Phocaea was at the forefront of long-range Greek TRADE and COLONIZATION, particularly in the western Mediterranean.

Located on the north shore of the entrance to the Bay of Smyrna, Phocaea (Greek: Phokaia) was established circa 1050 B.C. by IONIAN GREEKS who had emigrated from mainland Greece. The Phocaeans developed as brilliant seamen and began establishing anchorages and colonies to enlarge their shipping network. Probably one of their earliest colonies was Lampsacus, on the HELLESPONT (circa 650 B.C.).

Around 600 B.C., in a westward drive for raw tin, SILVER, IRON, and other resources, the Phocaeans founded their famous colony, MASSALIA (modern-day Marseille), at the mouth of the Rhone on the southern coast of France. This depot facilitated trade with the local Ligurians and CELTS. Farther west, the Phocaeans traded with the Celtic kingdom of Tartessus, in southern Spain. Such ventures brought Phocaean sailors into battle with the Carthaginians, a non-Greek people who had previously held a monopoly on maritime trade in the West.

The Greek historian HERODOTUS describes Phocaea's abrupt decline. Circa 545 B.C. the city was besieged by the invading Persians under King CYRUS (1). The Phocaeans, taking to their ships en masse, abandoned their home to the enemy. By way of an oath, they dropped an iron ingot into the sea and swore never to return until the metal should float up. (Nevertheless, half of them did decide to return.) The refugees sailed west. Rebuffed from Corsica by allied fleets of Carthaginians and ETRUSCANS, these Phocaeans eventually founded a new colony, Elea, on the southwestern coast of ITALY.

Phocaea, repopulated by the group that had returned, endured reduced fortunes under Persian rule. The city joined the doomed IONIAN REVOLT against the Persians

(499–493 B.C.). After Ionia was liberated at the end of the PERSIAN WARS (479 B.C.), Phocaea became part of the Athenian-controlled DELIAN LEAGUE.

Like the rest of Ionia, Phocaea passed to Spartan control at the end of the PELOPONNESIAN WAR (404 B.C.) and was handed back to the Persians by the terms of the KING'S PEACE (386 B.C.). Liberated by ALEXANDER THE GREAT'S conquest of the region in 334 B.C., Phocaea later passed to the Macedonian-Greek SELEUCID EMPIRE, then to the kingdom of PERGAMUM, and finally, in 129 B.C., to the domain of ROME.

(See also BRONZE; CARTHAGE; SHIPS AND SEAFARING.)

Phocis   Small region in the mountains of central Greece. The Phocians' significance was their intermittent control of the important sanctuary of the god APOLLO at DELPHI, located within Phocian territory. Their designs on Delphi brought them into periodic conflict with Apollo's priests and administrators there, who sometimes went so far as to declare holy war against Phocis.

Settlement in Phocis centered on two separate valleys: in the west, the Crisaean valley, adjoining the northern Corinthian Gulf; in the east, the valley of the middle Cephissus River, bordering BOEOTIA. Between the two Phocian valleys stood the traversible southern spur of Mt. Parnassus, where Delphi was located.

Coveted for its envelopment of Delphi and for its communication routes through mountainous central Greece, Phocis was surrounded by hostile Greek neighbors: Boeotia (to the east), West LOCRIS and Doris (to the west), and East Locris (to the north). Farther north lay another occasional enemy, THESSALY. In response to these pressures, the Phocians had by 600 B.C. become a unified, warlike people. They extended their power north of their home territory, even walling up the distant pass of THERMOPYLAE against southward attacks by the Thessalians. The Phocians' major conflicts with their neighbors were the three "Sacred Wars," so named because each involved Delphi's declaration of war. The Sacred Wars were consequential in allowing opportunities for intervention to such powerful states as ATHENS, SPARTA, THEBES, and MACEDON.

In the First Sacred War (circa 590 B.C.), Phocis was overrun by an alliance of Sicyonians, Thessalians, Locrians, and Athenians. The Phocians were deprived of the fertile Crisaean plain (which was henceforth left as an uncultivated offering to Apollo), and control of Delphi was handed over to a league of neighboring states (including Phocis) known as the AMPHICTYONIC LEAGUE. At regular meetings of the League delegates, the Phocians had two of the 12 votes.

In 457 B.C. Delphi was recaptured by the Phocians, by now in alliance with the powerful city of Athens. There followed the bloodless Second Sacred War (448 B.C.), in which the Spartans marched into Phocis and restored Delphi to the Amphictyonic League. As soon as the Spartans had withdrawn, the Athenians—on PERICLES' direction— marched to Delphi and returned it to the Phocians. But sometime in the next decades, the Phocians lost Delphi again and ceased to be Athenian allies. In the PELOPONNESIAN WAR (431–404 B.C.) and years following, Phocis was allied with SPARTA against Athens.

The Third Sacred War (355–346 B.C.) involved most of mainland Greece and gave the Macedonian king PHILIP II a chance for serious involvement in Greek affairs. The war began as a struggle between Phocis and the powerful Boeotian city of Thebes. At Thebes' prompting, Delphi declared war against certain Phocians who were impiously growing food on the Crisaean plain. But the Phocians seized Delphi and used funds stolen from the sanctuary to hire a mercenary army. With this force the Phocians fought back a coalition of Boeotians, Locrians, and Thessalians. A Phocian army even invaded Thessaly, where the Phocian commander was killed in battle against Macedonian Philip, who had entered the war as a Theban-Thessalian ally (352 B.C.). But in 346 B.C. an exhausted Phocis, its stolen funds depleted, surrendered to Philip. He garrisoned the region, occupied Delphi, and personally took over Phocis' two votes on the Amphictyonic council. The Phocians were made to pay an indemnity against their theft of Apollo's treasure, and the Crisaean plain was once more left fallow for the god.

In the 200s B.C. Phocis ceased to be important in Greek politics, as new Greek powers, such as AETOLIA, arose to dominate Delphi.

(See also CLEISTHENES (2).)

Phoenicia   Region corresponding roughly to modern-day Lebanon, located on the southern part of the east Mediterranean coast of ancient Syria. The Phoenicians were non-Greeks of Semitic race. They called themselves the Kinanu, and it is their ancestors who are known as Canaanites in the Bible. In the late second millennium B.C. these Canaanites were battered by foreign invaders, including the Jewish tribes led by Joshua (circa 1230 B.C.) and the Philistines (who may have been Mycenaean Greeks displaced by turmoil at home, circa 1190 B.C.). But by 1000 B.C. the Canaanites had compensated for lost territory by becoming the greatest seafarers of the ancient world.

Our word *Phoenician* comes from the Greeks, who called these people Phoinikes, "red men"—probably a reference to skin color (but not, apparently, to the purple murex dye, *porphura,* which was among the Phoenicians' most precious commodities). By about 900 B.C. Phoenician sea traders were visiting Greece as part of their middleman's network that stretched throughout the ancient Near East to the Red Sea and Persian Gulf, touching on EGYPT, Assyria, Babylonia, and the kingdoms of ASIA MINOR. Also by 900 B.C. (as ARCHAEOLOGY reveals) the Phoenicians had begun exploring the western Mediterranean, seeking out suppliers of SILVER, tin, and other coveted metals. To facilitate western TRADE, the Phoenicians founded colonies, notably CARTHAGE (Kart Hadasht, "new city,") in what is now Tunisia, and Gaddir ("walled place," modern Cadiz) in southern Spain.

At home, the three Phoenician seaports of BYBLOS, Sidon, and Tyre became wealthy commercial and manufacturing centers. It was probably at Tyre prior to 1000 B.C. that the Phoenicians invented their 22-character Semitic ALPHABET, a vast improvement over their prior, cuneiform script. When the Greeks learned to adapt this Phoenician alphabet to represent the Greek language (before 750 B.C.), Greek literature was born.

In other ways, the brilliant Phoenicians deeply affected the formative Greek culture of 900–700 B.C. Near Eastern textiles and bronzework, brought to Greece by Phoenician traders, helped to spark the Greek artistic revolution of the "Orientalizing period" (circa 730–630 B.C.). Certain oddities of Greek MYTH—such as the filial violence of both CRONUS and ZEUS, the flood of Deucalion, and the dying ADONIS—are best understood as being Near Eastern ideas that made their way to Greece via the Phoenicians (before 700 B.C.). Other Phoenician exports to Greece include the chicken (first domesticated in India) and the Semitic custom of reclining at meals, which the Greeks adapted to their drinking party known as the SYMPOSIUM.

The Greeks copied Phoenician shipbuilding techniques, circa 900–700 B.C. And it was surely in emulation of the Phoenicians that the Greeks embarked on their own seaborne trade and COLONIZATION in the 800s–500s B.C. Inevitably the Greeks came into conflict with their former mentors, in competition for western trade privileges. Circa 600 B.C., related to the founding of a Phocaean-Greek colony at MASSALIA (in southern France), a first sea battle was fought between Greeks and Carthaginian Phoenicians. Many such conflicts were to follow, particularly between Greeks and Carthaginians in SICILY.

Phoenicia surrendered to the conquering Persian king CYRUS (1) circa 540 B.C., and henceforth the Phoenicians, like other subject peoples, supplied levies for the Persian armed forces. Phoenician ships and crews were the pride of the Persian navy. Phoenician squadrons fought fiercely against the Greeks in the PERSIAN WARS, in 499–480 B.C.

Phoenicia revolted unsuccessfully from Persian rule in the 300s B.C. When the Macedonian king ALEXANDER THE GREAT invaded the Persian Empire, the Phoenicians surrendered readily except for the Phoenician fortress of Tyre, which fell only after a monumental siege (332 B.C.).

In the HELLENISTIC AGE (300–150 B.C.), Phoenicia became a province of the SELEUCID EMPIRE, providing fleets and revenues. By the time of the Roman conquest (63 B.C.), Phoenicia had ceased to exist as a separate entity.

(See also AL MINA; CADMUS; CYPRUS; SHIPS AND SEAFARING; THALES; WARFARE, NAVAL.)

phratry   See KINSHIP.

Phrygia   See ASIA MINOR.

Phrynichus   Athenian tragic playwright of the late 500s and early 400s B.C. A pioneer of early Athenian tragedy, Phrynichus was reportedly the first to introduce female characters onstage. He exerted a strong influence on the younger tragedian AESCHYLUS, particularly with his innovative use of recent events from the PERSIAN WARS. Only a few fragments of Phrynichus' work survive.

Competing at the annual drama festival known as the City Dionysia, Phrynichus won his first victory in about 510 B.C. In around 492 B.C. he presented *The Capture of Miletus*, based on the recent ravaging of that Greek city by the Persians after the failed IONIAN REVOLT. The Athenian audience burst into tears at the rendition of their ally's fate, but Phrynichus was fined 1,000 drachmas for re-

minding the Athenians of their woes, and future performance of the play was banned.

In around 476 B.C. he presented another historical drama, *The Phoenician Women*, giving the tragic, Persian viewpoint of the recent Greek naval victory at SALAMIS (1). This play surely inspired Aeschylus' tragedy *The Persians* (472 B.C.).

(See also MILETUS; THEATER.)

phyle   See KINSHIP.

Pindar   Greatest choral poet of ancient Greece. Born near the central Greek city of THEBES but often traveling, Pindar (Pindaros) lived circa 518–438 B.C. He apparently claimed kinship with the noble Aegid clan, which had branches at SPARTA and other Greek cities, and much of his poetry beautifully conveys the old-fashioned values of the aristocratic class, whose power was waning throughout Greece during Pindar's own lifetime. Pindar wrote various types of LYRIC POETRY, most of it choral poetry—that is, intended for public performance by a chorus. He is best remembered for his victory odes (*epinikia,*) composed in honor of various patrons' triumphs at such important Greek sports festivals as the OLYMPIC GAMES. In ornate language that often attains magnificence, Pindar combines a sense of joyous occasion with a sublime religious piety and a sadness over human transience or human injustice.

Forty-five of Pindar's victory odes survive whole, having come down to us numbered and arranged by a later ancient editor into four categories—Olympians, Pythians, Nemeans, and Isthmians. These categories reflect the four major sports festivals that occasioned nearly all the poems. The odes' honorees include winners in WRESTLING, PANKRATION, footraces, and the four-horse chariot race.

In Pindar's day, international athletes were always rich and male, and were usually members of the traditional noble class, for whom the games were part of a distinctive, aristocratic identity. Pindar's patrons, from many Greek cities, included three rulers: King Arcesilas IV of CYRENE (1), the tyrant Theron of ACRAGAS, and the tyrant HIERON (1) of SYRACUSE, the most powerful individual in the Greek world. Pindar wrote four odes for Hieron; of these, two are usually considered the poet's masterpieces—Olympian 1 and Pythian 1.

Perhaps invented by the poet IBYCUS (circa 530 B.C.), the Greek victory ode was composed for a fee paid by the victor or his family, and was performed by a chorus of men or boys at the athlete's home city sometime after the sports event. Like most choral verse, Pindar's odes are written in a Doric Greek dialect (although Pindar himself would have spoken a different dialect, Aeolic). The meters and lengths vary; most of Pindar's odes run between about 45 and 120 lines.

Employing an associative flow of ideas and images, the typical Pindaric ode salutes the athlete's home region, praises the inherited superiority of the upper class, and broods over the precariousness of human happiness. Many odes recount one or another Greek MYTH in such a way as to compare the mythical hero with the poem's patron. Beyond flattery, these comparisons convey the message that victory—for hero or athlete—is proof of divine favor.

Pindar's frequent use of myth makes him one of our important sources for these stories; among many examples are Pythian 9's tale of APOLLO and the nymph CYRENE (2), and Nemean 10's tale of CASTOR AND POLYDEUCES. But the reverent Pindar always minimizes or bowdlerizes the gods' cruelty and injustice, in comparison with other extant versions of such myths.

We know little of Pindar's life. His parents' names are variously reported. As a boy he studied poetry at ATHENS when that city had recently emerged as the world's first DEMOCRACY (508 B.C.). Significantly, Pindar's Athenian contacts came from right-wing, aristocratic circles; these friends included Megacles of the noble Alcmaeonid clan and Melesias, a champion wrestler who was the future father of the conservative Athenian politician THUCYDIDES (2). Both Megacles and Melesias represented Athenian upper-class forces that were soon destined to come to grief against the radical policies of the democracy. Megacles was ostracized by the Athenians in 486 B.C., and Pindar's ode Pythian 7—written to celebrate Megacles' chariot victory at the PYTHIAN GAMES at DELPHI that same year—is partly a consolation for his friend's political misfortune.

Pindar's fateful connection with Greek SICILY was forged in 490 B.C., when he wrote ode Pythian 6 for Xenocrates, a nobleman of the Sicilian Greek city of Acragas. (Xenocrates' brother was Theron, who was destined to become dictator of Acragas and one of Pindar's patrons.) Xenocrates had won the chariot race, which was the most prestigious event and one of the few in which contestants did not have to compete personally: The drivers were usually professionals, and the official contestants were considered to be the chariots' owners.

Xenocrates had a teenage son named Thrasybulus. Pythian 6 is remarkable for its frank expression of Pindar's infatuation with Thrasybulus. Homosexual attachments were a prominent part of Greek aristocratic life in that era, and Thrasybulus was probably the sort of glamorous youth who created a sensation amid the games' intense social atmosphere. We know that Pindar also wrote a drinking song for Thrasybulus and a later ode, Isthmian 2, for a chariot victory by him (circa 470 B.C.). Pindar's tone in these verses—confident, personal, not at all subservient in addressing his eminent employers—is a sign of his high social status.

But Pindar's comfortable world was soon shaken, when Persian armies invaded mainland Greece in 490 and 480 B.C. The allied Greeks' victory over the Persians brought on changes that eventually swept away the aristocrats' traditional way of life. Pindar's Thebes fell into deep disgrace for having collaborated with the occupying Persian forces, and the two cities that had led the Greek defense—Sparta and Athens—emerged as rival leaders of Greece.

Meanwhile, in Sicily, the Greek city of Syracuse had become the foremost power of the western Mediterranean, having led the Sicilian Greeks in defeating a Carthaginian invasion (480 B.C.). And it was at Syracuse that Pindar would deliver his greatest poetry.

He visited in 476 B.C., probably at the invitation of the dictator Hieron, who had won the horse-and-rider race at that summer's Olympic Games. (Like the chariot race, the horse race was officially won by the owner, not the rider.) Pindar's victory ode for Hieron that year is perhaps the most famous lyric poem of antiquity, placed by ancient editors at the beginning of the Pindaric corpus, as Olympian 1:

> Best of things is water. And gold,
> Like a fire at night, outshines all other wealth.
> And if you wish to sing of glory in the Games,
> Look no further in the daytime sky for any star
> More warming than the sun,
> Nor any contest grander than at Olympia. (lines 1–8)

The 116-line ode celebrates Hieron's victory by telling the legend of PELOPS, who supposedly established the first Olympic Games. Just as Pelops was beloved by the god POSEIDON, so is Hieron blessed by some god.

While in Sicily, Pindar may have indulged in a feud with the poet SIMONIDES of Ceos and his poet-nephew BACCHYLIDES, who were likewise guests at Hieron's court. Pindar's Olympian 2 contains a mysterious reference to a pair of crows who "chatter vainly against the sacred eagle of ZEUS"—which is sometimes interpreted as the poet's rebuke to his rivals.

Pindar left Sicily in around 475 B.C. In 474 B.C. Hieron defeated an Etruscan invasion of Greek southern ITALY, at the sea battle of CUMAE. Soon thereafter Hieron founded a new Sicilian city, Aetna, near volcanic Mt. Etna. Then, in 470 B.C., Hieron won the chariot race at the Pythian Games and commissioned an ode from Pindar.

Pindar's resultant poem, Pythian 1, is probably his best. The 100-line ode successfully interweaves the above-mentioned events with a worldview of harmony versus discord, justice versus evil. The poem contains an unforgettable description of the mythical monster TYPHON, the god Zeus' defeated enemy, who now lies shackled beneath Mt. Etna and whose fire-breathing fury supposedly causes the volcano's eruptions. Cleverly linked to local geography, Pindar's Typhon serves as a cosmic symbol for the ETRUSCANS, Carthaginians, and other defeated enemies of Hieron and Greek civilization. Yet the poem ends with a bold warning to Hieron not to abuse his power, lest he too provoke the gods' anger.

In 468 B.C. Hieron, then dying, crowned his achievements with a chariot victory at the Olympic Games. But he hired Bacchylides, not Pindar, to commemorate this triumph. The ruler may have been offended by Pythian 1's presumptuous warning.

Pindar's last years saw the final erosion of his familiar world, with Greece divided by hostility between Sparta and the imperialist Athens of PERICLES. One prominent victim of Athenian aggression was the Greek island of AEGINA. The anti-Athenian anger of Pindar's friends is suggested in the anxious tone of Pindar's last surviving victory ode, Pythian 8, written in 446 B.C. for a boy wrestler of Aegina. The poem also contains (lines 95–97) a poignant expression of Pindar's aristocratic religious outlook, which saw victory in SPORT as a god-given gift providing a brief triumph over the doom that awaits all mortals:

> We are creatures of a day. What is someone? What is he not?
> A human being is a shadow in a dream.

But when a god grants a brightness,
Then humans have a radiant splendor and their life is sweet.

Pindar lived until age 80. Supposedly he had a wife and three children, but we hear also of another young male friend, Theoxenus of Tenedos, in whose arms the poet is said to have died at the city of ARGOS. Pindar's memory was revered for centuries. Supposedly the priests of APOLLO at Delphi would close the temple daily with the words "May Pindar join the gods at dinner." And when the Macedonian king ALEXANDER THE GREAT captured and destroyed Thebes in 335 B.C., he is said to have spared the house that had been Pindar's.

(See also ALCMAEONIDS; ARISTOCRACY; CATANA; CHARIOTS; FATE; GREEK LANGUAGE; HIMERA; HOMOSEXUALITY; ISTHMIAN GAMES; NEMEAN GAMES; PERSIAN WARS.)

Piraeus Main port and naval base of ATHENS, located about four miles southwest of the city.

The westward-jutting peninsula known as the Piraeus (Greek: Peiraieus) was neglected in early Athenian times, when ships sheltered at the more southerly beachfront of Phalerum. In 510 B.C. the Athenian tyrant HIPPIAS (1) built a citadel on the Piraeus' hilltop, which was called Munychia. But it was the Athenian statesman THEMISTOCLES who, in 492 B.C., began developing the harbor for a growing Athenian war fleet.

By the mid-400s B.C. a walled town proper had arisen, laid out on a rectilinear grid by the city planner HIPPODAMUS. The naval base contained dockyards, arsenals, and ship sheds (covered drydocking for individual ships, the pride of Athenian naval technology). In addition, the port was a bustling commercial center, home to many of the METICS (resident aliens) who conducted Athens' import-export TRADE. It was also home to working-class Athenian citizens of left-wing loyalties, staunch supporters of such democratic politicians as PERICLES.

Piraeus had three harbors, whose entrances—flanked by half-immersed stone walls—could be sealed completely by the raising of massive chains. By 448 B.C. the town and harborfront were enclosed on north and south by the LONG WALLS, running up to Athens. The Long Walls made Piraeus and Athens into a single, linked fortress; even the Spartan land invasions in the PELOPONNESIAN WAR could not breach these defenses or obstruct Athens' communications with its naval base. The Long Walls were torn down after Athens' defeat in the war (404 B.C.), but were rebuilt by the Athenian general CONON (393 B.C.).

From 322 until 196 B.C. the citadel at Munychia held a Macedonian garrison—one of the Macedonians' four "fetters" of Greece. Today Piraeus remains vital as the harbor of modern Athens.

(See also GREECE, GEOGRAPHY OF; MACEDON; THIRTY TYRANTS, THE; WARFARE, NAVAL.)

Pirithous See CENTAURS; THESEUS.

Pisistratus Dictator or *turannos* who ruled ATHENS from 546 to 527 B.C. Like other Greek TYRANTS of that era, Pisistratus seized supreme power in his city with the common people's support, at the expense of the aristocrats.

His benevolent reign removed the nobles' grip on Athenian government and marked a step in the city's difficult progress from ARISTOCRACY to DEMOCRACY in the 500s B.C.

Pisistratus (Greek: Peisistratos, "adviser of the army") was himself born into an Athenian aristocratic family, circa 600 B.C. Before taking power, the dynamic and affable Pisistratus came to the political fore when he led an army against Athens' enemy, the city of MEGARA (1). He then set out to make himself dictator, by exploiting the volatile atmosphere of Athenian class tensions.

Despite the prior democratic reforms of the statesman SOLON (594 B.C.), Athenian politics and adjudication at this time were still dominated by the nobility. This dominance was resented by the middle class, which supplied the backbone of the army. When Pisistratus entered politics, the countryside held two opposing factions, divided along lines of region and class: the right-wing party of the plain, led by a certain Lycurgus; and the more democratic party of the coast, led by Megacles of the Alcmaeonid clan.

Pisistratus created a third, more left-wing party, by organizing a peasant following and styling himself as the champion of the common people. He then convinced the Athenian ASSEMBLY to vote him an escort of bodyguards. Legend claims that the aged Solon, who was Pisistratus' kinsman, warned the Athenians against the man's intentions, but Pisistratus, using his bodyguards, was able to seize the ACROPOLIS and become tyrant (circa 561 B.C.).

Soon pushed into exile by the combined effort of the two other factions, he was reinstated with the help of his former rival Megacles. But when this alliance broke down, Pisistratus fled once more (circa 556 B.C.). In around 546 B.C. he was back, leading a Greek mercenary army hired with a fortune he had made in SILVER-mining ventures in THRACE. Sailing from ERETRIA, the invaders landed near the town of MARATHON, where Pisistratus' family had its regional following. There Pisistratus was joined by other Athenians who wanted an end to aristocratic strife. On the road to Athens, Pisistratus defeated a government army at the Battle of Pallene, and the city was his again.

Pisistratus ruled shrewdly and moderately for 19 more years. He took aristocratic hostages but indulged in no vendettas or confiscations. He maintained Solon's laws and certain trappings of democratic government (although he also made sure that his own supporters held the top posts). He taxed reasonably—his tax on farm revenue, for instance, was 5 percent—and in return, he provided lavish building programs and public relief.

Like other dictators, Pisistratus aimed to make his city great. He assured Athens of a food supply by developing military outposts along the HELLESPONT, at Sigeum and in the Thracian CHERSONESE. These guarded the shipping lane for precious grain from the BLACK SEA. He elaborated Athens' national festivals, such as the PANATHENAEA, and encouraged industry and commerce. Under him, Athenian black-figure POTTERY reached its artistic peak and dominated the Mediterranean market. In SCULPTURE, ARCHITECTURE, and primitive THEATER, Pisistratus' Athens moved toward its amazing achievements in the next century. He fostered an Athenian patriotic culture that was partly intended to compete with cities such as CORINTH and partly intended to erase the old, local, aristocratic factionalism at home. His reign brought Athens forward in its progress from being a second-tier power in the 600s B.C. to being the capital of the Greek world in the 400s B.C.

Pisistratus died peacefully in 527 B.C. He was succeeded by his eldest son, HIPPIAS (1), whose reign was more troubled.

(See also ALCMAEONIDS.)

Pithecusae Greek trading colony located on a volcanic island now called Ischia, off the western coast of southern ITALY, near modern-day Naples. First occupied by Greeks from CHALCIS and ERETRIA circa 775 B.C., Pithecusae (Greek: Pithekoussai, "monkey island") was apparently the earliest substantial Greek settlement in the West. It was the first landfall in a wave of westward Greek COLONIZATION that swept across southern Italy and SICILY over the following two centuries.

Most knowledge of Pithecusae comes from modern AR-CHAEOLOGY on the site. The first Greeks found the island unoccupied. Their settlement site—atop a tall peninsula with two harbors, on the island's north shore—was chosen with an eye toward local seafaring. The island provided a safe base from which Greek merchant ships could sail the six miles to the Italian mainland, to conduct TRADE with the non-Greek peoples there. Foremost of these Italian trading partners were the ETRUSCANS, whose home region lay far to the north but who had an outpost at Capua, inland of the Bay of Naples.

What brought Greek traders from mainland Greece to western Italy was the lure of raw metals—IRON, SILVER, tin (for making BRONZE), and others—that would fetch high prices in the Greeks' home cities of Chalcis and Eretria. The Etruscans mined large amounts of iron ore on the island of Elba, off the western coast of northern Italy, and at least some of this valuable ore was then traded to the Greeks at Pithecusae. Archaeology at Pithecusae has yielded the remnants of ancient Greek foundries, probably indicating that the acquired iron was refined and forged into ingots by Greek smiths right there, before being shipped to Greece.

In exchange for Italian metals, the Pithecusae Greeks supplied WINE, painted POTTERY, worked metal, and other luxury goods. Modern discoveries at Pithecusae of ceramics and metalwork from ancient EGYPT and ASIA MINOR suggest that some of the Greeks' luxury items originated in the non-Greek Near East, traveling west via Greek trade networks.

By about 750 B.C. Pithecusae was crowded and prosperous. Affluence is suggested by the quantities of imported Near Eastern artifacts found in Greek tombs there. The most important archaeological item from Pithecusae is the so-called cup of Nestor, made between about 750 and 700 B.C. This is a Greek clay drinking cup, painted in Geometric style and inscribed with one of the earliest surviving examples of Greek alphabetic WRITING.

Pithecusae was largely abandoned in the late 700s B.C. One cause may have been the start of the LELANTINE WAR, which pitted Chalcidean Greeks against Eretrian Greeks as enemies. At least some of the Pithecusans—possibly the Chalcidean party—founded a new colony, CUMAE, on the Italian mainland opposite Pithecusae.

(See also ALPHABET.)

Pittacus Constitutional dictator of the city of MYTILENE, on the island of LESBOS circa 580 B.C. Among his enemies was the poet ALCAEUS, whose loathing of Pittacus has been immortalized in verse. Nevertheless, Pittacus seems to have governed well, calming the civil strife at Mytilene between the nobles and the middle class. His rule supplied an alternative to the brutal TYRANTS who in those years were arising throughout the Greek world to wrest power from the ARISTOCRACY. Pittacus later was listed among the SEVEN SAGES.

Born of noble blood, he was an illustrious soldier, having once killed an enemy champion in single combat during Mytilene's war against Athenian settlers at Sigeum, in northwestern ASIA MINOR, circa 600 B.C. Later Pittacus was elected as dictator for a term of 10 years. Like his contemporary, SOLON of ATHENS, he made new laws that loosened the aristocrats' monopoly on political power. By turning his back on the partisan interests of his own noble class, Pittacus earned the hatred of old-fashioned aristocrats such as Alcaeus.

Plataea Town of southern BOEOTIA, in central Greece. Plataea lies just north of Mt. Cithaeron, by the River Asopus. Due to fear of its powerful neighbor THEBES, Plataea formed an alliance with ATHENS circa 519 B.C. and remained a staunch ally during the PERSIAN WARS and PELOPONNESIAN WAR. Alone among the Greek states, Plataea sent soldiers to fight alongside the Athenians against the Persians at the Battle of MARATHON (490 B.C.). During the invasion of the Persian king XERXES, Plataea was occupied and sacked by the Persians (480 B.C.). Near the town in 479 B.C. the allied Greeks won the BATTLE OF PLATAEA, which destroyed the Persian threat in Greece.

At the outbreak of the Peloponnesian War in 431 B.C., Plataea was unsuccessfully attacked by Thebes. Plataea's civilians were soon evacuated to Athens. A garrison, left to defend Plataea, surrendered and was massacred after a grueling, two-year siege by the Spartans and Thebans (427 B.C.). Later the empty town was razed to the ground.

Rebuilt, Plataea was again destroyed by Thebes in 373 B.C. After conquering Greece in 338 B.C., the Macedonian king PHILIP II rebuilt Plataea in order to humiliate Thebes, and the town survived down to Roman times.

Plataea, battle of Climactic land battle in the Greeks' defense of their homeland in the PERSIAN WARS. This hard-fought Greek victory occurred in the late summer of 479 B.C., on the plain near the Boeotian town of PLATAEA. Coming about a year after the Greek naval triumph at SALAMIS (1), the Battle of Plataea destroyed the last remaining Persian force in Greece and ended the Persian king XERXES' dream of conquering Greece.

After his unexpected defeat at Salamis, Xerxes led most of his troops home in 480 B.C. To pursue the subjugation of Greece, he left behind his able general (and brother-in-law) Mardonius, with an army numbering perhaps 60,000. Mardonius' army included Greek soldiers from the mainland regions of THESSALY and BOEOTIA, whose cities were actively collaborating with the Persians.

After wintering in Thessaly, Mardonius and his army marched south in the spring of 479 B.C. On receiving news of a Greek allied army's approach from the south, Mardonius withdrew toward friendly THEBES, the chief city of Boeotia. About five miles south of Thebes, just north of Plataea, where the level terrain favored his CAVALRY, Mardonius and his army awaited the Greeks.

The Greek army, commanded by the Spartan general Pausanias, contained about 38,700 HOPLITES—armored infantry—with perhaps as many light-armed troops. The largest hoplite contingents came from SPARTA, ATHENS, and CORINTH. The town of Plataea fought on the Greek side, as a staunch ally of Athens and longtime enemy of Thebes.

The allied Greeks had the advantage of heavier armor and the high morale of a patriotic army. The Persians' advantage was their mighty cavalry, drawn from the Iranian plateau, from Scythia, and from Thessaly and Boeotia. In action against foot soldiers, these horsemen would ride up, send out arrows and javelins, then spur quickly out of reach, only to wheel and attack elsewhere. On flat land, they proved highly effective against the allied Greeks, who had no horse soldiers.

The battle was preceded by 12 days of waiting and skirmishing. The two main armies faced each other, separated by the Asopus River; they stood in battle array in the hot summer sun all day and retired to camps at night. Trying to provoke an attack, Mardonius repeatedly sent his horsemen to harass the Greeks and raid their arriving supply wagons. Both sides suffered from lack of food and fresh water. Pausanias was unwilling to risk an attack, yet each passing day made his army's position more difficult.

At dawn on the 13th day the Persians saw that the entire Greek force had withdrawn toward Plataea. The Greeks were marching in some disarray, having split into three different groups in the dark. Seizing his chance, Mardonius led a full-scale attack across the plain.

The Persian army rushed upon the two nearer Greek contingents. One of these numbered about 11,500 hoplites and contained the best soldiers in Greece, the Spartans, commanded by Pausanias. The other embattled Greek contingent was from Athens and (fittingly enough) Plataea; it numbered about 8,600 and was commanded by the Athenian general ARISTIDES. The Battle of Plataea consisted of two simultaneous actions, separated by perhaps a mile. Meanwhile the third Greek division, farthest away, began marching back toward the fighting.

In attacking so eagerly, Mardonius evidently underestimated the Greeks' discipline and defensive strength. Near a rural temple of DEMETER, the Spartans and their allies formed into battle order, while Mardonius, atop a white charger, brought on his elite Persian infantry. Equipped as archers, the Persians halted a distance from the Greeks and shot volleys of arrows. Greek hoplites, although armored, tended to be vulnerable at the neck, legs, and groin, and many were felled by the arrows before the command finally came to charge.

The hoplites pushed forward into the Persian ranks. The melee wore on until Mardonius was killed, hit in the head with a stone thrown by a Spartan. Seeing their general fall, the Persians broke ranks and fled. At the other end of the battlefield, the Athenians were locked in combat with Boeotian Greeks serving on the Persian side. But at the

sight of their Persian masters running away, the Boeotians retreated toward Thebes.

Many Persians ran to a large wooden palisade that Mardonius had built before the battle. This fort became a slaughterhouse, as the pursuing Greeks broke through and massacred the fugitives within. Although a sizable Persian contingent may have escaped back to Asia, the battle was a stunning Greek victory.

But Plataea marked a fleeting glory. Before the battle, the Greeks had sworn an oath of camaraderie—the famous Oath of Plataea—pledging, among other things, never to harm Athens, Sparta, or the town of Plataea. Yet within a few years hostilities had erupted between Athens and Sparta, and in the PELOPONNESIAN WAR Plataea itself was destroyed by the Spartans.

(See also WARFARE, LAND.)

Plato   Athenian philosopher who lived circa 427–347 B.C. and was one of the most influential thinkers in world history. Plato founded the Western world's first important

The authoritarianism in Plato's political philosophy is suggested in this stern portrait of the great thinker, from a marble bust of the Roman era. The likeness was probably copied from a Greek bronze statue of the 300s B.C.

institution of higher learning, the ACADEMY, which helped shape the course of PHILOSOPHY for the next 1,000 years. Among Plato's teachings, his theory of Forms provided a revolutionary concept of reality—a concept that he tied into corresponding theories of ethics and human knowledge, in a way that seemed to explain the universe.

Most prior Greek philosophers had searched for the universe's primal elements by examining the physical world. Plato, however, influenced by the earlier Greek philosopher PARMENIDES, theorized that true reality is not to be found in the visible, physical world but in an ideal world of eternal Forms. Our earthly phenomena are just imperfect copies of these Forms, Plato said. Such earthly copies include not only material items such as beds and tables, but human virtues such as justice and knowledge. Above all other Forms stands the Form of the Good, whose central and nutritive role in the universe is analogous to that of the sun in the visible world.

To later generations, Plato's concept of a supreme Form of the Good seemed to anticipate Christian monotheism, and so Plato's teachings enjoyed great prestige in the Christian-influenced Roman Empire, circa A.D. 100–400, when a movement called Neoplatonism arose. Probably because of his posthumous association with Christianity, Plato is almost unique among ancient Greek authors in that every major written work by him has been preserved.

For his philosophical writings, Plato chose a style that was unusual in his day—prose dialogue. We have about 30 of these. (A few, ascribed to Plato, are probably forgeries.) Each dialogue typically presents a philosophical discussion among several characters, many of whom are fictionalized versions of real-life personages of fifth-century-B.C. ATHENS. The characters debate a specific topic, such as the nature of courage. The beauty of Plato's technique is that it draws the reader directly into the inquiry. Of the dialogues, Plato's masterpiece is the *Republic* (Greek: Politeia, better translated as "form of government"), written circa 385–370 B.C. The *Republic* first addresses the question "What is justice?" and proceeds to sketch Plato's ideal government, run by a philosopher-king. Plato's theory of Forms receives its classic explanation in the lengthy *Republic*.

The scanty knowledge of Plato's life comes partly from information in the works of later ancient writers, such as the biographers Diogenes Laertius (A.D. 200s) and PLUTARCH (circa A.D. 100). Also preserved are a number of epistles supposedly written by Plato. Although these letters seem to be forgeries, many scholars today accept as accurate the autobiographical information in the famous "Seventh Letter."

Plato (Greek: Platon, "wide," perhaps a nickname) was born into a rich and aristocratic Athenian clan with strong political links; his mother's family claimed descent from the lawgiver SOLON. Plato grew up entirely during the epic PELOPONNESIAN WAR (431–404 B.C.). Like other wealthy Athenian youths, Plato probably did military service in the CAVALRY, perhaps on patrol in the countryside around Athens, circa 409–404 B.C.

Possibly through association with his slightly older kinsmen CRITIAS and Charmides, the young Plato joined the circle of the dynamic Athenian philosopher SOCRATES (circa 469–399 B.C.). Socrates' ethical-political inquiries proved to be the single greatest influence on Plato's thought. In most of his dialogues, written after Socrates' death, Plato immortalized his old mentor by making him the central speaker. Plato's choice of the dialogue form was undoubtedly an attempt to convey the mental excitement of Socrates' discussions.

The convulsive end of the Peloponnesian War changed Plato's life. After Athens' defeat (404 B.C.), Critias and Charmides became central figures in the dictatorship of the THIRTY TYRANTS. Invited to join them, Plato hesitated, and meanwhile the Thirty fell from power; Critias and Charmides were killed (403 B.C.). In 399 B.C., under the restored Athenian DEMOCRACY, Socrates was tried and executed—partly because of his association with Critias. As Socrates' disciple, Plato himself may have been in danger; in any case he seems to have been sickened by the bloody excesses of both right wing and left wing, and he rejected any plan of a political career, choosing instead a life of travel and reflection. But Socratic-type questions about politics and human nature remained foremost in Plato's mind, and he concluded (to use his later words) that there could be no end to human troubles until "either philosophers should become kings, or else those now ruling should become inspired to pursue philosophical wisdom" (*Republic* 473c).

Between about 399 and 387 B.C. Plato probably wrote many or all of the dialogues that modern scholars classify as "early." This group begins with the *Euthyphro*, *Apology*, and *Crito*—concerning Socrates' arrest, trial, and imprisonment—and culminates in the *Protagoras* and *Gorgias*. The early dialogues mainly explore ethical issues, of which the most important is the *Protagoras*' question, "Is moral virtue teachable?"

During these years Plato traveled to Greek southern ITALY, where he met followers of the early philosopher PYTHAGORAS. Aspects of Plato's views on death and the soul's immortality seem influenced by Pythagorean belief. Also, historical memories of a Pythagorean ruling class—which in prior centuries had existed in certain Italian-Greek cities—may have helped to shape Plato's goal to create a government of philosophers.

Circa 387 B.C., at about age 40, Plato made his fateful first voyage to SICILY, to the important Greek city of SYRACUSE, where the ruthless dictator DIONYSIUS (1) reigned. There Plato became both friend and teacher of a kinsman of the ruler—a handsome young nobleman named Dion. In keeping with prevalent Greek upper-class mores of the day, this male relationship was probably sexual: Plato's homosexual feelings are evident in surviving verses that he wrote for Dion. At the same time, the friendship may have suggested to Plato that he could win followers from among the Syracusan ruling circle and thus create his philosophers' government.

According to one story, the Sicilian visit became a fiasco when the cruel Dionysius, tiring of Plato's criticism, arranged for the philosopher to be sold into slavery; Plato's friends bought back his freedom. In any case, Plato soon returned to Athens and bought land and buildings in a park sacred to the mythical hero Academus, about a mile north of Athens' Dipylon Gate, circa 386 B.C. This holding became Plato's Academy, a university dedicated to philosophical inquiry and the preparation of future leaders of Athens and other Greek cities.

Plato devoted nearly the next 20 years to supervising the Academy and to writing the dialogues that modern scholars assign to his "middle" period. In addition to the *Republic*, these include the *Phaedo* (which combines an account of Socrates' last days in prison with Plato's introduction of his own theory of Forms) and the *Symposium* (on the nature of sexual love). The modern, popular notion of nonsexual, "Platonic" love is a distortion of Plato's views: Specifically addressing the aristocratic, male HOMOSEXUALITY of his own day, Plato believed that such feelings could ennoble and educate those partners who channeled their passion into a spiritual, not just a sexual, union. There is no evidence that Plato ever married, and from his writings he seems to have been of an abstemious homosexual nature.

Much of Plato's thought surely developed from life at the Academy—especially from his informal discussions and disputations with his colleagues (a procedure that derived from Socratic technique). This method of question-and-answer disputation was called dialectic (Greek: *dialektikē*, "discussion" or "dialogue"). In Plato's dialogues, Socrates employs dialectic to refute his opponents' preconceptions, by showing that certain conclusions logically drawn from those preconceptions are false. Plato saw dialectic as the premier philosophical technique, "the coping-stone set atop the other types of learning" (*Republic* 534e).

Only through dialectic can people hope to gain some knowledge of reality—that is, of the Forms. Otherwise we are in the dark. In the most famous allegory in all philosophical literature, the *Republic* compares nonphilosophical inquiry with the plight of prisoners trying to learn about reality by observing shadows in firelight on the wall of a cave.

In 367 B.C. Dionysius of Syracuse died and was succeeded by his son, Dionysius II. The 60-year-old Plato—at the invitation of his old friend Dion—again voyaged to Syracuse, in the hope of making a philosophical convert of the new ruler. But the younger Dionysius proved to be a weak and decadent leader, and was eventually ejected in a coup led by Dion himself (357 B.C.). Plato meanwhile returned to Athens (circa 361 B.C.), having despaired of ever creating an ideal state in Sicily.

The third and last stage of Plato's career covers the years circa 360 B.C. until his death, in 347 B.C. These years at the Academy saw the advancement of the brilliant "graduate student" ARISTOTLE, whom some considered to be Plato's likely successor as head of the school. Plato's dialogues of this period include the *Timaeus* (presenting Plato's picture of the physical cosmos) and the *Laws* (his longest and probably final work). Like the *Republic*, the *Laws* portray an ideal Platonic city-state, but in this later dialogue Plato's authoritarian bent is more pronounced—the laws of his utopian "Cretan city" include generous use of the death penalty—and the *Laws* can be read as evidence of an old man's embitterment. Plato died at age 80, pen in hand (according to legend). He was succeeded as head of the Academy by his nephew Speusippus.

(See also AFTERLIFE; ATLANTIS; EDUCATION; GORGIAS; MATHEMATICS; PROTAGORAS; SOPHISTS; SYMPOSIUM.)

Plutarch Greek biographer and moral essayist who lived during the early Roman Empire, circa A.D. 50–125. Born in the central Greek region of BOEOTIA, Plutarch (Greek: Ploutarchos) lived in Greece but was also a Roman citizen, and may have held the local Roman post of procurator. He traveled, read, and wrote widely, and for the last 30 years of his life was a priest at the god APOLLO's shrine at DELPHI. Plutarch embodies the Greek-Roman assimilation under the Roman Empire.

Although he did not exactly write history, he happens to be a major source of information for events in the Greek world circa 600–200 B.C. This is due to one of his writings, known by the title *Parallel Lives of the Noble Greeks and Romans*, which in its surviving form consists of 22 paired short biographies ("Lives"), one of a Greek leader, one of a Roman. Most of these biographical subjects lived in the great centuries of classical Greece and the Roman Republic, prior to Plutarch's time. Among his more important Greek Lives are those of SOLON, PERICLES, ALCIBIADES, and ALEXANDER THE GREAT. Although often sloppy about chronology, the Lives are full of historical and personal detail.

polis Ancient Greek city-state. The *polis* (from which word is derived the English word "politics") was the basic political unit of the classical Greeks. Between about 800 B.C. and 300 B.C., the map of Greece was a patchwork of autonomous city-states, some linked together by alliance or kinship, and some vying to dominate their neighbors, but each one capable of ruling itself as a self-contained political entity. Beginning as aristocracies, city-states developed as democracies or oligarchies in the 500s–400s B.C. The most important DEMOCRACY WAS ATHENS.

Two factors contributed to the city-state's emergence. One was the geography of Greece: mountains, islands, and small farming valleys naturally created discrete, small population centers, many with their own dialects and religious cults. The second reason, more peculiar to the 900s–800s B.C., has to do with the rejection of kingship in Greece during that era. A king may strive to unite various peoples under his single rule, because he is the government. But members of an aristocratic clan—who may rely for their power on local lands and on local religious cults, for which they supply the priesthood—might be prone to concentrate their rule in a smaller, more homogeneous area.

In population, the polis consisted of: full citzens (usually males over age 18, born of citizen parents); female and children citizens (protected by the law but without any voice in government); second-class citizens, such as PERIOECI; resident aliens or METICS; and SLAVES.

The age of the polis ended with the Macedonian conquest of Greece (338 B.C.), the campaigns of ALEXANDER THE GREAT (334–323 B.C.), and the subsequent rise of rich and powerful Greco-Macedonian kingdoms in the eastern Mediterranean and Near East. In the HELLENISTIC AGE (300–150 B.C.), individual Greek cities were not strong enough to survive independently.

(See also APOLLO; ARISTOCRACY; GREECE, GEOGRAPHY OF; HELOTS; HOPLITE; LAWS AND LAW COURTS; OLIGARCHY; TYRANTS; WOMEN.)

Polyclitus Prominent sculptor of the mid- and late 400s B.C., known for his idealized view of the male body.

Polyclitus (Greek: Polukleitos) was born at ARGOS. Working mainly in BRONZE, he established his reputation with a number of commissioned statues of victorious Olympian athletes, displayed at OLYMPIA. Two famous nudes by Polyclitus have survived in the form of later marble copies—the Doryphorus (spear carrier, in the Naples Museum), representing a muscular athlete; and the more sensual Diadumenus (ribbon binder, in the Athens National Museum and elsewhere), showing a youth tying an athlete's ribbon around his head. The Doryphorus brought to perfection the "counterpoise" stance: The man is shown resting his weight on one leg, knee locked, with the other leg drawn back. This stance became a standard pose for statuary down through Renaissance times.

Polyclitus' masterpiece was considered to be the giant GOLD and ivory cult statue of the goddess HERA, sculpted for Hera's temple at Polyclitus' native Argos (circa 420 B.C.). The statue, now lost, showed the goddess seated, with a scepter in one hand and a pomegranate in the other. The statue was in ancient times compared with the colossal statue of Olympian ZEUS sculpted by Polyclitus' rival, PHIDIAS. The Roman geographer Strabo considered the Hera to be the more lovely in technique, although smaller and less magnificent than the Zeus.

(See also SCULPTURE.)

Polycrates Dynamic tyrant of the Greek island of SAMOS. Reigning circa 540–522 B.C., Polycrates was the last great Greek ruler in the eastern Aegean, in the path of the advancing Persians. Usurping power at Samos with the help of his two brothers, Polycrates soon ruled alone, executing one brother and banishing the other. He made an alliance with the Egyptian pharaoh Amasis and built a navy of 100 longboats, arming them with bowmen. Under him, Samos became the preeminent eastern Greek state—replacing Samos' nearby rival, MILETUS, now under Persian rule.

Polycrates was a visionary of naval power. The historian HERODOTUS (the main source for Polycrates' story) called him "the first Greek to plan an empire by sea," and the historian THUCYDIDES (1) saw his domain as a forerunner of the Athenian sea empire of the 400s B.C.

True to his name ("ruling much"), Polycrates led his fleet east and west across the Aegean, mainly against his fellow Greeks. Many of his ventures were pure piracy, but he probably hoped to drive the Persians from the ASIA MINOR coast and capture the Greek cities there for himself. He attacked the territory of Persian-controlled Miletus and defeated a fleet coming to Miletus' aid from the Greek island of LESBOS; he put the prisoners to work as SLAVES on Samos. Polycrates also took over the holy island of DELOS and enlarged the prestigious Delian Games, in the god APOLLO's honor. Like other dictators, Polycrates glorified his capital city (also named Samos), while providing employment and amenities for the common people. He completed Samos' grand temple of the goddess HERA, which was one of the largest existing Greek temples at that time. He developed Samos' harbor and built or completed a tunneled aqueduct, an engineering marvel of its day, bringing water to Samos city.

Polycrates kept the most magnificent court of the Greek world. With generous retaining fees, he attracted the distinguished physician Democedes of CROTON and the famous lyric poets IBYCUS of RHEGIUM and ANACREON of Teos. In keeping with the upper-class tastes of the era, Polycrates (although married and a father) was an enthusiastic lover of boys and young men. It was probably for Polycrates that Anacreon wrote many of his extant poems of homosexual content.

Eventually the shadow of disaster fell across this glamorous despot, as he became caught between (on one side) the unstoppable Persian advance and (on the other) the antityrant elements of the Greek world. Circa 525 B.C. the skippers of a Samian war fleet rebelled and attacked Polycrates on Samos. They were defeated, but the survivors sailed west to get help from SPARTA, the foremost antityrant state of mainland Greece. The Spartans sent out an overseas expedition, their first in recorded history. However, Polycrates held out against their six-week siege of Samos city, and they went home.

Polycrates did not survive for long. Circa 522 B.C. the Persian governor of western Asia Minor lured Polycrates in person to the mainland, then had him seized and killed. A few years later the Persian king DARIUS (1) conquered Samos and installed, as his vassal, Polycrates' surviving brother, Syloson ("preserver of booty"). Samos' sea holdings were lost, and henceforth Samos' warships served in the Persian navy.

Among the folktales that sprang up around Polycrates, one has been retold in variations over the centuries. According to Herodotus, Polycrates, at the height of success, was warned that—because the gods are jealous of human happiness—he should offset the danger by throwing away whatever he valued most in the world. He decided to discard his priceless signet ring, an engraved emerald set in GOLD. Aboard one of his ships, he threw the ring into the sea. But a few days later, as Polycrates sat down to dinner, his servants brought in that very ring; it had been found in the belly of a fine fish being prepared for the ruler's meal. The gods had rejected Polycrates' offering, and he was doomed to die a miserable death.

(See also FATE; HOMOSEXUALITY; TYRANTS; WARFARE, NAVAL.)

Polygnotus See PAINTING.

Polynices See SEVEN AGAINST THEBES.

Polyphemus In HOMER's epic poem the *Odyssey* (written down circa 750 B.C.) Polyphemus ("much fame") is a CYCLOPS—one of a race of gigantic, savage, one-eyed creatures who inhabit a legendary region in the West, vaguely associated with SICILY. Polyphemus is the son of the god POSEIDON and the nymph Thoösa. In what is probably the *Odyssey*'s best-known episode, Polyphemus captures the Greek hero ODYSSEUS (book 9).

Odysseus discovers the monster's cave after landing his ship in the Cyclopes' territory on his voyage home from the TROJAN WAR. With 12 of his men, Odysseus enters the deserted cave to steal supplies, then willfully remains inside the cave, hoping to have a look at the Cyclops. But when Polyphemus returns with his goats and sheep, he immediately traps the Greeks by sealing the cave mouth with a slab of rock, too large for humans to drag away.

Then, catching two of the Greeks in his hands, Polyphemus beats out their brains and eats them raw for dinner. Stretching out amid his sheep, he falls asleep. The Greeks do not attack him, realizing that they can never escape without Polyphemus to move that massive boulder.

The next morning Polyphemus eats two more men for breakfast, leaves the Greeks penned inside as he pastures his flock and herd for another day, and then eats two more men for supper. By then Odysseus has a plan: He makes Polyphemus drunk by giving him a goatskin-full of WINE. When the Cyclops falls into a drunken sleep, Odysseus and his men ram a sharpened post into the monster's one eyeball, blinding him forever. The other Cyclopes, hearing Polyphemus' cries, gather outside the cave and ask what's wrong; Polyphemus shouts out that "Nobody" is hurting him—Nobody (Outis) being the false name by which Odysseus had identified himself to the giant. The other Cyclopes wander off, assuming that Polyphemus' affliction must come from the gods. Then the blind Polyphemus drags away the door boulder and, squatting at the cave mouth, waits furiously for the Greeks to try to run out.

Again Odysseus has a scheme. He ties each of his six remaining men to the underbellies of three sheep abreast, and himself takes hold of the fleecy underbelly of a big ram. As the Cyclops' sheeps and goats move out of the cave for grazing, the Greeks ride concealed among them. The giant strokes each animal's back as it passes, but he misses the Greeks hiding underneath.

The Greeks escape back to their beached ship and shipmates. Once safely aboard and headed out to sea, Odysseus cannot resist the impulse to shout out taunts to Polyphemus and tell him his own true name and lineage. Guided by the voice, Polyphemus hurls a boulder that nearly smashes the departing ship, then he prays to his divine father, Poseidon, to curse Odysseus. Thereafter, in the *Odyssey*, Poseidon is Odysseus' enemy, working to disrupt the hero's voyage home.

Centuries later, in the HELLENISTIC AGE, Polyphemus became the subject of poetic elaboration. The Sicilian Greek poet THEOCRITUS (circa 275 B.C.) wrote two poems describing how, prior to the Odysseus episode, Polyphemus loved and wooed the sea nymph Galatea. The idea of turning a savage, ugly giant into a romantic hero is typical of the frivolous intellectual tastes of the Hellenistic era.

Unrelated to the Cyclops story, the name *Polyphemus* also belonged to one of the Argonauts who sailed with the Thessalian hero JASON (1) in search of the Golden Fleece.

Poseidon   Greek god of the sea and all other bodies of water, as well as of earthquakes and horses. According to the Greek creation legend, Poseidon received the sea as his domain when he and his brothers, ZEUS and HADES, drew lots for their portions of the world, after deposing their royal father, CRONUS. Poseidon was variously described as younger or older than Zeus, but he was subordinate to Zeus in wisdom and moral significance.

A patron god of mariners and fishermen, Poseidon had important temples in certain Greek coastal regions in historical times. These sanctuaries included Cape MYCALE, on the central west coast of ASIA MINOR, and Cape Sunium, in the Athenian territory of ATTICA. The god was commemorated in the names of seaport cities such as POSEIDONIA

Poseidon aims his trident, on a silver coin known as a *statēr*, minted by the Greek city of Poseidonia (Paestum), in southern Italy, circa 540 B.C. The figure may represent a bronze statue of the god that stood in that prosperous city named after him. The Greek lettering reads POS.

and POTIDAEA. But his most famous shrine was at the Isthmus of CORINTH, in south-central Greece, where the biennial ISTHMIAN GAMES were held in his honor.

Poseidon was especially associated with storms at sea, which posed a deadly hazard to ancient shipping. In the god's sea-front temples, voyagers prayed for safe passage, and survivors of shipwrecks gave thanks for deliverance. When storms off the Thessalian coast in 480 B.C. destroyed part of the invading Persian fleet during the PERSIAN WARS, the Greeks paid thanks to Poseidon.

The name Poseidon clearly contains the Greek root *pos-*, "lord," but the rest of the meaning is lost to us. In Greek RELIGION, Poseidon (like Zeus) probably dates back to the earliest Greek-speaking tribesmen who arrived in Greece circa 2100 B.C. Poseidon's earliest form was probably not that of a sea god. (According to modern scholarship, the Greeks originally came from a landlocked region, the Danube basin, and at first lacked a word for "sea.") The prehistoric Poseidon may have been a deity of lakes, rivers, and mountain torrents. He became associated with earthquakes because they were thought to be caused by underground watercourses. In historical times, Poseidon's most common titles were "shaker of the earth" (Ennosigaios) and "holder of the earth" (Gaieochos).

Probably before or during the years of MYCENAEAN CIVILIZATION (circa 1600–1200 B.C.), Poseidon took on his familiar role as the sea god. This change surely occurred as the early Greeks learned shipbuilding and ventured out to sea for food supply, war, and TRADE. Poseidon is mentioned frequently in the surviving LINEAR B tablets from the late Mycenaean era, circa 1400–1200 B.C., but his precise roles in those years are unclear.

In taking over the sea, Poseidon's cult seems to have displaced the worship of certain pre-Greek sea deities.

Apparently some of these survived in Greek MYTH in the guise of lesser sea gods, such as NEREUS and his mermaid daughters, the Nereids. Poseidon was said to have married a Nereid named Amphitritē, after subduing her by pursuit and abduction. She bore him a son, Triton. The sound *trit*, which the names of mother and son share, is considered by scholars to be non-Greek and may represent a prehellenic word for "sea."

Poseidon's function as the horse god may date back to his earliest roots among prehistoric Greek-speaking tribes, who evidently arrived in mainland Greece with domesticated horses. In historical times there was a widespread cult of Poseidon Hippios (of the horse); images in some shrines showed the god as a horseman. According to one myth, the first horse had been born out of a rock where Poseidon's semen had fallen. Poseidon oversaw the horse- and chariot-racing at all Greek sports festivals, not just at his own Isthmian Games. He was said to have been the mentor of the hero PELOPS, founder of the OLYMPIC GAMES.

Mythologically, Poseidon embodied the raw power found in his elements—the sea, earthquakes, and horses. Legends portrayed him as strong, yet brutish and not clever, sometimes in contrast with the wise goddess ATHENA. Poseidon was an important god for the Athenians—he was associated with the hero THESEUS, and he had a shrine inside the patriotic-cult building known as the Erectheum, on the ACROPOLIS. But he was said to have lost a contest with Athena over who would become the chief Athenian deity. HOMER's *Odyssey* echoed this competition: Poseidon failed to destroy the wily hero ODYSSEUS, who was aided by Athena. Other legends described Athena as a civilizing influence over Poseidon's realm—she invented shipbuilding and the bridle, to bring humans to the sea and to tame the horse.

Poseidon's myths tended to associate him with monsters. He was the lover of the demon MEDUSA, and lover or father of the HARPIES. His beautiful mistress SCYLLA was transformed into a hideous sea monster by the jealous Amphitritē. His liaison with the earth goddess GAEA (who was technically his grandmother) produced the fierce giant Antaeus, later killed by HERACLES. With the nymph Thoösa, Poseidon was father of the brutal POLYPHEMUS. Less grotesquely, the Elean princess Tyro bore Poseidon twin sons, Neleus (later king of PYLOS and father of NESTOR) and Pelias (later king of Iolcus, in THESSALY, and enemy of JASON [1]).

In Greek art, Poseidon supplied a study in pure strength. He was often shown holding his trident—a three-pronged fish spear used by Greek fishermen. He resembled Zeus in his beard, physique, and mature age. A superb classical Greek statue found in the sea off Cape ARTEMISIUM in the A.D. 1920s, now identified as a likeness of Zeus, was for decades thought to represent Poseidon.

(See also CHARIOTS; GREEK LANGUAGE; SHIPS AND SEAFARING.)

Poseidonia Greek city on the southwestern coast of ITALY, on the Bay of Salerno. Located at the edge of a wide plain, Poseidonia was founded by colonists from the Italian Greek city of SYBARIS in the last quarter of the 600s B.C. and was named for the sea god POSEIDON. It was a sprawling city, encircled by a wall nearly three miles long but without natural defenses, and it thrived from FARMING and seaborne TRADE. Captured circa 390 B.C. by neighboring Italian tribesmen, the Lucanii, the city passed in 273 B.C. into the hands of the Romans, who called it Paestum.

The site now boasts the standing remnants of three Doric-style temples, built between about 550 and 450 B.C., which are among the best-preserved examples of Greek monumental ARCHITECTURE. Of these, the famous "Temple of Neptune" (actually a temple of HERA) rivals the Athenian PARTHENON in magnificence. The site also contains tombs of the 400s and 300s B.C. that have provided some of the few surviving Greek wall paintings.

(See also PAINTING.)

Potidaea Important Corinthian colony, founded circa 600 B.C. on the northern Aegean seacoast known as CHALCIDICĒ. Located on the narrow neck of the westernmost of Chalcidicē's three peninsulas, Potidaea had a fine harbor, facing west into the Thermaic Gulf and protected in later years by walls and by moles across the harbor mouth. Potidaea (Greek: *Poteidaia*, a form of the name of the god POSEIDON) was the only major Chalcidic city founded by Greeks who were not IONIAN GREEKS; it was also the only Corinthian colony located in the Aegean rather than in the Adriatic or farther west. Potidaea's location made it a gateway from Greece to the north Aegean. Exporting to Corinth, it was a source for TIMBER, SILVER, and SLAVES, available from nearby THRACE.

Potidaea became a foremost Chalcidic city, later surpassed only by its neighbor OLYNTHUS. With Thracian silver, Potidaea was minting coins by about 550 B.C. It led the local Greek resistance to the invading Persians of 480–479 B.C., its formidable defenses withstanding a Persian siege. After the Persian retreat, Potidaea joined the Athenian-led DELIAN LEAGUE. However, the city maintained close ties with Corinth—for instance, it received its chief political executives from Corinth every year.

As Athenian-Corinthian relations worsened in the mid-400s B.C., Potidaea became the object of increasing Athenian mistrust. In 434 B.C. the Athenians raised Potidaea's Delian annual tribute from six to 15 TALENTS. In resistance to Athenian demands, Potidaea rebelled, appealing to Corinth for help. Soon ATHENS and Corinth were in collision over Potidaea. At the Battle of Potidaea, fought outside the city (432 B.C.), an Athenian expeditionary force of 2,000 HOPLITES defeated a slightly smaller Corinthian-led army. This episode—like similar Athenian-Corinthian hostilities at the Adriatic city of CORCYRA—was a major cause in igniting the PELOPONNESIAN WAR.

Potidaea finally was captured by the Athenians in early 429 B.C., after an immense siege in which the Athenian army suffered from plague and the starving defenders were reduced to cannibalism. The Athenians allowed the Potidaeans to evacuate, and the city was repopulated with Athenian colonists. But after Athens' defeat in the Peloponnesian War (404 B.C.), the local Chalcidic Greeks took over the city.

As the second city (after Olynthus) of the powerful Chalcidic League, Potidaea again became a target of Athenian imperialism, in the early 300s B.C. It was captured by

Athenian troops in 363 B.C. The Athenians were expelled by the Macedonian king PHILIP II in 358 B.C., but Potidaea probably was destroyed by Philip in 348 B.C., in his war against Olynthus.

In around 316 B.C. the city was refounded as the capital of MACEDON under the ruler CASSANDER; the city's new name was Cassandreia. In the HELLENISTIC AGE (300–150 B.C.) it remained a major port of the north Aegean, and passed into the empire of ROME after the Romans dismantled the Macedonian kingdom in the mid-100s B.C.

(See also PERIANDER; PERSIAN WARS.)

pottery   The term *Greek pottery* refers mainly to ceramic vessels—storage jars, drinking cups, mixing bowls, plates, and the like, shaped from wet clay on a potter's wheel and fired to brittleness in a kiln heated to about 1,000 degrees Fahrenheit. In addition to hardening the pot, the baking permanently sets any glaze or paint, often with predictable changes of color.

Items of ceramic (Greek: *keramos*, "potter's clay") obviously played a large role in the TRADE and daily life of all ancient cultures. However, for modern archaeologists, pottery takes on a paramount importance well beyond its importance in ancient times, due to one simple fact: Pottery is nearly indestructible. A fired clay pot may shatter easily, but the sherds will last for 10,000 years once buried in the ground. Pottery is thus different from such perishable materials as textiles or wood, which also served major uses in ancient times but which have disappeared with barely a trace. Pottery provides the most available material link to the ancient world. Not every ancient site has a PARTHENON to offer, but almost every site has a buried wealth of potsherds, full of data for the archaeologist.

For example, Greek pottery is the best-surviving medium for ancient Greek PAINTING. The painted scenes that once adorned walls, wood panels, and canvases in ancient Greece are lost, but the scenes on pottery, baked onto the clay, have survived. Such vase painting suggests the developing techniques of all Greek painting, and the rendered scenes also provide priceless information about life in ancient Greece—for RELIGION, FARMING, warfare, and the like. Many of the illustrations in this book depict scenes that were originally painted on pottery.

More generally, the geographic distribution of pottery remnants can reveal ancient trade routes and waves of migrations. Mycenaean potsherds found in southern CYPRUS reveal that the early Greeks probably settled there circa 1300 B.C. Similarly, the huge troves of Corinthan and Athenian pottery found at sites in former Etruria (in northern ITALY, a region where the Greeks never settled) are a dramatic sign of the insatiable market for Greek goods that existed among the ETRUSCANS, circa 600–500 B.C.

The invention of pottery predates by thousands of years the dawn of Greek civilization, starting with pots shaped without a potter's wheel in Mesopotamia and eastern China, circa 5000 B.C. By 2500 B.C. pottery was being produced in the pre-Greek CYCLADES and by the Minoan people of CRETE. Technique was improved by introduction of the potter's wheel circa 2100 B.C.

Mainland Greece, with its soil rich in clay, was producing pottery before the Greeks arrived. The invading Greeks'

Athenian amphora of the mid-700s B.C., 16 inches tall, painted in the style known as Geometric. Geometric vase painting provides the first surviving sign of an awakened Greek artistic genius after the Dark Age (1100–900 B.C.). This impressive example shows how painters began to break up their abstract designs with renderings of live figures—in this case, deer (above), goats (middle), and geese (below). The repeated key pattern, sometimes called a meander, remained a familiar design on Greek textiles and other goods throughout antiquity.

conquest of Greece (circa 2100 B.C. and after) is probably signified by the disappearance of prior styles of pottery and the abrupt emergence (circa 1900 B.C.) of a new type, known to modern archaeologists as Minyan ware. Distinguished by its unadorned gray or gray-yellow glazing and

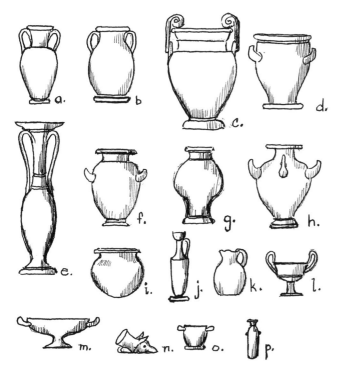

A sampling of Greek pottery designs from the 500s and 400s B.C. Several of these types had their specialized use at the upper-class drinking party known as the symposium. (a) Amphora, meaning "two handled"—the classic Greek wine-storage vessel; sealed with a lid or plug, the amphora was used for domestic storage and (in a modified shape) for seaborne export. (b) Pelike, another type of storage vessel. (c) Krater ("mixer"), used for mixing wine and water at a meal or symposium; this particular design is known as a volute (curl-handled) krater. (d) Bell Krater. (e) Loutrophoros, a ceremonial water vase, used at Athens for the special bath on the wedding day and for the funerary washing of a corpse. (f) Stamnos, a jar for storing wine or oil. (g) Psykter ("cooler"), used to cool undiluted wine prior to mixing and serving; the psykter's shape allowed it to fit partly inside a bowl of cold water. (h) Hydria, a water jug. (i) Lebes, a basin used like a krater, for mixing wine and water. (j) Lekythos, an oil flask. (k) Oinochoë, a "wine-pourer," for dispensing wine from the krater. (l) Kantharos, a type of wine cup. (m) Kylix, another type of wine cup: The shallow bowl was suited to a drinker reclining at a symposium. (n) Rhyton, a decorative form of wine cup, ending, at bottom, in an animal's head or other shape. (o) Skyphos, another wine cup. (p) Alabastron, a jar for perfumes and unguents; the jar acquired its name, in ancient times, because many specimens were carved from alabaster.

its shape that resembles vessels of beaten metal, Minyan ware was discovered and named in A.D. 1880 by the great archaeologist Heinrich Schliemann. This pottery has been found throughout Greece and is the earliest surviving vestige of those people whose descendants would create the MYCENAEAN CIVILIZATION (circa 1600–1200 B.C.).

More imaginative pottery shapes and painted designs emerge at the height of Mycenaean culture (circa 1400–1200 B.C.). Clearly influenced by Minoan pottery, Mycenaean ware included many shapes—such as the kylix (drinking bowl), pitcher (lekythos), and storage jar (amphora)—that would see use in later centuries in classical Greece.

Mycenaean artists typically painted brown on pale yellow. Designs include concentric bands and floral patterns, but a few discovered Mycenaean pots show animal figures such as birds and an octopus. One spectacular vase shows a column of soldiers.

The downfall of Mycenaean society (by 1100 B.C.) was followed by the gradual emergence of new pottery shapes and painted designs, recognizably derived from Mycenaean types. This new style, emerging specifically at ATHENS circa 1050 B.C., is today called Protogeometric—the forerunner to the Geometric style that developed circa 900 B.C. Protogeometric vase painting is distinguished by certain brown or black designs: concentric circles or half circles (drawn with a compass on the vase's side), triangles, crosshatching, and the earliest appearance of the typically Greek right-angle pattern known as the meander.

The Geometric period (circa 900–700 B.C.) saw new methods of decoration, along with certain new pottery shapes. On a background of tan or light yellow, the best Geometric ware shows a density of design, including (after about 800 B.C.) silhouette stick figures representing animal life or human activity, at first confined to horizontal panels around the vase. One of the most characteristic Geometric human scenes (circa 750 B.C.) is on the Dipylon amphora, named because it was discovered near Athens' ancient Dipylon Gate. Amid bands of meanders and diamond shapes, a panel shows triangular-torsoed stick figures in postures of mourning around a skirted figure on a bier. Appropriately, the five-foot-tall vase was deposited as an offering in an aristocratic tomb. This vase's artist, named the Dipylon painter by scholars, is considered to be the first recognizably individual Greek artist.

Geometric style spread to workshops throughout Greece, but by about 730 B.C. artists at the mighty commercial center of CORINTH had abandoned Geometric in favor of new designs, strongly influenced by artwork imported from the Near East. The first stage of this development—part of a larger Orientalizing movement in all Greek art—is known as the Protocorinthian style (730–625 B.C.). Next came the ripe Corinthian style (circa 625–550 B.C.). Widely popular, Corinthian-made potteries monopolized the market, both inside the Greek world and beyond, for nearly 150 years.

Protocorinthian and Corinthian vase painting shows mainly human or animal scenes, with some intervening decorations (typically floral, copied from Near Eastern motifs). Gone are the dark stick figures of Geometric art; there is wide use of color and detail, with attempts at foreshortening. One of the most admired such works is the Chigi Vase (circa 650 B.C.), illustrated with a warfare scene. Under influence of the emerging Athenian black-figure technique, the full Corinthian style (circa 560 B.C.) combined color with solemnity in a packed, friezelike scene. New pottery shapes also came into use in this era.

In imitation of Corinthian technique, the region of SPARTA produced a school of pottery painting in the early 500s B.C.. A number of lively scenes survive on the inside bottoms of Spartan cups.

Meanwhile Athenian black-figure pottery, destined to oust Corinthian ware from the world market, was being produced by 600 B.C. Early black figure saw the vast reduction of decorative designs in favor of scenes from MYTH; later scenes from daily life—farming, seafaring—become more common. The name *black figure* refers to the fact that men's skin was painted dark and rendered into black by glazing and firing; in contrast, women's flesh was shown as white. Incisions helped create detail and fine lines. One early epitome of the style is the Athenian-made Francoise Vase (circa 570 B.C.), discovered in what used to be Etruscan territory and now in the Florence archaeological museum. The small vase (a volute krater, or form of mixing bowl) uses a light background to present lively mythological scenes in miniature. The black-figure technique reached its zenith in about 550–525 B.C., with the work of such Athenian artists as the Amasis Painter.

But by 520 B.C. a new Athenian style had emerged, now called red figure. Red figure reversed the black-figure technique: The human and animal figures were left open to the clay's red color, and it was now the background that was painted and fired black. This represented a breakthrough for creating realistic figures and (through incision) precise detail. The style reached perfection in the work of the artist now known as the Berlin Painter (circa 480 B.C.), who was able to show beautifully realistic musculature and detail.

Another popular Athenian style of the 400s B.C. was the white-ground technique, which rendered figures in delicate lines against a white background. This style was used particularly for *lekythoi* (or *lekuthoi*, pitchers) deposited as offerings in tombs.

The decline of Athenian red-figure style circa 400 B.C. marks the end of the great phases of Greek vase painting. But red-figure continued—the painted scenes become more crowded, the figures more florid—into the 300s B.C., especially in Greek south Italy and SICILY (where Athenian-influenced vase-painting movements arose in the 440s B.C.). By 300 B.C. the new trade routes of the Hellenistic world were making metal pots and vessels more accessible, thereby removing any rich market for vase painting. But the 200s B.C. saw a renaissance of sorts, in the polychrome ware from Greek Sicily.

(See also ARCHAEOLOGY; FUNERAL CUSTOMS; MINOAN CIVILIZATION.)

Praxiteles The most important Greek sculptor of the 300s B.C. An Athenian, Praxiteles worked in the mid-century and was famous for his gorgeous, idealized nudes. He developed marble (as opposed to BRONZE) as a medium for free-standing statuary, and in keeping with the emerging tastes of the 300s B.C., he promoted the female nude as an object of artistic reverence, alongside the male.

Praxiteles' most celebrated work was his cult statue of APHRODITE for that goddess' temple at CNIDUS, in ASIA MINOR (circa 364 B.C.). Considered the most beautiful statue in the world, it showed the goddess standing naked, having disrobed for her bath, with one hand imperfectly covering her vulva. A later copy of this statue (not itself awe-inspiring) now stands in the Vatican Museum. Another well-known statue by Praxiteles, circa 340 B.C., showed the god HERMES holding the divine infant DIONYSUS; a statue of this type standing in the museum at OLYMPIA is now considered by modern scholars to be Praxiteles' original, rather than a later ancient copy.

(See also SCULPTURE.)

Priam In MYTH, the last king of TROY, killed in the Greeks' capture of the city in the TROJAN WAR. Priam was the son and successor of King Laomedon, and the husband of HECUBA, by whom he had 17 sons, including HECTOR and PARIS. Between wife and concubines, Priam had 50 sons and 12 daughters in all.

In HOMER's epic poem the *Iliad*, which describes events of the 10th year of the war, Priam appears as an important secondary character, elderly and wise. He disapproves of Paris' abduction of the Spartan princess HELEN, which has caused the war, but he shows nobility in his courtesy to her and his determination not to give her back to the Greeks against her will. Near the poem's end, after Hector has been slain by the Greek hero ACHILLES, Priam journeys, with the supernatural aid of the gods ZEUS and HERMES, to the Greek camp. There he successfully pleads with Achilles for the privilege of ransoming back Hector's corpse. The old king's grief and suppliant posture clearly foreshadow the destruction of his city.

The story of Priam's death was originally told in a Greek epic poem titled *The Capture of Troy* (*Iliou Persis*), which has not survived, but whose general plot is known. During the Greeks' sack of the city, Priam took refuge at an altar of Zeus but was impiously slain there by Achilles' son, NEOPTOLEMUS. For the classical Greeks, Priam represented the extremes of good and bad fortune that can visit a man in his lifetime.

Priam's name is apparently not Greek, and he may represent a memory of a real-life Trojan king of the historical Trojan War, circa 1220 B.C..

(See also EPIC POETRY; HELEN OF TROY.)

Priapus East Greek fertility god and guardian of orchards and gardens, often pictured as a puny or misshapen man with an enormous, erect penis. Priapus originated in the non-Greek regions of ASIA MINOR, and in classical times (400s B.C.) his worship among the Greeks was confined mainly to Lampsacus, on the east shore of the HELLESPONT. However, Priapus' cult spread during the HELLENISTIC AGE (circa 300–150 B.C.), becoming popular at ALEXANDRIA (1) and later at ROME. His worship involved the non-Greek custom of sacrificing donkeys. He seems to have had no MYTH, being described merely as the son of APHRODITE and DIONYSUS. His name is commemorated in the modern English adjective *priapic*.

Priapus' monstrous organ signified his guardian powers as well as his fertilizing aspect, for the Greeks had a notion that trespassers and thieves (male or female) are liable to rape by the rightful owners of a territory.

(See also HERMES.)

Procrustes In MYTH, a sadistic brigand, son of the god POSEIDON who preyed upon travelers in ATTICA until slain by the Athenian hero THESEUS.

Meeting a wayfarer on the road, Procrustes ("hammerer-out") would deceptively offer him lodging for the night, then force the victim to lie down on one of two beds. Tall men he would place on the shorter bed, and saw off their lower legs, to fit; he placed short men on the longer bed and stretched them out (for the bed was a rack).

The modern English adjective *procrustean* describes an attempt to fit facts or individual examples into a preconceived, inappropriate theory or rule.

Prometheus In MYTH, a benevolent Titan (demigod) who championed the cause of humankind against the gods. Although several characters in Greek myth defied or challenged the great god ZEUS, Prometheus is the only one who appears in a heroic light. Classical Greeks saw him as an example of the wily "trickster"—a type also represented by the god HERMES, the hero ODYSSEUS, and the real-life Athenian statesman THEMISTOCLES. Appropriately enough, *Promētheus* means "having forethought."

According to one version of the myth, it was Prometheus himself who created the first humans, shaping them out of clay while the goddess ATHENA breathed life into them. Zeus then called on Prometheus to decide the question of how humans should sacrifice to the gods—that is, which part of the sacrificed beast should be assigned as the gods' portion and which part should be eaten by mortals. In this arbitration, Prometheus chose to deceive and humiliate Zeus. As described in HESIOD's epic poem *Theogony* (circa 700 B.C.), Prometheus butchered an ox and disguised the bones in a wrapping of fat, but hid the ox's flesh and edible organs inside the stomach. He showed both items to Zeus, inviting him to choose between a bundle glistening with fat or a tripe. The king of the gods chose the inviting-looking bundle, and that is why (according to the legend) humans customarily keep the best sacrificial portions for themselves to eat, while burning only the animal's bones and hide for the gods.

Zeus retaliated by withholding the use of fire from humankind. But Prometheus stole fire from heaven inside a fennel stalk, to bestow on mortals.

Enraged at Prometheus' defiance, Zeus plotted revenge against him and his race of mortals (who up until then had included only one gender, the male). Zeus ordered the craftsman god HEPHAESTUS and other gods to create a new kind of human, lovely but treacherous: This was Pandora ("all gifts"), the first woman. She was sent to earth, where she married Prometheus' simple-minded brother, Epimetheus (having afterthought). Pandora brought with her, or otherwise acquired, a supernatural clay jar containing pain, toil, illness, and other evils not yet known in the world. When, out of curiosity, she opened the jar, the hateful contents flew out and propagated, leaving only hope inside the jar, to delude mortals.

For Prometheus himself, Zeus had a more terrible punishment. Hephaestus chained Prometheus to a lonely mountain crag, where every day an eagle would fly up and tear out the prisoner's liver, devouring it; but Prometheus never died, and every night his liver grew back, so that the torture could be repeated daily. Eventually the hero HERACLES released Prometheus, with Zeus' permission.

Prometheus' punishment is the subject of the stage tragedy *Prometheus Bound* (circa 455 B.C.), which traditionally has been ascribed to the Athenian playwright AESCHYLUS, but which lately is thought not to be his work.

(See also RELIGION; TITANS; WOMEN.)

prophecy and divination Because the ancient Greeks believed that the gods had a hand in human events, they also believed that humans could interpret the gods' wishes. While prayer and sacrifice sought to win the gods' favor, the interpretation of divine will was the goal of certain other activities, which the Greeks called *manteia*, denoting prophecy and other forms of divination. Prophets and soothsayers abounded in Greek MYTH, but they also remained important among living Greeks right down to the end of the ancient world.

In its strictest English-language use, the word *prophecy* refers to a form of soothsaying in which the god speaks directly through the mouth of the human prophet (Greek: *prophētēs*). "Divination," a more general word, includes less inspirational forms of future-telling. The most famous ancient Greek example of inspirational prophecy was the Delphic oracle. The oracle was a priestess of APOLLO, at the god's most holy sanctuary at DELPHI in central Greece. In response to an inquirer's question, she would go into a trance and deliver frenzied utterings, which a male priest would interpret and write down in verse. Similar oracles of Apollo were found at CUMAE and elsewhere.

Apollo's father, ZEUS, was the other main god of prophecy. At Zeus' very ancient sanctuary at DODONA, in northwestern Greece, the priests gave answers after listening for Zeus' prompting in the rustle of leaves in the god's sacred oak tree.

Other modes of divination were practiced by a seer (*mantis*), whose methods might involve observation of animal life, such as the flights and cries of passing birds—a practice now called augury. But the most common of the seer's duties was extispicy—that is, examining the removed organs (particularly the liver) of a newly sacrificed animal: A healthy-looking liver meant that the gods approved of the inquirer's intended course of action; an unhealthy one meant disapproval.

Still other forms of divination sought to find messages in dreams, in random consultations of texts of HOMER's poems, or in events of everyday life. A sneeze, for instance, was considered a good omen, applicable to whatever was being discussed or thought of at the moment. One peculiar rite, known as *kledon*, was practiced at various shrines of the god HERMES. After whispering a question into the ear of the god's cult statue, the worshipper would immediately stop up his or her own ears and leave the temple; outside, beyond the town marketplace, the worshipper would unplug his or her ears and listen for the god's answer delivered in the chance conversations of passersby.

Despite the scorn of intellectuals such as the philosopher XENOPHANES (circa 520 B.C.) and the historian THUCYDIDES (1) (circa 410 B.C.), a belief in divination ran deep and wide in Greek society. City governments employed official seers, and in times of crisis populations sought guidance in oracles and published prophecies. A Greek army on the march brought along a detail of seers, to determine divine will for military purposes; they usually did this by examining the entrails of sacrificial birds. The poet SIMONIDES wrote an epitaph for a Spartan seer, Megistias, killed at the Battle of THERMOPYLAE (480 B.C.). Supposedly Megistias foresaw the Spartans' annihilation, but chose to remain beside his fighting king.

(See also CASSANDRA; RELIGION.)

Propontis See MARMARA, SEA OF.

Propylaea See ACROPOLIS.

prostitutes A society such as ancient Greece—with its slave population, discrimination against female employment, and lack of charitable relief for the poor—was bound to have many prostitutes of both genders, drawn from a needy underclass. There was also a strong demand for prostitutes because the unromantic nature of most ancient Greek marriages combined with the enforced seclusion of citizen WOMEN, led many married men to seek sexual adventure outside the home. A market for teenage boys was fueled by the homosexual atmosphere of upper-class life in the 600s–300s B.C.

As with any aspect of ancient Greek society, much evidence for prostitution comes from ATHENS in the 400s–300s B.C. References to male or female prostitution occur in such Athenian writers as the comic playwright ARISTOPHANES (circa 410 B.C.) and the orator AESCHINES (circa 340 B.C.). The practice was legal at Athens, so long as the prostitute was not an Athenian citizen. If a male citizen was proved to have prostituted himself, he would basically lose his citizenship rights; a prostituted female citizen might suffer a similar penalty, while criminal charges could be brought against her male guardian. Prostitutes at Athens typically came from two social classes: METICS (resident aliens, meaning any Greeks not born of Athenian parents) and SLAVES. The slaves would be women or boys, sent out by their owners to earn money, which the owner then took.

No doubt many prostitutes lived the grim life of the streetwalker. But among females, the Greeks distinguished this lower-class type of prostitute, whom they called a *pornē*, from the high-class *hetaira* (literally "companion," often translated as "courtesan"). The *hetairai* were women skilled in flute-playing, singing, dancing, or acrobatics, skilled also in flirtation and sexual techniques, artful in conversation, and better educated than most citizen women. A number of extant Athenian vase paintings of the 400s B.C. give an idea of the *hetaira*'s typical appearance and activities, while a later Greek writer, Athenaeus (circa A.D. 200), has preserved a collection of the sayings and exploits of famous Athenian *hetairai*.

Hetairai were often organized into "houses" run by managers, who would protect the women and supervise their schedules and fees. Such houses flourished at all commercial centers of the Greek world, including CORINTH and NAUCRATIS, in addition to Athens. Only SPARTA—austere, anticommercial, and with notoriously promiscuous citizen women—was said to have no prostitutes at all.

The natural setting for *hetairai* was the upper-class drinking party known as the SYMPOSIUM. Hired for the night, or perhaps owned as slaves of the host or guests, the *hetairai* might dance or provide flute music, but they were also obliged to give sexual favors to the guests, all of whom

would be men. Certain verses of the poet ANACREON convey his infatuation with one or another *hetaira* at a symposium.

Such women were the only ones permitted to socialize with men informally outside of their homes. In one speech, the Athenian orator DEMOSTHENES (1) maintains that a certain woman cannot be an Athenian citizen, for she has been observed dining and drinking in male company, "just as a *hetaira* would do."

The hazard of *hetairai*—from the established male viewpoint—was that they might captivate rich young men, who would squander entire fortunes on them. From the *hetaira*'s viewpoint, the best she could hope for was to win a rich man's heart, so that he would purchase her freedom (if she were a slave) and set her up as his wife or mistress. At Athens, where a *hetaira* could not legally marry a citizen, she might at least become a common-law wife, or *pallakē*. Among the best-known *hetairai* were Thaïs, who was the mistress and mother of three children of PTOLEMY (1), and Phryne, who served as the sculptor PRAXITELES' model for his celebrated statue of APHRODITE.

Crude forms of birth control were no doubt an essential skill of the prostitute; unfortunately, little is known about this aspect. Ancient medical writers mention vaginal suppositories of woven fabric, used for contraception, and herbal potions thought to suppress fertility. But probably more significant are the several ancient references to abortion. A clause of the ancient Hippocratic Oath, for example, forbade physicians from inducing abortion, while the Roman poet Ovid (circa 20 B.C.) wrote a short poem lamenting his mistress' serious illness following a self-induced abortion.

(See also ASPASIA; HIPPOCRATES; HOMOSEXUALITY; MARRIAGE; MUSIC; SAPPHO.)

Protagoras The first and most famous of the SOPHISTS. Born at the Greek city of ABDERA, Protagoras lived circa 485–415 B.C. In his long and lucrative career, he traveled from city to city, instructing students in intellectual skills such as disputation and the arts of reasoning. Protagoras was the first thinker to acquire the title *sophistēs*, "professor" or "disputer." One of Protagoras' doctrines—that any question could admit two contradicting but equally true answers—became the hallmark of later sophists, who were notorious for their ability to debate a controversy from either side.

Visiting ATHENS, Protagoras may have influenced the thinking of the young SOCRATES. Protagoras was also a friend of the Athenian statesman PERICLES, who once spent a whole day arguing with him over a famous criminal case, in which a boy at a GYMNASIUM had been accidentally killed by a thrown javelin: Was the javelin-thrower guilty in the death, or was the javelin itself guilty? In 443 B.C., when the Athenians organized a colonizing expedition to Thurii, in southern ITALY, Protagoras was chosen, no doubt on Pericles' influence, to draw up the law code for the departing colonists.

Protagoras taught an extreme form of moral relativism—that is, that there are no absolute truths, neither morally nor scientifically. Rather, he said, human convention and opinion determine completely what we can call "right" and "wrong" or "hot" and "cold." By Protagoras' teaching, perception *is* reality, for reality is determined by what we perceive. This was the background of Protagoras' best-known statement, which opened his treatise titled *Truth:* "Man is the measure of all things." Rather than extolling the beauty of the human spirit (as it is sometimes interpreted), the statement simply introduces the idea that there are no absolute truths outside of what humans agree upon.

Protagoras' writings survive only as quotations or references by later authors. He appears as a fictionalized character in PLATO's dialogue *Protagoras,* where he debates with Socrates the question, "Can moral virtue (*aretē*) be taught?"

(See also ANTIPHON; LAWS AND LAW COURTS.)

Psyche   See EROS.

Ptolemy (1)   Macedonian general and, eventually, founder of a ruling dynasty in EGYPT; he lived circa 367–283 B.C. Ptolemy (*Ptolemaios,* "warlike") was born the son of a Macedonian nobleman at the court of King PHILIP II; Ptolemy's mother, Arsinoë, was probably at some point the king's mistress. The young Ptolemy was a friend of Philip's son, later known as ALEXANDER THE GREAT; after Alexander became king (336 B.C.), Ptolemy was appointed to his bodyguard and general staff. He accompanied Alexander on the conquest of the Persian Empire (334–323 B.C.) and later published a campaign memoir that, although now lost, served as a major source for the surviving account written by Arrian.

At Alexander's death (323 B.C.), Ptolemy was serving as his governor of Egypt. In the ensuing turmoil, Ptolemy emerged as one of the DIADOCHI (Successors), who seized great chunks of Alexander's empire for themselves. Ptolemy's dangerous adversary in these years was the Macedonian general ANTIGONUS (1), who, from his base in ASIA MINOR, sought to reknit the old empire under his own rule. By 315 B.C. Ptolemy had joined the coalition of secessionist Diadochi—SELEUCUS (1), CASSANDER, and LYSIMACHUS—against Antigonus and his son, DEMETRIUS POLIORCETES.

At the sea battle of Cyprian Salamis (306 B.C.), Ptolemy suffered a major defeat at Demetrius' hands, thereby losing control of the AEGEAN SEA. But his Egyptian fleets remained strong in the southeastern Mediterranean. In 304 B.C. Ptolemy officially declared himself king of Egypt—modern scholars denote him henceforth as Ptolemy I. In 301 B.C. his army joined in the massive campaign that destroyed his enemy Antigonus. Ptolemy took the honorary title *Sotēr* (savior) to commemorate this victory. The next 15 years saw Ptolemy pushing to acquire or recapture territory in Antigonus' old empire, including CYPRUS, Asia Minor, and the Aegean.

Ruling from his capital at ALEXANDRIA (1), Ptolemy began organizing his Egyptian kingdom along lines that would last 700 years, down through the Roman Empire. He instituted such important "Ptolemaic" features as careful registration and taxation throughout the land, and established military colonies for the breeding of the vital Macedonian soldiering class.

In his private life, Ptolemy was devoted to his mistress, Thaïs, who bore him three children. More officially, his third wife, Berenice—who was also his half sister—bore

him two children: a girl, Arsinoë, and a boy, PTOLEMY (2) II (born 308 B.C.). In 285 B.C. the younger Ptolemy became joint ruler with his ailing father, who died in 283 or 282 B.C. Ptolemy's descendants would rule Egypt until 30 B.C.

(See also CLEOPATRA; COLONIZATION; WARFARE, NAVAL.)

Ptolemy (2) II   Macedonian-descended king of EGYPT who reigned circa 283–246 B.C., after succeeding his father, PTOLEMY (1). Under the younger Ptolemy's rule, the land of Egypt—already rich, powerful, and ancient—became home to a great Greek culture. Ptolemy turned his capital at ALEXANDRIA (1) into the cultural center of the Hellenistic world, a magnet for Greek thinkers, poets, and craftsmen, far surpassing contemporary ATHENS. It was Ptolemy II who built, at Alexandria, those famous institutions of learning and patronage, the Library and Museum, and who sponsored such "Alexandrian" poets as CALLIMACHUS and APOLLONIUS. It was also Ptolemy who built the lighthouse at Pharos, considered one of the SEVEN WONDERS OF THE WORLD.

Ptolemy developed his father's administrative system for the kingdom, creating an efficient bureaucracy to tax, govern, and police Egypt. Militarily, Ptolemy clashed with the SELEUCID EMPIRE, with which Egypt shared a shifting, uneasy border in the Levant. The resulting First Syrian War (circa 274–271 B.C.) ended with important Ptolemaic gains in Syria and ASIA MINOR. The Second Syrian War (266–253 B.C.), less conclusive, ended with the MARRIAGE of Ptolemy's daughter Berenice to the Seleucid ruler Antiochus II.

Ptolemy's first wife (circa 289 B.C.) was Arsinoë, daughter of LYSIMACHUS, the Macedonian-born king of THRACE. She bore Ptolemy three children but was eventually banished for plotting to kill him. Ptolemy's second wife (circa 276 B.C.) was his own full sister, also named Arsinoë. This Arsinoë II died within a few years, but her influence was great and her reign marked the most brilliant years of the Alexandrian court. It was this marriage to his sister that earned Ptolemy the title—which he bore with pride—of Philadelphos (lover of his sibling). Brother-sister marriage among members of the royal family was an Egyptian tradition dating back to the pharaohs, and many subsequent Ptolemies embraced this custom so as to minimize the range for court intrigues and problems of succession.

Ptolemy died in 246 B.C. and was succeeded by Ptolemy III, his son by the first Arsinoë.

(See also HELLENISTIC AGE; JEWS; WOMEN.)

Ptolemy (3)   See ASTRONOMY.

Pygmalion   Mythical king of CYPRUS who sculpted an ivory statue of the goddess APHRODITE and then fell in love with it. The goddess herself, taking pity on him, breathed life into the statue, which became a living woman named Galatea. She and Pygmalion had a daughter, Paphos, after whom the Cypriot city was named.

Pylos   Coastal area of the southwest PELOPONNESE, encompassing the Bay of Navarino. In historical times, Pylos ("the gate") was part of the larger region known as MESSENIA.

Sheltered by the mile-and-a-half-long island of Sphacteria ("wasp place"), the bay provided an excellent harbor, one of the few of the entire Peloponnese. North and east of the harbor stretched the narrow but fertile coastal plain of western Messenia.

In Greek MYTH, Pylos was the home of King NESTOR, a Greek hero of the TROJAN WAR. Modern ARCHAEOLOGY confirms that the Pylos area was a western center of MYCENAEAN CIVILIZATION: Excavations in the foothills eight miles northeast of the bay have uncovered the foundations of an elaborate palace complex, built circa 1400 B.C. and second in size only to the Mycenaean citadels at MYCENAE and TIRYNS. Archaeologists have named this fortress the Palace of Nestor.

The palace consisted of three buildings, each two stories in height; their construction slightly resembled Tudor-style half-timbering. Their outer walls were made of limestone; interior elements included wooden columns and wainscoting. The complex was destroyed by fire circa 1200 B.C., probably amid the internecine wars that ended Mycenaean society.

Among the items recovered from the ruins by archaeologists are more than 1,200 clay tablets inscribed in LINEAR B script. A few of these tablets (datable to a spring season just before the palace's destruction) record the inhabitants' keeping of a coastal watch, presumably against enemy naval attack. It was probably just such an attack that demolished the palace.

Despite Pylos' rich natural resources and geographical features, the region remained unoccupied for centuries after the Mycenaean era (although it retained its name). In 425 B.C., during the PELOPONNESIAN WAR, the Athenian general DEMOSTHENES (2) fortified and garrisoned Pylos' harbor, which lay only 50 miles west of the enemy city of SPARTA. The ensuing fighting between Athenians and Spartans at Pylos, ending in an Athenian victory, was one of the most consequential events of the war.

After the battle, the Athenian fort at Pylos remained garrisoned until 409 B.C. Thereafter, the harbor front was uninhabited for the rest of antiquity. The general region is now home to the modern Greek city of Pylos, also called Navarino, located at the south end of Navarino Bay.

(See also CLEON.)

Pyrrhon See SKEPTICISM.

Pyrrhus King of the Molossian tribes in the northwestern Greek region called EPIRUS, reigning 297–272 B.C. With his magnetic but erratic personality, Pyrrhus brought his kingdom to a brief preeminence in Greek affairs. However, his ambition of conquering a western empire was destined to fail against the Romans' growing power in ITALY. Pyrrhus was the first Greek commander ever to fight the Romans, and is best remembered for his "Pyrrhic victory" at Ausculum, in southeastern Italy, where he lost so many men that he supposedly declared, "Another such victory and we are ruined."

Pyrrhus ("fiery") was born in 319 B.C., the son of the Molossian king Aiacides. The family was related by blood to the recently deceased Macedonian king ALEXANDER THE GREAT; they also claimed descent from the mythical hero NEOPTOLEMUS. When Aiacides was desposed and banished in 317 B.C., Pyrrhus began an exile that eventually involved him with some of the greatest personages of the day. As a teenager, Pyrrhus served as an officer for the Macedonian dynast DEMETRIUS POLIORCETES, who had married Pyrrhus' sister Deidameia. At about age 20, Pyrrhus married Antigone, the stepdaughter of King PTOLEMY (1) of EGYPT. With Ptolemy's support, Pyrrhus in 297 B.C. became a joint king of the Molossians. (He never officially called himself king of Epirus, although he is always referred to as such.) His partner-king was his kinsman Neoptolemus, whom he soon had murdered.

Discontented with the confines of humble Epirus, Pyrrhus spent much of his reign trying to conquer—or otherwise win the throne of—the nearby kingdom of MACEDON, the great realm to the northeast. In these years Macedon was ruled by Pyrrhus' former friend (now enemy) Demetrius, and later by Demetrius' son, Antigonus II.

At the same time, Pyrrhus sought to expand his domain westward toward the Adriatic and Italy. After his first wife died, Pyrrhus married Lanissa, the daughter of the Syracusan dictator AGATHOCLES (295 B.C.). Lanissa's dowry included the important Adriatic-Greek city of CORCYRA, to which Pyrrhus soon added other northwestern acquisitions, such as AMBRACIA, which he made into his capital. But when he polygamously took two more wives, Lanissa left him and married Demetrius (who soon claimed Corcyra).

In late 281 B.C. Pyrrhus received an appeal from the Greek city of TARAS, on the "heel" of Italy. The Tarentines had been contacting various Greek rulers, requesting military aid against the encroaching Romans. Lured by the hope of conquest, Pyrrhus sailed to Italy with 22,500 men and 20 Indian war elephants. In the spring of 280 B.C. the fighting began at Heraclea, on the Italian south-coast "instep," not far from Taras. The Romans—terrified by Pyrrhus' elephants (beasts they had never seen before)—retreated before the massed pikes of the Greek PHALANX. The Romans lost 7,000 men, with another 1,800 captured; but Pyrrhus lost 4,000.

Pyrrhus marched his army north to within 40 miles of ROME, but—discouraged by the lack of local Italian help—turned back south. The next year he experienced the costly victory at Ausculum that inspired his famous lament. After abruptly removing his forces to SICILY, to defend the Greek cities there against the Carthaginians, Pyrrhus returned to Italy to renew his Roman war. But he was defeated in battle near Capua, at Beneventum (where the Romans stampeded his elephants by shooting fire arrows), and he sailed back to Epirus, reportedly having lost two-thirds of his army (275 B.C.).

In a war against Antigonus II, Pyrrhus nearly conquered Macedon (274–273 B.C.). But he abruptly abandoned his advantage, marching far south to invade the territory of SPARTA. There the Spartans and Macedonians combined against him. Battling his way back northward, he encountered street-fighting in ARGOS and was killed—supposedly when an old woman threw a roof tile down onto his head.

A romantic figure and brilliant battlefield commander, Pyrrhus lacked the perseverance and long-range judgment to win an empire.

(See also DODONA; WARFARE, LAND.)

Pythagoras Greek philosopher, mathematican, and reputed miracle-worker, circa 570–500 B.C. No writings by Pythagoras exist, and his life and work are clouded in mystery; most information comes from statements by later writers.

Pythagoras apparently pioneered the study of MATHEMATICS in the Western world. Today he is best remembered for the discovery, made by him or one of his disciples, that the square of the length of the hypotenuse of a right triangle is equal to the sum of the squares of the two other sides' lengths—the famous Pythagorean theorem.

Born on the island of SAMOS, Pythagoras emigrated west in around 531 B.C., reportedly to escape the power of the Samian tyrant POLYCRATES. Settling at CROTON, the foremost Greek colony of south ITALY, Pythagoras founded a religious-philosophical society. He is said to have risen to great local influence. He wrote new laws for Croton and in effect governed the town with his followers, who numbered 300 and came from the Crotonian elite. This OLIGARCHY eventually provoked a people's revolution, and Pythagoras and his followers withdrew to the nearby Greek colony of Metapontum, where Pythagoras died and where his tomb stood in later days.

Two reported tenets of his teachings stand out. One is that the universe's secrets can be learned through the study of numbers. Pythagoras apparently drew great significance from his discovery that musical notes comprise an arithmetic progression; if a lyre string is stopped at its halfway point, the note produced by plucking the half-string is exactly one octave above the whole string's note. Since the Greeks associated MUSIC with the civilizing power of the god APOLLO, this discovery was thought to reveal part of the gods' scheme for the world. Similarly, Pythagoras saw the number 10 as a key to knowledge, mainly because 10 is the sum of the first four integers: 1 + 2 + 3 + 4. The Pythagoreans were known to observe a mystic symbol of 10 dots arranged in rows forming an equilateral triangle:

The second major Pythagorean concept was the transmigration of souls, meaning that the individual soul does not descend to the Underworld at the body's death (as taught by Greek RELIGION) but passes into a new bodily form. In this, Pythagoras may have been influenced by the mystical cult of ORPHISM, which was popular in Greek southern Italy. Like the Orphics, the Pythagoreans viewed the human soul as a fallen divinity, imprisoned in the body and condemned to a series of reincarnations as human, animal, or even vegetable life, until the soul's impurities are cleansed away. As the soul becomes purified in successive lives, it is rewarded with higher forms of incarnation, such as statesmen or poets, until at last it wins some kind of release among the gods; more often, however, the soul might descend from human to beast. One tale records how Pythagoras saw a man beating a puppy and asked him to stop, because (said Pythagoras) he recognized in the puppy's howls the human voice of a deceased friend.

To purify the soul, Pythagoras developed elaborate rules of right conduct for his followers; several such rules are described by later writers. The eating of certain meats was prohibited, since an animal might house a human soul. (It is unclear, however, whether Pythagoras was a complete vegetarian.) The most famous Pythagorean taboo, forbidding the eating of beans probably involved a belief that beans contain little souls (whose attempts to escape are the cause of the bean-eater's flatulence).

The list of Pythagorean commandments strikes a modern reader as odd. The emphasis is not on ethical conduct but on what might be called superstition. Disciples were forbidden to wear rings, to stir a fire with IRON, or to speak of Pythagorean matters in the dark. On rising from bed, a Pythagorean was obliged to smooth away the body's impression in the bedclothes. These and other precepts apparently involved an attempt to avert the "evil eye" or similar bad luck.

Such superstitions reveal how primitive Pythagoras' society was, and how great were his achievements in mathematics and other areas. Pythagoras was said to admit WOMEN to his following on an equal footing with men—which, if true, represents a revolutionary notion in sixth-century-B.C. Greek society. The Pythagorean doctrine of transmigration influenced the Sicilian Greek philosopher EMPEDOCLES (mid-400s B.C.) and the Athenian philosopher PLATO (early 300s B.C.).

After Pythagoras' death (circa 500 B.C.), his followers split into two sects: Of these, the group known as the Mathematicians became a force in early Greek ASTRONOMY. It was these Pythagoreans who first concluded that the earth is spherical, not flat. They also originated the famous theory of the music of the spheres, which imagined that the earth, sun, moon, and planets all revolve around a central fire and create a harmonious sound by virtue of their synchronized speeds; the music is inaudible to us humans.

The political history of the Pythagoreans was unhappy. In the late 400s B.C. they were nearly exterminated in a civil war at Metapontum. The next century saw the school reorganize itself at the Italian-Greek city of TARAS, before dying out by 300 B.C.

(See also AFTERLIFE.)

Pythia See DELPHI.

Pythian Games The second most important of the great religious sports festivals of the ancient Greek world, after the OLYMPIC GAMES. The Pythian Games were held every four years in honor of the god APOLLO, at Apollo's holiest shrine, DELPHI. The name refers to Delphi's inner sanctum, known as Pytho.

An eight-yearly festival of Apollo at Delphi, featuring competitions of the lyre and choruses, was the ancient forerunner of these games. In 582 B.C., after an allied Greek coalition had broken the local power of PHOCIS during the First Sacred War, the Pythian Games were reorganized in their classic form and were placed under the administration of the AMPHICTYONIC LEAGUE. But the games never lost their original aspect—celebrating Apollo as the god of MUSIC.

Unlike other panhellenic fetes, these games included competitions in singing and drama. The footraces were held in a stadium on the slopes of Mt. Parnassus, near Delphi; the horse and chariot races took place far below, on the Crisaean plain. The prize for victors was a garland of bay leaves. (The laurel tree was sacred to Apollo.)

The four-yearly Pythian Games fell halfway along the four-yearly Olympic cycle; that is, in the second spring after the last Olympics, and two years before the next one.

religion The ancient Greeks believed that their gods had a guiding hand in human affairs. Events as diverse as the annual sprouting of crops, disease epidemics, victory or defeat in war, and individual victories in sports events—which modern observers might assign to scientific causes—were seen by the Greeks as proof of the gods' involvement in human events great and small. Greek MYTH told of gods in "the old days" who had actually descended to earth to make love or war among mortals. Although the Greeks of 400 B.C. did not expect to meet a god in the street, they did expect their gods to be invisibly present—at least in major, public events.

This made the Greeks generally more pious than we are today, but also more legalistic and practical-minded in religious matters. Feeling close to their gods, the Greeks expected the gods to be influenced by prayers and sacrifices of animals or crops. The Greeks expected *results* from their faith. A farmer sowed his seeds in the field, then prayed to the gods to send the proper weather. To the ancient Greeks, both activities seemed equally practical.

Modern American society views religion as a purely private concern, legally separated from civic life. But the Greeks had the opposite view: Religion was a public matter. The goodwill of the gods toward the community was something to be carefully maintained (like the water supply or military defense). Therefore, Greek states kept certain priesthoods and temples at public expense, to oversee festivals, offer sacrifices, and generally safeguard the gods' benefaction. Relatedly, the crime of impiety (*asebeia*) was a serious one. When an Athenian jury condemned the philosopher SOCRATES to death on charges including impiety (399 B.C.), it was because some Athenians believed that Socrates' philosophical questioning had genuinely offended the gods and had threatened their care of ATHENS.

As is well known, Greek religion was polytheistic. This means that—unlike the JEWS, but like most other ancient peoples—the Greeks generally worshipped more than one god. Their gods were anthropomorphic, that is, they were pictured in human shape, male or female. The major gods were 12 in number; each was imagined as having power over one or more important aspects of human life. ATHENA was the goddess of handicraft and wisdom, but in "wisdom" she might overlap somewhat with APOLLO, whose province included reason and intellect, while in "handicraft" she might overlap with HEPHAESTUS, god of metallurgy. The king of the gods was ZEUS, sky-father, lord of justice and of universal order. Below him were HERA, POSEIDON, Athena, Apollo, ARTEMIS, APHRODITE, DIONYSUS, DEMETER, HERMES, Hephaestus, and ARES. Because they

A snake goddess or priestess of Minoan Crete, in an 11.6-inch-tall figurine made of a type of glazed earthenware known as faience, circa 1550 B.C., from Cnossus. The figure wears the Minoan aristocratic costume of a flounced skirt and open bodice that reveals the breasts. She holds two snakes, and atop her headdress sits a feline, perhaps a wildcat. The worship of a supreme nature goddess, or family of goddesses, was a central element of pre-Greek Aegean religion. The conquering Greek tribesmen of 2100 B.C. and later cautiously incorporated such elements into their own worship. Certain classical Greek female deities, such as Hera, Athena, and Artemis, are probably survivals of the ancient, non-Hellenic goddesses.

To the music of the flute and lyre, a lamb is brought to the altar for sacrifice, in this scene painted circa 500 B.C. on a wooden plaque found near Corinth. The priest carries on his head a tray with the utensils of sacrifice.

were imagined as dwelling in Zeus' palace in the sky atop MT. OLYMPUS, these 12 were known as the Olympian gods.

But the Greeks recognized a second important group of gods, called the chthonians (from the Greek word *chthon*, earth). These were deities of the earth and the Underworld, chief of whom were HADES and PERSEPHONE, the Underworld's king and queen. In SICILY and other parts of the western Greek world, Persephone became an important object of worship, as a guardian and judge of souls after death. The Greeks worshipped chthonian deities with rites different from those used for Olympian gods.

Aside from the Olympians and chthonians, there were lesser gods, demigods, and heroes, who might have shrines and cults confined to specific locales (such as the cult of HYACINTHUS in the region of SPARTA). Heroes were usually worshipped as the spirits of extraordinary, deceased mortal men. A few major heroes, such as HERACLES, were worshipped throughout the Greek world and were imagined as having turned into immortal gods after their death.

Worship of the gods involved the use of idols—manmade images, usually statues of wood, clay, stone, or BRONZE. With the advances in Greek marble carving in the 600s B.C. and the discovery of improved bronze-casting techniques in the 500s B.C., the representation of the gods became the highest calling of ancient Greek SCULPTURE. A god's idol might be miniature, life-size, or colossal (from the Greek *kolossos*, "a giant statue"). The god inhabited the idol, but not exclusively; a god could be everywhere at once, and many official statues of the same god might stand within a single city. All gods were portrayed as physically beautiful. They had been born but would never die; they lived forever in young adulthood or vigorous middle age. They were

omniscient regarding human affairs and were able to assume any shape. Every city probably kept a shrine to each of the major gods, but certain deities had special care of specific places: For instance, Athena was the guardian of Athens, Poseidon of CORINTH, and Hera of ARGOS and SAMOS.

The gods were imagined as concentrations of energy: The faces of their statues might be painted red, signifying life and power. But the gods were also capable of human flaws and emotions, such as anger, cruelty, and brutal sexual desire. In the latter 500s B.C. the philosopher XENOPHANES complained that conventional religion had ascribed to the gods "everything that is shameful and disgraceful among human beings—thieving and adultery and deceiving one another." One reason why Greek mythology has cast a spell over readers through the centuries is its vivid, endearing character portrayal of the gods, particularly in the poems of HOMER.

Modern scholars believe that the religious system presented by Homer (circa 750 B.C.) and HESIOD (circa 700 B.C.) arose gradually, starting with a fusion of two or three different cultures after 2100 B.C. But the facts are lost in prehistory, and scholars must rely on scant evidence from ARCHAEOLOGY in order to trace this process of religious fusion.

The Greeks were an Indo-European people, with a language and social patterns akin to other Indo-European peoples, such as the ancient Romans, Germans, and Persians. Almost certainly the Greek-speaking tribesmen who first invaded mainland Greece circa 2100 B.C. brought a patriarchal, warrior religion whose chief god was the sky-father, Zeus (a figure that scholars recognize under other names in other Indo-European mythologies). Over time, the Greek conquerors absorbed certain religious elements from the subjugated, pre-Greek inhabitants of the land (an agrarian, non-Indo-European folk with possible origins in ASIA MINOR). This people, like the non-Greeks of Asia Minor in later centuries, may have worshipped a central mother goddess or family of goddesses. Similarly, the non-Greek MINOAN CIVILIZATION of the Aegean islands, which strongly influenced the emerging Greek culture circa 2000–1400 B.C., evidently had important female deities. The prime result of these influences was the introduction into Greek religion of a few major goddesses not originally Greek; eventually they became known by the names Hera, Athena, Artemis, and Persephone.

By 1600 B.C. the mainland Greeks had organized themselves into the warlike and technologically advanced MYCENAEAN CIVILIZATION. It is this era that offers modern scholars the first clear glimpse at Greek religion, in the LINEAR B tablets excavated at Mycenaean CNOSSUS (circa 1400 B.C.) and PYLOS (circa 1200 B.C.). In their lists of religious ceremonies performed for various gods, the tablets reveal that the Mycenaean faith was surprisingly similar to the Greek religion of 700 or 1,000 years later. The tablets mention Zeus and Poseidon as important gods, as well as Hera, Athena, Dionysus, and other deities (some with names unfamiliar).

Of the 12 gods later to be known as the Olympians, only Apollo and Aphrodite definitely are not mentioned on the existing Linear B tablets. Scholars believe that these two latecomers entered Greek religion during the DARK AGE, after the collapse of Mycenaean society circa 1200 B.C. Both deities probably reached Greece from the Near East, via the mercantile island of CYPRUS. In general, Greek religion remained open to foreign deities, particularly goddesses; later in ancient times various Greek locales had important cults of the Egyptian goddess Isis, the Anatolian goddess Cybele, and the Thracian goddess Bendis.

By the 600s B.C., the principal gods and forms of worship were well established. Statues of the gods were housed in temples of simple design but elaborate craftsmanship. Probably descended from the design of Mycenaean palaces, the Greek temple was basically a glorified roof, suspended by pillars over a rectangular foundation, with a smaller area walled off within. There the idol of the god stood or sat. In the 600s B.C. wooden temples began to be replaced by buildings largely of stone, either limestone or marble. The famous Greek "orders" of ARCHITECTURE—the Doric, Ionic, and Corinthian—were invented as styles of temples.

A Greek temple was the sanctuary of a particular god; rarely would more than one god be worshipped in the same building. Unlike a modern church, a temple was not a place where congregations gathered. It had no rows of pews for multitudes of worshippers. Rather, the temple received individuals or small groups for prayer or sacrifice. The temple's priests or priestesses might be minor functionaries under the direction of a high priest who typically came from an aristocratic family associated with that specific sanctuary. At Athens, for example, a noble family named the Eteoboutadai traditionally supplied the priest and priestess for the two major cults on the ACROPOLIS. But Greek religion was noteworthy for the lack of power exercised by its priests (one important exception being

Apollo's priesthood at DELPHI). Technically, prayer and sacrifice could be conducted by worshippers without any priest.

The rite of sacrifice—practiced in various forms by many ancient religions—probably dates back to Stone Age hunters' rituals. The Greeks saw it as the essential way to win divine favor: "Gifts persuade the gods," a proverb ran. Although "bloodless" sacrifices of grain, cakes, or fruit might be offered, a more significant act involved the slaughter of a domestic animal. The noblest victim was a bull; the most common was a sheep. Goats, pigs, and chickens were also used. To the best modern knowledge, human sacrifice was extremely rare among the Greeks; one of the few such cases was the sacrifice of Persian prisoners during the national emergency of the PERSIAN WARS.

Greek sacrifice involved the notion that the god somehow fed upon the victim's blood—an essential rite was the splashing of the blood onto the altar (typically located outside the temple). But first the live animal would be led to the altar in a procession featuring the priest and the person who was providing the animal. The worshippers would be dressed in clean finery, their heads decked with garlands of leaves appropriate to the god: oak for Zeus, laurel for Apollo, and so on. A flute-player would provide stately background MUSIC.

Preliminary rituals included pelting the altar and the live animal with barley grains. Then the priest would cut the animal's throat. The carcass would be skinned and gutted. The inedible organs and bones would be burned at the altar for the god, the smoke ascending to the god's heavenly realm (for Olympian gods; for chthonian deities, the sacrifice emphasized the act of pouring blood into the ground). Then the edible flesh would be cooked as a meal for the celebrants. Large-scale animal sacrifices, sometimes at public expense, might provide community feasts for holidays.

The sacrifice thus benefited the human participants at least as much as it benefited the god. The god received blood, bones, and innards; the human participants received edible meat. Only in rare rituals would the meat be intentionally burned up or thrown away. The element of human self-interest in this rite was not lost upon certain thoughtful Greeks: In time there arose the explanatory myth of how the demigod PROMETHEUS had tricked Zeus into accepting the sacrifice's inferior portions for the gods.

Sacrifices, with their associated processions and feasts, defined the religious year. Although the religious calendar differed in its particulars from place to place, the calendar for any city or region told mainly which kind of offering was to go to which god or hero on which day of the year. At Athens, important holidays included the City Dionysia, an early spring festival of Dionysus, featuring the major annual stage-drama competition, and the PANATHENAEA, celebrating the goddess Athena's birthday in midsummer.

A society as intellectual as the Greeks' was bound to produce religious doubt. Independent thinkers such as the sophist PROTAGORAS (mid-400s B.C.) and the tragedian EURIPIDES (later 400s B.C.) were rumored to be atheists, disbelieving in the gods. Certain philosophers, such as Xenophanes, PLATO (early 300s B.C.), and ARISTOTLE (mid-300s B.C.), believed in a single, universal god. During the

HELLENISTIC AGE (circa 300–150 B.C.), the rise of giant Greco-Macedonian kingdoms created a new social order, and traditional religion—expressing as it did the public life of the old-fashioned Greek city-state—came in for questioning and modification. The philosophy STOICISM inclined toward a monotheism built around Zeus; the rival school of EPICUREANISM claimed that multiple gods existed but had no concern for humankind. Meanwhile there arose fringe "mystery" religions (Greek: *mustēria*, from *mustēs*, "an initiate"), offering to their followers a more personal faith and the promise of a happy life after death.

The great society of the Roman Empire saw further religious percolation, including the rise of Christianity among Greek communities of the eastern Mediterranean in the first century A.D. But many Greeks were still worshipping the old-fashioned gods in A.D. 391, when the Byzantine emperor Theodosius closed all pagan temples throughout his realm and decreed Christianity to be the only legal religion.

(See also ADONIS; AFTERLIFE; ASCLEPIUS; CASTOR AND POLYDEUCES; ELEUSINIAN MYSTERIES; FATE; FUNERAL CUSTOMS; FURIES; HECATĒ; HESTIA; NYMPHS; OLYMPIA; ORPHISM; PAN; PARTHENON; PHILOSOPHY; PHOENICIA; PINDAR; POLIS; PROPHECY AND DIVINATION; ROME; SAMOTHRACE; THEATER.)

Rhadamanthys   See AFTERLIFE; MINOS.

rhapsodes   See EPIC POETRY.

Rhegium   Greek seaport on the west coast of the "toe" of ITALY, commanding the Straits of Messina. Rhegium (Greek: *Rhēgion*) was founded in the late 700s B.C. by Ionian Greek colonists from CHALCIS; a second group founded a sister-city, ZANCLĒ, across the straits in SICILY. With its strategic position, Rhegium became one of the most important Ionian Greek cities of the West. Accordingly, it was an enemy of such powerful Dorian Greek cities as SYRACUSE, in eastern Sicily, and LOCRI, on the eastern Italian "toe."

Rhegium's most famous citizen was the lyric poet IBYCUS (circa 535 B.C.). The city reached a peak under its energetic tyrant Anaxilas (reigned 494–476 B.C.), who fortified the straits against Etruscan war fleets and enlarged the city population with refugees from the Spartan-dominated, Peloponnesian region of MESSENIA. In 433 B.C. Rhegium made an alliance with ATHENS for protection against Syracuse and Locri. In 415–413 B.C., during the PELOPONNESIAN WAR, Rhegium served as a naval base for the Athenians' campaign against Syracuse.

Circa 387 B.C. the Syracusan tyrant DIONYSIUS (1) captured Rhegium and destroyed it. Rebuilt, it eventually passed into the hands of ROME. The site is now the Italian seaport of Reggio.

(See DORIAN GREEKS; ETRUSCANS; IONIAN GREEKS.)

rhetoric   The art of public speaking. Although the word in modern English often implies empty or overblown verbiage, for the ancient Greeks the skill of *rhētorikē* meant nothing less than the ability to communicate in public. Rhetoric (also called oratory) was considered an essential skill for any active citizen in a DEMOCRACY because, in an age before telecommunications, newspapers, or advertising, the one way for an individual citizen to help shape public policy was to stand up before the citizen ASSEMBLY and make a speech. Although speeches might also play decision-making roles in an OLIGARCHY such as SPARTA or CORINTH, it was at the democratic cities—and particularly at the greatest of these, ATHENS—that the best speakers emerged. Among the more famous Athenian orators were DEMOSTHENES (1) (circa 340 B.C.) and the statesman PERICLES (circa 440 B.C.). The philosopher ARISTOTLE (circa 340 B.C.) wrote an extant treatise on rhetoric, one of many produced in antiquity.

The roots of Greek rhetoric go back to the war councils of the Mycenaean princes (circa 1200 B.C.) and the civic councils of the early Greek aristocratic states (circa 800 B.C.). The poems of HOMER (written down circa 750 B.C.) show the assembled Greek heroes delivering speeches in turn at war council. There an artful speaker such as ODYSSEUS is admired for his ability to persuade his fellow captains toward a certain course of action.

It was in the 400s B.C. that rhetoric emerged as an organized study, under the impetus of the developing Greek democracies. The strong poetic traditions of SICILY also may have played a role, because it was a Sicilian Greek who first brought advanced rhetoric to Athens—GORGIAS of Leontini who amazed the sophisticated Athenians with his spoken "display pieces" (427 B.C.).

Aside from the assembly, the other great arena for public speaking was the law courts, where citizens spoke as defendants or prosecutors in front of juries usually numbering several hundred. Some of the best-known surviving speeches are Athenian courtroom speeches from the 300s B.C., delivered by such powerful arguers as LYSIAS. Other men became rich as speechwriters, who did not usually deliver speeches personally but who would, for a fee, compose a courtroom speech for a client. Among the best known of these was ISOCRATES, who opened a school of rhetoric (the first one known) in around 390 B.C.

(See also AESCHINES; ANTIPHON; EDUCATION; ISAEUS; LAWS AND LAW COURTS; SOPHISTS.)

Rhetra   See LYCURGUS (1).

Rhodes   Large, diamond-shape Greek island, located off the southwest coast of ASIA MINOR. By 300 B.C. Rhodes had emerged as one of the foremost maritime powers of the Hellenistic world—wealthy from commerce and with a strong navy.

ARCHAEOLOGY has shown that the island engaged in Mycenaean-Greek TRADE and perhaps received settlers circa 1300 B.C., following an era of Minoan domination. But the main inhabitants at this time were most likely of non-Greek, Carian stock. Rhodes' recorded history begins circa 1000 B.C., with the arrival of DORIAN GREEKS, who sailed east from mainland Greece and founded three cities on the island—Ialysus, at the northern tip, Camirus (Kameiros), on the northwest coast, and Lindus (the largest of the three), in the east. These three cities formed part of a local Dorian federation that included the nearby island of Cos and the city of CNIDUS, on the Asia Minor coast. The Greek name *Rhodos* probably did not refer to the roses (*rhodai*) that still grow throughout the island, but rather came from some Carian name.

The Rhodian Greeks acquired certain non-Greek traditions from the East, chief of which was their national cult of the sun god, worshipped under his Greek name, HELIOS. Rhodes was unique among early Greek states in having a sun cult. According to one myth, Helios had chosen Rhodes as his own kingdom and had populated it with the descendants of his seven sons.

Rhodes' location made it a natural middleman between East and West, although early on it was overshadowed by other east Greek sea powers, such as MILETUS and SAMOS. The Rhodians took part in the trading and colonizing expansion in the 700s–500s B.C.: Rhodian Geometric POTTERY is among the main types found by archaeologists at the ancient Greek trade depot at AL MINA (on the Syrian coast), and it is known that Rhodians helped establish colonies in SICILY and south ITALY. By the 500s B.C. popular dictators known as TYRANTS had arisen in the Rhodian cities, as elsewhere in the Greek world. One such ruler, Cleobulus of Lindus, was later reckoned among the SEVEN SAGES of Greece.

After being conquered by the Persians in around 494 B.C., Rhodes was liberated circa 478 B.C. by the mainland Greeks, at the end of the PERSIAN WARS. Over the next 65 years the three Rhodian cities were members of the Athenian-dominated DELIAN LEAGUE. By then the Rhodian cities were governed as Athenian-inspired democracies—as they would remain, with one interruption, until the Roman conquest. But in 411–407 B.C., during the PELOPONNESIAN WAR, the Rhodian cities joined the widespread Delian revolt and won independence from ATHENS.

Then the three Rhodian cities agreed to combine their individual governments and form a union, with a new capital city, Rhodos. This Rhodes city—a seaport at the island's northern tip, near Ialysus—eventually became one of the most spectacular Greek cities, thriving from the shipping of grain and other essentials across the eastern Mediterranean.

Rhodes was recaptured by the Persians in 355 B.C. but regained independence and DEMOCRACY after the Macedonian king ALEXANDER THE GREAT invaded the Persian Empire (334 B.C.). Amid the turmoil following Alexander's death (323 B.C.), the island became a tempting target for the various DIADOCHI (Successors) who battled over Alexander's vast empire, and in 305–304 B.C. Rhodes city withstood a huge siege by the Macedonian prince DEMETRIUS POLIORCETES.

In the burst of self-confidence following this defense, the Rhodians erected a giant BRONZE statue of the sun god in around 290 B.C. that became known as the Colossus of Rhodes.

Designed by one Chares of Lindus, the 90-foot-tall Colossus (Greek: *Kolossos*, "giant statue") was counted as one of the SEVEN WONDERS OF THE WORLD. Although it collapsed in an earthquake after only about 65 years, its general appearance is known from ancient coins and an ancient writer's description. The Colossus stood alongside the city's harbor (but not bestriding the harbor entrance, legs apart, with ships passing between, as is sometimes claimed). The statue consisted of bronze plating over a frame with an interior staircase. The god, with a halo of sun rays, was shown holding up a torch, which workmen inside operated as a beacon fire to aid ships at sea. The design of the Colossus helped to inspire New York City's Statue of Liberty, erected in A.D. 1886.

The Rhodians were among the first east Greek allies of the emerging power of ROME; it was complaints from Rhodes and PERGAMUM that brought the Romans into the Second Macedonian War against the Macedonian king PHILIP V (200–197 B.C.). But Rhodes lost its prosperity when the Romans—seeking to punish Rhodes for its insufficient help in the Third Macedonian War (172–167 B.C.)—promoted the Aegean island of DELOS as the new center of east Mediterranean trade. Rhodes later became a Roman subject.

(See also COLONIZATION; SHIPS AND SEAFARING; WARFARE, NAVAL.)

Rome   Non-Greek city of west-central ITALY, located in Latium, about 16 miles from the Tyrrhenian coast. After being strongly influenced by the Greeks in such cultural rudiments as its ALPHABET and RELIGION, Rome emerged in the 300s B.C. as a dynamic military force, destined to conquer an empire that (by A.D. 117) stretched from Britain to Mesopotamia. The Romans subdued many peoples, but their conquest of the Greek world, accomplished between about 338 and 30 B.C., deeply affected their own civilization by introducing them to the material and literary splendors of the Greeks. As the Roman poet Horace wrote (circa 19 B.C.): "Captive Greece took mighty Rome captive, forcing culture onto rustic folk." Eventually Roman culture became largely Greek, even though the Romans spoke Latin, not Greek. As the Romans pushed their empire into Europe, they introduced Greek-originated ideas and customs to new places. It is because of the Romans that ancient Greek culture has colored European society, and hence American society.

The Romans belonged to the Latin ethnic group, but early on they were dominated by another Italian people, the ETRUSCANS. The name *Roma* may be of Etruscan origin; its meaning is unclear. Supposedly founded in 753 B.C. Rome surely partook of the pro-Greek atmosphere of Etruscan society in the 600s B.C. Greek goods and customs made their way north from Greek colonies of Campania, such as CUMAE and Neapolis (Naples). The Roman alphabet, used today for English as well as for most other European languages, was first adapted by the Romans (perhaps circa 600 B.C.) from an Etruscan model, itself based on the Greek alphabet. By the 400s B.C. the Romans were vigorously assimilating Greek gods, heroes, and MYTH into their own religion.

Rome supposedly expelled its last Etruscan king in 510 B.C. The next 150 years saw political strife between rich and poor: eventually Rome's government took shape as a republic, in which a wide citizen base was governed by a well-born elite, sitting on a legislative body called the senate.

These centuries also saw the spread of Roman control through Italy. First came the unification of Latium. By the mid-300s B.C. the Romans had reached Campania, where they entered into their first diplomatic relations with Italian-Greek cities. There followed an accelerated helleniza-tion of Roman culture, including the first Roman copying of Greek ARCHITECTURE, COINAGE, and POTTERY.

After subduing the central-Italian people known as the Samnites, the Romans had by 290 B.C. won all of central Italy. In the extreme south, conflict with the Greek city of TARAS brought the Romans into their first war against a Greek army, led by the Epirote king PYRRHUS. The hard-won defeat of Pyrrhus (275 B.C.) left the Romans masters of all Italy south of the Po River. Rome had become a world power, comparable with the Hellenistic kingdoms of the eastern Mediterranean or—closer to home—the African-Phoenician city of CARTHAGE. By this era the formidable Roman army could draw on a manpower pool of several million men, due to Rome's policy of re-quiring military service from its Italian-ethnic subject peoples.

Rome now became drawn into conflict with other great Mediterranean powers—particularly Carthage and MACEDON—which led eventually to the Romans' subjuga-tion of the Mediterranean Greek world. First came two bitter wars with Carthage. The First Punic (Carthaginian) War was fought from 264 to 241 B.C. It turned Rome into a naval power and left the victorious Romans in possession of SICILY, with the important Sicilian-Greek city of SYRACUSE as a Roman ally.

The Second Punic War (218–201 B.C.) saw Rome fighting for its life on Italian soil against the brilliant Carthaginian general Hannibal. Syracuse, siding this time with Carthage, was captured, plundered, and annexed by the vengeful Romans (211 B.C.). Another ally of Carthage was the dy-namic Macedonian king PHILIP V, whose domain included most of mainland Greece. In a Punic War sideshow, the Romans invaded Philip's Greek territories—their first ad-vance into Greece (214 B.C.). This First Macedonian War ended in a truce (205 B.C.), but Philip was by then a hated enemy of an increasingly eastward-looking Rome.

After the Second Punic War ended with Carthage's de-feat, the Romans turned a receptive ear to complaints from their eastern Greek allies PERGAMUM and RHODES against Philip's adventuring. The Second Macedonian War (200–197 B.C.) saw the second Roman invasion of mainland Greece. With the help of Greek allies from AETOLIA, the Roman general Titus Quinctius Flamininus in 197 B.C. won a total victory against Philip's army at the Battle of Cynocephalae ("dog heads"), in THESSALY. The battle broke Macedon's imperial power forever and incidentally estab-lished the superiority of Roman battle tactics against the more cumbersome Macedonian PHALANX. But by the terms of a lenient treaty, Macedon was allowed to continue as an independent state.

It was at the ISTHMIAN GAMES of 196 B.C.—after Macedo-nian control of Greece had been removed—that Flamininus issued his famous proclamation declaring freedom for the mainland Greeks: Neither Macedon nor Rome would be overlord there. In 194 B.C. Roman troops evacuated Greece.

But even unwillingly, the Romans found themselves enbroiled in Greek events. The Aetolians, disgruntled over Rome's light treatment of their defeated enemy Philip, invited the Seleucid king ANTIOCHUS (2) III to invade Greece as a new liberator. Reentering Greece in response, Roman forces defeated Antiochus at THERMOPYLAE (191 B.C.) and pursued him to ASIA MINOR—the Romans' first campaign in Asia—where they defeated him again, at the Battle of Magnesia (190 B.C.). This victory foreshadowed the future Roman conquest of the entire Greek East. By the treaty of Apamea (188 B.C.), Antiochus lost most of Asia Minor, which the Romans parceled to their allies Pergamum and Rhodes. In Greece, meanwhile, the Aetolians were de-feated and their power curtailed.

The accession of Philip's ambitious son Perseus brought new Macedonian adventurism, and another Roman inva-sion. The Third Macedonian War (172–167 B.C.) ended with Perseus' complete defeat at the Battle of Pydna, in southern Macedon. Deposed, Perseus died in a Roman prison. His kingdom was dismantled and was later an-nexed as a Roman province called Macedonia. In Greece, the League of ACHAEA was punished for its pro-Macedo-nianism with the removal of 1,000 well-born hostages to Rome. (One of these, named Polybius, was destined to become the last great Greek historian; so impressed was he by Roman power that he chose to write, in Greek, about the rise of Rome.)

When the Achaean League resumed anti-Roman agita-tion in 148 B.C., Roman patience with Greece was at an end. Marching south from the province of Macedonia, a Roman army defeated the Achaeans in battle. CORINTH, the league's major city, was besieged and captured. By order of the Roman senate the city was leveled to the ground as a warning to the Greeks (146 B.C.).

Greece's freedom was at an end, and the country was then incorporated into the Roman province of Macedonia. In the following centuries the Greek East fell. Much of Asia Minor was legally bequeathed to the Romans by the last Pergamene king, Attalus III, in 133 B.C. In 64 B.C. the Roman general Pompey annexed Syria, once the Seleucid heartland. And in 30 B.C. the Romans captured the last remaining Hellenistic kingdom, EGYPT.

(See also AENEAS; CASTOR AND POLYDEUCES; CLEOPATRA; HELLENISTIC AGE; HERACLES; SELEUCID EMPIRE.)

Roxane See ALEXANDER THE GREAT.

S

Sacred Wars See DELPHI; PHOCIS.

sacrifice See RELIGION.

Salamis (1) Inshore island in the northern Saronic Gulf, in central Greece. Salamis' eastern side is separated from the coast of ATTICA by a curved channel less than a mile wide in the narrows, and it was there that the crucial sea battle of the PERSIAN WARS was fought in late summer of 480 B.C., after the invading army of the Persian king XERXES had overrun northern and central Greece. The allied Greek victory at Salamis broke the Persian navy and stopped the invasion, opening the way for the Greeks' final land-victory, at PLATAEA, the following summer.

Well before the sea battle, Salamis was an important place, associated with the mythical hero AJAX (1). Salamis lay close offshore between the territories of ATHENS and MEGARA (1), and not far from the powerful island state of AEGINA. In the 600s B.C. the island was a bone of contention among those three warring states. By the mid-500s B.C. it was held by Athens.

In 480 B.C. the Persians and their allies overran a Spartan-led force at the Battle of THERMOPYLAE and swept southward into central Greece. The Persian navy sailed in escort. Reaching Athens, Xerxes' army found the city abandoned. The noncombatants had been evacuated, while most of Athens' fighting men were in the allied Greek navy, stationed inside Salamis' eastern channel.

Rounding the southern tip of Attica, the Persian navy sailed north into the Saronic Gulf. Not literally "Persian," this navy was manned largely by subject peoples—Phoenicians, Egyptians, and even IONIAN GREEKS—who were fighting for Xerxes. The Phoenician crews seem to have been superior seamen to any Greeks. Battered by prior fighting and storms, the Persians' warships by then probably numbered less than 400. Waiting at Salamis was a combined Greek fleet of perhaps 300 warships, with Athens supplying the largest contingent, almost 180 ships. The fleet's commander was Eurybiades, a Spartan.

The Greek plan to make a stand at Salamis probably originated with the brilliant Athenian soldier-statesman THEMISTOCLES, who foresaw that the way to offset the Persians' advantage in numbers and faster ships would be to lure them to battle inside the narrow straits. By vision and force of personality, he convinced Eurybiades and the other Greek sea commanders to stay together and risk a major fight, rather than dispersing to defend their homes.

According to two fifth-century-B.C. writers—the Athenian playwright AESCHYLUS (who may have fought at Sala-

mis) and the historian HERODOTUS—Themistocles tricked the Persian king by sending him an untruthful message, couched as a plea for personal favor, saying that the Greeks were ready to disband and flee. Supposedly Xerxes actually believed this report from an enemy commander. Whether this is true or not, we do know that Xerxes, overconfident and impatient, ordered his fleet to enter the Salamis channel at night and attack the Greeks at dawn. What followed was one of the most dramatic events in world history—the saving of Greece from foreign domination.

At dawn the Greeks sang their battle paean and launched their fleet in a concave crescent formation embracing the entire width of the channel. With the Persian ships rowing into this Greek "net," the two fleets collided.

As Themistocles had expected, the Persians' advantage in numbers and skill was lost in the crowded narrows. Whereas boarding tactics had been the rule at the recent sea battle at ARTEMISIUM, Salamis saw more use of Greek ramming techniques. The Persian ships—jammed together by the Greek enveloping tactic, and not as able to withstand ramming as the heavier ships of the Greeks—soon began to get the worst of it. Aeschylus gives a vivid description, presented from the Persian viewpoint, in his tragedy *The Persians* (472 B.C.): "The Greek ships, arrayed in a circle around us, bore in and rammed: our ships capsized, the sea was hidden in wrecks and corpses, and all the shores and reefs were heaped with dead . . ."

After most of the elite Phoenician squadron had been destroyed, the other Persian allies broke off and rowed away, pursued by the Greeks. At this point an amphibious landing of Athenian HOPLITES, under the general ARISTIDES, overran a contingent of Persian infantry on the nearby islet of Psytalleia.

The battle was over. The Greek side had lost 40 ships, the Persians (probably) several times that number. Among the Persian dead was the admiral Ariabignes, a brother of Xerxes.

Xerxes himself had watched the battle from a throne set up on the shore, and was so alarmed by his fleet's ruin that he soon started overland for home. Then vulnerable by sea, he was anxious to reach the HELLESPONT before the Greek navy could destroy his bridges there, to trap him inside Europe. Along with most of his huge invasion force, Xerxes crossed back into Asia, leaving behind a small army under his general Mardonius to continue the campaign on land. This army, in turn, was later destroyed by the Greeks at Plataea.

For the Athenians, who had abandoned their city yet still supplied the bulk of the Greek navy, the Battle of

Salamis was a patriotic triumph and a springboard to Athens' naval empire in the following decades. For the Persians, Salamis marked the turning point on their western frontier; no longer would they be purely the aggressors, as they had been for 70 years. With their naval power crippled for a decade, the Persians would subsequently go on the defensive—first against Athenian naval action in the eastern Mediterranean, and later, in the 300s B.C., against Spartan and Macedonian invasions of Asia. The process that would end with ALEXANDER THE GREAT's conquest of PERSIA (334–323 B.C.) began in the narrows at Salamis.

(See also WARFARE, NAVAL.)

Salamis (2)   See CYPRUS; DEMETRIUS POLIORCETES.

Samos   Fertile and mountainous Aegean island, 190 square miles in area, located two miles off the central west coast of ASIA MINOR. Samos was among the most prosperous and powerful Greek states of the 600s–400s B.C. The Samians were known for their seafaring, their domestic WINE, and their national cult of HERA, patron goddess of the island.

Occupied before 1200 B.C. by Greek colonists of the MYCENAEAN CIVILIZATION, Samos was resettled circa 1100–1000 B.C. by a new wave of immigrants—IONIAN GREEKS, fleeing from the Dorian invasion of mainland Greece. A capital city, also named Samos, arose in the southeast part of the island, opposite the Asia Minor coast.

Samos island became a foremost state of the eastern Greek region called IONIA, and played a major role in the Greek world's expansion through seaborne TRADE and COLONIZATION in the 800s–500s B.C. The principal Samian colony was at the north Aegean island of SAMOTHRACE (which, as the name suggests, served as an anchorage between Samos and THRACE). To buy luxury goods and raw metals, the Samians exported domestic products such as wine and woolens, along with Thracian SLAVES and SILVER acquired via Samothrace. Markets included both mainland Greece and Near Eastern kingdoms such as EGYPT. One Samian sea captain made a famously lucrative trading stop at the Celtic kingdom of Tartessus, in southern Spain, circa 650 B.C.

In this era Samos was governed as an ARISTOCRACY, with a ruling class known as *geomoroi*, "landholders." The island was home to famous craftsmen such as the architect Rhoecus, whose temple of Hera, begun circa 560 B.C., was the largest Greek building of its day (320 × 160 feet). All that survives of that proud structure today is a single Ionic column, standing at the site of ancient Samos city.

The island reached its zenith under POLYCRATES, who held supreme power circa 540–522 B.C. Polycrates carved out an Aegean sea empire and glorified Samos with engineering marvels such as an underground aqueduct and sea walls to defend the harbor entrance. His lavish court attracted poets, craftsmen, and other talents. But Samos' most important citizen, the philosopher and mathematician PYTHAGORAS, fled Polycrates' rule and emigrated to south ITALY.

After Polycrates was captured and murdered by the encroaching Persians under King DARIUS (1), Samos became a Persian holding, ruled by Greek puppet rulers. Joining the IONIAN REVOLT against the Persians, Samos contributed its entire navy, 60 ships. But at the revolt's climactic sea battle of Lade, fought near Samos, the Samian ships deserted for Persian offers of preferential treatment (494 B.C.). In 480 B.C., under Persian command, Samian crews fought against their fellow Greeks in King XERXES' invasion of Greece. The following year, Samos was liberated by the victorious Greek counteroffensive.

Then under the heavy influence of ATHENS, Samos became a DEMOCRACY and a foremost member of the Athenian-run DELIAN LEAGUE. Samos was one of the few Delian states to supply warships and crews in lieu of silver tribute. However, in 440 B.C. the island rebelled against Athens. The nine-month-long revolt amounted to an Athenian crisis: An attacking Athenian fleet was defeated by 70 ships under the Samian leader Melissus, but eventually Samos fell to an Athenian siege and blockade commanded by PERICLES himself. The island was purged of anti-Athenian elements and reduced to subject status, paying a yearly tribute as well as a huge war penalty of 1,400 TALENTS.

Samos remained a staunch Athenian ally—although no longer contributing warships—throughout the PELOPONNESIAN WAR (431–404 B.C.). After 413 B.C. the island was a major Athenian naval base. During the right-wing coup of the FOUR HUNDRED at Athens, the Samians supported the Athenian democratic government-in-exile, formed by the Athenian fleet and troops at Samos (411 B.C.). For their loyalty, the Samians received the unique honor of being made Athenian citizens (405 B.C.). But in the following year, at the war's end, Samos was captured by SPARTA.

In the next decades an independent Samos again fell into conflict with the Athenians, who captured the island and repopulated it with Athenian colonists (365 B.C.). Later the native population was allowed to return, by edict of the Macedonian conqueror ALEXANDER THE GREAT (effected posthumously, in 321 B.C.).

A bone of contention among the warring DIADOCHI (Successors) who divided up Alexander's empire after his death, Samos eventually passed to King PTOLEMY (2) II of Egypt (circa 281 B.C.). By then Samos was waning as an east Greek sea power, surpassed by the growing commercial might of RHODES. In 127 B.C. Samos was annexed by the Romans.

(See also BRONZE; PERSIAN WARS; SHIPS AND SEAFARING; TYRANTS.)

Samothrace   Small Greek island of the northeast AEGEAN SEA. Samothrace sits 25 miles south of the northern coast, which in antiquity belonged to the non-Greek land of THRACE. The oval island, only 68 square miles, consists of a single mountain, rising to 5,250 feet and heavily forested in ancient times. The pre-Greek inhabitants were probably of Thracian stock. Sometime between 700 and 550 B.C., Samothrace was conquered by seafaring Greeks, who sought it as a safe anchorage from which to make the trading run to the Thracian coast. (The mainland Thracians were a potentially hostile folk who nevertheless could offer such precious goods as SILVER and SLAVES.) As suggested by the name Samothrakē (Samos' Thrace), the island was probably occupied by Greeks from the island state of

SAMOS; possibly the Samians were preceded by Greeks from another (and nearer) Greek island state, LESBOS. Samothrace seems to have been a mixed settlement, where Greek and Thracian inhabitants lived in harmony; modern archaeological excavations have uncovered a cemetery that (as inscriptions reveal) was used by both peoples in the latter 500s B.C.

In classical times, Samothrace was known for its unique Greek religious cult, colored by the place's pre-Greek origins. This cult worshiped the Cabiri (Kabeiroi)—male deities, originally non-Greek, possibly from Thrace or Phrygia (in ASIA MINOR). The Cabiri varied between two and four in number and were patron gods of smiths and sailors as well as of fertility. Also called the Great Gods, they were sometimes identified with the Greek deities CASTOR AND POLYDEUCES. If the name Kabeiroi is Greek (which is uncertain), it probably means "the burners," in reference to the forge fire of an ancient smithy.

The Cabiri had cults at several shrines in the Greek world, but their main sanctuary was at Samothrace city, high in the island's hills. There were held the Mysteries of Samothrace—a secret cult, open only to initiates. These Mysteries (Greek: *mustēriai*, from *mustēs*, "an initiate") are often mentioned but rarely discussed by ancient writers, who feared to divulge secret details lest they provoke the Cabiri's wrath. The cult became popular in the HELLENISTIC AGE (300–150 B.C.): Men and WOMEN flocked to Samothrace for initiation and worship. King PTOLEMY (2) II of EGYPT (circa 270 B.C.) was an important patron of the sanctuary; remnants of the buildings that he financed can be seen today. Among the costly offerings that adorned the site was the wonderful second-century-B.C. marble statue known as the Winged Victory of Samothrace, excavated by French archaeologists in A.D. 1863 and now displayed at the Louvre.

Politically Samothrace became an ally of ATHENS, serving as a member of the Athenian-controlled DELIAN LEAGUE (478–404 B.C.) and the subsequent SECOND ATHENIAN LEAGUE (378–338 B.C.). Attractive as a naval base and propagandistic pawn, in Hellenistic times the island changed hands frequently, among the Macedonians, Ptolemaic Egyptians, and Seleucids.

(See also AFTERLIFE; SCULPTURE.)

Sappho Lyric poet of MYTILENE on the island of LESBOS, born in about 630 B.C. Her verses, of which perhaps two short poems and some 150 fragments survive today, have been much admired since ancient times for their directness of expression and personal honesty. Most of Sappho's poems were composed for solo recitation to the accompaniment of the lyre. She wrote in the Aeolic dialect and vernacular of Lesbos, a place whose strong lyric tradition had already produced such poets as TERPANDER, ARION, and Sappho's contemporary ALCAEUS. Sappho is one of the few female poets whom we know of from ancient Greece.

The typical subject of Sappho's verse is her physical love for several of the young WOMEN who were her comembers in a local cult of the goddess APHRODITE. The clearly homosexual nature of Sappho's poetry (and the presumably homosexual atmosphere of her female circle on Lesbos) gave rise in the 19th century A.D. to the word *lesbian*,

meaning "a woman who has sexual feelings for other women."

The only reliable facts about Sappho's life are those that can be gleaned from her verses. She was born into an aristocratic family of Mytilene. Although her poetry does not deal with politics, she apparently shared in the privileges and hazards of upper-class life. She went into brief exile in SICILY during a period of civil unrest on Lesbos circa 600 B.C., and she writes of happier days when her brother Larichus served as cupbearer at aristocratic drinking parties in Mytilene. Sappho had another brother, Charaxus, who sailed as a trader to NAUCRATIS, in EGYPT, where he squandered a fortune on a courtesan named Doricha; in certain verses Sappho rails against Doricha and wishes Charaxus a return to sanity. Later writers say that Sappho was married—as most women of her social class were bound to be—and we know that she had a daughter, Cleïs; in one poem Sappho expresses joy at the beauty of her daughter's face. In later centuries there arose a ridiculous tale that Sappho had died by jumping off a cliff for hopeless love of a sturdy ferryman.

Modern writers often assume that Sappho was a music teacher or similar, but her poems show only that she led a group of younger women, probably teenagers, in an official worship of Aphrodite—specifically, in observing the goddess' religious calendar and in maintaining various shrines in and around Mytilene. Some of Sappho's loveliest verses show the women venturing into the countryside or plaiting wildflower garlands to adorn a cult statue or altar. Sappho's sexual involvements with her young colleagues would seem to grow out of this sincere devotion to the goddess of sex and nature. The one poem by Sappho that definitely survives in its complete form is a hymn to Aphrodite in which the goddess, appearing before Sappho, promises to help the poet woo her beloved.

Not that love always means happiness in Sappho's poetry. "O Atthis," she writes to a girlfriend, "you have come to hate me in your mind, and flee to Andromeda!" (Andromeda was a rival of Sappho's, possibly the leader of a rival cult.) There was also the melancholy fact that, no matter what liaisons might arise between the girls in Sappho's group, all of these aristocratic maidens were destined to be married off by their families. (In ancient Greece girls were usually married by age 18; often by 14 or even 12.) Several of Sappho's poems are wedding songs, while other verses show distress within the group as members depart for MARRIAGE. Sappho's most famous poem, which survives nearly complete, records her own unabated desire as she watches a girlfriend talk with a handsome young man—probably the girl's husband and evidently at their wedding feast. He looks calm and godlike, the poet writes, "but whenever *I* even glance at you, my voice is lost, my tongue is stopped, and a delicate fire has spread under my skin . . ."

In another poem, which survives only as a mutilated fragment, Sappho recalls how she consoled a girlfriend who was departing from the group. " 'Honestly I want to die,' " the fragment begins, with the other woman talking. " 'Sappho, I swear it's against my will that I'm leaving you.' And I answered her thus: 'Go gladly and remember me, for you know how we cherished you. Or let me remind you, if you forget, what blissful days we enjoyed—how at my side you often put on garlands of violets and roses and crocuses . . . and on soft mattresses gently you satisfied desire. And there was . . . nothing sacred from which we kept away . . .' "

In the modern age of commercialized sex and sexual politics, it takes some effort to imagine the innocence of Sappho's world. Although Sappho has been appropriated as a feminist forerunner since the 1970s, the poems suggest a lady with an open and honest heart living in a land of private emotion, completely removed—as almost all women of those days were—from politics or grand social concerns. The religious and homosexual involution of C.: Certain verses by the male Spartan poet ALCMAN (circa 630 B.C.) seem to commemorate similar emotional tides within a Spartan girls' chorus.

The private world of Sappho and her lovers represents the female response to an outside world created by men— a world of politics, war, and male HOMOSEXUALITY, which excluded even upper-class women except as wives and mothers. One of Sappho's best-known poems involves a gentle rejection of male militarism: "Some say that a host of cavalry is the loveliest sight on earth, some say foot soldiers, some say warships. But I say it is the person you love."

What makes such verses great, despite their self-absorption, is their apparent honesty and unselfconsciousness. Sappho's eroticism is miles apart from the amused, detached male homosexuality in the poetry of IBYCUS or ANACREON (mid- and late 500s B.C.). Sappho's genius lies in her ability to convey her sincere feelings, in simple lyric meters and vocabulary. Her metaphors and similes are beautifully apt and often come from the natural world, as when she writes of a maiden who outshines other women as the moon outshines the stars. Among Sappho's admirers in later generations was the philosopher PLATO, who called her the tenth Muse. The Roman poet Catullus (circa 55 B.C.) imitated her verses in at least two of his poems.

(See also ARISTOCRACY; LYRIC POETRY; MUSES.)

satyrs The *saturai* were mythical spirits of the countryside and wilderness, personifying fertility and sexual desire. Satyrs were semihuman in form, with the ears, tail, legs, and hooves of a goat (or, in some earlier versions, of a horse). They were favorite subjects in vase painting and other art, and were always shown naked, with beards and pug noses, often with bald heads, and usually with erect penises. Satyrs were typically shown getting drunk or pursuing sexual gratification with their female counterparts, the NYMPHS.

The satyrs were associated with DIONYSUS, the god of WINE and fertility. Processions honoring Dionysus and the satyrs were held in the Athenian countryside (among other places), and these bawdy ceremonies apparently supplied the origin for Athenian stage comedy. Classical Athens retained a form of stage farce, less elaborate than formal comedy, known as the satyr play (*saturikon*).

(See also THEATER.)

science The ancient Greeks were the first Europeans to make inquiries into ASTRONOMY, physics, and biology, yet they were slow to recognize a separate discipline that might be called science. Greek scientific studies in the 500s and 400s B.C. were bound up in the concept of *philosophia* (love of arcane knowledge) and the search for philosophical truth. Such studies were conducted in imitation of MATHEMATICS, with emphasis on theory rather than on experiment and observation. It took the categorizing genius of ARISTOTLE (384–322 B.C.) to identify a separate discipline that he called *phusikē* (natural studies), in contrast to mathematics or PHILOSOPHY.

The origin of Western science is associated with the philosopher THALES of MILETUS (circa 585 B.C.). Thales was the first Greek to try to explain the universe without appeal to MYTH or RELIGION, but rather with reference to a primary physical substance (which he identified as water). Further crude theorizing in physics came from Thales' disciples ANAXIMANDER (circa 560 B.C.) and ANAXIMENES (circa 550 B.C.), and from such later philosophers as HERACLITUS (circa 500 B.C.), PARMENIDES (circa 490 B.C.), EMPEDOCLES (who originated the concept of four primal elements— earth, fire, water, and air, circa 450 B.C.), and the atomists LEUCIPPUS (circa 440 B.C.) and DEMOCRITUS (circa 430 B.C.). These thinkers all addressed, in highly original ways, the basic problem of physical change: How is it that matter apparently comes into being and grows out of nothing? (For instance, how is it that trees sprout leaves?)

But the single most important figure in ancient science was Aristotle, whose voluminous inquiries into all existing branches of knowledge served to create certain scientific categories still recognized today, such as physics and biology. Aristotle introduced the practice of studying by empirical observation, purely for the sake of scientific knowledge—a practice that became traditional at his Athenian school of higher learning, the LYCEUM (founded circa 335 B.C.).

In zoology specifically, Aristotle's inquiries were simply epoch-making. His zoological writings make reference to over 500 animal species, including about 120 kinds of fish and 60 kinds of insects, observed either in their natural habitat or in dissection. Although not the first Greek to practice animal dissection, he was the first to do so with a wide variety of species and to document his researches.

Aristotle tackled the problem of change with typical restraint and common sense, suggesting that a thing can come to be both what it already is and what it is not—in the sense that it fulfills an inborn potential, or *telos*. For example, the acorn "is" an oak tree, in terms of its *telos*. In physics, Aristotle pioneered the study of motion, which is now called dynamics.

Aristotle's work dominated Western science, for better or worse, for over 2,000 years, down to the 1600s A.D. His acceptance of Empedocles' notion of four elements guaranteed that that theory would remain for centuries the authoritative theory of chemistry. Even Aristotle's grosser errors became canonical: His treatise *On the Heavens* contains the notoriously inaccurate statement that a heavier object will fall at a proportionately faster speed than a lighter one. Although questioned in late antiquity, this false law of physics was taught in European schools until it was finally disproved by Galileo in around A.D. 1590.

Aristotle's successor at the Lyceum, THEOPHRASTUS (president circa 323–287 B.C.), carried on the master's studies in biology and physics, as did Theophrastus' successor, Straton (president circa 287–269 B.C.). Meanwhile, the Athenian philosophical movement of STOICISM popularized the Empedoclean-Aristotelian theory of four elements, while the rival school of EPICUREANISM adapted Leucippus' and Democritus' concept of atoms (*atomoi,* "indivisible particles"). But by the early 200s B.C. the center of scientific research was shifting from Athens to the wealthy court of the Ptolemies at ALEXANDRIA (1), where the Library and Museum attracted scholars from around the Greek world.

The main scientific advances at Alexandria were in the previously neglected field of mechanical engineering. From at least the 500s B.C. onward, the Greeks had possessed such simple mechanical devices as the lever, pulley, and winch, but in the 200s B.C. Alexandrian scientists explored new constructions and mechanical theories. Fueling this advance was the Ptolemies' eager financing of military technology, specifically the creation of bigger and better siege artillery, powered by torsion. The Alexandrian inventor Ctesibius (circa 270 B.C.) is credited with improvements in catapult design as well as with such peaceful inventions as the water pump and the water clock (which measured the passing hours through the controlled dripping of water). Ctesibius also pioneered the study of pneumatics (the action of air under pressure).

The century's greatest Greek scientist—and one of its greatest mathematicians—was ARCHIMEDES (active 260–212 B.C.), a lifelong citizen of the Sicilian-Greek city of SYRACUSE. Among his many accomplishments, Archimedes invented the branch of physics known today as hydrostatics (the study of properties of standing water). Archimedes added one important invention to Mediterranean technology—the Archimedes screw, still used today to draw water uphill continually, without suction.

One technological breakthrough of the late Greek world (circa 100 B.C.) was the simple expedient of powering a millstone by means of a donkey tethered to a pole, treading a circle. Further advances appeared under the Roman Empire, including the Alexandrian inventor Hero's famous experiment with steam power (circa A.D. 60). The slow advance of Greek technology can be explained partly by the availability of slave labor and by a lack of government funding for nonmilitary research.

(See also MEDICINE; SLAVES; WARFARE, SIEGE.)

Scopas   See SCULPTURE.

sculpture   The greatest artistic success of the ancient Greeks came in their shaping of marble (by carving) or metal (by casting) to represent the human body. By the mid-400s B.C. Greek sculptors, working mainly at the preeminent city of ATHENS, had achieved a realism in portraying human forms that went unequaled for nearly the next 2,000 years, until the Italian Renaissance. Modern art historians refer to the years 480–330 B.C. as the Greek "classical" era, meaning that that was when the Greeks established standards of excellence—in sculpture and in ARCHITECTURE—that helped to define those art forms for all time.

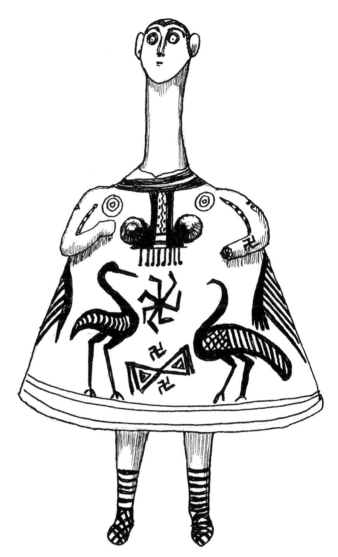

Clay female figurine from Boeotia, in central Greece, later 700s B.C. The bell-shape body was molded on a potter's wheel, then painted and baked in a kiln; the legs were made separately and fastened inside. The painted swastikas probably represent the good magic of rotating wheels or drills for making fire; it would be another 2,600 years before the Nazis appropriated that symbol as an emblem of hate.

Most Greek sculpture has not survived from ancient days. During late antiquity and the Middle Ages, priceless works were lost to neglect or plundered for their raw material. Today many ancient sculptures exist only as Roman copies of vanished Greek originals. Most of these Roman copies are marble, even if the Greek originals were BRONZE.

Classical Greek sculpture took several forms. A statue is by definition self-standing; Greek statues typically showed a human figure, male or female, representing a person or deity. Among the most important statues were idols of gods or goddesses set up in the deities' temples. Aside from statuary, there was architectural sculpture: Figures were carved in limestone or (usually) marble to be fastened to panels on the outside of temples and other public buildings. Typically showing scenes from MYTH, architectural as bronze or GOLD, the earliest extant Greek sculpture is in stone—specifically, the soft limestone that lies plentifully in the mountains of mainland Greece. The best-known early item of carved limestone is the bas-relief slab at the gateway of MYCENAE (circa 1250 B.C.). No doubt such sculpture developed from primitive traditions of wood-carving—however, almost no wooden items have survived from the distant past, since wood is such a perishable material. The earliest Greeks surely used wood for carving their statues of the gods; even as late as the 400s B.C.

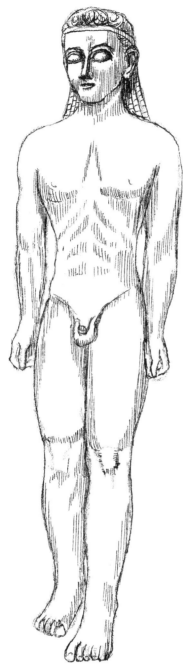

A marble *kouros* (male youth) from the temple of Poseidon at Cape Sunium, in Athenian territory, circa 585 B.C. This nearly 10-foot-tall statue has the typical *kouros* pose: face front, arms stiffly at sides, left leg advanced. The pose closely resembles that of earlier Egyptian statuary, and was probably originally copied by Greek artists who visited Egypt or who based their work on imported Egyptian figurines.

sculptures might be carved completely in the round (as though they were statues) or they might be left partly uncarved, where the figure would be fixed to the building. Another type of architectural sculpture was bas-relief, which was a panel of stone carved so that the figure (or, often, group of figures) would rise part-way out of the stone. Apart from buildings, bas-reliefs were also used for gravestones and public plaques.

Although classical sculptors could work in metals such

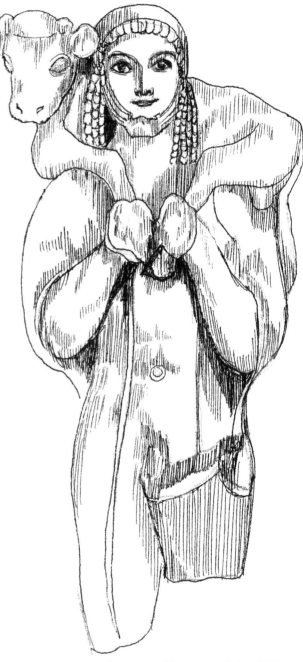

The Calf-bearer, an Athenian marble statue of about 560 B.C. Sculpted at just over life size, the man is bringing a bull calf to sacrifice. The artistic stylization of prior decades is still apparent, including the typical "Archaic smile" and *kouros*like leg positions, but the general pose is more daring, with a confident sense of motion. Here the Greeks have begun to make sculpture breathe life.

The celebrated Discobolus (Discus Thrower), a life-size marble statue of the Roman imperial age, copied from a bronze original (now lost) sculpted by Myron of Athens, circa 450 B.C. Intended to be seen only from this angle, the statue is unrealistically flattened in pose; nevertheless, the original work represented a landmark in the Greeks' quest to portray the human body in action.

wooden idols were still being worshipped at certain shrines.

After the collapse of MYCENAEAN CIVILIZATION, the grim DARK AGE (circa 1100–900 B.C.) saw artistic impoverishment throughout the Greek world. Sculpture may have been reduced largely to wood-carving and the shaping of clay. But sculpture in stone reemerged at the start of what scholars call the "archaic" era (circa 650–480 B.C.). Influenced by the monumental statuary carved in granite and porphyry that could be seen in EGYPT, Greek sculptors turned to the most suitable hard stone found in Greece—marble (a compacted form of limestone). Likewise, the Greeks copied the typical pose of Egyptian statuary—arms stiffly at sides, left leg advanced. A number of these early Greek statues survive, with the earliest dating from soon after 600 B.C. They are the figures known today as *kouroi* ("male youths"); typically they show a naked young man,

slightly larger than life size. The *kouroi* and their somewhat later female counterparts, the clothed figures known as *korai* ("maidens"), were works commissioned by wealthy families to stand at a young person's grave or as a dedicatory offering in a god's temple. Most extant *kouroi* and *korai* come from the region of Athens.

A comparison of individual *kouroi* reveals the advances made by Greek sculptors during the 500s B.C. Aided partly by a development in metal-carving tools, sculptors of the later part of the century were giving their figures more realistic musculature and facial expressions; Greek statues begin to show the famous "archaic smile"—intended to represent, not happiness necessarily, but the vividness of a living human face.

The defeat of the invading Persians in 480–479 B.C. opened a new phase in Greek art—the classical period—in which artistic confidence, religious piety, and prosperity all combined to produce one of the greatest cultural eras in world history. From about 450 to 404 B.C. the rich, imperial Athens—with its lavish public-building program begun by the statesman PERICLES—dominated the field. Greek sculptors abandoned the styles of prior generations and broke through to a new realism in showing the posture and proportions of the human form, particularly the male form; this breakthrough is embodied in the Athenian marble statue that modern scholars call the "Critius boy," carved circa 480 B.C. By the mid-400s B.C. the Argive sculptor POLYCLITUS had perfected the proportions of the standing male figure, with the statue known as the Doryphorus (spear carrier).

Meanwhile came an improved ability to show bodies in movement or in emotional poses. The best known surviving examples are the marble sculptures of gods and heroes that adorned the temple of ZEUS at OLYMPIA (circa 457 B.C.) and the Athenian PARTHENON (circa 440 B.C.).

The art was also revolutionized by improvements in metal casting: the early 400s B.C. saw the invention of a method of casting molten bronze around a wooden core, thus creating a hollow metal figure rather than (as in prior technology) a solid one. The new method allowed for

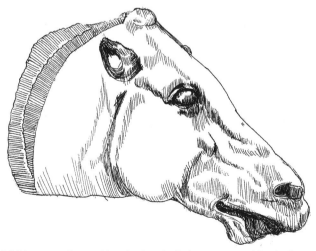

Vivid texture in marble: the head of the Moon goddess's chariot horse, from the east pediment of the Parthenon, circa 435 B.C. The sculpture is now among the Elgin Marbles in the British Museum in London.

The famous Winged Victory of Samothrace, now in the Louvre Museum in Paris. Carved in marble circa 190 B.C., the nearly eight-foot-tall statue is thought to commemorate a Rhodian naval victory over the Seleucids: The goddess Nike (Victory), now minus head and arms, is shown alighting on the prow of a ship. In ancient times the statue stood in an outdoor shrine on the holy island of Samothrace. The marble's gorgeous effects of movement and windswept drapery make this one of the most admired statues in the world.

greater realism in the rendering of musculature, clothing, and such fine details as beards and hair. Among the most famous Greek bronze statues surviving today are the Charioteer of DELPHI (circa 474 B.C.) and the Artemisium ZEUS (circa 455 B.C.). The new metal-casting method helped to create the century's two most famous statues—the colossal gold and ivory idols of ATHENA and Zeus, both sculpted by the Athenian PHIDIAS in the 430s B.C. but now lost.

Athens' defeat in the PELOPONNESIAN WAR (404 B.C.) saw the start of a less confident, less public-minded, but still innovative artistic era. In place of the former patriotism and piety, tastes in sculpture during the 300s B.C. turned to more personal subjects. Portraiture of living people became a growing field. In mythological subjects, the prior era's fascination with the male form was at last counterbalanced by an interest in the female form (usually shown clothed, sometimes nude). Relatedly, sculptors worked for better realism in the rendering of flowing clothing fabric and facial expression. Marble carving became the foremost sculptural discipline at this time, ahead of metal casting. The spirit and values of the mid-300s B.C. were embodied in the work of the Athenian sculptor PRAXITELES.

The conquests of the Macedonian king ALEXANDER THE GREAT (334–323 B.C.) recreated the map of the Greek world, with new centers of Greek culture and patronage arising in the East. Although the classical era had passed, the subsequent HELLENISTIC AGE (circa 300–150 B.C.) saw some advanced realism in specific effects, such as in conveying emotion or movement through pose, facial gesture, and clothing. Among the most admired extant Hellenistic sculptures are the bas-reliefs carved on the Great Altar of Zeus from the eastern Greek city of PERGAMUM, circa 180 B.C. Now housed in a Berlin museum, the Pergamene carvings show a favorite mythological scene—the primeval battle between the gods and GIANTS, representing the triumph of civilization over savagery.

Whether marble or metal, most ancient Greek sculpture was painted: The figure's flesh, clothing, and other features were vividly colored so as to project from a distance. On buildings, architectural sculptures stood out against dark painted backgrounds of red or blue. Today the paint on most of the surviving Greek sculpture has faded away through time.

(See also AEGINA; FUNERAL CUSTOMS; LACOÖN; MELOS; RELIGION; RHODES; SAMOTHRACE; SEVEN WONDERS OF THE WORLD; WOMEN.)

Scylla   A mythical, female sea monster, semihuman in shape, described in HOMER's epic poem the *Odyssey* (book 12). Scylla lived in a narrow sea channel sometimes identified as the Straits of Messina, between SICILY and the "toe" of ITALY. Opposite her cave, the whirlpool Charybdis operated; ships moving through the channel had to risk encountering either Scylla at one side or Charybdis at the other. Scylla (Mangler) had six heads and 12 legs, and normally fed herself on fish; but she would attack any passing ship, grabbing men from shipboard and devouring them.

In the *Odyssey* the hero ODYSSEUS, having been forewarned by the witch CIRCE, steers his ship close by Scylla to avoid the worse danger, Charybdis. Scylla, snatches six of Odysseus' men overboard, but the ship passes by, rowed by the surviving crew. Odysseus' adventure has given rise to the expression "Steering on Scylla to avoid Charybdis," meaning to choose the lesser of two evils.

Later writers described Scylla as having genitalia consisting of ferocious, snarling dogs' heads. This deformity was said to be the result of a magic spell. Scylla had once been a beautiful woman who became the lover of the sea god POSEIDON and was then hideously bewitched in a magical bath prepared by Poseidon's jealous wife, Amphitritē.

Scythia   See BLACK SEA.

Second Athenian League   The name given by modern historians to the alliance formed by ATHENS and various other Greek states in 378 B.C. and modeled after the DELIAN LEAGUE, which had been founded exactly 100 years before. The league arose in response to Spartan repression in the generation following SPARTA's victory in the PELOPONNESIAN WAR (404 B.C.) and particularly as a result of the unpopular Spartan-Persian treaty known as the KING's PEACE (386 B.C.).

In order to entice Greek allies into the league, Athens had to promise to avoid its previous imperialistic excesses of the Delian League, including the levying of tribute and the imposition of Athenian garrison colonies; the latter promise was not kept scrupulously. The most important ally in the league was THEBES. But, after defeating the Spartans at the Battle of LEUCTRA in 371 B.C., Thebes led other central Greek states in seceding from the league. Indeed, the rapid decline of Sparta after 371 B.C. removed much of the league's original reason for being.

The league received another blow in 357 B.C., when three major east Greek member states—CHIOS, RHODES, and BYZANTIUM—revolted. The league was dissolved by the Macedonian king PHILIP II, after he had seized mainland Greece in 338 B.C.

Seleuceia See ANTIOCH; SELEUCID EMPIRE.

Seleucid Empire The largest of several Hellenistic kingdoms that emerged from the Eastern conquests of the Macedonian king ALEXANDER THE GREAT (died 323 B.C.). At its greatest secure extent, circa 280 B.C., the Seleucid Empire occupied much of what had been Alexander's domain, from ASIA MINOR and the Levant to Afghanistan, but excluding EGYPT. The Seleucids' governing and soldiering classes came from a Greek and Macedonian minority based in the western cities and in military colonies throughout; these people ruled a population of Syrians, Persians, Phoenicians, Babylonians, JEWS, and others. Unfortunately, this unwieldy domain lacked a unifying identity and had trouble replenishing its Greco-Macedonian soldiery. For most of its 250-year history, the empire was losing territory.

The term *Seleucid Empire* is a translation of the kingdom's Greek name, Seleukis, which was derived from the name of the founding dynast, SELEUCUS (1), who lived circa 358–281 B.C. Seleucus began his career as a Macedonian nobleman and soldier under Alexander. After Alexander's death, he was one of several Macedonian generals, known collectively as the DIADOCHI (Successors), who seized portions of the conqueror's domain; Seleucus' share was by far the largest.

In 300 B.C. Seleucus built the Syrian coastal city of ANTIOCH, which later became the Seleucid capital, superseding the prior capital of Seleuceia-on-the-Tigris, in Mesopotamia. Brilliantly located to be a gateway between the Mediterranean and the East, Antioch became the seat of a lively Syro-Greek culture, personified for us by the poet and anthologist Meleager of Gadara (circa 100 B.C.). Antioch was named for Seleucus' son and eventual successor, Antiochus, who reigned 281–261 B.C. The names Antiochus and Seleucus (also Demetrius) became traditional for the Seleucid royal house.

Despite their Eastern holdings, the Seleucid rulers tended to look westward, being preoccupied with territorial conflicts against Ptolemaic Egypt. The six Syrian Wars (274–168 B.C.), fought over the two empires' shared frontier in the Levant, involved use of war elephants and huge phalanxes of heavy infantry. Meanwhile the city of PERGAMUM, revolting from Seleucid rule, established a rival domain in Asia Minor (circa 262 B.C.). Farther east, the hellenized kingdom of BACTRIA rebelled (circa 250 B.C.), and the non-Greek kingdom of Parthia emerged to deprive the Seleucids of territories in PERSIA and Mesopotamia (mid-200s–mid-100s B.C.).

ANTIOCHUS (2) III (reigned 223–187 B.C.), the greatest Seleucid king, re-created much of the old empire; but he came to grief against the overseas armies of the Italian city of ROME. The Romans defeated Antiochus in battle in Greece (191 B.C.) and Asia Minor (190 B.C.), and forced him to relinquish his holdings in Asia Minor (188 B.C.). Weakened by palace factions and by the Maccabean Revolt of the Palestinian Jews (167 B.C.), the Seleucid realm gradually shrank into the region that was called Syria (including modern Syria and Lebanon). The Seleucid kingship was dismantled in 64 B.C., when Syria became a Roman province.

The Seleucid Empire's significance for history is its role in hellenizing the Levant. This effect can be seen today in Greek-style ruins at such sites as Palmyra, in the Syrian desert. Long after the end of the Seleucid Empire, Greek was still the language of commerce, government, and higher learning in the Near East. Under Roman rule, in the first century A.D., the Bible's New Testament was written in Greek—rather than in Hebrew, Aramaic, or Latin—as a result of the Seleucid legacy.

(See also CELTS; HELLENISTIC AGE; WARFARE, LAND.)

Seleucus (1) Founder and first king of the SELEUCID EMPIRE, which was named after him. Living circa 358–281 B.C., Seleucus was one of the most capable of the DIADOCHI (Successors), who fought over the dismembered empire of ALEXANDER THE GREAT. The kingdom Seleucus received, which was the largest of any Successor's, extended from the Levant to the Indus River valley.

This Oriental despot was born—neither Oriental nor royal—in MACEDON, as the son of a Macedonian nobleman in the service of the Macedonian king PHILIP II. At about age 24, Seleucus (Shining one) accompanied Philip's son and successor, Alexander, on his invasion of the Persian Empire (334 B.C.). Although not mentioned as a prominent commander, Seleucus may have been a close friend of the young king, who was his near contemporary.

Following Alexander's spectacular conquests and his death, Seleucus became the Macedonian governor at Babylon (321 B.C.). Coming into conflict with ANTIGONUS (1), who was based in ASIA MINOR, in 316 B.C. Seleucus fled to the protection of PTOLEMY (1), in EGYPT, but in 312 B.C. he returned to Babylon with an army and recaptured the city. Later Seleucus conquered eastward to seize an empire that nearly compared with Alexander's. Soon Seleucus, like the other Diadochi, was calling himself "king" (*basileus*). In 301 B.C. he and the Macedonian dynast LYSIMACHUS destroyed Antigonus at the Battle of Ipsus, in central Asia Minor. An essential element in the victory was Seleucus' 480 trained Indian war elephants, acquired from an Indian king in exchange for Seleucid territorial concessions in the Indus region.

The Ipsus victory left Seleucus with secure gains in Syria and southeastern Asia Minor. He then sought westward outlets; no doubt he saw that his domain could survive only by remaining a part of the Greek world and by attracting Greek immigration. Having established an eastern capital, Seleuceia, on the River Tigris (circa 312 B.C.), he then founded a western capital, the great city of ANTIOCH, near the Syrian seacoast (300 B.C.).

Antioch was named for Seleucus' son and eventual successor, Antiochus. Prince Antiochus' mother was Apama, a noblewoman of BACTRIA whom Seleucus had married under Alexander's supervision, in 324 B.C. In 298 B.C., without divorcing, the 60-year-old Seleucus took a second wife—Stratonicē, the 20-year-old daughter of his prior enemy, DEMETRIUS POLIORCETES (Antigonus' son). Later, when the dynamic but unstable Demetrius invaded Asia Minor, Seleucus captured him (285 B.C.) and chivalrously provided a comfortable captivity in which Demetrius ate and drank himself to death.

In 281 B.C. Seleucus took aim at what would be his last great acquisition—Asia Minor, most of which was held by his former ally, Lysimachus, now king of Macedon. For Seleucus, Asia Minor represented the crucial westward link to the central Greek world. Seleucus defeated Lysimachus at the Battle of Corupedium, near Sardis. Lysimachus's death in the fighting inspired the 77-year-old Seleucus with dreams of seizing the vacant Macedonian throne. Seleucus invaded Macedon and was on the point of capturing the kingdom when he was murdered by one of his followers, a son of Ptolemy who wanted the throne for himself. Seleucus was succeeded by Antiochus, who had been reigning as his father's partner since 292 B.C.

The Seleucid Empire continued under Seleucus' descendants for nearly another 220 years, but the vast conquests of Seleucus I gradually were eroded.

(See also WARFARE, LAND.)

Seleucus (2): see Seleucid Empire

Selinus Westernmost Greek city in SICILY. The site of ancient Selinus, now uninhabited, boasts one of the grandest collections of Greek temple ruins from the 400s B.C.

Founded circa 628 B.C. by Dorian-Greek colonists from the city of Megara Hyblaea (on Sicily's east coast), Selinus occupied a terrace of land overlooking the southwestern Sicilian coastline, just south of modern Castelvetrano. According to one tale, Selinus was named for local growths of wild celery (Greek: *selinon*). Selinus enjoyed access to rich coastal farmland and sea trade, but it also lay within the territory of the Greeks' traditional enemies, the Carthaginians, who occupied western Sicily. This menace was decreased by the immense Greek victory over the Carthaginians at the Battle of HIMERA (480 B.C.), after which the Selinuntines became allies of the preeminent Sicilian-Greek city, SYRACUSE.

It was during these years—late 500s–early 400s B.C.—that the prosperous Selinuntines constructed the eight Doric-style temples whose remnants grace the site today. Of these, the building now known as Temple G (probably a shrine of APOLLO) must have been planned as one of the largest buildings in the Greek world; it was left incomplete, circa 480 B.C. The temples all seem to have been toppled by earthquake in ancient times.

A traditional enmity existed between Selinus and the northwestern city of Segesta (inhabited by a non-Greek native people, the Elymi). This conflict took on international proportions in 415 B.C., when the Segestans appealed for help to the powerful city of ATHENS, in mainland Greece. During the subsequent, ill-fated Athenian invasion of Sicily, Selinuntine troops fought alongside their Syracusan allies, against the Athenians (415–413 B.C.). But soon Segesta had found a new ally, CARTHAGE, and in 409 B.C. Selinus was destroyed by the Carthaginians in their drive to reconquer Sicily.

Reinhabited by refugees, Selinus probably remained a Carthaginian vassal until 250 B.C., when it was again destroyed in the course of the First Punic War, between Romans and Carthaginians.

(See also ARCHITECTURE; DORIAN GREEKS; PELOPONNESIAN WAR; ROME.)

Semele See DIONYSUS.

Semonides Early Greek lyric poet, probably of the late 600s B.C. Semonides came from the island of SAMOS but emigrated—possibly because of social unrest—to Amorgos, an island of the CYCLADES.

Semonides' work survives in one apparently complete poem and one fragment, both written in iambic meter, a form associated with satire. The poem—a social satire on the nature of WOMEN—is typical of the misogyny of Greek male culture of the 600s–400s B.C. The gods (the poem explains) made women from different materials or animals: from mud, the changeable sea, the weasel, the donkey, and so forth. According to the poem, only one type is good to marry: the one made from the quiet, industrious bee.

Semonides should not be confused with the more famous SIMONIDES of Ceos, who lived a century later.

(See also LYRIC POETRY.)

Sestos Small Greek city on the European shore of the HELLESPONT, midway along the strait, at the narrowest section. Sestos was the fortress of the Hellespont, with a hilltop citadel whose walls ran down to flank the best harbor in the region. Ideally situated to control shipping in the channel, Sestos served as a naval base for ATHENS in the 400s–300s B.C., protecting Athenian imports of precious grain from the BLACK SEA.

Sestos was founded by colonists from LESBOS in the 600s B.C. Around 512 B.C. Sestos was annexed by the Persian king DARIUS (1). The Persian king XERXES, in his campaign to conquer Greece, marched his vast army across the Hellespont on two pontoon bridges that reached the European shore near Sestos (480 B.C.). After the Persian defeat in Greece, Sestos was immediately captured and occupied by Athenian forces (479 B.C.). For nearly the next 75 years, Sestos was a member of the Athenian-led DELIAN LEAGUE.

Athenian-held Sestos played a prominent role in naval campaigns for the Hellespont in the later PELOPONNESIAN

WAR. After Athens' defeat (404 B.C.), Sestos was taken over by SPARTA. Sestos became Athens' subject again in 365 B.C. but soon rebelled. The Athenians crushed the rebellion, captured the city, sold the inhabitants as SLAVES, and repopulated the place with Athenian colonists (357 B.C.). In 334 B.C. the Macedonian king ALEXANDER THE GREAT brought his army to Sestos for his shipborne invasion of Persian territory across the Hellespont.

On the Asian shore opposite Sestos, one mile away, stood the Greek city of Abydos. In later Greek times there arose the legend of Hero and Leander, about a man of Abydos (Leander) who swam the Hellespont ever night to visit his girlfriend in Sestos.

(See also EUROPE AND ASIA; PERSIAN WARS.)

Seven Against Thebes, the   Well-known MYTH, part of the so-called Theban Cycle, about the legendary woes of the royal house of THEBES (a city in central Greece). By the 600s B.C. the myths of Thebes had been collected in an epic poem, now lost, called the *Thebaïd*. Like other episodes in the Theban Cycle, the tale of the Seven was presented onstage by the great Athenian playwright AESCHYLUS, in his tragedy *Seven Against Thebes* (467 B.C.).

According to the legend, the Theban king OEDIPUS had two sons, Eteocles (True Glory) and Polynices (Greek: Poluneikes, Much Strife). The brothers' destiny was not happy. After Oedipus' death, they argued as to who should be king—Eteocles was the elder, but the law of primogeniture did not apply among the Greeks—and they finally agreed to reign in alternate years. Eteocles drew the lot allowing him to rule first, and Polynices withdrew to the city of ARGOS, awaiting his turn.

With the year's passing it became obvious that Eteocles would not relinquish his throne. So Polynices enlisted the aid of Adrastus, king of Argos, whose daughter (Argeia) he had married. Polynices and Adrastus collected a mighty army, to be led by Polynices and six other champions, usually identified as Tydeus of Calydon; Amphiaraus of Argos; Capaneus of Argos; Hippomedon of Argos; Parthenopaeus of ARCADIA (son of the huntress ATALANTA); and, in most versions, King Adrastus.

Foremost among the Seven was Tydeus, a small but valiant man who had gone to Argos after being banished from Calydon for homicide. Like Polynices, Tydeus had married one of Adrastus' daughters, Deipyle. Their baby son, DIOMEDES, was destined to outshine even his father, as a hero in the TROJAN WAR.

Also notable among the Seven was Amphiaraus, a seer who foresaw that the expedition would be a disaster and that he himself would never return if he accompanied it. Amphiaraus knew that his wife, Eriphyle, would be approached to convince him to join the campaign, and he forbade her to accept any gift from Polynices. She disobeyed, and was bribed by the fabulous necklace that had once belonged to Polynices' great-grandmother Harmonia (wife of the hero CADMUS). As a result, Eriphyle successfully convinced Amphiaraus to join the doomed expedition—an odd concession by him, given his foreknowledge, but so runs the myth. In departing, Amphiaraus compelled his son ALCMAEON (1) to pledge to avenge his death by killing Eriphyle (i.e., Alcmaeon's own mother); he also

made the boy vow to launch his own attack on Thebes someday.

The tale of Amphiaraus was a favorite among the Greeks, who liked stories, mythical or otherwise, about people choosing a course of certain death. Amphiaraus' fateful departure from Argos is the subject of one of the most beautiful surviving Corinthian-style vase paintings, circa 560 B.C.

The Seven's attack on Thebes was, as Amphiaraus had predicted, a failure. At each of the seven gates of Thebes, one of the Seven fought a Theban champion in single combat. Polynices and Eteocles killed each other, and the rest of the Seven died, with the exception of King Adrastus, who escaped on his superlative horse. Amphiaraus, attempting to flee in his chariot, was swallowed up in the earth.

In the next generation, Amphiaraus' son Alcmaeon, true to his vow, marched to Thebes with six other heroes, including Tydeus' son, Diomedes; this time the city fell to the attackers. Alcmaeon's exploit was known as the Expedition of the Epigoni (Descendants) and was described in a now-lost epic poem titled *Epigonoi* (circa 600 B.C.).

Because Greek myths often provide distorted reflections of events in the distant Mycenaean age (circa 1600–1200 B.C.), this tale may contain a kernel of historical truth. Thebes and Argos represent the two heartlands of Mycenaean Greek civilization, with Argos equivalent to the royal city of MYCENAE. The legends of the Seven and the Epigoni surely commemorate repeated attempts by the king of Mycenae to capture Thebes and its fertile plain, perhaps circa 1350 B.C.

(See also ANTIGONE; MYCENAEAN CIVILIZATION.)

Seven Sages, the   A list of seven men said to be the wisest in ancient Greece. Each of the sages (Greek: *sophoi*) was a real person, typically a statesman, active in the 500s or late 600s B.C. One or two were benign TYRANTS. The list first appears in PLATO's dialogue *Protagoras*, written circa 390 B.C.; but the tradition probably dates back another 200 years. The seven were: (1) THALES, a philosopher and scientist of MILETUS; (2) PITTACUS, an elected ruler of MYTILENE, on the island of LESBOS (3) Bias, a judge and diplomat of Priënē (in IONIA); (4) SOLON, an elected ruler and lawgiver at ATHENS; (5) Cleobulus, tyrant of Lindus, on the island of RHODES; (6) CHILON, an EPHOR (chief official) at SPARTA; and (7) Myson of Chenae (in Trachis in central Greece), a rustic sage.

In other versions, the obscure Myson is replaced by PERIANDER, tyrant of CORINTH.

Seven Wonders of the World, the   By a popular tradition in the HELLENISTIC AGE (circa 300–150 B.C.), the Greeks counted seven monuments as preeminent in size or grandeur among all man-made works. In the most usual list, five of the seven wonders belonged to the Greek world; the other two—the oldest—were built by non-Greek civilizations. The seven were:

1. the pyramids of EGYPT, built between about 2700 and 1800 B.C.

2. the hanging gardens of Babylon, dating from the early 500s B.C.

3. the temple of the goddess ARTEMIS at EPHESUS, constructed over the period 550–430 B.C.

4. the giant cult statue of ZEUS fashioned circa 430 B.C. by the Athenian sculptor PHIDIAS for the god's temple at OLYMPIA.

5. the Mausoleum, or tomb of the ruler Mausolus, at HALICARNASSUS, built circa 350 B.C.

6. the Colossus of RHODES—a statue of the sun god, HELIOS, erected beside the harbor at Rhodes city circa 285 B.C.

7. the lighthouse at ALEXANDRIA (1), built on the isle of Pharos, outside Alexandria harbor, circa 270 B.C.

Of all these works, only the pyramids of Egypt survive today.

ships and seafaring "We live around a sea like frogs around a pond," the Athenian philosopher SOCRATES is reported to have said (circa 410 B.C.), summarizing the ancient Greeks' reliance on seaborne TRADE and transport. Like the Phoenicians, but unlike many other Mediterranean peoples, the Greeks had a genius for seamanship. Seafaring was proverbially dangerous in ancient times, but it was also far faster, more efficient, and more lucrative than overland freighting; mule teams or ox-drawn carts could never compare with the speed and efficency of ships carrying cargo (or troops) under sail or oar. New cities were established regularly to take advantage of nearby natural harbors or waterways. The great expansion of the Greek world in 800–500 B.C. was largely the story of seaborne trade, COLONIZATION, and conquest, along ever-enlarging shipping routes. On the military side, sea power came to mean (by the mid-500s B.C.) something like what air power means to modern generals—the fastest way to bring harm to enemies at various points, including the ability to stop their imports and starve them. Sea power became the necessary condition for imperial strength among classical Greek states. It was by sea that ATHENS became great in the 400s B.C.

Seafaring was encouraged by the geography of mainland Greece—with its mountains, scarce farmland, and immensely long overall coastline, nearly 2,000 miles total. Many Greeks, such as the people of the island AEGINA, took to the sea to compensate for their home region's lack of space or fertility.

Aside from Athens and Aegina, major GREEK seagoing states in various eras of history included CHALCIS, CORINTH, MILETUS, PHOCAEA, SAMOS, CHIOS, LESBOS, and RHODES.

The Mediterranean Sea—including its northeastern corner, the Greeks' AEGEAN SEA—offered inducements to seafaring. Unlike the Atlantic Ocean, the Mediterranean has no tides and is broken by many islands and peninsulas, with a reliable "window" of calm weather in summer. Against these natural advantages, the ancient Greeks were burdened by primitive means of navigation. Lacking the compass and sextant, they depended on visible reckonings, such as coastal landmarks and position of the stars. This need for landmarks—along with a fear of straying too

A pirate ship, at right, intercepts a merchant vessel, in a black-figure scene on an Athenian cup, late 500s B.C. The painter has clearly distinguished the two ships' designs: the low-riding pirate bireme, built for speed, powered mainly by oars, carrying a bronze ram; and the heavier, taller, deeper-hulled freight ship, built to hold cargo. Note the merchantman's thick mast. This ship would rely mostly on its sail for power. The horizontal lines atop the merchant's hull probably represent a defensive fence.

far from shelter—explains the ancient seafarers' persistent reliance on coastal routing. A Greek colony such as CORCYRA, on the Adriatic coast, long remained an important anchorage between Greece and ITALY, even though on a map it looks to be a northern detour.

Ancient vessels were made from TIMBER, and relied on two forms of power—wind and oars. Most ships carried a single mast, capable of hoisting one square sail typically made of patchwork linen (from the flax plant) or canvas (from Egyptian cotton). Unlike modern sailboats, ancient square-riggers could make direct progress only if the wind was fully or nearly astern; a wind from the side (abeam) required the ship to sail close-hauled, on a zigzag course of tacking; a wind forward of the beam tended to be unusable. Still, heavier ships, such as the larger merchant vessels, had to rely mostly on wind power. Smaller craft, such as warships, could make progress under oars alone if that was preferable to using the sail. War squadrons always rowed when moving into battle, for better control and uniformity.

Despite changes in ship design during the Greek epoch of 1600–100 B.C., these two basic categories persisted: merchant vessel and war vessel. The merchantmen were by far the larger type, designed to carry freight; such ships had deep drafts, wide beams, and heavy masts; their hulls were usually closed at top by a deck. They carried only a small crew. A war craft, by contrast, was designed for speed and was slender, shallow-drafted, and low-riding. Like a canoe or modern racing shell, the warship's hull was typically undecked (or only partly decked), and almost every foot of space was taken by oarsmen or shipboard soldiers.

The major difference in handling between warships and merchantmen was that war crews, sitting at their crowded benches, could not spend the night aboard ship easily. To rest or cook, they needed to pitch camp ashore. Merchant freighters, on the other hand, could sail day and night.

Sheltered under the top decking, the off-duty crew could rest on mattresses and could feed from supplies aboard. Under proper conditions, a freighter might leave the shore behind and course out over the open sea; with a favoring wind, such a ship might reach between four and six knots, traveling 100 or more miles in 24 hours.

Because a merchant ship was capable of long-range travel, ancient Greek merchant voyages tended to be ambitious and dangerous. Shipwrecks were common—as suggested by vase paintings, written references, and modern underwater ARCHAEOLOGY. Many harbors had temples where mariners went to pray and sacrifice for a safe voyage. The main patron gods of seafaring were POSEIDON, ATHENA, APHRODITE, and CASTOR AND POLYDEUCES.

For safety, shipping and naval operations normally were confined to the season from May to mid-September, when Mediterranean weather tends to be sunny and calm. But even this period seemed too long to HESIOD (circa 700 B.C.), whose epic poem *Works and Days* mentions that seafaring is safe for only the 50 days following the summer solstice—roughly June 21 to mid-August. Only the extremes of war or greed could induce ships to put to sea during autumn or winter. The hazard of winter seafaring lay partly in violent storms, but also in reduced visibility due to rain, fog, and short daylight. The apostle Paul's shipwreck of A.D. 57 (described in the New Testament book *Acts of the Apostles*, Chapter 27) occurred because the skipper had risked sailing from CRETE after the season's end.

(See also CYPRUS; GREECE, GEOGRAPHY OF; MASSALIA; MINOAN CIVILIZATION; MYCENAEAN CIVILIZATION; PHOENICIA; WARFARE, NAVAL.)

Sicily   Largest (9,860 square miles) island of the Mediterranean, lying just west of the "toe" of ITALY. Sicily is shaped roughly like an isosceles triangle, with its base facing east.

Renowned in ancient times for its fertile coastlands, Sicily attracted waves of Greek settlers at the outset of the great colonizing epoch (beginning after 750 B.C.). The island became a major part of the Greek world, producing a dozen or so noteworthy Greek cities, particularly the Corinthian colony of SYRACUSE, which grew to be the greatest Greek city of the West (400s–200s B.C.). Syracuse and other Sicilian-Greek cities supplied precious grain for CORINTH and other populous mainland Greek cities.

Despite prosperity, Greek Sicily had a tragic history, full of social unrest within cities and conflicts between cities inhabited by DORIAN GREEKS and those inhabited by IONIAN GREEKS. But worst of all were the repeated wars with the non-Greek, Carthaginian settlers of western Sicily.

The name Sicily (Greek: Sikelia) commemorates the Sicels—a native, non-Greek people of the island's east coast, whose prime land at Syracuse and on the plain of Catania was seized by the early Greek colonists. Other native peoples included the Sicani, in the island's west-central region; and the Elymi, in the west. The relatively unsophisticated level of these peoples no doubt helped to attract early Greek settlers. (The Greeks tended to plant colonies where they knew they could subdue local inhabitants.)

Sicily's earliest Greek colonies—NAXOS (2), Syracuse, CATANA, and Leontini—arose circa 734–728 B.C. along the island's 200-mile-long eastern coast. Among the coast's distinctive geographic features are volcanic Mt. Etna (over 11,000 feet tall) and the plain of Catania (whose fertility is due in part to volcanic lime, washed down from Etna's slopes). Toward the coast's southern end is the promontory of Syracuse, with its superb natural harbor and nearby farmland.

Most of the eastern coast was Ionian Greek, with the exception of Dorian Syracuse. But a Dorian bloc soon arose on Sicily's southern coast, with the founding of GELA (circa 688 B.C.) and ACRAGAS (circa 580 B.C.). After Syracuse, Acragas became the second most important city of Greek Sicily (400s B.C.).

Sicily's western side, meanwhile, had been occupied by settlers from the African-Phoenician city of CARTHAGE (which lay only about 135 miles southwest of Sicily). By about 500 B.C. Greeks and Carthaginians had settled into a bitter conflict that lasted over 250 years and monopolized the energies of Greek Sicily. The constant Carthaginian threat largely shaped the authoritarian nature of Sicilian Greek politics. Long after the age of TYRANTS had passed away in mainland Greece, Sicilian Greek cities were still relying on military tyrants who could promise protection against the hated Carthaginians. Foremost among such powerful rulers were the Syracusan tyrants: GELON (reigned circa 490–478 B.C.), his brother HIERON (1) (478–467 B.C.), DIONYSIUS (1) (circa 400–367 B.C.), and AGATHOCLES (317–289 B.C.). Under these men, Syracuse dominated the other Greek cities and several times came close to driving the Carthaginians from the island. At other times leaders from mainland Greece were called in to help the beleaguered Sicilian Greeks. For example, the Corinthian commander TIMOLEON (345–340 B.C.) defeated the Carthaginians in battle and gave the Greek cities new governments. King PYRRHUS of EPIRUS arrived and beat the Carthaginians (279 B.C.), then hurried away before exploiting his advantage.

The conquest of Sicily remained an elusive and dangerous dream for Syracuse, for Carthage, and for the mainland Greek city of ATHENS, whose attempt to capture Syracuse during the PELOPONNESIAN WAR proved a disaster (415–413 B.C.). It took the emerging might of Italian ROME to conquer Greeks and Carthaginians alike. Sicily became a battlefield between Rome and Carthage in the First Punic War (264–241 B.C.), before it passed to Roman control. A defiant Syracuse was besieged and captured by the Romans in 213–211 B.C., during the Second Punic War. Sicily was then made a province of the Roman Empire.

(See also COLONIZATION; CYCLOPS; HIMERA; PERSEPHONE; SELINUS; ZANCLĒ.)

Sicyon   Important city of the northeast PELOPONNESE, about 15 miles west of CORINTH. Sicyon was located two miles inland, on a fertile coastal plain beside the Corinthian Gulf, at the foot of a lofty hill that served as the city's ACROPOLIS. The city had a mixed Dorian and Achaean population, having been founded (probably) from Dorian ARGOS, circa 1000 B.C. Supposedly Sicyon was named for its local growths of cucumbers (*sikuai*).

Sicyon was one of the first Greek cities to be seized by TYRANTS. The dynasty created circa 660 B.C. by the tyrant Orthagoras lasted for over a century and made Sicyon into a great power. Orthogoras' descendant CLEISTHENES (2) ruled circa 600–570 B.C., leading the First Sacred War against PHOCIS and later celebrating his daughter's wedding with Homeric magnificence. This era saw construction of the Sicyonians' treasury house at DELPHI, prominently occupying the start of the Sacred Way up to APOLLO's temple.

Under Cleisthenes, wealthy Sicyon was a center for manufacturing and art, particularly SCULPTURE and PAINTING. A Sicyonian poet, Egigenes, is said to have been a pioneer in the writing of "tragic choruses" in the early 500s B.C. The city's schools of art remained important for centuries; Sicyon's most famous artist was the sculptor Lysippus (circa 330 B.C.).

The Sicyonian tyranny was extinguished circa 550 B.C. by the Spartans. Sicyon became an OLIGARCHY and a Spartan ally. Sicyonians fought alongside the Spartans in the PERSIAN WARS (480–479 B.C.) and PELOPONNESIAN WAR (431–404 B.C.). In the mid-200s B.C. Sicyon's most famous son was the statesman and commander Aratus.

(See also ACHAEA; DORIAN GREEKS; THEATER.)

siege warfare See WARFARE, SIEGE.

Sigeum See HELLESPONT.

Silenus Mythical semihuman creature, distinguished by his horse ears, horse tail, pot belly, shaggy hair, bald head, and pug nose. Like the SATYRS, whom he resembled, Silenus was associated with the wine god DIONYSUS. On vase paintings of the 500s–400s B.C., Silenus was portrayed cavorting in the god's retinue or lustily pursuing the NYMPHS. Sometimes he was shown as several creatures (called by the plural, Silenoi or Sileni). A widespread Athenian joke of the late 400s B.C. claimed that Silenus resembled the living philosopher SOCRATES.

silver Precious metal in both ancient and modern times. Silver traditionally has been less prized than GOLD, partly because silver ore is more common and less malleable for shaping.

Throughout antiquity, silverwork was used for luxury items such as plates, goblets, statuary, and jewelry; it was particularly favored for cups and plates with repoussé work. More important, silver was the prime substance for Greek COINAGE (first appearing just after 600 B.C.) Even before coinage, silver was being used in ingot form for high-level exchange, in the monetary unit known as the TALENT.

Unlike gold, silver existed as a natural mineral deposit in mainland Greece. The publicly owned silver mines at Laurium, 15 miles south of ATHENS, played a large role in Athenian greatness in the 500s–400s B.C. The Laurium mines helped finance naval fleets and the social programs of the DEMOCRACY.

But the main supplier of raw silver for the Greek world was the north Aegean coast (in the non-Greek land of THRACE) and particularly Mt. Pangaeus, near the lower Strymon River. From there—where silver seems to have been available by panning in local rivers—the precious ore

reached Greek hands by TRADE and eventually by conquest. In the 600s–500s B.C., Greek traders began operating out of THASOS, SAMOTHRACE, ABDERA, and AENUS, acquiring raw silver from local Thracian kings. From the Thracian shore, this silver—typically in the form of large, bowtie-shape ingots—would be shipped to markets in the Greek world or to foreign markets such as EGYPT and PHOENICIA. In this early era (650–480 B.C.), raw silver was one of the few goods that Greek traders could offer to the sophisticated Near Eastern markets.

During the mid-400s B.C., Mt. Pangaeus seems to have come under control of imperial Athens, through the Athenian colony at AMPHIPOLIS. In the mid-300s B.C. the region fell to the cash-hungry Macedonian king PHILIP II. Other sources of raw silver included non-Greek Asian kingdoms such as LYDIA and Phrygia, fabled for their wealth. Another source, circa 600–500 B.C., was the Celtic kingdom in southern Spain known as Tartessus (the Tarshish of the Bible). To this distant domain sailed Greek traders from SAMOS and PHOCAEA.

In the HELLENISTIC AGE (300–150 B.C.), silver became more plentiful among the Greeks, due to the formerly Persian-owned mines and treasure hordes acquired by ALEXANDER THE GREAT. The silver mines of Thrace, Asia Minor, and Mesopotamia made possible the beautiful, large-denomination coins of the Hellenistic Greek kings.

Simonides Lyric poet (circa 556–468 B.C.) from the island of Ceos, in the CYCLADES. Over his long lifetime, Simonides enjoyed success with many genres of poetry. He was famous for his epitaphs written to commemorate the Greek dead in the PERSIAN WARS (490–479 B.C.). He developed the choral song known as the victory ode (*epinikion*), thereby in effect clearing the way for his younger and greater contemporary, the Theban poet PINDAR. The only poems by Simonides that survive complete are epigrams—short poems in elegiac verse, written to be carved in stone. Fragments of his victory odes and his dirges exist, but other categories are largely lost, including Simonides' dithyrambs—narrative poems on mythological topics, performed by choruses.

Simonides' work was admired in the ancient world for its clarity, word choice, and sympathy for human transcience and suffering. His simple language went right to the heart—particularly in its expressions of sorrow—and was considered more effective than the grandiose phrasings of other great poets such as AESCHYLUS.

Simonides was also the first poet to work for hire—that is, to compose poetry for a one-time fee from successive patrons rather than in a permanent position in the retinue of some wealthy man. Simonides was said to be fond of money and shrewd in business dealings. He seems to have had an affable, worldly personality.

Ceos—where Simonides, son of Leoprepes, was born into an upper-class family—was an Ionian-ethnic island, 15 miles off the Athenian coast. After gaining repute for his victory odes, Simonides was summoned to ATHENS (circa 525 B.C.) by Hipparchus, the brother and cultural minister of the Athenian tyrant HIPPIAS (1). There Simonides enjoyed high payments from Hipparchus, whose majestic court hosted another famous poet, ANACREON.

At Athens in these years, Simonides surely wrote some of his many dithyrambs, now lost. Late in life Simonides set up an inscription commemorating his 56 victories in dithyramb contests at Athens; the inscription mentions that he himself taught the choruses, which were composed of men.

Simonides left Athens around 514–510 B.C., as the tyranny crumbled. He traveled to THESSALY, and there found employ with the wealthy Aleaud and Scopad clans. But his stay ended after the Scopads were decimated by the collapsing roof of a banquet hall. The disaster later inspired Simonides to write a poem containing one of his much-admired reflections on the brevity of human happiness: "Since you are mortal, never say what tomorrow will bring nor how long a man may be happy. For the darting of the dragonfly is not so swift as change of fortune."

By the early 490s B.C., Simonides, then approaching 60, was back in Athens, his former intimacy with the tyrants apparently pardoned. Amid the great events of the Persian Wars he found a new calling as the poet laureate of Greece. After the Athenian victory at MARATHON (490 B.C.), the Athenians honored their dead with an epitaph by Simonides, chosen over one submitted by the young poet Aeschylus (himself an Athenian who had fought at the battle). According to the story, the Athenians decided that Simonides' poem conveyed a keener sympathy.

In 479 B.C., after the invasion by the Persian king XERXES had been repelled, Simonides was again commissioned to write patriotic verse. He produced choral odes about the sea battles at ARTEMISIUM and SALAMIS (1) and several much-admired epitaphs for the Greek dead.

In 476 B.C., at age 80, Simonides left Athens again, this time for the grand court of the tyrant HIERON (1) of SYRACUSE, in SICILY. Simonides' departure might have been related to the political downfall at Athens of his friend THEMISTOCLES. At Syracuse, Simonides—with his nephew and protégé, the poet BACCHYLIDES—is said to have indulged in a feud with Pindar, who was also in Sicily at that time. Another tale claims that Hieron's wife once asked Simonides which is better, to acquire wisdom or wealth. "Wealth," the poet replied. "For I see the wise sitting at the doorsteps of the rich."

Simonides died in Sicily in 468 B.C. and was buried at ACRAGAS.

(See also IONIAN GREEKS; LYRIC POETRY; THEATER; THERMOPYLAE.)

Sinope   Prosperous Greek city on the BLACK SEA. Founded circa 625 B.C. by colonists from MILETUS, Sinope (modern Sinop, in Turkey) was located halfway along the north coast of ASIA MINOR, where the coast bulges advantageously northward. The city was built at the neck of a lofty promontory with a fine double harbor—the only first-rate harbor on the Black Sea's southern shore.

Isolated by mountains inland, Sinope depended for its prosperity on seaborne TRADE. The city was probably an export center for goods shipped in from other Greek ports of the Black Sea. Such goods would include the metals of Asia Minor—GOLD, SILVER, IRON, and copper and tin for BRONZE-making—as well as "Sinopic earth" (cinnabar, or red mercuric sulfide, used for making a prized red dye).

Sinope lay 220 miles due south of the eastern Crimean peninsula, which was home to PANTICAPAEUM and other Greek seaports; the shipping run between Sinope and the Crimea was a vital link in the Greek navigation of the Black Sea.

Sinope founded its own colonies, farther east on the Black Sea coast. Chief among these daughter cities was Trapezus ("the table," named for a mountain landmark), which is now the Turkish city of Trabzon. Unlike mountain-girt Sinope, Trapezus had access to the Asia Minor interior and the land route to the Near East.

Sinope probably offered nominal submission to the Persian kings who had conquered Asia Minor by 545 B.C. The city was being ruled a Greek dictator in the mid-400s B.C., when an Athenian fleet under the commander PERICLES arrived on a Black Sea expedition. The Athenians established DEMOCRACY at Sinope and settled Athenian colonists there. The city may have henceforth had some trading agreement with ATHENS, but it never became a member of the DELIAN LEAGUE.

Occupied briefly by Persian troops in around 375 B.C., Sinope was liberated by ALEXANDER THE GREAT's conquest of the Persian Empire (334–323 B.C.). The city seems to have remained independent of the several Hellenistic kingdoms that subsequently dominated Asia Minor, but was captured by the Iranian-ethnic kingdom of Pontus in 183 B.C. and became the Pontic capital. After the Romans' conquest of Pontus (63 B.C.), Sinope became a city of the Roman Empire.

Sinope was a center of Greek culture in the East, with grand temples, markets, and a tradition of learning. Its most famous citizen was the Cynic philosopher DIOGENES (mid-300s B.C.).

Sirens   In Greek MYTH the Seirēnes were sea witches who lured sailors to their deaths. The sailors, hearing the Sirens' enchanting song, would land their ship and go ashore, where the Sirens would kill them.

The Sirens are first mentioned in book 12 of HOMER's epic poem the *Odyssey* (written down circa 750 B.C.). There they are two in number; their appearance is never described. The seafaring hero ODYSSEUS defies the Sirens by plugging his crewmen's ears with beeswax beforehand. However, wishing to hear the song himself, he has his men bind him to the ship's mast. As the crew rows past the Sirens' island, Odysseus is bewitched by the song and pleads, unheeded, to be untied. Eventually the ship passes out of earshot. The Sirens, whose name may mean "Scorchers" or "Binders," are among several female threats whom Odysseus enounters on his voyage home.

Later generations of Greeks named 11 Sirens and pictured them as half woman, half bird. These Sirens resemble other winged, malevolent, female creatures of Greek myth—the HARPIES and the FURIES.

Sisyphus   One of the three infamous sinners—betrayers of the gods' friendship and thus eternally damned—encountered by the hero ODYSSEUS in book 11 of HOMER's epic poem the *Odyssey*. In his visit to the Underworld, Odysseus observes Sisyphus' punishment, which is to roll a great boulder up a hill but, on nearing the top, to have it slip backward past him, back down the slope. Sisyphus must pursue this frustrating and exhausting labor forever. In theory, his punishment would end if he could ever reach the summit and push the boulder down the *far* slope, but the vindictive gods will never allow this to happen.

The *Odyssey* does not mention Sisyphus' original offense, but later writers describe him as the conniving founder of the city of CORINTH. In exchange for receiving a perpetual water source for his city's citadel, Sisyphus informed the god Asopus that the great god ZEUS had abducted Asopus' daughter. When angry Zeus sent the god HADES to fetch Sisyphus to the Underworld, the hero outwitted Hades and tied him up, thus unnaturally stopping all death in the world. Eventually the gods captured Sisyphus.

Sisyphus was a hero of the "trickster" type—a category that included the god HERMES, the Titan PROMETHEUS, and Odysseus himself (sometimes said to be Sisyphus' illegitimate son).

(See also AFTERLIFE; TANTALUS.)

Skepticism Greek school of thought which believed that no positive knowledge is possible, since everything is in a constant state of flux. This notion appears early in Greek philosophy, in statements of HERACLITUS (circa 500 B.C.) and in the relativism of the SOPHISTS (latter 400s B.C.), who maintained that any argument could be opposed by an equally true counterargument. The Athenian philosopher SOCRATES (469–399 B.C.) was famous for his statement "All I know is that I know nothing." Socrates' younger contemporary Cratylus became convinced that all communication is impossible, since the original meanings of words are inevitably altered by changes in the speaker, the listener, and the words themselves. According to legend, Cratylus therefore refused to discuss anything, preferring to wiggle his finger when addressed.

The founding of a formal school of Skepticism was the work of Pyrrhon (circa 363–273 B.C.), a native of ELIS (in the western PELOPONNESE) who had accompanied ALEXANDER THE GREAT on his conquest of the Persian Empire (334–323 B.C.). Centered at Elis, the Skeptics (Skeptikoi, from the word Greek *skepsis*: "perception" or "doubt") withheld judgment on all matters, stating no positive doctrines but accepting tentatively the outward appearances of things for the sake of day-to-day living. This outlook was supposed to bring the Skeptic to the desired state of imperturbability (*ataraxia*).

Alongside STOICISM and EPICUREANISM, Skepticism became one of the major philosophies of the HELLENISTIC AGE (circa 300–150 B.C.). It had its greatest influence at the ACADEMY, the prestigious philosophical school founded at ATHENS by PLATO circa 385 B.C. Skepticism was introduced as a study at the Academy in the mid-200s B.C., and for nearly the next two centuries it provided the basis of much Academic thought. Rejecting many of their founder Plato's doctrines, the Academics of these years saw Skepticism as connecting back to Socrates' statement about knowing nothing. They sought to prove the impossibility of positive knowledge and also used Skepticism to attack their rivals the Stoics, whose beliefs placed faith in sensory perception.

slaves Slave labor was a fundamental part of the ancient Greek economy. No less a thinker than ARISTOTLE (circa 330 B.C.) believed that the use of slaves, although lamentable, was unavoidable; to him, a slave was "a living tool." Only the mystical PLATO (circa 360 B.C.) imagined a society where slavery would be abolished (and where WOMEN would be equal citizens). Unfortunately, we know little about ancient Greek slavery; no extant ancient Greek writer discusses it at length, and modern scholars must rely on a patchwork of references in ancient Greek poetry, written history, stage comedy, and courtroom and political speeches. Archaeological evidence, such as tombstone inscriptions, add valuable details. Inevitably—as in all studies of ancient Greek society—the available information focuses on the rich and literate city of ATHENS, especially from the 300s B.C. onward.

Generally, a slave is a person without civil rights, owned as a possession, usually by another individual. The owner might have the power of life and death over the slave. (However, at Athens an owner could not legally kill a slave at will, but needed state permission.) To his or her master, the slave owes all labor and service—including, if required, sexual favors. In exchange, the master shelters, clothes, and feeds his slaves, just as he would be expected to care for his dogs or cattle. The origins of this brutal institution are lost in prehistory; certainly slavery was not confined to the Greeks. In HOMER's epic poem the *Iliad* (written down circa 750 B.C.), slaves are shown to be war captives; the proud Trojans' preordained fate is death or slavery at the Greeks' hands.

War was a major source of slaves. A defeated people would face slavery at the victors' discretion. These enslaved captives could be either Greeks or non-Greeks. To take one of many examples: After subduing the defiant Greek island of MELOS in 416 B.C., during the PELOPONNESIAN WAR, the Athenians put to death the entire adult male population and sold the women and children into slavery. Rather than being transported back to the conquerors' home city, such captives probably were sold on the spot to slave traders, who followed every Greek army. These traders were businessmen, with access to cash and transport, who would bring their newly purchased slaves to some major market for resale. The east Greek states of CHIOS and EPHESUS were important slave markets in the 400s B.C.; other commercial centers, such as Athens and CORINTH, played their roles.

The fate of war captives was surely ghastly. Families must have been broken up. The demand for adult male slaves (purchased for heavy labor) was far less than the demand for women, teenagers, and children (purchased for domestic duties). Children were especially prized as being pliant, physically pleasing, and long-term investments, but the mothers surely were often left behind on the auction block.

In times of peace, slaves might originate outside the Greek world. References in Homer, along with other documents, show Greek pirates of the 700s B.C. swooping down on the coast of ASIA MINOR or THRACE to carry away captives. Thracian slaves are mentioned regularly in Greek literature, and the Greek colonies founded on Thrace's Aegean coast—such as ABDERA and AMPHIPOLIS probably were active in acquiring slaves. This was done with the help of the feuding Thracian chieftains of the interior, who would bring their war captives to the Greek town, to sell.

In Greek cities the "exposure" of unwanted infants, particularly girls, was a legal, common practice. Exposed infants—that is, babies abandoned outdoors by their parents—automatically assumed slave status and would become the property of whoever rescued them. However, Greek cities typically had laws forbidding, within the city, the enslavement of any recognized citizen, whether child or adult. The Athenian lawgiver SOLON (circa 594 B.C.) wrote famous legislation prohibiting the enslavement of any Athenian for debt.

Except for SPARTA—with its large numbers of HELOTS, or serfs—Greek cities tended to depend heavily on slave labor. The Athenian state-run SILVER mines at Laurium may have employed as many as 30,000 slaves in the 400s–300s B.C. Modern ARCHAEOLOGY has revealed the hellish work conditions at Laurium, where slaves, including children, worked 10-hour shifts in black crawl-tunnels with poor ventilation; it seems to have been cheaper to let the workers sicken and die than to improve conditions. Smaller and less gruesome were privately owned factories, where skilled slaves produced beds, knives, and other household goods. The largest such operation now known of was an Athenian shield factory, owned by the family of the orator LYSIAS, which employed 120 slaves.

Some Athenian slaves probably worked semiindependently of their owners. A slave might practice a trade or keep a shop, delivering the profits to his master; a slave might be bailiff of a country estate, overseeing the slave farmhands and selling produce for the absentee owner. Slaves might be educated men, serving as clerks or tutors. Such trusted workers might look forward to the day when their master would free them (perhaps in his legal last will). At Athens, a freed slave took on the status of a metic, or resident alien.

Slave-population numbers are difficult to guess. Modern estimates suggest that, at the outbreak of the PELOPONNE-SIAN WAR (431 B.C.), Athenian territory was home to slightly over 100,000 slaves; this figure is probably equivalent to the number of nonslaves living there at that time. The ratio of 1:1 implies that slave ownership went fairly far down the social ladder. At the bottom of the scale might be (for example) a cobbler owning one slave to assist him in the shop and to do domestic work in the evening. Purely domestic slaves, bringing in little or no income, were a sign of wealth; the richest Athenian families might have 50 household slaves. No doubt many Athenians were too poor to afford even one slave, but this was viewed as a hardship.

With a 1:1 ratio, the Athenian slave population seems ominously large; yet there never was a slave revolt comparable to that led by Spartacus against the Romans in 73 B.C. The closest we come to this is the Athenian historian THUCYDIDES' (1) statement that 30,000 skilled slaves ran away from Athens during the latter part of the Peloponnesian War (413–404 B.C.). Reasons for this relative lack of unrest include the diverse origins and languages of the slaves, and their piecemeal segregation by owners and duties. In fact, when a Greek army marched to war, each HOPLITE had a servant (probably a slave) to carry armor and supplies. This would seem to provide a golden opportunity for revolt or escape, but the slaves lacked the motive and organization.

(See also METICS; NICIAS; PROSTITUTES.)

Smyrna   See ARCHAEOLOGY; IONIA.

Socrates   Athenian philosopher who lived 469–399 B.C. Although Socrates left no writings, he is of landmark importance in Western thought. By his example, PHILOSO-PHY was turned away from its prior emphasis on natural SCIENCE and became directed more toward questions of ethics—that is, the right conduct of life. In the words of the Roman thinker Cicero (circa 50 B.C.), "Socrates was the first to call philosophy down from the sky."

Socrates profoundly influenced his pupil PLATO. Socrates' central conclusions—that happiness depends solely on living a moral life, and that moral virtue is equivalent to knowledge and is therefore teachable—became the spring-

Socrates' unhandsome appearance was a source of jokes among his contemporary Athenians, who compared him with the grotesque mythical figure Silenus. This portrait comes from a marble bust of the Roman era, thought to be copied from a bronze statue made by the renowned Athenian sculptor Lysippus, circa 330 B.C., 70 years after the philosopher's death.

board for Plato's elaborate theory of reality and system of ethics. Socrates' equation of virtue with knowledge led to the daring corollary that evil is ignorance and hence unintentional. According to Socrates, the evildoer acts mistakenly in harming his or her own soul.

Socrates himself taught no doctrine, claiming simply, "I know nothing." This professed ignorance was denoted by the Greek word *eironeia:* that is, the famous Socratic irony. Unlike other philosophers, he never founded a school or charged fees for lessons. At ATHENS he attracted a following of young, aristocratic men, with whom he would dispute ethical or political issues. Since he took no money, his goal was evidently to prepare these students for public office by teaching them to think for themselves. His "Socratic method" consisted of asking questions, particularly in pursuit of definitions. For example: "You say that this man is a better citizen than his opponent? Let us consider therefore what we mean by 'good citizen.' " This question-and-answer process forced the respondent to examine his own preconceptions, in a search for general truths.

Socrates' disputations and skeptical outlook led him to be associated—unfairly, from his viewpoint—with those well-paid teachers of intellectual skills, the SOPHISTS. One famous sophist, PROTAGORAS of ABDERA, who taught at Athens circa 455–415 B.C., is recorded as believing that expertise (Greek: *aretē*) in most fields is teachable. Protagoras surely inspired Socrates' similar but more profound view that moral virtue (likewise denoted by the Greek word *aretē*) is teachable.

Among Socrates' well-heeled students were two destined for great importance in Athenian politics: the brilliant but erratic ALCIBIADES and the extreme right-winger CRITIAS (who was killed in 403 B.C. while trying to abolish Athenian DEMOCRACY). The link with Critias was probably responsible for Socrates' death. In 399 B.C. the 70-year-old philosopher was prosecuted for corrupting the youth (a charge implicitly referring to Critias and Alcibiades) and for impiety. After making a flamboyantly nonconciliatory defense speech, Socrates was found guilty by the jury of 501 Athenians and was sentenced to die by being given poison. The famous prison scene, in which Socrates discusses the soul's immortality with his visitors before drinking the fatal hemlock, is recounted in Plato's dialogue *Phaedo.*

Information about Socrates comes mainly from two sources. Plato's fictional dialogues contain much biographical information about Socrates while also foisting onto him Plato's own more complex philosophical theories. Here the insoluble question arises: What aspects of Plato's Socrates show us the real Socrates, and how much is just a mask for Plato's own thought? Probably a more biographical Socrates is portrayed in Plato's early work (such as the *Crito* or *Euthyphro,*) while the Socrates in the *Republic* and later dialogues is a more fictionalized character, who discusses concepts that the real Socrates never explored. The other source is XENOPHON, a stolid Athenian soldier and historian who was one of Socrates' disciples and who wrote about him in three nonfiction memoirs, *Memorabilia,* *Symposium,* and *Apology.* (The latter two titles also belong to works by Plato.) That Socrates could be the beloved mentor of two such different people as Plato and Xenophon is itself revealing. Other information comes from ARISTOPH-

ANES' stage comedy *Clouds* (performed 423 B.C.), in which a crackpot scientist named Socrates operates a "Thinking Shop." Clearly the real Socrates was a dynamic personality who fascinated friend and foe alike. When one contemporary Athenian asked the oracle at DELPHI who was the wisest of all men, the answer came back "Socrates."

He was born into the Athenian middle class, the son of a stonecutter or sculptor named Sophroniscus. Socrates came of age during the political primacy of PERICLES, when Athens reached its zenith as a radical democracy and imperial naval power—twin developments that alienated many conservative Athenians. As a youth, Socrates supposedly studied under a disciple of the philosopher ANAXAGORAS, but became discouraged by the emphasis on physics and cosmology rather than ethics.

During the PELOPONNESIAN WAR between Athens and SPARTA (431–404 B.C.), Socrates served as foot soldier in the POTIDAEA campaign (432–430 B.C.), at the disastrous Battle of Delium (424 B.C.), and at AMPHIPOLIS (422 B.C.), by which time he would have been about 47 years old. On campaign, he distinguished himself by his physical endurance and courage. One anecdote describes Socrates striding calmly amid the Athenian retreat from Delium, defying the pursuing enemy CAVALRY.

In later life he repeatedly resisted the political hysteria of Athens in crisis. In 406 B.C., while serving a citizen's normal duty on the council panel to prepare the agenda for the Athenian ASSEMBLY, Socrates refused to go along with an illegal motion ordering a group trial for six commanders charged with negligence after the sea battle of Arginusae. In 403 B.C. Socrates defied his old pupil Critias, who had seized power in the coup of the THIRTY TYRANTS: Brought before the Thirty, Socrates refused to help them arrest a certain intended victim. No doubt Socrates himself would have become a victim of the Thirty, had not their reign of terror soon ended. Instead, Socrates was denounced by prominent accusers after the restoration of democracy.

That Socrates should have been prosecuted under the democracy is significant. An original thinker who defied convention, Socrates was known for his antidemocratic sentiments. As Xenophon's memoirs make clear, he criticized the democracy's inability to entrust its government to the most apt and expert people. Without advocating revolution or dictatorship, Socrates seems to have favored a meritocracy, with power entrusted to a worthy ruling class. Such beliefs—although purely theoretical—must have seemed damning after Critias' right-wing coup. Following the Thirty's downfall, the Athenians' wrath turned against Socrates.

Across the centuries, Socrates' eccentric personality communicates itself. Although magnetic in character, he was physically unattractive in middle age—balding, pug-nosed, and paunchy. His contemporaries humorously compared him to the mythical figure SILENUS. Socrates was famous for his austerity: Impervious to cold and fatigue, he almost always went barefoot, even over frozen winter ground at the siege of Potidaea.

He had three sons by his wife, Xanthippe (whom several sources describe as a bad-tempered shrew). Socrates may have practiced a stonecutter's trade, like his father, but he evidently spent most of his time at the Athenian sports grounds where educated men of leisure congregated. There he would conduct semipublic debates with his followers or rivals.

Socrates' disciples (all male) included several of the most glamorous youths of the day in the homosexual society of aristocratic Athens. A homosexual ambiance pervades many of Plato's dialogues, such as the *Symposium*, where the topic is love and where the mutual attraction between Socrates and Alcibiades is discussed. But as so much is said about his indifference to physical needs, it remains unclear whether Socrates actually was intimate with these young men.

(See also GYMNASIUM; HOMOSEXUALITY; LAWS AND LAW COURTS; SKEPTICISM.)

Solon Athenian statesman who in the early 500s B.C. drafted a new code of law, which averted a revolution at ATHENS and laid the foundations for Athenian DEMOCRACY. Solon's democratic reforms came at a time of major class tension: between the nobles and the increasingly confident middle class, as well as between the country landowners and the tenant farmers who paid rent under onerous conditions. Solon alleviated the burdens of the poor farmers and, at the upper end of the social scale, broke the nobles' traditional monopoly on political power.

Lawgiver, poet, and patriot, Solon saved Athens from civil war and dictatorship, and laid the foundations for Athenian democracy, circa 594 B.C. His likeness here is based on a marble bust of the Roman era.

Later generations revered Solon as one of the SEVEN SAGES of Greece. The texts of his laws, carved into wooden tablets, were kept on display in the Athenian AGORA. These laws supplied the code at Athens for nearly 200 years, until the legal revisions of 410–400 B.C.; even afterward, the Athenians still called their code "the laws of Solon."

Solon was also a poet. A number of his verses are preserved in two later written works—ARISTOTLE's *Constitution of Athens* (circa 340 B.C.) and PLUTARCH's biography of Solon (circa A.D. 100). But because Solon is the earliest important figure in Athenian history, much of his life and work remain unclear to us.

Born around 630 B.C. into an aristocratic Athenian family, Solon first distinguished himself in a war against MEGARA (1). His patriotism won the respect of his fellow citizens, and they asked him to serve as arbitrator and lawgiver, to avert civil war. It was probably in 594 B.C., the year in which he served as chef ARCHON, that Solon instituted his famous reforms.

The most grievous social problem facing Athens at that time was the plight of the rural poor in ATTICA, the 1,000-square-mile district surrounding Athens. It seems that the rich landowners were crushing the tenant farmers who paid part of their produce as rent. The law allowed loans to be secured on the person of the borrower; that is, defaulting debtors were liable to be sold as SLAVES. Tenant farmers who forfeited on their rent (in bad harvest years, e.g.) were faced with catastrophe, for both themselves and their families.

Meanwhile, in the city, the better-off commoners had their own grievances, directed against the government. Citizens of the middle class—who supplied the backbone of Athens' army, serving as HOPLITE infantry—were chafing at being excluded from government. At this time, Athens was still an ARISTOCRACY. All political power was held by a few ancient, upper-class families (of which Solon was a member). Only aristocrats could hold executive posts or be admitted to the powerful decision-making council and supreme court known as the AREOPAGUS. All of the law courts, overseen by aristocrats, showed a strong upper-class bias; the existing law code, devised around 620 B.C. by the nobleman DRACO, was mainly repressive.

By the early 500s B.C. there was fear at Athens that these two aggrieved classes—the peasantry and the hoplite class—would join forces and place a dictator, or *turannos*, on the throne. TYRANTS had seized power at CORINTH and Megara, and Athens itself had seen a failed coup by a would-be tyrant named CYLON in around 620 B.C. In this volatile atmosphere Athens turned to Solon, whom all parties trusted. He could have exploited the situation to make himself tyrant; but he was a greater man than that.

Solon rescued the tenant farmers and other poor Athenians by forbidding that loans be secured on the borrower's person—henceforth no Athenian could be enslaved for debt—Solon also cancelled all such existing debts. "I plucked up the marker-stones," he declares in one poem, referring to the hated signs standing at repossessed farms. He also freed Athenians who had been enslaved, bringing home many who had been sold abroad. Solon's poverty-relief measures were known as the *seisachtheia*, "the shaking off of burdens."

Solon overhauled the crude Draconian law code and created political changes that broke the aristocrats' monopoly on power. He devised a new system of social ranking, with four citizen classes based entirely on income level. This new emphasis on income worked against the traditional emphasis on noble birth. The upper classes might still be full of rich aristocrats, but now other men—commoners—could be admitted to these classes. Since a man's social rank determined his eligibility for various political offices, this was a significant democratic change.

Solon enfranchised the lowest income class, the Athenian peasants, admitting them as citizens and allowing them to vote in the ASSEMBLY. He created a new deliberative COUNCIL, of 400 members, open to the upper three income classes; the council members served for one year, after being chosen by lot. This new council, very influential in that it helped prepare the agenda for the assembly, served to counterbalance the old aristocratic Areopagus. Also, it was probably Solon who set up the Athenian "people's court," to which any citizen had the right of appeal; this assured that a defendant could be heard by a jury of his peers rather than by a court consisting of nobles.

The principal beneficiaries of Solon's reforms were the rich commoners—nonaristocratic landowners, traders, and manufacturers who could now take their place in Athens' upper classes and enjoy greater political voice. Solon was not a complete democrat; he was more of a timocrat. As his poetry makes clear, he believed that the populace should be governed by a worthy, fair-minded ruling class.

Certain embittered verses by Solon indicate that his reforms came as a disappointment to both extremes: The nobles resented their losses, and the poorer classes were angry that Solon had stopped short of redistributing the land. Having made the Athenians promise to obey his laws unchanged for 10 years, Solon left Athens to travel (some say) to EGYPT and LYDIA. Returning home, he lived long enough to see his kinsman PISISTRATUS make himself tyrant in Athens (circa 561 B.C.)—the very outcome that Solon had tried to prevent. But Solon's laws and political apparatus survived in Athens (the tyrants did not) and set the stage for CLEISTHENES' (1) reforms (508 B.C.) and the radical democracy of the 400s B.C.

(See also LAWS AND LAW COURTS.)

sophists *Sophistai* (experts or professors) were itinerant teachers who, in the latter 400s B.C., went from city to city in the Greek world, tutoring young men in disputation and other intellectual skills. These lessons, for which the sophists charged hefty fees, were designed to help students achieve practical success as lawyers, politicians, speechwriters and so on. The visiting sophists caused a cultural sensation at the wealthy, intellectual city of ATHENS, but they also earned a reputation for amoral cleverness; the hallmark of a sophist was the ability to argue any question from two opposing sides. The modern English word *sophistry* (meaning "clever but false reasoning") retains the negative associations of this rationalistic movement.

Yet not all sophists were amoral. The best were genuine thinkers, breaking new intellectual ground and answering a need for "college-level" EDUCATION in their day—the sophist Hippias of ELIS, for instance, taught a method

of memory training, among many offered courses (circa 430 B.C.).

The earliest and most famous sophist was PROTAGORAS of ABDERA (circa 445 B.C.); another prominent one was Prodicus of Ceos (circa 430 B.C.). The orator GORGIAS of Leontini (circa 427 B.C.) was associated with the sophists, due to the ingenuity of the arguments he used in public speaking. And in the eyes of many Athenians, the philosopher SOCRATES (469–399 B.C.) was a sophist, on account of his question-and-answer method of inquiry.

Sophocles Athenian tragic playwright who lived circa 496–406 B.C. Sophocles is reckoned among the three great classical tragedians (the other two being AESCHYLUS and EURIPIDES). His work has been seen as Greek tragedy's high point, wherein the separate requirements of plot and character are fused most successfully. He is said to have written 123 plays, of which only seven survive today;

Sophocles was—to modern taste, at least—the finest of the Athenian tragedians. In his day he was also a well-liked figure, elected several times by his countrymen to high office, including two generalships. His human insight and sophistication are suggested in this likeness, from a bronze bust of Hellenistic or Roman date.

among these is the play that the philosopher and critic ARISTOTLE admired most of any Greek tragedy, *Oedipus the King*.

SOPHOCLES ("famed for wisdom") was born the son of a rich manufacturer at the Athenian DEME of Colonus—the beauty of which he would commemorate in a famous choral ode in his tragedy *Oedipus at Colonus*. His life spanned the near century of Athenian greatness. Growing up during the PERSIAN WARS, he partook of the spirit of Athenian confidence following the miraculous victory over the Persians. After the Greek sea victory at SALAMIS (1) (480 B.C.), the teenage Sophocles was honored with the assignment of dancing and singing, naked and anointed, at the trophy monument. To win this commission he must have been a performer of considerable talent and beauty. Supposedly, in his earlier plays he occasionally appeared in roles onstage, but gave that up due to a weak voice.

At about age 28 Sophocles won his first tragedy-competition, at the major annual drama festival known as the City Dionysia (468 B.C.). We know that one of the playwrights whom he defeated this year was the well-established Aeschylus, but the centuries have not preserved the three Sophoclean tragedies (and one satyr-play) that comprised this winning entry. Sophocles went on to win a total of 24 victories in over 65 more years of writing. Of the seven times (only) when he failed to win first prize, he always took second prize, never third. This remarkable record shows that Sophocles was the most successful tragedian of his day; Euripides, by contrast, won first place only five times. Relatedly, Sophocles was highly popular as a social figure. Several ancient writings testify to his charming personality and many friends—who included the historian HERODOTUS and the Athenian statesman CIMON.

In the development of Greek THEATER, Sophocles is credited by ancient writers with making several changes in the way tragedy was performed. Most important, he raised the number of speaking actors from two to three—which greatly increased the possible number of speaking parts, since an actor could play multiple roles. Sophocles supposedly ended the custom of presenting tragedies as linked trilogies, in favor of presenting three tragedies on unrelated topics. And he is said to have increased the size of the chorus from 12 to 15 men and to have developed the use of painted backdrops as scenery.

Outside the theater, Sophocles played a substantial role in Athenian public life prior to and during the PELOPONNESIAN WAR (431–404 B.C.). He was devoted to ATHENS, and—unlike Aeschylus, Euripides, and other artists—he refused all lucrative invitations to visit other cities of the Greek world. In 443 B.C. Sophocles was elected to the office of treasurer of the DELIAN LEAGUE. In 441 B.C., reportedly in recognition of his play *Antigone*, he was elected for the following year as a general of the Athenian armed forces. Under PERICLES' senior command, General Sophocles served in the campaign against the rebellious island of SAMOS. Possibly (we do not know for sure) Sophocles was later elected as general a second time.

In 420 B.C. Sophocles was active in developing the god ASCLEPIUS' cult at Athens, which may mean that he helped establish a public hospital. In 413 B.C., amid the state emergency following the Athenian military disaster at SYR-

ACUSE, the Athenians appointed Sophocles, age 83, to a special executive post.

Sophocles married twice and we hear of two sons. One of them became a successful writer of tragedies, as did one of Sophocles' grandsons. Like other Greek men of his era, Sophocles may have pursued a private life that we would call bisexual. An anecdote in the memoirs of the writer ION (2) describes Sophocles relaxing at a SYMPOSIUM during the Samian military campaign: Sophocles coaxed a kiss from a handsome slave boy, then commented to the other guests, "So you see, I am not so bad a strategist as Pericles believes." Another famous quotation is preserved in the philosopher PLATO's treatise *The Republic*: When Sophocles in old age was asked if he was still able to have sex with a woman, he replied that he had happily escaped from his sex urge, "as from a cruel and insane master."

In about 406 B.C., after Euripides died, the 90-year-old Sophocles presented his chorus dressed in mourning, to commemorate his dead rival. Soon afterward Sophocles died, and he was remembered in ARISTOPHANES' comedy *Frogs* (405 B.C.), which is set in the Underworld: "He was a good fellow up there and now he's a good fellow down here."

The first performance dates for most of Sophocles' seven existing tragedies are unknown. But the following sequence is plausible: *Ajax* (performed perhaps circa 450–445 B.C.), *Antigone* (perhaps circa 442 B.C.), *The Women of Trachis* (perhaps circa 429–420 B.C.), *Oedipus the King* (perhaps circa 429–420 B.C.), *Electra* (perhaps circa 418–410 B.C.), *Philoctetes* (409 B.C.), and *Oedipus at Colonus* (performed in 401 B.C., after Sophocles' death). These tragedies survived because they were chosen in later antiquity to be taught in schools. Of the lost plays, the titles of many are known and there exist some fragments of text; apparently about a third of them were drawn from legends of the TROJAN WAR. All of Sophocles' plays, in keeping with the custom of the day, retold tales already familiar from EPIC POETRY and other sources.

A brief description of a few of the plays must suffice here. *The Women of Trachis* presents the horrible death of the hero HERACLES, poisoned by a venom-smeared robe given to him by his innocent wife, Deineira (who wrongly believes the robe to be a magic charm to reclaim her husband's love). The play is named for the characters portrayed by the chorus: local women in the region of central Greece where Heracles legendarily met his end. The action emphasizes Heracles' agony in death and the tragic irony that the world's mightiest hero, vanquisher of monsters, should die at the unintending hands of his gentle, loving, naive wife.

Oedipus the King and *Antigone* are today among the most accessible and widely read of all existing Greek tragedies. *Oedipus* is often misunderstood as the tragedy of a man undone by arrogance, or *hubris*. But, like much of Sophocles' work, the play is really about the limits of human knowledge and power. The Theban king OEDIPUS, at the play's outset, seems as blessed as a human can be: young, wise, powerful, just, beloved by his people. By the play's end he is blind, accursed, and homeless; his prior happiness has been revealed as an illusion. Unlike certain other Greek tragic heroes, Oedipus onstage does not do anything evil or foolhardy to bring on the gods' wrath. Although he does show pride and anger early in the play, these do not lead to bad action—rather, Oedipus' destructive actions lie buried, unbeknownst to him, in the past. *Oedipus the King* presents his gradual discovery of his own calamity; ironically, it is Oedipus' intelligence and perseverance that bring about this discovery.

The tightly plotted play opens with King Oedipus' city of THEBES ravaged by plague and famine; the oracle at DELPHI has revealed that the land lies under a curse because the murderer of the prior king, Laius, has not been punished. Since the killer's identity is unknown, Oedipus vows to discover it. The blind seer TIRESIAS is summoned for questioning. The ensuing scene is perhaps the finest in all extant Greek tragedy. As Oedipus grows puzzled and enraged, Tiresias at first refuses to speak, then angrily replies that Oedipus himself is the defiler of the land; Oedipus himself is the murderer. Oedipus ignores the stated message and shrewdly assumes that Tiresias is trying to discredit him as king, perhaps in league with the ambitious nobleman Creon.

Tiresias, alone of all the characters, understands that Oedipus slew King Laius years ago, in a quarrel at a crossroads far from Thebes. Oedipus, traveling from CORINTH (where he had grown up as the son of the king and queen there), did not know the identity of the arrogant older man whom he killed. Worse, he did not know that Laius was his own father. Worst of all, Oedipus has since married Laius' widow, Jocasta, and had children by her—who is really Oedipus' mother. He had been taken from her at his birth, and she did not recognize him as a young man.

The tragedy moves toward its climax as Oedipus, like an ideal courtroom judge, gradually uncovers the truth by interrogating witnesses, including Jocasta and a Theban shepherd who is the sole survivor of Laius' fatal fight. At one point Oedipus realizes that it was he who killed Laius, but he holds onto the hope that at least Laius was not his father (and Jocasta his mother). When finally the whole truth is publicly revealed, Oedipus in horror gouges out his own eyes and Jocasta hangs herself (both offstage). The final scenes show the blind Oedipus departing into self-imposed exile, cursed by his revealed parricide and incest. The tragedy ends with the statement—appearing in the writings of Sophocles' friend Herodotus—that no human being should be called happy until he is safely dead. Sophocles' *Oedipus at Colonus* presents the hero's final days, when he finds sanctuary and divine forgiveness under protection of the mythical Athenian king THESEUS.

Antigone presents the classic conflict of public duty versus private conscience. In the aftermath of the destructive war of the SEVEN AGAINST THEBES, Oedipus' daughter ANTIGONE chooses to disobey an edict of Creon (who is now ruler of Thebes) in favor of her religious and familial obligations. Creon has forbidden anyone on pain of death to grant funeral rites to the slain attacker Polynices, Antigone's brother. In breaking this law, Antigone discards all happiness, including her betrothal to Creon's son, Haemon. But death comes as a triumph for her and a curse for Creon and his family. It is Antigone who has upheld the higher, unwritten law of social obligation.

The Sophoclean hero or heroine tends to be a splendid individualist, set apart from other people by a refusal to compromise. These protagonists are larger than life—stronger, wiser, or more pious than others—yet they inevitably succumb to circumstance or FATE. This two-sided view is essential to Sophocles' work: The hero may be called "godlike," but usually he fails due to his human flaws, such as anger or ignorance. Ignorance in particular interests Sophocles: He portrays Oedipus and Heracles each as violently undone by an imperfect, human understanding of the facts.

Above human beings stand the gods. Where Aeschylus' surviving tragedies convey a faith in divine wisdom and Euripides' show gods who are malicious and capricious, Sophocles' work falls somewhere in between. In Sophocles, the great god ZEUS controls mortal destiny, but Zeus' purpose may be mysterious to the onstage characters and to the audience. "All of these events are Zeus," the chorus announces at the end of *The Women of Trachis*, amid Heracles' gruesome death. Zeus' purpose seems needlessly cruel, except that Heracles will win blessed immortality after his suffering. Sophocles refuses to clarify or justify the workings of the gods. Rather, his plays mirror the religious questions of real life.

(See also AJAX [1]; DEMOCRACY; ELECTRA; HOMOSEXUALITY; HUBRIS; MYTH; ORESTES; PHILOCTETES; RELIGION.)

Sparta One of the greatest city-states of mainland Greece, often in rivalry with ATHENS. Located in the southern Greek region known as the PELOPONNESE, Sparta dominated southern Greece after 600 B.C. In 404 B.C. Sparta became the supreme Greek city, having defeated Athens in the PELOPONNESIAN WAR. But a declining birthrate among Sparta's citizens, poor relations with its subject cities, a constant fear of serf revolt, and a resistance to change all made the failure of the Spartan Empire inevitable. In 371–369 B.C. Sparta was unexpectedly defeated and crippled (although not captured) by its erstwhile ally THEBES. Thereafter Sparta confined itself mainly to intrigues in the Peloponnese, often in defense against armies of the kingdom of MACEDON. In 222 B.C., amid internal political convulsions, Sparta was defeated and captured by the Macedonian king Antigonus III. Within the next 80 years, however, both Macedon and Sparta surrendered to the armies of ROME.

Sparta was unique among Greek cities, because of its program of patriotic indoctrination and full-time military service for males. This program arose in the mid 600s B.C., probably at the hands of a Spartan statesman named LYCURGUS (1), who was revered as a hero by later generations of Spartans. Today nothing is known for sure about Lycurgus' life, but he evidently shaped Sparta into an authoritarian state, devoted to maintaining the bravest and most disciplined of all HOPLITE armies. Between about 600 B.C. and 371 B.C. the Spartans could justifiably claim to be the best infantry in the Greek world (also superior to the one foreign infantry they encountered, the Persians). Eventually, however, the Spartans' reliance on traditional battle tactics left them vulnerable to new tactics, developed by the Thebans and Macedonians.

The grim world of Sparta was the "other" Greece, in contrast to the exuberant life and literature of classical

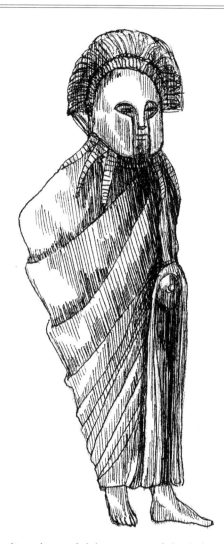

The austerity, valor, and sinister aspect of classical Sparta are apparent in this bronze figurine of a cloaked Spartan soldier, circa 400 B.C. Note the plaited hair beneath the helmet. Spartans were known for their long tresses. On the morning before their heroic last stand at Thermopylae (480 B.C.), the Spartan soldiers combed their hair.

Athens. The Spartans were secretive and anti-intellectual. They rarely let in visitors and left behind few writings. Most modern knowledge of Sparta comes from information preserved by ancient Greek writers who were not themselves Spartan—the historians HERODOTUS, THUCYDIDES (1), and XENOPHON, the philosopher ARISTOTLE, and the biographer PLUTARCH. Still, much remains mysterious about the Spartans today.

Although militaristic and austere, Sparta was not necessarily repulsive to all other Greeks. In a Greek world where ideals of DEMOCRACY (rule by the many) vied against those of OLIGARCHY (rule by the few), Sparta was the model oligarchy. It maintained its dominance over other Greek cities by supporting the ruling classes of those cities. In so doing, Sparta counterbalanced the appeal of Athens, which was trying to export democracy to the lower-income classes everywhere. In southern Greece, aristocrats and other wealthy people welcomed Spartan protection and leader-

ship. And many other Greeks admired Sparta's army and stable government.

Even Athens itself contained pro-Spartan individuals. Typically these Athenians were aristocrats. The best remembered of them is the gentleman-soldier-historian Xenophon (circa 428–354 B.C.). Similarly, Xenophon's contemporary, the Athenian philosopher PLATO, used Spartan government as inspiration for the authoritarian utopias imagined in his writings *The Republic* and *The Laws*.

Sparta ("sown land," or possibly "broom shrub") was located in the region called LACONIA, or Lacedaemon, in the southeastern Peloponnese. Nestled between lofty mountain ranges on the west and the east, the city sat at the northern tip of the Laconian plain, which fans out triangularly 25 miles southward to the sea. Just west of the city, the River Eurotas flowed southward through the widening plain. Nowadays a Greek provincial capital, Sparti, stands near the ancient site.

Long before the city of Sparta arose, the fertile and rain-nourished Laconian plain was home to pre-Greek inhabitants in the third millennium B.C. and to Mycenaean Greeks in the second millennium B.C. According to legend, this was the kingdom ruled by MENELAUS at the time of the TROJAN WAR (circa 1220 B.C.). Archaeological excavations reveal that the town of Amyclae, just south of the future site of Sparta, was a civic center containing a sanctuary of a pre-Greek god named HYACINTHUS. Later Amyclae became an important shrine of the Greek god APOLLO, one of the patron gods of the Spartan region.

Sparta itself was founded by descendants of the DORIAN GREEKS who overran most of the Peloponnese circa 1100–1000 B.C., after the collapse of MYCENAEAN CIVILIZATION. Circa 950 B.C. four local Dorian villages became amalgamated around a shared civic center. This multipart beginning probably explains the odd Spartan institution of dual kingship, whereby Sparta always had two coequal kings, drawn from two royal familes, the Agiads and Eurypontids. One reason the double kingship survived for the next 700 years was that Spartan kings were primarily military leaders, commanding armies in the field. Certainly by the 600s B.C.—and probably long before—the kings' power at home was being curtailed by other governmental branches, specifically the EPHORS (Greek: *ephorai*, "overseers") and the COUNCIL.

In appearance Sparta remained a sprawling village, interspersed with trees. Disdaining wealth and pomp, the Spartans of later centuries continued to make their houses out of wood and to build very few public stone monuments. Sparta at its height of power had no perimeter wall; the Spartans preferred to trust in their army and their protective mountains. Not until 318 B.C. did an enclosing wall finally go up, against the threat of a Macedonian invasion. Similarly, the Spartan ACROPOLIS was a puny, unwalled mound (very unlike the lofty citadels of other Greek cities), whose main job was to hold the temple of the city's patron deity, ATHENA. The building itself was known as the temple of Athena Chalkioikos ("Athena in Bronze"), so named because its walls were covered in BRONZE plates showing mythical scenes.

This temple and that of ARTEMIS Orthia, which stood to the west, by the Eurotas' bank, were Sparta's two main monuments. Thucydides, writing during the Spartan heyday (circa 400 B.C.), contrasted such physical simplicity with the elaborate grandeur of Athens. In a famous passage, Thucydides commented that if Sparta were to become deserted and only its empty structures viewed by future generations, people would never believe that the city had been so great a power.

The earliest events of Spartan history saw Sparta still governed as an old-style ARISTOCRACY, prior to Lycurgus' reforms. By about 700 B.C. this aristocratic Sparta had conquered Laconia (including the seaport of Gytheum, about 25 miles south of Sparta, to serve as the Spartans' maritime link).

The capture of Laconia and its inhabitants created three distinct social levels. At the top were the original Spartans, called Spartiatai (or "Spartiates"). Below them were the PERIOECI (Greek: *perioikoi*, "dwellers around"), second-class citizens who were descendants of the earliest-subdued Laconian villagers. As their name suggests, the perioeci lived in towns that roughly formed a circle around the Spartiates' prime land on the central plain. The perioeci were obligated to pay taxes and serve alongside the Spartiates in the army; they enjoyed legal protections, rights of ownership, and so on but were excluded from the political process. Spartiates and perioeci together comprised the people known as the Lacedaemonians.

Below these was another group, far inferior in status. They were the serfs, or HELOTS (Greek: *helotai*), supposedly named for the Laconian coastal village of Helos ("marsh"). It seems that the Spartans, at some point in their expansion, changed their policy toward the villagers they subdued: instead of admitting them as lesser citizens, they began to subjugate them as serfs. This brutal serf system existed elsewhere in the Greek world, but the Spartans used it the most.

The helot population increased enormously after the Spartan capture of the fertile Greek region called MESSENIA, in the southwestern Peloponnese. Invading across the high Tagytus mountain range that borders Sparta to the west, Spartan armies descended into the Messenian plain and fought a long war (circa 735–715 B.C.) to seize the farmlands and helotize the people. Sparta's victory in this First Messenian War is associated with a Eurypontid king, Theompus, the earliest reliably real figure in Spartan history.

The annexation of Messenia made Sparta rich, with control of two-fifths of the Peloponnese. In a land where famine was a constant threat, Sparta alone had solved the food problem, with supplies from Messenia abundant to its needs: no other Greek state ever managed to subdue an entire Greek region and population. Yet the helots in Messenia and Laconia were also a source of fear. They outnumbered their Spartan masters—by the mid 400s B.C. the ratio would be seven to one. Fear of helot uprisings, which occurred periodically, kept Spartan armies close to home and created a mass neurosis in the Spartan mentality. Every year the ephors would formally declare war on the helots, thus allowing them to be killed with impunity; Spartan youths were enrolled in the Krypteia, or Secret Society, whose job it was to murder subversive helots. Conversely, the helots became a magnet for intrigues by Sparta's external enemies, such as Athens in the 400s B.C.

But helot freedom had to wait until 369 B.C., when the triumphant Theban leader EPAMINONDAS liberated the helots of Messenia (although not those of Laconia), setting up an independent Messenian state.

Meanwhile, Sparta came into conflict with the one southern Greek power that could challenge Spartan ambition—the city of ARGOS, in the northeastern Peloponnese. In 669 B.C. a Spartan army was defeated by an Argive army at Hysiai, on the road to Argos. Modern scholars believe that the battle involved the victory of organized hoplite tactics, as perfected by the Argive king PHEIDON, over the old-fashioned, individualistic fighting of the aristocratic Spartan army. Scholars further believe that the defeat fully discredited the aristocratic leadership at Sparta and brought on drastic changes in government, society, and military organization, through the reforms of Lycurgus.

Lycurgus changed Sparta from a narrow aristocracy to a broad-based oligarchy. That is, he brought middle-class Spartan men into the government as voters and soldiers, creating an enlarged citizen body of about 9,000 adult males. Henceforth, perhaps once a month, all citizens over age 30 gathered at the *apella*, or ASSEMBLY, to vote on proposals framed by a 30-man council called the *gerousia* ("the elders"), which included the two kings. The assembly could vote only "yes" or "no" and not initiate any business. But its voting included such duties as declaring war or peace, ratifying the appointment of military leaders, and electing ephors and council members aside from the kings. Council members had to be 60 or more years old; they held office for the rest of their lives. Although elected annually, ephors had immense power as chief legal officials and as monitors controlling the kings' conduct.

In theory any citizen could be elected to the council or ephorate, but in practice these offices probably tended to be filled by members of a few rich and prestigious families. Both the council and the ephorate were probably survivals of the old aristocratic government, now adapted to the Lycurgan government.

Most significant, Lycurgus' system created a large, full-time hoplite army, supported by Sparta's captured territories. The 9,000 full-rank citizens—the Spartiates—henceforth were barred from practicing any profession aside from soldiering; crafts and TRADE became solely the province of the perioeci. Instead, every male Spartiate (excepting perhaps the rich) received an allotment of public land in Messenia or Laconia. Farmed by helots overseen by state officers, the allotment produced food that supported the citizen and his family; the citizen was thus free to devote himself to lifelong military drill and campaigning.

Of all Greek city-states, only Sparta had a permanent soldiering class, free of the need to earn a livelihood as farmers or craftsmen. Spartan armies, for example, could campaign through late summer and autumn, when other states' farmer-soldiers might feel obliged to return home for their various sowings and harvests.

The Spartiates—the full-rank male Spartan citizens—sometimes called themselves *homoioi*, "equals." They were equal in that each man had a land allotment (although rich families undoubtedly kept their traditional estates in Laconia) and each had one vote in the assembly. But *homoioi* also means "those who are all the same," and this nuance reveals the essence of Lycurgus' reforms. Sparta became a city where concepts of the individual and the family were erased, in favor of a grim system of conformity and state service.

An individual Spartiate reached full citizenship by meeting three requirements: legitimate birth to Spartiate-rank parents, completion of a 13-year-long boyhood training program called the *agogē*, and election at age 20 to one of the communal dining messes that were the center of Spartan adult male life. The *agogē* process is described in the writings of Plutarch and Xenophon. At birth, boys (and girls) were examined for health and were accepted or rejected by state officials—not by the family father, as elsewhere in the Greek world. Those rejected were killed. At age seven, boys were taken from their mothers, to be brought up in "packs" under supervision of older boys who reported to a state officer. By age 12, boys lived in barracks, sleeping on mattresses of river rushes that they gathered themselves. Year round they went shoeless and were allowed only one outer garment, a kind of cloak. Their staple diet was a kind of pork-boiled porridge—the famous "Spartan broth"—which they were encouraged to supplement by stealing food from local storages and farms. If caught stealing, they were beaten for their clumsiness.

Boys received schooling in traditional Spartan musical art forms, such as competitive choral singing and dancing, but book-learning was minor. Their EDUCATION was mainly physical—combat-type sports and games. They exercised naked outdoors in all weather. They were subjected to routine hazing and abuse by the older boys, intended to toughen them. Homosexual pairings between older and younger boys were officially encouraged, as being helpful to patriotic esprit.

Foremost, boys were taught to obey orders and endure hardship without complaint. Scrutinized and judged by their elders, they grew up fearing disgrace more than death. Plutarch recounts how a Spartan boy somehow stole a fox cub and hid it under his cloak, only to have the animal start biting out his insides. Rather than cry out or otherwise reveal his theft, the child collapsed and died without a whimper.

The training program was designed to prepare boys for a soldiering life and to weed out anyone unsuitable for full citizenship. For each young male the reckoning came at age 20, when he sought admittance to one of the Spartan dining messes (Greek: *sussitia*). Each *sussition* probably consisted of about 15 men, who voted among themselves whether to admit or reject a candidate. Once selected, a new member remained part of the mess for life, barring expulsion, and he contributed a portion of his land's produce to help feed the group. (Their diet seems to have been far more ample and varied than the boyhood fare.) A man's mess-mates—not his family—were the center of his life and identity.

Thus, at Sparta, a man's most intense feelings were normally directed at other males. Men were still expected to marry citizen-rank WOMEN and breed children, but MARRIAGE was not a deeply emotional affair. Probably until age 30 a male Spartan continued to live with his mess-mates, visiting his wife only in secret. And Spartan law even allowed a husband to arrange for his wife to sleep with another suitable man, possibly at her request—a custom that other Greeks found outrageous.

The Spartan screening system inevitably created classes of citizens in between the Spartiates and perioeci. Spartiates who failed to win admittance to a sussition—or who were expelled because of cowardice or inability to provide a share of food—ceased to be Spartiates and descended to a rank known as Inferiors. Conversely, individual perioeci boys might be selected for the rugged Spartiate training program, but it is unclear whether such boys could ever attain full citizenship. A tendency to exclude groups on grounds of inferior birth was a deeply ingrained aspect of Spartan mentality: as early as 700 B.C. (probably before Lycurgus' reforms) a rejected group called "the sons of virgins" sailed off to found Sparta's only major colony, TARAS, in southern ITALY.

Sparta's extreme social conditions created an unusual life for women, as well. Sparta was unique among ancient Greek cities in the freedom it granted to its female citizens—this was partly due to reliance on women to manage estates while men were away at war. Somewhat similarly to boys, Spartan girls received training in choral singing, dancing, and gymnastics. Scandalously (as it seemed to other Greeks), Spartan girls exercised in public, naked or scantily clad. Like boys, Spartan girls apparently might experience homosexual love affairs in the course of their upbringing—as suggested in extant verses by the poet ALCMAN (circa 620 B.C.). And Spartan women had a unique reputation for heterosexual initiative and promiscuity. At least some of these conditions were intended to mold Spartan females into healthy and capable child-bearers, to supply future soldiers for the state.

Yet for reasons that are not clear, Sparta eventually faced a dwindling of its Spartiate population (as opposed to the perioeci or helots). Perhaps Spartan marriages were unproductive, with the problem worsened by the brutal screening system. The shrinkage reached crisis proportions after 464 B.C., when a catastrophic earthquake wiped out large numbers of Spartiates in the city center. By 425 B.C. the Spartiates numbered no more than 3,000—whereas Lycurgus in the mid-600s B.C. had made arrangements based on a population of 9,000.

In theory, Sparta's population problem could have been eased by promoting large numbers of perioeci up to Spartiate status. Although this drastic measure was proposed amid Sparta's political turmoil of the 200s B.C., it was never carried out. Spartiates viewed themselves as an ethnically distinct elite, raised from birth to dominate "lesser" peoples. Their mentality did not allow for mass enrollments of new citizens.

Another long-term result of Lycurgus' reforms was loss of trade and arts. Although the late 600s B.C. saw Sparta's cultural heyday—with verses being written by choral poets such as Alcman and TYRTAEUS, and Laconian perioeci bronzesmiths producing a valued metalwork for export—in the 500s B.C. Sparta's crafts declined, even as the city thrived. A major reason for this was Sparta's decision not to adopt COINAGE, an invention that revolutionized Greek trade soon after 600 B.C. Instead of using coins, Sparta retained a primitive, awkward currency of IRON rods, thereby assuring itself, intentionally, of economic and cultural isolation.

The 500s B.C. saw Sparta dominate the Peloponnese, to become the foremost mainland Greek city. After an attempt to conquer ARCADIA met with defeat (circa 570 B.C.), Sparta changed its policy toward its Peloponnesian neighbors: instead of seeking to crush and helotize them (which was certain to inspire fierce resistance), Sparta began offering them a place in a Spartan-led alliance that modern scholars call the Peloponnesian League. This "league policy" is sometimes associated with the Spartan ephor CHILON, circa 556 B.C. Eventually the league embraced most Peloponnesian states except Argos, which remained Sparta's enemy but was badly defeated in wars circa 546 and 494 B.C. Sparta's staunchest allies in the league included TEGEA, CORINTH, and ELIS.

Sparta reached its peak under its most capable king, the first CLEOMENES (1), who reigned circa 520–490 B.C. and tried to organize all of Greece against the coming Persian invasion. During the PERSIAN WARS (490–479 B.C.), Sparta led the land-army defense of mainland Greece, alongside the Athenian-led defense at sea. The final victory brought Sparta immense prestige—which it soon lost to Athens.

The years 478–431 B.C. saw a steady worsening of Sparta's relations with Athens. Thucydides relates that the Spartans feared the growth of Athenian power—specifically Athens' naval empire organized through the DELIAN LEAGUE. But Sparta also feared the appeal and aggressive export of Athenian democracy, so dangerous to Sparta's oligarchic friends in its allied cities. Spartan-Athenian hostility culminated in the huge Peloponnesian War (431–404 B.C.). A hard-fought victory left Sparta dominant over an exhausted Greece.

Sparta had by now acquired a navy and an important foreign ally, the empire of PERSIA, but Sparta's turn as an imperial power was destined to last barely a generation. By harsh rule in Greece and adventuring in ASIA MINOR, Sparta managed to unite several of its former allies and enemies, including Persia, Corinth, Thebes, Athens, and Argos, in a war against itself. This CORINTHIAN WAR ended with Spartan hypocrisy exposed by the shameful terms of the KING'S PEACE (386 B.C.): far from seeking to liberate the Greek cities of Asia Minor from Persian rule as they claimed, the Spartans just wanted to retain power in Greece.

Growing anti-Spartan anger fueled the emergence of a challenger state, Thebes. The Spartan king AGESILAUS, despite personal virtues, proved blind to the danger posed by the Theban leader Epaminondas. At the Battle of LEUCTRA (371 B.C.) the age-old myth of Spartan invincibility was shattered: a Spartan army was destroyed and Sparta thrown forever on the defensive. Invading Laconia for the first time in its history, Epaminondas decided not to attack Sparta itself. But he proceeded to dismantle the Spartan empire, by freeing Messenia and Arcadia from Spartan domination (369–368 B.C.). Messenia's liberation in particular was devastating to Sparta, since it removed the city's source of surplus food. With these events, the Peloponnesian League more or less dissolved.

Sparta remained resistant to Thebes and then to the Macedonians, who conquered Greece in 338 B.C. While the Macedonian king ALEXANDER THE GREAT campaigned through distant Persia, the Spartan king Agis III led a few Peloponnesian states in revolt. But he was defeated and killed at the Battle of Megalopolis, at the hands of the Macedonian regent ANTIPATER (331 B.C.).

By the mid-200s B.C. Sparta was a relatively poor state, its citizen body shrunken to a mere 700 Spartiates, vastly outnumbered by perioeci and helots. In foreign policy, Sparta lay on the defensive against the Achaean League, which was now the dominant Peloponnesian power. This era saw the final significant moments of Spartan history. Hoping to revive the city's prior greatness, the young king Agis IV tried to abolish debts, redistribute land, and enroll perioeci as new citizens. Opposed by conservative Spartiates, Agis died violently (241 B.C.), but his revolutionary plan inspired a new king, Cleomenes III.

After a few years on the throne, Cleomenes discarded constitutional kingship and seized absolute power (226 B.C.). He abolished the ephorate, redistributed the land into 4,000 lots, and enrolled perioeci as full citizens. Next, seeking to recover Sparta's military supremacy in the Peloponnese, he waged war with the armies of the Achaean League. But when the Achaean League leader Aratus invited the powerful Macedonian king Antigonus III to intervene, Sparta's revival was doomed. At the Battle of Sellasia, just north of Sparta, Antigonus destroyed a Spartan army (222 B.C.). Cleomenes fled into exile, soon to die, while Antigonus seized Sparta and canceled Cleomenes' reforms.

Of the Spartan kings who came later, Nabis was a noteworthy adventurer during and after the Second Macedonian War (220–197 B.C.). By 146 B.C., however, Sparta and all of Greece had passed into Roman hands. During the first century A.D., under the Roman Empire, Sparta enjoyed a revival as a tourist attraction. Spectators were regaled by the sight of such age-old endurance contests as Spartan boys running the gauntlet of whips at the altar of Artemis Orthia.

(See also ACHAEA; AGIAD CLAN; CIMON; EURYPONTID CLAN; HOMOSEXUALITY; LYSANDER; MUSIC; TYRANTS; WARFARE, LAND.)

Speusippus   See ACADEMY; PLATO.

sport   The ancient Greeks of the 700s B.C. and later revered competitive sports, especially for men and boys. Sport was seen as a character-building facet of a boy's EDUCATION and as a way of demonstrating the gods' favor for individual competitors. The Greeks associated sport with RELIGION. Their major competitions took place at festivals in the gods' honor; the most famous of these were the OLYMPIC GAMES, sacred to the great god ZEUS.

The Greeks overwhelmingly favored individual contests over team sports. Typical events at public competitions included foot races (200 yards, 400 yards, long distance, and the race in armor), field events (long jump, discus throw, and javelin throw), combative events (WRESTLING, BOXING, and PANKRATION), and the horse races and chariot races.

WOMEN and girls were not completely barred from sports. For instance, SPARTA maintained a vigorous program of girls' gymnastics, and the Olympic Games included a girls' foot race (held outside the main sanctuary where the males competed). However, the Greeks' love of sport specifically involved a glorification of male strength and beauty; in most public events, males competed naked or nearly so, their bodies anointed with oil. Such display was considered inappropriate for Greek women. (In the few events open to them, female competitors wore tunics.)

Every Greek city-state had its own schedule of festivals, where local athletes might compete in honor of the city's patron gods. In every city, the GYMNASIUM was an important local institution, where the wealthier men might pass the day in socializing and political discussions, as well as sports practice. The Greeks' serious approach to sports was an aspect of Greek culture that attracted many non-Greek peoples—including the ETRUSCANS, Romans, and certain of the JEWS—who came into contact with the Greeks in the centuries before Christ.

(See also CHARIOTS; HOMOSEXUALITY; ISTHMIAN GAMES; NEMEAN GAMES; PENTATHLON; PINDAR; PYTHIAN GAMES.)

statues See SCULPTURE

Stesichorus Important Greek lyric poet, active in the first half of the 500s B.C. and usually associated with the Sicilian-Greek city of HIMERA. Stesichoros, meaning "chorus master," may have been a title; according to one story, the man's real name was Tisias. Prolific and inventive, Stesichoras was the first west Greek literary celebrity, and he strongly influenced later generations of Greek poets. He was known for his long narrative poems on mythological subjects, perhaps 26 in number. Of these, much has been lost; the remnants survive mainly in quotations by later writers and in some recovered ancient papyrus fragments, published in A.D. 1973, containing portions of Stesichorus' poem *Geryoneis* (about the mythical creature Geryon, slain by HERACLES).

Writing in the Doric Greek dialect, Stesichorus told tales inspired by HOMER's verses and by other poems of the epic cycle. In the *Wooden Horse*, the *Capture of Troy*, and *Homecomings*, he described the TROJAN WAR and its aftermath. His *Oresteia*—about the murder of AGAMEMNON and the vengeance taken by his son, ORESTES—almost certainly later influenced the *Oresteia* stage trilogy by the Athenian playwright AESCHYLUS.

Stesichorus was both clever and verbose. The *Geryoneis* ran at least 1,300 lines (probably much longer) and featured a sympathetic portrait of the three-bodied Geryon, who traditionally had been presented as a fierce monster.

It is not known how Stesichorus' poems were performed. He usually is called a choral poet, meaning that his work typically would have been sung or chanted by a trained chorus at a public event. However, his more recently discovered verses seem to lack certain hallmarks of choral song, leading certain scholars to believe that at least some of Stesichorus' poems were written for a solo singer accompanying himself on the lyre. Perhaps this performer was the poet himself.

Stesichorus is traditionally credited with inventing the "triad"—a three-stanza metrical grouping that became an essential element in lyric verse and Athenian stage drama. The triad consisted of the *strophē*, or "turn," the *antistrophē*, "counterturn," and the *epodos*, "after-song." This sequence was repeated throughout the poem. The strophe and antistrophe had to be metrically identical; the epode was different. Supposedly the names denote the chorus' motions in singing the parts—dancing rightward on the strophe, leftward on the antistrophe, and standing still for the epode.

One well-known story claims that Stesichorus wrote two different poems dealing with the mythical HELEN OF TROY. The first was a conventional treatment of Helen's adulterous elopement with the Trojan prince PARIS. The second poem, entitled the *Palinodia* (retraction), claimed that the first version was untrue: Helen had never left her Spartan husband, MENELAUS; rather, the gods had sent a phantom-Helen to TROY, so that the city's doom might be fulfilled.

The daring contrivance of this palinode was much admired in the ancient world and gave rise to a legend: Supposedly, after writing the first Helen poem, Stesichorus was struck blind. Only after publishing the retraction did the poet regain his sight.

(See also LYRIC POETRY; MUSIC.)

Stoicism School of Greek PHILOSOPHY founded at ATHENS *circa* 300 B.C. by Zeno, a Cypriot of Phoenician-Greek descent. With its ideal of a virtuous life impervious to misfortune, and its assurance of an ordered universe in which the individual person played a role, Stoicism addressed certain feelings of change and doubt that accompanied the start of the HELLENISTIC AGE.

During Hellenistic times (circa 300–150 B.C.), Stoicism remained a modest influence, about equal to its major rival philosophy, EPICUREANISM. But once introduced at the non-Greek city of ROME circa 144 B.C., Stoicism captured the Roman mentality (more so than the Greek) and grew into the major intellectual movement of the early Roman Empire. Unlike Epicureanism, Stoicism encouraged its followers to engage in the public life of government, and many ruling-class Romans were Stoics, such as the statesman Seneca the Younger (circa A.D. 60) and the emperor Marcus Aurelius (circa A.D. 170).

Stoicism's long history is divided by scholars into three phases, called the Early, Middle, and Late Stoa. The Middle Stoa, circa 144–30 B.C., saw Stoicism transplanted to Rome, where it flourished as the Late Stoa, circa 30 B.C.–mid-200s A.D. Although finally submerged by the spread of Christianity, Stoicism helped to shape the ethical outlook of early Christian thinkers. Most modern knowledge of Stoicism comes from extant Greek and Latin writings of the late period. The early Stoic writings survive only in brief quotations or summaries by later authors.

Zeno, the Stoic founder, immigrated to Athens as a young man and probably studied the teachings of the Athenian CYNICS. By about age 33 he had created his own philosophy, borrowing Cynic ethical theories of self-sufficiency along with select teachings of PLATO and others. Zeno presented his new ideas in public lectures and disputations in a part of the Athenian AGORA known as the Stoa Poikilē, or Painted Colonnade. This locale gave the new movement its name.

Zeno died around 261 B.C., bequeathing his school to a disciple named Cleanthes. Cleanthes' successor was Chrysippus (president 232–207 B.C.), a highly important thinker: His writings (not extant today) clarified Zeno's teachings and fused Stoic ethics, scientific theories, and theories of knowledge into a coherent, interconnected system.

Although most ancient Greeks worshipped many gods, the Stoics inclined toward belief in a single divinity. They pictured a universe imbued with a divine purpose or intelligence, variously referred to as reason (Greek: *logos*), fire, breath, or FATE. Every human being, they believed, contained a small version of the divine purpose, which is demonstrated in the human ability to reason, plan and give speech (*logos*).

This idea of a shared divine spark is background to one of Stoicism's most appealing teachings: the essential brotherhood of all people. The Stoic was taught to ignore distinctions of weath and birth and to see all humans, including WOMEN and SLAVES, as being spiritually the same material.

Stoicism recognized only two classes of people: the virtuous and the wicked. Like SOCRATES and Plato, the Stoics described virtue (Greek: *aretē*) as a form of knowledge or wisdom. Courage, moderation, and other virtues were evidence of this wisdom; greed, fear, sensuality were symptoms of ignorance. Wisdom produced virtue and virtue produced happiness; the greedy or dishonest person could not, by definition, be happy. In life, virtue was to be exercised, not in the cloistered setting that Epicureanism favored, but in areas of social responsibility: earning a living, raising a family, holding public office.

Contrary to some popular belief, the Stoics did not seek out pain. But they believed that pain, poverty, and death were not to be feared, since those ills could not harm a person's virtue. Thus the Stoic could find consolation in adversity: Happiness dwelt within, unaffected by external misfortune or external success. "Only the wise man is rich," a Stoic proverb ran. The Stoic tried to emulate the calm and grandeur of the universe, by accepting all events in life with a serene mind. This state of mind the Stoics called *apatheia*, "absence of passion."

Stoicism involved many other beliefs in fields such as logic and SCIENCE, but it was its noble, reassuring ethical system that made Stoicism so popular in later antiquity. Today we still use the adjectives "stoic" and "stoical" to denote an indifference to pain or misfortune.

Styx See AFTERLIFE.

Sybaris Affluent Greek city on the Gulf of Taranto, on the "instep" of the southern coast of ITALY. Founded by colonists from ACHAEA in around 720 B.C., Sybaris grew rich from its fertile farmland and its TRADE with the ETRUSCANS of the north. The city's reputation for luxury has produced the English word *sybarite*, meaning "someone excessively devoted to pleasure."

Sybaris enlarged its territory at the expense of local non-Greeks and founded Italian colonies of its own, including POSEIDONIA (Paestum). But rivalry with the nearby Greek city of CROTON led to a war, in which Sybaris was defeated and obliterated (510 B.C.). Nearly 70 years later, the Athenians established the colony Thurii near the site of Sybaris.

symposium Type of all-male, after-dinner drinking party. The symposium (Greek: *sumposion*, "drinking together"; plural: *sumposia*) played a vital role in ancient Greek aristocratic life. Only the rich could afford to host symposia; and the symposium's expense, its male homosexual or bisexual aspect, and its intellectual games were all part of a separate aristocratic identity.

The feasting of kings and their retainers goes back to prehistoric times. The poems of HOMER, written down circa 750 B.C. describe aristocratic warriors sitting at banquets. By the 600s B.C. the symposium had arisen, probably through Phoenician influence, as a new form of social gathering. Snacks such as sesame cakes might be served, but the symposium's central activity was the semiritualistic drinking of WINE. All the drinkers were men, of equal noble status. The servers and entertainers were typically young female and male SLAVES chosen for their beauty.

The symposium was the natural setting for the type of woman known as a *hetaira*, or courtesan; such women might be dancers or flute-girls, but they would also be obliged to give sexual favors. Similarly, according to ancient Greek aristocratic taste, the symposium would be the scene of various homosexual attractions, whether between guests or involving a guest and slave. One anecdote tells how the great Athenian playwright SOPHOCLES coaxed a kiss from a slave boy at a symposium (circa 440 B.C.). A surviving tomb painting from POSEIDONIA (circa 480 B.C.) shows a youthful guest warding off a caress from his couchmate, a slightly older man.

The symposium was governed by rules. The typical "symposiast" (reveler) did not sit but rather reclined on a couch, propped up on his left elbow, Phoenician style. There were usually between seven and 15 couches, with two men to each. One of the drinkers was appointed "king," or master of ceremonies, to decide on the sequence of activities.

Symposia had their own equipment, including many of the forms of Greek POTTERY that are known from ARCHAE-

A symposium scene, from an Athenian red-figure cup, circa 485 B.C. Tended by slave boys, the symposiasts recline on couches, drinking wine from the kind of shallow cup known as the kylix. The setting is a special room of the host's house, where walls are hung with wine pitchers and kylixes.

OLOGY. The men drank from cups such as the *kulix* (or kylix)—wide and shallow, for sipping on the recline. According to Greek custom, the wine was diluted with water, usually one part wine to two or three water. The mixture was prepared in a large bowl, the *kratēr*, and from there distributed by slaves to the drinkers' cups. During the festivities the wine-to-water ratio might be altered according to the wishes of the "king," but drinking undiluted wine was considered very unhealthy. Games included *kottabos* (in which wine dregs were flicked from the cup at a target) and competitive singing by individual symposiasts to flute accompaniment—the origin of the Greek verse form known as the elegy.

Undoubtedly events could grow wild as men got drunk—for example, more than one vase painting shows *hetairai* climbing onto couches to embrace guests. The party might end in a *komos*, a drunken torchlight procession in honor of DIONYSUS, the wine god.

But symposia were also the setting for philosophical and, especially, political discussion. The symposiasts were usually men of right-wing views, and the symposium was designed to foster a sense of exclusivity in an atmosphere of male bonding. At ATHENS and other democracies, such a strong minority could be subversive. The right-wing conspiracies that threatened democratic Athens in 457, 411, and 404 B.C. were surely hatched in gatherings such as symposia.

The symposium's dual intellectual-sexual nature is conveyed in idealized terms by the philosopher PLATO, in his dialogue titled the *Symposium*. In this fictional account, SOCRATES, ARISTOPHANES, ALCIBIADES, and other notable fifth-century-B.C. Athenians meet to drink and discuss the nature of sexual love.

(See also ALCAEUS; ANACREON; ARISTOCRACY; HOMOSEXUALITY; LYRIC POETRY; MUSIC; OLIGARCHY; PHOENICIA; POTTERY; PROSTITUTES.)

Syracuse Preeminent Greek city of SICILY, and one of the grandest and most violent cities of Greek history. Under a series of military dictators in the 400s–200s B.C., Syracuse led the Sicilian Greeks in their constant struggle against the Carthaginians, who occupied Sicily's western corner. Syracuse played a major role in the downfall of ATHENS during the PELOPONNESIAN WAR, when the Syracusans totally destroyed an Athenian invasion force (413 B.C.). By the first half of the 300s B.C., Syracuse vied with North African CARTHAGE for control of the western Mediterranean, before falling to the expansionism of ROME (211 B.C.).

Syracuse (now the Sicilian city of Siracusa) was one of the earliest Greek colonies. Located on the south part of Sicily's east coast, it was founded circa 733 B.C. by shipborne Corinthian settlers who subdued the local native Sicels. The site combined fertile coastal farmland with the best harbor of eastern Sicily—a double harbor, formed by the inshore island Ortygia ("quail island"). The Greeks fortified Ortygia and the mainland opposite, eventually linking them with a causeway and bridge.

The Syracuse colony was undoubtedly intended as a breadbasket for its mother city, CORINTH. Syracuse's surplus harvest of barley and wheat was probably offered first for sale to the Corinthians.

Like Corinth, Syracuse in the 700s–500s B.C. was governed as an ARISTOCRACY. The descendants of the first settlers became a ruling class known as the Gamoroi, "landholders," who lived in the city and whose farms occupied the best land of the plain. Lower on the social ladder were the descendants of later Greek arrivals, who might be craftsmen in the city or have smallholdings in the hills. Lowest of all were the Sicel-ethnic serfs, who tilled the land for their Greek masters (and whose condition resembled that of the Spartan HELOTS). These distinctive social layers created class hatreds, which—when added to the external menace of the Carthaginians and native Sicilian tribes—were destined to have important political consequences. By the 400s B.C. Syracuse saw TYRANTS (*turannoi*, "dictators") arising, as champions of the common people and as promisers of military security.

The first tyrant was GELON; ruler of the nearby Greek city of GELA, who seized Syracuse under pretext of aiding the beleaguered Gamoroi (485 B.C.). Gelon transferred his capital to Syracuse and from there ruled as the most powerful individual in the Greek world. He led Greek Sicily to victory against the Carthaginians at the Battle of HIMERA (480 B.C.), and he extended the Syracusan empire westward, subjugating Carthaginians, Greeks, and natives across two-thirds of Sicily.

Gelon was succeeded by his brother HIERON (1), who ruled 478–467 B.C. and extended Syracusan power to Greek ITALY with his defeat of the Etruscans at the sea battle of CUMAE (474 B.C.). Under Hieron, Syracuse became a cultural capital, attracting literary artists of old Greece, such as AESCHYLUS, PINDAR, SIMONIDES, and BACCHYLIDES. The Greek amphitheater, ancient Syracuse's best-known surviving remnant, was first built under Hieron, quite possibly for the performance of Aeschylus' play *Women of Etna* (476 B.C.). A Sicilian Greek playwright Epicharmus, probably a Syracusan, is said to have pioneered the writing of stage comedy in the early 400s B.C.

After Hieron's death, the tyranny was overthrown (466 B.C.). A Syracusan DEMOCRACY, based on the Athenian model, lasted for about 60 years. During this time Syracuse remained the foremost Dorian-Greek city of Sicily, and was hostile to such local Ionian-Greek cities as Leontini. Eventually Syracuse's enemies appealed to the great city of Athens for help. The ambitious Athenians dreamed of capturing Syracuse's wheatfields and other riches, and in 415 B.C. they dispatched an ill-fated armada that eventually totaled perhaps 50,000 troops and 173 warships. The attackers besieged Syracuse but, after two years' fighting, were annihilated. The architects of this Syracusan triumph were the Spartan general Gylippus and the Syracusan statesman-general Hermocrates.

In 410 B.C. a new threat emerged, as the Carthaginians began a series of eastward campaigns in Sicily. Following so close after the Athenian attack, this crisis threw the Syracusan democracy into turmoil, and circa 405 B.C. there emerged a new dictator, DIONYSIUS (1), a military officer and former follower of Hermocrates (since killed in civil war). Seizing power, Dionysius married Hermocrates' daughter and embarked on a war that rolled the Carthaginians back from the very walls of Syracuse to Sicily's western corner. The ensuing peace (397 B.C.) left Greek Sicily securely under Syracusan protection and rule.

The prosperous years until Dionysius' death (367 B.C.) saw Syracuse as the magnificent capital of the western Greeks. Dionysius made Ortygia island his private citadel and fortified Syracuse's inland heights, called the Epipolae ("above the city"); still standing are the remains of one of his forts, now called the Castle of Euryalus. Like Hieron, Dionysius sought to attract poets and thinkers to his court. In 387 B.C. the Athenian philosopher PLATO made the first of his ill-fated visits to Syracuse. (The philosopher was unimpressed, complaining that in Sicily everyone overate twice a day and never slept alone.)

Dionysius' son and successor, Dionysius II, lacked his father's ability to rule successfully, and was ousted after 10 years. Greek Sicily relapsed into civil strife, and Syracuse's power declined.

The reorganization of Greek Sicily by the Corinthian commander TIMOLEON (345–circa 340 B.C.) left Syracuse with a moderate OLIGARCHY on the Corinthian pattern. Yet, once again, a dictator emerged: AGATHOCLES, who ruled 317–287 B.C. The hectic events of his reign saw Syracuse again besieged by the Carthaginians and again leading the defense of Greek Sicily.

After Agathocles' death, a new autocrat emerged from the ranks of military officers: Hieron II, who in 269 B.C. declared himself king and whose 54-year reign brought Syracuse to its final peak of glory. Hieron wisely made an alliance with the emerging power of Rome.

Under him, Syracuse was one of the three great Greek cities of the HELLENISTIC AGE, along with ALEXANDRIA (1) and ANTIOCH. Syracuse's most famous citizen of this era was the mathematician-scientist ARCHIMEDES.

Syracuse's downfall came amid the conflict between Rome and Carthage. After Hieron's death (215 B.C.), his successor made a foolhardy alliance with Carthage that soon brought Syracuse under attack by Roman forces. From 213–211 B.C. the city's huge fortifications withstood the Roman siege: According to legend, the aged Archimedes invented a giant glass lens, designed to focus sun rays on Roman warships in the harbor and set them afire. But eventually the city's outer wall fell to night assault and its inner wall to treason. The Romans captured Syracuse amid slaughter and looting. Among the booty shipped back to Rome were statues and other art objects that spurred the Romans' fascination with Greek culture. Later, Syracuse became the chief city of Roman Sicily and eventually lost its Greek identity.

(See also THEATER.)

T

talent Large monetary unit, equivalent to a certain weight (Greek: *talanton*) of SILVER, usually about 58 pounds. The talent measure was in use well before the beginnings of Greek COINAGE (circa 595 B.C.), and no talent coin existed. In ATHENS in the 400s to 300s B.C., a talent was considered equivalent to 6,000 of the silver coins known as drachmae.

Tantalus In HOMER's epic poem the *Odyssey*, the hero ODYSSEUS visits the Underworld (book 11), and sees three great sinners there who are suffering eternal punishment for having betrayed the friendship of the gods. The sinners are the giant Tityus and the mortals SISYPHUS and Tantalus. Described as an old man, Tantalus is tortured by unrelieved thirst and hunger. He stands in a cool pond up to his neck, with the boughs of fruit trees hanging above his head; but when he lowers his head to drink, the water drains away instantly, and when he reaches up for fruit, the boughs are lifted away by the wind. Tantalus' fate has given rise to our English word *tantalize*, meaning "to cause torment by showing but withholding something desirable." In Greek the name Tantalus may have meant "sufferer."

Like certain other characters in Greek MYTH, Tantalus was described as a non-Greek; he was a wealthy king in ASIA MINOR and a son of the great god ZEUS. But Tantalus' good fortune brought on insane arrogance (HUBRIS). To test the omniscience of the gods, he invited them to a banquet, then he killed his own young son, PELOPS, and cooked him as the dish. But only the goddess DEMETER—distracted by grief for her lost daughter, PERSEPHONE—ate a few bites, from the shoulder. Zeus and the other gods detected the trick and brought Pelops back to life, replacing his shoulder's missing part with ivory. Tantalus was sent off to an eternal pain that mirrors his crime somewhat. Through Pelops, Tantalus was the ancestor of the mythical kings who eventually ruled MYCENAE, in Greece. This family of the Pelopidae included ATREUS, AGAMEMNON, and ORESTES; the misfortunes of this royal house can be understood partly as originating with Tantalus' crime.

(See also AFTERLIFE.)

Taras Greek city founded by Spartan colonists in the late 700s B.C. on the "instep" of the southern coast of ITALY. Often known by its Latin name, Tarentum, this city is now the Italian seaport of Taranto, located beside the Gulf of Taranto.

As the only colony established by SPARTA, Taras maintained a Spartan form of government until switching to DEMOCRACY, circa 475 B.C. Taras thrived from FARMING, manufacturing (including POTTERY and jewelry), and Adriatic TRADE, despite the hostility of local Italian tribes. By the mid-400s B.C. it was the foremost Greek city of south Italy. The city's peak came in the 300s B.C., when it was producing such luminaries as the philosopher-engineer Archytas and the philospher–musical theorist Aristoxenus. But by midcentury Taras was being threatened by its Italian enemies—the nearby Lucanii and the more distant, expansionist city of ROME.

For help, Taras appealed to other Greek states. The last and best known of these episodes came in 280 B.C., when the dynamic King PYRRHUS of EPIRUS arrived with his army to fight the Romans. After early victories gave way to defeat, Pyrrhus withdrew.

Taras was captured by the Romans and became a Roman subject state (272 B.C.). Later it was an important port city of Roman Italy.

Tarentum See TARAS.

Tartarus See AFTERLIFE.

Tegea City in the southeastern part of the plain of ARCADIA, lying along the main route between SPARTA, to the south, and ARGOS, to the northeast. Tegea was one of the two original Arcadian cities. (The other was MANTINEA, Tegea's rival.)

Tegea arose circa 600 B.C. from an amalgamation of local villages, to resist Spartan aggression. Renowned as fighters, the Tegeans defeated a Spartan invasion circa 585 B.C., but in the mid-500s B.C. they made peace and became staunch Spartan allies. Tegean troops had the traditional privilege of occupying the left wing of the battle line in a Spartan-led army.

Tegean troops played a prominent role alongside the Spartans in the PELOPONNESIAN WAR (431–404 B.C.). But after Sparta's defeat by THEBES in 371 B.C., Tegea turned against the crippled Sparta and fought on the Theban side at the Battle of Mantinea (362 B.C.). Later, however, enmity with Mantinea brought Tegea back to a treaty with Sparta.

The glory of ancient Tegea was its temple of the goddess ATHENA. Destroyed by fire in 390 B.C., this shrine was rebuilt as the largest and grandest temple in southern Greece. The site, three miles outside Tegea, was excavated by French archaeologists in the early 1900s A.D.; the later temple's foundations and column fragments are visible today.

Teiresias See TIRESIAS.

Telemachus In MYTH, the son of King ODYSSEUS and PENELOPE. Telemachus' name, "distant battle," refers to his father's part in the TROJAN WAR, which began just when the boy was born. A subplot in HOMER's epic poem the *Odyssey* traces Telemachus' passage to maturity. At the *Odyssey's* opening, Telemachus, about 20 years old, is a dutiful but timid prince, overawed by the confident nobles who crowd his absent father's palace at ITHACA and compete for Penelope's hand in MARRIAGE. Unable to make them leave, Telemachus voyages by ship to the Peloponnesian courts of NESTOR and MENELAUS, to ask about the whereabouts of the father he has not seen since he was a baby. With the goddess ATHENA's help, Telemachus sails back to Ithaca (evading an ambush by the suitors en route) and encounters his father, who has returned in disguise. By now the boy is showing considerable confidence and initiative. In the climactic scene, where Odysseus slaughters the suitors (book 22), Telemachus fights bravely alongside his father, sustaining a wrist wound.

Other legends add a few details. Supposedly Telemachus later married his father's former paramour, CIRCE.

Teos See IONIA.

Terpander Lyric poet of the mid-600s B.C. Terpander came from the island of LESBOS but composed and performed at SPARTA. His name means "delight of man," and legend claims that his poetry soothed the Spartans at a time of civil crisis. Only a few lines ascribed to Terpander survive, and even these were probably not really written by him. He was apparently an important early figure in creating choral poetry as an art form and in making Sparta a center for that art. Almost certainly, he influenced the younger Spartan choral poet ALCMAN.

Terpander may have pioneered certain techniques in playing the lyre. However, the story crediting him with inventing the seven-stringed lyre (in place of four strings) is not true, as the seven-stringed type dates back to the second millennium B.C.

(See also LYRIC POETRY; MUSIC.)

Thales The first Greek thinker to try to explain the universe in nonreligious, rationalistic terms. As such, Thales can be considered both the first scientist and the first philosopher in the West. He lived between about 610 and 540 B.C. in MILETUS, in the flourishing Greek region called IONIA, in ASIA MINOR. Although he left no writings, he is best remembered for his theory that water is the fundamental constituent of all matter.

Thales and two of his followers, ANAXIMANDER and ANAXIMENES, comprise the first known philosophical movement: the Milesian School of natural philosophers. These men used physical observations to inquire about a primary substance in the universe. The significance of such thinkers is that they were the first to look for answers without reference to the traditional myths of the gods. Later generations considered Thales to be one of the SEVEN SAGES of Greece. He was famous in his lifetime, and took part in great events. In around 545 B.C., when Ionia was under attack by the Persian king CYRUS (1), Thales proposed the creation of a federated union of Ionian cities.

He established—probably on the basis of Babylonian or Egyptian learning—the studies of ASTRONOMY and geometry among the Greeks; he was believed to be the first Greek to accurately predict the year of a solar eclipse (probably 585 B.C.). Like other Greek wise men, he is said to have traveled to EGYPT, where he supposedly calculated a pyramid's height by measuring its shadow simultaneously with that of his walking stick and calculating the ratios.

He was of partial Phoenician descent, and Near Eastern influences may lurk behind his emphasis on water and his theory that the earth floats like a log in a cosmic watery expanse. (This resembles the Semitic myth of watery creation in the biblical book of Genesis.) But in Thales' mind such ideas became scientific.

(See also MATHEMATICS; PHILOSOPHY; PHOENICIA; SCIENCE.)

Thasos Wealthy island of the northern AEGEAN SEA, lying six miles offshore of the non-Greek mainland of THRACE. Thasos is a round island, about 13 miles in diameter, rising to a central peak. The Greek poet ARCHILOCHUS sourly described it as sticking up like a donkey's back, topped with forests (circa 670 B.C.). The main city was a seaport, also named Thasos (site of the modern Greek town of Limen); it lay at the north of the island, opposite the mainland.

As one of the very few Aegean sites to offer mineral deposits of GOLD, Thasos attracted Greek colonists at an early date, circa 700 B.C. These colonists—of whom Archilochus eventually was one—came from the island of Paros in the CYCLADES. Although Thasos' mines were soon exhausted, the island served as a base from which to pursue mining and panning on the nearby Thracian mainland. The most important mainland prospecting area lay on the eastern slopes of Mt. Pangaeus, at a site called Skaptē Hulē (dug-out wood), about 25 miles due northwest of Thasos.

But the Pangaeus fields existed amid a hostile native people—the warlike Thracians. Much of Archilochus' surviving poetry describes fierce fighting against the "mophaired men of Thrace," and what these verses seem to portray is Greek military protection for miners and for convoys of raw gold and SILVER, moving from the slopes of Pangaeus to the safe anchorage at Thasos.

The island grew rich from its exports of precious ore, TIMBER, local marble, and, probably, SLAVES acquired as Thracian war captives.

Thasos had a renowned school of SCULPTURE in the 600s and 500s B.C. and was the birthplace of the painter Polygnotus (circa 475 B.C.).

The island surrendered to the advancing Persians under King DARIUS (1) in 491 B.C. After the Persian defeat in 479 B.C., Thasos joined the Athenian-controlled DELIAN LEAGUE, and, like most other Delian allies, reorganized its government as a DEMOCRACY.

But in 465 B.C. Thasos revolted from the Delian League, over Athenian interference in its crucial mining operations on the mainland. An Athenian war fleet defeated the Thasians in a sea battle, and besieged and captured the capital city (465–463 B.C.). Relatedly, the Athenians began colonizing the west foot of Mt. Pangaeus, at the site later known as AMPHIPOLIS.

The Athenians apparently now took over Thasos' mainland mines, at least temporarily. We know that in 454 B.C. Thasos was paying only three TALENTS of Delian tribute a year—a tiny assessment for that rich state, unless it was no longer rich. But by 446 B.C. the Thasian tribute had jumped to 30 talents a year, which may signify the Thasians' recovery of their mainland possessions.

In 411 B.C., amid the widespread Delian revolt against a beleaguered ATHENS in the PELOPONNESIAN WAR, Thasos rebelled again and went over to the Spartan side. But the common people resisted this defection, and the next years saw civil war on Thasos, followed by recapture by the Athenians (407 B.C.). In 404 B.C., at the war's end, the exhausted island's Athenian loyalists were massacred by the Spartan general LYSANDER.

In 389 B.C. the island joined the SECOND ATHENIAN LEAGUE, of which it remained a faithful member. Thasos was seized by the Macedonian king PHILIP II in 340 B.C.; by about 300 B.C. he and his successors had depleted the Pangaeus mines. Thasos remained a part of the Macedonian kingdom until 196 B.C., when it was freed by the Romans in their liberation of Greece.

As befits a rich capital, Thasos city boasted a number of grand buildings. Among the remnants visible today are parts of the AGORA and of a temple of HERACLES (400s B.C.). Most impressive is the partly surviving ancient city wall, dating back to the Thasians' preparation for revolt in 411 B.C.

(See also PAINTING; PERSIAN WARS; ROME.)

theater   Many cultures have produced some original form of public performance that might be called drama or theater. It was the ancient Greeks, specifically the Athenians of the 400s B.C., who perfected the two genres known as tragedy and comedy, which defined European theater for centuries following. Shakespeare (circa A.D. 1600) wrote tragedies and comedies because those were playwriting's two principal forms, inherited from the Greeks.

Greek tragedy drew on tales from MYTH. This meant that in most cases the entire audience would be familiar with the story's plot beforehand (similar to an audience at a modern Easter pageant). The playwright's skill would lie in shaping the material so as to communicate a particular world view or character portrait.

Most Greek tragedy presents the downfall of a hero or other lofty protagonist. In the more sophisticated tragedies, this downfall is shown to be due to HUBRIS (arrogance): Having attracted the anger of the gods, the hero is destroyed by a disastrously arrogant decision of his own. The clearest example of this pattern is in AESCHYLUS' *Agamemnon*, where the hero displays his hubris by walking atop a priceless tapestry.

The philosopher ARISTOTLE, in a famous passage in his literary-critical *Poetics*, explains that tragedy's artistic goal is to arouse the audience's emotions of pity and fear in a way that purges these feelings and provides relief; this purging is called *katharsis*. Greek tragedy also strives to examine the nature—or absence—of divine justice, and the role of humankind in the universe.

Comedy, on the other hand, was always meant to be funny and riotous, with characters less noble and compli-

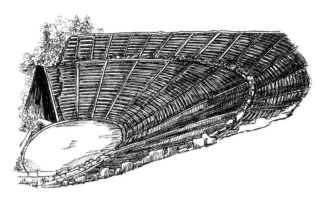

The theater at Epidaurus, in the Peloponnese, as it appears today. Built into the side of a hill, this structure dates from the mid-300s B.C. and is the best preserved of any ancient Greek theater, with excellent acoustics. Its 55 rows of limestone benches could seat 12,000. Actors and chorus would perform mainly on the 66-foot-wide, open disk of beaten earth, called the *orchēstra* (dancing floor). Behind the orchestra, at far left, there would have been a stage building, or *skēnē*, which provided storage, changing rooms, and a means of backdrop display. Also, characters such as gods could make an entrance by being winched down from the skene roof.

cated than those in tragedy or even real life. ARISTOPHANES' 11 surviving comedies, dating between 425 and 388 B.C., are farcical, fantastical works that celebrate the city's political life, while mocking specific individuals and institutions.

Ancient Greek theater was a form of poetry, with all dialogue being spoken or sung in verse. Theater was also mass culture, intended for an audience that included several social classes, and it thrived in democracies such as ATHENS (although it is unclear whether WOMEN, even female citizens, were allowed to attend). Because the theater audiences comprised much the same constituency as the democratic ASSEMBLY, Greek drama often conveyed strong political messages.

Tragedy apparently was invented at Athens in the 500s B.C. (although the city of SICYON claimed this honor, too). Comedy was developed at various Greek states, including SYRACUSE and other cities of SICILY, before being instituted most grandly at Athens. Both genres probably emerged from the public performances of choral singers at religious festivals in honor of DIONYSUS, the god of WINE and fertility, who was a favorite deity of the masses. Greek comedy, with its farce and obscenity, seems clearly derived from the riotous Dionysian procession known as the *komos*, held at many Greek states; *komoidia* means "song of the *komos*." Vase paintings show the *komos* as a parade of masked men costumed as SATYRS and carrying a log carved as a huge penis. The *komos* was probably an occasion for rude jokes exchanged with onlookers; out of this interplay there may have developed the *parabasis* (a direct choral address to the audience that was a hallmark of Athenian stage comedy). The belief that Dionysus could spiritually possess his worshippers probably led to individual speaking or singing roles in the god's character.

The origins of tragedy are less clear than those of comedy. According to Aristotle's *Poetics*, tragedy emerged from the narrative choral song known as the dithyramb. Like the *komos*, the dithyramb was performed at certain festivals of Dionysus, but was more solemn and rehearsed than the *komos*. The meaning of the word *dithyramb* is unclear now, but it is known, from extant dithyrambs written by the poet BACCHYLIDES (mid-400s B.C.), that these songs told tales from Greek myth, not necessarily relating to Dionysus. The chorus would sing or chant the story and probably dance interpretively.

Assuming that what Aristotle said was accurate, there must have come a time at Athens when certain types of performance broke away from the rules governing the dithyramb's poetic meter, size of chorus, and so on. Most significantly, individual performers began stepping out of the chorus, to sing or speak roles in the character of a mythical figure in the story. This step is traditionally credited to a shadowy genius named Thespis (circa 535 B.C.), who also may have introduced the single actor's wearing of a mask to signify character. Thespis' actor apparently delivered set speeches in between the chorus' presentations; the actor may have changed masks to assume multiple roles.

This new Athenian art form was called *tragoidia*, "goat song," perhaps referring to its presentation at festivals of Dionysus. (The sacrifice of goats was part of the god's cult.) By about 534 B.C., under the enlightened tyrant PISISTRATUS, tragedy performance was installed officially at the important Athenian early springtime festival called the City Dionysia. By 500 B.C. this festival had developed into the annual, publicly funded, three-way competition that was the famous occasion for all performances of new tragedy. Performances were held in a theater—first of wood or earth, later of marble—at the south base of the Athenian ACROPOLIS.

The Dionysia contest typically presented three authors' new work—in each case, a tragic trilogy and a satyr play (*saturikon*), for comic relief. The satyr play, an offshoot of comedy, featured a clownish chorus always represented as satyrs. In a typical plot, this gang would "wander into" one of the great myths, such as that of PERSEUS (1). Today only one complete satyr play survives: the *Cyclops* of EURIPIDES (circa 410 B.C.).

The three groups of plays performed at the Dionysia would have been selected previously, in written form, by an official who judged the applicants. Qualifying entrants were assigned a paying sponsor (*chorēgos*), who paid part of the cost out of pocket as a form of state service; the balance came directly out of state funds. At the festival a panel of judges, consisting of ordinary citizens chosen by lottery, awarded first, second, and third prize to the three playwrights. Sophocles is said to have won first prize 24 times, second prize seven times, and never third.

Comedy, less esteemed than tragedy, was installed as a competition at the City Dionysia only around 488 B.C. The entries were single plays, one by each playwright. By 440 B.C. a second contest had been introduced, at the Lenaea (a Dionysian festival in midwinter). Normally five comedies competed. The two earliest comedy-writers about whom any information exists are CRATINUS and Crates (mid-400s B.C.).

Comedy employed a chorus of 24 members; tragedy used 12 at first, later 15. Comic choruses tended to be more important to the play's action and message than were their tragic counterparts—at least by the time of Aeschylus (525–456 B.C.), who was remembered for having reduced the tragic chorus' speaking role. All Athenian actors and stage choruses were male, although the roles might be female.

The first Athenian tragedian about whom any information exists is PHRYNICHUS (circa 540–475 B.C.). His tragedies may have been a form of costumed oratorio, with a simple narrative. The story would unfold around a single speaking actor (the *protagonistēs*), who might assume multiple roles by changing masks and who would deliver his speeches in alternation with the chorus.

Stagecraft, plotting, and characterization were improved by Aeschylus, whose *Persians* (472 B.C.) is the earliest extant tragedy. Emphasizing individual roles over the chorus, Aeschylus introduced a second speaking actor, which allowed for better-developed conflicts onstage and which greatly increased the number of available roles. Sophocles (circa 496–406 B.C.) introduced a third speaking actor, and he emphasized characterizations of his protagonists that tended to remove the chorus from the plot. His younger rival Euripides (circa 485–406 B.C.) produced innovations in plotting and in characterization of mythical characters that often seem intended to disturb the audience. These three tragedians—along with a fourth, AGATHON (late 400s B.C.)—were recognized in their own day as being the greatest practitioners of the art.

With the deaths of Euripides and Sophocles (both in 406 B.C.) and Athens' defeat in the PELOPONNESIAN WAR (404 B.C.), the great age of Athenian theater had passed. New tragedies were still performed—the Syracusan tyrant DIONYSIUS (1) wrote one that won a single-play prize at Athens in 367 B.C.—but in the absence of steady new talent, a tradition arose of restaging the plays of the three classic tragedians.

Comedy at Athens enjoyed a resurgence with MENANDER, Philemon, and other writers of "New Comedy," in the late 300s B.C. This socially satirical comedy of manners—so different from the directly political invective of Aristophanes or EUPOLIS—went on to influence the work of the Roman comedy-writers Plautus and Terence (100s B.C.).

Outside Athens, monumental theater buildings had arisen before circa 470 B.C., when the Syracusan tyrant HIERON (1) constructed a theater for the production of a tragedy by the visiting Aeschylus. By the HELLENISTIC AGE (300–150 B.C.), every major Greek city had a marble theater, seating as many as 24,000, where performances might consist largely of restaged plays of the three great Athenian tragedians. Today admirable ancient theaters have been reconstructed at EPIDAURUS, DODONA, PERGAMUM, and EPHESUS, among other sites.

Thebes   Major city of central Greece. Set in the eastern plain of BOEOTIA, seven-gated Thebes (Greek: Thēbai) enjoyed wide farmlands and control of overland routes both north and south. Today the ancient city's central fortress, the Cadmea (named for Thebes' legendary founder, King CADMUS) lies buried directly under the modern Greek town of Thivai.

Thebes was known for its first-rate army, which included a strong CAVALRY arm, and throughout its history the city usually dominated all of Boeotia. In the 300s B.C. Thebes became the foremost Greek power for a brief time, after defeating and displacing its former ally SPARTA.

Along with its neighbor and rival ORCHOMENUS, Thebes seems to have been preeminent in the earlier centuries of MYCENAEAN CIVILIZATION, circa 1600–1350 B.C. Archaeological excavations at Thebes, hampered by the site's modern town, have yielded portions of a Mycenaean palace as well as LINEAR B tablets and Near Eastern seal stones (which may add credence to the legend that King Cadmus originally came from the Levant). Thebes' importance in this era is reflected in the city's large role in Greek MYTH. The ruling caste's fortunes and misfortunes are recounted in the tales of HERACLES, OEDIPUS, and the SEVEN AGAINST THEBES. The legend of the Seven—with its sequel, the tale of the Epigoni—possibly commemorates Thebes' downfall at the hands of the rival Greek kingdom of MYCENAE, circa 1350 B.C.

By the 500s B.C. the tales of Thebes had been organized into three epic poems, the *Oedipodia*, the *Thebaïd*, and the *Epigoni*, collectively known as the Theban Cycle. These stories later became favorite material for the Athenian tragic playwrights.

In historical times Thebes' hegemony over the other Boeotian cities took the form of a federation—the Boeotian League, wherein Thebes supplied two or more of the 11 delegates, and every other represented state supplied one. Thebes shared an ill-defined, mountainous frontier with ATHENS, which lay about 40 miles southeast. The two cities were enemies in the late 500s–early 300s B.C., due largely to Athens' alliance with Thebes' defiant neighbor PLATAEA. The creation of the Athenian DEMOCRACY (508 B.C.) further alarmed the Thebans, whose government was an OLIGARCHY.

During the Persian king XERXES' invasion of Greece (480–479 B.C.), Thebes was the major Greek city to submit to the Persians, and it contributed soldiers and a base of operations for the Persian war effort. After the Persian defeat, Thebes fell into deep discredit. Stripped of its Boeotian hegemony by the vindictive Spartans, the city remained a minor power until the mid-400s B.C., when the Spartans revived it as an ally against their common enemy, Athens. As a staunch Spartan ally, Thebes again became the chief city of Boeotia. The Thebans' attack on Plataea precipitated the PELOPONNESIAN WAR (431–404 B.C.). In 424 B.C. the Thebans defeated an Athenian invasion at the Battle of Delium.

Victory in the war added to Thebes' prestige and territory, gained at the expense of its small Boeotian neighbors. But, alienated by Spartan arrogance, Thebes soon made alliance with Athens, ARGOS, and CORINTH against Sparta, in the CORINTHIAN WAR (394–386 B.C.).

The following decades saw the acme of Theban power, under the city's greatest statesman and general, EPAMINONDAS. He led his countrymen against the Spartans—supposedly the greatest soldiers in Greece—and defeated them at the Battle of LEUCTRA (371 B.C.). The Thebans then dismantled Sparta's empire, marching into the PELOPONNESE and freeing MESSENIA and ARCADIA from Spartan control. The Theban army of these years was distinguished by the Sacred Band (Hieros Lochos), an elite corps of 300 HOPLITES consisting of paired male lovers. Classical Thebes

in general was known for its military male-homosexual society.

In the mid-300s B.C. Thebes was confronted with the emerging power of MACEDON. Initially siding with the Macedonian king PHILIP II, Thebes switched sides to join the Athenian-led alliance that opposed Philip. After taking part in the disastrous Greek defeat at CHAERONEA (338 B.C.), Thebes was stripped of power and garrisoned by Philip's vengeful troops. A Theban revolt after Philip's death brought the city to complete destruction at the hands of Philip's successor, ALEXANDER THE GREAT: 6,000 Thebans were supposedly killed and 30,000 sold as SLAVES (335 B.C.). Thebes was rebuilt in 316 B.C. by the Macedonian ruler CASSANDER but was never again a great power. By the first century B.C. it was little more than a village.

Greek Thebes had nothing to do with the magnificent ancient Egyptian city called Thebes (modern Luxor and Karnak). Thēbai was a hellenized name, given to the Egyptian site by Greek visitors. One ancient Egyptian name for the city was Apet; to the Greek ear, this apparently sounded enough like Thēbai for the two names to become assimilated.

(See also ANTIGONE; DIONYSUS; EPIC POETRY; HOMOSEXUALITY; PERSIAN WARS; WARFARE, LAND.)

Themistocles   Athenian statesman and general who lived circa 528–463 B.C. In the history of ATHENS, Themistocles is second in importance only to PERICLES (who was his political successor). By urging the Athenians to create a powerful navy in the early 480s B.C., Themistocles provided his city and all of Greece with an effective defense against invasion in the PERSIAN WARS (480 B.C.). This Athenian navy then became the means by which Athens acquired and held its sea empire of 479–404 B.C. In domestic politics, Themistocles was a radical left-winger who helped broaden the base of the Athenian DEMOCRACY, in opposition to the rich and noble-born. He was also an inveterate enemy of SPARTA, foreseeing, as few Athenians did in the 470s B.C., that Sparta would be Athens' next great foe.

Themistocles remained a controversial figure long after his death. His deeds are described by the historian HERODOTUS (circa 435 B.C.), by the biographer PLUTARCH (circa A.D. 100), and briefly by the historian THUCYDIDES (1) (circa 410 B.C.). Herodotus and Plutarch show signs of having used anti-Themistoclean source material, which condemns the man for his greed and flawed patriotism. But Thucydides praises Themistocles' foresight and decisiveness, summarizing him with these words: "With his inborn genius and speed of action, he was the best for doing at a moment's notice exactly what the emergency called for" (book 1, 103).

Themistocles' father Neocles was of an aristocratic Athenian family. Themistocles' mother is sometimes described as a non-Athenian—which, if true, would have placed the future statesman in a relatively humble social category. Themistocles rose quickly as a left-wing politician in the newly created democracy, being elected chief ARCHON at about age 35 and organizing the development of PIRAEUS as Athens' harbor and naval base (493 B.C.).

The years after the Battle of MARATHON (490 B.C.) saw political turmoil at Athens, with at least five prominent

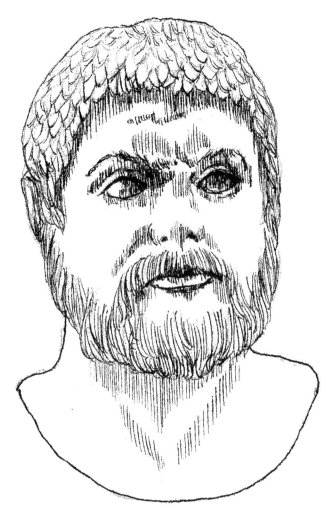

The hero of the Persian Wars. This portrait of Themistocles comes from a marble bust, titled with his name and found at Ostia, near Rome. The likeness is thought to be based on an original Greek portrait statue of circa 460 B.C. The figure's blunt, pugnacious features recall the fact that, however brilliant he may have been, Themistocles was basically a soldier and left-wing politician. In this face can be seen the hearty, outspoken man of the people.

of staff. (He was not, however, the top-ranking Greek admiral; that honor went to a Spartan.) To Themistocles can be ascribed the strategy of confronting the southward-sailing Persian fleet inside certain narrow channels, where the invaders' superior numbers and seamanship would be canceled out. After moderate success at the sea battle of ARTEMISIUM, this strategy resulted in a total Greek victory at SALAMIS (1), inside a mile-wide channel not far from Athens, specifically chosen by Themistocles long before the event.

Even before the war was over, Themistocles foresaw that Sparta would be Athens' next enemy. Although the Spartans respected him, Themistocles began to steer his city against them. The walls of Athens, destroyed by the Persians, were rebuilt despite Spartan disapproval (479 B.C.), and alliances were formed with ARGOS and THESSALY, two enemies of Sparta.

But Themistocles' downfall soon followed. The creation of the DELIAN LEAGUE and the naval counteroffensive against the Persians brought new Athenian soldier-politicians to the fore, particularly the young, conservative, pro-Spartan CIMON. Circa 471 B.C. the Athenians voted to ostracize Themistocles, who was by then nearly 60. He spent the next years traveling around the PELOPONNESE, where his activities seem connected with the establishment of anti-Spartan democracies at Argos, at MANTINEA, and elsewhere. The alarmed Spartans appealed to Athens, claiming to have proof that Themistocles was conducting treasonous intrigues with the Persian king. Indicted by the Athenians for treason, Themistocles fled Greece for ASIA MINOR (circa 467 B.C.). Eventually making his way to the Persian court at Susa, he became a valued adviser to King Artaxerxes I (464 B.C.). For about the last year of his life, the hero of Salamis served as a local Persian governor at the city of Magnesia-on-the-Maeander, in Asia Minor. He died, probably of natural causes, at age 65.

Themistocles was married twice, to Athenian women; his second wife, Archippe, accompanied or followed him to Asia with the youngest of their children. Two of Themistocles' daughters had the unusual names Italia and Sybaris, which may indicate yet another of his foreign-policy visions—namely, the extension of Athenian influence westward to the Greek cities of southern ITALY.

(See also AESCHYLUS; ARISTIDES; ATTICA; SYBARIS; WARFARE, NAVAL; XANTHIPPUS.)

politicians banished by OSTRACISM—but not Themistocles, who emerged as the city's leading statesman. It was then he made his great contribution to Athens' future. In 483 B.C. the publicly owned SILVER mines at Laurium, outside Athens, produced a bonanza yield. Although it was customary for such surpluses to be distributed as cash to each citizen, Themistocles convinced the Athenian ASSEMBLY to use the revenue to build new warships. His ostensible reason was Athens' current war against the nearby Greek state of AEGINA, but his real reason seems to have been his expectation of the Persian invasion. Within three years Athens had acquired the biggest navy in the Greek world—going from 70 warships to 200—and all its new vessels were the superior type of warship known as the trireme.

When the Greek states allied against the attack of the Persian king XERXES (480 B.C.), Themistocles served as a fleet commander and as one of the strategy-planning chiefs

Theocritus   One of the best-known poets of the HELLENISTIC AGE. Born circa 300 B.C. in the Sicilian Greek city of SYRACUSE, Theocritus flourished at Egyptian ALEXANDRIA (1), at the wealthy and literary court of king PTOLEMY (2) II Philadelphus; there Theocritus seems to have been a friend of the great Alexandrian poet CALLIMACHUS. Theocritus wrote in several poetic genres, but his most influential verses were his innovative pastoral poems. It was Theocritus who bequeathed to future generations of Roman and European Renaissance poets the pretty literary convention of Sicilian shepherds and shepherdesses, pining for love or playing contentedly on panpipes. Of his work, 31 poems survive today; in addition to bucolic poems, these include mimes (Greek: *mimoi*, realistic or satirical scenes rendered into hexameters) and epyllia (little epics)—hexameter poems of several hundred lines presenting a heroic MYTH or love story.

Theognis Lyric poet of the Greek city of MEGARA (1), of the mid-500s B.C. Theognis' work survives as a collection of nearly 700 elegiac couplets, many of which apparently were written by anonymous later poets in imitation of Theognis' poems. Within this collection (known as the Theognidia), modern scholars have tentatively identified a core of about 300 poems, which seem to reflect a consistent poetic personality and a composition date in the mid-500s B.C.

One reason why Theognis' poetry attracted imitators is that he spoke for an entire ancient Greek social class—the ARISTOCRACY, who by the 500s B.C. had lost their monopoly on wealth and political power, and whose very lives were threatened, in some cities, by the rise of violent popular leaders known as TYRANTS (*turannoi*). Theognis' verses angrily bewail the fact that noble families are intermarrying with wealthy families of lower social status. "Now blood is diluted by base wealth," the poet typically laments. Most of the poems of the Theognidia are drinking songs (*skolia*), written for musical recitation at the upper-class drinking party known as the SYMPOSIUM. These poems heartily reflect the symposium's intellectual and homosexual atmosphere. Nearly all of the authentic Theognis poems are addressed to a certain Cyrnus, who was the poet's boyfriend; in certain verses Theognis speaks candidly of his love and passion for Cyrnus.

(See also HOMOSEXUALITY; LYRIC POETRY.)

Theophrastus Greek philosopher and essayist who lived circa 370–287 B.C. Born on the island of LESBOS, he was active at ATHENS, where he succeeded ARISTOTLE as head of the philosophical school known as the LYCEUM. Theophrastus maintained the school's broad range of study, and he eventually moved the location to a larger building complex, called the Peripatos.

Theophrastus wrote on a variety of scientific and philosophical subjects, but is best remembered today for his extant work titled *Characters*, a collection of 30 sketches of eccentric or abnormal personality types.

Thera Greek island of the mid AEGEAN SEA, at the southern edge of the island group known as the CYCLADES. The island's alternative, Italian-derived name, Santorini, commemorates St. Irene of Salonika. Covering only 29 square miles, Thera consists of an irregular, westward-opening crescent that rises dramatically from the sea. This odd shape marks the remnant of an ancient volcanic cone, which—as geology and ARCHAEOLOGY show—erupted in about 1480 B.C., blasting itself in half.

Before the eruption, Thera was home to a branch of the pre-Greek MINOAN CIVILIZATION. Modern excavations at Akrotiri, in southern Thera, have uncovered the most important Minoan site known to us outside of CRETE, with remnants of streets and villas, and some fine frescoes. This settlement was preserved for millennia by being buried in lava after its inhabitants had fled.

Thera's cataclysmic eruption of 1480 B.C. probably resembled that of the Indonesian island Krakatoa in A.D. 1883.

The resultant tidal waves battered other islands, including Crete, 70 miles south. The disaster may have been the direct cause of a sharp social decline observable in Crete at that time, including depopulation of Crete's coastlines. Thera's destruction may possibly lie behind the later Greek legend of ATLANTIS.

Circa 1000 B.C., Thera was occupied by DORIAN GREEKS from the mainland, who founded Thera city, in the south mid-island. An impressive temple of Carnean APOLLO, datable to the 500s B.C., is among the city's early monuments revealed by excavation. Scratched onto rocks nearby are ancient Greek graffiti, many with jocular homosexual messages, which represent some of the earliest surviving examples of Greek alphabetic WRITING (600s B.C.).

Despite prosperity, early Dorian Thera was badly enough afflicted by drought to send out large numbers of its young men to found a colony in North Africa, circa 630 B.C. This colony was CYRENE (1), which became one of the foremost cities of the Greek world within just a few decades.

Although Dorian-Greek and not Ionian, Thera was a subject ally of ATHENS in the DELIAN LEAGUE (400s B.C.). In the 200s B.C. the island became a naval base for Ptolemaic EGYPT, before being annexed by the Romans in the first century B.C. Most of Thera city's physical remains, including the AGORA, colonnades, and temple of the god DIONYSUS, date from the Ptolemaic and Roman occupations.

(See also COLONIZATION; HOMOSEXUALITY.)

Theramenes Athenian commander and right-wing politician of the late PELOPONNESIAN WAR and its aftermath. Theramenes was instrumental in arranging the Athenian surrender to SPARTA at the war's end (404 B.C.). More important, he was involved in both of the oligarchic coups at ATHENS in these years—that of the FOUR HUNDRED (411 B.C.) and that of the THIRTY TYRANTS (404–403 B.C.). As leader of the moderate faction within the Thirty, Theramenes lost his life for opposing the extremist leader, CRITIAS.

The contemporary Athenian historian XENOPHON, who knew Theramenes, portrays him as a cynical opportunist. But later writers (including ARISTOTLE in his *Constitution of Athens*) have seen Theramenes as a true moderate and patriot, trying to steer his city through disaster.

(See also OLIGARCHY.)

Thermopylae Seaside mountain pass, connecting the frontiers of THESSALY and East LOCRIS, at the eastern edge of the Mt. Oeta range, in northeast-central Greece. In 480 B.C., during the PERSIAN WARS, Thermopylae was the site of a famous three-day battle in which 5,000 Greek HOPLITES, led by the Spartan king Leonidas, blocked a Persian army of perhaps 200,000, marching south under the command of King XERXES.

Long before the battle, Thermopylae was recognized as the strategic weak point on the main route into central Greece. Just under four miles long and less than 50 feet wide at its narrowest point, the pass was hemmed (to the west) by a wall of cliffs and (to the east) by the waters of the Gulf of Malis. Circa 600 B.C. the pass was further closed by a wall, built by inhabitants of the central Greek region of PHOCIS to block the southward incursions of their enemies, the Thessalians.

North of the pass, the River Spercheios flows into the gulf. The river's alluvial deposits have changed the coastline since ancient times, resulting in a seaward widening of the pass. No longer is Thermopylae the grim "catwalk" it once was. The name Thermopulai, "hot gates," refers to local sulfur springs. Ancient Thermopylae also served as a sanctified meeting place for the local peoples of the AMPHICTYONIC LEAGUE.

The Battle of Thermopylae was fought in the late summer of 480 B.C., simultaneous with the sea battle of ARTEMISIUM, 40 miles away. On the Greek side, the object of both battles was to stop the enemy at narrow points on land and sea, north of the Greek heartland. Unfortunately, the Thermopylae effort was undermined by the indecision of leaders at SPARTA, who sent an advance guard under Leonidas, but then declined to send the main army—in part because their religious festival of the Carneia offically forbade the dispatching of troops during that time. So Leonidas was left to hold Thermopylae with his small force.

The Persian infantry entered the northern mouth of the pass and charged against the Greek hoplites, who had formed up alongside and behind the stone wall. The Greek historian HERODOTUS, the main source for information on the Persian Wars, describes how the Persians and their countrymen, the Medes—unable to use their superior numbers in the narrow space—fell back repeatedly with heavy losses. Even Xerxes' elite legion of Immortals failed to push through. The invaders were disadvantaged by having shorter thrusting spears and lighter armor than the Greeks, and perhaps by having inferior morale and discipline. But, after two days, with the help of a traitorous local Greek named Ephialtes, the Persians found a mountain footpath that brought their troops down behind the Greek lines.

Leonidas, learning that he was being outflanked, sent most of the army south to safety. The battle's climax, on the third day, was the heroic last stand of Leonidas with his royal guard of 300 Spartans and 1,100 other Greek troops, from BOEOTIA. The Persian troops, lashed on by the whips of their commanders, attacked from both ends of the pass; in the crush, some fell into the sea or were trampled underfoot. The Greeks fought until their spears were broken, then fought on with their swords. Leonidas was killed, as were all his 300 Spartans—except for one man, who had been sent home ill. The Boeotians died, too, or surrendered. On the Persian side, the dead included two of Xerxes' brothers. But the invading hordes marched through, into central Greece.

Although in military terms Thermopylae was a Greek failure—with more men and better organization they might have seriously stalled the Persian land invasion—the battle became very significant in emotional and patriotic terms. Like other suicidal exploits, the Spartan defense at Thermopylae inspired strong reverence in the Greek mind. Monuments to the slain were erected at Sparta and in the pass itself, where the defenders were buried. The poet SIMONIDES wrote one of the epitaphs:

> Here is the tomb of that famous Megistias
> Whom the Mede slew crossing the River Spercheios.

> A seer, he clearly saw the goddesses of death approach,
> Yet could not desert his Spartan king.

But the most enduring Thermopylae epitaph is an anonymous elegiac couplet, from a commemorative column in the pass:

> Go tell the Spartans, O passerby,
> That here we lie, complying with their orders.

The Thermopylae pass remained a strategic bone of contention throughout Greek history. In 352 B.C. it was occupied by the Macedonian king PHILIP II, in a military-diplomatic maneuver against ATHENS. In 279 B.C. a united Greek army defended the pass—again unsuccessfully—against the southward invasion of the CELTS.

(See also LYRIC POETRY; PROPHECY AND DIVINATION.)

Theron See ACRAGAS.

Theseus Mythical Athenian hero and king. Whereas many Greek legendary heroes were connected with the cities of THEBES or MYCENAE, relatively few figures were attached to ATHENS. Theseus therefore came to be viewed as an Athenian national hero—not only a great warrior and lover, but a civilizer, who rid the countryside of brigands and who unified the villages of ATTICA into a single federation, centered at Athens. In historical times, he was honored at an annual festival, the Theseia, held in the fall.

Being an Athenian cultural treasure, the Theseus MYTH was reworked by generations of poets and storytellers. Several of Theseus' adventures seem specifically modeled on those of HERACLES, in an attempt to make Theseus as important as that greatest Greek hero. Some of this nationalistic myth-making perhaps occurred in the 500s B.C., under the Athenian tyrants PISISTRATUS and HIPPIAS (1).

The mythical Theseus probably commemorates some actual Athenian king of the Mycenaean era, perhaps circa 1300 B.C., whose lasting achievement was to consolidate the 1,000-square-mile region of Attica under his rule. Possibly this king's name really was Theseus (Settler). Alternatively, the king's accomplishments were eventually ascribed to an existing legendary hero. The historical essence of the Theseus legend appealed to PLUTARCH (circa A.D. 100), who wrote a biography of Theseus among his biographies of more strictly historical Greeks and Romans.

According to the myth, the Athenian king Aegeus, being childless, consulted the god APOLLO's oracle at DELPHI. He received the advice to return home but that, until arriving, he should not "loosen the wineskin's jutting foot." In fact this was a warning, not against alcohol, but against sexual intercourse; however, Aegeus, failing to understand, dallied with King Pittheus' daughter, Aethra, at the city of Troezen (in the northeastern PELOPONNESE). Departing from Troezen, Aegeus left behind a sword and a pair of sandals underneath a huge boulder; he secretly instructed Aethra that when their future son—whom Aethra was confident she had conceived—was strong enough to shift the boulder and recover the items of proof, he should go to Athens.

The son was born at Troezen and named Theseus. King Pittheus, to protect his daughter's reputation, claimed that

Theseus' father was the god POSEIDON, who supposedly had seduced Aethra. On reaching manhood, Theseus easily pushed the boulder away from the sandals and sword. Taking these tokens of identity, he went to find his father at Athens. Rather than use the quick sea route across the Saronic Gulf, Theseus chose to travel by land, around the top of the gulf, for adventure's sake.

On the journey, he encountered a series of sadistic brigands and slew them all. Reaching Athens, Theseus was acknowledged by his father, King Aegeus. Soon, however, Theseus came into danger from Aegaeus' paramour, the Colchian princess MEDEA, who feared Theseus' influence and right of succession. At Medea's prompting, Aegeus sent Theseus against the Bull of MARATHON, which was ravaging the countryside. (This adventure is clearly based on Heracles' Seventh Labor, against the Cretan Bull.) Theseus captured the animal and sacrificed it to Apollo.

Next came the hero's most famous adventure. Athens (according to the myth) was at this time a subject city of Crete, having been subdued by the Cretan king MINOS. The Athenians were compelled to pay a yearly tribute of seven youths and seven maidens, who were sent by Athenian ship to Crete to be locked up inside the maze of the LABYRINTH. There the young people would be devoured by the ferocious Minotaur (bull of Minos), who was the monstrous half-man, half-bull offspring of Minos' wife, Queen Pasiphaë. Theseus, deciding to put a stop to this oppression, volunteered to sail to Crete amid the next batch of young people.

Reaching Crete, he met the princess Ariadne, daughter of Minos and Pasiphaë. She fell in love with him and, to guide him through the dark Labyrinth, gave him a thread, one end of which he fastened near the entrance. Then he slew the Minotaur, escaped from the Labyrinth, and fled from Crete with Ariadne aboard the waiting Athenian tribute ship. But Theseus callously abandoned Ariadne on the island of Naxos, where she was later found and wed by the god DIONYSUS.

Theseus' ship sailed on toward Athens. The hero had arranged with his father that, if he were successful on Crete, he would substitute the tribute ship's black sail for another one, colored white. In the excitement of the return Theseus forgot this agreement, and Aegeus, seeing the ship approach under a black sail, assumed that his son had been killed. The remorseful Aegeus immediately threw himself into the sea and drowned. Supposedly, from this, the AEGEAN SEA was named.

Succeeding his father as king, Theseus brought about the unification of Attica. He then departed on a campaign against the distant-dwelling AMAZONS, the tribe of fierce female warriors. One version says that Theseus accompanied Heracles' Amazon expedition, another that Theseus went on his own, but in either case he brought back to Athens, as his captive, the Amazon queen, Hippolyta. In pursuit, the mass of Amazons marched to Athens and besieged the city, but were repulsed. (In classical times, the Amazonomachy, or Battle with the Amazons, was a favorite subject of monumental Athenian SCULPTURE and PAINTING.) Hippolyta died after bearing Theseus a son, HIPPOLYTUS. Later Theseus married Phaedra, the younger sister of Ariadne, and this young wife conceived an illicit passion for her stepson, resulting in the tragic deaths of both.

One of Theseus' last adventures was his visit to the Underworld with his friend Pirithous, king of the Lapiths. Their ambitious goal was to carry off the goddess PERSEPHONE, to be Pirithous' wife. However, Persephone's husband, HADES, captured the two heroes by tricking them into sitting on an enchanted bench, which held them fast. Theseus eventually was rescued by Heracles, who managed to rip him out of his chair; but Pirithous remained in the nether world.

The end of Theseus' life was unhappy. Ousted from Athens by a rebellion, he voyaged to the island of Scyros, where he was murdered by the king, Lycomedes, who disputed Theseus' claim to a local estate.

The myth has a historical aftermath. In about 476 B.C. the Athenian commander CIMON brought back from Scyros a skeleton and relics said to be those of Theseus. The hero was reburied in a monumental tomb in the middle of Athens. The writer Plutarch mentions that the tomb became a sanctuary for runaway SLAVES and the needy, because, in his lifetime, Theseus was the champion of the oppressed.

(See also MYCENAEAN CIVILIZATION; PROCRUSTES.)

Thespis See THEATER.

Thessaly Large, northeasternly region of Greece. Thessaly, whose best-known landmark is Mt. OLYMPUS, consists of two wide and fertile plains, enclosed on four sides by mountains. To the south rises the range of Mt. Oeta, which—in ancient Greek geographical terms—separated Thessaly from PHOCIS and the rest of central Greece; the principal route through this barrier was the pass of THERMOPYLAE. On Thessaly's western and southwestern frontiers stood the Pindus mountain range, beyond which lay the Greek states of EPIRUS and AETOLIA. On the north, more mountains divided Thessaly from the kingdom of MACEDON. And, in the east, the north-to-south line of Mt. Olympus, Mt. Ossa, and Mt. Pelion occupied the Aegean coast and rendered the shoreline harborless and unusable for shipping.

The only suitable harbor in all Thessaly was located at the top of the Gulf of Pagasae (the modern Gulf of Volos), nestled behind the southward-jutting peninsula called Magnesia. There two seaports emerged in different phases of history: First was Iolcus (modern Volos), a major city of the MYCENAEAN CIVILIZATION (circa 1600–1200 B.C.), fabled as the home of the hero JASON (1). Later came nearby Pagasae, which flourished in the 400s and 300s B.C.

The name Thessaly commemorates the Thessaloi, a Greek people of the Aeolian ethnic group who subjugated the area circa 1100 B.C., soon after the Mycenaean collapse. The inhabitants retained an Aeolic dialect in ancient times. Thessaly's wealth lay in its wide plains, which provided grain, cattle, and—the region's hallmark—horses. The Thessalians were famous horsemen; their armies, unlike most Greek forces, consisted mainly of CAVALRY.

Isolated by its mountain perimeter, Thessaly developed as a self-contained, horse-ranching ARISTOCRACY. During the great centuries of Greek history (800–300 B.C.) it remained politically and culturally backward, with power and ownership concentrated in the hands of a few immensely rich families. One such family was the Aleudae, based at the city of Larissa, in the northern plain. These aristocrats resisted both kingship and DEMOCRACY, and it was not until about 374 B.C. that Thessaly became united under one leader, the tyrant JASON (2) of the city of Pherae.

In the 350s B.C. Thessaly fell under the sway of the Macedonian king PHILIP II, and Thessalian cavalry did important service in the army of Philip and of his son ALEXANDER THE GREAT. Valuable as a doorway into central Greece, Thessaly received a Macedonian garrison, which was housed (circa 293 B.C.) in a newly built fortress called Demetrias, adjoining Pagasae. The region remained under Macedonian control until the Romans broke Macedon's power, in 168 B.C.

(See also AEOLIAN GREEKS; GREECE, GEOGRAPHY OF; GREEK LANGUAGE; ROME; TYRANTS.)

Thetis   See ACHILLES; PELEUS.

Thirty Tyrants, the   Name commonly given to the 30-man committee that ruled in ATHENS as a dictatorial puppet government for the Spartans immediately after Athens' surrender in the PELOPONNESIAN WAR (404 B.C.). The Thirty ruled for over a year, from spring 404 until early autumn 403 B.C., in a reign of terror.

The 30 men were right-wing Athenians, elected by the Athenians at a rigged meeting of the ASSEMBLY under the watchful eye of the occupying Spartan general, LYSANDER. Elected ostensibly to draft a new law code, they seized the government, with Lysander's approval, and dismantled the Athenian DEMOCRACY. They disbanded the democratic COUNCIL, abolished the people's law courts, and convinced Lysander to install a Spartan garrison on the ACROPOLIS. With their private police force of 300 "whip-bearers," the Thirty conducted arrests and executions to eradicate opposition and raise funds by confiscation. As many as 1,500 citizens and METICS (resident aliens) are said to have lost their lives; many others fled.

Eventually there arose division within the Thirty, between the extremist CRITIAS and the more moderate THERAMENES. At length, Critias had Theramenes put to death.

When a resistance band under the exiled Athenian soldier Thrasybulus seized a fortress in the Attic countryside, the Thirty began preparing a refuge for themselves at the town of Eleusis, outside Athens. This led to their worst atrocity—the execution of 300 men of Eleusis—perhaps the town's entire adult male population.

But the despots' days were numbered. When Thrasybulus' rebels seized Athens' harbor town of PIRAEUS, the Thirty brought out an army against them and were defeated (winter 404–403 B.C.). Critias was killed in the fighting, as was his nephew and lieutenant Charmides. The remnants of the Thirty were deposed, and full democracy was restored by the autumn of 403 B.C. Individual members of the Thirty were executed thereafter.

Remarkably, the Spartans had made little effort to save their puppet government. This inactivity owed much to a political rift at Sparta, where King Pausanias was repudiat-

ing Lysander's policies. Pausanias soon came to terms with the new Athenian government.

The Thirty represent the worst of three different oligarchic attempts to subvert the Athenian democracy during the 400s B.C. The first of these was the plot to hand over the LONG WALLS to the Spartans in 457 B.C.; the second was the coup d'etat of the FOUR HUNDRED in 411 B.C. All three attempts were inspired by an upper-class desire to rid the government of its lowest-income class of citizens, the *thetes*. Such plots failed because they inevitably did harm to the middle class—the HOPLITE class—as well as causing patriotic outrage.

(See also OLIGARCHY; PLATO; SOCRATES; SYMPOSIUM.)

Thrace   Non-Greek land extending for about 300 miles across the top of the AEGEAN SEA and the Sea of MARMARA, with a further coastline on the southwestern BLACK SEA, south of the Danube River. This region is now divided among the modern nations of Greece, Turkey, and Bulgaria. Relevant geographical features include three large rivers that empty southward into the Aegean; these are (from west to east) the Strymon, Nestus, and Hebrus. The largely mountainous interior rises to the Rhodope chain, which runs east–west along southern Bulgaria and the modern Greek border.

The ancient Greek name Thraikē probably renders the sound of some native name for the homeland. The Thracians spoke a non-Greek language and possessed northern European physical features, such as red hair. They brewed beer but for WINE they depended on Greek imports; they craved many Greek goods. The Thracians were renowned for their ferocity in battle (they used heavy slashing swords) and for their drunken devotions to the god DIONYSUS (a deity who may have originally been a Thracian beer god).

Throughout most of their history the Thracians remained grouped into various warlike tribes, chief of which were the Odrysae, in the east. Only in the mid-400s B.C. did a united Thracian kingdom emerge, briefly, under the Odrysian king Sitalces and his successors. The Thracians lived in villages, with no cities until Roman times (after the mid-100s B.C.).

Although Greece and Thrace had no common border—being buffered by the kingdom of MACEDON, northeast of Greece—Greek traders and settlers gravitated to Thrace's Aegean coast early on, to acquire the land's precious resources. GOLD and SILVER deposits at Mt. Pangaeus brought the Greek colonists to nearby THASOS (circa 700 B.C.) and Maroneia (mid-600s B.C.), in defiance of hostile local tribes. Other Greek colonies in Thrace included BYZANTIUM (mid-600s B.C.), AENUS (circa 600 B.C.), Perinthus (circa 600 B.C.), the CHERSONESE region (mid-500s B.C.), ABDERA (545 B.C.), and AMPHIPOLIS (437 B.C.).

Besides resources for mining, Thrace offered wheat, TIMBER, and SLAVES to the Greeks. The slaves were often freeborn Thracian war captives, acquired in internal Thracian feuds and then sold to local Greek traders. Thracians might even sell their own children (according to the Greek historian HERODOTUS). Thracian slaves, particularly females, are mentioned periodically in writings from ancient Greece.

The internal history of Thrace consists mainly of the rise of the Odrysian state in the 400s B.C., followed by its collapse in the 300s B.C. The Macedonian kings PHILIP II and ALEXANDER THE GREAT invaded Thrace repeatedly (340s–330s B.C.), annexing the Strymon valley and other locales and subjugating the entire land. After Alexander's death, Thrace briefly became a sovereign power under the Macedonian dynast LYSIMACHUS (reigned 323–281 B.C.), before returning to Macedonian control. After the Roman conquest of Macedon in 167 B.C., Thrace became an uneasy section of the Roman province of Macedonia.

(See also COLONIZATION; ORPHEUS.)

Thucydides (1)   Athenian historian who left behind a detailed but unfinished account of the PELOPONNESIAN WAR (431–404 B.C.), which he had lived through. Thucydides' work is one of the most valuable and impressive writings surviving from ancient Greece.

Although he was not the first Greek to write history—his main predecessor was HERODOTUS—Thucydides was in many ways the first real historian, the first author to apply rationalistic standards in inquiring about the past. Unlike Herodotus, Thucydides keeps strict chronology and avoids fables, biographical anecdotes, and (usually) digressions, instead focusing austerely on the most important events. In Herodotus' work, war and politics can result from the gods, FATE, or sexual passion. In Thucydides there are no such fairy-tale causes of action; Thucydides, like a modern political scientist, explains events in terms of a fundamental human drive for power and self-advantage.

In modern terms, Thucydides was as much a journalist as a historian, because much of what he reported was current or very recent as he was writing. He was no mere library scholar; already an adult when the Peloponnesian War began, he served (briefly and unsuccessfully) as an Athenian general. He states in his opening sentence that he began writing as soon as fighting broke out, since he believed that the war would be bigger and more worth recording that any prior conflict. He says that, for all events described, he either witnessed them firsthand or carefully interviewed participants afterward. Also (implicitly contrasting himself with the crowd-pleasing Herodotus), Thucydides announces that his opus will be the last word on the subject, and that it contains only strictly relevant information: "My work is written not as a display piece for an immediate audience, but as a prize to last forever" (book 1)

Born in around 460–455 B.C., Thucydides belonged to a very distinguished Athenian family with wealth from GOLD- and SILVER-mining privileges in the Mt. Pangaeus region of THRACE. His father, although Athenian, bore the Thracian royal name Olorus. According to a plausible reconstruction of Thucydides' family tree, the eminent Athenian soldier and statesman CIMON was Thucydides' granduncle, while Thucydides' maternal grandfather was the right-wing politician THUCYDIDES (2), in whose honor the future historian was named. Yet the younger Thucydides at some point deserted his family's politics and followed the radical democrat PERICLES, the bitter enemy of both Cimon and the elder Thucydides. In his writing, Thucyd-ides the historian (normally sparing of praise) extolls Pericles' intelligence and leadership.

The young Thucydides probably heard the historian Herodotus give public readings in ATHENS, circa 445–425 B.C. The earliest episode in Thucydides' life that is known today comes from a rare autobiographical mention in his history. Early in the Peloponnesian War, probably in 430 B.C., Thucydides caught the deadly plague that was sweeping through Athens. Unlike many Athenians, he survived, and in one passage he describes with scientific detachment the bodily symptoms of the disease and its devastation of the city (book 2).

By 424 B.C. he had been elected as one of Athens' 10 annual generals. He was assigned to the strategically vital coast of Thrace, where he already wielded influence as a mining mogul, but there (as he coolly relates in his history) he failed to prevent the Spartan commander BRASIDAS from capturing the Athenian colony of AMPHIPOLIS. For this failure, Thucydides was banished by the Athenians and lived away for 20 years, returning only when the exiles were recalled in 404 B.C., after Athens' final defeat.

Ironically, it was exile that allowed him to produce so fine a history. As an independently wealthy member of the international ARISTOCRACY in the days before general suspicion of espionage, Thucydides was able to travel through the theaters of war. "Because of my exile, I saw what was happening on both sides, particularly on the Peloponnesian side," he relates (book 5). He may have made the Thracian mining region his base, but he journeyed to SPARTA, whose paltry physical monuments he comments on (book 1), and to SYRACUSE, which looms large in his written work. He must have been a relentless seeker of truth—observing battles, interviewing soldiers and politicians, consulting archives, collating notes, and rejecting items in a drive for accuracy. Much of his information is amazing in its detail, but he rarely allows the details to submerge his story.

According to a later writer, Thucydides was married and had children. Apparently he died in around 400 B.C., only a few years after his return to Athens. He never completed his history, which breaks off in midsentence describing events of 411 B.C. (The last seven years of the war are recounted by the historian XENOPHON, in his *Hellenica*.)

Thucydides' history was organized by later scholars into eight "books," but in content it can be broken down into four unequal sections: (1) books 1–5 (partial), on the Archidamian War and Peace of NICIAS, 431–421 B.C.; (2) book 5 (remainder), on the troubled years of temporary peace, 421–415 B.C.; (3) books 6 and 7, a beautifully polished section describing the calamitous Athenian invasion of Syracuse, 415–413 B.C.; and (4) book 8, describing naval events in the eastern Aegean and the right-wing coup of the FOUR HUNDRED in Athens, 412–411 B.C. Both book 5 and the unfinished book 8 are clearly unrevised drafts. The work's well-known set pieces include Pericles' Funeral Oration over the Athenian dead (book 2), the description of civil war and moral breakdown at the city of CORCYRA (book 3), and the Melian Dialogue, in which notions of Might vesus Right are debated prior to the Athenians' devastation of the helpless island of MELOS (book 5). Writing in his native Attic prose, Thucydides uses a uniquely clipped and condensed style, which is sometimes difficult to understand. He seems to want to fit as much meaning into as few words as possible.

The writer's intelligence and mastery of his material tend to disguise certain flaws or quirks. It may be that Thucydides overrates Pericles as a soldier and glosses over a major Periclean military setback, the failed siege of the Peloponnesian city of EPIDAURUS in 430 B.C. Conversely, Thucydides surely underrates the motives and abilities of the Athenian politician CLEON; Thucydides seems to have disliked him personally—perhaps Cleon proposed the edict banishing General Thucydides in 424 B.C. Also, Thucydides shows himself totally uninterested in finance, despite its importance for the Athenian empire and war effort. His treatment of Greek-Persian affairs is skimpy, insofar as it was PERSIA's funding of the Spartan navy, 413–404 B.C., that enabled Sparta to defeat Athens. But some of these shortcomings might have been corrected had the historian lived to complete his work.

Another idiosyncrasy is Thucydides' way of ignoring questions of emotional motive, such as patriotism or religious piety, as not being true causes of events. Modern historians believe that emotional issues were important in such gatherings as the Athenian democratic ASSEMBLY, where demagogues might whip up the citizens to get them to vote a certain way. But in Thucydides' world it is only the victims, such as the doomed Melians in the Melian Dialogue, who make emotional appeals to the gods or to justice; the winners always speak in rational terms of self-advantage and expediency.

Nevertheless, Thucydides stands as a monumental early figure in history-writing. The personal impression he creates is that of an Olympian intellectual, patterned after his hero, Pericles. An atheist and a loner, enamored of intelligence and power, Thucydides could combine icy objectivity with a fascination for human affairs. He was a cynical genius, well suited to recording the suicidal event that toppled the classical Greek world.

Thucydides (2)   Athenian politician who led the right-wing opposition to the radical democrat PERICLES in the 440s B.C. and was ostracized by Pericles' efforts. Thucydides' defeat brought on a 14-year period of Periclean supremacy in the Athenian DEMOCRACY. This Thucydides probably was the maternal grandfather and namesake of the historian THUCYDIDES (1).

The future politician was born circa 500 B.C. His father, Melesias, an Athenian nobleman and celebrated wrestler, was a friend of the poet PINDAR. Thucydides grew up in the cosmopolitan world of Greek aristocrats and married a sister of the great Athenian soldier CIMON. After Cimon's death (circa 450 B.C.), Thucydides succeeded him as leader of the conservative opposition. He organized a system whereby all his followers sat together in the ASSEMBLY, in order to create more noise and effect during public debate.

An able orator, in around 448 B.C. Thucydides attacked Pericles' public-building policy, whereby allied DELIAN LEAGUE funds, contributed for defense against PERSIA, were to be used instead to erect grand monuments in ATHENS (such as the PARTHENON). Thucydides may have been sin-

cerely offended by Athens' domination of other Greek states, as well as by Periclean domestic policies that opened up political opportunities to the Athenian lower-income classes.

But Pericles' support was overwhelming, and the Athenians voted to ostracize Thucydides (circa 443 B.C.). He seems to have returned to Athens after his term of exile ended, circa 433 B.C., but we hear of no more political activity.

(See also OSTRACISM; WRESTLING.)

Thurii   See SYBARIS.

timber   In ancient mainland Greece, thick forests flourished in the foothills and mountains, which were cooler and far less heavily farmed than the arable plains. Trees carpeted mountainous regions such as ARCADIA and the Ossa-Pelion massif of southern THESSALY. The wood of these trees was a crucial construction material for public buildings, private homes, furnishings, machinery, and ships.

Foothills and lower mountain slopes supported a natural growth of deciduous trees; the most common was the oak, in several varieties. The oak's strong, heavy wood was an important source of timber for building beams and furniture. A tall, majestic tree, the "king" of the Greek forests, the oak was from an early period associated with ZEUS, king of the gods.

Other timber-producing trees of the middle-altitude region might include cypress, elm, and ash. Ash provides a tough, flexible wood that early on was used for military purposes. In HOMER's poems, the warriors' spears are described as being of ash.

Above 4,000 feet over sea level, the mountain regions produced the tallest conifers—fir, juniper, and cedar—which as a group supplied much of the best building timber in the ancient Mediterranean. Cedars produce large, durable beams, prized for big constructions such as a temple's roof supports. The fir tree, which grew in several varieties, was the seagoing tree; its lengthy, flexible, light-weight wood was well suited to the building of warships. Alternatively, the shipwright could use pine or cypress, which grew at lower altitudes than fir and were more accessible throughout the Greek world, but which gave the disadvantage of somewhat heavier planking. Merchant ships usually were made of pine.

Like grain, timber was a crucial import for the populous, powerful cities of central and southern Greece in the classical period. For the imperial ATHENS of the 400s B.C., with its need to maintain a large war fleet, foreign policy was partly a matter of ensuring friendship with—or else conquering—those states that had access to valuable forests of fir trees (such as MACEDON or the CHALCIDICĒ region).

In RELIGION, the Greeks associated certain trees with the Olympian gods. As mentioned, the oak was linked to Zeus; similarly, the olive "belonged" to ATHENA, the laurel to APOLLO, the cedar and nut trees to ARTEMIS, and the myrtle to APHRODITE. DIONYSUS, the god of WINE and vegetation in general, was associated with the pine, the plane tree, and most fruit-bearing trees as well as with the grape and ivy vines. Despite these divine associations, such trees generally were not immune from being cut down. In most

Greek communities, however, there existed sacred groves dedicated to a god or local hero, where the trees were sacrosanct.

(See also DODONA; FARMING; SHIPS AND SEAFARING.)

Timoleon Corinthian-born commander and statesman, who in the 340s B.C. became the political savior of Greek SICILY. Liberating the various cities from their dictatorial TYRANTS, Timoleon led the Greeks in resistance to Carthaginian invaders.

Timoleon was a soldier and aristocrat of middle age, living in CORINTH in 345 B.C., when the Corinthians sent him with a small mercenary force to Sicily, in response to a request for help from the aristocrats of SYRACUSE. (Syracuse was the foremost Sicilian Greek city, and a Corinthian colony and ally.) The Syracusans were seeking aid against the tyrant Dionysius II, who, after a period of exile, had reinstated himself as master of the city. By combining military strategy with diplomacy, Timoleon ousted Dionysius. Then he reorganized the Syracusan government, creating a liberal OLIGARCHY on the Corinthian model. Thereafter he began a campaign to oust the tyrants of the other Sicilian-Greek cities and reorganize those governments.

Amid these events, in 341 B.C., a seaborne Carthaginian invasion force landed in Sicily. Timoleon won an initial victory over the invaders at the Crimisus River, in northwestern Sicily, but he was forced to sue for peace with the Carthaginians, after the Greek tyrants joined the war against him. Having made this peace, he was then free to defeat the tyrants separately (circa 340–337 B.C.) and to complete his reorganization of the cities' governments. He spent his final few years as a revered private citizen at Syracuse and died in 334 B.C. His achievements were short-lived: Within a generation chaos had returned to Greek Sicily, and there arose a new Syracusan tyrant, AGATHOCLES.

(See also CARTHAGE.)

tin See BRONZE.

Tiresias In MYTH, a blind Theban seer, renowned for his wisdom and longevity; he was said to have lived for seven generations. In HOMER's epic poem the *Odyssey* (book II), the hero ODYSSEUS visits the Underworld to consult the ghost of Tiresias, who prophesies as to what FATE awaits the hero at home and mentions certain precautions to observe on the return voyage.

Tiresias was not born blind; the best-known story about him tells how he lost his sight. Out walking one day, he encountered two snakes mating and hit them with his walking stick, killing the female. Immediately he was transformed into a woman, and remained so until, some time later, he again encountered two snakes mating. This time he killed the male and so regained his prior gender. Because Tiresias had had this unique experience, the god ZEUS and his wife, HERA, asked Tiresias to settle their argument as to who gets more pleasure from sexual intercourse, men or WOMEN. Zeus claimed it was women, but Hera modestly insisted it was men. When Tiresias declared that women get nine times more pleasure than men, Hera

angrily blinded him. But Zeus, to compensate, gave him long life and the inner sight of prophecy.

(See also PROPHECY AND DIVINATION; SOPHOCLES.)

Tiryns Ancient Greek city in the Argolid region of the northeastern PELOPONNESE, near modern Nauplion. With its spectacular fortifications, first excavated by archaeologist Heinrich Schliemann in A.D. 1884, Tiryns provides the best surviving example of a fortress city of the MYCENAEAN CIVILIZATION (circa 1300 B.C.).

Tiryns was ideally located, atop a rocky hill in the southeast Argive plain, beside the sea (which now lies a mile away, due to coastal changes since ancient times). The site was inhabited long before the Greek era; the name Tiruns was not originally Greek for its ending resembles the distinctive *nth* sound found in such other pre-Greek names as CORINTH and OLYNTHUS. Archaeological evidence suggests that Tiryns was among the first sites in Greece to be taken over by Greek-speaking invaders. The Greeks' arrival probably is indicated in the destruction of the pre-Greek palace known as the House of the Tiles, at nearby Lerna (circa 2100 B.C.).

In the Mycenaean age (1600–1200 B.C.) Tiryns served as the port for the overlord city of MYCENAE, nine miles to the north. Tiryns received its famous encircling walls, at two separate levels of its terraced hill, after 1400 B.C. These fortifications, made of huge limestone blocks weighing as much as 14 tons each, rose about 65 feet, with a typical thickness of 20–30 feet. In the east there stood a tall gateway, protected by an enfilading wall, similar to the Lion Gate at Mycenae. Greeks of later centuries ascribed these works to the mythical giant creatures known as Cyclopes, and they coined the adjective *Cyclopean*, which still is used today to denote Mycenaean grand masonry.

Within the walls stood a city complex, ascending to the summit's two royal halls (east and west), built in different eras. These two palaces were made of limestone and wood; a bathhouse and remnants of frescoes throughout suggest a primitive luxury. Tiryns' prominence in this epoch is reflected in Greek MYTH, which claimed that the hero HERACLES was born there or that his family came from there.

Like Mycenae and other cities, Tiryns shows clear signs of having been destroyed by fire circa 1200 B.C., amid the collapse of Mycenaean society. Reinhabited on a reduced scale, the town survived into historical times. But by then the nearby city of ARGOS was the dominant power of the plain, and in around 470 B.C. the Argives destroyed Tiryns.

(See also ARCHAEOLOGY; CYCLOPS; GREEK LANGUAGE.)

Titans In MYTH, the Titans were a race of primeval gods who preceded the Olympian gods. They were the offspring of the goddess GAEA (Mother Earth) and the god Uranus (Greek: Ouranos, Sky); the most important of their number was CRONUS, who ruled the universe prior to his son ZEUS.

The Titans represent the brutal era before Zeus' civilizing reign. As recounted in HESIOD's epic poem the *Theogony* (circa 700 B.C.), Zeus became king after leading his fellow gods in a 10-year war against Cronus and the other Titans. (But two Titans, PROMETHEUS and Themis, fought beside the gods, as did other supernatural beings—the Cyclopes and the Hekatoncheires, or "hundred-handed ones.") Defeated, the enemy Titans now lie shackled forever in the lowest level of the Underworld—all except ATLAS, whose immense strength is employed in holding up the sky.

If the name Titan (plural: Titanes) is actually a Greek word, it may mean "honored one." More probably it is not Greek, but derives from the lost language of the pre-Greek inhabitants of Greece. The legend of the defeated Titans may distortedly commemorate the Greeks' conquest of a native people, circa 2100 B.C.

(See also AFTERLIFE; CYCLOPS; GIANTS; OLYMPUS, MT.; ORPHISM; RELIGION.)

trade The Greeks share with the Phoenicians the distinction of being the greatest traders and seafarers of the ancient Mediterranean. One important reason why the Greeks were able to make such a strong impression on other ancient peoples—ETRUSCANS, CELTS, Romans, Scythians—was that they sailed far and wide in search for new markets to trade in.

For the archaeologist, ancient Greek trade routes often are suggested by the presence, in non-Greek lands, of surviving Greek POTTERY and metalwork. Written references by such authors as the Greek historian HERODOTUS or the Roman geographer Strabo also help fill in the picture. It is known that as early as about 1300 B.C., long before the invention of COINAGE, Mycenaean-Greek merchants were spanning the Mediterranean, trading by barter. Archaeological evidence shows that these early Greeks had trade depots in the Lipari Islands (near SICILY) as well as in CYPRUS, western ASIA MINOR, and the northern Levant. During the Greek trading and colonizing expansion of circa 800–500 B.C., Greek merchants visited southern France and Spain and the farthest corners of the BLACK SEA. Circa 308 B.C. a Greek skipper, Pytheas of MASSALIA, explored the Atlantic seaboard, sailing around Britain and perhaps reaching Norway in a search for new trade routes.

What lured the Greeks along these amazing distances was the need to acquire certain valuable resources or luxury goods (depending on the locale). Southern Spain, for example, offered raw tin, a necessary component of the alloy BRONZE, a metal on which Greek society of all centuries strongly depended. In the elaborate trade network of the ancient Mediterranean, tin was quarried by Celtic peoples in British Cornwall and (possibly) in northwest Spain, and was brought to the Greek market, overland to the Mediterranean, by Celtic middlemen. Other prizes of the western Mediterranean included SILVER ore and lead. In exchange, the Greeks might offer metalwork, textiles, or WINE, brought from mainland Greece or from the Greek cities of Asia Minor.

Other lands offered more sophisticated commodities and different rules of trade. To acquire the carved ivories, metalwork, or textiles of EGYPT, the Greeks of 800–500 B.C. might have to offer SLAVES, silver ore, or other basic resources that the wealthy Egyptians lacked. Perhaps most desirable of all Egyptian goods was the grain surplus of the Nile valley, which Greek merchants could ship to hungry Greece for lucrative resale. Grain, along with TIMBER, was also a major attraction of Sicily, ITALY, THRACE, and the northern Black Sea coast. Among valuable Greek

Trade-borne influences from the Near East are embodied in the griffin, a dragonlike beast of Mesopotamian religion whose image the Greeks appropriated for their own art in the 700s and 600s B.C., during the "Orientalizing period." The Greeks saw griffins and other Eastern motifs on imported metalwork, textiles, and carved ivory, and Greek craftsmen then adapted these designs. This drawing shows a bronze cauldron attachment, cast on the Greek island of Rhodes, circa 600 B.C.

exports were the painted pottery of CORINTH (600s B.C.) and ATHENS (500s B.C.), and the woolens of MILETUS and other cities of IONIA.

The conquests of the Macedonian king ALEXANDER THE GREAT (334–323 B.C.) changed the trading map of the ancient world, flooding the Greek market with luxury goods. For instance, the availability of metalwork now snuffed out most of the demand for ceramic pottery.

(See also AEGINA; AL MINA; CARTHAGE; CHALCIDICĒ; CHALCIS; CHIOS; COLONIZATION; CRETE; GOLD; ILLYRIS; PITHECUSAE; PHOENICIA; RHODES; SAMOS; SHIPS AND SEAFARING; TROY.)

tragedy   See THEATER.

trees   See TIMBER.

tribes   See KINSHIP.

trireme   See WARFARE, NAVAL.

Troilus   Son of the mythical Trojan king PRIAM and his wife, HECUBA. As a handsome youth in the early days of the TROJAN WAR, Troilus was slain by the Greek champion ACHILLES.

Although Troilus had only a small role in Greek MYTH, he became important later, in the medieval European legend of Troilus and Cressida.

Trojan War   The legend of the Trojan War is today the best-known story from Greek MYTH. Amazingly rich in its many characters and events, the Trojan War legend describes how the allied Greeks—under the command of King AGAMEMNON of MYCENAE—sailed from Greece and laid siege to the non-Greek city of TROY (also called Ilium), located outside the mouth of the HELLESPONT sea channel, in northwest ASIA MINOR. The Greeks had come at Agamemnon's behest to avenge the abduction of the beautiful Helen—wife of Agamemnon's brother MENELAUS—by the Trojan prince PARIS. The Trojans, then sheltering Paris and the compliant Helen within their walls, had refused the Greeks' demand for Helen's return. This was the official reason for the war, although the Greeks also craved Troy's wealth.

The siege, consisting mainly of battlefield fighting on the plains outside the city, lasted for 10 years and saw the death of many heroes of both sides, including the Trojan prince HECTOR and the Greek champion ACHILLES. Finally Troy fell to a stratagem devised by the wily Greek hero ODYSSEUS. Pretending to abandon the siege, the Greeks sailed away, leaving behind a huge wooden horse they had constructed—the so-called Trojan Horse. Hidden within this hollow monument was a picked force of Greek "commandoes," but the unsuspecting Trojans, believing the horse to be a Greek offering to the gods, brought it into the city. After nightfall, the Greeks emerged from the horse and opened Troy's gates to the waiting Greek army (which had hurried back under cover of darkness).

Destruction of the rich and proud city followed, with the inhabitants all massacred or captured as SLAVES. However, the Greeks' arrogant and impious behavior at the

sack of Troy angered the gods, who decreed that many of the surviving Greek heroes would be killed on the voyage home.

The immense saga—of which HOMER's *Iliad* and *Odyssey* present only a portion—was the product of Greek oral poetic tradition over several centuries, approximately 1200–550 B.C. The legend produced a body of EPIC POETRY, describing major episodes of the war. The earliest and greatest of the heroic epics were the two poems written down circa 750 B.C. and ascribed to the poet Homer—namely, the *Iliad*, or *Tale of Ilium*, recounting the "passion" of the Greek hero Achilles during the war's 10th year; and the *Odyssey*, describing the homecoming of the Greek hero Odysseus, with several back references to the last days of the war.

These two Homeric poems, however, were not the only epic poems dealing with the Trojan War. The classical Greeks (400s B.C.) knew at least six other Trojan-related epics, not ascribed to Homer. These poems did not survive antiquity, but their story plots are summarized by later writers. The lost Trojan War epics were: (1) the *Cypria* ("Tales from CYPRUS"), describing the war's causes and outset; (2) the *Aethiopis*, recounting Achilles' slaying of the Ethiopian king MEMNON (a Trojan ally) and Achilles' own death in battle; (3) *The Little Iliad*, describing the madness of the Greek hero AJAX (1) and the episode of the Wooden Horse; (4) the *Capture of Troy* (*Iliou Persis*), recording the Greeks' bloody capture of the city; (5) the *Homecomings*, recording the calamitous homeward voyages of the surviving Greek heroes other than Odysseus; and (6) the *Telegonia*, describing Odysseus' last days and the related adventures of Telegonus, who was his son by the witch CIRCE.

In A.D. 1870–1890 the pioneering German archaeologist Heinrich Schliemann discovered and excavated the ancient site of Troy, which—amid its various ancient strata—showed clear signs of having been destroyed by a great fire circa 1220 B.C. Since this epoch-making discovery, scholars have universally come to agree that the Trojan War of legend was based on some genuine event of the late Mycenaean age. Certainly the Mycenaean Greeks possessed the wealth, manpower, and organization to launch such an expedition, and they probably had sufficient cause; even if we dismiss the abduction of Spartan Helen as a fiction, there remains that fact that ancient Troy must have controlled the Hellespont waterway, and was probably preying on the Mycenaeans' vital imports of raw metals.

(See also AJAX (2); ARCHAEOLOGY; ANDROMACHE; BRONZE; CASSANDRA; DIOMEDES; HECUBA; HELEN OF TROY; LAOCOÖN; NEOPTOLEMUS; NESTOR; PHILOCTETES; PRIAM; TRADE; WARFARE, LAND; WARFARE, SIEGE.)

Troy   Ancient city of ASIA MINOR, usually described as being inhabited by a non-Greek people. In the late BRONZE AGE (circa 1300 B.C.), Troy was evidently one of the most powerful cities of the eastern Mediterranean. The site of ancient Troy—at the village of Hissarlik, in modern-day Turkey—was discovered by the great German archaeologist Heinrich Schliemann in A.D. 1870. But Troy had been famous for 3,000 years prior, immortalized in the ancient Greek epic cycle of the TROJAN WAR, of which the two greatest poems are HOMER's *Iliad* and *Odyssey*. Schliemann's epoch-making discovery revealed that the historical Troy had achieved affluence in the early second millennium B.C., before being destroyed in an immense fire, at the archaeological level known as Troy VIIa (circa 1220 B.C.). Modern scholars tend to believe that this destruction marks the same event commemorated in the mythical Trojan War.

The name *Troy* comes to us from the poet Homer and other Greek writers who called the city Troia and Ilion. These names seem to have no meaning in the Greek language, and may simply preserve the ancient Trojans' names for their city. Greek MYTH mentions two Trojan founding fathers—Tros and his son Ilus—who gave the city its two names; but this sounds like a later rationalization. The first founder of Troy, according to Greek myth, was Tros' grandfather, named Dardanus.

Troy is situated at the very northwestern corner of Asia Minor, about four miles south of the western mouth of the HELLESPONT waterway and about four miles inland of the AEGEAN SEA. Set on a height commanding the seaside plain, ancient Troy could raid or toll the merchant shipping in the Hellespont without itself being blockaded inside the strait; its inland site left it safe from surprise raids by sea. Troy's control of the Hellespont must have been a crucial factor in the city's prosperity and was probably the reason why the real Trojan War was fought: The Mycenaean Greeks of the mid-1200s B.C. may have decided to destroy Troy because it was blocking their badly needed imports of metals from Asia Minor.

Schliemann's excavations uncovered no less than nine levels of habitation at Troy, dating from circa 2500 B.C. to just after the birth of Christ. The significant levels for our purpose begin with Troy VI (about 1900–1300 B.C.), which marks the arrival of certain invaders—an accomplished people, with horses and superior building techniques, who enlarged and refortified the city. These Trojans may have been Luwians, displaced from interior Asia Minor by the Hittites. Or these Trojans may have been kinfolk of the first Greeks, who in the same centuries were descending through the Greek peninsula, overrunning or assimilating the prior inhabitants there.

This impressive Troy VI was devastated by an earthquake circa 1300 B.C. Troy VIIa represents the survivors' rebuilding in the wreckage, along the plan of the preceding city. This Troy was destroyed by fire circa 1220 B.C., in what may have been the Trojan War.

Scholars have always noted that the small size of Schliemann's Troy VIIa seems at odds with Homer's picture of a mighty and defiant fortress city. Until archaeologists can find convincing evidence of wider ancient city limits, it must be assumed that the Trojan War legends greatly exaggerate the grandeur of the real Troy of the 1200s B.C.

(See also ARCHAEOLOGY.)

Tydeus See SEVEN AGAINST THEBES.

Typhon One of several mythical monsters that arose to challenge the might of ZEUS soon after the beginning of the world. Like the rebellious GIANTS, Typhon (or Typhoeus) was a son of the earth goddess, GAEA. The epic poet HESIOD, in his *Theogony* (circa 700 B.C.), vividly describes Typhon as having a vaguely human form with 100 serpents' heads that sparked fire and made the deafening noise of many different beasts. This grotesque creature tried to take over the world, but was incinerated by Zeus' thunderbolts. Being immortal, Typhon could not die, and was imprisoned for eternity in Tartarus (the lowest abyss of the Underworld).

Like other hybrid creatures described in Hesiod, Typhon probably represents a mythological borrowing from the Near East. Greek legend often associated Typhon with the southeastern ASIA MINOR region called Cilicia, and perhaps Typhon was originally a spirit of Cilician volcanoes.

As Greek exploration opened up SICILY and southern ITALY in about 800–600 B.C., Typhon became "reassigned" to that new frontier (as did other Greek mythical figures). At that time he was said to be imprisoned under Mt. Etna, in Sicily, where his fire and rage were the cause of that volcano's eruptions.

In surviving artwork, Typhon is sometimes shown as a snake-footed man in combat with Zeus. For the Greeks, the battle provided a picturesque emblem of savagery versus civilization.

(See also PINDAR.)

tyrants The Greek tyrants were dictators who first arose at various Greek cities on a wave of middle-class anger in the mid-600s B.C. and seized the government from the aristocrats who had hitherto ruled. The Greek title tyrant (*turannos*) did not at first denote the cruel abuse of power, as it does now; rather, it had a neutral meaning—"usurper with supreme power." The word *turannos* apparently was imported to Greece from the Near East, where it had been used to describe the usurping Lydian king Gyges (circa 670 B.C.). The point of this newly minted Greek word was that, for the first time in Greek memory, these despots were coming from outside the narrow ruling circles.

What brought the first tyrants to power was a growing Greek middle class that was ready for political and social privileges that the aristocrats had denied them. By the mid-600s B.C., CORINTH and other cities had been active in seaborne TRADE for probably 200 years. This trade had created a class of merchants and manufacturers with wealth but no prestige or political voice. When a revolution in military tactics brought these middle-class men into the army to serve as HOPLITES, their potential power increased greatly. There followed, throughout the greater Greek cities, a wave of tyrants' revolutions.

The first Greek *turannos* now known is PHEIDON of ARGOS (circa 670 B.C.). Other famous tyrants include CYPSELUS of Corinth (and his son PERIANDER), CLEISTHENES (2) of SICYON, PISISTRATUS of ATHENS (and his son HIPPIAS (1)), and POLYCRATES of SAMOS. Tyrants also held power at MEGARA (1), MILETUS, and many other Greek cities. These rulers spanned the period from about 670–510 B.C. Their dynasties rarely lasted for more than two generations, since it was impossible to maintain or repeat constructively the popular discontent that had raised the tyrant to power in the first place.

Like other popular dictators throughout history, the tyrants sought to bring amenities to the poor and to relieve unemployment with public works. Tyrants such as Polycrates were known for improving local water supplies as well as for their monumental building programs, which glorified themselves and their cities. Tyrants might also be known for their unbridled passions and sexuality.

The enemy of the tyrants was the panhellenic class of the ARISTOCRACY, who had most to fear from revolution. But the tyrants also had an enemy in the Greek city of SPARTA, which by the 500s B.C. had developed into a kind of Greek police force, sending out troops in response to appeals from displaced aristocrats of Samos, Athens, and elsewhere. In 510 B.C. the Spartans ejected the last important tyrant of mainland Greece, Hippias of Athens; in his place, the Athenians soon established the first Greek DEMOCRACY.

The great supporter of the tyrants was the Eastern kingdom of PERSIA, which used Greek tyrants as puppet rulers in its conquest of Greek ASIA MINOR. The Persians also negotiated in secret with independent tyrants in mainland Greece, in hopes of winning collaborators.

Tyrants played a major role in the history of Greek SICILY, particularly at the preeminent city of SYRACUSE. This was due in part to bitter class divisions within Sicilian-Greek society. But another factor was the abiding external menace of the Carthaginians, who inhabited the western corner of Sicily and who periodically launched massive campaigns to subjugate the island's Greek cities. There thus arose a demand for capable Greek military dictators who could organize defense. The early 400s B.C. saw a series of grand Sicilian tyrants, culminating in GELON of Syracuse (circa 480 B.C.), his brother HIERON (1) (circa 470 B.C.), and Theron of ACRAGAS (circa 480 B.C.), all of whom were known for their wealth and magnificent court.

Although Syracuse saw periods of democracy, conditions there remained ripe for tyrants down through the 200s B.C., long after tyranny had faded away elsewhere in the Greek world. The most famous or notorious of the later Syracusan tyrants were DIONYSIUS (1) (circa 390 B.C.) and AGATHOCLES (circa 300 B.C.); Agathocles was the first tyrant to assume the title king.

(See also PERSIAN WARS.)

Tyre See PHOENICIA.

Tyrtaeus Spartan lyric poet of the later 600s B.C. who wrote verses encouraging his countrymen in their difficult fight in the Second Messenian War. According to tradition, Tyrtaeus was a Spartan general and his poems were sung by soldiers in competitions at the evening meal. There is also a late, absurd tale claiming that Tyrtaeus was originally a lame Athenian schoolmaster.

Only fragments of Tyrtaeus' work survive today, in several dozen lines of elegiac verse. (The elegy is a form intended for recital to flute accompaniment.) Tyrtaeus' poems were patriotic and instructional—reminding his listeners, for example, how beautiful it is to die for one's country, and urging them to adhere to the specific tactics of HOPLITE warfare. His poetry of propaganda belonged to a larger program by which SPARTA transformed itself in those years into a militaristic society, with the best army in Greece.

Tyrtaeus is said to have composed marching songs (now lost) in the Spartan Doric dialect, but he wrote his elegies in Ionic Greek, following the literary convention of his day. His verses contain epic-style language with many echoes of HOMER. Depending on how late in the 600s B.C. Tyrtaeus lived, he may have been influenced by the Ionian poet CALLINUS of EPHESUS, who also developed the elegy's use for patriotic themes.

(See also GREEK LANGUAGE; LYRIC POETRY; MESSENIA.)

U

Ulysses   See ODYSSEUS.

Underworld   See AFTERLIFE; HADES.

Uranus   See CRONUS; GAEA.

vase-painting   See POTTERY.

vines and viticulture   See FARMING; WINE.

voting   See DEMOCRACY.

W

warfare, land The famous land battles of ancient Greece were primarily collisions of heavy infantry, who met on open ground by mutual consent. CAVALRY remained secondary, partly due to the mountainous Greek terrain and partly due to technical matters, such as the small size of Greek horses and the absence of stirrups. It took the tactical genius of the Macedonian king PHILIP II and his son ALEXANDER THE GREAT (mid-300s B.C.) to make cavalry at least partly effective against infantry.

It is less obvious why light-armed foot soldiers were underemployed until the 400s B.C. Commanders in mountainous Greece might have been expected to develop guerrilla tactics using projectile troops—javelin men, archers, and slingers. However, for social and psychological reasons, such tactics did not arise among the major Greek states until the PELOPONNESIAN WAR (431–404 B.C.). Instead, the Greeks (a sporting people) tended to think of war as a manly and consensual contest of strength. "Whenever the Greeks declare war on one another, they go find the most smoothe and level piece of ground and fight their battle there," states an observer cited by the historian HERODOTUS (circa 435 B.C.). In classical Greece, as in Europe of the 18th century A.D., brutal warfare was conducted along certain gentlemanly guidelines.

Greek military history falls into three major phases: (1) Mycenaean and post-Mycenaean warfare, circa 1600–750 B.C., (2) the age of the HOPLITE, or heavily armed infantryman, circa 650–320 B.C.; and (3) the era of the Macedonian-developed PHALANX, perfected by Alexander the Great and employed by his successors. The HELLENISTIC AGE (circa 300–150 B.C.) saw larger armies, commanded by absolute monarchs, battling for vast tracts of land in the eastern Mediterranean and using such innovative war "machines" as elephants. In the end, phalanx and elephant fell to the superior maneuverability of the legions of Republican ROME (100s B.C.).

To understand Mycenaean warfare, modern scholars rely on such archaeological evidence as arms and armor from burial sites and artistic depictions of soldiering on surviving artwork. Also helpful if approached cautiously are battle descriptions in HOMER's *Iliad* (written down circa 750 B.C. but purporting to describe the TROJAN WAR, fought circa 1220 B.C.).

Mycenaean warfare (somewhat like medieval European warfare) revolved around armored noblemen who would take the field at the head of their kinsmen and retainers. These warlords are commemorated in such Homeric heroes as ACHILLES and DIOMEDES. In battle, such champions would seek out their social equals on the enemy side, to duel it out with javelins, jabbing spears, and stabbing swords. As IRON was yet unknown, the choice metal for war was BRONZE. The most dramatic Mycenaean armament—the body-covering shield of oxhide, stretched on a wooden frame, such as AJAX (1) carries in the *Iliad*—was obsolete by about 1450 B.C., replaced by wooden shields faced with bronze. CHARIOTS were used for war—perhaps to convey champions across the plain to the fighting, or perhaps for chariot battles, on the model of contemporary Egyptian and Hittite warfare. The rough terrain of Greece does not generally lend itself to chariots, but the plains of ARGOS and BOEOTIA would have sufficed.

The depressed years of the DARK AGE (1100–900 B.C.) continued the Mycenaean pattern on an impoverished scale. War was still a job for aristocrats and their followers. Bronze armor became scarcer then, but iron—more plentiful than bronze—made its first appearance, in swords and spear-points.

The emergence of the hoplite (after 700 B.C.) banished the old, individualistic approach to war. Henceforth battles were decided not by dueling champions but by large con-

This battle scene, painted on a Corinthian wine pitcher circa 650 B.C., is the best surviving illustration of the hoplite tactics that dominated Greek warfare between 700 and 350 B.C. The armies collide in orderly rows, with the front-rank men jabbing overhand with spears. (An extra spear is held in the shield hand.) The presence of a flute-player to keep the men in step may identify the left-hand army as Spartan. The scene accurately shows the hoplites' armor and round shields. The battle ranks, cleverly foreshortened by the anonymous ancient painter, would have in fact contained several hundred men each.

centrations of armored hoplites, arrayed in tight rows, who fought by pushing with their spears and shields in an organized effort to break the enemy formation. The hoplite era includes the most famous conflicts of mainland Greece: the PERSIAN WARS and Peloponnesian Wars (400s B.C.), and the wars of SPARTA and THEBES (300s B.C.).

The precise origins of hoplite warfare are unclear. It is known that in 669 B.C. the army of ARGOS decisively defeated a Spartan force at Hysiae in the eastern PELOPONNESE; evidence suggests that the Argives at this battle were the first Greeks to use large-scale hoplite tactics, against Spartans who were still fighting in the disorganized old manner.

The hoplite's name came from his new-style shield, the *hoplon*, made of bronze-faced wood. The *hoplon* was round, wide (about three feet in diameter), heavy (about 16 pounds), and deeply concave; in battle it enclosed the soldier's torso while extending toward his neighbor on the left. Other hoplite armament included a bronze helmet and breastplate, a six- to eight-foot-long jabbing spear, and a stabbing or slashing sword of forged iron, for use after the spear had broken. A hoplite's armor, weighing about 60 pounds, would not be worn on the march, but would be carried in wagons or by personal SLAVES or servants.

One difference between the hoplite and previous Greek warriors was that the hoplite was nearly helpless on his own. He remained an effective fighter so long as he stayed grouped with his fellows. Should soldiers leave the battle formation—whether by turning and fleeing from the rear ranks or by breaking ranks in front to chase an enemy— the whole structure was endangered. This obligation to stand fast in battle shaped the classical Greek values of steadfastness, calmness in danger, and the ability to endure fatigue and pain.

Another difference between the hoplite and the armored warrior of prior centuries was that the hoplite was probably not an aristocrat. (Aristocrats comprised the cavalry at this time.) The hoplite was a middle-class citizen, levied in defense of his city-state and fighting against analogous citizens in the enemy army. Cities henceforth relied for their defense on large numbers of uniformly equipped citizens, not on a few noble heroes; the coming of the hoplite thus saw the democratization of warfare.

Ancient writers use the word *othismos*, the "push," to describe hoplite battle. Down to the 300s B.C., the battle line typically was formed six to eight rows deep, with a front of several hundred men. The men in the first three rows advanced with spears leveled; behind them, the soldiers held their spears pointing upward. As the two armies collided, sometimes at a full run, the front ranks of both sides met the enemy spear-points while thrusting overhand with their own spears. But at the same time, the rear ranks kept pushing forward—each man actually leaning his shield into the back of his countryman ahead and shoving him toward the enemy. The armies tried to push each other apart.

For the front hoplites, battle must have been hellish and exhausting. There are tales of soldiers dying of suffocation, pinned upright by the press of men. More usually, of course, soldiers died of wounds—possibly to the neck or genitals (areas not always protected by shield or breast-plate), or possibly to the head or chest (the armor splitting under the spear point). When a hoplite fell, the man behind him took his place with leveled spear, and most men who went down alive would be very lucky not to be trampled to death.

The heart of the battle, the excruciating *othismos*, lasted probably less than an hour. As one army started to get the worst of it, soldiers on that side would begin fleeing (rear rows first), and suddenly the entire losing army might be reduced to a chaotic rout as every man dropped his shield and spear, running in panic. Then the cavalry might come into play, either in slaughtering a fleeing enemy or in covering its own side's retreat.

Unlike modern armies, the Greeks did not usually make war all year long. The campaign season lasted from March until October, when weather conditions were right for encamping and moving large numbers of men. Furthermore, since most Greek soldiers were farmers by trade, the end of summer usually saw a consensus to stop campaigning and go home for the autumn sowings and vintage. One reason for the military superiority of SPARTA in the 500s to 400s B.C. was that its citizens were professional soldiers; they had no job but war. (Their fields at home were worked by a society of serfs, the HELOTS.)

In strategy, hoplite warfare was very simple, and relied on an element of cooperation between antagonists. The aggressor army would march into enemy territory, and the defending army would march out to meet it. Defenders did not like to stay confined behind their city walls—the Athenian statesman PERICLES' strategy for the Peloponnesian War was unique in its passive defense. Rather, defending armies would come out to protect the farms of the countryside, upon which the city's food supply might depend.

The size of hoplite armies could vary greatly. At the Battle of PLATAEA, against the Persians (479 B.C.), the allied Greek force was said to number over 38,000 hoplites. At the Battle of MANTINEA (418 B.C.), the opposing armies had about 3,600 hoplites each, plus light-armed troops. Other armies, fielded in local disputes, might be much smaller.

By the late 400s B.C. military minds had begun adjusting hoplite tactics. It had long been observed that an army's right wing tended to outperform its left wing—partly because the better troops were stationed on the right, but partly because the battle line itself would drift rightward on the approach to battle, as each man leaned right to gain the protection of his neighbor's shield; this rightward drift meant that each army advantageously outflanked the other on the right. To increase this advantage, the Boeotians at the Battle of Delium used a fortified right wing, 25 rows deep, and defeated an Athenian army (424 B.C.).

Similarly, the Spartans' superior tactics involved speedily gaining the upper hand on their right wing and then wheeling the right wing leftward, rolling up the enemy from the side. However, at the Battle of LEUCTRA (371 B.C.), the Theban general EPAMINONDAS beat the "invincible" Spartans by stacking his *left* wing 50 men deep and crushing the 12-man-deep Spartan right wing before it could destroy him.

Theban triumphs partly inspired the innovations of King Philip II of MACEDON (reigned 359–336 B.C.), who brought land warfare into its third stage with his development of the formation that modern scholars call the phalanx. To make his Macedonian peasantry effective against Greek hoplites, Philip forged a new-style heavy infantry—lighter-armored than hoplites and equipped with an innovative, 13- to 14-foot-long pike, the *sarissa*. The phalanx formation consisted of about 9,000 such infantry, arrayed 16 rows deep, with the first five rows presenting their pikes forward. On the model of Spartan organization, Philip's phalanx was articulated into 1,500-man battalions, for maneuverability. A well-trained phalanx enjoyed advantages over a hoplite army in terms of its weapons' reach and probably its level of fatigue. (The phalanx men were not crowded together as closely as the hoplites, nor were they so heavily armored.)

In battle Philip employed the phalanx mainly for defense—to halt and punish an enemy attack—while using his cavalry to charge against any gap that opened in the enemy formation. The phalanx was the anvil; the cavalry was the hammer. The superiority of Philip's tactics was proven at the Battle of CHAERONEA (338 B.C.), where his troops totally defeated a hoplite army of allied Greeks. The hoplite age was passing, 330 years after the Battle of Hysiae.

Philip's innovations helped produce the spectacular victories won by his royal son, Alexander the Great (reigned 336–323 B.C.). In military history, Alexander opened the door to further development of cavalry, under Persian influence and using larger Persian horse breeds. More important, Alexander's conquests changed the social-political basis of ancient warfare, by creating several large and wealthy Greco-Macedonian kingdoms around the eastern Mediterranean. The incessant warfare of these Hellenistic kingdoms—particularly Ptolemaic EGYPT against the Levantine SELEUCID EMPIRE (274–168 B.C.)—saw the use of huge, unwieldy phalanxes, upward of 20,000 men. These kingdoms were constantly trying to perpetuate a Greco-Macedonian soldiering class, so as to reduce reliance on mercenaries.

The most dramatic aspect of Hellenistic warfare was its use of war elephants, which Alexander first encountered in his Indus Valley campaigns (327–325 B.C.). Later generations of Greeks acquired elephants from Africa as well from India. Typically ridden by a handler and an archer or javelin man, the war elephant provided a kind of live "tank," used especially by the early Seleucid armies. At the Battle of Ipsus (301 B.C.), the 480 elephants of King SELEUCUS (1) helped to destroy his enemy, ANTIGONUS (1). Deployed usually to screen the phalanx and charge ahead, elephants were effective mainly as weapons of terror, to scare horses and soldiers not accustomed to them. But use of these unmaneuverable beasts had diminished by the mid-200s B.C., as commanders learned to deflect their charge with arrow or javelin volleys, or to let them pass harmlessly through by having the troops open corridors in their path.

(See also CRETE; DEMOSTHENES (2); LELANTINE WAR; PHEIDON; PHILIP V; PYRRHUS.)

warfare, naval   In the earliest centuries of Greek history (circa 1600–700 B.C.), warships served primarily as troop

transports—the troops in question were the oarsmen, who would leave their benches to go ashore as infantry. Eventually the sea itself became the scene of fighting, as defending states learned to send out ships to oppose an oncoming enemy fleet on the water. The historian THUCYDIDES (1) writes that the earliest remembered sea battle was fought circa 664 B.C. between the navies of two Greek states, CORINTH and CORCYRA. Thucydides mentions another early naval battle, circa 600 B.C., between Phocaean Greeks and Carthaginians in the western Mediterranean.

Such early sea fights, tactically crude, were probably conducted as infantry battles on water—that is, the opposing fleets would crowd together, allowing boarding parties of spearmen to fight it out hand to hand. But naval warfare was changed by two developments circa 700–430 B.C.: the invention and gradual spread of the superior Greek warship type known as the trireme; and the acceptance of ramming tactics over boarding tactics, as a more effective way to destroy the enemy without sustaining harm oneself. The famous Greek sea battles of the 400s B.C.—such as the Battle of SALAMIS (1) (480 B.C.)—featured trireme warships maneuvering to ram and sink the enemy.

The trireme was invented in response to a specific challenge: how to fit as many oarsmen as possible into a vessel that would still be seaworthy. Throughout Greek history, the oar was the main power source for any warship. Most war vessels carried one midship mast, capable of hoisting a single square sail; but the mast was not always used and typically could be removed. It was the oarsmen who powered the ship through battle and in passage through contrary winds or calms.

The more oarsmen, the greater the power—up to a point. The practical limit in length for an undecked wooden ship is about 100 feet; beyond that, the hull is liable to break apart. In a simple ship design, this 100-foot limit confined the total number of rowers to about 50—whence the Greek term *pentekontoros*, or penteconter, meaning a 50-oared, single-leveled vessel. Images in vase paintings and other surviving art show penteconters as resembling Viking longboats, with the oarsmen seated along one level in the open air. There was no deck—that is, no layer of planking across the hull's top. The rowers sat in double file—two men side by side on each bench, each man handling his own oar, which projected out (left or right) through thole-pins above the gunwale.

Circa 900–700 B.C., Greek shipwrights—probably copying Phoenician designs—learned to fit more oarsmen into a 100-foot hull. They did so by laying the benches at two different levels (still without any deck) and providing oarports below the gunwales for the lower row of oars.

This two-leveled type of Greek warship is known by modern scholars as the bireme. But the real breakthrough came circa 700 B.C., with the Greeks' development of a three-leveled warship (again, probably based on a Phoenician model). This was the ship that the Greeks called the *triērēs* (three-rowing or three-fitted) and that is now called the trireme—the ancient Greek warship par excellence.

However, the trireme caught on slowly. Penteconters, cheaper and easier to build, were still being used by some Greek states as late as the PERSIAN WARS (490–479 B.C.). The universal change to trireme navies came after 479 B.C.

A trireme typically carried 160 to 170 oarsmen, far more than prior designs could hold. The oars emerged in three rows on each side—one above the gunwale and two below. The added power meant greater speed, hence greater maneuverability in battle. The ample power also meant that triremes could take on more weight; therefore, they carried such array as partial decking fore and aft, vertical defensive screens, and an overhead central catwalk running the length of the vessel. These structures created room for a contingent of archers and HOPLITES, who did not row but served purely as ship's soldiers, to aim projectiles in battle and fend off enemy boarding attempts. By the time of the Peloponnesian War, the number of shipboard hoplites had settled at 10. Including crew, officers, and soldiers, the trireme might carry 215 men in all.

With its partial decking, the trireme hull could be built to about 120 feet long. The design was narrow, with a width at the gunwales of only 12 to 16 feet. There was scant room for passengers or any cargo aside from drinking water—most food supplies had to come from shore or from supply ships. A trireme had no accommodations for eating or resting. Confined to their crowded benches, the crew could not easily spend a night at sea; to cook and sleep, they had to camp ashore every night.

This basic restriction applied to all ancient warships and influenced all Greek naval strategy. A navy could conduct operations only within range of a friendly harbor or beaching place. To fight an offensive naval war, a state would have to create and maintain forward naval bases (as SPARTA did against Athenian interests in western ASIA MINOR in the later PELOPONNESIAN WAR, circa 413–404 B.C.). Relatedly, naval blockades of enemy harbors tended to be faulty, since fleets could not ride off the coast day and night.

Ramming tactics favored the faster and more skillful ship or fleet. The Greek ship's ram was a wooden post—sheathed in BRONZE, often styled as a boar's head—jutting forward of the prow at water level. The challenge in ramming was to avoid a head-on collision (which could sink both ships) and instead strike the enemy vessel abeam or astern. The classic ramming maneuver was called *diekplous*, "going through and out." When two opposing fleets moved to battle, each side would typically deploy in rows abreast. In *diekplous*, a fleet would suddenly shift from row-abreast to single-file formation, with ships prow to stern, and then veer inward, perhaps by squadrons, to penetrate the enemy's row-abreast formation. Proceeding through the enemy line, the attackers would come around to ram from behind.

Because the *diekplous* required sea room, a commander wishing to defend against this tactic could seek to offer battle inside a confined area—as the Greeks did at Salamis, against a superior Persian fleet. A way to equalize terms further was to try to compel a battle of boarding tactics—as the Syracusans did in their three victories over the Athenians inside SYRACUSE's harbor (414–413 B.C.). A commander intent on boarding tactics might outfit his ships beforehand with raised platforms to hold more soldiers. Generally, the superior-skilled fleets in Greek naval history sought to force battle on the open sea, where there was room for ramming by *diekplous*.

A fleet's skill and speed depended on crews' training and ships' maintenance (since wooden hulls need periodic drying out, caulking, etc.). The top Greek naval power of the 400s–early 300s B.C. was the city of ATHENS, with its standing fleet of 200 triremes (circa 431 B.C.) and its elaborate shipbuilding and drydocking facilities at PIRAEUS. The creator of this navy was the Athenian statesman THEMISTOCLES, circa 488 B.C. The heir to Themistocles' program was the statesman PERICLES (active circa 462–429 B.C.). Unlike many Greek states that employed foreign mercenary oarsmen, Athens employed its lower-income citizens as rowers. This military reliance on large number of poorer citizens effectively increased their political power in the DEMOCRACY.

Athens aside, the important navies in Greek history down to the late 300s B.C. belonged to Corinth, Corcyra, SAMOS, PHOCAEA, AEGINA, CHIOS, LESBOS, and Syracuse. Sparta, not originally a sea power, developed a navy with Persian funding in order to defeat Athens in the Peloponnesian War. Among non-Greek peoples, the Phoenicians (ruled by PERSIA by the mid-500s B.C.) were the Greeks' most dangerous enemies at sea. Farther west, the Greeks' naval opponents included the ETRUSCANS and those Phoenician colonists, the Carthaginians.

New navies and naval tactics arose in the later 300s B.C. Partly as a result of ALEXANDER THE GREAT's studious contact with powerful Phoenician fleets (332 B.C.), Greek shipwrights began building warships larger than the trireme. There are types called "four-rowing" (*tetrērēs*), "five-rowing" (*pentērēs*), and others, eventually up to "16," and even "30" and "40." Modern scholars believe that these names describe three-leveled ships containing multiple oarsmen per oar at one or more levels. The "five" type, usually called the quinquereme, may have employed two men per oar on the upper two rows, with one man per oar on the lowest row. The quinquereme was favored by Greek and Roman navies of the HELLENISTIC AGE (300–150 B.C.).

The late 300s–200s B.C. saw two Hellenistic empires battling for control of the eastern Mediterranean—Ptolemaic EGYPT and the dynasty of ANTIGONUS (1) (based first in Asia Minor, then in MACEDON). The best-known sea battle of this era is also called the Battle of Salamis, fought near a different Salamis, on the island of CYPRUS, in 306 B.C. There a fleet of 108 warships under Antigonus' son DEMETRIUS POLIORCETES destroyed 140 ships of PTOLEMY (1). This era also saw the rise of RHODES as a naval and mercantile power.

Ramming remained the primary battle tactic, but a more remote means of attack emerged in the form of shipborne catapults, shooting arrows. Too bulky for a trireme or quinquereme, a catapult might be fitted onto a merchant ship or onto two warships lashed together. Historians report the first use of such naval artillery at Alexander the Great's siege of Tyre. Demetrius Poliorcetes used naval arrow-catapults in his Battle of Cyprian Salamis and at his siege of Rhodes (305–304 B.C.). Later navies of the Hellenistic era probably also employed rock-throwing catapults.

The deadly use of fire against wooden ships is at least as old as the Peloponnesian War, when the Syracusans sent out a fire-ship against Athenian vessels in Syracuse harbor (413 B.C.). Later, circa 190 B.C., Rhodian warships carried fire-pots—containers of pitch, set ablaze and then slung by long poles onto nearby enemy ships. But the more fiendish tactic of shooting burning naphtha ("Greek fire") across the water at enemy vessels was not invented until the late 600s A.D., by the medieval Byzantines.

In the second half of the 200s B.C., naval control of the Mediterranean was won by the Romans, who had copied and adapted the designs and tactics of Hellenistic and Carthaginian navies.

(See also AEGEAN SEA; AEGOSPOTAMI; ARTEMISIUM; BOSPORUS; CARTHAGE; CIMON; CONON; CORINTHIAN WAR; CYZICUS; DELIAN LEAGUE; HELLESPONT; HIERON (1); IONIAN REVOLT; LYSANDER; PHOENICIA; POLYCRATES; ROME; SHIPS AND SEAFARING; TIMBER; WARFARE, SIEGE.

warfare, siege   Siege warfare entails the capture of forts and fortified cities. For much of Greek history, this process remained crude. An attacking army might try to take a site by storm—that is, by massed assault using scaling ladders. Alternatively, there was siege—the attackers would encircle the enemy walls with earthworks (a procedure known as circumvallation) and wait to starve out the defenders or to win treasonous help from within. The legend of a 10-year-long siege of TROY surely reflects the frustrations of primitive siege warfare.

By the late 700s B.C. the armies of Near Eastern empires (well in advance of the Greeks) were using machines and scientific methods to reduce enemy fortifications. The Greeks, impressed by Persian siegecraft in the era 546–479 B.C., began copying these tactics. These included use of battering rams to tear open an enemy wall, siege mounds to bring attackers to the top of a wall, and miners tunneling under walls. Yet the PELOPONNESIAN WAR (431–404 B.C.) saw only a few successful sieges, such as at PLATAEA or BYZANTIUM. Siege was still largely a waiting game, and a fortress city with a harbor (such as ATHENS) could almost never be captured without being completely blockaded, by both warships at sea and troops on land.

A breakthrough came circa 399 B.C., with the Greeks' development of the catapult. Credit goes to the engineers of the Syracusan dictator DIONYSIUS (1), who probably were copying the machines used by their enemies, the Carthaginians. The earliest type of Greek catapult resembled a giant medieval crossbow, set atop a stand. This wooden device shot a single, six-foot-long arrow and was typically aimed at troops visible at the parapet of the enemy wall. Although unable to knock down walls, the arrow could easily pierce a man's armor, hence the Greek name *katapeltēs*, "shield-piercer."

Volleys from multiple catapults provided valuable "covering fire" for troops conducting siege operations. Typically the besiegers' catapults would be perched inside of siege towers. The siege tower was a wooden structure, built as tall as the enemy wall and used as a kind of multistory armored car, with catapults set on internal parapets and sighted through portholes. Pushed forward on wheels, the tower approached the enemy wall gradually, while engineers prepared a roadbed ahead. Some towers were apparently designed to be pushed right up to the city wall, in order to bring up attacking troops on internal ladders or to employ a ground-level battering ram.

The first recorded use of Greek catapults and siege towers was at Dionysius' capture of the Carthaginian stronghold of Moyta, in western SICILY (397 B.C.). Dionysius' dramatic successes brought swift imitation and improvement by other Greek commanders. A more powerful type of arrow-shooting catapult arose, perhaps under the Thessalian dictator Jason of Pherae (reigned circa 385–370 B.C.). This new design used the principle of torsion. Instead of a horizontal, bending wooden bow, the torsion catapult featured two vertical cylinders of taut skeining (composed of human hair or animal sinew), set on either side of the front of a long wooden stock. Into each skein was slotted one wooden arm, extending outward; the arms' outer ends were attached to a bowstring. Winched back, the bowstring brought great tension from the twisting skeins to the arrow waiting in the groove of the stock.

Siegecraft in the Greek world reached its mature stage under the Macedonian king PHILIP II (reigned 359–336 B.C.), who successsfully coordinated the use of tower, catapult, and battering ram. Philip's catapults were torsion-powered arrow-shooters. It was Philip's son ALEXANDER THE GREAT who (among other achievements) pioneered the use of rock-firing catapults. Alexander's "stone-throwers" (*lithoboloi* or *petroboloi*) could hurl 170-pound boulders more than 190 yards, to batter apart enemy walls. Mounted aboard ships, these deadly machines helped Alexander to capture the island fortress of Tyre, in one of history's monumental sieges, conducted simultaneously by land and sea (332 B.C.).

Stone-throwing catapults were made of wood and were torsion-powered, looking much like the foregoing arrow-shooting type. This design remained the basic form for Greek catapults. (A different design, with one vertical arm, has been popularized by medieval European artwork and now represents what most people think of as a catapult. Although probably invented by the Greeks, this type was used later, by Roman armies.)

Alexander and his successors of the HELLENISTIC AGE (circa 300–150 B.C.) were the most ambitious besiegers the world had yet seen. The Macedonian prince DEMETRIUS POLIORCETES won his surname (Poliorkētēs means "city besieger") from his huge—but unsuccessful—siege of RHODES in 305–304 B.C. Demetrius' armament included two massive battering rams (iron-clad treetrunks, perhaps 130 feet long, hung inside mobile sheds and worked by hundreds of men) and an armor-plated siege tower called the Hēlēpolis, or "taker of cities," containing nine levels of stone-throwers and arrow-catapults.

Tactics of defense also improved in this era—with catapults on city walls shooting flaming projectiles against wooden siege machines—but generally the advantage had now swung to the besiegers. The Romans of the late 200s B.C. adopted Hellenistic siege tactics and brought them to new technical heights in conquering the Mediterranean. In 213–211 B.C. the Romans besieged and captured the Greek fortress city of SYRACUSE, which 200 years before had defied and destroyed a besieging Athenian army.

(See also ROME; SCIENCE; THESSALY; WARFARE, LAND; WARFARE, NAVAL.)

wine The making of wine—the alcoholic beverage produced by controlled fermentation of grapes—was practiced in the eastern Mediterranean long before the first Greek-speakers arrived, circa 2100 B.C. Greeks probably learned wine-making from the pre-Hellenic inhabitants of mainland Greece. By the Mycenaean era (circa 1600–1200 B.C.), wine played an important role in the economy and recreations of the Greeks. The Greeks' development of wine-making skills may be reflected in legends describing the arrival of a newcomer god, DIONYSUS, lord of viticulture and wine.

By the 500s B.C., the Greeks had perfected their wine production to the point where Greek wines were an important export item, prized throughout the Mediterranean. The Greeks exported wine to such advanced Eastern kingdoms as EGYPT, while Greek traders in THRACE, southern Gaul, and elsewhere used wine to exploit the native peoples and to acquire local goods cheaply (much as European fur traders used brandy on the northeast American Indians in the 1600s A.D.).

The hilly terrain of Greece was (and is) apt for vine-growing, and renowned wines were produced in particular by some of the islands, including THASOS, CHIOS, LESBOS, RHODES, Lemnos, and Cos. But wine was made throughout the Greek world and was cherished by both the peasantry (as a mode of riotous release, on festival days sacred to Dionysus) and by the upper classes (as an object of ritualized consumption, at the type of male drinking party known as the SYMPOSIUM).

In Greece, grapes were harvested in September. Juice was produced by workers first treading the grapes in vats, then crushing the remaining pulp in wine presses. The juice was stored indoors in tubs and allowed to ferment for six months; then it was filtered through cloth and sealed inside storage jars (*amphorai*). Unlike modern wine-makers, the Greeks could not halt the natural fermentation. Therefore, Greek wines had to be consumed within three or four years. These wines had more natural sugar than modern wines, and hence more alcohol, perhaps 18 percent (versus 12.5 percent for most modern consumer wines).

Not surprisingly, the ancient Greeks customarily mixed water into their sweet, potent wine just before drinking—the typical ratio was two or three parts water to one part wine. Such distinctive Greek POTTERY forms as the *psuktēr* (cooler) and *kratēr* (mixer) were developed expressly for the chilling and diluting of wine. Once cooled, the straight wine would be poured into a krater, mixed with water, and ladled out. Undiluted wine was considered unhealthy, but watered wine, if used moderately, was viewed as medicinal—as wine is still viewed today in some Mediterranean countries.

(See also CELTS; FARMING; MASSALIA.)

women The inferior role of women in ancient Greek society reminds us that the Greeks remained in some ways a primitive people. Even in classical ATHENS, at its height of democratic enlightenment (mid-400s B.C.), women were considered second-class citizens at best. A female Athenian citizen, protected by law, was far better off than a slave or a metic (resident alien); but the citizen woman also was forbidden to own much property, to inherit in her own name, to vote, or to attend political debate in the ASSEMBLY.

Women might practice humble trades such as street-cleaning, but this was a sign of hardship, and very few skilled or lucrative trades were open to women. Instead, a woman was considered to be the ward of a man—her husband, her father, or her guardian (such as a brother).

Female SLAVES might run errands in public, but a respectable woman's place was in the home—specifically, in the house's interior and secluded "women's chambers." There the wife or mature daughter was expected to perform—or, if the household was wealthy, to oversee—such domestic duties as child care, cleaning, and spinning and weaving. The only crucial public role of citizen women was to give birth to young citizens for the state.

Of all the major Greek city-states, only warlike SPARTA granted its citizen women any degree of freedom, allowing them to own quantities of property and encouraging such male-imitative pastimes as girls' gymnastics. The unusual freedom of Spartan women probably arose from their roles as estate managers during the absences of their soldiering husbands and kinsmen. Spartan women also had a unique reputation for promiscuity. The philosopher ARISTOTLE (mid-300s B.C.) criticized Spartan society for allowing its women too much freedom and influence. Of all Greek thinkers, only PLATO, in his dialogue *The Republic* (circa 360 B.C.), could imagine a utopian society where women and men would be equally eligible for roles in government.

As with all other social topics from ancient Greece, the existing evidence for women's lives comes overwhelmingly from Athens in the 400s–300s B.C. This evidence includes extant stage drama, courtroom speeches, and archaeological remains such as tombstone inscriptions.

Woman as worker. A woman places clothing or bedding in a storeroom chest, in a terra-cotta relief from the Greek city of Locri, in southern Italy, circa 460 B.C. This woman is probably a slave or servant, but even upper-class Greek ladies might spend their days on certain domestic duties, such as spinning and weaving.

Woman as entertainment. A young woman plays a double flute, in a Greek marble bas-relief, circa 460 B.C. A flute-girl would generally be the kind of slave known as a *hetaira* (companion). She would provide music at male drinking parties, often accompanying the guests' poetry recitals, and would also be obliged to give sexual favors. This carving appears on one side of the marble Ludovisi Throne—probably in fact an altar to the goddess Aphrodite—now in a museum in Rome.

Nothing is known for certain about the role of women during the era of MYCENAEAN CIVILIZATION (circa 1600–1200 B.C.). But a cautious reading of Greek MYTH and particularly HOMER's poems suggests that Mycenaean baronial society did allow considerable power to a few highborn ladies. For example, the queens PENELOPE and CLYTAEMNESTRA in the *Odyssey* are each shown as managing a kingdom in her husband's absence.

The segregation and devaluation of women in Greece began sometime in the era 1100–700 B.C. and may be connected to two related changes: the emergence of the Greek city-state, with its wide, citizen-based military organization; and the idealization of male strength and beauty, as expressed most fully in institutionalized male HOMOSEXUALITY. The misogyny expressed by the poets HESIOD (circa 700 B.C.) and SEMONIDES (late 600s B.C.)—in contrast with Homer's often-charming portrayal of women—surely indicates this social shift.

In archaic and classical Greece, circa 800–300 B.C., women were seen as the opposites or subjugated opponents of men. Male Greek society valued rational discourse, military courage, and physical endurance and self-restraint. Women were believed to be irrational, fearful, and ruled by physical desires. One reason why women were kept secluded at home is that they were thought liable to sexual seduction (or other mischief) if they ventured out unescorted. "The highest honor for a woman is to be least talked about by men, whether in praise or criticism," declared the Athenian statesman PERICLES in his famous Funeral Oration of 431 B.C.

Only in RELIGION were Greek women allowed to participate in public life to any degree. Holy days, such as the autumn sowing festival of the Thesmophoria, provided occasions when women were expected to leave the house and congregate for religious ceremonies. Also, certain priesthoods were reserved for women, such as the office of Pythia or priestess of APOLLO at DELPHI. These religious duties probably date back to a prehistoric belief in the magical properties of female fertility. Similarly, the large

Woman as ideal beauty. A statue of an aristocratic maiden (*korē*), carved in marble circa 515 B.C. and set up as an offering on the Athenian acropolis. Such *korai* statues were the female counterpart to the male *kouroi* of this era.

role played in Greek mythology by such goddesses as ATHENA or HERA may represent vestiges of a non-Greek religion, adopted by the first arriving Greek tribes circa 2100 B.C.

There is some evidence that female infanticide was regularly practiced at classical Athens and elsewhere. A father could choose whether to officially accept a child or not, and many men might prefer to raise sons, who could help with the farmwork, inherit the estate, and so on.

Girls had a narrow upbringing. A girl of the middle and upper classes grew up in the women's chambers of the house, perhaps rarely seen by her father. At age seven, she might start attending a girls' school, but only long enough to acquire literacy and numeracy. In various Greek cities, she might be considered ready for MARRIAGE by age 14 or even 12. The poet SAPPHO (circa 580 B.C.) has some poignant verses on the plight of an aristocratic girl who must go from the beautiful security of childhood to the strangeness and anxiety of marriage.

Marrying young, women bore children young and died young. Medical examination of extant skeletons from ancient Greece suggest that women typically died around age 36, having borne four children. The estimated average age for men at death was 45, but (since husbands might be 20 years older than their wives) many wives must have survived their husbands and remarried. At classical Athens and elsewhere, a woman's inheritance would legally pass to her new husband's hands at her remarriage, and so heiresses and widows were in great demand.

The oppression of women seems to have lightened somewhat in the changed world of the HELLENISTIC AGE (circa 300–150 B.C.). The royal courts at ALEXANDRIA (1), ANTIOCH, and PELLA created a breed of rich, ruling-class women who were able to influence public and cultural life. These royal examples probably created new opportunities for other women—at least for wealthy citizen women. A surviving public inscription from CYMĒ honors a certain Archippe for her public donations (100s B.C.), while papyrus documents from Ptolemaic EGYPT show women to have been active as lessors, creditors, and debtors much more so than at Athens in prior centuries.

Relatedly, Hellenistic culture placed a new, greater value on female intelligence and sexuality. The Athenian philosopher Epicurus (circa 300 B.C.) is said to have admitted women to his school on equal terms with men. Female beauty reemerged as an ideal for painters and sculptors—most famously in the Athenian sculptor PRAXITELES' celebrated nude statue of the goddess APHRODITE (mid-300s B.C.). In literature, poets such as APOLLONIUS (circa 270 B.C.) showed a new interest in female psychology and male-female romance.

(See also ALCMAN; AMAZONS; EPICUREANISM; ERINNA; EURIPIDES; FUNERAL CUSTOMS; MAENADS; METICS; POLIS; PROSTITUTES; SCULPTURE.)

wrestling

Among the sports-loving Greeks, wrestling was a highly popular and well-organized event. Less brutal than BOXING and PANKRATION, the other two major Greek combat sports, wrestling was considered a valuable element in a boy's EDUCATION and military training as well as

Two wrestlers, from a bas-relief scene on a marble Athenian statue base, circa 510 B.C. The man on the right grips his opponent's left wrist and forearm, perhaps working for an arm-drag. Note the two men's difference in size: The ancient Greeks observed no weight classes, and size could be a distinct advantage. This statue base was probably part of a young aristocrat's gravestone, showing one of his favorite pastimes in life.

an appropriate exercise for well-bred adult men. Most sizable Greek towns of the classical age contained at least one *palaistra* (enclosed wrestling court).

Wrestling was among the most prestigious events at great athletic festivals such as the OLYMPIC or PYTHIAN GAMES. In addition to separate competitions for men and boys, there were wrestling contests that served as the final round in the five-event PENTATHLON.

The patron deity of wrestlers was the god HERMES, who was said to have invented the sport. In fact, wrestling predates the Greek world, having been practiced in the ancient Near East. The Greeks adopted the sport early in their history, maybe in the Mycenaean era (1600–1200 B.C.). Wrestling is described in HOMER's poems—in the *Iliad*; for example, the heroes ODYSSEUS and AJAX (1) engage in a friendly match (book 23). Other mythical wrestlers included the hero HERACLES, who in the course of his 12th Labor had to grapple the giant Antaeus to death.

In its rules, the classical Greek sport was much like modern collegiate free-style wrestling, only tougher. There were no weight classes for the ancient Greeks—advantage favored size and weight. Punching was forbidden, but other harmful moves, such as strangleholds, shoulder-throws, or breaking an opponent's finger, were allowed.

Like other Greek athletes, wrestlers competed naked and were typically slicked down with olive oil (which prevented abrasions but also had a ritualistic significance). They competed atop soft ground; practice sessions were generally held in a mud pit, and formal matches took place on the *skamma*, a special bed of fine sand. At grand events, the matches might be held in a stadium full of spectators.

The bout began with the two competitors standing, facing each other, and ended with a "fall," when one man was forced down in such a way that his back, his shoulders, or his chest and stomach touched the ground. (Alternatively, a man could lose by being thrown off the wrestling ground.) To win in an official contest, a man had to gain three falls against his adversary; therefore, a single match could run to as many as five bouts.

(See also SPORT.)

writing　The use of written symbols to convey words or ideas predates the Greeks by many centuries. In the third millennium B.C., well before the start of Greek civilization, the Sumerians and other Mesopotamian peoples had developed a script called cuneiform. The Egyptians of this era had several forms of writing, including the famous hieroglyphic pictograms that can still be seen on inner pyramid walls and the like. The Mycenaean Greeks had by 1400 B.C. adapted a form of writing from Minoan CRETE. Modern scholars call this Mycenaean script LINEAR B.

The use of Linear B ended with the collapse of MYCENAEAN CIVILIZATION circa 1200 B.C. Thereafter, the invention of the versatile Greek ALPHABET, circa 775 B.C., brought widespread literacy to the Greek world and allowed the Greeks to produce a great literary culture. The emergence of the first Greek literature—HOMER's epic poems *Iliad* and *Odyssey* is clearly a result of the invention of the alphabet, which allowed such orally developed stories to be written down (circa 750 B.C.).

The earliest extant specimen of Greek alphabetic script is a verse inscription scratched onto a Geometric cup, found in a grave at the Greek colony of PITHECUSAE, in western ITALY, and dated to about 725 B.C.: *I am the cup of Nestor and this I say: whoever drinks from me will instantly be smitten by lovely Aphrodite.* This inscription, coming from what was then the Greek world's western frontier, is striking evidence of the spread of the alphabet among the Greeks within one or two generations. General literacy also is indicated by extant graffiti and public inscriptions of the latter 600s B.C., from THERA, CHIOS, and elsewhere. But the most dramatic specimen of early Greek writing is the graffito scratched onto the left leg of a colossal statue of the Egyptian pharaoh Ramses II at Abu Simbel, 700 miles up the Nile. Written in about 591 B.C. by Greek mercenary soldiers in the service of Pharaoh Psamtik II, the message, still visible today, records the names of the expedition's leaders. It is the earliest known announcement that "Kilroy was here."

More typical of early Greek writing materials were clay tablets or the skins of animals, which could be pumiced smooth and stitched together to form a roll. For scratching in clay, a stylus was used, but animal skins took pen and ink. (The pen was merely a quill or the like, and the ink often was a vegetable mixure.) By the 600s B.C., papyrus—cheaper and plentiful—had replaced animal skins for Greek writing. The papyrus plant, which grew only in the Egyptian delta, had a pith that, when cut into strips, could be arranged and pressed so as to create a sheet of paperlike material. These sheets would then be glued together to form a continuous roll, normally of 30–35 feet. When two wooden pins were fastened to the ends of the roll, the result was a scroll, which could be rolled back and forth between the pins. The papyrus scroll was the ancient Greeks' "book" (*biblion*), the premier tool for reading and writing. Long works were produced in multiple scrolls, or "books." Most lengthy writings from the Greco-Roman world—such as those of Homer or Vergil—are still divided editorially into different books, even if the whole work is now contained in one modern volume.

The drawback of papyrus was that it came exclusively from Egypt. After the Greek-Egyptian king Ptolemy V discontinued papyrus exports circa 200 B.C.—supposedly to hamper the rival book-collecting of King Eumenes II of PERGAMUM—the Pergamenes responded by inventing parchment (Greek: *pergamēnē*). Parchment (or vellum, as it is now known) was made from cattle, goat, or sheep skin processed in an improved way so as to smooth it down to papyruslike quality. Today important documents, such as college diplomas, still are printed on vellum, hence the nickname "sheepskin."

(See also BYBLOS, EPIC POETRY.)

X

Xanthippus Athenian soldier and politician who lived circa 525–475 B.C. and was active in the PERSIAN WARS.

Xanthippus (Yellow Horse) was born to an aristocratic family at ATHENS. In the days after CLEISTHENES' (1) democratic reforms (508 B.C.), Xanthippus began his political career by marrying Cleisthenes' niece Agariste. Uncle and niece both belonged to the noble and ambitious Alcmaeonid clan. Agariste and Xanthippus had three children, the youngest of whom was the future left-wing leader PERICLES, born in 494 B.C.

With family wealth and connections, Xanthippus rose to political prominence in 489 B.C., when he prosecuted the soldier-statesman MILTIADES on a charge of deceiving the people. With Miltiades' downfall, Xanthippus briefly became the foremost politician in Athens, but was himself soon on the defensive, probably at the hands of the left-wing leader THEMISTOCLES. In 484 B.C. Xanthippus was ostracized by the Athenians—one of five important OSTRACISMS in the 480s B.C.

Although the normal term of banishment in an ostracism was 10 years, Xanthippus was recalled early, in the state emergency created by the Persian king XERXES' invasion of Greece (480 B.C.). Elected as a general for the year 479 B.C., Xanthippus commanded the Athenian contingent at the land battle of MYCALE, where the Greeks liberated IONIA. Elected general again, for 478 B.C., he led a Greek fleet against the Persian-held fortress of SESTOS, on the HELLESPONT; after a grueling siege, Xanthippus captured the city and crucified the hated Persian governor. Xanthippus returned to great acclaim at Athens, but died within a few years, perhaps from injuries received at Sestos. By 472 B.C. the young Pericles had inherited his father's estate (the other son had died) and was embarked on his own political career.

(See also ALCMAEONIDS.)

Xenophanes Greek poet and philosopher, born circa 570 B.C. at Colophon, in ASIA MINOR, but active in Greek SICILY and southern ITALY. There he traveled from city to city, presenting his PHILOSOPHY in verses (some of which have survived as quotations in later writers' work). Xenophanes apparently lived into his 90s. He is said to have founded the philosophical school at the Italian-Greek city of Elea and to have tutored there the student destined to be the most famous Eleatic philosopher, PARMENIDES.

It is known that Xenophanes left Colophon at age 25—probably to escape the invading Persians under King CYRUS (1). Like his contemporary PYTHAGORAS, Xenophanes brought to the western Greeks the rationalism and sophistication of his native IONIA, and he seems to have caused an intellectual sensation with his attack on traditional Homeric theology. "HOMER and HESIOD," Xenophanes wrote, "have attributed to the gods everything that is shameful and disgraceful among human beings—theft and adultery and deceiving each other." Xenophanes also expressed the enlightened theory that humans have created gods in their own image: The Ethiopians say that their gods are black and snub-nosed; the Thracians make their gods blue-eyed and red-haired. "But if cattle or horses or lions had hands," Xenophanes wrote, "then horses would portray their gods as horses, and cattle as cattle." Rather, said Xenophanes, there is one god, who is in no way like humans, either in form or thought. Xenophanes' concept of God—as a single, unmoved mover—may have inspired Parmenides' more elaborate concept of a unified, motionless Reality.

Remarkably, Xenophanes' also intuited the nature of fossils. Observing the impressions of fish and seaweed in rock quarries near SYRACUSE, Xenophanes supposedly guessed that that area had once been under water and that the earth's land in general is created from, and returns to, the sea.

A brilliant, dilettantish observer rather than a deep thinker, Xenophanes stands as an embodiment of Ionian rationalism and as an intellectual "pollenizer" for Greek Sicily and Italy. His lifetime saw the violent end of the Ionian Enlightenment and the emergence of new intellectual centers, among the mainland and western Greeks. One poignant fragment by Xenophanes gives a brief self-portrait of a wanderer, dispossessed by the Persians' conquest of his native land: "In winter, as you lie by your fire after dinner, drinking wine and cracking nuts, speak thus to the wayfarer at your door: 'What is your name, my friend? Where do you live? What is your age? How old were you when the Persians came?' "

(See also PERSIAN WARS; RELIGION.)

Xenophon Athenian historian, soldier, and gentleman, who lived circa 428–circa 354 B.C. Xenophon wrote several histories and treatises, but is best remembered for producing the most exciting memoir from the ancient world—the *Anabasis* ("expedition to the interior"), describing his adventures in leading 10,000 beleaguered Greek mercenary soldiers out of the heart of the Persian Empire (401–400 B.C.). Among Xenophon's other surviving work is his valuable but biased history titled *Hellenica* (Greek history). Beginning at the point where THUCYDIDES' (1) history of the PELOPONNESIAN WAR abruptly breaks off, with the events of 411 B.C., Xenophon's *Hellenica* recounts the turbulent events of his own lifetime, from the later years of the war to the rise of THEBES and the destruction of Spartan hegemony in Greece (411–362 B.C.). In terms of breadth, objectivity, and analysis, Xenophon as a historian is far inferior to his predecessor Thucydides.

Born into a rich, aristocratic family at ATHENS, Xenophon was a contemporary of the philosopher PLATO. Like Plato, the young Xenophon, although not a deep thinker, became a disciple of the Athenian philosopher SOCRATES. Today our most important source of information about the real-life Socrates (as opposed to the Socrates of Plato's scheme) is Xenophon's memoir titled *Memorabilia*.

The Peloponnesian War was being waged throughout Xenophon's entire childhood and youth, and—like other Athenian aristocrats—Xenophon was to show strong pro-Spartan sympathies during his life. He probably partici-pated in the war's last years—perhaps in the naval battle at Arginusae (406 B.C.), which looms large in his *Hellenica*, and perhaps in CAVALRY patrols outside the city. Also in these years, he married an Athenian woman, Philesia, and had two sons, Gryllus (named for Xenophon's father) and Diodorus.

In 401 B.C., after Athens' defeat in the war and the disastrous reign of the THIRTY TYRANTS, Xenophon left his native city for what was fated to be an absence of 35 years. As a member of the discredited Socratic circle, he probably felt uncomfortable in the restored DEMOCRACY. (And, in fact, within two years the Athenians would con-demn and execute Socrates.) Xenophon went to ASIA MI-NOR as one of about 10,000 Greek mercenary HOPLITE soldiers in the service of the Persian prince Cyrus but, as it turned out, Cyrus used this force in his attempt to seize the Persian throne from his brother, Artaxerxes II. After Cyrus was killed and his army defeated at the Battle of Cunaxa, near Babylon (401 B.C.), the frightened and demor-alized surviving Greek troops elected Xenophon to lead them the 1,500 miles out of hostile territory. Four months later, the Greeks finally emerged from the Persian interior, at the Greek colony of Trapezus, on the southeast coast of the BLACK SEA. These exploits are recounted in the *Anabasis*.

Circa 399 B.C., during Xenophon's absence, the Atheni-ans declared him exiled forever; the probable causes were Xenophon's pro-Spartanism and prior friendship with Soc-rates. The following years saw Xenophon doing military service under the Spartan king AGESILAUS. During the CORINTHIAN WAR (in which Sparta and Athens were ene-mies, 395–386 B.C.), Xenophon probably served against his fellow Athenians. Xenophon remained a close associate of Agesilaus, whose deeds are recounted in the *Hellenica* and in Xenophon's laudatory biography *Agesilaus*.

In around 388 B.C. Xenophon and his family were settled by their Spartan friends at an estate near OLYMPIA. There Xenophon devoted himself to writing and to the pursuits addressed in his treatises *On Hunting* and *On Household Management*. In a spirit of gratitude toward Sparta, he wrote a laudatory account of the Spartan governmental system, the *Lakedaimonion politeia*. Other works of the 380s and 370s B.C. include the treatise *On Horsemanship* (our only surviving full account from ancient Greece on this subject) and the *Education of Cyrus*, a historical novel that imagines the ideal tutoring received by the old-time Persian

ruler CYRUS (1) (not the Cyrus of the *Anabasis*). Some of these works seem to have developed out of Xenophon's notes toward the EDUCATION of his two sons.

In 371 B.C., after Sparta's defeat at Theban hands at the Battle of LEUCTRA, the people of ELIS (near Olympia) rose in defiance of Sparta and reclaimed local lands, including Xenophon's estate. Xenophon, then nearly 60, was eventu-ally allowed by the Athenians to return.

At Athens he lived his last decade or so, finishing the *Hellenica* (which ends with a description of the Theban victory over the Spartans at the second Battle of MANTINEA in 362 B.C.). Xenophon's own son Gryllus was killed at this battle, fighting on the Spartan side. On receiving the tragic news, Xenophon is said to have replied, "I knew my son was mortal."

Xenophon rewards modern study because he is the com-plete Athenian gentleman. He embodied many of the val-ues and interests of his social class—soldiering, horsemanship, estate management, pro-Spartanism (exag-gerated, in Xenophon's case), respect for education, and so on. He had verbal skill (his prose is plain and workman-like) and a strong interest in right and wrong. However, his youthful dabbling in PHILOSOPHY clearly had more to do with Socrates' dynamic personality and the intellectual milieu of late-fifth-century B.C. Athens than with any ge-nius of Xenophon's.

His flaws as a historian are apparent in his failure to present a full, unbiased account in his magnum opus, the *Hellenica*.

Xerxes Persian king who led the failed invasion of GREECE in 480 B.C. Our knowledge of Xerxes comes mainly from the ancient Greek historian HERODOTUS, whose view-point is hostile; he portrays the king as sometimes reflective and aesthetic-minded, but also intoxicated by power and capable of great lust, cruelty, anger, and cowardice. "Xerxes" is the Greek form of the name, which in Persian sounded like Khshah-yar-shan and meant "king of kings." In the biblical book of Esther, Xerxes is called King Aha-suerus and is pictured favorably.

As son and heir of the brilliant king DARIUS (1), Xerxes was about 32 years old when he came to the throne in late 486 B.C. Soon he turned his attention westward, to resume his father's conflict with the mainland Greeks. Various Persian inscriptions of Xerxes' reign make it clear that he was a pious worshipper of Ahura Mazda, the Zoroastrian supreme god, and he may have thought he was on a divine mission to conquer Greece.

After four years' preparation, including the spanning of the HELLESPONT with elaborate twin bridges of boats, Xerxes led a mighty host—perhaps 200,000 troops and 600 warships—to Europe. Descending through THRACE and MACEDON, he subdued northern and central Greece, but his navy was first eroded due to storms and a sea battle, and then destroyed when Xerxes overconfidently decided to attack the Greek fleet inside SALAMIS (1) channel. After the Salamis defeat in late summer 480 B.C., Xerxes hurried back to Asia with most of his force, leaving behind a Persian army, which the Greeks smashed at PLATAEA the following summer.

Little is known of Xerxes' life afterward. He apparently

devoted himself to construction of royal buildings at Persepolis, the empire's summer capital, and was assassinated in a palace intrigue around 465 B.C. He was succeeded by his son Artaxerxes I.

To the Greeks, Xerxes supplied the prime living example of the sin of HUBRIS—insane pride that leads to divinely prompted self-destruction. The abiding picture of him in Western tradition (whether true or false) is a vainglorious emperor who, after a Hellespont storm had wrecked his boat-bridges, ordered his men to punish the channel by lashing its waters with whips.

(See also PERSIA; PERSIAN WARS.)

Z

Zanclē Greek city of northeast SICILY, located beside the three-mile-wide Strait of Messina, which separates Sicily from the "toe" of ITALY. Located on a crescent-shaped peninsula enclosing a fine harbor, Zanclē (now the modern Sicilian city of Messina), was founded circa 725 B.C. by Greek settlers from CUMAE and CHALCIS. Its sister city was RHEGIUM, across the strait.

Zanclē's original setters were of the Ionian-Greek ethnic group, but circa 490 B.C., Zanclē was repopulated by settlers of Dorian-Greek identity, commanded by the powerful tyrant Anaxilas of Rhegium. Chief among these newcomers were refugees from the Peloponnesian region of MESSENIA, which was suffering under Spartan domination. Anaxilas, whose ancestral home was Messenia, then renamed Zanclē as Messana (in the Doric dialect) or Messene (in Ionic form).

Zanclē-Messana eventually became an ally of Syracuse against ATHENS and the Carthaginians in the latter 400s and early 300s B.C. In 264 B.C. the city was captured by the Carthaginians—an episode that eventually brought on the titanic First Punic War between CARTHAGE and ROME (264–241 B.C.). After the Roman victory, Messana thrived as a Roman ally.

(See also DORIAN GREEKS, IONIAN GREEKS.)

Zeno (1) See STOICISM.

Zeno (2) See PARMENIDES.

Zeus King of the gods, spiritual father of gods and men, and dispenser of divine justice in the universe, according to Greek RELIGION and MYTH. Zeus' name appears on Mycenaean Greek LINEAR B tablets dated to 1400 B.C., but his cult is probably much older than that. Unlike many of the classical Greek gods, Zeus was truly Greek in origin. Probably he was originally a Sky Father—a weather god and chief deity—worshipped by the Greek tribes who first came southward from the Balkans into the land of Greece circa 2100 B.C. In name and function, this primitive Zeus seems related to celestial father gods of other ancient Indo-European peoples, such as the Roman Jupiter (*Zeu-pater*) and the Germanic Tiu. The name *Zeus* is also related to the Greek word *dios*, "bright." Throughout antiquity Zeus never lost his function as a weather god. "Sometimes Zeus is clear, sometimes he rains," writes the poet THEOCRITUS (circa 265 B.C.). Zeus is portrayed as sending thunderstorms against his enemies in HOMER's epic poem the *Iliad* (written down circa 750 B.C.), and in art he is often shown hurling a thunderbolt.

But it was Zeus' moral dimension as orderer of the universe that made him so important to Greek civilization. During the Mycenaean age (1600–1200 B.C.) Zeus was the supreme king over a human society organized around kings; from his cloud-shrouded palace atop Mt. OLYMPUS, in Thessaly, Zeus ruled the subordinate gods and oversaw human events in the world below.

As the age of Greek kings passed away during the 1100s–900s B.C., Zeus retained his preeminence, developing into a kind of chief judge, peacemaker, and civic god. Myths describe how he had destroyed various primeval monsters and forged peace in place of violence. The poet HESIOD (circa 700 B.C.) invokes Zeus as the lord of justice, and the god's other epithets include Kosmetas (orderer), Polieos (overseer of the *polis*—the city), Soter (savior), and Eleutherios (guarantor of political freedoms). He maintained the laws, protected suppliants, instituted festivals, gave prophecies (such as at his important oracle at DODONA), and in general oversaw the fruits of civilized life. This high moral

The Artemisium Zeus: a 6.5-foot-tall bronze statue, sculpted circa 455 B.C. and recovered in the 1920s from the sea-channel off Cape Artemisium, at the northern tip of Euboea. This most magnificent of surviving Greek bronzes stands in the National Museum in Athens. Previously identified as Poseidon, the figure is now thought to show Zeus hurling a thunderbolt (since lost). With such bolts, Zeus vanquished the Giants, the monster Typhon, and other enemies of civilization and peace.

The head of the Artemisium Zeus, with its plaited hair and beard. The eye sockets, which in ancient times would have been inlaid with onyx or another material, now stand empty, revealing the statue's hollow core. This work, beautifully conveys the power and grandeur of the king of the gods.

conception of Zeus reaches its epitome in certain choral odes in the tragedies of the great Athenian playwright AESCHYLUS—for instance, in *The Suppliants* (performed 472 B.C.) and *The Eumenides* (458 B.C.). By the 200s B.C. the Stoic philosophers tended to see Zeus as the single, universal deity.

Zeus' central myths—as recorded particularly by Hesiod in his *Theogony*—tell how the young Zeus usurped the heavenly kingdom from his father, CRONUS, then reordered this kingdom and defended it against attacks by monstrous demigods. In the myth, Cronus, king of the race of TITANS, always ate each of his newborn children, but baby Zeus was spirited away by his mother, Rhea, and brought up on the island of CRETE. Returning, Zeus led a revolt against his father, banished him, and installed himself as king of a new race of immortals, many of whom he fathered. As king of the gods, Zeus ruled the sky; his brothers, POSEIDON and HADES, ruled the sea and Underworld, respectively.

In defending their kingdom, Zeus and the other gods withstood three separate assaults by the violent offspring of GAEA (Mother Earth). The three rebellious parties were the GIANTS, the multiserpent-headed TYPHON, and the twin brothers called the Aloadae, who piled Mt. Ossa atop Olympus and Mt. Pelion atop Ossa in their attempt to storm heaven. Originally such mythical battles may have commemorated the first Greeks' conquest of mainland Greece and the subjugation of the prior inhabitants' religion by the invaders' cult of Zeus.

During the TROJAN WAR (as described in Homer's *Iliad*), Zeus sided with the aggrieved Greek hero ACHILLES and brought misfortune to the Greeks, until Achilles' quarrel with AGAMEMNON was resolved. Yet even Zeus had to yield to FATE, and so was unable to save his mortal son Sarpedon from preordained death in the fighting at TROY (*Iliad,* book 16).

However, Zeus had a second, more grotesque mythological dimension, which certain Greek thinkers found puzzling or repugnant. For "the father of gods and men" (as the poet Homer calls him) was quite literally that, insofar as his numerous love affairs produced many offspring, both human and immortal. His wife and sister, the goddess HERA, bore him the gods ARES, HEPHAESTUS, and Hebe, but Zeus' liaisons with other goddesses and mortal women (often involving rape or deception) produced the deities ATHENA, APOLLO, ARTEMIS, HERMES, and DIONYSUS, the hero HERACLES, and others. Among the humans whom Zeus raped or seduced were: the Spartan queen Leda (whom he approached in the form of a swan and who gave birth to two pairs of twins—CASTOR AND POLYDEUCES and HELEN OF TROY and CLYTAEMNESTRA), the Argive princess Danaë (whom he visited as a shower of gold and who bore him the hero PERSEUS [1]), the Phoenician princess EUROPA (whom he abducted as a bull and who bore him MINOS and Rhadamanthys), and the Trojan prince GANYMEDE (who was abducted by an eagle, perhaps Zeus in disguise, to become Zeus' lover and cupbearer atop Olympus).

The legends of Zeus' sexual exploits with human women probably arose individually throughout Greece during the Mycenaean age, in an attempt to connect the great god with some local noble family's genealogy. But the unsavory discrepancy in these myths—the god of justice descending to earth to commit rape and adultery—was commented on by intellectuals such as the Ionian philosopher XENOPHANES (latter 500s B.C.) and was exploited in the tragedies of the innovative Athenian playwright EURIPIDES (latter 400s B.C.).

In art, Zeus was often shown as a bearded, kingly figure in vigorous middle age. His weapons were the thunderbolt and the aegis (a magic cloak or shield, originally imagined as a stormcloud). Of his innumerable shrines throughout the Greek world, probably the most magnificent was his Doric-style temple built in the mid-400s B.C. at OLYMPIA, at the site of the great sports festival held in his honor— the games of Olympian Zeus. Inside the temple was an immense GOLD-and-ivory statue of the enthroned god, sculpted by PHIDIAS and reckoned as one of the SEVEN WONDERS OF THE WORLD for its size and grandeur.

(See also OLYMPIC GAMES; PROMETHEUS; STOICISM.)

Zeuxis   See PAINTING.

BIBLIOGRAPHY

Adcock, F. E. *The Greek and Macedonian Art of War*. Berkeley: University of California Press, 1957.

Anderson, J. K. *Xenophon*. New York: Charles Scribner's Sons, 1974.

Andrewes, Anthony. *Greek Tyrants*. London: Hutchinson University Library, 1974.

Austin, M. M., ed. *The Hellenistic World from Alexander to the Roman Conquest: A Selection of Ancient Sources in Translation*. New York: Cambridge University Press, 1981.

Avery, Catherine B., ed. *The New Century Classical Handbook*. New York: Appleton-Century-Crofts, 1962.

Barber, Robin. *Blue Guide: Greece*. New York: W. W. Norton, 1987.

Barnes, Jonathan. *Aristotle*. New York: Oxford University Press, 1982.

Bickerman, E. J. *Chronology of the Ancient World*. New York: Thames and Hudson, 1968.

Biers, William R. *The Archaeology of Ancient Greece: an Introduction*. Ithaca, N.Y.: Cornell University Press, 1980.

Blanchard, Paul. *Blue Guide: Southern Italy: From Rome to Calabria*. New York: W. W. Norton, 1990.

Boardman, John. *Greek Art*. New York: Thames and Hudson, 1987.

———. *The Greek Sculpture: The Classical Period*. New York: Thames and Hudson, 1985.

———. *The Greeks Overseas*. Rev. ed. New York: Thames and Hudson, 1980.

Boardman, John, and Hammond, N. G. L., eds. *The Cambridge Ancient History: The Expansion of the Greek World, Eighth to Sixth Centuries B.C.*, vol. 3, part 3. 2nd ed. New York: Cambridge University Press, 1982.

Boardman, John, et al., eds. *The Cambridge Ancient History: Persia, Greece, and the Western Mediterranean c. 525–479 B.C.*, vol. 4. 2nd ed. New York: Cambridge University Press, 1988.

Boardman, John; Griffin, Jasper; and Murray, Oswyn, eds. *The Oxford History of the Classical World*, chapts. 1–15. New York: Oxford University Press, 1986.

Bowra, C. M. *Greek Lyric Poetry: From Alcman to Simonides*. New York: Oxford University Press, 1936.

———. *Homer*. London: Gerald Duckworth and Company, 1972.

———. *Pindar*. New York: Oxford University Press, 1964.

Bright, John. *A History of Israel*. 2nd ed. Philadelphia: Westminster Press, 1972.

Buck, Carl Darling. *Comparative Grammar of Greek and Latin*. Chicago: University of Chicago Press, 1933.

Burkert, Walter. *Ancient Mystery Cults*. Cambridge, Mass.: Harvard University Press, 1987.

———. *Greek Religion*. Trans. John Raffan. Cambridge, Mass.: Harvard University Press, 1985.

Burn, Andrew Robert. *The Lyric Age of Greece*. New York: St. Martin's, 1960.

———. *The Penguin History of Greece*. Rev. ed. New York: Penguin Books, 1985.

———. *Persia and the Greeks*. London: Edward Arnold, 1970.

Bury, J. B., Cook, S. A., and Adcock, F. E., eds. *The Cambridge Ancient History: Athens 478–401 B.C.*, vol. 5. 1st ed. New York: Cambridge University Press, 1964.

———. *The Cambridge Ancient History: Macedon: 401–301 B.C.*, vol. 6. 1st ed. New York: Cambridge University Press, 1964.

Casson, Lionel. *Ships and Seamanship in the Ancient World*. Princeton, N.J.: Princeton University Press, 1971.

Cawkwell, George. *Philip of Macedon*. Boston: Faber & Faber, 1978.

———. Lectures presented at Oxford University ("Thucydides" and "Persia and Greece"), 1977–1978.

Ceram, C. W. *Gods, Graves, and Scholars: The Story of Archaeology*. Trans. E. B. Garside. New York: Alfred A. Knopf, 1952.

Childe, Gordon. *What Happened in History*. Rev. ed. Foreword and footnotes Grahame Clarke. Baltimore: Penguin Books, 1964.

Cook, R. M. *Greek Painted Pottery*. London: Methuen and Co., 1960.

Cotterell, Arthur, ed. *The Encyclopedia of Ancient Civilizations*. New York: Mayflower Books, 1980.

Davies, J. K. *Democracy and Classical Greece*. Atlantic Highlands, N.J.: Humanities Press, 1978.

Dover, K. J. *Aristophanic Comedy*. Berkeley: University of California Press, 1972.

———. *Greek Homosexuality*. Cambridge, Mass.: Harvard University Press, 1978.

Easterling, P. E., et al., eds. *The Cambridge History of Classical Literature*. New York: Cambridge University Press, 1989.

Edwards, I. E. S., et al., eds. *The Cambridge Ancient History: The Middle East and the Aegean Region, circa 1380–1000 B.C.*, vol. 2, part 2. 3rd ed. New York: Cambridge University Press, 1975.

Edwards, Paul, ed. *The Encyclopedia of Philosophy*, vol. 1–8. New York: Macmillan, 1967.

Ehrenberg, Victor. *Sophocles and Pericles*. Oxford: Basil Blackwell, 1954.

Ellis, Walter M. *Alcibiades*. New York: Routledge, 1989.

Evans, J. A. S. *Herodotus*. Boston: Twayne Publishers, 1982.

Finley, M. I. *Early Greece: The Bronze and Archaic Ages*. Rev. ed. London: Chatto and Windus, 1981.

Fornara, Charles W., ed. and trans. *Archaic Times to the End of the Peloponnesian War*. Baltimore: Johns Hopkins University Press, 1977.

Fornara, Charles W., and Samons, Loren J. *Athens from Cleisthenes to Pericles*. Berkeley: University of California Press, 1991.

Forrest, W. G. *The Emergence of Greek Democracy*. New York: World University Library, 1966.

——. *A History of Sparta: 950–192* B.C. London: Hutchinson University Library, 1968.

Fox, Robin Lane. *The Search for Alexander*. Boston: Little, Brown; 1980.

Fuller, J. F. C. *The Generalship of Alexander the Great*. New Brunswick, N.J.: Rutgers University Press, 1960.

Gagarin, Michael, *Early Greek Law*. Berkeley: University of California Press, 1986.

Gomme, A. W. *A Historical Commentary on Thucydides*, vol. 1. New York: Oxford University Press, 1945.

Gomme, A. W., Andrewes, A., and Dover, K. J. *A Historical Commentary on Thucydides*, vol. 4. New York: Oxford University Press, 1970.

Graves, Robert. *The Greek Myths*, vols. 1 and 2. Baltimore: Penguin, 1955.

Green, Peter. *Alexander of Macedon, 356–323 B.C.* Berkeley: University of California Press, 1991.

——. *A Concise History of Ancient Greece*. London: Thames and Hudson, 1979.

Griffin, Jasper. *Homer on Life and Death*. New York: Oxford University Press, 1980.

Guthrie, W. K. C. *Socrates*. New York: Cambridge University Press, 1971.

Halperin, David. *One Hundred Years of Homosexuality*. New York: Routledge, 1990.

Hamilton, J. R. *Alexander the Great*. Pittsburgh: University of Pittsburgh Press, 1974.

——. *Plutarch: Alexander: A Commentary*. New York: Oxford University Press, 1969.

Hammond, N. G. L., and Scullard, H. H., eds. *The Oxford Classical Dictionary*. New York: Oxford University Press, 1970.

Hare, R. M. *Plato*. New York: Oxford University Press, 1982.

Harris, H. A. *Greek Athletes and Athletics*. Bloomington: Indiana University Press, 1966.

Hignett, C. *A History of the Athenian Constitution*. New York: Oxford University Press, 1952.

Hornblower, Simon. *Thucydides*. Baltimore: Johns Hopkins University Press, 1987.

How, W. W. and Wells, J. *A Commentary on Herodotus*, vols. 1 and 2. New York: Oxford University Press, 1975.

Hussey, Edward. *The Presocratics*. London: Gerald Duckworth and Company, 1972.

Janson, H. W. *History of Art*. Englewood Cliffs, N.J.: Prentice-Hall, 1969.

Jones, A. H. M. *Athenian Democracy*. Oxford: Basil Blackwell, 1975.

——. *Sparta*. Cambridge, Mass.: Harvard University Press, 1967.

Kirk, G. S. *Homer and the Oral Tradition*. New York: Cambridge University Press, 1976.

——. *The Songs of Homer*. New York: Cambridge University Press, 1962.

Kirk, G. S., and Raven, J. E. *The Presocratic Philosophers*. New York: Cambridge University Press, 1957.

Knox, Bernard. *The Oldest Dead White European Males and Other Reflections on the Classics*. New York: W. W. Norton, 1993.

Lefkowitz, Mary R. *The Lives of the Greek Poets*. Baltimore: Johns Hopkins University Press, 1981.

Levi, Peter. *Atlas of the Greek World*. New York: Facts On File, 1984.

Lewis, David, et al., eds. *The Cambridge Ancient History: The Fifth Century*, vol. 5. 2nd ed. New York: Cambridge University Press, 1992.

Liddell, H. G., ed. *An Intermediate Greek-English Lexicon* (abridgement of the 7th ed. of Liddell and Scott's Greek-English Lexicon). New York: Oxford University Press, 1889.

Lloyd, G. E. R. *Aristotle: The Growth and Structure of His Thought*. New York: Cambridge University Press, 1968.

——. *Early Greek Science: Thales to Aristotle*. London: Chatto and Windus, 1970.

——. *Greek Science after Aristotle*. London: Chatto and Windus, 1973.

Long, A. A. *Hellenistic Philosophy: Stoics, Epicureans, Sceptics*. London: Gerald Duckworth and Company, 1974.

Luce, J. V. *An Introduction to Greek Philosophy*. New York: Thames and Hudson, 1992.

Macadam, Alta, ed. *The Blue Guide: Sicily*. New York: W. W. Norton, 1988.

MacDowell, Douglas M. *The Law in Classical Athens*. Ithaca, N.Y.: Cornell University Press, 1978.

MacKendrick, Paul. *The Greek Stones Speak: The Story of Archaeology in Greek Lands*. New York: W. W. Norton, 1962.

Martin, Roland. *Greek Architecture*. New York: Electa/Rizzoli, 1988.

McDonagh, Bernard. *Blue Guide: Turkey: The Aegean and Mediterranean Coasts*. New York: W. W. Norton, 1989.

McLeish, Kenneth. *The Theatre of Aristophanes*. New York: Taplinger Publishing Company, 1980.

Meiggs, Russell. *The Athenian Empire*. New York: Oxford University Press, 1972.

——. *Trees and Timber in the Ancient Mediterranean World*. New York: Oxford University Press, 1982.

Meiggs, Russell, and Lewis, David, eds. *A Selection of Greek Historical Inscriptions*. New York: Oxford University Press, 1969.

Morrison, J. R., and Williams, R. T. *Greek Oared Ships: 900–322 B.C.* New York: Cambridge University Press, 1968.

Murray, Gilbert. *Euripides and His Age*. New York: Henry Holt and Company, 1913.

Murray, Oswyn. *Early Greece*. 2nd ed. Cambridge, Mass.: Harvard University Press, 1993.

——. Lectures presented at Oxford University ("Greek Poetry and Greek History"), 1978–1979.

———. "Sweet Feasts of the Grape." *The Times Higher Education Supplement,* July 6, 1990, pp. 13–17.

Murray, Oswyn, and Price, Simon, eds. *The Greek City from Homer to Alexander.* New York: Oxford University Press, 1990.

New York Times Atlas of the World, The. New York: Times Books, 1993.

Ostwald, Martin. *From Popular Sovereignty to the Sovereignty of Law: Law, Society, and Politics in Fifth-Century Athens.* Berkeley: University of California Press, 1986.

Pickard-Cambridge, A. W. *Dithyramb, Tragedy, and Comedy.* New York: Oxford University Press, 1962.

Podlecki, A. J. *The Life of Themistocles.* Montreal: McGill-Queen's University Press, 1975.

Poliakoff, Michael B. *Combat Sports in the Ancient World.* New Haven: Yale University Press, 1987.

Politt, J. J. *Art in the Hellenistic Age.* New York: Cambridge University Press, 1986.

Pomeroy, Sarah B. *Goddesses, Whores, Wives, and Slaves: Women in Classical Antiquity.* New York: Schocken Books, 1975.

Race, William H. *Pindar.* Boston: Twayne Publishers, 1986.

Renault, Mary. *The Nature of Alexander.* New York: Pantheon Books, 1975.

Rhodes, P. J. *A Commentary on the Aristotelian Athenaion Politeia.* New York: Oxford University Press, 1981.

Richter, Gisela, M. A. *A Handbook of Greek Art.* London: Phaidon Press, 1959.

Robertson, Alec, and Stevens, Denis, eds. *The Pelican History of Music,* vol. 1. Baltimore: Penguin Books, 1960.

Robertson, Martin. *A History of Greek Art,* vols. 1 and 2. New York: Cambridge University Press, 1975.

Rose, H. J. *A Handbook of Greek Literature.* New York: E. P. Dutton, 1934, 1961.

———. *A Handbook of Greek Mythology.* New York: E. P. Dutton, 1959.

———. *Religion in Greece and Rome.* New York: Harper and Row, 1959.

Ross, W. D. *Aristotle.* London: Methuen and Co., 1949.

de Ste. Croix, G. E. M. *The Origins of the Peloponnesian War.* London: Gerald Duckworth and Company, 1972.

Salmon, J. B. *Wealthy Corinth: A History of the City to 338 B.C..* New York: Oxford University Press, 1984.

Scodel, Ruth. *Sophocles.* Boston: Twayne Publishers, 1984.

Sherratt, Andrew, ed. *The Cambridge Encyclopedia of Archaeology.* New York: Crown Publishers/Cambridge University Press, 1980.

Singer, Charles. *Greek Biology & Greek Medicine.* New York: Oxford University Press, 1922.

Snodgrass, A. M. *Arms and Armour of the Greeks.* Ithaca, N.Y.: Cornell University Press, 1967.

Snowden, Frank M., Jr. *Blacks in Antiquity: Ethiopians in the Greco-Roman Experience.* Cambridge, Mass.: Belknap Press/Harvard University Press, 1970.

Stewart, Andrew. *Greek Sculpture: An Exploration,* vols. 1 and 2. New Haven: Yale University Press, 1990.

Stillwell, Richard, ed. *The Princeton Encylcopedia of Classical Sites.* Princeton, N.J.: Princeton University Press, 1976.

Stockton, David. *The Classical Athenian Democracy.* New York: Oxford University Press, 1990.

Stone, I. F. *The Trial of Socrates.* Boston: Little, Brown; 1988.

Tarn, W. W. *Hellenistic Military and Naval Developments.* New York: Cambridge University Press, 1930.

Taylour, Lord William. *The Mycenaeans.* 2nd ed. New York: Thames and Hudson, 1983.

Wade-Gery, H. T. "Thucydides the son of Melesias." *Journal of Hellenic Studies,* vol. 52, 1932, pp. 205–227.

Walbank, F. W. et al., eds. *The Cambridge Ancient History: The Hellenistic World,* vol. 7, part 1. 2nd ed. New York: Cambridge University Press, 1984.

———. *The Cambridge Ancient History: The Rise of Rome to 220 B.C.,* vol. 7, part 2. 2nd ed. New York: Cambridge University Press, 1989.

———. *The Cambridge Ancient History: Rome and the Mediterranean to 133 B.C.,* vol. 8, 2nd ed. New York: Cambridge University Press, 1989.

Walker, Steven F. *Theocritus.* Boston: Twayne Publishers, 1980.

Weidman, Thomas. *Greek and Roman Slavery.* Baltimore: Johns Hopkins University Press, 1981.

Wellesz, Egon. *The New Oxford History of Music. Vol. 1: Ancient and Oriental Music.* New York: Oxford University Press, 1957.

Winnington-Ingram, R. P. *Sophocles: An Interpretation.* New York: Cambridge University Press, 1980.

Wood, Michael. *In Search of the Trojan War.* New York: Facts On File, 1985.

Zimmerman, Bernhard: *Greek Tragedy: An Introduction.* Trans. Thomas Marier. Baltimore: Johns Hopkins University Press, 1991.

Ancient authors in Greek and in translation

Aeschylus. *Agamemnon, Libation-bearers, Eumenides, Fragments.* Greek text, trans. Herbert Weir Smyth, ed. Hugh Lloyd-Jones. Cambridge, Mass.: Harvard University Press (Loeb Classical Library), 1971.

———. *Prometheus Bound, The Suppliants, Seven Against Thebes, The Persians.* Trans. and intro. Philip Vellacott. Baltimore: Penguin Books, 1961.

———. *Septem quae supersunt tragoediae.* Greek text, 2nd rev. ed. Denys Page. New York: Oxford University Press, 1972.

Archilochus et al., *Greek Lyric Poetry.* Ed. and trans. Willis Barnstone. New York: Schocken Books, 1967.

———. *Greek Lyrics.* Ed. and trans. Richmond Lattimore. Chicago: University of Chicago Press, 1981.

———. *The Penguin Book of Greek Verse.* Greek text, ed. and trans. by Constantine A. Trypanis. New York: Penguin Books, 1971.

Aristophanes. *The Complete Plays of Aristophanes.* Ed. and intro. by Moses Hadas. New York: Bantam Books, 1962.

Aristotle. *The Athenian Constitution* et al., Greek text, trans. H. Rackham. Cambridge, Mass.: Harvard University Press (Loeb Classical Library), 1952.

———. *Ethica Nicomachea.* Greek text, ed. I. Bywater. New York: Oxford University Press, 1890.

———. *Nicomachean Ethics.* Trans., intro., and notes Martin Ostwald. Indianapolis: Bobbs-Merrill Company, 1962.

—————. *On the Art of Poetry.* Ed. and trans. Lane Cooper. Ithaca, N.Y.: Cornell University Press, 1947.

—————. *Politics.* Ed. and trans. Ernest Baker. New York: Oxford University Press, 1946.

Arrian. *The Campaigns of Alexander.* Trans. Aubrey de Sélincourt, intro., and notes J. R. Hamilton. New York: Penguin Books, 1971.

Euripides. *Bacchae.* Greek text, ed. and notes E. R. Dodds. New York: Oxford University Press, 1960.

—————. *The Complete Greek Tragedies,* vols. 3 and 4. Trans. David Grene and Richmond Lattimore. Chicago: University of Chicago Press, 1959.

—————. *Hippolytos.* Greek text, ed. and notes W. S. Barrett. New York: Oxford University Press, 1964.

—————. *Medea.* Greek text, ed. and notes Denys L. Page. New York: Oxford University Press, 1938.

Herodotus. *Historiae,* vols. 1 and 2. Greek text, ed. Charles Hude. New York: Oxford University Press, 1927.

—————. *The Histories.* Trans. Aubrey de Sélincourt, intro. and notes A. R. Burn. New York: Penguin Books, 1972.

Hesiod. *Theogony* Trans. Apostolos N. Athanassakis. Baltimore: Johns Hopkins University Press, 1983.

—————. *Works and Days.* Trans. Apostolos N. Athanassakis. Baltimore: Johns Hopkins University Press, 1983.

Homer. *The Iliad,* vols. 1 and 2. Greek text, ed. and trans. A. T. Murray, Cambridge, Mass.: Harvard University Press (Loeb Classical Library), 1939, 1942.

—————. *The Iliad,* trans. Alston Hurd Chase and William G. Perry, Jr. New York: Bantam Books, 1960.

—————. *The Iliad.* Trans. Robert Fitzgerald. Garden City, N.Y.: Anchor Press/Doubleday, 1975.

—————. *The Odyssey,* vols. 1 and 2. Greek text, ed. W. B. Stanford. New York: St. Martin's Press, 1965.

—————. *The Odyssey.* Trans. Robert Fitzgerald. Garden City, N.Y.: Anchor Press/Doubleday, 1961.

Menander. *The Dyskolos of Menander.* Greek text, ed. and intro. E. W. Handley. Cambridge, Mass.: Harvard University Press, 1965.

Pausanias. *Guide to Greece,* vols. 1 and 2. Trans. and intro. Peter Levi. New York: Penguin Books, 1979.

Pindar. *The Odes of Pindar.* Trans. Richmond Lattimore. Chicago: University of Chicago Press, 1947.

—————. *Pindari carmina cum fragmentis.* Greek text, ed. C. M. Bowra. New York: Oxford University Press, 1947.

Plato. *Euthyphro, Apology, Crito.* Trans. and intro. F. J. Church. Indianapolis: Bobbs-Merrill Company, 1948.

—————. *Gorgias.* Greek text, ed. and notes E. R. Dodds. New York: Oxford University Press, 1959.

—————. *Gorgias.* Trans. and intro. W. C. Helmbold. Indianapolis: Bobbs-Merrill, 1952.

—————. *Protagoras.* Trans. and intro. Gregory Vlastos. Indianapolis: Bobbs-Merrill Company, 1956.

—————. *The Republic of Plato.* Trans. and intro. Francis Macdonald Cornford. New York: Oxford University Press, 1941.

—————. *Res Publica,* in vol. 4 of *Opera* (complete works). Greek text, ed. John Burnet. New York: Oxford University Press, 1903.

—————. *The Symposium.* Trans. and intro. Walter Hamilton. Baltimore: Penguin Books, 1951.

Plutarch. *The Age of Alexander.* Trans. Ian Scott-Kilvert, intro. G. T. Griffith. New York: Penguin Books, 1973.

—————. *Plutarch's Lives,* vols. 1–7, 9. Greek text, trans. Bernadotte Perrin. Cambridge, Mass.: Harvard University Press (Loeb Classical Library), 1967.

—————. *Plutarch on Sparta.* Trans., intro., and notes Richard J. A. Talbert. New York: Penguin Books, 1988.

—————. *The Rise and Fall of Athens.* Trans. and intro. Ian Scott-Kilvert. New York: Penguin Books, 1960.

Sappho, et al., *Greek Lyric,* vol. 1. Greek text, ed. and trans. David Campbell. Cambridge, Mass.: Harvard University Press (Loeb Classical Library), 1982.

Sophocles. *The Complete Plays of Sophocles.* Trans. Richard Claverhouse Jebb, intro. Moses Hadas. New York: Bantam Books, 1971.

—————. *Oedipus Rex.* Greek text, ed. R. D. Dawe. New York: Cambridge University Press, 1982.

Stesichorus, et al., *Greek Lyric,* vol. 3. Greek text, ed. and trans. David Campbell. Cambridge, Mass.: Harvard University Press (Loeb Classical Library), 1991.

Thucydides. *Historiae,* vols. 1 and 2. Greek text, ed. Henry Stuart Jones and Enoch Powell. New York: Oxford University Press, 1960, 1963.

—————. *The Peloponnesian War.* Trans. and intro. Rex Warner. Baltimore: Penguin Books, 1954.

Xenophon. *Conversations of Socrates: Socrates' Defence, Memoirs of Socrates, The Dinner Party, The Estate-manager.* Trans. Hugh Tredennick and Robin Waterfield, ed. Robin Waterfield. New York: Penguin Books, 1990.

—————. *A History of My Times (Hellenica).* Trans. Rex Warner, intro. and notes George Cawkwell. New York: Penguin Books, 1979.

—————. *The Persian Expedition.* Trans. Rex Warner, intro. and notes George Cawkwell. New York: Penguin Books, 1972.

INDEX

Main essays are indicated by **bold** page numbers.